Rhodes State College

College Prep Mathematics

Charles P. McKeague

*xyz*textbooks

*xyz*textbooks

Rhodes State College
College Prep Mathematics
Charles P. McKeague

Editorial Director: Patrick McKeague

Sales Manager: Amy Jacobs

Production Manager: Matthew Hoy

Project Manager: Staci Truelson

Text Designer: Devin Christ

Cover Image: ©2014 Beth Basista

AIE Edition: ISBN-13: 978-1-630980-17-7/ ISBN-10: 1-630980-17-X
Student Edition: ISBN-13: 978-1-630980-16-0/ ISBN-10: 1-630980-16-1

For product information and technology assistance, contact us at
XYZ Textbooks, 1-877-745-3499

For permission to use material from this text or product,
e-mail: **info@mathtv.com**

XYZ Textbooks
1339 Marsh St.
San Luis Obispo, CA 93401
USA

For your course and learning solutions, visit **www.xyztextbooks.com**

Printed in the United States of America

Brief Contents

0 Whole Numbers 1

1 Real Numbers and Algebraic Expressions 79

2 Fractions and Mixed Numbers 147

3 Decimals 233

4 Unit Analysis 291

5 Linear Equations and Inequalities in One Variable 313

6 Exponents and Polynomials 397

7 Factoring Polynomials 469

8 Rational Expressions, Equations, and Functions 533

9 Graphing Linear Equations and Inequalities in Two Variables 603

10 Systems of Linear Equations 683

11 Inequalities, Graphs,and Functions 739

12 Rational Exponents and Radicals 827

13 Quadratic Equations and Functions 897

14 Exponential and Logarithmic Functions 925

Contents

0 Whole Numbers 1

0.1 Place Value and Names for Numbers 2
0.2 Addition with Whole Numbers, and Perimeter 12
0.3 Subtraction with Whole Numbers 24
0.4 Rounding Numbers, Estimating Answers, and Displaying Information 34
0.5 Multiplication with Whole Numbers, and Area 42
0.6 Division with Whole Numbers 57
0.7 Exponents and Order of Operations 67
Chapter 0 Summary 76
Chapter 0 Test 78

1 Real Numbers and Algebraic Expressions 79

1.1 Real Numbers 80
1.2 Addition and Subtraction of Real Numbers 94
1.3 Multiplication of Real Numbers 105
1.4 Division of Real Numbers 114
1.5 Properties of Real Numbers 125
1.6 Simplifying Expressions 135
Chapter 1 Summary 144
Chapter 1 Test 146

2 Fractions and Mixed Numbers 147

2.1 The Meaning and Properties of Fractions 148
2.2 Prime Numbers, Factors, and Reducing to Lowest Terms 159
2.3 Multiplication with Fractions, and the Area of a Triangle 167
2.4 Division with Fractions 177
2.5 Addition and Subtraction with Fractions 184
2.6 Mixed-Number Notation 197
2.7 Multiplication and Division with Mixed Numbers 204
2.8 Addition and Subtraction with Mixed Numbers 211
2.9 Order of Operations and Complex Fractions 220
Chapter 2 Summary 228
Chapter 2 Test 232

3 Decimals 233

3.1 Decimal Notation, Place Value, and Rounding 234
3.2 Addition and Subtraction with Decimals 243
3.3 Multiplication with Decimals 251
3.4 Division with Decimals 259
3.5 Fractions and Decimals 270
3.6 Square Roots and The Pythagorean Theorem 278
Chapter 3 Summary 288
Chapter 3 Test 290

4 Unit Analysis 291

4.1 Introduction to Dimensional Analysis 292
4.2 Conversions in the Metric System 297
4.3 Distance and Length 300
4.4 Volume and Capacity 303
4.5 Weight and Mass 306
4.6 Temperature Conversions 308
Chapter 4 Test 311

5 Linear Equations and Inequalities in One Variable 313

5.1 Addition Property of Equality 314
5.2 Multiplication Property of Equality 323
5.3 Solving Linear Equations in One Variable 332
5.4 Formulas 343
5.5 Number, Age, Perimeter, Coin Applications 355
5.6 Integer, Interest, Triangle, Miscellaneous Applications 366
5.7 Linear Inequalities in One Variable 378
Chapter 5 Summary 393
Chapter 5 Test 395

6 Exponents and Polynomials 397

6.1 Multiplication with Exponents and Scientific Notation 398
6.2 Division with Exponents and Negative Exponents 409
6.3 Operations with Monomials 423
6.4 Addition and Subtraction of Polynomials 431
6.5 Multiplication of Polynomials 437
6.6 Special Products 446
6.7 Dividing a Polynomial by a Monomial 453
6.8 Dividing a Polynomial by a Polynomial 459
Chapter 6 Summary 465
Chapter 6 Test 467

7 Factoring Polynomials 469

7.1 The Greatest Common Factor, Factoring by Grouping 470
7.2 Factoring Trinomials of the Form $x^2 + bx + c$ 478
7.3 Factoring Trinomials of the Form $ax^2 + bx + c$ 485
7.4 The Difference of Two Squares and Perfect Square Trinomials 493
7.5 The Sum and Difference of Two Cubes 501
7.6 Factoring: A General Review 505
7.7 Solving Quadratic Equations by Factoring 511
7.8 Applications of Quadratic Equations 519
Chapter 7 Summary 530
Chapter 7 Test 532

8 Rational Expressions, Equations, and Functions 533

8.1 Basic Properties and Reducing to Lowest Terms 534
8.2 Multiplication and Division of Rational Expressions 545
8.3 Addition and Subtraction of Rational Expressions 556
8.4 Complex Rational Expressions 568
8.5 Rational Equations 576
8.6 Applications 588
Chapter 8 Summary 599
Chapter 8 Test 601

9 Graphing Linear Equations and Inequalities in Two Variables 603

9.1 The Rectangular Coordinate System 604
9.2 Solutions and Graphs of Linear Equations 614
9.3 Graphing Linear Equations Using Intercepts 632
9.4 Graphing Linear Equations Using Slope 644
9.5 Finding the Equation of a Line 657
9.6 Graphing Linear Inequalities in Two Variables 670
Chapter 9 Summary 679
Chapter 9 Test 681

10 Systems of Linear Equations 683

10.1 Solving Linear Systems by Graphing 684
10.2 The Substitution Method 694
10.3 The Elimination Method 702
10.4 Systems of Linear Equations in Three Variables 713
10.5 Systems of Linear Inequalities and Applications 722
Chapter 10 Summary 736
Chapter 10 Test 738

11 Inequalities, Graphs, and Functions 739

11.1 Interval Notation and Linear Inequalities 740
11.2 Sets and Compound Inequalities 751
11.3 Absolute Value Equations 761
11.4 Absolute Value Inequalities 768
11.5 Graphs of Equations 777
11.6 Introduction to Functions 797
11.7 Function Notation 812
Chapter 11 Summary 824
Chapter 11 Test 826

12 Rational Exponents and Radicals 827

12.1 Rational Exponents 828
12.2 Simplifying Radicals 842
12.3 Addition and Subtraction of Radical Expressions 854
12.4 Multiplication and Division of Radical Expressions862 862
12.5 Radical Equations and Functions 872
12.6 Complex Numbers 884
Chapter 12 Summary 893
Chapter 12 Test 895

13 Quadratic Equations and Functions 897

13.1 Completing the Square 898
13.2 The Quadratic Formula 911
Chapter 13 Summary 923
Chapter 13 Test 924

14 Exponential and Logarithmic Functions 925

14.1 Exponential Functions and Applications 926
14.2 Inverse Functions 940
14.3 Logarithmic Functions and Applications 954
14.4 Properties of Logarithms 966
14.5 Common and Natural Logarithms with Applications 975
14.6 Exponential Equations, Change of Base, and Applications 985
Chapter 14 Summary 994
Chapter 14 Test 996

Preface

XYZ Textbooks is on a mission to improve the quality and affordability of course materials for mathematics. This custom book brings a new level of technological innovation, which is certain to improve student proficiency. This book is presented in a format in which each section is organized by easy-to-follow learning objectives. Each objective includes helpful example problems, as well as additional practice problems in the margins, and are referenced in each exercise set. To help students relate what they are studying to the real world, each chapter and section begin with a real-life application of key concepts. You and your students will encounter extensions of these opening applications in the exercise sets.

The following features will appear in your custom book and are designed to help your students succeed.

QR Codes

Unique technology embedded in each exercise set allows students with a smart phone or other internet-connected mobile device to view our video tutorials without being tied to a computer.

Video Examples Section 1.1

A The Real Number Line

To represent negative numbers in algebra, we use what is called the *real number line*. Here is how we construct a real number line: We first draw a straight line and label a convenient point on the line with 0. Then we mark off equally spaced distances in both directions from 0. Label the points to the right of 0 with the numbers 1, 2, 3, . . . (the dots mean "and so on"). The points to the left of 0 we label in order, −1, −2, −3, Here is what it looks like:

Vocabulary Review

A list of fill-in-the-blank sentences appears at the beginning of each exercise set to help students better comprehend and verbalize concepts.

Vocabulary Review

Choose the correct words to fill in the blanks below.

 prime composite remainder lowest terms

1. A _____ number is any whole number greater than 1 that has exactly two divisors: 1 and the number itself.
2. A number is a divisor of another number if it divides it without a _____.
3. Any whole number greater than 1 that is not a prime number is called a _____ number.
4. A fraction is said to be in _____ if the numerator and the denominator have no factors in common other than the number 1.

Getting Ready For Class

Simple writing exercises for students to answer after reading each section increases their understanding of key concepts.

GETTING READY FOR CLASS

After reading through the preceding section, respond in your own words and in complete sentences.

A. What is a prime number?

B. Why is the number 22 a composite number?

C. Factor 120 into a product of prime factors.

D. How would you reduce a fraction to lowest terms?

Key words

A list of important vocabulary appears at the beginning of each section to help students prepare for new concepts. These words can also be found in blue italics within the sections.

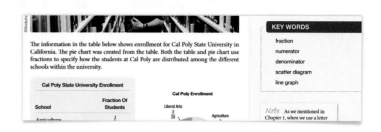

Find the Mistake

Complete sentences that include common mistakes appear in each exercise set to help students identify errors and aid their comprehension of the concepts presented in that section.

Landmark Review

A review appears in the middle of each chapter for students to practice key skills, check progress, and address difficulties.

End of Chapter Tests

Problems at the end of each chapter provide students with a comprehensive review of each chapter's concepts.

Supplements

Instructor

Name: _____

Phone: _____

Email: _____

Office Hours: _____

Teacher Assistant

Name: _____

Phone: _____

Email: _____

Office Hours: _____

On-Campus Tutoring

Location: _____

Hours: _____

Study Group

Name: _____

Phone: _____

Email: _____

Name: _____

Phone: _____

Email: _____

Name: _____

Phone: _____

Email: _____

For the Instructor

Online Homework XYZ Homework provides powerful online instructional tools for faculty and students. Randomized questions provide unlimited practice and instant feedback with all the benefits of automatic grading. Tools for instructors can be found at *www.xyzhomework.com* and include the following:

- Quick setup of your online class
- More than 3,000 randomized questions, similar to those in the textbook, for use in a variety of assessments, including online homework, quizzes, and tests
- Text and videos designed to supplement your instruction
- Automated grading of online assignments
- Flexible gradebook
- Message boards and other communication tools, enhanced with calculator-style input for proper mathematics notation

MathTV.com With more than 8,000 videos, MathTV.com provides the instructor with a useful resource to help students learn the material. MathTV.com features videos of most examples in the book, explained by the author and a variety of peer instructors. If a problem can be completed more than one way, the peer instructors often solve it by different methods. Instructors can also use the site's Build a Playlist feature to create a custom list of videos for posting on their class blog or website.

For the Student

Online Homework XYZ Homework provides powerful online instruction and homework practice for students. Benefits for the student can be found at *www.xyzhomework.com* and include the following:

- Unlimited practice with problems similar to those in the text
- Online quizzes and tests for instant feedback on performance
- Online video examples
- Convenient tracking of class programs

MathTV.com Students have access to math instruction 24 hours a day, seven days a week, on MathTV.com. Assistance with any problem or subject is never more than a few clicks away.

Online Book This text is available online for both instructors and students. Tightly integrated with MathTV.com, students can read the book and watch videos of the author and peer instructors explaining most examples. Access to the online book is available free with the purchase of a new book.

Additional Chapter Features

Learning Objectives

We have provided a list of the section's important concepts at the beginning of each section near the section title, noted with an orange capital letter, which is reiterated beside each learning objective header in the section.

Example and Practice Problems

Example problems, with work, explanations, and solutions are provided for every learning objective. Beside each example problem, we have also provided a corresponding practice problem and its answer in the margin.

Margin Notes

In many sections, you will find yellow notes in the margin that correspond to the section's current discussion. These notes contain important reminders, fun facts, and helpful advice for working through problems.

Colored Boxes

Throughout each section, we have highlighted important definitions, properties, rules, how to's, and strategies in colored boxes for easy reference.

Blueprint for Problem Solving

We provide students with a step-by-step guide for solving application problems in a clear and efficient manner.

Facts from Geometry

Throughout the book, we highlight specific geometric topics and show students how they apply to the concepts they are learning in the course.

Using Technology

Throughout the book, we provide students with step-by-step instructions for how to work some problems on a scientific or graphing calculator.

Descriptive Statistics

In some sections, we call out to the use of descriptive statistics as they are applied to the concepts students are learning, and thus, applied to real life.

Vocabulary Review

At the beginning of every exercise set, these fill-in-the-blank sentences help students comprehend and verbalize concepts from the corresponding section.

Problems

Each section contains problems ranging in difficulty that allow students to practice each learning objective presented in the section. The majority of the sections will also include problems in one or more of the following categories:

- Applying the Concepts
- Extending the Concepts or One Step Further
- Improving your Quantitative Literacy
- Getting Ready for the Next Section
- Extending the Concepts

Chapter Summary

At the end of each chapter, we provide a comprehensive summary of the important concepts from the chapter. Beside each concept, we also provide additional example problems for review.

Answers to Odd-Numbered Questions

At the end of the book, we have supplied answers for all odd-numbered problems that appear in the following features:

- Landmark Reviews
- Vocabulary Reviews
- Exercise Sets
- Find the Mistake Exercises

Whole Numbers

0.1 Place Value and Names for Numbers

0.2 Addition with Whole Numbers, and Perimeter

0.3 Subtraction with Whole Numbers

0.4 Rounding Numbers, Estimating Answers, and Displaying Information

0.5 Multiplication with Whole Numbers, and Area

0.6 Division with Whole Numbers

0.7 Exponents and Order of Operations

Washington D.C., USA
Gray Buildings © 2008 Sanborn
Gray Buildings © District of Columbia (DC GIS) & CyberCity
Gray Buildings © 2010 CyberCity

Just before the beginning of the United States' involvement in World War II, our military was in great need of office space in and around Washington, DC. Coming up with a solution to this problem fell to Brigadier General Brehon B. Somervell who worked in the construction division. He summoned one of his subordinates and a civilian architect to his office on Thursday July 17, 1941, and informed them that he wanted plans for a structure to house 40,000 workers by Monday. After what was described as a "very busy weekend," the Monday deadline was met and the idea for the Pentagon was born. Construction began September 11, 1941 and was completed January 15, 1943 at a cost of $85 million. Still serving as headquarters of the Department of Defense, approximately 25,000 civilian and military employees report to work there each day. These workers use 131 stairways, 13 elevators, or 19 escalators to reach their offices on one of 7 floors. Despite its large size, the building is still regarded as one of the most efficient in the world. It takes no more than 7 minutes to walk between any two points.

921 feet

Suppose you wanted to know the length of the Pentagon's perimeter. After examining the illustration above, we can tell that the length of one outside wall of the Pentagon measures 921 feet. We can then use this measurement to calculate the total perimeter. In this chapter, we will work problems similar to this one.

OBJECTIVES

A Identify the place value for digits.

B Write numbers in expanded form.

C Write numbers in words.

KEY WORDS

place value

expanded form

set

counting numbers

whole numbers

number line

**Video Examples
Section 0.1**

Practice Problems

1. Give the place value of each digit in the number 17,045.

Answers

1. 1 = ten thousands, 7 = thousands, 0 = hundreds, 4 = tens, 5 = ones

0.1 Place Value and Names for Numbers

April 3rd is International Pillow Fight Day. In 2010, the 5th Pillow Fight NYC took place in Union Square, New York City. People from all over the city filled the square and wielded pillows. When the fight began, people swung their pillows from side to side, hoping it would land on the nearest victim before getting hit themselves. The city's inaugural pillow fight took place in 2006 and had 500 participants. By 2010, the event garnered at least 5,000 people.

In this section, we will learn about place value and its relation to decimals. Comparing the number of participants in the 2006 pillow fight to the 2010 pillow fight, we can see that the 5 moves from the hundreds place value to the thousands place value.

A Place Value

Our number system is based on the number 10 and is therefore called a "base 10" number system. We write all numbers in our number system using the digits 0, 1, 2, 3, 4, 5, 6, 7, 8, and 9. The positions of the digits in a number determine the values of the digits. For example, the 5 in the number 251 has a different value from the 5 in the number 542.

The place values in our number system are as follows: The first digit on the right is in the ones column. The next digit to the left of the ones column is in the tens column. The next digit to the left is in the hundreds column. For a number like 542, the digit 5 is in the hundreds column, the 4 is in the tens column, and the 2 is in the ones column.

If we keep moving to the left, the columns increase in value. The following table shows the name and value of each of the first seven columns in our number system. Also, notice how the number 1 is located in the appropriate place value column.

Table 1

Millions Column	Hundred Thousands Column	Ten Thousands Column	Thousands Column	Hundreds Column	Tens Column	Ones Column
1,000,000	100,000	10,000	1,000	100	10	1

EXAMPLE 1 Give the place value of each digit in the number 305,964.

Solution Starting with the digit at the right, we have 4 in the ones column, 6 in the tens column, 9 in the hundreds column, 5 in the thousands column, 0 in the ten thousands column, and 3 in the hundred thousands column.

Large Numbers

The photograph shown here was taken by the Hubble telescope in April 2002. The object in the photograph is called the Cone Nebula. In astronomy, distances to objects like the Cone Nebula are given in light-years—the distance light travels in a year. If we assume light travels 186,000 miles in one second, then a light-year is 5,865,696,000,000 miles; that is,

<center>5 trillion, 865 billion, 696 million miles</center>

To find the place value of digits in large numbers, we can use Table 2. Note how the Ones, Thousands, Millions, Billions, and Trillions categories are each broken into Ones, Tens, and Hundreds. Note also that we have written the digits for our light-year in the last row of the table.

NASA

Table 2

Trillions			Billions			Millions			Thousands			Ones		
Hundreds	Tens	Ones	Hundreds	Tens	Ones	Hundreds	Tens	Ones	Hundreds	Tens	Ones	Hundreds	Tens	Ones
		5	8	6	5	6	9	6	0	0	0	0	0	0

EXAMPLE 2 Give the place value of each digit in the number 73,890,672,540.

Solution The following diagram shows the place value of each digit.

Ten Billions	Billions	Hundred Millions	Ten Millions	Millions	Hundred Thousands	Ten Thousands	Thousands	Hundreds	Tens	Ones
7	3,	8	9	0,	6	7	2,	5	4	0

Note When writing numbers with four or more digits, we use a comma to separate every three digits, starting from the ones column.

2. Give the place value of each digit in the number 4,572,106,890.

B Expanded Form

We can use the idea of place value to write numbers in expanded form. For example, the number 542 is written in condensed (standard) form. We can also write it in expanded form as

$$542 = 500 + 40 + 2$$

because the 5 is in the hundreds column, the 4 is in the tens column, and the 2 is in the ones column.

Here are more examples of numbers written in expanded form:

EXAMPLE 3 Write 5,478 in expanded form.

Solution $5,478 = 5,000 + 400 + 70 + 8$

3. Write 23,187 in expanded form.

Answers

Billions	Hundred Millions	Ten Millions	Millions	Hundred Thousands	Ten Thousands	Thousands	Hundreds	Tens	Ones

2. 4, 5 7 2, 1 0 6, 8 9 0
3. 20,000 + 3,000 + 100 + 80 + 7

We can use money to make the results from Example 3 more intuitive. Suppose you have $5,478 in cash as follows:

Thousands	Hundreds	Tens	Ones
$5,000	$400	$70	$8

Using this diagram as a guide, we can write

$$\$5,478 = \$5,000 + \$400 + \$70 + \$8$$

which shows us that our work writing numbers in expanded form is consistent with our intuitive understanding of the different denominations of money.

EXAMPLE 4 Write 56,094 in expanded form.

Solution Notice that there is a 0 in the hundreds column. This means we have 0 hundreds. In expanded form, we have

$$56,094 = 50,000 + 6,000 + 90 + 4$$

↑
*Note that we don't have
to include the 0 hundreds.*

Now let's reverse the process. In the next two examples, we will take an expression in expanded form and condense it back to standard form.

EXAMPLE 5 Write $10,000 + 7,000 + 500 + 60 + 9$ in condensed (standard) form.
Solution $10,000 + 7,000 + 500 + 60 + 9 = 17,569$

EXAMPLE 6 Write $5,000,000 + 70,000 + 600 + 3$ in condensed (standard) form.
Solution Note that the digit is 0 in the hundred thousands, thousands and tens place. The answer is

$$5,070,603$$

C Writing Numbers in Words

The idea of place value and expanded form can be used to help write the names for numbers. Naming numbers and writing them in words takes some practice. Let's begin by looking at the names of some two-digit numbers. Table 3 lists a few. Notice that the two-digit numbers that are greater than twenty and do not end in 0 have two parts. These parts are separated by a hyphen.

Table 3

Number	In English	Number	In English
25	*Twenty-five*	30	*Thirty*
47	*Forty-seven*	62	*Sixty-two*
93	*Ninety-three*	77	*Seventy-seven*
88	*Eighty-eight*	50	*Fifty*

4. Write 1,049,580 in expanded form.

5. Write
$60,000 + 8,000 + 700 + 90 + 3$
in condensed (standard) form.

6. Write $4,000,000 + 200,000 + 6,000 + 40 + 5$ in condensed (standard) form.

Answers
4. $1,000,000 + 40,000 + 9,000 + 500 + 80$
5. 68,793
6. 4,206,045

The following examples give the names for some larger numbers. In each case, the names are written according to the place values given in Table 2.

EXAMPLE 7　Write each number in words.

a. 452　　**b.** 397　　**c.** 608

Solution

a. Four hundred fifty-two

b. Three hundred ninety-seven

c. Six hundred eight

EXAMPLE 8　Write each number in words.

a. 3,561　　**b.** 53,662　　**c.** 547,801

Solution

a. Three thousand, five hundred sixty-one

↑
Notice how the comma separates the thousands from the hundreds.

b. Fifty-three thousand, six hundred sixty-two

c. Five hundred forty-seven thousand, eight hundred one

EXAMPLE 9　Write each number in words.

a. 507,034,005　　**b.** 739,600,075　　**c.** 5,003,007,006

Solution

a. Five hundred seven million, thirty-four thousand, five

b. Seven hundred thirty-nine million, six hundred thousand, seventy-five

c. Five billion, three million, seven thousand, six

The next examples show how we write a number given in words as a number written with digits.

EXAMPLE 10　Write five thousand, six hundred forty-two, using digits instead of words.

Solution　Five thousand, six hundred forty-two

| 5 | 6 | 42 | → | 5,642 |

EXAMPLE 11　Write each number using digits instead of words.

a. Three million, fifty-one thousand, seven hundred

b. Two billion, five

c. Seven million, seven hundred seven

Solution

a. 3,051,700

b. 2,000,000,005

c. 7,000,707

7. Write each number in words.
a. 752
b. 386
c. 507

8. Write each number in words.
a. 4,865
b. 51,508
c. 408,623

9. Write each number in words.
a. 406,058,210
b. 3,200,486,404

10. Write eight thousand, two hundred thirty-nine using digits instead of words.

11. Write each number using digits instead of words.
a. Seven million, forty-two thousand, six hundred.
b. Twelve million, four hundred.
c. Six hundred thousand, eight.

Answers
7. a. Seven hundred fifty-two
b. Three hundred eighty-six
c. Five hundred seven
8. a. Four thousand, eight hundred sixty-five.
b. Fifty-one thousand, five hundred eight.
c. Four hundred eight thousand, six hundred twenty three.
9. a. Four hundred six million, fifty-eight thousand, two hundred ten.
b. Three billion, two hundred million, four hundred eighty-six thousand, four hundred four.
10. 8,239
11. a. 7,042,600　**b.** 12,000,400
c. 600,008

Sets and the Number Line

In mathematics, a collection of numbers is called a set. In this chapter, we have been working with the set of counting numbers (also called natural numbers) and the set of whole numbers, which are defined as follows:

$$\text{Counting numbers} = \{1, 2, 3, \ldots\}$$

$$\text{Whole numbers} = \{0, 1, 2, 3, \ldots\}$$

The dots mean "and so on," and the braces { } are used to group the numbers in the set together.

Another way to visualize the whole numbers is with a number line. To draw a number line, we simply draw a straight line and mark off equally spaced points along the line, as shown in Figure 1. We label the point at the left with 0 and the rest of the points, in order, with the numbers 1, 2, 3, 4, 5, and so on.

FIGURE 1

The arrow on the right indicates that the number line can continue in that direction forever.

GETTING READY FOR CLASS

Each section of the book will end with some problems and questions like the ones below. They are for you to answer after you have read through the section but before you go to class. All of them require that you give written responses in complete sentences. Writing about mathematics is a valuable exercise. As with all problems in this course, approach these writing exercises with a positive point of view. You will get better at giving written responses to questions as the course progresses. Even if you never feel comfortable writing about mathematics, just attempting the process will increase your understanding and ability in this course.

After reading through the preceding section, respond in your own words and in complete sentences.

A. Give the place value of the 9 in the number 305,964.

B. Write the number 742 in expanded form.

C. Place a comma and a hyphen in the appropriate place so that the number 2,345 is written correctly in words below:

Two thousand three hundred forty five

D. Is there a largest whole number?

Vocabulary Review

Choose the correct words to fill in the blanks below.

comma set expanded hyphen counting condensed whole place value

1. _____ is the value of a digit in a number determined by its position.

2. The number 619 is written in _____ form, but is written in _____ form when it appears as 600 + 10 + 9.

3. The name of a two-digit number that is greater then twenty and does not end in 0 has two parts separated by a _____.

4. When writing a number with four or more digits, use a _____ to separate every three digits.

5. A collection of numbers is called a _____.

6. The set of _____ numbers is defined as {1, 2, 3,...}.

7. The set of _____ numbers is defined as {0, 1, 2, 3,...}.

Problems

A Give the place value of each digit in the following numbers.

1. 78	**2.** 93	**3.** 45	**4.** 79
5. 348	**6.** 789	**7.** 608	**8.** 450
9. 2,378	**10.** 6,481	**11.** 273,569	**12.** 768,253

Give the place value of the 5 in each of the following numbers.

13. 458,992	**14.** 75,003,782	**15.** 507,994,787	**16.** 320,906,050
17. 267,894,335	**18.** 234,345,678,789	**19.** 4,569,000	**20.** 50,000

B Write each of the following numbers in expanded form.

21. 658	**22.** 479	**23.** 68	**24.** 71
25. 4,587	**26.** 3,762	**27.** 32,674	**28.** 54,883

29. 3,462,577

30. 5,673,524

31. 407

32. 508

33. 30,068

34. 50,905

35. 3,004,008

36. 20,088,060

Write each of the following numbers in condensed (standard) form.

37. 500 + 40 + 7

38. 900 + 60 + 5

39. 6,000 + 300 + 20 + 3

40. 5,000 + 700 + 90 + 9

41. 20,000 + 7,000 + 30 + 5

42. 40,000 + 8,000 + 40 + 9

43. 70,000 + 900 + 7

44. 90,000 + 700 + 4

45. 100,000 + 80,000 + 600 + 40 +3

46. 700,000 + 40,000 + 7,000 + 50 + 7

47. 400,000 + 6,000 + 8

48. 700,000 + 40 + 5

C Write each of the following numbers in words.

49. 29

50. 75

51. 40

52. 90

53. 573

54. 895

55. 707

56. 405

57. 770

58. 450

59. 23,540

60. 56,708

61. 3,004

62. 5,008

63. 3,040

64. 5,080

65. 104,065,780

66. 637,008,500

67. 5,003,040,008

68. 7,050,800,001

69. 2,546,731

70. 6,998,454

71. 20,432,000

72. 300,508,100

Write each of the following numbers with digits instead of words.

73. Three hundred twenty-five

74. Forty-eight

75. Five thousand, four hundred thirty-two

76. One hundred twenty-three thousand, sixty-one

77. Eighty-six thousand, seven hundred sixty-two

78. One hundred million, two hundred thousand, three hundred

79. Two million, two hundred

80. Two million, two

81. Two million, two thousand, two hundred

82. Two billion, two hundred thousand, two hundred two

Applying the Concepts

83. Hot Air Balloon The first successful crossing of the Atlantic in a hot air balloon was made in August 1978 by Maxie Anderson, Ben Abruzzo, and Larry Newman of the United States. The 3,100-mile trip took approximately 140 hours. What is the place value of the 3 in the distance covered by the balloon?

84. Seating Arrangements The number of different ways in which 10 people can be seated at a table with 10 places is 3,628,800. What is the place value of the 3 in this number?

85. Seating Capacity The Rose Bowl, in Pasadena, CA, has a seating capacity of 106,721. Write this number in expanded form.

86. Astronomy The average distance from the sun to the earth is 92,897,416 miles. Write this number in expanded form.

87. Baseball Salaries Salary studies by mlbfanhouse.com show Major League Baseball's 2010 average player salary was $3,014,572, representing an increase of 0.6% from the previous season's average. Write 3,014,572 in words.

88. Top Disney Movies The illustration shows the average income for some of Disney's top grossing movies.

TOP GROSSING ANIMATED DISNEY MOVIES
www.destican.com

Snow White and the Seven Dwarfs (*1937*)	$782,620,000
101 Dalmatians (*1961*)	$717,405,900
Fantasia (*1941*)	$596,252,200
The Lion King (*1994*)	$554,524,300
The Jungle Book (*1967*)	$529,021,800

From the chart, write the following gross incomes in words.

a. Snow White and the Seven Dwarfs

b. The Jungle Book

Extending the Concepts

Many of the problem sets in this book will end with a few problems like the ones below. These problems challenge you to extend your knowledge of the material in the problem set. In most cases, there are no examples in the text similar to these problems. You should approach these problems with a positive point of view; even though you may not always work them correctly, just the process of attempting them will increase your knowledge and ability in mathematics.

The numbers on the number line below are each 1 inch apart. As with all number lines, the arrow indicates that the number line continues to the right indefinitely. Use this number line to answer Problems 91–100.

89. How far apart are the numbers 1 and 4?

90. How far apart are the numbers 2 and 5?

91. How far apart are the numbers 0 and 5?

92. How far apart are the numbers 0 and 10?

93. What number is 4 inches to the right of 2?

94. What number is 4 inches to the right of 3?

95. What number is 4 inches to the left of 7?

96. What number is 4 inches to the left of 12?

97. If 1 foot is 12 inches in length, what number is 1 foot from 0?

98. What number is 2 feet from 0?

Find the Mistake

Each sentence below contains a mistake. Circle the mistake and write the correct word(s) or numbers(s) on the line provided.

1. The place value of the 7 in the number 562,472 is hundreds. _____

2. The number 12,789 written in expanded form is 1,200 + 700 + 80 + 9. _____

3. The number 9,023,627,003 written in words is nine and twenty-three million, six and twenty-seven thousand, three.

4. Writing forty million, three hundred forty-eight thousand, thirteen in digits gives 4,034,813. _____

OBJECTIVES

A Solve addition problems without carrying.

B Solve addition problems involving carrying.

C Use the properties of addition.

D Solve addition equations.

E Solve perimeter problems.

KEY WORDS

carrying

sum

addition property of 0

commutative property of addition

associative property of addition

equation

variable

solution

inspection

polygon

perimeter

Video Examples
Section 0.2

Note Table 1 is a summary of the addition facts that you *must* know in order to make a successful start in your study of basic mathematics. You *must* know how to add any pair of numbers that come from the list. You *must* be fast and accurate. You don't want to have to think about the answer to 7 + 9. You should know it's 16. Memorize these facts now. Don't put it off until later.

When it comes to taking care of your houseplants, there is no more need to guess when they need water. A new device measures the water content of the plant's soil and sends a text message to the owner. The message can notify the owner of her thirsty plant, or even if she has overwatered her plant. Each device sells for $99. If we wanted to buy the device for two house plants, we would need to add 99 and 99 to find how much we will pay (before sales tax). In this section, we will learn how to do addition with whole numbers, such as adding 99 and 99. Let's begin by visualizing addition on the number line.

Facts of Addition

Using lengths to visualize addition can be very helpful. In mathematics, we generally do so by using the number line. For example, we add 3 and 5 on the number line like this: Start at 0 and move to 3, as shown in Figure 1. From 3, move 5 more units to the right. This brings us to 8. Therefore, $3 + 5 = 8$.

FIGURE 1

If we do this kind of addition on the number line with all combinations of the numbers 0 through 9, we get the results summarized in Table 1.

We call the information in Table 1 our basic addition facts. Your success with the examples and problems in this section depends on knowing the basic addition facts.

Table 1

Addition Table

	0	1	2	3	4	5	6	7	8	9
0	0	1	2	3	4	5	6	7	8	9
1	1	2	3	4	5	6	7	8	9	10
2	2	3	4	5	6	7	8	9	10	11
3	3	4	5	6	7	8	9	10	11	12
4	4	5	6	7	8	9	10	11	12	13
5	5	6	7	8	9	10	11	12	13	14
6	6	7	8	9	10	11	12	13	14	15
7	7	8	9	10	11	12	13	14	15	16
8	8	9	10	11	12	13	14	15	16	17
9	9	10	11	12	13	14	15	16	17	18

Suppose we want to use the Table 1 to find the answer to $3 + 5$. We locate the 3 in the column on the left and the 5 in the row at the top. We read *across* from the 3 and *down* from the 5. The entry in the table that is across from 3 and below 5 is 8.

A Adding Whole Numbers

To add whole numbers, we add digits with the same place value. First we add the digits in the ones place, then the tens place, then the hundreds place, and so on.

EXAMPLE 1 Add: $43 + 52$.

Solution This type of addition is best done vertically—aligning the digits with the same place value. First, we add the digits in the ones place.

$$\begin{array}{r} 43 \\ + 52 \\ \hline 5 \end{array}$$

Then we add the digits in the tens place.

$$\begin{array}{r} 43 \\ + 52 \\ \hline 95 \end{array}$$

EXAMPLE 2 Add: $165 + 801$.

Solution Writing the sum vertically, we have

$$\begin{array}{r} 165 \\ + 801 \\ \hline 966 \end{array}$$

← ———— Add ones place.

———— Add tens place.

———— Add hundreds place.

B Addition with Carrying

In Examples 1 and 2, the sums of the digits with the same place value were always 9 or less. There are many times when the sum of the digits with the same place value will be a number larger than 9. In these cases, we have to do what is called *carrying* in addition. The following examples illustrate this process.

EXAMPLE 3 Add: $197 + 213 + 324$.

Solution We write the sum vertically and add digits with the same place value.

$$\begin{array}{r} \overset{1}{197} \\ 213 \\ + 324 \\ \hline 4 \end{array}$$

When we add the ones, we get $7 + 3 + 4 = 14$. We write the 4 and carry the 1 to the tens column.

$$\begin{array}{r} \overset{1\,1}{197} \\ 213 \\ + 324 \\ \hline 34 \end{array}$$

We add the tens, including the 1 that was carried over from the last step. We get 13, so we write the 3 and carry the 1 to the hundreds column.

$$\begin{array}{r} \overset{1\,1}{197} \\ 213 \\ + 324 \\ \hline 734 \end{array}$$

We add the hundreds, including the 1 that was carried over from the last step.

Practice Problems

1. Add: $23 + 45$.

> *Note* To show *why* we add digits with the same place value, we can write each number showing the place value of the digits:
>
> $$\begin{array}{r} 43 = 4 \text{ tens} + 3 \text{ ones} \\ + 52 = 5 \text{ tens} + 2 \text{ ones} \\ \hline 9 \text{ tens} + 5 \text{ ones} \end{array}$$

2. Add: $243 + 526$.

3. Add: $238 + 432 + 186$.

Answers

1. 68
2. 769
3. 856

4. Add: 17,386 + 5,978 + 468.

EXAMPLE 4 Add: 46,789 + 2,490 + 864.

Solution We write the sum vertically and then use the shorthand form of addition.

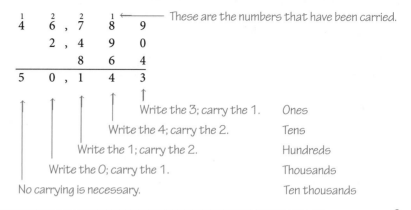

Adding numbers as we are doing here takes some practice. Most people don't make mistakes in carrying. Most mistakes in addition are made in adding the numbers in the columns. That is why it is so important that you maintain accuracy with the basic addition facts given in this chapter.

Vocabulary

The word we use to indicate addition is *sum*. If we say "the sum of 3 and 5 is 8," what we mean is $3 + 5 = 8$. The word sum always indicates addition. We can state this fact in symbols by using the letters a and b to represent numbers.

> **DEFINITION** sum
>
> If a and b are any two numbers, then the *sum* of a and b is $a + b$. To find the sum of two numbers, we add them.

Note When mathematics is used to solve everyday problems, the problems are almost always stated in words. The translation of English to symbols is a very important part of mathematics.

Table 2 gives some phrases and sentences in English and their mathematical equivalents written in symbols.

Table 2

In English	In Symbols
The sum of 4 and 1	$4 + 1$
4 added to 1	$1 + 4$
8 more than m	$m + 8$
x increased by 5	$x + 5$
The sum of x and y	$x + y$
The sum of 2 and 4 is 6.	$2 + 4 = 6$

Answer

4. 23,832

C Properties of Addition

Once we become familiar with addition, we may notice some facts about addition that are true regardless of the numbers involved. The first of these facts involves the number 0 (zero).

Whenever we add 0 to a number, the result is the original number. For example,

$$7 + 0 = 7 \quad \text{and} \quad 0 + 3 = 3$$

Because this fact is true no matter what number we add to 0, we call it a property of 0

> **PROPERTY** Addition Property of 0
>
> If we let a represent any number, then it is always true that
>
> $$a + 0 = a \quad \text{and} \quad 0 + a = a$$
>
> *In words*: Adding 0 to any number leaves that number unchanged.

A second property we notice by becoming familiar with addition is that the order of two numbers in a sum can be changed without changing the result.

$$3 + 5 = 8 \quad \text{and} \quad 5 + 3 = 8$$
$$4 + 9 = 13 \quad \text{and} \quad 9 + 4 = 13$$

This fact about addition is true for *all* numbers. The order in which you add two numbers doesn't affect the result. We call this fact the *commutative property of addition*, and we write it in symbols as follows:

> **PROPERTY** Commutative Property of Addition
>
> If a and b are any two numbers, then it is always true that
>
> $$a + b = b + a$$
>
> *In words*: Changing the order of two numbers in a sum doesn't change the result.

EXAMPLE 5 Use the commutative property of addition to rewrite each sum.

a. $4 + 6$ **b.** $5 + 9$ **c.** $3 + 0$ **d.** $7 + n$

Solution The commutative property of addition indicates that we can change the order of the numbers in a sum without changing the result. Applying this property we have

a. $4 + 6 = 6 + 4$
b. $5 + 9 = 9 + 5$
c. $3 + 0 = 0 + 3$
d. $7 + n = n + 7$

Notice that we did not actually add any of the numbers. The instructions were to use the commutative property, and the commutative property involves only the order of the numbers in a sum.

The last property of addition we will consider here has to do with sums of more than two numbers. Suppose we want to find the sum of 2, 3, and 4. We could add 2 and 3 first, and then add 4 to what we get.

$$(2 + 3) + 4 = 5 + 4 = 9$$

Or, we could add the 3 and 4 together first and then add the 2.

$$2 + (3 + 4) = 2 + 7 = 9$$

The result in both cases is the same. If we try this with any other numbers, the same thing happens.

Note When we use letters to represent numbers, as we do when we say, "If a and b are any two numbers," then a and b are called variables, because the values they take on vary. We use the variables a and b in the definitions and properties on this page because we want you to know that the definitions and properties are true for all numbers that you will encounter in this book.

5. Use the commutative property of addition to rewrite each sum.
a. $5 + 9$
b. $12 + 13$
c. $17 + 0$
d. $x + 3$

Note This discussion is here to show why we write the next property the way we do. Sometimes it is helpful to look ahead to the property itself (in this case, the associative property of addition) to see what it is that is being justified.

Answers
5. a. $9 + 5$ **b.** $13 + 12$
c. $0 + 17$ **d.** $3 + x$

We call this fact about addition the *associative property of addition*, and we write it in symbols as follows:

> **PROPERTY** Associative Property of Addition
>
> If a, b, and c represent any three numbers, then
>
> $$(a + b) + c = a + (b + c)$$
>
> *In words*: Changing the grouping of three or more numbers in a sum doesn't change the result.

6. Use the associative property of addition to rewrite each sum.
 a. $(2 + 5) + 6$
 b. $(x + 3) + 5$
 c. $5 + (4 + 7)$
 d. $11 + (8 + n)$

EXAMPLE 6 Use the associative property of addition to rewrite each sum.

a. $(5 + 6) + 7$ **b.** $(3 + 9) + 1$ **c.** $6 + (8 + 2)$ **d.** $4 + (9 + n)$

Solution The associative property of addition indicates that we are free to regroup the numbers in a sum without changing the result.

a. $(5 + 6) + 7 = 5 + (6 + 7)$

b. $(3 + 9) + 1 = 3 + (9 + 1)$

c. $6 + (8 + 2) = (6 + 8) + 2$

d. $4 + (9 + n) = (4 + 9) + n$

The commutative and associative properties of addition tell us that when adding whole numbers, we can use any order and grouping. When adding several numbers, it is sometimes easier to look for pairs of numbers whose sums are 10, 20, and so on.

7. Add: $8 + 7 + 2 + 3 + 6$.

EXAMPLE 7 Add: $9 + 3 + 2 + 7 + 1$

Solution

$$9 + 3 + 2 + 7 + 1 = (9 + 1) + (3 + 7) + 2 \quad \textit{Associative property of addition}$$
$$= 10 + 10 + 2$$
$$= 22$$

D Solving Equations

We can use the addition table to help solve some simple equations. If n is used to represent a number, then the equation

$$n + 3 = 5$$

will be true if n is 2. The number 2 is therefore called a *solution* to the equation, because, when we replace n with 2, the equation becomes a true statement.

$$2 + 3 = 5$$

Note The letter n as we are using it here is a variable that is a solution to an equation. We will do more work with equations later in the book.

Equations like this are really just puzzles, or questions. When we say, "Solve the equation $n + 3 = 5$," we are asking the question, "What number do we add to 3 to get 5?"

When we solve equations by reading the equation to ourselves and then stating the solution, as we did with the equation above, we are solving the equation by *inspection*.

8. Find the solution to each equation by inspection.
 a. $n + 4 = 9$
 b. $n + 7 = 15$
 c. $3 + n = 12$
 d. $15 = n + 6$

EXAMPLE 8 Find the solution to each equation by inspection.

a. $n + 5 = 9$ **b.** $n + 6 = 12$ **c.** $4 + n = 5$ **d.** $13 = n + 8$

Solution We find the solution to each equation by using the addition facts given in Table 1.

a. The solution to $n + 5 = 9$ is 4 because $4 + 5 = 9$.

b. The solution to $n + 6 = 12$ is 6 because $6 + 6 = 12$.

c. The solution to $4 + n = 5$ is 1 because $4 + 1 = 5$.

d. The solution to $13 = n + 8$ is 5 because $13 = 5 + 8$.

Answers

6. a. $2 + (5 + 6)$
 b. $x + (3 + 5)$
 c. $(5 + 4) + 7$
 d. $(11 + 8) + n$
7. 26
8. a. 5 **b.** 8 **c.** 9 **d.** 9

E Perimeter

We end this section with an introduction to perimeter. Let's start with the definition of a polygon.

> **DEFINITION** polygon
>
> A *polygon* is a closed geometric figure, with at least three sides, in which each side is a straight line segment.

> **FACTS FROM GEOMETRY** Perimeter
>
> The most common polygons are squares, rectangles, and triangles. Examples of these are shown in Figure 2.
>
>
>
> FIGURE 2
>
> In the square, *s* is the length of the side, and each side has the same length. In the rectangle, *l* stands for the length, and *w* stands for the width. The width is usually the lesser of the two. In the triangle, *h* stands for height.

> **DEFINITION** perimeter
>
> The *perimeter* of any polygon is the sum of the lengths of the sides, and it is denoted with the letter P.

To find the perimeter of a polygon, we add all the lengths of the sides together.

EXAMPLE 9 Find the perimeter of each geometric figure.

Solution In each case, we find the perimeter by adding the lengths of all the sides.

a. The figure is a square. Because the length of each side in the square is the same, the perimeter is

$$P = 15 + 15 + 15 + 15 = 60 \text{ inches}$$

b. In the rectangle, two of the sides are 24 feet long, and the other two are 37 feet long. The perimeter is the sum of the lengths of the sides.

$$P = 24 + 24 + 37 + 37 = 122 \text{ feet}$$

c. For this polygon, we add the lengths of the sides together. The result is the perimeter.

$$P = 36 + 23 + 24 + 12 + 24 = 119 \text{ yards}$$

9. Find the perimeter of each geometric figure.

a.

b.

c.

USING TECHNOLOGY Calculators

From time to time we will include some notes like this one, which show how a calculator can be used to assist us with some of the calculations in the book. Most calculators on the market today fall into one of two categories: those with algebraic logic and those with function logic. Calculators with algebraic logic have a key with an equals sign on it. Calculators with function logic do not have an equals key. Instead they have a key labeled ENTER or EXE (for execute). Scientific calculators use algebraic logic, and graphing calculators, such as the TI-83/84, use function logic.

Here are the sequences of keystrokes to use to work the problem shown in Part c of Example 9.

Scientific Calculator: 36 $\boxed{+}$ 23 $\boxed{+}$ 24 $\boxed{+}$ 12 $\boxed{+}$ 24 $\boxed{=}$

Graphing Calculator: 36 $\boxed{+}$ 23 $\boxed{+}$ 24 $\boxed{+}$ 12 $\boxed{+}$ 24 $\boxed{\text{ENT}}$

GETTING READY FOR CLASS

After reading through the preceding section, respond in your own words and in complete sentences.

A. What number is the sum of 6 and 8?

B. Make up an addition problem using the number 456 that involves carrying from the ones column to the tens column only.

C. Explain the difference between the commutative property of addition and the associative property of addition.

D. What is the perimeter of a polygon?

Vocabulary Review

Choose the correct words to fill in the blanks below.

sum	associative	inspection	place value	carrying
commutative	addition	solution	perimeter	zero

1. Learning the basic _____ facts is essential to your study of mathematics.

2. To add whole numbers, add digits of the same _____.

3. If the sum of the digits with the same place value is larger than 9, use the process called _____ to complete the problem.

4. The _____ of a and b is $a + b$ if a and b are any two numbers.

5. The addition property of _____ states that adding zero to any number leaves that number unchanged.

6. The _____ property of addition states that changing the order of two numbers in a sum doesn't change the result.

7. The _____ property of addition states that changing the grouping of three numbers in a sum doesn't change the result.

8. When we solve equations by reading the equation to ourselves and then stating the _____, we are solving the equation by _____.

9. To find the _____ of a polygon, we add all the lengths of the sides together.

Problems

A Find each of the following sums. (Add.)

1. $3 + 5 + 7$　　　　**2.** $2 + 8 + 6$　　　　**3.** $1 + 4 + 9$　　　　**4.** $2 + 8 + 3$

5. $5 + 9 + 4 + 6$　　**6.** $8 + 1 + 6 + 2$　　**7.** $1 + 2 + 3 + 4 + 5$　　**8.** $5 + 6 + 7 + 8 + 9$

9. $9 + 1 + 8 + 2$　　**10.** $7 + 3 + 6 + 4$

Add each of the following. (There is no carrying involved in these problems.)

11. $\begin{array}{r} 43 \\ +25 \\ \hline \end{array}$　　　　**12.** $\begin{array}{r} 56 \\ +23 \\ \hline \end{array}$　　　　**13.** $\begin{array}{r} 81 \\ +17 \\ \hline \end{array}$

14. $\begin{array}{r} 37 \\ +22 \\ \hline \end{array}$　　　　**15.** $\begin{array}{r} 4,281 \\ +3,016 \\ \hline \end{array}$　　　　**16.** $\begin{array}{r} 2,749 \\ +1,250 \\ \hline \end{array}$

17. 3,482
 + 3,005

18. 2,496
 +7,503

19. 32
 21
 +43

20. 521
 340
 +135

21. 6,245
 203
 + 1,001

22. 27
 4,510
 + 342

B Add each of the following. (All problems involve carrying in at least one column.)

23. 49
 +16

24. 85
 +29

25. 74
 +28

26. 36
 +46

27. 682
 +193

28. 439
 +270

29. 638
 +191

30. 444
 +595

31. 4,963
 +5,428

32. 8,291
 +7,489

33. 6,205
 +9,999

34. 8,888
 +9,999

35. 56,789
 +98,765

36. 45,678
 +87,654

37. 52,468
 +58,642

38. 13,579
 +97,531

39. 4,296
 8,720
 +4,375

40. 5,637
 481
 +7,899

41. 4,994
 449
 +9,449

42. 6,824
 371
 +4,857

43. 12
 34
 56
 +78

44. 21
 43
 65
 +87

45. 999
 444
 555
 +222

46. 646
 464
 525
 +252

47. 9,245
 672
 8,341
 + 27

48. 45
 9,876
 54
 + 6,789

49. 123
 469
 87
 + 95

50. 835
 84
 107
 + 98

51. 5,894
 256
 + 1,045

52. 8,265
 648
 + 2,384

Complete the following tables.

53.

First Number a	Second Number b	Their Sum a + b
61	38	
63	36	
65	34	
67	32	

54.

First Number a	Second Number b	Their Sum a + b
10	45	
20	35	
30	25	
40	15	

55.

First Number a	Second Number b	Their Sum a + b
9	16	
36	64	
81	144	
144	256	

56.

First Number a	Second Number b	Their Sum a + b
25	75	
24	76	
23	77	
22	78	

C Rewrite each of the following using the commutative property of addition.

57. $5 + 9$ **58.** $2 + 1$ **59.** $3 + 8$ **60.** $9 + 2$ **61.** $6 + 4$ **62.** $1 + 7$

Rewrite each of the following using the associative property of addition.

63. $(1 + 2) + 3$ **64.** $(4 + 5) + 9$ **65.** $(2 + 1) + 6$ **66.** $(2 + 3) + 8$

67. $1 + (9 + 1)$ **68.** $2 + (8 + 2)$ **69.** $(4 + n) + 1$ **70.** $(n + 8) + 1$

D Find a solution for each equation by inspection.

71. $n + 6 = 10$ **72.** $n + 4 = 7$ **73.** $n + 8 = 13$ **74.** $n + 6 = 15$

75. $4 + n = 12$ **76.** $5 + n = 7$ **77.** $17 = n + 9$ **78.** $13 = n + 5$

79. $15 = n + 6$ **80.** $23 = n + 14$ **81.** $16 = 12 + n$ **82.** $24 = 16 + n$

Write each of the following expressions in words. Use the word *sum* in each case.

83. 4 + 9

84. 9 + 4

85. 8 + 1

86. 9 + 9

87. 2 + 3 = 5

88. 8 + 2 = 10

Write each of the following in symbols.

89. a. The sum of 5 and 2

b. 3 added to 8

90. a. The sum of a and 4

b. 6 more than x

91. a. m increased by 1

b. The sum of m and n

92. a. The sum of 4 and 8 is 12.

b. The sum of a and b is 6.

E Find the perimeter of each figure. (Note that we have abbreviated the units on each figure to save space. The abbreviation for feet is ft, inches is in., and yards is yd.) The first four figures are squares.

93.

3 in.

94.

9 in.

95.

4 ft

96.

2 ft

97.

3 yd

10 yd

98.

1 yd

5 yd

99.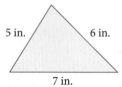

5 in. 6 in.

7 in.

100.

4 in. 10 in.

12 in.

Applying the Concepts

The application problems that follow are related to addition of whole numbers. Read each problem carefully to determine exactly what you are being asked to find. Don't assume that just because a number appears in a problem you have to use it to solve the problem. Sometimes you do, and sometimes you don't.

101. Gallons of Gasoline Tim bought gas for his economy car twice last month. The first time he bought 18 gallons and the second time he bought 16 gallons. What was the total amount of gasoline Tim bought last month?

102. Tallest Mountain The world's tallest mountain is Mount Everest. On May 5, 1999, it was found to be 7 feet taller than it was previously thought to be. Before this date, Everest was thought to be 29,028 feet high. That height was determined by B. L. Gulatee in 1954. What is the current height of Mount Everest?

103. Checkbook Balance On Monday, Bob had a balance of $241 in his checkbook. On Tuesday, he made a deposit of $108, and on Thursday, he wrote a check for $24. What was the balance in his checkbook on Wednesday?

104. Number of Passengers A plane flying from Los Angeles to New York left Los Angeles with 67 passengers on board. The plane stopped in Bakersfield and picked up 28 passengers, and then it stopped again in Dallas where 57 more passengers came on board. How many passengers were on the plane when it landed in New York?

Check No.	Date	Description of Transaction	Payment/Debit (-)	Deposit/Credit (+)	Balance
					$241 00
	6/5	ATM Deposit		$108 00	?
1251	6/7	Electric Bill	$24 00		

Find the Mistake

Each sentence below contains a mistake. Circle the mistake and write the correct word(s) or numbers(s) on the line provided.

1. To find the sum of 786 and 49, add the ones place by writing the 1 and carrying the 5. _____

2. The problem $(12 + 7) + 3 = 12 + (7 + 3)$ uses the commutative property of addition. _____

3. The solution for the equation $47 = x + 29$ is $x = 8$. _____

4. The perimeter of a rectangle with a length of 35 inches and a width of 15 inches is 50 inches. _____

A Using vocabulary for subtraction.

B Subtracting whole numbers without borrowing.

C Subtract whole numbers with borrowing.

D Solve subtraction equations.

difference

borrowing

**Video Examples
Section 0.3**

0.3 Subtraction with Whole Numbers

Image © sxc.hu, dafalias, 2005

The Excalibur Hotel and Casino in Las Vegas, Nevada has a new attraction called the Virtusphere. The plastic sphere is ten feet in diameter and spins on 45 small casters embedded in its frame. A person wearing knee and elbow pads and a virtual gaming headset encloses himself in the sphere. As a video game plays on the headset's screen, the person can interact with the game by walking, running, and rolling without leaving the safety of the sphere.

Let's say the person paid to play a video game inside the Virtusphere for 20 minutes. If the person played for 13 minutes, how many minutes does he have left? To answer this question, we need to understand a little more about subtraction.

Subtraction is the opposite operation of addition. If you understand addition and can work simple addition problems quickly and accurately, then subtraction shouldn't be difficult for you.

A Vocabulary

The word *difference* always indicates subtraction. We can state this in symbols by letting the letters *a* and *b* represent numbers.

> **DEFINITION** difference
>
> The **difference** of two numbers *a* and *b* is
>
> $$a - b$$

Table 1 gives some word statements involving subtraction and their mathematical equivalents written in symbols.

Table 1

In English	In Symbols
The difference of 9 and 1	$9 - 1$
The difference of 1 and 9	$1 - 9$
The difference of *m* and 4	$m - 4$
The difference of *x* and *y*	$x - y$
3 subtracted from 8	$8 - 3$
2 subtracted from *t*	$t - 2$
The difference of 7 and 4 is 3.	$7 - 4 = 3$
The difference of 9 and 3 is 6.	$9 - 3 = 6$

B The Meaning of Subtraction

When we want to subtract 3 from 8, we write

$$8 - 3, \qquad 8 \text{ subtract } 3, \qquad \text{or} \qquad 8 \text{ minus } 3$$

The number we are looking for here is the difference between 8 and 3, or the number we add to 3 to get 8; that is,

$$8 - 3 = ? \qquad \text{is the same as} \qquad ? + 3 = 8$$

In both cases, we are looking for the number we add to 3 to get 8. The number we are looking for is 5. We have two ways to write the same statement:

Subtraction		Addition
$8 - 3 = 5$	or	$5 + 3 = 8$

For every subtraction problem, there is an equivalent addition problem. Table 2 lists some examples.

Table 2

Subtraction		Addition
$7 - 3 = 4$	because	$4 + 3 = 7$
$9 - 7 = 2$	because	$2 + 7 = 9$
$10 - 4 = 6$	because	$6 + 4 = 10$
$15 - 8 = 7$	because	$7 + 8 = 15$

To subtract numbers with two or more digits, we align the numbers vertically and subtract in columns.

EXAMPLE 1 Subtract: $376 - 241$.

Solution We write the problem vertically, aligning digits with the same place value. Then we subtract in columns.

$$
\begin{array}{r}
376 \\
- \ 241 \\
\hline
135
\end{array}
$$
← Subtract the bottom number in each column from the number above it.

We can visualize Example 1 using money.

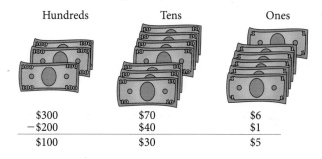

Hundreds	Tens	Ones
$300	$70	$6
−$200	$40	$1
$100	$30	$5

EXAMPLE 2 Subtract 503 from 7,835.

Solution In symbols, this statement is equivalent to

$$7,835 - 503$$

To subtract, we write 503 below 7,835 and then subtract in columns.

$$
\begin{array}{r}
7,835 \\
- \ \ \ 503 \\
\hline
7,332
\end{array}
$$

$5 - 3 = 2$ Ones
$3 - 0 = 3$ Tens
$8 - 5 = 3$ Hundreds
$7 - 0 = 7$ Thousands

Practice Problems

1. Subtract: $485 - 123$.

2. Subtract 320 from 1,865.

Answers

1. 362
2. 1,545

As you can see, subtraction problems like the ones in Examples 1 and 2 are fairly simple. We write the problem vertically, lining up the digits with the same place value, and subtract in columns. We always subtract the bottom number from the top number.

C Subtraction with Borrowing

Subtraction must involve *borrowing* when the bottom digit in any column is larger than the digit above it. In one sense, borrowing is the reverse of the carrying we did in addition.

3. Subtract: $87 - 29$.

EXAMPLE 3 Subtract: $92 - 45$.

Solution We write the problem vertically with the place values of the digits showing.

$$92 = 9 \text{ tens} + 2 \text{ ones}$$
$$- \ 45 = 4 \text{ tens} + 5 \text{ ones}$$

Look at the ones column. We cannot subtract immediately, because 5 is larger than 2. Instead, we borrow 1 ten from the 9 tens in the tens column. We can rewrite the number 92 as

$$9 \text{ tens} + 2 \text{ ones}$$
$$= 8 \text{ tens} + 1 \text{ ten} + 2 \text{ ones}$$
$$= 8 \text{ tens} + 12 \text{ ones}$$

Note The discussion here shows why borrowing is necessary and how we go about it. To understand borrowing, you should pay close attention to this discussion.

Now we are in a position to subtract.

$$92 = 9 \text{ tens} + 2 \text{ ones} = 8 \text{ tens} + 12 \text{ ones}$$
$$- \ 45 = 4 \text{ tens} + 5 \text{ ones} = 4 \text{ tens} + \ \ 5 \text{ ones}$$
$$4 \text{ tens} + \ \ 7 \text{ ones}$$

The result is 4 tens + 7 ones, which can be written in standard form as 47.

Writing the problem out in this way is more trouble than is actually necessary. The shorthand form of the same problem looks like this:

$$\overset{8}{\cancel{9}} \ \overset{12}{\cancel{2}}$$
$$- \ 4 \ 5$$
$$\overline{4 \ 7}$$

⟵ This shows we have borrowed 1 ten to go with the 2 ones.

$12 - 5 = 7$ *Ones*

$8 - 4 = 4$ *Tens*

This shortcut form shows all the necessary work involved in subtraction with borrowing. We will use it from now on.

The borrowing that changed 9 tens + 2 ones into 8 tens + 12 ones can be visualized with money.

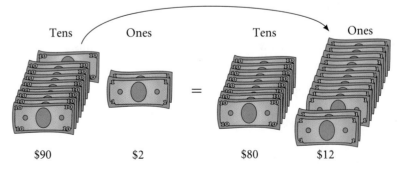

Tens	Ones		Tens	Ones
$90	$2	=	$80	$12

EXAMPLE 4 Find the difference of 549 and 187.

Solution In symbols, the difference of 549 and 187 is written

$$549 - 187$$

Writing the problem vertically so that the digits with the same place value are aligned, we have

$$
\begin{array}{r}
549 \\
- \, 187 \\
\end{array}
$$

The top number in the tens column is smaller than the number below it. This means that we will have to borrow from the next larger column.

$$
\begin{array}{r}
\overset{4}{\cancel{5}}\,\overset{14}{\cancel{4}}\,9 \\
- \, 1 \ 8 \ 7 \\
\hline
3 \ 6 \ 2 \\
\end{array}
$$

← Borrow 1 hundred to go with the 4 tens.

$9 - 7 = 2$ Ones
$14 - 8 = 6$ Tens
$4 - 1 = 3$ Hundreds

The actual work we did in borrowing looks like this:

5 hundreds + 4 tens + 9 ones

= 4 hundreds + 1 hundred + 4 tens + 9 ones

= 4 hundreds + 14 tens + 9 ones

EXAMPLE 5 Jo Ann has $742 in her checking account. If she writes a check for $615 to pay the rent, how much is left in her checking account?

Check No.	Date	Description of Transaction	Payment/Debit (-)	Deposit/Credit (+)	Balance
					$742 00
1630	5/3	Rent	$615 00		?

Solution To find the amount left in the account after she has written the rent check, we subtract.

$$
\begin{array}{r}
\$7 \,\overset{3}{\cancel{4}}\,\overset{12}{\cancel{2}} \\
- \ 6 \ 1 \ 5 \\
\hline
\$1 \ 2 \ 7 \\
\end{array}
$$

She has $127 left in her account after writing a check for the rent.

USING TECHNOLOGY Calculators

Here is how we would work the problem shown in Example 5 on a calculator:

Scientific Calculator: 742 ⊟ 615 ⊟

Graphing Calculator: 742 ⊟ 615 [ENT]

4. Find the difference of 268 and 179.

5. Joshua has $615 in his checking account. If he writes a check for $289 to make his car payment, how much is left?

Answers
4. 89
5. $326

D Solving Equations

We can solve equations involving subtraction as we did with addition, by inspection. For example, the equation

$$n - 4 = 8$$

has a solution of 12, since $12 - 4 = 8$. Later, we will learn other methods for solving these equations.

EXAMPLE 6 Find the solution to each equation by inspection.

a. $n - 3 = 7$ **b.** $n - 4 = 9$ **c.** $12 - n = 5$ **d.** $7 = 18 - n$

Solution

a. The solution to $n - 3 = 7$ is 10, because $10 - 3 = 7$.

b. The solution to $n - 4 = 9$ is 13, because $13 - 4 = 9$.

c. The solution to $12 - n = 5$ is 7, because $12 - 7 = 5$.

d. The solution to $7 = 18 - n$ is 11, because $7 = 18 - 11$.

6. Find the solution to each equation by inspection.
 a. $n - 5 = 9$
 b. $n - 1 = 7$
 c. $19 - n = 15$
 d. $4 = 12 - n$

Estimating

One way to estimate the answer to the problem shown in Example 5 is to round 742 to 700 and 615 to 600 and then subtract 600 from 700 to obtain 100, which is an estimate of the difference. Making a mental estimate in this manner will help you catch some of the errors that will occur if you press the wrong buttons on your calculator. We will cover estimating and rounding more thoroughly in the next section.

GETTING READY FOR CLASS

After reading through the preceding section, respond in your own words and in complete sentences.

A. Which sentence below describes the problem in Example 1?

 a. The difference of 241 and 376 is 135.

 b. The difference of 376 and 241 is 135.

B. Write a subtraction problem using the number 234 that involves borrowing from the tens column to the ones column.

C. Write a subtraction problem using the number 234 in which the answer is 111.

D. Describe how you would subtract the number 56 from the number 93.

Answers

6. a. $n = 14$ **b.** $n = 8$
 c. $n = 4$ **d.** $n = 8$

Vocabulary Review

Choose the correct words to fill in the blanks below.

columns borrowing difference estimation

1. The _____ of two numbers a and b is $a - b$, and is the result of subtraction.

2. To subtract numbers with two or more digits, we align the number vertically and subtract in _____.

3. Subtraction involves _____ when the bottom digit in any column is larger than the digit above it.

4. _____ by rounding is a good way to catch errors.

Problems

A Perform the indicated operation.

1. Subtract 24 from 56.

2. Subtract 71 from 89.

3. Subtract 23 from 45.

4. Subtract 97 from 98.

5. Find the difference of 29 and 19.

6. Find the difference of 37 and 27.

7. Find the difference of 126 and 15.

8. Find the difference of 348 and 32.

Write each of the following expressions in words. Use the word *difference* in each case.

9. $10 - 2$

10. $9 - 5$

11. $a - 6$

12. $7 - x$

13. $8 - 2 = 6$

14. $m - 1 = 4$

Write each of the following expressions in symbols.

15. The difference of 8 and 3

16. The difference of x and 2

17. 9 subtracted from y

18. a subtracted from b

19. The difference of 3 and 2 is 1.

20. The difference of 10 and y is 5.

B Work each of the following subtraction problems.

21.
$$\begin{array}{r} 975 \\ -\ 663 \\ \hline \end{array}$$

22.
$$\begin{array}{r} 480 \\ -\ 260 \\ \hline \end{array}$$

23.
$$\begin{array}{r} 904 \\ -\ 501 \\ \hline \end{array}$$

24.
$$\begin{array}{r} 657 \\ -\ 507 \\ \hline \end{array}$$

25.
$$\begin{array}{r} 9,876 \\ -\ 8,765 \\ \hline \end{array}$$

26.
$$\begin{array}{r} 5,008 \\ -\ 3,002 \\ \hline \end{array}$$

27.
$$\begin{array}{r} 7,976 \\ -\ 3,432 \\ \hline \end{array}$$

28.
$$\begin{array}{r} 6,980 \\ -\ \ \ 470 \\ \hline \end{array}$$

C Find the difference in each case. (These problems all involve borrowing.)

29. $52 - 37$

30. $65 - 48$

31. $70 - 37$

32. $90 - 21$

33. $74 - 69$

34. $31 - 28$

35. $51 - 18$

36. $64 - 58$

37. $329 - 234$

38. $518 - 492$

39. $348 - 196$

40. $759 - 661$

41. $\begin{array}{r} 932 \\ -\ 658 \\ \hline \end{array}$

42. $\begin{array}{r} 895 \\ -\ 597 \\ \hline \end{array}$

43. $\begin{array}{r} 647 \\ -\ 159 \\ \hline \end{array}$

44. $\begin{array}{r} 842 \\ -\ 199 \\ \hline \end{array}$

45. $\begin{array}{r} 905 \\ -\ 367 \\ \hline \end{array}$

46. $\begin{array}{r} 804 \\ -\ 238 \\ \hline \end{array}$

47. $\begin{array}{r} 600 \\ -\ 437 \\ \hline \end{array}$

48. $\begin{array}{r} 800 \\ -\ 342 \\ \hline \end{array}$

49. $\begin{array}{r} 4{,}583 \\ -\ 2{,}973 \\ \hline \end{array}$

50. $\begin{array}{r} 7{,}849 \\ -\ 2{,}957 \\ \hline \end{array}$

51. $\begin{array}{r} 79{,}040 \\ -\ 32{,}957 \\ \hline \end{array}$

52. $\begin{array}{r} 86{,}492 \\ -\ 78{,}506 \\ \hline \end{array}$

Complete the following tables.

53.

First Number a	Second Number b	The Difference of a and b $a - b$
25	15	
24	16	
23	17	
22	18	

54.

First Number a	Second Number b	The Difference of a and b $a - b$
90	79	
80	69	
70	59	
60	49	

55.

First Number a	Second Number b	The Difference of a and b $a - b$
400	256	
400	144	
225	144	
225	81	

56.

First Number a	Second Number b	The Difference of a and b $a - b$
100	36	
100	64	
25	16	
25	9	

D Find the solution to each equation by inspection.

57. $n - 1 = 5$

58. $n - 4 = 11$

59. $n - 5 = 14$

60. $n - 6 = 9$

61. $12 - n = 3$

62. $15 - n = 10$

63. $20 - n = 7$

64. $25 - n = 24$

65. $10 = n - 15$ **66.** $12 = n - 5$ **67.** $8 = n - 3$ **68.** $10 = n - 4$

69. $5 = 14 - n$ **70.** $4 = 20 - n$ **71.** $5 = 6 - n$ **72.** $8 = 12 - n$

Applying the Concepts

Not all of the following application problems involve only subtraction. Some involve addition as well. Be sure to read each problem carefully.

73. Checkbook Balance Diane has $504 in her checking account. If she writes five checks for a total of $249, how much does she have left in her account?

74. Checkbook Balance Larry has $763 in his checking account. If he writes a check for each of the three bills listed, how much will he have left in his account?

Item	Amount
Rent	$418
Phone	$25
Car repair	$117

75. Tallest Mountain The world's tallest mountain is Mount Everest. On May 5, 1999, it was found to be 7 feet taller than it was previously thought to be. Before this date, Everest was thought to be 29,028 feet high. That height was determined by B. L. Gulatee in 1954. The first measurement of Everest was in 1856. At that time, the height was thought to be 29,002 feet. What is the difference between the current height of Everest and the height measured in 1856?

76. Home Prices In 1985, Mr. Hicks paid $137,500 for his home. He sold it in 2000 for $260,600. What is the difference between what he sold it for and what he bought it for?

77. Enrollment Six years ago, there were 567 students attending Smith Elementary School. Today the same school has an enrollment of 399 students. How much of a decrease in enrollment has there been in the last six years at Smith School?

78. Oil Spills In March 1977, an oil tanker hit a reef off Taiwan and spilled 3,134,500 gallons of oil. In March 1989, an oil tanker hit a reef off Alaska and spilled 10,080,000 gallons of oil. How much more oil was spilled in the 1989 disaster?

Checkbook Balance On Monday, Gil has a balance of $425 in his checkbook. On Tuesday, he deposits $149 into the account. On Wednesday, he writes a check for $37, and on Friday he writes a check for $188. Use this information to answer Problems 79–82.

Check No.	Date	Description of Transaction	Payment/Debit (-)	Deposit/Credit (+)	Balance
					$425 00
	9/13	Deposit		$149 00	
1622	9/14	Water Bill	$37 00		
1622	9/16	Car Payment	$188 00		

79. Find Gil's balance after he makes the deposit on Tuesday.

80. What is his balance after he writes the check on Wednesday?

81. To the nearest ten dollars, what is his balance at the end of the week?

82. To the nearest ten dollars, what is his balance before he writes the check on Friday?

83. iPad Sales The bar chart shows the projects sales of iPads from 2010–2012.

Source: Data from emarketer.com

Year	Projected Sales (Millions)
2010	
2011	
	30.1

a. Use the information in the bar chart to fill in the missing entries in the table.

b. What is the difference in projected iPad sales between 2012 and 2010?

84. U.S. Mobile Gaming Revenues The bar chart shows the projected revenue for paid mobile games.

Source: emarketer.com

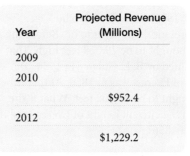

Year	Projected Revenue (Millions)
2009	
2010	
	$952.4
2012	
	$1,229.2

a. Use the chart to fill in the missing entries in the table.

b. What is the difference in projected revenue between 2012 and 2009?

Improving Your Quantitative Literacy

85. Beijing Olympic Venues The bar chart below shows the seating capacity for several Olympic venues in Beijing.

BEIJING OLYMPIC VENUE CAPACITY
en.beijing2008.cn

Venue		Capacity
The National Stadium		91,000
The National Aquatics Center		17,000
The National Indoor Stadium		18,000
Beijing Olympic Green Tennis Courts		17,400
Olympic Sport Center Stadium		38,000

Venue	Capacity
National Stadium	
National Aquatic Center	
	18,000
Beijing Olympic Green Tennis Courts	
	38,000

a. Use the information in the bar chart to fill in the missing entries in the table.

b. How many more people can attend a competition in the National Stadium than in the National Aquatic Center?

Find the Mistake

Each sentence below contains a mistake. Circle the mistake and write the correct word(s) or numbers(s) on the line provided.

1. Translating "The difference of 22 and 3 is 19" into symbols gives us $3 - 22 = 19$. _____

2. To subtract 50 from 290, vertically align the tens column of the first number with the hundreds column of the second number. _____

3. To find the difference of 85 and 27, begin by subtracting 7 from 5. _____

4. The solution for the equation $108 = 130 - y$ is $y = 238$. _____

Landmark Review: Checking Your Progress

This feature is intended as a review of key skills from the preceding sections in the chapter. Each review will give you the opportunity to utilize important concepts and check your progress as you practice problems of different types. Take note of any problems in this review that you find difficult. There is no better time than now to revisit and master those difficult concepts in preparation for any upcoming exams or subsequent chapters.

Write each of the following numbers in expanded form.

1. 549

2. 1,493

3. 60,243

4. 30,403,005

Perform the indicated operations.

5. $4 + 2 + 12$

6. $4 + 3 + 8 + 9 + 5 + 2$

7. $\begin{array}{r} 52 \\ + 65 \\ \hline \end{array}$

8. $\begin{array}{r} 2,435 \\ + 5,215 \\ \hline \end{array}$

9. $\begin{array}{r} 14,253 \\ 25,489 \\ 53,503 \\ + 10,456 \\ \hline \end{array}$

10. $\begin{array}{r} 145,358 \\ 256,789 \\ + \quad 15 \\ \hline \end{array}$

11. $\begin{array}{r} 957 \\ - 427 \\ \hline \end{array}$

12. $\begin{array}{r} 6,492 \\ - 5,257 \\ \hline \end{array}$

13. Subtract 15 from 32.

14. Subtract 5 from 27.

15. Find the difference of 37 and 25.

16. Find the difference of 142 and 40.

Find a solution for each equation by inspection.

17. $5 + x = 10$

18. $1 + n = 15$

19. $12 = 5 + a$

20. $29 = y + 9$

21. $x - 1 = 15$

22. $7 - n = 3$

23. $10 - y = 5$

24. $34 - x = 33$

KEY WORDS

rounding

estimate

bar chart

horizontal axis

vertical axis

**Video Examples
Section 0.4**

Note After you have used the steps listed here to work a few problems, you will find that the procedure becomes almost automatic.

0.4 Rounding Numbers, Estimating Answers, and Displaying Information

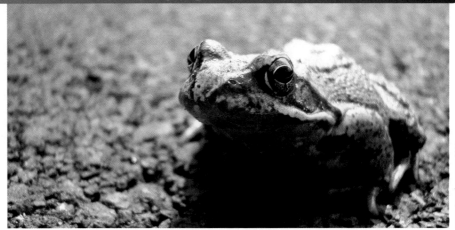

Image © sxc.hu, romos, 2009

In Greece, millions of frogs decided to migrate all at once across a busy highway. Scientists believe the frogs left a nearby lake in search of food, but needless to say, the amphibian influx caused a major traffic jam. Greek officials shut down the highway for two hours out of concern for driver safety. Previously, three motorists had skidded off the road trying to avoid the frogs. It was nearly impossible to count all the frogs crossing the highway, so scientists relied on their estimation and rounding skills to come up with the quantity of millions.

In this section, we will learn how to round numbers to a specific place value and then apply that skill to estimating answers for math problems. First, let's consider another example that uses rounded numbers.

When we talk about numbers, it is sometimes helpful to use numbers that have been *rounded off*, rather than exact numbers. For example, the city where I live has a population of 45,119. But when I tell people how large the city is, I usually say, "The population is about 45,000." The number 45,000 is the original number rounded to the nearest thousand. The number 45,119 is closer to 45,000 than it is to 46,000, so it is rounded to 45,000. We can visualize this situation on the number line.

A Rounding Numbers

The steps used in rounding numbers are given below.

> **HOW TO** Steps for Rounding Whole Numbers
>
> **1.** Locate the digit just to the right of the place you are to round to.
>
> **2.** If that digit is less than 5, replace it and all digits to its right with zeros.
>
> **3.** If that digit is 5 or more, replace it and all digits to its right with zeros, and add 1 to the digit to its left.

You can see from these steps that in order to round a number you must be told what column (or place value) to round to.

EXAMPLE 1 Round 5,382 to the nearest hundred.

Solution The 3 is in the hundreds column. We look at the digit just to its right, which is 8. Because 8 is greater than 5, we add 1 to the 3, and we replace the 8 and 2 with zeros.

1. Round 2,849 to the nearest hundred.

EXAMPLE 2 Round 94 to the nearest ten.

Solution The 9 is in the tens column. To its right is 4. Because 4 is less than 5, we simply replace it with 0.

2. Round 1,427 to the nearest ten.

EXAMPLE 3 Round 973 to the nearest hundred.

Solution We have a 9 in the hundreds column. To its right is 7, which is greater than 5. We add 1 to 9 to get 10, and then replace the 7 and 3 with zeros.

3. Round 955 to the nearest hundred.

EXAMPLE 4 Round 47,256,344 to the nearest million.

Solution We have 7 in the millions column. To its right is 2, which is less than 5. We simply replace all the digits to the right of 7 with zeros to get

47,256,344 is 47,000,000 to the nearest million
 ↑ ↑ ↑
 Less than 5 Leave as is. Replaced with zeros

4. Round 479,580 to nearest thousand.

Table 1 gives more examples of rounding.

Table 1

| Original Number | Rounded to the Nearest | | |
	Ten	Hundred	Thousand
6,914	6,910	6,900	7,000
8,485	8,490	8,500	8,000
5,555	5,560	5,600	6,000
1,234	1,230	1,200	1,000

RULE Calculating and Rounding

If we are doing calculations and are asked to round our answer, we do all our arithmetic first and then round the result. That is, the last step is to round the answer; we don't round the numbers first and then do the arithmetic.

Answers
1. 2,800
2. 1,430
3. 1,000
4. 480,000

5. Using the pie chart from Example 5, answer the following questions.
 a. To the nearest hundred dollars, what is the total amount spent on taxes and house payments?
 b. To the nearest thousand dollars, how much of their income is spent on items other than house payments?

EXAMPLE 5 The pie chart below shows how a family earning $36,913 a year spends their money.

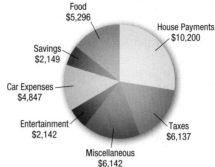

Food $5,296
House Payments $10,200
Savings $2,149
Car Expenses $4,847
Entertainment $2,142
Taxes $6,137
Miscellaneous $6,142

a. To the nearest hundred dollars, what is the total amount spent on food and entertainment?

b. To the nearest thousand dollars, how much of their income is spent on items other than taxes and savings?

Solution In each case, we add the numbers in question and then round the sum to the indicated place.

a. We add the amounts spent on food and entertainment and then round that result to the nearest hundred dollars.

Food	$5,296
Entertainment	+ 2,142
Total	$7,438 = $7,400 to the nearest hundred dollars

b. We add the numbers for all items except taxes and savings.

House payments	$10,200
Food	5,296
Car expenses	4,847
Entertainment	2,142
Miscellaneous	+ 6,142
Total	$28,627 = $29,000 to the nearest thousand dollars

B Estimating Sums and Differences

When we *estimate* the answer to a problem, we simplify the problem so that an approximate answer can be found quickly. There are a number of ways of doing this. One common method is to use rounded numbers to simplify the arithmetic necessary to arrive at an approximate answer, as our next example shows.

6. Estimate the answer to the following addition problem by rounding each number to the nearest thousand.

 15,465
 1,564
 893
 + 11,989

EXAMPLE 6 Estimate the answer to the following addition problem by rounding each number to the nearest thousand.

 4,872
 1,691
 777
 + 6,124

Solution We round each of the four numbers in the sum to the nearest thousand. Then we add the rounded numbers.

4,872	rounds to	5,000
1,691	rounds to	2,000
777	rounds to	1,000
+ 6,124	rounds to	+ 6,000
		14,000

We estimate the answer to this problem to be approximately 14,000. The actual answer, found by adding the original unrounded numbers, is 13,464.

EXAMPLE 7 Estimate the answer to the following subtraction problem by rounding each number to the nearest hundred.

$$5,423 - 1,856$$

Solution Round each number to the nearest hundred, then subtract.

5,432	rounds to	5,400
1,856	rounds to	− 1,900
		3,500

We estimate the answer to be approximately 3,500. The actual answer, found by subtracting the original numbers, is 3,567.

The method used in Examples 6 and 7 above does not conflict with the rule we stated before Example 5. In Examples 6 and 7, we are asked to *estimate* an answer, so it is okay to round the numbers in the problem before adding them. In Example 5, we are asked for a rounded answer, meaning that we are to find the exact answer to the problem and then round to the indicated place. In this case, we must not round the numbers in the problem before adding.

C Bar Charts

DESCRIPTIVE STATISTICS

Below are two representations for the amount of caffeine in five different drinks, one numeric and one visual.

Table 2

Beverage (6-Ounce Cup)	Caffeine (In Milligrams)
Brewed Coffee	100
Instant Coffee	70
Tea	50
Cocoa	5
Decaffeinated Coffee	4

FIGURE 1

The diagram in Figure 1 is called a *bar chart*. The horizontal line below which the drinks are listed is called the *horizontal axis*, while the vertical line that is labeled from 0 to 100 is called the *vertical axis*.

EXAMPLE 8 Refer to the bar chart shown in Figure 1. How many milligrams of caffeine are consumed in one serving of instant coffee?

Solution Find the instant coffee bar on the horizontal axis and read its corresponding value from the vertical axis.

The amount of caffeine in one serving of instant coffee is 70 milligrams.

GETTING READY FOR CLASS

After reading through the preceding section, respond in your own words and in complete sentences.

A. Describe the process you would use to round 5,382 to the nearest thousand.

B. Describe the process you would use to round 47,256,344 to the nearest ten thousand.

C. Find a number not containing the digit 7 that will round to 700 when rounded to the nearest hundred.

D. When I ask a class of students to round the number 7,499 to the nearest thousand, a few students will give the answer as 8,000. In what way are these students using the rule for rounding numbers incorrectly?

7. Estimate the answer to the following subtraction problem by rounding each number to the nearest hundred.

$$12,487 - 5,806$$

8. Using Figure 1, how much caffeine is consumed in one serving of decaffeinated coffee?

Answers

7. 6,700

8. 4 milligrams

EXERCISE SET 0.4

Vocabulary Review

Choose the correct words to fill in the blanks below.

more round estimate less horizontal axis vertical axis

1. When asked to _____ your answer, do all calculations first then round the result.

2. When rounding numbers, if the digit to the right of the place you are to round to is _____ than 5, replace it and all digits to its right with zeros.

3. When rounding numbers, if the digit to the right of the place you are to round to is 5 or _____ , replace it and all digits to its right with zeros, and add 1 to the digit to its left.

4. The horizontal line below the bars of a bar chart is called the _____ , whereas the vertical line to the left of the bars is called the _____ .

5. When asked to _____ your answer, use rounded numbers to arrive at an approximate answer.

Problems

A Round each of the numbers to the nearest ten.

1. 42	**2.** 44	**3.** 46	**4.** 48	**5.** 45	**6.** 73
7. 77	**8.** 75	**9.** 458	**10.** 455	**11.** 471	**12.** 680
13. 56,782	**14.** 32,807	**15.** 4,504	**16.** 3,897	**17.** 1,195	**18.** 10,998

Round each of the numbers to the nearest hundred.

19. 549	**20.** 954	**21.** 833	**22.** 604	**23.** 899	**24.** 988
25. 1090	**26.** 6,778	**27.** 5,044	**28.** 56,990	**29.** 39,603	**30.** 31,999

Round each of the numbers to the nearest thousand.

31. 4,670	**32.** 9,054	**33.** 9,760	**34.** 4,444
35. 978	**36.** 567	**37.** 657,892	**38.** 688,909
39. 608,433	**40.** 3,789,345	**41.** 5,744,500	**42.** 509,905

Complete the following table by rounding the numbers on the left as indicated by the headings in the table.

Original Number	Rounded to the Nearest		
	Ten	Hundred	Thousand
43. 7,821			
44. 5,945			
45. 5,999			
46. 4,353			
47. 10,985			
48. 11,108			
49. 99,999			
50. 95,505			

Applying the Concepts

51. Average Salary Salary studies by mlbfanhouse.com show that major league baseball's average player salary for the 2010 season was $3,014,572, representing an increase of 0.6% over the previous season's average. Round the 2010 average player salary to the nearest hundred thousand.

52. Tallest Mountain The world's tallest mountain is Mount Everest. On May 5, 1999, it was found to be 7 feet taller than it was previously thought to be. Before this date, Everest was thought to be 29,028 feet high. That height was determined by B. L. Gulatee in 1954. The first measurement of Everest was in 1856. At that time the height was given as 29,002 feet. Round the current height, the 1954 height, and the 1856 height of Mount Everest to the nearest thousand.

Fox Ratings The chart shows the number of viewers watching primetime shows on the Fox network in one week during 2009. Use the information to answer Problems 53-56.

53. What is the number of viewers watching these Fox shows during the week?

54. Using your answer from #53, is the statement "About 27 million viewers watched these shows" correct?

55. To the nearest hundred thousand, how many people watched "So You Think You Can Dance" during the week?

56. To the nearest ten thousand, how many people watched "Lie to Me" during the week?

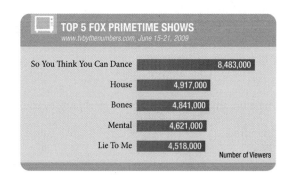

TOP 5 FOX PRIMETIME SHOWS
www.tvbythenumbers.com, June 15-21, 2009

So You Think You Can Dance	8,483,000
House	4,917,000
Bones	4,841,000
Mental	4,621,000
Lie To Me	4,518,000

Number of Viewers

Business Expenses The pie chart shows one year's worth of expenses for a small business. Use the chart to answer Problems 57–60.

57. To the nearest hundred dollars, how much was spent on postage and supplies?

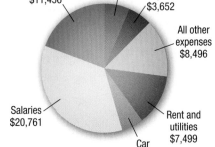

Supplies
$11,456

Postage
$3,792

Telephone
$3,652

All other
expenses
$8,496

Salaries
$20,761

Rent and
utilities
$7,499

Car
expenses
$3,205

58. Find the total amount spent, to the nearest hundred dollars, on rent, utilities and car expenses.

59. To the nearest thousand dollars, how much was spent on items other than salaries, rent and utilities?

60. To the nearest thousand dollars, how much was spent on items other than postage, supplies, and car expenses?

B Estimating Estimate the answer to each of the following addition problems by rounding each number to the indicated place value and then adding.

61. Hundred

750
275
+ 120

62. Thousand

1,891
765
+ 3,223

63. Hundred

472
422
536
+511

64. Hundred

399
601
744
+ 298

65. Thousand

25,399
7,601
18,744
+ 6,298

66. Thousand

9,999
8,888
7,777
+ 6,666

Estimate the answer to the following problems by rounding each number to the indicated place value and then subtracting.

67. Hundred

17,487 − 5,640

68. Hundred

986 − 249

69. Thousand

8,946 − 3,815

70. Thousand

18,955 − 12,049

71. Thousand

204,489 − 107,815

72. Thousand

49,899 − 29,846

C Working with Bar Charts

73. Exercise The following table lists the number of calories burned in 1 hour of exercise by a person who weighs 150 pounds. Construct a bar chart from the information in the table.

Calories Burned by a 150-Pound Person in One Hour	
Activity	**Calories**
Bicycling	374
Bowling	265
Handball	680
Jazzercise	340
Jogging	680
Skiing	544

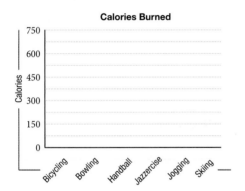

74. Fast Food The following table lists the number of calories consumed by eating some popular fast foods. Construct a bar chart from the information in the table.

Calories in Fast Food	
Food	**Calories**
McDonald's hamburger	270
Burger King hamburger	260
Jack in the Box hamburger	280
McDonald's Big Mac	510
Burger King Whopper	630
Jack in the Box Colossus burger	940

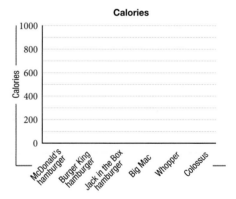

Find the Mistake

Each sentence below contains a mistake. Circle the mistake and write the correct word(s) or numbers(s) on the line provided.

1. Rounding 12,456 to the nearest hundred gives us 12,400. _____

2. To round 102,673 to the nearest ten, replace the 3 with a zero and add 1 to the 7. _____

3. To estimate the sum of 14,256 and 2,789 by rounding to the nearest hundred, add 14,200 and 2,800. _____

4. Using the bar chart below, Student C's home is 13 more miles from school than Student A's home.

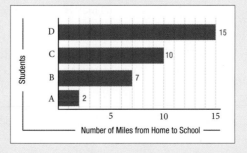

Practice Problems

1. Multiply 5 · 400 using repeated addition.

**Video Examples
Section 0.5**

0.5 Multiplication with Whole Numbers, and Area

Engineers are working to perfect a new night vision system based on the amazing vision of a bee. The system will hopefully be used as a heads-up display to help drivers better navigate dark roads. A bee's eyes are called compound eyes because they are made up of 6000 tiny lenses. Each lens is attached to a tube containing 8 light receptors that help the bee see large range of colors and levels of light, even on the darkest nights.

Considering the information about a bee's eyes, how many total light receptors does he have? To answer this problem, we need to understand multiplication with whole numbers. Multiplication is what we will cover in this section.

A Multiplication as Repeated Addition

To begin, we can think of multiplication as shorthand for repeated addition. That is, multiplying 3 times 4 can be thought of this way:

$$3 \text{ times } 4 = 4 + 4 + 4 = 12$$

Multiplying 3 times 4 means to add three 4s. We can write 3 times 4 as 3×4, or $3 \cdot 4$.

EXAMPLE 1 Multiply $4 \cdot 400$ using repeated addition.

Solution Using the definition of multiplication as repeated addition, we have

$$4 \cdot 400 = 400 + 400 + 400 + 400$$

$$= 1,600$$

One way to visualize this process is to think about a 4 by 400 meter relay team.

400 meters 400 meters 400 meters 400 meters

1,600 meters

Notice that if we had multiplied 4 and 4 to get 16 and then attached two zeros on the right, the result would have been the same.

Answer

1. 2,000

B Vocabulary and Properties

There are many ways to indicate multiplication. All the following statements are equivalent. They all indicate multiplication with the numbers 3 and 4.

$$3 \cdot 4, \quad 3 \times 4, \quad 3(4), \quad (3)4, \quad (3)(4), \quad \begin{array}{r} 4 \\ \underline{\times\ 3} \end{array}$$

If one or both of the numbers we are multiplying are represented by variables, we may also use the following notation:

$$5n \quad \text{means} \quad 5 \text{ times } n$$
$$ab \quad \text{means} \quad a \text{ times } b$$

We use the word *product* to indicate multiplication. If we say, "The product of 3 and 4 is 12," then we mean

$$3 \cdot 4 = 12$$

Both $3 \cdot 4$ and 12 are called the product of 3 and 4. The 3 and 4 are called *factors*.

DEFINITION factors

Factors are numbers that when multiplied together give a product.

Table 1 gives some word statements involving multiplication and their mathematical equivalents written in symbols.

Table 1

In English	In Symbols
The product of 2 and 5	$2 \cdot 5$
The product of 5 and 2	$5 \cdot 2$
The product of 4 and n	$4n$
The product of x and y	xy
The product of 9 and 6 is 54.	$9 \cdot 6 = 54$
The product of 2 and 8 is 16.	$2 \cdot 8 = 16$

EXAMPLE 2 Identify the products and factors in the statement
$$9 \cdot 8 = 72$$

Solution The factors are 9 and 8, and the products are $9 \cdot 8$ and 72.

EXAMPLE 3 Identify the products and factors in the statement
$$30 = 2 \cdot 3 \cdot 5$$

Solution The factors are 2, 3, and 5. The products are $2 \cdot 3 \cdot 5$ and 30.

Distributive Property

To develop an efficient method of multiplication, we need to use what is called the *distributive property*. To begin, consider the following two problems:

Problem 1	Problem 2
$3(4 + 5)$	$3(4) + 3(5)$
$= 3(9)$	$= 12 + 15$
$= 27$	$= 27$

Note The kind of notation we will use to indicate multiplication will depend on the situation. For example, when we are solving equations that involve letters, it is not a good idea to indicate multiplication with the symbol \times, since it could be confused with the variable x. The symbol we will use to indicate multiplication most often in this book is the multiplication dot.

2. Identify the products and factors in the statement $5 \cdot 7 = 35$.

3. Identify the products and factors in the statement $2 \cdot 5 \cdot 11 = 110$.

Answers
2. Factors: 5, 7
 Products: $5 \cdot 7$, 35
3. Factors: 2, 5, 11
 Products: $2 \cdot 5 \cdot 11$, 110

The result in both cases is the same number, 27. This indicates that the original two expressions must have been equal also; that is,

$$3(4 + 5) = 3(4) + 3(5)$$

This is an example of the distributive property. We say that multiplication *distributes* over addition.

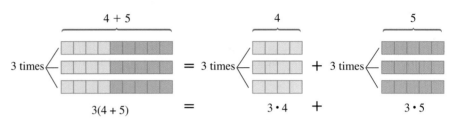

We can write this property in symbols using the letters a, b, and c to represent any three numbers.

> **PROPERTY** **Distributive Property**
>
> If a, b, and c represent any three numbers, then
>
> $$a(b + c) = a(b) + a(c)$$

More Properties of Multiplication

> **PROPERTY** **Multiplication Property of 0**
>
> If a represents any number, then
>
> $$a \cdot 0 = 0 \qquad \text{and} \qquad 0 \cdot a = 0$$
>
> *In words:* Multiplication by 0 always results in 0.

> **PROPERTY** **Multiplication Property of 1**
>
> If a represents any number, then
>
> $$a \cdot 1 = a \qquad \text{and} \qquad 1 \cdot a = a$$
>
> *In words:* Multiplying any number by 1 leaves that number unchanged.

> **PROPERTY** **Commutative Property of Multiplication**
>
> If a and b are any two numbers, then
>
> $$ab = ba$$
>
> *In words:* The order of the numbers in a product doesn't affect the result.

> **PROPERTY** **Associative Property of Multiplication**
>
> If a, b, and c represent any three numbers, then
>
> $$(ab)c = a(bc)$$
>
> *In words:* We can change the grouping of the numbers in a product without changing the result.

To visualize the commutative property, we can think of a dining room with 8 guests.

4 tables with 2 people each = 2 tables with 4 people each

EXAMPLE 4 Use the commutative property of multiplication to rewrite each of the following products.

a. $7 \cdot 9$ **b.** $4(6)$

Solution Applying the commutative property to each expression, we have

a. $7 \cdot 9 = 9 \cdot 7$

b. $4(6) = 6(4)$

EXAMPLE 5 Use the associative property of multiplication to rewrite each of the following products.

a. $(2 \cdot 7) \cdot 9$ **b.** $3 \cdot (8 \cdot 2)$

Solution Applying the associative property of multiplication, we regroup as follows:

a. $(2 \cdot 7) \cdot 9 = 2 \cdot (7 \cdot 9)$

b. $3 \cdot (8 \cdot 2) = (3 \cdot 8) \cdot 2$

C Multiplication with Whole Numbers

Suppose we want to find the product 7(65). By writing 65 as $60 + 5$ and applying the distributive property, we have

$$7(65) = 7(60 + 5) \qquad 65 = 60 + 5$$
$$= 7(60) + 7(5) \quad \text{Distributive property}$$
$$= 420 + 35 \qquad \text{Multiply first, then add.}$$
$$= 455$$

We can write the same problem vertically like this:

$$
\begin{array}{r}
60 + 5 \\
\times \qquad 7 \\
\hline
35 \\
+ \qquad 420 \\
\hline
455
\end{array}
$$

$\leftarrow 7(5) = 35$
$\leftarrow 7(60) = 420$

This saves some space in writing. But notice that we can cut down on the amount of writing even more if we write the problem this way:

Step 2: $7(6) = 42$; Add the → $\overset{3}{6}5$ Step 1: $7(5) = 35$; Write the 5
3 we carried to 42 to get 45. $\times\,7$ in the ones column, and then carry
 $\overline{455}$ ← the 3 to the tens column.

This shortcut notation takes some practice.

EXAMPLE 6 Multiply: 9(43).

Solution

Step 2: $9(4) = 36$; Add the → $\overset{2}{4}3$ Step 1: $9(3) = 27$; Write the 7
2 we carried to 36 to get 38. $\times\,9$ in the ones column, and then carry
 $\overline{387}$ ← the 2 to the tens column.

4. Use the commutative property of multiplication to rewrite each product.
 a. $5 \cdot 12$
 b. $6(x + y)$

5. Use the associative property of multiplication to rewrite each product.
 a. $(5 \cdot 8) \cdot 3$
 b. $3 \cdot (7 \cdot n)$

Note When using the distributive property, we multiply $a(b)$ and $a(c)$ before adding their products together because of a rule called order of operations. Part of this rule states that once operations inside parentheses have been worked, then we multiply before we add. We will study this rule more later on in this chapter.

6. Multiply: 7(59).

Answers
4. a. $12 \cdot 5$ **b.** $(x + y)(6)$
5. a. $5 \cdot (8 \cdot 3)$ **b.** $(3 \cdot 7) \cdot n$
6. 413

7. Multiply: 67(45).

EXAMPLE 7 Multiply: 52(37).

Solution This is the same as 52(30 + 7) or by the distributive property

$$52(30) + 52(7)$$

We can find each of these products by using the shortcut method.

$$
\begin{array}{r} 52 \\ \times\ 30 \\ \hline 1{,}560 \end{array}
\qquad
\begin{array}{r} \overset{1}{52} \\ \times\ 7 \\ \hline 364 \end{array}
$$

The sum of these two numbers is 1,560 + 364 = 1,924. Here is a summary of what we have so far:

$$
\begin{aligned}
52(37) &= 52(30 + 7) & & 37 = 30 + 7 \\
&= 52(30) + 52(7) & & \text{Distributive property} \\
&= 1{,}560 + 364 & & \text{Multiply first, then add.} \\
&= 1{,}924
\end{aligned}
$$

The shortcut form for this problem is

$$
\begin{array}{r}
52 \\
\times\ 37 \\
\hline
364 \\
+\ 1{,}560 \\
\hline
1{,}924
\end{array}
\qquad
\begin{aligned}
&\longleftarrow 7(52) = 364 \\
&\longleftarrow 30(52) = 1{,}560
\end{aligned}
$$

In this case, we have not shown any of the numbers we carried, simply because it becomes very messy.

8. Multiply: 124(375).

EXAMPLE 8 Multiply: 279(428).

Solution
$$
\begin{array}{r}
279 \\
\times\ 428 \\
\hline
2{,}232 \\
5{,}580 \\
+\ 111{,}600 \\
\hline
119{,}412
\end{array}
\qquad
\begin{aligned}
&\longleftarrow 8(279) = 2{,}232 \\
&\longleftarrow 20(279) = 5{,}580 \\
&\longleftarrow 400(279) = 111{,}600
\end{aligned}
$$

USING TECHNOLOGY Calculator

Here is how we would work the problem shown in Example 8 on a calculator:

Scientific Calculator: 279 $\boxed{\times}$ 428 $\boxed{=}$
Graphing Calculator: 279 $\boxed{\times}$ 428 $\boxed{\text{ENT}}$

Estimating

One way to estimate the answer to the problem shown in Example 8 is to round each number to the nearest hundred and then multiply the rounded numbers. Doing so would give us this:

$$300(400) = 120{,}000$$

Our estimate of the answer is 120,000, which is close to the actual answer, 119,412. Making estimates is important when we are using calculators; having an estimate of the answer will keep us from making major errors in multiplication.

Answers
7. 3,015
8. 46,500

D Applications

EXAMPLE 9 A supermarket orders 35 cases of a certain soft drink. If each case contains 12 cans of the drink, then how many cans were ordered?

Solution We have 35 cases and each case has 12 cans. The total number of cans is the product of 35 and 12, which is 35(12).

$$
\begin{array}{r}
12 \\
\times\ 35 \\
\hline
60 \\
+\ 360 \\
\hline
420
\end{array}
$$

 60 ←———— 5(12) = 60

 + 360 ←———— 30(12) = 360

There is a total of 420 cans of the soft drink.

EXAMPLE 10 Shirley earns $12 an hour for the first 40 hours she works each week. If she has $109 deducted from her weekly check for taxes and retirement, how much money will she take home if she works 38 hours this week?

Solution To find the amount of money she earned for the week, we multiply 12 and 38. From that total we subtract 109. The result is her take-home pay. Without showing all the work involved in the calculations, here is the solution:

$$38(\$12) = \$456 \qquad \textit{Her total weekly earnings}$$

$$\$456 - \$109 = \$347 \qquad \textit{Her take-home pay}$$

EXAMPLE 11 In 1993, the government standardized the way in which nutrition information is presented on the labels of most packaged food products. Figure 1 shows one of these standardized food labels. It is from a box of original Cheez-It baked snack crackers. Approximately how many crackers are in the box, and what is the total number of calories consumed if all the crackers in the box are eaten?

Nutrition Facts		
Serving Size 30 g. (About 27 crackers)		
Servings Per Container: 9		
Amount Per Serving		
Calories 150	Calories from fat 70	
		% Daily Value*
Total Fat 8g		12%
Saturated Fat 2g		10%
Trans Fat 0g		
Cholesterol 0mg		0%
Sodium 230mg		10%
Total Carbohydrate 17g		6%
Sugars 0g		
Protein 3g		
Vitamin A 2%	•	Vitamin C 0%
*Percent Daily Values are based on a 2,000 calorie diet		

FIGURE 1

Solution Reading toward the top of the label, we see that there are about 27 crackers in one serving, and approximately 9 servings in the box. Therefore, the total number of crackers in the box is

$$9(27) = 243 \text{ crackers}$$

This is an approximate number, because each serving is approximately 27 crackers.
Reading further we find that each serving contains 150 calories. Therefore, the total number of calories consumed by eating all the crackers in the box is

$$9(150) = 1{,}350 \text{ calories}$$

As we progress through the book, we will study more of the information in nutrition labels.

9. A bar orders 45 cases of beer. If each case contains 24 bottles of beer, how many bottles were ordered?

10. Rodrigo earns $9 per hour at his tutoring job. If he works 35 hours and has $42 deducted for taxes, how much will his paycheck be for the week?

11. From Example 11, how many calories from fat will you consume if you eat the entire box of crackers?

Note The letter g that is shown after some of the numbers in the nutrition label in Figure 1 stands for grams, a unit used to measure weight. The unit mg stands for milligrams, another smaller unit of weight. We will have more to say about these units later in the book.

Answers

9. 1,080 beers
10. $273
11. 630 calories from fat

12. Using the chart from Example 12 suppose a 150-pound person bicycles for 3 hours and goes jogging for 2 hours in a week. How many calories did that person burn in a week?

EXAMPLE 12 The table below lists the number of calories burned in 1 hour of exercise by a person who weighs 150 pounds. Suppose a 150-pound person goes bowling for 2 hours after having eaten the box of crackers mentioned in Example 11. Will he or she burn all the calories consumed from the crackers?

Activity	Calories Burned in 1 Hour by a 150-Pound Person
Bicycling	374
Bowling	265
Handball	680
Jazzercize	340
Jogging	680
Skiing	544

Solution Each hour of bowling burns 265 calories. If the person bowls for 2 hours, he or she will burn a total of

$$2(265) = 530 \text{ calories}$$

Because the box of crackers contained 1,350 calories, not all of them have been burned with 2 hours of bowling.

Area

> **DEFINITION** area
>
> The **area** of a flat object is a measure of the amount of surface the object has.

The rectangle in Figure 2 below has an area of 6 square inches, because that is the number of squares (each of which is 1 inch long and 1 inch wide) it takes to cover the rectangle.

Note To understand some of the notation we use for area, we need to talk about exponents. The 2 in the expression 3^2 is an exponent. The expression 3^2 is read "3 to the second power," or "3 squared," and it is defined this way:

$$3^2 = 3 \cdot 3 = 9$$

As you can see, the exponent 2 in the expression 3^2 tells us to multiply two 3s together. Here are some additional expressions containing the exponent 2.

$$4^2 = 4 \cdot 4 = 16$$
$$5^2 = 5 \cdot 5 = 25$$
$$11^2 = 11 \cdot 11 = 121$$

We will cover exponents in more detail later in this chapter.

FIGURE 2

It is no coincidence that the area of the rectangle in Figure 2 and the product of the length and the width are the same number. We can calculate the area of the rectangle in Figure 2 by simply multiplying the length and the width together.

$$\begin{aligned} \text{Area} &= (\text{length}) \cdot (\text{width}) \\ &= (3 \text{ inches}) \cdot (2 \text{ inches}) \\ &= (3 \cdot 2) \cdot (\text{inches} \cdot \text{inches}) \\ &= 6 \text{ square inches} \end{aligned}$$

The unit *square inches* can be abbreviated as *sq. in.* or *in².*

Answer

12. 2,482 calories

FACTS FROM GEOMETRY Area

Figure 3 shows two common geometric figures along with the formulas for their areas.

Area = (side)(side)
 = (side)2
 = s^2
Square

Area = (length)(width)
 = lw
Rectangle

FIGURE 3

EXAMPLE 13 Find the total area of the house and deck shown in Figure 4.

FIGURE 4

13. From the figure in Example 13, find the area of the two bedrooms.

Solution We begin by drawing an additional line (see Figure 5) so that the original figure is now composed of two rectangles. Next, we fill in the missing dimensions on the two rectangles (Figure 6).

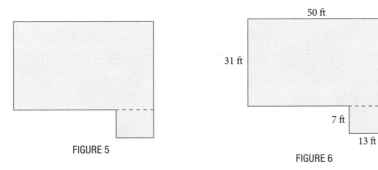

FIGURE 5

FIGURE 6

Finally, we calculate the area of the original figure by adding the areas of the individual figures.

$$
\begin{aligned}
\text{Area} &= \text{Area small rectangle} + \text{Area large rectangle} \\
&= \quad\quad 13 \cdot 7 \quad\quad + \quad\quad 50 \cdot 31 \\
&= \quad\quad\quad 91 \quad\quad\quad + \quad\quad\quad 1550 \\
&= 1641 \text{ square feet}
\end{aligned}
$$

Answer

13. 420 square feet

14. Find the solution to each of
the following equations by
inspection.
 a. $5 \cdot n = 30$
 b. $7 \cdot n = 56$
 c. $24 = 4 \cdot n$
 d. $33 = 3 \cdot n$

E Solving Equations

If n is used to represent a number, then the equation

$$4 \cdot n = 12$$

is read "4 times n is 12," or "The product of 4 and n is 12." This means that we are looking for the number we multiply by 4 to get 12. The number is 3. Because the equation becomes a true statement if n is 3, we say that 3 is the solution to the equation.

EXAMPLE 14 Find the solution to each of the following equations by inspection.

a. $6 \cdot n = 24$ **b.** $4 \cdot n = 36$ **c.** $15 = 3 \cdot n$ **d.** $21 = 3 \cdot n$

Solution

a. The solution to $6 \cdot n = 24$ is 4, because $6 \cdot 4 = 24$.

b. The solution to $4 \cdot n = 36$ is 9, because $4 \cdot 9 = 36$.

c. The solution to $15 = 3 \cdot n$ is 5, because $15 = 3 \cdot 5$.

d. The solution to $21 = 3 \cdot n$ is 7, because $21 = 3 \cdot 7$.

GETTING READY FOR CLASS

After reading through the preceding section, respond in your own words and in complete sentences.

A. Explain how multiplication is shorthand for repeated addition.

B. Use the variables a, b, and c to explain the distribution property.

C. Explain how you would find the area of a flat object.

D. What is the difference between the commutative property of multiplication and the associative property of multiplication?

Answers

14. a. 6 **b.** 8 **c.** 6 **d.** 11

Vocabulary Review

Choose the correct words to fill in the blanks below.

associative	distributive	addition	multiplication
unchanged	factors	area	commutative

1. Multiplication is repeated _____ .

2. _____ are numbers that when multiplied together give a product.

3. The _____ property states that $a(b + c) = a(b) + a(c)$.

4. The _____ property of 0 states that multiplication by 0 always results in 0.

5. The multiplication property of 1 states that multiplying any number by 1 leaves that number _____ .

6. The _____ property of multiplication states that the order of the numbers in a product doesn't affect the result.

7. The _____ property of multiplication states that $(ab)c = a(bc)$.

8. The _____ of a flat object is a measure of the amount of surface the object has.

Problems

A Multiply each of the following using repeated addition.

1. $3 \cdot 100$

2. $7 \cdot 100$

3. $3 \cdot 200$

4. $4 \cdot 200$

5. $6 \cdot 500$

6. $8 \cdot 400$

7. $5 \cdot 1{,}000$

8. $8 \cdot 1{,}000$

9. $3 \cdot 7{,}000$

10. $6 \cdot 7{,}000$

11. $9 \cdot 9{,}000$

12. $7 \cdot 7{,}000$

B Write each of the following expressions in words, using the word *product.*

13. $6 \cdot 7$

14. $9(4)$

15. $2 \cdot n$

16. $5 \cdot x$

17. $9 \cdot 7 = 63$

18. $(5)(6) = 30$

Write each of the following in symbols.

19. The product of 7 and n

20. The product of 9 and x

21. The product of 6 and 7 is 42.

22. The product of 8 and 9 is 72.

23. The product of 0 and 6 is 0.

24. The product of 1 and 6 is 6.

Identify the products in each statement.

25. $9 \cdot 7 = 63$

26. $2(6) = 12$

27. $4(4) = 16$

28. $5 \cdot 5 = 25$

Identify the factors in each statement.

29. $2 \cdot 3 \cdot 4 = 24$

30. $6 \cdot 1 \cdot 5 = 30$

31. $12 = 2 \cdot 2 \cdot 3$

32. $42 = 2 \cdot 3 \cdot 7$

Rewrite each of the following using the commutative property of multiplication.

33. $5(9)$

34. $4(3)$

35. $6 \cdot 7$

36. $8 \cdot 3$

Rewrite each of the following using the associative property of multiplication.

37. $2 \cdot (7 \cdot 6)$

38. $4 \cdot (8 \cdot 5)$

39. $3 \times (9 \times 1)$

40. $5 \times (8 \times 2)$

Use the distributive property to rewrite each expression, then simplify.

41. $7(2 + 3)$

42. $4(5 + 8)$

43. $9(4 + 7)$

44. $6(9 + 5)$

45. $3(x + 1)$

46. $5(x + 8)$

47. $2(x + 5)$

48. $4(x + 3)$

C Find each of the following products (multiply). In each case, use the shortcut method.

49.
$$\begin{array}{r} 25 \\ \times\, 4 \\ \hline \end{array}$$

50.
$$\begin{array}{r} 43 \\ \times\, 9 \\ \hline \end{array}$$

51.
$$\begin{array}{r} 38 \\ \times\, 6 \\ \hline \end{array}$$

52.
$$\begin{array}{r} 45 \\ \times\, 7 \\ \hline \end{array}$$

53.
$$\begin{array}{r} 18 \\ \times\, 2 \\ \hline \end{array}$$

54.
$$\begin{array}{r} 29 \\ \times\, 3 \\ \hline \end{array}$$

55.
$$\begin{array}{r} 72 \\ \times\, 20 \\ \hline \end{array}$$

56.
$$\begin{array}{r} 68 \\ \times\, 30 \\ \hline \end{array}$$

57.
$$\begin{array}{r} 19 \\ \times\, 50 \\ \hline \end{array}$$

58.
$$\begin{array}{r} 24 \\ \times\, 40 \\ \hline \end{array}$$

59.
$$\begin{array}{r} 69 \\ \times\, 25 \\ \hline \end{array}$$

60.
$$\begin{array}{r} 27 \\ \times\, 36 \\ \hline \end{array}$$

61.
$$\begin{array}{r} 11 \\ \times\, 11 \\ \hline \end{array}$$

62.
$$\begin{array}{r} 12 \\ \times\, 21 \\ \hline \end{array}$$

63.
$$\begin{array}{r} 97 \\ \times\, 16 \\ \hline \end{array}$$

64.
$$\begin{array}{r} 24 \\ \times\, 39 \\ \hline \end{array}$$

65. 168
× 25

66. 452
× 34

67. 728
× 91

68. 680
× 76

69. 698
× 400

70. 879
× 600

71. 111
× 111

72. 123
× 321

73. 532
× 200

74. 277
× 900

75. 856
× 232

76. 455
× 248

77. 976
× 628

78. 432
× 555

79. 2,468
× 135

80. 2,725
× 324

81. 24,563
× 735

82. 56,728
× 852

83. 44,777
× 5,888

84. 33,999
× 2,555

Complete the following tables.

85.

First Number a	Second Number b	Their Product ab
11	11	
11	22	
22	22	
22	44	

86.

First Number a	Second Number b	Their Product ab
25	15	
25	30	
50	15	
50	30	

87.

First Number a	Second Number b	Their Product ab
25	10	
25	100	
25	1,000	
25	10,000	

88.

First Number a	Second Number b	Their Product ab
11	111	
11	222	
22	111	
22	222	

89.

First Number a	Second Number b	Their Product ab
12	20	
36	20	
12	40	
36	40	

90.

First Number a	Second Number b	Their Product ab
10	12	
100	12	
1,000	12	
10,000	12	

D Find the area enclosed by each figure. (Note that some of the units on the figures come from the metric system. The abbreviations are as follows: Meter is abbreviated m, centimeter is cm, and millimeter is abbreviated mm. A meter is about 3 inches longer than a yard.)

91.

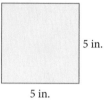

5 in.

5 in.

92.

10 in.

10 in.

93.

7 m

12 m

94.

3 m

12 m

95.

8 cm

8 cm

16 cm

8 cm

96.

6 in.

18 in.

8 in.

20 in.

E Find a solution for each equation.

97. $4 \cdot n = 12$

98. $3 \cdot n = 12$

99. $9 \cdot n = 81$

100. $6 \cdot n = 36$

101. $0 = n \cdot 5$

102. $6 = 1 \cdot n$

103. $48 = 4 \cdot n$

104. $50 = 5 \cdot n$

105. $n \cdot 6 = 66$

106. $n \cdot 5 = 45$

107. $81 = 9 \cdot n$

108. $49 = 7 \cdot n$

109. Planning a Trip A family decides to drive their compact car on their vacation. They figure it will require a total of about 130 gallons of gas for the vacation. If each gallon of gas will take them 22 miles, how long is the trip they are planning?

110. Rent A student pays $475 rent each month. How much money does she spend on rent in 2 years?

111. Reading House Plans Find the area of the floor of the house shown here if the garage is not included with the house, and if the garage is included with the house.

112. The Alphabet and Area The box shown here is called a typecase, or printer's tray. It was used in early typesetting to store the letters that would be used to lay out a page of type. Use the box to answer the questions below.

a. What letter is used most often in printed material?

b. Which letter is printed more often, the letter *i* or the letter *f*?

c. What is the relationship between area and how often a letter in the alphabet is printed?

Exercise and Calories The table below is an extension of the table we used in Example 12 of this section. It gives the amount of energy expended during 1 hour of various activities for people of different weights. The accompanying figure is a nutrition label from a bag of cheddar Goldfish crackers. Use the information from the table and the nutrition label to answer Problems 113–118.

Calories Burned Through Exercise

| | Calories Per Hour | | |
Activity	120 Pounds	150 Pounds	180 Pounds
Bicycling	299	374	449
Bowling	212	265	318
Handball	544	680	816
Jazzercise	272	340	408
Jogging	544	680	816
Skiing	435	544	653

Nutrition Facts

Serving Size 55 pieces
Servings Per Container About 4

Amount Per Serving

Calories 140 Calories from fat 45

	% Daily Value*
Total Fat 5g	8%
Saturated Fat 1g	5%
Cholesterol 5mg	2%
Sodium 250mg	10%
Total Carbohydrate 20g	7%
Dietary Fiber 1g	4%
Sugars 1g	
Protein 4g	

| Vitamin A 0% | • | Vitamin C 0% |
| Calcium 4% | • | Iron 2% |

*Percent Daily Values are based on a 2,000 calorie diet

113. Suppose you weigh 180 pounds. How many calories would you burn if you play handball for 2 hours and then ride your bicycle for 1 hour?

114. How many calories are burned by a 120-lb person who jogs for 1 hour and then goes bike riding for 2 hours?

115. How many calories would you consume if you ate the entire bag of crackers?

116. Approximately how many crackers are in the bag?

117. If you weigh 180 pounds, will you burn off the calories consumed by eating 3 servings of Goldfish crackers if you ride your bike 1 hour?

118. If you weigh 120 pounds, will you burn off the calories consumed by eating 3 servings of Goldfish crackers if you ride your bike for 1 hour?

Estimating

Mentally estimate the answer to each of the following problems by rounding each number to the indicated place and then multiplying.

119.
$$\begin{array}{r} 750 \\ \times\ 12 \end{array}$$
hundred
ten

120.
$$\begin{array}{r} 591 \\ \times\ 323 \end{array}$$
hundred
hundred

121.
$$\begin{array}{r} 3{,}472 \\ \times\ 511 \end{array}$$
thousand
hundred

122.
$$\begin{array}{r} 399 \\ \times\ 298 \end{array}$$
hundred
hundred

123.
$$\begin{array}{r} 2{,}399 \\ \times\ 698 \end{array}$$
thousand
hundred

124.
$$\begin{array}{r} 9{,}999 \\ \times\ 666 \end{array}$$
thousand
hundred

One Step Further: Number Sequences

A *geometric sequence* is a sequence of numbers in which each number is obtained from the previous number by multiplying by the same number each time. For example, the sequence 3, 6, 12, 24, . . . is a geometric sequence, starting with 3, in which each number comes from multiplying the previous number by 2. Find the next number in each of the following geometric sequences.

125. 5, 10, 20, . . .

126. 10, 50, 250, . . .

127. 2, 6, 18, . . .

128. 12, 24, 48, . . .

Find the Mistake

Each sentence below contains a mistake. Circle the mistake and write the correct word(s) or numbers(s) on the line provided.

1. Factors are numbers that when multiplied together give a sum. _____

2. The distributive property is used to show that $16(5 + 9) = 16(5) + 9$. _____

3. The first step when multiplying 73 and 4 is to multiply 4 and 3 by writing down the 1 and carrying the 2. _____

4. The area of a rectangle with a length of 35 inches and a width of 8 inches is 43 square inches. _____

0.6 Division with Whole Numbers

OBJECTIVES

A Understand the relationship between multiplication and division.

B Find quotients of whole numbers with no remainder.

C Find quotients of whole numbers with nonzero remainders.

KEY WORDS

quotient

dividend

divisor

long division

remainder

Image © sxc.hu, ColinBroug, 2010

In 2008, the *Alaska Ranger* crab-fishing ship sank leaving the 47-member crew floating in the near-freezing waters of the Bering Sea. The Coast Guard, 230 miles away, picked up the ship's mayday call before it sank and immediately set out to rescue the crew by helicopter. Other fishing vessels also picked up the ship's mayday and changed their courses toward the last known coordinates. One of the closest vessels was 250 miles away, and at top speed could travel at 35 miles per hour. How many hours did it take this vessel to reach the survivors? To answer this question, we need to divide 250 by 35. In this section, we will work with more division problems similar to this one.

A Notation

As was the case with multiplication, there are many ways to indicate division. All the following statements are equivalent. They all mean 10 divided by 5.

$$10 \div 5, \quad \frac{10}{5}, \quad 10/5, \quad 5\overline{)10}$$

The kind of notation we use to write division problems will depend on the situation. We will use the notation $5\overline{)10}$ mostly with the long division problems found in this chapter. The notation $\frac{10}{5}$ will be used in the chapter on fractions and in later chapters. The horizontal line used with the notation $\frac{10}{5}$ is called the fraction bar.

Vocabulary

The word *quotient* is used to indicate division. If we say, "The quotient of 10 and 5 is 2," then we mean

$$10 \div 5 = 2 \quad \text{or} \quad \frac{10}{5} = 2$$

The 10 is called the dividend, and the 5 is called the divisor. All the expressions, $10 \div 5$, $\frac{10}{5}$, and 2, are called the quotient of 10 and 5.

**Video Examples
Section 0.6**

Table 1

In English	In Symbols
The quotient of 15 and 3	$15 \div 3$, or $\frac{15}{3}$, or 15/3
The quotient of 3 and 15	$3 \div 15$, or $\frac{3}{15}$, or 3/15
The quotient of 8 and n	$8 \div n$, or $\frac{8}{n}$, or 8/n
x divided by 2	$x \div 2$, or $\frac{x}{2}$, or x/2
The quotient of 21 and 3 is 7.	$21 \div 3 = 7$, or $\frac{21}{3} = 7$

The Meaning of Division

One way to arrive at an answer to a division problem is by thinking in terms of multiplication. For example, if we want to find the quotient of 32 and 8, we may ask, "What do we multiply by 8 to get 32?"

$$32 \div 8 = ? \qquad \text{means} \qquad 8 \cdot ? = 32$$

Because we know from our work with multiplication that $8 \cdot 4 = 32$, it must be true that

$$32 \div 8 = 4$$

Table 2 lists some additional examples.

Table 2

Division		Multiplication
$18 \div 6 = 3$	because	$6 \cdot 3 = 18$
$32 \div 8 = 4$	because	$8 \cdot 4 = 32$
$10 \div 2 = 5$	because	$2 \cdot 5 = 10$
$72 \div 9 = 8$	because	$9 \cdot 8 = 72$

B Division by One-Digit Numbers

Consider the following division problem:

$$465 \div 5$$

We can think of this problem as asking the question, "How many fives can we subtract from 465?" To answer the question, we begin subtracting multiples of 5. One way to organize this process is shown below.

$$
\begin{array}{r}
90 \\
5\overline{)465} \\
-450 \\
\hline
15
\end{array}
$$

← We first guess that there are at least 90 fives in 465.

← 90(5) = 450

← 15 is left after we subtract 90 fives from 465.

What we have done so far is subtract 90 fives from 465 and found that 15 is still left. Because there are 3 fives in 15, we continue the process.

$$
\begin{array}{r}
3 \\
90 \\
5\overline{)465} \\
-450 \\
\hline
15 \\
-15 \\
\hline
0
\end{array}
$$

← There are 3 fives in 15.

← 3 · 5 = 15

← The difference is 0.

The total number of fives we have subtracted from 465 is

$$90 + 3 = 93 \qquad \text{The number of fives subtracted from 465}$$

We now summarize the results of our work.

$$465 \div 5 = 93 \qquad \text{Which we check}$$
$$\text{with multiplication} \rightarrow
\begin{array}{r}
\overset{1}{93} \\
\times \ 5 \\
\hline
465
\end{array}$$

Notation

The division problem just shown can be shortened by eliminating some of the information.

The shorthand form for this problem

$$
\begin{array}{r}
3 \\
90 \\
5\overline{)465} \\
-\,450 \\
\hline
15 \\
-\,15 \\
\hline
0
\end{array}
$$

looks like this.

$$
\begin{array}{r}
93 \\
5\overline{)465} \\
-\,45\!\downarrow \\
\hline
15 \\
-\,15 \\
\hline
0
\end{array}
$$

The arrow indicates that we bring down the 5 after we subtract.

The problem shown above on the right is the shortcut form of what is called *long division*. Here is an example showing this shortcut form of long division from start to finish.

EXAMPLE 1 Divide: 595 ÷ 7.

Solution Because 7(8) = 56, our first estimate of the number of sevens that can be subtracted from 595 is 80.

$$
\begin{array}{r}
8 \\
7\overline{)595} \\
-\,56\!\downarrow \\
\hline
35
\end{array}
$$

← The 8 is placed above the tens column, so we know our first estimate is 80.
← 8(7) = 56
← 59 − 56 = 3; then bring down the 5.

Since 7(5) = 35, we have

$$
\begin{array}{r}
85 \\
7\overline{)595} \\
-\,56\!\downarrow \\
\hline
35 \\
-\,35 \\
\hline
0
\end{array}
$$

← There are 5 sevens in 35.

← 5(7) = 35
← 35 − 35 = 0

Our result is 595 ÷ 7 = 85, which we can check with multiplication.

$$
\begin{array}{r}
\overset{3}{85} \\
\times\,7 \\
\hline
595
\end{array}
$$

Division by Two-Digit Numbers

EXAMPLE 2 Divide: 9,380 ÷ 35.

Solution In this case our divisor, 35, is a two-digit number. The process of division is the same. We still want to find the number of thirty-fives we can subtract from 9,380.

$$
\begin{array}{r}
2 \\
35\overline{)9380} \\
-\,70\!\downarrow \\
\hline
238
\end{array}
$$

← The 2 is placed above the hundreds column.

← 2(35) = 70
← 93 − 70 = 23; then bring down the 8.

We can make a few preliminary calculations to help estimate how many thirty-fives are in 238.

$$5 \times 35 = 175 \qquad 6 \times 35 = 210 \qquad 7 \times 35 = 245$$

Practice Problems

1. Divide: 1,272 ÷ 8.

2. Divide: 13,152 ÷ 24.

Answers
1. 159
2. 548

Because 210 is the closest to 238 without being larger than 238, we use 6 as our next estimate.

$$
\begin{array}{r}
26 \\
35\overline{)9380} \\
-70\downarrow \\
\hline
238 \\
-210\downarrow \\
\hline
280
\end{array}
$$

← 6 in the tens column means this estimate is 60.

← 6 (35) = 210

← 238 − 210 = 28; bring down the 0.

Because 35(8) = 280, we have

$$
\begin{array}{r}
268 \\
35\overline{)9380} \\
-70\downarrow \\
\hline
238 \\
-210\downarrow \\
\hline
280 \\
-280 \\
\hline
0
\end{array}
$$

← 8(35) = 280

← 280 − 280 = 0

We can check our result with multiplication.

$$
\begin{array}{r}
268 \\
\times\quad 35 \\
\hline
1{,}340 \\
+\ 8{,}040 \\
\hline
9{,}380
\end{array}
$$

EXAMPLE 3 Divide: 1,872 by 18.

Solution Here is the first step:

$$
\begin{array}{r}
1 \\
18\overline{)1872} \\
-18 \\
\hline
0
\end{array}
$$

← 1 is placed above hundreds column.

← Multiply 1(18) to get 18.

← Subtract to get 0.

The next step is to bring down the 7 and divide again.

$$
\begin{array}{r}
10 \\
18\overline{)1872} \\
-18\downarrow \\
\hline
07 \\
-0 \\
\hline
7
\end{array}
$$

← 0 is placed above tens column. 0 is the largest number we can multiply by 18 and not go over 7.

← Multiply 0(18) to get 0.

← Subtract to get 7.

Here is the complete problem:

$$
\begin{array}{r}
104 \\
18\overline{)1872} \\
-18\downarrow \\
\hline
07 \\
-0\downarrow \\
\hline
72 \\
-72 \\
\hline
0
\end{array}
$$

To show our answer is correct, we multiply.

$$18(104) = 1{,}872$$

3. Divide: 5,200 ÷ 16.

C Division with Remainders

Suppose Darlene is making bread using a five-pound (80 oz.) bag of flour. Her bread recipe calls for 30 ounces of flour for each loaf. To see how many ounces she could get from a 80-ounce bag, she would divide 80 by 30. If she did so, she would find that she could make two loaves, but she would have 20 ounces of flour left over. A diagram of this problem is shown in Figure 1.

20 ounces
of left-over flour

Flour

5 pounds (80 oz) of flour 30 ounces per loaf of bread
60 ounces total

FIGURE 1

Writing the results in the diagram as a division problem looks like this:

$$
\begin{array}{r}
2 \quad \leftarrow \text{Quotient} \\
\text{Divisor} \rightarrow 30\overline{)80} \quad \leftarrow \text{Dividend} \\
\underline{-\ 60} \\
20 \quad \leftarrow \text{Remainder}
\end{array}
$$

EXAMPLE 4 Divide: $1,690 \div 67$.

Solution Dividing as we have previously, we get

$$
\begin{array}{r}
25 \\
67\overline{)1690} \\
\underline{-\ 134{\downarrow}} \\
350 \\
\underline{-\ 335} \\
15 \quad \leftarrow \text{15 is left over.}
\end{array}
$$

We have 15 left, and because 15 is less than 67, no more sixty-sevens can be subtracted. In a situation like this, we call 15 the remainder and write

These indicate that the remainder is 15.

$$
\begin{array}{r}
25 \ \text{R}\ 15 \\
67\overline{)1690} \\
\underline{-\ 134{\downarrow}} \\
350 \\
\underline{-\ 335} \\
15
\end{array}
\qquad \text{or} \qquad
\begin{array}{r}
25\,\tfrac{15}{67} \\
67\overline{)1690} \\
\underline{-\ 134{\downarrow}} \\
350 \\
\underline{-\ 335} \\
15
\end{array}
$$

Both forms of notation shown above indicate that 15 is the remainder. The notation R 15 is the notation we will use in this chapter. The notation $\frac{15}{67}$ will be useful in the chapter on fractions.

To check a problem like this, we multiply the divisor and the quotient as usual, and then add the remainder to this result.

$$
\begin{array}{r}
67 \\
\times\ 25 \\
\hline
335 \\
+\ 1,340 \\
\hline
1,675 \quad \leftarrow \text{Product of divisor and quotient}
\end{array}
$$

$$1,675 + 15 = 1,690$$

Remainder Dividend

4. Divide: $1,572 \div 23$.

5. A family has an annual income of $34,320. How much is their average monthly income?

Note To estimate the answer to Example 5 quickly, we can replace 35,880 with 36,000 and mentally calculate

$$36,000 \div 12$$

which gives an estimate of 3,000. Our actual answer, 2,990, is close enough to our estimate to convince us that we have not made a major error in our calculation.

USING TECHNOLOGY Calculators

Here is how we would work the problem shown in Example 4 on a calculator:

Graphing Calculator: 1690 $\boxed{\div}$ 67 $\boxed{\text{ENT}}$

Scientific Calculator: 1690 $\boxed{\div}$ 67 $\boxed{=}$

In both cases the calculator will display 25.223881 (give or take a few digits at the end), which gives the remainder in decimal form. We will discuss decimals later in the book.

EXAMPLE 5 A family has an annual income of $35,880. How much is their average monthly income?

Solution Because there are 12 months in a year and the yearly (annual) income is $35,880, we want to know what $35,880 divided into 12 equal parts is. Therefore, we have

$$
\begin{array}{r}
2990 \\
12\overline{)35880} \\
-\,24 \\
\hline
118 \\
-\,108 \\
\hline
108 \\
-\,108 \\
\hline
00 \\
\end{array}
$$

Because $35{,}880 \div 12 = 2{,}990$, the monthly income for this family is $2,990.

Division by Zero

We cannot divide by 0. That is, we cannot use 0 as a divisor in any division problem. Here's why:

Suppose there was an answer to the problem

$$\frac{8}{0} = ?$$

That would mean that $0 \cdot ? = 8$. But we already know that multiplication by 0 always produces 0. There is no number we can use for the ? to make a true statement out of $0 \cdot ? = 8$. Because this was equivalent to the original division problem

$$\frac{8}{0} = ?$$

we have no number to associate with the expression $\frac{8}{0}$. It is undefined.

RULE Division by Zero

Division by 0 is undefined. Any expression with a divisor of 0 is undefined. We cannot divide by 0.

GETTING READY FOR CLASS

After reading through the preceding section, respond in your own words and in complete sentences.

A. Which sentence below describes the problem shown in Example 1?

 a. The quotient of 7 and 595 is 85.

 b. Seven divided by 595 is 85.

 c. The quotient of 595 and 7 is 85.

B. To find a solution to a division problem, how would you think in terms of multiplication?

C. Example 4 shows that $1{,}690 \div 67$ gives a quotient of 25 with a remainder of 15. If we were to divide 1,692 by 67, what would the remainder be?

D. Explain why division by 0 is undefined in mathematics.

Answer

5. $2,860

Vocabulary Review

Choose the correct words to fill in the blanks below.

remainder divisor quotient undefined

1. The word _____ is used to indicate division.
2. In the expression $12 \div 6$, the 12 is called the _____, and the 6 is called the _____.
3. When a number does not divide evenly into another number, we are left with a _____.
4. An expression with a divisor of zero is _____.

Problems

A Write each of the following in symbols.

1. The quotient of 6 and 3

2. The quotient of 3 and 6

3. The quotient of 45 and 9

4. The quotient of 12 and 4

5. The quotient of r and s

6. The quotient of s and r

7. The quotient of 20 and 4 is 5.

8. The quotient of 20 and 5 is 4.

Write a multiplication statement that is equivalent to each of the following division statements.

9. $6 \div 2 = 3$

10. $6 \div 3 = 2$

11. $\dfrac{36}{9} = 4$

12. $\dfrac{36}{4} = 9$

13. $\dfrac{48}{6} = 8$

14. $\dfrac{35}{7} = 5$

15. $28 \div 7 = 4$

16. $81 \div 9 = 9$

B Find each of the following quotients. (Divide.)

17. $25 \div 5$

18. $72 \div 8$

19. $40 \div 5$

20. $12 \div 2$

21. $9 \div 0$

22. $7 \div 1$

23. $360 \div 8$

24. $285 \div 5$

25. $\dfrac{138}{6}$

26. $\dfrac{267}{3}$

27. $5\overline{)7,650}$

28. $5\overline{)5,670}$

29. 5)6,750
30. 5)6,570
31. 3)54,000
32. 3)50,400

33. 3)50,040
34. 3)50,004
35. 4)96,000
36. 5)25,450

Divide. You shouldn't have any wrong answers because you can always check your results with multiplication.

37. 1,440 ÷ 32
38. 1,206 ÷ 67
39. $\dfrac{2,401}{49}$
40. $\dfrac{4,606}{49}$

41. 28)12,096
42. 28)96,012
43. 63)90,594
44. 45)17,595

45. 87)61,335
46. 79)48,032
47. 45)135,900
48. 56)227,920

C Divide. The following division problems all have remainders.

49. 6)370
50. 8)390
51. 3)271
52. 3)172

53. 26)345
54. 26)543
55. 71)16,620
56. 71)33,240

57. 23)9,250
58. 23)20,800
59. 169)5,950
60. 391)34,450

Complete the following tables.

61.

First Number	Second Number	The Quotient of *a* and *b*
a	*b*	$\dfrac{a}{b}$
100	25	
100	26	
100	27	
100	28	

62.

First Number	Second Number	The Quotient of *a* and *b*
a	*b*	$\dfrac{a}{b}$
100	25	
101	25	
102	25	
103	25	

Estimating

Work these problems mentally, without using a calculator.

63. The quotient 845 ÷ 93 is closest to which of the following numbers?

 a. 10 **b.** 100 **c.** 1,000 **d.** 10,000

64. The quotient 762 ÷ 43 is closest to which of the following numbers?

 a. 2 **b.** 20 **c.** 200 **d.** 2,000

65. The quotient 15,208 ÷ 771 is closest to which of the following numbers?

 a. 2 **b.** 20 **c.** 200 **d.** 2,000

66. The quotient 24,471 ÷ 523 is closest to which of the following numbers?

 a. 5 **b.** 50 **c.** 500 **d.** 5,000

Without a calculator give a one-digit estimate for each of the following quotients. That is, for each quotient, mentally estimate the answer using one of the digits 1, 2, 3, 4, 5, 6, 7, 8, or 9.

67. 316 ÷ 289

68. 662 ÷ 289

69. 728 ÷ 355

70. 728 ÷ 177

71. 921 ÷ 243

72. 921 ÷ 442

73. 673 ÷ 109

74. 673 ÷ 218

Applying the Concepts

The application problems that follow may involve more than division. Some may require addition, subtraction, or multiplication, whereas others may use a combination of two or more operations.

75. Monthly Income A family has an annual income of $22,200. How much is their monthly income?

76. Hourly Wages If a man works an 8-hour shift and is paid $96, how much does he make for 1 hour?

77. Price per Pound If 6 pounds of a certain kind of fruit cost 96¢, how much does 1 pound cost?

78. Cost of a Dress A dress shop orders 45 dresses for a total of $675. If they paid the same amount for each dress, how much was each dress?

79. Fitness Walking The guidelines for fitness indicate that a person who walks 10,000 steps daily is physically fit. According to experts, it takes just over 2,000 steps to walk one mile. If that is the case, how many miles do you need to walk in order to take 10,000 steps?

80. Filling Glasses How many 8-ounce glasses can be filled from three 32-ounce bottles of soda?

81. Filling Glasses How many 5-ounce glasses can be filled from a 32-ounce bottle of milk? How many ounces of milk will be left in the bottle when all the glasses are full?

82. Filling Glasses How many 3-ounce glasses can be filled from a 28-ounce bottle of milk? How many ounces of milk will be left in the bottle when all the glasses are filled?

83. Filling Glasses How many 32-ounce bottles of Coke will be needed to fill sixteen 6-ounce glasses?

84. Filling Glasses How many 28-ounce bottles of 7-Up will be needed to fill fourteen 6-ounce glasses?

85. Cost of Wine If a person paid $192 for 16 bottles of wine, how much did each bottle cost?

86. Miles per Gallon A traveling salesman kept track of his mileage for 1 month. He found that he traveled 1,104 miles and used 48 gallons of gas. How many miles did he travel on each gallon of gas?

87. Milligrams of Calcium Suppose one egg contains 25 milligrams of calcium, a piece of toast contains 40 milligrams of calcium, and a glass of milk contains 215 milligrams of calcium. How many milligrams of calcium are contained in a breakfast that consists of three eggs, two glasses of milk, and four pieces of toast?

88. Milligrams of Iron Suppose a glass of juice contains 3 milligrams of iron and a piece of toast contains 2 milligrams of iron. If Diane drinks two glasses of juice and has three pieces of toast for breakfast, how much iron is contained in the meal?

Calculator Problems

Find each of the following quotients using a calculator.

89. $305,026 \div 698$

90. $771,537 \div 949$

91. 18,436,466 divided by 5,678

92. 2,492,735 divided by 2,345

93. The quotient of 603,955 and 695

94. The quotient of 875,124 and 876

95. $4,903\overline{)27,868,652}$

96. $3,090\overline{)2,308,230}$

97. Gallons per Minute If a 79,768-gallon tank can be filled in 472 minutes, how many gallons enter the tank each minute?

98. Weight per Case A truckload of 632 crates of motorcycle parts weighs 30,968 pounds. How much does each of the crates weigh, if they each weigh the same amount?

Find the Mistake

Each sentence below contains a mistake. Circle the mistake and write the correct word(s) or numbers(s) on the line provided.

1. The division notation $\frac{10}{5}$ is equivalent to $10\overline{)5}$. _____

2. The quotient of 198 and 11 is 11. _____

3. To divide 1,640 by 12, first subtract 12 from 40. _____

4. Dividing 2,380 by 13 gives a remainder of 13. _____

0.7 Exponents and Order of Operations

Image © sxc.hu, Des¹gn, 2005

OBJECTIVES

A Evaluate exponential expressions with whole numbers.

B Use order of operations to evaluate expressions.

KEY WORDS

base

exponent

order of operations

In 1856, the steamboat *Arabia* got stuck in sunken woody debris in the Missouri River. One of the dead trees ripped a hole in the ship's hull, which subsequently filled with water and sank the ship. Over the years, the course of the river has shifted a half mile away, leaving the ship buried far below a Kansas City cornfield for more than a century.

In 1987, the local Hawley family set out to uncover the ship. After three weeks of digging using heavy equipment, the Hawleys finally reached the hull of the ship. The nutrient-rich mud had preserved much of the hull's cargo in pristine condition. The family uncovered artifacts such as elegant china, fancy clothes made of silk and beaver furs, medicines, perfumes, tools, weapons, eyeglasses, even jars of preserved food that are still edible nearly 150 years later!

The hole dug to find the Arabia was as large as a football field and was 45 feet deep. To calculate the volume of dirt excavated in order to unearth the Arabia, we need to multiply the square footage of a football field by the depth of the hole. A football field is 57,600 square feet (ft^2), therefore the hole's volume is 2,592,000 cubic feet (ft^3)! The notation for square feet and cubic feet uses exponents, which we will discuss further now.

A Exponents

Exponents are a shorthand way of writing repeated multiplication. In the expression 2^3, 2 is called the *base* and 3 is called the *exponent*. The expression 2^3 is read "2 to the third power" or "2 cubed." The exponent 3 tells us to use the base 2 as a multiplication factor three times.

$$2^3 = 2 \cdot 2 \cdot 2 \qquad \textit{2 is used as a factor three times.}$$

We can simplify the expression by multiplication.

$$2^3 = 2 \cdot 2 \cdot 2$$
$$= 4 \cdot 2$$
$$= 8$$

The expression 2^3 is equal to the number 8. We can summarize this discussion with the following definition:

> **DEFINITION** exponent
>
> An *exponent* is a whole number that indicates how many times the base is to be used as a factor. Exponents indicate repeated multiplication.

For example, in the expression 5^2, 5 is the base and 2 is the exponent. The meaning of the expression is

$$5^2 = 5 \cdot 5 \qquad \textit{5 is used as a factor two times.}$$
$$= 25$$

The expression 5^2 is read "5 to the second power" or "5 squared."

**Video Examples
Section 0.7**

1. Name the base and exponent.
 a. 3^4
 b. 5^{10}
 c. x^5

EXAMPLE 1 Name the base and the exponent.
a. 3^2 **b.** 3^3 **c.** 2^4

Solution

a. The base is 3, and the exponent is 2. The expression is read "3 to the second power" or "3 squared."

b. The base is 3, and the exponent is 3. The expression is read "3 to the third power" or "3 cubed."

c. The base is 2, and the exponent is 4. The expression is read "2 to the fourth power."

As you can see from this example, a base raised to the second power is also said to be *squared,* and a base raised to the third power is also said to be *cubed.* These are the only two exponents (2 and 3) that have special names. All other exponents are referred to only as "fourth powers," "fifth powers," "sixth powers," and so on.

The next example shows how we can simplify expressions involving exponents by using repeated multiplication.

2. Expand and multiply.
 a. 5^2
 b. 4^3
 c. 3^4

EXAMPLE 2 Expand and multiply.
a. 3^2 **b.** 4^2 **c.** 3^3 **d.** 2^4

Solution

a. $3^2 = 3 \cdot 3 = 9$

b. $4^2 = 4 \cdot 4 = 16$

c. $3^3 = 3 \cdot 3 \cdot 3 = 9 \cdot 3 = 27$

d. $2^4 = 2 \cdot 2 \cdot 2 \cdot 2 = 4 \cdot 4 = 16$

USING TECHNOLOGY Calculators

Here is how we use a calculator to evaluate exponents, as we did in Example 2d:

Scientific Calculator: 2 $\boxed{x^y}$ 4 $\boxed{=}$

Graphing Calculator: 2 $\boxed{\wedge}$ 4 $\boxed{\text{ENT}}$ or 2 $\boxed{x^y}$ 4 $\boxed{\text{ENT}}$
(depending on the calculator)

Finally, we should consider what happens when the numbers 0 and 1 are used as exponents.

RULE Exponent One

Any number raised to the first power is itself. that is, if we let the letter a represent any number, then
$$a^1 = a$$

RULE Exponent Zero

Any number other than 0 raised to the 0 power is 1. That is, if a represents any nonzero number, then it is always true that
$$a^0 = 1$$

3. Simplify.
 a. 8^1
 b. 10^0
 c. 16^1
 d. x^0

EXAMPLE 3 Simplify.
a. 5^1 **b.** 9^1 **c.** 4^0 **d.** 8^0

Solution

a. $5^1 = 5$

b. $9^1 = 9$

c. $4^0 = 1$

d. $8^0 = 1$

B Order of Operations

The symbols we use to specify operations, $+, -, \cdot, \div$, along with the symbols we use for grouping, () and [], serve the same purpose in mathematics as punctuation marks in English. They may be called the punctuation marks of mathematics.

Consider the following sentence:

Bob said John is tall.

It can have two different meanings, depending on how we punctuate it:

1. "Bob," said John, "is tall."
2. Bob said, "John is tall."

Without the punctuation marks we don't know which meaning the sentence has.

Now, consider the following mathematical expression:

$$4 + 5 \cdot 2$$

What should we do? Should we add 4 and 5 first, or should we multiply 5 and 2 first? There seem to be two different answers. In mathematics, we want to avoid situations in which two different results are possible. Therefore, we follow the rule for *order of operations*.

RULE Order of Operations

When evaluating mathematical expressions, we will perform the operations in the following order:

1. If the expression contains grouping symbols, such as parentheses (), brackets [], or a fraction bar, then we perform the operations inside the grouping symbols, or above and below the fraction bar, first.

2. Then we evaluate, or simplify, any numbers with exponents.

3. Then we do all multiplications and divisions in order, starting at the left and moving right.

4. Finally, we do all additions and subtractions, from left to right.

According to our rule, the expression $4 + 5 \cdot 2$ would have to be evaluated by multiplying 5 and 2 first, and then adding 4. The correct answer—and the only answer—to this problem is 14.

$$4 + 5 \cdot 2 = 4 + 10 \qquad \text{Multiply first, then add.}$$
$$= 14$$

Here are some more examples that illustrate how we apply the rule for order of operations to simplify (or evaluate) expressions.

EXAMPLE 4 Simplify: $4 \cdot 8 - 2 \cdot 6$.

Solution We multiply first and then subtract.

$$4 \cdot 8 - 2 \cdot 6 = 32 - 12 \qquad \text{Multiply first, then subtract.}$$
$$= 20$$

EXAMPLE 5 Simplify: $5 + 2(7 - 1)$.

Solution According to the rule for the order of operations, we must do what is inside the parentheses first.

$$5 + 2(7 - 1) = 5 + 2(6) \qquad \text{Work inside parentheses first.}$$
$$= 5 + 12 \qquad \text{Then multiply.}$$
$$= 17 \qquad \text{Then add.}$$

4. Simplify: $5 \cdot 12 - 3 \cdot 7$.

5. Simplify: $3 + 7(8 - 3)$.

Answers
4. 39
5. 38

6. Simplify: $4 \cdot 3^2 + 24 \div 2^3 - 6$.

EXAMPLE 6 Simplify: $9 \cdot 2^3 + 36 \div 3^2 - 8$.

Solution

$$9 \cdot 2^3 + 36 \div 3^2 - 8 = 9 \cdot 8 + 36 \div 9 - 8 \quad \text{Work exponents first.}$$
$$= 72 + 4 - 8 \qquad \text{Then multiply and divide, left to right.}$$
$$= 76 - 8 \quad\left.\vphantom{\begin{matrix}a\\b\end{matrix}}\right\}$$
$$= 68 \qquad \text{Add and subtract left to right,}$$

USING TECHNOLOGY Calculators

Here is how we use a calculator to work the problem shown in Example 5:

Scientific Calculator: $5 \boxed{+} 2 \boxed{\times} \boxed{(} \boxed{(} 7 \boxed{-} 1 \boxed{)} \boxed{=}$

Graphing Calculator: $5 \boxed{+} 2 \boxed{(} \boxed{(} 7 \boxed{-} 1 \boxed{)} \boxed{\text{ENT}}$

Example 6 on a calculator looks like this:

Scientific Calculator: $9 \boxed{\times} 2 \boxed{x^y} 3 \boxed{+} 36 \boxed{\div} 3 \boxed{x^y} 2 \boxed{-} 8 \boxed{=}$

Graphing Calculator: $9 \boxed{\times} 2 \boxed{\wedge} 3 \boxed{+} 36 \boxed{\div} 3 \boxed{\wedge} 2 \boxed{-} 8 \boxed{\text{ENT}}$

7. Simplify: $4 + 2[12 - 2(6 - 3)]$.

EXAMPLE 7 Simplify: $3 + 2[10 - 3(5 - 2)]$.

Solution The brackets, [], are used in the same way as parentheses. In a case like this, we move to the innermost grouping symbols first and begin simplifying.

$$3 + 2[10 - 3(5 - 2)] = 3 + 2[10 - 3(3)]$$
$$= 3 + 2[10 - 9]$$
$$= 3 + 2[1]$$
$$= 3 + 2$$
$$= 5$$

Table 1 lists some English expressions and their corresponding mathematical expressions written in symbols.

Table 1

In English	Mathematical Equivalent
5 times the sum of 3 and 8	$5(3 + 8)$
Twice the difference of 4 and 3	$2(4 - 3)$
6 added to 7 times the sum of 5 and 6	$6 + 7(5 + 6)$
The sum of 4 times 5 and 8 times 9	$4 \cdot 5 + 8 \cdot 9$
3 subtracted from the quotient of 10 and 2	$10 \div 2 - 3$

GETTING READY FOR CLASS

After reading through the preceding section, respond in your own words and in complete sentences.

A. What is an exponent?

B. Give a written description of how you would use the order of operations to simplify the expression below.

$$3 + 4(5 + 6)$$

C. What happens when 0 is used as an exponent?

D. Name the four steps in the order of operations.

Vocabulary Review

Choose the correct words to fill in the blanks below.

| zero | right | factor | left | grouping symbols |
| after | multiplication | exponent | base | before |

1. Exponents are a way of writing repeated _____.

2. An exponent is a whole number that indicates how many times the base is to be used as a _____.

3. In the expression 4^3, the 3 is called the _____ and the 4 is called the _____.

4. Any number other than zero raised to the _____ power is 1.

5. For the order of operations, we perform all operations inside the _____ before any other operations.

6. When performing operations, we always start at the _____ and move to the _____.

7. When following the order of operations, simplify any numbers with exponents _____ doing any multiplications or divisions.

8. When following the order of operations, we do all subtractions or additions _____ any multiplications or divisions.

Problems

A For each of the following expressions, name the base and the exponent.

1. 4^5 **2.** 5^4 **3.** 3^6 **4.** 6^3 **5.** 8^2

6. 2^8 **7.** 9^1 **8.** 1^9 **9.** 4^0 **10.** 0^4

Use the definition of exponents as indicating repeated multiplication to simplify each of the following expressions.

11. 6^2 **12.** 7^2 **13.** 2^3 **14.** 2^4

15. 1^4 **16.** 5^1 **17.** 9^0 **18.** 27^0

19. 9^2 **20.** 8^2 **21.** 10^1 **22.** 8^1

23. 12^1 **24.** 16^0 **25.** 45^0 **26.** 3^4

B Use the rule for the order of operations to simplify each expression.

27. $16 - 8 + 4$

28. $16 - 4 + 8$

29. $20 \div 2 \cdot 10$

30. $40 \div 4 \cdot 5$

31. $20 - 4 \cdot 4$

32. $30 - 10 \cdot 2$

33. $3 + 5 \cdot 8$

34. $7 + 4 \cdot 9$

35. $3 \cdot 6 - 2$

36. $5 \cdot 1 + 6$

37. $6 \cdot 2 + 9 \cdot 8$

38. $4 \cdot 5 + 9 \cdot 7$

39. $4 \cdot 5 - 3 \cdot 2$

40. $5 \cdot 6 - 4 \cdot 3$

41. $5^2 + 7^2$

42. $4^2 + 9^2$

43. $480 + 12(32)^2$

44. $360 + 14(27)^2$

45. $3 \cdot 2^3 + 5 \cdot 4^2$

46. $4 \cdot 3^2 + 5 \cdot 2^3$

47. $8 \cdot 10^2 - 6 \cdot 4^3$

48. $5 \cdot 11^2 - 3 \cdot 2^3$

49. $2(3 + 6 \cdot 5)$

50. $8(1 + 4 \cdot 2)$

51. $19 + 50 \div 5^2$

52. $9 + 8 \div 2^2$

53. $9 - 2(4 - 3)$

54. $15 - 6(9 - 7)$

55. $4 \cdot 3 + 2(5 - 3)$

56. $6 \cdot 8 + 3(4 - 1)$

57. $4[2(3) + 3(5)]$

58. $3[2(5) + 3(4)]$

59. $(7 - 3)(8 + 2)$

60. $(9 - 5)(9 + 5)$

61. $3(9 - 2) + 4(7 - 2)$

62. $7(4 - 2) - 2(5 - 3)$

63. $18 + 12 \div 4 - 3$

64. $20 + 16 \div 2 - 5$

65. $4(10^2) + 20 \div 4$

66. $3(4^2) + 10 \div 5$

67. $8 \cdot 2^4 + 25 \div 5 - 3^2$

68. $5 \cdot 3^4 + 16 \div 8 - 2^2$

69. $5 + 2[9 - 2(4 - 1)]$

70. $6 + 3[8 - 3(1 + 1)]$

71. $3 + 4[6 + 8(2 - 0)]$

72. $2 + 5[9 + 3(4 - 1)]$

73. $\dfrac{15 + 5(4)}{17 - 12}$

74. $\dfrac{20 + 6(2)}{11 - 7}$

Translate each English expression into an equivalent mathematical expression written in symbols. Then simplify.

75. 8 times the sum of 4 and 2

76. 3 times the difference of 6 and 1

77. Twice the sum of 10 and 3

78. 5 times the difference of 12 and 6

79. 4 added to 3 times the sum of 3 and 4

80. 25 added to 4 times the difference of 7 and 5

81. 9 subtracted from the quotient of 20 and 2

82. 7 added to the quotient of 6 and 2

83. The sum of 8 times 5 and 5 times 4

84. The difference of 10 times 5 and 6 times 2

Applying the Concepts

Nutrition Labels Use the three nutrition labels below to work Problems 85–88.

Pizza Dough (Crust)

Nutrition Facts

Serving Size 1/6 of package (65g)
Servings Per Container: 6

Amount Per Serving

Calories 160	Calories from fat 18

	% Daily Value*
Total Fat 2g	3%
Saturated Fat 0.5g	3%
Poly unsaturated Fat 0g	
Monounsaturated Fat 0g	
Cholesterol 0mg	0%
Sodium 470mg	20%
Total Carbohydrate 31g	10%
Dietary Fiber 1g	4%
Sugars 4g	
Protein 5g	

Vitamin A 0%	•	Vitamin C 0%
Calcium 0%	•	Iron 10%

*Percent Daily Values are based on a 2,000 calorie diet

Tomato Sauce

Nutrition Facts

Serving Size 1/4 cup (62g)
Servings Per Container: 5

Amount Per Serving

Calories 15	Calories from fat 0

	% Daily Value*
Total Fat 0g	0%
Saturated Fat 0g	0%
Cholesterol 0mg	0%
Sodium 360mg	15%
Potassium 0mg	0%
Total Carbohydrate 3g	1%
Dietary Fiber 1g	4%
Sugars 2g	
Protein 1g	

Vitamin A 4%	•	Vitamin C 8%
Calcium 0%	•	Iron 0%

*Percent Daily Values are based on a 2,000 calorie diet. Your daily values may be higher or lower depending on your calorie needs.

Shredded Mozzarella Cheese

Nutrition Facts

Serving Size 1 oz (28.3g)
Servings Per Container: 12

Amount Per Serving

Calories 72	Calories from fat 41

	% Daily Value*
Total Fat 4.5g	7%
Saturated Fat 2.9g	14%
Cholesterol 18mg	6%
Sodium 175mg	7%
Total Carbohydrate 0.8g	0%
Fiber 0g	0%
Sugars 0.3g	
Protein 6.9g	

Vitamin A 3%	•	Vitamin C 0%
Calcium 22%	•	Iron 0%

*Percent Daily Values (DV) are based on a 2,000 calorie diet

Find the total number of calories in each of the following meals.

85. Pizza Crust 1 serving
Tomato Sauce 1 serving
Cheese 1 serving

86. Pizza Crust 1 serving
Tomato Sauce 2 servings
Cheese 1 serving

87. Pizza Crust 2 servings
Tomato Sauce 1 serving
Cheese 1 serving

88. Pizza Crust 2 servings
Tomato Sauce 1 serving
Cheese 2 servings

The following table lists the number of calories consumed by eating some popular fast foods. Use the table to work Problems 89 and 90.

Calories in Food

Food	Calories
McDonald's hamburger	270
Burger King hamburger	260
Jack in the Box hamburger	280
McDonald's Big Mac	510
Burger King Whopper	630
Jack in the Box Colossus burger	940

89. Compare the total number of calories in the meal in Problem 85 with the number of calories in a McDonald's Big Mac.

90. Compare the total number of calories in the meal in Problem 88 with the number of calories in a Burger King hamburger.

One Step Further: Number Sequences

There is a relationship between the two sequences below. The first sequence is the *sequence of odd numbers*. The second sequence is called the *sequence of squares*.

$1, 3, 5, 7, \ldots$ The sequence of odd numbers

$1, 4, 9, 16, \ldots$ The sequence of squares

91. Add the first two numbers in the sequence of odd numbers.

92. Add the first three numbers in the sequence of odd numbers.

93. Add the first four numbers in the sequence of odd numbers.

94. Add the first five numbers in the sequence of odd numbers.

95. Compare the answers to Problems 91–94 with the numbers in the sequence of squares. In your own words, explain how the sequence of odd numbers can be used to obtain the sequence of squares.

96. Draw a diagram that shows how the sequence of odd numbers can be used to produce the sequence of squares. The diagram can contain numbers, addition symbols, equal signs, and arrows, but it should not contain any words. Your goal is to draw the diagram so that anyone who knows how to add whole numbers can look at it and see the relationship between the numbers in the two sequences.

Find the Mistake

Each sentence below contains a mistake. Circle the mistake and write the correct word(s) or numbers(s) on the line provided.

1. The expression 8^4 means that the exponent 4 is used as a factor 8 times. _____

2. The number 7 raised to the third power equals $3^7 = 2{,}187$. _____

3. Following the order of operations, work all additions and subtractions before evaluating any exponents.

4. Simplifying $(4 + 1)^3 - (9 + 1)^2$ gives 189. _____

Chapter 0 Summary

The numbers in brackets indicate the sections in which the topics were discussed.

EXAMPLES

The margins of the chapter summaries will be used for examples of the topics being reviewed, whenever it is convenient.

1. The number 42,103,045 written in words is "forty-two million, one hundred three thousand, forty-five." The number 5,745 written in expanded form is $5,000 + 700 + 40 + 5$.

Place Values for Whole Numbers [0.1]

The place values for the digits of any base 10 number are as follows:

Table 1

Trillions			Billions			Millions			Thousands			Ones		
Hundreds	Tens	Ones	Hundreds	Tens	Ones	Hundreds	Tens	Ones	Hundreds	Tens	Ones	Hundreds	Tens	Ones

2. The sum of 5 and 2 is $5 + 2$. The difference of 5 and 2 is $5 - 2$. The product of 5 and 2 is $5 \cdot 2$. The quotient of 10 and 2 is $10 \div 2$.

Vocabulary Associated with Addition, Subtraction, Multiplication, and Division [0.2, 0.3, 0.5, 0.6]

The word *sum* indicates addition.

The word *difference* indicates subtraction.

The word *product* indicates multiplication.

The word *quotient* indicates division.

3. a. $3 + 2 = 2 + 3$
b. $3 \cdot 2 = 2 \cdot 3$
c. $(x + 3) + 5 = x + (3 + 5)$
d. $(4 \cdot 5) \cdot 6 = 4 \cdot (5 \cdot 6)$
e. $3(4 + 7) = 3(4) + 3(7)$

Properties of Addition and Multiplication [0.2, 0.5]

If a, b, and c represent any three numbers, then the properties of addition and multiplication used most often are

Commutative property of addition: $a + b = b + a$

Commutative property of multiplication: $a \cdot b = b \cdot a$

Associative property of addition: $(a + b) + c = a + (b + c)$

Associative property of multiplication: $(a \cdot b) \cdot c = a \cdot (b \cdot c)$

Distributive property: $a(b + c) = a(b) + a(c)$

4. The perimeter of the rectangle below is
$P = 37 + 37 + 24 + 24$
$\quad = 122$ feet

24 ft

37 ft

Perimeter of a Polygon [0.2]

The *perimeter* of any polygon is the sum of the lengths of the sides, and it is denoted with the letter P.

Steps for Rounding Whole Numbers [0.4]

1. Locate the digit just to the right of the place you are to round to.
2. If that digit is less than 5, replace it and all digits to its right with zeros.
3. If that digit is 5 or more, replace it and all digits to its right with zeros, and add 1 to the digit to its left.

5. 5,482 to the nearest ten is 5,480. 5,482 to the nearest hundred is 5,500. 5,482 to the nearest thousand is 5,000.

Formulas for Area [0.5]

Below are two common geometric figures, along with the formulas for their areas.

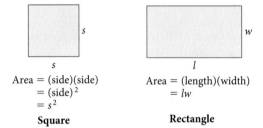

Area = (side)(side)
= (side)2
= s^2

Square

Area = (length)(width)
= lw

Rectangle

Division by 0 (Zero) [0.6]

Division by 0 is undefined. We cannot use 0 as a divisor in any division problem.

6. Each of the expressions below are undefined.

$5 \div 0$ $\quad \dfrac{7}{0} \quad$ 4/0

Order of Operations [0.7]

To simplify a mathematical expression:

1. We simplify the expression inside the grouping symbols first. Grouping symbols are parentheses (), brackets [], or a fraction bar.
2. Then we evaluate any numbers with exponents.
3. We then perform all multiplications and divisions in order, starting at the left and moving right.
4. Finally, we do all the additions and subtractions, from left to right.

7. $4 + 6(8 - 2)$
$= 4 + 6(6)$ *Work inside parentheses first,*
$= 4 + 36$ *Then multiply,*
$= 40$ *Then add.*

Exponents [0.7]

In the expression 2^3, 2 is the *base* and 3 is the *exponent*. An exponent is a shorthand notation for repeated multiplication. The exponent 0 is a special exponent. Any nonzero number to the 0 power is 1.

8. $2^3 = 2 \cdot 2 \cdot 2 = 8$
$5^0 = 1$
$3^1 = 3$

At the end of each chapter, we will provide a test that you should use as a helpful study tool for a chapter exam. Remember, it is still important to study every aspect of the material presented in the chapter, and not rely solely on the Chapter Test to study for an exam.

1. Write the number 30,652 in words. [0.1]

2. Write the number six million, seven thousand, twenty-nine with digits instead of words. [0.1]

3. Write the number 285,634 in expanded form. [0.1]

Identify each of the statements in Problems 4–7 as an example of one of the following properties. [0.2, 0.5]

 a. Addition property of 0

 b. Multiplication property of 0

 c. Multiplication property of 1

 d. Commutative property of addition

 e. Commutative property of multiplication

 f. Associative property of addition

 g. Associative property of multiplication

4. $8 + 9 = 9 + 8$

5. $(5 \cdot 7) \cdot 8 = 5 \cdot (7 \cdot 8)$

6. $4 \cdot 1 = 4$

7. $8 \cdot 0 = 0$

Find each of the following sums. (Add.) [0.2]

8.
$$\begin{array}{r} 237 \\ + 461 \\ \hline \end{array}$$

9.
$$\begin{array}{r} 3,821 \\ 467 \\ + 20,315 \\ \hline \end{array}$$

10. Find the perimeter. [0.2]

Find each of the following differences. (Subtract.) [0.3]

11.
$$\begin{array}{r} 846 \\ - 325 \\ \hline \end{array}$$

12.
$$\begin{array}{r} 8,502 \\ - 3,715 \\ \hline \end{array}$$

13. Round the number 238,543 to the nearest ten thousand. [0.4]

Find each of the following products. (Multiply.) [0.5]

14. 5(162)

15. 74(267)

Find each of the following quotients. (Divide.) [0.6]

16. $1,224 \div 17$

17. $512\overline{)11,776}$

Write the following in numbers, then simplify as much as possible.

18. Twice the sum of 13 and 4. [0.2, 0.3, 0.5, 0.6]

19. The quotient of 18 and 6 increased by 12. [0.2, 0.3, 0.5, 0.6]

Use the rule for the order of operations to simplify each expression as much as possible. [0.7]

20. $5(4)^2 - 5(2)^3$

21. $9 - 3(7 - 5)$

22. $5 + 3(29 - 4)$

23. $5(x - 4)$

The chart shows how many Facebook fans each of these have. Use the information to answer the following questions.

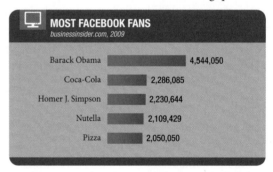

24. How many more fans does Barack Obama have than Homer J. Simpson? Write your answer in digits and words.

25. How many fewer fans does Pizza have than Coca-Cola? Write your answer in digits and words.

Real Numbers and Algebraic Expressions

1

1.1 Real Numbers

1.2 Addition and Subtraction of Real Numbers

1.3 Multiplication of Real Numbers

1.4 Division of Real Numbers

1.5 Properties of Real Numbers

1.6 Simplifying Expressions

A dam built on a large river can produce hydroelectric power. Gravity forces water through the propellers of a turbine near the bottom of the dam wall. The turbine rotates a metal shaft in a generator, producing electricity. The Three Gorges Dam on the Yangtze River in China is the world's largest hydroelectric power project. The total electricity generating capacity of the dam is estimated to be 18,000 megawatts of power. However, the environmental and human impact of the project is still a source of

debate. When the dam was completed in 2006, the 660-kilometer-long reservoir displaced approximately 1,300,000 people.

If 1,503 villages were submerged, how many people lived in each village before being relocated? (For the purpose of this problem, let's assume there were an equal number of people in each village.) We can find the answer to this question by solving this division problem:

$$\text{People relocated from each village} = \frac{\text{Total people relocated}}{\text{\# of villages}} = \frac{1,300,000}{1,503}$$

This type of division is one of the skills you will use as you work through this chapter.

OBJECTIVES

A Locate and label points on the real number line.

B Change a fraction to an equivalent fraction with a new denominator.

C Simplify expressions containing absolute value.

D Identify the opposite of a number.

E Translate between phrases written in English and expressions written in symbols.

F Find the perimeter and area of squares, rectangles, and triangles.

KEY WORDS

real number line

coordinate

equivalent fractions

absolute value

opposites

variable

algebraic expression

algebraic equation

algebraic inequality

area

perimeter

**Video Examples
Section 1.1**

Note If there is no sign (+ or −) in front of a non-zero number, the number is assumed to be positive (+). The number zero has no sign.

1.1 Real Numbers

©iStockphoto.com/KKIDD

Each winter, thousands of spectators come to a mountain resort in Wyoming to watch the World Championship Snow King Hill Climb. One by one, snowmobilers attempt to race up a 1,500-foot snow-covered slope on Snow King Mountain. At its steepest point, the slope measures 45 degrees! The steep slope often overpowers the snowmobile, which comes tumbling back down the hill, sometimes taking the rider with it.

The weather and the conditions of the snow may also play a part in the success of the snowmobilers. The table and bar chart shown here give the record low temperature, in degrees Fahrenheit, for each month of the year in the nearby city of Jackson, Wyoming. Notice that some of these temperatures are represented by negative numbers which we will begin working with in this section.

FIGURE 1

A The Real Number Line

To represent negative numbers in algebra, we use what is called the *real number line*. Here is how we construct a real number line: We first draw a straight line and label a convenient point on the line with 0. Then we mark off equally spaced distances in both directions from 0. Label the points to the right of 0 with the numbers 1, 2, 3, . . .(the dots mean "and so on"). The points to the left of 0 we label in order, −1, −2, −3, Here is what it looks like:

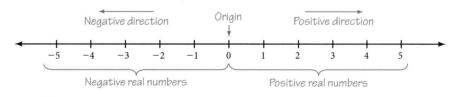

The numbers increase in value going from left to right. If we "move" to the right, we are moving in the positive direction. If we move to the left, we are moving in the negative direction. When we compare two numbers on the number line, the number on the left is always smaller than the number on the right. For instance, -3 is smaller than -1 because it is to the left of -1 on the number line.

EXAMPLE 1 Locate and label the points on the real number line associated with the numbers $-3.5, -1\frac{1}{4}, \frac{1}{2}, \frac{3}{4}, 2.5$.

Solution We draw a real number line from -4 to 4 and label the points in question.

Practice Problems

1. Locate and label the points associated with $-2, -\frac{1}{2}, 0, 1.5,$ 2.75.

> **DEFINITION** coordinate
>
> The number associated with a point on the real number line is called the *coordinate* of that point.

In the preceding example, the numbers $\frac{1}{2}, \frac{3}{4}, 2.5, -3.5,$ and $-1\frac{1}{4}$ are the coordinates of the points they represent.

> **DEFINITION** real numbers
>
> The numbers that can be represented with points on the real number line are called *real numbers*.

Note There are other numbers on the number line that you may not be as familiar with. They are irrational numbers such as $\pi, \sqrt{2}, \sqrt{3}$. We will introduce these numbers later in the book.

Real numbers include whole numbers, fractions, decimals, and other numbers that are not as familiar to us as these.

B Equivalent Fractions on the Number Line

The number line can be used to visualize fractions. Recall that for the fraction $\frac{a}{b}$, a is called the numerator and b is called the denominator. As long as $\frac{a}{b}$ is a proper fraction where $a < b$, the denominator b indicates the number of equal parts in the interval from 0 to 1 on the number line. The numerator a indicates how many of those parts we have. If we take that part of the number line from 0 to 1 and divide it into *three equal parts*, we say that we have divided it into *thirds* (Figure 2). Each of the three segments is $\frac{1}{3}$ (one third) of the whole segment from 0 to 1.

FIGURE 2

Two of these smaller segments together are $\frac{2}{3}$ (two thirds) of the whole segment. And three of them would be $\frac{3}{3}$ (three thirds), or the whole segment.

Let's do the same thing again with six equal divisions of the segment from 0 to 1 (Figure 3). In this case we say each of the smaller segments has a length of $\frac{1}{6}$ (one sixth).

FIGURE 3

The same point we labeled with $\frac{1}{3}$ in Figure 2 is now labeled with $\frac{2}{6}$. Likewise, the point we labeled earlier with $\frac{2}{3}$ is now labeled $\frac{4}{6}$. It must be true then that

$$\frac{2}{6} = \frac{1}{3} \qquad \text{and} \qquad \frac{4}{6} = \frac{2}{3}$$

Actually, there are many fractions that name the same point as $\frac{1}{3}$. If we were to divide the segment between 0 and 1 into 12 equal parts, 4 of these 12 equal parts $\left(\frac{4}{12}\right)$ would be the same as $\frac{2}{6}$ or $\frac{1}{3}$; that is,

$$\frac{4}{12} = \frac{2}{6} = \frac{1}{3}$$

Even though these three fractions look different, each names the same point on the number line, as shown in Figure 4. All three fractions have the same value because they all represent the same number.

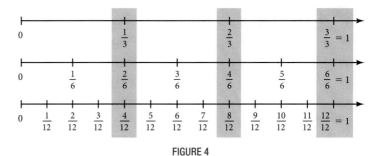

FIGURE 4

> **DEFINITION** equivalent fractions
>
> Fractions that represent the same number are said to be *equivalent*. Equivalent fractions may look different, but they must have the same value.

It is apparent that every fraction has many different representations, each of which is equivalent to the original fraction. The next two properties review a way of changing the terms of a fraction without changing its value.

> **PROPERTY** Equivalent Fractions Using Multiplication
>
> Multiplying the numerator and denominator of a fraction by the same nonzero number never changes the value of the fraction.

> **PROPERTY** Equivalent Fractions Using Division
>
> Dividing the numerator and denominator of a fraction by the same nonzero number never changes the value of the fraction.

EXAMPLE 2 Write $\frac{3}{4}$ as an equivalent fraction with denominator 20.

Solution The denominator of the original fraction is 4. The fraction we are trying to find must have a denominator of 20. We know that if we multiply 4 by 5, we get 20. The property for equivalent fractions using multiplication indicates that we are free to multiply the denominator by 5 as long as we do the same to the numerator.

$$\frac{3}{4} = \frac{3 \cdot 5}{4 \cdot 5} = \frac{15}{20}$$

The fraction $\frac{15}{20}$ is equivalent to the fraction $\frac{3}{4}$.

2. Write $\frac{5}{8}$ as an equivalent fraction with denominator 48.

C Absolute Values

Representing numbers on the number line lets us give each number two important properties: a direction from zero and a distance from zero. The direction from zero is represented by the sign in front of the number. (A number without a sign is understood to be positive.) The distance from zero is called the *absolute value* of the number, as the following definition indicates:

> **DEFINITION** absolute value
>
> The *absolute value* of a real number is its distance from zero on the number line. If x represents a real number, then the absolute value of x is written $|x|$.

Note To help you understand absolute value, consider the distance between two cities. If you were to travel from city A to city B, the distance would be a positive measurement. The same goes for traveling from city B back to city A. Just as with absolute value, the distance from one point to another is always positive, regardless of the direction.

EXAMPLE 3 Write each expression without absolute value symbols.

a. $|5|$ **b.** $|-5|$ **c.** $\left|-\frac{1}{2}\right|$ **d.** $-|-8|$

Solution

a. $|5| = 5$ The number 5 is 5 units from zero.

b. $|-5| = 5$ The number -5 is 5 units from zero.

c. $\left|-\frac{1}{2}\right| = \frac{1}{2}$ The number $-\frac{1}{2}$ is $\frac{1}{2}$ units from zero.

d. $-|-8| = -8$ The number -8 is 8 units from zero, but the preceding negative sign gives the value a direction of 8 units to the left of zero.

3. Write each expression without absolute value symbols.

a. $|7|$

b. $|-7|$

c. $\left|-\frac{3}{4}\right|$

d. $-|-12|$

Note The value of the expression in part d of Example 3 is much different than the value of the expression $-(-8)$, where we use the parentheses instead of absolute value symbols.

$-|-8| = -8$, whereas $-(-8) = 8$

The absolute value of a number is *never* negative. It is the distance the number is from zero without regard to which direction it is from zero. When working with the absolute value of sums and differences, we must simplify the expression inside the absolute value symbols first and then find the absolute value of the simplified expression.

EXAMPLE 4 Simplify each expression.

a. $|8 - 3|$ **b.** $|3 \cdot 2^3 + 2 \cdot 3^2|$ **c.** $|9 - 2| - |8 - 6|$

Solution

a. $|8 - 3| = |5| = 5$

b. $|3 \cdot 2^3 + 2 \cdot 3^2| = |3 \cdot 8 + 2 \cdot 9| = |24 + 18| = |42| = 42$

c. $|9 - 2| - |8 - 6| = |7| - |2| = 7 - 2 = 5$

4. Simplify each expression.

a. $|7 - 2|$

b. $|2 \cdot 3^2 + 5 \cdot 2^2|$

c. $|10 - 4| - |11 - 9|$

Answers

2. $\frac{30}{48}$

3. a. 7 **b.** 7 **c.** $\frac{3}{4}$ **d.** -12

4. a. 5 **b.** 38 **c.** 4

5. Give the opposite of each number.
 a. 8
 b. −5
 c. $-\dfrac{2}{3}$
 d. −4.2

D Opposites

Another important concept associated with numbers on the number line is that of *opposites*. Here is the definition:

> **DEFINITION** opposites
>
> Numbers the same distance from zero but in opposite directions from zero are called *opposites*.

EXAMPLE 5 Give the opposite of each number.

a. 5 **b.** −3 **c.** $\dfrac{1}{4}$ **d.** −2.3

Solution

	Number	Opposite	
a.	5	−5	5 and −5 are opposites.
b.	−3	3	−3 and 3 are opposites.
c.	$\dfrac{1}{4}$	$-\dfrac{1}{4}$	$\dfrac{1}{4}$ and $-\dfrac{1}{4}$ are opposites.
d.	−2.3	2.3	−2.3 and 2.3 are opposites.

Each negative number is the opposite of some positive number, and each positive number is the opposite of some negative number. The opposite of a negative number is a positive number. In symbols, if a represents a positive number, then

$$-(-a) = a$$

Opposites always have the same absolute value. And, when you add any two opposites, the result is always zero.

$$a + (-a) = 0$$

Also, you will see that understanding opposites is important to graphing subtraction of real numbers in the next section.

E Variables: An Intuitive Look

In the previous chapter, you probably noticed the use of letters instead of numbers in certain expressions. These letters are called *variables*, which we will now discuss in greater detail.

When you filled out the application for the school you are attending, there was a space to fill in your first name. "First name" is a variable quantity because the value it takes depends on who is filling out the application. For example, if your first name is Manuel, then the value of "First Name" is Manuel. However, if your first name is Christa, then the value of "First Name" is Christa.

If we denote "First Name" as F, "Last Name" as L, and "Whole Name" as W, then we take the concept of a variable further and write the relationship between the names this way:

$$F + L = W$$

We use the $+$ symbol loosely here to represent writing the names together with a space between them. This relationship we have written holds for all people who have only a first name and a last name. For those people who have a middle name, the relationship between the names is

$$F + M + L = W$$

where M represents a middle name.

A similar situation exists in algebra when we let a letter stand for a number or a group of numbers. For instance, if we say "let a and b represent numbers," then a and b are called variables because the values they take on vary. Here is the formal definition:

> **DEFINITION** variable
>
> A *variable* is a letter that stands for (represents) a mathematical quantity.

Furthermore, much of what we do in algebra involves comparison of quantities. We will begin by listing some symbols used to compare mathematical quantities. The comparison symbols fall into two major groups: equality symbols and inequality symbols. We use the variables a and b in the following lists so that the relationships shown there are true for all numbers that we will encounter in this book. By using variables, the following statements are general statements about all numbers, rather than specific statements about only a few numbers.

> **PROPERTY** Comparison Symbols
>
Equality:	$a = b$	a is equal to b
> | | $a \neq b$ | a is not equal to b |
> | *Inequality:* | $a < b$ | a is less than b |
> | | $a > b$ | a is greater than b |
> | | $a \geq b$ | a is greater than or equal to b |
> | | $a \leq b$ | a is less than or equal to b |

The symbols for inequality, $<$ and $>$, always point to the smaller of the two quantities being compared. For example, $3 < x$ means 3 is smaller than x. In this case, we can say "3 is less than x" or "x is greater than 3"; both statements are correct. Similarly, the expression $5 > y$ can be read as "5 is greater than y" or as "y is less than 5" because the inequality symbol is pointing to y, meaning y is the smaller of the two quantities.

Next, we consider the symbols used to represent the four basic operations: addition, subtraction, multiplication, and division.

> **PROPERTY** Operation Symbols
>
Addition:	$a + b$	The *sum* of a and b
> | *Subtraction:* | $a - b$ | The *difference* of a and b |
> | *Multiplication:* | $a \cdot b, (a)(b), a(b), (a)b, ab$ | The *product* of a and b |
> | *Division:* | $a \div b, a/b, \dfrac{a}{b}, b\overline{)a}$ | The *quotient* of a and b |

Note In the past, you may have used the notation 3×5 to denote multiplication. In algebra, it is best to avoid this notation if possible, because the multiplication symbol \times can be confused with the variable x when written by hand.

When we encounter the word *sum*, the implied operation is addition. To find the sum of two numbers, we simply add them. *Difference* implies subtraction, *product* implies multiplication, and *quotient* implies division. Notice also that there is more than one way to write the product or quotient of two numbers.

> **PROPERTY** Grouping Symbols
>
> Parentheses () and brackets [] are the symbols used for grouping numbers together. Occasionally, braces { } are also used for grouping, although they are usually reserved for solution set notation.

The following table illustrates the relationship between the symbols for comparing, operating, and grouping and the English language.

Mathematical Expression	Written Equivalent
$4 + 1 = 5$	The sum of 4 and 1 is 5.
$8 - 1 < 10$	The difference of 8 and 1 is less than 10.
$2(3 + 4) = 14$	Twice the sum of 3 and 4 is 14.
$3x \geq 15$	The product of 3 and x is greater than or equal to 15.
$\frac{y}{2} = y - 2$	The quotient of y and 2 is equal to the difference of y and 2.

Lastly, remember that a variable is a letter used to represent a number or a group of numbers. An expression that contains variables with other numbers and symbols is called an *algebraic expression*. Here is a formal definition:

> **DEFINITION** algebraic expression
>
> An *algebraic expression* is an expression that contains any combination of numbers, variables, operation symbols, and grouping symbols.

Each of the following is an algebraic expression:

$$3x + 5 \qquad 4t^2 - 9 \qquad x^2 - 6xy + y^2$$

$$-15x^2y^4z^5 \qquad \frac{a^2 - 9}{a - 3} \qquad \frac{(x - 3)(x + 2)}{4x}$$

You will encounter a variety of algebraic expressions in this book.

Furthermore, when we set an algebraic expression equal to a value or another expression, the result is an *algebraic equation*. Here are some examples of algebraic equations:

$$4x + 2 = 10 \qquad 2a^2 + 3 = -1 \qquad \frac{(y + 2)(y - 3)}{6y}$$

If an inequality symbol is used in place of an equals sign, then the result is called an *algebraic inequality*.

F Formulas for Area and Perimeter

FACTS FROM GEOMETRY Formulas for Area and Perimeter

A square, rectangle, and triangle are shown in the following figures. Note that we have labeled the dimensions of each with variables. The formulas for the perimeter and area of each object are given in terms of its dimensions.

Square
Perimeter = $4s$
Area = s^2

Rectangle
Perimeter = $2l + 2w$
Area = lw

Triangle
Perimeter = $a + b + c$
Area = $\frac{1}{2}bh$

The formula for perimeter gives us the distance around the outside of the object along its sides, whereas the formula for area gives us a measure of the amount of surface the object has.

Note The vertical line labeled h in the triangle is its height, or altitude. It extends from the top of the triangle down to the base, meeting the base at an angle of 90°. The altitude of a triangle is always perpendicular to the base. The small square shown where the altitude meets the base is used to indicate that the angle formed is 90°.

EXAMPLE 6 Find the perimeter and area of each figure.

a.
5 ft

b.
6 in.
8 in.

c.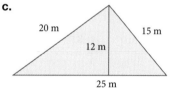
20 m 15 m
12 m
25 m

6. Find the perimeter and area of the figures in Example 6, with the following changes:

 a. $s = 4$ feet

 b. $l = 7$ inches
 $w = 5$ inches

 c. $a = 40$ meters
 $b = 50$ meters (base)
 $c = 30$ meters
 $h = 24$ meters

Solution We use the preceding formulas to find the perimeter and the area. In each case, the units for perimeter are linear units, whereas the units for area are square units.

a. Perimeter = $4s = 4 \cdot 5$ feet = 20 feet
Area = $s^2 = (5 \text{ feet})^2 = 25$ square feet

b. Perimeter = $2l + 2w = 2(8 \text{ inches}) + 2(6 \text{ inches}) = 28$ inches
Area = $lw = (8 \text{ inches})(6 \text{ inches}) = 48$ square inches

c. Perimeter = $a + b + c = (20 \text{ meters}) + (25 \text{ meters}) + (15 \text{ meters}) = 60$ meters
Area = $\frac{1}{2}bh = \frac{1}{2}(25 \text{ meters})(12 \text{ meters}) = 150$ square meters

GETTING READY FOR CLASS

After reading through the preceding section, respond in your own words and in complete sentences.

A. What is a real number?

B. What is an equivalent fraction of $\frac{2}{6}$? Explain.

C. What is a variable?

D. Explain how you find the perimeter and the area of a rectangle.

Answer

6. a. Perimeter = 16 feet
Area = 16 square feet

b. Perimeter = 24 inches
Area = 35 square inches

c. Perimeter = 120 meters
Area = 600 square meters

EXERCISE SET 1.1

Vocabulary Review

Choose the correct words to fill in the blanks below.

| absolute value | nonzero | real | algebraic expression | coordinate |
| equivalent | variables | opposites | real number line | |

1. The _____ is a straight line with points that represent positive and negative numbers and zero.
2. The number associated with a point on the number line is called a _____.
3. Numbers that can be represented with points on a number line are called _____ numbers.
4. Although _____ fractions look different, they have the same value.
5. Multiplying (or dividing) the numerator and denominator of a fraction by the same _____ number never changes the value of the fraction.
6. The _____ of a real number is its distance from zero on the number line.
7. Numbers with the same absolute value but have points in opposite directions from zero on the number line are called _____.
8. An _____ contains any combination of numbers, _____, operation symbols, and grouping symbols.

Problems

A Draw a number line that extends from -5 to 5. Label the points with the following coordinates.

1. 5

2. -2

3. -4

4. -3

5. 1.5

6. -1.5

7. $\dfrac{9}{4}$

8. $\dfrac{8}{3}$

B Write each of the following fractions as an equivalent fraction with denominator 24.

9. $\dfrac{3}{4}$

10. $\dfrac{5}{6}$

11. $\dfrac{1}{2}$

12. $\dfrac{1}{8}$

13. $\dfrac{5}{8}$

14. $\dfrac{7}{12}$

Write each fraction as an equivalent fraction with denominator 60.

15. $\dfrac{3}{5}$

16. $\dfrac{5}{12}$

17. $\dfrac{11}{30}$

18. $\dfrac{9}{10}$

19. $\dfrac{5}{6}$

20. $\dfrac{2}{3}$

C Simplify each expression.

21. $|8 - 2|$

22. $|6 - 1|$

23. $|5 \cdot 2^3 - 2 \cdot 3^2|$

24. $|2 \cdot 10^2 + 3 \cdot 10|$

25. $|7 - 2| - |4 - 2|$

26. $|10 - 3| - |4 - 1|$

27. $10 - |7 - 2(5 - 3)|$

28. $12 - |9 - 3(7 - 5)|$

29. $15 - |8 - 2(3 \cdot 4 - 9)| - 10$

30. $25 - |9 - 3(4 \cdot 5 - 18)| - 20$

31. $25 - 2|8 - 2 \cdot 3| - 4$

32. $30 - 4|6 - 4(3 - 2)| - 7 \cdot 2$

C D For each of the following numbers, give the opposite and the absolute value. (Assume all variables are nonzero.)

33. 10

34. 8

35. $\dfrac{3}{4}$

36. $\dfrac{5}{7}$

37. $\dfrac{11}{2}$

38. $\dfrac{16}{3}$

39. -3

40. -5

41. $-\dfrac{2}{5}$

42. $-\dfrac{3}{8}$

43. x

44. a

E Place one of the symbols $<$ or $>$ between each of the following to make the resulting statement true.

45. $-5 \quad -3$

46. $-8 \quad -1$

47. $-3 \quad -7$

48. $-6 \quad 5$

49. $|-4| \quad -|-4|$

50. $3 \quad -|-3|$

51. $7 \quad -|-7|$

52. $-7 \quad |-7|$

53. $-\dfrac{3}{4} \quad -\dfrac{1}{4}$

54. $-\dfrac{2}{3} \quad -\dfrac{1}{3}$

55. $-\dfrac{3}{2} \quad -\dfrac{3}{4}$

56. $-\dfrac{8}{3} \quad -\dfrac{17}{3}$

For each sentence below, write an equivalent algebraic equation or inequality.

57. The sum of x and 5 is 14.

58. The difference of x and 4 is 8.

59. The product of 5 and y is less than 30.

60. The product of 8 and y is greater than 16.

61. The product of 5 and y is greater than or equal to the difference of y and 16.

62. The product of 3 and y is less than or equal to the sum of y and 6.

63. The quotient of x and 3 is equal to the sum of x and 2.

64. The quotient of x and 2 is equal to the difference of x and 4.

F Find the perimeter and area of each figure.

65.

1 in.

1 in.

66.

15 mm

15 mm

67.

0.75 in.

1.5 in.

68.

1.5 cm

4.5 cm

69.

2.75 cm 3.5 cm

2.5 cm

4 cm

70.

1.8 in. 1.2 in.

1 in.

2 in.

Applying the Concepts

71. Football Yardage A football team gains 6 yards on one play and then loses 8 yards on the next play. To what number on the number line does a loss of 8 yards correspond? The total yards gained or lost on the two plays corresponds to what negative number?

72. Checking Account Balance A woman has a balance of $20 in her checking account. If she writes a check for $30, what negative number can be used to represent the new balance in her checking account?

Temperature In the United States, temperature is measured on the Fahrenheit temperature scale. On this scale, water boils at 212 degrees and freezes at 32 degrees. To denote a temperature of 32 degrees on the Fahrenheit scale, we write

32°F, which is read "32 degrees Fahrenheit"

Use this information for Problems 73 and 74.

73. Temperature and Altitude Marilyn is flying from Seattle to San Francisco on a Boeing 737 jet. When the plane reaches an altitude of 35,000 feet, the temperature outside the plane is 64 degrees below zero Fahrenheit. Represent the temperature with a negative number. If the temperature outside the plane gets warmer by 10 degrees, what will the new temperature be?

74. Temperature Change At 10:00 in the morning in White Bear Lake, Minnesota, John notices the temperature outside is 10 degrees below zero Fahrenheit. Write the temperature as a negative number. An hour later it has warmed up by 6 degrees. What is the temperature at 11:00 that morning?

Wind Chill The table to the right gives wind chill temperatures. The left column gives the air temperature, and the first row is wind speed in miles per hour. The numbers within the table indicate how cold the weather will feel. For example, if the thermometer reads 30°F and the wind is blowing at 15 miles per hour, the wind chill temperature is 9°F. Use the table to answer Problems 75 and 76.

75. Find the wind chill temperature if the thermometer reads 20°F and the wind is blowing at 25 miles per hour.

76. Which will feel colder: a day with an air temperature of 10°F with a 25-mile-per-hour wind, or a day with an air temperature of 25° F and a 10-mile-per-hour wind?

Wind Chill Temperatures					
Air Temperature (°F)	Wind Speed (mph)				
	10	15	20	25	30
30°	16°	9°	4°	1°	−2°
25°	10°	2°	−3°	−7°	−10°
20°	3°	−5°	−10°	−15°	−18°
15°	−3°	−11°	−17°	−22°	−25°
10°	−9°	−18°	−24°	−29°	−33°
5°	−15°	−25°	−31°	−36°	−41°
0°	−22°	−31°	−39°	−44°	−49°
−5°	−27°	−38°	−46°	−51°	−56°

77. Scuba Diving Steve is scuba diving near his home in Maui. At one point he is 100 feet below the surface. Represent this number with a negative number. If he descends another 5 feet, what negative number will represent his new position?

78. Pizza Sales The chart shows top sellers of frozen pizza in one year. Write a mathematical statement using one of the symbols < or > to compare the following:

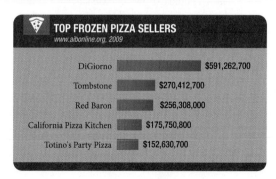

a. Digiorno to California Pizza Kitchen

b. Totino's Party Pizza to Tombstone

c. Red Baron to California Pizza Kitchen

79. Geometry Find the area and perimeter of an $8\frac{1}{2}$-by-11-inch piece of notebook paper.

80. Geometry Find the area and perimeter of an $8\frac{1}{2}$-by-$5\frac{1}{2}$-inch piece of paper.

Calories and Exercise The table here gives the amount of energy expended per hour for various activities for a person weighing 120, 150, or 180 pounds. Use the table to answer questions 81–84.

81. Suppose you weigh 120 pounds. How many calories will you burn if you play handball for 2 hours and then ride your bicycle for an hour?

82. How many calories are burned by a person weighing 150 pounds who jogs for $\frac{1}{2}$ hour and then goes bicycling for 2 hours?

83. Two people go skiing. One weighs 180 pounds and the other weighs 120 pounds. If they ski for 3 hours, how many more calories are burned by the person weighing 180 pounds?

Energy Expended from Exercising

Activity	Calories per Hour		
	120 lb	150 lb	180 lb
Bicycling	299	374	449
Bowling	212	265	318
Handball	544	680	816
Horseback trotting	278	347	416
Jazzercise	272	340	408
Jogging	544	680	816
Skiing (downhill)	435	544	653

84. Two people spend an evening bowling for 3 hours. If one weighs 120 pounds and the other weighs 150 pounds, how many more calories are burned during the evening by the person weighing 150 pounds?

85. Cookie Sales The chart shows cookie sales for one year. Use it to answer the following questions.

a. What was the revenue for Oreo cookie sales during the year?

b. True or false? The chart shows sales of private label cookies to be less than $400,000,000.

c. True or false? The chart shows sales of Lofthouse cookies to be more than $100,000,000.

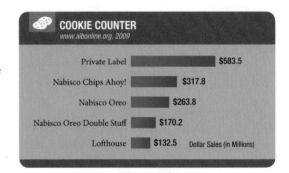

Improving Your Quantitative Literacy

86. Quantitative literacy is a subject discussed by many people involved in teaching mathematics. The person they are concerned with when they discuss it is you. We are going to work at improving your quantitative literacy, but before we do that we should answer the question, What is quantitative literacy? Lynn Arthur Steen, a noted mathematics educator, has stated that quantitative literacy is "the capacity to deal effectively with the quantitative aspects of life."

 a. Give a definition for the word *quantitative*.

 b. Give a definition for the word *literacy*.

 c. Are there situations that occur in your life that you find distasteful, or that you try to avoid, because they involve numbers and mathematics? If so, list some of them here. (For example, some people find the process of buying a car particularly difficult because they feel that the numbers and details of the financing are beyond them.)

Find the Mistake

Each sentence below contains a mistake. Circle the mistake and write the correct word(s) or numbers(s) on the line provided.

1. Writing the fraction $\frac{4}{11}$ as an equivalent fraction with a denominator of 33 gives us $\frac{4}{33}$. _____

2. A real number and its opposite will have different absolute values. _____

3. In the algebraic expression $2x - y$, the 2 is the variable. _____

4. The perimeter of a rectangle with a length of 32 and a width of 8 is 256. _____

Image © sxc.hu, palm3559, 2007

**Video Examples
Section 1.2**

If you have an extra $20,000 and a knack for poker, you may receive a coveted invitation to the National Heads-Up Poker Championship held annually in the United States. The money would be your buy-in to the single-elimination tournament, which hosts one-on-one matches in bracket form. The last player at the table is crowned champion and receives a prize of $500,000.

No invitation? Instead, suppose that you are playing a friendly game of poker with some friends, and you lose $3 on the first hand and $4 on the second hand. If you represent winning with positive numbers and losing with negative numbers, how can you translate this situation into symbols? Because you lost $3 and $4 for a total of $7, one way to represent this situation is with addition of negative numbers:

$$(-\$3) + (-\$4) = -\$7$$

From this equation, we see that the sum of two negative numbers is a negative number. To generalize addition with positive and negative numbers, we use the number line.

Because real numbers have both a distance from zero (absolute value) and a direction from zero (+ or − sign), we can think of addition of two numbers in terms of distance and direction from zero.

A Adding Positive and Negative Numbers

Let's look at a problem for which we know the answer. Suppose we want to add the numbers 3 and 4. The problem is written 3 + 4. To put it on the number line, we read the problem as follows:

1. The 3 tells us to "start at the origin and move 3 units in the positive direction."
2. The + sign is read "and then move."
3. The 4 means "4 units in the positive direction."

To summarize, 3 + 4 means to start at the origin, move 3 units in the positive direction, and then move an additional 4 units also in the positive direction.

We end up at 7, which is the answer to our problem: 3 + 4 = 7.

Let's try other combinations of positive and negative 3 and 4 on the number line.

Practice Problems

EXAMPLE 1 Add $3 + (-4)$.

Solution Starting at the origin, move 3 units in the positive direction and then 4 units in the negative direction.

We end up at -1; therefore, $3 + (-4) = -1$.

EXAMPLE 2 Add $-3 + 4$.

Solution Starting at the origin, move 3 units in the negative direction and then 4 units in the positive direction.

We end up at 1; therefore, $-3 + 4 = 1$.

EXAMPLE 3 Add $-3 + (-4)$.

Solution Starting at the origin, move 3 units in the negative direction and then 4 units in the negative direction.

We end up at -7; therefore, $-3 + (-4) = -7$. Here is a summary of what we have just completed:

$$3 + 4 = 7$$
$$3 + (-4) = -1$$
$$-3 + 4 = 1$$
$$-3 + (-4) = -7$$

If we look closely at the preceding addition problems, we can summarize our results with the following rule:

RULE Adding Real Numbers

To add two real numbers with

1. the *same* sign, simply add their absolute values and use the common sign. (If both numbers are positive, then the answer is positive. If both numbers are negative, then the answer is negative.)

2. *different* signs, subtract the smaller absolute value from the larger. The answer will have the sign of the number with the larger absolute value.

Practice Problems

1. Add $2 + (-5)$. You can use the number line in Example 1 if you like.

2. Add $-2 + 5$.

3. Add $-2 + (-5)$.

Note This rule is what we have been working towards. It is very important. Be sure that you understand it and can use it. The problems we have done up to this point have been done simply to justify this rule. Now that we have the rule, we no longer need to do our addition problems on the number line.

Answers

1. -3
2. 3
3. -7

This rule covers all possible combinations of addition with real numbers. You must memorize it. After you have worked a number of problems, it will seem almost automatic.

EXAMPLE 4 Add all combinations of positive and negative 10 and 13.

Solution Rather than work these problems on the number line, we use the rule for adding positive and negative numbers to obtain our answers.

$$10 + 13 = 23$$
$$10 + (-13) = -3$$
$$-10 + 13 = 3$$
$$-10 + (-13) = -23$$

4. Add all possible combinations of positive and negative 9 and 12.
a. $9 + 12$
b. $9 + (12)$
c. $-9 + 12$
d. $-9 + (12)$

B Subtracting Positive and Negative Numbers

Because we want to make as few rules as possible, we can define subtraction in terms of addition. By doing so, we can then use the rule for addition to solve our subtraction problems.

> **RULE** Subtracting Real Numbers
>
> To subtract one real number from another, simply add its opposite.
> Algebraically, the rule is written like this: If a and b represent two real numbers, then it is always true that
>
> $$\underbrace{a - b}_{\text{To subtract } b,} \quad = \quad \underbrace{a + (-b)}_{\text{add the opposite of } b.}$$

This is how subtraction is defined in algebra. This definition of subtraction will not conflict with what you already know about subtraction, but it will allow you to do subtraction using negative numbers.

EXAMPLE 5 Subtract all possible combinations of positive and negative 7 and 2.

Solution

$$\left.\begin{array}{l} 7 - 2 = 7 + (-2) = 5 \\ -7 - 2 = -7 + (-2) = -9 \end{array}\right\} \quad \text{Subtracting 2 is the same as adding } -2.$$

$$\left.\begin{array}{l} 7 - (-2) = 7 + 2 = 9 \\ -7 - (-2) = -7 + 2 = -5 \end{array}\right\} \quad \text{Subtracting } -2 \text{ is the same as adding 2.}$$

5. Subtract all possible combinations of positive and negative 4 and 7.
a. $7 - 4$
b. $-7 - 4$
c. $7 - (-4)$
d. $-7 - (-4)$

Notice that each subtraction problem is first changed to an addition problem. The rule for addition is then used to arrive at the answer.

EXAMPLE 6 Subtract all combinations of positive and negative 8 and 13.

Solution

$$\left.\begin{array}{l} 8 - 13 = 8 + (-13) = -5 \\ -8 - 13 = -8 + (-13) = -21 \end{array}\right\} \quad \text{Subtracting 13 is the same as adding } -13.$$

$$\left.\begin{array}{l} 8 - (-13) = 8 + 13 = 21 \\ -8 - (-13) = -8 + 13 = 5 \end{array}\right\} \quad \text{Subtracting } -13 \text{ is the same as adding 13.}$$

6. Subtract all combinations of positive and negative 4 and 6.
a. $6 - 4$
b. $-6 - 4$
c. $6 - (-4)$
d. $-6 - (-4)$

Answers

4. a. 21 **b.** -3 **c.** 3 **d.** -21
5. a. 3 **b.** -11 **c.** 11 **d.** -3
6. a. 2 **b.** -10 **c.** 10 **d.** -2

We have defined subtraction in terms of addition, and we still obtain answers consistent with the answers we are used to getting with subtraction. Moreover, we now can do subtraction problems involving both positive and negative numbers.

As you proceed through the following examples and the problem set, you will begin to notice shortcuts you can use in working the problems. You will not always have to change subtraction to addition of the opposite to be able to get answers quickly. Use all the shortcuts you wish as long as you consistently get the correct answers.

C Order of Operations

Recall our discussion about the order of operations in the previous chapter. Let's apply this rule to some problems involving real numbers.

EXAMPLE 7 Add $-3 + 2 + (-4)$.

Solution Applying the rule for order of operations, we add left to right.

$$-3 + 2 + (-4) = -1 + (-4)$$
$$= -5$$

7. Add $-5 + 2 + (-7)$.

EXAMPLE 8 Add $-8 + [2 + (-5)] + (-1)$.

Solution Adding inside the brackets first and then left to right, we have

$$-8 + [2 + (-5)] + (-1) = -8 + (-3) + (-1)$$
$$= -11 + (-1)$$
$$= -12$$

8. Add $-4 + [2 + (-3)] + (-1)$.

EXAMPLE 9 Simplify each expression as much as possible.

a. $7 + (-3) - 5$ **b.** $8 - (-2) - 6$ **c.** $-2 - (-3 + 1) - 5$

Solution

a. $7 + (-3) - 5 = 7 + (-3) + (-5)$ *Begin by changing all subtractions to*
$\qquad\qquad\quad = 4 + (-5)$ *additions. Then add left to right.*
$\qquad\qquad\quad = -1$

b. $8 - (-2) - 6 = 8 + 2 + (-6)$ *Begin by changing all subtractions to*
$\qquad\qquad\quad = 10 + (-6)$ *additions. Then add left to right.*
$\qquad\qquad\quad = 4$

c. $-2 - (-3 + 1) - 5 = -2 - (-2) - 5$ *Do what is in the parentheses first.*
$\qquad\qquad\qquad\quad = -2 + 2 + (-5)$
$\qquad\qquad\qquad\quad = -5$

9. Simplify each expression as much as possible.

a. $4 + (-2) - 3$

b. $9 - 4 - 5$

c. $-3 - (-5 + 1) - 6$

Note When writing lines of simplified calculations or algebraic expression (as we will encounter later in this chapter), it is important to remember to begin each line with an equals sign, as we have done here.

The next two examples involve multiplication and exponents as well as addition and subtraction. Remember, according to the rule for order of operations, we evaluate the numbers containing exponents and multiply before we subtract.

EXAMPLE 10 Simplify $2 \cdot 5 - 3 \cdot 8 - 4 \cdot 9$.

Solution First, we multiply left to right, and then we subtract.

$$2 \cdot 5 - 3 \cdot 8 - 4 \cdot 9 = 10 - 24 - 36$$
$$= -14 - 36$$
$$= -50$$

10. Simplify $3 \cdot 9 - 4 \cdot 10 - 5 \cdot 11$.

Answers
7. -10
8. -6
9. a. 1 **b.** 0 **c.** 5
10. -68

11. Simplify $4 \cdot 2^5 - 3 \cdot 5^2$.

EXAMPLE 11 Simplify $3 \cdot 2^3 + 2 \cdot 4^2$.

Solution We begin by evaluating each number that contains an exponent. Then we multiply before we subtract.

$$3 \cdot 2^3 - 2 \cdot 4^2 = 3 \cdot 8 + 2 \cdot 16$$
$$= 24 + 32$$
$$= 56$$

D Writing Words in Symbols

Just as we discussed in the previous section, it is important to be able to translate the English language into symbols for operations with real numbers. The next four examples will let you practice doing so.

12. Add 8 to -5.

EXAMPLE 12 Add 7 to -3.

Solution First, we write the problem in terms of addition, then we simplify.

$$-3 + 7 = 4$$

13. Add -3 to 8.

EXAMPLE 13 Add -5 to 2.

Solution $2 + (-5) = -3$

14. Find the difference of 8 and 6.

EXAMPLE 14 Find the difference of 9 and 2.

Solution Written in symbols, the problem looks like this:

$$9 - 2 = 7$$

The difference of 9 and 2 is 7.

15. Find the difference of 4 and -7.

EXAMPLE 15 Find the difference of 3 and -5.

Solution Subtracting -5 from 3 we have

$$3 - (-5) = 3 + 5$$
$$= 8$$

E Complementary and Supplementary Angles

FACTS FROM GEOMETRY Complementary and Supplementary Angles

If you have studied geometry at all, you know that there are 360° in a full rotation—the number of degrees swept out by the radius of a circle as it rotates once around the circle.

We can apply our knowledge of algebra to help solve some simple geometry problems. Before we do, however, we need to review some of the vocabulary associated with angles.

> **DEFINITION** complementary and supplementary angles
>
> In geometry, two angles that add to 90° are called *complementary angles*. In a similar manner, two angles that add to 180° are called *supplementary angles*.
>
>
>
> Complementary angles: $x + y = 90°$ Supplementary angles: $x + y = 180°$

EXAMPLE 16 Find x in each of the following diagrams.

a. **b.**

Solution We use subtraction to find each angle.

a. Because the two angles are complementary, we can find x by subtracting 30° from 90°.

$$x = 90° - 30° = 60°$$

We say 30° and 60° are complementary angles. The complement of 30° is 60°.

b. The two angles in the diagram are supplementary. To find x, we subtract 45° from 180°.

$$x = 180° - 45° = 135°$$

We say 45° and 135° are supplementary angles. The supplement of 45° is 135°.

16. Find x in each of the following diagrams.

a.

b.

Subtracting and Taking Away

For some people taking algebra for the first time, subtraction of positive and negative numbers can be a problem. These people may believe that $-5 - 9$ should be -4 or 4, not -14. If this is happening to you, you probably are thinking of subtraction in terms of taking one number away from another. Thinking of subtraction in this way works well with positive numbers if you always subtract the smaller number from the larger. In algebra, however, we encounter many situations other than this. The definition of subtraction, that $a - b = a + (-b)$, clearly indicates the correct way to use subtraction; that is, when working subtraction problems, you should think "addition of the opposite," not "take one number away from another." To be successful in algebra, you need to apply properties and definitions exactly as they are presented here.

> **GETTING READY FOR CLASS**
>
> *After reading through the preceding section, respond in your own words and in complete sentences.*
>
> **A.** Explain how you would add 3 and -5 on the number line.
>
> **B.** Define subtraction in terms of addition.
>
> **C.** Why is it important to understand absolute value in order to know how to add positive and negative numbers?
>
> **D.** What are complementary and supplementary angles?

Answers

16. a. 45° **b.** 120°

EXERCISE SET 1.2

Vocabulary Review

Choose the correct words to fill in the blanks below.

| opposite | different | same | complementary | supplementary |

1. To add two real numbers with the _____ sign, add their absolute values and use the common sign.
2. To add two real numbers with _____ signs, subtract the smaller absolute value from the larger and add the sign of the number with the larger absolute value.
3. To subtract one real number from another, add its _____.
4. Two angles that add to 90° are called _____ angles.
5. Two angles that add to 180° are called _____ angles.

Problems

A Work the following addition problems. You may want to begin by doing a few on the number line.

1. $6 + (-3)$
2. $7 + (-8)$
3. $13 + (-20)$
4. $15 + (-25)$

5. $18 + (-32)$
6. $6 + (-9)$
7. $-6 + 3$
8. $-8 + 7$

9. $-30 + 5$
10. $-18 + 6$
11. $-6 + (-6)$
12. $-5 + (-5)$

13. $-9 + (-10)$
14. $-8 + (-6)$
15. $-10 + (-15)$
16. $-18 + (-30)$

B Subtract.

17. $5 - 8$
18. $6 - 7$
19. $3 - 9$
20. $2 - 7$

21. $5 - 5$
22. $8 - 8$
23. $-8 - 2$
24. $-6 - 3$

25. $-4 - 12$
26. $-3 - 15$
27. $-6 - 6$
28. $-3 - 3$

29. $-8 - (-1)$ **30.** $-6 - (-2)$ **31.** $15 - (-20)$ **32.** $20 - (-5)$

33. $-4 - (-4)$ **34.** $-5 - (-5)$ **35.** $-3 - (-12)$ **36.** $-7 - (-20)$

C Work the following problems using the rule for addition of real numbers. You may want to refer back to the rule for order of operations.

37. $5 + (-6) + (-7)$ **38.** $6 + (-8) + (-10)$ **39.** $3 - 2 - 5$

40. $4 - 8 - 6$ **41.** $-22 + 4 - 10$ **42.** $-13 + 6 - 5$

43. $5 + [6 + (-2)] + (-3)$ **44.** $10 + [8 + (-5)] + (-20)$ **45.** $[6 + (-2)] + [3 + (-1)]$

46. $[18 + (-5)] + [9 + (-10)]$ **47.** $20 + (-6) + [3 + (-9)]$ **48.** $18 + (-2) + [9 + (-13)]$

49. $16 - [(4 - 5) - 1]$ **50.** $15 - [(4 - 2) - 3]$ **51.** $5 - [(2 - 3) - 4]$

52. $6 - [(4 - 1) - 9]$ **53.** $21 - [-(3 - 4) - 2] - 5$ **54.** $30 - [-(10 - 5) - 15] - 25$

55. $12 - [-(7 - 2) - 5] - 9$ **56.** $13 - [-(5 - 12) - 3] - 7$

The problems that follow involve some multiplication. Be sure that you work inside the parentheses first, then multiply, and finally, add left to right. Use the rule for order of operations to simplify each expression as much as possible.

57. $-5 + 2(-3 + 7)$

58. $-3 + 4(-2 + 7)$

59. $9 + 3(-8 + 10)$

60. $4 + 5(-2 + 6)$

61. $2 \cdot 8 - 3 \cdot 5$

62. $3 \cdot 4 - 6 \cdot 7$

63. $3 \cdot 5 - 2 \cdot 7$

64. $6 \cdot 10 - 5 \cdot 20$

65. $2 \cdot 3^2 + 5 \cdot 2^2$

66. $3 \cdot 7^2 + 2 \cdot 8^2$

67. $4 \cdot 3^3 - 5 \cdot 2^3$

68. $3 \cdot 6^2 - 2 \cdot 3^2 - 8 \cdot 6^2$

D Write the numerical expression that is equivalent to each of the following phrases and then simplify.

69. The sum of 5 and 9

70. The sum of 6 and -3

71. Four added to the sum of -7 and -5

72. Six added to the sum of -9 and 1

73. Subtract 4 from -7.

74. Subtract 5 from -19.

75. Subtract 17 from the sum of 4 and -5.

76. Subtract -6 from the sum of 6 and -3.

77. The difference of 8 and 5

78. The difference of 5 and 8

79. The difference of 8 and -5

80. The difference of 5 and -8

Answer the following questions.

81. What number do you add to -8 to get -5?

82. The sum of what number and -12 is 8?

83. What number do you subtract from 8 to get -2?

84. What number do you subtract from 1 to get 5?

E Find x in each of the following diagrams.

85.

86.

87.

88.

Applying the Concepts

89. Temperature Change The temperature inside a space shuttle is 73°F before reentry. During reentry the temperature inside the craft increases 10°. On landing it drops 8°F. Write an expression using the numbers 73, 10, and 8 to describe this situation. What is the temperature inside the shuttle on landing?

90. Temperature Change The temperature at noon is 23°F. Six hours later it has dropped 19°F, and by midnight it has dropped another 10°F. Write a subtraction problem that gives the temperature at midnight. What is the temperature at midnight?

91. Temperature Change The coldest temperature ever recorded in Antarctica was −129°F, while the highest recorded was 59°F. What is the absolute value of the difference between these two temperatures?

92. Temperature Change The annual mean temperature at the South Pole is −58°F, while that of the North Pole is 0°F. What is the absolute value of the difference between these two temperatures?

93. Temperature Change The highest temperature ever recorded on Earth was 136°F in Libya, while the coldest temperature ever recorded on Earth was −129°F in Antarctica. What is the absolute value of the difference between these two temperatures?

94. Temperature Change The highest North American temperature ever recorded was 134°F in Death Valley, CA, while the coldest North American temperature ever recorded was −87° in north Ice, Greenland. What is the absolute value of the difference between these two temperatures?

Profit Revenue and Costs In business, the difference of revenue and cost is profit, or $P = R - C$, where P is profit, R is revenue, and C is costs. Let's suppose the bar charts below show the cost and revenue for a children's shoe company called First Steps for a recent 5-year period. Use this information to answer the questions below.

95. What was the profit for the year 2007?

96. In what year was the largest increase in costs from the previous year? How much was the increase?

97. What was the only year the company had a loss? (a negative profit.) How much was the loss?

98. In which year was the profit the greatest? What was the profit in that year?

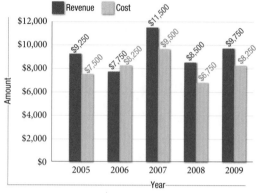

99. Health Clubs The chart shows the membership in US Health Clubs over a five-year period. Use it to answer the following questions.

a. How many more members were there in 2003 than 2001?

b. How many more members were there in 2004 than 2000?

c. If the trend from 2004 to 2005 continued, how many members were there be in 2006?

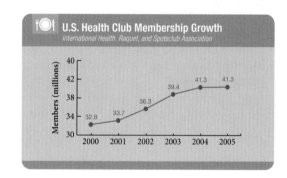

Find the Mistake

Each sentence below contains a mistake. Circle the mistake and write the correct word(s) or numbers(s) on the line provided.

1. The difference of -22 and -17 is -39. _____

2. To subtract one real number from another, subtract its opposite. _____

3. To work the problem $4^3 - (-8 + 4) - 2$ using the order of operations, we subtract the 2 before working what is inside the parentheses. _____

4. The supplement of an angle $88°$ is 2. _____

1.3 Multiplication of Real Numbers

OBJECTIVES

A Multiply any combination of positive and negative numbers.

B Simplify expressions using the rule for order of operations.

C Multiply positive and negative fractions.

D Multiply using the multiplication property of zero.

KEY WORDS

multiplication property of zero

Facebook made its Wall Street debut as a publicly traded company in May 2012. The stock opened at $42.05 per share, up 13% from its IPO pricing of $38. Then it hit $42.99 before falling back to $40 by the end of the first ten minutes of trading. Suppose you own 5 shares of Facebook stock during the time when the stock fell from $42.99 to $40, approximately a $3 drop per share. How much money have you lost? Intuitively, we know the loss is $15. Because it is a loss, we can express it as $-$15. To describe this situation with numbers, we would write

$$5 \text{ shares each lose } \$3. \rightarrow \quad 5(-3) = -15 \quad \leftarrow \text{ Total of } \$15$$

Reasoning in this manner, we conclude that the product of a positive number with a negative number is a negative number. Let's look at multiplication in more detail.

A Multiplication of Real Numbers

From our experience with counting numbers, we know that multiplication is simply repeated addition; that is, $3(5) = 5 + 5 + 5$. We will use this fact, along with our knowledge of negative numbers, to develop the rule for multiplication of any two real numbers. The following example illustrates multiplication with all of the possible combinations of positive and negative numbers.

**Video Examples
Section 1.3**

Practice Problems

1. Multiply.

 a. $4(2)$

 b. $4(-2)$

 c. $-4(2)$

 d. $-4(-2)$

EXAMPLE 1 Multiply.

a. $3(5)$ **b.** $3(-5)$ **c.** $-3(5)$ **d.** $-3(-5)$

Solution

a. Two positives: $3(5) = 5 + 5 + 5$

 $= 15$ *Positive answer*

b. One positive: $3(-5) = -5 + (-5) + (-5)$

 $= -15$ *Negative answer*

c. One negative: $-3(5) = -15$ *Negative answer*

To understand why multiplying a negative by a positive (or a positive by a negative) results in a negative answer, we consider the number line. Recall that, on the number line, the negative sign means to move to the left of the zero. If we think of -3 in the problem $-3(5)$ as moving 5 units to the left of the zero 3 times, then we have

Answers

1. a. 8 **b.** 8 **c.** 8 **d.** 8

d. Two negatives: $-3(-5) = ?$

With two negatives, $-3(-5)$, it is not possible to work the problem in terms of repeated addition. (It doesn't "make sense" to write -5 down a -3 number of times.) The answer is probably 15 (that's just a guess), but we need some justification for saying so. We will solve a different problem and in doing so get the answer to the problem $(-3)(-5)$.

Here is a problem to which we know the answer. We will work it two different ways.

$$-3[5 + (-5)] = -3(0) = 0$$

The answer is zero. We also can work the problem by distributing the -3 over both terms inside the parentheses. This distributive property is one of the properties we will cover more in depth later in the chapter.

$$-3[5 + (-5)] = -3(5) + (-3)(-5)$$
$$= -15 + ?$$

Because the answer to the problem is 0, our ? must be 15. (What else could we add to -15 to get 0? Only 15.)

> *Note* You may have to read the explanation for Example 1d several times before you understand it completely. The purpose of the explanation in Example 1d is simply to justify the fact that the product of two negative numbers is a positive number.

Here is a summary of the results we have obtained:

Original numbers have		The answer is
the same sign	$3(5) = 15$	positive
different signs	$3(-5) = -15$	negative
different signs	$-3(5) = -15$	negative
the same sign	$-3(-5) = 15$	positive

By examining Example 1 and the preceding table, we can use the information there to write the following rule. This rule tells us how to multiply any two real numbers.

> *Note* Students have trouble with the expression $-3(-5)$ because they want to subtract rather than multiply. Because we are very precise with the notation we use in algebra, the expression $-3(-5)$ has only one meaning: multiplication. A subtraction problem that uses the same numbers is $-3 - 5$. Compare the two following lists:
>
All Multiplication	No Multiplication
> | $5(4)$ | $5 + 4$ |
> | $-5(4)$ | $-5 + 4$ |
> | $5(-4)$ | $5 - 4$ |
> | $-5(-4)$ | $-5 - 4$ |

> **RULE** Multiplying Real Numbers
>
> To multiply any two real numbers, simply multiply their absolute values. The sign of the answer is
>
> **1.** *positive* if both numbers have the same sign (both + or both −).
>
> **2.** *negative* if the numbers have opposite signs (one +, the other −).

The following example illustrates how we use the preceding rule to multiply real numbers.

EXAMPLE 2 Multiply.

a. $-8(-3) = 24$

b. $-10(-5) = 50$ If the two numbers in the product have the same sign, the answer is positive.

c. $-4(-7) = 28$

d. $5(-7) = -35$

e. $-4(8) = -32$ If the two numbers in the product have different signs, the answer is negative.

f. $-6(10) = -60$

2. Multiply.

 a. $-4(-9)$

 b. $-11(-12)$

 c. $2(-7)$

 d. $4(-3)$

 e. $-2(10)$

 f. $-3(13)$

Answers

2. a. 36 **b.** 132 **c.** 14 **d.** -12
 e. -20 **f.** -39

B Using Order of Operations

In the following example, we combine the rule for order of operations with the rule for multiplication to simplify expressions. Remember, the rule for order of operations specifies that we are to work inside the parentheses first and then simplify numbers containing exponents. After this, we multiply and divide, left to right. The last step is to add and subtract, left to right.

EXAMPLE 3 Simplify as much as possible.

a. $-5(-3)(-4)$ **b.** $4(-3) + 6(-5) - 10$ **c.** $(-2)^3$
d. $-3(-2)^3 - 5(-4)^2$ **e.** $6 - 4(7 - 2)$

Solution

a. $-5(-3)(-4) = 15(-4)$

$$= -60$$

b. $4(-3) + 6(-5) - 10 = -12 + (-30) - 10$ Multiply.

$$= -42 - 10$$ Add.

$$= -52$$ Subtract.

c. $(-2)^3 = (-2)(-2)(-2)$ Definition of exponents

$$= -8$$ Multiply, left to right.

d. $-3(-2)^3 - 5(-4)^2 = -3(-8) - 5(16)$ Exponents first

$$= 24 - 80$$ Multiply.

$$= -56$$ Subtract.

e. $6 - 4(7 - 2) = 6 - 4(5)$ Inside parentheses first

$$= 6 - 20$$ Multiply.

$$= -14$$ Subtract.

EXAMPLE 4 Figure 1 gives the calories that are burned in 1 hour for a variety of exercise by a person weighing 150 pounds. Figure 2 gives the calories that are consumed by eating some popular fast foods. Find the net change in calories for a 150-pound person playing handball for 2 hours and then eating a Whopper.

Calories Burned in 1 Hour by a 150-Pound Person

FIGURE 1

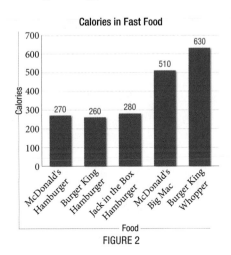

Calories in Fast Food

FIGURE 2

Solution The net change in calories will be the difference of the calories gained from eating and the calories lost from exercise.

$$\text{Net change in calories} = 630 - 2(680) = -730 \text{ calories}$$

3. Simplify as much as possible.

 a. $-7(-6)(-1)$

 b. $2(-8) + 3(-7) - 4$

 c. $(-3)^5$

 d. $-5(-2)^3 - 7(-3)^2$

 e. $7 - 3(4 - 9)$

Note It is important to pay close attention to the order of operations when simplifying problems involving even and odd exponents, and negative signs. Compare the following lists:

$$2^3 = 8 \qquad 2^4 = 16$$
$$(-2)^3 = -8 \qquad (-2)^4 = 16$$
$$-2^3 = -8 \qquad -2^4 = -16$$
$$-(-2)^3 = 8 \qquad -(-2)^4 = -16$$

4. Find the net change in calories for a 150-pound person jogging for 2 hours then eating a Big Mac.

C Multiplying Fractions

Previously, we mentioned that to multiply two fractions we multiply numerators and multiply denominators. We can apply the rule for multiplication of positive and negative numbers to fractions in the same way we apply it to other numbers. We multiply absolute values: The product is positive if both fractions have the same sign and negative if they have different signs. Here is an example:

EXAMPLE 5 Multiply.

a. $-\dfrac{3}{4}\left(\dfrac{5}{7}\right)$ **b.** $-6\left(\dfrac{1}{2}\right)$ **c.** $-\dfrac{2}{3}\left(-\dfrac{3}{2}\right)$

Solution

a. $-\dfrac{3}{4}\left(\dfrac{5}{7}\right) = -\dfrac{3 \cdot 5}{4 \cdot 7}$ *Different signs give a negative answer.*

$= -\dfrac{15}{28}$

b. $-6\left(\dfrac{1}{2}\right) = -\dfrac{6}{1}\left(\dfrac{1}{2}\right)$ *Different signs give a negative answer.*

$= -\dfrac{6}{2}$

$= -3$

c. $-\dfrac{2}{3}\left(-\dfrac{3}{2}\right) = \dfrac{2 \cdot 3}{3 \cdot 2}$ *Same signs give a positive answer.*

$= \dfrac{6}{6}$

$= 1$

D The Multiplication Property of Zero

As a reminder, we will cover properties of real numbers later in the chapter. For now we will consider the multiplication property of zero. It is of great importance, as you will see, throughout our study of algebra and deserves special attention.

> **RULE** Multiplication Property of Zero
>
> The product of any real number a and zero is 0.
>
> $a(0) = 0$ a is a real number.
>
> $-a(0) = 0$ a is a real number.
>
> $0(0) = 0$

EXAMPLE 6 Multiply.

a. $5(0)$ **b.** $0(6)$ **c.** $5{,}398{,}423(0)$ **d.** $(-1)(-6)(0)$

Solution

a. $5(0) = 0$

b. $0(6) = 0$

c. $5{,}398{,}423(0) = 0$

d. $(-1)(-6)(0) = 0$

5. Multiply.

a. $-\dfrac{2}{3}\left(\dfrac{5}{9}\right)$

b. $-8\left(\dfrac{1}{2}\right)$

c. $-\dfrac{3}{4}\left(-\dfrac{4}{3}\right)$

6. Multiply.
a. $8(0)$
b. $0(12)$
c. $8{,}4320(0)$
d. $(-4)(-3)(0)$

Answers

5. a. $-\dfrac{10}{27}$ **b.** -4 **c.** 1
6. a. 0 **b.** 0 **c.** 0 **d.** 0

Whenever you see a 0 as one of the numbers to be multiplied, you can simply write the answer down without any calculations. It doesn't matter what the other numbers multiply to be or if the numbers are positive, negative, fractions, very large, or very small.

EXAMPLE 7 Multiply.

a. $\dfrac{1}{12}(0)$ **b.** $25^3(0)$ **c.** $(-3)\left(\dfrac{9}{13}\right)(0)$ **d.** $(465,832)(0.00026)(0)$

Solution

a. $\dfrac{1}{12}(0) = 0$

b. $25^3(0) = 0$

c. $(-3)\left(\dfrac{9}{13}\right)(0) = 0$

d. $(465,832)(0.00026)(0) = 0$

7. Multiply.

a. $\dfrac{1}{2}(0)$

b. $4^3(0)$

c. $-14\left(\dfrac{8}{17}\right)(0)$

d. $(543,287)(0.42987)(0)$

Answers

7. a. 0 **b.** 0 **c.** 0 **d.** 0

Vocabulary Review

Choose the correct words to fill in the blanks below.

negative zero positive absolute values

1. To multiply two real numbers, first multiply their _____.
2. The sign of an answer to a multiplication problem involving two numbers will be _____ if both numbers have the same sign.
3. The sign of an answer to a multiplication problem involving two numbers will be _____ if both numbers have opposite signs.
4. The product of any real number and _____ is 0.

Problems

A Use the rule for multiplying two real numbers to find each of the following products.

1. $7(-6)$

2. $8(-4)$

3. $-8(2)$

4. $-16(3)$

5. $-3(-1)$

6. $-7(-1)$

7. $-11(-11)$

8. $-12(-12)$

B Use the rule for order of operations to simplify each expression as much as possible.

9. $-3(2)(-1)$

10. $-2(3)(-4)$

11. $-3(-4)(-5)$

12. $-5(-6)(-7)$

13. $-2(-4)(-3)(-1)$

14. $-1(-3)(-2)(-1)$

15. $(-7)^2$

16. $(-8)^2$

17. $(-3)^3$

18. $(-2)^4$

19. $-2(2-5)$

20. $-3(3-7)$

21. $-5(8-10)$

22. $-4(6-12)$

23. $(4-7)(6-9)$

24. $(3 - 10)(2 - 6)$

25. $(-3 - 2)(-5 - 4)$

26. $(-3 - 6)(-2 - 8)$

27. $-3(-6) + 4(-1)$

28. $-4(-5) + 8(-2)$

29. $2(3) - 3(-4) + 4(-5)$

30. $5(4) - 2(-1) + 5(6)$

31. $4(-3)^2 + 5(-6)^2$

32. $2(-5)^2 + 4(-3)^2$

33. $7(-2)^3 - 2(-3)^3$

34. $10(-2)^3 - 5(-2)^4$

35. $6 - 4(8 - 2)$

36. $7 - 2(6 - 3)$

37. $9 - 4(3 - 8)$

38. $8 - 5(2 - 7)$

39. $-4(3 - 8) - 6(2 - 5)$

40. $-8(2 - 7) - 9(3 - 5)$

41. $7 - 2[-6 - 4(-3)]$

42. $6 - 3[-5 - 3(-1)]$

43. $7 - 3[2(-4 - 4) - 3(-1 - 1)]$

44. $5 - 3[7(-2 - 2) - 3(-3 + 1)]$

45. $8 - 6[-2(-3 - 1) + 4(-2 - 3)]$

46. $4 - 2[-3(-1 + 8) + 5(-5 + 7)]$

47. $-5 - 2(-3 - 8)$

48. $-13 - 4(-2 - 6)$

49. $-5(2 - 3) - 7(4 - 9)$

50. $-8(-1 - 6) - 5(-2 - 3)$

C Multiply the following fractions.

51. $-\dfrac{2}{3} \cdot \dfrac{5}{7}$

52. $-\dfrac{6}{5} \cdot \dfrac{2}{7}$

53. $-8\left(\dfrac{1}{2}\right)$

54. $-12\left(\dfrac{1}{3}\right)$

55. $-\dfrac{3}{4}\left(-\dfrac{4}{3}\right)$

56. $-\dfrac{5}{8}\left(-\dfrac{8}{5}\right)$

57. $\left(-\dfrac{3}{4}\right)^2$

58. $\left(-\dfrac{2}{5}\right)^2$

D Find the following products.

59. $-2(0)$

60. $2x(0)$

61. $-7x(0)$

62. $\dfrac{1}{5}(0)$

63. $3x(-5)(0)$

64. $(16^5)\left(\dfrac{4}{19}\right)(0)(x)$

65. $(4{,}285)(x)(y)(0)$

66. $\left(\dfrac{1}{151}\right)(3)(z)(0)$

Simplify.

67. $2(5-5)+4$

68. $5x(2)+3x(0)+4$

69. $6(3+1)-5(0)$

70. $2(3)+4(5)-5(2)$

71. $\left(\dfrac{1}{2} \cdot 18\right)^2$

72. $\left[\dfrac{1}{2}(-10)\right]^2$

73. $\left(\dfrac{1}{2} \cdot 3\right)^2$

74. $\left(\dfrac{1}{2} \cdot 5\right)^2$

Applying the Concepts

75. Temperature Change The temperature is 25°F at 5:00 in the afternoon. If the temperature drops 6°F every hour after that, what is the temperature at 9:00 in the evening?

76. Investment Value Suppose you purchase $500 worth of a mutual fund and find that the value of your purchase doubles every 2 years. Write a sequence of numbers that gives the value of your purchase every 2 years for the first 10 years you own it. Is this sequence a geometric sequence?

77. Reading Charts Refer to the bar charts below to find the net change in calories for a 150-pound person who bowls for 3 hours and then eats a cheeseburger with onion and a strawberry shake.

78. Reading Charts Refer to the bar charts below to find the net change in calories for a 150-pound person who plays handball for 1.5 hours and then eats a Double-Double with onion and fries.

Source: www.in-n-out.com/nutritional info.asp

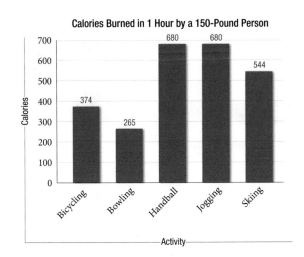

79. Using a Calculator You own 245 shares of a stock that rises $1.08 per share. How much did your investment gain in value?

80. Using a Calculator You own 548 shares of a stock that loses $2.54 per share. How much did your investment lose in value? (Hint: Answer should be a positive number.)

81. Using a Calculator You own 485 shares of stock x and 795 shares of stock y. During one week, x increases by $3.47 per share, while y loses $1.05 per share. How much did the value of your investment change over the week?

82. Using a Calculator You own 1,150 shares of stock A and 575 shares of stock B. During one week stock A increases by $2.57 per share, while stock B loses $6.40 per share. How much did the value of your investment change over the week?

Find the Mistake

Each sentence below contains a mistake. Circle the mistake and write the correct word(s) or numbers(s) on the line provided.

1. Multiplying a positive number by a negative number will give a positive answer. _____

2. Multiplying two numbers with the same sign will give a negative answer. _____

3. Simplifying the expression $(-15)(-3)(6)(-2)$ gives us 540. _____

4. The number -12 raised to the third power is equal to the number 12 raised to the third power. _____

A Divide any combination of positive and negative numbers.

B Divide positive and negative fractions.

C Simplify expressions using the rule for order of operations.

KEY WORDS

dividing with zero

**Video Examples
Section 1.4**

Note We are defining division this way simply so that we can use what we already know about multiplication to do division problems. We actually want as few rules as possible. Defining division in terms of multiplication allows us to avoid writing a separate rule for division.

1.4 Division of Real Numbers

Image © Ben Barczi, 2010

In 2010, a Peruvian inventor won the $200,000 grand prize for a competition that requested its entrants to suggest a new and ingenious way to save the earth. To battle global warming, the inventor proposed covering 173 acres of dry rocky land in the Andes Mountains with a mixture made from egg whites, lime, and water. The mixture creates a whitewash that when spread over the rocks would reflect sunlight and ideally reduce temperatures, thus slowing the melting of the area's glaciers.

Suppose it took 50 gallons to cover an acre of the mountaintop. Let's say each of a five-member team used a 2-gallon bucket to transport the whitewash mixture up the slope and pour it onto the rocks. How many trips up the slope would each team member have to make for the team to cover the acre? We can represent the answer to this question with division:

$$\frac{50}{2(5)} = \frac{50}{10} = 5$$

Here is a more detailed discussion of division with positive and negative numbers.

A Dividing Positive and Negative Numbers

The last of the four basic operations is division. We will use the same approach to define division as we used for subtraction; that is, we will define division in terms of rules we already know.

Recall that we developed the rule for subtraction of real numbers by defining subtraction in terms of addition. We changed our subtraction problems to addition problems and then added to get our answers. Because we already have a rule for multiplication of real numbers, and division is the inverse operation of multiplication, we will simply define division in terms of multiplication.

We know that division by the number 2 is the same as multiplication by its reciprocal $\frac{1}{2}$; that is, 6 divided by 2 is 3, which is the same as 6 times $\frac{1}{2}$. Similarly, dividing a number by 5 gives the same result as multiplying by $\frac{1}{5}$. We can extend this idea to all real numbers with the following rule:

> **RULE** Dividing Real Numbers
>
> If a and b represent any two real numbers (b cannot be 0), then it is always true that
>
> $$a \div b = \frac{a}{b} = a\left(\frac{1}{b}\right)$$

Notice that we are using parentheses in the above rule to imply multiplication. The expression $a\left(\frac{1}{b}\right)$ can also be written as $a \cdot \frac{1}{b}$. A common mistake is to forget the parentheses and write the expression as $a\frac{1}{b}$. This would instead imply a mixed number, written as $a + \frac{1}{b}$, and would be an error.

EXAMPLE 1 Write each division problem as an equivalent multiplication problem, and then multiply.

a. $\dfrac{6}{2}$ **b.** $\dfrac{6}{-2}$ **c.** $\dfrac{-6}{2}$ **d.** $\dfrac{-6}{-2}$

Solution

a. $\dfrac{6}{2} = 6\left(\dfrac{1}{2}\right) = 3$ *The product of two positives is positive.*

b. $\dfrac{6}{-2} = 6\left(-\dfrac{1}{2}\right) = -3$ ⎫
 The product of a positive and a negative is a negative.
c. $\dfrac{-6}{2} = -6\left(\dfrac{1}{2}\right) = -3$ ⎭

d. $\dfrac{-6}{-2} = -6\left(-\dfrac{1}{2}\right) = 3$ *The product of two negatives is positive.*

The second step in these examples is used only to show that we *can* write division in terms of multiplication. [In actual practice we wouldn't write $\frac{6}{2}$ as $6\left(\frac{1}{2}\right)$.] The answers, therefore, follow from the rule for multiplication; that is, like signs produce a positive answer, and unlike signs produce a negative answer.

Here are some examples. This time we will not show division as multiplication by the reciprocal. We will simply divide. If the original numbers have the same signs, the answer will be positive. If the original numbers have different signs, the answer will be negative.

EXAMPLE 2 Divide.

a. $\dfrac{12}{6} = 2$ *Like signs give a positive answer.*

b. $\dfrac{12}{-6} = -2$ *Unlike signs give a negative answer.*

c. $\dfrac{-12}{6} = -2$ *Unlike signs give a negative answer.*

d. $\dfrac{-12}{-6} = 2$ *Like signs give a positive answer.*

e. $\dfrac{15}{-3} = -5$ *Unlike signs give a negative answer.*

f. $\dfrac{-40}{-5} = 8$ *Like signs give a positive answer.*

g. $\dfrac{-14}{2} = -7$ *Unlike signs give a negative answer.*

B Division with Fractions

We can apply the definition of division with real numbers to fractions. Remember that dividing by a fraction is equivalent to multiplying by its reciprocal. For example, we can divide a number, even if it is another fraction, by the fraction $\frac{3}{4}$, by multiplying the number by the reciprocal of $\frac{3}{4}$, which is $\frac{4}{3}$.

$$\frac{2}{5} \div \frac{3}{4} = \frac{2}{5} \cdot \frac{4}{3} = \frac{8}{15}$$

Practice Problems

1. Change each problem to an equivalent multiplication problem, and then multiply to get your answer.

 a. $\dfrac{12}{4}$

 b. $\dfrac{12}{-4}$

 c. $\dfrac{-12}{4}$

 d. $\dfrac{-12}{-4}$

Note What we are saying here is that the work shown in Example 1 is shown simply to justify the answers we obtain. In the future, we won't show the middle step in these kinds of problems. Even so, we need to know that division is *defined* to be multiplication by the reciprocal.

2. Divide.

 a. $\dfrac{18}{3}$

 b. $\dfrac{18}{-3}$

 c. $\dfrac{-18}{3}$

 d. $\dfrac{-18}{-3}$

 e. $\dfrac{30}{-10}$

 f. $\dfrac{-50}{-25}$

 g. $\dfrac{-21}{3}$

3. Divide.

a. $\dfrac{3}{4} \div \dfrac{5}{7}$

b. $-\dfrac{7}{8} \div \dfrac{3}{5}$

c. $10 \div \left(-\dfrac{5}{6}\right)$

EXAMPLE 3 Divide.

a. $\dfrac{2}{3} \div \dfrac{5}{7}$ **b.** $-\dfrac{3}{4} \div \dfrac{7}{9}$ **c.** $8 \div \left(-\dfrac{4}{5}\right)$

Solution

a. $\dfrac{2}{3} \div \dfrac{5}{7} = \dfrac{2}{3} \cdot \dfrac{7}{5}$ *Rewrite as multiplication by the reciprocal.*

$\qquad\quad = \dfrac{14}{15}$ *Multiply.*

b. $-\dfrac{3}{4} \div \dfrac{7}{9} = -\dfrac{3}{4} \cdot \dfrac{9}{7}$ *Rewrite as multiplication by the reciprocal.*

$\qquad\qquad = -\dfrac{27}{28}$ *Multiply.*

c. $8 \div \left(-\dfrac{4}{5}\right) = \dfrac{8}{1}\left(-\dfrac{5}{4}\right)$ *Rewrite as multiplication by the reciprocal.*

$\qquad\qquad\quad = -\dfrac{40}{4}$ *Multiply.*

$\qquad\qquad\quad = -10$ *Divide 40 by 4 to simplify.*

C Using Order of Operations

The last step in each of the following examples involves reducing a fraction to lowest terms. Recall that to reduce a fraction to lowest terms, we divide the numerator and denominator by the largest number that divides each of them exactly. For example, to reduce $\dfrac{15}{20}$ to lowest terms, we divide 15 and 20 by 5 to get $\dfrac{3}{4}$.

4. Simplify.

a. $\dfrac{-6(5)}{9}$

b. $\dfrac{10}{-7 - 1}$

EXAMPLE 4 Simplify as much as possible.

a. $\dfrac{-4(5)}{6}$ **b.** $\dfrac{30}{-4 - 5}$

Solution

a. $\dfrac{-4(5)}{6} = \dfrac{-20}{6}$ *Simplify numerator.*

$\qquad\quad = -\dfrac{10}{3}$ *Reduce to lowest terms by dividing numerator and denominator by 2.*

b. $\dfrac{30}{-4 - 5} = \dfrac{30}{-9}$ *Simplify denominator.*

$\qquad\qquad = -\dfrac{10}{3}$ *Reduce to lowest terms by dividing numerator and denominator by 3.*

In the examples that follow, the numerators and denominators contain expressions that are somewhat more complicated than those we have seen thus far. To apply the rule for order of operations to these examples, we treat fraction bars the same way we treat grouping symbols; that is, fraction bars separate numerators and denominators so that each will be simplified separately.

Answers

3. a. $\dfrac{21}{20}$ **b.** $-\dfrac{35}{24}$ **c.** -12

4. a. $-\dfrac{10}{3}$ **b.** $-\dfrac{5}{4}$

EXAMPLE 5 Simplify.

a. $\dfrac{2(-3) + 4}{12}$ **b.** $\dfrac{5(-4) + 6(-1)}{2(3) - 4(1)}$

Solution

a. $\dfrac{2(-3) + 4}{12} = \dfrac{-6 + 4}{12}$ In the numerator, we multiply before we add.

$= \dfrac{-2}{12}$ Add.

$= -\dfrac{1}{6}$ Reduce to lowest terms by dividing the numerator and the denominator by 2.

b. $\dfrac{5(-4) + 6(-1)}{2(3) - 4(1)} = \dfrac{-20 + (-6)}{6 - 4}$ Multiply before adding.

$= \dfrac{-26}{2}$ Simplify numerator and denominator.

$= -13$ Divide -26 by 2.

We must be careful when we are working with expressions such as $(-5)^2$ and -5^2 that we include the negative sign with the base only when parentheses indicate we are to do so.

Unless there are parentheses to indicate otherwise, we consider the base to be only the number directly below and to the left of the exponent. If we want to include a negative sign with the base, we must use parentheses.

To simplify a more complicated expression, we follow the same rule. For example,

$$7^2 - 3^2 = 49 - 9$$ The bases are 7 and 3; the sign between the two terms is a subtraction sign.

For another example,

$$5^3 - 3^4 = 125 - 81$$ We simplify exponents first, then subtract.

EXAMPLE 6 Simplify.

a. $\dfrac{5^2 - 3^2}{-5 + 3}$ **b.** $\dfrac{(3 + 2)^2}{-3^2 - 2^2}$

Solution

a. $\dfrac{5^2 - 3^2}{-5 + 3} = \dfrac{25 - 9}{-2}$ Simplify numerator and denominator separately.

$= \dfrac{16}{-2}$

$= -8$

b. $\dfrac{(3 + 2)^2}{-3^2 - 2^2} = \dfrac{5^2}{-9 - 4}$ Simplify numerator and denominator separately.

$= \dfrac{25}{-13}$

$= -\dfrac{25}{13}$

5. Simplify.

a. $\dfrac{3(-4) + 9}{6}$

b. $\dfrac{6(-2) + 5(-3)}{5(4) - 11}$

6. Simplify.

a. $\dfrac{4^2 - 2^2}{-4 + 2}$

b. $\dfrac{(4 + 3)^2}{-4^2 - 3^2}$

Answers

5. a. $-\dfrac{1}{2}$ **b.** -3

6. a. -6 **b.** $-\dfrac{49}{25}$

Division with the Number 0

For every division problem there is an associated multiplication problem involving the same numbers. For example, the following two problems say the same thing about the numbers 2, 3, and 6:

Division Multiplication

$$\frac{6}{3} = 2 \qquad 6 = 2(3)$$

We can use this relationship between division and multiplication to clarify division involving the number 0.

First, dividing 0 by a number other than 0 is allowed and always results in 0. To see this, consider dividing 0 by 5. We know the answer is 0 because of the relationship between multiplication and division. This is how we write it:

$$\frac{0}{5} = 0 \qquad \text{because} \qquad 0 = 0(5)$$

However, dividing a nonzero number by 0 is not allowed in the real numbers. Suppose we were attempting to divide 5 by 0. We don't know if there is an answer to this problem, but if there is, let's say the answer is a number that we can represent with the letter n. If 5 divided by 0 is a number n, then

$$\frac{5}{0} = n \qquad \text{and} \qquad 5 = n(0)$$

This is impossible, because no matter what number n is, when we multiply it by 0 the answer must be 0. It can never be 5. In algebra, we say expressions like $\frac{5}{0}$ are undefined because there is no answer to them; that is, division by 0 is not allowed in the real numbers.

The only other possibility for division involving the number 0 is 0 divided by 0. We will treat problems like $\frac{0}{0}$ as if they were undefined also. Here is a rule that summarizes this information:

> **RULE** Dividing with Zero
>
> 1. Division by 0 is not defined.
> $$a \div 0 \text{ or } \frac{a}{0} \text{ is not defined, for all real numbers, } a.$$
> 2. Zero divided by any number is 0.
> $$a \div 0 \text{ or } \frac{0}{a} = 0 \text{ for all real numbers } a \text{ so long as } a \neq 0.$$

> **GETTING READY FOR CLASS**
>
> *After reading through the preceding section, respond in your own words and in complete sentences.*
>
> **A.** Why do you define division in terms of multiplication?
>
> **B.** How do you divide fractions?
>
> **C.** Why is it important to use the order of operations when dividing real numbers?
>
> **D.** Why is division by 0 not allowed with real numbers?

Vocabulary Review

Choose the correct words to fill in the blanks below.

 lowest terms zero negative reciprocal positive

1. We define division as the same as multiplication by the _____.
2. When dividing two real numbers with the same sign, the answer will be _____.
3. When dividing two real numbers with different signs, the answer will be _____.
4. Simplify the numerator and denominator of a fraction separately before reducing it to _____.
5. Dividing by _____ is not defined, whereas zero divided by any non-zero number is 0.

Problems

A Find the following quotients (divide).

1. $\dfrac{8}{-4}$ **2.** $\dfrac{10}{-5}$ **3.** $\dfrac{-48}{16}$ **4.** $\dfrac{-32}{4}$ **5.** $\dfrac{-7}{21}$ **6.** $\dfrac{-25}{100}$

7. $\dfrac{-39}{-13}$ **8.** $\dfrac{-18}{-6}$ **9.** $\dfrac{-6}{-42}$ **10.** $\dfrac{-4}{-28}$ **11.** $\dfrac{0}{-32}$ **12.** $\dfrac{0}{17}$

The following problems review all four operations with positive and negative numbers. Perform the indicated operations.

13. $-3 + 12$ **14.** $5 + (-10)$ **15.** $-3 - 12$ **16.** $5 - (-10)$

17. $-3(12)$ **18.** $5(-10)$ **19.** $-3 \div 12$ **20.** $5 \div (-10)$

B Divide and reduce all answers to lowest terms.

21. $\dfrac{4}{5} \div \dfrac{3}{4}$ **22.** $\dfrac{6}{8} \div \dfrac{3}{4}$ **23.** $-\dfrac{5}{6} \div \left(-\dfrac{5}{8}\right)$ **24.** $-\dfrac{7}{9} \div \left(-\dfrac{1}{6}\right)$

25. $\dfrac{10}{13} \div \left(-\dfrac{5}{4}\right)$ **26.** $\dfrac{5}{12} \div \left(-\dfrac{10}{3}\right)$ **27.** $-\dfrac{5}{6} \div \dfrac{5}{6}$ **28.** $-\dfrac{8}{9} \div \dfrac{8}{9}$

29. $-\dfrac{3}{4} \div \left(-\dfrac{3}{4}\right)$ **30.** $-\dfrac{6}{7} \div \left(-\dfrac{6}{7}\right)$ **31.** $-\dfrac{5}{12} \div \left(-\dfrac{3}{4}\right)$ **32.** $-\dfrac{4}{5} \div \dfrac{36}{25}$

C The following problems involve more than one operation. Simplify as much as possible.

33. $\dfrac{3(-2)}{-10}$

34. $\dfrac{4(-3)}{24}$

35. $\dfrac{-5(-5)}{-15}$

36. $\dfrac{-7(-3)}{-35}$

37. $\dfrac{-8(-7)}{-28}$

38. $\dfrac{-3(-9)}{-6}$

39. $\dfrac{27}{4-13}$

40. $\dfrac{27}{13-4}$

41. $\dfrac{20-6}{5-5}$

42. $\dfrac{10-12}{3-3}$

43. $\dfrac{2\cdot 5-10}{-3+9}$

44. $\dfrac{2\cdot 4-8}{2+8}$

45. $\dfrac{15(-5)-25}{2(-10)}$

46. $\dfrac{10(-3)-20}{5(-2)}$

47. $\dfrac{27-2(-4)}{-3(5)}$

48. $\dfrac{20-5(-3)}{10(-3)}$

49. $\dfrac{12-6(-2)}{12(-2)}$

50. $\dfrac{3(-4)+5(-6)}{10-6}$

51. $\dfrac{5^2-2^2}{-5+2}$

52. $\dfrac{7^2-4^2}{-7+4}$

53. $\dfrac{8^2-2^2}{8^2+2^2}$

54. $\dfrac{4^2-6^2}{4^2+6^2}$

55. $\dfrac{(5+3)^2}{-5^2-3^2}$

56. $\dfrac{(7+2)^2}{-7^2-2^2}$

57. $\dfrac{(8-4)^2}{8^2-4^2}$

58. $\dfrac{(6-2)^2}{6^2-2^2}$

59. $\dfrac{-4\cdot 3^2-5\cdot 2^2}{-8(7)}$

60. $\dfrac{-2\cdot 5^2+3\cdot 2^3}{-3(13)}$

61. $\dfrac{3\cdot 10^2+4\cdot 10+5}{345}$

62. $\dfrac{5\cdot 10^2+6\cdot 10+7}{567}$

63. $\dfrac{7-[(2-3)-4]}{-1-2-3}$

64. $\dfrac{2-[(3-5)-8]}{-3-4-5}$

65. $\dfrac{6(-4)-2(5-8)}{-6-3-5}$

66. $\dfrac{3(-4)-5(9-11)}{-9-2-3}$

67. $\dfrac{3(-5-3)+4(7-9)}{5(-2)+3(-4)}$

68. $\dfrac{-2(6-10)-3(8-5)}{6(-3)-6(-2)}$

69. $\dfrac{|3-9|}{3-9}$

70. $\dfrac{|4-7|}{4-7}$

71. $\dfrac{|-2-3|}{-2-3}$

72. $\dfrac{|4-12|}{4-12}$

73. Simplify each expression.

 a. $20 \div 4 \cdot 5$

 b. $-20 \div 4 \cdot 5$

 c. $20 \div (-4) \cdot 5$

 d. $20 \div 4(-5)$

 e. $-20 \div 4(-5)$

74. Simplify each expression.

 a. $32 \div 8 \cdot 4$

 b. $-32 \div 8 \cdot 4$

 c. $32 \div (-8) \cdot 4$

 d. $32 \div 8(-4)$

 e. $-32 \div 8(-4)$

75. Simplify each expression.

 a. $8 \div \dfrac{4}{5}$

 b. $8 \div \dfrac{4}{5} - 10$

 c. $8 \div \dfrac{4}{5}(-10)$

 d. $8 \div \left(-\dfrac{4}{5}\right) - 10$

76. Simplify each expression.

 a. $10 \div \dfrac{5}{6}$

 b. $10 \div \dfrac{5}{6} - 12$

 c. $10 \div \dfrac{5}{6}(-12)$

 d. $10 \div \left(-\dfrac{5}{6}\right) - 12$

77. What is the quotient of -12 and -4?

78. The quotient of -4 and -12 is what number?

79. What number do we divide by -5 to get 2?

80. What number do we divide by -3 to get 4?

81. Twenty-seven divided by what number is -9?

82. Fifteen divided by what number is -3?

83. If the quotient of -20 and 4 is decreased by 3, what number results?

84. If -4 is added to the quotient of 24 and -8, what number results?

Applying the Concepts

85. Investment Suppose that you and 3 friends bought equal shares of an investment for a total of $15,000 and then sold it later for only $13,600. How much did each person lose?

86. Investment If 8 people invest $500 each in a stamp collection and after a year the collection is worth $3,800, how much did each person lose?

87. Temperature Change Suppose that the temperature outside is dropping at a constant rate. If the temperature is 75°F at noon and drops to 61°F by 4:00 in the afternoon, by how much did the temperature change each hour?

88. Temperature Change In a chemistry class, a thermometer is placed in a beaker of hot water. The initial temperature of the water is 165°F. After 10 minutes the water has cooled to 72°F. If the water temperature drops at a constant rate, by how much does the water temperature change each minute?

89. Sewing If $\frac{6}{7}$ yard of material is needed to make a blanket, how many blankets can be made from 12 yards of material?

90. Manufacturing A clothing manufacturer is making scarves that require $\frac{3}{8}$ yard for material each. How many can be made from 27 yards of material?

91. Capacity Suppose a bag holds exactly $\frac{1}{4}$ pound of candy. How many of these bags can be filled from 12 pounds of candy?

92. Capacity A certain size bottle holds exactly $\frac{4}{5}$ pint of liquid. How many of these bottles can be filled from a 20-pint container?

93. Cooking A man is making cookies from a recipe that calls for $\frac{3}{4}$ teaspoon of oil. If the only measuring spoon he can find is a $\frac{1}{8}$ teaspoon, how many of these will he have to fill with oil in order to have a total of $\frac{3}{4}$ teaspoon of oil?

94. Cooking A cake recipe calls for $\frac{1}{2}$ cup of sugar. If the only measuring cup available is a $\frac{1}{8}$ cup, how many of these will have to be filled with sugar to make a total of $\frac{1}{2}$ cup of sugar?

95. Cartons of Milk If a small carton of milk holds exactly $\frac{1}{2}$ pint, how many of the $\frac{1}{2}$-pint cartons can be filled from a 14-pint container?

96. Pieces of Pipe How many pieces of pipe that are $\frac{2}{3}$ foot long must be laid together to make a pipe 16 feet long?

97. Internet Mailing Lists A company sells products on the Internet through an e-mail list. They predict that they sell one $50 product for every 25 people on their mailing list.

 a. What is their projected revenue if their list contains 10,000 e-mail addresses?

 b. What is their projected revenue if their list contains 25,000 e-mail addresses?

 c. They can purchase a list of 5,000 e-mail addresses for $5,000. Is this a wise purchase?

98. Internet Mailing Lists A new band has a following on the Internet. They sell their CDs through an e-mail list. They predict that they sell one $15 CD for every 10 people on their mailing list.

 a. What is their projected revenue if their list contains 5,000 e-mail addresses?

 b. What is their projected revenue if their list contains 20,000 e-mail addresses?

 c. If they need to make $45,000, how many people do they need on their e-mail list?

99. South Dakota holds two weather records: the fastest temperature rise and the fastest temperature drop. This change is caused by Chinook winds, which can warm temperatures rapidly. As the wind dies down, the resulting temperature plummets. The fastest temperature rise occurred in Spearfish, SD when the temperature rose 49° in 2 minutes, while the fastest temperature dropped 47° occurred in Rapid City, SD, when the temperature dropped 47° in 5 minutes. Find the change in temperature per minute for Spearfish and Rapid City.

100. Montana holds two weather records for the greatest decrease and the greatest increase in temperature in 24 hours. In Browning, Montana, the temperature dropped from 44°F to −56°F in a 24–hour period. In Loma, Montana, the temperature rose from −54°F to 48°F. Both changes were the result of arctic cold fronts (one entering, one leaving). Find the change in temperature per hour for Browing and Loma.

South Dakota

Montana

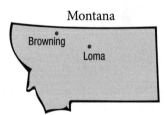

Find the Mistake

Each sentence below contains a mistake. Circle the mistake and write the correct word(s) or numbers(s) on the line provided.

1. The division problem $88 \div 4$ is the same as $4 \cdot \left(\frac{1}{88} \right)$. _____

2. Dividing two numbers with unlike signs gives a positive answer. _____

3. Dividing −55 by −5 will give us −11. _____

4. Simplifying $\frac{-3(6)}{9}$ gives us 2. _____

Landmark Review: Checking Your Progress

Give the opposite and absolute value for each of the following numbers.

1. $\dfrac{3}{4}$

2. -27

Simplify each expression.

3. $|15 - 4|$

4. $15 - 3|5 - 2 \cdot 3| - 3$

Write an equivalent algebraic expression in symbols.

5. The sum of y and 15 is 27.

6. The product of 3 and x is less than 24.

Add or subtract the following.

7. $-42 + 3.$

8. $-13 - 5$

9. $4 - 7$

10. $-52 + 37 - 15$

Which of the following equations matches with each illustration?

a.

b.

11. $x + y = 180$

12. $x + y = 90$

Simplify each of the following expressions.

13. $(-5)^2$

14. $-5(14 - 24)$

15. $\dfrac{15}{25 - 30}$

16. $\dfrac{52 - 13}{5 - 5}$

17. Thirty divided by what number is -10?

18. If the quotient of 8 and -4 is decreased by 2, what number results?

1.5 Properties of Real Numbers

Image © Katherine Heistand Shields, 2010

OBJECTIVES

A Rewrite expressions using the commutative and associative properties.

B Multiply using the distributive property.

C Identify properties used to rewrite an expression.

KEY WORDS

property

commutative

associative

grouping

distributive

identity

inverse

symmetric

In this section, we will list all the facts (properties) that you know from past experience are true about numbers in general. We will give each property a name so we can refer to it later in this book. Mathematics is very much like a game. The game involves numbers. The rules of the game are the properties we are developing in this chapter. The goal of the game is to extend the basic rules to as many situations as possible.

You know from past experience with numbers that it makes no difference in which order you add two numbers; that is, $3 + 5$ is the same as $5 + 3$. Imagine a surfer catching some waves on a beautiful sunny day. He is able to stand up on the first 3 waves but then immediately wipes out into the water. He has successful rides on the next 5 waves. Using our knowledge of addition, we know that the surfer attempted a total of 8 waves. Even if the surfer had successful rides on the first 5 waves and wiped out on the next three, the total waves he attempted would still be 8. This is an example of the *commutative property of addition*. We say addition is a commutative operation. Changing the order of the numbers does not change the answer. This fact about numbers is one of the properties of real numbers we will cover in this section.

Multiplication is another commutative operation because $3(5)$ is the same as $5(3)$. Changing the order of the two numbers you are multiplying does not change the answer.

A The Commutative and Associative Properties

For all properties listed in this section, *a*, *b*, and *c* represent real numbers.

> **PROPERTY** Commutative Property of Addition
>
> *In symbols:* $a + b = b + a$
>
> *In words:* Changing the *order* of the numbers in a sum will not change the result.

> **PROPERTY** Commutative Property of Multiplication
>
> *In symbols:* $a \cdot b = b \cdot a$
>
> *In words:* Changing the *order* of the numbers in a product will not change the result.

The statement $5 + 8 = 8 + 5$ is an example of the commutative property of addition. The statement $2 \cdot y = y \cdot 2$ is an example of the commutative property of multiplication.

**Video Examples
Section 1.5**

Practice Problems

1. Simplify $9 + y + 4$.

Note At this point, some students are confused by the expression $x + 8$; they feel that there is more to do, but they don't know what. At this point, there isn't any more that can be done with $x + 8$ unless we know what x is. So $x + 8$ is as far as we can go with this problem.

2. Simplify.

a. $3 + (7 + x)$

b. $3(4x)$

3. Simplify.

a. $\frac{1}{3}(3x)$

b. $2\left(\frac{1}{2}x\right)$

c. $16\left(\frac{3}{4}x\right)$

EXAMPLE 1 Simplify $5 + x + 3$.

Solution The expression $5 + x + 3$ can be simplified using the commutative property of addition.

$$5 + x + 3 = x + 5 + 3 \qquad \textit{Commutative property of addition}$$
$$= x + 8 \qquad \textit{Add.}$$

The other two basic operations, subtraction and division, are not commutative. The order in which we subtract or divide two numbers makes a difference in the answer.

Another property of numbers that you have used many times has to do with grouping. You know that when we add three numbers it makes no difference which two we add first. When adding $3 + 5 + 7$, we can add the 3 and 5 first and then the 7, or we can add the 5 and 7 first and then the 3. Mathematically, it looks like this: $(3 + 5) + 7 = 3 + (5 + 7)$. This property is true of multiplication as well. Operations that behave in this manner are called *associative* operations. The answer will not change when we change the association (or grouping) of the numbers.

> **PROPERTY** Associative Property of Addition
>
> *In symbols:* $a + (b + c) = (a + b) + c$
>
> *In words:* Changing the **grouping** of the numbers in a sum will not change the result.

> **PROPERTY** Associative Property of Multiplication
>
> *In symbols:* $a(bc) = (ab)c$
>
> *In words:* Changing the **grouping** of the numbers in a product will not change the result.

The following examples illustrate how the associative properties can be used to simplify expressions that involve both numbers and variables.

EXAMPLE 2 Simplify.

a. $4 + (5 + x)$ **b.** $5(2x)$

Solution

a. $4 + (5 + x) = (4 + 5) + x \qquad \textit{Associative property of addition}$
$$= 9 + x \qquad \textit{Add.}$$

b. $5(2x) = (5 \cdot 2)x \qquad \textit{Associative property of multiplication}$
$$= 10x \qquad \textit{Multiply.}$$

EXAMPLE 3 Simplify.

a. $\frac{1}{5}(5x)$ **b.** $3\left(\frac{1}{3}x\right)$ **c.** $12\left(\frac{2}{3}x\right)$

Solution

a. $\frac{1}{5}(5x) = \left(\frac{1}{5} \cdot 5\right)x \qquad \textit{Associative property of multiplication}$
$$= 1x \qquad \textit{Multiply.}$$
$$= x$$

b. $3\left(\frac{1}{3}x\right) = \left(3 \cdot \frac{1}{3}\right)x \qquad \textit{Associative property of multiplication}$
$$= 1x \qquad \textit{Multiply.}$$
$$= x$$

c. $12\left(\frac{2}{3}x\right) = \left(12 \cdot \frac{2}{3}\right)x \qquad \textit{Associative property of multiplication}$
$$= 8x \qquad \textit{Multiply.}$$

Answers

1. $13 + y$

2. a. $10 + x$ **b.** $12x$

3. a. x **b.** x **c.** $12x$

B The Distributive Property

The associative and commutative properties apply to problems that are either all multiplication or all addition. There is a third basic property that involves both addition and multiplication. It is called the *distributive property* and looks like this:

> **PROPERTY** Distributive Property
>
> *In symbols:* $a(b + c) = ab + ac$
>
> *In words:* Multiplication **distributes** over addition.

You will see as we progress through the book that the distributive property is used very frequently in algebra. We can give a visual justification to the distributive property by finding the areas of rectangles. Figure 1 shows a large rectangle that is made up of two smaller rectangles. We can find the area of the large rectangle two different ways.

Method 1

We can calculate the area of the large rectangle directly by finding its length and width. The width is 5 inches, and the length is $(3 + 4)$ inches.

$$\text{Area of large rectangle} = 5(3 + 4)$$
$$= 5(7)$$
$$= 35 \text{ square inches}$$

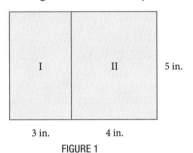

I II 5 in.

3 in. 4 in.

FIGURE 1

Method 2

Because the area of the large rectangle is the sum of the areas of the two smaller rectangles, we can find the area of each small rectangle and then add to find the area of the large rectangle.

$$\text{Area of large rectangle} = \text{Area of rectangle I} + \text{Area of rectangle II}$$
$$= \quad 5(3) \quad + \quad 5(4)$$
$$= \quad 15 \quad + \quad 20$$
$$= \quad 35 \text{ square inches}$$

In both cases the result is 35 square inches. Because the results are the same, the two original expressions must be equal. Stated mathematically, $5(3 + 4) = 5(3) + 5(4)$. We can either add the 3 and 4 first and then multiply that sum by 5, or we can multiply the 3 and the 4 separately by 5 and then add the products. The following examples illustrate how we use the distributive property with algebraic expressions.

EXAMPLE 4 Apply the distributive property to each expression, and then simplify.

a. $2(x + 3)$ **b.** $5(2x - 8)$ **c.** $5(2x + 4y)$ **d.** $4(2a + 3) + 8$ **e.** $-(a + 7)$

Solution

a. $2(x + 3) = 2(x) + 2(3)$ Distributive property
$= 2x + 6$ Multiply.

b. $5(2x - 8) = 5(2x) - 5(8)$ Distributive property
$= 10x - 40$ Multiply.

Notice in Part b that multiplication distributes over subtraction as well as addition.

c. $5(2x + 4y) = 5(2x) + 5(4y)$ Distributive property
$= 10x + 20y$ Multiply.

d. $4(2a + 3) + 8 = 4(2a) + 4(3) + 8$ Distributive property
$= 8a + 12 + 8$ Multiply.
$= 8a + 20$ Add.

e. $-(a + 7) = (-1)(a) + (-1)(1)$ Distributive property
$= -a - 7$ Multiply.

Note Because subtraction is defined in terms of addition, it is also true that the distributive property applies to subtraction as well as addition; that is, $a(b - c) = ab - ac$ for any three real numbers a, b, and c.

Note Students sometimes mistake $a(bc)$ as $ab \cdot ac$, which is not true. The expression $ab \cdot ac$ simplifies to a^2bc, which is not equal to the original expression. The distributive property states that multiplication distributes over addition, not multiplication.

4. Apply the distributive property to each expression, and then simplify the result.

a. $3(x + 2)$

b. $4(2x - 5)$

c. $3(7x - 6y)$

d. $5(7a - 3) + 2$

e. $-(b - 2)$

Note The negative sign in Example 4 part d is the same as distributing a -1 over the sum.

Answers

4. **a.** $3x + 6$ **b.** $8x - 20$
c. $21x - 18y$ **d.** $35a - 13$
e. $-b + 2$

5. Apply the distributive property to the expression, and then simplify the result.

$$\frac{1}{3}(3x + 6)$$

Next, we have some expressions to simplify that involve fractions.

EXAMPLE 5 Apply the distributive property to each expression, and then simplify the result.

a. $\frac{1}{2}(3x + 6)$ **b.** $3\left(\frac{1}{3}x + 5\right)$ **c.** $a\left(1 + \frac{1}{a}\right)$ **d.** $12\left(\frac{2}{3}x + \frac{1}{2}y\right)$

Solution

a. $\frac{1}{2}(3x + 6) = \frac{1}{2}(3x) + \frac{1}{2}(6)$ *Distributive property*

$\phantom{\frac{1}{2}(3x + 6)} = \frac{3}{2}x + 3$ *Multiply.*

b. $3\left(\frac{1}{3}x + 5\right) = 3 \cdot \frac{1}{3}x + 3 \cdot 5$ *Distributive property*

$\phantom{3\left(\frac{1}{3}x + 5\right)} = x + 15$ *Multiply.*

c. $a\left(1 + \frac{1}{a}\right) = a \cdot 1 + a \cdot \frac{1}{a}$ *Distributive property*

$\phantom{a\left(1 + \frac{1}{a}\right)} = a + 1$ *Multiply*

d. $12\left(\frac{2}{3}x + \frac{1}{2}y\right) = 12 \cdot \frac{2}{3}x + 12 \cdot \frac{1}{2}y$ *Distributive property*

$\phantom{12\left(\frac{2}{3}x + \frac{1}{2}y\right)} = 8x + 6y$ *Multiply.*

Special Numbers

In addition to the three properties mentioned so far, we want to include in our list two special numbers that have unique properties. They are the numbers 0 and 1.

> **PROPERTY** **Additive Identity Property**
>
> There exists a unique number 0 such that
>
> *In symbols:* $a + 0 = a$ and $0 + a = a$
>
> *In words:* Adding zero to any number does not change that value of the number.

> **PROPERTY** **Multiplicative Identity Property**
>
> There exists a unique number 1 such that
>
> *In symbols:* $a(1) = a$ and $(1)a = a$
>
> *In words:* Multiplying any number by 1 does not change the value of the number.

> **PROPERTY** **Additive Inverse Property**
>
> *In symbols:* $a + -a = 0$
>
> *In words:* Opposites add to 0.

> **PROPERTY** **Multiplicative Inverse Property**
>
> For every real number a, except 0, there exists a unique real number $\frac{1}{a}$ such that
>
> *In symbols:* $a\left(\frac{1}{a}\right) = 1$
>
> *In words:* Reciprocals multiply to 1.

Answer

5. $x + 2$

C Identifying Properties

A final property we would like to mention in this chapter is the *symmetric property*. This property allows us to move our variable in an equation from the right-hand side of the equals sign to the left-hand side once we have solved the equation.

PROPERTY Symmetric Property

In symbols: $a = b$, then $b = a$

In words: If one quantity equals another quantity, then the second quantity also equals the first.

Of all the basic properties listed, the commutative, associative, and distributive properties are the ones we will use most often. They are important because they will be used as justifications or reasons for many of the things we will do.

The following example illustrates how we use the preceding properties. Each sub-example contains an algebraic expression that has been changed in some way. The property that justifies the change is written to the right.

EXAMPLE 6 State the property that justifies the given statement.

a. $x + 5 = 5 + x$ Commutative property of addition

b. $(2 + x) + y = 2 + (x + y)$ Associative property of addition

c. $6(x + 3) = 6x + 18$ Distributive property

d. $2 + (-2) = 0$ Additive inverse property

e. $3\left(\dfrac{1}{3}\right) = 1$ Multiplicative inverse property

f. $(2 + 0) + 3 = 2 + 3$ Additive identity property

g. $(2 + 3) + 4 = 3 + (2 + 4)$ Commutative and associative properties of addition

h. $(x + 2) + y = (x + y) + 2$ Commutative and associative properties of addition

i. If $\dfrac{3}{4} = x$, then $x = \dfrac{3}{4}$. Symmetric property

As a final note on the properties of real numbers, we should mention that although some of the properties are stated for only two or three real numbers, they hold for as many numbers as needed. For example, the distributive property holds for expressions like $3(x + y + z + 5 + 2)$; that is,

$$3(x + y + z + 5 + 2) = 3x + 3y + 3z + 15 + 6$$

It is not important how many numbers are contained in the sum, only that it is a sum. Multiplication, you see, distributes over addition, whether there are two numbers in the sum or 200.

GETTING READY FOR CLASS

After reading through the preceding section, respond in your own words and in complete sentences.

A. What is the commutative property of addition?

B. Write the commutative property of multiplication in symbols and words.

C. Use the associative property of multiplication to simplify $4(3x)$.

D. How do you use the distributive property to find the area of a rectangle?

6. State the property that justifies the given statement.

a. $5 \cdot 4 = 4 \cdot 5$

b. $x + 7 = 7 + x$

c. $(3 + a) + b = 3 + (a + b)$

d. $6(x + 5) = 6x + 30$

e. $4\left(\dfrac{1}{4}\right) = 1$

f. $8 + (-8) = 0$

g. $(x + 2) + 3 = x + (2 + 3)$

h. $5(1) = 5$

i. If $\dfrac{2}{3} = y$, then $y = \dfrac{2}{3}$.

Answers

6. a. Commutative property of multiplication

b. Commutative property of addition

c. Associative property of addition

d. Distributive property

e. Multiplicative inverse

f. Additive inverse

g. Associative property of addition

h. Multiplicative identity

i. Symmetric property

EXERCISE SET 1.5

Vocabulary Review

Match the following definitions on the left with their property names on the right.

1. $a + b = b + a$
2. $a \times b = b \times a$
3. $a + (b + c) = (a + b) + c$
4. $a(bc) = (ab)c$
5. $a(b + c) = ab + ac$
6. $a + 0 = a$ and $0 + a = a$
7. $a(1) = a$ and $1(a) = a$
8. $a + (-a) = 0$
9. $a\left(\frac{1}{a}\right) = 1$
10. $a = b$ then $b = a$

a. Multiplicative inverse property
b. Commutative property of multiplication
c. Distributive property
d. Associative property of addition
e. Commutative property of addition
f. Additive identity property
g. Symmetric property
h. Associative property of multiplication
i. Additive inverse property
j. Multiplicative identity property

Problems

A Use the associative property to rewrite each of the following expressions, and then simplify the result.

1. $4 + (2 + x)$
2. $5 + (6 + x)$
3. $(x + 2) + 7$
4. $(x + 8) + 2$

5. $3(5x)$
6. $5(3x)$
7. $9(6y)$
8. $6(9y)$

9. $\frac{1}{2}(3a)$
10. $\frac{1}{3}(2a)$
11. $\frac{1}{3}(3x)$
12. $\frac{1}{4}(4x)$

13. $\frac{1}{2}(2y)$
14. $\frac{1}{7}(7y)$
15. $\frac{3}{4}\left(\frac{4}{3}x\right)$
16. $\frac{3}{2}\left(\frac{2}{3}x\right)$

17. $\frac{6}{5}\left(\frac{5}{6}a\right)$
18. $\frac{2}{5}\left(\frac{5}{2}a\right)$
19. $3(5xy)$
20. $12(3ab)$

21. $15(2xy)$
22. $5(4cd)$

B Apply the distributive property to each of the following expressions. Simplify when possible.

23. $8(x + 2)$

24. $5(x + 3)$

25. $8(x - 2)$

26. $5(x - 3)$

27. $4(y + 1)$

28. $4(y - 1)$

29. $3(6x + 5)$

30. $3(5x + 6)$

31. $2(3a + 7)$

32. $5(3a + 2)$

33. $9(6y - 8)$

34. $2(7y - 4)$

35. $\frac{1}{3}(3x + 6)$

36. $\frac{1}{2}(2x + 4)$

37. $-3(x - 4y)$

38. $-4(3a - 5b)$

39. $6(2x + 3y)$

40. $8(3x + 2y)$

41. $4(3a - 2b)$

42. $5(4a - 8b)$

43. $\frac{1}{2}(6x + 4y)$

44. $\frac{1}{3}(6x + 9y)$

45. $4(a + 4) + 9$

46. $6(a + 2) + 8$

47. $-2(2a - 4) + 3$

48. $-5(3x - 2) + 5$

49. $2(3x + 5) + 2$

50. $7(2x + 1) + 3$

51. $7(2x + 4) + 10$

52. $3(5x + 6) + 20$

53. $2(4x - 5) - 7$

54. $3(5x - 7) - 9$

Apply the distributive property and simplify, if possible.

55. $\frac{1}{2}(4x + 2)$

56. $\frac{1}{3}(6x + 3)$

57. $\frac{3}{4}(8x - 4)$

58. $\frac{2}{5}(5x + 10)$

59. $-\frac{1}{4}(8x - 16)$

60. $-\frac{3}{5}(15x + 5)$

61. $\frac{5}{6}(6x + 12)$

62. $\frac{2}{3}(9x - 3)$

63. $10\left(\dfrac{3}{5}x + \dfrac{1}{2}\right)$

64. $8\left(\dfrac{1}{4}x - \dfrac{5}{8}\right)$

65. $-12\left(\dfrac{1}{4}x + \dfrac{2}{3}\right)$

66. $-15\left(\dfrac{2}{3}a - \dfrac{3}{5}\right)$

67. $15\left(\dfrac{1}{3}x + \dfrac{2}{5}\right)$

68. $12\left(\dfrac{1}{12}m + \dfrac{1}{6}\right)$

69. $12\left(\dfrac{1}{2}m - \dfrac{5}{12}\right)$

70. $8\left(\dfrac{1}{8} + \dfrac{1}{2}m\right)$

71. $21\left(\dfrac{1}{3} + \dfrac{1}{7}x\right)$

72. $6\left(\dfrac{3}{2}y + \dfrac{1}{3}\right)$

73. $6\left(\dfrac{1}{2}x - \dfrac{1}{3}y\right)$

74. $12\left(\dfrac{1}{4}x + \dfrac{2}{3}y\right)$

75. $0.09(x + 2{,}000)$

76. $0.04(x + 7{,}000)$

77. $0.12(x + 500)$

78. $0.06(x + 800)$

79. $a\left(1 + \dfrac{1}{a}\right)$

80. $a\left(1 - \dfrac{1}{a}\right)$

81. $a\left(\dfrac{1}{a} - 1\right)$

82. $a\left(\dfrac{1}{a} + 1\right)$

C State the property or properties that justify the following.

83. $3 + 2 = 2 + 3$

84. $5 + 0 = 5$

85. $4\left(\dfrac{1}{4}\right) = 1$

86. $10(0.1) = 1$

87. $4 + x = x + 4$

88. $3(x - 10) = 3x - 30$

89. $2(y + 8) = 2y + 16$

90. $3 + (4 + 5) = (3 + 4) + 5$

91. $(3 + 1) + 2 = 1 + (3 + 2)$

92. $(5 + 2) + 9 = (2 + 5) + 9$

93. $(8 + 9) + 10 = (8 + 10) + 9$

94. $(7 + 6) + 5 = (5 + 6) + 7$

95. $3(x + 2) = 3(2 + x)$

96. $2(7y) = (7 \cdot 2)y$

97. If $\dfrac{4}{5} = a$, then $a = \dfrac{4}{5}$.

98. $a(5b) = 5(ab)$

99. $4(xy) = 4(yx)$

100. $3[2 + (-2)] = 3(0)$

101. $8[7 + (-7)] = 8(0)$

102. $7(1) = 7$

103. $\left(\dfrac{4}{3}\right)\left(\dfrac{3}{4}\right) = 1$

104. $\left(\dfrac{2}{3}\right)\left(\dfrac{3}{2}\right) = 1$

105. If $\dfrac{1}{8} = b$, then $b = \dfrac{1}{8}$.

106. $7 - 12 = -12 + 7$

Each of the following problems has a mistake in it. Correct the right-hand side.

107. $3(x + 2) = 3x + 2$

108. $5(4 + x) = 4 + 5x$

109. $9(a + b) = 9a + b$

110. $2(y + 1) = 2y + 1$

111. $3(0) = 3$

112. $5\left(\dfrac{1}{5}\right) = 5$

113. $3 + (-3) = 1$

114. $8(0) = 8$

115. $10(1) = 0$

116. $3 \cdot \dfrac{1}{3} = 0$

117. $\dfrac{2}{3}\left(\dfrac{3}{2}\right) = 0$

118. $6x - 6x = x$

Applying the Concepts

119. Getting Dressed While getting dressed for work, a man puts on his socks and puts on his shoes. Are the two statements "put on your socks" and "put on your shoes" commutative? That is, will changing the order of the events always produce the same result?

120. Skydiving A skydiver flying over the jump area is about to do two things: jump out of the plane and pull the rip cord. Are the two events "jump out of the plane" and "pull the rip cord" commutative?

121. Division Give an example that shows that division is not a commutative operation; that is, find two numbers for which changing the order of division gives two different answers.

122. Subtraction Simplify the expression $10 - (5 - 2)$ and the expression $(10 - 5) - 2$ to show that subtraction is not an associative operation.

123. Hours Worked Carlo works as a waiter. He works double shifts 4 days a week. The lunch shift is 2 hours and the dinner shift is 3 hours. Find the total number of hours he works per week using the numbers 2, 3, and 4. Do the calculation two different ways so that the results give further justification for the distributive property.

124. Jose earns $8 per hour regular pay and $12 per hour working overtime. He is paid overtime if he works more than 40 hours in a week. Find his pay if Jose works each of these weeks:

Week	Hours	Pay
1	36	
2	54	
3	60	

125. Quantitative Literacy and You What is quantitative literacy? Here is what the *Mathematical Association of America* has to say about quantitative literacy:

.... every college graduate should be able to apply simple mathematical methods to the solution of real-world problems. A quantitatively literate college graduate should be able to:

a. Interpret mathematical models such as formulas, graphs, tables, and schematics, and draw inferences from them.

b. Represent mathematical information symbolically, visually, numerically, and verbally.

c. Estimate and check answers to mathematical problems in order to determine reasonableness, identify alternatives, and select optimal results.

d. Recognize that mathematical and statistical methods have limits.

e. List two specific skills that you would like to acquire from this course that will improve your quantitative literacy. For example, you may want to better understand the financing options available for purchasing a new car.

Find the Mistake

Each sentence below contains a mistake. Circle the mistake and write the correct word(s) or numbers(s) on the line provided.

1. To simplify the problem $8\left(\frac{4}{7}x\right)$, we use the distributive property. _____

2. The distributive property states that multiplication distributes over division. _____

3. The additive inverse property states that $a(1) = a$ and $1(a) = a$. _____

4. The expression $35\left(\frac{2}{3}y + 18z\right)$ is equal to $35\left(\frac{2}{3}y\right) + 35(18z)$ because of the associative property of multiplication.

1.6 Simplifying Expressions

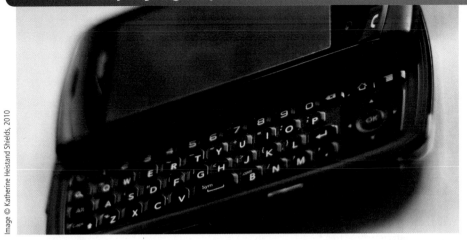

Image © Katherine Heistand Shields, 2010

OBJECTIVES

A Simplify expressions by combining similar terms.

B Simplify expressions involving parentheses.

C Calculate the value of an expression for a given value of the variable.

KEY WORDS

expression

term

similar terms

parentheses

value of an expression

If a cellular phone company charges $50 per month plus $30 for a data package and $0.10 for each text, then the amount of your monthly bill is given by the expression $50 + 30 + 0.10t$. To find the amount you will pay in one month if you send 100 texts, you substitute 100 for t and simplify the resulting expression. For example,

$$\text{Monthly bill} = 50 + 30 + 0.10(100)$$
$$90 = 50 + 30 + 10 \quad \text{in dollars}$$

In this section, we will learn how to simplify algebraic expressions similar to this.

A Combining Similar Terms

An *expression* is a number and/or variable combined with math operations used to represent a number. Examples include

$$-7y, \quad x + 7, \quad 2t - 9, \quad 4st - \frac{1}{4}s$$

Notice in the expressions above, "terms" are being added or subtracted to form the expression. A *term* is a number or a number and one or more variables multiplied together. Some terms in the expressions above include x, $2t$, and $-\frac{1}{4}s$.

> **DEFINITION** similar terms
>
> Two terms with the same variable part (same variables raised to the same powers) are called *similar* (or *like*) *terms*.

The terms $3x$ and $4x$ are similar because their variable parts are identical. Likewise, the terms $18y$, and $-10y$ are similar terms as are $3x^2$ and $-4x^2$.

To simplify an algebraic expression, we simply reduce the number of terms in the expression. We accomplish this by applying the distributive property along with our knowledge of addition and subtraction of positive and negative real numbers. The following examples illustrate the procedure.

Video Examples
Section 1.6

1. Simplify by combining similar terms.
 a. $5x + 2x$
 b. $12a - 7a$
 c. $6y - 8y + 5y$

EXAMPLE 1 Simplify by combining similar terms.

a. $3x + 4x$ **b.** $7a - 10a$ **c.** $18y - 10y + 6y$

Solution We combine similar terms by applying the distributive property.

a. $3x + 4x = (3 + 4)x$ *Distributive property*

 $= 7x$ *Add 3 and 4.*

b. $7a - 10a = (7 - 10)a$ *Distributive property*

 $= -3a$ *Add 7 and -10.*

c. $18y - 10y + 6y = (18 - 10 + 6)y$ *Distributive property*

 $= 14y$ *Add 18, -10, and 6.*

When the expression we intend to simplify is more complicated, we use the commutative and associative properties first.

2. Simplify each expression.

 a. $2x + 6 + 3x - 5$

 b. $8a - 2 - 4a + 5$

 c. $9x + 1 - x - 6$

EXAMPLE 2 Simplify each expression.

a. $3x + 5 + 2x - 3$ **b.** $4a - 7b - 2a + 3b$ **c.** $5x + \dfrac{1}{2}y - x - \dfrac{3}{2}y$

Solution We combine similar terms by applying the distributive property.

a. $3x + 5 + 2x - 3 = 3x + 2x + 5 - 3$ *Commutative property*

 $= (3x + 2x) + (5 - 3)$ *Associative property*

 $= (3 + 2)x + (5 - 3)$ *Distributive property*

 $= 5x + 2$

b. $4a - 7b - 2a + 3b = (4a - 2a) + (-7b + 3b)$ *Commutative and associative properties*

 $= (4 - 2)a + b(-7 + 3)$ *Distributive property*

 $= 2a - 4b$

c. $5x + \dfrac{1}{2}y - x - \dfrac{3}{2}y = (5x - x) + \dfrac{1}{2}y - \dfrac{3}{2}y$ *Commutative and associative properties*

 $= (5 - 1)x + \left(\dfrac{1}{2} - \dfrac{3}{2}\right)y$ *Distributive property*

 $= 4x - y$

Notice that in each case the result has fewer terms than the original expression. Because there are fewer terms, the resulting expression is said to be simpler than the original expression.

B Simplifying Expressions Involving Parentheses

If an expression contains parentheses, it is often necessary to apply the distributive property to remove the parentheses before combining similar terms.

3. Simplify $4(3x - 5) - 2$.

EXAMPLE 3 Simplify the expression $5(2x - 8) - 3$.

Solution We begin by distributing the 5 across $2x - 8$. We then combine similar terms.

$$5(2x - 8) - 3 = 10x - 40 - 3 \qquad \text{\textit{Distributive property}}$$

$$= 10x - 43 \qquad \text{\textit{Combine similar terms.}}$$

Answers

1. **a.** $7x$ **b.** $5a$ **c.** $3y$
2. **a.** $5x + 1$ **b.** $4a + 3$ **c.** $8x - 5$
3. $12x - 22$

EXAMPLE 4 Simplify $7 - 3(2y + 1)$.

Solution By the rule for order of operations, we must multiply before we add or subtract. For that reason, it would be incorrect to subtract 3 from 7 first. Instead, we multiply -3 and $2y + 1$ to remove the parentheses and then combine similar terms.

$$7 - 3(2y + 1) = 7 - 6y - 3 \qquad \text{Distributive property}$$
$$= -6y + 4 \qquad \text{Combine similar terms.}$$

EXAMPLE 5 Simplify $5(x - 2) - (3x + 4)$.

Solution We begin by applying the distributive property to remove the parentheses. The expression $-(3x + 4)$ can be thought of as $-1(3x + 4)$. Thinking of it in this way allows us to apply the distributive property.

$$-1(3x + 4) = -1(3x) + (-1)(4)$$
$$= -3x - 4$$

The complete solution looks like this:

$$5(x - 2) - (3x + 4) = 5x - 10 - 3x - 4 \qquad \text{Distributive property}$$
$$= 2x - 14 \qquad \text{Combine similar terms.}$$

As you can see from the explanation in Example 5, we use the distributive property to simplify expressions in which parentheses are preceded by a negative sign. In general, we can write

$$-(a + b) = -1(a + b)$$
$$= -a + (-b)$$
$$= -a - b$$

The negative sign outside the parentheses ends up changing the sign of each term within the parentheses. In words, we say "the opposite of a sum is the sum of the opposites."

C The Value of an Expression

An algebraic expression like $3x + 2$ has a certain value depending on what number we assign to x. For instance, when x is 4, $3x + 2$ becomes $3(4) + 2$, or 14. When x is -8, $3x + 2$ becomes $3(-8) + 2$, or -22. The value of an expression is found by replacing or substituting the variable with a given number.

EXAMPLE 6 Find the value of the following expressions by replacing the variable with the given number.

Expression	The Variable	Value of the Expression
a. $3x - 1$	$x = 2$	$3(2) - 1 = 6 - 1 = 5$
b. $-7a + 4$	$a = -3$	$-7(-3) + 4 = 21 + 4 = 25$
c. $2x - 3 + 4x$	$x = -1$	$2(-1) - 3 + 4(-1) = -2 - 3 + (-4)$ $= -9$
d. $2x - 5 - 8x$	$x = 5$	$2(5) - 5 - 8(5) = 10 - 5 - 40$ $= -35$
e. $y^2 - 6y + 9$	$y = 4$	$4^2 - 6(4) + 9 = 16 - 24 + 9 = 1$

4. Simplify $5 - 2(4y + 1)$.

5. Simplify $6(x - 3) - (7x + 2)$.

6. Find the following:
 a. The value of $4x - 7$ when $x = 3$
 b. The value of $2a + 4$ when $a = -5$
 c. The value of $2x - 5 + 6x$ when $x = -5$
 d. The value of $7x - 3 - 4x$ when $x = 10$
 e. The value of $y^2 - 10y + 25$ when $y = -2$

Answers
4. $-8y + 3$
5. $-x - 20$
6. a. 5 **b.** -6 **c.** -45 **d.** 27
 e. 49

Simplifying an expression should not change its value; that is, if an expression has a certain value when x is 5, then it will always have that value, no matter how much it has been simplified, as long as x is 5. If we were to simplify the expression in Example 6d first, it would look like

$$2x - 5 - 8x = -6x - 5$$

When x is 5, the simplified expression $-6x - 5$ is

$$-6(5) - 5 = -30 - 5 = -35$$

It has the same value as the original expression when x is 5.

We also can find the value of an expression that contains two variables if we know the values for both variables.

7. Find the value of $5a - 3b - 2$ when $a = 3$ and $b = 4$.

EXAMPLE 7 Find the value of the expression $2x - 3y + 4$ when $x = -5$ and $y = 6$.

Solution Substituting -5 for x and 6 for y, the expression becomes

$$2(-5) - 3(6) + 4 = -10 - 18 + 4$$
$$= -28 + 4$$
$$= -24$$

8. Find the value of $x^2 - 4xy - 4y^2$ when $x = 5$ and $y = 3$.

EXAMPLE 8 Find the value of the expression $x^2 - 2xy - y^2$ when $x = 3$ and $y = -4$.

Solution Replacing each x in the expression with the number 3 and each y in the expression with the number -4 gives us

$$3^2 - 2(3)(-4) - (-4)^2 = 9 - 2(3)(-4) - 16$$
$$= 9 - (-24) - 16$$
$$= 33 - 16$$
$$= 17$$

9. Find the value of $-7 - 5x + 9(x + y)$ when $x = 3$ and $y = -\dfrac{1}{3}$.

EXAMPLE 9 Find the value of the expression $-11 - 3x + 2(x + y)$ when $x = -3$ and $y = \dfrac{1}{2}$.

Solution Replacing x in the expression with -3, and replacing y with $\dfrac{1}{2}$ gives us.

$$-11 - 3(-3) + 2\left[(-3) + \dfrac{1}{2}\right] = -11 + 9 - 6 + 1$$
$$= -2 - 6 + 1$$
$$= -8 + 1$$
$$= -7$$

GETTING READY FOR CLASS

After reading through the preceding section, respond in your own words and in complete sentences.

A. What is a term?

B. What are similar terms?

C. Explain how you would use the distributive property to combine similar terms.

D. Explain how you would find the value of $5x + 3$ when x is 6.

Answers

7. -29

8. 1

9. 2

Vocabulary Review

Choose the correct words to fill in the blanks below.

parentheses similar expression value term variable

1. An _____ is a number and/or variable combined with math operations used to represent a number.
2. A _____ in an expression is a number or a number and one or more variables multiplied together.
3. Two monomials with the same variables raised to the same powers are called _____ terms.
4. The _____ of an expression is found by replacing or substituting the _____ with a given number.
5. To simplify an expression that contains _____ , it may be necessary to apply the distributive property to remove the parentheses before combining similar terms.

Problems

A Simplify the following expressions.

1. $3x - 6x$

2. $7x - 5x$

3. $-2a + a$

4. $3a - a$

5. $7x + 3x + 2y$

6. $8x - 2y - y$

7. $3a - 2b + 5a$

8. $7a + 3b + 6c$

9. $4x - 3 + \frac{1}{2}x$

10. $5x + 6 - 3x$

11. $\frac{3}{5}a + 5 + \frac{2}{5}a$

12. $6a + 7b + 8 - 3b$

13. $2x - 3 + 3x - 2$

14. $6x + 5 - 2x + 3 - 4x$

15. $3a - 1 + a + 3$

16. $-a + 2 + 8a - 7$

17. $-4x + 8 - 5x - 10$

18. $-\frac{1}{3}a - 1 - \frac{1}{2}a + 5$

19. $0.25b - 11 + 0.50b$

20. $8 - 0.75x - 15 - 2.30x$

21. $0.08x + 0.09x$

22. $0.04x + 0.05x$

23. $0.04x - 0.2x$

24. $0.08x - 0.3x$

B Simplify the following expressions.

25. $5(2x - 1) + 4$

26. $2(4x - 3) + 2$

27. $7(3y + 2) - 8$

28. $6(4y + 2) - 7$

29. $-3(2x - 1) + 5$

30. $-4(3x - 2) - 6x$

31. $5a - 2(a + 1)$

32. $7 - 8(2a + 3a)$

33. $6 - 4(x - 5)$

34. $12 - 3\left(\frac{1}{3}x - 2\right)$

35. $-9 - (2 - y) + 1$

36. $-10 - 2\left(\frac{1}{2} - \frac{3}{2}y\right) + 3y$

37. $-6x + 2(2 - 3x) + x$

38. $-7 - (3 + x) - 1$

39. $(4x - 7) - (2x + 5)$

40. $(7x - 3) - (4x + 2)$

41. $8(2a + 4) - (6a - 1)$

42. $9(3a + 5) - (8a - 7)$

43. $0.10(x - 2) + (x - 3)$

44. $2(2x + 1) - (x + 4) + x$

45. $4(2y - 8) - 2(y + 7)$

46. $5(y - 3) - y(4 - 2)$

47. $-(2x + 1) - \frac{1}{2}(x + 5)$

48. $-3(3x - 2) - (2x + 3)$

49. $(x + y) + (x - y)$

50. $2(x - 2) + 3(5x)$

51. $x + (x + 3)(-3)$

52. $x - 2(x + 2)$

53. $(x + 3y) + 3(2x - y)$

54. $(2x - y) - 2(x + 3y)$

55. $-6\left(\frac{1}{2}x - \frac{1}{3}y\right) + 12\left(\frac{1}{4}x + \frac{2}{3}y\right)$

56. $6\left(\frac{1}{3}x + \frac{1}{2}y\right) - 4\left(x + \frac{3}{4}y\right)$

57. $0.08x + 0.09(x + 2{,}000)$

58. $0.06x + 0.04(x + 7{,}000)$

59. $0.12x + 0.10(4{,}000 - x)$

60. $0.08x + 0.07(10{,}000 - x)$

C Evaluate the following expressions when $x = 2$. (Find the value of the expressions if x is 2.)

61. $3x - 1$

62. $4x + 3$

63. $-2x - 5$

64. $-3x + 6$

65. $x^2 - 8x + 16$

66. $x^2 - 10x + 25$

67. $(x - 4)^2$

68. $(x - 5)^2$

Evaluate the following expressions when $x = -5$. Then simplify the expression, and check to see that it has the same value for $x = -5$.

69. $7x - 4 - x - 3$

70. $3x + 4 + 7x - 6$

71. $5(2x + 1) + 4$

72. $2\left(\frac{1}{2}x - 10\right) + 5$

Evaluate the following expressions when $x = -3$ and $y = 5$.

73. $x^2 - 2xy + y^2$

74. $x^2 + 2xy + y^2$

75. $(x - y)^2$

76. $(y - x)^2$

77. $x^2 + 6xy + 9y^2$

78. $x^2 - 10xy + 25y^2$

79. $(x + 3y)^2$

80. $(x - 5y)^2$

Find the value of $b^2 - 4ac$ for the given values of a, b, and c. (You will see these types of problems later in the book.)

81. $a = 1, b = -5, c = -6$　　**82.** $a = 1, b = -6, c = 7$　　**83.** $a = 2, b = 4, c = -3$　　**84.** $a = 3, b = 4, c = -2$

Find the value of $12x - 3$ for each of the following values of x.

85. $\dfrac{1}{2}$　　　　　　**86.** $\dfrac{1}{3}$　　　　　　**87.** $\dfrac{1}{4}$　　　　　　**88.** $\dfrac{1}{6}$

89. $\dfrac{3}{2}$　　　　　　**90.** $\dfrac{2}{3}$　　　　　　**91.** $\dfrac{3}{4}$　　　　　　**92.** $\dfrac{5}{6}$

Applying the Concepts

93. Temperature and Altitude If the temperature on the ground is 70°F, then the temperature at A feet above the ground can be found from the expression $-0.0035A + 70$. Find the temperature at the following altitudes.

a. 8,000 feet

b. 12,000 feet

c. 24,000 feet

d. 29,035 feet (height of Mt. Everest)

94. Perimeter of a Rectangle The expression $2l + 2w$ gives the perimeter of a rectangle with length l and width w. Find the perimeter of the rectangles with the following lengths and widths.

a. Length = 8 meters, width = 5 meters

5 m

8 m

b. Length = 10 feet, width = 3 feet

3 ft

10 ft

95. Cellular Phone Rates A cellular phone company has a plan that charges $50 per month plus $0.10 for each text that you send. The expression $50 + 0.10t$ gives the amount of money you will pay for sending t text messages a month. Find the monthly bill.

a. 20 text messages in a month

b. 50 text messages in a month

c. 100 text messages in a month

96. Cost of Bottled Water A water bottling company charges $7.00 per month for their water dispenser and $1.10 for each gallon of water delivered. If you have g gallons of water delivered in a month, then the expression $7 + 1.1g$ gives the amount of your bill for that month. Find the monthly bill for each of the following deliveries.

a. 10 gallons

b. 20 gallons

c. 30 gallons

97. The total pressure as you dive below the surface of the ocean increases by the formula $P = 14.7 + 0.4x$, where x is the depth (in feet) and P is the pressure (in lb/in²). Find the total pressure for each of the following diving depths.

a. 15 feet

b. 25 feet

c. 50 feet

98. A water company charges a flat monthly rate of $7.50 plus $0.0058 per cubic foot of water used. Find the total charges for each of the following monthly usages.

a. 4,000 ft³

b. 12,000 ft³

c. 20,000 ft³

99. Taxes You work H hours a day and you pay 21% of your income for federal income tax and another 8% for miscellaneous deductions. This is represented by the expression $0.21H + 0.08H$. Simplify this expression. If you work 8 hours a day under these conditions, how many hours do you work to pay for your taxes and miscellaneous deductions?

100. Apartment Furniture The chart shows the cost of some furniture items. Suppose you and some roommates are moving in to an apartment. Write an expression to answer the following questions, then simplify that expression.

Item	Cost
Sofa	$180
Sphere Chair	$40
Bookcase	$60
Computer Desk	$140
Storage Drawers	$11

a. You are buying sphere chairs and will get $10 off each one you buy.

b. You are buying computer desks and will get $20 off each one. You will also get a $50 rebate on the total purchase.

c. You are buying 2 sofas and are getting a discount of $x off each one. You will also be getting a $100 rebate off the total purchase.

Perimeter If the perimeter of a figure is defined as the sum of the outside lengths, find the perimeter of the following figures. Simplify your answer.

101.

102.

$$3z - 3$$
$$2z + 5 \quad \frac{1}{3}z \quad \frac{1}{3}z \quad z \quad \frac{1}{3}z \quad z$$

Find the Mistake

Each sentence below contains a mistake. Circle the mistake and write the correct word(s) or numbers(s) on the line provided.

1. Simplifying the expression $12z + 9z - 3z$ by combining similar terms gives us $21z^2 - 3z$. _____

2. The value of the expression $8y + \frac{1}{3}y$ when $y = 3$ is $14\frac{1}{3}$. _____

3. To simplify the expression $-3(5x - 4) - 8x$, first subtract $8x$ from $5x$ to combine similar terms.

4. The value of the expression $(2x - 4)^2$ when $x = \frac{1}{2}$ is the same as the value when $x = 4$. _____

Absolute Value [1.1]

1. $|5| = 5$
 $|-5| = 5$

The *absolute value* of a real number is its distance from zero on the real number line. Absolute value is never negative.

Opposites [1.1]

2. The numbers 3 and -3 are opposites; their sum is 0.
 $3 + (-3) = 0$

Any two real numbers the same distance from zero on the number line but in opposite directions from zero are called *opposites*. Opposites always add to zero.

Symbols [1.1]

$a = b$	a is equal to b.
$a \neq b$	a is not equal to b.
$a < b$	a is less than b.
$a \nless b$	a is not less than b.
$a > b$	a is greater than b.
$a \ngtr b$	a is not greater than b.
$a \geq b$	a is greater than or equal to b.
$a \leq b$	a is less than or equal to b.

Addition of Real Numbers [1.2]

3. Add all combinations of positive and negative 10 and 13.
 $10 + 13 = 23$
 $10 + (-13) = -3$
 $-10 + 13 = 3$
 $-10 + (-13) = -23$

To add two real numbers with

1. the same sign, simply add their absolute values and use the common sign.

2. different signs, subtract the smaller absolute value from the larger absolute value. The answer has the same sign as the number with the larger absolute value.

Subtraction of Real Numbers [1.2]

4. Subtracting 2 is the same as adding -2.
 $7 - 2 = 7 + (-2) = 5$

To subtract one number from another, simply add the opposite of the number you are subtracting; that is, if a and b represent real numbers, then

$$a - b = a + (-b)$$

Multiplication of Real Numbers [1.3]

5. $3(5) = 15$
 $3(-5) = -15$
 $-3(5) = -15$
 $-3(-5) = 15$

To multiply two real numbers, simply multiply their absolute values. Like signs give a positive answer. Unlike signs give a negative answer.

Division of Real Numbers [1.4]

6. $-\frac{6}{2} = -6\left(\frac{1}{2}\right) = -3$
 $\frac{-6}{-2} = -6\left(-\frac{1}{2}\right) = 3$

Division by a number is the same as multiplication by its reciprocal. Like signs give a positive answer. Unlike signs give a negative answer.

Properties of Real Numbers [1.5]

	For Addition	*For Multiplication*
Commutative:	$a + b = b + a$	$a \cdot b = b \cdot a$
Associative:	$a + (b + c) = (a + b) + c$	$a \cdot (b \cdot c) = (a \cdot b) \cdot c$
Distributive:	$a(b + c) = ab + ac$	
Identity:	$a + 0 = a$	$a \cdot 1 = a$
Inverse:	$a + (-a) = 0$	$a\left(\dfrac{1}{a}\right) = 1$
Symmetric:	$a = b$ then $b = a$	

Similar Terms [1.6]

A *term* is a number or a number and one or more variables multiplied together. *Similar terms* are terms with the same variable part.

7. The terms $2x$, $5x$, and $-7x$ are all similar because their variable parts are the same.

Simplifying Expressions [1.6]

In this chapter, we simplified expressions that contained variables by using the distributive property to combine similar terms.

8. Simplify $3x + 4x$.
$$3x + 4x = (3 + 4)x$$
$$= 7x$$

COMMON MISTAKE

1. Interpreting absolute value as changing the sign of the number inside the absolute value symbols. $|-5| = +5, |+5| = -5$. (The first expression is correct; the second one is not.) To avoid this mistake, remember: Absolute value is a distance and distance is always measured in positive units.

2. Using the phrase "two negatives make a positive." This works only with multiplication and division. With addition, two negative numbers produce a negative answer. It is best not to use the phrase "two negatives make a positive" at all.

Translate into symbols. [1.1]

1. The difference of 15 and x is 12.

2. The product of 6 and a is 30.

3. The sum of 7 and a is greater than 15.

4. The quotient of 12 and n is less than or equal to 5.

Simplify according to the rule for order of operations. [1.1]

5. $|4 - 5| - |3 - 5|$

6. $15 - 3|2 - 4(3 - 1)| - 4 \cdot 2$

For each number, name the opposite and absolute value. [1.1]

7. -6

8. $\dfrac{5}{3}$

Add. [1.2]

9. $5 + (-8)$

10. $|-2 + 7| + |-3 + (-4)|$

Subtract. [1.2]

11. $-7 - 4$

12. $5 - (6 + 3) - 3$

Multiply. [1.3]

13. $7(-4)$

14. $3(-4)(-2)$

15. $9\left(-\dfrac{1}{3}\right)$

16. $\left(-\dfrac{3}{2}\right)^3$

Simplify. [1.1, 1.2, 1.3, 1.4]

17. $-2(3) - 7$

18. $2(3)^3 - 4(-2)^4$

19. $9 + 4(2 - 6)$

20. $5 - 3[-2(1 + 4) + 3(-3)]$

21. $\dfrac{-4(3) + 5(-2)}{-5 - 6}$

22. $\dfrac{4(3 - 5) - 2(-6 + 8)}{4(-2) + 10}$

Match each expression below with the letter of the property that justifies it. [1.5]

23. $4(2y) = (4 \cdot 2)y$

24. $5(x - 3) = 5x - 15$

25. $4 + x = x + 4$

26. $(a + 5) - 2 = a + (5 - 2)$

a. Commutative property of addition
b. Commutative property of multiplication
c. Associative property of addition
d. Associative property of multiplication
e. Distributive property

Simplify. [1.5, 1.6]

27. $5 + (7 + 3x)$

28. $3(-5y)$

29. $-5(2x - 3)$

30. $-\dfrac{1}{3}(6x + 12)$

31. $n - 5(n - 2)$

32. $(3a - b) - 5(a + 3b)$

33. $-10\left(\dfrac{1}{5}x - \dfrac{1}{2}y\right) + 12\left(\dfrac{1}{4}x - \dfrac{1}{3}y\right)$

34. $0.03n + 0.08(n + 3000)$

Use the illustration below to answer the following questions.

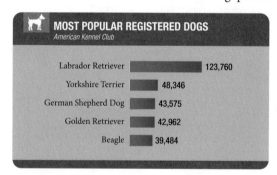

35. How many Labrador Retrievers are registered with the American Kennel Club?

36. What is the total number of Golden Retrievers and Beagles registered?

37. How many more Yorkshire Terriers are registered than German Shepherds?

Fractions and Mixed Numbers

2

2.1 The Meaning and Properties of Fractions

2.2 Prime Numbers, Factors, and Reducing to Lowest Terms

2.3 Multiplication with Fractions, and the Area of a Triangle

2.4 Division with Fractions

2.5 Addition and Subtraction with Fractions

2.6 Mixed-Number Notation

2.7 Multiplication and Division with Mixed Numbers

2.8 Addition and Subtraction with Mixed Numbers

2.9 Order of Operations and Complex Fractions

The Roman Colosseum in Rome, Italy is a testament to the advancements made in engineering and architecture by the ancient Roman Empire. Initiated by the Roman Emperor Vespasianin in the year 70 AD, it was completed in 80 AD after his death. The next Emperor, Titus, opened the Colosseum to the public. It is said the inaugural ceremony lasted for more than 100 days. The Roman Colosseum is an amphitheater that had the capacity to hold 50,000 people, a huge number at that time in history.

For comparison, the table and bar chart below shows the seating capacity for other coliseums.

Seating Capacity	
Los Angeles Coliseum	134,254
Oakland Coliseum	55,528
Roman Colosseum	50,000
Nassau Veteran's Memorial Coliseum	15,000
The Coliseum at Caesars Palace	4,100
The Coliseum, St. Petersburg, FL	2,000

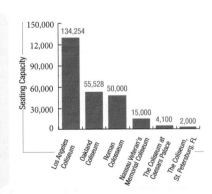

What fraction of the capacity of the Los Angeles Coliseum represents the capacity of the Roman Colosseum? Problems like this one require multiplying fractions, which is one of the topics of this chapter.

**Video Examples
Section 2.1**

Note As we mentioned in Chapter 1, when we use a letter to represent a number, or a group of numbers, that letter is called a variable. In the definition for a fraction, we are restricting the numbers that the variable b can represent to numbers other than 0. As you know we want to avoid writing an expression that would imply division by the number 0.

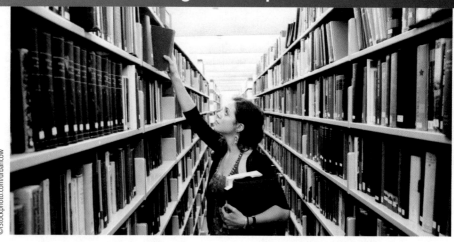

©iStockphoto.com/urbancow

The information in the table below shows enrollment for Cal Poly State University in California. The pie chart was created from the table. Both the table and pie chart use fractions to specify how the students at Cal Poly are distributed among the different schools within the university.

Cal Poly State University Enrollment	
School	**Fraction Of Students**
Agriculture	$\frac{2}{9}$
Architecture and Environmental Design	$\frac{1}{9}$
Business	$\frac{1}{9}$
Engineering	$\frac{5}{18}$
Liberal Arts	$\frac{3}{20}$
Science and Mathematics	$\frac{3}{20}$

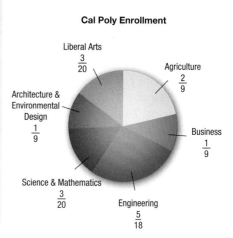

From the table, we see that $\frac{1}{9}$ (one-ninth) of the students are enrolled in the College of Business. This means one out of every nine students at Cal Poly is studying Business. The fraction $\frac{1}{9}$ tells us we have 1 part of 9 equal parts. That is, the students at Cal Poly could be divided into 9 equal groups, so that one of the groups contained all the business students and only business students.

Figure 1 shows a rectangle that has been divided into equal parts, four different ways. The shaded area for each rectangle is $\frac{1}{2}$ the total area.

Now that we have an intuitive idea of the meaning of fractions, here are the more formal definitions and vocabulary associated with fractions.

a. $\frac{1}{2}$ is shaded **b.** $\frac{2}{4}$ are shaded **c.** $\frac{3}{6}$ are shaded **d.** $\frac{4}{8}$ are shaded

FIGURE 1

A Identifying Parts of a Fraction

> **DEFINITION** fraction
>
> A *fraction* is any number that can be put in the form $\frac{a}{b}$ (also sometimes written *a/b*), where *a* and *b* are numbers and *b* is not 0.

Some examples of fractions are

$$\frac{1}{2} \qquad \frac{3}{4} \qquad \frac{7}{8} \qquad \frac{9}{5}$$

One-half Three-fourths Seven-eighths Nine-fifths

> **DEFINITION** terms
>
> For the fraction $\frac{a}{b}$, *a* and *b* are called the *terms* of the fraction. More specifically, *a* is called the *numerator,* and *b* is called the *denominator.*

EXAMPLE 1 Name the numerator and denominator for each fraction.

a. $\frac{3}{4}$ **b.** $\frac{a}{5}$ **c.** 7

Solution

a. The terms of the fraction $\frac{3}{4}$ are 3 and 4. The 3 is called the numerator, and the 4 is called the denominator.

b. The numerator of the fraction $\frac{a}{5}$ is *a*. The denominator is 5. Both *a* and 5 are called terms.

c. The number 7 may also be put in fraction form, because it can be written as $\frac{7}{1}$. In this case, 7 is the numerator and 1 is the denominator.

> **DEFINITION** proper and improper fraction
>
> A *proper fraction* is a fraction in which the numerator is less than the denominator. If the numerator is greater than or equal to the denominator, the fraction is called an *improper fraction.*

CLARIFICATION 1: The fractions $\frac{3}{4}$, $\frac{1}{8}$, and $\frac{9}{10}$ are all proper fractions, because in each case the numerator is less than the denominator.

CLARIFICATION 2: The numbers $\frac{9}{5}$, $\frac{10}{10}$, and 6 are all improper fractions, because in each case the numerator is greater than or equal to the denominator. (As we have seen, 6 can be written as $\frac{6}{1}$, in which case 6 is the numerator and 1 is the denominator.)

B Fractions on the Number Line

We can give meaning to the fraction $\frac{2}{3}$ by using a number line. If we take that part of the number line from 0 to 1 and divide it into *three equal parts*, we say that we have divided it into *thirds* (see Figure 2). Each of the three segments is $\frac{1}{3}$ (one third) of the whole segment from 0 to 1.

FIGURE 2

Practice Problems

1. Name the numerator and denominator for each fraction.

 a. $\frac{5}{6}$

 b. $\frac{1}{8}$

 c. $\frac{x}{3}$

Note There are many ways to give meaning to fractions like $\frac{2}{3}$ other than by using the number line. One popular way is to think of cutting a pie into three equal pieces, as shown below. If you take two of the pieces, you have taken $\frac{2}{3}$ of the pie.

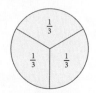

Two of these smaller segments together are $\frac{2}{3}$ (two thirds) of the whole segment. And three of them would be $\frac{3}{3}$ (three thirds), or the whole segment, as indicated in Figure 3.

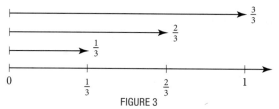

FIGURE 3

Let's do the same thing again with six and twelve equal divisions of the segment from 0 to 1 (see Figure 4).

The same point that we labeled with $\frac{1}{3}$ in Figure 3 is now labeled with $\frac{2}{6}$ and with $\frac{4}{12}$. It must be true then that

$$\frac{4}{12} = \frac{2}{6} = \frac{1}{3}$$

Although these three fractions look different, each names the same point on the number line, as shown in Figure 4. All three fractions have the same value, because they all represent the same number.

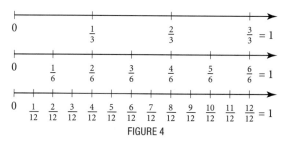

FIGURE 4

C Equivalent Fractions

DEFINITION equivalent fractions

Fractions that represent the same number are said to be *equivalent*. Equivalent fractions may look different, but they must have the same value.

It is apparent that every fraction has many different representations, each of which is equivalent to the original fraction. The next two properties give us a way of changing the terms of a fraction without changing its value.

PROPERTY Multiplication Property for Fractions

If a, b, and c are numbers and b and c are not 0, then it is always true that

$$\frac{a}{b} = \frac{a \cdot c}{b \cdot c}$$

In words: If the numerator and the denominator of a fraction are multiplied by the same nonzero number, the resulting fraction is equivalent to the original fraction.

2. Write $\frac{5}{6}$ as an equivalent fraction with a denominator of 30.

EXAMPLE 2 Write $\frac{3}{4}$ as an equivalent fraction with a denominator of 20.

Solution The denominator of the original fraction is 4. The fraction we are trying to find must have a denominator of 20. We know that if we multiply 4 by 5, we get 20. The multiplication property for fractions indicates that we are free to multiply the denominator by 5 so long as we do the same to the numerator.

$$\frac{3}{4} = \frac{3 \cdot 5}{4 \cdot 5} = \frac{15}{20}$$

The fraction $\frac{15}{20}$ is equivalent to the fraction $\frac{3}{4}$.

Answer

2. $\frac{25}{30}$

EXAMPLE 3 Write $\frac{3}{4}$ as an equivalent fraction with a denominator of $12x$.

Solution If we multiply 4 by $3x$, we will have $12x$.

$$\frac{3}{4} = \frac{3 \cdot 3x}{4 \cdot 3x} = \frac{9x}{12x}$$

3. Write $\frac{2}{3}$ as an equivalent fraction with a denominator of $15x$.

PROPERTY Division Property for Fractions

If a, b, and c are integers and b and c are not 0, then it is always true that

$$\frac{a}{b} = \frac{a \div c}{b \div c}$$

In words: If the numerator and the denominator of a fraction are divided by the same nonzero number, the resulting fraction is equivalent to the original fraction.

EXAMPLE 4 Write $\frac{10}{12}$ as an equivalent fraction with a denominator of 6.

Solution If we divide the original denominator 12 by 2, we obtain 6. The division property for fractions indicates that if we divide both the numerator and the denominator by 2, the resulting fraction will be equal to the original fraction.

$$\frac{10}{12} = \frac{10 \div 2}{12 \div 2} = \frac{5}{6}$$

4. Write $\frac{20}{25}$ as an equivalent fraction with a denominator of 5.

You know from previous examples that the following is true about fractions containing negative numbers.

If a and b are numbers and b is not equal to 0, then

$$\frac{-a}{b} = \frac{a}{-b} = -\frac{a}{b} \quad \text{and} \quad \frac{-a}{-b} = \frac{a}{b}$$

EXAMPLE 5 Write the following as equivalent fractions without a negative sign.

a. $\frac{-1}{-5}$ **b.** $\frac{-2}{-3}$ **c.** $\frac{-a}{-b}$ **d.** $\frac{-5}{-y}$

Solution Applying the rules for negative numbers, we have

a. $\frac{-1}{-5} = \frac{1}{5}$ **b.** $\frac{-2}{-3} = \frac{2}{3}$ **c.** $\frac{-a}{-b} = \frac{a}{b}$ **d.** $\frac{-5}{-y} = \frac{5}{y}$

5. Write the following as equivalent fractions without a negative sign.

a. $\frac{-6}{-7}$

b. $\frac{-11}{-3}$

c. $\frac{-x}{-y}$

d. $\frac{-a}{-4}$

EXAMPLE 6 Write the following as equivalent fractions by moving the negative sign in front of the fraction bar.

a. $\frac{-1}{4}$ **b.** $\frac{3}{-4}$ **c.** $\frac{-a}{b}$ **d.** $\frac{6}{-y}$

Solution Applying our rules for negative numbers, we have

a. $\frac{-1}{4} = -\frac{1}{4}$ **b.** $\frac{3}{-4} = -\frac{3}{4}$ **c.** $\frac{-a}{b} = -\frac{a}{b}$ **d.** $\frac{6}{-y} = -\frac{6}{y}$

6. Write the following as equivalent fractions by moving the negative sign in front of the fraction bar.

a. $\frac{-2}{9}$

b. $\frac{12}{-13}$

c. $\frac{x}{-y}$

d. $\frac{-8}{b}$

The Number 1 and Fractions

There are two situations involving fractions and the number 1 that occur frequently in mathematics. The first is when the denominator of a fraction is 1. In this case, if we let a represent any number, then

$$\frac{a}{1} = a$$

The second situation occurs when the numerator and the denominator of a fraction are the same nonzero number.

$$\frac{a}{a} = 1$$

7. Simplify each fraction.

a. $\frac{15}{1}$

b. $\frac{15}{15}$

c. $-\frac{45}{15}$

d. $-\frac{150}{15}$

EXAMPLE 7 Simplify each fraction.

a. $\frac{24}{1}$ b. $\frac{24}{24}$ c. $-\frac{48}{24}$ d. $-\frac{72}{24}$

Solution In each case, we divide the numerator by the denominator.

a. $\frac{24}{1} = 24$ b. $\frac{24}{24} = 1$ c. $-\frac{48}{24} = -2$ d. $-\frac{72}{24} = -3$

Comparing Fractions

We can compare fractions to see which is larger or smaller when they have the same denominator.

EXAMPLE 8 Write each fraction as an equivalent fraction with denominator 24. Then write them in order from smallest to largest.

$$\frac{5}{8} \qquad \frac{5}{6} \qquad \frac{3}{4} \qquad \frac{2}{3}$$

Solution We begin by writing each fraction as an equivalent fraction with denominator 24.

$$\frac{5}{8} = \frac{15}{24} \qquad \frac{5}{6} = \frac{20}{24} \qquad \frac{3}{4} = \frac{18}{24} \qquad \frac{2}{3} = \frac{16}{24}$$

Now that they all have the same denominator, the smallest fraction is the one with the smallest numerator and the largest fraction is the one with the largest numerator. Writing them in order from smallest to largest we have

$$\frac{15}{24} \; < \; \frac{16}{24} \; < \; \frac{18}{24} \; < \; \frac{20}{24}$$

or

$$\frac{5}{8} \; < \; \frac{2}{3} \; < \; \frac{3}{4} \; < \; \frac{5}{6}$$

8. Write each fraction as an equivalent fraction with denominator 36. Then write the original fractions in order from smallest to largest.

$$\frac{5}{9}, \; \frac{1}{3}, \; \frac{5}{6}, \; \frac{5}{12}$$

EXAMPLE 9 Write each fraction as an equivalent fraction with denominator 24.

$$-\frac{1}{4} \qquad -\frac{3}{8} \qquad -\frac{5}{6} \qquad -\frac{5}{8}$$

Solution We begin by writing each fraction as an equivalent fraction with denominator 24.

$$-\frac{1}{4} = -\frac{6}{24} \qquad -\frac{3}{8} = -\frac{9}{24} \qquad -\frac{5}{6} = -\frac{20}{24} \qquad -\frac{5}{8} = -\frac{15}{24}$$

Now we can write them in order from smallest to largest.

$$-\frac{20}{24} \; < \; -\frac{15}{24} \; < \; -\frac{9}{24} \; < \; -\frac{6}{24}$$

or

$$-\frac{5}{6} \; < \; -\frac{5}{8} \; < \; -\frac{3}{8} \; < \; -\frac{1}{4}$$

GETTING READY FOR CLASS

After reading through the preceding section, respond in your own words and in complete sentences.

A. What is a fraction?

B. Which term in the fraction $\frac{7}{8}$ is the numerator?

C. Is the fraction $\frac{3}{9}$ a proper fraction?

D. What word do we use to describe fractions such as $\frac{1}{5}$ and $\frac{4}{20}$, which look different, but have the same value?

Answers

7. a. 15 **b.** 1 **c.** −3 **d.** −10

8. $\frac{1}{3} < \frac{5}{12} < \frac{5}{9} < \frac{5}{6}$

Vocabulary Review

Choose the correct words to fill in the blanks below.

numerator equivalent proper fraction denominator improper

1. A _____ is any number that can be put in the form $\frac{a}{b}$, where a and b are numbers and b is not zero.

2. For the fraction $\frac{a}{b}$, the term a is called the _____, and the term b is called the

 _____.

3. A(n) _____ fraction is a fraction in which the numerator is less than the denominator.

4. A fraction is considered to be a(n) _____ fraction if the numerator is equal to or greater than the denominator.

5. _____ fractions are fractions that may look different but represent the same number.

Problems

A Name the numerator of each fraction.

1. $\frac{1}{3}$
2. $\frac{1}{4}$
3. $\frac{-2}{3}$
4. $\frac{2}{4}$
5. $\frac{x}{-8}$
6. $\frac{y}{10}$
7. $\frac{-a}{b}$
8. $\frac{x}{-y}$

Name the denominator of each fraction.

9. $\frac{2}{5}$
10. $\frac{3}{5}$
11. 6
12. 2
13. $\frac{a}{-12}$
14. $\frac{-b}{14}$

Complete the following tables.

15.

Numerator a	Denominator b	Fraction $\frac{a}{b}$
3	5	
-1		$-\frac{1}{7}$
	$-y$	$\frac{x}{y}$
$x+1$	x	

16.

Numerator a	Denominator b	Fraction $\frac{a}{b}$
2	9	
	-3	$-\frac{4}{3}$
-1		$\frac{1}{x}$
x		$\frac{x}{x+1}$

17. For the set of numbers $\left\{ \frac{3}{4}, \frac{6}{5}, \frac{12}{3}, \frac{1}{2}, \frac{9}{10}, \frac{20}{10} \right\}$, list all the proper fractions.

18. For the set of numbers $\left\{ \frac{1}{8}, \frac{7}{9}, \frac{6}{3}, \frac{18}{6}, \frac{3}{5}, \frac{9}{8} \right\}$, list all the improper fractions.

Indicate whether each of the following is *True* or *False*.

19. Every whole number greater than 1 can also be expressed as an improper fraction.

20. Some improper fractions are also proper fractions.

21. Adding the same number to the numerator and the denominator of a fraction will not change its value.

22. The fractions $\frac{3}{4}$ and $\frac{9}{16}$ are equivalent.

B Locate each of the following numbers on this number line.

23. $-\frac{3}{4}$

24. $-\frac{7}{8}$

25. $-\frac{15}{16}$

26. $-\frac{3}{8}$

27. $-\frac{1}{4}$

28. $-\frac{1}{16}$

29. $\frac{1}{2}$

30. $\frac{1}{4}$

31. $\frac{15}{16}$

32. $\frac{7}{8}$

33. Write each fraction as an equivalent fraction with denominator 100. Then write the original fractions in order from smallest to largest.

$$\frac{3}{10} \qquad \frac{1}{20} \qquad \frac{4}{25} \qquad \frac{2}{5}$$

34. Write each fraction as an equivalent fraction with denominator 30. Then write the original fractions in order from smallest to largest.

$$\frac{1}{15} \qquad \frac{5}{6} \qquad \frac{7}{10} \qquad \frac{1}{2}$$

Write each of the following fractions as an equivalent fraction with a denominator of 6.

35. $\frac{2}{3}$

36. $\frac{1}{2}$

37. $-\frac{55}{66}$

38. $-\frac{65}{78}$

Write each of the following fractions as an equivalent fraction with a denominator of 12.

39. $\frac{2}{3}$

40. $\frac{5}{6}$

41. $-\frac{56}{84}$

42. $-\frac{143}{156}$

Write each fraction as an equivalent fraction with a denominator of 12x.

43. $\dfrac{1}{6}$ **44.** $\dfrac{3}{4}$ **45.** $\dfrac{1}{2}$ **46.** $\dfrac{2}{3}$

Write each number as an equivalent fraction with a denominator of 8.

47. 2 **48.** 1 **49.** -5 **50.** -8

51. One-fourth of the first circle below is shaded. Use the other three circles to show three other ways to shade one-fourth of the circle.

52. The six-sided figures below are hexagons. One-third of the first hexagon is shaded. Shade the other three hexagons to show three other ways to represent one-third.

C Simplify by dividing the numerator by the denominator.

53. $\dfrac{3}{1}$ **54.** $\dfrac{3}{3}$ **55.** $-\dfrac{6}{3}$ **56.** $-\dfrac{12}{3}$ **57.** $\dfrac{-37}{-1}$ **58.** $\dfrac{-37}{-37}$

Divide the numerator and the denominator of each of the following fractions by 2.

59. $\dfrac{6}{8}$ **60.** $\dfrac{10}{12}$ **61.** $-\dfrac{86}{94}$ **62.** $-\dfrac{106}{142}$

Divide the numerator and the denominator of each of the following fractions by 3.

63. $\dfrac{12}{9}$ **64.** $\dfrac{33}{27}$ **65.** $\dfrac{-39}{51}$ **66.** $\dfrac{57}{-69}$

67. For each square below, what fraction of the area is given by the shaded region?

a.

b.

c.

d.

68. For each square below, what fraction of the area is given by the shaded region?

a.

b.

c.

d.

Applying the Concepts

69. Sending E-mail The pie chart below shows the number of workers who responded to a survey about sending non-work-related e-mail from the office. 100 people responded to the survey. Use the pie chart to fill in the table.

Workers sending personal e-mail from the office

Never 16
More than ten times a day 5
One to five times a day 47
Five to ten times a day 32

How often workers send non-work-related e-mail from the office	Fraction of respondents saying yes
Never	
1 to 5 times a day	
5 to 10 times a day	
More than 10 times a day	

70. Surfing the Internet The pie chart below shows the number of workers who responded to a survey about viewing non-work-related sites during working hours. 100 people responded to the survey. Use the pie chart to fill in the table.

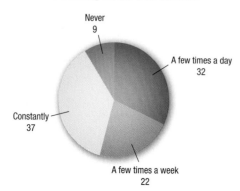

Workers surfing the net from the office

Never 9
A few times a day 32
Constantly 37
A few times a week 22

How often workers view non-work-related sites from the office	Fraction of respondents saying yes
Never	
A few times a week	
A few times a day	
Constantly	

71. Number of Children If there are 3 girls in a family with 5 children, then we say that $\frac{3}{5}$ of the children are girls. If there are 4 girls in a family with 5 children, what fraction of the children are girls?

72. Medical School If 3 out of every 7 people who apply to medical school actually get accepted, what fraction of the people who apply get accepted?

73. Number of Students Of the 43 people who started a math class meeting at 10:00 each morning, only 29 finished the class. What fraction of the people finished the class?

74. Number of Students In a class of 51 students, 23 are freshmen and 28 are juniors. What fraction of the students are freshmen?

75. Expenses If your monthly income is $1,791 and your house payment is $1,121, what fraction of your monthly income must go to pay your house payment?

76. Expenses If you spend $623 on food each month and your monthly income is $2,599, what fraction of your monthly income do you spend on food?

Use the chart on the right to answer questions 77 and 78.

77. Write the number of iPhones in Western Europe as an equivalent fraction with denominator 50.

78. Write the number of iPhones in Oceania as an equivalent fraction with denominator 75.

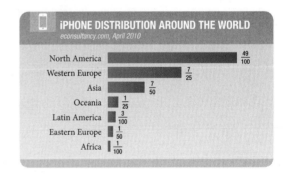

Estimating

79. Which of the following fractions is closest to the number 0?

a. $\frac{1}{2}$　　b. $\frac{1}{3}$　　c. $\frac{1}{4}$　　d. $\frac{1}{5}$

80. Which of the following fractions is closest to the number 1?

a. $\frac{1}{2}$　　b. $\frac{1}{3}$　　c. $\frac{1}{4}$　　d. $\frac{1}{5}$

81. Which of the following fractions is closest to the number 0?

a. $\frac{1}{8}$　　b. $\frac{3}{8}$　　c. $\frac{5}{8}$　　d. $\frac{7}{8}$

82. Which of the following fractions is closest to the number 1?

a. $\frac{1}{8}$　　b. $\frac{3}{8}$　　c. $\frac{5}{8}$　　d. $\frac{7}{8}$

Getting Ready for the Next Section

Multiply.

83. $2 \cdot 2 \cdot 3 \cdot 3 \cdot 3$ **84.** $2^2 \cdot 3^3$ **85.** $2^2 \cdot 3 \cdot 5$ **86.** $2 \cdot 3^2 \cdot 5$

Divide.

87. $12 \div 3$ **88.** $15 \div 3$ **89.** $20 \div 4$ **90.** $24 \div 4$

91. $42 \div 6$ **92.** $72 \div 8$ **93.** $102 \div 2$ **94.** $105 \div 7$

Find the Mistake

Each sentence below contains a mistake. Circle the mistake and write the correct word(s) or numbers(s) on the line provided.

1. For the fraction $\frac{21}{7}$, the numerator is 7. _____

2. The fraction $\frac{90}{15}$ is considered a proper fraction. _____

3. The multiplication property for fractions states that if the numerator and the denominator of a fraction are multiplied by the same nonzero number, the resulting fraction is improper to the original fraction. _____

4. If we divide the numerator and denominator of the fraction $\frac{8}{12}$ by 4, then we get the equivalent fraction $\frac{4}{3}$. _____

OBJECTIVES

A Identify prime factors of a composite number.

B Reduce fractions to lowest terms.

KEY WORDS

prime number

divisor

composite number

reduce

lowest terms

In 2006, Christian Schou earned the world record title for highest slackline walk at a height of 3,280 feet across a Norwegian fjord. A slackline is made of flat nylon webbing and anchored between two points, similar to a tightrope. But in contrast to a taut tightrope, the slackline is strung to allow it to stretch and bounce under the walker's feet.

Suppose the length of a walker's slackline is 30 feet. If the walker walks 15 feet to the middle of the rope, we can say the the walker has walked $\frac{15}{30}$ of the rope's length. We know that he is also standing in the middle of the rope, therefore, he has walked $\frac{1}{2}$ of the rope's length. The fraction $\frac{15}{30}$ is equivalent to the fraction $\frac{1}{2}$; that is, they both have the same value. The mathematical process we use to rewrite $\frac{15}{30}$ as $\frac{1}{2}$ is called *reducing to lowest terms*. Before we look at that process, we need to define some new terms.

A Prime and Composite Numbers

> **DEFINITION** prime number
>
> A *prime number* is any whole number greater than 1 that has exactly two divisors—itself and 1. (A number is a divisor of another number if it divides it without a remainder.)
>
> Prime numbers = {2, 3, 5, 7, 11, 13, 17, 19, 23, 29, 31, 37, . . . }

> **DEFINITION** composite number
>
> Any whole number greater than 1 that is not a prime number is called a *composite number*. A composite number always has at least one divisor other than itself and 1.

EXAMPLE 1 Identify each of the numbers below as either a prime number or a composite number. For those that are composite, give two divisors other than the number itself or 1.

a. 43 **b.** 12

Solution

a. 43 is a prime number, because the only numbers that divide it without a remainder are 43 and 1.

b. 12 is a composite number, because it can be written as $12 = 4 \cdot 3$, which means that 4 and 3 are divisors of 12. (These are not the only divisors of 12; other divisors are 1, 2, 6, and 12.)

**Video Examples
Section 2.2**

Note You may have already noticed that the word *divisor* as we are using it here means the same as the word *factor*. A divisor and a factor of a number are the same thing. A number can't be a divisor of another number without also being a factor of it.

Practice Problems

1. Identify each number as either prime or composite. For those that are composite, give two divisors other than the number itself or 1.
 a. 61
 b. 33

Answers

1. a. Prime **b.** 3, 11

Every composite number can be written as the product of prime factors. Let's look at the composite number 108. We know we can write 108 as $2 \cdot 54$. The number 2 is a prime number, but 54 is not prime. Because 54 can be written as $2 \cdot 27$, we have

$$108 = 2 \cdot 54$$
$$= 2 \cdot 2 \cdot 27$$

Now the number 27 can be written as $3 \cdot 9$ or $3 \cdot 3 \cdot 3$ (because $9 = 3 \cdot 3$), so

$$108 = 2 \cdot 54$$
$$108 = 2 \cdot 2 \cdot 27$$
$$108 = 2 \cdot 2 \cdot 3 \cdot 9$$
$$108 = 2 \cdot 2 \cdot 3 \cdot 3 \cdot 3$$

This last line is the number 108 written as the product of prime factors. We can use exponents to rewrite the last line.

$$108 = 2^2 \cdot 3^3$$

EXAMPLE 2 Factor 60 into a product of prime factors.

Solution We begin by writing 60 as $6 \cdot 10$ and continue factoring until all factors are prime numbers.

$$60 = 6 \cdot 10$$
$$= 2 \cdot 3 \cdot 2 \cdot 5$$
$$= 2^2 \cdot 3 \cdot 5$$

Notice that if we had started by writing 60 as $3 \cdot 20$, we would have achieved the same result:

$$60 = 3 \cdot 20$$
$$= 3 \cdot 2 \cdot 10$$
$$= 3 \cdot 2 \cdot 2 \cdot 5$$
$$= 2^2 \cdot 3 \cdot 5$$

B Reducing Fractions

We can use the method of factoring numbers into prime factors to help reduce fractions to lowest terms. Here is the definition for lowest terms:

> **DEFINITION** lowest terms
>
> A fraction is said to be in *lowest terms* if the numerator and the denominator have no factors in common other than the number 1.

Clarification 1: The fractions $\frac{1}{2}, \frac{1}{3}, \frac{2}{3}, \frac{1}{4}, \frac{3}{4}, \frac{1}{5}, \frac{2}{5}, \frac{3}{5}$, and $\frac{4}{5}$ are all in lowest terms, because in each case the numerator and the denominator have no factors other than 1 in common. That is, in each fraction, no number other than 1 divides both the numerator and the denominator exactly (without a remainder).

Note This process works by writing the original composite number as the product of any two of its factors and then writing any factor that is not prime as the product of any two of its factors. The process is continued until all factors are prime numbers.

2. Factor 80 into a product of prime factors.

Note There are some "shortcuts" to finding the divisors of a number. For instance, if a number ends in 0 or 5, then it is divisible by 5. If a number ends in an even number (0, 2, 4, 6, or 8), then it is divisible by 2. A number is divisible by 3 if the sum of its digits is divisible by 3. For example, 921 is divisible by 3 because the sum of its digits is $9 + 2 + 1 = 12$, which is divisible by 3.

Answer

2. $2^4 \cdot 5$

Clarification 2: The fraction $\frac{6}{8}$ is not written in lowest terms, because the numerator and the denominator are both divisible by 2. To write $\frac{6}{8}$ in lowest terms, we apply the division property for fractions and divide both the numerator and the denominator by 2.

$$\frac{6}{8} = \frac{6 \div 2}{8 \div 2} = \frac{3}{4}$$

The fraction $\frac{3}{4}$ is in lowest terms, because 3 and 4 have no factors in common except the number 1.

Reducing a fraction to lowest terms is simply a matter of dividing the numerator and the denominator by all the factors they have in common. We know from the division property for fractions that this will produce an equivalent fraction.

EXAMPLE 3　　Reduce the fraction $\frac{12}{15}$ to lowest terms by first factoring the numerator and the denominator into prime factors, and then dividing both the numerator and the denominator by the factor they have in common.

Solution　　The numerator and the denominator factor as follows:

$$12 = 2 \cdot 2 \cdot 3 \quad \text{and} \quad 15 = 3 \cdot 5$$

The factor they have in common is 3. The division property for fractions tells us that we can divide both terms of a fraction by 3 to produce an equivalent fraction.

$$\frac{12}{15} = \frac{2 \cdot 2 \cdot 3}{3 \cdot 5} \qquad \textit{Factor the numerator and the denominator completely.}$$

$$= \frac{2 \cdot 2 \cdot 3 \div 3}{3 \cdot 5 \div 3} \qquad \textit{Divide by 3.}$$

$$= \frac{2 \cdot 2}{5} = \frac{4}{5}$$

The fraction $\frac{4}{5}$ is equivalent to $\frac{12}{15}$ and is in lowest terms, because the numerator and the denominator have no factors other than 1 in common.

We can shorten the work involved in reducing fractions to lowest terms by using a slash to indicate division. For example, we can write the above problem this way:

$$\frac{12}{15} = \frac{2 \cdot 2 \cdot \cancel{3}}{\cancel{3} \cdot 5} = \frac{4}{5}$$

So long as we understand that the slashes through the 3s indicate that we have divided both the numerator and the denominator by 3, we can use this notation.

EXAMPLE 4　　Laura is having a party. She puts 4 six-packs of diet soda in a cooler for her guests. At the end of the party, she finds that only 4 sodas have been consumed. What fraction of the sodas are left? Write your answer in lowest terms.

Solution　　She had 4 six-packs of soda, which is $4(6) = 24$ sodas. Only 4 were consumed at the party, so 20 are left. The fraction of sodas left is

$$\frac{20}{24}$$

Factoring 20 and 24 completely and then dividing out both the factors they have in common gives us

$$\frac{20}{24} = \frac{\cancel{2} \cdot \cancel{2} \cdot 5}{\cancel{2} \cdot \cancel{2} \cdot 2 \cdot 3} = \frac{5}{6}$$

3. Reduce the fraction $\frac{30}{45}$ to lowest terms by using prime factors.

Note　The slashes in Example 4 indicate that we have divided both the numerator and the denominator by $2 \cdot 2$, which is equal to 4. With some fractions it is apparent at the start what number divides the numerator and the denominator. For instance, you may have recognized that both 20 and 24 in Example 4 are divisible by 4. We can divide both terms by 4 without factoring first. The division property for fractions guarantees that dividing both terms of a fraction by 4 will produce an equivalent fraction.

$$\frac{20}{24} = \frac{20 \div 4}{24 \div 4} = \frac{5}{6}$$

4. Laura is baking cupcakes for her party. She bakes two dozen cupcakes, and six are eaten. What fraction of the cupcakes are left?

Answers

3. $\frac{2}{3}$

4. $\frac{3}{4}$

5. Reduce $\frac{10}{35}$ to lowest terms.

EXAMPLE 5 Reduce $\frac{6}{42}$ to lowest terms.

Solution We begin by factoring both terms. We then divide through by any factors common to both terms.

$$\frac{6}{42} = \frac{2 \cdot 3}{2 \cdot 3 \cdot 7} = \frac{1}{7}$$

We must be careful in a problem like this to remember that the slashes indicate division. They are used to indicate that we have divided both the numerator and the denominator by $2 \cdot 3 = 6$. The result of dividing the numerator 6 by $2 \cdot 3$ is 1. It is a very common mistake to call the numerator 0 instead of 1 or to leave the numerator out of the answer.

6. Reduce $\frac{6}{30}$ to lowest terms.

EXAMPLE 6 Reduce $\frac{4}{40}$ to lowest terms.

$$\frac{4}{40} = \frac{2 \cdot 2 \cdot 1}{2 \cdot 2 \cdot 2 \cdot 5}$$

$$= \frac{1}{10}$$

7. Reduce $\frac{140}{30}$ to lowest terms.

EXAMPLE 7 Reduce $\frac{105}{30}$ to lowest terms.

$$\frac{105}{30} = \frac{3 \cdot 5 \cdot 7}{2 \cdot 3 \cdot 5}$$

$$= \frac{7}{2}$$

GETTING READY FOR CLASS

After reading through the preceding section, respond in your own words and in complete sentences.

A. What is a prime number?

B. Why is the number 22 a composite number?

C. Factor 120 into a product of prime factors.

D. How would you reduce a fraction to lowest terms?

Answers

5. $\frac{2}{7}$

6. $\frac{1}{5}$

7. $\frac{14}{3}$

Vocabulary Review

Choose the correct words to fill in the blanks below.

 prime composite remainder lowest terms

1. A _____ number is any whole number greater than 1 that has exactly two divisors: 1 and the number itself.

2. A number is a divisor of another number if it divides it without a _____.

3. Any whole number greater than 1 that is not a prime number is called a _____ number.

4. A fraction is said to be in _____ if the numerator and the denominator have no factors in common other than the number 1.

Problems

A Identify each of the numbers below as either a prime number or a composite number. For those that are composite, give at least one divisor (factor) other than the number itself or the number 1.

1. 11

2. 23

3. 105

4. 41

5. 81

6. 50

7. 13

8. 219

Factor each of the following into a product of prime factors.

9. 12

10. 8

11. 81

12. 210

13. 215

14. 75

15. 15

16. 42

B Reduce each fraction to lowest terms.

17. $\dfrac{5}{10}$

18. $\dfrac{3}{6}$

19. $\dfrac{4}{6}$

20. $\dfrac{4}{10}$

21. $\dfrac{8}{10}$

22. $\dfrac{6}{10}$

23. $\dfrac{36}{20}$

24. $\dfrac{32}{12}$

25. $\dfrac{42}{66}$

26. $\dfrac{36}{60}$

27. $\dfrac{24}{40}$

28. $\dfrac{50}{75}$

29. $\dfrac{14}{98}$

30. $\dfrac{12}{84}$

31. $\dfrac{70}{90}$

32. $\dfrac{80}{90}$

33. $\dfrac{42}{30}$

34. $\dfrac{18}{90}$

35. $\dfrac{150}{210}$

36. $\dfrac{110}{70}$

37. $\dfrac{45}{75}$

38. $\dfrac{180}{108}$

39. $\dfrac{60}{36}$

40. $\dfrac{105}{30}$

41. $\dfrac{96}{108}$ **42.** $\dfrac{66}{84}$ **43.** $\dfrac{126}{165}$ **44.** $\dfrac{210}{462}$ **45.** $\dfrac{102}{114}$ **46.** $\dfrac{255}{285}$

47. $\dfrac{294}{693}$ **48.** $\dfrac{273}{385}$

49. Reduce each fraction to lowest terms.
 a. $\dfrac{6}{51}$ **b.** $\dfrac{6}{52}$ **c.** $\dfrac{6}{54}$ **d.** $\dfrac{6}{56}$ **e.** $\dfrac{6}{57}$

50. Reduce each fraction to lowest terms.
 a. $\dfrac{6}{42}$ **b.** $\dfrac{6}{44}$ **c.** $\dfrac{6}{45}$ **d.** $\dfrac{6}{46}$ **e.** $\dfrac{6}{48}$

51. Reduce each fraction to lowest terms.
 a. $\dfrac{2}{90}$ **b.** $\dfrac{3}{90}$ **c.** $\dfrac{5}{90}$ **d.** $\dfrac{6}{90}$ **e.** $\dfrac{9}{90}$

52. Reduce each fraction to lowest terms.
 a. $\dfrac{3}{105}$ **b.** $\dfrac{5}{105}$ **c.** $\dfrac{7}{105}$ **d.** $\dfrac{15}{105}$ **e.** $\dfrac{21}{105}$

53. The answer to each problem below is wrong. Give the correct answer.
 a. $\dfrac{5}{15} = \dfrac{\cancel{5}}{3 \cdot \cancel{5}} = \dfrac{0}{3}$

 b. $\dfrac{5}{6} = \dfrac{3 + \cancel{2}}{4 + \cancel{2}} = \dfrac{0}{4}$

 c. $\dfrac{6}{30} = \dfrac{\cancel{2} \cdot \cancel{3}}{\cancel{2} \cdot \cancel{3} \cdot 5} = 5$

54. The answer to each problem below is wrong. Give the correct answer.
 a. $\dfrac{10}{20} = \dfrac{7 + \cancel{3}}{17 + \cancel{3}} = \dfrac{7}{17}$

 b. $\dfrac{9}{36} = \dfrac{\cancel{3} \cdot \cancel{3}}{2 \cdot 2 \cdot \cancel{3} \cdot \cancel{3}} = \dfrac{0}{4}$

 c. $\dfrac{4}{12} = \dfrac{\cancel{2} \cdot \cancel{2}}{\cancel{2} \cdot \cancel{2} \cdot 3} = 3$

55. Which of the fractions $\frac{6}{8}$, $\frac{15}{20}$, $\frac{9}{16}$, and $\frac{21}{28}$ does not reduce to $\frac{3}{4}$?

56. Which of the fractions $\frac{4}{9}$, $\frac{10}{15}$, $\frac{8}{12}$, and $\frac{6}{12}$ do not reduce to $\frac{2}{3}$?

The number line below extends from 0 to 2, with the segment from 0 to 1 and the segment from 1 to 2 each divided into 8 equal parts. Locate each of the following numbers on this number line.

57. $\dfrac{1}{2}$, $\dfrac{2}{4}$, $\dfrac{4}{8}$, and $\dfrac{8}{16}$

58. $\dfrac{3}{2}$, $\dfrac{6}{4}$, $\dfrac{12}{8}$, and $\dfrac{24}{16}$

59. $\dfrac{5}{4}$, $\dfrac{10}{8}$, and $\dfrac{20}{16}$

60. $\dfrac{1}{4}$, $\dfrac{2}{8}$, and $\dfrac{4}{16}$

Applying the Concepts

61. Income A family's monthly income is $2,400, and they spend $600 each month on food. Write the amount they spend on food as a fraction of their monthly income in lowest terms.

62. Hours and Minutes There are 60 minutes in 1 hour. What fraction of an hour is 20 minutes? Write your answer in lowest terms.

63. Final Exam Suppose 33 people took the final exam in a math class. If 11 people got an A on the final exam, what fraction of the students did not get an A on the exam? Write your answer in lowest terms.

64. Income Tax A person making $21,000 a year pays $3,000 in income tax. What fraction of the person's income is paid as income tax? Write your answer in lowest terms.

Nutrition The nutrition labels below are from two different snack crackers. Use them to work Problems 65–70.

Cheez-It Crackers

Nutrition Facts
Serving Size 30 g. (About 27 crackers)
Servings Per Container: 9

Amount Per Serving

Calories 150	Calories from fat 70

	% Daily Value*
Total Fat 8g	12%
Saturated Fat 2g	10%
Trans Fat 0g	
Polysaturated Fat 4g	
Monounsaturated Fat 2g	
Cholesterol 0mg	0%
Sodium 230mg	10%
Total Carbohydrate 17g	6%
Dietary Fiber less than 1g	3%
Sugars 0g	
Protein 3g	

Vitamin A 2%	•	Vitamin C 0%
Calcium 4%	•	Iron 6%

*Percent Daily Values are based on a 2,000 calorie diet

Goldfish Crackers

Nutrition Facts
Serving Size 55 pieces
Servings Per Container About 4

Amount Per Serving

Calories 140	Calories from fat 45

	% Daily Value*
Total Fat 5g	8%
Saturated Fat 1g	5%
Cholesterol 5mg	2%
Sodium 250mg	10%
Total Carbohydrate 20g	7%
Dietary Fiber 1g	4%
Sugars 1g	
Protein 4g	

Vitamin A 0%	•	Vitamin C 0%
Calcium 4%	•	Iron 2%

*Percent Daily Values are based on a 2,000 calorie diet

65. What fraction of the calories in Cheez-It crackers comes from fat?

66. What fraction of the calories in Goldfish crackers comes from fat?

67. For Cheez-It crackers, what fraction of the total fat is from saturated fat?

68. For Goldfish crackers, what fraction of the total fat is from saturated fat?

69. What fraction of the total carbohydrates in Cheez-It crackers is from sugar?

70. What fraction of the total carbohydrates in Goldfish crackers is from sugar?

Getting Ready for the Next Section

Multiply.

71. $1 \cdot 3 \cdot 1$

72. $2 \cdot 4 \cdot 5$

73. $3 \cdot 5 \cdot 3$

74. $1 \cdot 4 \cdot 1$

75. $5 \cdot 5 \cdot 1$

76. $6 \cdot 6 \cdot 2$

Factor into prime factors.

77. 60 **78.** 72 **79.** $15 \cdot 4$ **80.** $8 \cdot 9$

Expand and multiply.

81. 3^2 **82.** 4^2 **83.** 5^2 **84.** 6^2

Improving Your Quantitative Literacy

85. Wimbledon The graphic shown here gives the most Wimbledon Men's Champions by country. Which of the following is the closest to the fraction of champions from France as compared to Australia?

a. $\dfrac{1}{10}$ **b.** $\dfrac{1}{3}$ **c.** $\dfrac{3}{4}$ **d.** $\dfrac{1}{2}$

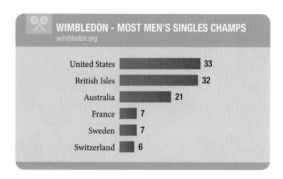

WIMBLEDON - MOST MEN'S SINGLES CHAMPS
wimbledon.org

United States 33
British Isles 32
Australia 21
France 7
Sweden 7
Switzerland 6

Find the Mistake

Each sentence below contains a mistake. Circle the mistake and write the correct word(s) or numbers(s) on the line provided.

1. The number 30 is a prime number because it has 10 as a divisor. _____

2. The number 70 factored into a product of primes is $7 \cdot 10$. _____

3. When reducing the fraction $\dfrac{32}{48}$ to lowest terms, we divide out the common factors 2 and 3 to get $\dfrac{2}{3}$.

4. Reducing the fraction $\dfrac{112}{14}$ to lowest terms gives us $\dfrac{7}{2}$. _____

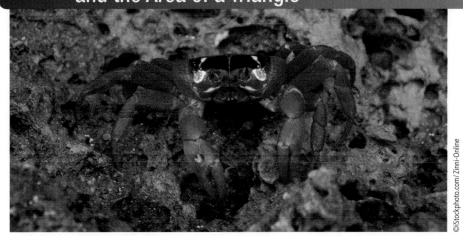

2.3 Multiplication with Fractions, and the Area of a Triangle

OBJECTIVES

A Multiply fractions with no common factors.

B Multiply and simplify fractions.

C Apply the distributive property and find the area of a triangle.

©iStockphoto.com/Zinni-Online

Once a year on Christmas Island, in Australia, millions of bright red crabs migrate from the rain forest for several miles to the beach. They spawn in the sea, where the eggs hatch almost immediately upon contact with the salt water. For days, the crabs swarm train tracks, highways, and other busy thoroughfares on the island. They even take over golf courses, and according to a special rule, a golfer must play a ball where it lies even if a crab knocks it to another spot.

Let's suppose a group of crabs are crawling down a railroad track when they hear a train coming. $\frac{3}{4}$ of the crabs scurry into the bushes on either side of the track, $\frac{1}{2}$ of the $\frac{3}{4}$ going to the right and $\frac{1}{2}$ to the left. What fraction of the original group of crabs crawls to the right of the track? This question can be answered by multiplying $\frac{1}{2}$ and $\frac{3}{4}$. Here is the problem written in symbols:

$$\frac{1}{2} \cdot \frac{3}{4} = \frac{3}{8}$$

If you analyze this example, you will discover that to multiply two fractions, we multiply the numerators and then multiply the denominators. We begin this section with the rule for multiplication of fractions.

KEY WORDS

simplify

area of a triangle

**Video Examples
Section 2.3**

A Multiplication of Fractions

> **RULE** Product of Two Fractions
>
> The product of two fractions is a fraction whose numerator is the product of the two numerators, and whose denominator is the product of the two denominators. We can write this rule in symbols as follows:
>
> If a, b, c, and d represent any numbers and b and d are not zero, then
>
> $$\frac{a}{b} \cdot \frac{c}{d} = \frac{a \cdot c}{b \cdot d}$$

EXAMPLE 1 Multiply: $\frac{3}{5} \cdot \frac{2}{7}$.

Solution Using our rule for multiplication, we multiply the numerators and multiply the denominators.

$$\frac{3}{5} \cdot \frac{2}{7} = \frac{3 \cdot 2}{5 \cdot 7} = \frac{6}{35}$$

The product of $\frac{3}{5}$ and $\frac{2}{7}$ is the fraction $\frac{6}{35}$. The numerator 6 is the product of 3 and 2, and the denominator 35 is the product of 5 and 7.

Practice Problems

1. Multiply: $\frac{5}{6} \cdot \frac{7}{8}$.

Answer

1. $\frac{35}{48}$

2. Multiply: $\frac{7}{12} \cdot 5$.

3. Multiply: $\frac{2}{3}\left(\frac{1}{3} \cdot \frac{5}{9}\right)$.

EXAMPLE 2 Multiply: $\frac{3}{8} \cdot 5$.

Solution The number 5 can be written as $\frac{5}{1}$. That is, 5 can be considered a fraction with numerator 5 and denominator 1. Writing 5 this way enables us to apply the rule for multiplying fractions.

$$\frac{3}{8} \cdot 5 = \frac{3}{8} \cdot \frac{5}{1}$$

$$= \frac{3 \cdot 5}{8 \cdot 1}$$

$$= \frac{15}{8}$$

EXAMPLE 3 Multiply: $\frac{1}{2}\left(\frac{3}{4} \cdot \frac{1}{5}\right)$.

Solution We find the product inside the parentheses first and then multiply the result by $\frac{1}{2}$.

$$\frac{1}{2}\left(\frac{3}{4} \cdot \frac{1}{5}\right) = \frac{1}{2}\left(\frac{3}{20}\right)$$

$$= \frac{1 \cdot 3}{2 \cdot 20}$$

$$= \frac{3}{40}$$

The properties of multiplication apply to fractions as well. That is, if a, b, and c are fractions, then

$$a \cdot b = b \cdot a$$ Multiplication with fractions is commutative.

$$a \cdot (b \cdot c) = (a \cdot b) \cdot c$$ Multiplication with fractions is associative.

To demonstrate the associative property for fractions, let's do Example 3 again, but this time we will apply the associative property first.

$$\frac{1}{2}\left(\frac{3}{4} \cdot \frac{1}{5}\right) = \left(\frac{1}{2} \cdot \frac{3}{4}\right) \cdot \frac{1}{5}$$ Associative property

$$= \left(\frac{1 \cdot 3}{2 \cdot 4}\right) \cdot \frac{1}{5}$$

$$= \left(\frac{3}{8}\right) \cdot \frac{1}{5}$$

$$= \frac{3 \cdot 1}{8 \cdot 5}$$

$$= \frac{3}{40}$$

The result is identical to that of Example 3.

B Multiplying and Simplifying Fractions

The answers to all the examples so far in this section have been in lowest terms. Let's see what happens when we multiply two fractions to get a product that is not in lowest terms.

4. Multiply: $\frac{9}{16} \cdot \frac{4}{3}$.

EXAMPLE 4 Multiply: $\frac{15}{8} \cdot \frac{4}{9}$.

Solution Multiplying the numerators and multiplying the denominators, we have

$$\frac{15}{8} \cdot \frac{4}{9} = \frac{15 \cdot 4}{8 \cdot 9}$$

$$= \frac{60}{72}$$

Answers

2. $\frac{35}{12}$

3. $\frac{10}{81}$

4. $\frac{3}{4}$

The product is $\frac{60}{72}$, which can be reduced to lowest terms by factoring 60 and 72 and then dividing out any factors they have in common.

$$\frac{60}{72} = \frac{2 \cdot 2 \cdot 3 \cdot 5}{2 \cdot 2 \cdot 2 \cdot 3 \cdot 3}$$

$$= \frac{5}{6}$$

We can actually save ourselves some time by factoring before we multiply. Here's how it is done:

$$\frac{15}{8} \cdot \frac{4}{9} = \frac{15 \cdot 4}{8 \cdot 9}$$

$$= \frac{(3 \cdot 5) \cdot (2 \cdot 2)}{(2 \cdot 2 \cdot 2) \cdot (3 \cdot 3)}$$

$$= \frac{3 \cdot 5 \cdot 2 \cdot 2}{2 \cdot 2 \cdot 2 \cdot 3 \cdot 3}$$

$$= \frac{5}{6}$$

The result is the same in both cases. Reducing to lowest terms before we actually multiply takes less time. Here are some additional examples. Problems like these will be useful when we solve equations.

EXAMPLE 5
$$\frac{9}{2} \cdot \frac{8}{18} = \frac{9 \cdot 8}{2 \cdot 18}$$

$$= \frac{(3 \cdot 3) \cdot (2 \cdot 2 \cdot 2)}{2 \cdot (2 \cdot 3 \cdot 3)}$$

$$= \frac{3 \cdot 3 \cdot 2 \cdot 2 \cdot 2}{2 \cdot 2 \cdot 3 \cdot 3}$$

$$= \frac{2}{1}$$

$$= 2$$

5. Multiply: $\frac{9}{5} \cdot \frac{10}{21}$.

Note Although $\frac{2}{1}$ is in lowest terms, it is still simpler to write the answer as just 2. We will always do this when the denominator is the number 1.

EXAMPLE 6
$$\frac{2}{3} \cdot \frac{6}{5} \cdot \frac{5}{8} = \frac{2 \cdot 6 \cdot 5}{3 \cdot 5 \cdot 8}$$

$$= \frac{2 \cdot (2 \cdot 3) \cdot 5}{3 \cdot 5 \cdot (2 \cdot 2 \cdot 2)}$$

$$= \frac{2 \cdot 2 \cdot 3 \cdot 5}{3 \cdot 5 \cdot 2 \cdot 2 \cdot 2}$$

$$= \frac{1}{2}$$

6. Multiply: $\frac{3}{4} \cdot \frac{4}{9} \cdot \frac{6}{7}$.

In Chapter 1, we did some work with exponents. We can extend our work with exponents to include fractions, as the following examples indicate.

EXAMPLE 7
$$\left(\frac{3}{4}\right)^2 = \frac{3}{4}\left(\frac{3}{4}\right)$$

$$= \frac{3 \cdot 3}{4 \cdot 4}$$

$$= \frac{9}{16}$$

7. Simplify: $\left(\frac{5}{6}\right)^2$.

Answers

5. $\frac{6}{7}$

6. $\frac{2}{7}$

7. $\frac{25}{36}$

8. Multiply: $\left(\frac{4}{5}\right)^2 \cdot \frac{5}{6}$.

EXAMPLE 8 $\left(\frac{5}{6}\right)^2 \cdot \frac{1}{2} = \frac{5}{6} \cdot \frac{5}{6} \cdot \frac{1}{2}$

$$= \frac{5 \cdot 5 \cdot 1}{6 \cdot 6 \cdot 2}$$

$$= \frac{25}{72}$$

9. Simplify each expression.

a. $\left(\frac{4}{5}\right)\left(-\frac{2}{5}\right)$

b. $\left(-\frac{3}{5}\right)\left(-\frac{7}{10}\right)$

EXAMPLE 9 Simplify each expression.

a. $\left(\frac{2}{3}\right)\left(-\frac{3}{5}\right)$ **b.** $\left(-\frac{7}{8}\right)\left(-\frac{5}{14}\right)$

Solution

a. $\left(\frac{2}{3}\right)\left(-\frac{3}{5}\right) = -\frac{6}{15} = -\frac{2}{5}$ *The rule for multiplication also holds for fractions.*

b. $\left(-\frac{7}{8}\right)\left(-\frac{5}{14}\right) = \frac{35}{112} = \frac{5}{16}$

The word *of* used in connection with fractions indicates multiplication. If we want to find $\frac{1}{2}$ of $\frac{2}{3}$, then what we do is multiply $\frac{1}{2}$ and $\frac{2}{3}$.

10. Find $\frac{3}{4}$ of $\frac{5}{6}$.

EXAMPLE 10 Find $\frac{1}{2}$ of $\frac{2}{3}$.

Solution Knowing the word *of*, as used here, indicates multiplication, we have

$$\frac{1}{2} \text{ of } \frac{2}{3} = \frac{1}{2} \cdot \frac{2}{3}$$

$$= \frac{1 \cdot 2}{2 \cdot 3} = \frac{1}{3}$$

This seems to make sense. Logically, $\frac{1}{2}$ of $\frac{2}{3}$ should be $\frac{1}{3}$, as Figure 1 shows.

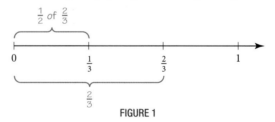

FIGURE 1

11. What is $\frac{7}{8}$ of 24?

Note As you become familiar with multiplying fractions, you may notice shortcuts that reduce the number of steps in the problems. It's okay to use these shortcuts if you understand why they work and are consistently getting correct answers. If you are using shortcuts and not consistently getting correct answers, then go back to showing all the work until you completely understand the process.

EXAMPLE 11 What is $\frac{3}{4}$ of 12?

Solution Again, the word *of* means multiply.

$$\frac{3}{4} \text{ of } 12 = \frac{3}{4}(12)$$

$$= \frac{3}{4}\left(\frac{12}{1}\right)$$

$$= \frac{3 \cdot 12}{4 \cdot 1}$$

$$= \frac{3 \cdot 2 \cdot 2 \cdot 3}{2 \cdot 2 \cdot 1}$$

$$= \frac{9}{1} = 9$$

Answers

8. $\frac{8}{15}$

9. a. $-\frac{4}{5}$ **b.** $\frac{21}{50}$

10. $\frac{5}{8}$

11. 21

C The Distributive Property and Area

The distributive property can also be applied to fractions. Recall that

$$a(b + c) = a \cdot b + a \cdot c \quad \text{or} \quad a(b - c) = a \cdot b - a \cdot c$$

The next example illustrates using the distributive property with fractions.

EXAMPLE 12 Apply the distributive property, then simplify.

a. $6\left(2 + \dfrac{1}{3}\right)$ **b.** $12\left(\dfrac{3}{4} - \dfrac{1}{2}\right)$ **c.** $-5\left(4 - \dfrac{2}{3}\right)$

Solution Apply the distributive property to each expression.

a. $6\left(2 + \dfrac{1}{3}\right) = 6 \cdot 2 + 6 \cdot \dfrac{1}{3} = 12 + \dfrac{6}{3} = 12 + 2 = 14$

b. $12\left(\dfrac{3}{4} - \dfrac{1}{2}\right) = 12 \cdot \dfrac{3}{4} - 12 \cdot \dfrac{1}{2} = \dfrac{12}{1} \cdot \dfrac{3}{4} - \dfrac{12}{1} \cdot \dfrac{1}{2} = 9 - 6 = 3$

c. $-5\left(4 - \dfrac{2}{3}\right) = -5 \cdot 4 - (-5)\dfrac{2}{3} = -20 + \dfrac{10}{3} = -\dfrac{60}{3} + \dfrac{10}{3} = -\dfrac{50}{3}$

> **12.** Apply the distributive property, then simplify.
>
> **a.** $5\left(4 + \dfrac{3}{5}\right)$
>
> **b.** $8\left(\dfrac{3}{4} - \dfrac{1}{2}\right)$
>
> **c.** $-3\left(2 - \dfrac{1}{3}\right)$

FACTS FROM GEOMETRY The Area of a Triangle

The formula for the area of a triangle is one application of multiplication with fractions. Figure 2 shows a triangle with base b and height h. Below the triangle is the formula for its area. As you can see, it is a product containing the fraction $\frac{1}{2}$.

Area = $\frac{1}{2}$ (base)(height)

$A = \frac{1}{2} bh$

FIGURE 2

EXAMPLE 13 Find the area of the triangle in Figure 3.

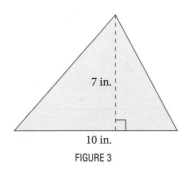

7 in.

10 in.

FIGURE 3

> **13.** Find the area of a triangle with base 8 in. and height 11 in.

Solution Applying the formula for the area of a triangle, we have

$$A = \frac{1}{2} bh = \frac{1}{2} \cdot 10 \cdot 7 = 5 \cdot 7 = 35 \text{ in}^2$$

Vocabulary Review

Choose the correct words to fill in the blanks below.

triangle multiplication fractions product

1. The _____ of two fractions is a fraction whose numerator is the product of the two numerators, and whose denominator is the product of the two denominators.

2. In symbols, the multiplication of _____ is written as $\frac{a}{b} \cdot \frac{c}{d} = \frac{a \cdot c}{b \cdot d}$.

3. The word *of* used in connection with fractions indicates _____.

4. The formula for the area of a _____ is $A = \frac{1}{2}bh$.

Problems

A Find each of the following products. (Multiply.)

1. $\frac{2}{3} \cdot \frac{4}{5}$

2. $\frac{5}{6} \cdot \frac{7}{4}$

3. $\frac{1}{2} \cdot \frac{7}{4}$

4. $\frac{3}{5} \cdot -\frac{4}{7}$

5. $\frac{5}{3} \cdot \frac{3}{5}$

6. $\frac{4}{7} \cdot -\frac{7}{4}$

7. $\frac{3}{4} \cdot 9$

8. $\frac{2}{3} \cdot -5$

9. $\frac{6}{7} \left(\frac{7}{6} \right)$

10. $-\frac{2}{9} \left(\frac{9}{2} \right)$

11. $\frac{1}{2} \cdot \frac{1}{3} \cdot \frac{1}{4}$

12. $\frac{2}{3} \cdot \frac{4}{5} \cdot \frac{1}{3}$

13. $\frac{2}{5} \cdot \frac{3}{5} \cdot \frac{4}{5}$

14. $\frac{1}{4} \cdot -\frac{3}{4} \cdot \frac{3}{4}$

15. $\frac{3}{2} \cdot \frac{5}{2} \cdot \frac{7}{2}$

16. $\frac{4}{3} \cdot -\frac{5}{3} \cdot \frac{7}{3}$

Complete the following tables.

17.

First Number x	Second Number y	Their Product xy
$\frac{1}{2}$	$\frac{2}{3}$	
$\frac{2}{3}$	$\frac{3}{4}$	
$\frac{3}{4}$	$\frac{4}{5}$	
$\frac{5}{a}$	$-\frac{a}{6}$	

18.

First Number x	Second Number y	Their Product xy
12	$\frac{1}{2}$	
12	$\frac{1}{3}$	
12	$\frac{1}{4}$	
12	$\frac{1}{6}$	

19.

First Number x	Second Number y	Their Product xy
$\frac{1}{2}$	30	
$\frac{1}{5}$	30	
$\frac{1}{6}$	30	
$\frac{1}{15}$	30	

20.

First Number x	Second Number y	Their Product xy
$\frac{1}{3}$	$\frac{3}{5}$	
$\frac{3}{5}$	$\frac{5}{7}$	
$\frac{5}{7}$	$\frac{7}{9}$	
$-\frac{7}{b}$	$\frac{b}{11}$	

B Multiply each of the following. Be sure all answers are written in lowest terms.

21. $\dfrac{9}{20} \cdot \dfrac{4}{3}$

22. $\dfrac{135}{16} \cdot -\dfrac{2}{45}$

23. $\dfrac{3}{4} \cdot 12$

24. $\dfrac{3}{4} \cdot 20$

25. $\dfrac{1}{3}\,(3)$

26. $\dfrac{1}{5}\,(5)$

27. $\dfrac{2}{5} \cdot -20$

28. $\dfrac{3}{5} \cdot 15$

29. $\dfrac{72}{35} \cdot -\dfrac{55}{108} \cdot \dfrac{7}{110}$

30. $\dfrac{32}{27} \cdot \dfrac{72}{49} \cdot \dfrac{1}{40}$

Expand and simplify each of the following.

31. $\left(\dfrac{2}{3}\right)^2$

32. $\left(\dfrac{3}{5}\right)^2$

33. $\left(\dfrac{3}{4}\right)^2$

34. $\left(\dfrac{2}{7}\right)^2$

35. $\left(\dfrac{1}{2}\right)^2$

36. $\left(-\dfrac{1}{3}\right)^2$

37. $\left(\dfrac{2}{3}\right)^3$

38. $\left(-\dfrac{3}{5}\right)^3$

39. $\left(\dfrac{3}{4}\right)^2 \cdot \dfrac{8}{9}$

40. $\left(\dfrac{5}{6}\right)^2 \cdot \dfrac{12}{15}$

41. $\left(\dfrac{1}{2}\right)^2 \left(\dfrac{3}{5}\right)^2$

42. $\left(\dfrac{3}{8}\right)^2 \left(-\dfrac{4}{3}\right)^2$

43. $\left(\dfrac{1}{2}\right)^2 \cdot 8 + \left(\dfrac{1}{3}\right)^2 \cdot 9$

44. $\left(\dfrac{2}{3}\right)^2 \cdot 9 + \left(\dfrac{1}{2}\right)^2 \cdot 4$

45. Find $\frac{3}{8}$ of 64.

46. Find $\frac{2}{3}$ of 18.

47. What is $\frac{1}{3}$ of the sum of 8 and 4?

48. What is $\frac{3}{5}$ of the sum of 8 and 7?

49. Find $\frac{1}{2}$ of $\frac{3}{4}$ of 24.

50. Find $\frac{3}{5}$ of $\frac{1}{3}$ of 15.

Find the mistakes in Problems 51–54. Correct the right-hand side of each one.

51. $\frac{1}{2} \cdot \frac{3}{5} = \frac{4}{10}$

52. $\frac{2}{7} \cdot \frac{3}{5} = \frac{5}{35}$

53. $-\frac{3}{4} \cdot -\frac{1}{3} = -\frac{1}{4}$

54. $-\frac{2}{5} \cdot -\frac{2}{3} = -\frac{4}{15}$

C Apply the distributive property, then simplify.

55. $4\left(3 + \frac{1}{2}\right)$

56. $4\left(2 - \frac{3}{4}\right)$

57. $-12\left(\frac{1}{2} + \frac{2}{3}\right)$

58. $-12\left(\frac{3}{4} - \frac{1}{6}\right)$

59. $9\left(\frac{2}{3} - \frac{1}{9}\right)$

60. $12\left(\frac{1}{2} - \frac{1}{3}\right)$

61. $-16\left(\frac{5}{8} - \frac{1}{4}\right)$

62. $-24\left(\frac{2}{3} - \frac{1}{6}\right)$

63. Find the area of the triangle with base 19 inches and height 14 inches.

64. Find the area of the triangle with base 13 inches and height 8 inches.

65. The base of a triangle is $\frac{4}{3}$ feet and the height is $\frac{2}{3}$ feet. Find the area.

66. The base of a triangle is $\frac{8}{7}$ feet and the height is $\frac{14}{5}$ feet. Find the area.

Find the area of each figure.

67.

3 yd

2 yd

68.

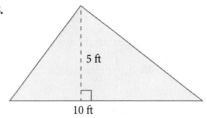

5 ft

10 ft

Applying the Concepts

Use the information in the pie chart to answer questions 69 and 70. Round to the nearest student.

69. Reading a Pie Chart If there are approximately 15,800 students attending Cal Poly, approximately how many of them are studying agriculture?

70. Reading a Pie Chart If there are exactly 15,828 students attending Cal Poly, how many of them are studying engineering? Round to the nearest student.

Cal Poly Enrollment

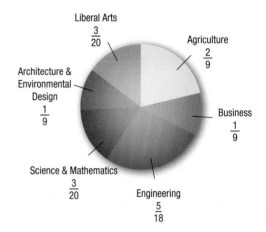

Liberal Arts $\frac{3}{20}$

Agriculture $\frac{2}{9}$

Architecture & Environmental Design $\frac{1}{9}$

Business $\frac{1}{9}$

Science & Mathematics $\frac{3}{20}$

Engineering $\frac{5}{18}$

71. Hot Air Balloon Aerostar International makes a hot air balloon called the Rally 105 that has a volume of 105,400 cubic feet. Another balloon, the Rally 126, was designed with a volume that is approximately $\frac{6}{5}$ the volume of the Rally 105. Find the volume of the Rally 126 to the nearest hundred cubic feet.

72. Health Care According to the Department of Veteran's Affairs, approximately $\frac{4}{25}$ of military veterans have diabetes. If there approximately are 3.1 million veterans, how many have diabetes? Round number to the nearest ten thousand.

73. Bicycle Safety The National Safe Kids Campaign and Bell Sports sponsored a study that surveyed 8,159 children ages 5 to 14 who were riding bicycles. Approximately $\frac{2}{5}$ of the children were wearing helmets, and of those, only $\frac{13}{20}$ were wearing the helmets correctly. About how many of the children were wearing helmets correctly?

74. Bicycle Safety From the information in Problem 73, how many of the children surveyed do not wear helmets?

Geometric Sequences A geometric sequence is a sequence in which each term comes from the previous term by multiplying by the same number each time. For example, the sequence $1, \frac{1}{2}, \frac{1}{4}, \frac{1}{8}, \ldots$ is a geometric sequence in which each term is found by multiplying the previous term by $\frac{1}{2}$. By observing this fact, we know that the next term in the sequence will be $\frac{1}{8} \cdot \frac{1}{2} = \frac{1}{16}$.

Find the next number in each of the geometric sequences below.

75. $1, \frac{1}{3}, \frac{1}{9}, \ldots$

76. $1, \frac{1}{4}, \frac{1}{16}, \ldots$

77. $\frac{3}{2}, 1, \frac{2}{3}, \frac{4}{9}, \ldots$

78. $\frac{2}{3}, 1, \frac{3}{2}, \frac{9}{4}, \ldots$

Estimating For each problem below, mentally estimate if the answer will be closest to 0, 1, 2 or 3. Make your estimate without using pencil and paper or a calculator.

79. $\frac{11}{5} \cdot \frac{19}{20}$

80. $\frac{3}{5} \cdot \frac{1}{20}$

81. $\frac{16}{5} \cdot \frac{23}{24}$

82. $\frac{9}{8} \cdot \frac{31}{32}$

Getting Ready for the Next Section

In the next section we will do division with fractions. As you already know, division and multiplication are closely related. These review problems are intended to let you see more of the relationship between multiplication and division.

Perform the indicated operations.

83. $8 \div 4$

84. $8 \cdot \frac{1}{4}$

85. $15 \div 3$

86. $15 \cdot \frac{1}{3}$

87. $18 \div 6$

88. $18 \cdot \frac{1}{6}$

For each number below, find a number to multiply it by to obtain 1.

89. $\frac{3}{4}$

90. $\frac{9}{5}$

91. $\frac{1}{3}$

92. $\frac{1}{4}$

93. 7

94. 2

Find the Mistake

Each sentence below contains a mistake. Circle the mistake and write the correct word(s) or numbers(s) on the line provided.

1. To find the product of two fractions, multiply the numerators and put them over the largest denominator.

2. To multiply $\frac{6}{7}$ by $\frac{12}{9}$, find the product of the numerators and divide it by the sum of the denominators to get $\frac{72}{63}$.

3. Simplifying $\left(\frac{5}{6}\right)^2 \cdot \frac{8}{9}$ gives $\frac{80}{54} = \frac{40}{27}$. _____

4. The area of a triangle with a height of 14 inches and a base of 32 inches is 448 square inches. _____

2.4 Division with Fractions

Image © Ben Barczi 2010

OBJECTIVES

A Divide fractions and simplify.

B Use division in order of operations problems.

KEY WORDS

reciprocal

In 2010, a Peruvian inventor won the $200,000 grand prize for a competition that requested its entrants to suggest a new and ingenious way to save the earth. To battle global warming, the inventor proposed covering 173 acres of dry rocky land in the Andes Mountains with a mixture made from egg whites, lime, and water. The mixture creates a whitewash that when spread over the rocks would reflect sunlight and ideally reduce temperatures, thus slowing the melting of the area's glaciers.

Suppose a truck transports the whitewash in a 32-gallon tank up to the mountaintop. Let's say the inventor used a bucket that only held $\frac{3}{4}$ of a gallon to pour the whitewash onto the rocks. How many times would the inventor need to fill up his bucket to empty the tank? In order to answer this question, we need to learn more about how to divide with fractions. But before we define division with fractions, we must first introduce the idea of *reciprocals*. Look at the following multiplication problems:

$$\frac{3}{4} \cdot \frac{4}{3} = \frac{12}{12} = 1 \qquad \frac{7}{8} \cdot \frac{8}{7} = \frac{56}{56} = 1$$

In each case the product is 1. Whenever the product of two numbers is 1, we say the two numbers are reciprocals.

**Video Examples
Section 2.4**

A Dividing Fractions

> **DEFINITION** reciprocals
>
> Two numbers whose product is 1 are said to be *reciprocals.* In symbols, the reciprocal of $\frac{a}{b}$ is $\frac{b}{a}$, because
>
> $$\frac{a}{b} \cdot \frac{b}{a} = \frac{a \cdot b}{b \cdot a} = \frac{a \cdot b}{a \cdot b} = 1 \qquad (a \neq 0, b \neq 0)$$

Every number has a reciprocal except 0. The reason 0 does not have a reciprocal is because the product of *any* number with 0 is 0. It can never be 1. Reciprocals of whole numbers are fractions with 1 as the numerator. For example, the reciprocal of 5 is $\frac{1}{5}$, because

$$5 \cdot \frac{1}{5} = \frac{5}{1} \cdot \frac{1}{5} = \frac{5}{5} = 1$$

Table 1 lists some numbers and their reciprocals.

Table 1

Number	Reciprocal	Reason
$\frac{3}{4}$	$\frac{4}{3}$	Because $\frac{3}{4} \cdot \frac{4}{3} = \frac{12}{12} = 1$
$\frac{9}{5}$	$\frac{5}{9}$	Because $\frac{9}{5} \cdot \frac{5}{9} = \frac{45}{45} = 1$
$\frac{1}{3}$	3	Because $\frac{1}{3} \cdot 3 = \frac{1}{3} \cdot \frac{3}{1} = \frac{3}{3} = 1$
7	$\frac{1}{7}$	Because $7 \cdot \frac{1}{7} = \frac{7}{1} \cdot \frac{1}{7} = \frac{7}{7} = 1$

Division with fractions is accomplished by using reciprocals. More specifically, we can define division by a fraction to be the same as multiplication by its reciprocal. Here is the precise definition:

> **DEFINITION** division by a fraction
>
> If a, b, c, and d are numbers and b, c, and d are all not equal to 0, then
>
> $$\frac{a}{b} \div \frac{c}{d} = \frac{a}{b} \cdot \frac{d}{c}$$

This definition states that dividing by the fraction $\frac{c}{d}$ is exactly the same as multiplying by its reciprocal $\frac{d}{c}$. Because we developed the rule for multiplying fractions in the last section, we do not need a new rule for division. We simply *replace the divisor by its reciprocal* and multiply. Here are some examples to illustrate the procedure:

EXAMPLE 1 Divide: $\frac{1}{2} \div \frac{1}{4}$.

Solution The divisor is $\frac{1}{4}$, and its reciprocal is $\frac{4}{1}$. Applying the definition of division for fractions, we have

$$\frac{1}{2} \div \frac{1}{4} = \frac{1}{2} \cdot \frac{4}{1}$$
$$= \frac{1 \cdot 4}{2 \cdot 1}$$
$$= \frac{1 \cdot 2 \cdot 2}{2 \cdot 1}$$
$$= \frac{2}{1}$$
$$= 2$$

The quotient of $\frac{1}{2}$ and $\frac{1}{4}$ is 2. Or, $\frac{1}{4}$ "goes into" $\frac{1}{2}$ two times. Logically, our definition for division of fractions seems to be giving us answers that are consistent with what we know about fractions from previous experience. Because 2 times $\frac{1}{4}$ is $\frac{2}{4}$ or $\frac{1}{2}$, it seems logical that $\frac{1}{2}$ divided by $\frac{1}{4}$ should be 2.

EXAMPLE 2 Divide: $-\frac{3}{8} \div \frac{9}{4}$.

Solution Dividing by $\frac{9}{4}$ is the same as multiplying by its reciprocal, which is $\frac{4}{9}$.

$$-\frac{3}{8} \div \frac{9}{4} = -\frac{3}{8} \cdot \frac{4}{9}$$
$$= -\frac{3 \cdot 2 \cdot 2}{2 \cdot 2 \cdot 2 \cdot 3 \cdot 3}$$
$$= -\frac{1}{6}$$

The quotient of $-\frac{3}{8}$ and $\frac{9}{4}$ is $-\frac{1}{6}$.

Note Defining division to be the same as multiplication by the reciprocal does make sense. If we divide 6 by 2, we get 3. On the other hand, if we multiply 6 by $\frac{1}{2}$ (the reciprocal of 2), we also get 3. Whether we divide by 2 or multiply by $\frac{1}{2}$, we get the same result.

Practice Problems

1. Divide: $\frac{3}{4} \div \frac{1}{6}$.

2. Divide: $-\frac{8}{15} \div \frac{4}{5}$.

Answers

1. $\frac{9}{2}$

2. $-\frac{2}{3}$

EXAMPLE 3 Divide: $\frac{2}{3} \div 2$.

Solution The reciprocal of 2 is $\frac{1}{2}$. Applying the definition for division of fractions, we have

$$\frac{2}{3} \div 2 = \frac{2}{3} \cdot \frac{1}{2}$$

$$= \frac{2 \cdot 1}{3 \cdot 2}$$

$$= \frac{1}{3}$$

EXAMPLE 4 Divide: $2 \div \frac{1}{3}$.

Solution We replace $\frac{1}{3}$ by its reciprocal, which is 3, and multiply.

$$2 \div \frac{1}{3} = 2(3)$$

$$= 6$$

Here are some further examples of division with fractions. Notice in each case that the first step is the only new part of the process.

EXAMPLE 5 $\frac{4}{27} \div \frac{16}{9}$

Solution We replace $\frac{16}{9}$ by its reciprocal and multiply.

$$= \frac{4}{27} \cdot \frac{9}{16}$$

$$= \frac{4 \cdot 9}{3 \cdot 9 \cdot 4 \cdot 4}$$

$$= \frac{1}{12}$$

In Example 5 we did not factor the numerator and the denominator completely in order to reduce to lowest terms because, as you have probably already noticed, it is not necessary to do so. We need to factor only enough to show what numbers are common to the numerator and the denominator. If we factored completely in the second step, it would look like this:

$$= \frac{2 \cdot 2 \cdot 3 \cdot 3}{3 \cdot 3 \cdot 3 \cdot 2 \cdot 2 \cdot 2 \cdot 2}$$

$$= \frac{1}{12}$$

The result is the same in both cases. From now on, we will factor numerators and denominators only enough to show the factors we are dividing out.

EXAMPLE 6 Divide.

a. $\frac{16}{35} \div 8$ **b.** $27 \div \frac{3}{2}$

Solution

a. $\frac{16}{35} \div 8 = \frac{16}{35} \cdot \frac{1}{8}$

$$= \frac{2 \cdot 8 \cdot 1}{35 \cdot 8}$$

$$= \frac{2}{35}$$

3. Divide: $\frac{5}{6} \div 3$.

4. Divide: $4 \div \frac{2}{3}$.

5. Divide: $\frac{9}{16} \div \frac{9}{4}$.

6. Divide.

 a. $\frac{15}{26} \div 5$

 b. $18 \div \frac{9}{10}$

Answers

3. $\frac{5}{18}$

4. 6

5. $\frac{1}{4}$

6. a. $\frac{3}{26}$ **b.** 20

b. $27 \div \dfrac{3}{2} = 27 \cdot \dfrac{2}{3}$

$$= \frac{\overset{9}{\cancel{3} \cdot 9} \cdot 2}{\cancel{3}}$$

$$= 18$$

B Order of Operations

The next two examples combine what we have learned about division of fractions with the rule for order of operations.

7. The quotient of $\frac{2}{3}$ and $\frac{1}{6}$ is decreased by 2. What number results?

EXAMPLE 7 The quotient of $\frac{8}{3}$ and $\frac{1}{6}$ is increased by 5. What number results?

Solution Translating to symbols, we have

$$\frac{8}{3} \div \frac{1}{6} + 5 = \left[\frac{8}{3} \cdot \frac{6}{1} \right] + 5$$

$$= 16 + 5$$

$$= 21$$

8. Simplify: $25 \div \left(\frac{5}{6}\right)^2 + 32 \div \left(\frac{4}{5}\right)^2$.

EXAMPLE 8 Simplify $32 \div \left(\frac{4}{3}\right)^2 + 75 \div \left(\frac{5}{2}\right)^2$.

Solution According to the rule for order of operations, we must first evaluate the numbers with exponents, then we divide, and finally we add.

$$\left[32 \div \left(\frac{4}{3}\right)^2 \right] + \left[75 \div \left(\frac{5}{2}\right)^2 \right] = 32 \div \frac{16}{9} + 75 \div \frac{25}{4}$$

$$= 32 \cdot \frac{9}{16} + 75 \cdot \frac{4}{25}$$

$$= 18 + 12$$

$$= 30$$

9. Repeat Example 9 if they have 16 yards of material, and each blanket requires $\frac{2}{3}$ yard of material.

EXAMPLE 9 A 4-H Club is making blankets to keep their lambs clean at the county fair. If each blanket requires $\frac{3}{4}$ yard of material, how many blankets can they make from 9 yards of material?

Solution To answer this question we must divide 9 by $\frac{3}{4}$.

$$9 \div \frac{3}{4} = 9 \cdot \frac{4}{3}$$

$$= 3 \cdot 4$$

$$= 12$$

They can make 12 blankets from the 9 yards of material.

GETTING READY FOR CLASS

After reading through the preceding section, respond in your own words and in complete sentences.

A. What do we call two numbers whose product is 1?

B. True or false? The quotient of $\frac{3}{5}$ and $\frac{3}{8}$ is the same as the product of $\frac{3}{5}$ and $\frac{8}{3}$.

C. How are multiplication and division of fractions related?

D. Dividing by $\frac{19}{9}$ is the same as multiplying by what number?

Answers

7. 2

8. 86

9. 24 blankets

Vocabulary Review

Choose the correct words to fill in the blanks below.

divisor product fraction reciprocal

1. Two numbers whose _____ is 1 are said to be reciprocals.

2. In symbols, the _____ of $\frac{a}{b}$ is $\frac{b}{a}$ if a and b are not equal to zero.

3. To divide by a fraction, replace the _____ by its reciprocal and multiply.

4. In symbols, division by a _____ is written as $\frac{a}{b} \div \frac{c}{d} = \frac{a}{b} \cdot \frac{d}{c}$.

Problems

A Find the quotient in each case by replacing the divisor by its reciprocal and multiplying.

1. $\frac{3}{4} \div \frac{1}{5}$

2. $\frac{1}{3} \div \frac{1}{2}$

3. $\frac{2}{3} \div \frac{1}{2}$

4. $\frac{5}{8} \div \frac{1}{4}$

5. $6 \div \frac{2}{3}$

6. $8 \div -\frac{3}{4}$

7. $20 \div \frac{1}{10}$

8. $16 \div \frac{1}{8}$

9. $\frac{3}{4} \div 2$

10. $\frac{3}{5} \div 2$

11. $\frac{7}{8} \div -\frac{7}{8}$

12. $\frac{4}{3} \div \frac{4}{3}$

13. $\frac{7}{8} \div \frac{8}{7}$

14. $\frac{4}{3} \div \frac{3}{4}$

15. $\frac{9}{16} \div \frac{3}{4}$

16. $\frac{25}{36} \div \frac{5}{6}$

17. $\frac{25}{46} \div \frac{40}{69}$

18. $-\frac{25}{24} \div \frac{15}{36}$

19. $\frac{13}{28} \div \frac{39}{14}$

20. $\frac{28}{125} \div \frac{5}{2}$

21. $\frac{27}{196} \div \frac{9}{392}$

22. $\frac{16}{135} \div \frac{2}{45}$

23. $\frac{25}{18} \div 5$

24. $-\frac{30}{27} \div 6$

25. $6 \div \frac{4}{3}$

26. $-12 \div \frac{4}{3}$

27. $\frac{4}{3} \div 6$

28. $\frac{4}{3} \div 12$

29. $\frac{3}{4} \div \frac{1}{2} \cdot 6$

30. $12 \div \frac{6}{7} \cdot 7$

31. $\frac{2}{3} \cdot \frac{3}{4} \div \frac{5}{8}$

32. $4 \cdot -\frac{7}{6} \div 7$

33. $\frac{35}{10} \cdot \frac{80}{63} \div -\frac{16}{27}$

34. $\frac{20}{72} \cdot \frac{42}{18} \div \frac{20}{16}$

B Simplify each expression as much as possible using order of operations.

35. $10 \div \left(\dfrac{1}{2}\right)^2$

36. $12 \div \left(\dfrac{1}{4}\right)^2$

37. $\dfrac{18}{35} \div \left(\dfrac{6}{7}\right)^2$

38. $\dfrac{48}{55} \div \left(-\dfrac{8}{11}\right)^2$

39. $\dfrac{4}{5} \div \dfrac{1}{10} + 5$

40. $\dfrac{3}{8} \div \dfrac{1}{16} + 4$

41. $10 + \dfrac{11}{12} \div \dfrac{11}{24}$

42. $15 + \dfrac{13}{14} \div \dfrac{13}{42}$

43. $24 \div \left(\dfrac{2}{5}\right)^2 + 25 \div \left(\dfrac{5}{6}\right)^2$

44. $18 \div \left(\dfrac{3}{4}\right)^2 + 49 \div \left(\dfrac{7}{9}\right)^2$

45. $100 \div \left(\dfrac{5}{7}\right)^2 + 200 \div \left(\dfrac{2}{3}\right)^2$

46. $64 \div \left(\dfrac{8}{11}\right)^2 + 81 \div \left(\dfrac{9}{11}\right)^2$

47. What is the quotient of $\frac{3}{8}$ and $\frac{5}{8}$?

48. Find the quotient of $\frac{4}{5}$ and $\frac{16}{25}$.

49. If the quotient of 18 and $\frac{3}{5}$ is increased by 10, what number results?

50. If the quotient of 50 and $\frac{5}{3}$ is increased by 8, what number results?

51. Show that multiplying 3 by 5 is the same as dividing 3 by $\frac{1}{5}$.

52. Show that multiplying 8 by $\frac{1}{2}$ is the same as dividing 8 by 2.

Applying the Concepts

Although many of the application problems that follow involve division with fractions, some do not. Be sure to read the problems carefully.

53. **Sewing** If $\frac{6}{7}$ yard of material is needed to make a blanket, how many blankets can be made from 12 yards of material?

54. **Manufacturing** A clothing manufacturer is making scarves that require $\frac{3}{8}$ yard of material each. How many can be made from 27 yards of material?

55. Capacity Suppose a bag of candy holds exactly $\frac{1}{4}$ pound of candy. How many of these bags can be filled from 12 pounds of candy?

56. Capacity A certain size bottle holds exactly $\frac{4}{5}$ pint of liquid. How many of these bottles can be filled from a 20-pint container?

57. Cooking Audra is making cookies from a recipe that calls for $\frac{3}{4}$ teaspoon of oil. If the only measuring spoon she can find is a $\frac{1}{8}$ teaspoon, how many of these will she have to fill with oil in order to have a total of $\frac{3}{4}$ teaspoon of oil?

58. Cooking A cake recipe calls for $\frac{1}{2}$ cup of sugar. If the only measuring cup available is a $\frac{1}{8}$ cup, how many of these will have to be filled with sugar to make a total of $\frac{1}{2}$ cup of sugar?

59. Student Population If 14 of every 32 students attending Cuesta College are female, what fraction of the students is female? (Simplify your answer.)

60. Population If 27 of every 48 residents of a small town are male, what fraction of the population is male? (Simplify your answer.)

61. Student Population If 14 of every 32 students attending Cuesta College are female, and the total number of students at the school is 4,064, how many of the students are female?

62. Population If 27 of every 48 residents of a small town are male, and the total population of the town is 17,808, how many of the residents are male?

63. Cartons of Milk If a small carton of milk holds exactly $\frac{1}{2}$ pint, how many of the $\frac{1}{2}$-pint cartons can be filled from a 14-pint container?

64. Pieces of Pipe How many pieces of pipe that are $\frac{2}{3}$ foot long must be laid together to make a pipe 16 feet long?

Find the Mistake

Each sentence below contains a mistake. Circle the mistake and write the correct word(s) or numbers(s) on the line provided.

1. Two numbers whose quotient is 1 are said to be reciprocals. _____

2. Dividing the fraction $\frac{12}{7}$ by $\frac{4}{9}$, is equivalent to $\frac{7}{12} \cdot \frac{9}{4}$. _____

3. To work the problem $\frac{22}{5} \div \frac{10}{3}$, multiply the first fraction by its reciprocal. _____

4. The quotient of $\frac{14}{11}$ and $\frac{32}{6}$ is $\frac{224}{33}$. _____

OBJECTIVES

A Add and subtract fractions with common denominators.

B Add and subtract fractions with unlike denominators.

KEY WORDS

least common denominator (LCD)

**Video Examples
Section 2.5**

Note Most people who have done any work with adding fractions know that you add fractions that have the same denominator by adding their numerators, but not their denominators. However, most people don't know why this works. The reason why we add numerators but not denominators is because of the distributive property. That is what the discussion here is all about. If you really want to understand addition of fractions, pay close attention to this discussion.

In Las Vegas, the Stratosphere's hotel and casino has opened another attraction for thrill-seekers. SkyJump is a death-defying controlled free-fall from the 108th floor of the hotel to the ground. Jump Package 1 includes the jump cost plus a DVD of the jump for $114.99. A jump without the DVD costs $99.99. Suppose you are part of a group of people that schedules an appointment for the SkyJump. $\frac{2}{7}$ of the group buys Jump Package 1 and $\frac{3}{7}$ buys the jump without the DVD. The remaining $\frac{1}{7}$ of the group decides not to jump and instead buys the "Chicken" shirt available for purchase at the SkyJump store. What fraction represents the amount of people in the group that actually jumped? To answer this question, we must be able to add fractions with a common denominator.

A Addition and Subtraction with Common Denominators

Adding and subtracting fractions is actually just another application of the distributive property. The distributive property looks like this:

$$a(b + c) = a(b) + a(c)$$

where a, b, and c may be whole numbers or fractions. We will want to apply this property to expressions like

$$\frac{2}{7} + \frac{3}{7}$$

But before we do, we must make one additional observation about fractions.

The fraction $\frac{2}{7}$ can be written as $2 \cdot \frac{1}{7}$, because

$$2 \cdot \frac{1}{7} = \frac{2}{1} \cdot \frac{1}{7} = \frac{2}{7}$$

Likewise, the fraction $\frac{3}{7}$ can be written as $3 \cdot \frac{1}{7}$, because

$$3 \cdot \frac{1}{7} = \frac{3}{1} \cdot \frac{1}{7} = \frac{3}{7}$$

In general, we can say that the fraction $\frac{a}{b}$ can always be written as $a \cdot \frac{1}{b}$, because

$$a \cdot \frac{1}{b} = \frac{a}{1} \cdot \frac{1}{b} = \frac{a}{b}$$

To add the fractions $\frac{2}{7}$ and $\frac{3}{7}$, we simply rewrite each of them as we have done above and apply the distributive property. Here is how it works:

$$\frac{2}{7} + \frac{3}{7} = 2 \cdot \frac{1}{7} + 3 \cdot \frac{1}{7} \qquad \textit{Rewrite each fraction.}$$

$$= (2 + 3) \cdot \frac{1}{7} \qquad \textit{Apply the distributive property.}$$

$$= 5 \cdot \frac{1}{7} \qquad \textit{Add 2 and 3 to get 5.}$$

$$= \frac{5}{7} \qquad \textit{Rewrite } 5 \cdot \frac{1}{7} \textit{ as } \frac{5}{7}.$$

We can visualize the process shown above by using circles that are divided into 7 equal parts.

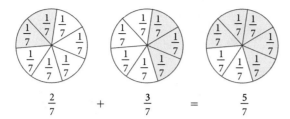

$$\frac{2}{7} \quad + \quad \frac{3}{7} \quad = \quad \frac{5}{7}$$

The fraction $\frac{5}{7}$ is the sum of $\frac{2}{7}$ and $\frac{3}{7}$. The steps and diagrams above show why we add numerators, *but do not add denominators*. Using this example as justification, we can write a rule for adding two fractions that have the same denominator.

> **RULE** Addition with Common Denominator
>
> To add two fractions that have the same denominator, we add their numerators to get the numerator of the answer. The denominator in the answer is the same denominator as in the original fractions.

What we have here is the sum of the numerators placed over the *common denominator*. In symbols, we have the following:

> **Addition and Subtraction of Fractions**
>
> If a, b, and c are numbers, and c is not equal to 0, then
> $$\frac{a}{c} + \frac{b}{c} = \frac{a + b}{c}$$
> This rule holds for subtraction as well. That is,
> $$\frac{a}{c} - \frac{b}{c} = \frac{a - b}{c}$$

EXAMPLE 1 Add or subtract.

a. $\frac{3}{8} + \frac{1}{8}$ **b.** $\frac{9}{5} - \frac{3}{5}$ **c.** $\frac{3}{7} + \frac{2}{7} + \frac{9}{7}$

Solution

a. $\frac{3}{8} + \frac{1}{8} = \frac{3 + 1}{8}$ Add numerators; keep the same denominator.

$= \frac{4}{8}$ The sum of 3 and 1 is 4.

$= \frac{1}{2}$ Reduce to lowest terms.

b. $\frac{9}{5} - \frac{3}{5} = \frac{9 - 3}{5}$ Subtract numerators; keep the same denominator.

$= \frac{6}{5}$ The difference of 9 and 3 is 6.

c. $\frac{3}{7} + \frac{2}{7} + \frac{9}{7} = \frac{3 + 2 + 9}{7}$

$= \frac{14}{7}$

$= 2$

B Addition and Subtraction with Unlike Denominators

As Example 1 indicates, addition and subtraction are simple, straightforward processes when all the fractions have the same denominator. We will now turn our attention to the process of adding fractions that have different denominators. In order to get started, we need the following definition:

DEFINITION least common denominator

The *least common denominator* (LCD) for a set of denominators is the smallest number that is exactly divisible by each denominator. (Note that, in some books, the least common denominator is also called the *least common multiple*.)

In other words, all the denominators of the fractions involved in a problem must divide into the least common denominator exactly. That is, they divide it without leaving a remainder.

2. Find the LCD for the fractions $\frac{7}{15}$ and $\frac{9}{25}$.

EXAMPLE 2 Find the LCD for the fractions $\frac{5}{12}$ and $\frac{7}{18}$.

Solution The least common denominator for the denominators 12 and 18 must be the smallest number divisible by both 12 and 18. We can factor 12 and 18 completely and then build the LCD from these factors. Factoring 12 and 18 completely gives us

$$12 = 2 \cdot 2 \cdot 3 \qquad 18 = 2 \cdot 3 \cdot 3$$

Now, if 12 is going to divide the LCD exactly, then the LCD must have factors of $2 \cdot 2 \cdot 3$. If 18 is to divide it exactly, it must have factors of $2 \cdot 3 \cdot 3$. We don't need to repeat the factors that 12 and 18 have in common.

$$\left. \begin{array}{l} 12 = 2 \cdot 2 \cdot 3 \\ 18 = 2 \cdot 3 \cdot 3 \end{array} \right\} \qquad \begin{array}{c} \text{12 divides the LCD.} \\ \downarrow \\ \text{LCD} = 2 \cdot 2 \cdot 3 \cdot 3 = 36 \\ \uparrow \\ \text{18 divides the LCD.} \end{array}$$

The LCD for 12 and 18 is 36. It is the smallest number that is divisible by both 12 and 18; 12 divides it exactly three times, and 18 divides it exactly two times.

Note The ability to find least common denominators is very important in mathematics. The discussion here is a detailed explanation of how to find an LCD.

We can visualize the results in Example 2 with the diagram below. It shows that 36 is the smallest number that both 12 and 18 divide evenly. As you can see, 12 divides 36 exactly 3 times, and 18 divides 36 exactly 2 times.

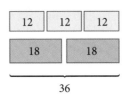

3. Add: $\frac{7}{15} + \frac{9}{25}$.

EXAMPLE 3 Add: $\frac{5}{12} + \frac{7}{18}$.

Solution We can add fractions only when they have the same denominators. In Example 2, we found the LCD for $\frac{5}{12}$ and $\frac{7}{18}$ to be 36. We change $\frac{5}{12}$ and $\frac{7}{18}$ to equivalent fractions that have 36 for a denominator by applying the multiplication property for fractions.

$$\frac{5}{12} = \frac{5 \cdot 3}{12 \cdot 3} = \frac{15}{36}$$

$$\frac{7}{18} = \frac{7 \cdot 2}{18 \cdot 2} = \frac{14}{36}$$

Answers

2. 75

3. $\frac{62}{75}$

The fraction $\frac{15}{36}$ is equivalent to $\frac{5}{12}$, because it was obtained by multiplying both the numerator and the denominator by 3. Likewise, $\frac{14}{36}$ is equivalent to $\frac{7}{18}$, because it was obtained by multiplying the numerator and the denominator by 2. All we have left to do is to add numerators.

$$\frac{15}{36} + \frac{14}{36} = \frac{29}{36}$$

The sum of $\frac{5}{12}$ and $\frac{7}{18}$ is the fraction $\frac{29}{36}$. Let's write the complete problem again step by step.

$$\frac{5}{12} + \frac{7}{18} = \frac{5 \cdot 3}{12 \cdot 3} + \frac{7 \cdot 2}{18 \cdot 2} \qquad \textit{Rewrite each fraction as an equivalent fraction with denominator 36.}$$

$$= \frac{15}{36} + \frac{14}{36}$$

$$= \frac{29}{36} \qquad \textit{Add numerators; keep the common denominator.}$$

EXAMPLE 4 Find the LCD for $\frac{3}{4}$ and $\frac{1}{6}$.

Solution We factor 4 and 6 into products of prime factors and build the LCD from these factors.

$$\left.\begin{array}{l} 4 = 2 \cdot 2 \\ 6 = 2 \cdot 3 \end{array}\right\} \quad \text{LCD} = 2 \cdot 2 \cdot 3 = 12$$

The LCD is 12. Both denominators divide it exactly; 4 divides 12 exactly 3 times, and 6 divides 12 exactly 2 times.

EXAMPLE 5 Add: $\frac{3}{4} + \frac{1}{6}$.

Solution In Example 4, we found that the LCD for these two fractions is 12. We begin by changing $\frac{3}{4}$ and $\frac{1}{6}$ to equivalent fractions with denominator 12.

$$\frac{3}{4} = \frac{3 \cdot 3}{4 \cdot 3} = \frac{9}{12}$$

$$\frac{1}{6} = \frac{1 \cdot 2}{6 \cdot 2} = \frac{2}{12}$$

The fraction $\frac{9}{12}$ is equal to the fraction $\frac{3}{4}$, because it was obtained by multiplying the numerator and the denominator of $\frac{3}{4}$ by 3. Likewise, $\frac{2}{12}$ is equivalent to $\frac{1}{6}$, because it was obtained by multiplying the numerator and the denominator of $\frac{1}{6}$ by 2. To complete the problem, we add numerators.

$$\frac{9}{12} + \frac{2}{12} = \frac{11}{12}$$

The sum of $\frac{3}{4}$ and $\frac{1}{6}$ is $\frac{11}{12}$. Here is how the complete problem looks:

$$\frac{3}{4} + \frac{1}{6} = \frac{3 \cdot 3}{4 \cdot 3} + \frac{1 \cdot 2}{6 \cdot 2} \qquad \textit{Rewrite each fraction as an equivalent fraction with denominator 12.}$$

$$= \frac{9}{12} + \frac{2}{12}$$

$$= \frac{11}{12} \qquad \textit{Add numerators; keep the same denominator.}$$

4. Find the LCD for $\frac{7}{12}$ and $\frac{3}{8}$.

5. Add: $\frac{7}{12} + \frac{3}{8}$.

Note We can visualize the work in Example 5 using circles and shading.

6. Subtract: $\frac{3}{4} - \frac{5}{12}$.

EXAMPLE 6 Subtract: $\frac{7}{15} - \frac{3}{10}$.

Solution Let's factor 15 and 10 completely and use these factors to build the LCD.

15 divides the LCD.

$$\left.\begin{array}{l} 15 = 3 \cdot 5 \\ 10 = 2 \cdot 5 \end{array}\right\} \quad \text{LCD} = 2 \cdot 3 \cdot 5 = 30$$

10 divides the LCD.

Changing to equivalent fractions and subtracting, we have

$$\frac{7}{15} - \frac{3}{10} = \frac{7 \cdot 2}{15 \cdot 2} - \frac{3 \cdot 3}{10 \cdot 3} \quad \text{Rewrite as equivalent fractions with the LCD for the denominator.}$$

$$= \frac{14}{30} - \frac{9}{30}$$

$$= \frac{5}{30} \qquad\qquad \text{Subtract numerators; keep the LCD.}$$

$$= \frac{1}{6} \qquad\qquad \text{Reduce to lowest terms.}$$

As a summary of what we have done so far, and as a guide to working other problems, we now list the steps involved in adding and subtracting fractions with different denominators.

> **HOW TO** **To Add or Subtract Any Two Fractions**
>
> **Step 1:** Factor each denominator completely, and use the factors to build the LCD. (Remember, the LCD is the smallest number divisible by each of the denominators in the problem.)
>
> **Step 2:** Rewrite each fraction as an equivalent fraction with the LCD. This is done by multiplying both the numerator and the denominator of the fraction in question by the appropriate whole number.
>
> **Step 3:** Add or subtract the numerators of the fractions produced in Step 2. This is the numerator of the sum or difference. The denominator of the sum or difference is the LCD.
>
> **Step 4:** Reduce the fraction produced in Step 3 to lowest terms if it is not already in lowest terms.

The idea behind adding or subtracting fractions is really very straight-forward. We can only add or subtract fractions that have the same denominators. If the fractions we are trying to add or subtract do not have the same denominators, we rewrite each of them as an equivalent fraction with the LCD for a denominator.

Here are some additional examples of sums and differences of fractions:

7. Subtract: $\frac{1}{2} - \frac{3}{10}$.

EXAMPLE 7 Subtract: $\frac{3}{5} - \frac{1}{6}$.

Solution The LCD for 5 and 6 is their product, 30. We begin by rewriting each fraction with this common denominator.

$$\frac{3}{5} - \frac{1}{6} = \frac{3 \cdot 6}{5 \cdot 6} - \frac{1 \cdot 5}{6 \cdot 5}$$

$$= \frac{18}{30} - \frac{5}{30}$$

$$= \frac{13}{30}$$

Answers

6. $\frac{1}{3}$

7. $\frac{1}{5}$

EXAMPLE 8 Add: $\frac{1}{6} + \frac{1}{8} + \frac{1}{4}$.

Solution We begin by factoring the denominators completely and building the LCD from the factors that result.

$$
\left.\begin{array}{l}
6 = 2 \cdot 3 \\
8 = 2 \cdot 2 \cdot 2 \\
4 = 2 \cdot 2
\end{array}\right\} \qquad
LCD = 2 \cdot 2 \cdot 2 \cdot 3 = 24
$$

8 divides the LCD.
4 divides the LCD. 6 divides the LCD.

We then change to equivalent fractions and add as usual.

$$
\frac{1}{6} + \frac{1}{8} + \frac{1}{4} = \frac{1 \cdot 4}{6 \cdot 4} + \frac{1 \cdot 3}{8 \cdot 3} + \frac{1 \cdot 6}{4 \cdot 6} = \frac{4}{24} + \frac{3}{24} + \frac{6}{24} = \frac{13}{24}
$$

EXAMPLE 9 Subtract: $3 - \frac{5}{6}$.

Solution The denominators are 1 (because $3 = \frac{3}{1}$) and 6. The smallest number divisible by both 1 and 6 is 6.

$$
3 - \frac{5}{6} = \frac{3}{1} - \frac{5}{6} = \frac{3 \cdot 6}{1 \cdot 6} - \frac{5}{6} = \frac{18}{6} - \frac{5}{6} = \frac{13}{6}
$$

Here are three examples that involve addition with negative numbers.

EXAMPLE 10 Add: $\frac{3}{8} + \left(-\frac{1}{8}\right)$.

Solution We subtract absolute values. The answer will be positive, because $\frac{3}{8}$ is positive.

$$
\frac{3}{8} + \left(-\frac{1}{8}\right) = \frac{2}{8}
$$
$$
= \frac{1}{4} \qquad \textit{Reduce to lowest terms.}
$$

EXAMPLE 11 Add: $\frac{1}{10} + \left(-\frac{4}{5}\right) + \left(-\frac{3}{20}\right)$.

Solution To begin, change each fraction to an equivalent fraction with an LCD of 20.

$$
\frac{1}{10} + \left(-\frac{4}{5}\right) + \left(-\frac{3}{20}\right) = \frac{1 \cdot 2}{10 \cdot 2} + \left(-\frac{4 \cdot 4}{5 \cdot 4}\right) + \left(-\frac{3}{20}\right)
$$
$$
= \frac{2}{20} + \left(-\frac{16}{20}\right) + \left(-\frac{3}{20}\right)
$$
$$
= -\frac{14}{20} + \left(-\frac{3}{20}\right)
$$
$$
= -\frac{17}{20}
$$

EXAMPLE 12 Find the difference of $-\frac{3}{5}$ and $\frac{2}{5}$.

Solution

$$
-\frac{3}{5} - \frac{2}{5} = -\frac{3}{5} + \left(-\frac{2}{5}\right)
$$
$$
= -\frac{5}{5}
$$
$$
= -1
$$

8. Add: $\frac{1}{3} + \frac{3}{8} + \frac{1}{4}$.

9. Subtract: $4 - \frac{5}{7}$.

10. Add: $\frac{4}{5} + \left(-\frac{2}{5}\right)$.

11. Add: $\frac{5}{12} + \left(-\frac{5}{6}\right) + \left(-\frac{5}{24}\right)$.

12. Find the difference of $-\frac{5}{8}$ and $\frac{3}{8}$.

Answers

8. $\frac{23}{24}$

9. $\frac{23}{7}$

10. $\frac{2}{5}$

11. $-\frac{25}{24}$

12. $-\frac{8}{8} = -1$

13. Write the fractions from smallest to largest.

$$\frac{3}{4}, \frac{7}{12}, \frac{5}{8}, \frac{13}{16}$$

Comparing Fractions

As we have shown previously, we can compare fractions to see which is larger or smaller when they have the same denominator. Now that we know how to find the LCD for a set of fractions, we can use the LCD to write equivalent fractions with the intention of comparing them.

EXAMPLE 13 Find the LCD for the fractions below, then write each fraction as an equivalent fraction with the LCD for a denominator. Then write them in order from smallest to largest.

$$\frac{5}{8} \qquad \frac{5}{16} \qquad \frac{3}{4} \qquad \frac{1}{2}$$

Solution The LCD for the four fractions is 16. We begin by writing each fraction as an equivalent fraction with denominator 16.

$$\frac{5}{8} = \frac{10}{16} \qquad \frac{5}{16} = \frac{5}{16} \qquad \frac{3}{4} = \frac{12}{16} \qquad \frac{1}{2} = \frac{8}{16}$$

Now that they all have the same denominator, the smallest fraction is the one with the smallest numerator, and the largest fraction is the one with the largest numerator. Writing them in order from smallest to largest we have

$$\frac{5}{16} \quad < \quad \frac{8}{16} \quad < \quad \frac{10}{16} \quad < \quad \frac{12}{16}$$

$$\frac{5}{16} \quad < \quad \frac{1}{2} \quad < \quad \frac{5}{8} \quad < \quad \frac{3}{4}$$

GETTING READY FOR CLASS

After reading through the preceding section, respond in your own words and in complete sentences.

A. When adding two fractions with the same denominators, we always add their _____, but we never add their _____.

B. What does the abbreviation LCD stand for?

C. What is the first step when finding the LCD for the fractions $\frac{5}{12}$ and $\frac{7}{18}$?

D. When adding fractions, what is the last step?

Answers

13. $\frac{7}{12} < \frac{5}{8} < \frac{3}{4} < \frac{13}{16}$

Vocabulary Review

The following is a list of steps for adding and subtracting fractions. Choose the correct words to fill in the blanks below.

numerators least common denominator lowest terms equivalent

Step 1: Factor each denominator completely, and use the factors to build the _____.

Step 2: Rewrite each fraction as an _____ fraction with the LCD.

Step 3: Add or subtract the _____ of the fractions produced in Step 2.

Step 4: Reduce the fraction produced in Step 3 to _____ if it is not already.

Problems

A Find the following sums and differences, and reduce to lowest terms. (Add or subtract as indicated.)

1. $\dfrac{3}{6} + \dfrac{1}{6}$

2. $\dfrac{2}{5} + \dfrac{3}{5}$

3. $\dfrac{5}{8} - \dfrac{3}{8}$

4. $\dfrac{6}{7} - \dfrac{1}{7}$

5. $\dfrac{3}{4} - \dfrac{1}{4}$

6. $-\dfrac{7}{9} + \dfrac{4}{9}$

7. $\dfrac{2}{3} + \left(-\dfrac{1}{3}\right)$

8. $\dfrac{9}{8} - \dfrac{1}{8}$

9. $\dfrac{1}{4} + \dfrac{2}{4} + \dfrac{3}{4}$

10. $\dfrac{2}{5} + \dfrac{3}{5} + \dfrac{4}{5}$

11. $\dfrac{x+7}{2} + \left(-\dfrac{1}{2}\right)$

12. $\dfrac{x+5}{4} - \dfrac{3}{4}$

13. $\dfrac{1}{10} + \dfrac{3}{10} + \dfrac{4}{10}$

14. $\dfrac{3}{20} + \dfrac{1}{20} + \dfrac{4}{20}$

15. $\dfrac{1}{3} + \dfrac{4}{3} + \dfrac{5}{3}$

16. $\dfrac{5}{4} + \dfrac{4}{4} + \dfrac{3}{4}$

B Complete the following tables.

17.

First Number a	Second Number b	The Sum of a and b $a + b$
$\dfrac{1}{2}$	$\dfrac{1}{3}$	
$\dfrac{1}{3}$	$\dfrac{1}{4}$	
$\dfrac{1}{4}$	$\dfrac{1}{5}$	
$\dfrac{1}{5}$	$\dfrac{1}{6}$	

18.

First Number a	Second Number b	The Sum of a and b $a + b$
1	$\dfrac{1}{2}$	
1	$\dfrac{1}{3}$	
1	$\dfrac{1}{4}$	
1	$\dfrac{1}{5}$	

19.

First Number a	Second Number b	The Sum of a and b $a + b$
$\dfrac{1}{12}$	$\dfrac{1}{2}$	
$\dfrac{1}{12}$	$\dfrac{1}{3}$	
$\dfrac{1}{12}$	$\dfrac{1}{4}$	
$\dfrac{1}{12}$	$\dfrac{1}{6}$	

20.

First Number a	Second Number b	The Sum of a and b $a + b$
$\dfrac{1}{8}$	$\dfrac{1}{2}$	
$\dfrac{1}{8}$	$\dfrac{1}{4}$	
$\dfrac{1}{8}$	$\dfrac{1}{16}$	
$\dfrac{1}{8}$	$\dfrac{1}{24}$	

Find the LCD for each of the following, then use the methods developed in this section to add or subtract as indicated.

21. $\dfrac{4}{9} + \dfrac{1}{3}$

22. $\dfrac{1}{2} + \dfrac{1}{4}$

23. $2 + \dfrac{1}{3}$

24. $3 + \dfrac{1}{2}$

25. $\dfrac{3}{4} + 1$

26. $\dfrac{3}{4} + 2$

27. $\dfrac{1}{2} + \dfrac{2}{3}$

28. $\dfrac{1}{8} + \dfrac{3}{4}$

29. $\dfrac{1}{4} + \dfrac{1}{5}$

30. $\dfrac{1}{3} + \dfrac{1}{5}$

31. $\dfrac{1}{2} + \dfrac{1}{5}$

32. $\dfrac{1}{2} - \dfrac{1}{5}$

33. $\dfrac{5}{12} + \dfrac{3}{8}$

34. $\dfrac{9}{16} + \dfrac{7}{12}$

35. $\dfrac{8}{30} - \dfrac{1}{20}$

36. $\dfrac{9}{40} - \dfrac{1}{30}$

37. $\dfrac{3}{10} + \dfrac{1}{100}$

38. $\dfrac{9}{100} + \dfrac{7}{10}$

39. $\dfrac{10}{36} + \dfrac{9}{48}$

40. $\dfrac{12}{28} + \dfrac{9}{20}$

41. $\dfrac{17}{30} + \dfrac{11}{42}$

42. $\dfrac{19}{42} + \dfrac{13}{70}$

43. $\dfrac{25}{84} - \left(-\dfrac{41}{90}\right)$

44. $\dfrac{23}{70} + \dfrac{29}{84}$

45. $\dfrac{13}{126} - \dfrac{13}{180}$

46. $\dfrac{17}{84} + \left(-\dfrac{17}{90}\right)$

47. $\dfrac{3}{4} + \dfrac{1}{8} + \dfrac{5}{6}$

48. $\dfrac{3}{8} + \dfrac{2}{5} + \dfrac{1}{4}$

49. $\dfrac{3}{10} + \dfrac{5}{12} + \dfrac{1}{6}$

50. $\dfrac{5}{21} + \dfrac{1}{7} + \dfrac{3}{14}$

51. $\dfrac{1}{2} + \dfrac{1}{3} + \dfrac{1}{4} + \dfrac{1}{6}$

52. $\dfrac{1}{8} + \dfrac{1}{4} - \left(-\dfrac{1}{5}\right) + \dfrac{1}{10}$

53. $10 - \dfrac{2}{9}$

54. $9 - \dfrac{3}{5}$

55. $4 - \dfrac{2}{3}$

56. $5 + \left(-\dfrac{3}{4}\right)$

57. $\dfrac{1}{10} + \dfrac{4}{5} - \dfrac{3}{20}$

58. $\dfrac{1}{2} + \dfrac{3}{4} - \dfrac{5}{8}$

59. $\dfrac{1}{4} - \dfrac{1}{8} - \left(-\dfrac{1}{2}\right) - \dfrac{3}{8}$

60. $\dfrac{7}{8} + \left(-\dfrac{3}{4}\right) + \dfrac{5}{8} - \dfrac{1}{2}$

61. $-\dfrac{1}{6} - \dfrac{5}{6}$

62. $-\dfrac{4}{7} - \dfrac{3}{7}$

63. $-\dfrac{13}{70} - \dfrac{23}{42}$

64. $-\dfrac{17}{60} - \dfrac{17}{90}$

There are two ways to work the problems below. You can combine the fractions inside the parentheses first and then multiply, or you can apply the distributive property first, then add.

65. $15\left(\dfrac{2}{3} + \dfrac{3}{5}\right)$

66. $15\left(\dfrac{4}{5} - \dfrac{1}{3}\right)$

67. $4\left(\dfrac{1}{2} + \dfrac{1}{4}\right)$

68. $6\left(\dfrac{1}{3} + \dfrac{1}{2}\right)$

69. $4\left(3 - \dfrac{3}{4}\right)$

70. $6\left(5 - \dfrac{2}{3}\right)$

71. $9\left(\dfrac{1}{3} + \dfrac{1}{9}\right)$

72. $12\left(\dfrac{1}{3} + \dfrac{1}{4}\right)$

73. Write the fractions in order from smallest to largest.

$$\dfrac{3}{4} \qquad \dfrac{3}{8} \qquad \dfrac{1}{2} \qquad \dfrac{1}{4}$$

74. Write the fractions in order from smallest to largest.

$$\dfrac{1}{2} \qquad \dfrac{1}{6} \qquad \dfrac{1}{4} \qquad \dfrac{1}{3}$$

75. Find the sum of $\dfrac{3}{7}$, 2, and $\dfrac{1}{9}$.

76. Find the sum of 6, $\dfrac{6}{11}$, and 11.

77. Give the difference of $\dfrac{7}{8}$ and $\dfrac{1}{4}$.

78. Give the difference of $\dfrac{9}{10}$ and $\dfrac{1}{100}$.

Arithmetic Sequences An arithmetic sequence is a sequence in which each term comes from the previous term by adding the same number each time. For example, the sequence $1, \frac{3}{2}, 2, \frac{5}{2}, \ldots$ is an arithmetic sequence that starts with the number 1. Then each term after that is found by adding $\frac{1}{2}$ to the previous term. By observing this fact, we know that the next term in the sequence will be $\frac{5}{2} + \frac{1}{2} = \frac{6}{2} = 3$.

Find the next number in each arithmetic sequence below.

79. $1, \dfrac{4}{3}, \dfrac{5}{3}, 2, \ldots$

80. $1, \dfrac{5}{4}, \dfrac{3}{2}, \dfrac{7}{4}, \ldots$

81. $\dfrac{3}{2}, 2, \dfrac{5}{2}, \ldots$

82. $\dfrac{2}{3}, 1, \dfrac{4}{3}, \ldots$

Applying the Concepts

Some of the application problems below involve multiplication or division, while others involve addition or subtraction.

83. Capacity One carton of milk contains $\frac{1}{2}$ pint while another contains 4 pints. How much milk is contained in both cartons?

84. Baking A recipe calls for $\frac{2}{3}$ cup of flour and $\frac{3}{4}$ cup of sugar. What is the total amount of flour and sugar called for in the recipe?

85. Budget A family decides that they can spend $\frac{5}{8}$ of their monthly income on house payments. If their monthly income is \$2,120, how much can they spend for house payments?

86. Savings A family saves $\frac{3}{16}$ of their income each month. If their monthly income is \$1,264, how much do they save each month?

Reading a Pie Chart The pie chart below shows how the students at Cal Poly State University are distributed among the different schools at the university. Use the information in the pie chart to answer questions 79 and 80.

87. If the students in the Schools of Engineering and Business are combined, what fraction results?

88. What fraction of the university's students are enrolled in the Schools of Agriculture, Engineering, and Business combined?

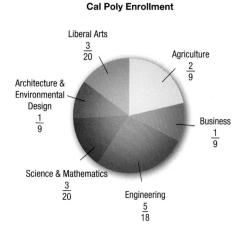

Cal Poly Enrollment

Liberal Arts $\frac{3}{20}$

Agriculture $\frac{2}{9}$

Architecture & Environmental Design $\frac{1}{9}$

Business $\frac{1}{9}$

Science & Mathematics $\frac{3}{20}$

Engineering $\frac{5}{18}$

89. Final Exam Grades The table gives the fraction of students in a class of 40 that received grades of A, B, or C on the final exam. Fill in all the missing parts of the table.

Grade	Number of Students	Fraction of Students
A		$\frac{1}{8}$
B		$\frac{1}{5}$
C		$\frac{1}{2}$
Below C		
Total	40	1

90. Flu During a flu epidemic a company with 200 employees has $\frac{1}{10}$ of their employees call in sick on Monday and another $\frac{3}{10}$ call in sick on Tuesday. What is the total number of employees calling in sick during this 2-day period?

91. Subdivision A 6-acre piece of land is subdivided into $\frac{3}{5}$-acre lots. How many lots are there?

92. Cutting Wood A 12-foot piece of wood is cut into shelves. If each is $\frac{3}{4}$ foot in length, how many shelves are there?

Find the perimeter of each figure.

93.

$\frac{3}{8}$ in.

$\frac{3}{8}$ in.

94.

$\frac{3}{8}$ in.

$\frac{3}{4}$ in.

95.

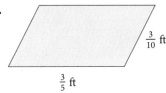

$\frac{3}{10}$ ft

$\frac{3}{5}$ ft

96.

$\frac{1}{3}$ ft \quad $\frac{1}{3}$ ft

$\frac{3}{5}$ ft

- - -

Getting Ready for the Next Section

Simplify.

97. $9 \cdot 6 + 5$

98. $4 \cdot 6 + 3$

99. Write 2 as a fraction with denominator 8.

100. Write 2 as a fraction with denominator 4.

101. Write 1 as a fraction with denominator 8.

102. Write 5 as a fraction with denominator 4.

Add.

103. $\frac{8}{4} + \frac{3}{4}$ \qquad **104.** $\frac{16}{8} + \frac{1}{8}$ \qquad **105.** $2 + \frac{1}{8}$ \qquad **106.** $2 + \frac{3}{4}$ \qquad **107.** $1 + \frac{1}{8}$ \qquad **108.** $5 + \frac{3}{4}$

Divide.

109. $11 \div 4$ \qquad **110.** $10 \div 3$ \qquad **111.** $208 \div 24$ \qquad **112.** $207 \div 26$

Find the Mistake

Each sentence below contains a mistake. Circle the mistake and write the correct word(s) or numbers(s) on the line provided.

1. The fractions $\frac{a}{c}$ and $\frac{b}{c}$ can be added to become $\frac{a+b}{c}$ because they have different denominators. _____

2. Subtracting $\frac{12}{21}$ from $\frac{18}{21}$, gives us $\frac{30}{21}$. _____

3. The least common denominator for a set of denominators is the smallest number that is exactly divisible by each numerator. _____

4. The LCD for the fractions $\frac{4}{6}$, $\frac{2}{8}$ and $\frac{3}{4}$ is 12. _____

2.6 Mixed-Number Notation

Pogopalooza is an annual world championship for stunt pogo stick athletes. Pogo jumpers compete in various events, such as the most or the least jumps in a minute, the highest jump, and numerous exhibitions of acrobatic stunts. Imagine you are competing in the High Jump competition. The crowd cheers as you bounce a remarkable $8\frac{1}{4}$ feet into the air! Being a skilled mathematician, you realize that your score of $8\frac{1}{4}$ feet, or eight and one-fourth feet, is actually a *mixed number*. It is the sum of a whole number and a proper fraction. With mixed-number notation, we leave out the addition sign.

Mixed-Number Notation

Here are some further examples of mixed number notation:

$$2\frac{1}{8} = 2 + \frac{1}{8}, \quad 6\frac{5}{9} = 6 + \frac{5}{9}, \quad 11\frac{2}{3} = 11 + \frac{2}{3}$$

The notation used in writing mixed numbers (writing the whole number and the proper fraction next to each other) must always be interpreted as addition. It is a mistake to read $5\frac{3}{4}$ as meaning 5 times $\frac{3}{4}$. If we want to indicate multiplication, we must use parentheses or a multiplication symbol. That is,

$$5\frac{3}{4} \text{ is \textbf{not} the same as } 5\left(\frac{3}{4}\right).$$

This implies addition.

These imply multiplication.

$$5\frac{3}{4} \text{ is \textbf{not} the same as } 5 \cdot \frac{3}{4}.$$

**Video Examples
Section 2.6**

A Changing Mixed Numbers to Improper Fractions

To change a mixed number to an improper fraction, we write the mixed number with the + sign showing and then add the two numbers, as we did earlier.

EXAMPLE 1 Change $2\frac{3}{4}$ to an improper fraction.

Solution
$$2\frac{3}{4} = 2 + \frac{3}{4}$$ Write the mixed number as a sum.

$$= \frac{2}{1} + \frac{3}{4}$$ Show that the denominator of 2 is 1.

$$= \frac{4 \cdot 2}{4 \cdot 1} + \frac{3}{4}$$ Multiply the numerator and the denominator of $\frac{2}{1}$ by 4 so both fractions will have the same denominator.

Practice Problems

1. Change $4\frac{2}{3}$ to an improper fraction.

Answer

1. $\frac{14}{3}$

$$= \frac{8}{4} + \frac{3}{4}$$

$$= \frac{11}{4} \qquad \textit{Add the numerators; keep the common denominator.}$$

The mixed number $2\frac{3}{4}$ is equal to the improper fraction $\frac{11}{4}$. The diagram that follows further illustrates the equivalence of $2\frac{3}{4}$ and $\frac{11}{4}$.

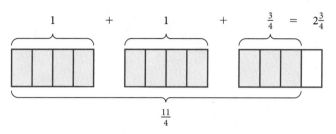

2. Change $3\frac{1}{5}$ to an improper fraction.

EXAMPLE 2 Change $2\frac{1}{8}$ to an improper fraction.

Solution $\quad 2\frac{1}{8} = 2 + \frac{1}{8} \qquad \textit{Write as addition.}$

$$= \frac{2}{1} + \frac{1}{8} \qquad \textit{Write the whole number 2 as a fraction.}$$

$$= \frac{8 \cdot 2}{8 \cdot 1} + \frac{1}{8} \qquad \textit{Change } \tfrac{2}{1} \textit{ to a fraction with denominator 8.}$$

$$= \frac{16}{8} + \frac{1}{8}$$

$$= \frac{17}{8} \qquad \textit{Add the numerators.}$$

If we look closely at Examples 1 and 2, we can see the following shortcut that will let us change a mixed number to an improper fraction without so many steps.

 Shortcut: To change a mixed number to an improper fraction, simply multiply the whole number by the denominator of the fraction, and add the result to the numerator of the fraction. The result is the numerator of the improper fraction we are looking for. The denominator is the same as the original denominator.

3. Change $5\frac{7}{8}$ to an improper fraction.

EXAMPLE 3 Use the shortcut to change $5\frac{3}{4}$ to an improper fraction.

Solution

 1. First, we multiply 4×5 to get 20.

 2. Next, we add 20 to 3 to get 23.

 3. The improper fraction equal to $5\frac{3}{4}$ is $\frac{23}{4}$.

Here is a diagram showing what we have done:

Mathematically, our shortcut is written like this:

$$5\frac{3}{4} = \frac{(4 \cdot 5) + 3}{4} = \frac{20 + 3}{4} = \frac{23}{4}$$

The result will always have the same denominator as the original mixed number.

The shortcut shown in Example 3 works because the whole-number part of a mixed number can always be written with a denominator of 1. Therefore, the LCD for a whole number and fraction will always be the denominator of the fraction. That is why we multiply the whole number by the denominator of the fraction.

$$5\frac{3}{4} = 5 + \frac{3}{4} = \frac{5}{1} + \frac{3}{4} = \frac{4 \cdot 5}{4 \cdot 1} + \frac{3}{4} = \frac{4 \cdot 5 + 3}{4} = \frac{23}{4}$$

EXAMPLE 4 Change $6\frac{5}{9}$ to an improper fraction.

Solution Using the first method, we have

$$6\frac{5}{9} = 6 + \frac{5}{9} = \frac{6}{1} + \frac{5}{9} = \frac{9 \cdot 6}{9 \cdot 1} + \frac{5}{9} = \frac{54}{9} + \frac{5}{9} = \frac{59}{9}$$

Using the shortcut method, we have

$$6\frac{5}{9} = \frac{(9 \cdot 6) + 5}{9} = \frac{54 + 5}{9} = \frac{59}{9}$$

B Changing Improper Fractions to Mixed Numbers

To change an improper fraction to a mixed number, we divide the numerator by the denominator. The result is used to write the mixed number.

EXAMPLE 5 Change $\frac{11}{4}$ to a mixed number.

Solution Dividing 11 by 4 gives us

$$\begin{array}{r} 2 \\ 4\overline{)11} \\ \underline{-8} \\ 3 \end{array}$$

We see that 4 goes into 11 two times with 3 for a remainder. We write this result as

$$\frac{11}{4} = 2 + \frac{3}{4} = 2\frac{3}{4}$$

The improper fraction $\frac{11}{4}$ is equivalent to the mixed number $2\frac{3}{4}$.

One easy way to visualize the results in Example 5 is to imagine running 13 laps on a $\frac{1}{4}$ mile track. Your 13 laps are equivalent to $\frac{13}{4}$ miles. In miles, your laps are equal to 3 miles plus 1 quarter-mile, or $3\frac{1}{4}$ miles.

EXAMPLE 6 Write as a mixed number.

a. $\frac{10}{3}$ **b.** $\frac{208}{24}$

Solution

a.
$$\begin{array}{r} 3 \\ 3\overline{)10} \\ \underline{9} \\ 1 \end{array}$$ so $\frac{10}{3} = 3 + \frac{1}{3} = 3\frac{1}{3}$

b.
$$\begin{array}{r} 8 \\ 24\overline{)208} \\ \underline{192} \\ 16 \end{array}$$ so $\frac{208}{24} = 8 + \frac{16}{24} = 8 + \frac{2}{3} = 8\frac{2}{3}$

Reduce to lowest terms.

4. Change $8\frac{6}{7}$ to an improper fraction.

Calculator Note The sequence of keys to press on a calculator to obtain the numerator in Example 4 looks like this:

$$9 \;\boxed{\times}\; 6 \;\boxed{+}\; 5 \;\boxed{=}$$

5. Change $\frac{13}{3}$ to a mixed number.

Note This division process shows us how many ones are in $\frac{11}{4}$ and, when the ones are taken out, how many fourths are left.

6. Write as a mixed number.

a. $\frac{27}{4}$

b. $\frac{76}{8}$

Answers

4. $\frac{62}{7}$

5. $4\frac{1}{3}$

6. a. $6\frac{3}{4}$ **b.** $9\frac{1}{2}$

Long Division, Remainders, and Mixed Numbers

Mixed numbers give us another way of writing the answers to long division problems that contain remainders. Here is how we divide 1,690 by 67:

$$
\begin{array}{r}
25 \text{ R } 15 \\
67\overline{)1690} \\
\underline{134\downarrow} \\
350 \\
\underline{335} \\
15
\end{array}
$$

The answer is 25 with a remainder of 15. Using mixed numbers, we can now write the answer as $25\frac{15}{67}$. That is,

$$
\frac{1,690}{67} = 25\frac{15}{67}
$$

The quotient of 1,690 and 67 is $25\frac{15}{67}$.

GETTING READY FOR CLASS

After reading through the preceding section, respond in your own words and in complete sentences.

A. What is a mixed number?

B. The expression $5\frac{3}{4}$ is equivalent to what addition problem?

C. The improper fraction $\frac{11}{4}$ is equivalent to what mixed number?

D. Why is $\frac{13}{5}$ an improper fraction, but $\frac{3}{5}$ is not an improper fraction?

Vocabulary Review

Choose the correct words to fill in the blanks below.

| addition | denominator | proper fraction |
| improper fraction | remainder | mixed number |

1. A mixed number is the sum of a whole number and a _____.
2. The notation used in writing mixed numbers is always interpreted as _____.
3. We leave out the addition sign when writing a number as a _____.
4. To change a mixed number to an _____, write the mixed number with the addition sign showing and then add the two numbers.
5. To change an improper fraction to a mixed number, divide the numerator by the _____.
6. An answer to a long division problem that contains a _____ can be rewritten as a mixed number.

Problems

A Change each mixed number to an improper fraction.

1. $4\frac{2}{3}$ 2. $3\frac{5}{8}$ 3. $5\frac{1}{4}$ 4. $7\frac{1}{2}$ 5. $1\frac{5}{8}$ 6. $1\frac{6}{7}$

7. $15\frac{2}{3}$ 8. $-\left(17\frac{3}{4}\right)$ 9. $4\frac{20}{21}$ 10. $-\left(5\frac{18}{19}\right)$ 11. $12\frac{31}{33}$ 12. $14\frac{29}{31}$

B Change each improper fraction to a mixed number.

13. $\frac{9}{8}$ 14. $\frac{10}{9}$ 15. $-\frac{19}{4}$ 16. $\frac{23}{5}$ 17. $\frac{29}{6}$ 18. $\frac{7}{2}$

19. $\frac{13}{4}$ 20. $\frac{41}{15}$ 21. $\frac{109}{27}$ 22. $\frac{319}{23}$ 23. $-\frac{428}{15}$ 24. $\frac{769}{27}$

Applying the Concepts

25. iPad Applications The chart shows the number of each type of applications compatible with the iPad. Use the information to answer the following questions.

a. The number of game apps is what fraction of the education apps?

b. The number of productivity apps is what fraction of the entertainment apps?

26. NBA Finals The chart shows the number of NBA Finals appearances by several teams. Use the information to answer the following questions.

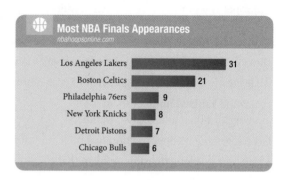

a. The number of Lakers appearances is what fraction of the Bulls appearances?

b. The number of Pistons appearances is what fraction of the Celtics appearances?

27. Stocks Suppose a stock is selling on a stock exchange for $5\frac{1}{4}$ dollars per share. If the price increases $\frac{3}{4}$ dollar per share, what is the new price of the stock?

28. Stocks Suppose a stock is selling on a stock exchange for $5\frac{1}{4}$ dollars per share. If the price increases 2 dollars per share, what is the new price of the stock?

29. Height If a man is 71 inches tall, then in feet his height is $5\frac{11}{12}$ feet. Change $5\frac{11}{12}$ to an improper fraction.

30. Height If a woman is 63 inches tall, then her height in feet is $\frac{63}{12}$. Write $\frac{63}{12}$ as a mixed number.

31. Gasoline Prices The price of unleaded gasoline is $305\frac{1}{5}$¢ per gallon. Write this number as an improper fraction.

32. Gasoline Prices Suppose the price of gasoline is $308\frac{1}{5}$¢ if purchased with a credit card, but 5¢ less if purchased with cash. What is the cash price of the gasoline?

Getting Ready for the Next Section

Change to improper fractions.

33. $2\dfrac{3}{4}$ **34.** $3\dfrac{1}{5}$ **35.** $4\dfrac{5}{8}$ **36.** $1\dfrac{3}{5}$ **37.** $2\dfrac{4}{5}$ **38.** $5\dfrac{9}{10}$

Find the following products. (Multiply.)

39. $\dfrac{3}{8} \cdot \dfrac{3}{5}$ **40.** $\dfrac{11}{4} \cdot \dfrac{16}{5}$ **41.** $-\dfrac{2}{3}\left(\dfrac{9}{16}\right)$ **42.** $\dfrac{7}{10}\left(\dfrac{5}{21}\right)$

Find the quotients. (Divide.)

43. $\dfrac{4}{5} \div \dfrac{7}{8}$ **44.** $\dfrac{3}{4} \div \dfrac{1}{2}$ **45.** $-\dfrac{8}{5} \div \dfrac{14}{5}$ **46.** $\dfrac{59}{10} \div -2$

Improving Your Quantitative Literacy

Starbucks Stores The chart shows the number of Starbucks stores in each state. Use the information to answer the following questions.

47. According to the chart, which of the following statements is closer to the truth?

 a. There are five times as many stores in California as there are in Illinois.

 b. There are twice as many stores in Washington as there are in New York.

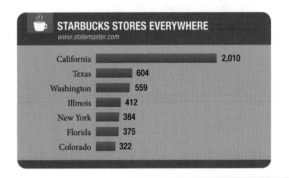

STARBUCKS STORES EVERYWHERE
www.statemaster.com

California	2,010
Texas	604
Washington	559
Illinois	412
New York	384
Florida	375
Colorado	322

48. According to the chart, which fraction best represents the fraction of stores in the chart that are described below. Choose between $\dfrac{1}{2}, \dfrac{1}{3}, \dfrac{1}{4}$, and $\dfrac{1}{5}$.

 a. Illinois as compared to California

 b. Colorado as compared to Texas

Find the Mistake

Each sentence below contains a mistake. Circle the mistake and write the correct word(s) or numbers(s) on the line provided.

 1. For mixed-number notation, writing the whole number next to the fraction implies multiplication.

 2. A shortcut for changing a mixed number to an improper fraction is to multiply the whole number by the numerator of the fraction, and then add the result to the denominator of the fraction.

 3. Changing $6\dfrac{4}{5}$ to an improper fraction gives us $\dfrac{29}{5}$. _____

 4. Writing $\dfrac{70}{12}$ as a mixed number gives us $6\dfrac{1}{6}$. _____

2.7 Multiplication and Division with Mixed Numbers

©iStockphoto.com/negaprion

Fire tornadoes are a difficult and dangerous element of firefighting. The visible part of the fire tornado is called the core, where hot ash and combustible gases ignite in flame and rise hundreds of feet to the sky. The outer layer of the tornado is the circulating air that fuels the core with fresh oxygen. As the fire tornado creeps across the land, it sets anything in its path ablaze, hurls debris into the air, and spins so fast it can knock down trees. The funnel of flames can create wind speeds of more than 100 miles per hour, which is as strong as a category 2 hurricane.

Suppose a fire tornado created a wind speed of 100 miles per hour. What if this speed was $2\frac{1}{2}$ times the wind speed outside the tornado? How fast is the wind speed outside the tornado? Recalling our work from the previous section, we know that the number $2\frac{1}{2}$ is a *mixed number*. To answer the question, we must be able to divide 100 by $2\frac{1}{2}$. Division with mixed numbers is one of the topics we will cover in this section.

A Multiplication with Mixed Numbers

EXAMPLE 1 Multiply: $2\frac{3}{4} \cdot 3\frac{1}{5}$.

Solution We begin by changing each mixed number to an improper fraction.

$$2\frac{3}{4} = \frac{11}{4} \quad \text{and} \quad 3\frac{1}{5} = \frac{16}{5}$$

Using the resulting improper fractions, we multiply as usual. (That is, we multiply numerators and multiply denominators.)

$$\frac{11}{4} \cdot \frac{16}{5} = \frac{11 \cdot 16}{4 \cdot 5}$$

$$= \frac{11 \cdot 4 \cdot 4}{4 \cdot 5}$$

$$= \frac{44}{5} \quad \text{or} \quad 8\frac{4}{5}$$

EXAMPLE 2 Multiply: $3 \cdot 4\frac{5}{8}$.

Solution Writing each number as an improper fraction, we have

$$3 = \frac{3}{1} \quad \text{and} \quad 4\frac{5}{8} = \frac{37}{8}$$

**Video Examples
Section 2.7**

Practice Problems

1. Multiply: $1\frac{1}{4} \cdot 2\frac{3}{5}$.

Note As you can see, once you have changed each mixed number to an improper fraction, you multiply the resulting fractions the same way you did in Section 2.3.

2. Multiply: $6 \cdot 2\frac{1}{8}$.

Answers

1. $3\frac{1}{4}$

2. $12\frac{3}{4}$

The complete problem looks like this:

$$3 \cdot 4\frac{5}{8} = \frac{3}{1} \cdot \frac{37}{8} \qquad \text{Change to improper fractions.}$$

$$= \frac{111}{8} \qquad \text{Multiply numerators and multiply denominators.}$$

$$= 13\frac{7}{8} \qquad \text{Write the answer as a mixed number.}$$

B Division with Mixed Numbers

Dividing mixed numbers also requires that we change all mixed numbers to improper fractions before we actually do the division.

EXAMPLE 3 Divide: $1\frac{3}{5} \div 2\frac{4}{5}$.

3. Divide: $3\frac{3}{4} \div 2\frac{1}{4}$.

Solution We begin by rewriting each mixed number as an improper fraction.

$$1\frac{3}{5} = \frac{8}{5} \quad \text{and} \quad 2\frac{4}{5} = \frac{14}{5}$$

We then divide using the same method we used in Section 2.4. We multiply by the reciprocal of the divisor. Here is the complete problem:

$$1\frac{3}{5} \div 2\frac{4}{5} = \frac{8}{5} \div \frac{14}{5} \qquad \text{Change to improper fractions.}$$

$$= \frac{8}{5} \cdot \frac{5}{14} \qquad \text{To divide by } \frac{14}{5}, \text{ multiply by } \frac{5}{14}.$$

$$= \frac{8 \cdot 5}{5 \cdot 14} \qquad \text{Multiply numerators and multiply denominators.}$$

$$= \frac{4 \cdot 2 \cdot 5}{5 \cdot 2 \cdot 7} \qquad \text{Divide out factors common to the numerator and denominator.}$$

$$= \frac{4}{7} \qquad \text{Answer in lowest terms.}$$

EXAMPLE 4 Divide: $5\frac{9}{10} \div 2$.

4. Divide: $8\frac{3}{4} \div 5$.

Solution We change to improper fractions and proceed as usual.

$$5\frac{9}{10} \div 2 = \frac{59}{10} \div \frac{2}{1} \qquad \text{Write each number as an improper fraction.}$$

$$= \frac{59}{10} \cdot \frac{1}{2} \qquad \text{Write division as multiplication by the reciprocal.}$$

$$= \frac{59}{20} \qquad \text{Multiply numerators and multiply denominators.}$$

$$= 2\frac{19}{20} \qquad \text{Change to a mixed number.}$$

GETTING READY FOR CLASS

After reading through the preceding section, respond in your own words and in complete sentences.

A. What is the first step when multiplying or dividing mixed numbers?

B. What is the reciprocal of $2\frac{4}{5}$?

C. Dividing $5\frac{9}{10}$ by 2 is equivalent to multiplying $5\frac{9}{10}$ by what number?

D. Find $4\frac{5}{8}$ of 3.

Answers

3. $1\frac{2}{3}$

4. $1\frac{3}{4}$

Vocabulary Review

Choose the correct words to fill in the blanks below.

improper fraction mixed number divisor numerators

1. To multiply mixed numbers, first change each _____ to an improper fraction.

2. To multiply improper fractions, multiply _____ and multiply denominators.

3. Division of mixed numbers requires that we change all mixed numbers to _____ before actually doing the dividing.

4. To divide improper fractions, multiply by the reciprocal of the _____.

Problems

Write your answers as proper fractions or mixed numbers, not as improper fractions.

A Find the following products. (Multiply.)

1. $3\frac{2}{5} \cdot 1\frac{1}{2}$

2. $2\frac{1}{3} \cdot 6\frac{3}{4}$

3. $5\frac{1}{8} \cdot 2\frac{2}{3}$

4. $1\frac{5}{6} \cdot 1\frac{4}{5}$

5. $2\frac{1}{10} \cdot 3\frac{3}{10}$

6. $4\frac{7}{10} \cdot 3\frac{1}{10}$

7. $1\frac{1}{4} \cdot 4\frac{2}{3}$

8. $3\frac{1}{2} \cdot 2\frac{1}{6}$

9. $2 \cdot 4\frac{7}{8}$

10. $-10 \cdot 1\frac{1}{4}$

11. $\frac{3}{5} \cdot 5\frac{1}{3}$

12. $\frac{2}{3} \cdot 4\frac{9}{10}$

13. $2\frac{1}{2} \cdot 3\frac{1}{3} \cdot 1\frac{1}{2}$

14. $3\frac{1}{5} \cdot 5\frac{1}{6} \cdot 1\frac{1}{8}$

15. $\frac{3}{4} \cdot 7 \cdot 1\frac{4}{5}$

16. $\frac{7}{8} \cdot 6 \cdot \left(-1\frac{5}{6}\right)$

B Find the following quotients. (Divide.)

17. $3\frac{1}{5} \div 4\frac{1}{2}$

18. $1\frac{4}{5} \div 2\frac{5}{6}$

19. $6\frac{1}{4} \div 3\frac{3}{4}$

20. $8\frac{2}{3} \div 4\frac{1}{3}$

21. $10 \div 2\frac{1}{2}$

22. $12 \div 3\frac{1}{6}$

23. $8\frac{3}{5} \div 2$

24. $-12\frac{6}{7} \div 3$

25. $\left(\frac{3}{4} \div 2\frac{1}{2}\right) \div 3$

26. $\frac{7}{8} \div \left(1\frac{1}{4} \div -4\right)$

27. $\left(8 \div 1\frac{1}{4}\right) \div -2$

28. $8 \div \left(1\frac{1}{4} \div 2\right)$

29. $2\frac{1}{2} \cdot \left(3\frac{2}{5} \div 4\right)$

30. $4\frac{3}{5} \cdot \left(2\frac{1}{4} \div 5\right)$

31. Find the product of $2\frac{1}{2}$ and 3.

32. Find the product of $\frac{1}{5}$ and $3\frac{2}{3}$.

33. What is the quotient of $2\frac{3}{4}$ and $3\frac{1}{4}$?

34. What is the quotient of $1\frac{1}{5}$ and $2\frac{2}{5}$?

Applying the Concepts

35. Cooking A certain recipe calls for $2\frac{3}{4}$ cups of sugar. If the recipe is to be doubled, how much sugar should be used?

36. Cooking If a recipe calls for $3\frac{1}{2}$ cups of flour, how much flour will be needed if the recipe is tripled?

37. Cooking If a recipe calls for $2\frac{1}{2}$ cups of sugar, how much sugar is needed to make $\frac{1}{3}$ of the recipe?

38. Cooking A recipe calls for $3\frac{1}{4}$ cups of flour. If Diane is using only half the recipe, how much flour should she use?

39. Number Problem Find $\frac{3}{4}$ of $1\frac{7}{9}$. (Remember that *of* means multiply.)

40. Number Problem Find $\frac{5}{6}$ of $2\frac{4}{15}$.

41. Cost of Gasoline If a gallon of gas costs $305\frac{1}{5}$¢, how much does 8 gallons cost?

42. Cost of Gasoline If a gallon of gas costs $308\frac{1}{5}$¢, how much does $\frac{1}{2}$ gallon cost?

43. Distance Traveled If a car can travel $32\frac{3}{4}$ miles on a gallon of gas, how far will it travel on 5 gallons of gas?

44. Distance Traveled If a new car can travel $20\frac{3}{10}$ miles on 1 gallon of gas, how far can it travel on $\frac{1}{2}$ gallon of gas?

45. Sewing If it takes $1\frac{1}{2}$ yards of material to make a pillow cover, how much material will it take to make 3 pillow covers?

46. Sewing If the material for the pillow covers in Problem 45 costs $2 a yard, how much will it cost for the material for the 3 pillow covers?

Find the area of each figure.

47.

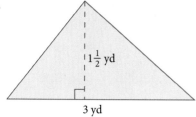

$1\frac{1}{2}$ yd

3 yd

48.

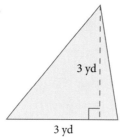

3 yd

3 yd

49. Write the numbers in order from smallest to largest.

$$2\frac{1}{8} \qquad \frac{5}{4} \qquad \frac{3}{4} \qquad 1\frac{1}{2}$$

50. Write the numbers in order from smallest to largest.

$$1\frac{3}{8} \qquad \frac{7}{8} \qquad \frac{7}{4} \qquad 1\frac{11}{16}$$

Nutrition The figure below shows nutrition labels for two different cans of corn.

Can 1

Nutrition Facts

Serving Size 1 cup
Servings Per Container About 2 ½

Amount Per Serving

Calories 133 Calories from fat 15

	% Daily Value*
Total Fat 3g	3%
Saturated Fat 1g	1%
Cholesterol 0mg	0%
Sodium 530mg	22%
Total Carbohydrate 30g	10%
Dietary Fiber 3g	13%
Sugars 4g	
Protein 4g	

Vitamin A 0%	•	Vitamin C 23%
Calcium 1%	•	Iron 8%

*Percent Daily Values are based on a 2,000 calorie diet

Can 2

Nutrition Facts

Serving Size 1 cup
Servings Per Container About 1 ½

Amount Per Serving

Calories 184 Calories from fat 10

	% Daily Value*
Total Fat 2g	2%
Saturated Fat 1g	1%
Cholesterol 0mg	0%
Sodium 730mg	30%
Total Carbohydrate 46g	15%
Dietary Fiber 3g	12%
Sugars 6g	
Protein 4g	

Vitamin A 0%	•	Vitamin C 20%
Calcium 1%	•	Iron 5%

*Percent Daily Values are based on a 2,000 calorie diet

51. Compare the total number of calories in the two cans of corn.

52. Compare the total amount of sugar in the two cans of corn.

53. Compare the total amount of sodium in the two cans of corn.

54. Compare the total amount of protein in the two cans of corn.

Getting Ready for the Next Section

55. Write as equivalent fractions with denominator 15.

 a. $\dfrac{2}{3}$ **b.** $\dfrac{1}{5}$ **c.** $\dfrac{3}{5}$ **d.** $\dfrac{1}{3}$

56. Write as equivalent fractions with denominator 12.

 a. $\dfrac{3}{4}$ **b.** $\dfrac{1}{3}$ **c.** $\dfrac{5}{6}$ **d.** $\dfrac{1}{4}$

57. Write as equivalent fractions with denominator 20.

 a. $\dfrac{1}{4}$ **b.** $\dfrac{3}{5}$ **c.** $\dfrac{9}{10}$ **d.** $\dfrac{1}{10}$

58. Write as equivalent fractions with denominator 24.

 a. $\dfrac{3}{4}$ **b.** $\dfrac{7}{8}$ **c.** $\dfrac{5}{8}$ **d.** $\dfrac{3}{8}$

Add or subtract the following fractions, as indicated.

59. $\dfrac{2}{3} + \dfrac{1}{5}$

60. $\dfrac{3}{4} + \dfrac{5}{6}$

61. $\dfrac{2}{3} + \dfrac{8}{9}$

62. $\dfrac{1}{4} + \dfrac{3}{5} + \dfrac{9}{10}$

63. $\dfrac{9}{10} - \dfrac{3}{10}$

64. $\dfrac{7}{10} - \dfrac{3}{5}$

65. $\dfrac{5}{12} - \dfrac{1}{6}$

66. $\dfrac{2}{3} - \dfrac{1}{4} - \dfrac{1}{6}$

One Step Further

To find the square of a mixed number, we first change the mixed number to an improper fraction, and then we square the result. For example, $\left(2\dfrac{1}{2} \right)^2 = \left(\dfrac{5}{2} \right)^2 = \dfrac{25}{4}$. If we are asked to write our answer as a mixed number, we write it as $6\dfrac{1}{4}$.

Find each of the following squares, and write your answers as mixed numbers.

67. $\left(1\dfrac{1}{2} \right)^2$

68. $\left(3\dfrac{1}{2} \right)^2$

69. $\left(1\dfrac{3}{4} \right)^2$

70. $\left(2\dfrac{3}{4} \right)^2$

71. The length of one side of a square is $1\dfrac{1}{2}$ ft. Write the area of the square as a mixed number.

72. The length of one side of a square is $2\dfrac{1}{2}$ ft. Write the area of the square as a mixed number.

73. The length of one side of a cube is $1\dfrac{1}{2}$ ft. Write the volume of the cube as a mixed number.

74. The length of one side of a cube is $1\dfrac{1}{3}$ ft. Write the volume of the cube as a mixed number.

Find the Mistake

Each sentence below contains a mistake. Circle the mistake and write the correct word(s) or numbers(s) on the line provided.

1. To multiply two mixed numbers, multiply the whole numbers and multiply the fractions.

2. Multiplying $4\frac{3}{8}$ and $9\frac{2}{7}$ gives us the mixed number $45\frac{1}{3}$. _____

3. To divide the mixed numbers $12\frac{2}{5}$ and $3\frac{12}{5}$, change the mixed numbers to improper fractions and then multiply numerators and denominators. _____

4. The answer to the division problem $3\frac{9}{14} \div 2$ written as a mixed number is $\frac{51}{28}$.

Landmark Review: Checking Your Progress

Find the following sums and differences and reduce to lowest terms.

1. $\frac{7}{10} + \frac{3}{10}$

2. $\frac{2}{5} - \frac{1}{5}$

3. $\frac{2}{3} + \frac{3}{5}$

4. $\frac{1}{2} + \frac{3}{4}$

5. $\frac{3}{5} + \frac{2}{3} - \frac{4}{7}$

6. $\frac{4+x}{3} - \frac{2}{3}$

7. Find the sum of $\frac{2}{3}, \frac{1}{5}$, and $\frac{1}{2}$.

8. Find the difference of $\frac{7}{10}$ and $\frac{3}{5}$.

Change each mixed number to an improper fraction.

9. $3\frac{5}{8}$

10. $4\frac{2}{3}$

11. $10\frac{1}{2}$

12. $1\frac{1}{4}$

Change each improper fraction to a mixed number.

13. $\frac{14}{3}$

14. $\frac{23}{5}$

15. $\frac{7}{2}$

16. $\frac{42}{17}$

Perform the indicated operations.

17. $4\frac{1}{4} \cdot 5\frac{1}{2}$

18. $3\frac{1}{3} \cdot 2\frac{5}{6}$

19. $5\frac{2}{3} \cdot 6\frac{5}{8}$

20. $5\frac{1}{4} \div 4\frac{3}{8}$

21. $3\frac{7}{10} \div 1\frac{3}{5}$

22. $4\frac{5}{8} \div 2\frac{2}{3}$

Dock diving is a sport where dogs compete to see who can jump the highest or the farthest off the end of a dock into a body of water. Many of the dogs' handlers use a chase object to lure the dog as far as possible off the dock. The handler commands the dog to begin a running start and then throws the chase object out over the water. The dog leaps after the object, and once he splashes into the water, an event official measures his jump distance. The official measures the dog's final distance from the end of the dock to the point where the base of the dog's tail first hit the water.

In this section, we will work addition and subtraction problems that involve mixed numbers. Let's say that the top two dogs jumped 21 feet $10\frac{1}{4}$ inches and 21 feet $11\frac{7}{8}$ inches. How much further did the latter dog jump? To solve this problem, we need to know how to subtract $10\frac{1}{4}$ from $11\frac{7}{8}$. After reading this section, you will be more comfortable working with mixed numbers such as these.

The notation we use for mixed numbers is especially useful for addition and subtraction. When adding and subtracting mixed numbers, we will assume you recall from section 3.1 how to go about finding a least common denominator (LCD).

**Video Examples
Section 2.8**

A Addition with Mixed Numbers

EXAMPLE 1 Add: $3\frac{2}{3} + 4\frac{1}{5}$.

Solution We begin by writing each mixed number showing the $+$ sign. We then apply the commutative and associative properties to rearrange the order and grouping.

$$3\frac{2}{3} + 4\frac{1}{5} = 3 + \frac{2}{3} + 4 + \frac{1}{5} \qquad \text{Expand each number to show the + sign.}$$

$$= 3 + 4 + \frac{2}{3} + \frac{1}{5} \qquad \text{Commutative property}$$

$$= (3 + 4) + \left(\frac{2}{3} + \frac{1}{5}\right) \quad \text{Associative property}$$

$$= 7 + \left(\frac{5 \cdot 2}{5 \cdot 3} + \frac{3 \cdot 1}{3 \cdot 5}\right) \quad \begin{array}{l}\text{Add } 3 + 4 = 7\text{; then multiply to get}\\ \text{the LCD.}\end{array}$$

$$= 7 + \left(\frac{10}{15} + \frac{3}{15}\right) \qquad \text{Write each fraction with the LCD.}$$

$$= 7 + \frac{13}{15} \qquad \text{Add the numerators.}$$

$$= 7\frac{13}{15} \qquad \begin{array}{l}\text{Write the answer in mixed-number}\\ \text{notation.}\end{array}$$

Practice Problems

1. Add: $2\frac{1}{3} + 3\frac{3}{4}$.

Answer

1. $6\frac{1}{12}$

As you can see, we obtain our result by adding the whole-number parts ($3 + 4 = 7$) and the fraction parts $\left(\frac{2}{3} + \frac{1}{5} = \frac{13}{15} \right)$ of each mixed number. Knowing this, we can save ourselves some writing by doing the same problem in columns.

$$3\frac{2}{3} = 3\frac{2 \cdot 5}{3 \cdot 5} = 3\frac{10}{15}$$

$$+\, 4\frac{1}{5} = 4\frac{1 \cdot 3}{5 \cdot 3} = 4\frac{3}{15} \qquad \textit{Write each fraction with LCD 15}$$

$$7\frac{13}{15} \qquad \textit{Add whole numbers, then add fractions.}$$

The second method shown above requires less writing and lends itself to mixed-number notation. We will use this method for the rest of this section.

EXAMPLE 2 Add: $5\frac{3}{4} + 9\frac{5}{6}$.

Solution The LCD for 4 and 6 is 12. Writing the mixed numbers in a column and then adding looks like this:

$$5\frac{3}{4} = 5\frac{3 \cdot 3}{4 \cdot 3} = 5\frac{9}{12}$$

$$+\, 9\frac{5}{6} = 9\frac{5 \cdot 2}{6 \cdot 2} = 9\frac{10}{12}$$

$$14\frac{19}{12}$$

The fraction part of the answer is an improper fraction. We rewrite it as a whole number and a proper fraction.

$$14\frac{19}{12} = 14 + \frac{19}{12} \qquad \textit{Write the mixed number with a + sign.}$$

$$= 14 + 1\frac{7}{12} \qquad \textit{Write } \tfrac{19}{12} \textit{ as a mixed number.}$$

$$= 15\frac{7}{12} \qquad \textit{Add 14 and 1.}$$

EXAMPLE 3 Add: $5\frac{2}{3} + 6\frac{8}{9}$.

Solution

$$5\frac{2}{3} = 5\frac{2 \cdot 3}{3 \cdot 3} = 5\frac{6}{9}$$

$$+\, 6\frac{8}{9} = 6\frac{8}{9} = 6\frac{8}{9}$$

$$11\frac{14}{9} = 12\frac{5}{9}$$

The last step involves writing $\frac{14}{9}$ as $1\frac{5}{9}$ and then adding 11 and 1 to get 12.

EXAMPLE 4 Add: $3\frac{1}{4} + 2\frac{3}{5} + 1\frac{9}{10}$.

Solution The LCD is 20. We rewrite each fraction as an equivalent fraction with denominator 20 and add.

$$3\frac{1}{4} = 3\frac{1 \cdot 5}{4 \cdot 5} = 3\frac{5}{20}$$

$$2\frac{3}{5} = 2\frac{3 \cdot 4}{5 \cdot 4} = 2\frac{12}{20}$$

$$+\, 1\frac{9}{10} = 1\frac{9 \cdot 2}{10 \cdot 2} = 1\frac{18}{20}$$

$$6\frac{35}{20} = 7\frac{15}{20} = 7\frac{3}{4} \qquad \textit{Reduce to lowest terms.}$$

$$\frac{35}{20} = 1\frac{15}{20} \qquad \textit{Change to a mixed number.}$$

2. Redo Practice Problem 1 using method 2.

Note Once you see how to change from a whole number and an improper fraction to a whole number and a proper fraction, you will be able to do this step without showing any work.

3. Add: $4\frac{3}{4} + 5\frac{3}{8}$.

4. Add: $2\frac{2}{3} + 3\frac{1}{4} + 5\frac{5}{6}$.

Answers

2. $6\frac{1}{12}$

3. $10\frac{1}{8}$

4. $11\frac{3}{4}$

We should note here that we could have worked each of the first four examples in this section by first changing each mixed number to an improper fraction and then adding as we did earlier in this chapter. To illustrate, if we were to work Example 4 this way, it would look like this:

$$3\frac{1}{4} + 2\frac{3}{5} + 1\frac{9}{10} = \frac{13}{4} + \frac{13}{5} + \frac{19}{10} \qquad \text{Change to improper fractions.}$$

$$= \frac{13 \cdot 5}{4 \cdot 5} + \frac{13 \cdot 4}{5 \cdot 4} + \frac{19 \cdot 2}{10 \cdot 2} \qquad \text{LCD is 20.}$$

$$= \frac{65}{20} + \frac{52}{20} + \frac{38}{20} \qquad \text{Equivalent fractions}$$

$$= \frac{155}{20} \qquad \text{Add numerators.}$$

$$= 7\frac{15}{20} = 7\frac{3}{4} \qquad \text{Change to a mixed number, and reduce.}$$

As you can see, the result is the same as the result we obtained in Example 4.

There are advantages to both methods. The method just shown works well when the whole-number parts of the mixed numbers are small. The vertical method shown in Examples 1–4 works well when the whole-number parts of the mixed numbers are large.

B Subtraction with Mixed Numbers

Subtraction with mixed numbers is very similar to addition with mixed numbers.

EXAMPLE 5 Subtract: $3\frac{9}{10} - 1\frac{3}{10}$.

Solution Because the denominators are the same, we simply subtract the whole numbers and subtract the fractions.

$$\begin{array}{r} 3\frac{9}{10} \\ -\ 1\frac{3}{10} \\ \hline 2\frac{6}{10} = 2\frac{3}{5} \end{array} \qquad \text{Reduce to lowest terms.}$$

An easy way to visualize the results in Example 5 is to imagine 3 dollar bills and 9 dimes in your pocket. If you spend 1 dollar and 3 dimes, you will have 2 dollars and 6 dimes left.

Ones Dimes

EXAMPLE 6 Subtract: $12\frac{7}{10} - 8\frac{3}{5}$.

Solution The common denominator is 10. We must rewrite $8\frac{3}{5}$ as an equivalent fraction with denominator 10.

$$12\frac{7}{10} = \quad 12\frac{7}{10} \quad = \quad 12\frac{7}{10}$$

$$-\ 8\frac{3}{5} = -\ 8\frac{3 \cdot 2}{5 \cdot 2} = -\ 8\frac{6}{10}$$

$$\rule{2cm}{0.4pt}$$

$$4\frac{1}{10}$$

5. Subtract: $4\frac{5}{8} - 1\frac{1}{8}$.

6. Subtract: $8\frac{1}{2} - 3\frac{1}{6}$.

Answers

5. $3\frac{1}{2}$

6. $5\frac{1}{3}$

7. Subtract: $12 - 8\frac{5}{9}$.

Note Convince yourself that 10 is the same as $9\frac{7}{7}$. The reason we choose to write the 1 we borrowed as $\frac{7}{7}$ is that the fraction we eventually subtracted from $\frac{7}{7}$ was $\frac{2}{7}$. Both fractions must have the same denominator, 7, so that we can subtract.

8. Subtract: $12\frac{1}{6} - 5\frac{5}{6}$.

9. Subtract: $9\frac{1}{3} - 3\frac{3}{4}$.

EXAMPLE 7 Subtract: $10 - 5\frac{2}{7}$.

Solution In order to have a fraction from which to subtract $\frac{2}{7}$, we borrow 1 from 10 and rewrite the 1 we borrow as $\frac{7}{7}$. The process looks like this:

$$10 = \quad 9\frac{7}{7} \quad \longleftarrow \text{We rewrite 10 as } 9 + 1, \text{ which is } 9 + \frac{7}{7} = 9\frac{7}{7}.$$

$$\underline{- \; 5\frac{2}{7} = - \; 5\frac{2}{7}} \quad \text{Then we can subtract as usual.}$$

$$4\frac{5}{7}$$

EXAMPLE 8 Subtract: $8\frac{1}{4} - 3\frac{3}{4}$.

Solution Because $\frac{3}{4}$ is larger than $\frac{1}{4}$, we again need to borrow 1 from the whole number. The 1 that we borrow from the 8 is rewritten as $\frac{4}{4}$, because 4 is the denominator of both fractions.

$$8\frac{1}{4} = \quad 7\frac{5}{4} \quad \longleftarrow \text{Borrow 1 in the form } \frac{4}{4}; \text{ then } \frac{4}{4} + \frac{1}{4} = \frac{5}{4}.$$

$$\underline{- \; 3\frac{3}{4} = - \; 3\frac{3}{4}}$$

$$4\frac{2}{4} = 4\frac{1}{2} \quad \text{Reduce to lowest terms.}$$

EXAMPLE 9 Subtract: $4\frac{3}{4} - 1\frac{5}{6}$.

Solution This is about as complicated as it gets with subtraction of mixed numbers. We begin by rewriting each fraction with the common denominator 12.

$$4\frac{3}{4} = \quad 4\frac{3 \cdot 3}{4 \cdot 3} = \quad 4\frac{9}{12}$$

$$\underline{- \; 1\frac{5}{6} = - \; 1\frac{5 \cdot 2}{6 \cdot 2} = - \; 1\frac{10}{12}}$$

Because $\frac{10}{12}$ is larger than $\frac{9}{12}$, we must borrow 1 from 4 in the form $\frac{12}{12}$ before we subtract.

$$4\frac{9}{12} = \quad 3\frac{21}{12} \quad \longleftarrow 4 = 3 + 1 = 3 + \frac{12}{12}, \text{ so } 4\frac{9}{12} = \left(3 + \frac{12}{12}\right) + \frac{9}{12}$$

$$\underline{- \; 1\frac{10}{12} = - \; 1\frac{10}{12}} \qquad\qquad\qquad = 3 + \left(\frac{12}{12} + \frac{9}{12}\right)$$

$$2\frac{11}{12} \qquad\qquad\qquad\qquad\qquad = 3 + \frac{21}{12}$$

$$\qquad\qquad\qquad\qquad\qquad\qquad\qquad = 3\frac{21}{12}$$

GETTING READY FOR CLASS

After reading through the preceding section, respond in your own words and in complete sentences.

A. Is it necessary to "borrow" when subtracting $1\frac{3}{10}$ from $3\frac{9}{10}$?

B. To subtract $1\frac{2}{7}$ from 10, it is necessary to rewrite 10 as what mixed number?

C. To subtract $11\frac{20}{30}$ from $15\frac{3}{30}$, it is necessary to rewrite $15\frac{3}{30}$ as what mixed number?

D. Rewrite $14\frac{19}{12}$ so that the fraction part is a proper fraction instead of an improper fraction.

Answers

7. $3\frac{4}{9}$

8. $6\frac{1}{3}$

9. $5\frac{7}{12}$

Vocabulary Review

Choose the correct words to fill in the blanks below.

proper columns addition sign improper LCD borrow

1. To add mixed numbers, rewrite each mixed number showing the _____, and then apply the commutative and associative properties.

2. Adding mixed numbers in _____ requires less writing than the method that uses the addition sign and grouping.

3. When adding mixed numbers in columns, (1) add the whole numbers, (2) write each fraction with the _____, and (3) add the fractions.

4. If your answer to a mixed-number addition problem is a(n) _____ fraction, rewrite it as a whole number and a _____ fraction.

5. When subtracting mixed numbers, if the fraction in the second mixed number is larger than the fraction in the first mixed number, _____ 1 from the whole number in the first mixed number.

Problems

A B Add and subtract the following mixed numbers as indicated.

1. $2\frac{1}{5} + 3\frac{3}{5}$

2. $8\frac{2}{9} + 1\frac{5}{9}$

3. $4\frac{3}{10} + 8\frac{1}{10}$

4. $5\frac{2}{7} + 3\frac{3}{7}$

5. $6\frac{8}{9} - 3\frac{4}{9}$

6. $12\frac{5}{12} + \left(-7\frac{1}{12}\right)$

7. $9\frac{1}{6} + 2\frac{5}{6}$

8. $9\frac{1}{4} - \left(-5\frac{3}{4}\right)$

9. $3\frac{5}{8} - 2\frac{1}{4}$

10. $7\frac{9}{10} - 6\frac{3}{5}$

11. $11\frac{1}{3} + 2\frac{5}{6}$

12. $1\frac{5}{8} + 2\frac{1}{2}$

13. $7\frac{5}{12} - 3\frac{1}{3}$

14. $7\frac{3}{4} - 3\frac{5}{12}$

15. $6\frac{1}{3} - 4\frac{1}{4}$

16. $5\frac{4}{5} - 3\frac{1}{3}$

17. $10\frac{5}{6} + 15\frac{3}{4}$

18. $11\frac{7}{8} - \left(-9\frac{1}{6}\right)$

19. $18\frac{1}{8} + \left(-6\frac{3}{4}\right)$

20. $10\frac{1}{3} - 4\frac{1}{6}$

21. $5\dfrac{2}{3}$

$+\ 6\dfrac{1}{3}$

22. $8\dfrac{5}{6}$

$+\ 9\dfrac{5}{6}$

23. $10\dfrac{13}{16}$

$-\ 8\dfrac{5}{16}$

24. $17\dfrac{7}{12}$

$-\ 9\dfrac{5}{12}$

25. $6\dfrac{1}{2}$

$+\ 2\dfrac{5}{14}$

26. $9\dfrac{11}{12}$

$+\ 4\dfrac{1}{6}$

27. $1\dfrac{5}{8}$

$+\ 1\dfrac{3}{4}$

28. $7\dfrac{6}{7}$

$+\ 2\dfrac{3}{14}$

29. $4\dfrac{2}{3}$

$+\ 5\dfrac{3}{5}$

30. $9\dfrac{4}{9}$

$+\ 1\dfrac{1}{6}$

31. $5\dfrac{4}{10}$

$-\ 3\dfrac{1}{3}$

32. $12\dfrac{7}{8}$

$-\ 3\dfrac{5}{6}$

A Find the following sums. (Add.)

33. $1\dfrac{1}{4} + 2\dfrac{3}{4} + 5$

34. $6 + 5\dfrac{3}{5} + 8\dfrac{2}{5}$

35. $7\dfrac{1}{10} + 8\dfrac{3}{10} + 2\dfrac{7}{10}$

36. $5\dfrac{2}{7} + 8\dfrac{1}{7} + 3\dfrac{5}{7}$

37. $\dfrac{3}{4} + 8\dfrac{1}{4} + 5$

38. $\dfrac{5}{8} + 1\dfrac{1}{8} + 7$

39. $3\dfrac{1}{2} + 8\dfrac{1}{3} + 5\dfrac{1}{6}$

40. $4\dfrac{1}{5} + 7\dfrac{1}{3} + 8\dfrac{1}{15}$

41. $8\dfrac{2}{3}$

$9\dfrac{1}{8}$

$+\ 6\dfrac{1}{4}$

42. $7\dfrac{3}{5}$

$8\dfrac{2}{3}$

$+\ 1\dfrac{1}{5}$

43. $6\dfrac{1}{7}$

$9\dfrac{3}{14}$

$+\ 12\dfrac{1}{2}$

44. $1\dfrac{5}{6}$

$2\dfrac{3}{4}$

$+\ 5\dfrac{1}{2}$

45. $10\dfrac{1}{20}$

$11\dfrac{4}{5}$

$+\ 15\dfrac{3}{10}$

46. $18\dfrac{7}{12}$

$19\dfrac{3}{16}$

$+\ 10\dfrac{2}{3}$

47. $10\dfrac{3}{4}$

$12\dfrac{5}{6}$

$+\ 9\dfrac{5}{8}$

48. $4\dfrac{5}{9}$

$9\dfrac{2}{3}$

$+\ 8\dfrac{5}{6}$

B The following problems all involve the concept of borrowing. Subtract in each case.

49. $8 - 1\frac{3}{4}$ 　　　**50.** $5 - 3\frac{1}{3}$ 　　　**51.** $15 - 5\frac{3}{10}$ 　　　**52.** $24 - 10\frac{5}{12}$

53. $8\frac{1}{4} - 2\frac{3}{4}$ 　　　**54.** $12\frac{3}{10} - 5\frac{7}{10}$ 　　　**55.** $9\frac{1}{3} - 8\frac{2}{3}$ 　　　**56.** $7\frac{1}{6} - 6\frac{5}{6}$

57. $4\frac{1}{4} - 2\frac{1}{3}$ 　　　**58.** $6\frac{1}{5} - 1\frac{2}{3}$ 　　　**59.** $9\frac{2}{3} - 5\frac{3}{4}$ 　　　**60.** $12\frac{5}{6} - 8\frac{7}{8}$

61. $16\frac{3}{4} - 10\frac{4}{5}$ 　　　**62.** $18\frac{5}{12} - 9\frac{3}{4}$ 　　　**63.** $10\frac{3}{10} - 4\frac{4}{5}$ 　　　**64.** $9\frac{4}{7} - 7\frac{2}{3}$

65. $13\frac{1}{6} - 12\frac{5}{8}$ 　　　**66.** $21\frac{2}{5} - 20\frac{5}{6}$ 　　　**67.** $19\frac{1}{4} - 8\frac{5}{6}$ 　　　**68.** $22\frac{7}{10} - 18\frac{4}{5}$

69. Find the difference between $6\frac{1}{5}$ and $2\frac{7}{10}$. 　　　**70.** Give the difference between $5\frac{1}{3}$ and $1\frac{5}{6}$.

71. Find the sum of $3\frac{1}{8}$ and $2\frac{3}{5}$. 　　　**72.** Find the sum of $1\frac{5}{6}$ and $3\frac{4}{9}$.

Applying the Concepts

73. Building Two pieces of molding $5\frac{7}{8}$ inches and $6\frac{3}{8}$ inches long are placed end to end. What is the total length of the two pieces of molding together?

74. Jogging A jogger runs $2\frac{1}{2}$ miles on Monday, $3\frac{1}{4}$ miles on Tuesday, and $2\frac{2}{5}$ miles on Wednesday. What is the jogger's total mileage for this 3-day period?

75. Horse Racing If a racehorse runs at both Churchill Downs and Keeneland racetracks, she will run $1\frac{3}{8}$ miles at Churchill Downs, and $1\frac{1}{2}$ miles at Keeneland. How much further will she run at Keeneland?

76. Triple Crown The three races that constitute the Triple Crown in horse racing are shown in the table.

　　a. Write the distances in order from smallest to largest.

　　b. How much longer is the Belmont Stakes race than the Preakness Stakes?

Race	Distance (miles)
Kentucky Derby	$1\frac{1}{4}$
Preakness Stakes	$1\frac{3}{16}$
Belmont Stakes	$1\frac{1}{2}$

Source: ESPN.com

77. Length of Jeans A pair of jeans is $32\frac{1}{2}$ inches long. How long are the jeans after they have been washed if they shrink $1\frac{1}{3}$ inches?

78. Manufacturing A clothing manufacturer has two rolls of cloth. One roll is $35\frac{1}{2}$ yards, and the other is $62\frac{5}{8}$ yards. What is the total number of yards in the two rolls?

Area and Perimeter The diagrams below show the dimensions of playing fields for the National Football League (NFL), the Canadian Football League, and arena football.

Football Fields

Arena — $28\frac{1}{3}$ yd — 50 yd

NFL — $53\frac{1}{3}$ yd — 100 yd

Canadian — 65 yd — 110 yd

79. Find the perimeter of each football field.

80. Find the area of each football field.

Stock Prices In March 1995, rumors that Michael Jordan would return to basketball sent stock prices for the companies whose products he endorses higher. The table below gives some of the details of those increases. Use the table to work Problems 81-84.

81. a. Find the difference in the price of Nike stock between March 13 and March 8.

b. If you owned 100 shares of Nike stock, how much more are the 100 shares worth on March 13 than on March 8?

82. a. Find the difference in price of General Mills stock between March 13 and March 8.

b. If you owned 1,000 shares of General Mills stock on March 8, how much more would they be worth on March 13?

Stock Prices for Companies with Michael Jordan Endorsements			
Company	Product Endorsed	Stock Price (Dollars) 3/8/95	3/13/95
Nike	Air Jordans	$74\frac{7}{8}$	$77\frac{3}{8}$
Quaker Oats	Gatorade	$32\frac{1}{4}$	$32\frac{5}{8}$
General Mills	Wheaties	$60\frac{1}{2}$	$63\frac{3}{8}$
McDonald's		$32\frac{7}{8}$	$34\frac{3}{8}$

83. If you owned 200 shares of McDonald's stock on March 8, how much more would they be worth on March 13?

84. If you owned 100 shares of McDonald's stock on March 8, how much more would they be worth on March 13?

Getting Ready for the Next Section

Multiply or divide as indicated.

85. $\dfrac{11}{8} \cdot \dfrac{29}{8}$

86. $\dfrac{3}{4} \div \dfrac{5}{6}$

87. $\dfrac{7}{6} \cdot \dfrac{12}{7}$

88. $10\dfrac{1}{3} \div 8\dfrac{2}{3}$

Combine.

89. $\dfrac{3}{4} + \dfrac{5}{8}$

90. $\dfrac{1}{2} + \dfrac{2}{3}$

91. $2\dfrac{3}{8} + 1\dfrac{1}{4}$

92. $3\dfrac{2}{3} + 4\dfrac{1}{3}$

Improving Your Quantitative Literacy

93. Horse Racing A column on horse racing in the *Daily News* in Los Angeles reported that the horse Action This Day ran 3 furlongs in $35\dfrac{1}{5}$ seconds and another horse, Halfbridled, went two-fifths of a second faster. How many seconds did it take Halfbridled to run 3 furlongs?

Find the Mistake

Each sentence below contains a mistake. Circle the mistake and write the correct word(s) or numbers(s) on the line provided.

1. To begin adding $3\dfrac{9}{14}$ and $5\dfrac{1}{3}$, write each mixed number with the addition sign and then apply the commutative and associative properties, such that $\left(3 + \dfrac{2}{5}\right) + \left(5 + \dfrac{1}{3}\right)$. _____

2. The final answer for the problem $5\dfrac{2}{3} + 6\dfrac{7}{9}$ is $11\dfrac{13}{9}$. _____

3. The first step when subtracting $8 - 3\dfrac{2}{7} = 4\dfrac{5}{7}$, is to borrow 1 from 8 in the form of $\dfrac{8}{8}$. _____

4. The work for the subtraction problem $5\dfrac{2}{3} - 2\dfrac{4}{5}$ is shown below. Circle the mistake and write the correct work on the lines provided.

$$5\dfrac{2}{3} = 5\dfrac{2 \cdot 5}{3 \cdot 5} = 5\dfrac{10}{15} = 5\left(\dfrac{10}{15} + \dfrac{15}{15}\right) = 5\dfrac{25}{15} \quad \underline{\hspace{5cm}}$$

$$-\ 2\dfrac{4}{5} = 2\dfrac{4 \cdot 3}{5 \cdot 3} = 2\dfrac{12}{15} \qquad\qquad \rightarrow \qquad 2\dfrac{12}{15} \quad \underline{\hspace{5cm}}$$

$$\overline{} \qquad 3\dfrac{13}{15} \quad \underline{\hspace{5cm}}$$

OBJECTIVES

A Simplify expressions involving fractions using the order of operations.

B Simplify complex fractions.

KEY WORDS

complex fraction

**Video Examples
Section 2.9**

Practice Problems

1. Simplify the expression.

$$4 + \left(1\frac{1}{3}\right)\left(4\frac{1}{2}\right)$$

Answer

1. 10

Each year, the coastal waters of Qingdao, China are infiltrated with a vibrant green algae bloom that covers more than 150 square miles. The algae thrive in high temperatures and in polluted waters caused by runoff from agriculture land and fish farms. The bloom sucks oxygen out of the water, threatening the local marine life. As it washes onshore and dries in the hot sun, it gives off a noxious odor that smells like rotten eggs. Soldiers and other volunteers work tirelessly to clear the water and the beaches of the algae before the bloom drastically damages the local ecosystem.

Suppose you have volunteered to help clean up a portion of the algae-affected coast. The area of the beach you'll be responsible for has a width of $5\frac{1}{4}$ feet and a length of $10\frac{2}{3}$ feet. You also need to clean part of the water that measures 20 square feet. You wonder how large the total area is for which you are responsible. The total area is given by the following expression:

$$20 + \left(5\frac{1}{4}\right)\left(10\frac{2}{3}\right)$$

In this section, we will learn how to simplify expressions, such as the one above.

A Order of Operations

Let's use the order of operations to simplify expressions that contain both fractions and mixed numbers.

EXAMPLE 1 Simplify the expression: $5 + \left(2\frac{1}{2}\right)\left(3\frac{2}{3}\right)$.

Solution The rule for order of operations indicates that we should multiply $2\frac{1}{2}$ times $3\frac{2}{3}$ and then add 5 to the result.

$$5 + \left(2\frac{1}{2}\right)\left(3\frac{2}{3}\right) = 5 + \left(\frac{5}{2}\right)\left(\frac{11}{3}\right) \qquad \textit{Change the mixed numbers to improper fractions.}$$

$$= 5 + \frac{55}{6} \qquad \textit{Multiply the improper fractions.}$$

$$= \frac{30}{6} + \frac{55}{6} \qquad \textit{Write 5 as } \frac{30}{6} \textit{ so both numbers have the same denominator.}$$

$$= \frac{85}{6} \qquad \textit{Add fractions by adding their numerators.}$$

$$= 14\frac{1}{6} \qquad \textit{Write the answer as a mixed number.}$$

EXAMPLE 2 Simplify: $\left(\frac{3}{4} + \frac{5}{8}\right)\left(2\frac{3}{8} + 1\frac{1}{4}\right)$.

Solution We begin by combining the numbers inside the parentheses.

$$\frac{3}{4} + \frac{5}{8} = \frac{3 \cdot 2}{4 \cdot 2} + \frac{5}{8} \qquad \text{and} \qquad 2\frac{3}{8} = \quad 2\frac{3}{8} \quad = \quad 2\frac{3}{8}$$

$$= \frac{6}{8} + \frac{5}{8} \qquad\qquad\qquad +1\frac{1}{4} = +1\frac{1 \cdot 2}{4 \cdot 2} = +1\frac{2}{8}$$

$$= \frac{11}{8} \qquad\qquad\qquad\qquad\qquad\qquad 3\frac{5}{8}$$

Now that we have combined the expressions inside the parentheses, we can complete the problem by multiplying the results.

$$\left(\frac{3}{4} + \frac{5}{8}\right)\left(2\frac{3}{8} + 1\frac{1}{4}\right) = \left(\frac{11}{8}\right)\left(3\frac{5}{8}\right)$$

$$= \frac{11}{8} \cdot \frac{29}{8} \qquad \text{Change } 3\tfrac{5}{8} \text{ to an improper fraction.}$$

$$= \frac{319}{64} \qquad \text{Multiply fractions.}$$

$$= 4\frac{63}{64} \qquad \text{Write the answer as a mixed number.}$$

EXAMPLE 3 Simplify: $\frac{3}{5} + \frac{1}{2}\left(3\frac{2}{3} + 4\frac{1}{3}\right)^2$.

Solution We begin by combining the expressions inside the parentheses.

$$\frac{3}{5} + \frac{1}{2}\left(3\frac{2}{3} + 4\frac{1}{3}\right)^2 = \frac{3}{5} + \frac{1}{2}(8)^2 \quad \text{The sum inside the parentheses is 8.}$$

$$= \frac{3}{5} + \frac{1}{2}(64) \quad \text{The square of 8 is 64.}$$

$$= \frac{3}{5} + 32 \qquad \tfrac{1}{2} \text{ of 64 is 32.}$$

$$= 32\frac{3}{5} \qquad\quad \text{The result is a mixed number.}$$

B Complex Fractions

DEFINITION complex fraction

A *complex fraction* is a fraction in which the numerator and/or the denominator are themselves fractions or combinations of fractions.

Each of the following is a complex fraction:

$$\frac{\frac{3}{4}}{\frac{5}{6}}, \qquad \frac{3 + \frac{1}{2}}{2 - \frac{3}{4}}, \qquad \frac{\frac{1}{2} + \frac{2}{3}}{\frac{3}{4} - \frac{1}{6}}$$

2. Simplify: $\left(\frac{2}{3} + \frac{5}{6}\right)\left(3\frac{1}{4} - 1\frac{1}{2}\right)$.

3. Simplify: $3\frac{5}{6} + \frac{1}{3}\left(4\frac{1}{3} + 2\frac{2}{3}\right)^2$.

Answers

2. $2\frac{5}{8}$

3. $20\frac{1}{6}$

4. Simplify: $\dfrac{\frac{2}{3}}{\frac{5}{6}}$.

EXAMPLE 4 Simplify: $\dfrac{\frac{3}{4}}{\frac{5}{6}}$.

Solution This is actually the same as the problem $\frac{3}{4} \div \frac{5}{6}$, because the bar between $\frac{3}{4}$ and $\frac{5}{6}$ indicates division. Therefore, it must be true that

$$\frac{\frac{3}{4}}{\frac{5}{6}} = \frac{3}{4} \div \frac{5}{6}$$

$$= \frac{3 \cdot 6}{4 \cdot 5}$$

$$= \frac{18}{20}$$

$$= \frac{9}{10}$$

As you can see, we continue to use properties we have developed previously when we encounter new situations. In Example 4, we use the fact that division by a number and multiplication by its reciprocal produce the same result. We are taking a new problem, simplifying a complex fraction, and thinking of it in terms of a problem we have done previously, division by a fraction.

5. Simplify: $\dfrac{\frac{1}{4} + \frac{1}{3}}{\frac{1}{2} - \frac{1}{3}}$.

EXAMPLE 5 Simplify: $\dfrac{\frac{1}{2} + \frac{2}{3}}{\frac{3}{4} - \frac{1}{6}}$.

Note We are going to simplify this complex fraction by two different methods. This is the first method.

Solution Let's decide to call the numerator of this complex fraction the *top* of the fraction and its denominator the *bottom* of the complex fraction. It will be less confusing if we name them this way. The LCD for all the denominators on the top and bottom is 12, so we can multiply the top and bottom of this complex fraction by 12 and be sure all the denominators will divide it exactly. This will leave us with only whole numbers on the top and bottom.

$$\frac{\frac{1}{2} + \frac{2}{3}}{\frac{3}{4} - \frac{1}{6}} = \frac{12\left(\frac{1}{2} + \frac{2}{3}\right)}{12\left(\frac{3}{4} - \frac{1}{6}\right)} \qquad \text{Multiply the top and bottom by the LCD.}$$

$$= \frac{12 \cdot \frac{1}{2} + 12 \cdot \frac{2}{3}}{12 \cdot \frac{3}{4} - 12 \cdot \frac{1}{6}} \qquad \text{Distributive property}$$

$$= \frac{6 + 8}{9 - 2} \qquad \text{Multiply each fraction by 12.}$$

$$= \frac{14}{7} \qquad \text{Add on the top and subtract on the bottom.}$$

$$= 2 \qquad \text{Reduce to lowest terms.}$$

The problem can be worked in another way also. We can simplify the top and bottom of the complex fraction separately. Simplifying the top, we have

$$\frac{1}{2} + \frac{2}{3} = \frac{1 \cdot 3}{2 \cdot 3} + \frac{2 \cdot 2}{3 \cdot 2} = \frac{3}{6} + \frac{4}{6} = \frac{7}{6}$$

Simplifying the bottom, we have

$$\frac{3}{4} - \frac{1}{6} = \frac{3 \cdot 3}{4 \cdot 3} - \frac{1 \cdot 2}{6 \cdot 2} = \frac{9}{12} - \frac{2}{12} = \frac{7}{12}$$

We now write the original complex fraction again using the simplified expressions for the top and bottom. Then we proceed as we did in Example 4.

$$\frac{\dfrac{1}{2} + \dfrac{2}{3}}{\dfrac{3}{4} - \dfrac{1}{6}} = \frac{\dfrac{7}{6}}{\dfrac{7}{12}}$$

$$= \frac{7}{6} \div \frac{7}{12} \qquad \textit{The divisor is } \tfrac{7}{12}.$$

$$= \frac{7}{6} \cdot \frac{12}{7} \qquad \textit{Replace } \tfrac{7}{12} \textit{ by its reciprocal and multiply.}$$

$$= \frac{7 \cdot 2 \cdot 6}{6 \cdot 7} \qquad \textit{Divide out common factors.}$$

$$= 2$$

Note The fraction bar that separates the numerator of the complex fraction from its denominator works like parentheses. If we were to rewrite this problem without it, we would write it like this:

$$\left(\frac{1}{2} + \frac{2}{3}\right) \div \left(\frac{3}{4} - \frac{1}{6}\right)$$

That is why we simplify the top and bottom of the complex fraction separately and then divide.

EXAMPLE 6 Simplify: $\dfrac{3 + \dfrac{1}{2}}{2 - \dfrac{3}{4}}$.

Solution The simplest approach here is to multiply both the top and bottom by the LCD for all fractions, which is 4.

$$\frac{3 + \dfrac{1}{2}}{2 - \dfrac{3}{4}} = \frac{4\left(3 + \dfrac{1}{2}\right)}{4\left(2 - \dfrac{3}{4}\right)} \qquad \textit{Multiply the top and bottom by 4.}$$

$$= \frac{4 \cdot 3 + 4 \cdot \dfrac{1}{2}}{4 \cdot 2 - 4 \cdot \dfrac{3}{4}} \qquad \textit{Distributive property}$$

$$= \frac{12 + 2}{8 - 3} \qquad \textit{Multiply each number by 4.}$$

$$= \frac{14}{5} \qquad \textit{Add on the top and subtract on the bottom.}$$

$$= 2\frac{4}{5}$$

6. Simplify: $\dfrac{4 - \dfrac{1}{3}}{2 + \dfrac{1}{2}}$.

EXAMPLE 7 Simplify: $\dfrac{10\dfrac{1}{3}}{8\dfrac{2}{3}}$.

Solution The simplest way to simplify this complex fraction is to think of it as a division problem.

$$\frac{10\dfrac{1}{3}}{8\dfrac{2}{3}} = 10\frac{1}{3} \div 8\frac{2}{3} \qquad \textit{Write with a } \div \textit{ symbol.}$$

$$= \frac{31}{3} \div \frac{26}{3} \qquad \textit{Change to improper fractions.}$$

$$= \frac{31}{3} \cdot \frac{3}{26} \qquad \textit{Write in terms of multiplication.}$$

$$= \frac{31 \cdot 3}{3 \cdot 26} \qquad \textit{Divide out the common factor 3.}$$

$$= \frac{31}{26} = 1\frac{5}{26} \qquad \textit{Answer as a mixed number.}$$

7. Simplify: $\dfrac{8\dfrac{3}{4}}{5\dfrac{1}{2}}$.

Answers

6. $1\dfrac{7}{15}$

7. $1\dfrac{13}{22}$

GETTING READY FOR CLASS

After reading through the preceding section, respond in your own words and in complete sentences.

A. What is a complex fraction?

B. Rewrite $\dfrac{\frac{5}{6}}{\frac{1}{3}}$ as a multiplication problem.

C. True or false? The rules for order of operations tell us to work inside parentheses first.

D. True or false? We find the LCD when we add or subtract fractions, but not when we multiply them. Explain.

Vocabulary Review

Choose the correct words to fill in the blanks below.

LCD simplify complex fraction

whole numbers bottom top

1. A _____ is a fraction in which the numerator and/or the denominator are themselves fractions or combinations of fractions.

2. We call the numerator of a complex fraction the _____, and the denominator the _____.

3. One method to simplify a complex fraction is to find the _____ of all the denominators in the top and bottom of the complex fraction.

4. Multiplying the top and bottom of a complex fraction by the LCD will leave only _____ on the top and the bottom.

5. A second method to _____ a complex fraction is to simplify the top and bottom of the complex fraction separately.

Problems

A Use the rule for order of operations to simplify each of the following.

1. $3 + \left(1\frac{1}{2}\right)\left(2\frac{2}{3}\right)$

2. $7 - \left(1\frac{3}{5}\right)\left(2\frac{1}{2}\right)$

3. $8 - \left(\frac{6}{11}\right)\left(1\frac{5}{6}\right)$

4. $10 + \left(2\frac{4}{5}\right)\left(\frac{5}{7}\right)$

5. $\frac{2}{3}\left(1\frac{1}{2}\right) + \frac{3}{4}\left(1\frac{1}{3}\right)$

6. $\frac{2}{5}\left(2\frac{1}{2}\right) + \frac{5}{8}\left(3\frac{1}{5}\right)$

7. $2\left(1\frac{1}{2}\right) + 5\left(6\frac{2}{5}\right)$

8. $4\left(5\frac{3}{4}\right) + 6\left(3\frac{5}{6}\right)$

9. $\left(\frac{3}{5} + \frac{1}{10}\right)\left[\frac{1}{2} - \left(-\frac{3}{4}\right)\right]$

10. $\left(\frac{2}{9} + \frac{1}{3}\right)\left(\frac{1}{5} + \frac{1}{10}\right)$

11. $\left(2 + \frac{2}{3}\right)\left(3 + \frac{1}{8}\right)$

12. $\left(3 - \frac{3}{4}\right)\left(3 + \frac{1}{3}\right)$

13. $\left(1 + \frac{5}{6}\right)\left(1 - \frac{5}{6}\right)$

14. $\left(2 - \frac{1}{4}\right)\left(2 + \frac{1}{4}\right)$

15. $\frac{2}{3} + \frac{1}{3}\left(2\frac{1}{2} + \frac{1}{2}\right)^2$

16. $\frac{3}{5} + \frac{1}{4}\left(2\frac{1}{2} - \frac{1}{2}\right)^3$

17. $2\frac{3}{8} + \frac{1}{2}\left(\frac{1}{3} + \frac{5}{3}\right)^3$

18. $8\frac{2}{3} + \frac{1}{3}\left(\frac{8}{5} + \frac{7}{5}\right)^2$

19. $2\left(\frac{1}{2} + \frac{1}{3}\right) + 3\left(\frac{2}{3} + \frac{1}{4}\right)$

20. $5\left(\frac{1}{5} + \frac{3}{10}\right) + 2\left(\frac{1}{10} + \frac{1}{2}\right)$

B Simplify each complex fraction as much as possible.

21. $\dfrac{\frac{2}{3}}{\frac{3}{4}}$

22. $\dfrac{\frac{5}{6}}{\frac{3}{12}}$

23. $\dfrac{\frac{2}{3}}{\frac{4}{3}}$

24. $\dfrac{\frac{7}{9}}{\frac{5}{9}}$

25. $\dfrac{\frac{11}{20}}{\frac{5}{10}}$

26. $\dfrac{\frac{9}{16}}{\frac{3}{4}}$

27. $\dfrac{\frac{1}{2} + \frac{1}{3}}{\frac{1}{2} - \frac{1}{3}}$

28. $\dfrac{\frac{1}{4} + \frac{1}{5}}{\frac{1}{4} - \frac{1}{5}}$

29. $\dfrac{\frac{5}{8} - \frac{1}{4}}{\frac{1}{8} + \frac{1}{2}}$

30. $\dfrac{\frac{3}{4} + \frac{1}{3}}{\frac{2}{3} + \frac{1}{6}}$

31. $\dfrac{\frac{9}{20} - \frac{1}{10}}{\frac{1}{10} + \frac{9}{20}}$

32. $\dfrac{\frac{1}{2} + \frac{2}{3}}{\frac{3}{4} + \frac{5}{6}}$

33. $\dfrac{1 + \frac{2}{3}}{1 - \frac{2}{3}}$

34. $\dfrac{5 - \frac{3}{4}}{2 + \frac{3}{4}}$

35. $\dfrac{2 + \frac{5}{6}}{5 - \frac{1}{3}}$

36. $\dfrac{9 - \frac{11}{5}}{3 + \frac{13}{10}}$

37. $\dfrac{3 + \frac{5}{6}}{1 + \frac{5}{3}}$

38. $\dfrac{10 + \frac{9}{10}}{5 + \frac{4}{5}}$

39. $\dfrac{\frac{1}{3} + \frac{3}{4}}{2 - \frac{1}{6}}$

40. $\dfrac{3 + \frac{5}{2}}{\frac{5}{6} + \frac{1}{4}}$

41. $\dfrac{\frac{5}{6}}{3 + \frac{2}{3}}$

42. $\dfrac{9 - \frac{3}{2}}{\frac{7}{4}}$

43. $\dfrac{8 + \frac{3}{4}}{\frac{5}{8}}$

44. $\dfrac{\frac{5}{6}}{3 - \frac{1}{3}}$

Simplify each of the following complex fractions.

45. $\dfrac{2\frac{1}{2} + \frac{1}{2}}{3\frac{3}{5} - \frac{2}{5}}$

46. $\dfrac{5\frac{3}{8} + \frac{5}{8}}{4\frac{1}{4} + 1\frac{3}{4}}$

47. $\dfrac{2 + 1\frac{2}{3}}{3\frac{5}{6} - 1}$

48. $\dfrac{5 + 8\frac{3}{5}}{2\frac{3}{10} + 4}$

49. $\dfrac{3\frac{1}{4} - 2\frac{1}{2}}{5\frac{3}{4} + 1\frac{1}{2}}$

50. $\dfrac{9\frac{3}{8} + 2\frac{5}{8}}{6\frac{1}{2} + 7\frac{1}{2}}$

51. $\dfrac{3\frac{1}{4} + 5\frac{1}{6}}{2\frac{1}{3} + 3\frac{1}{4}}$

52. $\dfrac{8\frac{5}{6} + 1\frac{2}{3}}{7\frac{1}{3} + 2\frac{1}{4}}$

53. $\dfrac{6\frac{2}{3} + 7\frac{3}{4}}{8\frac{1}{2} + 9\frac{7}{8}}$

54. $\dfrac{3\frac{4}{5} - 1\frac{9}{10}}{6\frac{5}{6} - 2\frac{3}{4}}$

55. $\dfrac{4\frac{1}{3} - 1\frac{2}{3}}{8\frac{1}{4} - 5\frac{1}{2}}$

56. $\dfrac{3\frac{1}{2} + 2\frac{1}{6}}{2 - \frac{5}{6}}$

57. What is twice the sum of $2\frac{1}{5}$ and $\frac{3}{6}$?

58. Find 3 times the difference of $1\frac{7}{9}$ and $\frac{2}{9}$.

59. Add $5\frac{1}{4}$ to the sum of $\frac{3}{4}$ and 2.

60. Subtract $\frac{7}{8}$ from the product of 2 and $3\frac{1}{2}$.

Applying the Concepts

61. Manufacturing A dress manufacturer usually buys two rolls of cloth, one of $32\frac{1}{2}$ yards and the other of $25\frac{1}{3}$ yards, to fill his weekly orders. If his orders double one week, how much of the cloth should he order? (Give the total yardage.)

62. Body Temperature Suppose your normal body temperature is $98\frac{3}{5}°$ Fahrenheit. If your temperature goes up $3\frac{1}{5}°$ on Monday and then down $1\frac{4}{5}°$ on Tuesday, what is your temperature on Tuesday?

Find the Mistake

Each sentence below contains a mistake. Circle the mistake and write the correct word(s) or numbers(s) on the line provided.

1. The first step to solving the problem $\frac{1}{4} - \left(2\frac{3}{8} - 1\frac{5}{8}\right)^2$ is to rewrite the problem so it looks like $\frac{1}{4} + \left(-2\frac{3}{8} + -1\frac{5}{8}\right)^2$.

2. To simplify $5\frac{2}{3} + \left(10\frac{1}{3} \cdot \frac{2}{3}\right)$ you must first change $10\frac{1}{3}$ into the reciprocal $\frac{31}{3}$ before multiplying by $\frac{2}{3}$.

3. A complex fraction is a fraction in which a fraction or combination of fractions appear in the denominator of the original fraction.

4. To simplify the complex fraction $\dfrac{\frac{1}{6} + \frac{2}{3}}{\frac{5}{6} + \frac{5}{12}}$, multiply the top and bottom of the fraction by 6.

Chapter 2 Summary

EXAMPLES

1. Each of the following is a fraction:
$$\frac{1}{2}, \quad \frac{3}{4}, \quad \frac{8}{1}, \quad \frac{7}{3}$$

Definition of Fractions [2.1]

A fraction is any number that can be written in the form $\frac{a}{b}$, where a and b are numbers and b is not 0. The number a is called the *numerator*, and the number b is called the *denominator*.

Properties of Fractions [2.1]

2. Change $\frac{3}{4}$ to an equivalent fraction with denominator 12.
$$\frac{3}{4} = \frac{3 \cdot 3}{4 \cdot 3} = \frac{9}{12}$$

Multiplying the numerator and the denominator of a fraction by the same nonzero number will produce an equivalent fraction. The same is true for dividing the numerator and denominator by the same nonzero number. In symbols the properties look like this:

If a, b, and c are numbers and b and c are not 0, then

Multiplication property for fractions $\quad \frac{a}{b} = \frac{a \cdot c}{b \cdot c}$

Division property for fraction $\quad \frac{a}{b} = \frac{a \div c}{b \div c}$

Fractions and the Number 1 [2.1]

3. $\frac{5}{1} = 5, \frac{5}{5} = 1$

If a represents any number, then
$$\frac{a}{1} = a \quad \text{and} \quad \frac{a}{a} = 1 \quad \text{(where } a \text{ is not 0)}$$

Reducing Fractions to Lowest Terms [2.2]

4. $\dfrac{90}{588} = \dfrac{2 \cdot 3 \cdot 3 \cdot 5}{2 \cdot 2 \cdot 3 \cdot 7 \cdot 7}$

$\qquad = \dfrac{3 \cdot 5}{2 \cdot 7 \cdot 7}$

$\qquad = \dfrac{15}{98}$

To reduce a fraction to lowest terms, factor the numerator and the denominator, and then divide both the numerator and denominator by any factors they have in common.

Multiplying Fractions [2.3]

5. $\dfrac{3}{5} \cdot \dfrac{4}{7} = \dfrac{3 \cdot 4}{5 \cdot 7} = \dfrac{12}{35}$

To multiply fractions, multiply numerators and multiply denominators.

The Area of a Triangle [2.3]

6. If the base of a triangle is 10 inches and the height is 7 inches, then the area is

$A = \dfrac{1}{2}bh$

$\quad = \dfrac{1}{2} \cdot 10 \cdot 7$

$\quad = 5 \cdot 7$

$\quad = 35$ square inches

The formula for the area of a triangle with base b and height h is

$$A = \frac{1}{2}\, bh$$

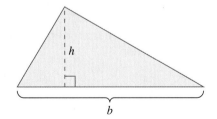

Reciprocals [2.4]

Any two numbers whose product is 1 are called *reciprocals*. The reciprocal of a is $\frac{1}{a}$ because their product is 1.

7. The reciprocal of $\frac{2}{3}$ is $\frac{3}{2}$.
$$\frac{2}{3} \cdot \frac{3}{2} = 1$$

Division with Fractions [2.4]

To divide by a fraction, you must multiply by its reciprocal. That is, the quotient of two fractions is defined to be the product of the first fraction with the reciprocal of the second fraction (the divisor).

8. $\frac{3}{8} \div \frac{1}{3} = \frac{3}{8} \cdot \frac{3}{1} = \frac{9}{8}$

COMMON MISTAKE

1. A common mistake made with division of fractions occurs when we multiply by the reciprocal of the first fraction instead of the reciprocal of the divisor. For example,

$$\frac{2}{3} \div \frac{5}{6} \neq \frac{3}{2} \cdot \frac{5}{6}$$

Remember, we perform division by multiplying by the reciprocal of the divisor (the fraction to the right of the division symbol).

2. If the answer to a problem turns out to be a fraction, that fraction should always be written in lowest terms. It is a mistake not to reduce to lowest terms.

Least Common Denominator (LCD) [2.5]

The *least common denominator* (LCD) for a set of denominators is the smallest number that is exactly divisible by each denominator.

Addition and Subtraction of Fractions [2.5]

To add (or subtract) two fractions with a common denominator, add (or subtract) numerators and use the common denominator. *In symbols:* If a, b, and c are numbers with c not equal to 0, then

$$\frac{a}{c} + \frac{b}{c} = \frac{a+b}{c} \quad \text{and} \quad \frac{a}{c} - \frac{b}{c} = \frac{a-b}{c}$$

9. $\frac{1}{8} + \frac{3}{8} = \frac{1+3}{8}$
$$= \frac{4}{8}$$
$$= \frac{1}{2}$$

Additional Facts about Fractions

1. In some books, fractions are called *rational numbers*.

2. Every whole number can be written as a fraction with a denominator of 1.

3. The commutative, associative, and distributive properties are true for fractions.

4. The word *of* as used in the expression "$\frac{2}{3}$ *of* 12" indicates that we are to multiply $\frac{2}{3}$ and 12.

5. Two fractions with the same value are called *equivalent fractions*.

Mixed-Number Notation [2.6]

A mixed number is the sum of a whole number and a fraction. The $+$ sign is not shown when we write mixed numbers; it is implied. The mixed number $4\frac{2}{3}$ is actually the sum $4 + \frac{2}{3}$.

Changing Mixed Numbers to Improper Fractions [2.6]

10. $4\frac{2}{3} = \dfrac{3 \cdot 4 + 2}{3} = \dfrac{14}{3}$

↑ Mixed number ↑ Improper fraction

To change a mixed number to an improper fraction, we write the mixed number showing the $+$ sign and add as usual. The result is the same if we multiply the denominator of the fraction by the whole number and add what we get to the numerator of the fraction, putting this result over the denominator of the fraction.

Changing an Improper Fraction to a Mixed Number [2.6]

11. Change $\dfrac{14}{3}$ to a mixed number.

To change an improper fraction to a mixed number, divide the denominator into the numerator. The quotient is the whole-number part of the mixed number. The fraction part is the remainder over the divisor.

Multiplication and Division with Mixed Numbers [2.7]

12. $2\frac{1}{3} \cdot 1\frac{3}{4} = \dfrac{7}{3} \cdot \dfrac{7}{4} = \dfrac{49}{12} = 4\frac{1}{12}$

To multiply or divide two mixed numbers, change each to an improper fraction and multiply or divide as usual.

Addition and Subtraction with Mixed Numbers [2.8]

13.
$$3\frac{4}{9} = 3\frac{4}{9} \qquad = 3\frac{4}{9}$$
$$+\ 2\frac{2}{3} = 2\frac{2 \cdot 3}{3 \cdot 3} = 2\frac{6}{9}$$
$$5\frac{10}{9} = 6\frac{1}{9}$$

Common denominator ↑ Add fractions.
Add whole numbers.

To add or subtract two mixed numbers, add or subtract the whole-number parts and the fraction parts separately. This is best done with the numbers written in columns.

Borrowing in Subtraction with Mixed Numbers [2.8]

14.
$$4\frac{1}{3} = \quad 4\frac{2}{6} = \quad 3\frac{8}{6}$$
$$-\ 1\frac{5}{6} = -1\frac{5}{6} = -1\frac{5}{6}$$
$$2\frac{3}{6} = 2\frac{1}{2}$$

It is sometimes necessary to borrow when doing subtraction with mixed numbers. We always change to a common denominator before we actually borrow.

Complex Fractions [2.9]

A fraction that contains a fraction in its numerator or denominator is called a *complex fraction*.

> **COMMON MISTAKE**
>
> 1. The most common mistake when working with fractions occurs when we try to add two fractions without using a common denominator. For example,
>
> $$\frac{2}{3} + \frac{4}{5} \neq \frac{2+4}{3+5}$$
>
> If the two fractions we are trying to add don't have the same denominators, then we *must* rewrite each one as an equivalent fraction with a common denominator. *We never add denominators when adding fractions.*
> **Note** We do *not* need a common denominator when multiplying fractions.
>
> 2. A common mistake when working with mixed numbers is to confuse mixed-number notation for multiplication of fractions. The notation $3\frac{2}{5}$ does *not* mean 3 *times* $\frac{2}{5}$. It means 3 *plus* $\frac{2}{5}$.
>
> 3. Another mistake occurs when multiplying mixed numbers. The mistake occurs when we don't change the mixed number to an improper fraction before multiplying and instead try to multiply the whole numbers and fractions separately.
>
> $$2\frac{1}{2} \cdot 3\frac{1}{3} = (2 \cdot 3) + \left(\frac{1}{2} \cdot \frac{1}{3}\right) \qquad \text{Mistake}$$
>
> $$= 6 + \frac{1}{6}$$
>
> $$= 6\frac{1}{6}$$
>
> Remember, the correct way to multiply mixed numbers is to first change to improper fractions and then multiply numerators and multiply denominators. This is correct:
>
> $$2\frac{1}{2} \cdot 3\frac{1}{3} = \frac{5}{2} \cdot \frac{10}{3} = \frac{50}{6} = 8\frac{2}{6} = 8\frac{1}{3} \qquad \text{Correct}$$

15.
$$\frac{4 + \dfrac{1}{3}}{2 - \dfrac{5}{6}} = \frac{6\left(4 + \dfrac{1}{3}\right)}{6\left(2 - \dfrac{5}{6}\right)}$$

$$= \frac{6 \cdot 4 + 6 \cdot \dfrac{1}{3}}{6 \cdot 2 - 6 \cdot \dfrac{5}{6}}$$

$$= \frac{24 + 2}{12 - 5}$$

$$= \frac{26}{7} = 3\frac{5}{7}$$

Reduce to lowest terms. [2.2]

1. $\dfrac{25}{125}$

2. $\dfrac{32}{128}$

Perform the indicated operations. Reduce all answers to lowest terms. [2.2, 2.3, 2.4]

3. $3 \cdot \dfrac{5}{8}$

4. $\dfrac{3}{16}\left(\dfrac{8}{9}\right)$

5. $\dfrac{3}{8} \cdot -\dfrac{12}{15}$

6. $\dfrac{7}{5}\left(-\dfrac{1}{14}\right)$

7. $\dfrac{1}{2} \cdot -\dfrac{3}{8} \cdot -\dfrac{4}{5}$

8. $\dfrac{7}{5} \cdot -\dfrac{12}{7} \cdot -\dfrac{1}{3}$

9. $\dfrac{8}{15} \div -\dfrac{2}{3}$

10. $\dfrac{3}{5} \div -9$

11. $16 \div \dfrac{8}{3}$

12. $\dfrac{36}{18} \div \dfrac{9}{2}$

Simplify each of the following as much as possible. [2.2, 2.3, 2.4]

13. $\dfrac{2}{3} \cdot \dfrac{1}{2} \div \dfrac{5}{6}$

14. $\dfrac{10}{7} \div 5 \cdot \dfrac{3}{4}$

15. $\dfrac{6}{11} \div \dfrac{3}{5} \div \dfrac{1}{2}$

16. $\dfrac{7}{8} \div \left(\dfrac{3}{2}\right)^2$

17. $\left(\dfrac{5}{18}\right)\left(\dfrac{4}{5}\right) \div \dfrac{10}{9}$

18. $\left(\dfrac{6}{5} \div \dfrac{9}{8}\right)\left(\dfrac{5}{8}\right)$

19. Find the area of the triangle. [2.3]

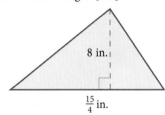

8 in.

$\dfrac{15}{4}$ in.

20. Sewing A dress requires $\dfrac{6}{5}$ yards of material to make. If you have 12 yards of material, how many dresses can you make? [2.4]

Perform the indicated operations. Reduce all answers to lowest terms. [2.5, 2.6]

21. $\dfrac{3}{8} + \dfrac{1}{8}$

22. $\dfrac{9}{14} - \dfrac{3}{14}$

23. $5 + \dfrac{5}{8}$

24. $\dfrac{5}{6} - \left(-\dfrac{2}{3}\right)$

25. $\dfrac{3}{8} + \dfrac{7}{12} + \dfrac{1}{3}$

26. $5\dfrac{1}{5} + 4\dfrac{3}{4}$

27. $\dfrac{4}{7} - \dfrac{3}{14}$

28. $\dfrac{5}{6} + \dfrac{2}{3} + 2$

29. $\dfrac{7}{18} - \dfrac{1}{6}$

30. $1\dfrac{5}{8} + \left(-2\dfrac{1}{4}\right)$

31. $9\dfrac{5}{6} + 6\dfrac{11}{18}$

32. $3\dfrac{5}{8} + 4\dfrac{1}{6} - 2\dfrac{4}{9}$

33. Sewing A dress that is $18\dfrac{5}{8}$ inches long is shortened by $2\dfrac{3}{4}$ inches. What is the new length of the dress? [2.5]

34. Find the perimeter of the triangle below. [2.5]

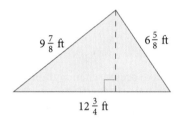

$9\dfrac{7}{8}$ ft $6\dfrac{5}{8}$ ft

$12\dfrac{3}{4}$ ft

Simplify each of the following as much as possible. [2.7, 2.8, 2.9]

35. $5 + 2\left(4\dfrac{3}{4}\right)$

36. $\left(4\dfrac{3}{4} + \dfrac{1}{2}\right)\left(6\dfrac{5}{6} - 4\dfrac{1}{3}\right)$

37. $\dfrac{\dfrac{1}{12} + \dfrac{3}{4}}{\dfrac{5}{6} - \dfrac{1}{3}}$

38. $6 - 3\left(1\dfrac{4}{5}\right)$

39. $\dfrac{\dfrac{9}{16}}{5\dfrac{1}{4}}$

40. $\dfrac{3\dfrac{1}{6} + \dfrac{1}{3}}{5\dfrac{5}{12} - 2\dfrac{1}{6}}$

The illustration shows what new iPhone buyers had as their previous phone.

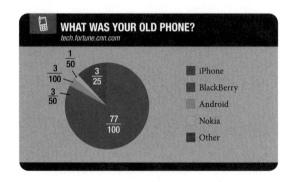

WHAT WAS YOUR OLD PHONE?
tech.fortune.cnn.com

$\dfrac{1}{50}$

$\dfrac{3}{100}$

$\dfrac{3}{50}$

$\dfrac{3}{25}$

$\dfrac{77}{100}$

■ iPhone
■ BlackBerry
■ Android
□ Nokia
■ Other

41. Write the fraction of iPhone buyers who had a Blackberry as their previous phone as an equivalent fraction with a denominator of 100.

42. Write the fraction of iPhone buyers who previously had a Nokia phone as an equivalent fraction with a denominator of 200.

Decimals

3

3.1 Decimal Notation,
Place Value, and Rounding

3.2 Addition and Subtraction
with Decimals

3.3 Multiplication with
Decimals

3.4 Division with Decimals

3.5 Fractions and Decimals

3.6 Square Roots and
The Pythagorean Theorem

The ancient Egyptian culture required that the body of a ruler be housed in a large tomb; the more impressive the structure, the smoother that ruler's journey into the afterlife would be. Thus explains the spectacular architectural wonders that are the Egyptian Pyramids. The largest of Egypt's famous pyramids, the Great Pyramid of Giza is believed to be the tomb of King Khufu and is recognized as one of the seven wonders of the ancient world. Khufu began construction on his pyramid soon after taking his throne, completing it approximately 20 years later around 2560 BC.

The current height of the Great Pyramid measures 138.8 meters. Historians believe that erosion has shortened the pyramid throughout its 4,000-year history, and that its original height was 146.5 meters. Suppose you wanted to calculate how much volume was lost due to this erosion. You would need to know that each side s of the pyramid measures 230.4 meters in length and that the area of the base B of a pyramid can be found using the formula

$$B = s \cdot s$$

You would also need to know that volume V of a pyramid can be found with the formula

$$V = \frac{1}{3}B \cdot h$$

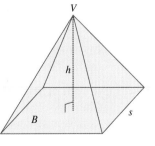

where h is the height of the pyramid at its apex. If you wanted to find the volume of the Great Pyramid when it was built and its volume now, you will need to know how to multiply and subtract numbers containing decimals. Decimal numbers are what we will explore in this chapter.

Video Examples Section 3.1

Image © sxc.hu, shho, 2009

A person's handwriting is extremely difficult to mimic. To make copying signatures of the world's elite businessmen and women even more impossible, a company based in Germany has produced a pen with plant DNA embedded in the ink. Each pen sold has a distinctively different type of DNA, making words written with the pen distinguishable from words written with other pens. A forensic test is needed to view the plant DNA, but uniqueness of each pen gives peace of mind that a forgery is less likely to happen.

Each pen containing plant DNA sells for $15,850. If this pen sold in San Luis Obispo, California, the sales tax would be an additional $1,386.88. In this chapter, we will focus our attention on decimals. Anyone who has used money in the United States has worked with decimals already. For example, the sales tax on the pen contains a decimal:

$$\$1386.88$$
Decimal point

What is interesting and useful about decimals is their relationship to fractions and to powers of ten. The work we have done up to now—especially our work with fractions—can be used to develop the properties of decimal numbers.

A Decimal Notation and Place Value

Previously you have used the idea of place value for the digits in a whole number. The name and the place value of each of the first seven columns in our number system is shown here:

Millions Column	Hundred Thousands Column	Ten Thousands Column	Thousands Column	Hundreds Column	Tens Column	Ones Column
1,000,000	100,000	10,000	1,000	100	10	1

As we move from right to left, we multiply by 10 each time. The value of each column is 10 times the value of the column on its right, with the rightmost column being 1. Up until now we have always looked at place value as increasing by a factor of 10 each time we move one column to the left.

Ten Thousands		Thousands		Hundreds		Tens		Ones
10,000	←	1,000	←	100	←	10	←	1
	Multiply by 10		Multiply by 10		Multiply by 10		Multiply by 10	

To understand the idea behind decimal numbers, we notice that moving in the opposite direction, from left to right, we *divide* by 10 each time.

Ten Thousands		Thousands		Hundreds		Tens		Ones
10,000	→	1,000	→	100	→	10	→	1
	Divide by 10		Divide by 10		Divide by 10		Divide by 10	

If we keep going to the right, the next column will have to be

$$1 \div 10 = \frac{1}{10} \qquad \text{Tenths}$$

The next one after that will be

$$\frac{1}{10} \div 10 = \frac{1}{10} \cdot \frac{1}{10} = \frac{1}{100} \qquad \text{Hundredths}$$

After that, we have

$$\frac{1}{100} \div 10 = \frac{1}{100} \cdot \frac{1}{10} = \frac{1}{1,000} \qquad \text{Thousandths}$$

We could continue this pattern as long as we wanted. We simply divide by 10 to move one column to the right. (And remember, dividing by 10 gives the same result as multiplying by $\frac{1}{10}$.)

To show where the ones column is, we use a *decimal point* between the ones column and the tenths column.

Thousands	Hundreds	Tens	Ones		Tenths	Hundredths	Thousandths	Ten Thousandths	Hundred Thousandths
1,000	100	10	1	↑ Decimal point	$\frac{1}{10}$	$\frac{1}{100}$	$\frac{1}{1,000}$	$\frac{1}{10,000}$	$\frac{1}{100,000}$

The ones column can be thought of as the middle column, with columns larger than 1 to the left and columns smaller than 1 to the right. The first column to the right of the ones column is the tenths column, the next column to the right is the hundredths column, the next is the thousandths column, and so on. The decimal point is always written between the ones column and the tenths column.

We can use the place value of decimal fractions to write them in expanded form.

EXAMPLE 1 Write 423.576 in expanded form.

Solution $423.576 = 400 + 20 + 3 + \dfrac{5}{10} + \dfrac{7}{100} + \dfrac{6}{1,000}$

EXAMPLE 2 Write each number in words.

a. 0.4 **b.** 0.04 **c.** 0.004

Solution

a. 0.4 is "four tenths."

b. 0.04 is "four hundredths."

c. 0.004 is "four thousandths."

Note Because the digits to the right of the decimal point have fractional place values, numbers with digits to the right of the decimal point are called *decimal fractions*. In this book, we will also call them *decimal numbers*, or simply *decimals* for short.

Practice Problems

1. Write 18.439 in expanded form.

2. Write each number in words.
 a. 0.1
 b. 0.01
 c. 0.001

Answers

1. $10 + 8 + \dfrac{4}{10} + \dfrac{3}{100} + \dfrac{9}{1000}$

2. **a.** One tenth
 b. One hundredth
 c. One thousandth

3. Write each number in words.
 a. 4.2
 b. 4.12
 c. 4.012

4. Write 10.805 in words.

5. Write 14.0836 in words.

6. Write each number as a fraction or mixed number. Do not reduce to lowest terms.
 a. 0.08

 b. 4.206

 c. 12.5814

When a decimal number contains digits to the left of the decimal point, we use the word "and" to indicate where the decimal point is when writing the number in words.

EXAMPLE 3 Write each number in words.
a. 5.4 **b.** 5.04 **c.** 5.004
Solution
a. 5.4 is "five and four tenths."
b. 5.04 is "five and four hundredths."
c. 5.004 is "five and four thousandths."

EXAMPLE 4 Write 3.64 in words.
Solution The number 3.64 is read "three and sixty-four hundredths." The place values of the digits are as follows:

$$
\begin{array}{ccc}
3 & . & 6 \quad\quad 4 \\
\uparrow & & \uparrow \quad\quad\ \ \nwarrow \\
3\text{ ones} & & 6\text{ tenths} \quad 4\text{ hundredths}
\end{array}
$$

We read the decimal part as "sixty-four hundredths" because

$$6\text{ tenths} + 4\text{ hundredths} = \frac{6}{10} + \frac{4}{100} = \frac{60}{100} + \frac{4}{100} = \frac{64}{100}$$

EXAMPLE 5 Write 25.4936 in words.
Solution Using the idea given in Example 4, we write 25.4936 in words as "twenty-five and four thousand, nine hundred thirty-six ten thousandths."

In order to understand addition and subtraction of decimals in the next section, we need to be able to convert decimal numbers to fractions or mixed numbers.

EXAMPLE 6 Write each number as a fraction or a mixed number. Do not reduce to lowest terms.
a. 0.004 **b.** 3.64 **c.** 25.4936
Solution
a. Because 0.004 is 4 thousandths, we write

$$0.004 = \frac{4}{1{,}000}$$

Three digits after the decimal point ↑ ↖ Three zeros

b. Looking over the work in Example 4, we can write

$$3.64 = 3\frac{64}{100}$$

Two digits after the decimal point ↑ ↖ Two zeros

c. From the way in which we wrote 25.4936 in words in Example 5, we have

$$25.4936 = 25\frac{4936}{10{,}000}$$

Four digits after the decimal point ↑ ↑ Four zeros

B Rounding Decimal Numbers

The rule for rounding decimal numbers is similar to the rule for rounding whole numbers. If the digit in the column to the right of the one we are rounding to is 5 or more, we add 1 to the digit in the column we are rounding to; otherwise, we leave it alone. We then replace all digits to the right of the column we are rounding to with zeros if they are to the left of the decimal point; otherwise, we simply delete them. Table 1 illustrates the procedure.

Table 1

Rounded to the Nearest

Number	Whole Number	Tenth	Hundredth
24.785	25	24.8	24.79
2.3914	2	2.4	2.39
0.98243	1	1.0	0.98
14.0942	14	14.1	14.09
0.545	1	0.5	0.55

EXAMPLE 7 Round 9,235.492 to the nearest hundred.

Solution The number next to the hundreds column is 3, which is less than 5. We change all digits to the right to 0, and we can drop all digits to the right of the decimal point, so we write

9,200

EXAMPLE 8 Round 0.00346 to the nearest ten thousandth.

Solution Because the number to the right of the ten thousandths column is more than 5, we add 1 to the 4 and get

0.0035

GETTING READY FOR CLASS

After reading through the preceding section, respond in your own words and in complete sentences.

A. Write 754.326 in expanded form.

B. Write $400 + 70 + 5 + \frac{1}{10} + \frac{3}{100} + \frac{7}{1,000}$ in decimal form.

C. Write seventy-two and three tenths in decimal form.

D. How many places to the right of the decimal point is the hundredths column?

7. Round 8,456.085 to the nearest hundred.

8. Round 8,456.085 to the nearest hundredth.

EXERCISE SET 3.1

Vocabulary Review

Choose the correct words to fill in the blanks below.

| hundredths | left | decimal point | fractional |
| tenths | right | thousandths | place value |

1. Digits to the right of a decimal point have _____ place values.
2. A _____ is used to separate the ones column and the tenths column.
3. In the decimal number 0.036, the 6 is in the _____ column.
4. In the decimal number 4.169, the 1 is in the _____ column.
5. In the decimal number 10.0977, the 9 is in the _____ column.
6. We use _____ of decimal fractions to write them in expanded form.
7. When a decimal number contains digits to the _____ of the decimal point, we use the word *and* to indicate where the decimal point is when writing the number in words.
8. When rounding decimal numbers, if the digit in the column to the _____ of the one we are rounding to is 5 or more, we add 1 to the digit in the column to which we are rounding.

Problems

A Write out the name of each number in words.

1. 0.3 **2.** 0.03 **3.** 0.015 **4.** 0.0015

5. 3.4 **6.** 2.04 **7.** 52.7 **8.** 46.8

Write each number as a fraction or a mixed number. Do not reduce your answers.

9. 405.36 **10.** 362.78 **11.** 9.009 **12.** 60.06

13. 1.234 **14.** 12.045 **15.** 0.00305 **16.** 2.00106

Give the place value of the 5 in each of the following numbers.

17. 458.327 **18.** 327.458 **19.** 29.52 **20.** 25.92 **21.** 0.00375

22. 0.00532 **23.** 275.01 **24.** 0.356 **25.** 539.76 **26.** 0.123456

Write each of the following as a decimal number.

27. Fifty-five hundredths

28. Two hundred thirty-five ten thousandths

29. Six and nine tenths

30. Forty-five thousand and six hundred twenty-one thousandths

31. Eleven and eleven hundredths

32. Twenty-six thousand, two hundred forty-five and sixteen hundredths

33. One hundred and two hundredths

34. Seventy-five and seventy-five hundred thousandths

35. Three thousand and three thousandths

36. One thousand, one hundred eleven and one hundred eleven thousandths

B Complete the following table.

	Rounded to the Nearest			
Number	Whole Number	Tenth	Hundredth	Thousandth
37. 47.5479				
38. 100.9256				
39. 0.8175				
40. 29.9876				
41. 0.1562				
42. 128.9115				
43. 2,789.3241				
44. 0.8743				
45. 99.9999				
46. 71.7634				

Applying the Concepts

47. Penny Weight If you have a penny dated anytime from 1959 through 1982, its original weight was 3.11 grams. If the penny has a date of 1983 or later, the original weight was 2.5 grams. Write the two weights in words.

48. 100 Meters The chart shows some close finishes at the 2010 Winter Olympics in Vancouver. Use the information to answer the following questions.

a. What is the place value of the 6 in the finish of the Men's Slalom?

b. Write the the difference in times for the Defago/Svindal finish in words.

49. Speed of Light The speed of light is 186,282.3976 miles per second. Round this number to the nearest hundredth.

50. Halley's Comet Halley's comet was seen from the earth during 1986. It will be another 76.1 years before it returns. Write 76.1 in words.

51. Nutrition A 50-gram egg contains 0.15 milligram of riboflavin. Write 0.15 in words.

52. Nutrition One medium banana contains 0.64 milligram of Vitamin B6. Write 0.64 in words.

53. Gasoline Prices The bar chart below was created from a survey by the U.S. Department of Energy's Energy Information Administration during four weeks in 2011. It gives the average price of regular gasoline for the United States on each Monday throughout the four week period. Use the information in the chart to fill in the table.

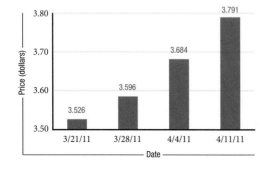

Price of 1 Gallon of Regular Gasoline	
Date	Price (Dollars)
3/21/11	
3/28/11	
4/4/11	
4/11/11	

54. Speed and Time The bar chart below was created from data given by *Car and Driver* magazine. It gives the minimum time in seconds for a Toyota Echo to reach various speeds from a complete stop. Use the information in the chart to fill in the table.

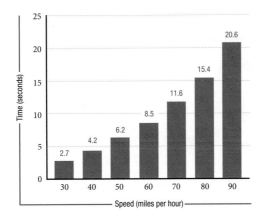

Speed (miles per hour)	Time (seconds)
30	
40	
50	
60	
70	
80	
90	

For each pair of numbers, place the correct symbol, < or >, between the numbers.

55. a. 0.02 0.2

 b. 0.3 0.032

56. a. 0.45 0.5

 b. 0.5 0.56

57. Write the following numbers in order from smallest to largest.

 0.02 0.05 0.025 0.052 0.005 0.002

58. Write the following numbers in order from smallest to largest.

 0.2 0.02 0.4 0.04 0.42 0.24

59. Which of the following numbers will round to 7.5?

 7.451 7.449 7.54 7.56

60. Which of the following numbers will round to 3.2?

 3.14999 3.24999 3.279 3.16111

Change each decimal to a fraction, and then reduce to lowest terms.

61. 0.25

62. 0.75

63. 0.125

64. 0.375

65. 0.625

66. 0.0625

67. 0.875

68. 0.1875

Estimating For each pair of numbers, choose the number that is closest to 10.

69. 9.9 and 9.99

70. 8.5 and 8.05

71. 10.5 and 10.05

72. 10.9 and 10.99

Estimating For each pair of numbers, choose the number that is closest to 0.

73. 0.5 and 0.05

74. 0.10 and 0.05

75. 0.01 and 0.02

76. 0.1 and 0.01

Getting Ready for the Next Section

In the next section we will do addition and subtraction with decimals. To understand the process of addition and subtraction, we need to understand the process of addition and subtraction with mixed numbers.

Find each of the following sums and differences. (Add or subtract.)

77. $4\dfrac{3}{10} + 2\dfrac{1}{100}$

78. $5\dfrac{35}{100} + 2\dfrac{3}{10}$

79. $8\dfrac{5}{10} - 2\dfrac{4}{100}$

80. $6\dfrac{3}{100} - 2\dfrac{125}{1,000}$

81. $5\dfrac{1}{10} + 6\dfrac{2}{100} + 7\dfrac{3}{1,000}$

82. $4\dfrac{27}{100} + 6\dfrac{3}{10} + 7\dfrac{123}{1,000}$

Find the Mistake

Each sentence below contains a mistake. Circle the mistake and write the correct word(s) or numbers(s) on the line provided.

1. To move a place value from the tens column to the hundreds column, you must divide by ten. _____

2. The decimal 0.09 can be written as the fraction $\dfrac{9}{1,000}$. _____

3. The decimal 142.9643 written as a mixed number is $142\dfrac{9,643}{1,000}$. _____

4. Rounding the decimal 0.06479 to the nearest tenth gives us 0.065. _____

The chart shows the top finishing times for the women's 400-meter race during the 2009 World Track and Field Championship in Berlin. In order to analyze the different finishing times, it is important that you are able to add and subtract decimals, and that is what we will cover in this section.

Runner	Time (seconds)
Sanya Richards (USA)	49.00
Shericka Williams (Jamaica)	49.32
Antonia Krivoshapka (Russia)	49.71
Novlene Williams-Mills (Jamaica)	49.77

A Addition of Decimals

Suppose you are earning $8.50 an hour and you receive a raise of $1.25 an hour. Your new hourly rate of pay is

$$\begin{array}{r} \$8.50 \\ +\ \$1.25 \\ \hline \$9.75 \end{array}$$

To add the two rates of pay, we align the decimal points, and then add in columns.

To see why this is true in general, we can use mixed-number notation.

$$8.50 = 8\frac{50}{100}$$
$$+\ 1.25 = 1\frac{25}{100}$$
$$9\frac{75}{100} = 9.75$$

We can visualize the mathematics above by thinking in terms of money.

**Video Examples
Section 3.2**

Practice Problems

1. Add by first changing to fractions: $16.47 + 5.9 + 45.842$.

EXAMPLE 1 Add by first changing to fractions: $25.43 + 2.897 + 379.6$.

Solution We first change each decimal to a mixed number. We then write each fraction using the least common denominator and add as usual.

$$25.43 = 25\frac{43}{100} = 25\frac{430}{1,000}$$

$$2.897 = 2\frac{897}{1,000} = 2\frac{897}{1,000}$$

$$+\ 379.6 = 379\frac{6}{10} = 379\frac{600}{1,000}$$

$$406\frac{1,927}{1,000} = 407\frac{927}{1,000} = 407.927$$

Again, the result is the same if we just line up the decimal points and add as if we were adding whole numbers.

$$\begin{array}{r} 25.430 \\ 2.897 \\ +\ 379.600 \\ \hline 407.927 \end{array}$$

Notice that we can fill in zeros on the right to help keep the numbers in the correct columns. Doing this does not change the value of any of the numbers.

Note: The decimal point in the answer is directly below the decimal points in the problem.

B Subtraction of Decimals

The same thing would happen if we were to subtract two decimal numbers. We can use these facts to write a rule for addition and subtraction of decimal numbers.

> **RULE** Addition (or Subtraction) of Decimal Numbers
>
> To add (or subtract) decimal numbers, we line up the decimal points and add (or subtract) as usual. The decimal point in the result is written directly below the decimal points in the problem.

We will use this rule for the rest of the examples in this section.

2. Subtract: $42.809 - 13.658$.

EXAMPLE 2 Subtract: $39.812 - 14.236$.

Solution We write the numbers vertically, with the decimal points lined up, and subtract as usual.

$$\begin{array}{r} 39.812 \\ -\ 14.236 \\ \hline 25.576 \end{array}$$

3. Add: $9 + 1.05 + 8.7 + 3.86$.

EXAMPLE 3 Add: $8 + 0.002 + 3.1 + 0.04$.

Solution To make sure we keep the digits in the correct columns, we can write zeros to the right of the rightmost digits.

$$8 = 8.000$$
$$3.1 = 3.100$$
$$0.04 = 0.040$$

Writing the extra zeros here is really equivalent to finding a common denominator for the fractional parts of the original four numbers—now we have a thousandths column in all the numbers.

Answers

1. $68\frac{212}{1,000} = 68.212$

2. 29.151

3. 22.61

This doesn't change the value of any of the numbers, and it makes our task easier. Now we have

$$
\begin{array}{r}
8.000 \\
0.002 \\
3.100 \\
+\ 0.040 \\
\hline
11.142
\end{array}
$$

EXAMPLE 4 Subtract: $5.9 - 3.0814$.

Solution In this case, it is very helpful to write 5.9 as 5.9000, since we will have to borrow in order to subtract.

$$
\begin{array}{r}
5.9000 \\
-\ 3.0814 \\
\hline
2.8186
\end{array}
$$

EXAMPLE 5 Subtract 3.09 from the sum of 9 and 5.472.

Solution Writing the problem in symbols, we have

$$
(9 + 5.472) - 3.09 = 14.472 - 3.09
$$
$$
= 11.382
$$

EXAMPLE 6 Add: $-4.75 + (-2.25)$.

Solution Because both signs are negative, we add absolute values. The answer will be negative.

$$
-4.75 + (-2.25) = -7.00
$$

EXAMPLE 7 Add: $3.42 + (-6.89)$.

Solution The signs are different, so we subtract the smaller absolute value from the larger absolute value. The answer will be negative, because 6.89 is larger than 3.42 and the sign in front of 6.89 is $-$.

$$
3.42 + (-6.89) = -3.47
$$

4. Subtract: $3.7 - 1.9034$.

5. Subtract 8.92 from the sum of 12 and 5.01.

6. Add: $-3.5 + (-1.25)$.

7. Add: $6.58 + (-10.47)$.

GETTING READY FOR CLASS

After reading through the preceding section, respond in your own words and in complete sentences.

A. When adding numbers with decimals, why is it important to line up the decimal points?

B. Write 379.6 in mixed-number notation.

C. Why do we line up the decimals when we add or subtract vertically?

D. How many quarters does the decimal 0.75 represent?

Answers
4. 1.7966
5. 8.09
6. -4.75
7. -3.89

EXERCISE SET 3.2

Vocabulary Review

Choose the correct words to fill in the blanks below.

decimal point columns zeros money value

1. It is helpful to think in terms of _____ to add and subtract decimal numbers.
2. To add two decimal numbers, align the decimal points and then add in _____ .
3. When adding or subtracting decimal numbers, the _____ in the answer is directly below the decimals in the problems.
4. It is important to write _____ to the right of the rightmost digits to keep the digits in the correct columns.
5. Writing zeros to the right of the rightmost digits does not change the _____ of any of the numbers.

Problems

A Find each of the following sums. (Add.)

1. $2.91 + 3.28$

2. $8.97 + 2.04$

3. $0.04 + 0.31 + 0.78$

4. $0.06 + 0.92 + 0.65$

5. $-3.89 + 2.4$

6. $7.65 + 3.8$

7. $4.532 + 1.81 + 2.7$

8. $9.679 + 3.49 + 6.5$

9. $-0.081 + (-5) + (-2.94)$

10. $0.396 + 7 + 3.96$

11. $5.0003 + 6.78 + 0.004$

12. $27.0179 + 7.89 + 0.009$

13.
$$\begin{array}{r} 7.123 \\ 8.12 \\ + 9.1 \\ \hline \end{array}$$

14.
$$\begin{array}{r} 5.432 \\ 4.32 \\ + 3.2 \\ \hline \end{array}$$

15.
$$\begin{array}{r} 9.001 \\ 8.01 \\ + 7.1 \\ \hline \end{array}$$

16.
$$\begin{array}{r} 6.003 \\ 5.02 \\ + 4.1 \\ \hline \end{array}$$

17.
$$\begin{array}{r} 89.7854 \\ 3.4 \\ 65.35 \\ + 100.006 \\ \hline \end{array}$$

18.
$$\begin{array}{r} 57.4698 \\ 9.89 \\ 32.032 \\ + 572.0079 \\ \hline \end{array}$$

19.
$$\begin{array}{r} 543.21 \\ + 123.45 \\ \hline \end{array}$$

20.
$$\begin{array}{r} 987.654 \\ + 456.789 \\ \hline \end{array}$$

B Find each of the following differences. (Subtract.)

21. $99.34 - 88.23$

22. $47.69 - 36.58$

23. $5.97 - 2.4$

24. $9.87 - 1.04$

25. $6.3 - 2.08$

26. $7.5 - 3.04$

27. $-28.96 - (-149.37)$

28. $796.45 - 32.68$

29. $45 - 0.067$

30. $48 - 0.075$

31. $8 - 0.327$

32. $-0.962 - (-12)$

33. $765.432 - 234.567$

34. $654.321 - 123.456$

35. $100.42 - 56.87$

36. $12 - 1.93$

37. $-4.082 - (-10)$

38. $20 - 5.86$

Subtract.

39. $\begin{array}{r} 34.07 \\ -\ 6.18 \\ \hline \end{array}$
40. $\begin{array}{r} 25.008 \\ -\ 3.119 \\ \hline \end{array}$
41. $\begin{array}{r} 40.04 \\ -\ 4.4 \\ \hline \end{array}$
42. $\begin{array}{r} 50.05 \\ -\ 5.5 \\ \hline \end{array}$
43. $\begin{array}{r} 768.436 \\ -\ 356.998 \\ \hline \end{array}$
44. $\begin{array}{r} 495.237 \\ -\ 247.668 \\ \hline \end{array}$

Add and subtract as indicated.

45. $(7.8 - 4.3) + 2.5$

46. $(8.3 - 1.2) + 3.4$

47. $7.8 - (4.3 + 2.5)$

48. $8.3 - (1.2 + 3.4)$

49. $(9.7 - 5.2) - 1.4$

50. $(7.8 - 3.2) - 1.5$

51. $9.7 - (5.2 - 1.4)$

52. $7.8 - (3.2 - 1.5)$

53. $5.9 - (4.03 - 2.3)$

54. $10 - (3 - 1.06)$

55. $12.2 - (9.1 + 0.3)$

56. $15.6 - (4.9 + 3.87)$

57. $-9.01 - 2.4$

58. $-8.23 - 5.4$

59. $-0.89 - 1.01$

60. $-0.42 - 2.04$

61. $-3.4 - 5.6 - 8.5$

62. $-2.1 - 3.1 - 4.1$

63. Subtract 5 from the sum of 8.2 and 0.072.

64. Subtract 8 from the sum of 9.37 and 2.5.

65. What number is added to 0.035 to obtain 4.036?

66. What number is added to 0.043 to obtain 6.054?

Applying the Concepts

67. Shopping A family buying school clothes for their two children spends $25.37 at one store, $39.41 at another, and $52.04 at a third store. What is the total amount spent at the three stores?

68. Expenses A 4-H Club member is raising a lamb to take to the county fair. If she spent $75 for the lamb, $25.60 for feed, and $35.89 for shearing tools, what was the total cost of the project?

69. Take-Home Pay A college professor making $2,105.96 per month has deducted from her check $311.93 for federal income tax, $158.21 for retirement, and $64.72 for state income tax. How much does the professor take home after the deductions have been taken from her monthly income?

70. Take-Home Pay A cook making $1,504.75 a month has deductions of $157.32 for federal income tax, $58.52 for Social Security, and $45.12 for state income tax. How much does the cook take home after the deductions have been taken from his check?

71. Rectangle The logo on a business letter is rectangular. The rectangle has a width of 0.84 inches and a length of 1.41 inches. Find the perimeter.

72. Rectangle A small sticky note is a rectangle. It has a width of 21.4 millimeters and a length of 35.8 millimeters. Find the perimeter.

73. Change A person buys $4.57 worth of candy. If he pays for the candy with a $10 bill, how much change should he receive?

74. Checking Account A checking account contains $342.38. If checks are written for $25.04, $36.71, and $210, how much money is left in the account?

Check No.	Date	Description of Transaction	Payment/Debit (-)		Deposit/Credit (+)		Balance	
1630	7/3	Deposit			$342	38	$342	38
	7/5	Albertsons	$25	04				
	7/5	Rite Aid	$36	71				
	7/7	Gas Company	$210	00			?	

Downhill Skiers The chart shows the times for several downhill skiers in the 2010 Vancouver Olympics. Use the information to answer the following questions.

75. How much faster was Elisabeth Goergl than Andrea Fischbacher?

76. How much faster was Lindsey Vonn than Julia Mancuso?

FASTEST DOWNHILL SKIER - VANCOUVER 2010
www.vancouver2010.com

Lindsey Vonn	1:44:19
Julia Mancuso	1:44:75
Elisabeth Goergl	1:45:65
Andrea Fischbacher	1:45:68

hours:minutes:seconds

77. Mac Users The chart shows the increase in the number of Mac users each year. Use it to answer the following questions.

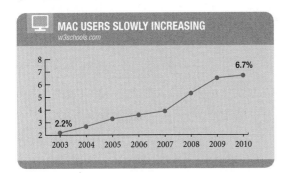

a. Was the increase in users during 2007 more or less than 5% of total users?

b. Was the increase in users during 2009 more or less than 5%?

c. Estimate the percent of total users increase from 2007 to 2009.

78. Movie Tickets The chart shows the increase in movie ticket prices. Use the information to answer the following questions.

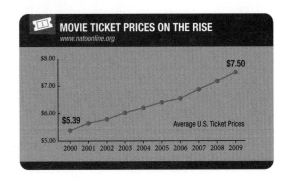

a. Were movie ticket prices above $6 in 2001?

b. Were ticket prices above $6 in 2007?

c. Were ticket prices in 2008 below $7?

79. Geometry A rectangle has a perimeter of 9.5 inches. If the length is 2.75 inches, find the width.

80. Geometry A rectangle has a perimeter of 11 inches. If the width is 2.5 inches, find the length.

81. Change Suppose you eat dinner in a restaurant and the bill comes to $16.76. If you give the cashier a $20 bill and a penny, how much change should you receive? List the bills and coins you should receive for change.

82. Change Suppose you buy some tools at the hardware store and the bill comes to $37.87. If you give the cashier two $20 bills and 2 pennies, how much change should you receive? List the bills and coins you should receive for change.

Sequences Find the next number in each sequence.

83. 2.5, 2.75, 3, . . .

84. 3.125, 3.375, 3.625, . . .

Getting Ready for the Next Section

To understand how to multiply decimals, we need to understand multiplication with whole numbers, fractions, and mixed numbers. The following problems review these concepts.

85. $\dfrac{1}{10} \cdot \dfrac{3}{10}$

86. $\dfrac{5}{10} \cdot \dfrac{6}{10}$

87. $\dfrac{3}{100} \cdot \dfrac{17}{100}$

88. $\dfrac{7}{100} \cdot \dfrac{31}{100}$

89. $5\left(\dfrac{3}{10}\right)$

90. $7 \cdot \dfrac{7}{10}$

91. $56 \cdot 25$

92. $39(48)$

93. $\dfrac{5}{10} \cdot \dfrac{3}{10}$

94. $\dfrac{5}{100} \cdot \dfrac{3}{1{,}000}$

95. $2\dfrac{1}{10} \cdot \dfrac{7}{100}$

96. $3\dfrac{5}{10} \cdot \dfrac{4}{100}$

97. $305(436)$

98. $403(522)$

99. $5(420 + 3)$

100. $3(550 + 2)$

Improving Your Quantitative Literacy

101. Facebook Users The chart shows the number of active Facebook users over several years. Use the information to answer the following questions.

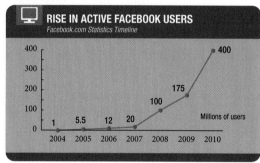

a. How many more users were there in 2007 than in 2006?

b. Between which years was the increase the most?

c. If the trend from 2009 to 2010 continues, how many users will there be in 2011?

Find the Mistake

Each sentence below contains a mistake. Circle the mistake and write the correct word(s) or numbers(s) on the line provided.

1. To add 32.69 and 4.837, align the rightmost digit of each number and add in columns. _____

2. To add $0.004 + 5.06 + 32$ by first changing each decimal to a fraction would give us the problem $\dfrac{4}{1{,}000} + \dfrac{5}{600} + 32$.

3. When subtracting $8.7 - 2.0163$, we make sure to keep the digits in the correct columns by writing 8.7 as 8.0007. _____

4. Subtracting 4.367 from the sum of 12.1 and 0.036 gives us 11.3333. _____

3.3 Multiplication with Decimals

The mystery of the roving rocks in Racetrack Playa of Death Valley, California has been puzzling scientists for decades. Water from winter rains wash down the slopes of the surrounding mountains and form a small lake in the Playa. During the summer, the lake evaporates and the muddy bed dries up, cracking into a mosaic of hexagonal clay tiles. Scientists have been trying to discover what causes large stones to move across the lakebed and leave behind tracks. One hypothesis suggests that as night temperatures fall below freezing, ice forms around each tile. Frequent winds up to 90 miles per hour can move this ice, allowing for a conveyor belt of sorts to move the rocks. To date, no one has ever witnessed the rocks moving, but scientists believe they may move at half the jogging speed of a human. Let's suppose a person is jogging at 8.5 miles per hour. If a rock in Racetrack Playa was moving at half the jogger's rate, it would be moving at 4.25 miles per hour. Therefore, it must be true that

$$\frac{1}{2} \text{ of 8.5 is 4.25}$$

But because $\frac{1}{2}$ can be written as 0.5 and *of* translates to multiply, we can write this problem again as

$$0.5 \times 8.5 = 4.25$$

If we were to ignore the decimal points in this problem and simply multiply 5 and 85, the result would be 425.

A Multiplication with Decimals

Multiplication with decimal numbers is similar to multiplication with whole numbers. The difference lies in deciding where to place the decimal point in the answer. To find out how this is done, we can use fraction notation.

EXAMPLE 1 Change each decimal to a fraction and multiply.

$$0.5 \times 0.3$$

Solution Changing each decimal to a fraction and multiplying, we have

$$0.5 \times 0.3 = \frac{5}{10} \times \frac{3}{10} \qquad \text{Change to fractions.}$$

$$= \frac{15}{100} \qquad \text{Multiply numerators and multiply denominators.}$$

$$= 0.15 \qquad \text{Write the answer in decimal form.}$$

The result is 0.15, which has two digits to the right of the decimal point.

**Video Examples
Section 3.3**

Practice Problems

1. Change each decimal to a fraction and multiply.

$$0.4 \times 0.08$$

Note To indicate multiplication we are using a × sign here instead of a dot so we won't confuse the decimal points with the multiplication symbol.

Answer

1. 0.032

What we want to do now is find a shortcut that will allow us to multiply decimals without first having to change each decimal number to a fraction. Let's look at another example.

EXAMPLE 2 Change each decimal to a fraction and multiply: 0.05×0.003.

Solution

$$0.05 \times 0.003 = \frac{5}{100} \times \frac{3}{1,000} \quad \text{Change to fractions.}$$

$$= \frac{15}{100,000} \quad \text{Multiply numerators and multiply denominators.}$$

$$= 0.00015 \quad \text{Write the answer in decimal form.}$$

The result is 0.00015, which has a total of five digits to the right of the decimal point.

Looking over these first two examples, we can see that the digits in the result are just what we would get if we simply forgot about the decimal points and multiplied; that is, $3 \times 5 = 15$. The decimal point in the result is placed so that the total number of digits to its right is the same as the total number of digits to the right of both decimal points in the original two numbers. The reason this is true becomes clear when we look at the denominators after we have changed from decimals to fractions.

EXAMPLE 3 Multiply: 2.1×0.07.

Solution

$$2.1 \times 0.07 = 2\frac{1}{10} \times \frac{7}{100} \quad \text{Change to fractions.}$$

$$= \frac{21}{10} \times \frac{7}{100}$$

$$= \frac{147}{1,000} \quad \text{Multiply numerators and multiply denominators.}$$

$$= 0.147 \quad \text{Write the answer as a decimal.}$$

Again, the digits in the answer come from multiplying $21 \times 7 = 147$. The decimal point is placed so that there are three digits to its right, because that is the total number of digits to the right of the decimal points in 2.1 and 0.07.

We summarize this discussion with the following rule:

> **RULE** Multiplication with Decimal Numbers
>
> To multiply two decimal numbers, follow these steps:
>
> 1. Multiply as you would if the decimal points were not there.
> 2. Place the decimal point in the answer so that the number of digits to its right is equal to the total number of digits to the right of the decimal points in the original two numbers in the problem.

EXAMPLE 4 How many digits will be to the right of the decimal point in the following product?

$$2.987 \times 24.82$$

Solution There are three digits to the right of the decimal point in 2.987 and two digits to the right in 24.82. Therefore, there will be $3 + 2 = 5$ digits to the right of the decimal point in their product.

2. Change each decimal to a fraction and multiply.

0.08×0.009.

3. Multiply: 3.2×0.09.

4. How many digits will be to the right of the decimal point in the following product?

4.632×0.0008

Answers
2. 0.00072
3. 0.288
4. Seven

EXAMPLE 5 Multiply: 3.05×4.36.

Solution We can set this up as if it were a multiplication problem with whole numbers. We multiply and then place the decimal point in the correct position in the answer.

$$
\begin{array}{r}
3.05 \quad \longleftarrow \text{2 digits to the right of decimal point} \\
\times\ 4.36 \quad \longleftarrow \text{2 digits to the right of decimal point} \\
\hline
1830 \\
915 \\
+\ 12\ 20 \\
\hline
13.2980 \\
\uparrow
\end{array}
$$

The decimal point is placed so that there are $2 + 2 = 4$ digits to its right.

As you can see, multiplying decimal numbers is just like multiplying whole numbers, except that we must place the decimal point in the result in the correct position.

Estimating

Look back to Example 5. We could have placed the decimal point in the answer by rounding the two numbers to the nearest whole number and then multiplying them. Because 3.05 rounds to 3 and 4.36 rounds to 4, and the product of 3 and 4 is 12, we estimate that the answer to 3.05×4.36 will be close to 12. Then, we place the decimal point in the product 132980 between the 3 and the 2 in order to make it into a number close to 12.

EXAMPLE 6 Estimate the answer to each of the following products.

a. 29.4×8.2 **b.** 68.5×172 **c.** $(6.32)^2$

Solution

a. Because 29.4 is approximately 30 and 8.2 is approximately 8, we estimate this product to be about $30 \times 8 = 240$. (If we were to multiply 29.4 and 8.2, we would find the product to be exactly 241.08.)

b. Rounding 68.5 to 70 and 172 to 170, we estimate this product to be $70 \times 170 = 11,900$. (The exact answer is 11,782.) Note here that we do not always round the numbers to the nearest whole number when making estimates. The idea is to round to numbers that will be easy to multiply.

c. Because 6.32 is approximately 6 and $6^2 = 36$, we estimate our answer to be close to 36. (The actual answer is 39.9424.)

B Order of Operations with Decimals

We can use the rule for order of operations to simplify expressions involving decimal numbers and addition, subtraction, and multiplication.

EXAMPLE 7 Perform the indicated operations: $0.05(4.2 + 0.03)$.

Solution We begin by adding inside the parentheses.

$$0.05(4.2 + 0.03) = 0.05(4.23) \quad \text{Add.}$$
$$= 0.2115 \quad \text{Multiply.}$$

Notice that we could also have used the distributive property first, and the result would be unchanged.

$$0.05(4.2 + 0.03) = 0.05(4.2) + 0.05(0.03) \quad \text{Distributive property}$$
$$= 0.210 + 0.0015 \quad \text{Multiply.}$$
$$= 0.2115 \quad \text{Add.}$$

5. Multiply: 2.16×3.05.

6. Estimate the answer to each of the following products.
 a. 39.6×9.1
 b. 58.7×141.3
 c. $(9.84)^2$

7. Perform the indicated operations.
 $0.07(3.9 + 6.05)$

Answers
5. 6.588
6. a. 360 **b.** 8,400 **c.** 100
7. 0.6965

8. Simplify: $9.5 + 11(2.3)^2$.

EXAMPLE 8 Simplify: $4.8 + 12(3.2)^2$.

Solution According to the rule for order of operations, we must first evaluate the number with an exponent, then multiply, and finally add.

$$4.8 + 12(3.2)^2 = 4.8 + 12(10.24) \quad (3.2)^2 = 10.24$$
$$= 4.8 + 122.88 \quad \text{Multiply.}$$
$$= 127.68 \quad \text{Add.}$$

9. Simplify each expression.
 a. $(-4)(1.2)$
 b. $(-0.2)(-0.5)$

EXAMPLE 9 Simplify each expression.

a. $(-5)(3.4)$ **b.** $(-0.4)(-0.8)$

Solution
a. $(-5)(3.4) = -17.0$ *The rule for multiplication also holds for decimals.*
b. $(-0.4)(-0.8) = 0.32$

Applications

10. How much will Sally from Example 10 make if she works 48 hours in one week?

EXAMPLE 10 Sally earns $6.32 for each of the first 36 hours she works in one week, and $9.48 in overtime pay for each additional hour she works in the same week. How much money will she make if she works 42 hours in one week?

Solution The difference between 42 and 36 is 6 hours of overtime pay. The total amount of money she will make is

Pay for the first Pay for the
36 hours next 6 hours
$$6.32(36) + 9.48(6) = 227.52 + 56.88$$
$$= 284.40$$

She will make $284.40 for working 42 hours in one week.

Note To estimate the answer to Example 10 before doing the actual calculations, we would do the following:

$6(40) + 9(6) = 240 + 54 = 294$

GETTING READY FOR CLASS

After reading through the preceding section, respond in your own words and in complete sentences.

 A. If you multiply 34.76 and 0.072, how many digits will be to the right of the decimal point in your answer?

 B. To simplify the expression $0.053(9) + 67.42$, what would be the first step according to the rule for order of operations?

 C. What is the purpose of estimating?

 D. What are some applications of decimals that we use in our everyday lives?

Answers
8. 67.69
9. a. -4.8 **b.** 0.10
10. $341.28

Vocabulary Review

Choose the correct words to fill in the blanks in the paragraph below.

answer decimal points multiply digits

To multiply two decimal numbers, first _____ as you would if the decimal points were not there.
Then place the decimal point in the _____ so that the number of _____ to its
right is equal to the total number of digits to the right of the _____ in the original two numbers
in the problem.

Problems

A Find each of the following products. (Multiply.)

1. 0.7
 × 0.4

2. 0.8
 × 0.3

3. 0.07
 × 0.4

4. 0.8
 × 0.03

5. 0.03
 × 0.09

6. 0.07
 × 0.002

7. 2.6(0.3)

8. 8.9(0.2)

9. 0.9
 × 0.88

10. 0.8
 × 0.99

11. 3.12
 × 0.005

12. 4.69
 × 0.006

13. 4.003
 × 6.07

14. 7.0001
 × 3.04

15. 5(0.006)

16. 7(0.005)

17. 75.14
 × 2.5

18. 963.8
 × 0.24

19. 0.1
 × 0.02

20. 0.3
 × 0.02

21. 2.796(10)

22. 97.531(100)

23. 0.0043
 × 100

24. 12.345
 × 1,000

25. 49.94
 × 1,000

26. 157.02
 × 10,000

27. 987.654
 × 10,000

28. 1.23
 × 100,000

B Perform the following operations according to the rule for order of operations.

29. $2.1(3.5 - 2.6)$

30. $5.4(9.9 - 6.6)$

31. $0.05(0.02 + 0.03)$

32. $-0.04(0.07 + 0.09)$

33. $2.02(0.03 + 2.5)$

34. $-4.04(0.05 + 6.6)$

35. $(2.1 + 0.03)(3.4 + 0.05)$

36. $(9.2 + 0.01)(3.5 + 0.03)$

37. $(2.1 - 0.1)(2.1 + 0.1)$

38. $(9.6 - 0.5)(9.6 + 0.5)$

39. $3.08 - 0.2(5 + 0.03)$

40. $4.09 + 0.5(6 + 0.02)$

41. $4.23 - 5(0.04 + 0.09)$

42. $7.89 - 2(0.31 + 0.76)$

43. $2.5 + 10(4.3)^2$

44. $3.6 + 15(2.1)^2$

45. $100(1 + 0.08)^2$

46. $500(1 + 0.12)^2$

47. $(1.5)^2 + (2.5)^2 + (3.5)^2$

48. $(1.1)^2 + (2.1)^2 + (3.1)^2$

Applying the Concepts

Solve each of the following word problems. Note that not all of the problems are solved by simply multiplying the numbers in the problems. Many of the problems involve addition and subtraction as well as multiplication.

49. Number Problem What is the product of 6 and the sum of 0.001 and 0.02?

50. Number Problem Find the product of 8 and the sum of 0.03 and 0.002.

51. Number Problem What does multiplying a decimal number by 100 do to the decimal point?

52. Number Problem What does multiplying a decimal number by 1,000 do to the decimal point?

53. Home Mortgage On a certain home mortgage, there is a monthly payment of $9.66 for every $1,000 that is borrowed. What is the monthly payment on this type of loan if $143,000 is borrowed?

54. Caffeine Content If 1 cup of regular coffee contains 105 milligrams of caffeine, how much caffeine is contained in 3.5 cups of coffee?

55. Long-Distance Charges If a phone company charges $0.45 for the first minute and $0.35 for each additional minute for a long-distance call, how much will a 20-minute long-distance call cost?

56. Price of Gasoline If gasoline costs $3.05 per gallon when you pay with a credit card, but $0.06 per gallon less if you pay with cash, how much do you save by filling up a 12-gallon tank and paying for it with cash?

57. Car Rental Suppose it costs $15 per day and $0.12 per mile to rent a car. What is the total bill if a car is rented for 2 days and is driven 120 miles?

58. Car Rental Suppose it costs $20 per day and $0.08 per mile to rent a car. What is the total bill if the car is rented for 2 days and is driven 120 miles?

59. Wages A man earns $5.92 for each of the first 36 hours he works in one week and $8.88 in overtime pay for each additional hour he works in the same week. How much money will he make if he works 45 hours in one week?

60. Wages A student earns $8.56 for each of the first 40 hours she works in one week and $12.84 in overtime pay for each additional hour she works in the same week. How much money will she make if she works 44 hours in one week?

61. Rectangle A rectangle has a width of 33.5 millimeters and a length of 254 millimeters. Find the area.

62. Rectangle A rectangle has a width of 2.56 inches and a length of 6.14 inches. Find the area rounded to the nearest hundredth.

63. Rectangle The logo on a business letter is rectangular. The rectangle has a width of 0.84 inches and a length of 1.41 inches. Find the area rounded to the nearest hundredth.

64. Rectangle A small sticky note is a rectangle. It has a width of 21.4 millimeters and a length of 35.8 millimeters. Find the area.

Getting Ready for the Next Section

To get ready for the next section, which covers division with decimals, we will review multiplication and division with whole numbers and fractions.

Perform the following operations.

65. 3,758 ÷ 2

66. 9,900 ÷ 22

67. 50,032 ÷ 33

68. 90,902 ÷ 5

69. 20)5,960

70. 30)4,620

71. 4 × 8.7

72. 5 × 6.7

73. 27 × 1.848

74. 35 × 32.54

75. 38)31,350

76. 25)377,800

Improving Your Quantitative Literacy

77. Containment System Holding tanks for hazardous liquids are often surrounded by containment tanks that will hold the hazardous liquid if the main tank begins to leak. We see that the center tank has a height of 16 feet and a radius of 6 feet. The outside containment tank has a height of 4 feet and a radius of 8 feet. The formula to calculate the volume of a tank is $V = \pi r^2 h$. Use 3.14 as an estimate for π. If the center tank is full of heating fuel and develops a leak at the bottom, will the containment tank be able to hold all the heating fuel that leaks out?

Find the Mistake

Each sentence below contains a mistake. Circle the mistake and write the correct word(s) or numbers(s) on the line provided.

1. To multiply 18.05 by 3.5, multiply as if the numbers were whole numbers and then place the decimal in the answer with two digits to its right. _____

2. To estimate the answer for 24.9 × 7.3, round 24.9 to 20 and 7.3 to 7. _____

3. To simplify (8.43 + 1.002) − (0.05)(3.2), first subtract the product of 0.05 and 3.2 from 1.002 before adding 8.43.

4. Lucy pays $1.52 a pound for the first three pounds of candy she buys at a candy store, and pays $3.27 for each additional pound. To find how much she will pay if she buys 5.2 pounds of candy, we must solve the problem 1.52(3) + 3.27(5.2).

3.4 Division with Decimals

OBJECTIVES

A Divide decimal numbers with exact answers.

B Divide decimal numbers with rounded answers.

A coffee maker company announced the development of their new home espresso machine that scans your fingerprint to take an order. The new device can distinguish between six different housemates based on a scan of each fingerprint and its matching preferences (e.g., strength, froth, etc.). The current estimated selling price for the espresso machine is $3,200.

Suppose you and your three roommates have decided to purchase one of these espresso machines selling for $3,199.99. If you decide to split the bill equally, how much does each person owe? To find out, you will have to divide 3199.99 by 4. In this section, we will find out how to do division with any combination of decimal numbers.

A Dividing Decimal Numbers (Exact Answers)

EXAMPLE 1 Divide: 5,974 ÷ 20.

Solution

```
        298
   20)5974
      40↓
      ───
      197
      180↓
      ───
      174
      160
      ───
       14
```

In the past, we have written this answer as $298\frac{14}{20}$ or, after reducing the fraction, $298\frac{7}{10}$. Because $\frac{7}{10}$ can be written as 0.7, we could also write our answer as 298.7. This last form of our answer is exactly the same result we obtain if we write 5,974 as 5,974.0 and continue the division until we have no remainder. Here is how it looks:

```
        298.7
   20)5974.0
      40↓ │
      ───
      197 │
      180↓│
      ───
      174 │
      160↓
      ───
      140
      140
      ───
        0
```

Notice that we place the decimal point in the answer directly above the decimal point in the problem.

Video Examples Section 3.4

Practice Problems

1. Divide: 4,870 ÷ 50.

Note We can estimate the answer to Example 1 by rounding 5,974 to 6,000 and dividing by 20:

$$\frac{6,000}{20} = 300$$

Answer

1. 97.4

Let's try another division problem. This time one of the numbers in the problem will be a decimal.

EXAMPLE 2 Divide: $34.8 \div 4$.

Solution We can use the ideas from Example 1 and divide as usual. The decimal point in the answer will be placed directly above the decimal point in the problem.

$$
\begin{array}{r}
8.7 \\
4)\overline{34.8} \\
\underline{32} \\
28 \\
\underline{28} \\
0
\end{array}
\qquad
\begin{array}{rl}
\textit{Check:} & 8.7 \\
\times & 4 \\
\hline & 34.8
\end{array}
$$

The answer is 8.7.

We can use these facts to write a rule for dividing decimal numbers.

> **RULE** Division with Decimal Numbers
>
> To divide a decimal by a whole number, we do the usual long division as if there were no decimal point involved. The decimal point in the answer is placed directly above the decimal point in the problem.

Here are some more examples to illustrate the procedure:

EXAMPLE 3 Divide: $49.896 \div 27$.

Solution

$$
\begin{array}{r}
1.848 \\
27)\overline{49.896} \\
\underline{27} \\
22\,8 \\
\underline{21\,6} \\
1\,29 \\
\underline{1\,08} \\
216 \\
\underline{216} \\
0
\end{array}
\qquad
\begin{array}{rl}
\textit{Check:} & 1.848 \\
\times & 27 \\
\hline & 12936 \\
& 36\,96 \\
\hline & 49.896
\end{array}
$$

We can write as many zeros as we choose after the rightmost digit in a decimal number without changing the value of the number. For example,

$$6.91 = 6.910 = 6.9100 = 6.91000$$

There are times when this can be very useful, as Example 4 shows.

EXAMPLE 4 Divide: $1,138.9 \div 35$.

Solution

$$
\begin{array}{r}
32.54 \\
35)\overline{1138.90} \\
\underline{105} \\
88 \\
\underline{70} \\
18\,9 \\
\underline{17\,5} \\
1\,40 \\
\underline{1\,40} \\
0
\end{array}
$$

Write 0 after the 9. It doesn't change the original number, but it gives us another digit to bring down.

$$
\begin{array}{rl}
\textit{Check:} & 32.54 \\
\times & 35 \\
\hline & 162\,70 \\
& 976\,2 \\
\hline & 1,138.90
\end{array}
$$

2. Divide: $56.9 \div 5$.

> *Note* We never need to make a mistake with division, because we can always check our results with multiplication.

3. Divide: $54.632 \div 25$.

4. Divide: $1,105.8 \div 15$.

Answers
2. 11.38
3. 2.18528
4. 73.72

Until now we have considered only division by whole numbers. Extending division to include division by decimal numbers is a matter of knowing what to do about the decimal point in the divisor.

EXAMPLE 5 Divide: $31.35 \div 3.8$.

Solution In fraction form, this problem is equivalent to

$$\frac{31.35}{3.8}$$

If we want to write the divisor as a whole number, we can multiply the numerator and the denominator of this fraction by 10:

$$\frac{31.35 \times 10}{3.8 \times 10} = \frac{313.5}{38}$$

So, since this fraction is equivalent to the original fraction, our original division problem is equivalent to

$$
\begin{array}{r}
8.25 \\
38\overline{)313.50} \\
\underline{304} \\
95 \\
\underline{76} \\
190 \\
\underline{190} \\
0
\end{array}
$$ *Put 0 after the last digit.*

We can summarize division with decimal numbers by listing the following points, as illustrated in the first five examples.

> **Summary of Division with Decimals**
>
> 1. We divide decimal numbers by the same process used to divide whole numbers. The decimal point in the answer is placed directly above the decimal point in the dividend.
>
> 2. We are free to write as many zeros after the last digit in a decimal number as we need.
>
> 3. If the divisor is a decimal, we can change it to a whole number by moving the decimal point to the right as many places as necessary so long as we move the decimal point in the dividend the same number of places.

B Dividing Decimal Numbers (Rounded Answers)

EXAMPLE 6 Divide, and round the answer to the nearest hundredth.

$$0.3778 \div 0.25$$

Solution First, we move the decimal point two places to the right.

$$0.25.\overline{)37.78}$$

5. Divide: $46.354 \div 4.9$.

> *Note* We do not always use the rules for rounding numbers to make estimates. For example, to estimate the answer in Example 5, $31.35 \div 3.8$, we can get a rough estimate of the answer by reasoning that 3.8 is close to 4 and 31.35 is close to 32. Therefore, our answer will be approximately $32 \div 4 = 8$.

6. Divide, and round the answer to the nearest hundredth.

$$0.3849 \div 0.49$$

> *Note* Moving the decimal point two places in both the divisor and the dividend is justified like this:
>
> $$\frac{0.3778 \times 100}{0.25 \times 100} = \frac{37.78}{25}$$

Answers

5. 9.46
6. 0.79

Then we divide, using long division.

$$
\begin{array}{r}
1.5112 \\
25\overline{)37.7800} \\
\underline{25} \\
12\,7 \\
\underline{12\,5} \\
28 \\
\underline{25} \\
30 \\
\underline{25} \\
50 \\
\underline{50} \\
0
\end{array}
$$

Rounding to the nearest hundredth, we have 1.51. We actually did not need to have this many digits to round to the hundredths column. We could have stopped at the thousandths column and rounded off.

EXAMPLE 7 Divide and round to the nearest tenth: $17 \div 0.03$.

Solution Because we are rounding to the nearest tenth, we will continue dividing until we have a digit in the hundredths column. We don't have to go any further to round to the tenths column.

$$
\begin{array}{r}
5\,66.66 \\
0.03.\overline{)17.00.00} \\
\underline{15} \\
2\,0 \\
\underline{1\,8} \\
20 \\
\underline{18} \\
20 \\
\underline{18} \\
20 \\
\underline{18} \\
2
\end{array}
$$

Rounding to the nearest tenth, we have 566.7.

Applications

EXAMPLE 8 If a man earning $5.26 an hour receives a paycheck for $170.95 before deductions, how many hours did he work?

Solution To find the number of hours the man worked, we divide $170.95 by $5.26.

$$
\begin{array}{r}
32.5 \\
5.26.\overline{)170.95.0} \\
\underline{157\,8} \\
13\,15 \\
\underline{10\,52} \\
2\,63\,0 \\
\underline{2\,63\,0} \\
0
\end{array}
$$

The man worked 32.5 hours.

7. Divide, and round to the nearest tenth.

$19 \div 0.07$

8. If you earn $8.30 per hour and receive a paycheck for $311.25 before deductions, how many hours did you work?

EXAMPLE 9 A telephone company charges $0.43 for the first minute and then $0.33 for each additional minute for a long-distance call. If a long-distance call costs $3.07, how many minutes was the call?

Solution To solve this problem we need to find the number of additional minutes for the call. To do so, we first subtract the cost of the first minute from the total cost, and then we divide the result by the cost of each additional minute. Without showing the actual arithmetic involved, the solution looks like this:

$$\text{The number of additional minutes} = \frac{\overset{\text{Total cost of the call}}{3.07} - \overset{\text{Cost of the first minute}}{0.43}}{\underset{\substack{\text{Cost of each} \\ \text{additional minute}}}{0.33}} = \frac{2.64}{0.33} = 8$$

The call was 9 minutes long. (The number 8 is the number of additional minutes past the first minute.)

9. Repeat Example 9 if the long distance call costs $13.63.

DESCRIPTIVE STATISTICS Grade Point Average

I have always been surprised by the number of my students who have difficulty calculating their grade point average (GPA). During her first semester in college, my daughter, Amy, earned the following grades:

Class	Units	Grade
Algebra	5	B
Chemistry	4	C
English	3	A
History	3	B

When her grades arrived in the mail, she told me she had a 3.0 grade point average, because the A and C grades averaged to a B. I told her that her GPA was a little less than a 3.0. What do you think? Can you calculate her GPA? If not, you will be able to after you finish this section.

When you calculate your grade point average (GPA), you are calculating what is called a *weighted average*. To calculate your grade point average, you must first calculate the number of grade points you have earned in each class that you have completed. The number of grade points for a class is the product of the number of units the class is worth times the value of the grade received. The table below shows the value that is assigned to each grade.

Grade	Value
A	4
B	3
C	2
D	1
F	0

If you earn a B in a 4-unit class, you earn $4 \times 3 = 12$ grade points. A grade of C in the same class gives you $4 \times 2 = 8$ grade points. To find your grade point average for one term (a semester or quarter), you must add your grade points and divide that total by the number of units. Round your answer to the nearest hundredth.

Answer

9. 41 minutes

10. Calculate the grade-point average.

Class	Units	Grade
Algebra	4	A
Biology	3	B
English	3	B
History	3	C
PE	1	A

EXAMPLE 10 Calculate Amy's grade point average using the previous information.

Solution We begin by creating two more columns: one for the value of each grade (4 for an A, 3 for a B, 2 for a C, 1 for a D, and 0 for an F), and another for the grade points earned for each class. To fill in the grade points column, we multiply the number of units by the value of the grade.

Class	Units	Grade	Value	Grade Points
Algebra	5	B	3	$5 \times 3 = 15$
Chemistry	4	C	2	$4 \times 2 = 8$
English	3	A	4	$3 \times 4 = 12$
History	3	B	3	$3 \times 3 = 9$
Total Units	15			Total Grade Points: 44

To find her grade point average, we divide 44 by 15 and round (if necessary) to the nearest hundredth.

$$\text{Grade point average} = \frac{44}{15} = 2.93$$

GETTING READY FOR CLASS

After reading through the preceding section, respond in your own words and in complete sentences.

A. The answer to the division problem in Example 1 is $298\frac{14}{20}$. Write this number in decimal notation.

B. In Example 4, we place a 0 at the end of a number without changing the value of the number. Why is the placement of this 0 helpful?

C. The expression $0.3778 \div 0.25$ is equivalent to the expression $37.78 \div 25$ because each number was multiplied by what?

D. Briefly explain how to divide with decimals.

Answer

10. 3.14

Vocabulary Review

Choose the correct words to fill in the blanks below.

divisor above long division right last

1. To divide decimal numbers, use _____ as if the decimal point was not there.

2. In a long division problem that involves a decimal point, write the decimal point in the answer directly _____ the decimal point in the problem.

3. Writing zeros after the _____ digit in a decimal number will not change the value.

4. If the _____ is a decimal, we can change it to a whole number by moving the decimal point to the _____ as many places as necessary so long as we move the decimal point in the dividend the same number of places.

Problems

A Perform each of the following divisions.

1. $394 \div 20$

2. $486 \div 30$

3. $248 \div 40$

4. $372 \div 80$

5. $5\overline{)26}$

6. $8\overline{)36}$

7. $25\overline{)276}$

8. $50\overline{)276}$

9. $28.8 \div 6$

10. $15.5 \div 5$

11. $77.6 \div 8$

12. $31.48 \div 4$

13. $35\overline{)92.05}$

14. $26\overline{)146.38}$

15. $45\overline{)190.8}$

16. $55\overline{)342.1}$

17. $86.7 \div 34$

18. $411.4 \div 44$

19. $29.7 \div 22$

20. $488.4 \div 88$

21. $4.5\overline{)29.25}$

22. $3.3\overline{)21.978}$

23. $0.11\overline{)1.089}$

24. $0.75\overline{)2.40}$

25. $2.3\overline{)0.115}$

26. $6.6\overline{)0.198}$

27. $0.012\overline{)1.068}$

28. $0.052\overline{)0.23712}$

29. $1.1\overline{)2.42}$

30. $2.2\overline{)7.26}$

31. $0.014\overline{)0.0644}$

32. $0.38\overline{)9.652}$

B Carry out each of the following divisions only so far as needed to round the results to the nearest hundredth.

33. $26\overline{)35}$

34. $18\overline{)47}$

35. $3.3\overline{)56}$

36. $4.4\overline{)75}$

37. $0.1234 \div 0.5$

38. $-0.543 \div 2.1$

39. $19 \div 7$

40. $16 \div (-6)$

41. $0.059\overline{)0.69}$

42. $0.048\overline{)0.49}$

43. $1.99 \div 0.5$

44. $0.99 \div 0.5$

45. $2.99 \div 0.5$

46. $3.99 \div (-0.5)$

47. $-3.82 \div 0.9$

48. $1.79 \div 0.08$

Applying the Concepts

49. Hot Air Balloon Since the pilot of a hot air balloon can only control the balloon's altitude, he relies on the winds for travel. To ride on the jet streams, a hot air balloon must rise as high as 12 kilometers. Convert this to miles by dividing by 1.61. Round your answer to the nearest tenth of a mile.

50. Hot Air Balloon December and January are the best times for traveling in a hot-air balloon because the jet streams in the Northern Hemisphere are the strongest. They reach speeds of 400 kilometers per hour. Convert this to miles per hour by dividing by 1.61. Round to the nearest whole number.

51. Women's Golf The table below gives the top five money earners for the Ladies' Professional Golf Association (LPGA) in 2010. Fill in the last column of the table by finding the average earning per tournament for each golfer. Round your answers to the nearest ten dollars.

Rank	Name	Number of Tournaments	Total Earnings	Average per Tournament
1.	Na Yeon Choi	23	$1,871,165.50	
2.	Jiyai Shin	19	$1,783,127.00	
3.	Cristie Kerr	21	$1,601,551.75	
4.	Yani Tseng	19	$1,573,529.00	
5.	Suzann Pettersen	19	$1,557,174.50	

52. Men's Golf The table below gives the top five earners for the men's Professional Golf Association (PGA) in 2010. Fill in the last column of the table by finding the average earnings per tournament for each golfer. Round your answers to the nearest hundred dollars.

Rank	Name	Number of Tournaments	Total Earnings	Average per Tournament
1.	Matt Kuchar	26	$4,910,477	
2.	Jim Furyk	21	$4,809,622	
3.	Ernie Els	20	$4,558,861	
4.	Dustin Johnson	23	$4,473,122	
5.	Steve Stricker	19	$4,190,235	

53. Wages If a woman earns $33.90 for working 6 hours, how much does she earn per hour?

54. Wages How many hours does a person making $6.78 per hour have to work in order to earn $257.64?

55. Gas Mileage If a car travels 336 miles on 15 gallons of gas, how far will the car travel on 1 gallon of gas?

56. Gas Mileage If a car travels 392 miles on 16 gallons of gas, how far will the car travel on 1 gallon of gas?

57. Wages Suppose a woman earns $6.78 an hour for the first 36 hours she works in a week and then $10.17 an hour in overtime pay for each additional hour she works in the same week. If she makes $294.93 in one week, how many hours did she work overtime?

58. Wages Suppose a woman makes $286.08 in one week. If she is paid $5.96 an hour for the first 36 hours she works and then $8.94 an hour in overtime pay for each additional hour she works in the same week, how many hours did she work overtime that week?

59. Phone Bill Suppose a telephone company charges $0.41 for the first minute and then $0.32 for each additional minute for a long-distance call. If a long-distance call costs $2.33, how many minutes was the call?

60. Phone Bill Suppose a telephone company charges $0.45 for each of the first three minutes and then $0.29 for each additional minute for a long-distance call. If a long-distance call costs $3.67, how many minutes was the call?

Grade Point Average The following grades were earned by Steve during his first term in college. Use these data to answer Problems 61–64.

61. Calculate Steve's GPA.

Class	Units	Grade
Basic mathematics	3	A
Health	2	B
History	3	B
English	3	C
Chemistry	4	C

62. If his grade in chemistry had been a B instead of a C, by how much would his GPA have increased?

63. If his grade in health had been a C instead of a B, by how much would his grade point average have dropped?

64. If his grades in both English and chemistry had been Bs, what would his GPA have been?

Calculator Problems Work each of the following problems on your calculator. If rounding is necessary, round to the nearest hundred thousandth.

65. $7 \div 9$

66. $11 \div 13$

67. $243 \div 0.791$

68. $67.8 \div 37.92$

69. $0.0503 \div 0.0709$

70. $429.87 \div 16.925$

Getting Ready for the Next Section

In the next section, we will consider the relationship between fractions and decimals in more detail. The problems below review some of the material that is necessary to make a successful start in the next section.

Reduce to lowest terms.

71. $\dfrac{75}{100}$

72. $\dfrac{220}{1,000}$

73. $\dfrac{12}{18}$

74. $\dfrac{15}{30}$

75. $\dfrac{75}{200}$

76. $\dfrac{220}{2,000}$

77. $\dfrac{38}{100}$

78. $\dfrac{75}{1,000}$

Write each fraction as an equivalent fraction with denominator 100.

79. $\dfrac{3}{5}$

80. $\dfrac{1}{2}$

81. $\dfrac{5}{1}$

82. $\dfrac{17}{20}$

Write each fraction as an equivalent fraction with denominator 15.

83. $\dfrac{4}{5}$

84. $\dfrac{2}{3}$

85. $\dfrac{4}{1}$

86. $\dfrac{2}{1}$

87. $\dfrac{6}{5}$

88. $\dfrac{7}{3}$

Divide.

89. $3 \div 4$

90. $3 \div 5$

91. $7 \div 8$

92. $3 \div 8$

Find the Mistake

Each sentence below contains a mistake. Circle the mistake and write the correct word(s) or numbers(s) on the line provided.

1. The answer to the problem $25\overline{)70.75}$ will have a decimal point placed with four digits to its right. _____

2. To work the problem $27.468 \div 8.4$, multiply 8.4 by 10 and then divide. _____

3. To divide 0.6778 by 0.54, multiply both numbers by 10 to move the decimal point two places to the right. _____

4. Samantha earns $10.16 an hour as a cashier. She received a paycheck for $309.88. To find out how many hours she worked, you must solve the problem $10.16 \div 309.88$. _____

Landmark Review: Checking Your Progress

Write each of the following in words.

1. 1.15

2. 45.08

3. 0.005

4. 245.157

Write each of the following as a decimal number.

5. Sixty-seven ten thousandths

6. Five and six tenths

7. Twenty-three and fourteen thousandths

8. Two thousand thirteen and fifteen hundredths

Find each of the following sums and differences.

9. $24.13 + 4.15$

10. $6.000014 + 3.15$

11. $100.00001 + 24.1583$

12. $5.387 + 6.412$

13. $8.3 - 5.2$

14. $14.2 - 7.13$

15. $27.57 - 14.24$

16. $92.42 - 14.05$

Perform each of the following operations.

17. $4.735(10)$

18. $0.075(0.03)$

19. $1.4 \div 0.07$

20. $0.24 \div 0.6$

Perform the following operations according to the rule for order of operations.

21. $4.3(3.8 - 2.6)$

22. $(2.85 - 1.7)(5.67 + 4.2)$

23. $5.5 + 2.2(14 - 12.5)$

24. $(1.3)^2 + (5.1)^2 + (2.4)^2$

OBJECTIVES

A Convert between fractions and decimals.

B Simplify expressions involving both fractions and decimals.

KEY WORDS

convert

Video Examples Section 3.5

Practice Problems

1. Write $\frac{5}{8}$ as a decimal.

2. Write $\frac{5}{7}$ as a decimal rounded to the thousandths place.

If you are shopping for clothes and a store has a sale advertising $\frac{1}{3}$ off the regular price, how much can you expect to pay for a pair of pants that normally sells for $31.95? If the sale price of the pants is $22.30, have they really been marked down by $\frac{1}{3}$? To answer questions like these, we need to know how to solve problems that involve fractions and decimals together.

We begin this section by showing how to convert back and forth between fractions and decimals.

A Converting Fractions to Decimals

You may recall that the notation we use for fractions can be interpreted as implying division. That is, the fraction $\frac{3}{4}$ can be thought of as meaning "3 divided by 4." We can use this idea to convert fractions to decimals.

EXAMPLE 1 Write $\frac{3}{4}$ as a decimal.

Solution Dividing 3 by 4, we have

$$
\begin{array}{r}
.75 \\
4\overline{)3.00} \\
28 \\
\hline
20 \\
20 \\
\hline
0
\end{array}
$$

The fraction $\frac{3}{4}$ is equal to the decimal 0.75.

EXAMPLE 2 Write $\frac{7}{12}$ as a decimal rounded to the thousandths place.

Solution Because we want the decimal to be rounded to the thousandths place, we divide to the ten thousandths place and round off to the thousandths place.

$$
\begin{array}{r}
.5833 \\
12\overline{)7.0000} \\
60 \\
\hline
100 \\
96 \\
\hline
40 \\
36 \\
\hline
40 \\
36 \\
\hline
4
\end{array}
$$

Answers

1. 0.625

2. 0.714

Rounding off to the thousandths place, we have 0.583. Because $\frac{7}{12}$ is not exactly the same as 0.583, we write

$$\frac{7}{12} \approx 0.583$$

where the symbol \approx is read "is approximately equal to."

If we wrote more zeros after 0.583 in Example 2, the pattern of 3s would continue for as many places as we choose to divide. When we get a sequence of digits that repeat like this, 0.58333 . . . , we can indicate the repetition by writing

$$0.58\overline{3} \qquad \text{The bar over the 3 indicates that the 3 repeats from there on.}$$

EXAMPLE 3 Write $\frac{3}{11}$ as a decimal.

Solution Dividing 3 by 11, we have

$$
\begin{array}{r}
.272727 \\
11\overline{)3.000000} \\
22 \\
\hline
80 \\
77 \\
\hline
30 \\
22 \\
\hline
80 \\
77 \\
\hline
30 \\
22 \\
\hline
80 \\
77 \\
\hline
3 \\
\end{array}
$$

No matter how long we continue the division, the remainder will never be 0, and the pattern will continue. We write the decimal form of $\frac{3}{11}$ as $0.\overline{27}$, where

$$0.\overline{27} = 0.272727\ldots \qquad \text{The dots mean "and so on."}$$

Converting Decimals to Fractions

To convert decimals to fractions, we take advantage of the place values we assigned to the digits to the right of the decimal point.

EXAMPLE 4 Write 0.38 as a fraction in lowest terms.

Solution 0.38 is 38 hundredths, or

$$0.38 = \frac{38}{100}$$

$$= \frac{19}{50} \qquad \text{Divide the numerator and the denominator by 2 to reduce to lowest terms.}$$

The decimal 0.38 is equal to the fraction $\frac{19}{50}$.

We could check our work here by converting $\frac{19}{50}$ back to a decimal. We do this by dividing 19 by 50. That is,

$$
\begin{array}{r}
.38 \\
50\overline{)19.00} \\
150 \\
\hline
400 \\
400 \\
\hline
0 \\
\end{array}
$$

3. Write $\frac{5}{6}$ as a decimal.

Note The bar over the 2 and the 7 in $0.\overline{27}$ is used to indicate that the pattern repeats itself indefinitely.

4. Write 0.76 as a fraction in lowest terms.

Answers

3. $0.8\overline{3}$

4. $\frac{19}{25}$

5. Convert 0.045 to a fraction.

EXAMPLE 5 Convert 0.075 to a fraction.

Solution We have 75 thousandths, or

$$0.075 = \frac{75}{1,000}$$

$$= \frac{3}{40} \qquad \textit{Divide the numerator and the denominator by 25 to reduce to lowest terms.}$$

6. Write 12.08 as a mixed number.

EXAMPLE 6 Write 15.6 as a mixed number.

Solution Converting 0.6 to a fraction, we have

$$0.6 = \frac{6}{10} = \frac{3}{5} \qquad \textit{Reduce to lowest terms.}$$

Since $0.6 = \frac{3}{5}$, we have $15.6 = 15\frac{3}{5}$.

B Problems Containing Both Fractions and Decimals

We continue this section by working some problems that involve both fractions and decimals.

7. Simplify: $\frac{12}{25}$ (1.41 − 0.56).

EXAMPLE 7 Simplify: $\frac{19}{50}$ (1.32 + 0.48).

Solution In Example 4, we found that $0.38 = \frac{19}{50}$. Therefore, we can rewrite the problem as

$$\frac{19}{50}(1.32 + 0.48) = 0.38(1.32 + 0.48) \quad \textit{Convert all numbers to decimals.}$$

$$= 0.38(1.80) \qquad \textit{Add: 1.32 + 0.48.}$$

$$= 0.684 \qquad \textit{Multiply: 0.38 × 1.80.}$$

8. Simplify: $\frac{3}{4} + \frac{2}{3}$ (0.66).

EXAMPLE 8 Simplify: $\frac{1}{2} + (0.75)\left(\frac{2}{5}\right)$.

Solution We could do this problem one of two different ways. First, we could convert all fractions to decimals and then simplify.

$$\frac{1}{2} + (0.75)\left(\frac{2}{5}\right) = 0.5 + 0.75(0.4) \quad \textit{Convert to decimals.}$$

$$= 0.5 + 0.300 \qquad \textit{Multiply: 0.75 × 0.4.}$$

$$= 0.8 \qquad \textit{Add.}$$

Or, we could convert 0.75 to $\frac{3}{4}$ and then simplify.

$$\frac{1}{2} + (0.75)\left(\frac{2}{5}\right) = \frac{1}{2} + \frac{3}{4}\left(\frac{2}{5}\right) \qquad \textit{Convert decimals to fractions.}$$

$$= \frac{1}{2} + \frac{3}{10} \qquad \textit{Multiply: } \frac{3}{4} \times \frac{2}{5}.$$

$$= \frac{5}{10} + \frac{3}{10} \qquad \textit{The common denominator is 10.}$$

$$= \frac{8}{10} \qquad \textit{Add numerators.}$$

$$= \frac{4}{5} \qquad \textit{Reduce to lowest terms.}$$

The answers are equivalent. That is, $0.8 = \frac{8}{10} = \frac{4}{5}$. Either method can be used with problems of this type.

EXAMPLE 9 Simplify: $\left(\frac{1}{2}\right)^3 (2.4) + \left(\frac{1}{4}\right)^2 (3.2)$.

Solution This expression can be simplified without any conversions between fractions and decimals. To begin, we evaluate all numbers that contain exponents. Then we multiply. After that, we add.

$$\left(\frac{1}{2}\right)^3 (2.4) + \left(\frac{1}{4}\right)^2 (3.2) = \frac{1}{8}\,(2.4) + \frac{1}{16}\,(3.2) \qquad \text{Evaluate exponents.}$$

$$= 0.3 + 0.2 \qquad \text{Multiply by } \tfrac{1}{8} \text{ and } \tfrac{1}{16}.$$

$$= 0.5 \qquad \text{Add.}$$

9. Simplify: $\left(\frac{1}{3}\right)^2 (0.81) + \left(\frac{1}{2}\right)^3 (1.2)$.

Applications

EXAMPLE 10 If a shirt that normally sells for $27.99 is on sale for $\frac{1}{3}$ off, what is the sale price of the shirt?

Solution To find out how much the shirt is marked down, we must find $\frac{1}{3}$ of 27.99. That is, we multiply $\frac{1}{3}$ and 27.99, which is the same as dividing 27.99 by 3.

$$\frac{1}{3}(27.99) = \frac{27.99}{3} = 9.33$$

The shirt is marked down $9.33. The sale price is $9.33 less than the original price.

$$\text{Sale price} = 27.99 - 9.33 = 18.66$$

The sale price is $18.66. We also could have solved this problem by simply multiplying the original price by $\frac{2}{3}$, since, if the shirt is marked $\frac{1}{3}$ off, then the sale price must be $\frac{2}{3}$ of the original price. Multiplying by $\frac{2}{3}$ is the same as dividing by 3 and then multiplying by 2. The answer would be the same.

10. A pair of jeans that normally sell for $38.60 are on sale for $\frac{1}{4}$ off. What is the sale price of the jeans?

> **GETTING READY FOR CLASS**
>
> *After reading through the preceding section, respond in your own words and in complete sentences.*
>
> **A.** To convert fractions to decimals, do we multiply or divide the numerator by the denominator?
>
> **B.** The decimal 0.13 is equivalent to what fraction?
>
> **C.** Write 36 thousandths in decimal form and in fraction form.
>
> **D.** Explain how to write the fraction $\frac{84}{1,000}$ in lowest terms.

Answers

9. 0.24

10. $28.95

EXERCISE SET 3.5

Vocabulary Review

Choose the correct words to fill in the blanks below.

 repeats place values division reduce

1. To convert a fraction to a decimal, think of the fraction in terms of _____ .

2. A bar over the right most digit(s) in a decimal number indicates that digit _____ indefinitely.

3. To convert a decimal to a fraction, use the _____ of the digits and _____ to lowest terms.

Problems

A Each circle below is divided into 8 equal parts. The number below each circle indicates what fraction of the circle is shaded. Convert each fraction to a decimal.

1.

$\dfrac{1}{8}$

2.

$\dfrac{3}{8}$

3.

$\dfrac{5}{8}$

4.

$\dfrac{7}{8}$

Complete the following tables by converting each fraction to a decimal.

5.

Fraction	$\frac{1}{5}$	$\frac{2}{5}$	$\frac{3}{5}$	$\frac{4}{5}$	$\frac{5}{5}$
Decimal					

6.

Fraction	$\frac{1}{6}$	$\frac{2}{6}$	$\frac{3}{6}$	$\frac{4}{6}$	$\frac{5}{6}$	$\frac{6}{6}$
Decimal						

Convert each of the following fractions to a decimal.

7. $\dfrac{1}{2}$

8. $-\dfrac{12}{25}$

9. $\dfrac{14}{25}$

10. $\dfrac{14}{32}$

11. $\dfrac{18}{32}$

12. $\dfrac{9}{16}$

13. $-\dfrac{13}{16}$

14. $\dfrac{13}{8}$

Write each fraction as a decimal rounded to the hundredths place.

15. $\dfrac{12}{13}$

16. $\dfrac{17}{19}$

17. $\dfrac{3}{11}$

18. $-\dfrac{5}{11}$

19. $\dfrac{2}{23}$

20. $-\dfrac{3}{28}$

21. $\dfrac{12}{43}$

22. $\dfrac{15}{51}$

Complete the following table by converting each decimal to a fraction.

23.

Decimal	0.125	0.250	0.375	0.500	0.625	0.750	0.875
Fraction							

24.

Decimal	0.1	0.2	0.3	0.4	0.5	0.6	0.7	0.8	0.9
Fraction									

Write each decimal as a fraction in lowest terms.

25. 0.15 **26.** 0.45 **27.** 0.08 **28.** 0.06 **29.** 0.375 **30.** 0.475

Write each decimal as a mixed number.

31. 5.6 **32.** 8.4 **33.** 5.06 **34.** 8.04 **35.** 1.22 **36.** 2.11

B Simplify each of the following as much as possible, and write all answers as decimals.

37. $\frac{1}{2}(2.3 + 2.5)$ **38.** $\frac{3}{4}(1.8 + 7.6)$ **39.** $\dfrac{1.99}{\frac{1}{2}}$ **40.** $\dfrac{2.99}{\frac{1}{2}}$

41. $3.4 - \frac{1}{2}(0.76)$ **42.** $6.7 - \frac{1}{5}(0.45)$ **43.** $\frac{2}{5}(0.3) + \frac{3}{5}(0.3)$ **44.** $\frac{1}{8}(0.7) + \frac{3}{8}(0.7)$

45. $6\left(\frac{3}{5}\right)(0.02)$ **46.** $8\left(\frac{4}{5}\right)(0.03)$ **47.** $\frac{5}{8} + 0.35\left(\frac{1}{2}\right)$ **48.** $\frac{7}{8} + 0.45\left(\frac{3}{4}\right)$

49. $\left(\frac{1}{3}\right)^2(5.4) + \left(\frac{1}{2}\right)^3(3.2)$ **50.** $\left(\frac{1}{5}\right)^2(7.5) + \left(\frac{1}{4}\right)^2(6.4)$ **51.** $(0.25)^2 + \left(\frac{1}{4}\right)^2(3)$ **52.** $(0.75)^2 + \left(\frac{1}{4}\right)^2(7)$

Applying the Concepts

53. Price of Beef If each pound of beef costs $2.59, how much does $3\frac{1}{4}$ pounds cost?

54. Price of Gasoline What does it cost to fill a $15\frac{1}{2}$-gallon gas tank if the gasoline is priced at 305.2¢ per gallon? Convert your answer to dollars.

55. Sale Price A dress that costs $57.99 is on sale for $\frac{1}{3}$ off. What is the sale price of the dress?

56. Sale Price A suit that normally sells for $121 is on sale for $\frac{1}{4}$ off. What is the sale price of the suit?

57. Perimeter of the Sierpinski Triangle The diagram shows one stage of what is known as the Sierpinski triangle. Each triangle in the diagram has three equal sides. The large triangle is made up of 4 smaller triangles. If each side of the large triangle is 2 inches, and each side of the smaller triangles is 1 inch, what is the perimeter of the shaded region?

58. Perimeter of the Sierpinski Triangle The diagram shows another stage of the Sierpinski triangle. Each triangle in the diagram has three equal sides. The largest triangle is made up of a number of smaller triangles. If each side of the large triangle is 2 inches, and each side of the smallest triangles is 0.5 inch, what is the perimeter of the shaded region?

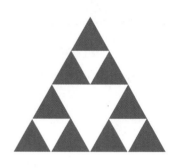

59. Average Gain in Stock Price The table below shows the amount of gain each day of one week in 2011 for the price of Verizon stock. Complete the table by converting each fraction to a decimal and rounding to the nearest hundredth if necessary.

	Change in Stock Price	
Day	Gain ($)	As a Decimal ($) (to the nearest hundredth)
Monday	$\frac{3}{5}$	
Tuesday	$\frac{1}{2}$	
Wednesday	$\frac{1}{25}$	
Thursday	$\frac{1}{5}$	
Friday	$\frac{1}{10}$	

60. Average Gain in Stock Price The table below shows the amount of gain each day of one week in 2011 for the price of AT&T stock. Complete the table by converting each fraction to a decimal and rounding to the nearest hundredth, if necessary.

	Change in Stock Price	
Day	Gain	As a Decimal ($) (to the nearest hundredth)
Monday	$\frac{3}{10}$	
Tuesday	$\frac{3}{50}$	
Wednesday	$\frac{2}{25}$	
Thursday	$\frac{1}{10}$	
Friday	0	

61. Nutrition If 1 ounce of ground beef contains 50.75 calories and 1 ounce of halibut contains 27.5 calories, what is the difference in calories between a $4\frac{1}{2}$-ounce serving of ground beef and a $4\frac{1}{2}$-ounce serving of halibut?

62. Nutrition If a 1-ounce serving of baked potato contains 48.3 calories and a 1-ounce serving of chicken contains 24.6 calories, how many calories are in a meal of $5\frac{1}{4}$ ounces of chicken and a $3\frac{1}{3}$-ounce baked potato?

Taxi Ride The Texas Junior College Teachers Association annual conference was held in Austin. At that time, a taxi ride in Austin was $1.25 for the first $\frac{1}{5}$ of a mile and $0.25 for each additional $\frac{1}{5}$ of a mile. The charge for a taxi to wait is $12.00 per hour. Use this information for Problems 63 through 66.

63. If the distance from one of the convention hotels to the airport is 7.5 miles, how much will it cost to take a taxi from that hotel to the airport?

64. If you were to tip the driver of the taxi in Problem 63 $1.50, how much would it cost to take a taxi from the hotel to the airport?

65. Suppose the distance from one of the hotels to one of the western dance clubs in Austin is 12.4 miles. If the fare meter in the taxi gives the charge for that trip as $16.50, is the meter working correctly?

66. Suppose that the distance from a hotel to the airport is 8.2 miles, and the ride takes 20 minutes. Is it more expensive to take a taxi to the airport or to just sit in the taxi?

Getting Ready for the Next Section

Expand and simplify.

67. 6^2

68. 8^2

69. 5^2

70. 10^2

71. 5^3

72. 2^5

73. 3^2

74. 2^3

75. $\left(\frac{1}{3}\right)^4$

76. $\left(\frac{3}{4}\right)^3$

77. $\left(\frac{5}{6}\right)^2$

78. $\left(\frac{3}{5}\right)^3$

79. $(0.5)^2$

80. $(0.1)^3$

81. $(1.2)^2$

82. $(2.1)^2$

83. $3^2 + 4^2$

84. $5^2 + 12^2$

85. $6^2 + 8^2$

86. $2^2 + 3^2$

Find the Mistake

Each sentence below contains a mistake. Circle the mistake and write the correct word(s) or numbers(s) on the line provided.

1. The correct way to write $\frac{6}{11}$ as a decimal is 0.54545454.... _____

2. Writing 14.3 as a fraction gives us $14 \cdot \frac{3}{10} = \frac{42}{10} = \frac{21}{5}$. _____

3. The simplified answer to the problem $\frac{12}{45(0.256 + 0.14)}$ contains both fractions and decimals. _____

4. Simplifying the problem $\left(\frac{3}{2}\right)(0.5) + \left(\frac{1}{2}\right)^2(6.7)$ by first converting all decimals to fractions gives us $\left(\frac{3}{2}\right)\left(\frac{1}{2}\right) + \left(\frac{1}{2}\right)^2\left(\frac{67}{100}\right)$.

OBJECTIVES

A Simplify expressions involving square roots.

B Use a calculator to find square roots.

C Use the Pythagorean theorem to solve problems.

KEY WORDS

square root

radical sign

perfect squares

irrational number

Pythagorean theorem

spiral of roots

Video Examples
Section 3.6

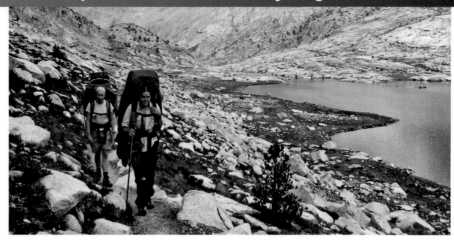

Imagine you are a seasoned backpacker and decide to hike through Evolution Valley in Kings Canyon National Park, California. You hike directly east along one leg of a trail for 6 miles, then turn north and hike for 8 more miles. If you were to draw a straight line from where you began to where you are now, what would be the distance? There is a formula in geometry that gives the distance d:

$$d = \sqrt{6^2 + 8^2}$$

where $\sqrt{}$ is called the square root symbol. If we simplify what is under the square root symbol, we have

$$d = \sqrt{36 + 64}$$
$$= \sqrt{100}$$

The expression $\sqrt{100}$ stands for the number we square to get 100. Because $10 \cdot 10 = 100$, that number is 10. Therefore, the straight-line distance from the starting point of your hike to where you are now is 10 miles. We'll discuss square roots in detail this section.

A Square Roots

Previously, we did some work with exponents. In particular, we spent some time finding squares of numbers. For example, we considered expressions like this:

$$5^2 = 5 \cdot 5 = 25$$
$$7^2 = 7 \cdot 7 = 49$$
$$x^2 = x \cdot x$$

We say that "the square of 5 is 25" and "the square of 7 is 49." To square a number, we multiply it by itself. When we ask for the *square root* of a given number, we want to know what number we square in order to obtain the given number. We say that the square root of 49 is 7, because 7 is the number we square to get 49. Likewise, the square root of 25 is 5, because $5^2 = 25$. The symbol we use to denote square root is $\sqrt{}$, which is also called a *radical sign*. Here is the precise definition of square root:

> **DEFINITION** square root
>
> The **square root** of a positive number a, written \sqrt{a}, is the number we square to get a.
>
> $$\text{If} \quad \sqrt{a} = b \quad \text{then} \quad b^2 = a.$$

Note The square root we are describing here is actually the principal or positive square root. There is another square root that is a negative number. We won't see it in this book, but, if you go on to take an algebra course, you will see it there.

We list some common square roots in Table 1.

Table 1		
Statement	In Words	Reason
$\sqrt{0} = 0$	The square root of 0 is 0	Because $0^2 = 0$
$\sqrt{1} = 1$	The square root of 1 is 1	Because $1^2 = 1$
$\sqrt{4} = 2$	The square root of 4 is 2	Because $2^2 = 4$
$\sqrt{9} = 3$	The square root of 9 is 3	Because $3^2 = 9$
$\sqrt{16} = 4$	The square root of 16 is 4	Because $4^2 = 16$
$\sqrt{25} = 5$	The square root of 25 is 5	Because $5^2 = 25$

Numbers like 1, 9, and 25, whose square roots are whole numbers, are called *perfect squares*. To find the square root of a perfect square, we look for the whole number that is squared to get the perfect square. The following examples involve square roots of perfect squares.

EXAMPLE 1 Simplify: $7\sqrt{64}$.

Solution The expression $7\sqrt{64}$ means 7 times $\sqrt{64}$. To simplify this expression, we write $\sqrt{64}$ as 8 and multiply.

$$7\sqrt{64} = 7 \cdot 8 = 56$$

We know $\sqrt{64} = 8$, because $8^2 = 64$.

EXAMPLE 2 Simplify: $\sqrt{9} + \sqrt{16}$.

Solution We write $\sqrt{9}$ as 3 and $\sqrt{16}$ as 4. Then we add.

$$\sqrt{9} + \sqrt{16} = 3 + 4 = 7$$

EXAMPLE 3 Simplify: $\sqrt{\dfrac{25}{81}}$.

Solution We are looking for the number we square to get $\frac{25}{81}$. We know that when we multiply two fractions, we multiply the numerators and multiply the denominators. Because $5 \cdot 5 = 25$ and $9 \cdot 9 = 81$, the square root of $\frac{25}{81}$ must be $\frac{5}{9}$.

$$\sqrt{\frac{25}{81}} = \frac{5}{9} \quad \text{because} \quad \left(\frac{5}{9}\right)^2 = \frac{5}{9} \cdot \frac{5}{9} = \frac{25}{81}$$

In Examples 4–6, we simplify each expression as much as possible.

EXAMPLE 4 Simplify: $12\sqrt{25} = 12 \cdot 5 = 60$.

EXAMPLE 5 Simplify: $\sqrt{100} - \sqrt{36} = 10 - 6 = 4$.

EXAMPLE 6 Simplify: $\sqrt{\dfrac{49}{121}} = \dfrac{7}{11}$ because $\left(\dfrac{7}{11}\right)^2 = \dfrac{7}{11} \cdot \dfrac{7}{11} = \dfrac{49}{121}$.

Practice Problems

1. Simplify: $5\sqrt{81}$.

2. Simplify: $\sqrt{25} + \sqrt{36}$.

3. Simplify: $\sqrt{\dfrac{36}{49}}$.

4. Simplify: $25\sqrt{9}$.

5. Simplify: $\sqrt{144} - \sqrt{36}$.

6. Simplify: $\sqrt{\dfrac{81}{144}}$.

Answers
1. 45
2. 11
3. $\frac{6}{7}$
4. 75
5. 6
6. $\frac{3}{4}$

So far in this section we have been concerned only with square roots of perfect squares. The next question is, "What about square roots of numbers that are not perfect squares, like $\sqrt{7}$, for example?" We know that

$$\sqrt{4} = 2 \quad \text{and} \quad \sqrt{9} = 3$$

And because 7 is between 4 and 9, $\sqrt{7}$ should be between $\sqrt{4}$ and $\sqrt{9}$. That is, $\sqrt{7}$ should be between 2 and 3. But what is it exactly? The answer is, we cannot write it exactly in decimal or fraction form. Because of this, it is called an *irrational number*. We can approximate it with a decimal, but we can never write it exactly with a decimal. Table 2 gives some decimal approximations for $\sqrt{7}$. The decimal approximations were obtained by using a calculator. We could continue the list to any accuracy we desired. However, we would never reach a number in decimal form whose square was exactly 7.

Table 2

Approximations for the Square Root of 7

Accurate to the nearest	The square root of 7 is	Check by squaring
Tenth	$\sqrt{7} = 2.6$	$(2.6)^2 = 6.76$
Hundredth	$\sqrt{7} = 2.65$	$(2.65)^2 = 7.0225$
Thousandth	$\sqrt{7} = 2.646$	$(2.646)^2 = 7.001316$
Ten thousandth	$\sqrt{7} = 2.6458$	$(2.6458)^2 = 7.00025764$

B Using Calculators to Find Square Roots

7. Give a decimal approximation for the expression $6\sqrt{10}$ that is accurate to the nearest thousandth.

EXAMPLE 7 Give a decimal approximation for the expression $5\sqrt{12}$ that is accurate to the nearest ten thousandth.

Solution Let's agree not to round to the nearest ten thousandth until we have first done all the calculations. Using a calculator, we find $\sqrt{12} \approx 3.4641016$. Therefore,

$$5\sqrt{12} \approx 5(3.4641016) \qquad \sqrt{12} \text{ on calculator}$$
$$= 17.320508 \qquad \text{Multiply.}$$
$$= 17.3205 \qquad \text{Round to the nearest ten thousandth.}$$

8. Approximate $\sqrt{245} - \sqrt{131}$ to the nearest thousandth.

EXAMPLE 8 Approximate $\sqrt{301} + \sqrt{137}$ to the nearest hundredth.

Solution Using a calculator to approximate the square roots, we have

$$\sqrt{301} + \sqrt{137} \approx 17.349352 + 11.704700 = 29.054052$$

To the nearest hundredth, the answer is 29.05.

9. Approximate $\sqrt{\dfrac{9}{26}}$ to the nearest ten thousandth.

EXAMPLE 9 Approximate $\sqrt{\dfrac{7}{11}}$ to the nearest thousandth.

Solution Because we are using calculators, we first change $\frac{7}{11}$ to a decimal and then find the square root.

$$\sqrt{\frac{7}{11}} \approx \sqrt{0.6363636} \approx 0.7977240$$

To the nearest thousandth, the answer is 0.798.

Answers
7. 18.974
8. 4.207
9. 0.5883

C The Pythagorean Theorem

FACTS FROM GEOMETRY Pythagorean Theorem

A *right triangle* is a triangle that contains a 90° (or right) angle. The longest side in a right triangle is called the *hypotenuse*, and we use the letter c to denote it. The two shorter sides are denoted by the letters a and b. The Pythagorean theorem states that the hypotenuse is the square root of the sum of the squares of the two shorter sides. Here is the previous statement in symbols:

$$c = \sqrt{a^2 + b^2}$$

EXAMPLE 10 Find the length of the hypotenuse in each right triangle.

a.

3 m
4 m

b.

5 in.
7 in.

Solution We apply the formula given above.

a. When $a = 3$ and $b = 4$,

$$c = \sqrt{3^2 + 4^2}$$
$$= \sqrt{9 + 16}$$
$$= \sqrt{25}$$
$$= 5 \text{ meters}$$

b. When $a = 5$ and $b = 7$,

$$c = \sqrt{5^2 + 7^2}$$
$$= \sqrt{25 + 49}$$
$$= \sqrt{74}$$
$$\approx 8.60 \text{ inches}$$

In part a, the solution is a whole number, whereas in part b, we must use a calculator to get 8.60 as an approximation to $\sqrt{74}$.

EXAMPLE 11 A ladder is leaning against a barn to reach a hay loft 6-feet above the barn floor. If the bottom of the ladder is 8 feet from the wall, how long is the ladder?

Solution A picture of the situation is shown in Figure 2. We let c denote the length of the ladder. Applying the Pythagorean theorem, we have

$$c = \sqrt{6^2 + 8^2}$$
$$= \sqrt{36 + 64}$$
$$= \sqrt{100}$$
$$= 10 \text{ feet}$$

The ladder is 10 feet long.

6 ft
8 ft

FIGURE 2

10. Find the length of the hypotenuse in a right triangle with the following leg lengths.

 a. 6 ft, 8 ft

 b. 9 m, 14 m

11. A rope is tied from the top of a 12-foot pole to the ground 5 feet from the base of the pole. If there is no slack in the rope, how long is the rope?

x
12 ft
5 ft

FIGURE 1

Answers

10. a. 10 ft **b.** 16.64 m

11. 13 ft

FACTS FROM GEOMETRY The Spiral of Roots

The diagram showed here is called the *spiral of roots*. It is constructed using the Pythagorean theorem and it gives us a way to visualize positive square roots. The table below gives us the decimal equivalents (some of which are approximations) of the first 10 square roots in the spiral. The line graph can be constructed from the table or from the spiral.

Approximate length of diagonals

Number	Positive Square Root
1	1
2	1.41
3	1.73
4	2
5	2.24
6	2.45
7	2.65
8	2.83
9	3
10	3.16

To visualize the square roots of the counting numbers, we can construct the spiral of roots another way. To begin, we draw two line segments, each of length 1, at right angles to each other. Then we use the Pythagorean theorem to find the length of the diagonal. Figure 3 illustrates:

$$x = \sqrt{1^2 + 1^2}$$
$$x = \sqrt{2}$$

FIGURE 3

$$x = \sqrt{(\sqrt{2})^2 + 1^2}$$
$$= \sqrt{2 + 1}$$
$$= \sqrt{3}$$

FIGURE 4

$$x = \sqrt{(\sqrt{3})^2 + 1^2}$$
$$= \sqrt{3 + 1}$$
$$= \sqrt{4}$$
$$= 2$$

FIGURE 5

Next, we construct a second triangle by connecting a line segment of length 1 to the end of the first diagonal so that the angle formed is a right angle. We find the length of the second diagonal using the Pythagorean theorem. Figure 4 illustrates this procedure. As we continue to draw new triangles by connecting line segments of length 1 to the end of each previous diagonal, so that the angle formed is a right angle, the spiral of roots begins to appear (see Figure 5).

GETTING READY FOR CLASS

After reading through the preceding section, respond in your own words and in complete sentences.

A. Which number is larger, the square of 10 or the square root of 10?

B. Give a definition for the square root of a number.

C. What two numbers will the square root of 20 fall between?

D. What is the Pythagorean theorem?

Vocabulary Review

Choose the correct words to fill in the blanks below.

spiral of roots hypotenuse irrational radical sign

Pythagorean theorem square root whole numbers right

1. The _____ of a positive number a is the number we square to get a.

2. The symbol we use to denote a square root is called a _____.

3. Numbers whose square roots are _____ are called perfect squares.

4. An _____ number, such as $\sqrt{7}$, is a number that can be approximated with a decimal number but can never be written exactly as a decimal or fraction.

5. A _____ triangle is a triangle that contains a 90 degree angle.

6. The longest side of a right triangle is called the _____.

7. The _____ states that the hypotenuse is the square root of the sum of the squares of the two shorter sides.

8. The _____ is constructed using the Pythagorean theorem and is a way to visualize positive square roots of the counting numbers.

Problems

A Find each of the following square roots without using a calculator.

1. $\sqrt{64}$ **2.** $\sqrt{100}$ **3.** $\sqrt{81}$ **4.** $\sqrt{49}$

5. $\sqrt{36}$ **6.** $\sqrt{144}$ **7.** $\sqrt{25}$ **8.** $\sqrt{169}$

Simplify each of the following expressions without using a calculator.

9. $3\sqrt{25}$ **10.** $9\sqrt{49}$ **11.** $6\sqrt{64}$ **12.** $11\sqrt{100}$

13. $15\sqrt{9}$ **14.** $8\sqrt{36}$ **15.** $16\sqrt{9}$ **16.** $9\sqrt{16}$

17. $\sqrt{49} + \sqrt{64}$ **18.** $\sqrt{1} + \sqrt{0}$ **19.** $\sqrt{16} - \sqrt{9}$ **20.** $\sqrt{25} - \sqrt{4}$

21. $3\sqrt{25} + 9\sqrt{49}$ **22.** $6\sqrt{64} + 11\sqrt{100}$ **23.** $15\sqrt{9} - 9\sqrt{16}$ **24.** $7\sqrt{49} - 2\sqrt{4}$

25. $\sqrt{\dfrac{16}{49}}$

26. $\sqrt{\dfrac{100}{121}}$

27. $\sqrt{\dfrac{36}{64}}$

28. $\sqrt{\dfrac{81}{144}}$

29. $\sqrt{\dfrac{4}{9}}$

30. $\sqrt{\dfrac{25}{16}}$

31. $\sqrt{\dfrac{1}{16}}$

32. $\sqrt{\dfrac{1}{100}}$

Indicate whether each of the statements in Problems 33–36 is *True* or *False*.

33. $\sqrt{4} + \sqrt{9} = \sqrt{4 + 9}$

34. $\sqrt{\dfrac{16}{25}} = \dfrac{\sqrt{16}}{\sqrt{25}}$

35. $\sqrt{25 \cdot 9} = \sqrt{25} \cdot \sqrt{9}$

36. $\sqrt{100} - \sqrt{36} = \sqrt{100 - 36}$

B Calculator Problems

Use a calculator to work problems 37 through 56.

Approximate each of the following square roots to the nearest ten thousandth.

37. $\sqrt{1.25}$

38. $\sqrt{12.5}$

39. $\sqrt{125}$

40. $\sqrt{1250}$

Approximate each of the following expressions to the nearest hundredth.

41. $2\sqrt{3}$

42. $3\sqrt{2}$

43. $5\sqrt{5}$

44. $5\sqrt{3}$

45. $\dfrac{\sqrt{3}}{3}$

46. $\dfrac{\sqrt{2}}{2}$

47. $\sqrt{\dfrac{1}{3}}$

48. $\sqrt{\dfrac{1}{2}}$

Approximate each of the following expressions to the nearest thousandth.

49. $\sqrt{12} + \sqrt{75}$

50. $\sqrt{18} + \sqrt{50}$

51. $\sqrt{87}$

52. $\sqrt{68}$

53. $2\sqrt{3} + 5\sqrt{3}$

54. $3\sqrt{2} + 5\sqrt{2}$

55. $7\sqrt{3}$

56. $8\sqrt{2}$

C Find the length of the hypotenuse in each right triangle. Round to the nearest hundredth, if rounding is necessary.

57.

58.

59.

60.

61.

62.

63.

64.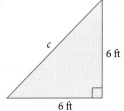

Applying the Concepts

65. Geometry One end of a wire is attached to the top of a 24-foot pole; the other end of the wire is anchored to the ground 18 feet from the bottom of the pole. If the pole makes an angle of 90° with the ground, find the length of the wire.

66. Geometry Two children are trying to cross a stream. They want to use a log that goes from one bank to the other. If the left bank is 5 feet higher than the right bank and the stream is 12 feet wide, how long must a log be to just barely reach?

67. Geometry A ladder is leaning against the top of a 15-foot wall. If the bottom of the ladder is 20 feet from the wall, how long is the ladder?

68. Geometry A wire from the top of a 24-foot pole is fastened to the ground by a stake that is 10 feet from the bottom of the pole. How long is the wire?

69. Spiral of Roots Construct your own spiral of roots by using a ruler. Draw the first triangle by using two 1-inch lines. The first diagonal will have a length of $\sqrt{2}$ inches. Each new triangle will be formed by drawing a 1-inch line segment at the end of the previous diagonal so that the angle formed is 90°. Draw your spiral until you have at least six right triangles.

70. Spiral of Roots Construct a spiral of roots by using line segments of length 2 inches. The length of the first diagonal will be $2\sqrt{2}$ inches. The length of the second diagonal will be $2\sqrt{3}$ inches. What will be the length of the third diagonal?

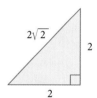

Use a calculator to work problems 71 and 72.

71. Lighthouse Problem The higher you are above the ground, the farther you can see. If your view is unobstructed, then the distance in miles that you can see from h feet above the ground is given by the formula

$$d = \sqrt{\frac{3h}{2}}$$

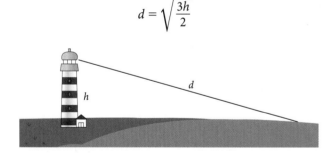

The following figure shows a lighthouse with a door and windows at various heights. The preceding formula can be used to find the distance to the ocean horizon from these heights. Use the formula and a calculator to complete the following table. Round your answers to the nearest whole number.

Height h (feet)	Distance d (miles)
10	
50	
90	
130	
170	
190	

72. Pendulum Problem The time (in seconds) it takes for the pendulum on a clock to swing through one complete cycle is given by the formula

$$T = \frac{11}{7}\sqrt{\frac{L}{2}}$$

where L is the length (in feet) of the pendulum. Use this formula and a calculator to complete the following table. Round your answers to the nearest hundredth.

Length L (feet)	Time T (seconds)
1	
2	
3	
4	
5	
6	

Improving Your Quantitative Literacy

73. Millionaire Households The chart shows the number of millionaire households over a ten year period. Use the information to answer the following questions.

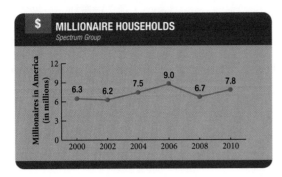

a. If the trend from 2008 to 2010 continues, how many millionaire households will there be in 2012?

b. Estimate the number of millionaire households in 2005.

Find the Mistake

Each sentence below contains a mistake. Circle the mistake and write the correct word(s) or numbers(s) on the line provided.

1. The square root of a positive number x is the number we square to get \sqrt{x}. _____

2. The square root of 225 is 15, and can be written in symbols as $\sqrt{15} = 225$. _____

3. Simplifying the radical $\sqrt{\frac{196}{36}}$ gives us $\frac{98}{36}$ because $\left(\frac{98}{36}\right)^2 = \frac{98}{36} + \frac{98}{36} = \frac{196}{36}$. _____

4. The Pythagorean theorem states that $b = \sqrt{a^2 + c^2}$. _____

Chapter 3 Summary

EXAMPLES

1. The number 4.123 in words is "four and one hundred twenty-three thousandths."

Place Value [3.1]

The place values for the first five places to the right of the decimal point are

Decimal Point	Tenths	Hundredths	Thousandths	Ten Thousandths	Hundred Thousandths
.	$\frac{1}{10}$	$\frac{1}{100}$	$\frac{1}{1,000}$	$\frac{1}{10,000}$	$\frac{1}{100,000}$

Rounding Decimals [3.1]

2. 357.753 rounded to the nearest
Tenth: 357.8
Ten: 360

If the digit in the column to the right of the one we are rounding to is 5 or more, we add 1 to the digit in the column we are rounding to; otherwise, we leave it alone. We then replace all digits to the right of the column we are rounding to with zeros if they are to the left of the decimal point; otherwise, we simply delete them.

Addition and Subtraction with Decimals [3.2]

3.
```
   3.400
  25.060
+  0.347
--------
  28.807
```

To add (or subtract) decimal numbers, we align the decimal points and add (or subtract) as if we were adding (or subtracting) whole numbers. The decimal point in the answer goes directly below the decimal points in the problem.

Multiplication with Decimals [3.3]

4. If we multiply 3.49×5.863, there will be a total of $2 + 3 = 5$ digits to the right of the decimal point in the answer.

To multiply two decimal numbers, we multiply as if the decimal points were not there. The decimal point in the product has as many digits to the right as there are total digits to the right of the decimal points in the two original numbers.

Division with Decimals [3.4]

5.
```
           1.39
    2.5.)3.4.75
        25
        --
         97
         75
        ---
        225
        225
        ---
          0
```

To begin a division problem with decimals, we make sure that the divisor is a whole number. If it is not, we move the decimal point in the divisor to the right as many places as it takes to make it a whole number. We must then be sure to move the decimal point in the dividend the same number of places to the right. Once the divisor is a whole number, we divide as usual. The decimal point in the answer is placed directly above the decimal point in the dividend.

Changing Fractions to Decimals [3.5]

To change a fraction to a decimal, we divide the numerator by the denominator.

6. $\dfrac{4}{15} = 0.2\overline{6}$ because

$$
\begin{array}{r}
.266 \\
15\overline{)4.000} \\
\underline{30}\downarrow \\
100 \\
\underline{90}\downarrow \\
100 \\
\underline{90} \\
10
\end{array}
$$

Changing Decimals to Fractions [3.5]

To change a decimal to a fraction, we write the digits to the right of the decimal point over the appropriate power of 10.

7. $0.781 = \dfrac{781}{1{,}000}$

Square Roots [3.6]

The square root of a positive number a, written \sqrt{a}, is the number we square to get a.

8. $\sqrt{49} = 7$ because
$7^2 = 7 \cdot 7 = 49$

Pythagorean Theorem [3.6]

In any right triangle, the length of the longest side (the hypotenuse) is equal to the square root of the sum of the squares of the two shorter sides.

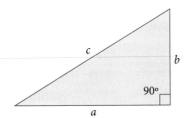

$c = \sqrt{a^2 + b^2}$

1. Write the decimal number 11.819 in words. [3.1]

2. Give the place value of the 8 in the number 61.8276. [3.1]

3. Write seventy-three and forty-six ten thousandths as a decimal number. [3.1]

4. Round 100.9052 to the nearest hundredth. [3.1]

Perform the following operations. [3.2, 3.3, 3.4]

5. $6 + 0.8 + 0.22$

6. $28.332 - 16.608$

7. $6.9(2.40)$

8. $96.4768 \div 16.52$

9. $6.8 + 3.3$

10. $16.47 - 8.58$

11. $0.5(3.7)(-1.8)$

12. $11.616 \div 52.8$

13. A person purchases $11.39 worth of goods at a drugstore. If a $10 bill and a $5 bill were used to pay for the purchases, how much change is received? [3.2]

14. If coffee sells for $3.29 per pound, how much will 5.5 pounds of coffee cost? [3.3]

15. If a person earns $489 for working 60 hours, what is the person's hourly wage? [3.4]

16. Write $\frac{17}{25}$ as a decimal. [3.5]

17. Write 0.38 as a fraction in lowest terms. [3.5]

18. $6.8(3.7 + 0.08)$

19. $6.1 - 4(0.93)$

20. $41.901 - 7(3.11 + 0.462)$ **21.** $\frac{7}{8}(0.11) + \frac{5}{6}(0.45)$

22. $23.4 - 8(6.01 - 4.2)$ **23.** $\frac{1}{5}(0.17) + 8(6.13 - 2.8)$

Find each of the following square roots without using a calculator. [3.6]

24. $\sqrt{100}$

25. $\sqrt{9}$

Simplify each of the following without using a calculator. [3.6]

26. $3\sqrt{25}$

27. $10\sqrt{9}$

Find the length of the hypotenuse in each right triangle. Round to the nearest hundredth if necessary. [3.6]

28.

29.

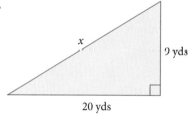

The diagram shows the annual sales for different magazines. Use the information to answer the following questions.

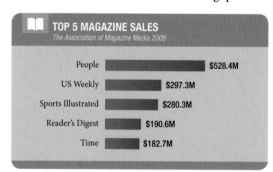

30. What is the average monthly sales for People Magazine?

31. What are the average monthly sales for US Weekly and Time Magazine combined?

Unit Analysis

4.1 Introduction to Dimensional Analysis

4.2 Conversions in the Metric System

4.3 Distance and Length

4.4 Volume and Capacity

4.5 Weight and Mass

4.6 Temperature Conversions

Machu Picchu, Peru
Image © 2010 DigitalGlobe
Image © 2010 GeoEye

Unit Analysis Supplement

Rhodes State College
Developed by Rodney Null

The ruins of Machu Picchu stand high in the Andes mountains of Peru and are among the most beautiful and mysterious ancient building sites in the world. Most scholars believe that the Incas began construction on the site around 1400 AD. Archaeological excavations have found temples, baths, palaces, storage rooms, and 150 houses. Builders used blocks weighing up to 50 tons to create the structures. Despite the primitive technology of the time, the blocks form tight mortar-less joints that the blade of a knife cannot fit between. Alignment of some of the structures also shows remarkable knowledge of astronomy still being studied today. And the surrounding agricultural terraces and local springs are believed to have produced enough food and water to make the small city self-supporting.

The ruins are also very well preserved, adding to their archaeological value. At 2,350 meters above sea level, they sit on the edge of a sheer cliff and are invisible from the river that flows through the Urubamba Valley 600 meters below. Though the Spanish conquerors reportedly searched for the location a century after its construction, they never found it. As a result, it is one of the only Incan settlements that was not plundered by the conquering armies.

Suppose you want to convert the above heights to the more familiar units of feet and miles. You could use a table like the one shown below and a process called unit analysis, which is one of the topics of this chapter. Unit analysis can be used to compare any units of measure, such as those for length, weight, temperature, and volume. Once you have worked through this chapter, come back and convert these distances so you can begin planning your trip to Machu Picchu!

Conversion Factors Between the Metric and U.S. Systems of Measurement	
Length	
inches and centimeters	2.54 cm = 1 in.
feet and meters	1 m = 3.28 ft
miles and kilometers	1.61 km = 1 mi

Farmers Co-op Swap Meet

A regional cooperative of farmers has organized a distribution center where growers can exchange farm goods of equal value. This gives someone who, for example, only raises potatoes the opportunity to trade some of their potatoes for tomatoes, zucchini, apples, etc. The farmers like this because it gives them the ability to offer customers a variety of products which they themselves did not grow. They also appreciate that it is a swapping of goods and requires no money.

To eliminate individual haggling over fair exchange rates, prior to each swap meet, the cooperative establishes the exchange rates for the day's business. These rates change from one meeting to the next based upon a number of factors such as season, rain fall, demand, etc.

On one trading day some of the exchange rates agreed upon are as follows:

3 apples	=	5 tomatoes
2 potatoes	=	1 tomato
2 ears (of corn)	=	7 zucchinis
4 tomatoes	=	3 ears (of corn)

To make the calculations easier, each farmer gets a deck of conversion cards based on the equivalent values for that trading day. For example the apples to tomatoes card for the exchange rate above looks like:

one side	flip side
3 apples	5 tomatoes
5 tomatoes	3 apples

Here is how they use the cards.

EXAMPLE 1 If Sunshine Farms has 600 tomatoes they wish to exchange for apples they would arrange the following calculation using a card.

Solution Create a conversion card

600 tomatoes	3 apples
1	5 tomatoes

And then perform the calculation:

$$600 \times \frac{3}{5} = 360 \text{ apples.}$$

EXAMPLE 2 If Evergreen Farms has 1500 apples they wish to exchange for tomatoes they would arrange the following calculation using a card.

Solution Create a conversion card:

1500 apples	5 tomatoes
1	3 apples

And then perform the calculation:

$$1500 \times \frac{5}{3} = 2500 \text{ tomatoes}$$

**Video Examples
Section 4.1**

EXAMPLE 3 If Veggie Best Farms has 420 zucchinis they wish to trade for tomatoes they would arrange the following calculation using cards.

Solution Create a conversion card:

420 zucchinis	2 ears	4 tomatoes
1	7 zucchinis	3 ears

And then perform the calculation:

$$420 \times \frac{2}{7} \times \frac{4}{3} = 160 \text{ tomatoes}$$

EXAMPLE 4 If Only Taters Farms has 3600 potatoes they wish to swap for zucchini they would arrange the following calculation using cards.

Solution Create a conversion card:

3600 potatoes	1 tomato	3 ears	7 zucchinis
1	2 potatoes	4 tomatoes	2 ears

And then perform the calculation:

$$3600 \times \frac{1}{2} \times \frac{3}{4} \times \frac{7}{2} = 4725 \text{ zucchinis}$$

GETTING READY FOR CLASS

After reading through the preceding section, respond in your own words and in complete sentences.

A. Explain how the cards are chosen and arranged.

B. Explain how a calculation is done from an arrangement of cards.

Check with your instructor or classmates

Problems

Using the conversion table, show the arrangement of cards and results for the following exchanges of produce.

3 apples	=	5 tomatoes
2 potatoes	=	1 tomato
2 ears (of corn)	=	7 zucchinis
4 tomatoes	=	3 ears (of corn)

1. 450 apples swapped for tomatoes

2. 240 ears of corn traded for tomatoes

3. 126 zucchinis exchanged for ears of corn

4. 10 dozen ears of corn swapped for zucchinis

5. 427 tomatoes traded for apples

6. 1 gross of potatoes exchanged for tomatoes

7. 700 potatoes traded for apples

8. 640 tomatoes swapped for zucchinis

9. 225 potatoes exchanged for ears of corn

10. 1500 zucchinis swapped for potatoes

11. 75 apples traded for ears of corn

12. 500 apples exchanged for zucchinis

Developing Dimensional Analysis Techniques

You have some expertise in converting zucchinis to apples, etc. at the swap meet. Let's discuss the mathematical concepts at work when you use an arrangement of conversion cards to determine the appropriate calculations involved in a swap. There are two.

(1) Since the fractions displayed on each card are based on produce of equal value, we can say both the numerator and the denominator of the fraction have equal value. Based on your experience, when the numerator and denominator of a fraction have equal value what is the overall value of the fraction?

For Example: $\dfrac{5}{5} = ?$ or $\dfrac{19}{19} = ?$ or $\dfrac{846}{846} = ?$

That's right! They have a value of 1

(2) Your arrangements of cards represented multiplying the initial amount of produce by fractions whose values are 1. Based on your experience, what happens to the value of something when you multiply it by 1?

For Example: $63 \times 1 = ?$ or $1 \times 375 = ?$ or $916 \times 1 = ?$

That's right! The product is the same as the value of the number being multiplied by the 1.

> *Note* This concept is formally known as the *Identity Property of Multiplication* and 1 is referred to as the *Multiplicative Identity*. It is often stated symbolically as follows:
> $$1 \times a = a \times 1 = a$$

So at the swap meet, when a farmer made an arrangement of cards to calculate a swap, like:

3600 potatoes	1 tomato	3 ears corn	7 zucchinis
1	2 potatoes	4 tomatoes	2 ears corn

The mathematical calculation behind it is:

$$3600 \text{ potatoes} = \frac{3600 \text{ potatoes}}{1} \times \frac{1 \text{ tomato}}{2 \text{ potatoes}} \times \frac{3 \text{ ears corn}}{4 \text{ tomatoes}} \times \frac{7 \text{ zucchinis}}{2 \text{ ears corn}}$$

The 3600 potatoes are getting multiplied by three ones. Very strange looking ones, but since the numerator and denominator have equal value ones none the less. And because 1 is the multiplicative identity the initial value that the 3600 potatoes represent is preserved, but that value has been changed into 4725 zucchinis.

Now let's apply the same ideas to covert units in a measurement.

EXAMPLE 5 Suppose you want to convert a length of 2 feet to an equivalent length only in inches.

Solution First you need to know the equivalence (aka conversion factor) between feet and inches.

That is: 1 ft = 12 in

This could be arranged to make one of those "fancy looking" ones two different ways:
$$\frac{1 \text{ ft}}{12 \text{ in}} \text{ or } \frac{12 \text{ in}}{1 \text{ ft}}$$

From your experience at the swap meet you can probably guess it is the second one you want (Can you explain why?) and the calculation will look as follows:

$$2 \text{ ft} = \frac{2 \text{ ft}}{1} \times \frac{12 \text{ in}}{1 \text{ ft}} = 24 \text{ in}$$

> *Note* This can be thought of as simplifying out the "fancy one"
> $$\frac{\text{ft}}{\text{ft}} = 1$$

EXAMPLE 6 If you had a punch bowl filled with 64 cups of punch and you wanted to pour it into 1 gallon jugs for storage, how many one gallon jugs would you need?

Solution To answer this you essentially want to figure out,

$$64 \text{ cups} = ? \text{ gallons}$$

The calculation, using fancy ones, you set up to solve this will depend on the conversion factors you have for the units involved. Here are two possibilities.

Know conversion factors	Know conversion factors
1 pt = 2 c	1 c = 8 fl oz
1 qt = 2 pt	1 qt = 32 fl oz
1 gal = 4 qt	1 gal = 4 qt
Set up:	**Set up:**
$64c = \dfrac{64c}{1} \times \dfrac{1pt}{2c} \times \dfrac{1qt}{2pt} \times \dfrac{1gal}{4qt}$	$64c = \dfrac{64c}{1} \times \dfrac{8floz}{1c} \times \dfrac{1qt}{32floz} \times \dfrac{1\,gal}{4qt}$
And then with the calculator do:	**And then with the calculator do:**
64/2/2/4 to get 4 gallons	64×8/32/4 to get 4 gallons

Notice several things:
- We did not have a conversion factor that would take us directly from cups to gallons, so more than one conversion factor and corresponding "fancy one" was required.
- In each example if you can cancel/simplify out all of the units involved except for gallons.
- In typing calculations into the calculator you do not need to worry about the 1's, Remember that Identity Property?
- For the values other than one, if they are in the numerator you are multiplying by them, if they are in the denominator you are dividing by them. Does that make sense?

●

Using all of the conversion factor given in the punch bowl problem, set up two different sets of "fancy ones" calculations to convert 15 gallons to cups. Show the corresponding calculations you would type into your calculator. Be sure all units except cups eventually cancel out.

Note Some sources refer to the "fancy ones" as unit fractions.

Note The technique you are using here is referred to by names such as: Dimensional Analysis, Unit Analysis, or Factor Labeling Method. Regardless of the name, it is a powerful way to change units!

Check with your instructor or a classmate

4.2 Conversions in the Metric System

Just like we have done in the U.S. Customary System, using equivalences (a.k.a. conversion factors) and applying the techniques of dimensional analysis can be used to do unit conversions in the Metric System. That said, because of the special structure of the Metric System, there is a very convenient shortcut which most people that work with the metric system use for basic conversions to avoid difficult calculations and get quick answers without even using a calculator. Interested? If so, read on. Otherwise, feel free to do it the "hard way".

Unlike equivalences in the U.S. Customary System, all of the equivalences between units, in the Metric System, are based on powers of ten (e.g. $.001 = 10^{-3}$, $1 = 10^0$, $100 = 10^2$).

For example:

In the U.S. Customary System	In the Metric System
1 lb = 16 oz	1 cm = .01 m
1 gal = 4 qt	1 L = 1000 mL
1 mi = 5280 ft	1 dag = 10 g

This use of powers of ten in the metric system, and the fact that we work in a base ten numeration system, gives us the opportunity for a short cut. Let's look at some metric conversions using dimensional analysis.

$$15 \text{ cm} = \frac{15 \text{ cm}}{1} \times \frac{10 \text{ mm}}{1 \text{ cm}} = 150 \text{ mm}$$

or

$$2.6 \text{ g} = \frac{2.6 \text{ g}}{1} \times \frac{1 \text{ hg}}{100 \text{ g}} = .026 \text{ hg}$$

or

$$57600 \text{ mL} = \frac{57600 \text{ mL}}{1} \times \frac{1 \text{ L}}{1000 \text{ mL}} = 57.6 \text{ L}$$

or

$$.082 \text{ km} = \frac{.082 \text{ km}}{1} \times \frac{1000 \text{ m}}{1 \text{ km}} = 82 \text{ m}$$

Video Examples
Section 4.2

Notice that the arrangement of digits in the resulting measurement is the same as the arrangement of the digits in the original measurement, with some creative use of zero as place holders. What has changed is the placement of the decimal point in the arrangement of digits. So to do a basic metric conversion we really just need to figure out two things:

1. Which direction (left or right) to move the decimal point.
2. How far to move it.

Both of these questions can be answered from knowing how the prefix (beginning part) of all but the base units of measure (i.e. meter, gram, liter, etc.) are ordered. See the following chart to get started.

Prefix Structure in the Metric System (aka SI prefixes)							
Prefix	kilo	hecto	deca	base unit	deci	centi	milli
Abbreviation	k	h	da	varies	d	c	m

Using this chart let's revisit the 4 examples we just looked at.

1. **Converting 15 cm to millimeters.**
 * Since we are going to be moving the decimal point, it is important to know where it is currently located. With no decimal point shown, it can be assumed to be at the end of the number (i.e. to the right of the 5) so we are really looking at 15. cm
 * Now find the c (centi) and the m (milli) prefixes in the chart.
 * Because we are starting with cm, start at centi and jump one block at a time to milli, noting the direction and number of jumps required to get to milli.
 * The direction was to the **right** and the number of jumps is **1**. These are the directions for move the decimal point. To move the decimal point one place to the right requires putting a 0 on the 15. cm to make it 15.0 cm and the result is 150. mm or just 150 mm

Summarizing: To convert 15 cm to millimeters move the decimal point one place to the right and you get 150 mm.

2. **Converting 2.6 g to hectograms.**
 * In this problem the location of the decimal point is clear, but there is no prefix on the unit. That is because this measurement of mass is using the base unit of grams. So, start at the base unit in the chart and jump one block at a time till you get to h (hecto) again noting the direction and number of jumps required.
 * The direction was to the **left** and the number of jumps is **2**. These are the directions for moving the decimal point. Once again we need to insert 0's to allow for this move, thinking of 2.6 g as 02.6 g and then getting the resulting .026 hg after moving the decimal point

Summarizing: To convert 2.6 g to hectograms move the decimal point two places to the left and you get .026 hg.

Now it's your turn. Explain (in words) how to convert the remaining two examples with this shortcut which some folks call "sliding the decimal point".

3. Converting 57600 mL to Liters.

4. Converting .082 km to meters.

Check with your instructor or classmates

Chapter 4 Unit Analysis Supplement © Rhodes State College, Rodney Null

Prefix Structure in the Metric System (a.k.a. SI prefixes)

Prefix	kilo	hecto	deca	base unit	deci	centi	milli
Abbreviation	k	h	da	varies	d	c	m
Translation	1000	100	10	1	.1	.01	.001
Distance	kilometer	hectometer	decameter	meter	decimeter	centimeter	millimeter
Abbreviation	km	hm	dam	m	dm	cm	mm
Example Equivalences	.004 km	.04 hm	.4 dam	4 m	40 dm	400 cm	4000 mm
Volume	kiloliter	hectoliter	decaliter	liter	deciliter	centiliter	milliliter
Abbreviation	kL	hL	daL	L	dL	cL	mL
Example Equivalences	7 kL	70 hL	700 daL	7000 L	70000 dL	700000 cL	7000000 mL
Mass	kilogram	hectogram	decagram	gram	decigram	centigram	milligram
Abbreviation	kg	hg	dag	g	dg	cg	mg
Example Equivalences	.000008 kg	.00008 hg	.0008 dag	.008 g	.08 dg	.8 cg	8 mg

4.3 Distance and Length

Video Examples
Section 4.3

Note A dollar bill is a little more than 6 inches long or a little bigger than $\frac{1}{2}$ foot.

Note In some situations **m** is used to represent miles as in 30 mpg = 30 miles per gallon or 55 mph = 55 miles per hour. In the metric system, **m** is the abbreviation for meter(s). When measuring time, **m** or **min** is sometimes used to abbreviate minutes such as 6 h 15 m = 6 hours 15 minutes. This can be confusing! When you see **m** used in a problem, how will you determine what it represents? In this class, **m** will only be used to represent meters.

Attribute you are measuring

Distance/Length

Examples of tools used to measure it

Ruler	Odometer	Micrometer
Tape Measure	Trundle Wheel	Caliper

Common units used in the U.S. Customary System

- **Names**
Inches	Feet
Yards	Miles

- **References for estimation**
 1 inch — a U.S. quarter is a little less than an inch in diameter
 1 foot — a little longer than this piece of paper
 1 yard — the width of the doorway to this classroom
 1 mile — the distance your car travels in one minute when you are going 60 mph

- **Abbreviations used**
Inch (es): in or ”	(Example: 8 inches = 8 in = 8”)
Feet (foot): ft or ’	(Example: 5 feet = 5 ft = 5’)
Yard(s): yd(s)	(Example: 6 yards = 6 yds)
Mile(s): mi	(Example: 4 miles = 4 mi)

Units used in the Metric System

Names	kilometer	hectometer	decameter	meter	decimeter	centimeter	millimeter
Abbreviation	km	hm	dam	m	dm	cm	mm
Example Equivalences	.004 km	.04 hm	.4 dam	4 m	40 dm	400 cm	4000 mm

- **References for estimation**
 1 millimeter — about the thickness of a U.S. dime
 1 centimeter —
 1 meter —
 1 kilometer —

Equivalences used in U.S. Customary System for conversions

1 foot = 12 inches
1 yard = 3 feet
1 mile = 5280 feet

- **Dimensional analysis examples using these equivalences**

$$4 \text{ yds} = \frac{4 \text{ yds}}{1} \times \frac{3 \text{ ft}}{1 \text{yd}} = 12 \text{ ft}$$

$$\frac{3}{10} \text{ mi} = \frac{3 \text{ mi}}{10} \times \frac{5280 \text{ ft}}{1 \text{ mi}} = 1584 \text{ ft}$$

$$7 \text{ mi} = \frac{7 \text{ mi}}{1} \times \frac{5280 \text{ ft}}{1 \text{ mi}} \times \frac{1 \text{ yd}}{3 \text{ ft}} = 12320 \text{ yds}$$

Chapter 4 Unit Analysis Supplement © Rhodes State College, Rodney Null

Equivalences between U.S. Customary System and Metric System for conversions

1 inch = 2.54 centimeters
1 mile ≈ 1.61 kilometers

- **Dimensional analysis examples using these equivalences**

$$2\frac{1}{2}\text{ ft} = \frac{5}{2}\text{ ft} = \frac{5\text{ ft}}{2} \times \frac{12\text{ in}}{1\text{ft}} \times \frac{2.54\text{ cm}}{1\text{ in}} = 76.2\text{ cm}$$

$$8500\text{ m} = 8.5\text{ km} = \frac{8.5\text{ km}}{1} \times \frac{1\text{ mi}}{1.61\text{ km}} \approx 5.28\text{ mi}$$

U.S. Customary

Convert each of the following. Show work.

1. 60 in to feet

2. 8 yd 1 ft to feet

3. 6.2 mi to yards

4. 82 in to feet and inches
(To whole feet first with leftover inches.)

5. 184,800 ft to miles

6. $\frac{2}{3}$ ft to inches

7. $9\frac{1}{2}$ yd to feet

8. 5' 10" to inches

9. 10,560 yd to miles

10. 13 ft to yards

Metric

Convert each of the following. You can use dimensional analysis or slide the decimal point.

11. 280 mm to centimeters

12. 8.3 m to centimeters

13. 400 m to kilometers

14. .12 hm to decimeters

15. 750 cm to millimeters

16. 39 km to decameters

17. 5.18 dm to hectometers

18. .007 km to millimeters

19. 63,000 cm to meters

20. 21 km to meters

Converting between the Two Systems

Convert each of the following. If necessary, round to 3 significant digits.

21. 10 km to miles

22. $4\frac{1}{2}$ in to centimeters

23. 75 yd to meters

24. 480 cm to feet

25. 5 ft 7 in to centimeters

26. 73 mm to inches

27. 48 mi to meters

28. 6.3 dam to feet

29. 4,000 ft to kilometers

30. 95,000 cm to yards

4.4 Volume and Capacity

Attribute you are measuring

Volume/Capacity

Examples of tools used to measure it

Measuring Cups Gas Gauge Pipette

Graduated Cylinder Measuring Spoons Syringe

Common Units used in the U.S. Customary System

- **Names**

 Fluid Ounces Cups Pints Quarts

 Tablespoons Gallons Teaspoons

Note *Fluid Ounces* for measuring volume are *not* the same as *Ounces* for measuring weight!

- **References for estimation**
 1 Teaspoon —
 1 Tablespoon —
 1 Fluid Ounce —
 1 Cup —
 Pint —
 1 Quart —
 1 Gallon —

- **Abbreviations used**

 Teaspoon(s): tsp (Example: 5 teaspoons = 5 tsp)
 Tablespoon(s): tbsp (Example: 3 tablespoons = 3 tbsp)
 Fluid Ounce(s): fl oz (Example: 7 fluid ounces = 7 fl oz)
 Cup(s): c (Example: 2 cups = 2 c)
 Pint(s): pt (Example: 4 pints = 4 pt)
 Quart(s): qt (Example: 6 quarts = 6 qt)
 Gallon(s): gal (Example: 9 gallons = 9 gal)

Video Examples Section 4.4

Units used in the Metric System

Names	kiloliter	hectoliter	decaliter	liter	deciliter	centiliter	milliliter
Abbreviation	kL	hL	daL	L	dL	cL	mL
Example Equivalences	.009 kL	.09 hL	.9 daL	9 L	90 dL	900 cL	9000 mL

Note Milliliters are equivalent to cubic centimeters and are used interchangeably:

$15 \text{ mL} = 15 \text{ cm}^3 = 15 \text{ cc}$

- **References for estimation**
 1 milliliter — 1 liter —
 1 deciliter — 1 kiloliter —

Equivalences used in U.S. Customary System for conversions

 1 gallon = 4 quarts 1 cup = 8 fluid ounces
 1 quart = 2 pints 1 fluid ounce = 2 tablespoons
 1 pint = 2 cups 1 tablespoon = 3 teaspoons

- **Dimensional analysis examples using these equivalences**

$$62\text{qt} = \frac{62 \text{ qt}}{1} \times \frac{1 \text{ gal}}{4 \text{ qt}} = \frac{31 \text{ gal}}{2} = 15\frac{1}{2} \text{ gal}$$

$$2\text{ c} = \frac{2 \text{ c}}{1} \times \frac{8 \text{ fl oz}}{1 \text{ c}} \times \frac{2 \text{ tbsp}}{1 \text{ fl oz}} = 32 \text{ tbsp}$$

Equivalences between U.S. Customary System and Metric System for conversions

$1 \text{ liter} \approx 1.06 \text{ quarts}$

- **Dimensional analysis examples using this equivalence**

$$12 \text{ pt} = \frac{12 \text{ pt}}{1} \times \frac{1 \text{ qt}}{2 \text{ pt}} \times \frac{1 \text{ L}}{1.06 \text{ qt}} \approx 5.66 \text{ L}$$

$$50000 \text{ mL} = 50 \text{ L} = \frac{50 \text{ L}}{1} \times \frac{1.06 \text{ qt}}{1 \text{ L}} \times \frac{1 \text{ gal}}{4 \text{ qt}} \approx 13.3 \text{ gal}$$

U.S. Customary

Convert each of the following. Show work.

1. 4 c to fluid ounces

2. 7 gal to pints

3. 56 fl oz to quarts

4. $3\frac{1}{2}$ qt to cups

5. 19 pt to quarts and pints

6. 13 gal 3 qt to quarts

7. 36 c to gallons

8. $\frac{2}{3}$ gal to fluid ounces

9. 43 fl oz to cups and fluid ounces

10. 5 pt to fluid ounces

Metric

Convert each of the following. You can use dimensional analysis or slide the decimal point. (Metric)

11. 2 L to milliliters

12. 425 cc to deciliters

13. .025 kL to hectoliters

14. 75 mL to cubic centimeters

15. 4000 cL to liters

16. 3.9 daL to milliliters

17. 580 L to kiloliters

18. 670 cm³ to liters

19. .29 L to decaliters

20. 95 dL to centiliters

Converting between the Two Systems

Convert each of the following. If necessary, round to 3 significant digits. Show work.

21. 12 fl oz to milliliters

22. 2 L to cups

23. $3\frac{1}{2}$ gal to liters

24. 500 cc to pints

25. 9.8 kL to gallons

26. 5 qt 1 pt to liters

27. 8700 cm³ to quarts

28. .3 hL to fluid ounces

29. $5\frac{1}{3}$ c to centiliters

30. 5 pt to cubic centimeters

4.5 Weight and Mass

Note Weight and Mass are actually two different things - mass is the "stuff" something is made of and weight is the force of gravity pulling on this stuff. On earth, we don't really bother to distinguish them, but if you went to the moon you would certainly notice the difference. On the moon, you still have the same amount of stuff making you, but your weight would be different, since the moon's gravitational pull is different than the earth's. Technically, converting between mass and weight is wrong - they are two different things, like apples and peanuts.

Note *Ounces* for measuring weight are *not* the same as *Fluid Ounces* for measuring volume!

Video Examples
Section 4.5

Attribute you are measuring

Weight/Mass

Examples of tools used to measure it

Triple Beam Balance Analytic Balance Scales (e.g. bathroom scales, postal scales)

Common units used in the U.S. Customary System

- **Names**

 Ounce Pound Ton

- **References for estimation**
 1 ounce —
 1 pound —
 1 ton —
- **Abbreviations used**
 Ounce(s): oz (Example: 13 ounces = 13 oz)
 Pound(s): lb or # (Example: 8 pounds = 8 lb = 8 #)

Units used in the Metric System

Names	kilogram	hectogram	decagram	gram	decigram	centigram	milligram
Abbreviation	kg	hg	dag	g	dg	cg	mg
Example Equivalences	.004 kg	.04 hg	.4 dag	4 g	40 dg	400 cg	4000 mg

- References for estimation
 1 milligram —
 1 gram —
 1 kilogram —

Equivalences used in U.S. Customary System for conversions

 1 pound = 16 ounces
 1 ton = 2000 pounds

- **Dimensional analysis examples using these equivalences**

$$5\frac{3}{4}\text{ lb} = \frac{23}{4}\text{ lb} = \frac{23\text{ lb}}{4} \times \frac{16\text{ oz}}{1\text{ lb}} = 92\text{ oz}$$

$$800\text{ lb} = \frac{800\text{ lb}}{1} \times \frac{1\text{ ton}}{2000\text{ lb}} = \frac{2\text{ ton}}{5} = \frac{2}{5}\text{ ton}$$

$$\frac{5}{8}\text{ ton} = \frac{5\text{ ton}}{8} \times \frac{2000\text{ lb}}{1\text{ ton}} \times \frac{16\text{ oz}}{1\text{ lb}} = 20000\text{ oz}$$

Equivalences between U.S. Customary System and Metric System for conversions

 1 kilogram ≈ 2.20 pounds
 1 ounce ≈ 28.3 grams

- **Dimensional analysis examples using these equivalences**

$$3\text{ ton} = \frac{3\text{ ton}}{1} \times \frac{2000\text{ lb}}{1\text{ ton}} \times \frac{1\text{ kg}}{2.20\text{ lb}} \approx 2730\text{ kg}$$

$$14000\text{ mg} = 14\text{g} = \frac{14\text{g}}{1} \times \frac{1\text{ oz}}{28.3\text{ g}} \approx .495\text{ oz}$$

$$134\text{ lb} = \frac{134\text{ lb}}{1} \times \frac{1\text{ kg}}{2.20\text{ lb}} \approx 60.9\text{ kg}$$

U.S. Customary

Convert each of the following. Show work. (U.S. Customary)

1. 112 oz to pounds

2. $4 \frac{1}{2}$ ton to pounds

3. 12 lb 13 oz to ounces

4. .025 ton to ounces

5. 5.6 lb to pounds and ounces
(To whole pounds with leftover ounces)

6. 97 oz to pounds and ounces
(To whole pounds with leftover ounces)

7. 4 oz to pounds

8. $\frac{3}{4}$ ton to pounds

9. 2 ton 300 lb to pounds

10. 19000 lbs to tons and pounds
(To whole tons with leftover pounds)

Metric

Convert each of the following. You can use dimensional analysis or slide the decimal point.

11. 25 g to decigrams

12. 500 mg to grams

13. .3 cg to decagrams

14. 73.6 kg to grams

15. 9.2 g to kilograms

16. 43.21 dg to hectograms

17. 9.3 g to milligrams

18. 83.9 mg to kilograms

19. .0023 hg to centigrams

20. .0191 dag to kilograms

Converting between the Two Systems

Convert each of the following. If necessary, round to 3 significant digits. Show work.

21. 180 lbs to kilograms

22. 700 g to ounces

23. 1400 kg to tons

24. $\frac{1}{4}$ lb to grams

25. 500 mg to ounces

26. 95 hg to pounds

27. 5 lb 7 oz to kilograms

28. 1.5 kg to ounces

29. 85 oz to decigrams

30. $\frac{3}{4}$ ton to kilograms

The two systems (or scales) we use to discuss temperature measurements are Fahrenheit and Celsius (a.k.a. Centigrade). There are others, for example Rankin and Kelvin scales.

Converting between Fahrenheit temperature measurements and Celsius temperature measurements requires a different technique than we have used for conversions of distance, volume, and weight. There are several reasons for this.

1. In each scale we will only use one unit, the degree. The symbol for degree looks like this: 90° read as ninety degrees.

2. Degrees in the Fahrenheit scale are not the same size as degrees in the Celsius scale. You can see this if you look at two points of reference, the boiling point and freezing point of pure water.

Measurement Scale	Freezing Point of Water	Boiling Point of Water
Fahrenheit	32°	212°
Celsius	0°	100°

Which would have to be bigger, degrees in the Fahrenheit scale or degrees in the Celsius scale? Explain your reasoning.

3. The zero degree mark on each scale is not the same and neither one mean zero (or no) temperature. So it is possible to have a temperature less than zero degrees, in other words you can have negative temperatures, on both scales.

Hey, nobody said this was going to be easy!

The method we will use to convert between temperatures measured on the Fahrenheit and Celsius scales is different than dimensional analysis and the sliding the decimal point strategies we have used so far. It involves using one of two versions of a formula. The version you use depends on which scale you have the temperature measured on initially.

If you start with a temperature measured in degrees on the Fahrenheit scale, you will want to use this version of the formula:

$$C = \frac{5}{9}(F - 32)$$

EXAMPLE 1 What is 77° F in Celsius?

Solution Substitute 77 into the formula for F to get

$$C = \frac{5}{9}(77 - 32)$$

Doing the calculation on the right side of the equation it simplifies to:

$$C = 25$$

So 77° F = 25° C

**Video Examples
Section 4.6**

If you start with a temperature measured in degrees on the Celsius scale, you will want to use this version of the formula:

$$F = \frac{9}{5}C + 32$$

EXAMPLE 2 What is $-15°$ C in Fahrenheit?

Solution Substitute -15 into the formula for the C to get:

$$F = \frac{9}{5}(-15) + 32$$

Doing the calculation on the right side of the equation it simplifies to:

$$F = 5$$

So $-15°$ C $= 5°$ F.

Fahrenheit-Celsius/Centigrade

Convert each of the following. Show use of the appropriate formula. If necessary, round to 3 significant digits.

1. 98° F to Celsius

2. 100° C to Fahrenheit

3. 23° C to Fahrenheit

4. 0° F to Centigrade

5. 23° F to Celsius

6. −10° F to Celsius

7. 350° F to Centigrade

8. −40° C to Fahrenheit

9. 0° C to Fahrenheit

10. 55° F to Celsius

Convert each of the following. Show appropriate work. If necessary, round to 3 significant digits.

1. 95 mm to centimeters

2. $3\frac{1}{2}$ lb to ounces

11. 425° F to Celsius

12. 4600 mL to kilometers

3. 28° C to Fahrenheit

4. 2000 cc to cups

13. 6'2" to meters

14. 13.7 gal to liters

5. 9 yd 1 ft to kilograms

6. 20 mi to yards

15. 5500 cm to hectometers

16. 64 fl oz to gallons

7. 48 fl oz to quarts

8. 17 mg to grams

17. 16 ton 400 lb to pounds

18. $5\frac{2}{3}$ lb to kilograms

9. 63,000 cm to meters

10. 21 hg to ounces

19. 10 km to miles

20. 230 cm³ to milliliters

Linear Equations and Inequalities in One Variable

5

5.1 Addition Property of Equality

5.2 Multiplication Property of Equality

5.3 Solving Linear Equations in One Variable

5.4 Formulas

5.5 Number, Age, Perimeter, Coin Applications

5.6 Integer, Interest, Triangle, Miscellaneous Applications

5.7 Linear Inequalities in One Variable

Estádio Jornalista Mário Filho Stadium
Rio de Janiero, Brazil
© 2010 Google

The honor of hosting the 2014 FIFA World Cup for soccer has fallen to Brazil, making it just the fifth country to host two World Cup tournaments. Expected to be one of the main venues of the competition, Estádio Jornalista Mário Filho in Rio de Janiero is the sixth largest stadium in the world. The stadium was originally constructed for the 1950 World Cup, where it acquired the nickname "Maracana" after a nearby river. The Maracana hosted the 1950 tournament final where Uruguay upset the home team Brazil in a stunning 2-1 win. At that time, the crowd was reported to have been 200,000 soccer fans, including those who illegally stormed the gates and entered after the start of the match. That crowd is still the largest ever to watch a soccer match. Due to an accident in 1992 resulting in two deaths, renovations greatly reduced Maracana's capacity, which is currently listed at 103,045.

Suppose the 2014 FIFA World Cup lists their ticket prices in the same manner as the 2010 World Cup in South Africa. The following chart shows the 2010 prices by seating category in US dollars.

	Category 1	Category 2	Category 3	Wheelchair
Opening Match	$450.00	$300.00	$200.00	$70.00
Group Stage	$160.00	$120.00	$80.00	$20.00
Round of 16	$200.00	$150.00	$100.00	$50.00
Quarter Finals	$300.00	$200.00	$150.00	$75.00
Semi-Finals	$600.00	$400.00	$250.00	$100.00
3rd/4th Place	$300.00	$200.00	$150.00	$75.00
The Final	$900.00	$600.00	$400.00	$150.00

Suppose the Stadium Maracana sells out for the final game with 100,000 paid tickets. If the seating prices hold, and there are 50,000 Category 2 tickets sold, 15,000 Category 3 tickets sold, and 5,000 Wheelchair tickets sold, how many Category 1 tickets will be sold? An equation to answer this question looks like this:

$$100,000 = 50,000 + 15,000 + 5,000 + x$$

Solving the equation above requires the use of the addition property of equality, one of the topics we will cover in this chapter.

©iStockphoto.com/IvicaNS

Video Examples
Section 5.1

Note The symbol { } is used in roster notation to represent a list of numbers in a set.

Practice Problems

1. Is 4 a solution to $2x + 3 = 7$?

Note We use a question mark over the equal signs to show that we don't know yet whether the two sides of the equation are equal.

Answer

1. No

Did you know you can roughly calculate the outside temperature by counting the number of times a cricket chirps? Researchers at the Library of Congress suggest that you focus on one cricket and count the number of chirps in 15 seconds. Then add 37 to that number and you'll get the approximate outside temperature in degrees Fahrenheit. We can also use this information to write the following linear equation:

$$t = n + 37$$

where t is the outside temperature in degrees Fahrenheit and n is the number of cricket chirps in 15 seconds. Suppose the temperature outside is 70 degrees Fahrenheit. After plugging this value into the equation, we can solve for n to find the number of chirps; that is, we must find all replacements for the variable that make the equation a true statement. To do so, we must first understand how to use the addition property of equality, which we will cover in this section.

A Solutions to Equations

DEFINITION solution set

The *solution set* for an equation is the set of all numbers that when used in place of the variable make the equation a true statement

For example, the equation $x + 2 = 5$ has the solution set {3} because when x is 3 the equation becomes the true statement $3 + 2 = 5$, or $5 = 5$.

EXAMPLE 1 Is 5 a solution to $2x - 3 = 7$?

Solution We substitute 5 for x in the equation, and then simplify to see if a true statement results. A true statement means we have a solution; a false statement indicates the number we are using is not a solution.

$$\begin{aligned}
\text{When} \rightarrow \qquad & x = 5 \\
\text{the equation} \rightarrow \qquad & 2x - 3 = 7 \\
\text{becomes} \rightarrow \qquad & 2(5) - 3 \overset{?}{=} 7 \\
& 10 - 3 \overset{?}{=} 7 \\
& 7 = 7 \qquad \text{A true statement}
\end{aligned}$$

Because $x = 5$ turns the equation into the true statement $7 = 7$, we know 5 is a solution to the equation.

EXAMPLE 2 Is -2 a solution to $8 = 3x + 4$?

Solution Substituting -2 for x in the equation, we have

$$8 \overset{?}{=} 3(-2) + 4$$

$$8 \overset{?}{=} -6 + 4$$

$$8 = -2 \qquad \text{A false statement}$$

$$x \neq -2$$

Substituting -2 for x in the equation produces a false statement. Therefore, $x = -2$ is not a solution to the equation.

2. Is $\dfrac{4}{3}$ a solution to $8 = 3x + 4$?

Note The symbol \neq means "does not equal".

The important thing about an equation is its solution set. Therefore, we make the following definition to classify together all equations with the same solution set.

> **DEFINITION** equivalent equation
>
> Two or more equations with the same solution set are said to be *equivalent equations*.

Equivalent equations may look different but must have the same solution set.

EXAMPLE 3

a. $x + 2 = 5$ and $x = 3$ are equivalent equations because both have solution set {3}.

b. $a - 4 = 3$, $a - 2 = 5$, and $a = 7$ are equivalent equations because they all have solution set {7}.

c. $y + 3 = 4$, $y - 8 = -7$, and $y = 1$ are equivalent equations because they all have solution set {1}.

3. Answer the following.
 a. Are the equations $x + 3 = 9$ and $x = 6$ equivalent?
 b. Are the equations $a - 5 = 3$ and $a = 6$ equivalent?
 c. Are $y + 2 = 5$, $y - 1 = 2$, and $y = 3$ equivalent equations?

B Addition Property of Equality

If two numbers are equal and we increase (or decrease) both of them by the same amount, the resulting quantities are also equal. We can apply this concept to equations. Adding the same amount to both sides of an equation always produces an equivalent equation—one with the same solution set. This fact about equations is called the *addition property of equality* and can be stated more formally as follows:

> **PROPERTY** Addition Property of Equality
>
> For any three algebraic expressions A, B, and C,
>
> *In symbols*: if $A = B$
>
> then $A + C = B + C$
>
> *In words*: Adding the same quantity to both sides of an equation will not change the solution set.

Note We will use this property many times in the future. Be sure you understand it completely by the time you finish this section.

This property is just as simple as it seems. We can add any amount to both sides of an equation and always be sure we have not changed the solution set.

Consider the equation $x + 6 = 5$. We want to solve this equation for the value of x. To solve for x means we want to end up with x on one side of the equal sign and a number on the other side. Because we want x by itself, we will add -6 to both sides.

$$x + 6 + (-6) = 5 + (-6) \qquad \text{Addition property of equality}$$

$$x + 0 = -1$$

$$x = -1$$

All three equations say the same thing about x. They all say that x is -1. All three equations are equivalent. The last one is just easier to read.

Answers
2. Yes
3. **a.** Yes **b.** No **c.** Yes

Here are some further examples of how the addition property of equality can be used to solve equations.

4. Solve $x - 3 = 10$ for x.

EXAMPLE 4 Solve the equation $x - 5 = 12$ for x.

Solution Because we want x alone on the left side, we choose to add 5 to both sides.

$$x - 5 + 5 = 12 + 5 \qquad \textit{Addition property of equality}$$

$$x + 0 = 17$$

$$x = 17$$

To check our solution to Example 4, we substitute 17 for x in the original equation:

$$\text{When} \rightarrow \qquad\qquad x = 17$$

$$\text{the equation} \rightarrow \qquad x - 5 = 12$$

$$\text{becomes} \rightarrow \qquad 17 - 5 \overset{?}{=} 12$$

$$12 = 12 \qquad \textit{A true statement}$$

As you can see, our solution checks. The purpose for checking a solution to an equation is to catch any mistakes we may have made in the process of solving the equation.

5. Solve $a + \dfrac{2}{3} = -\dfrac{1}{6}$ for a.

EXAMPLE 5 Solve $a + \dfrac{3}{4} = -\dfrac{1}{2}$ for a.

Solution Because we want a by itself on the left side of the equal sign, we add the opposite of $\frac{3}{4}$ to each side of the equation.

$$a + \frac{3}{4} + \left(-\frac{3}{4}\right) = -\frac{1}{2} + \left(-\frac{3}{4}\right) \qquad \textit{Addition property of equality}$$

$$a + 0 = -\frac{1}{2} \cdot \frac{2}{2} + \left(-\frac{3}{4}\right) \qquad \textit{LCD on the right side is 4.}$$

$$a = -\frac{2}{4} + \left(-\frac{3}{4}\right) \qquad \textit{$\frac{2}{4}$ is equivalent to $\frac{1}{2}$.}$$

$$a = -\frac{5}{4} \qquad \textit{Add fractions.}$$

The solution is $a = -\frac{5}{4}$. To check our result, we replace a with $-\frac{5}{4}$ in the original equation. The left side then becomes $-\frac{5}{4} + \frac{3}{4}$, which reduces to $-\frac{1}{2}$, so our solution checks.

6. Solve $2.7 + x = 8.1$ for x.

EXAMPLE 6 Solve $7.3 + x = -2.4$ for x.

Solution Again, we want to isolate x, so we add the opposite of 7.3 to both sides.

$$7.3 + (-7.3) + x = -2.4 + (-7.3) \qquad \textit{Addition property of equality}$$

$$0 + x = -9.7$$

$$x = -9.7$$

Sometimes it is necessary to simplify each side of an equation before using the addition property of equality. The reason we simplify both sides first is that we want as few terms as possible on each side of the equation before we use the addition property of equality. The following examples illustrate this procedure.

7. Solve $-2x + 3 + 3x = 6 + 8$ for x.

EXAMPLE 7 Solve $-x + 2 + 2x = 7 + 5$ for x.

Solution We begin by combining similar terms on each side of the equation. Then we use the addition property to solve the simplified equation.

$$x + 2 = 12 \qquad \textit{Simplify both sides first.}$$

$$x + 2 + (-2) = 12 + (-2) \qquad \textit{Addition property of equality}$$

$$x + 0 = 10$$

$$x = 10$$

Answers

4. 13

5. $-\dfrac{5}{6}$

6. 5.4

7. 11

EXAMPLE 8 Solve $4(2a - 3) - 7a = 2 - 5$.

Solution We must begin by applying the distributive property to separate terms on the left side of the equation. Following that, we combine similar terms and then apply the addition property of equality.

$$4(2a - 3) - 7a = 2 - 5 \qquad \text{Original equation}$$
$$8a - 12 - 7a = 2 - 5 \qquad \text{Distributive property}$$
$$a - 12 = -3 \qquad \text{Simplify each side.}$$
$$a - 12 + 12 = -3 + 12 \qquad \text{Add 12 to each side.}$$
$$a = 9$$

To check our solution, we replace a with 9 in the original equation.

$$4(2 \cdot 9 - 3) - 7 \cdot 9 \overset{?}{=} 2 - 5$$
$$4(15) - 63 \overset{?}{=} -3$$
$$60 - 63 \overset{?}{=} -3$$
$$-3 = -3 \qquad \text{A true statement}$$

We can also add a term involving a variable to both sides of an equation.

EXAMPLE 9 Solve $3x - 5 = 2x + 7$.

Solution We can solve this equation in two steps. First, we add $-2x$ to both sides of the equation. When this has been done, x appears on the left side only. Second, we add 5 to both sides.

$$3x + (-2x) - 5 = 2x + (-2x) + 7 \qquad \text{Add } -2x \text{ to both sides.}$$
$$x - 5 = 7 \qquad \text{Simplify each side.}$$
$$x - 5 + 5 = 7 + 5 \qquad \text{Add 5 to both sides.}$$
$$x = 12 \qquad \text{Simplify each side.}$$

8. Solve $5(3a - 4) - 14a = 25$.

Note Remember that terms cannot be combined until the parentheses are removed.

Note Again, we place a question mark over the equal sign because we don't know yet whether the expressions on the left and right side of the equal sign will be equal.

9. Solve $5x - 4 = 4x + 6$.

A Note on Subtraction

Although the addition property of equality is stated for addition only, we can subtract the same number from both sides of an equation as well. Because subtraction is defined as addition of the opposite, subtracting the same quantity from both sides of an equation does not change the solution.

$$x + 2 = 12 \qquad \text{Original equation}$$
$$x + 2 - 2 = 12 - 2 \qquad \text{Subtract 2 from each side.}$$
$$x = 10$$

Note In my experience teaching algebra, I find that students make fewer mistakes if they think in terms of addition rather than subtraction. So, you are probably better off if you continue to use the addition property just the way we have used it in the examples in this section. But, if you are curious as to whether you can subtract the same number from both sides of an equation, the answer is yes.

GETTING READY FOR CLASS

After reading through the preceding section, respond in your own words and in complete sentences.

A. What is a solution set for an equation?

B. What are equivalent equations?

C. Explain in words the addition property of equality.

D. How do you check a solution to an equation?

Answers
8. 45
9. 10

Vocabulary Review

Choose the correct words to fill in the blanks below.

equivalent addition subtraction solution set

1. The set of numbers used in place of the variable in an equation are called the _____ for that equation.

2. Two or more equations with the same solution set are said to be_____ equations.

3. The _____ property of equality states that adding the same quantity to both sides of an equation will not change the solution set.

4. Since subtraction is defined as addition of the opposite, we can apply the addition property of equality to both sides of a _____ problem and the solution will not change.

Problems

A Check to see if the number to the right of each of the following equations is the solution to the equation.

1. $2x + 1 = 5; 2$

2. $4x + 3 = 7; 1$

3. $3x + 4 = 19; 5$

4. $3x + 8 = 14; 2$

5. $2x - 4 = 2; 4$

6. $5x - 6 = 9; 3$

7. $2x + 1 = 3x + 3; -2$

8. $4x + 5 = 2x - 1; -6$

9. $x - 4 = 2x + 1; -4$

10. $x - 8 = 3x + 2; -5$

11. $x - 2 = 3x + 1; \dfrac{-3}{2}$

12. $x + 4 = 4x - 1; \dfrac{5}{3}$

B Solve the following equations.

13. $x - 3 = 8$

14. $x - 2 = 7$

15. $x + 2 = 6$

16. $x + 5 = 4$

17. $a + \dfrac{1}{2} = -\dfrac{1}{4}$

18. $a + \dfrac{1}{3} = -\dfrac{5}{6}$

19. $x + 2.3 = -3.5$

20. $x + 7.9 = 23.4$

21. $y + 11 = -6$

22. $y - 3 = -1$

23. $x - \dfrac{5}{8} = -\dfrac{3}{4}$

24. $x - \dfrac{2}{5} = -\dfrac{1}{10}$

25. $m - 6 = -10$

26. $m - 10 = -6$

27. $6.9 + x = 3.3$

28. $7.5 + x = 2.2$

29. $5 = a + 4$

30. $12 = a - 3$

31. $-\dfrac{5}{9} = x - \dfrac{2}{5}$

32. $-\dfrac{7}{8} = x - \dfrac{4}{5}$

Simplify both sides of the following equations as much as possible, and then solve.

33. $4x + 2 - 3x = 4 + 1$

34. $5x + 2 - 4x = 7 - 3$

35. $8a - \dfrac{1}{2} - 7a = \dfrac{3}{4} + \dfrac{1}{8}$

36. $9a - \dfrac{4}{5} - 8a = \dfrac{3}{10} - \dfrac{1}{5}$

37. $-3 - 4x + 5x = 18$

38. $10 - 3x + 4x = 20$

39. $-11x + 2 + 10x + 2x = 9$

40. $-10x + 5 - 4x + 15x = 0$

41. $-2.5 + 4.8 = 8x - 1.2 - 7x$

42. $-4.8 + 6.3 = 7x - 2.7 - 6x$

43. $2y - 10 + 3y - 4y = 18 - 6$

44. $15 - 21 = 8x + 3x - 10x$

The following equations contain parentheses. Apply the distributive property to remove the parentheses, then simplify each side before using the addition property of equality.

45. $2(x + 3) - x = 4$

46. $5(x + 1) - 4x = 2$

47. $-3(x - 4) + 4x = 3 - 7$

48. $-2(x - 5) + 3x = 4 - 9$

49. $5(2a + 1) - 9a = 8 - 6$

50. $4(2a - 1) - 7a = 9 - 5$

51. $-(x + 3) + 2x - 1 = 6$

52. $-(x - 7) + 2x - 8 = 4$

53. $4y - 3(y - 6) + 2 = 8$

54. $7y - 6(y - 1) + 3 = 9$

55. $-3(2m - 9) + 7(m - 4) = 12 - 9$

56. $-5(m - 3) + 2(3m + 1) = 15 - 8$

Solve the following equations by the method used in Example 9 in this section. Check each solution in the original equation.

57. $4x = 3x + 2$

58. $6x = 5x - 4$

59. $8a = 7a - 5$

60. $9a = 8a - 3$

61. $2x = 3x + 1$

62. $4x = 3x + 5$

63. $3y + 4 = 2y + 1$

64. $5y + 6 = 4y + 2$

65. $2m - 3 = m + 5$

66. $8m - 1 = 7m - 3$

67. $4x - 7 = 5x + 1$

68. $3x - 7 = 4x - 6$

69. $5x - \dfrac{2}{3} = 4x + \dfrac{4}{3}$

70. $3x - \dfrac{5}{4} = 2x + \dfrac{1}{4}$

71. $8a - 7.1 = 7a + 3.9$

72. $10a - 4.3 = 9a + 4.7$

Applying the Concepts

73. Movie Tickets A movie theater has a total of 300 seats. For a special Saturday night preview, they reserve 20 seats to give away to their VIP guests at no charge. If x represents the number of tickets they can sell for the preview, then x can be found by solving the equation $x + 20 = 300$.

 a. Solve the equation for x.

 b. If tickets for the preview are $7.50 each, what is the maximum amount of money they can make from ticket sales?

74. Movie Tickets A movie theater has a total of 250 seats. For a special Saturday night preview, they reserve 45 seats to give away to their VIP guests at no charge. If x represents the number of tickets they can sell for the preview, then x can be found by solving the equation $x + 45 = 250$.

 a. Solve the equation for x.

 b. If tickets for the preview are $8.00 each, what is the maximum amount of money they can make from ticket sales?

75. Geometry Two angles are complementary angles. If one of the angles is 23°, then solving the equation $x + 23° = 90°$ will give you the other angle. Solve the equation.

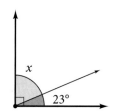

76. Geometry Two angles are supplementary angles. If one of the angles is 23°, then solving the equation $x + 23° = 180°$ will give you the other angle. Solve the equation.

77. Theater Tickets The El Portal Center for the Arts in North Hollywood, California, holds a maximum of 400 people. The two balconies hold 86 and 89 people each; the rest of the seats are at the stage level. Solving the equation $x + 86 + 89 = 400$ will give you the number of seats on the stage level.

 a. Solve the equation for x.

 b. If tickets on the stage level are $30 each, and tickets in either balcony are $25 each, what is the maximum amount of money the theater can bring in for a show?

78. Geometry The sum of the angles in the triangle on the swing set is 180°. Use this fact to write an equation containing x. Then solve the equation.

79. Using the equation $t = n + 40$ given in the introduction of this section, complete the following table.

n	t
32	
	90
44	
	136

80. Using the equation $T = 40 + O$ which relates overtime hours to total hours, complete the following table.

O	T
10	
	51
17	
	62

81. Geometry The three angles shown in the triangle at the front of the tent in the following figure add up to 180°. Use this fact to write an equation containing x, and then solve the equation to find the number of degrees in the angle at the top of the triangle.

82. Suppose Ava and Kendra are bridesmaids in a friend's wedding. Answer the following questions about the money they spend.

a. Suppose Kendra spent $210 for a bridesmaid dress, then bought new shoes to go with the dress. If the total bill was $289, does the equation $210 + x = 289$ describe this situation? If so, what does x represent?

b. Suppose Ava buys a bridesmaid dress on sale and then spends $120 on her makeup, nails, and hair. If the total bill is $219, and x represents a positive number, does the equation $219 + x = 120$ describe the situation?

Getting Ready for the Next Section

To understand all of the explanations and examples in the next section you must be able to work the problems below.

Simplify.

83. $\frac{3}{2}\left(\frac{2}{3}y\right)$

84. $\frac{5}{2}\left(-\frac{2}{5}y\right)$

85. $\frac{1}{5}(5x)$

86. $-\frac{1}{4}(-4a)$

87. $\frac{1}{5}(30)$

88. $-\frac{1}{4}(24)$

89. $\frac{3}{2}(4)$

90. $\frac{1}{26}(13)$

91. $12\left(-\frac{3}{4}\right)$

92. $12\left(\frac{1}{2}\right)$

93. $\frac{3}{2}\left(-\frac{5}{4}\right)$

94. $\frac{5}{3}\left(-\frac{6}{5}\right)$

95. $13 + (-5)$

96. $-13 + (-5)$

97. $-\dfrac{3}{4} + \left(-\dfrac{1}{2}\right)$

98. $-\dfrac{7}{10} + \left(-\dfrac{1}{2}\right)$

99. $7x + (-4x)$

100. $5x + (-2x)$

101. $3x + (-12x)$

102. $5x - 9x$

Find the Mistake

Each sentence below contains a mistake. Circle the mistake and write the correct word or expression on the line provided.

1. The number 2 is a solution to the equation $3x - 9 = 8x + 1$. _____

2. Saying that the number 12 is a solution to the equation $\dfrac{1}{6}y + 3 = 5$ is a false statement. _____

3. The addition property of equality states that $A(C) = B(C)$ if $A = B$. _____

4. To solve the equation $\dfrac{5}{6}b + 6 = 11$ for b, begin by adding -11 to both sides. _____

Image © Vandenwyngaert Stephane

5.2 Multiplication Property of Equality

OBJECTIVES

A Use the multiplication property of equality to solve an equation.

B Use the addition and multiplication properties of equality together to solve an equation.

KEY WORDS

multiplication property of equality

Henry Ford made famous the mass production assembly line in an effort to streamline his automobile manufacturing. He realized that an assembler was taking 8.5 hours to complete his entire task because the assembler had to retrieve parts on his own as needed. Ford asked that parts be delivered to an assembler, which reduced the time it took to complete the task to 2.5 minutes! Later, Ford installed the moving conveyor belt, reducing the completion time even further.

Imagine yourself working on an assembly line. If it takes you 2.5 minutes to assemble your item, how many items can you assemble in an hour's time? To answer this question, we must use the given information to set up the following equation:

$$60 = 2.5x$$

Notice we converted an hour into 60 minutes to match the units for your assembly time. In this section, we will introduce you to the multiplication property of equality. You will use this property to solve linear equations, such as the one for the assembly line.

A Multiplication Property of Equality

In the previous section, we found that adding the same number to both sides of an equation never changed the solution set. The same idea holds for multiplication by numbers other than zero. We can multiply both sides of an equation by the same nonzero number and always be sure we have not changed the solution set. (The reason we cannot multiply both sides by zero will become apparent later.) This fact about equations is called the *multiplication property of equality*, which can be stated formally as follows:

> **PROPERTY** Multiplication Property of Equality
>
> For any three algebraic expressions A, B, and C, where $C \neq 0$,
>
> *In symbols:* if $A = B$
>
> then $AC = BC$
>
> *In words:* Multiplying both sides of an equation by the same nonzero number will not change the solution set.

Video Examples Section 5.2

Note This property is also used many times throughout the book. Make sure you understand it completely.

Suppose we want to solve the equation $5x = 30$. We have $5x$ on the left side but would like to have just x. We choose to multiply both sides by $\frac{1}{5}$ which allows us to divide out the fives and isolate the variable. The solution follows.

$$5x = 30$$

$$\frac{1}{5}(5x) = \frac{1}{5}(30) \qquad \text{Multiplication property of equality}$$

$$\left(\frac{1}{5} \cdot 5\right)x = \frac{1}{5}(30) \qquad \text{Associative property of multiplication}$$

$$1x = 6 \qquad\qquad \text{Note: } 1x \text{ is the same as } x.$$

$$x = 6$$

We chose to multiply by $\frac{1}{5}$ because it is the reciprocal of 5 and $\left(\frac{1}{5}\right)(5) = 1$. We can see that multiplication by any number except zero will not change the solution set. If, however, we were to multiply both sides by zero, the result would always be $0 = 0$. Although the statement $0 = 0$ is true, we have lost our variable and cannot solve the equation. This is the only restriction of the multiplication property of equality. We are free to multiply both sides of an equation by any number except zero in an effort to isolate the variable.

Here are some more examples that use the multiplication property of equality.

EXAMPLE 1 Solve $-4a = 24$ for a.

Solution Because we want the variable a alone on the left side, we choose to multiply both sides by $-\frac{1}{4}$.

$$-\frac{1}{4}(-4a) = -\frac{1}{4}(24) \qquad \text{Multiplication property of equality}$$

$$\left[-\frac{1}{4}(-4)\right]a = -\frac{1}{4}(24) \qquad \text{Associative property}$$

$$a = -6$$

EXAMPLE 2 Solve $-\frac{t}{3} = 5$ for t.

Solution Because division by 3 is the same as multiplication by $\frac{1}{3}$, we can write $-\frac{t}{3}$ as $-\frac{1}{3}t$. To solve the equation, we multiply each side by the reciprocal of $-\frac{1}{3}$, which is -3.

$$-\frac{t}{3} = 5 \qquad\qquad \text{Original equation}$$

$$-\frac{1}{3}t = 5 \qquad\qquad \text{Dividing by 3 is equivalent to multiplying by } \frac{1}{3}.$$

$$-3\left(-\frac{1}{3}t\right) = -3(5) \qquad \text{Multiply each side by } -3.$$

$$t = -15$$

EXAMPLE 3 Solve $\frac{2}{3}y = \frac{4}{9}$.

Solution We can multiply both sides by $\frac{3}{2}$ and have $1y$ on the left side.

$$\frac{3}{2}\left(\frac{2}{3}y\right) = \frac{3}{2}\left(\frac{4}{9}\right) \qquad \text{Multiplication property of equality}$$

$$\left(\frac{3}{2} \cdot \frac{2}{3}\right)y = \frac{3}{2}\left(\frac{4}{9}\right) \qquad \text{Associative property}$$

$$y = \frac{2}{3} \qquad\qquad \text{Simplify } \frac{3}{2}\left(\frac{4}{9}\right) = \frac{12}{18} = \frac{2}{3}.$$

EXAMPLE 4 Solve $5 + 8 = 10x + 20x - 4x$.

Solution Our first step will be to simplify each side of the equation.

$$13 = 26x \qquad\qquad \text{Simplify by combining similar terms.}$$

$$\frac{1}{26}(13) = \frac{1}{26}(26x) \qquad \text{Multiplication property of equality}$$

$$\frac{13}{26} = x \qquad\qquad \text{Reduce to lowest terms.}$$

$$\frac{1}{2} = x$$

$$x = \frac{1}{2} \qquad\qquad \text{Symmetric property}$$

Practice Problems

1. Solve $3a = -27$ for a.

2. Solve $\frac{t}{4} = 6$ for t.

3. Solve $\frac{2}{5}y = \frac{3}{4}$.

Note Notice in Examples 1 through 3 that if the variable is being multiplied by a number like -4 or $\frac{2}{3}$, we always multiply by the number's reciprocal, $-\frac{1}{4}$ or $\frac{3}{2}$, to end up with just the variable on one side of the equation.

4. Solve $3 + 7 = 6x + 8x - 9x$.

Note Remember the symmetric property allows us to move our variable from the right to the left-hand side once we have solved an equation. For any real numbers, if $a = b$, then $b = a$.

Answers

1. -9

2. 24

3. $\frac{15}{8}$

4. 2

B Solving Equations Using Addition and Multiplication

In the next three examples, we will use both the addition property of equality and the multiplication property of equality.

EXAMPLE 5 Solve $6x + 5 = -13$ for x.

Solution We begin by adding -5 to both sides of the equation.

$$6x + 5 + (-5) = -13 + (-5) \qquad \text{Add } -5 \text{ to both sides.}$$
$$6x = -18 \qquad \text{Simplify.}$$
$$\frac{1}{6}(6x) = \frac{1}{6}(-18) \qquad \text{Multiply both sides by } \frac{1}{6}.$$
$$x = -3$$

5. Solve $4x + 3 = -13$ for x.

EXAMPLE 6 Solve $5x = 2x + 12$ for x.

Solution We begin by adding $-2x$ to both sides of the equation.

$$5x + (-2x) = 2x + (-2x) + 12 \qquad \text{Add } -2x \text{ to both sides.}$$
$$3x = 12 \qquad \text{Simplify.}$$
$$\frac{1}{3}(3x) = \frac{1}{3}(12) \qquad \text{Multiply both sides by } \frac{1}{3}.$$
$$x = 4 \qquad \text{Simplify.}$$

6. Solve $7x = 4x + 15$ for x.

Note Notice that in Example 6 we used the addition property of equality first to combine all the terms containing x on the left side of the equation. Once this had been done, we used the multiplication property to isolate x on the left side.

EXAMPLE 7 Solve $3x - 4 = -2x + 6$ for x.

Solution We begin by adding $2x$ to both sides.

$$3x + 2x - 4 = -2x + 2x + 6 \qquad \text{Add } 2x \text{ to both sides.}$$
$$5x - 4 = 6 \qquad \text{Simplify.}$$

Now we add 4 to both sides.

$$5x - 4 + 4 = 6 + 4 \qquad \text{Add 4 to both sides.}$$
$$5x = 10 \qquad \text{Simplify.}$$
$$\frac{1}{5}(5x) = \frac{1}{5}(10) \qquad \text{Multiply by } \frac{1}{5}.$$
$$x = 2 \qquad \text{Simplify.}$$

7. Solve $2x - 5 = -3x + 10$ for x.

The next example involves fractions. You will see that the properties we use to solve equations containing fractions are the same as the properties we used to solve the previous equations. Also, the LCD that we used previously to add fractions can be used with the multiplication property of equality to simplify equations containing fractions.

EXAMPLE 8 Solve $\frac{2}{3}x + \frac{1}{2} = -\frac{3}{4}$.

Solution We can solve this equation by applying our properties and working with the fractions, or we can begin by eliminating the fractions.

Method 1 Working with the fractions.

$$\frac{2}{3}x + \frac{1}{2} + \left(-\frac{1}{2}\right) = -\frac{3}{4} + \left(-\frac{1}{2}\right) \qquad \text{Add } -\frac{1}{2} \text{ to each side.}$$
$$\frac{2}{3}x = -\frac{5}{4} \qquad \text{Note that } -\frac{3}{4} + \left(-\frac{1}{2}\right) = -\frac{3}{4} + \left(-\frac{2}{4}\right).$$
$$\frac{3}{2}\left(\frac{2}{3}x\right) = \frac{3}{2}\left(-\frac{5}{4}\right) \qquad \text{Multiply each side by } \frac{3}{2}.$$
$$x = -\frac{15}{8}$$

8. Solve $\frac{3}{5}x + \frac{1}{2} = -\frac{7}{10}$.

Answers
5. -4
6. 5
7. 3
8. -2

Note Our original equation has denominators of 3, 2, and 4. The LCD for these three denominators is 12, and it has the property that all three denominators will divide it evenly. Therefore, if we multiply both sides of our equation by 12, each denominator will divide into 12 and we will be left with an equation that does not contain any denominators other than 1.

Method 2 Eliminating the fractions in the beginning.

$$12\left(\frac{2}{3}x + \frac{1}{2}\right) = 12\left(-\frac{3}{4}\right)$$ Multiply each side by the LCD 12.

$$12\left(\frac{2}{3}x\right) + 12\left(\frac{1}{2}\right) = 12\left(-\frac{3}{4}\right)$$ Distributive property on the left side

$$8x + 6 = -9$$ Multiply.

$$8x = -15$$ Add -6 to each side.

$$x = -\frac{15}{8}$$ Multiply each side by $\frac{1}{8}$.

As the third line in Method 2 indicates, multiplying each side of the equation by the LCD eliminates all the fractions from the equation.

As you can see, both methods yield the same solution.

A Note on Division

Because *division* is defined as multiplication by the reciprocal, multiplying both sides of an equation by the same number is equivalent to dividing both sides of the equation by the reciprocal of that number; that is, multiplying each side of an equation by $\frac{1}{3}$ and dividing each side of the equation by 3 are equivalent operations. If we were to solve the equation $3x = 18$ using division instead of multiplication, the steps would look like this:

$$3x = 18$$ Original equation

$$\frac{3x}{3} = \frac{18}{3}$$ Divide each side by 3.

$$x = 6$$

Using division instead of multiplication on a problem like this may save you some writing. However, with multiplication, it is easier to explain "why" we end up with just one x on the left side of the equation. (The "why" has to do with the associative property of multiplication.) Continue to use multiplication to solve equations like this one until you understand the process completely. Then, if you find it more convenient, you can use division instead of multiplication.

GETTING READY FOR CLASS

After reading through the preceding section, respond in your own words and in complete sentences.

A. Explain in words the multiplication property of equality.

B. If an equation contains fractions, how do you use the multiplication property of equality to clear the equation of fractions?

C. Why is it okay to divide both sides of an equation by the same nonzero number?

D. Give an example of using an LCD with the multiplication property of equality to simplify an equation.

Vocabulary Review

Choose the correct words to fill in the blanks below.

multiplication division nonzero reciprocal

1. Multiplying both sides of an equation by the same _____ number will not change the solution set.

2. $AC = BC$, where $C \neq 0$ is the _____ property of equality written in symbols.

3. _____ is defined as multiplication by the reciprocal.

4. Dividing both sides of the equation by the same number is equivalent to multiplying both sides of the equation by the _____ of that number.

Problems

A Solve the following equations using the multiplication property of equality. Be sure to show your work.

1. $5x = 10$

2. $6x = 12$

3. $7a = 28$

4. $4a = 36$

5. $-8x = 4$

6. $-6x = 2$

7. $8m = -16$

8. $5m = -25$

9. $-3x = -9$

10. $-9x = -36$

11. $-7y = -28$

12. $-15y = -30$

13. $2x = 0$

14. $7x = 0$

15. $-5x = 0$

16. $-3x = 0$

17. $\dfrac{x}{3} = 2$

18. $\dfrac{x}{4} = 3$

19. $-\dfrac{m}{5} = 10$

20. $-\dfrac{m}{7} = 1$

21. $-\dfrac{x}{2} = -\dfrac{3}{4}$

22. $-\dfrac{x}{3} = \dfrac{5}{6}$

23. $\dfrac{2}{3}a = 8$

24. $\dfrac{3}{4}a = 6$

25. $-\dfrac{3}{5}x = \dfrac{9}{5}$

26. $-\dfrac{2}{5}x = \dfrac{6}{15}$

27. $-\dfrac{5}{8}y = -20$

28. $-\dfrac{7}{2}y = -14$

Simplify both sides as much as possible, and then solve.

29. $-4x - 2x + 3x = 24$

30. $7x - 5x + 8x = 20$

31. $4x + 8x - 2x = 15 - 10$

32. $5x + 4x + 3x = 4 + 8$

33. $-3 - 5 = 3x + 5x - 10x$

34. $10 - 16 = 12x - 6x - 3x$

35. $18 - 13 = \frac{1}{2}a + \frac{3}{4}a - \frac{5}{8}a$

36. $20 - 14 = \frac{1}{3}a + \frac{5}{6}a - \frac{2}{3}a$

37. $\frac{1}{2}x - \frac{1}{3}x = 2 - 4$

38. $\frac{1}{3}x - \frac{1}{4}x = \frac{1}{2} + 2$

39. $\frac{1}{2}x + \frac{1}{3}x = \frac{1}{2} - 3$

40. $\frac{1}{4}y + \frac{1}{3}y = 2 - 16$

Solve the following equations by multiplying both sides by -1.

41. $-x = 4$

42. $-x = -3$

43. $-x = -4$

44. $-x = 3$

45. $15 = -a$

46. $-15 = -a$

47. $-y = \frac{1}{2}$

48. $-y = -\frac{3}{4}$

B Solve each of the following equations using the addition and multiplication properties where necessary.

49. $3x - 2 = 7$

50. $2x - 3 = 9$

51. $2a + 1 = 3$

52. $5a - 3 = 7$

53. $\frac{1}{8} + \frac{1}{2}x = \frac{1}{4}$

54. $\frac{1}{3} + \frac{1}{7}x = -\frac{8}{21}$

55. $6x = 2x - 12$

56. $8x = 3x - 10$

57. $2y = -4y + 18$

58. $3y = -2y - 15$

59. $-7x = -3x - 8$

60. $-5x = -2x - 12$

61. $8x + 4 = 2x - 5$

62. $5x + 6 = 3x - 6$

63. $x + \dfrac{1}{2} = \dfrac{1}{4}x - \dfrac{5}{8}$

64. $\dfrac{1}{3}x + \dfrac{2}{5} = \dfrac{1}{5}x - \dfrac{2}{5}$

65. $6m - 3 = m + 2$

66. $6m - 5 = m + 5$

67. $\dfrac{1}{2}m - \dfrac{1}{4} = \dfrac{1}{12}m + \dfrac{1}{6}$

68. $\dfrac{1}{2}m - \dfrac{5}{12} = \dfrac{1}{12}m + \dfrac{5}{12}$

69. $9y + 2 = 6y - 4$

70. $6y + 14 = 2y - 2$

71. $\dfrac{3}{2}y + \dfrac{1}{3} = y - \dfrac{2}{3}$

72. $\dfrac{3}{2}y + \dfrac{7}{2} = \dfrac{1}{2}y - \dfrac{1}{2}$

Applying the Concepts

73. Break-Even Point Movie theaters pay a certain price for the movies that you and I see. Suppose a theater pays $1,500 for each showing of a popular movie. If they charge $7.50 for each ticket they sell, then the equation

$$7.5x = 1,500$$

gives the number of tickets they must sell to equal the $1,500 cost of showing the movie. This number is called the break-even point. Solve the equation for x to find the break-even point.

74. Basketball Laura plays basketball for her community college. In one game she scored 13 points total, with a combination of free throws, field goals, and three-pointers. Each free throw is worth 1 point, each field goal is 2 points, and each three-pointer is worth 3 points. If she made 1 free throw and 3 field goals, then solving the equation

$$1 + 3(2) + 3x = 13$$

will give us the number of three-pointers she made. Solve the equation to find the number of three-point shots Laura made.

75. Taxes Suppose 21% of your monthly pay is withheld for federal income taxes and another 8% is withheld for Social Security, state income tax, and other miscellaneous items. If you are left with $987.50 a month in take-home pay, then the amount you earned before the deductions were removed from your check is given by the equation

$$G - 0.21G - 0.08G = 987.5$$

Solve this equation to find your gross income.

76. Rhind Papyrus The *Rhind Papyrus* is an ancient document that contains mathematical riddles. One problem asks the reader to find a quantity such that when it is added to one-fourth of itself the sum is 15. The equation that describes this situation is

$$x + \dfrac{1}{4}x = 15$$

Solve this equation.

Angles The sum of the measures of the interior angles in a triangle is 180°. Given the following angle measures, set up an equation and solve for x.

77.

78.

Angles Solve for x.

79.

80.

81. Music Downloads Suppose you wish to purchase music through the internet for a personal playlist. It costs $10 to join and $1.50 per song. You have $40 in credit to use toward joining and making your playlist. How many songs can you purchase without going over budget?

82. Music Downloads Ellis and Celia wish to purchase music through the internet. Ellis finds a site that costs $20 to join and $1.25 per song thereafter. Celia finds one that costs $10 to join and $1.50 per song thereafter. How many songs must they purchase so that they have spent the same amount?

83. Recall from section 1.6 the formula $P = 14.7 + 0.4x$, which gives the pressure (in lb/in²) at a diving depth of x feet. Use this formula to find the diving depths which would result in the following pressures:

 a. 22.7 lb/in²

 b. 28.7 lb/in²

 c. 44.7 lb/in²

84. Suppose the total bill from a water company is given by $B = 7.5 + 0.0058w$, where B is the total bill and w is the water usage (in cubic feet). Use this formula to find the water usage resulting in the following monthly water bills.

 a. $56.80

 b. $94.50

 c. $167.00

85. The formula for Franco's pay is $P = 320 + 12(h - 40)$, where P is the pay (in $) and h is the hours worked. How many hours must Franco work to earn $500 in one week?

86. The formula for Esmeralda's weekly pay is $P = 400 + 15(h - 40)$, where P is the pay (in $) and h is the hours worked. How many hours must Esmeralda work to earn $520 in one week?

Getting Ready for the Next Section

To understand all of the explanations and examples in the next section you must be able to work the problems below.

Solve each equation.

87. $2x = 4$

88. $3x = 24$

89. $30 = 5x$

90. $0 = 5x$

91. $0.17x = 510$

92. $0.1x = 400$

93. $0.12x = 78$

94. $0.35x = 420$

Apply the distributive property and then simplify if possible.

95. $3(x - 5) + 4$

96. $5(x - 3) + 2$

97. $0.09(x + 2{,}000)$

98. $0.04(x + 7{,}000)$

99. $7 - 3(2y + 1)$

100. $4 - 2(3y + 1)$

101. $3(2x - 5) - (2x - 4)$

102. $4(3x - 2) - (6x - 5)$

Find the Mistake

Each sentence below contains a mistake. Circle the mistake and write the correct word or number on the line provided.

1. The multiplication property of equality states that multiplying one side of an equation by the same nonzero number will not change the solution set. _____

2. To solve $-\frac{2b}{9} = 3$ for b, we multiply both sides of the equation by the opposite of $-\frac{2}{9}$. _____

3. To begin solving the equation $\frac{5}{6}x + \frac{1}{2} = -\frac{1}{4}$, we can either add $\frac{5}{6}$ to each side of the equation or multiply both sides by the LCD 12. _____

4. To solve the equation $-\frac{8}{9}y = -3$, we begin by multiplying both sides of the equation by the LCD of 9 and 3. _____

5.3 Solving Linear Equations in One Variable

©iStockphoto.com/mevans

**Video Examples
Section 5.3**

In the spring of 2010, over 500 robots took over a California town for the seventh annual RoboGames, held at the San Mateo County Expo Center. Robots of all shapes and sizes competed in events, such as stair climbing, obstacle running, kung-fu, sumo wrestling, basketball, and even combat battles. Advance tickets to the games were $20, and entry fees for most events were $35 each. Suppose you purchased an advance ticket and entered a robot into x number of events, spending $125 total. We could write the following equation based on the above information, and then solve for x to determine how many events you entered.

$$35x + 20 = 125$$

In this section, we will use the material we have developed in the first two sections of this chapter to build a method for solving any linear equation such as the one above.

A Solving Linear Equations

The equations we have worked with in the first two sections of this chapter are called *linear equations in one variable*. Let's begin with a formal definition.

> **DEFINITION** linear equation in one variable
>
> A *linear equation in one variable* is any equation that can be put in the form $Ax + B = C$, where A, B, and C are real numbers and A is not zero.

Each of the equations we will solve in this section is a linear equation in one variable. The steps we use to solve a linear equation in one variable are listed here:

> **HOW TO** Strategy for Solving Linear Equations in One Variable
>
> **Step 1a:** Use the distributive property to separate terms, if necessary.
>
> **1b:** If fractions are present, consider multiplying both sides by the LCD to eliminate the fractions. If decimals are present, consider multiplying both sides by a power of 10 to clear the equation of decimals.
>
> **1c:** Combine similar terms on each side of the equation.
>
> **Step 2:** Use the addition property of equality to get all variable terms on one side of the equation and all constant terms on the other side. A variable term is a term that contains the variable (for example, $5x$). A constant term is a term that does not contain the variable (the number 3, for example).
>
> **Step 3:** Use the multiplication property of equality to get x (that is, $1x$) by itself on one side of the equation.
>
> **Step 4:** Check your solution in the original equation to be sure that you have not made a mistake in the solution process.

Note You may have some previous experience solving equations. Even so, you should solve the equations in this section using the method developed here. Your work should look like the examples in the text. If you have learned shortcuts or a different method of solving equations somewhere else, you can always go back to them later. What is important now is that you are able to solve equations by the methods shown here.

As you work through the examples in this section, it is not always necessary to use all four steps when solving equations. The number of steps used depends on the equation. In Example 1, there are no fractions or decimals in the original equation, so step 1b will not be used. Likewise, after applying the distributive property to the left side of the equation in Example 1, there are no similar terms to combine on either side of the equation, making step 1c also unnecessary.

EXAMPLE 1 Solve $2(x + 3) = 10$.

Solution To begin, we apply the distributive property to the left side of the equation to separate terms.

Step 1a:	$2x + 6 = 10$	Distributive property
Step 2:	$2x + 6 + (-6) = 10 + (-6)$	Addition property of equality
	$2x = 4$	
Step 3:	$\dfrac{1}{2}(2x) = \dfrac{1}{2}(4)$	Multiply each side by $\dfrac{1}{2}$.
	$x = 2$	The solution is 2.

The solution to our equation is 2. We check our work (to be sure we have not made either a mistake in applying the properties or an arithmetic mistake) by substituting 2 into our original equation and simplifying each side of the result separately.

Step 4:	When \rightarrow	$x = 2$
	the equation \rightarrow	$2(x + 3) = 10$
	becomes \rightarrow	$2(2 + 3) \overset{?}{=} 10$
		$2(5) \overset{?}{=} 10$
		$10 = 10$ A true statement

Our solution checks.

The general method of solving linear equations is actually very simple. We can add any number to both sides of the equation and multiply both sides by any nonzero number. The equation may change in form, but the solution set will not.

The examples that follow show a variety of equations and their solutions. When you have finished this section and worked the problems in the problem set, the steps in the solution process should be a description of how you operate when solving equations.

EXAMPLE 2 Solve $3(x - 5) + 4 = 13$ for x.

Solution Our first step will be to apply the distributive property to the left side of the equation.

Step 1a:	$3x - 15 + 4 = 13$	Distributive property
Step 1c:	$3x - 11 = 13$	Simplify the left side.
Step 2:	$3x - 11 + 11 = 13 + 11$	Add 11 to both sides.
	$3x = 24$	
Step 3:	$\dfrac{1}{3}(3x) = \dfrac{1}{3}(24)$	Multiply both sides by $\dfrac{1}{3}$.
	$x = 8$	The solution is 8.

Step 4:	When \rightarrow	$x = 8$
	the equation \rightarrow	$3(x - 5) + 4 = 13$
	becomes \rightarrow	$3(8 - 5) + 4 \overset{?}{=} 13$
		$3(3) + 4 \overset{?}{=} 13$
		$9 + 4 \overset{?}{=} 13$
		$13 = 13$ A true statement

3. Solve.

$7(x - 3) + 5 = 4(3x - 2) - 8$

EXAMPLE 3 Solve $5(x - 3) + 2 = 5(2x - 8) - 3$.

Solution In this case, we first apply the distributive property to each side of the equation.

Step 1a:	$5x - 15 + 2 = 10x - 40 - 3$	*Distributive property*
Step 1c:	$5x - 13 = 10x - 43$	*Simplify each side.*
Step 2:	$5x + (-5x) - 13 = 10x + (-5x) - 43$	*Add −5x to both sides.*
	$-13 = 5x - 43$	
	$-13 + 43 = 5x - 43 + 43$	*Add 43 to both sides.*
	$30 = 5x$	
Step 3:	$\dfrac{1}{5}(30) = \dfrac{1}{5}(5x)$	*Multiply both sides by $\frac{1}{5}$.*
	$6 = x$	*The solution is 6.*

Step 4: Replacing x with 6 in the original equation, we have

$$5(6 - 3) + 2 \overset{?}{=} 5(2 \cdot 6 - 8) - 3$$

$$5(3) + 2 \overset{?}{=} 5(12 - 8) - 3$$

$$5(3) + 2 \overset{?}{=} 5(4) - 3$$

$$15 + 2 \overset{?}{=} 20 - 3$$

$$17 = 17 \qquad \textit{A true statement}$$

Note It makes no difference on which side of the equal sign x ends up. Most people prefer to have x on the left side because we read from left to right, and it seems to sound better to say x is 6 rather than 6 is x. Both expressions, however, have exactly the same meaning.

4. Solve the equation.

$0.06x + 0.04(x + 7,000) = 680$

EXAMPLE 4 Solve the equation $0.08x + 0.09(x + 2,000) = 690$.

Solution We can solve the equation in its original form by working with the decimals, or we can eliminate the decimals first by using the multiplication property of equality and solving the resulting equation. Both methods follow.

Method 1
Working with the decimals.

	$0.08x + 0.09(x + 2,000) = 690$	*Original equation*
Step 1a:	$0.08x + 0.09x + 0.09(2,000) = 690$	*Distributive property*
Step 1c:	$0.17x + 180 = 690$	*Simplify the left side.*
Step 2:	$0.17x + 180 + (-180) = 690 + (-180)$	*Add −180 to each side.*
	$0.17x = 510$	
Step 3:	$\dfrac{0.17x}{0.17} = \dfrac{510}{0.17}$	*Divide each side by 0.17.*
	$x = 3,000$	

Note that we divided each side of the equation by 0.17 to obtain the solution. This is still an application of the multiplication property of equality because dividing by 0.17 is equivalent to multiplying by $\frac{1}{0.17}$.

Method 2
Eliminating the decimals in the beginning.

	$0.08x + 0.09(x + 2,000) = 690$	*Original equation*
Step 1a:	$0.08x + 0.09x + 180 = 690$	*Distributive property*
Step 1b:	$100(0.08x + 0.09x + 180) = 100(690)$	*Multiply both sides by 100.*
	$8x + 9x + 18,000 = 69,000$	

Step 1c: $17x + 18,000 = 69,000$ Simplify the left side.

Step 2: $17x = 51,000$ Add $-18,000$ to each side.

Step 3: $\dfrac{17x}{17} = \dfrac{51,000}{17}$ Divide each side by 17.

$x = 3,000$

Substituting 3,000 for x in the original equation, we have

Step 4: $0.08(3,000) + 0.09(3,000 + 2,000) \overset{?}{=} 690$

$0.08(3,000) + 0.09(5,000) \overset{?}{=} 690$

$240 + 450 \overset{?}{=} 690$

$690 = 690$ A true statement

EXAMPLE 5 Solve $7 - 3(2y + 1) = 16$.

Solution We begin by multiplying -3 times the sum of $2y$ and 1.

Step 1a: $7 - 6y - 3 = 16$ Distributive property

Step 1c: $-6y + 4 = 16$ Simplify the left side.

Step 2: $-6y + 4 + (-4) = 16 + (-4)$ Add -4 to both sides.

$-6y = 12$

Step 3: $-\dfrac{1}{6}(-6y) = -\dfrac{1}{6}(12)$ Multiply both sides by $-\dfrac{1}{6}$.

$y = -2$

There are two things to notice about the example that follows. First, the distributive property is used to remove parentheses that are preceded by a negative sign. Second, the addition property and the multiplication property are not shown in as much detail as in the previous examples.

EXAMPLE 6 Solve $3(2x - 5) - (2x - 4) = 6 - (4x + 5)$.

Solution When we apply the distributive property to remove the grouping symbols and separate terms, we have to be careful with the signs. Remember, we can think of $-(2x - 4)$ as $-1(2x - 4)$, so that

$$-(2x - 4) = -1(2x - 4) = -2x + 4$$

It is not uncommon for students to make a mistake with this type of simplification and write the result as $-2x - 4$, which is incorrect. Here is the complete solution to our equation:

$3(2x - 5) - (2x - 4) = 6 - (4x + 5)$ Original equation

$6x - 15 - 2x + 4 = 6 - 4x - 5$ Distributive property

$4x - 11 = -4x + 1$ Simplify each side.

$8x - 11 = 1$ Add $4x$ to each side.

$8x = 12$ Add 11 to each side.

$x = \dfrac{12}{8}$ Multiply each side by $\dfrac{1}{8}$.

$x = \dfrac{3}{2}$ Reduce to lowest terms.

The solution, $\frac{3}{2}$, checks when replacing x in the original equation.

5. Solve $4 - 2(3y + 1) = -16$.

6. Solve.
$4(3x - 2) - (6x - 5) = 6 - (3x + 1)$

Answers
5. 3
6. $\frac{8}{9}$

7. Solve $2(t + 2) - 2 = 2(t + 3) - 4$.

EXAMPLE 7 Solve $-2(3t - 2) = 2(4 - 3t) - 4$.

Solution We apply the distributive property to simplify both sides of the equation.

$$-2(3t - 2) = 2(4 - 3t) - 4 \qquad \textit{Original equation}$$
$$-6t + 4 = 8 - 6t - 4 \qquad \textit{Distributive property}$$
$$-6t + 4 = -6t + 4 \qquad \textit{Simplify each side.}$$
$$-6t = -6t \qquad \textit{Add } -4 \textit{ to each side.}$$
$$0 = 0 \qquad \textit{Add } 6t \textit{ to each side.}$$

We have eliminated the variable and are left with a true statement. This means that any real number that we use in place of the variable will be a solution to the equation. The solution is all real numbers.

8. Solve $-2(3y - 2) = 3(3 - 2y)$.

EXAMPLE 8 Solve $4(y - 5) + 3 = 2(2y - 3)$.

Solution Again, we apply the distributive property.

$$4(y - 5) + 3 = 2(2y - 3) \qquad \textit{Original equation}$$
$$4y - 20 + 3 = 4y - 6 \qquad \textit{Distributive property}$$
$$4y - 17 = 4y - 6 \qquad \textit{Simplify each side.}$$
$$-17 = -6 \qquad \textit{Add } -4y \textit{ to each side.}$$

We have eliminated the variable and are left with a false statement. This means that no real numbers used in place of the variable will be a solution to the equation. There is no solution.

GETTING READY FOR CLASS

After reading through the preceding section, respond in your own words and in complete sentences.

A. What is the first step in solving a linear equation containing parentheses?

B. When solving a linear equation, why should you get all variable terms on one side and all constant terms on the other before using the multiplication property of equality?

C. What is the last step in solving a linear equation?

D. If an equation contains decimals, what can you do to eliminate the decimals?

Answers

7. All real numbers

8. No solution

Vocabulary Review

The following is a list of steps for solving a linear equation in one variable. Choose the correct words to fill in the blanks below.

solution	addition	multiplication	distributive
LCD	decimals	similar	constant

Step 1a: Use the _____ property to separate terms.

 1b: To clear any fractions, multiply both sides of the equation by the _____ . To clear _____ , multiply both sides by a power of 10.

 1c: Combine _____ terms.

Step 2: Use the _____ property of equality to get all variable terms on one side of the equation and all _____ terms on the other side.

Step 3: Use the _____ property of equality to get x by itself on one side of the equation.

Step 4: Check your _____ in the original equation.

Problems

A Solve each of the following equations using the four steps shown in this section.

1. $2(x + 3) = 12$

2. $3(x - 2) = 6$

3. $6(x - 1) = -18$

4. $4(x + 5) = 16$

5. $2(4a + 1) = -6$

6. $3(2a - 4) = 12$

7. $14 = 2(5x - 3)$

8. $-25 = 5(3x + 4)$

9. $-2(3y + 5) = 14$

10. $-3(2y - 4) = -6$

11. $-5(2a + 4) = 0$

12. $-3(3a - 6) = 0$

13. $1 = \frac{1}{2}(4x + 2)$

14. $1 = \frac{1}{3}(6x + 3)$

15. $3(t - 4) + 5 = -4$

16. $5(t - 1) + 6 = -9$

Solve each equation.

17. $4(2x + 1) - 7 = 1$

18. $6(3y + 2) - 8 = -2$

19. $\frac{1}{2}(x - 3) = \frac{1}{4}(x + 1)$

20. $\frac{1}{3}(x - 4) = \frac{1}{2}(x - 6)$

21. $-0.7(2x - 7) = 0.3(11 - 4x)$

22. $-0.3(2x - 5) = 0.7(3 - x)$

23. $-2(3y + 3) = 3(1 - 2y) - 9$

24. $-5(4y - 3) = 2(2 - 10y) + 11$

25. $\frac{3}{4}(8x - 4) + 3 = \frac{2}{5}(5x + 10) - 1$

26. $\frac{5}{6}(6x + 12) + 1 = \frac{2}{3}(9x - 3) + 5$

27. $0.06x + 0.08(100 - x) = 6.5$

28. $0.05x + 0.07(100 - x) = 6.2$

29. $6 - 5(2a - 3) = 1$

30. $-8 - 2(3 - a) = 0$

31. $0.2x - 0.5 = 0.5 - 0.2(2x - 13)$

32. $0.4x - 0.1 = 0.7 - 0.3(6 - 2x)$

33. $2(t - 3) + 3(t - 2) = 28$

34. $-3(t - 5) - 2(2t + 1) = -8$

35. $5(x - 2) - (3x + 4) = 3(6x - 8) + 10$

36. $3(x - 1) - (4x - 5) = 2(5x - 1) - 7$

37. $2(5x - 3) - (2x - 4) = 5 - (6x + 1)$

38. $3(4x - 2) - (5x - 8) = 8 - (2x + 3)$

39. $-(3x + 1) + 7 = 4 - (3x + 2)$

40. $-(6x + 2) = 8 - 2(3x + 1)$

41. $x + (2x - 1) = 2$

42. $x + (5x + 2) = 20$

43. $x - (3x + 5) = -3$

44. $x - (4x - 1) = 7$

45. $15 = 3(x - 1)$

46. $12 = 4(x - 5)$

47. $4x - (-4x + 1) = 5$

48. $-2x - (4x - 8) = -1$

49. $5x - 8(2x - 5) = 7$

50. $3x + 4(8x - 15) = 10$

51. $2(4x - 1) = 4(2x + 3)$

52. $2(3y - 3) = 3(2y + 4)$

53. $0.2x + 0.5(12 - x) = 3.6$

54. $0.3x + 0.6(25 - x) = 12$

55. $0.5x + 0.2(18 - x) = 5.4$

56. $0.1x + 0.5(40 - x) = 32$

57. $x + (x + 3)(-3) = x - 3$

58. $x - 2(x + 2) = x - 2$

59. $5(x + 2) + 3(x - 1) = -9$

60. $4(x + 1) + 3(x - 3) = 2$

61. $3(x - 3) + 2(2x) = 5$

62. $2(x - 2) + 3(5x) = 30$

63. $-5(y + 3) = -5y - 15$

64. $3(4y - 1) - 5 = 4(3y - 2)$

65. $3x + 2(x - 2) = 6$

66. $5x - (x - 5) = 25$

67. $50(x - 5) = 30(x + 5)$

68. $34(x - 2) = 26(x + 2)$

69. $0.08x + 0.09(x + 2{,}000) = 860$

70. $0.11x + 0.12(x + 4{,}000) = 940$

71. $0.10x + 0.12(x + 500) = 214$

72. $0.08x + 0.06(x + 800) = 104$

73. $5x + 10(x + 8) = 245$

74. $5x + 10(x + 7) = 175$

75. $5x + 10(x + 3) + 25(x + 5) = 435$

76. $5(x + 3) + 10x + 25(x + 7) = 390$

The next two problems are intended to give you practice reading, and paying attention to, the instructions that accompany the problems you are working. Working these problems is an excellent way to get ready for a test or a quiz.

77. Work each problem according to the instructions.

 a. Solve $4x - 5 = 0$.

 b. Solve $4x - 5 = 25$.

 c. Add $(4x - 5) + (2x + 25)$.

 d. Solve $4x - 5 = 2x + 25$.

 e. Multiply $4(x - 5)$.

 f. Solve $4(x - 5) = 2x + 25$.

78. Work each problem according to the instructions.

 a. Solve $3x + 6 = 0$.

 b. Solve $3x + 6 = 4$.

 c. Add $(3x + 6) + (7x + 4)$.

 d. Solve $3x + 6 = 7x + 4$.

 e. Multiply $3(x + 6)$.

 f. Solve $3(x + 6) = 7x + 4$.

79. Solve the equation $35x + 20 = 125$, given at the introduction of this section, to find the number of robot events that were entered.

80. A traveling circus with thrill rides charges $6 for admission plus $1.50 for each ride you take. If you pay $16.50 total, how many rides did you take?

81. A stock trading company charges a commission of $15 per trade plus $0.025 per dollar value of stock traded. If your commission is $120, what was the value of the stock you traded?

82. Brad earns a base monthly salary of $2,400 plus $0.018 commission on his total sales. If his monthly pay is $3,246, what were his sales that month?

Getting Ready for the Next Section

To understand all of the explanations and examples in the next section you must be able to work the problems below.

Solve each equation.

83. $40 = 2x + 12$ **84.** $80 = 2x + 12$ **85.** $12 + 2y = 6$ **86.** $3x + 18 = 6$

87. $24x = 6$ **88.** $45 = 0.75x$ **89.** $70 = x \cdot 210$ **90.** $15 = x \cdot 80$

Apply the distributive property.

91. $\dfrac{1}{2}(-3x + 6)$ **92.** $\dfrac{2}{3}(-6x + 9)$ **93.** $-\dfrac{1}{4}(-5x + 20)$ **94.** $-\dfrac{1}{3}(-4x + 12)$

95. $12\left(\dfrac{1}{3}x - \dfrac{1}{4}\right)$ **96.** $-8\left(-\dfrac{1}{4}x + \dfrac{1}{8}\right)$ **97.** $-24\left(-\dfrac{1}{3}x - \dfrac{1}{8}\right)$ **98.** $-15\left(\dfrac{3}{5}x - \dfrac{2}{3}\right)$

Find the Mistake

Each sentence below contains a mistake. Circle the mistake and write the correct word(s) or equation on the line provided.

1. A linear equation in one variable is any equation that can be put in the form of $y = Ax + B$. _____

2. $12 + 6y - 4 = 3y$ is not a linear equation in one variable because it cannot be simplified to the form $Ax + B = C$.

3. To solve the linear equation $3(2x + 7) = 2(x - 1)$, begin by dividing both sides of the equation by $(x - 1)$.

4. To solve a linear equation, make sure to keep similar terms on opposite sides of the equation.

5.4 Formulas

Image © sxc.hu, connect7, 2008

OBJECTIVES

A Find the value of a variable in a formula given replacements for the other variables.

B Solve a formula for one of its variables.

C Continue to find the complement and supplement of an angle.

D Solve basic percent problems.

KEY WORDS

formula

solving for a variable

Recall our discuss from Section 1 about finding the outside temperature by listening to a cricket chirp. What if we needed to find the number of chirps based on a Celsius temperature? The formula to find the outside temperature in Celsius is

$$t = \left(\frac{n}{3}\right) + 4$$

where t is the outside temperature in degrees Celsius and n is the number of cricket chirps in 25 seconds. In this section, we will learn about formulas and how to solve a formula for one of its variables. For instance, we will learn how the Celsius temperature formula could be solved for n and become

$$n = 3(t - 4)$$

Then we could provide a numerical replacement for t and easily calculate a value for n.

A Solving Formulas

In this section, we continue solving equations by working with formulas. Here is the definition of a formula.

> **DEFINITION** formula
>
> In mathematics, a **formula** is an equation that contains more than one variable.

The equation $P = 2l + 2w$, which tells us how to find the perimeter of a rectangle, is an example of a formula.

To begin our work with formulas, we will consider some examples in which we are given numerical replacements for all but one of the variables.

**Video Examples
Section 5.4**

Practice Problems

1. Suppose the livestock pen in Example 1 has a perimeter of 80 feet. If the width is still 6 feet, what is the new length?

EXAMPLE 1 The perimeter P of a rectangular livestock pen is 40 feet. If the width w is 6 feet, find the length.

Solution First, we substitute 40 for P and 6 for w in the formula $P = 2l + 2w$. Then we solve for l.

When → $P = 40$ and $w = 6$

the formula → $P = 2l + 2w$

becomes → $40 = 2l + 2(6)$

$40 = 2l + 12$ Multiply 2 and 6.

$28 = 2l$ Add -12 to each side.

$14 = l$ Multiply each side by $\frac{1}{2}$.

To summarize our results, if a rectangular pen has a perimeter of 40 feet and a width of 6 feet, then the length must be 14 feet.

2. Find x when y is 9 in the formula $3x + 2y = 6$.

EXAMPLE 2 Find y when $x = 4$ in the formula $3x + 2y = 6$.

Solution We substitute 4 for x in the formula and then solve for y.

When → $x = 4$

the formula → $3x + 2y = 6$

becomes → $3(4) + 2y = 6$

$12 + 2y = 6$ Multiply 3 and 4.

$2y = -6$ Add -12 to each side.

$y = -3$ Multiply each side by $\frac{1}{2}$.

B Solving for a Variable

In the next examples we will solve a formula for one of its variables without being given numerical replacements for the other variables.

Consider the formula for the area of a triangle:

$$A = \tfrac{1}{2}bh$$

where A = area, b = length of the base, and h = height of the triangle.

Suppose we want to solve this formula for h. What we must do is isolate the variable h on one side of the equal sign. We begin by multiplying both sides by 2, because it is the reciprocal of $\frac{1}{2}$.

$$2 \cdot A = 2 \cdot \frac{1}{2}bh$$

$$2A = bh$$

Then we divide both sides by b.

$$\frac{2A}{b} = \frac{bh}{b}$$

$$h = \frac{2A}{b}$$

The original formula $A = \frac{1}{2}bh$ and the final formula $h = \frac{2A}{b}$ both give the same relationship among A, b, and h. The first one has been solved for A and the second one has been solved for h.

> **RULE** Solving for a Variable
>
> To solve a formula for one of its *variables*, we must isolate that variable on either side of the equal sign. All other variables and constants will appear on the other side.

EXAMPLE 3 Solve $3x + 2y = 6$ for y.

Solution To solve for y, we must isolate y on the left side of the equation. To begin, we use the addition property of equality to add $-3x$ to each side.

$$3x + 2y = 6 \qquad \text{Original formula}$$

$$3x + (-3x) + 2y = (-3x) + 6 \qquad \text{Add } -3x \text{ to each side.}$$

$$2y = -3x + 6 \qquad \text{Simplify the left side.}$$

$$\frac{1}{2}(2y) = \frac{1}{2}(-3x + 6) \qquad \text{Multiply each side by } \frac{1}{2}.$$

$$y = -\frac{3}{2}x + 3$$

3. Solve $5x - 4y = 20$ for y.

EXAMPLE 4 Solve $h = vt - 16t^2$ for v.

Solution Let's begin by interchanging the left and right sides of the equation. That way, the variable we are solving for, v, will be on the left side.

$$vt - 16t^2 = h \qquad \text{Symmetric property}$$

$$vt - 16t^2 + 16t^2 = h + 16t^2 \qquad \text{Add } 16t^2 \text{ to each side.}$$

$$vt = h + 16t^2$$

$$\frac{vt}{t} = \frac{h + 16t^2}{t} \qquad \text{Divide each side by } t.$$

$$v = \frac{h + 16t^2}{t}$$

We know we are finished because we have isolated the variable we are solving for on the left side of the equation and it does not appear on the other side.

4. Solve $P = 2w + 2l$ for l.

EXAMPLE 5 Solve $\dfrac{y - 1}{x} = \dfrac{3}{2}$ for y.

Solution Although we will do more extensive work with formulas of this form later in the book, we need to know how to solve this particular formula for y in order to understand some things in the next chapter. We begin by multiplying each side of the formula by x. Doing so will simplify the left side of the equation, and make the rest of the solution process simple.

$$\frac{y - 1}{x} = \frac{3}{2} \qquad \text{Original formula}$$

$$x \cdot \frac{y - 1}{x} = \frac{3}{2} \cdot x \qquad \text{Multiply each side by } x.$$

$$y - 1 = \frac{3}{2}x \qquad \text{Simplify each side.}$$

$$y = \frac{3}{2}x + 1 \qquad \text{Add 1 to each side.}$$

This is our solution. If we look back to the first step, we can justify our result on the left side of the equation this way: Dividing by x is equivalent to multiplying by its reciprocal $\frac{1}{x}$. Here is what it looks like when written out completely:

$$x \cdot \frac{y - 1}{x} = x \cdot \frac{1}{x}(y - 1) = 1(y - 1) = (y - 1)$$

5. Solve $\dfrac{y - 2}{x} = \dfrac{4}{3}$ for y.

Answers

3. $y = \frac{5}{4}x - 5$

4. $l = \frac{P - 2w}{2}$

5. $y = \frac{4}{3}x + 2$

C Complementary and Supplementary Angles

> **FACTS FROM GEOMETRY** More on Complementary
> and Supplementary Angles
>
> In Chapter 1 we defined complementary angles as angles that add to 90°; that is, if x and y are complementary angles, then
>
> $$x + y = 90°$$
>
> If we solve this formula for y, we obtain a formula equivalent to our original formula:
>
> $$y = 90° - x$$
>
> Because y is the complement of x, we can generalize by saying that the complement of angle x is the angle $90° - x$. By a similar reasoning process, we can say that the supplement of angle x is the angle $180° - x$. To summarize, if x is an angle, then
>
> The complement of x is $90° - x$, and
>
> The supplement of x is $180° - x$
>
> If you go on to take a trigonometry class, you will see this formula again.

6. Find the complement and the supplement of 35°.

EXAMPLE 6 Find the complement and the supplement of 25°.

Solution We can use the formulas above with $90° - x$ and $180° - x$.

The complement of 25° is $90° - 25° = 65°$.

The supplement of 25° is $180° - 25° = 155°$.

D Basic Percent Problems

The last examples in this section show how basic percent problems can be translated directly into equations. To understand these examples, you must recall that *percent* means "per hundred" that is, 75% is the same as $\frac{75}{100}$, 0.75, and, in reduced fraction form, $\frac{3}{4}$. Likewise, the decimal 0.25 is equivalent to 25%. To change a decimal to a percent, we move the decimal point two places to the right and write the % symbol. To change from a percent to a decimal, we drop the % symbol and move the decimal point two places to the left. The table that follows gives some of the most commonly used fractions and decimals and their equivalent percents.

Fraction	Decimal	Percent
$\frac{1}{2}$	0.5	50%
$\frac{1}{4}$	0.25	25%
$\frac{3}{4}$	0.75	75%
$\frac{1}{3}$	$0.33\overline{3}$	$33\frac{1}{3}\%$
$\frac{2}{3}$	0.667	$66\frac{2}{3}\%$
$\frac{1}{5}$	0.2	20%
$\frac{2}{5}$	0.4	40%

Answer

6. 55°; 145°

EXAMPLE 7 What number is 25% of 60?

7. What number is 25% of 74?

Solution To solve a problem like this, we let $x =$ the number in question (that is, the number we are looking for). Then, we translate the sentence directly into an equation by using an equal sign for the word "is" and multiplication for the word "of." Here is how it is done:

$$\text{What number is 25\% of 60?}$$
$$x \qquad = 0.25 \cdot 60$$
$$= 15$$

Notice that we must write 25% as a decimal in order to do the arithmetic in the problem. The number 15 is 25% of 60.

EXAMPLE 8 What percent of 24 is 6?

8. What percent of 84 is 21?

Solution Translating this sentence into an equation, as we did in Example 7, we have

$$\text{What percent of 24 is 6?}$$
$$x \qquad \cdot \ 24 = 6$$
$$24x = 6$$

Next, we multiply each side by $\frac{1}{24}$. (This is the same as dividing each side by 24.)

$$\frac{1}{24}(24x) = \frac{1}{24}(6)$$
$$x = \frac{6}{24}$$
$$= \frac{1}{4}$$
$$= 0.25, \text{ or } 25\%$$

25% of 24 is 6, or in other words, the number 6 is 25% of 24.

EXAMPLE 9 45 is 75% of what number?

9. 35 is 40% of what number?

Solution Again, we translate the sentence directly.

$$\text{45 is 75\% of what number?}$$
$$45 = 0.75 \cdot \qquad x$$

Next, we multiply each side by $\frac{1}{0.75}$ (which is the same as dividing each side by 0.75).

$$\frac{1}{0.75}(45) = \frac{1}{0.75}(0.75x)$$
$$\frac{45}{0.75} = x$$
$$60 = x$$

The number 45 is 75% of 60.

Answers

7. 18.5
8. 25%
9. 87.5

10. In another pizza ingredient, pepperoni, 75% of the calories are from fat. If one serving of pepperoni contains 132 calories, how many of those calories are from fat?

EXAMPLE 10 The American Dietetic Association (ADA) recommends eating foods in which the calories from fat are less than 30% of the total calories. The nutrition labels from two ingredients you may use for a pizza are shown in Figure 1. For each ingredient, what percent of the total calories come from fat?

PIZZA DOUGH

Nutrition Facts

Serving Size 1/6 of package (65g)
Servings Per Container: 6

Amount Per Serving

Calories 160 Calories from fat 18

	% Daily Value*
Total Fat 2g	3%
Saturated Fat 0.5g	3%
Poly unsaturated Fat 0g	
Monounsaturated Fat 0g	
Cholesterol 0mg	0%
Sodium 470mg	20%
Total Carbohydrate 31g	10%
Dietary Fiber 1g	4%
Sugars 4g	
Protein 5g	

| Vitamin A 0% | • | Vitamin C 0% |
| Calcium 0% | • | Iron 10% |

*Percent Daily Values are based on a 2,000 calorie diet

MOZZARELLA CHEESE

Nutrition Facts

Serving Size 1 oz (28.3g)
Servings Per Container: 12

Amount Per Serving

Calories 72 Calories from fat 41

	% Daily Value*
Total Fat 4.5g	7%
Saturated Fat 2.9g	14%
Cholesterol 18mg	6%
Sodium 175mg	7%
Total Carbohydrate 0.8g	0%
Fiber 0g	0%
Sugars 0.3g	
Protein 6.9g	

| Vitamin A 3% | • | Vitamin C 0% |
| Calcium 22% | • | Iron 0% |

*Percent Daily Values (DV) are based on a 2,000 calorie diet

FIGURE 1

Solution The information needed to solve this problem is located towards the top of each label. Each serving of the pizza dough contains 160 calories, of which 18 calories come from fat. To find the percent of total calories that come from fat, we must answer this question:

18 is what percent of 160?

For the mozzarella cheese, one serving contains 72 calories, of which 41 calories come from fat. To find the percent of total calories that come from fat, we must answer this question:

41 is what percent of 72?

Translating each equation into symbols, we have

18 is what percent of 160 41 is what percent of 72

$$18 = x \cdot 160 \qquad\qquad 41 = x \cdot 72$$

$$x = \frac{18}{160} \qquad\qquad\qquad x = \frac{41}{72}$$

$x = 0.11$ to the nearest hundredth $x = 0.57$ to the nearest hundredth

$x = 11\%$ $x = 57\%$

Comparing the two ingredients, 11% of the calories in the pizza dough are fat calories, whereas 57% of the calories in the mozzarella cheese are fat calories. According to the ADA, pizza dough is the healthier ingredient.

GETTING READY FOR CLASS

After reading through the preceding section, respond in your own words and in complete sentences.

A. What is a formula?

B. How do you solve a formula for one of its variables?

C. What are complementary angles?

D. Translate the following question into an equation: what number is 40% of 20?

Answer

10. 99 calories

Vocabulary Review

Choose the correct words to fill in the blanks below.

| isolate | variable | complementary | formula | supplement |

1. A formula is an equation with more than one _____.

2. $P = 2l + 2w$ is the _____ for finding the perimeter of a rectangle.

3. To solve for a variable in a formula, _____ that variable on either side of the equals sign.

4. If x and y are _____ angles, then $x + y = 90°$.

5. The _____ of angle x is the angle $180° - x$.

Problems

A Use the formula $P = 2l + 2w$ to find the length l of a rectangular lot if

1. The width w is 50 feet and the perimeter P is 300 feet.

2. The width w is 75 feet and the perimeter P is 300 feet.

Use the formula $2x + 3y = 6$ to find y when

3. $x = 3$ **4.** $x = -2$ **5.** $x = 0$ **6.** $x = -3$

Use the formula $2x - 5y = 20$ to find x when

7. $y = 2$ **8.** $y = -4$ **9.** $y = 0$ **10.** $y = -6$

Use the equation $y = (x + 1)^2 - 3$ to find the value of y when

11. $x = -2$ **12.** $x = -1$ **13.** $x = 1$ **14.** $x = 2$

15. Use the formula $y = \dfrac{20}{x}$ to find y when

 a. $x = 10$

 b. $x = 5$

16. Use the formula $y = 2x^2$ to find y when

 a. $x = 5$

 b. $x = -6$

17. Use the formula $y = Kx$ to find K when

 a. $y = 15$ and $x = 3$

 b. $y = 72$ and $x = 4$

18. Use the formula $y = Kx^2$ to find K when

 a. $y = 32$ and $x = 4$

 b. $y = 45$ and $x = 3$

B Solve each of the following for the indicated variable.

19. $A = lw$ for l

20. $d = rt$ for r

21. $V = lwh$ for h

22. $PV = nRT$ for P

23. $P = a + b + c$ for a

24. $P = a + b + c$ for b

25. $x - 3y = -1$ for x

26. $x + 3y = 2$ for x

27. $-3x + y = 6$ for y

28. $2x + y = -17$ for y

29. $2x + 3y = 6$ for y

30. $4x + 5y = 20$ for y

31. $y - 3 = -2(x + 4)$ for y

32. $y + 5 = 2(x + 2)$ for y

33. $y - 3 = -\dfrac{2}{3}(x + 3)$ for y

34. $y - 1 = -\dfrac{1}{2}(x + 4)$ for y

35. $P = 2l + 2w$ for w

36. $P = 2l + 2w$ for l

37. $h = vt + 16t^2$ for v

38. $h = vt - 16t^2$ for v

39. $A = \pi r^2 + 2\pi rh$ for h

40. $A = 2\pi r^2 + 2\pi rh$ for h

41. $I = prt$ for t

42. $PV = nrT$ for T

43. Solve for y.

a. $\dfrac{y - 1}{x} = \dfrac{3}{5}$

b. $\dfrac{y - 2}{x} = \dfrac{1}{2}$

c. $\dfrac{y - 3}{x} = 4$

44. Solve for y.

a. $\dfrac{y + 1}{x} = -\dfrac{3}{5}$

b. $\dfrac{y + 2}{x} = -\dfrac{1}{2}$

c. $\dfrac{y + 3}{x} = -4$

45. Solve for y.

a. $\dfrac{y + 2}{x} = -\dfrac{1}{2}$

b. $\dfrac{y - 2}{x} = -3$

c. $\dfrac{y + 1}{x} = -\dfrac{1}{3}$

46. Solve for y.

a. $\dfrac{y - 1}{x} = -\dfrac{3}{4}$

b. $\dfrac{y + 5}{x} = -7$

c. $\dfrac{y + 2}{x} = -3$

Solve each formula for y.

47. $\dfrac{x}{7} - \dfrac{y}{3} = 1$

48. $\dfrac{x}{5} - \dfrac{y}{9} = 1$

49. $-\dfrac{1}{4}x + \dfrac{1}{8}y = 1$

50. $-\dfrac{1}{9}x + \dfrac{1}{3}y = 1$

C Find the complement and the supplement of each angle.

51. 30°

52. 60°

53. 45°

54. 15°

D Translate each of the following into an equation, and then solve that equation.

55. What number is 25% of 40?

56. What number is 75% of 40?

57. What number is 12% of 2,000?

58. What number is 9% of 3,000?

59. What percent of 28 is 7?

60. What percent of 28 is 21?

61. What percent of 40 is 14?

62. What percent of 20 is 14?

63. 32 is 50% of what number?

64. 16 is 50% of what number?

65. 240 is 12% of what number?

66. 360 is 12% of what number?

Applying the Concepts

More About Temperatures As we mentioned in Chapter 1, in the U.S. system, temperature is measured on the Fahrenheit scale. In the metric system, temperature is measured on the Celsius scale. On the Celsius scale, water boils at 100 degrees and freezes at 0 degrees. To denote a temperature of 100 degrees on the Celsius scale, we write 100°C, which is read "100 degrees Celsius."

Table 1 is intended to give you an intuitive idea of the relationship between the two temperature scales. Table 2 gives the formulas, in both symbols and words, that are used to convert between the two scales.

Table 1

Situation	Temperature	
	Fahrenheit	Celsius
Water freezes	32°F	0°C
Room temperature	68°F	20°C
Normal body temperature	98.6°F	37°C
Water boils	212°F	100°C

Table 2

To convert from	Formula in symbols	Formula in words
Fahrenheit to Celsius	$C = \frac{5}{9}(F - 32)$	Subtract 32, then multiply by $\frac{5}{9}$.
Celsius to Fahrenheit	$F = \frac{9}{5}C + 32$	Multiply by $\frac{9}{5}$, then add 32.

67. Let $F = 212$ in the formula $C = \frac{5}{9}(F - 32)$, and solve for C. Does the value of C agree with the information in Table 1?

68. Let $C = 100$ in the formula $F = \frac{9}{5}C + 32$, and solve for F. Does the value of F agree with the information in Table 1?

69. Let $F = 68$ in the formula $C = \frac{5}{9}(F - 32)$, and solve for C. Does the value of C agree with the information in Table 1?

70. Let $C = 37$ in the formula $F = \frac{9}{5}C + 32$, and solve for F. Does the value of F agree with the information in Table 1?

71. Solve the formula $F = \frac{9}{5}C + 32$ for C.

72. Solve the formula $C = \frac{5}{9}(F - 32)$ for F.

Nutrition Labels The nutrition label in Figure 2 is from a caffe latte. The label in Figure 3 is from a nonfat caffe latte. Use the information on these labels for problems 73–76. Round your answers to the nearest tenth of a percent.

CAFFE LATTE

Nutrition Facts	
Serving Size 16 fl oz (453.0g)	
Servings Per Container: 1	
Amount Per Serving	
Calories 272	Calories from fat 126
	% Daily Value*
Total Fat 14.0g	**22%**
Saturated Fat 9.0g	**45%**
Cholesterol 60mg	**20%**
Sodium 208mg	**9%**
Total Carbohydrate 22.0g	**7%**
Fiber 0g	**0%**
Sugars 19.0g	
Protein 14.0g	
Vitamin A 12% • Vitamin C 0%	
Calcium 48% • Iron 0%	
*Percent Daily Values (DV) are based on a 2,000 calorie diet	

FIGURE 2

NONFAT CAFFE LATTE

Nutrition Facts	
Serving Size 16 fl oz (453.0g)	
Servings Per Container: 1	
Amount Per Serving	
Calories 168	Calories from fat 0
	% Daily Value*
Total Fat 0.3g	**0%**
Saturated Fat 0.0g	**0%**
Cholesterol 8mg	**3%**
Sodium 232mg	**10%**
Total Carbohydrate 24.0g	**8%**
Fiber 0g	**0%**
Sugars 21.0g	
Protein 16.0g	
Vitamin A 16% • Vitamin C 0%	
Calcium 48% • Iron 0%	
*Percent Daily Values (DV) are based on a 2,000 calorie diet	

FIGURE 3

73. What percent of the calories in the caffe latte are calories from fat?

74. What percent of the calories in the nonfat caffe latte are calories from fat?

75. One nonfat caffe latte weighs 453 grams, of which 24 grams are carbohydrates. What percent are carbohydrates?

76. One caffe latte weighs 453 grams. What percent is sugar?

Circumference The circumference of a circle is given by the formula $C = 2\pi r$. Find r using the following values. (The values given for π are approximations.)

77. The circumference C is 44 meters and π is $\frac{22}{7}$.

78. The circumference C is 176 meters and π is $\frac{22}{7}$.

79. The circumference is 9.42 inches and π is 3.14.

80. The circumference is 12.56 inches and π is 3.14.

Volume The volume of a cylinder is given by the formula $V = \pi r^2 h$. Find the height h if

81. The volume V is 42 cubic feet, the radius is $\frac{7}{22}$ feet, and π is $\frac{22}{7}$.

82. The volume V is 84 cubic inches, the radius is $\frac{7}{11}$ inches, and π is $\frac{22}{7}$.

83. The volume is 6.28 cubic centimeters, the radius is 3 centimeters, and π is 3.14.

84. The volume is 12.56 cubic centimeters, the radius is 2 centimeters, and π is 3.14.

Baseball Major league baseball has various player awards at the end of the year. One of them is the Rolaids Relief Man of the Year. To compute the Relief Man standings, points are gained or taken away based on the number of wins, losses, saves, and blown saves a relief pitcher has at the end of the year. The pitcher with the most Rolaids Points is the Rolaids Relief Man of the Year. The formula $P = 3s + 2w + t - 2l - 2b$ gives the number of Rolaids points a pitcher earns, where s = saves, w = wins, t = tough saves, l = losses, and b = blown saves. Use this formula to complete the following tables.

85. National League 2010

Pitcher, Team	W	L	Saves	Tough Saves	Blown Saves	Rolaids Points
Heath Bell, San Diego	6	1	47	3	5	
Brian Wilson, San Francisco	3	3	48	5	7	
Carlos Marmol, Chicago	2	3	38	5	5	
Billy Wagner, Atlanta	7	2	37	7	0	
Francisco Cordero, Cincinati	6	5	40	8	0	

86. American League 2010

Pitcher, Team	W	L	Saves	Tough Saves	Blown Saves	Rolaids Points
Rafael Soriano, Tampa Bay	3	2	45	0	3	
Joakim Soria, Kansas City	1	2	43	0	3	
Neftali Feliz, Texas	4	3	40	1	3	
Jonathan Papelbon, Boston	5	7	37	1	8	
Kevin Gregg, Toronto	2	6	37	1	6	

Getting Ready for the Next Section

To understand all of the explanations and examples in the next section you must be able to work the problems below.

Write an equivalent expression in English. Include the words *sum* and *difference* when possible.

87. $4 + 1 = 5$

88. $7 + 3 = 10$

89. $6 - 2 = 4$

90. $8 - 1 = 7$

91. $x - 15 = -12$

92. $2x + 3 = 7$

93. $x + 3 = 4(x - 3)$

94. $2(2x - 5) = 2x - 34$

For each of the following expressions, write an equivalent equation.

95. Twice the sum of 6 and 3 is 18.

96. Four added to the product of 5 and -1 is -1.

97. The sum of twice 5 and 3 is 13.

98. Twice the difference of 8 and 2 is 12.

99. The sum of a number and five is thirteen.

100. The difference of ten and a number is negative eight.

101. Five times the sum of a number and seven is thirty.

102. Five times the difference of twice a number and six is negative twenty.

Find the Mistake

Each sentence below contains a mistake. Circle the mistake and write the correct word(s) or equation on the line provided.

1. A formula contains only one variable. _____

2. To solve for a variable in a formula, make sure the variable appears on both sides of the equation.

3. To solve the formula $2ab = a + 8$ for b, multiply each side of the equation by $2a$. _____

4. The question "What percent of 108 is 9?" can be translated into the equation $\frac{x}{9} = 108$. _____

Landmark Review: Checking Your Progress

Solve each of the following equations.

1. $x + 3 = 15$

2. $x - 5 = 20$

3. $5.6 + y = 10.8$

4. $-\frac{1}{5} + b = \frac{2}{3}$

5. $4x = 16$

6. $\frac{1}{5}y = 2$

7. $-5x + 3 + 3x + 5 = 0$

8. $3y + 4 - 6y = -10 + 5$

9. $4(x + 3) = 24$

10. $-4 - 3(x + 2) - 3 = -25$

Solve each of the following for the indicated variable.

11. $A = lw$ for l

12. $P = a + b + c$ for c

13. $x - 4 + 3y = 14$ for x

14. $\frac{y - 3}{x} = -\frac{4}{3}$ for y

5.5 Number, Age, Perimeter, Coin Applications

An artist in Vermont built a sculpture of a dinosaur called Vermontasaurus that is 25 feet tall. The artist used found junk and scrap wood to construct his controversial work of art. The controversy lies in the structure's adherence to state building codes. The state's fire safety board has demanded that Vermontasaurus visitors not walk under the structure because of its questionable structural integrity. The area has been roped off, pending approval of a building permit.

Suppose you were tasked with constructing a permanent rectangular fence around the legs of the Vermontasaurus. You know that the length of the fence is 4 feet more than three times the width, and the perimeter is 168 feet. How would you find the length and width? In this section, we will use our knowledge of linear equations to solve similar application problems.

Problem Solving

To begin this section, we list the steps used in solving application problems. We call this strategy the *Blueprint for Problem Solving*. It is an outline that will overlay the solution process we use on all application problems.

**Video Examples
Section 5.5**

BLUEPRINT FOR PROBLEM SOLVING

Step 1: *Read* the problem, and then mentally *list* the items that are known and the items that are unknown. Make sure to include units of measurement where appropriate.

Step 2: *Assign a variable* to one of the unknown items. (In most cases this will amount to letting $x =$ the item that is asked for in the problem.) Then *translate* the other *information* in the problem to expressions involving the variable. Draw a diagram or create a table, where appropriate, to help convey this data.

Step 3: *Reread* the problem, and then *write an equation*, using the items and variables listed in steps 1 and 2, that describes the situation.

Step 4: *Solve the equation* found in step 3.

Step 5: *Write* your *answer* using a complete sentence.

Step 6: *Reread* the problem, and *check* your solution with the original words in the problem.

There are a number of substeps within each of the steps in our blueprint. For instance, with steps 1 and 2 it is always a good idea to draw a diagram or picture if it helps visualize the relationship between the items in the problem. In other cases a table helps organize the information. As you gain more experience using the blueprint to solve application problems, you will find additional techniques that expand the blueprint.

To help with problems of the type shown next in Examples 1 and 2, here are some common English words and phrases and their mathematical translations.

English	Algebra
The sum of a and b	$a + b$
The difference of a and b	$a - b$
The product of a and b	$a \cdot b$
The quotient of a and b	$\frac{a}{b}$
of	\cdot (multiply)
is	$=$ (equals)
A number	x
4 more than x	$x + 4$
4 times x	$4x$
4 less than x	$x - 4$

A Number Problems

EXAMPLE 1 The sum of twice a number and three is seven. Find the number.

Solution Using the Blueprint for Problem Solving as an outline, we solve the problem as follows:

Step 1: *Read* the problem, and then mentally *list* the items that are known and the items that are unknown.

> *Known items:* The numbers 3 and 7

> *Unknown items:* The number in question

Step 2: *Assign a variable* to one of the unknown items. Then *translate* the other *information* in the problem to expressions involving the variable.

> Let $x =$ the number asked for in the problem, then "The sum of twice a number and three" translates to $2x + 3$.

Step 3: *Reread* the problem, and then *write an equation,* using the items and variables listed in steps 1 and 2, that describes the situation.
With all word problems, the word *is* translates to $=$.

> The sum of twice x and 3 is 7.
> $$\underbrace{2x + 3}\qquad = 7$$

Step 4: *Solve the equation* found in step 3.

$$2x + 3 = 7$$
$$2x + 3 + (-3) = 7 + (-3)$$
$$2x = 4$$
$$\frac{1}{2}(2x) = \frac{1}{2}(4)$$
$$x = 2$$

Step 5: *Write* your *answer* using a complete sentence.

 The number is 2.

Step 6: *Reread* the problem, and *check* your solution with the original words in the problem.

 The sum of twice 2 and 3 is 7; a true statement.

You may find some examples and problems in this section that you can solve without using algebra or our blueprint. It is very important that you solve these problems using the methods we are showing here. The purpose behind these problems is to give you experience using the blueprint as a guide to solving problems written in words. Your answers are much less important than the work that you show to obtain your answer. You will be able to condense the steps in the blueprint later in the course. For now, though, you need to show your work in the same detail that we are showing in the examples in this section.

EXAMPLE 2 One number is three more than twice another; their sum is eighteen. Find the numbers.

Solution

Step 1: *Read and list.*
 Known items: Two numbers that add to 18. One is 3 more than twice the other.
 Unknown items: The numbers in question.

Step 2: *Assign a variable, and translate information.*
 Let $x =$ the first number. The other is $2x + 3$.

Step 3: *Reread and write an equation.*

 Their sum is 18.

 $x + (2x + 3) = 18$

Step 4: *Solve the equation.*

$$x + (2x + 3) = 18$$
$$3x + 3 = 18$$
$$3x + 3 + (-3) = 18 + (-3)$$
$$3x = 15$$
$$x = 5$$

Step 5: *Write the answer.*
 The first number is 5. The other is $2 \cdot 5 + 3 = 13$.

Step 6: *Reread and check.*
 The sum of 5 and 13 is 18, and 13 is 3 more than twice 5.

B Age Problems

Remember as you read through the solutions to the examples in this section, step 1 is done mentally. Read the problem, and then mentally list the items that you know and the items that you don't know. The purpose of step 1 is to give you direction as you begin to work application problems. Finding the solution to an application problem is a process; it doesn't happen all at once. The first step is to read the problem with a purpose in mind. That purpose is to mentally note the items that are known and the items that are unknown.

2. If three times the difference of a number and two were decreased by six, the result would be three. Find the number.

> *Note* Although these applications contain more than one variable quantity, we are choosing to write our equation in terms of a single variable to make the process of finding solutions easier.

Answer

2. 5

3. Pete is five years older than Mary. Five years ago Pete was twice as old as Mary. Find their ages now.

EXAMPLE 3 Bill is 6 years older than Tom. Three years ago Bill's age was four times Tom's age. Find the age of each boy now.

Solution Applying the Blueprint for Problem Solving, we have

Step 1: *Read and list.*

Known items: Bill is 6 years older than Tom. Three years ago Bill's age was four times Tom's age.

Unknown items: Bill's age and Tom's age

Step 2: *Assign a variable, and translate information.*

Let T = Tom's age now. That makes Bill $T + 6$ years old now. A table like the one shown here can help organize the information in an age problem. Notice how we placed the T in the box that corresponds to Tom's age now.

	Three Years Ago	Now
Bill		$T + 6$
Tom		T

If Tom is T years old now, 3 years ago he was $T - 3$ years old. If Bill is $T + 6$ years old now, 3 years ago he was $T + 6 - 3 = T + 3$ years old. We use this information to fill in the remaining spaces in the table.

	Three Years Ago	Now
Bill	$T + 3$	$T + 6$
Tom	$T - 3$	T

Step 3: *Reread and write an equation.*

Reading the problem again, we see that 3 years ago Bill's age was four times Tom's age. Writing this as an equation, we have

Bill's age 3 years ago = 4 · (Tom's age 3 years ago):

$$T + 3 = 4(T - 3)$$

Step 4: *Solve the equation.*

$$T + 3 = 4(T - 3)$$
$$T + 3 = 4T - 12$$
$$T + (-x) + 3 = 4T + (-T) - 12$$
$$3 = 3T - 12$$
$$3 + 12 = 3T - 12 + 12$$
$$15 = 3T$$
$$T = 5$$

Step 5: *Write the answer.*

Tom is 5 years old. Bill is 11 years old.

Step 6: *Reread and check.*

If Tom is 5 and Bill is 11, then Bill is 6 years older than Tom. Three years ago Tom was 2 and Bill was 8. At that time, Bill's age was four times Tom's age. As you can see, the answers check with the original problem.

C Perimeter Problems

To understand Example 4 completely, you need to recall that the perimeter of a rectangle is the sum of the lengths of the sides. The formula for the perimeter is $P = 2l + 2w$.

EXAMPLE 4 The length of a rectangle is 5 inches more than twice the width. The perimeter is 34 inches. Find the length and width.

Solution When working problems that involve geometric figures, a sketch of the figure helps organize and visualize the problem.

Step 1: *Read and list.*

> *Known items:* The figure is a rectangle. The length is 5 inches more than twice the width. The perimeter is 34 inches.
>
> *Unknown items:* The length and the width

Step 2: *Assign a variable, and translate information.*

> Because the length is given in terms of the width (the length is 5 more than twice the width), we let x = the width of the rectangle. The length is 5 more than twice the width, so it must be $2x + 5$. The diagram below is a visual description of the relationships we have listed so far.

x

$2x + 5$

Step 3: *Reread and write an equation.*

> The equation that describes the situation is

Twice the length + twice the width is the perimeter

$$2(2x + 5) \quad + \quad 2x \quad = \quad 34$$

Step 4: *Solve the equation.*

$2(2x + 5) + 2x = 34$	*Original equation*
$4x + 10 + 2x = 34$	*Distributive property*
$6x + 10 = 34$	*Add 4x and 2x.*
$6x = 24$	*Add −10 to each side.*
$x = 4$	*Divide each side by 6.*

Step 5: *Write the answer.*

> The width x is 4 inches. The length is $2x + 5 = 2(4) + 5 = 13$ inches.

Step 6: *Reread and check.*

> If the length is 13 and the width is 4, then the perimeter must be $2(13) + 2(4) = 26 + 8 = 34$, which checks with the original problem.

D Coin Problems

EXAMPLE 5 Jennifer has $2.45 in dimes and nickels. If she has 8 more dimes than nickels, how many of each coin does she have?

Solution

Step 1: *Read and list.*

> *Known items:* The type of coins, the total value of the coins, and that there are 8 more dimes than nickels.
>
> *Unknown items:* The number of nickels and the number of dimes

4. The length of a rectangle is three more than twice the width. The perimeter is 36 inches. Find the length and the width.

5. Amy has $1.75 in dimes and quarters. If she has 7 more dimes than quarters, how many of each coin does she have?

Answers

4. Width is 5; length is 13

5. 3 quarters, 10 dimes

Step 2: *Assign a variable, and translate information.*

If we let N = the number of nickels, then $N + 8$ = the number of dimes. Because the value of each nickel is 5 cents, the amount of money in nickels is $5N$. Similarly, because each dime is worth 10 cents, the amount of money in dimes is $10(N + 8)$. Here is a table that summarizes the information we have so far:

	Nickels	Dimes
Number	N	$N + 8$
Value (in cents)	$5N$	$10(N + 8)$

Step 3: *Reread and write an equation.*

Because the total value of all the coins is 245 cents, the equation that describes this situation is

Amount of money in nickels		Amount of money in dimes		Total amount of money
$5N$	+	$10(N + 8)$	=	245

Step 4: *Solve the equation.*

To solve the equation, we apply the distributive property first.

$5N + 10N + 80 = 245$	Distributive property
$15N + 80 = 245$	Add $5N$ and $10N$.
$15N = 165$	Add -80 to each side.
$N = 11$	Divide each side by 15.

Step 5: *Write the answer.*

The number of nickels is $N = 11$.
The number of dimes is $N + 8 = 11 + 8 = 19$.

Step 6: *Reread and check.*

To check our results

11 nickels are worth $5(11) =$	55 cents
19 dimes are worth $10(19) =$	190 cents

The total value is 245 cents = \$2.45

When you begin working the problems in the problem set that follows, there are a few things to remember. The first is that you may have to read the problems a number of times before you begin to see how to solve them. The second thing to remember is that word problems are not always solved correctly the first time you try them. Sometimes it takes a few attempts and some wrong answers before you can set up and solve these problems correctly. Don't give up.

> ### GETTING READY FOR CLASS
>
> *After reading through the preceding section, respond in your own words and in complete sentences.*
>
> **A.** Why is the first step in the Blueprint for Problem Solving done mentally?
>
> **B.** What is the last thing you do when solving an application problem?
>
> **C.** Why is it important to still use the blueprint and show your work even if you can solve a problem without using algebra?
>
> **D.** Write an application problem whose solution depends on solving the equation $4x + 2 = 10$.

Vocabulary Review

The following is a list of steps for the Blueprint for Problem Solving. Choose the correct words to fill in the blanks below.

| equation | variable | known | sentence | unknown | problem |

Step 1: Read the problem and mentally list the _____ and unknown items.

Step 2: Assign a variable to one of the _____ items, and then translate the rest of the information into expressions involving the variable.

Step 3: Reread the problem and write an equation using the _____ from step 2.

Step 4: Solve the _____.

Step 5: Write your answer using a complete _____.

Step 6: Reread the _____ and check your solution.

Problems

Solve the following word problems using the steps given in the Blueprint for Problem Solving.

A Number Problems

1. The sum of a number and five is thirteen. Find the number.

2. The difference of ten and a number is negative eight. Find the number.

3. The sum of twice a number and four is fourteen. Find the number.

4. The difference of four times a number and eight is sixteen. Find the number.

5. Five times the sum of a number and seven is thirty. Find the number.

6. Five times the difference of twice a number and six is negative twenty. Find the number.

7. One number is two more than another. Their sum is eight. Find both numbers.

8. One number is three less than another. Their sum is fifteen. Find the numbers.

9. One number is four less than three times another. If their sum is increased by five, the result is twenty-five. Find the numbers.

10. One number is five more than twice another. If their sum is decreased by ten, the result is twenty-two. Find the numbers.

B Age Problems

11. Shelly is 3 years older than Michele. Four years ago the sum of their ages was 67. Find the age of each person now.

	Four Years Ago	Now
Shelly	$x - 1$	$x + 3$
Michele	$x - 4$	x

12. Cary is 9 years older than Dan. In 7 years the sum of their ages will be 93. Find the age of each man now. (Begin by filling in the table.)

	Now	In Seven Years
Cary	$x + 9$	
Dan	x	$x + 7$

13. Cody is twice as old as Evan. Three years ago the sum of their ages was 27. Find the age of each boy now.

	Three Years Ago	Now
Cody		
Evan	$x - 3$	x

14. Justin is 2 years older than Ethan. In 9 years the sum of their ages will be 30. Find the age of each boy now.

	Now	In Nine Years
Justin		
Ethan	x	

15. Fred is 4 years older than Barney. Five years ago the sum of their ages was 48. How old are they now?

	Five Years Ago	Now
Fred		
Barney		x

16. Tim is 5 years older than JoAnn. Six years from now the sum of their ages will be 79. How old are they now?

	Now	Six Years From Now
Tim		
JoAnn	x	

17. Jack is twice as old as Lacy. In 3 years the sum of their ages will be 54. How old are they now?

18. John is 4 times as old as Martha. Five years ago the sum of their ages was 50. How old are they now?

19. Pat is 20 years older than his son Patrick. In 2 years Pat will be twice as old as Patrick. How old are they now?

20. Diane is 23 years older than her daughter Amy. In 6 years Diane will be twice as old as Amy. How old are they now?

C Perimeter Problems

21. The perimeter of a square is 36 inches. Find the length of one side.

22. The perimeter of a square is 44 centimeters. Find the length of one side.

23. The perimeter of a square is 60 feet. Find the length of one side.

24. The perimeter of a square is 84 meters. Find the length of one side.

25. One side of a triangle is three times the shortest side. The third side is 7 feet more than the shortest side. The perimeter is 62 feet. Find all three sides.

26. One side of a triangle is half the longest side. The third side is 10 meters less than the longest side. The perimeter is 45 meters. Find all three sides.

27. One side of a triangle is half the longest side. The third side is 12 feet less than the longest side. The perimeter is 53 feet. Find all three sides.

28. One side of a triangle is 6 meters more than twice the shortest side. The third side is 9 meters more than the shortest side. The perimeter is 75 meters. Find all three sides.

29. The length of a rectangle is 5 inches more than the width. The perimeter is 34 inches. Find the length and width.

$x + 5$

30. The width of a rectangle is 3 feet less than the length. The perimeter is 10 feet. Find the length and width.

31. The length of a rectangle is 7 meters more than twice the width. The perimeter is 68 meters. Find the length and width.

32. The length of a rectangle is 4 inches more than three times the width. The perimeter is 72 inches. Find the length and width.

33. The length of a rectangle is 6 feet more than three times the width. The perimeter is 36 feet. Find the length and width.

34. The length of a rectangle is 3 centimeters less than twice the width. The perimeter is 54 centimeters . Find the length and width.

D Coin Problems

35. Marissa has $4.40 in quarters and dimes. If she has 5 more quarters than dimes, how many of each coin does she have?

	Dimes	Quarters
Number	x	$x + 5$
Value (in cents)	$10(x)$	$25(x + 5)$

36. Kendra has $2.75 in dimes and nickels. If she has twice as many dimes as nickels, how many of each coin does she have? (Completing the table may help you get started.)

	Nickels	Dimes
Number	x	$2x$
Value (in cents)	$5(x)$	

37. Tanner has $4.35 in nickels and quarters. If he has 15 more nickels than quarters, how many of each coin does he have?

	Nickels	Quarters
Number	$x + 15$	x
Value (in cents)		

38. Connor has $9.00 in dimes and quarters. If he has twice as many quarters as dimes, how many of each coin does he have?

	Dimes	Quarters
Number	x	$2x$
Value (in cents)		

39. Sue has $2.10 in dimes and nickels. If she has 9 more dimes than nickels, how many of each coin does she have?

	Dimes	Nickels
Number		x
Value (in cents)		

40. Mike has $1.55 in dimes and nickels. If he has 7 more nickels than dimes, how many of each coin does he have?

	Dimes	Nickels
Number	x	
Value (in cents)		

41. Katie has a collection of nickels, dimes, and quarters with a total value of $4.35. There are 3 more dimes than nickels and 5 more quarters than nickels. How many of each coin is in her collection? (*Hint:* Let N = the number of nickels.)

	Nickels	Dimes	Quarters
Number	N		
Value			

42. Mary Jo has $3.90 worth of nickels, dimes, and quarters. The number of nickels is 3 more than the number of dimes. The number of quarters is 7 more than the number of dimes. How many of each coin does she have? (*Hint:* Let D = the number of dimes.)

	Nickels	Dimes	Quarters
Number			
Value			

43. Corey has a collection of nickels, dimes, and quarters with a total value of $2.55. There are 6 more dimes than nickels and twice as many quarters as nickels. How many of each coin is in her collection?

	Nickels	Dimes	Quarters
Number	x		
Value			

44. Kelly has a collection of nickels, dimes, and quarters with a total value of $7.40. There are four more nickels than dimes and twice as many quarters as nickels. How many of each coin is in her collection?

	Nickels	Dimes	Quarters
Number			
Value			

Getting Ready for the Next Section

To understand all of the explanations and examples in the next section you must be able to work the problems below. Simplify the following expressions.

45. $x + 2x + 2x$

46. $x + 2x + 3x$

47. $x + 0.075x$

48. $x + 0.065x$

49. $0.09(x + 2{,}000)$

50. $0.06(x + 1{,}500)$

Find the Mistake

Each application problem below is followed by a corresponding equation. Write true or false if the equation accurately represents the problem. If false, write the correct equation on the line provided.

1. The sum of 7 and the quotient of m and n is 9 more than n.

Equation: $\dfrac{7}{(m + n)} = n + 9$ _____

2. Danny is 30 years older than his daughter Emma. Four years ago, Danny's age was six times Emma's age. Find Danny and Emma's age.

Equation: $x + 26 = 6(x - 4)$ _____

3. The width of a rectangle is 6 centimeters more than three times the length. The perimeter is 42 centimeters. Find the length and width.

Equation: $3x + 6 + x = 42$ _____

4. Amelie has $3.10 in nickels and dimes. If the number of dimes she has is 4 less than twice the number of nickels, how many of each coin does she have?

Equation: $2x - 4 + x = 310$ _____

Apply the Blueprint for Problem Solving to solve:

A Consecutive Integer Problems

B Interest Problems

C Triangle Angle Problems

D Miscellaneous Problems

KEY WORDS

consecutive integers

vertices

**Video Examples
Section 5.6**

Note When assigning a variable or an expression to two consecutive integers, we could choose the approach shown here, or we could assign x to the larger integer and $x - 1$ to the smaller integer. Either case would yield the same answers. But for the sake of simplicity, we will continue to assign x to the first and smaller of the two consecutive integers.

5.6 Integer, Interest, Triangle, Miscellaneous Applications

©iStockphoto.com/polygrafix

In Las Vegas, the Stratosphere's hotel and casino has opened another attraction for thrill-seekers. SkyJump is a death-defying controlled free-fall from the 108th floor of the hotel to the ground. Jump Package 1 includes the jump cost plus a DVD of the jump for $114.99. A jump without the DVD costs $99.99. Suppose in one weekend, SkyJump had 15 jumpers, and receipts for that weekend's jumps totalled $1574.85. In this section, we will practice more application problems using linear equations, such as the one needed to determine how many jumpers purchased Jump Package 1 and how many paid just for the jump.

A Consecutive Integer Problems

Our first example involves consecutive integers. Here is the definition:

> **DEFINITION** consecutive integers
>
> When we ask for *consecutive integers*, we mean integers that are next to each other on the number line, like 5 and 6, or 13 and 14, or -4 and -3.

In the dictionary, consecutive is defined as following one another in uninterrupted order. If we ask for consecutive odd integers, then we mean odd integers that follow one another on the number line. For example, 3 and 5, 11 and 13, and -9 and -7 are consecutive odd integers. As you can see, to get from one odd integer to the next consecutive odd integer we add 2.

If we are asked to find two consecutive integers and we let x equal the first integer, the next one must be $x + 1$, because consecutive integers always differ by 1. Likewise, if we are asked to find two consecutive odd or even integers, and we let x equal the first integer, then the next one will be $x + 2$ because consecutive even or odd integers always differ by 2. Here is a table that summarizes this information.

In Words	Using Algebra	Example
Two consecutive integers	$x, x + 1$	The sum of two consecutive integers is 15. $x + (x + 1) = 15$ or $7 + 8 = 15$
Three consecutive integers	$x, x + 1, x + 2$	The sum of three consecutive integers is 24. $x + (x + 1) + (x + 2) = 24$ or $7 + 8 + 9 = 24$
Two consecutive odd integers	$x, x + 2$	The sum of two consecutive odd integers is 16. $x + (x + 2) = 16$ or $7 + 9 = 16$
Two consecutive even integers	$x, x + 2$	The sum of two consecutive even integers is 18. $x + (x + 2) = 18$ or $8 + 10 = 18$

EXAMPLE 1 The sum of two consecutive odd integers is 28. Find the two integers.

Solution

Step 1: *Read and list.*

Known items: Two consecutive odd integers. Their sum is equal to 28.
Unknown items: The numbers in question.

Step 2: *Assign a variable, and translate information.*

If we let $x =$ the first of the two consecutive odd integers, then $x + 2$ is the next consecutive one.

Step 3: *Reread and write an equation.*

Their sum is 28.

$$x + (x + 2) = 28$$

Step 4: *Solve the equation.*

$$2x + 2 = 28 \qquad \textit{Simplify the left side.}$$
$$2x = 26 \qquad \textit{Add } -2 \textit{ to each side.}$$
$$x = 13 \qquad \textit{Multiply each side by } \frac{1}{2}.$$

Step 5: *Write the answer.*

The first of the two integers is 13. The second of the two integers will be two more than the first, which is 15.

Step 6: *Reread and check.*

Suppose the first integer is 13. The next consecutive odd integer is 15. The sum of 15 and 13 is 28.

B Interest Problems

EXAMPLE 2 Suppose you invest a certain amount of money in an account that earns 8% in annual interest. At the same time, you invest $2,000 more than that in an account that pays 9% in annual interest. If the total interest from both accounts at the end of the year is $690, how much is invested in each account? (Hint: The formula for simple interest is $I = Prt$, where I is interest, P is principal amount invested, r is interest rate, and t is time in years.)

Solution

Step 1: *Read and list.*

Known items: The interest rates, the total interest earned, and how much more is invested at 9%
Unknown items: The amounts invested in each account.

Step 2: *Assign a variable, and translate information.*

Let $P =$ the amount of money invested at 8%. From this, $P + 2,000 =$ the amount of money invested at 9%. The interest earned on P dollars invested at 8% is $0.08P$. The interest earned on $P + 2,000$ dollars invested at 9% is $0.09(P + 2,000)$.

Here is a table that summarizes this information:

	Dollars invested at 8%	Dollars invested at 9%
Number of	P	$P + 2,000$
Interest on	$0.08P$	$0.09(P + 2,000)$

Practice Problems

1. The sum of two consecutive integers is 27. What are the two integers?

Note Now that you have worked through a number of application problems using our blueprint, you probably have noticed that step 3, in which we write an equation that describes the situation, is the key step. Anyone with experience solving application problems will tell you that there will be times when your first attempt at step 3 results in the wrong equation. Remember, mistakes are part of the process of learning to do things correctly.

2. On July 1, 1992, Alfredo opened a savings account that earned 4% in annual interest. At the same time, he put $3,000 more than he had in his savings into a mutual fund that paid 5% annual interest. One year later, the total interest from both accounts was $510. How much money did Alfredo put in his savings account?

Answers
1. 13 and 14
2. $4,000

Step 3: *Reread and write an equation.*

Because the total amount of interest earned from both accounts is $690, the equation that describes the situation is

Interest earned at 8%	+	Interest earned at 9%	=	Total interest earned
0.08P	+	0.09(P + 2,000)	=	690

Step 4: *Solve the equation.*

$$0.08P + 0.09(P + 2,000) = 690$$

$0.08P + 0.09P + 180 = 690$ *Distributive property*

$0.17P + 180 = 690$ *Add 0.08P and 0.09P.*

$0.17P = 510$ *Add −180 to each side.*

$P = 3,000$ *Divide each side by 0.17.*

Step 5: *Write the answer.*

The amount of money invested at 8% is $3,000, whereas the amount of money invested at 9% is $P + 2,000 = 3,000 + 2,000 = \$5,000$.

Step 6: *Reread and check.*

The interest at 8% is 8% of 3,000 = 0.08(3,000) = $240
The interest at 9% is 9% of 5,000 = 0.09(5,000) = $450

The total interest = $690

C Triangle Angle Problems

FACTS FROM GEOMETRY Labeling Triangles and the Sum of the Angles

One way to label the important parts of a triangle is to label the vertices with capital letters and the sides with small letters, as shown in Figure 1.

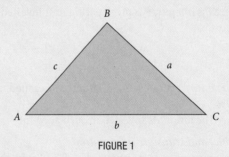

FIGURE 1

In Figure 1, notice that side a is opposite vertex A, side b is opposite vertex B, and side c is opposite vertex C. Also, because each vertex is the vertex of one of the angles of the triangle, we refer to the three interior angles as A, B, and C.

In any triangle, the sum of the interior angles is 180°. For the triangle shown in Figure 1, the relationship is written

$$A + B + C = 180°$$

EXAMPLE 3 The angles in a triangle are such that one angle is twice the smallest angle, whereas the third angle is three times as large as the smallest angle. Find the measure of all three angles.

Solution

Step 1: *Read and list.*

Known items: The sum of all three angles is 180°, one angle is twice the smallest angle, the largest angle is three times the smallest angle.

Unknown items: The measure of each angle

Step 2: *Assign a variable, and translate information.*

Let A be the smallest angle, then $2A$ will be the measure of another angle and $3A$ will be the measure of the largest angle.

Step 3: *Reread and write an equation.*

When working with geometric objects, drawing a generic diagram sometimes will help us visualize what it is that we are asked to find. In Figure 2, we draw a triangle with angles A, B, and C.

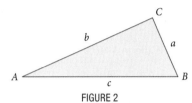

FIGURE 2

We can write the value of angle B in terms of A as $B = 2A$, and the value of angle C as $C = 3A$. We know that the sum of angles A, B, and C will be 180°, so our equation becomes

$$A + 2A + 3A = 180°$$

Step 4: *Solve the equation.*

$$A + 2A + 3A = 180°$$
$$6A = 180°$$
$$A = 30°$$

Step 5: *Write the answer.*

The smallest angle A measures 30°
Angle B measures $2A$, or $2(30°) = 60°$
Angle C measures $3A$, or $3(30°) = 90°$

Step 6: *Reread and check.*

The angles must add to 180°.
$$A + B + C = 180°$$
$$30° + 60° + 90° \stackrel{?}{=} 180°$$
$$180° = 180° \quad \textit{Our answers check.}$$

D Miscellaneous Problems

EXAMPLE 4 Tickets to a school play cost $6 for adults and $4 for children. A total of 200 tickets were sold, and $940 was collected in receipts. How many adult tickets and how many child tickets were sold?

Solution

Step 1: *Read and list.*

Known items: 200 tickets were sold, $940 was collected
Unknown items: Number of adult tickets, number of child tickets

3. The angles in a triangle are such that one angle is three times the smallest angle, whereas the largest angle is five times the smallest angle. Find the measure of all three angles.

4. Tickets to a community theater cost $10 for adults and $6 for children. A total of $680 was collected for one evening performance. If 20 more adults attended than children, how many children and how many adults attended the theater?

Answers

3. 20°, 60°, and 100°
4. 50 adults, 30 children

Step 2: *Assign a variable, and translate information.*

Let x be the number of adult tickets sold, then $200 - x$ will be the number of child tickets sold.

Step 3: *Reread and write an equation.*

Price of adult \times # Adult $+$ Price of child \times # Child $=$ Total collected

$$6 \cdot (x) \quad\quad + \quad 4 \cdot (200 - x) \quad = \quad 940$$

Step 4: *Solve the equation.*

$$6(x) + 4(200 - x) = 940$$
$$6x + 800 - 4x = 940$$
$$2x + 800 = 940$$
$$2x = 140$$
$$x = 70$$

Step 5: *Write the answer.*

There were 70 adult tickets sold, and $200 - 70 = 130$ child tickets sold.

Step 6: *Reread and check.*

The total receipts must add to $940.

$$6 \cdot (70) + 4 \cdot (130) \overset{?}{=} 940$$
$$420 + 520 \overset{?}{=} 940$$
$$940 = 940$$

GETTING READY FOR CLASS

After reading through the preceding section, respond in your own words and in complete sentences.

A. Write an equation for the sum of two consecutive integers being 13.

B. Write an application problem whose solution depends on solving the equation $x + 0.075x = 500$.

C. How do you label triangles?

D. What rule is always true about the three angles in a triangle?

Vocabulary Review

Choose the correct words to fill in the blanks below.

even sum consecutive integers

1. _____ integers are integers that are next to each other on the number line.

2. Two consecutive _____ integers would be given by x and $x + 2$.

3. Two consecutive _____ would be given by x and $x + 1$.

4. In any triangle, the _____ of the interior angles is 180°, and written by $A + B + C = 180°$.

A Consecutive Integer Problems

1. The sum of two consecutive integers is 11. Find the numbers.

2. The sum of two consecutive integers is 15. Find the numbers.

3. The sum of two consecutive integers is −9. Find the numbers.

4. The sum of two consecutive integers is −21. Find the numbers.

5. The sum of two consecutive odd integers is 28. Find the numbers.

6. The sum of two consecutive odd integers is 44. Find the numbers.

7. The sum of two consecutive even integers is 106. Find the numbers.

8. The sum of two consecutive even integers is 66. Find the numbers.

9. The sum of two consecutive even integers is −30. Find the numbers.

10. The sum of two consecutive odd integers is −76. Find the numbers.

11. The sum of three consecutive odd integers is 57. Find the numbers.

12. The sum of three consecutive odd integers is −51. Find the numbers.

13. The sum of three consecutive even integers is 132. Find the numbers.

14. The sum of three consecutive even integers is −108. Find the numbers.

B Interest Problems

15. Suppose you invest money in two accounts. One of the accounts pays 8% annual interest, whereas the other pays 9% annual interest. If you have $2,000 more invested at 9% than you have invested at 8%, how much do you have invested in each account if the total amount of interest you earn in a year is $860? (Begin by completing the following table.)

	Dollars invested at 8%	Dollars invested at 9%
Number of	x	
Interest on		

16. Suppose you invest a certain amount of money in an account that pays 11% interest annually, and $4,000 more than that in an account that pays 12% annually. How much money do you have in each account if the total interest for a year is $940?

	Dollars invested at 11%	Dollars invested at 12%
Number of	x	
Interest on		

17. Tyler has two savings accounts that his grandparents opened for him. The two accounts pay 10% and 12% in annual interest; there is $500 more in the account that pays 12% than there is in the other account. If the total interest for a year is $214, how much money does he have in each account?

18. Travis has a savings account that his parents opened for him. It pays 6% annual interest. His uncle also opened an account for him, but it pays 8% annual interest. If there is $800 more in the account that pays 6%, and the total interest from both accounts is $104, how much money is in each of the accounts?

19. A stockbroker has money in three accounts. The interest rates on the three accounts are 8%, 9%, and 10%. If she has twice as much money invested at 9% as she has invested at 8%, three times as much at 10% as she has at 8%, and the total interest for the year is $280, how much is invested at each rate? (*Hint:* Let $x =$ the amount invested at 8%.)

20. An accountant has money in three accounts that pay 9%, 10%, and 11% in annual interest. He has twice as much invested at 9% as he does at 10% and three times as much invested at 11% as he does at 10%. If the total interest from the three accounts is $610 for the year, how much is invested at each rate? (*Hint:* Let $x =$ the amount invested at 10%.)

c Triangle Angle Problems

21. Two angles in a triangle are equal and their sum is equal to the third angle in the triangle. What are the measures of each of the three interior angles?

22. One angle in a triangle measures twice the smallest angle, whereas the largest angle is six times the smallest angle. Find the measures of all three angles.

23. The smallest angle in a triangle is $\frac{1}{5}$ as large as the largest angle. The third angle is twice the smallest angle. Find the three angles.

24. One angle in a triangle is half the largest angle but three times the smallest. Find all three angles.

25. A right triangle has one 37° angle. Find the other two angles.

26. In a right triangle, one of the acute angles is twice as large as the other acute angle. Find the measure of the two acute angles.

27. One angle of a triangle measures 20° more than the smallest, while a third angle is twice the smallest. Find the measure of each angle.

28. One angle of a triangle measures 50° more than the smallest, while a third angle is three times the smallest. Find the measure of each angle.

D Miscellaneous Problems

29. Ticket Prices Miguel is selling tickets to a barbecue. Adult tickets cost $6.00 and children's tickets cost $4.00. He sells six more children's tickets than adult tickets. The total amount of money he collects is $184. How many adult tickets and how many children's tickets did he sell?

	Adult	Child
Number	x	$x + 6$
Income	$6(x)$	$4(x + 6)$

30. Working Two Jobs Maggie has a job working in an office for $10 an hour and another job driving a tractor for $12 an hour. One week she works in the office twice as long as she drives the tractor. Her total income for that week is $416. How many hours did she spend at each job?

Job	Office	Tractor
Hours Worked	$2x$	x
Wages Earned	$10(2x)$	$12x$

31. Phone Bill The cost of a long-distance phone call is $0.41 for the first minute and $0.32 for each additional minute. If the total charge for a long-distance call is $5.21, how many minutes was the call?

32. Phone Bill Danny, who is 1 year old, is playing with the telephone when he accidentally presses one of the buttons his mother has programmed to dial her friend Sue's number. Sue answers the phone and realizes Danny is on the other end. She talks to Danny, trying to get him to hang up. The cost for a call is $0.23 for the first minute and $0.14 for every minute after that. If the total charge for the call is $3.73, how long did it take Sue to convince Danny to hang up the phone?

33. Hourly Wages JoAnn works in the publicity office at the state university. She is paid $12 an hour for the first 35 hours she works each week and $18 an hour for every hour after that. If she makes $492 one week, how many hours did she work?

34. Hourly Wages Diane has a part-time job that pays her $8.50 an hour. During one week she works 26 hours and is paid $230.10. She realizes when she sees her check that she has been given a raise. How much per hour is that raise?

35. Office Numbers Professors Wong and Gil have offices in the mathematics building at Miami Dade College. Their office numbers are consecutive odd integers with a sum of 14,660. What are the office numbers of these two professors?

36. Cell Phone Numbers Diana and Tom buy two cell phones. The phone numbers assigned to each are consecutive integers with a sum of 11,109,295. If the smaller number is Diana's, what are their phone numbers?

37. Age Marissa and Kendra are 2 years apart. Their ages are two consecutive even integers. Kendra is the younger of the two. If Marissa's age is added to twice Kendra's age, the result is 26. How old is each girl?

38. Age Justin's and Ethan's ages form two consecutive odd integers. What is the difference of their ages?

39. Arrival Time Jeff and Carla Cole are driving separately from San Luis Obispo, California, to the north shore of Lake Tahoe, a distance of 425 miles. Jeff leaves San Luis Obispo at 11:00 a.m. and averages 55 miles per hour on the drive, Carla leaves later, at 1:00 p.m. but averages 65 miles per hour. Which person arrives in Lake Tahoe first?

40. Piano Lessons Tyler's parents pay him $0.50 to do the laundry and $1.25 to mow the lawn. In one month, he does the laundry 6 more times than he mows the lawn. If his parents pay him $13.50 that month, how many times did he mow the lawn?

At one time, the Texas Junior College Teachers Association annual conference was held in Austin. At that time a taxi ride in Austin was $1.25 for the first $\frac{1}{5}$ of a mile and $0.25 for each additional $\frac{1}{5}$ of a mile. Use this information for Problems 41 and 42.

41. Cost of a Taxi Ride If the distance from one of the convention hotels to the airport is 7.5 miles, how much will it cost to take a taxi from that hotel to the airport?

42. Cost of a Taxi Ride Suppose the distance from one of the hotels to one of the western dance clubs in Austin is 12.4 miles. If the fare meter in the taxi gives the charge for that trip as $16.50, is the meter working correctly?

43. Geometry The length and width of a rectangle are consecutive even integers. The perimeter is 44 meters. Find the length and width.

44. Geometry The length and width of a rectangle are consecutive odd integers. The perimeter is 128 meters. Find the length and width.

45. Geometry The angles of a triangle are three consecutive integers. Find the measure of each angle.

46. Geometry The angles of a triangle are three consecutive even integers. Find the measure of each angle.

47. Boards A 78-centimeter-long board is cut into 3 pieces. The longest piece is twice as long as the shortest piece. The other piece is 6 cm longer than the shortest piece. How long is each piece?

48. Rope A 52-meter-long rope is cut into three segments. The shortest piece is 6 meters shorter than the next piece, which is 13 meters shorter than the longest piece. How long is each piece?

Ike and Nancy give western dance lessons at the Elk's Lodge on Sunday nights. The lessons cost $3.00 for members of the lodge and $5.00 for nonmembers. Half of the money collected for the lesson is paid to Ike and Nancy. The Elk's Lodge keeps the other half. One Sunday night Ike counts 36 people in the dance lesson. Use this information to work Problems 47 through 50.

49. Dance Lessons What is the least amount of money Ike and Nancy will make?

50. Dance Lessons What is the largest amount of money Ike and Nancy will make?

51. Dance Lessons At the end of the evening, the Elk's Lodge gives Ike and Nancy a check for $80 to cover half of the receipts. Can this amount be correct?

52. Dance Lessons Besides the number of people in the dance lesson, what additional information does Ike need to know to always be sure he is being paid the correct amount?

53. Solve the problem given at the introduction to this section to find how many jumpers purchased package 1.

54. Solve the same problem as in problem 51, but this time assume Sky Jump had 30 jumpers with total receipts of $3,269.70.

Getting Ready for the Next Section

To understand all the explanations and examples in the next section you must be able to work the problems below.

Solve the following equations.

55. a. $x - 3 = 6$

b. $x + 3 = 6$

c. $-x - 3 = 6$

d. $-x + 3 = 6$

56. a. $x - 7 = 16$

b. $x + 7 = 16$

c. $-x - 7 = 16$

d. $-x + 7 = 16$

57. a. $\dfrac{x}{4} = -2$

b. $-\dfrac{x}{4} = -2$

c. $\dfrac{x}{4} = 2$

d. $-\dfrac{x}{4} = 2$

58. a. $3a = 15$

b. $3a = -15$

c. $-3a = 15$

d. $-3a = -15$

59. $2.5x - 3.48 = 4.9x + 2.07$

60. $2(1 - 3x) + 4 = 4x - 14$

61. $3(x - 4) = -2$

62. $-2(x - 3) = 8$

63. Solve $2x - 3y = 6$ for y.

64. Solve $3x + 2y = 7$ for y.

65. Solve $3x - 2y = 5$ for x.

66. Solve $2x - 5y = -10$ for x.

Find the Mistake

Each application problem below is followed by a corresponding equation. Write true or false if the equation accurately represents the problem. If false, write the correct equation on the line provided. (Hint: Some equations may appear in a simplified form.)

1. The sum of three consecutive integers divided by 2 is 36.

Equation: $\dfrac{(3x + 3)}{2} = 36$ _____

2. You have money in two savings accounts that pay 7% and 8% in annual interest. If you have four times as much money in the account that pays 7%, and the total interest of both accounts is $396, how much money is in each account?

Equation: $0.07x + 0.08(4x) = 396$ _____

3. One angle of a triangle measures 4 degrees more than the smallest angle. The largest angle is 16 degrees less than twice the smallest angle. Find the measure of each angle.

Equation: $4x - 12 = 180$ _____

4. Tickets for a dinner fundraiser cost $15 per plate with dessert and $12 per plate without dessert. 150 tickets were sold for a total of $2070. How many of each ticket was sold?

Equation: $15x + 12(150 - x) = 2070$ _____

OBJECTIVES

A Use the addition property for inequalities to solve an inequality and graph the solution set.

B Use the multiplication property for inequalities to solve inequalities.

C Use both the addition and multiplication properties to solve and graph inequalities.

D Solve and graph compound inequalities.

E Translate and solve application problems involving inequalities.

KEY WORDS

linear inequality

addition property for inequalities

multiplication property for inequalities

compound inequality

continued inequality

double inequality

**Video Examples
Section 5.7**

Image © sxc.hu/ckforjc, 2004

Suppose we know that the state of New Mexico has no more than 10 organic stores. We can write the following inequality to represent this information:

$$x \leq 10$$

If we know for certain of three stores in New Mexico, we can add to the original inequality and say $x + 3 \leq 10$. Up to now we have solved equalities. For example, when we say $x + 3 = 10$, we know x must be 7 since the statement is only true when $x = 7$. But if we say that $x + 3 \leq 10$, then there are many numbers that make this statement true. If x is 3, $3 + 3$ is less than 10. If x is 2.5, $2.5 + 3$ is less than 10. If x is any real number less than or equal to 7, then the statement $x + 3 \leq 10$ is true.

We also know that any number greater than 7 mark the statement false. If x is 8, then $x + 3$ is not less than 10.

When we solve an inequality, we look for all possible solutions—and there will usually be many of them. Any number that makes an inequality true is a *solution* to that inequality. All the numbers that make the inequality true are called the *solution set*.

We can express a solution set symbolically. If the solution set to an inequality is all the real numbers less than 7, we say the solution is $x < 7$. If the solution set to this inequality is all numbers less than or equal to 7, we say the solution is $x \leq 7$.

We can also express the solution set graphically using the number line. When a solution set is all numbers less than a certain number, we make that number with an open circle.

If we want to include the number (7, in this case) in the solution set, we mark that number with a closed circle.

Linear inequalities are solved by a method similar to the one used in solving linear equations. The only real differences between the methods are in the multiplication property for inequalities and in graphing the solution set.

A Addition Property for Inequalities

An inequality differs from an equation only with respect to the comparison symbol between the two quantities being compared. In place of the equal sign, we use < (less than), ≤ (less than or equal to), > (greater than), or ≥ (greater than or equal to). The addition property for inequalities is almost identical to the addition property for equality.

> **PROPERTY** Addition Property for Inequalities
>
> For any three algebraic expressions A, B, and C,
>
> $$\text{if} \qquad A < B$$
> $$\text{then} \qquad A + C < B + C$$
>
> *In words*: Adding the same quantity to both sides of an inequality will not change the solution set.

It makes no difference which inequality symbol we use to state the property. Adding the same amount to both sides always produces an inequality equivalent to the original inequality. Also, because subtraction can be thought of as addition of the opposite, this property holds for subtraction as well as addition.

EXAMPLE 1 Solve $x + 5 < 7$ and graph the solution.

Solution To isolate x, we add -5 to both sides of the inequality

$$x + 5 < 7$$
$$x + 5 + (-5) < 7 + (-5) \qquad \text{Addition property for inequalities}$$
$$x < 2$$

We can go one step further here and graph the solution set. The solution set is all real numbers less than 2. To graph this set, we simply draw a straight line and label the center 0 (zero) for reference. Then we label the 2 on the right side of zero and extend an arrow beginning at 2 and pointing to the left. We use an open circle at 2 because it is not included in the solution set. Here is the graph.

EXAMPLE 2 Solve $x - 6 \le -3$ and graph the solution.

Solution Adding 6 to each side will isolate x on the left side

$$x - 6 \le -3$$
$$x - 6 + 6 \le -3 + 6 \qquad \text{Add 6 to both sides.}$$
$$x \le 3$$

The graph of the solution set is

Notice that the dot at the 3 in Example 2 is darkened because 3 is included in the solution set. In this book we will use open circles on the graphs of solution sets with $<$ or $>$ and closed (darkened) circles on the graphs of solution sets with \le or \ge.

B Multiplication Property for Inequalities

To see the idea behind the multiplication property for inequalities, we will consider three true inequality statements and explore what happens when we multiply both sides by a positive number and then what happens when we multiply by a negative number.
 Consider the following three true statements:

$$3 < 5 \qquad -3 < 5 \qquad -5 < -3$$

Now multiply both sides by the positive number 4.

$$4(3) < 4(5) \qquad 4(-3) < 4(5) \qquad 4(-5) < 4(-3)$$
$$12 < 20 \qquad \quad -12 < 20 \qquad \quad -20 < -12$$

Practice Problems

1. Solve $x + 3 < 5$ and graph the solution.

2. Solve $x - 4 \le 1$ and graph the solution.

Answers

1. $x < 2$
2. $x \le 5$

Note This discussion is intended to show why the multiplication property for inequalities is written the way it is. You may want to look ahead to the property itself and then come back to this discussion if you are having trouble making sense out of it.

In each case, the inequality symbol in the result points in the same direction it did in the original inequality. We say the "sense" of the inequality doesn't change when we multiply both sides by a positive quantity.

Notice what happens when we go through the same process but multiply both sides by -4 instead of 4:

$$3 < 5 \qquad\qquad -3 < 5 \qquad\qquad -5 < -3$$

$$-4(3) > -4(5) \qquad -4(-3) > -4(5) \qquad -4(-5) > -4(-3)$$

$$-12 > -20 \qquad\qquad 12 > -20 \qquad\qquad 20 > 12$$

In each case, we have to change the direction in which the inequality symbol points to keep each statement true. Multiplying both sides of an inequality by a negative quantity always reverses the sense of the inequality. Our results are summarized in the multiplication property for inequalities.

PROPERTY **Multiplication Property for Inequalities**

For any three algebraic expressions A, B, and C,

> if $\qquad\qquad A < B$
> then $\qquad\quad AC < BC$ \qquad when C is positive
> and $\qquad\quad AC > BC$ \qquad when C is negative

In words: Multiplying both sides of an inequality by a positive number does not change the solution set. When multiplying both sides of an inequality by a negative number, it is necessary to reverse the inequality symbol to produce an equivalent inequality.

Note Because division is defined in terms of multiplication, this property is also true for division. We can divide both sides of an inequality by any nonzero number we choose. If that number happens to be negative, we must also reverse the direction of the inequality symbol.

We can multiply both sides of an inequality by any nonzero number we choose. If that number happens to be negative, we must also reverse the sense of the inequality.

EXAMPLE 3 Solve $3a < 15$ and graph the solution.

Solution We begin by multiplying each side by $\frac{1}{3}$. Because $\frac{1}{3}$ is a positive number, we do not reverse the direction of the inequality symbol.

$$3a < 15$$

$$\frac{1}{3}(3a) < \frac{1}{3}(15) \qquad \text{Multiply each side by } \frac{1}{3}.$$

$$a < 5$$

3. Solve $4a < 12$ and graph the solution.

EXAMPLE 4 Solve $-3a \le 18$ and graph the solution.

Solution We begin by multiplying both sides by $-\frac{1}{3}$. Because $-\frac{1}{3}$ is a negative number, we must reverse the direction of the inequality symbol at the same time that we multiply by $-\frac{1}{3}$.

$$-3a \le 18$$

$$-\frac{1}{3}(-3a) \ge -\frac{1}{3}(18) \qquad \text{Multiply both sides by } -\frac{1}{3} \text{ and reverse the direction of the inequality symbol.}$$

$$a \ge -6$$

4. Solve $-2x \le 10$ and graph the solution.

Answers

3. $a < 3$
4. $x \ge -5$

EXAMPLE 5 Solve $-\dfrac{x}{4} > 2$ and graph the solution.

Solution To isolate x, we multiply each side by -4. Because -4 is a negative number, we also must reverse the direction of the inequality symbol.

$$-\frac{x}{4} > 2$$

$$-4\left(-\frac{x}{4}\right) < -4(2) \qquad \text{Multiply each side by } -4, \text{ and reverse the}$$
$$\text{direction of the inequality symbol.}$$

$$x < -8$$

-8 0

5. Solve $-\dfrac{x}{3} > 2$ and graph the solution.

C Solving Linear Inequalities in One Variable

To solve more complicated inequalities, we use the following steps:

> **HOW TO** Solving Linear Inequalities in One Variable
>
> **Step 1a:** Use the distributive property to separate terms, if necessary.
>
> **1b:** If fractions are present, consider multiplying both sides by the LCD to eliminate the fractions. If decimals are present, consider multiplying both sides by a power of 10 to clear the inequality of decimals.
>
> **1c:** Combine similar terms on each side of the inequality.
>
> **Step 2:** Use the addition property for inequalities to get all variable terms on one side of the inequality and all constant terms on the other side.
>
> **Step 3:** Use the multiplication property for inequalities to get x by itself on one side of the inequality.
>
> **Step 4:** Graph the solution set.

EXAMPLE 6 Solve and graph: $2.5x - 3.48 < -4.9x + 2.07$.

Solution We have two methods we can use to solve this inequality. We can simply apply our properties to the inequality the way it is currently written and work with the decimal numbers, or we can eliminate the decimals to begin with and solve the resulting inequality.

Method 1 Working with the decimals.

$$2.5x - 3.48 < -4.9x + 2.07 \qquad \text{Original inequality}$$

$$2.5x + 4.9x - 3.48 < -4.9x + 4.9x + 2.07 \qquad \text{Add } 4.9x \text{ to each side.}$$

$$7.4x - 3.48 < 2.07$$

$$7.4x - 3.48 + 3.48 < 2.07 + 3.48 \qquad \text{Add } 3.48 \text{ to each side.}$$

$$7.4x < 5.55$$

$$\frac{7.4x}{7.4} < \frac{5.55}{7.4} \qquad \text{Divide each side by } 7.4.$$

$$x < 0.75$$

6. Solve and graph:
$4.5x + 2.31 > 6.3x - 4.89.$

Answers
5. $x < -6$
6. $x < 4$

Method 2 Eliminating the decimals in the beginning.

Because the greatest number of places to the right of the decimal point in any of the numbers is 2, we can multiply each side of the inequality by 100 and we will be left with an equivalent inequality that contains only whole numbers.

$2.5x - 3.48 < -4.9x + 2.07$	Original inequality
$100(2.5x - 3.48) < 100(-4.9x + 2.07)$	Multiply each side by 100.
$100(2.5x) - 100(3.48) < 100(-4.9x) + 100(2.07)$	Distributive property
$250x - 348 < -490x + 207$	Multiply.
$740x - 348 < 207$	Add 490x to each side.
$740x < 555$	Add 348 to each side.
$\dfrac{740x}{740} < \dfrac{555}{740}$	Divide each side by 740.
$x < 0.75$	

The solution by either method is $x < 0.75$. Here is the graph:

$$\xleftarrow{\hspace{2cm}} \begin{array}{cc} \underset{0}{|} & \overset{\ominus}{\underset{0.75}{|}} \end{array} \xrightarrow{\hspace{2cm}}$$

7. Solve and graph
$2(x - 5) \geq -6.$

$$\xleftarrow{\hspace{4cm}} \xrightarrow{\hspace{4cm}}$$

EXAMPLE 7 Solve and graph: $3(x - 4) \geq -2$.

Solution

$3x - 12 \geq -2$	Distributive property
$3x\ \ 12 + 12 \geq -2 + 12$	Add 12 to both sides.
$3x \geq 10$	
$\dfrac{1}{3}(3x) \geq \dfrac{1}{3}(10)$	Multiply both sides by $\dfrac{1}{3}$.
$x \geq \dfrac{10}{3}$	

$$\xleftarrow{\hspace{2cm}} \begin{array}{cc} \underset{0}{|} & \overset{\bullet}{\underset{\frac{10}{3}}{|}} \end{array} \xrightarrow{\hspace{2cm}}$$

8. Solve and graph
$3(1 - 2x) + 5 < 3x - 1.$

$$\xleftarrow{\hspace{4cm}} \xrightarrow{\hspace{4cm}}$$

EXAMPLE 8 Solve and graph: $2(1 - 3x) + 4 < 4x - 14$.

Solution

$2 - 6x + 4 < 4x - 14$	Distributive property
$-6x + 6 < 4x - 14$	Simplify.
$-6x + 6 + (-6) < 4x - 14 + (-6)$	Add -6 to both sides.
$-6x < 4x - 20$	
$-6x + (-4x) < 4x + (-4x) - 20$	Add $-4x$ to both sides.
$-10x < -20$	
$\left(-\dfrac{1}{10}\right)(-10x) > \left(-\dfrac{1}{10}\right)(-20)$	Multiply by $-\dfrac{1}{10}$, reverse the direction of the inequality.
$x > 2$	

$$\xleftarrow{\hspace{2cm}} \begin{array}{cc} \underset{0}{|} & \overset{\ominus}{\underset{2}{|}} \end{array} \xrightarrow{\hspace{2cm}}$$

Answers

7. $x \geq 2$

8. $x > 1$

EXAMPLE 9 Solve $2x - 3y < 6$ for y.

Solution We can solve this formula for y by first adding $-2x$ to each side and then multiplying each side by $-\frac{1}{3}$. When we multiply by $-\frac{1}{3}$ we must reverse the direction of the inequality symbol. Because this is a formula, we will not graph the solution.

$$2x - 3y < 6 \qquad \text{Original formula}$$
$$2x + (-2x) - 3y < (-2x) + 6 \qquad \text{Add } -2x \text{ to each side.}$$
$$-3y < -2x + 6$$
$$-\frac{1}{3}(-3y) > -\frac{1}{3}(-2x + 6) \qquad \text{Multiply each side by } -\frac{1}{3}.$$
$$y > \frac{2}{3}x - 2$$

9. Solve $2x - 3y < 6$ for x.

D Solving Compound Inequalities

Two inequalities considered together and connected by the words *and* or *or* are called *compound inequalities*.

EXAMPLE 10 Solve the compound inequality.
$$3t + 7 \leq -4 \qquad \text{or} \qquad 3t + 7 \geq 4$$

Solution We solve each half of the compound inequality separately, then we graph the solution set.

$$3t + 7 \leq -4 \qquad \text{or} \quad 3t + 7 \geq 4$$
$$3t \leq -11 \qquad \text{or} \qquad 3t \geq -3 \qquad \text{Add } -7.$$
$$t \leq -\frac{11}{3} \qquad \text{or} \qquad t \geq -1 \qquad \text{Multiply by } \frac{1}{3}.$$

The solution set can be graphed in the following way:

10. Solve and graph the solution for the compound inequality
$$3t - 6 \leq -3 \text{ or } 3t - 6 \geq 3.$$

Sometimes compound inequalities that use the word *and* as the connecting word can be written in a shorter form. For example, the compound inequality $-3 \leq x$ and $x \leq 4$ can be written $-3 \leq x \leq 4$. Inequalities of the form $-3 \leq x \leq 4$ are called *continued inequalities* or *double inequalities*. The graph of $-3 \leq x \leq 4$ is

EXAMPLE 11 Solve and graph $-3 \leq 2x - 5 \leq 3$.

Solution We can extend our properties for addition and multiplication to cover this situation. If we add a number to the middle expression, we must add the same number to the outside expressions. We do the same for multiplication, remembering to reverse the direction of the inequality symbols if we multiply by a negative number.

$$-3 \leq 2x - 5 \leq 3$$
$$2 \leq 2x \leq 8 \qquad \text{Add 5 to all three members.}$$
$$1 \leq x \leq 4 \qquad \text{Multiply through by } \frac{1}{2}.$$

Here is how we graph this solution set:

11. Solve and graph
$$-7 \leq 2x + 1 \leq 7.$$

Answers
9. $x < \frac{3}{2}y + 3$
10. $t \leq -\frac{11}{3}$ or $t \geq -1$
11. $1 \leq x \leq 4$

E Applications

When working application problems that involve inequalities, the phrases "at least" and "at most" translate as follows:

In Words	In Symbols
x is at least 30	$x \geq 30$
x is at most 20	$x \leq 20$

Our next example is similar to an example done earlier in this chapter. This time it involves an inequality instead of an equation.

We can modify our Blueprint for Problem Solving to solve application problems whose solutions depend on writing and then solving inequalities.

12. The sum of two consecutive integers is at least 27. What are the possibilities for the first of the two integers?

EXAMPLE 12 The sum of two consecutive odd integers is at most 28. What are the possibilities for the first of the two integers?

Solution When we use the phrase "their sum is at most 28," we mean that their sum is less than or equal to 28.

Step 1: *Read and list.*

Known items: Two consecutive odd integers. Their sum is less than or equal to 28.
Unknown items: The numbers in question.

Step 2: *Assign a variable, and translate information.*

If we let $x =$ the first of the two consecutive odd integers, then $x + 2$ is the next consecutive one.

Step 3: *Reread and write an inequality.*

Their sum is at most 28.

$$x + (x + 2) \leq 28$$

Step 4: *Solve the inequality.*

$$2x + 2 \leq 28 \qquad \textit{Simplify the left side.}$$

$$2x \leq 26 \qquad \textit{Add } -2 \textit{ to each side.}$$

$$x \leq 13 \qquad \textit{Multiply each side by } \frac{1}{2}.$$

Step 5: *Write the answer.*

The first of the two integers must be an odd integer that is less than or equal to 13. The second of the two integers will be two more than whatever the first one is.

Step 6: *Reread and check.*

Suppose the first integer is 13. The next consecutive odd integer is 15. The sum of 15 and 13 is 28. If the first odd integer is less than 13, the sum of it and the next consecutive odd integer will be less than 28.

13. Re-work Example 13 if she gets $700 or more in interest.

EXAMPLE 13 Monica wants to invest money into an account that pays $6\frac{1}{4}\%$ annual interest. She wants to be able to count on at least $500 in interest at the end of the year. How much money should she invest to get $500 or more at the end of the year?

Solution The phrase "or more" tells us this is an inequality problem. Once we find out the amount needed to get $500 interest, we can solve the problem and answer the question.

Step 1: *Read and list.*

Known items: The interest rate if $6\frac{1}{4}\%$. The minimum acceptable interest amount is $500.
Unknown items: The investment amount.

Answers
12. The first integer is at least 13;
 $x \geq 13$
13. $x \geq 11{,}200$

Step 2: *Assign a variable, and translate information.*

We will let $x =$ the amount to invest.

In words: $6\frac{1}{4}\%$ of x must be at least \$500.

Step 3: *Reread and write an inequality.*

Our sentence from step 2 can be translated directly into the inequality.

$$0.0625x \geq 500$$

Step 4: *Solve the inequality.*

$$0.0625x \geq 500 \qquad \textit{Original inequality}$$
$$x \geq 8{,}000 \qquad \textit{Divide both sides by 0.0625.}$$

Step 5: *Write the answer.*

We now know that if Monica invests \$8,000, her interest will be \$500. Any more than \$8,000 will give her more than \$500 so the answer is that Monica must invest at least \$8,000.

Step 6: *Reread and check.*

Put in any numbers greater than or equal to 8,000 in the inequality and check to make sure the result is at least 500.

$$x = 9{,}000$$
$$0.0625(9{,}000) = 562.5, \text{ which is} > \$500.$$

GETTING READY FOR CLASS

After reading through the preceding section, respond in your own words and in complete sentences.

A. State the addition property for inequalities.

B. When do you darken the circle on the graph of a solution set?

C. How is the multiplication property for inequalities different from the multiplication property of equality?

D. When do you reverse the direction of an inequality symbol?

Vocabulary Review

Choose the correct words to fill in the blanks below.

> graph compound positive inequalities negative compound

1. The addition property for _____ states that $A + C < B + C$ if $A < B$.

2. The multiplication property for inequalities states that multiplying both sides of an inequality by a _____ number does not change the solution set.

3. When multiplying both sides of an inequality by a _____ number, it is necessary to reverse the inequality symbol.

4. The last step to solving a linear inequality in one variable is to _____ the solution set.

5. A _____ inequality is more than one inequality considered together using the words *and* or *or*.

Problems

A Solve the following inequalities using the addition property of inequalities. Graph each solution set.

1. $x - 5 < 7$

2. $x + 3 < -5$

3. $a - 4 \leq 8$

4. $a + 3 \leq 10$

5. $x - 4.3 > 8.7$

6. $x - 2.6 > 10.4$

7. $y + 6 \geq 10$

8. $y + 3 \geq 12$

9. $2 < x - 7$

10. $3 < x + 8$

11. $-4 > x + 2$

12. $-5 \geq x - 2$

B Solve the following inequalities using the multiplication property of inequalities. If you multiply both sides by a negative number, be sure to reverse the direction of the inequality symbol. Graph the solution set.

13. $3x < 6$

14. $2x < 14$

15. $5a \leq 25$

16. $4a \leq 16$

17. $\frac{x}{3} > 5$

18. $\frac{x}{7} > 1$

19. $-2x > 6$

20. $-3x \geq 9$

21. $-3x \geq -18$

22. $-8x \geq -24$

23. $-\frac{x}{5} \leq 10$

24. $-\frac{x}{9} \geq -1$

25. $-\frac{2}{3}y > 4$

26. $-\frac{3}{4}y > 6$

27. $-\frac{x}{4} > 1$

28. $-\frac{1}{3}x \leq -2$

29. $-\frac{5}{6}x < 10$

30. $-\frac{2}{3}x < -6$

C Solve the following inequalities. Graph the solution set in each case.

31. $2x - 3 < 9$

32. $3x - 4 < 17$

33. $-\frac{1}{5}y - \frac{1}{3} \leq \frac{2}{3}$

34. $-\frac{1}{6}y - \frac{1}{2} \leq \frac{2}{3}$

35. $-7.2x + 1.8 > -19.8$

36. $-7.8x - 1.3 > 22.1$

37. $\frac{2}{3}x - 5 \leq 7$

38. $\frac{3}{4}x - 8 \leq 1$

39. $-\frac{2}{5}a - 3 > 5$

40. $-\dfrac{4}{5}a - 2 > 10$

41. $5 - \dfrac{3}{5}y > -10$

42. $4 - \dfrac{5}{6}y > -11$

43. $0.3(a + 1) \leq 1.2$

44. $0.4(a - 2) \leq 0.4$

45. $2(5 - 2x) \leq -20$

46. $7(8 - 2x) > 28$

47. $3x - 5 > 8x$

48. $8x - 4 > 6x$

49. $\dfrac{1}{3}y - \dfrac{1}{2} \leq \dfrac{5}{6}y + \dfrac{1}{2}$

50. $\dfrac{7}{6}y + \dfrac{4}{3} \leq \dfrac{11}{6}y - \dfrac{7}{6}$

51. $-2.8x + 8.4 < -14x - 2.8$

52. $-7.2x - 2.4 < -2.4x + 12$

53. $3(m - 2) - 4 \geq 7m + 14$

54. $2(3m - 1) + 5 \geq 8m - 7$

55. $3 - 4(x - 2) \leq -5x + 6$

56. $8 - 6(x - 3) \leq -4x + 12$

57. $4 - 3(x - 2) \geq -x + 4$

58. $5 - 2(2x - 1) \geq -2x + 4$

59. $-3x + 9 < 4 - 6(2x - 5)$

60. $-2x + 4 < 5 - 3(2x - 6)$

Solve each of the following formulas for *y*.

61. $3x + 2y < 6$

62. $-3x + 2y < 6$

63. $2x - 5y > 10$

64. $-2x - 5y > 5$

65. $-3x + 7y \leq 21$

66. $-7x + 3y \leq 21$

67. $2x - 4y \geq -4$

68. $4x - 2y \geq -8$

The next two problems are intended to give you practice reading, and paying attention to, the instructions that accompany the problems you are working.

69. Work each problem according to the instructions given.

a. Evaluate $-5x + 3$ when $x = 0$.

b. Solve: $-5x + 3 = -7$.

c. Is 0 a solution to $-5x + 3 < -7$?

d. Solve: $-5x + 3 < -7$.

70. Work each problem according to the instructions given.

a. Evaluate $-2x - 5$ when $x = 0$.

b. Solve: $-2x - 5 = 1$.

c. Is 0 a solution to $-2x - 5 > 1$?

d. Solve: $-2x - 5 > 1$.

For each graph below, write an inequality whose solution is the graph.

71.

72.

73.

74.

D Solve and graph the solution sets for the following compound inequalities.

75. $x + 5 \le -2$ or $x + 5 \ge 2$

76. $3x + 2 < -3$ or $3x + 2 > 3$

77. $5y + 1 \le -4$ or $5y + 1 \ge 4$

78. $7y - 5 \le -2$ or $7y - 5 \ge 2$

79. $2x + 5 < 3x - 1$ or $x - 4 > 2x + 6$

80. $3x - 1 > 2x + 4$ or $5x - 2 < 3x + 4$

81. $3x + 1 < -8$ or $-2x + 1 \le -3$

82. $2x - 5 \le -1$ or $-3x - 6 < -15$

Solve the following continued inequalities. Use a line graph to write each solution set.

83. $-2 \le m - 5 \le 2$

84. $-3 \le m + 1 \le 3$

85. $-40 < 20a + 20 < 20$

86. $-60 < 50a - 40 < 60$

87. $0.5 \leq 0.3a - 0.7 \leq 1.1$

88. $0.1 \leq 0.4a + 0.1 \leq 0.3$

89. $3 < \dfrac{1}{2}x + 5 < 6$

90. $5 < \dfrac{1}{4}x + 1 < 9$

91. $4 < 6 + \dfrac{2}{3}x < 8$

92. $3 < 7 + \dfrac{4}{5}x < 15$

93. $-2 < -\dfrac{1}{2}x + 1 \leq 1$

94. $-3 \leq -\dfrac{1}{3}x - 1 < 2$

95. $-\dfrac{1}{2} \leq \dfrac{3x + 1}{2} \leq \dfrac{1}{2}$

96. $-\dfrac{5}{6} \leq \dfrac{2x + 5}{3} \leq \dfrac{5}{6}$

97. $-1.5 \leq \dfrac{2x - 3}{4} \leq 3.5$

98. $-1.25 \leq \dfrac{2x + 3}{4} \leq .75$

99. $-\dfrac{3}{4} \leq \dfrac{4x - 3}{2} \leq 1.5$

100. $-\dfrac{5}{6} \leq \dfrac{5x + 2}{3} \leq \dfrac{5}{6}$

Applying the Concepts

101. Consecutive Integers The sum of two consecutive integers is at least 583. What are the possibilities for the first of the two integers?

102. Consecutive Integers The sum of two consecutive integers is at most 583. What are the possibilities for the first of the two integers?

103. Number Problems The sum of twice a number and six is less than ten. Find all solutions.

104. Number Problems Twice the difference of a number and three is greater than or equal to the number increased by five. Find all solutions.

105. Number Problems The product of a number and four is greater than the number minus eight. Find the solution set.

106. Number Problems The quotient of a number and five is less than the sum of seven and two. Find the solution set.

107. Geometry Problems The length of a rectangle is 3 times the width. If the perimeter is to be at least 48 meters, what are the possible values for the width? (If the perimeter is at least 48 meters, then it is greater than or equal to 48 meters.)

108. Geometry Problems The length of a rectangle is 3 more than twice the width. If the perimeter is to be at least 51 meters, what are the possible values for the width? (If the perimeter is at least 51 meters, then it is greater than or equal to 51 meters.)

109. Geometry Problems The numerical values of the three sides of a triangle are given by three consecutive even integers. If the perimeter is greater than 24 inches, what are the possibilities for the shortest side?

110. Geometry Problems The numerical values of the three sides of a triangle are given by three consecutive odd integers. If the perimeter is greater than 27 inches, what are the possibilities for the shortest side?

111. Car Heaters If you have ever gotten in a cold car early in the morning you know that the heater does not work until the engine warms up. This is because the heater relies on the heat coming off the engine. Write an equation using an inequality sign to express when the heater will work if the heater works only after the engine is 100°F.

112. Exercise When Kate exercises, she either swims or runs. She wants to spend a minimum of 8 hours a week exercising, and she wants to swim 3 times the amount she runs. What is the minimum amount of time she must spend doing each exercise?

113. Profit and Loss Movie theaters pay a certain price for the movies that you and I see. Suppose a theater pays $1,500 for each showing of a popular movie. If they charge $7.50 for each ticket they sell, then they will lose money if ticket sales are less than $1,500. However, they will make a profit if ticket sales are greater than $1,500. What is the range of tickets they can sell and still lose money? What is the range of tickets they can sell and make a profit?

114. Stock Sales Suppose you purchase x shares of a stock at $12 per share. After 6 months you decide to sell all your shares at $20 per share. Your broker charges you $15 for the trade. If your profit is at least $3,985, how many shares did you purchase in the first place?

Maintaining Your Skills

The problems that follow review some of the more important skills you have learned in previous sections and chapters. You can consider the time you spend working these problems as time spent studying for exams.

Answer the following percent problems.

115. What number is 25% of 32?

116. What number is 15% of 75?

117. What number is 20% of 120?

118. What number is 125% of 300?

119. What percent of 36 is 9?

120. What percent of 16 is 9?

121. What percent of 50 is 5?

122. What percent of 140 is 35?

123. 16 is 20% of what number?

124. 6 is 3% of what number?

125. 8 is 2% of what number?

126. 70 is 175% of what number?

Simplify each expression.

127. $-|-5|$

128. $\left(-\dfrac{2}{3}\right)^3$

129. $-3 - 4(-2)$

130. $2^4 + 3^3 \div 9 - 4^2$

131. $5\,|3 \quad 8|\,-6\,|2 - 5|$

132. $7 - 3(2 - 6)$

133. $5 - 2[-3(5 - 7) - 8]$

134. $\dfrac{5 + 3(7 - 2)}{2(-3) - 4}$

135. Find the difference of -3 and -9.

136. If you add -4 to the product of -3 and 5, what number results?

137. Apply the distributive property to $\dfrac{1}{2}(4x - 6)$.

138. Use the associative property to simplify $-6\left(\dfrac{1}{3}x\right)$.

Find the Mistake

Each sentence below contains a mistake. Circle the mistake and write the correct word(s) or number on the line provided.

1. If the solution set to an inequality is all the real numbers greater than 9, the solution is $x \le 9$. _____

2. To include the number 14 in the solution to the inequality $x \ge 14$, we mark the number on the graph with an open circle.

3. The last step in solving an inequality is to use the multiplication property for inequalities. _____

4. Upon solving the inequality $-5y < 4x + 9$ for y, the inequality symbol remains the same. _____

Chapter 5 Summary

Solution Set [5.1]

The *solution set* for an equation (or inequality) is all the numbers that, when used in place of the variable, make the equation (or inequality) a true statement.

1. The solution set for the equation $x + 2 = 5$ is $\{3\}$ because when x is 3 the equation is $3 + 2 = 5$, or $5 = 5$.

Equivalent Equations [5.1]

Two equations are called *equivalent* if they have the same solution set.

2. The equation $a - 4 = 3$ and $a - 2 = 5$ are equivalent because both have solution set $\{7\}$.

Addition Property of Equality [5.1]

When the same quantity is added to both sides of an equation, the solution set for the equation is unchanged. Adding the same amount to both sides of an equation produces an equivalent equation.

3. Solve $x - 5 = 12$.
$$x - 5\,(+5) = 12\,(+5)$$
$$x + 0 = 17$$
$$x = 17$$

Multiplication Property of Equality [5.2]

If both sides of an equation are multiplied by the same nonzero number, the solution set is unchanged. Multiplying both sides of an equation by a nonzero quantity produces an equivalent equation.

4. Solve $3x = 18$.
$$\frac{1}{3}(3x) = \frac{1}{3}(18)$$
$$x = 6$$

Linear Equations [5.3]

A linear equation in one variable is any equation that can be put in the form $Ax + B = C$, where A, B and C are real numbers and a is not zero.

Strategy for Solving Linear Equations in One Variable [5.3]

Step 1a: Use the distributive property to separate terms, if necessary.

1b: If fractions are present, consider multiplying both sides by the LCD to eliminate the fractions. If decimals are present, consider multiplying both sides by a power of 10 to clear the equation of decimals.

1c: Combine similar terms on each side of the equation.

Step 2: Use the addition property of equality to get all variable terms on one side of the equation and all constant terms on the other side. A variable term is a term that contains the variable (for example, $5x$). A constant term is a term that does not contain the variable (the number 3, for example).

Step 3: Use the multiplication property of equality to get x (that is, $1x$) by itself on one side of the equation.

Step 4: Check your solution in the original equation to be sure that you have not made a mistake in the solution process.

5. Solve $2(x + 3) = 10$.
$$2x + 6 = 10$$
$$2x + 6 + (-6) = 10 + (-6)$$
$$2x = 4$$
$$\frac{1}{2}(2x) = \frac{1}{2}(4)$$
$$x = 2$$

Formulas [5.4]

6. Solving $P = 2l + 2w$ for l, we have
$$P - 2w = 2l$$
$$\frac{P - 2w}{2} = l$$

A formula is an equation with more than one variable. To solve a formula for one of its variables, we use the addition and multiplication properties of equality to move everything except the variable in question to one side of the equal sign so the variable in question is alone on the other side.

Blueprint for Problem Solving [5.5, 5.6]

Step 1: *Read* the problem, and then mentally *list* the items that are known and the items that are unknown.

Step 2: *Assign a variable* to one of the unknown items. (In most cases this will amount to letting $x =$ the item that is asked for in the problem.) Then *translate* the other *information* in the problem to expressions involving the variable.

Step 3: *Reread* the problem, and then *write an equation,* using the items and variables listed in steps 1 and 2, that describes the situation.

Step 4: *Solve the equation* found in step 3.

Step 5: *Write* your *answer* using a complete sentence.

Step 6: *Reread* the problem, and *check* your solution with the original words in the problem.

Addition Property for Inequalities [5.7]

7. Solve $x + 5 < 7$.
$$x + 5 + (-5) < 7 + (-5)$$
$$x < 2$$

Adding the same quantity to both sides of an inequality produces an equivalent inequality, one with the same solution set.

Multiplication Property for Inequalities [5.7]

8. Solve $-3a \leq 18$.
$$-\frac{1}{3}(-3a) \geq -\frac{1}{3}(18)$$
$$a \geq -6$$

Multiplying both sides of an inequality by a positive number never changes the solution set. If both sides are multiplied by a negative number, the sign of the inequality must be reversed to produce an equivalent inequality.

Strategy for Solving Linear Inequalities in One Variable [5.7]

9. Solve $3(x - 4) \geq -2$.
$$3x - 12 \geq -2$$
$$3x - 12\, (+12) \geq -2\, (+12)$$
$$3x \geq 10$$
$$\frac{1}{3}(3x) \geq \frac{1}{3}(10)$$
$$x \geq \frac{10}{3}$$

Step 1a: Use the distributive property to separate terms, if necessary.

 1b: If fractions are present, consider multiplying both sides by the LCD to eliminate the fractions. If decimals are present, consider multiplying both sides by a power of 10 to clear the inequality of decimals.

 1c: Combine similar terms on each side of the inequality.

Step 2: Use the addition property for inequalities to get all variable terms on one side of the inequality and all constant terms on the other side.

Step 3: Use the multiplication property for inequalities to get x by itself on one side of the inequality.

Step 4: Graph the solution set.

Is 4 a solution to each of the following equations? [5.1]

1. $5x + 4 = 8$ **2.** $3y - 5 = 7$

Is 7 a solution to each of the following equations? [5.1]

3. $4n + 3 = 31$ **4.** $6a - 6 = 5$

Solve the following equations. [5.1]

5. $x - 7 = 10$

6. $y + 2 = 7$

7. $n + 4.2 = 9.4$

Solve the following equations. [5.2, 5.3]

8. $3x - 2 = 7$ **9.** $4y + 15 = y$

10. $\frac{1}{4}x - \frac{1}{12} = \frac{1}{3}x + \frac{1}{6}$ **11.** $-3(3 - 2x) - 7 = 8$

12. $3x - 9 = -6$

13. $0.05 + 0.07(100 - x) = 3.2$

14. $4(t - 3) + 2(t + 4) = 2t - 16$

15. $4x - 2(3x - 1) = 2x - 8$

16. $5(x + 2) - 3x + 3(x + 4) = 32$

17. $3(a + 7) + 6a - 3(a + 2) = 45$

18. If $3x - 4y = 16$, find y when $x = 4$. [5.4]

19. If $3x - 4y = 16$, find x when $y = 2$. [5.4]

20. Solve $2x + 6y = 12$ for y. [5.4]

21. Solve $x^2 = v^2 + 2ad$ for a. [5.4]

Solve each word problem. [5.5, 5.6]

22. Age Problem Paul is twice as old as Becca. Five years ago the sum of their ages was 44. How old are they now?

23. Geometry The length of a rectangle is 5 centimeters less than 3 times the width. The perimeter is 150 centimeters. What are the length and width?

24. Coin Problem A man has a collection of dimes and nickels with a total value of $1.70. If he has 8 more dimes than he has nickels, how many of each coin does he have?

25. Investing A woman has money in two accounts. One account pays 6% annual interest, whereas the other pays 12% annual interest. If she has $500 more invested at 12% than she does at 6% and her total interest for a year is $186, how much does she have in each account?

Solve each inequality, and graph the solution. [5.7]

26. $\frac{1}{2}x - 2 > 3$ **27.** $-6y \le 24$

28. $0.3 - 0.2x < 1.1$ **29.** $3 - 2(n - 1) \ge 9$

30. $5x - 3 < 2x$ **31.** $-3 \le 2x - 7$

Write an inequality whose solution is the given graph. [5.7]

32.

```
←————————⊕————————————→
        -3          0
```

33.

```
←————————●————————————→
        -1          0
```

34.

```
←————————⊕————————————→
        -2          0
```

The snapshot shows the number of hours per day the average commute is for several cities throughout the United States. Use the information to determine the percentage of each day (24 hours) are spent commuting in a car for the following cities. Round to the nearest tenth of a percent.

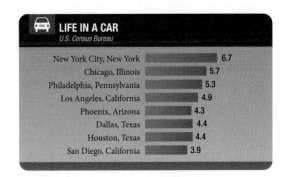

LIFE IN A CAR
U.S. Census Bureau

City	Hours
New York City, New York	6.7
Chicago, Illinois	5.7
Philadelphia, Pennsylvania	5.3
Los Angeles, California	4.9
Phoenix, Arizona	4.3
Dallas, Texas	4.4
Houston, Texas	4.4
San Diego, California	3.9

35. New York City **36.** Los Angeles

Exponents and Polynomials

6.1 Multiplication with Exponents and Scientific Notation

6.2 Division with Exponents and Negative Exponents

6.3 Operations with Monomials

6.4 Addition and Subtraction of Polynomials

6.5 Multiplication of Polynomials

6.6 Special Products

6.7 Dividing a Polynomial by a Monomial

6.8 Dividing a Polynomial by a Polynomial

Lake Baikal, Siberia.
Image © TerraMetrics
© 2010 Cnes/ Spot Image

Lake Baikal is the largest natural lake in the world. Located in the southeastern part of Siberia, it is also the deepest freshwater lake on Earth and contains the largest amount of fresh surface water. Experts estimate that nearly 20% of the world's fresh water lies within its banks. With its beginnings dating back 31 million years, it is also considered one of the oldest lakes on the planet. Despite its age, Lake Baikal is still expanding at a rate of 2 centimeters per year. Experts believe that in several million years this expansion will result in a new great ocean covering Asia.

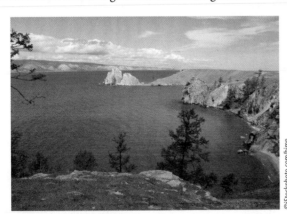

©iStockphoto.com/fiimn

Lake Baikal has 26 islands of varying sizes. Suppose one of these islands was the shape of a rectangle; one side was x^2 meters long, and the other side was x^3 meters long. How would you find the surface area of this island? The solution to this problem requires multiplication of exponents, which is one of the topics we will discuss in this chapter.

OBJECTIVES

A Use the definition of integer exponents to evaluate expressions containing exponents.

B Use the product property of exponents to simplify expressions.

C Use the power property of exponents to simplify expressions.

D Use the distributive property of exponents to simplify expressions.

E Write numbers greater than 10 in scientific notation and standard form.

KEY WORDS

product property of exponents

power property of exponents

distributive property of exponents

volume

rectangular solid

scientific notation

Practice Problems

1. Simplify.
 a. 5^3
 b. -2^4
 c. $(-3)^4$
 d. $\left(-\dfrac{2}{3}\right)^2$

During a hike up Bishop's Peak in San Luis Obispo, CA, a thirty-year-old hiker's heart beat 145 times per minute. If he hiked up the mountain and back down again in 4.5 hours, his heart beat approximately 39,150 times. In this section, we will learn how to write this large number in scientific notation. For instance, 39,150 can be written as 3.915×10^4. But first, we must learn to evaluate exponents and work with properties of exponents to further understand their meaning in this notation.

A Evaluating Exponents

Recall that an *exponent* is a number written just above and to the right of another number, which is called the *base*. In the expression 5^2, for example, the exponent is 2 and the base is 5. the expression 5^2 is read "5 to the second power" or "5 squared." The meaning of the expression is

$$5^2 = 5 \cdot 5 = 25$$

In the expression 5^3, the exponent is 3 and the base is 5. The expression 5^3 is read "5 to the third power" or "5 cubed." The meaning of the expression is

$$5^3 = 5 \cdot 5 \cdot 5 = 125$$

Here are some further examples:

EXAMPLE 1 Write each expression as a single number.

a. 4^3 **b.** -3^4 **c.** $(-2)^5$ **d.** $\left(-\dfrac{3}{4}\right)^2$

Solution

a. $4^3 = 4 \cdot 4 \cdot 4 = 16 \cdot 4 = 64$ Exponent 3, base 4

b. $-3^4 = -3 \cdot 3 \cdot 3 \cdot 3 = -81$ Exponent 4, base 3

c. $(-2)^5 = (-2)(-2)(-2)(-2)(-2) = -32$ Exponent 5, base -2

d. $\left(-\dfrac{3}{4}\right)^2 = \left(-\dfrac{3}{4}\right)\left(-\dfrac{3}{4}\right) = \dfrac{9}{16}$ Exponent 2, base $-\dfrac{3}{4}$

Question: In what way are $(-5)^2$ and -5^2 different?

Answer: In the first case, the base is -5. In the second case, the base is 5. The answer to the first is 25. The answer to the second is -25. Can you tell why? Would there be a difference in the answers if the exponent in each case were changed to 3? To clarify whether the base of an exponential expression contains a negative, it is helpful to enclose the base in parenthesis as we have done with $(-5)^2$.

Answers

1. **a.** 125 **b.** -16 **c.** 81 **d.** $\dfrac{4}{9}$

We can simplify our work with exponents by developing some properties of exponents. We want to list the things we know are true about exponents and then use these properties to simplify expressions that contain exponents.

B Product Property of Exponents

The first property of exponents applies to products with the same base. We can use the definition of exponents, as indicating repeated multiplication, to simplify expressions like $7^4 \cdot 7^2$.

$$7^4 \cdot 7^2 = (7 \cdot 7 \cdot 7 \cdot 7)(7 \cdot 7)$$
$$= (7 \cdot 7 \cdot 7 \cdot 7 \cdot 7 \cdot 7)$$
$$= 7^6 \qquad \text{Note: } 4 + 2 = 6.$$

As you can see, multiplication with the same base resulted in addition of exponents. We can summarize this result with the following property.

> **PROPERTY** Product Property of Exponents
>
> If a is any real number and r and s are integers, then
>
> $$a^r \cdot a^s = a^{r+s}$$
>
> *In words:* To multiply two expressions with the same base, add exponents and use the common base.

Here is an example using the product property.

EXAMPLE 2 Use the product property to simplify the following expressions. Leave your answers in terms of exponents.

a. $5^3 \cdot 5^6$ **b.** $x^7 \cdot x^8$ **c.** $3^4 \cdot 3^8 \cdot 3^5$

Solution

a. $5^3 \cdot 5^6 = 5^{3+6} = 5^9$

b. $x^7 \cdot x^8 = x^{7+8} = x^{15}$

c. $3^4 \cdot 3^8 \cdot 3^5 = 3^{4+8+5} = 3^{17}$

C Power Property of Exponents

Another common type of expression involving exponents is one in which an expression containing an exponent is raised to another power. The expression $(5^3)^2$ is an example.

$$(5^3)^2 = (5^3)(5^3)$$
$$= 5^{3+3}$$
$$= 5^6 \qquad \text{Note: } 3 \cdot 2 = 6.$$

This result offers justification for the power property of exponents.

> **PROPERTY** Power Property of Exponents
>
> If a is any real number and r and s are integers, then
>
> $$(a^r)^s = a^{rs}$$
>
> *In words:* A power raised to another power is the base raised to the product of the powers.

**Video Examples
Section 6.1**

2. Simplify, leaving your answers in terms of exponents.
 a. $5^4 \cdot 5^5$
 b. $x^3 \cdot x^7$
 c. $4^5 \cdot 4^2 \cdot 4^6$

Note In Example 2, notice that each base in the original problem is the same base that appears in the answer and that it is written only once in the answer. A very common mistake that people make when they first begin to use the product property is to write a 2 in front of the base in the answer. For example, people making this mistake would get $2x^{15}$ or $(2x)^{15}$ as the result in Example 2b. To avoid this mistake, you must be sure you understand the meaning of the product property exactly as it is written.

Answers

2. a. 5^9 **b.** x^{10} **c.** 4^{13}

3. Simplify.
 a. $(3^4)^5$
 b. $(x^7)^2$

EXAMPLE 3 Simplify the following expressions:
a. $(4^5)^6$ **b.** $(x^3)^5$
Solution

a. $(4^5)^6 = 4^{5 \cdot 6} = 4^{30}$

b. $(x^3)^5 = x^{3 \cdot 5} = x^{15}$

D Distributive Property of Exponents

The third property of exponents applies to expressions in which the product of two or more numbers or variables is raised to a power. Let's look at how the expression $(2x)^3$ can be simplified:

$$(2x)^3 = (2x)(2x)(2x)$$

$$= (2 \cdot 2 \cdot 2)(x \cdot x \cdot x)$$

$$= 2^3 \cdot x^3 \qquad \text{Note: The exponent 3 distributes over the product } 2x.$$

$$= 8x^3$$

We can generalize this result into a third property of exponents.

> **PROPERTY** **Distributive Property of Exponents**
>
> If a and b are any two real numbers and r is an integer, then
>
> $$(ab)^r = a^r b^r$$
>
> *In words:* The power of a product is the product of the powers.

Note A common mistake that students make when using this property is to not distribute the exponent to both parts of the base as in

$$(2x)^3 \neq 2x^3$$

Be careful to follow the property correctly, as we have shown here.

Here are some examples using the distributive property of exponents to simplify expressions.

4. Simplify.

 a. $\left(-\dfrac{1}{3}x^3y^2\right)^2$

 b. $(x^5)^2(x^4)^3$

 c. $(3y)^2(2y^3)$

 d. $(3x^3y^2)^2(2xy^4)^3$

EXAMPLE 4 Simplify the following expressions:

a. $\left(-\dfrac{1}{4}x^2y^3\right)^2$ **b.** $(x^4)^3(x^2)^5$ **c.** $(2y)^3(3y^2)$ **d.** $(2x^2y^5)^3(3x^4y)^2$

Solution

a. $\left(-\dfrac{1}{4}x^2y^3\right)^2 = \left(-\dfrac{1}{4}\right)^2(x^2)^2(y^3)^2$ Distributive property of exponents

$$= \dfrac{1}{16}x^4y^6 \qquad\qquad\qquad \text{Power property of exponents}$$

b. $(x^4)^3(x^2)^5 = x^{12} \cdot x^{10}$ Power property of exponents

$$= x^{22} \qquad\qquad\qquad \text{Product property of exponents}$$

c. $(2y)^3(3y^2) = 2^3y^3(3y^2)$ Distributive property of exponents

$$= 8 \cdot 3(y^3 \cdot y^2) \qquad \text{Commutative and associative properties}$$

$$= 24y^5 \qquad\qquad\quad \text{Product property of exponents}$$

d. $(2x^2y^5)^3(3x^4y)^2 = 2^3(x^2)^3(y^5)^3 \cdot 3^2(x^4)^2y^2$ Distributive property of exponents

$$= 8x^6y^{15} \cdot 9x^8y^2 \qquad\qquad \text{Power property of exponents}$$

$$= (8 \cdot 9)(x^6x^8)(y^{15}y^2) \qquad \text{Commutative and associative properties}$$

$$= 72x^{14}y^{17} \qquad\qquad\qquad \text{Product property of exponents}$$

Answers

3. a. 3^{20} **b.** x^{14}

4. a. $\dfrac{1}{9}x^6y^4$ **b.** x^{22} **c.** $18y^5$

 d. $72x^9y^{16}$

FACTS FROM GEOMETRY Volume of a Rectangular Solid

It is easy to see why the phrase "five squared" is associated with the expression 5^2. Simply find the area of the square shown in Figure 1 with a side of 5.

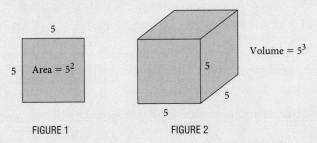

FIGURE 1 FIGURE 2

To see why the phrase "five cubed" is associated with the expression 5^3, we have to find the *volume* of a cube for which all three dimensions are 5 units long. The volume of a cube is a measure of the space occupied by the cube. To calculate the volume of the cube shown in Figure 2, we multiply the three dimensions together to get $5 \cdot 5 \cdot 5 = 5^3$.

The cube shown in Figure 2 is a special case of a general category of three dimensional geometric figures called *rectangular solids*. Rectangular solids have rectangles for sides, and all connecting sides meet at right angles. The three dimensions are length, width, and height. To find the volume of a rectangular solid, we find the product of the three dimensions, as shown in Figure 2.

> *Note* If we include units with the dimensions of the diagrams, then the units for the area will be square units and the units for volume will be cubic units. More specifically,
>
> If a square has a side 5 inches long, then its area will be
> $A = (5 \text{ inches})^2 = 25 \text{ inches}^2$
> where the unit inches2 stands for square inches.
>
> If a cube has a single side 5 inches long, then its volume will be
> $V = (5 \text{ inches})^3 = 125 \text{ inches}^3$
> where the unit inches3 stands for cubic inches.
>
> If a rectangular solid has a length of 5 inches, a width of 4 inches, and a height of 3 inches, then its volume is
> $V = (5 \text{ inches})(4 \text{ inches})(3 \text{ inches})$
> $= 60 \text{ inches}^3$

E Scientific Notation for Large Numbers

Many branches of science require working with very large numbers. In astronomy, for example, distances commonly are given in light-years. A light-year is the distance light travels in a year. It is approximately

$$5,880,000,000,000 \text{ miles}$$

This number is difficult to use in calculations because of the number of zeros it contains. Scientific notation provides a way of writing very large numbers in a more manageable form.

DEFINITION scientific notation

A number is in *scientific notation* when it is written as the product of a number between 1 and 10 and an integer power of 10. A number written in scientific notation has the form

$$n \times 10^r$$

where $1 \le n < 10$ and $r =$ an integer.

EXAMPLE 5 Write 376,000 in scientific notation.

Solution We must rewrite 376,000 as the product of a number between 1 and 10 and a power of 10. To do so, we move the decimal point 5 places to the left so that it appears between the 3 and the 7. Then we multiply this number by 10^5. The number that results has the same value as our original number and is written in scientific notation.

$$376,000 = 3.76 \times 10^5$$

Moved 5 places.

Decimal point was originally here.

Keeps track of the 5 places we moved the decimal point.

5. Write 4,810,000 in scientific notation.

Answer

5. 4.81×10^6

6. Write 5,473,000 in scientific notation.

EXAMPLE 6 Write 4,758,000 in scientific notation.

Solution As with the previous example, we must move the decimal to the left but in this case, we move it 6 places. It now appears between the 4 and the 7. We round to the nearest thousandth and multiply our result by 10^6.

$$4,758,000 = 4.758 \times 10^6$$

Moved 6 places.

Decimal point was originally here.

Keeps track of the 6 places we moved the decimal point.

Our answer has the same value as our original number and is written in scientific notation.

7. Write 3.05×10^5 in expanded form.

EXAMPLE 7 Write 4.52×10^3 in expanded form.

Solution Since 10^3 is 1,000, we can think of this as simply a multiplication problem; that is,

$$4.52 \times 10^3 = 4.52 \times 1,000 = 4,520$$

On the other hand, we can think of the exponent 3 as indicating the number of places we need to move the decimal point to write our number in expanded form. Since our exponent is positive 3, we move the decimal point three places to the right.

$$4.52 \times 10^3 = 4,520$$

In the next section, we will see how a negative power of 10 affects our answers in scientific notation.

GETTING READY FOR CLASS

After reading through the preceding section, respond in your own words and in complete sentences.

A. Explain the difference between -5^2 and $(-5)^2$.

B. How do you multiply two expressions containing exponents when they each have the same base?

C. Explain in words how you would use the product property and the distributive property of exponents to simplify $(2x^3)^2$.

D. How would you write 5,190,000 in scientific notation?

Answers

6. 5.473×10^6

7. 305,000

Vocabulary Review

Choose the correct words to fill in the blanks below.

> power distributive scientific notation
>
> base product exponent

1. In the expression 7^3, the 7 is called the _____ and the 3 is called the _____ .

2. The _____ property for exponents states that you should add exponents and use the common base to multiply two expressions with the same base.

3. The _____ property for exponents states that a power raised to another power is the base raised to the product of the powers.

4. The _____ property for exponents states that the power of a product is the product of the powers.

5. _____ is a way of writing very large numbers that is more manageable and takes the form $n \times 10^r$.

Problems

A Name the base and exponent in each of the following expressions. Then use the definition of exponents as repeated multiplication to simplify.

1. 4^2

2. 6^2

3. $(0.3)^2$

4. $(0.03)^2$

5. 4^3

6. 10^3

7. $(-5)^2$

8. -5^2

9. -2^3

10. $(-2)^3$

11. 3^4

12. $(-3)^4$

13. $\left(\dfrac{2}{3}\right)^2$

14. $\left(\dfrac{2}{3}\right)^3$

15. $\left(\dfrac{1}{2}\right)^4$

16. $\left(\dfrac{4}{5}\right)^2$

17. $\left(-\dfrac{3}{4}\right)^2$

18. $\left(-\dfrac{2}{3}\right)^4$

19. $-\left(\dfrac{1}{6}\right)^2$

20. $-\left(\dfrac{1}{9}\right)^2$

21. a. Complete the following table.

Number x	1	2	3	4	5	6	7
Square x^2							

 b. Using the results of part a, fill in the blank in the following statement: For numbers larger than 1, the square of the number is _____ than the number.

22. a. Complete the following table.

Number x	$\frac{1}{2}$	$\frac{1}{3}$	$\frac{1}{4}$	$\frac{1}{5}$	$\frac{1}{6}$	$\frac{1}{7}$	$\frac{1}{8}$
Square x^2							

 b. Using the results of part a, fill in the blank in the following statement: For numbers between 0 and 1, the square of the number is _____ than the number.

B Use the product property of exponents to simplify the following expressions.

23. $x^4 \cdot x^5$

24. $x^7 \cdot x^3$

25. $y^{10} \cdot y^{20}$

26. $y^{30} \cdot y^{30}$

27. $2^5 \cdot 2^4 \cdot 2^3$

28. $4^2 \cdot 4^3 \cdot 4^4$

29. $x^4 \cdot x^6 \cdot x^8 \cdot x^{10}$

30. $x^{20} \cdot x^{18} \cdot x^{16} \cdot x^{14}$

C Use the power property of exponents to write each of the following problems with a single exponent. (Assume all variables are positive numbers.)

31. $(x^2)^5$

32. $(x^5)^2$

33. $(5^4)^3$

34. $(5^3)^4$

35. $(y^3)^3$

36. $(y^2)^2$

37. $(2^5)^{10}$

38. $(10^5)^2$

39. $(a^3)^x$

40. $(a^5)^x$

41. $(b^x)^y$

42. $(b^r)^s$

D Use the distributive property of exponents to simplify each of the following expressions.

43. $(4x)^2$

44. $(2x)^4$

45. $(2y)^5$

46. $(5y)^2$

47. $(-3x)^4$

48. $(-3x)^3$

49. $(0.5ab)^2$

50. $(0.4ab)^2$

51. $(4xyz)^3$

52. $(5xyz)^3$

53. $(-4xy)^3$

54. $(-3xy)^4$

Simplify the following expressions by using the properties of exponents.

55. $(2x^4)^3$

56. $(3x^5)^2$

57. $(4a^3)^2$

58. $(5a^2)^2$

59. $(x^2)^3(x^4)^2$

60. $(x^5)^2(x^3)^5$

61. $(a^3)^1(a^2)^4$

62. $(a^4)^1(a^1)^3$

63. $(2x)^3(2x)^4$

64. $(3x)^2(3x)^3$

65. $(3x^2)^3(2x)^4$

66. $(3x)^3(2x^3)^2$

67. $(4x^2y^3)^2$

68. $(9x^3y^5)^2$

69. $\left(\dfrac{2}{3}a^4b^5\right)^3$

70. $\left(\dfrac{3}{4}ab^7\right)^3$

71. Complete the following table, and then construct a line graph of the information in the table

Number x	-3	-2	-1	0	1	2	3
Square x^2							

72. Complete the table, and then construct a line graph of the information in the table.

Number x	-3	-2	-1	0	1	2	3
Cube x^3							

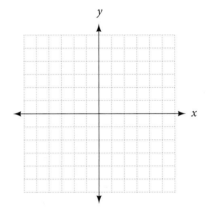

73. Complete the table. When you are finished, notice how the points in this table could be used to refine the line graph you created in Problem 71.

Number x	-2.5	-1.5	-0.5	0	0.5	1.5	2.5
Square x^2							

74. Complete the following table. When you are finished, notice that this table contains exactly the same entries as the table from Problem 73. This table uses fractions, whereas the table from Problem 73 uses decimals.

Number x	$-\dfrac{5}{2}$	$-\dfrac{3}{2}$	$-\dfrac{1}{2}$	0	$\dfrac{1}{2}$	$\dfrac{3}{2}$	$\dfrac{5}{2}$
Square x^2							

E Write each number in scientific notation.

75. 43,200

76. 432,000

77. 570

78. 5,700

79. 238,000

80. 2,380,000

81. 10,000,000,000

82. 20

Write each number in standard form.

83. 2.49×10^3 **84.** 2.49×10^4 **85.** 3.52×10^2 **86.** 3.52×10^5

87. 2.8×10^4 **88.** 2.8×10^3 **89.** -3.1×10^5 **90.** -4.2×10^6

Applying the Concepts

91. The five largest lakes in the world (by area) are listed below in the table. Complete the table by writing each area in scientific notation.

Name	Location	Area (sq. mi.)	Scientific Notation
Caspian Sea	5 countries	152,239	
Superior	US, Canada	31,820	
Victoria	Tanzania, Uganda	26,828	
Huron	US, Canada	23,010	
Michigan	US	22,400	

92. The table below lists the five outer planets (from Earth), and their average distance to Earth in miles. Write each distance in scientific notation.

Planet	Average Distance	Scientific Notation
Mars	48.60 million	
Jupiter	390.60 million	
Saturn	777.00 million	
Uranus	1.69 billion	
Neptune	2.70 billion	

93. Volume of a Cube Find the volume of a cube if each side is 3 inches long.

94. Volume of a Cube Find the volume of a cube if each side is 3 feet long.

95. Volume of a Cube A bottle of perfume is packaged in a box that is in the shape of a cube. Find the volume of the box if each side is 2.5 inches long. Round to the nearest tenth.

96. Volume of a Cube A television set is packaged in a box that is in the shape of a cube. Find the volume of the box if each side is 18 inches long.

97. Volume of a Box A DVD box set of your favorite TV series is in the shape of a rectangular solid. Find the volume of the container if the height is 8 inches, the width is 4.5 inches, and the depth is 1 inch.

98. Volume of a Box Your textbook is in the shape of a rectangular solid. Find the volume in cubic inches.

99. Volume of a Box If a box has a volume of 42 cubic feet, is it possible for you to fit inside the box? Explain your answer.

100. Volume of a Box A box has a volume of 45 cubic inches. Will a can of soup fit inside the box? Explain your answer.

101. Age in seconds If you are 21 years old, you have been alive for more than 650,000,000 seconds. Write the number of seconds in scientific notation.

102. Distance Around the Earth The distance around the Earth at the equator is more than 130,000,000 feet. Write the number of feet in scientific notation.

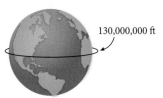

130,000,000 ft

103. Lifetime Earnings If you earn at least $12 an hour and work full-time for 30 years, you will make at least 7.4×10^5 dollars. Write your total earnings in expanded form.

104. Heart Beats per Year If your pulse is 72, then in one year your heart will beat at least 3.78×10^7 times. Write the number of heart beats in expanded form.

105. Investing If you put $1,000 into a savings account every year from the time you are 25 years old until you are 55 years old, you will have more than 1.8×10^5 dollars in the account when you reach 55 years of age (assuming 10% annual interest). Write 1.8×10^5 in expanded form.

106. Investing If you put $20 into a savings account every month from the time you are 20 years old until you are 30 years old, you will have more than 3.27×10^3 dollars in the account when you reach 30 years of age (assuming 6% annual interest compounded monthly). Write 3.27×10^3 in expanded form.

107. The value of an investment of $8,000 growing at 7% per year is given by the formula $V = 8000(1.07)^t$, where t represents the number of years the money is invested. Find the value of the investment after 20 years.

108. The value on an investment of $15,000 growing at 5% per year is given by the formula $V = 15000(1.05)^t$, where t represents the number of years the money is invested. Find the value of this investment after 25 years.

Displacement The displacement, in cubic inches, of a car engine is given by the formula

$$d = \pi \cdot s \cdot c \cdot \left(\frac{1}{2} \cdot b \right)^2$$

where s is the stroke, b is the bore, and c is the number of cylinders. (Note: The bore is the diameter of a cylinder in a piston engine and the stroke is the distance the cylinder travels.)

Calculate the engine displacement for each of the following cars. Use 3.14 to approximate π and round to the nearest cubic inch.

109. Ferrari Modena 8 cylinders, 3.35 inches of bore, 3.11 inches of stroke

110. Audi A8 8 cylinders, 3.32 inches of bore, 3.66 inches of stroke

111. Mitsubishi Eclipse 6 cylinders, 3.59 inches of bore, 2.99 inches of stroke

112. Porsche 911 GT3 6 cylinders, 3.94 inches of bore, 3.01 inches of stroke

Getting Ready for the Next Section

Subtract.

113. $4 - 7$

114. $-4 - 7$

115. $4 - (-7)$

116. $-4 - (-7)$

117. $15 - 20$

118. $15 - (-20)$

119. $-15 - (-20)$

120. $-15 - 20$

Simplify each expression.

121. $2(3) - 4$

122. $5(3) - 10$

123. $4(3) - 3(2)$

124. $-8 - 2(3)$

125. $2(5 - 3)$

126. $2(3) - 4 - 3(-4)$

127. $5 + 4(-2) - 2(-3)$

128. $2(3) + 4(5) - 5(2)$

Find the Mistake

Each sentence below contains a mistake. Circle the mistake and write the correct word(s) on the line provided.

1. Simplifying the expression $\left(-\frac{12}{32}\right)^3$ will give a positive answer. _____

2. You would use the distributive property for exponents to simplify the expression $x^8 \cdot x^4 \cdot x^{10}$.

3. To simplify the expression $(x^3 \cdot y)^4 \, (y \cdot z^2)^8$, we must first use the power property of exponents.

4. 5.83×10^5 in an example of a number written in expanded form. _____

OBJECTIVES

A Convert from negative to positive exponents.

B Use the quotient property of exponents to simplify expressions.

C Use the expanded distributive property of exponents to simplify expressions.

D Simplify expressions involving exponents of 0 and 1.

E Simplify expressions using combinations of the properties of exponents.

F Use negative exponents when writing numbers less than 1 in scientific notation and standard form.

KEY WORDS

negative exponent property

quotient property of exponents

expanded distributive property of exponents

Imagine you are a sculpture artist specializing in glass blowing. Your most demanded pieces are multicolored glass balls used as Christmas ornaments or sun-catchers hung in windows. Each glass ball fits into a cube-shaped box with a side of x inches long. Suppose you need to figure out how many small cube-shaped boxes will fit into a larger shipping box, also shaped like a cube but with a side of $5x$ inches long. First, you would need to find the volume of each box.

$$\text{Volume for the smaller box: } x \cdot x \cdot x = x^3$$
$$\text{Volume for the larger box: } 5x \cdot 5x \cdot 5x = (5x)^3$$

To find how many smaller boxes fit in the larger box, we use division.

$$\frac{(5x)^3}{x^3}$$

In this section, we will learn a property of exponents under division that will help us simplify this problem.

Previously, we found that multiplication with the same base results in addition of exponents; that is, $a^r \cdot a^s = a^{r+s}$. Since division is the inverse operation of multiplication, we can expect division with the same base to result in subtraction of exponents.

A Negative Exponents

To develop the properties for exponents under division, we again apply the definition of exponents.

$$\frac{x^5}{x^3} = \frac{x \cdot x \cdot x \cdot x \cdot x}{x \cdot x \cdot x} \qquad \frac{2^4}{2^7} = \frac{2 \cdot 2 \cdot 2 \cdot 2}{2 \cdot 2 \cdot 2 \cdot 2 \cdot 2 \cdot 2 \cdot 2}$$

$$= \frac{x \cdot x \cdot x}{x \cdot x \cdot x} (x \cdot x) \qquad = \frac{2 \cdot 2 \cdot 2 \cdot 2}{2 \cdot 2 \cdot 2 \cdot 2} \cdot \frac{1}{2 \cdot 2 \cdot 2}$$

$$= 1(x \cdot x) \qquad = \frac{1}{2 \cdot 2 \cdot 2}$$

$$= x^2 \quad \text{Note: } 5 - 3 = 2. \qquad = \frac{1}{2^3} \quad \text{Note: } 7 - 4 = 3.$$

In both cases, division with the same base resulted in subtraction of the smaller exponent from the larger. The problem is deciding whether the answer is a fraction. The problem is resolved easily by the following property.

**Video Examples
Section 6.2**

> **PROPERTY** Negative Exponent Property
>
> If r is a positive integer, then $a^{-r} = \dfrac{1}{a^r} = \left(\dfrac{1}{a}\right)^r$ where $(a \neq 0)$.

1. Write each expression with a positive exponent and then simplify.

 a. 3^{-2}
 b. 4^{-3}
 c. $5x^{-4}$

The following examples illustrate how we use this property to simplify expressions that contain negative exponents.

EXAMPLE 1 Write each expression with a positive exponent and then simplify.

a. 2^{-3} **b.** 5^{-2} **c.** $3x^{-6}$

Solution

a. $2^{-3} = \dfrac{1}{2^3} = \dfrac{1}{8}$ *Note: Negative exponents do not indicate negative numbers. they indicate reciprocals.*

b. $5^{-2} = \dfrac{1}{5^2} = \dfrac{1}{25}$

c. $3x^{-6} = 3 \cdot \dfrac{1}{x^6} = \dfrac{3}{x^6}$ *Note: Be careful not to apply the negative exponents to the 3.*

B Quotient Property of Exponents

Now let us look back to our original problem and try to work it again with the help of a negative exponent. We know that $\frac{2^4}{2^7} = \frac{1}{2^3}$. Let us decide now that with division of the same base, we will always subtract the exponent in the denominator from the exponent in the numerator and see if this conflicts with what we know is true.

$$\frac{2^4}{2^7} = 2^{4-7} \qquad \textit{Subtract the bottom exponent from the top exponent.}$$

$$= 2^{-3} \qquad \textit{Subtract.}$$

$$= \frac{1}{2^3} \qquad \textit{Definition of negative exponents}$$

Subtracting the exponent in the denominator from the exponent in the numerator and then using the negative exponent property gives us the same result we obtained previously. We can now continue the list of properties of exponents we started in Section 6.1.

PROPERTY Quotient Property of Exponents

If a is any real number and r and s are integers, then

$$\frac{a^r}{a^s} = a^{r-s} \qquad (a \neq 0)$$

In words: To divide with the same base, subtract the exponent in the denominator from the exponent in the numerator and raise the base to the exponent that results.

The following examples show how we use the quotient property of exponents and the negative exponent property to simplify expressions involving division.

2. Simplify each expression.

 a. $\dfrac{x^{10}}{x^4}$

 b. $\dfrac{x^5}{x^7}$

 c. $\dfrac{2^{21}}{2^{25}}$

EXAMPLE 2 Simplify the following expressions.

a. $\dfrac{x^9}{x^6}$ **b.** $\dfrac{x^4}{x^{10}}$ **c.** $\dfrac{2^{15}}{2^{20}}$

Solution

a. $\dfrac{x^9}{x^6} = x^{9-6} = x^3$

b. $\dfrac{x^4}{x^{10}} = x^{4-10} = x^{-6} = \dfrac{1}{x^6}$

c. $\dfrac{2^{15}}{2^{20}} = 2^{15-20} = 2^{-5} = \dfrac{1}{2^5} = \dfrac{1}{32}$

Answers

1. **a.** $\frac{1}{9}$ **b.** $\frac{1}{64}$ **c.** $\frac{5}{x^4}$

2. **a.** x^6 **b.** $\frac{1}{x^2}$ **c.** $\frac{1}{16}$

C Expanded Distributive Property of Exponents

Our final property of exponents is similar to the distributive property of exponents from Section 6.1, but it involves division instead of multiplication. After we have stated the property, we will give a proof of it. The proof shows why this property is true.

> **PROPERTY** Expanded Distributive Property of Exponents
>
> If a and b are any two real numbers ($b \neq 0$) and r is an integer, then
>
> $$\left(\frac{a}{b}\right)^r = \frac{a^r}{b^r}$$
>
> *In words:* A quotient raised to a power is the quotient of the powers.

Note This property may also lead to division by zero (undefined) if a is zero and r is less than zero.

Proof

$$\left(\frac{a}{b}\right)^r = \left(a \cdot \frac{1}{b}\right)^r \qquad \text{Definition of division}$$

$$= a^r \cdot \left(\frac{1}{b}\right)^r \qquad \text{Distributive property of exponents}$$

$$= a^r \cdot b^{-r} \qquad \text{Negative exponent property}$$

$$= a^r \cdot \frac{1}{b^r} \qquad \text{Negative exponent property}$$

$$= \frac{a^r}{b^r} \qquad \text{Multiply.}$$

EXAMPLE 3 Simplify the following expressions:

a. $\left(\dfrac{x}{2}\right)^3$ **b.** $\left(\dfrac{5}{y}\right)^2$ **c.** $\left(\dfrac{2}{3}\right)^4$

Solution

a. $\left(\dfrac{x}{2}\right)^3 = \dfrac{x^3}{2^3} = \dfrac{x^3}{8}$

b. $\left(\dfrac{5}{y}\right)^2 = \dfrac{5^2}{y^2} = \dfrac{25}{y^2}$

c. $\left(\dfrac{2}{3}\right)^4 = \dfrac{2^4}{3^4} = \dfrac{16}{81}$

3. Simplify each expression.

a. $\left(\dfrac{x}{5}\right)^2$

b. $\left(\dfrac{2}{a}\right)^3$

c. $\left(\dfrac{3}{4}\right)^3$

D Zero and One as Exponents

Recall the definition of 0 and 1 as exponents from Chapter 0. We will review the concept here and justify why we use them as defined.

To obtain an expression for x^1, we will solve a problem two different ways:

$$\left.\begin{array}{l} \dfrac{x^3}{x^2} = \dfrac{x \cdot x \cdot x}{x \cdot x} = x \\[2mm] \dfrac{x^3}{x^2} = x^{3-2} = x^1 \end{array}\right\} \qquad \text{Hence } x^1 = x.$$

Stated generally, this rule says that $a^1 = a$. This seems reasonable and we will use it since it is consistent with our property of division using the same base.

We use the same procedure to obtain an expression for x^0

$$\left.\begin{array}{l} \dfrac{5^2}{5^2} = \dfrac{25}{25} = 1 \\[2mm] \dfrac{5^2}{5^2} = 5^{2-2} = 5^0 \end{array}\right\} \qquad \text{Hence } 5^0 = 1.$$

Answers

3. a. $\dfrac{x^2}{25}$ **b.** $\dfrac{8}{a^3}$ **c.** $\dfrac{27}{64}$

It seems, therefore, that the best definition of x^0 is 1 for all bases x except $x = 0$. In the case of $x = 0$, we have 0^0, which we will not define. This definition will probably seem awkward at first. Most people would like to define x^0 as 0 when they first encounter it. Remember, the zero in this expression is an exponent, so x^0 does not mean to multiply by zero. Thus, we can make the general statement that $a^0 = 1$ for all real numbers except $a = 0$.

Here are some examples involving the exponents 0 and 1.

EXAMPLE 4 Simplify the following expressions.

a. 8^0 **b.** 8^1 **c.** $4^0 + 4^1$ **d.** $(2x^2y)^0$

Solution

a. $8^0 = 1$

b. $8^1 = 8$

c. $4^0 + 4^1 = 1 + 4 = 5$

d. $(2x^2y)^0 = 1$

E Combinations of Properties

Here is a summary of the definitions and properties of exponents we have developed so far. For each definition or property in the list, a and b are real numbers, and r and s are integers.

Definitions of Exponents	Properties of Exponents	
$a^1 = a$	**1.** $a^r \cdot a^s = a^{r+s}$	Product property of exponents
$a^0 = 1 \quad a \neq 0$	**2.** $(a^r)^s = a^{rs}$	Power property of exponents
	3. $(ab)^r = a^r b^r$	Distributive property of exponents
	4. $a^{-r} = \dfrac{1}{a^r} = \left(\dfrac{1}{a}\right)^r \quad a \neq 0$	Negative exponent property
	5. $\dfrac{a^r}{a^s} = a^{r-s} \quad a \neq 0$	Quotient property of exponents
	6. $\left(\dfrac{a}{b}\right)^r = \dfrac{a^r}{b^r} \quad b \neq 0$	Expanded distributive property of exponents

Here are some additional examples. These examples use a combination of the preceding properties and definitions.

EXAMPLE 5 Simplify each expression. Write all answers with positive exponents.

a. $\dfrac{(5x^3)^2}{x^4} = \dfrac{25x^6}{x^4}$ *Power and distributive properties of exponents*

$\qquad = 25x^2$ *Quotient property of exponents*

b. $\dfrac{x^{-8}}{(x^2)^3} = \dfrac{x^{-8}}{x^6}$ *Power property of exponents*

$\qquad = x^{-8-6}$ *Quotient property of exponents*

$\qquad = x^{-14}$ *Subtract.*

$\qquad = \dfrac{1}{x^{14}}$ *Negative exponent property*

4. Simplify each expression.

 a. 10^0

 b. 10^1

 c. $6^1 + 6^0$

 d. $(3x^5y^2)^0$

5. Simplify, and write your answers with positive exponents only.

 a. $\dfrac{x^{-6}}{(x^3)^4}$

 b. $\dfrac{(2x^3)^2}{x^4}$

 c. $\left(\dfrac{y^8}{y^3}\right)^2$

 d. $(2x^4)^{-2}$

 e. $x^{-6} \cdot x^2$

 f. $\dfrac{a^5(a^{-2})^4}{(a^{-3})^2}$

Answers

4. a. 1 **b.** 10 **c.** 7 **d.** 1

5. a. $\dfrac{1}{x^{18}}$ **b.** $4x^2$ **c.** y^{10} **d.** $\dfrac{1}{4x^8}$

 e. $\dfrac{1}{x^4}$ **f.** a^3

c. $\left(\dfrac{y^5}{y^3}\right)^2 = \dfrac{(y^5)^2}{(y^3)^2}$ *Expanded distributive property of exponents*

 $= \dfrac{y^{10}}{y^6}$ *Power property of exponents*

 $= y^4$ *Quotient property of exponents*

Notice in Example 5c that we could have simplified inside the parentheses first and then raised the result to the second power:

$$\left(\frac{y^5}{y^3}\right)^2 = (y^{5-3})^2 = (y^2)^2 = y^4$$

d. $(3x^5)^{-2} = \dfrac{1}{(3x^5)^2}$ *Negative exponent property*

 $= \dfrac{1}{9x^{10}}$ *Power and distributive properties of exponents*

e. $x^{-8} \cdot x^5 = x^{-8+5}$ *Product property of exponents*

 $= x^{-3}$ *Add.*

 $= \dfrac{1}{x^3}$ *Negative exponent property*

f. $\dfrac{(a^3)^2 a^{-4}}{(a^{-4})^3} = \dfrac{a^6 a^{-4}}{a^{-12}}$ *Power property of exponents*

 $= \dfrac{a^2}{a^{-12}}$ *Product property of exponents*

 $= a^{14}$ *Quotient property of exponents*

In the next two examples we use division to compare the area and volume of geometric figures.

EXAMPLE 6 Suppose you have two squares, one of which is larger than the other. If the length of a side of the larger square is 3 times as long as the length of a side of the smaller square, how many of the smaller squares will it take to cover up the larger square?

Solution If we let x represent the length of a side of the smaller square, then the length of a side of the larger square is $3x$. The area of each square, along with a diagram of the situation, is given in Figure 1.

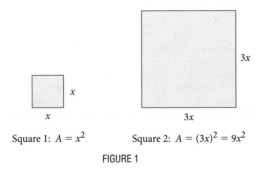

Square 1: $A = x^2$ Square 2: $A = (3x)^2 = 9x^2$

FIGURE 1

To find out how many smaller squares it will take to cover up the larger square, we divide the area of the larger square by the area of the smaller square.

$$\frac{\text{Area of square 2}}{\text{Area of square 1}} = \frac{9x^2}{x^2} = 9$$

It will take 9 of the smaller squares to cover the larger square.

6. Suppose the side of the larger square in Example 6 is 4 times as long as the side of the smaller square. How many smaller squares will it take to cover the larger square?

Answer

6. It will take 16 smaller squares to cover the larger square.

7. Suppose the side of the larger box in Example 7 is 4 times as long as the side of the smaller box. How many of the smaller boxes can be fit into the larger box?

EXAMPLE 7 Suppose you have two boxes, each of which is a cube. If the length of a side in the second box is 3 times as long as the length of a side of the first box, how many of the smaller boxes will fit inside the larger box?

Solution If we let x represent the length of a side of the smaller box, then the length of a side of the larger box is $3x$. The volume of each box, along with a diagram of the situation, is given in Figure 2.

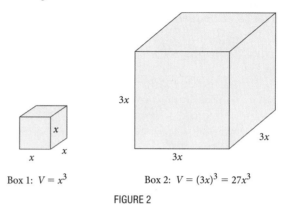

Box 1: $V = x^3$ Box 2: $V = (3x)^3 = 27x^3$

FIGURE 2

To find out how many smaller boxes will fit inside the larger box, we divide the volume of the larger box by the volume of the smaller box.

$$\frac{\text{Volume of box 2}}{\text{Volume of box 1}} = \frac{27x^3}{x^3} = 27$$

We can fit 27 of the smaller boxes inside the larger box.

F Scientific Notation for Small Numbers

Now that we have completed our list of definitions and properties of exponents, we can expand the work we did previously with scientific notation.

Recall that a number is in scientific notation when it is written in the form

$$n \times 10^r$$

where $1 \leq n < 10$ and r is an integer.

Since negative exponents give us reciprocals, we can use negative exponents to write very small numbers in scientific notation. For example, the number 0.00057, when written in scientific notation, is equivalent to 5.7×10^{-4}. Here's why:

$$5.7 \times 10^{-4} = 5.7 \times \frac{1}{10^4} = 5.7 \times \frac{1}{10,000} = \frac{5.7}{10,000} = 0.00057$$

The table below lists some other numbers in both scientific notation and expanded form.

Number written the long way		Same number written in scientific notation
376,000	=	3.76×10^5
49,500	=	4.95×10^4
3,200	=	3.2×10^3
591	=	5.91×10^2
46	=	4.6×10^1
8	=	8×10^0
0.47	=	4.7×10^{-1}
0.093	=	9.3×10^{-2}
0.00688	=	6.88×10^{-3}
0.0002	=	2×10^{-4}
0.000098	=	9.8×10^{-5}

Answer

7. The larger box will hold 64 smaller boxes.

Notice that in each case, when the number is written in scientific notation, the decimal point in the first number is placed so that the number is between 1 and 10. The exponent on 10 in the second number keeps track of the number of places we moved the decimal point in the original number to get a number between 1 and 10.

$$376,000 = 3.76 \times 10^5$$

Moved 5 places.

Decimal point was originally here.

Keeps track of the 5 places we moved the decimal point.

$$0.00688 = 6.88 \times 10^{-3}$$

Moved 3 places.

Keeps track of the 3 places we moved the decimal point.

EXAMPLE 8 Write 0.000584 in scientific notation.

Solution We move the decimal point 4 place to the right so that 0.000584 will be a product of a number between 1 and 10 and a power of 10. The decimal point appears between the 5 and the 8, and we multiply the result by 10^{-4}.

$$0.000584 = 5.84 \times 10^{-4}$$

Decimal point was originally here.

Moved 4 places.

Keeps track of the 4 places we moved the decimal point.

8. Write 0.00273 in scientific notation.

GETTING READY FOR CLASS

After reading through the preceding section, respond in your own words and in complete sentences.

A. How do you divide two expressions containing exponents when they each have the same base?

B. Explain the difference between 3^2 and 3^{-2}.

C. Explain what happens when you use 0 as an exponent.

D. What does a negative exponent mean in scientific notation?

Vocabulary Review

Choose the correct words to fill in the blanks below.

exponents quotient expanded negative

1. The _____ exponent property states that $a^{-r} = \frac{1}{a^r} = \left(\frac{1}{a}\right)^r$ if r is a positive integer and a does not equal zero.

2. The _____ property for exponents allows you to divide with the same base by subtracting the exponent in the denominator from the exponent in the numerator and raise the base to the exponent that results.

3. The _____ distributive property for exponents states that $\left(\frac{1}{a}\right)^r = \frac{a}{b^r}$ if a and b are any two real numbers, b does not equal zero, and r is an integer.

4. Negative _____ used in scientific notation represent very small numbers.

Problems

A Write each of the following with positive exponents, and then simplify, when possible.

1. 3^{-2} **2.** 3^{-3} **3.** 6^{-2} **4.** 2^{-6}

5. 8^{-2} **6.** 3^{-4} **7.** 5^{-3} **8.** 9^{-2}

9. $2x^{-3}$ **10.** $5x^{-1}$ **11.** $(2x)^{-3}$ **12.** $(5x)^{-1}$

13. $(5y)^{-2}$ **14.** $5y^{-2}$ **15.** 10^{-2} **16.** 10^{-3}

17. Complete the following table.

Number x	Square x^2	Power of 2 2^x
-3		
-2		
-1		
0		
1		
2		
3		

18. Complete the following table.

Number x	Cube x^3	Power of 3 3^x
-3		
-2		
-1		
0		
1		
2		
3		

B Use the quotient property of exponents to simplify each of the following expressions. Write all answers that contain exponents with positive exponents only.

19. $\dfrac{5^1}{5^3}$

20. $\dfrac{7^6}{7^8}$

21. $\dfrac{x^{10}}{x^4}$

22. $\dfrac{x^4}{x^{10}}$

23. $\dfrac{4^3}{4^0}$

24. $\dfrac{4^0}{4^3}$

25. $\dfrac{(2x)^7}{(2x)^4}$

26. $\dfrac{(2x)^4}{(2x)^7}$

27. $\dfrac{6^{11}}{6}$

28. $\dfrac{8^7}{8}$

29. $\dfrac{6}{6^{11}}$

30. $\dfrac{8}{8^7}$

31. $\dfrac{2^{-5}}{2^3}$

32. $\dfrac{2^{-5}}{2^{-3}}$

33. $\dfrac{2^5}{2^{-3}}$

34. $\dfrac{2^{-3}}{2^{-5}}$

35. $\dfrac{(3x)^{-5}}{(3x)^{-8}}$

36. $\dfrac{(2x)^{-10}}{(2x)^{-15}}$

37. $\dfrac{(-4x)^{10}}{(-4x)^8}$

38. $\dfrac{(-3x)^4}{(-3x)^7}$

C Use the expanded distributive property of exponents to simplify each of the following expressions. Write all answers with positive exponents only.

39. $\left(\dfrac{x}{3}\right)^3$

40. $\left(\dfrac{x}{2}\right)^4$

41. $\left(\dfrac{4}{y}\right)^2$

42. $\left(\dfrac{-6}{y}\right)^2$

43. $\left(-\dfrac{5}{x}\right)^3$

44. $\left(\dfrac{-4}{y}\right)^2$

45. $\left(\dfrac{-8}{y}\right)^2$

46. $\left(\dfrac{-3}{a}\right)^4$

D Simplify the following expressions.

47. $(-27x)^0$

48. $(-9y^2)^0$

49. $(-2y)^1$

50. $(-8y^2)^1$

51. $(-3ab^2)^1$

52. $(-4x^2y)^1$

53. $(-2xy^2)^0$

54. $\left(\dfrac{x^3}{y^4}\right)^0$

E Simplify the following expressions. Any answers that contain exponents should contain positive exponents only.

55. $(3xy)^4$

56. $(4xy)^3$

57. 10^0

58. 10^1

59. $(2a^2b)^1$

60. $(2a^2b)^0$

61. $(7y^3)^{-2}$

62. $(5y^4)^{-2}$

63. $x^{-3}x^{-5}$

64. $x^{-6} \cdot x^8$

65. $y^7 \cdot y^{-10}$

66. $y^{-4} \cdot y^{-6}$

67. $\dfrac{(x^2)^3}{x^4}$

68. $\dfrac{(x^5)^3}{x^{10}}$

69. $\dfrac{(a^4)^3}{(a^3)^2}$

70. $\dfrac{(a^5)^3}{(a^5)^2}$

71. $\dfrac{y^7}{(y^2)^8}$

72. $\dfrac{y^2}{(y^3)^4}$

73. $\left(\dfrac{y^7}{y^2}\right)^8$

74. $\left(\dfrac{y^2}{y^5}\right)^4$

75. $\dfrac{(x^{-2})^3}{x^{-5}}$

76. $\dfrac{(x^2)^{-3}}{x^{-5}}$

77. $\left(\dfrac{x^{-2}}{x^{-5}}\right)^3$

78. $\left(\dfrac{x^2}{x^{-5}}\right)^{-3}$

79. $\left(\dfrac{2a^{-3}b^2}{4a^{-5}b^6}\right)^{-2}$

80. $\left(\dfrac{3a^{-4}b^3}{2a^{-3}b^4}\right)^{-3}$

81. $\dfrac{(a^3)^2(a^4)^5}{(a^5)^2}$

82. $\dfrac{(a^4)^8(a^2)^5}{(a^3)^4}$

83. $\dfrac{(a^{-2})^3(a^4)^2}{(a^{-3})^{-2}}$

84. $\dfrac{(a^{-5})^{-3}(a^7)^{-1}}{(a^{-3})^5}$

85. $\dfrac{(3x^{-2}y^{-4})(4x^3y^{-2})^{-2}}{(3x^3y^{-2})^{-2}}$

86. $\dfrac{(4x^{-7}y^{-3})(2x^4y^{-3})^{-1}}{(2x^2y^{-2})^{-2}}$

87. Complete the following table, and then construct a line graph of the information in the table.

Number x	Power of x 2^x
-3	
-2	
-1	
0	
1	
2	
3	

88. Complete the following table, and then construct a line graph of the information in the table.

Number x	Power of x 3^x
-3	
-2	
-1	
0	
1	
2	
3	

F Write each of the following numbers in scientific notation.

89. 0.0048

90. 0.000048

91. 25

92. 35

93. 0.000009

94. 0.0009

95. -0.000087

96. -0.00000514

Write each of the following numbers in standard form.

97. 4.23×10^{-3}

98. 4.23×10^{3}

99. 8×10^{-5}

100. 8×10^{5}

101. 4.2×10^{0}

102. 4.2×10^{1}

103. -3.6×10^{-3}

104. -2.05×10^{-4}

105. Complete the following table.

Standard Form	Scientific Notation $n \times 10^r$
0.000357	3.57×10^{-4}
0.00357	
0.0357	
0.357	
3.57	
35.7	
357	
3,570	
35,700	

106. Complete the following table.

Standard Form	Scientific Notation $n \times 10^r$
0.000123	1.23×10^{-4}
	1.23×10^{-3}
	1.23×10^{-2}
	1.23×10^{-1}
	1.23×10^{0}
	1.23×10^{1}
	1.23×10^{2}
	1.23×10^{3}
	1.23×10^{4}

Applying the Concepts

107. Some home computers can do a calculation in 2×10^{-3} seconds. Write this number in standard form.

108. Some of the cells in the human body have a radius of 3×10^{-5} inches. Write this number in standard form.

109. **Margin of Victory** The chart below shows some close finishes from the 2008 Beijing Olympic games. Write each number in scientific notation.

110. Some cameras used in scientific research can take one picture every 0.000000167 second. Write this number in scientific notation.

111. The number 25×10^3 is not in scientific notation because 25 is larger than 10. Write 25×10^3 in scientific notation.

112. The number 0.25×10^3 is not in scientific notation because 0.25 is less than 1. Write 0.25×10^3 in scientific notation.

113. The number 23.5×10^4 is not in scientific notation because 23.5 is not between 1 and 10. Rewrite 23.5×10^4 in scientific notation.

114. The number 375×10^3 is not in scientific notation because 375 is not between 1 and 10. Rewrite 375×10^3 in scientific notation.

115. The number 0.82×10^{-3} is not in scientific notation because 0.82 is not between 1 and 10. Rewrite 0.82×10^{-3} in scientific notation.

116. The number 0.93×10^{-2} is not in scientific notation because 0.93 is not between 1 and 10. Rewrite 0.93×10^{-2} in scientific notation.

Comparing Areas Suppose you have two squares, one of which is larger than the other. Suppose further that the side of the larger square is twice as long as the side of the smaller square.

117. If the length of the side of the smaller square is 10 inches, give the area of each square. Then find the number of smaller squares it will take to cover the larger square.

118. How many smaller squares will it take to cover the larger square if the length of the side of the smaller square is 1 foot?

119. If the length of the side of the smaller square is x, find the area of each square. Then find the number of smaller squares it will take to cover the larger square.

120. Suppose the length of the side of the larger square is 1 foot. How many smaller squares will it take to cover the larger square?

Comparing Volumes Suppose you have two boxes, each of which is a cube. Suppose further that the length of a side of the second box is twice as long as the length of a side of the first box.

121. If the length of a side of the first box is 6 inches, give the volume of each box. Then find the number of smaller boxes that will fit inside the larger box.

122. How many smaller boxes can be placed inside the larger box if the length of a side of the second box is 1 foot?

123. If the length of a side of the first box is x, find the volume of each box. Then find the number of smaller boxes that will fit inside the larger box.

124. Suppose the length of a side of the larger box is 12 inches. How many smaller boxes will fit inside the larger box?

125. The radius of Earth is 6,378 km, while the radius of Jupiter is 71,472 km. How many Earths would fit inside of Jupiter? Answer this by comparing volume. The volume can be calculated using the formula $\frac{4}{3}\pi r^3$.

126. The radius of Jupiter is 71,472 km, while the radius of the sun is 695,000 km. How many Jupiters would fit inside the sun? Answer this by comparing their volumes. The volume can be calculated using the formula $\frac{4}{3}\pi r^3$.

Getting Ready for the Next Section

Simplify.

127. $3(4.5)$

128. $\dfrac{1}{2} \cdot \dfrac{5}{7}$

129. $\dfrac{4}{5}(10)$

130. $\dfrac{9.6}{3}$

131. $6.8(3.9)$

132. $9 - 20$

133. $-3 + 15$

134. $2x \cdot x \cdot \dfrac{1}{2}x$

135. $x^5 \cdot x^3$

136. $y^2 \cdot y$

137. $\dfrac{x^3}{(x^2)}$

138. $\dfrac{x^2}{x}$

139. $\dfrac{y^3}{y^5}$

140. $\dfrac{x^2}{x^5}$

141. $\dfrac{x^4}{(x^2)^2}$

142. $\dfrac{(y^2)^3}{y^6}$

Write in expanded form.

143. 3.4×10^2

144. 6.0×10^{-4}

145. 3.5×10^{-5}

146. -2.1×10^4

Find the Mistake

Each sentence below contains a mistake. Circle the mistake and write the correct expression on the line provided.

1. The negative exponent property says that $x^{-y} = \dfrac{y}{x}$. _____

2. When using the quotient property of exponents to simplify the expression $\dfrac{y^{12}}{y^8}$, we get y^{20}. _____

3. When using the expanded distributive property of exponents to simplify the expression $\left(\dfrac{x}{3}\right)^4$, we get $\dfrac{x^4}{3}$. _____

4. The number 0.00238 written in scientific notation is 2.38×10^{-2}. _____

6.3 Operations with Monomials

OBJECTIVES

A Use rules of exponents to multiply monomials.

B Use rules of exponents to divide monomials.

C Multiply and divide numbers written in scientific notation.

D Add and subtract monomials.

KEY WORDS

monomial

numerical coefficient

terms

similar terms

A preschool teacher needs to cover the wood floors in two of her home's bedrooms with carpet to provide a softer play surface for her students. She hires a carpet installer who creates a formula to know how much carpet two bedrooms in the house needs. The first room is in the shape of a square with a wall the length of x. The second room is the shape of a rectangle, with one wall the same length as the first room but a second wall twice as long. Using our knowledge of area, we can write two monomials that each represent the area of the rooms.

Square room area: $x \cdot x = x^2$

Rectangular room area: $x \cdot 2x = 2x^2$

Later in this section, we can add these monomials to determine how much carpet the installer will need. But first, we will discuss multiplication and division of monomials.

> **DEFINITION** monomial
>
> A *monomial* is a one-term expression that is either a constant (number) or the product of a constant and one or more variables raised to whole number exponents.

The following are examples of monomials:

$$-3 \qquad 15x \qquad -23x^2y \qquad 49x^4y^2z^4 \qquad \frac{3}{4}a^2b^3$$

The numerical part of each monomial is called the *numerical coefficient*, or just *coefficient*. Monomials are also called *terms*.

A Multiplying Monomials

There are two basic steps involved in the multiplication of monomials. First, we rewrite the products using the commutative and associative properties. Then, we simplify by multiplying coefficients and adding exponents of like bases.

EXAMPLE 1 Multiply each expression.

a. $(-3x^2)(4x^3)$ **b.** $\left(\frac{4}{5}x^5 \cdot y^2\right)(10x^3 \cdot y)$

Solution

a. $(-3x^2)(4x^3) = (-3 \cdot 4)(x^2 \cdot x^3)$ *Commutative and associative properties*

$= -12x^5$ *Multiply coefficients, add exponents.*

b. $\left(\frac{4}{5}x^5 \cdot y^2\right)(10x^3 \cdot y) = \left(\frac{4}{5} \cdot 10\right)(x^5 \cdot x^3)(y^2 \cdot y)$ *Commutative and associative properties*

$= 8x^8y^3$ *Multiply coefficients, add exponents.*

**Video Examples
Section 6.3**

Practice Problems

1. Multiply.

 a. $(-2x^3)(5x^2)$

 b. $(4x^6y^3)(2xy^2)$

Answers

1. **a.** $-10x^5$ **b.** $8x^7y^5$

You can see that in each case the work was the same—multiply coefficients and add exponents of the same base.

B Dividing Monomials

We can expect division of monomials to proceed in a similar way. Since our properties are consistent, division of monomials will result in division of coefficients and subtraction of exponents of like bases.

EXAMPLE 2 Divide each expression.

a. $\dfrac{15x^3}{3x^2}$ **b.** $\dfrac{39x^2y^3}{3xy^5}$

Solution

a. $\dfrac{15x^3}{3x^2} = \dfrac{15}{3} \cdot \dfrac{x^3}{x^2}$ *Write as separate fractions.*

$= 5x$ *Divide coefficients, subtract exponents.*

b. $\dfrac{39x^2y^3}{3xy^5} = \dfrac{39}{3} \cdot \dfrac{x^2}{x} \cdot \dfrac{y^3}{y^5}$ *Write as separate fractions.*

$= 13x \cdot \dfrac{1}{y^2}$ *Divide coefficients, subtract exponents.*

$= \dfrac{13x}{y^2}$ *Write answer as a single fraction.*

In Example 2b, the expression $\frac{y^3}{y^5}$ simplifies to $\frac{1}{y^2}$ because of the quotient property of exponents and the negative exponent property. If we were to show all the work in this simplification process, it would look like this:

$$\dfrac{y^3}{y^5} = y^{3-5}$$ *Quotient property of exponents*

$$= y^{-2}$$ *Subtract.*

$$= \dfrac{1}{y^2}$$ *Negative exponent property*

The point of this explanation is this: Even though we may not show all the steps when simplifying an expression involving exponents, the result we obtain still can be justified using the properties of exponents. We have not introduced any new properties in Example 2; we have just not shown the details of each simplification.

EXAMPLE 3 Divide $25a^5b^3$ by $50a^2b^7$.

Solution

$$\dfrac{25a^5b^3}{50a^2b^7} = \dfrac{25}{50} \cdot \dfrac{a^5}{a^2} \cdot \dfrac{b^3}{b^7}$$ *Write as separate fractions.*

$$= \dfrac{1}{2} \cdot a^3 \cdot \dfrac{1}{b^4}$$ *Divide coefficients, subtract exponents.*

$$= \dfrac{a^3}{2b^4}$$ *Write answer as a single fraction.*

Notice in Example 3 that dividing 25 by 50 results in $\frac{1}{2}$. This is the same result we would obtain if we reduced the fraction $\frac{25}{50}$ to lowest terms, and there is no harm in thinking of it that way. Also, notice that the expression $\frac{b^3}{b^7}$ simplifies to $\frac{1}{b^4}$ by the quotient property of exponents and the negative exponent property, even though we have not shown the steps involved in doing so.

2. Divide.

a. $\dfrac{27x^4y^3}{9xy^2}$

b. $\dfrac{25x^4}{5x}$

3. Divide $\dfrac{13a^6b^2}{39a^4b^7}$.

Answers

2. a. $3x^3y$ **b.** $5x^3$

3. $\dfrac{a^2}{3b^5}$

C Multiplication and Division of Numbers Written in Scientific Notation

We multiply and divide numbers written in scientific notation using the same steps we used to multiply and divide monomials.

EXAMPLE 4 Multiply $(4 \times 10^7)(2 \times 10^{-4})$.

Solution Since multiplication is commutative and associative, we can rearrange the order of these numbers and group them as follows:

$$(4 \times 10^7)(2 \times 10^{-4}) = (4 \times 2)(10^7 \times 10^{-4})$$
$$= 8 \times 10^3$$

Notice that we add exponents, $7 + (-4) = 3$, when we multiply with the same base.

EXAMPLE 5 Divide $\dfrac{9.6 \times 10^{12}}{3 \times 10^4}$.

Solution We group the numbers between 1 and 10 separately from the powers of 10 and proceed as we did in Example 4:

$$\frac{9.6 \times 10^{12}}{3 \times 10^4} = \frac{9.6}{3} \times \frac{10^{12}}{10^4}$$
$$= 3.2 \times 10^8$$

Notice that the procedure we used in both of these examples is very similar to multiplication and division of monomials, for which we multiplied or divided coefficients and added or subtracted exponents.

D Addition and Subtraction of Monomials

Addition and subtraction of monomials will be almost identical since subtraction is defined as addition of the opposite. With multiplication and division of monomials, the key was rearranging the numbers and variables using the commutative and associative properties. With addition, the key is application of the distributive property. We sometimes use the phrase *combine monomials* to describe addition and subtraction of monomials.

> **DEFINITION** similar terms
>
> Two monomials with the same variable part (same variables raised to the same powers) are called *similar* (or *like*) *terms*.

You can add only similar terms. This is because the distributive property (which is the key to addition of monomials) cannot be applied to terms that are not similar.

EXAMPLE 6 Combine the following monomials.
a. $-3x^2 + 15x^2$ **b.** $9x^2y - 20x^2y$ **c.** $5x^2 + 8y^2$

solution
a. $-3x^2 + 15x^2 = (-3 + 15)x^2$ Distributive property
$\qquad\qquad\quad = 12x^2$ Add coefficients.

b. $9x^2y - 20x^2y = (9 - 20)x^2y$ Distributive property
$\qquad\qquad\quad\ \ = -11x^2y$ Add coefficients.

c. $5x^2 + 8y^2$ In this case, we cannot apply the distributive property, so we cannot add the monomials.

4. Multiply $(3 \times 10^6)(3 \times 10^{-8})$.

5. Divide $\dfrac{4.8 \times 10^{20}}{2 \times 10^{12}}$.

6. Combine if possible.

 a. $-4x^2 + 9x^2$

 b. $12x^2y - 15x^2y$

 c. $6x^2 + 9x^3$

Answers
4. 9×10^{-2}
5. 2.4×10^8
6. a. $5x^2$ **b.** $-3x^2y$
 c. Cannot be combined

The next examples show how we simplify expressions containing monomials when more than one operation is involved.

7. Apply the distributive property.

a. $a^2\left(2 - \dfrac{4}{a}\right)$

b. $xy\left(\dfrac{3}{x} + \dfrac{2}{y}\right)$

EXAMPLE 7 Apply the distributive property to simplify.

a. $x^2\left(1 - \dfrac{6}{x}\right)$ **b.** $ab\left(\dfrac{1}{b} - \dfrac{1}{a}\right)$

Solution

a. $x^2\left(1 - \dfrac{6}{x}\right) = x^2 \cdot 1 - x^2 \cdot \dfrac{6}{x} = x^2 - \dfrac{6x^2}{x} = x^2 - 6x$

b. $ab\left(\dfrac{1}{b} - \dfrac{1}{a}\right) = ab \cdot \dfrac{1}{b} - ab \cdot \dfrac{1}{a} = \dfrac{ab}{b} - \dfrac{ab}{a} = a - b$

8. Simplify $\dfrac{(10x^3y^2)(6xy^4)}{12x^2y^3}$.

EXAMPLE 8 Simplify $\dfrac{(6x^4y)(3x^7y^5)}{9x^5y^2}$.

Solution We begin by multiplying the two monomials in the numerator.

$$\dfrac{(6x^4y)(3x^7y^5)}{9x^5y^2} = \dfrac{18x^{11}y^6}{9x^5y^2} \qquad \text{Simplify numerator.}$$

$$= 2x^6y^4 \qquad \text{Divide.}$$

9. Simplify $\dfrac{(1.2 \times 10^6)(6.3 \times 10^{-5})}{6 \times 10^{-10}}$.

EXAMPLE 9 Simplify $\dfrac{(6.8 \times 10^5)(3.9 \times 10^{-7})}{7.8 \times 10^{-4}}$.

Solution We group the numbers between 1 and 10 separately from the powers of 10.

$$\dfrac{(6.8)(3.9)}{7.8} \times \dfrac{(10^5)(10^{-7})}{10^{-4}} = 3.4 \times 10^{5+(-7)-(-4)}$$

$$= 3.4 \times 10^2$$

10. Simplify $\dfrac{24x^7}{3x^2} + \dfrac{14x^9}{7x^4}$.

EXAMPLE 10 Simplify $\dfrac{14x^5}{2x^2} + \dfrac{15x^8}{3x^5}$.

Solution Simplifying each expression separately and then combining similar terms gives

$$\dfrac{14x^5}{2x^2} + \dfrac{15x^8}{3x^5} = 7x^3 + 5x^3 \qquad \text{Divide.}$$

$$= 12x^3 \qquad \text{Add.}$$

> **GETTING READY FOR CLASS**
>
> *After reading through the preceding section, respond in your own words and in complete sentences.*
>
> **A.** What is a monomial?
>
> **B.** Describe how you would multiply $3x^2$ and $5x^2$.
>
> **C.** Describe how you would add $3x^2$ and $5x^2$.
>
> **D.** Describe how you would multiply two numbers written in scientific notation.

Answers

7. a. $2a^2 - 4a$ **b.** $3y + 2x$

8. $5x^2y^3$

9. 1.26×10^{11}

10. $10x^5$

Vocabulary Review

Choose the correct words to fill in the blanks below.

 terms coefficient subtraction variable constant

1. A monomial is a one-term expression that is either a _____ or the product of a constant and one or more variables raised to whole number exponents.

2. Monomials are also called _____.

3. The numerical _____ of $7xy^2$ is 7.

4. Two monomials with the same _____ part are called similar terms.

5. Division of monomials will result in division of coefficients and _____ of exponents of like bases.

Problems

A Multiply each expression and simplify.

1. $(3x^4)(4x^3)$ **2.** $(6x^5)(-2x^2)$ **3.** $(-2y^4)(8y^7)$ **4.** $(5y^{10})(2y^5)$

5. $(8x)(4x)$ **6.** $(7x)(5x)$ **7.** $(10a^3)(10a)(2a^2)$ **8.** $(5a^4)(10a)(10a^4)$

9. $(6ab^2)(-4a^2b)$ **10.** $(-5a^3b)(4ab^4)$ **11.** $(4x^2y)(3x^3y^3)(2xy^4)$ **12.** $(5x^6)(-10xy^4)(-2x^2y^6)$

B Divide each expression and simplify. Write all answers with positive exponents only.

13. $\dfrac{15x^3}{5x^2}$ **14.** $\dfrac{25x^5}{5x^4}$ **15.** $\dfrac{18y^9}{3y^{12}}$ **16.** $\dfrac{24y^4}{8y^7}$

17. $\dfrac{32a^3}{64a^4}$ **18.** $\dfrac{25a^5}{75a^6}$ **19.** $\dfrac{21a^2b^3}{-7ab^5}$ **20.** $\dfrac{32a^5b^6}{8ab^5}$

21. $\dfrac{3x^3y^2z}{27xy^2z^3}$ **22.** $\dfrac{5x^5y^4z}{30x^3yz^2}$ **23.** $\dfrac{-4x^2y^3z}{16x^5yz^2}$ **24.** $\dfrac{-16x^3yz^2}{24xy^7z}$

25. Fill in the table.

a	b	ab	$\dfrac{a}{b}$	$\dfrac{b}{a}$
10	$5x$			
$20x^3$	$6x^2$			
$25x^5$	$5x^4$			
$3x^{-2}$	$3x^2$			
$-2y^4$	$8y^7$			

26. Fill in the table.

a	b	ab	$\dfrac{a}{b}$	$\dfrac{b}{a}$
$10y$	$2y^2$			
$10y^2$	$2y$			
$5y^3$	15			
5	$15y^3$			
$4y^{-3}$	$4y^3$			

Simplify. Write all answers with positive exponents only.

27. $\dfrac{(3x^2)(8x^5)}{6x^4}$

28. $\dfrac{(7x^3)(6x^8)}{14x^5}$

29. $\dfrac{(9a^2b)(2a^3b^4)}{18a^5b^7}$

30. $\dfrac{(21a^5b)(2a^8b^4)}{14ab}$

31. $\dfrac{(4x^3y^2)(9x^4y^{10})}{(3x^5y)(2x^6y)}$

32. $\dfrac{(5x^4y^4)(10x^3y^3)}{(25xy^5)(2xy^7)}$

33. $\dfrac{(-2xy^2)(-4x^3y)}{(-6xy^3)(xy)}$

34. $\dfrac{(-12a^3b^2)(3ab^3)}{(-6a^2b^2)(ab)}$

Find each product. Write all answers in scientific notation.

35. $(3 \times 10^3)(2 \times 10^5)$

36. $(4 \times 10^8)(1 \times 10^6)$

37. $(3.5 \times 10^4)(5 \times 10^{-6})$

38. $(7.1 \times 10^5)(2 \times 10^{-8})$

39. $(5.5 \times 10^{-3})(2.2 \times 10^{-4})$

40. $(3.4 \times 10^{-2})(4.5 \times 10^{-6})$

Find each quotient. Write all answers in scientific notation.

41. $\dfrac{8.4 \times 10^5}{2 \times 10^2}$

42. $\dfrac{9.6 \times 10^{20}}{3 \times 10^6}$

43. $\dfrac{6 \times 10^8}{2 \times 10^{-2}}$

44. $\dfrac{8 \times 10^{12}}{4 \times 10^{-3}}$

45. $\dfrac{2.5 \times 10^{-6}}{5 \times 10^{-4}}$

46. $\dfrac{4.5 \times 10^{-8}}{9 \times 10^{-4}}$

47. $\dfrac{-4.0 \times 10^{-6}}{8.0 \times 10^{-3}}$

48. $\dfrac{-1.5 \times 10^{-3}}{-6.0 \times 10^2}$

Simplify each expression, and write all answers in scientific notation.

49. $\dfrac{(6 \times 10^8)(3 \times 10^5)}{9 \times 10^7}$

50. $\dfrac{(8 \times 10^4)(5 \times 10^{10})}{2 \times 10^7}$

51. $\dfrac{(5 \times 10^3)(4 \times 10^{-5})}{2 \times 10^{-2}}$

52. $\dfrac{(7 \times 10^6)(4 \times 10^{-4})}{1.4 \times 10^{-3}}$

53. $\dfrac{(2.8 \times 10^{-7})(3.6 \times 10^4)}{2.4 \times 10^3}$

54. $\dfrac{(5.4 \times 10^2)(3.5 \times 10^{-9})}{4.5 \times 10^6}$

Combine by adding or subtracting similar terms as indicated.

55. $3x^2 + 5x^2$

56. $4x^3 + 8x^3$

57. $8x^5 - 19x^5$

58. $75x^6 - 50x^6$

59. $2a + a - 3a$

60. $5a + a - 6a$

61. $10x^3 - 8x^3 + 2x^3$

62. $7x^5 + 8x^5 - 12x^5$

63. $20ab^2 - 19ab^2 + 30ab^2$

64. $18a^3b^2 - 20a^3b^2 + 10a^3b^2$

65. $-3x^2y - 7x^2y - x^2y$

66. $-2a^3b + a^3b - 12a^3b$

67. Fill in the table.

a	b	ab	a + b
$5x$	$3x$		
$4x^2$	$2x^2$		
$3x^3$	$6x^3$		
$2x^4$	$-3x^4$		
x^5	$7x^5$		

68. Fill in the table.

a	b	ab	a − b
$2y$	$3y$		
$-2y$	$3y$		
$4y^2$	$5y^2$		
y^3	$-3y^3$		
$5y^4$	$7y^4$		

Apply the distributive property to simplify each expression.

69. $xy\left(x + \dfrac{1}{y}\right)$

70. $xy\left(y + \dfrac{1}{x}\right)$

71. $xy\left(\dfrac{1}{y} + \dfrac{1}{x}\right)$

72. $xy\left(\dfrac{1}{x} - \dfrac{1}{y}\right)$

73. $x^2\left(1 - \dfrac{4}{x^2}\right)$

74. $x^2\left(1 - \dfrac{9}{x^2}\right)$

75. $x^2\left(1 - \dfrac{1}{x} - \dfrac{6}{x^2}\right)$

76. $x^2\left(1 - \dfrac{5}{x} + \dfrac{6}{x^2}\right)$

77. $x^2\left(1 - \dfrac{5}{x}\right)$

78. $x^2\left(1 - \dfrac{3}{x}\right)$

79. $x^2\left(1 - \dfrac{8}{x}\right)$

80. $x^2\left(1 - \dfrac{6}{x}\right)$

Simplify.

81. $\dfrac{18x^4}{3x} + \dfrac{21x^7}{7x^4}$

82. $\dfrac{24x^{10}}{6x^4} + \dfrac{32x^7}{8x}$

83. $\dfrac{45a^6}{9a^4} - \dfrac{50a^8}{2a^6}$

84. $\dfrac{16a^9}{4a} - \dfrac{28a^{12}}{4a^4}$

85. $\dfrac{6x^7y^4}{3x^2y^2} + \dfrac{8x^5y^8}{2y^6}$

86. $\dfrac{40x^{10}y^{10}}{8x^2y^5} + \dfrac{10x^8y^8}{5y^3}$

Getting Ready for the Next Section

Simplify.

87. $3 - 8$

88. $-5 + 7$

89. $-1 + 7$

90. $1 - 8$

91. $3(5)^2 + 1$

92. $3(-2)^2 - 5(-2) + 4$

93. $2x^2 + 4x^2$

94. $3x^2 - x^2$

95. $-5x + 7x$

96. $x - 2x$

97. $-(2x + 9)$

98. $-(4x^2 - 2x - 6)$

99. Find the value of $2x + 3$ when $x = 4$.

100. Find the value of $(3x)^2$ when $x = 3$.

101. Find the value of $(-2x)^2$ when $x = -3$.

102. Find the value of $(-5y)^3$ when $y = -1$.

Find the Mistake

Each sentence below contains a mistake. Circle the mistake and write the correct word(s) or expression on the line provided.

1. The coefficient of the monomial $12x^4y^2z$ is 4. _____

2. Multiplying the monomials $-15x^3y$ and $2x^2$ gives us $30x^6y$. _____

3. When dividing $28x^6y^2$ by $4x^2y$, write the problem as the separate fractions $\frac{28x^6}{4x^2} \cdot \frac{y^2}{y}$. _____

4. Monomials with similar terms have the same coefficients but different variable parts.

Landmark Review: Checking Your Progress

Simplify the following expressions. Assume all variables are positive numbers and any answers containing exponents should contain positive exponents only.

1. $x^5 \cdot x^3$

2. $y^{15} \cdot y^{12}$

3. $(x^3)^2$

4. $(a^x)^y$

5. $(3a)^3$

6. $(2y)^2(4y)^3$

7. $\dfrac{5^4}{5^2}$

8. $\dfrac{(3x)^5}{(3x)^2}$

9. $\dfrac{2^{-4}}{2^{-2}}$

10. $\dfrac{(a^2)^3}{(a^4)^2}$

11. $(4xy)^1$

12. $(14x^ny^m)^0$

13. $\dfrac{9 \times 10^5}{3 \times 10^3}$

14. $\dfrac{(3 \times 10^5)(6 \times 10^4)}{(9 \times 10^6)}$

15. $(6x^2y)(3x^3y^3)$

16. $\dfrac{(3a^3b^2c)}{(6abc^4)}$

6.4 Addition and Subtraction of Polynomials

OBJECTIVES

A Add and subtract polynomials.

B Find the value of a polynomial for a given value of the variable.

KEY WORDS

polynomial

binomial

trinomial

degree

Suppose we left the television and the ceiling fan on when we left the house for the day. The amount of energy wasted by the ceiling fan was 5 less watts per hour than the television that wasted x watts per hour. The next day, we left on the television, the ceiling fan, and the stereo, which wasted three times the amount of the television. How much total energy did we waste over the two days?

To help us solve this problem, we will arrange all our given information into a table.

| | Energy Use (Watts/Hour) | | |
	Television	Ceiling fan	Stereo
Day 1	x	$x - 5$	
Day 2	x	$x - 5$	$3x$

Next we need to learn how to add these two polynomials together to get our answer. Once you have read through this section, return to this problem and try to solve it.

In this section we will extend what we learned in Section 6.3 to expressions called polynomials. We begin this section with the definition of a polynomial.

> **DEFINITION** polynomial
>
> A *polynomial* is a finite sum of monomials.

Here are some examples of polynomials:

$$3x^2 + 2x + 1 \qquad 15x^2y + 21xy^2 - y^2 \qquad 3a - 2b + 4c - 5d$$

Polynomials can be further classified by the number of terms they contain. A polynomial with two terms is called a binomial. If it has three terms, it is a trinomial. As stated before, a monomial has only one term.

> **DEFINITION** degree
>
> The degree of a monomial is the sum of the exponents of the variables; whereas the *degree* of a polynomial in one variable is the highest power to which the variable is raised.

Various degrees of polynomials:

$$3x^5 + 2x^3 + 1 \qquad \text{A trinomial of degree 5}$$
$$2x + 1 \qquad \text{A binomial of degree 1}$$
$$3x^2 + 2x + 1 \qquad \text{A trinomial of degree 2}$$
$$3x^5 \qquad \text{A monomial of degree 5}$$
$$-9 \qquad \text{A monomial of degree 0}$$

**Video Examples
Section 6.4**

A Adding and Subtracting Polynomials

There are no new rules for adding one or more polynomials. We rely only on our previous knowledge. Here are some examples.

1. Add
$(2x^2 + 3x + 4) + (5x^2 - 6x - 3)$.

EXAMPLE 1 Add $(2x^2 - 5x + 3) + (4x^2 + 7x - 8)$.

Solution We use the commutative and associative properties to group similar terms together and then apply the distributive property to add

$$(2x^2 - 5x + 3) + (4x^2 + 7x - 8)$$
$$= (2x^2 + 4x^2) + (-5x + 7x) + (3 - 8) \qquad \text{Commutative and associative properties}$$
$$= (2 + 4)x^2 + (-5 + 7)x + (3 - 8) \qquad \text{Distributive property}$$
$$= 6x^2 + 2x - 5 \qquad \text{Add.}$$

The results here indicate that to add two polynomials, we add coefficients of similar terms.

2. Add $x^2 + 7x + 4x + 28$.

EXAMPLE 2 Add $x^2 + 3x + 2x + 6$.

Solution The only similar terms here are the two middle terms. We combine them as usual to get

$$x^2 + 3x + 2x + 6 = x^2 + 5x + 6$$

The opposite of a polynomial is the opposite of each of its terms. When you subtract one polynomial from another you subtract each of its terms.

3. Subtract
$(5x^2 + x + 2) - (x^2 + 3x + 7)$.

EXAMPLE 3 Subtract $(3x^2 + x + 4) - (x^2 + 2x + 3)$.

Solution To subtract $x^2 + 2x + 3$, we change the sign of each of its terms and add. If you are having trouble remembering why we do this, remember that we can think of $-(x^2 + 2x + 3)$ as $-1(x^2 + 2x + 3)$. If we distribute the -1 across $x^2 + 2x + 3$, we get $-x^2 - 2x - 3$.

$$(3x^2 + x + 4) - (x^2 + 2x + 3)$$
$$= 3x^2 + x + 4 - x^2 - 2x - 3 \qquad \text{Take the opposite of each term in the second polynomial.}$$
$$= (3x^2 - x^2) + (x - 2x) + (4 - 3)$$
$$= 2x^2 - x + 1$$

4. Subtract $-4x^2 - 3x + 8$ from $x^2 + x + 1$.

EXAMPLE 4 Subtract $-4x^2 + 5x - 7$ from $x^2 - x - 1$.

Solution The polynomial $x^2 - x - 1$ comes first, then the subtraction sign, and finally the polynomial $-4x^2 + 5x - 7$ in parentheses.

$$(x^2 - x - 1) - (-4x^2 + 5x - 7)$$
$$= x^2 - x - 1 + 4x^2 - 5x + 7 \qquad \text{Take the opposite of each term in the second polynomial.}$$
$$= (x^2 + 4x^2) + (-x - 5x) + (-1 + 7)$$
$$= 5x^2 - 6x + 6$$

B Finding the Value of a Polynomial

The last topic we want to consider in this section is finding the value of a polynomial for a given value of the variable.

To find the value of the polynomial $3x^2 + 1$ when x is 5, we replace x with 5 and simplify the result.

When → $x = 5$

the polynomial → $3x^2 + 1$

becomes → $3(5)^2 + 1 = 3(25) + 1$

$$= 75 + 1$$

$$= 76$$

EXAMPLE 5 Find the value of $3x^2 - 5x + 4$ when $x = -2$.

Solution

When → $x = -2$

the polynomial → $3x^2 - 5x + 4$

becomes → $3(-2)^2 - 5(-2) + 4 = 3(4) + 10 + 4$

$$= 12 + 10 + 4$$

$$= 26$$

5. Find the value of $2x^2 - x + 3$ when $x = -3$.

EXAMPLE 6 Find the value of $2x^2 + 3x + 5$ when $x = 3$.

Solution

When → $x = 3$

the polynomial → $2x^2 + 3x + 5$

becomes → $2(3)^2 + 3(3) + 5 = 2(9) + 9 + 5$

$$= 18 + 9 + 5$$

$$= 32$$

6. Find the value of $4x^2 + 2x - 3$ when $x = 5$.

GETTING READY FOR CLASS

After reading through the preceding section, respond in your own words and in complete sentences.

A. What is the difference between a binomial and a trinomial?

B. What is the degree of a polynomial?

C. Describe how you would subtract one polynomial from another.

D. How you would find the value of $3x^2 - 5x + 4$ when x is -2?

Answers

5. 24

6. 107

Vocabulary Review

Choose the correct words to fill in the blanks below.

degree similar polynomial opposite

1. A _____ is a finite sum of monomials.

2. The _____ of a polynomial in one variable is the highest power to which the variable is raised.

3. To add and subtract polynomials, first group _____ terms.

4. The _____ of a polynomial is the opposite of each of its terms.

Problems

Identify each of the following polynomials as a trinomial, binomial, or monomial, and give the degree in each case.

1. $2x^3 - 3x^2 + 1$

2. $4x^2 - 4x + 1$

3. $5 + 8a - 9a^3$

4. $6 + 12x^3 + x^4$

5. $2x - 1$

6. $4 + 7x$

7. $45x^2 - 1$

8. $3a^3 + 8$

9. $7a^2$

10. $90x$

11. -4

12. 56

A Perform the following additions and subtractions.

13. $(2x^2 + 3x + 4) + (3x^2 + 2x + 5)$

14. $(x^2 + 5x + 6) + (x^2 + 3x + 4)$

15. $(3a^2 - 4a + 1) + (2a^2 - 5a + 6)$

16. $(5a^2 - 2a + 7) + (4a^2 - 3a + 2)$

17. $x^2 + 4x + 2x + 8$

18. $x^2 + 5x - 3x - 15$

19. $6x^2 - 3x - 10x + 5$

20. $10x^2 + 30x - 2x - 6$

21. $x^2 - 3x + 3x - 9$

22. $x^2 - 5x + 5x - 25$

23. $3y^2 - 5y - 6y + 10$

24. $y^2 - 18y + 2y - 12$

25. $(6x^3 - 4x^2 + 2x) + (9x^2 - 6x + 3)$

26. $(5x^3 + 2x^2 + 3x) + (2x^2 + 5x + 1)$

27. $\left(\frac{2}{3}x^2 - \frac{1}{5}x - \frac{3}{4}\right) + \left(\frac{4}{3}x^2 - \frac{4}{5}x + \frac{7}{4}\right)$

28. $\left(\frac{3}{8}x^3 - \frac{5}{7}x^2 - \frac{2}{5}\right) + \left(\frac{5}{8}x^3 - \frac{2}{7}x^2 + \frac{7}{5}\right)$

29. $(a^2 - a - 1) - (-a^2 + a + 1)$

30. $(5a^2 - a - 6) - (-3a^2 - 2a + 4)$

31. $\left(\frac{5}{9}x^3 + \frac{1}{3}x^2 - 2x + 1\right) - \left(\frac{2}{3}x^3 + x^2 + \frac{1}{2}x - \frac{3}{4}\right)$

32. $\left(4x^3 - \frac{2}{5}x^2 + \frac{3}{8}x - 1\right) - \left(\frac{9}{2}x^3 + \frac{1}{4}x^2 - x + \frac{5}{6}\right)$

33. $(4y^2 - 3y + 2) + (5y^2 + 12y - 4) - (13y^2 - 6y + 20)$

34. $(2y^2 - 7y - 8) - (6y^2 + 6y - 8) + (4y^2 - 2y + 3)$

35. Subtract $10x^2 + 23x - 50$ from $11x^2 - 10x + 13$.

36. Subtract $2x^2 - 3x + 5$ from $4x^2 - 5x + 10$.

37. Subtract $3y^2 + 7y - 15$ from $11y^2 + 11y + 11$.

38. Subtract $15y^2 - 8y - 2$ from $3y^2 - 3y + 2$.

39. Add $50x^2 - 100x - 150$ to $25x^2 - 50x + 75$.

40. Add $7x^2 - 8x + 10$ to $-8x^2 + 2x - 12$.

41. Subtract $2x + 1$ from the sum of $3x - 2$ and $11x + 5$.

42. Subtract $3x - 5$ from the sum of $5x + 2$ and $9x - 1$.

B Find the value of the following polynomials for the given variables.

43. Find the value of the polynomial $x^2 - 2x + 1$ when x is 3. **44.** Find the value of the polynomial $(x - 1)^2$ when x is 3.

Applying the Concepts

45. Packaging A crystal ball with a diameter of 6 inches is being packaged for shipment. If the crystal ball is placed inside a circular cylinder with radius 3 inches and height 6 inches, how much volume will need to be filled with padding? Leave your answer in terms of π. (The volume of a sphere with radius r is $\frac{4}{3}\pi r^3$, and the volume of a right circular cylinder with radius r and height h is $\pi r^2 h$.)

46. Packaging Suppose the circular cylinder of Problem 45 has a radius of 4 inches and a height of 7 inches. How much volume will need to be filled with padding? Leave your answer in terms of π.

Getting Ready for the Next Section

Simplify each expression.

47. $(-5)(-1)$

48. $3(-4)$

49. $(-1)(6)$

50. $(-7) \cdot 8$

51. $(5x)(-4x)$

52. $(3x)(2x)$

53. $3x(-7)$

54. $3x(-1)$

55. $5x + (-3x)$

56. $-3x - 10x$

57. $7x - 15x$

58. $-5x - x$

Multiply.

59. $3(2x - 6)$

60. $5(2xy - 7)$

61. $-3(xy - 4)$

62. $-4(a^2 - 9)$

Find the Mistake

Each sentence below contains a mistake. Circle the mistake and write the correct expression on the line provided.

1. The degree of the polynomial $x^2 + 6 + 4x - x^7$ is 2. _____

2. Adding $x^3 + 4x^2 - 8 + x^2$ gives us $6x^5 - 8$. _____

3. The first step when subtracting $(x - 5)^2$ from $(x + 5)^2$ is to use the distributive property on the operation symbols so the problem looks like $(x + 5)^2 + (x + 5)^2$. _____

4. The value of the polynomial $7a^2 + 5a - 10$ when a is 4 is 794. _____

6.5 Multiplication of Polynomials

©iStockphoto.com/Nikada

OBJECTIVES

A Multiply a monomial with a polynomial.

B Multiply two binomials using the FOIL method.

C Multiply two polynomials using the Column method.

KEY WORDS

FOIL method

column method

revenue

Video Examples
Section 6.5

Swamp soccer is an exhilarating team sport played just like a standard game of soccer, except the field is a muddy swamp or bog. By the end of the game, most if not all players are soaked and covered head to toe in mud. Suppose you are a coach of your local swamp soccer team. You have coached the team for x years and plan to spend two more years coaching before retiring. If the amount of money you spend on uniforms each year can be given by the binomial $30x + 10$, how much will you have spent by the end of your coaching career?

We can set up this problem as the multiplication of two binomials. Therefore, if $x + 2$ equals the number of total coaching years, then we have

$$(x + 2)(30x + 10)$$

Before we explore the multiplication of two binomials further, let's look at an example of multiplying monomials with polynomials.

A Multiplying Monomials with Polynomials

We begin our discussion of multiplication of polynomials by finding the product of a monomial and a trinomial.

EXAMPLE 1 Multiply $3x^2(2x^2 + 4x + 5)$.

Solution Applying the distributive property gives us

$$3x^2(2x^2 + 4x + 5) = 3x^2(2x^2) + 3x^2(4x) + 3x^2(5) \qquad \text{Distributive property}$$
$$= 6x^4 + 12x^3 + 15x^2 \qquad \text{Multiply.}$$

B Multiplying Binomials

The distributive property is the key to multiplication of polynomials. We can use it to find the product of any two polynomials. There are some shortcuts we can use in certain situations, however. Let's look at an example that involves the product of two binomials.

EXAMPLE 2 Multiply $(3x - 5)(2x - 1)$.

Solution
$$(3x - 5)(2x - 1) = 3x(2x - 1) - 5(2x - 1)$$
$$= 3x(2x) + 3x(-1) + (-5)(2x) + (-5)(-1)$$
$$= 6x^2 - 3x - 10x + 5$$
$$= 6x^2 - 13x + 5$$

Practice Problems

1. Multiply $2x^3(4x^2 - 5x + 3)$.

2. Multiply $(2x - 4)(3x + 2)$.

Answers
1. $8x^5 - 10x^4 + 6x^3$
2. $6x^2 - 8x - 8$

If we look closely at the second and third lines of work in this example, we can see that the terms in the answer come from all possible products of terms in the first binomial with terms in the second binomial. This result is generalized as follows.

> **RULE** Multiplication of Two Polynomials
>
> To multiply any two polynomials, multiply each term in the first with each term in the second.

There are two ways we can put this rule to work, the first of which is called the FOIL method.

FOIL Method

If we look at the original problem in Example 2 and then at the answer, we see that the first term in the answer came from multiplying the first terms in each binomial.

$$3x \cdot 2x = 6x^2 \qquad \textit{First}$$

The middle term in the answer came from adding the products of the two outside terms with the two inside terms in each binomial.

$$
\begin{aligned}
3x(-1) &= -3x \qquad \textit{Outside}\\
-5(2x) &= -10x \qquad \textit{Inside}\\
\hline
&= -13x
\end{aligned}
$$

The last term in the answer came from multiplying the two last terms.

$$-5(-1) = 5 \qquad \textit{Last}$$

To summarize the FOIL method, we will multiply another two binomials.

EXAMPLE 3 Multiply $(2x + 3)(5x - 4)$.

Solution $(2x + 3)(5x - 4) = \underline{2x(5x)} + \underline{2x(-4)} + \underline{3(5x)} + \underline{3(-4)}$

$$\qquad\qquad\qquad\qquad\quad \textit{First} \quad\ \textit{Outside} \quad \textit{Inside} \quad\ \textit{Last}$$

$$= 10x^2 - 8x + 15x - 12$$
$$= 10x^2 + 7x - 12$$

With practice, adding the products of the outside and inside terms ($-8x + 15x = 7x$) can be done mentally.

C Column Method

The FOIL method can be applied only when multiplying two binomials. To find products of polynomials with more than two terms, we use what is called the Column method.

The column method of multiplying two polynomials is very similar to long multiplication with whole numbers. It is just another way of finding all possible products of terms in one polynomial with terms in another polynomial.

EXAMPLE 4 Multiply $(2x + 3)(3x^2 - 2x + 1)$.

Solution

$$
\begin{array}{r}
3x^2 - 2x + 1\\
2x + 3\\
\hline
6x^3 - 4x^2 + 2x\\
9x^2 - 6x + 3\\
\hline
6x^3 + 5x^2 - 4x + 3
\end{array}
$$

$\leftarrow 2x(3x^2 - 2x + 1)$
$\leftarrow 3(3x^2 - 2x + 1)$
Add similar terms.

It will be to your advantage to become very fast and accurate at multiplying polynomials. You should be comfortable using either method. The following examples illustrate the three types of multiplication.

3. Multiply $(3x + 5)(2x - 1)$.

4. Multiply $(3x - 2)(2x^2 - 3x + 4)$.

Answers

3. $6x^2 + 7x - 5$

4. $6x^3 - 13x^2 + 18x - 8$

EXAMPLE 5 Multiply.

a. $4a^2(2a^2 - 3a + 5)$

$$\begin{array}{r} 2a^2 - 3a + 5 \\ 4a^2 \\ \hline 8a^4 - 12a^3 + 20a \end{array}$$

b. $(x - 2)(y + 3) = x(y) + x(3) + (-2)(y) + (-2)(3)$

$$\qquad\qquad\qquad\quad\; F \qquad O \qquad\quad I \qquad\quad L$$

$$= xy + 3x - 2y - 6$$

c. $(x + y)(a - b) = x(a) + x(-b) + y(a) + y(-b)$

$$\qquad\qquad\qquad\quad\; F \qquad O \qquad\quad I \qquad\quad L$$

$$= xa - xb + ya - yb$$

d. $(5x - 1)(2x + 6) = 5x(2x) + 5x(6) + (-1)(2x) + (-1)(6)$

$$\qquad\qquad\qquad\qquad\; F \qquad O \qquad\qquad I \qquad\qquad L$$

$$= 10x^2 + 30x + (-2x) + (-6)$$

$$= 10x^2 + 28x - 6$$

Applications

EXAMPLE 6 The length of a rectangle is 3 more than twice the width. Write an expression for the area of the rectangle.

Solution We begin by drawing a rectangle and labeling the width with x. Since the length is 3 more than twice the width, we label the length with $2x + 3$.

Since the area A of a rectangle is the product of the length and width, we write our formula for the area of this rectangle as

$$A = x(2x + 3)$$

$$A = 2x^2 + 3x \qquad \text{Multiply.}$$

Suppose that a store sells x items at p dollars per item. The total amount of money obtained by selling the items is called the *revenue*. It can be found by multiplying the number of items sold, x, by the price per item, p. For example, if 100 items are sold for $6 each, the revenue is $100(6) = \$600$. Similarly, if 500 items are sold for $8 each, the total revenue is $500(8) = \$4,000$. If we denote the revenue with the letter R, then the formula that relates R, x, and p is

$$\text{Revenue} = (\text{number of items sold})(\text{price of each item})$$

In symbols: $R = xp$.

5. Multiply.

a. $5a^2(3a^2 - 2a + 4)$

b. $(x + 3)(y - 5)$

c. $(x + 2y)(a + b)$

d. $(4x + 1)(3x + 5)$

6. The length of a rectangle is 5 more than three times the width. Write an expression for the area.

Answers

5. a. $15a^4 - 10a^3 + 20a^2$

b. $xy - 5x + 3y - 15$

c. $ax + bx + 2ay + 2by$

d. $12x^2 + 23x + 5$

6. $3x^2 + 5x$

7. Suppose the store in Example 7 can sell x flash drives for p dollars according to the equation $x = 700 - 110p$. Find the revenue.

EXAMPLE 7 A store selling flash drives for home computers knows from past experience that it can sell x flash drives each day at a price of p dollars each, according to the equation $x = 800 - 100p$. Write a formula for the daily revenue that involves only the variables R and p.

Solution From our previous discussion we know that the revenue R is given by the formula

$$R = xp$$

But, since $x = 800 - 100p$, we can substitute $800 - 100p$ for x in the revenue equation to obtain

$$R = (800 - 100p)p$$

$$R = 800p - 100p^2$$

This last formula gives the revenue R in terms of the price p.

GETTING READY FOR CLASS

After reading through the preceding section, respond in your own words and in complete sentences.

A. Describe how you would use the Column method to multiply two polynomials.

B. Describe how the distributive property is used to multiply a monomial and a polynomial.

C. Describe how you would use the FOIL method to multiply two binomials.

D. Show how the product of two binomials can be a trinomial.

Vocabulary Review

Choose the correct words to fill in the blanks below.

term distributive products column binomials

1. To multiply a monomial with a polynomial, apply the _____ property.

2. To multiply any two polynomials, multiply each _____ in the first with each term in the second.

3. The FOIL method allows us to multiply any two _____ by finding the sum of the products of the first, outside, inside, and last two terms.

4. Use the column method to find the _____ of polynomials with more than two terms.

5. The _____ method of multiplying two polynomials is similar to long multiplication with whole numbers.

Problems

A Multiply the following by applying the distributive property.

1. $2x(3x + 1)$

2. $4x(2x - 3)$

3. $2x^2(3x^2 - 2x + 1)$

4. $5x(4x^3 - 5x^2 + x)$

5. $2ab(a^2 - ab + 1)$

6. $3a^2b(a^3 + a^2b^2 + b^3)$

7. $y^2(3y^2 + 9y + 12)$

8. $5y(2y^2 - 3y + 5)$

9. $4x^2y(2x^3y + 3x^2y^2 + 8y^3)$

10. $6xy^3(2x^2 + 5xy + 12y^2)$

B C Multiply the following binomials. You should do about half the problems using the FOIL method and the other half using the column method. Remember, you want to be comfortable using both methods.

11. $(x + 3)(x + 4)$

12. $(x + 2)(x + 5)$

13. $(x + 6)(x + 1)$

14. $(x + 1)(x + 4)$

15. $\left(x + \dfrac{1}{2}\right)\left(x + \dfrac{3}{2}\right)$

16. $\left(x + \dfrac{3}{5}\right)\left(x + \dfrac{2}{5}\right)$

17. $(a + 5)(a - 3)$

18. $(a - 8)(a + 2)$

19. $(x - a)(y + b)$

20. $\left(y + \dfrac{5}{6}\right)\left(y - \dfrac{5}{6}\right)$ **21.** $\left(y - \dfrac{4}{7}\right)\left(y + \dfrac{4}{7}\right)$ **22.** $(2x - 3)(x - 4)$ **23.** $(3x - 5)(x - 2)$

24. $(a + 2)(2a - 1)$ **25.** $(a - 6)(3a + 2)$ **26.** $(2x - 5)(3x - 2)$ **27.** $(3x + 6)(2x - 1)$

28. $(2x + 3)(a + 4)$ **29.** $(2x - 3)(a - 4)$ **30.** $(5x - 4)(5x + 4)$ **31.** $(6x + 5)(6x - 5)$

32. $\left(2x - \dfrac{1}{2}\right)\left(x + \dfrac{3}{2}\right)$ **33.** $\left(4x - \dfrac{3}{2}\right)\left(x + \dfrac{1}{2}\right)$ **34.** $(1 - 2a)(3 - 4a)$ **35.** $(1 - 3a)(3 + 2a)$

For each of the following problems, fill in the area of each small rectangle and square, and then add the results together to find the indicated product. Also, show that you get the same answer by multiplying the binomials and simplifying the product.

36. $(x + 2)(x + 3)$

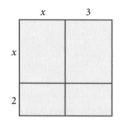

37. $(x + 4)(x + 5)$

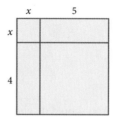

38. $(x + 1)(2x + 2)$

39. $(2x + 1)(2x + 2)$

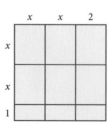

Multiply the following polynomials, and simplify your answer.

40. $(a - 3)(a^2 - 3a + 2)$

41. $(a + 5)(a^2 + 2a + 3)$

42. $(x + 2)(x^2 - 2x + 4)$

43. $(x + 3)(x^2 - 3x + 9)$

44. $(2x + 1)(x^2 + 8x + 9)$

45. $(3x - 2)(x^2 - 7x + 8)$

46. $(5x^2 + 2x + 1)(x^2 - 3x + 5)$

47. $(2x^2 + x + 1)(x^2 - 4x + 3)$

48. $(x^2 + 3)(2x^2 - 5)$

49. $(4x^3 - 8)(5x^3 + 4)$

50. $(3a^4 + 2)(2a^2 + 5)$

51. $(7a^4 - 8)(4a^3 - 6)$

52. $(x + 3)(x + 4)(x + 5)$

53. $(x - 3)(x - 4)(x - 5)$

Simplify the following expressions.

54. $(x - 3)(x - 2) + 2$

55. $(2x - 5)(3x + 2) - 4$

56. $(2x - 3)(4x + 3) + 4$

57. $(3x + 8)(5x - 7) + 52$

58. $(x + 4)(x - 5) + (-5)(2)$

59. $(x + 3)(x - 4) + (-4)(2)$

60. $2(x - 3) + x(x + 2)$

61. $5(x + 3) + 1(x + 4)$

62. $3x(x + 1) - 2x(x - 5)$

63. $4x(x - 2) - 3x(x - 4)$

64. $x(x + 2) - 3$

65. $2x(x - 4) + 6$

66. $a(a - 3) + 6$

67. $a(a - 4) + 8$

Applying the Concepts

68. Area The length of a rectangle is 5 units more than twice the width. Write an expression for the area of the rectangle.

69. Area The length of a rectangle is 2 more than three times the width. Write an expression for the area of the rectangle.

70. Area The width and length of a rectangle are given by two consecutive integers. Write an expression for the area of the rectangle.

71. Area The width and length of a rectangle are given by two consecutive even integers. Write an expression for the area of the rectangle.

72. A rectangular swimming pool has a length which is three times as long as its width. A five foot pathway is built to surround the entire pool.

 a. Write and simplify an expression for the total area of the pool plus the pathway.

 b. Write and simplify an expression for the area of the pathway alone.

73. A circular garden has a radius of r feet. A 2 foot wide rock border is placed around the entire garden.

 a. Write and simplify an expression in terms of π for the total area of the garden plus the rock border.

 b. Write and simplify an expression for the area of the rock border alone.

74. A rectangular piece of cardboard has a length that is 1 inch more than twice the width w. Squares of sides 2 inches are cut from the corners, and the resulting flaps are folded up to create the sides of a box.

a. Write and simplify an expression for the volume of the box.

b. Write and simplify an expression for the surface area of the box. Assume the box is open (with no top).

75. Revenue A store selling hair ribbons knows that the number of ribbons it can sell each week, x, is related to the price per ribbon, p, by the equation $x = 1,200 - 100p$. Write an expression for the weekly revenue that involves only the variables R and p. (*Remember:* The equation for revenue is $R = xp$.)

76. Revenue A store selling headphones knows from past experience that the number of headphones it can sell each week, x, is related to the price per set, p, by the equation $x = 1,300 - 100p$. Write an expression for the weekly revenue that involves only the variables R and p.

Getting Ready for the Next Section

Simplify each expression.

77. $13 \cdot 13$

78. $3x \cdot 3x$

79. $2(x)(-5)$

80. $2(2x)(-3)$

81. $6x + (-6x)$

82. $3x + (-3x)$

83. $(2x)(-3) + (2x)(3)$

84. $(2x)(-5y) + (2x)(5y)$

Multiply the polynomials and simplify your answer.

85. $-4(3x - 4)$

86. $-2x(2x + 7)$

87. $(x - 1)(x + 2)$

88. $(x + 5)(x - 6)$

89. $(x + 3)(x + 3)$

90. $(3x - 2)(3x - 2)$

91. $(5x + 3)(5x - 3)$

92. $(4x - 2)(4x + 2)$

Find the Mistake

Each sentence below contains a mistake. Circle the mistake and write the correct word(s) or expression on the line provided.

1. To find the product of two binomials, multiply the first terms and then multiply the second terms.

2. When using the FOIL method to multiply the binomials $(9x - 2)$ and $(5x + 3)$, the product of the two outside terms is -6.

3. For the problem $(x^2 - 4)(3x^2 + 6)$, adding the products of the two outside terms with the two inside terms gives us $4x^2 + 2$. _____

4. For the problem $4x^2(3x^2 + 4x + 9)$, the degree of the answer is 2. _____

A Find the square of a binomial.

B Multiply expressions of the form $(a + b)(a - b)$.

square of a binomial

difference of two squares

6.6 Special Products

©iStockphoto.com/ mountainberryphoto

Suppose an extreme whitewater kayaker decides to spend a weekend making runs down the Class III Gorge run of the South Fork American River in California. The kayaker considers a successful run one where he paddles the length of the river without flipping over. If the number of successful runs for the kayaker during the weekend equals $(x + 2)^2$, how would we square this binomial to help find a solution? Let's explore this concept further.

A Squaring Binomials

In this section, we will combine the results of the last section with our definition of exponents to find some special products.

Practice Problems

1. Find the square of $(4x - 5)$.

EXAMPLE 1 Find the square of $(3x - 2)$.

Solution To square $(3x - 2)$, we multiply it by itself.

$$
\begin{aligned}
(3x - 2)^2 &= (3x - 2)(3x - 2) & \textit{Definition of exponents} \\
&= 9x^2 - 6x - 6x + 4 & \textit{FOIL method} \\
&= 9x^2 - 12x + 4 & \textit{Combine similar terms.}
\end{aligned}
$$

Notice that the first and last terms in the answer are the square of the first and last terms in the original problem and that the middle term is twice the product of the two terms in the original binomial.

2. Multiply.

a. $(x + y)^2$

b. $(x - y)^2$

EXAMPLE 2 Expand and multiply.

a. $(a + b)^2$ **b.** $(a - b)^2$

Solution

a. $(a + b)^2 = (a + b)(a + b)$
$$= a^2 + 2ab + b^2$$

b. $(a - b)^2 = (a - b)(a - b)$
$$= a^2 - 2ab + b^2$$

Answers

1. $16x^2 - 40x + 25$

2. a. $x^2 + 2xy + y^2$

b. $x^2 - 2xy + y^2$

Binomial squares having the form of Example 2 occur very frequently in algebra. It will be to your advantage to memorize the following rule for squaring a binomial.

> **RULE** Square of a Binomial
>
> The square of a binomial is the sum of the square of the first term, the square of the last term, and twice the product of the two original terms. In symbols this rule is written as follows:
>
> $$(a + b)^2 \quad = \quad a^2 \quad + \quad 2ab \quad + \quad b^2$$
>
> $(a + b)(a + b) \;=\;$ Square of first term \qquad Twice product of the two terms \qquad Square of last term

This rule also applies to squares of differences, such as $(a - b)^2$, because the difference can be thought of as the addition of a negative.

EXAMPLE 3 Multiply using the preceding rule.

		First term squared	Twice their product	Last term squared		Answer
a. $(x - 5)^2$	$=$	x^2	$+ \; 2(x)(-5)$	$+ \quad 25$	$=$	$x^2 - 10x + 25$
b. $(x + 2)^2$	$=$	x^2	$+ \quad 2(x)(2)$	$+ \quad 4$	$=$	$x^2 + 4x + 4$
c. $(2x - 3)^2$	$=$	$4x^2$	$+ \; 2(2x)(-3)$	$+ \quad 9$	$=$	$4x^2 - 12x + 9$
d. $(5x - 4)^2$	$=$	$25x^2$	$+ \; 2(5x)(-4)$	$+ \quad 16$	$=$	$25x^2 - 40x + 16$

B More Special Products

Another special product that occurs frequently is $(a + b)(a - b)$. The only difference in the two binomials is the sign between the two terms. The interesting thing about this type of product is that the middle term is always zero. Here are some examples.

EXAMPLE 4 Multiply using the FOIL method.

a. $(2x - 3)(2x + 3) = 4x^2 + 6x - 6x - 9$ \qquad *FOIL method*

$\qquad\qquad\qquad\qquad = 4x^2 - 9$

b. $(x - 5)(x + 5) = x^2 + 5x - 5x - 25$ \qquad *FOIL method*

$\qquad\qquad\qquad\quad = x^2 - 25$

c. $(3x - 1)(3x + 1) = 9x^2 + 3x - 3x - 1$ \qquad *FOIL method*

$\qquad\qquad\qquad\qquad = 9x^2 - 1$

Notice that in each case the middle term is zero and therefore doesn't appear in the answer. The answers all turn out to be the difference of two squares. Here is a rule to help you memorize the result:

> **RULE** Difference of Two Squares
>
> When multiplying two binomials that differ only in the sign between their terms, subtract the square of the last term from the square of the first term.
>
> $$(a - b)(a + b) = a^2 - b^2$$

3. Multiply using the method shown in Example 3.

 a. $(x - 3)^2$

 b. $(x + 5)^2$

 c. $(3x - 2)^2$

 d. $(4x + 5)^2$

4. Multiply using the FOIL method.

 a. $(3x - 5)(3x + 5)$

 b. $(x - 4)(x + 4)$

 c. $(4x - 1)(4x + 1)$

**Video Examples
Section 6.6**

Answers

 3. a. $x^2 - 6x + 9$

 b. $x^2 + 10x + 25$

 c. $9x^2 - 12x + 4$

 d. $16x^2 + 40x + 25$

 4. a. $9x^2 - 25$ **b.** $x^2 - 16$

 c. $16x^2 - 1$

Here are some problems that result in the difference of two squares:

EXAMPLE 5 Multiply using the preceding rule.

a. $(x + 3)(x - 3) = x^2 - 9$

b. $(a + 2)(a - 2) = a^2 - 4$

c. $(9a + 1)(9a - 1) = 81a^2 - 1$

d. $(2x - 5y)(2x + 5y) = 4x^2 - 25y^2$

e. $(3a - 7b)(3a + 7b) = 9a^2 - 49b^2$

5. Multiply using the formula
$(a + b)(a - b) = a^2 - b^2$.
a. $(x + 2)(x - 2)$
b. $(a + 7)(a - 7)$
c. $(6a + 1)(6a - 1)$
d. $(8x - 2y)(8x + 2y)$
e. $(4a - 3b)(4a + 3b)$

Although all the problems in this section can be worked correctly using the methods in the previous section, they can be done much faster if the two rules are *memorized*. Here is a summary of the two rules:

$$\left.\begin{array}{l} (a + b)^2 = (a + b)(a + b) = a^2 + 2ab + b^2 \\ (a - b)^2 = (a - b)(a - b) = a^2 - 2ab + b^2 \end{array}\right\} \text{Binomial Squares}$$

$$(a - b)(a + b) = a^2 - b^2 \qquad \text{Difference of two squares}$$

6. Expand and simplify.
$x^2 + (x - 1)^2 + (x + 1)^2$

EXAMPLE 6 Write an expression in symbols for the sum of the squares of three consecutive even integers. Then, simplify that expression.

Solution If we let $x =$ the first of the even integers, then $x + 2$ is the next consecutive even integer, and $x + 4$ is the one after that. An expression for the sum of their squares is

$$x^2 + (x + 2)^2 + (x + 4)^2 \qquad \text{Sum of squares}$$

$$= x^2 + (x^2 + 4x + 4) + (x^2 + 8x + 16) \qquad \text{Expand squares.}$$

$$= 3x^2 + 12x + 20 \qquad \text{Add similar terms.}$$

GETTING READY FOR CLASS

After reading through the preceding section, respond in your own words and in complete sentences.

A. Describe how you would square the binomial $(a + b)$.

B. Explain why $(x + 3)^2$ cannot be $x^2 + 9$.

C. Why is the middle term in the product of $(a + b)$ and $(a - b)$ equal to zero?

D. When multiplied out, how will $(x + 3)^2$ and $(x - 3)^2$ differ?

Answers
5. a. $x^2 - 4$ **b.** $a^2 - 49$ **c.** $36a^2 - 1$
d. $64x^2 - 4y^2$ **e.** $16a^2 - 9b^2$
6. $3x^2 + 2$

Vocabulary Review

Choose the correct words to fill in the blanks below.

subtract sum square of a binomial difference of two squares

1. The square of a binomial is the _____ of the square of the first term, the square of the last term, and twice the product of the two original terms.

2. The _____ is given by $(x + y)^2 = x^2 + 2xy + y^2$.

3. When multiplying two binomials that differ only in the sign between their terms, _____ the square of the last term from the square of the first term.

4. The _____ is given by $(a - b)(a + b) = a^2 - b^2$.

Problems

A Perform the indicated operations for each binomial.

1. $(x - 2)^2$

2. $(x + 2)^2$

3. $(a + 3)^2$

4. $(a - 3)^2$

5. $(x - 5)^2$

6. $(x - 4)^2$

7. $\left(a - \dfrac{1}{2}\right)^2$

8. $\left(a + \dfrac{1}{2}\right)^2$

9. $(x + 10)^2$

10. $(x - 10)^2$

11. $(a + 0.8)^2$

12. $(a - 0.4)^2$

13. $(2x - 1)^2$

14. $(3x + 2)^2$

15. $(4a + 5)^2$

16. $(4a - 5)^2$

17. $(3x - 2)^2$

18. $(2x - 3)^2$

19. $(3a + 5b)^2$

20. $(5a - 3b)^2$

21. $(4x - 5y)^2$

22. $(5x + 4y)^2$

23. $(7m + 2n)^2$

24. $(2m - 7n)^2$

25. $(6x - 10y)^2$

26. $(10x + 6y)^2$

27. $(x^2 + 5)^2$

28. $(x^2 + 3)^2$

29. $(a^2 + 1)^2$

30. $(a^2 - 2)^2$

31. $(x^3 - 4)^2$

32. $(y^3 + 2)^2$

Comparing Expressions The following problems are meant to show that $(x + y)^2 \neq x^2 + y^2$ and $(x - y)^2 \neq x^2 - y^2$. Fill in each table.

33.

x	$(x + 3)^2$	$x^2 + 9$	$x^2 + 6x + 9$
1			
2			
3			
4			

34.

x	$(x - 5)^2$	$x^2 - 25$	$x^2 - 10x + 25$
1			
2			
3			
4			

35.

a	1	3	3	4
b	1	5	4	5
$(a + b)^2$				
$a^2 + b^2$				
$a^2 + ab + b^2$				
$a^2 + 2ab + b^2$				

36.

a	2	5	2	4
b	1	2	5	3
$(a - b)^2$				
$a^2 - b^2$				
$a^2 - 2ab + b^2$				

B Multiply the following binomials.

37. $(a + 5)(a - 5)$

38. $(a - 6)(a + 6)$

39. $(y - 1)(y + 1)$

40. $(y - 2)(y + 2)$

41. $(9 + x)(9 - x)$

42. $(10 - x)(10 + x)$

43. $(2x + 5)(2x - 5)$

44. $(3x + 5)(3x - 5)$

45. $\left(4x + \dfrac{1}{3}\right)\left(4x - \dfrac{1}{3}\right)$

46. $\left(6x + \dfrac{1}{4}\right)\left(6x - \dfrac{1}{4}\right)$

47. $(2a + 7)(2a - 7)$

48. $(3a + 10)(3a - 10)$

49. $(6 - 7x)(6 + 7x)$

50. $(7 - 6x)(7 + 6x)$

51. $(x^2 + 3)(x^2 - 3)$

52. $(x^2 + 2)(x^2 - 2)$

53. $(a^2 + 4)(a^2 - 4)$

54. $(a^2 + 9)(a^2 - 9)$

55. $(5y^4 - 8)(5y^4 + 8)$

56. $(7y^5 + 6)(7y^5 - 6)$

Simplify the following expressions.

57. $(x + 3)(x - 3) + (x - 5)(x + 5)$

58. $(x - 7)(x + 7) + (x - 4)(x + 4)$

59. $(2x + 3)^2 - (4x - 1)^2$

60. $(3x - 5)^2 - (2x + 3)^2$

61. $(a + 1)^2 - (a + 2)^2 + (a + 3)^2$

62. $(a - 1)^2 + (a - 2)^2 - (a - 3)^2$

63. $(2x + 3)^3$

64. $(3x - 2)^3$

65. $(x + 5)^2 - (x - 5)^2$

66. $(2x + 1)^2 - (2x - 1)^2$

67. $(x - y)^2 - (x + y)^2$

68. $(a + b)^2 - (a - b)^2$

69. $(x - 1)^2 + x^2 + (x + 1)^2$

70. $x^2 + (x + 1)^2 + (x + 2)^2$

71. $(2a - 1)^3$

72. $(4a - 3)^3$

Applying the Concepts

73. Shortcut The formula for the difference of two squares can be used as a shortcut to multiplying certain whole numbers if they have the correct form. Use the difference of two squares formula to multiply 49(51) by first writing 49 as $(50 - 1)$ and 51 as $(50 + 1)$.

74. Shortcut Use the difference of two squares formula to multiply 101(99) by first writing 101 as $(100 + 1)$ and 99 as $(100 - 1)$.

75. Comparing Expressions Evaluate the expression $(x + 3)^2$ and the expression $x^2 + 6x + 9$ for $x = 2$.

76. Comparing Expressions Evaluate the expression $x^2 - 25$ and the expression $(x - 5)(x + 5)$ for $x = 6$.

77. Number Problem Write an expression for the sum of the squares of two consecutive integers. Then, simplify that expression.

78. Number Problem Write an expression for the sum of the squares of two consecutive odd integers. Then, simplify that expression.

79. Number Problem Write an expression for the sum of the squares of three consecutive integers. Then, simplify that expression.

80. Number Problem Write an expression for the sum of the squares of three consecutive odd integers. Then, simplify that expression.

81. Area We can use the concept of area to further justify our rule for squaring a binomial. The length of each side of the square shown in the figure is $a + b$. (The longer line segment has length a and the shorter line segment has length b.) The area of the whole square is $(a + b)^2$. However, the whole area is the sum of the areas of the two smaller squares and the two smaller rectangles that make it up. Write the area of the two smaller squares and the two smaller rectangles and then add them together to verify the formula $(a + b)^2 = a^2 + 2ab + b^2$.

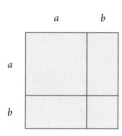

82. Area The length of each side of the large square shown in the figure is $x + 5$. Therefore, its area is $(x + 5)^2$. Find the area of the two smaller squares and the two smaller rectangles that make up the large square, then add them together to verify the formula $(x + 5)^2 = x^2 + 10x + 25$.

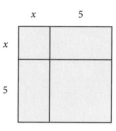

83. You have an x foot square available to plant a lawn. You want a 2 foot border around the lawn. Write and simplify an expression for the actual amount of grass to plant.

84. You have a lawn which is 20 feet by 20 feet square. You want to plant x feet of flowers around the lawn. Write an expression for the total area of lawn plus flowers, then simplify.

Getting Ready for the Next Section

Simplify.

85. $\dfrac{10x^3}{5x}$

86. $\dfrac{15x^2}{5x}$

87. $\dfrac{3x^2}{3}$

88. $\dfrac{4x^2}{2}$

89. $\dfrac{9x^2}{3x}$

90. $\dfrac{3x^4}{3x^2}$

91. $\dfrac{24x^3y^2}{8x^2y}$

92. $\dfrac{4x^2y^3}{8x^2y}$

Find the Mistake

Each sentence below contains a mistake. Circle the mistake and write the correct word(s) or expression on the line provided.

1. To find the square of the binomial $(5a^2 - 9a)$, we multiply $5a^2$ by $-9a$. _____

2. The second term in a binomial square is the square of the last term. _____

3. Simplifying $(7x - 3)^2$ gives us $7x^2 - 21x + 9$. _____

4. The problem $(4a - 6)(4a + 6) = 16a^2 - 36$ is an example of a binomial square. _____

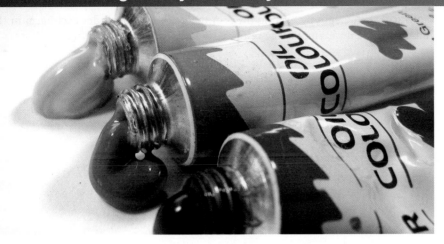

The number of tubes an artist used to paint his masterpieces in one month may be given by $3x^2 + 2x$. If $3x$ represents the number of paintings he completed in that month, how would we find the number of tubes he used per painting? We would set this up as a division problem. To solve it, we need to understand how to divide a polynomial by a monomial.

A Dividing a Polynomial by a Monomial

To divide a polynomial by a monomial, we will use the definition of division and apply the distributive property. Follow the steps in this example closely.

EXAMPLE 1　Divide $10x^3 - 15x^2$ by $5x$.

Solution

$$\frac{10x^3 - 15x^2}{5x} = (10x^3 - 15x^2)\frac{1}{5x}$$
Division by $5x$ is the same as multiplication by $\frac{1}{5x}$.

$$= 10x^3\left(\frac{1}{5x}\right) - 15x^2\left(\frac{1}{5x}\right)$$
Distribute $\frac{1}{5x}$ to both terms.

$$= \frac{10x^3}{5x} - \frac{15x^2}{5x}$$
Multiplication by $\frac{1}{5x}$ is the same as division by $5x$.

$$= 2x^2 - 3x$$
Divide monomials.

If we were to leave out the first steps in Example 1, the problem would look like this:

$$\frac{10x^3 - 15x^2}{5x} = \frac{10x^3}{5x} - \frac{15x^2}{5x}$$

$$= 2x^2 - 3x$$

The problem is much shorter and clearer this way. You may leave out the first two steps from Example 1 when working problems in this section. They are part of Example 1 only to help show you why the following rule is true.

> **RULE**　Division of a Polynomial by a Monomial
>
> To divide a polynomial by a monomial, simply divide each term in the polynomial by the monomial.

On the next page are some further examples using our rule for division of a polynomial by a monomial.

**Video Examples
Section 6.7**

Practice Problems

1. Divide $8x^3 - 12x^2$ by $4x$.

2. Divide $\dfrac{5x^2 - 10}{5}$.

3. Divide $\dfrac{8x^2 - 2}{2}$.

4. Find the quotient of $15x^4 - 10x^3$ and $5x$.

5. Divide $\dfrac{21x^3y^2 + 14x^2y^2 - 7x^2y^3}{7x^2y}$.

EXAMPLE 2 Divide $\dfrac{3x^2 - 6}{3}$.

Solution We begin by writing the 3 in the denominator under each term in the numerator. Then we simplify the result.

$$\frac{3x^2 - 6}{3} = \frac{3x^2}{3} - \frac{6}{3} \qquad \text{Divide each term in the numerator by 3.}$$

$$= x^2 - 2 \qquad \text{Simplify.}$$

EXAMPLE 3 Divide $\dfrac{4x^2 - 2}{2}$.

Solution Dividing each term in the numerator by 2, we have

$$\frac{4x^2 - 2}{2} = \frac{4x^2}{2} - \frac{2}{2} \qquad \text{Divide each term in the numerator by 2.}$$

$$= 2x^2 - 1 \qquad \text{Simplify.}$$

EXAMPLE 4 Find the quotient of $27x^3 - 9x^2$ and $3x$.

Solution We again are asked to divide the first polynomial by the second one.

$$\frac{27x^3 - 9x^2}{3x} = \frac{27x^3}{3x} - \frac{9x^2}{3x} \qquad \text{Divide each term by 3x.}$$

$$= 9x^2 - 3x \qquad \text{Simplify.}$$

EXAMPLE 5 Divide $\dfrac{24x^3y^2 + 16x^2y^2 - 4x^2y^3}{8x^2y}$.

Solution Writing $8x^2y$ under each term in the numerator and then simplifying, we have

$$\frac{24x^3y^2 + 16x^2y^2 - 4x^2y^3}{8x^2y} = \frac{24x^3y^2}{8x^2y} + \frac{16x^2y^2}{8x^2y} - \frac{4x^2y^3}{8x^2y}$$

$$= 3xy + 2y - \frac{y^2}{2}$$

From the examples in this section, it is clear that to divide a polynomial by a monomial, we must divide each term in the polynomial by the monomial. Often, students taking algebra for the first time will make the following mistake:

$$\frac{x + 2}{2} = x + 1 \qquad \text{Mistake}$$

The mistake here is in not dividing both terms in the numerator by 2. The correct way to divide $x + 2$ by 2 looks like this:

$$\frac{x + 2}{2} = \frac{x}{2} + \frac{2}{2} = \frac{x}{2} + 1 \qquad \text{Correct}$$

GETTING READY FOR CLASS

After reading through the preceding section, respond in your own words and in complete sentences.

A. What property of real numbers is the key to dividing a polynomial by a monomial?

B. Describe how you would divide the polynomial $x^2 + 10x$ by $5x$.

C. Describe how you would divide a polynomial by a negative monomial.

D. In your own words, explain the mistake in the problem $\dfrac{x + 2}{2} = x + 1$.

Answers

2. $x^2 - 2$

3. $4x^2 - 1$

4. $3x^3 - 2x^2$

5. $3xy + 2y - y^2$

Problems

A Divide the following polynomials by $5x$.

1. $5x^2 - 10x$

2. $10x^3 - 15x$

3. $15x - 10x^3$

4. $50x^3 - 20x^2$

5. $25x^2y - 10xy$

6. $15xy^2 + 20x^2y$

7. $35x^5 - 30x^4 + 25x^3$

8. $40x^4 - 30x^3 + 20x^2$

9. $50x^5 - 25x^3 + 5x$

10. $75x^6 + 50x^3 - 25x$

11. $-20x^3 + 15x^2 - 5x$

12. $100x^5 - 80x^4 - 40x^3$

Divide the following by $-2a$.

13. $8a^2 - 4a$

14. $a^3 - 6a^2$

15. $16a^5 + 24a^4$

16. $30a^6 + 20a^3$

17. $8ab + 10a^2$

18. $6a^2b - 10ab^2$

19. $12a^3b - 6a^2b^2 + 14ab^3$

20. $4ab^3 - 16a^2b^2 - 22a^3b$

21. $a^2 + 2ab + b^2$

22. $a^2b - 2ab^2 + b^3$

23. $-3a^4 - 8a^3 + 12a^2$

24. $-5a^5 - 4a^4 - 16a^2$

Find the following quotients.

25. $\dfrac{6x + 8y}{2}$

26. $\dfrac{9x - 3y}{3}$

27. $\dfrac{7y - 21}{-7}$

28. $\dfrac{14y - 12}{2}$

29. $\dfrac{10xy - 8x}{2x}$

30. $\dfrac{12xy^2 - 18x}{-6x}$

31. $\dfrac{x^2y - x^3y^2}{x}$

32. $\dfrac{x^2y - x^3y^2}{x^2}$

33. $\dfrac{x^2y - x^3y^2}{-x^2y}$

34. $\dfrac{ab + a^2b^2}{ab}$

35. $\dfrac{a^2b^2 - ab^2}{-ab^2}$

36. $\dfrac{a^2b^2c - ab^2c^2}{abc}$

37. $\dfrac{x^3 - 3x^2y + xy^2}{x}$

38. $\dfrac{x^2 - 3xy^2 + xy^3}{x}$

39. $\dfrac{10a^2 - 15a^2b + 25a^2b^2}{5a^2}$

40. $\dfrac{11a^2b^2 - 33ab}{-11ab}$

41. $\dfrac{26x^2y^2 - 13xy}{-13xy}$

42. $\dfrac{6x^2y^2 - 3xy}{6xy}$

43. $\dfrac{4x^2y^2 - 2xy}{4xy}$

44. $\dfrac{6x^2a + 12x^2b - 6x^2c}{36x^2}$

45. $\dfrac{5a^2x - 10ax^2 + 15a^2x^2}{20a^2x^2}$

46. $\dfrac{12ax - 9bx + 18cx}{6x^2}$

47. $\dfrac{16x^5 + 8x^2 + 12x}{12x^3}$

48. $\dfrac{27x^2 - 9x^3 - 18x^4}{-18x^3}$

Divide. Assume all variables represent positive numbers.

49. $\dfrac{9a^{5m} - 27a^{3m}}{3a^{2m}}$

50. $\dfrac{26a^{3m} - 39a^{5m}}{13a^{3m}}$

51. $\dfrac{10x^{5m} - 25x^{3m} + 35x^m}{5x^m}$

52. $\dfrac{18x^{2m} + 24x^{4m} - 30x^{6m}}{6x^{2m}}$

Simplify each numerator, and then divide.

53. $\dfrac{2x^3(3x + 2) - 3x^2(2x - 4)}{2x^2}$

54. $\dfrac{5x^2(6x - 3) + 6x^3(3x - 1)}{3x}$

55. $\dfrac{(x + 2)^2 - (x - 2)^2}{2x}$

56. $\dfrac{(x - 3)^2 - (x + 3)^2}{3x}$

57. $\dfrac{(x + 5)^2 + (x + 5)(x - 5)}{2x}$

58. $\dfrac{(x - 4)^2 + (x + 4)(x - 4)}{2x}$

59. Comparing Expressions Evaluate the expression $\dfrac{10x + 15}{5}$ and the expression $2x + 3$ when $x = 2$.

60. Comparing Expressions Evaluate the expression $\dfrac{6x^2 + 4x}{2x}$ and the expression $3x + 2$ when $x = 5$.

61. Comparing Expressions Show that the expression $\dfrac{3x + 8}{2}$ is not the same as the expression $3x + 4$ by replacing x with 10 in both expressions and simplifying the results.

62. Comparing Expressions Show that the expression $\dfrac{x + 10}{x}$ is not equal to 10 by replacing x with 5 and simplifying.

Getting Ready for the Next Section

Divide.

63. $27\overline{)3,962}$

64. $13\overline{)18,780}$

65. $\dfrac{2x^2 + 5x}{x}$

66. $\dfrac{7x^5 + 9x^3 + 3x^7}{x^3}$

Multiply each expression.

67. $x(x - 3)$

68. $(x - 3)(-2)$

69. $2x^2(x - 5)$

70. $10x(x - 5)$

71. $x^2(2x - 3)$

72. $2x(4x - 1)$

73. $-3(x - 7)$

74. $-5(x + 2)$

Subtract each expression.

75. $(x^2 - 5x) - (x^2 - 3x)$

76. $(2x^3 + 0x^2) - (2x^3 - 10x^2)$

77. $(-2x + 8) - (-2x + 6)$

78. $(4x - 14) - (4x - 10)$

79. $(-4x + 3) - (-4x - 2)$

80. $(-8x - 12) - (-8x + 3)$

81. $(-4x^2 - 2x) - (-4x^2 + 3x)$

82. $(-x^2 + 8x) - (-x^2 + 12x)$

Find the Mistake

Each sentence below contains a mistake. Circle the mistake and write the correct word(s) or expression on the line provided.

1. To divide a polynomial by a monomial, divide only the first term of the polynomial by the monomial. _____

2. Dividing $15x^3 - 6x$ by $-3x$ gives us $15x^3 - 2$. _____

3. The quotient of $12x^2y + 4x^4y^2 - x^3y^2$ and x^2y gives us $12xy + 4x^2y - 1$. _____

4. If you divide $-9x^2 - 12x$ by $-3x$, the answer is $-3x - 4$. _____

6.8 Dividing a Polynomial by a Polynomial

Image © Katherine Heistand Shields, 2010

KEY WORDS

long division of polynomials

remainder

Suppose you earn an annual salary of $42,500. You are considering buying a luxurious house and need to know your monthly gross income to further calculate if you can afford the mortgage payments. To determine your monthly income, we need to divide 42,500 by 12. How would you set up a long division problem to find this answer? This section will cover how to do so and then transition to *long division of polynomials*.

A Dividing a Polynomial by a Polynomial

Since long division for polynomials is very similar to long division with whole numbers, we will begin by reviewing a division problem with whole numbers. You may realize when looking at Example 1 that you don't have a very good idea why you proceed as you do with long division. What you do know is that the process always works. We are going to approach the explanations in this section in much the same manner; that is, we won't always be sure why the steps we will use are important, only that they always produce the correct result.

EXAMPLE 1 Divide $27\overline{)3,962}$.

Solution

$$
\begin{array}{r}
1 \\
27\overline{)3962} \\
27 \\
\hline
12
\end{array}
$$

Estimate 27 into 39.

Multiply $1 \times 27 = 27$.

Subtract $39 - 27 = 12$.

$$
\begin{array}{r}
1 \\
27\overline{)3962} \\
27\downarrow \\
\hline
126
\end{array}
$$

Bring down the 6.

These are the four basic steps in long division. Estimate, multiply, subtract, and bring down the next term. To finish the problem, we simply perform the same four steps again.

$$
\begin{array}{r}
14 \\
27\overline{)3962} \\
27\downarrow\downarrow \\
\hline
126 \\
108\downarrow \\
\hline
182
\end{array}
$$

4 is the estimate.

Multiply to get 108.

Subtract to get 18, then bring down the 2.

Practice Problems

1. Divide $35\overline{)4281}$.

**Video Examples
Section 6.8**

Answer

1. $122\frac{11}{35}$

One more time.

$$\begin{array}{r} 146 \\ 27\overline{)3962} \\ \underline{27} \\ 126 \\ \underline{108} \\ 182 \\ 162 \\ \underline{} \\ 20 \end{array}$$ *6 is the estimate.*

 Multiply to get 162.

 Subtract to get 20.

Since there is nothing left to bring down, we have our answer.

$$\frac{3962}{27} = 146 + \frac{20}{27} \qquad \text{or} \qquad 146\frac{20}{27}$$

To check out our answer, we evaluate the product of 146 and 27, then add the remainder.

$$(146)(27) + 20 = 3{,}942 + 20 = 3{,}962$$

Here is how it works with polynomials:

EXAMPLE 2 Divide $\dfrac{x^2 - 5x + 8}{x - 3}$.

Solution

$$\begin{array}{r} x \phantom{{}-5x+8} \\ x - 3\overline{)x^2 - 5x + 8} \\ \underline{\mp\, x^2 \pm 3x} \phantom{{}+8} \\ -\,2x \phantom{{}+8} \\ -\,2x + 8 \end{array}$$

Estimate $x^2 \div x = x$.

Multiply $x(x - 3) = x^2 - 3x$.

Subtract $(x^2 - 5x) - (x^2 - 3x) = -2x$.

Bring down the 8.

Notice that to subtract one polynomial from another, we add its opposite. That is why we change the signs on $x^2 - 3x$ and add what we get to $x^2 - 5x$. (To subtract the second polynomial, simply change the signs and add.)

We perform the same four steps again.

$$\begin{array}{r} x - 2 \\ x - 3\overline{)x^2 - 5x + 8} \\ \underline{\mp\, x^2 \pm 3x} \phantom{{}+8} \\ -\,2x + 8 \\ \underline{\pm\, 2x \mp 6} \\ 2 \end{array}$$

-2 is the estimate $(-2x \div x = -2)$.

Multiply $-2(x - 3) = -2x + 6.$.

Subtract $(-2x + 8) - (-2x + 6) = 2$.

Since there is nothing left to bring down, we have our answer.

$$\frac{x^2 - 5x + 8}{x - 3} = x - 2 + \frac{2}{x - 3}$$

To check our answer, we multiply $(x - 3)(x - 2)$ to get $x^2 - 5x + 6$. Then, adding on the remainder 2 we have $x^2 - 5x + 8$.

EXAMPLE 3 Divide $\dfrac{6x^2 - 11x - 14}{2x - 5}$.

Solution

$$\begin{array}{r} 3x + 2 \\ 2x - 5\overline{)6x^2 - 11x - 14} \\ \underline{\mp\, 6x^2 \pm 15x} \phantom{{}-14} \\ +\,4x - 14 \\ \underline{\mp\, 4x \pm 10} \\ -\,4 \end{array}$$

$$\frac{6x^2 - 11x - 14}{2x - 5} = 3x + 2 + \frac{-4}{2x - 5}$$

To check our answer, we multiply $(2x - 5)(3x + 2)$ to get $6x^2 - 11x - 10$. Then, adding on the remainder -4, we have $6x^2 - 11x - 14$.

2. Divide $\dfrac{x^2 - 5x + 8}{x - 2}$.

3. Divide $\dfrac{8x^2 - 6x - 5}{2x - 3}$.

Answers

2. $x - 3 + \dfrac{2}{x - 2}$

3. $4x + 3 + \dfrac{4}{2x - 3}$

One last step is sometimes necessary. The two polynomials in a division problem must both be in descending powers of the variable and cannot skip any powers from the highest power down to the constant term. It is also beneficial to keep similar powers vertically aligned as we divide.

EXAMPLE 4 Divide $\dfrac{2x^3 - 3x + 2}{x - 5}$.

4. Divide $\dfrac{3x^3 - 2x + 1}{x - 3}$.

Solution The problem will be much less confusing if we write $2x^3 - 3x + 2$ as $2x^3 + 0x^2 - 3x + 2$. Adding $0x^2$ does not change our original problem.

$$
\begin{array}{r}
2x^2 \\
x - 5 \overline{)\, 2x^3 + 0x^2 - 3x + 2} \\
\mp\, 2x^3 \pm 10x^2 \\
\hline
+\, 10x^2 - 3x
\end{array}
$$

Estimate $2x^3 \div x = 2x^2$.

Multiply $2x^2(x - 5) = 2x^3 - 10x^2$.

Subtract
$(2x^3 + 0x^2) - (2x^3 - 10x^2) = 10x^2$.
Bring down the next term.

Adding the term $0x^2$ gives us a column in which to write $10x^2$. (Remember, you can add and subtract only similar terms.)

Here is the completed problem:

$$
\begin{array}{r}
2x^2 + 10x + 47 \\
x - 5 \overline{)\, 2x^3 + 0x^2 - 3x + 2} \\
\mp\, 2x^3 \pm 10x^2 \\
\hline
+\, 10x^2 - 3x \\
\mp\, 10x^2 \pm 50x \\
\hline
+\, 47x + 2 \\
\mp\, 47x \pm 235 \\
\hline
237
\end{array}
$$

Our answer is $\dfrac{2x^3 - 3x + 2}{x - 5} = 2x^2 + 10x + 47 + \dfrac{237}{x - 5}$

You can check with multiplication that this quotient is correct.

As you can see, long division with polynomials is a mechanical process. Once you have done it correctly a couple of times, it becomes very easy to produce the correct answer.

> **GETTING READY FOR CLASS**
>
> *After reading through the preceding section, respond in your own words and in complete sentences.*
>
> **A.** What are the four steps used in long division with whole numbers?
>
> **B.** How is division of two polynomials similar to long division with whole numbers?
>
> **C.** How would you check your answer when dividing polynomials?
>
> **D.** How do you use 0 when dividing the polynomial $2x^3 - 3x + 2$ by $x - 5$?

Answer

4. $3x^2 + 9x + 25 + \dfrac{76}{x - 3}$

Vocabulary Review

Choose the correct words to fill in the blanks below.

opposite estimate powers variable long division

1. Dividing a polynomial by a polynomial is very similar to _____ with whole numbers.
2. The four basic steps in long division are _____, multiply, subtract, and bring down the next term.
3. To subtract one polynomial from another, add its _____.
4. The two polynomials in a division problem must both be in descending powers of the _____.
5. The two polynomials in a division problem must also not skip any _____ from the highest power down to the constant term.

Problems

A Divide the following polynomials.

1. $\dfrac{x^2 - 5x + 6}{x - 3}$

2. $\dfrac{x^2 - 5x + 6}{x - 2}$

3. $\dfrac{a^2 + 9a + 20}{a + 5}$

4. $\dfrac{a^2 + 9a + 20}{a + 4}$

5. $\dfrac{x^2 - 6x + 9}{x - 3}$

6. $\dfrac{x^2 + 10x + 25}{x + 5}$

7. $\dfrac{2x^2 + 5x - 3}{2x - 1}$

8. $\dfrac{4x^2 + 4x - 3}{2x - 1}$

9. $\dfrac{2a^2 - 9a - 5}{2a + 1}$

10. $\dfrac{4a^2 - 8a - 5}{2a + 1}$

11. $\dfrac{x^2 + 5x + 8}{x + 3}$

12. $\dfrac{x^2 + 5x + 4}{x + 3}$

13. $\dfrac{a^2 + 3a + 2}{a + 5}$

14. $\dfrac{a^2 + 4a + 3}{a + 5}$

15. $\dfrac{x^2 + 2x + 1}{x - 2}$

16. $\dfrac{x^2 + 6x + 9}{x - 3}$

17. $\dfrac{x^2 + 5x - 6}{x + 1}$

18. $\dfrac{x^2 - x - 6}{x + 1}$

19. $\dfrac{a^2 + 3a + 1}{a + 2}$

20. $\dfrac{a^2 - a + 3}{a + 1}$

21. $\dfrac{2x^2 - 2x + 5}{2x + 4}$

22. $\dfrac{15x^2 + 19x - 4}{3x + 8}$

23. $\dfrac{6a^2 + 5a + 1}{2a + 3}$

24. $\dfrac{4a^2 + 4a + 3}{2a + 1}$

25. $\dfrac{6a^3 - 13a^2 - 4a + 15}{3a - 5}$

26. $\dfrac{2a^3 - a^2 + 3a + 2}{2a + 1}$

27. $\dfrac{8x^3 + 2x^2 - 4x - 5}{2x - 1}$

28. $\dfrac{9x^3 - 3x^2 + x - 2}{3x + 1}$

29. $\dfrac{2x^3 - 9x^2 + 11x - 6}{2x^2 - 3x + 2}$

30. $\dfrac{6x^3 + 7x^2 - x + 3}{3x^2 - x + 1}$

31. $\dfrac{x^4 - 3x^3 + 2x^2 - 3x - 9}{x^2 - 2x - 3}$

32. $\dfrac{x^3 + 3x^2 - 4x - 12}{x^2 + x - 6}$

33. $\dfrac{x^4 - x^3 - 4x^2 - x + 1}{x^2 - 3x + 1}$

34. $\dfrac{x^4 - x^3 - x + 1}{x^2 - 2x + 1}$

Fill in the missing terms in the numerator, and then use long division to find the quotients (see Example 4).

35. $\dfrac{x^3 + 4x + 5}{x + 1}$

36. $\dfrac{x^3 + 4x^2 - 8}{x + 2}$

37. $\dfrac{x^3 - 1}{x - 1}$

38. $\dfrac{x^3 + 1}{x + 1}$

39. $\dfrac{x^3 - 8}{x - 2}$

40. $\dfrac{x^3 + 27}{x + 3}$

41. $\dfrac{x^3 + 64}{x + 4}$

42. $\dfrac{x^3 - 27}{x - 3}$

Applying the Concepts

Long Division The chart shows the number of goals scored in the World Cup by the highest scoring countries of all time. If there have been 18 World Cup tournaments since the inaugural games in 1930, how many goals per World Cup have the following countries scored? Round to the nearest tenth.

43. Argentina

44. Spain

45. Brazil

46. Italy

FIFA WORLD CUP - MOST POINTS OF ALL-TIME
www.fifa.com

Brazil	212
Germany	187
Italy	153
Argentina	121
England	94
France	86
Spain	81

Maintaining Your Skills

47. $6(3 + 4) + 5$

48. $[(1 + 2)(2 + 3)] + (4 \div 2)$

49. $1^2 + 2^2 + 3^2$

50. $(1 + 2 + 3)^2$

51. $5(6 + 3 \cdot 2) + 4 + 3 \cdot 2$

52. $(1 + 2)^3 + [(2 \cdot 3) + (4 \cdot 5)]$

53. $(1^3 + 2^3) + [(2 \cdot 3) + (4 \cdot 5)]$

54. $[2(3 + 4 + 5)] \div 3$

55. $(2 \cdot 3 + 4 + 5) \div 3$

56. $10^4 + 10^3 + 10^2 + 10^1$

57. $6 \cdot 10^3 + 5 \cdot 10^2 + 4 \cdot 10^1$

58. $5 \cdot 10^3 + 2 \cdot 10^2 + 8 \cdot 10^1$

59. $1 \cdot 10^3 + 7 \cdot 10^2 + 6 \cdot 10^1 + 0$

60. $4(2 - 1) + 5(3 - 2)$

61. $4 \cdot 2 - 1 + 5 \cdot 3 - 2$

62. $2^3 + 3^2 \cdot 4 - 5$

63. $(2^3 + 3^2) \cdot 4 - 5$

64. $2(2^2 + 3^2) + 3(3^2)$

Find the Mistake

Each sentence below contains a mistake. Circle the mistake and write the correct word(s) or expression on the line provided.

1. For a division of polynomials problem, you have found the answer when you are left with only a whole number to bring down. _____

2. To begin the problem $\frac{x^2 - 8x + 4}{x - 4}$, we first estimate $8x \div x = 8$. _____

3. The quotient of $x^3 - 9x + 2$ and $x - 1$ has no remainder. _____

4. When dividing $9x^3 - 7x - 4$ by $3x - 2$, we first write the problem as $3x - 2 \overline{)9x^3 - 7x - 4}$.

Chapter 6 Summary

Exponents: Definition and Properties [6.1, 6.2]

Integer exponents indicate repeated multiplications.

EXAMPLES

$$a^r \cdot a^s = a^{r+s} \qquad \text{Product property for exponents}$$

$$\frac{a^r}{a^s} = a^{r-s} \qquad \text{Quotient property for exponents}$$

$$(ab)^r = a^r \cdot b^r \qquad \text{Distributive property for exponents}$$

$$\left(\frac{a}{b}\right)^r = \frac{a^r}{b^r} \qquad \text{Expanded distributive property for exponents}$$

$$(a^r)^s = a^{r \cdot s} \qquad \text{Power property for exponents}$$

$$a^{-r} = \frac{1}{a^r} \qquad \text{Negative exponents property}$$

EXAMPLES

1. a. $2^3 = 2 \cdot 2 \cdot 2 = 8$

b. $x^5 \cdot x^3 = x^{5+3} = x^8$

c. $\frac{x^5}{x^3} = x^{5-3} = x^2$

d. $(3x)^2 = 3^2 \cdot x^2 = 9x^2$

e. $\left(\frac{2}{3}\right)^3 = \frac{2^3}{3^3} = \frac{8}{27}$

f. $(x^5)^3 = x^{5 \cdot 3} = x^{15}$

g. $3^{-2} = \frac{1}{3^2} = \frac{1}{9}$

Multiplication of Monomials [6.3]

To multiply two monomials, multiply coefficients and add exponents.

2. $(5x^2)(3x^4) = 15x^6$

Division of Monomials [6.3]

To divide two monomials, divide coefficients and subtract exponents.

3. $\frac{12x^9}{4x^5} = 3x^4$

Scientific Notation [6.1, 6.2, 6.3]

A number is in scientific notation when it is written as the product of a number between 1 and 10 and an integer power of 10.

4. $768,000 = 7.68 \times 10^5$
$0.00039 = 3.9 \times 10^{-4}$

Addition of Polynomials [6.4]

To add two polynomials, add coefficients of similar terms.

5. $(3x^2 - 2x + 1) + (2x^2 + 7x - 3)$
$= 5x^2 + 5x - 2$

Subtraction of Polynomials [6.4]

To subtract one polynomial from another, add the opposite of the second to the first.

6. $(3x + 5) - (4x - 3)$
$= 3x + 5 - 4x + 3$
$= -x + 8$

Multiplication of Polynomials [6.5]

7. a. $2a^2(5a^2 + 3a - 2)$
 $= 10a^4 + 6a^3 - 4a^2$

b. $(x + 2)(3x - 1)$
 $= 3x^2 - x + 6x - 2$
 $= 3x^2 + 5x - 2$

c.
$$
\begin{array}{r}
2x^2 - 3x + 4 \\
3x - 2 \\
\hline
6x^3 - 9x^2 + 12x \\
- 4x^2 + 6x - 8 \\
\hline
6x^3 - 13x^2 + 18x - 8
\end{array}
$$

To multiply a polynomial by a monomial, we apply the distributive property. To multiply two binomials we use the FOIL method. In other situations we use the Column method. Each method achieves the same result: To multiply any two polynomials, we multiply each term in the first polynomial by each term in the second polynomial.

Special Products [6.6]

8. $(x + 3)^2 = x^2 + 6x + 9$
 $(x - 3)^2 = x^2 - 6x + 9$
 $(x + 3)(x - 3) = x^2 - 9$

$\left.\begin{array}{l} (a + b)^2 = a^2 + 2ab + b^2 \\[6pt] (a - b)^2 = a^2 - 2ab + b^2 \end{array}\right\}$ *Binomial squares*

$(a + b)(a - b) = a^2 - b^2$ *Difference of two squares*

Dividing a Polynomial by a Monomial [6.7]

9. $\dfrac{12x^3 - 18x^2}{6x} = 2x^2 - 3x$

To divide a polynomial by a monomial, divide each term in the polynomial by the monomial.

Long Division with Polynomials [6.8]

10. $x^2 - 5x - 8 \div x - 3$

 $= x - 2 + \dfrac{2}{x - 3}$

$$
\begin{array}{r}
x - 2 \\
x - 3 \overline{)\, x^2 - 5x + 8} \\
\underline{\mp x^2 \pm 3x} \downarrow \\
- 2x + 8 \\
\underline{\pm 2x \mp 6} \\
2
\end{array}
$$

Division with polynomials is similar to long division with whole numbers. The steps in the process are estimate, multiply, subtract, and bring down the next term. The divisors in all the long-division problems in this chapter were binomials.

Simplify each of the following expressions. [6.1]

1. $(-3)^5$

2. $\left(\dfrac{2}{5}\right)^2$

3. $(2x^2)^3(3x^2)^2$

4. Write 2.47×10^5 in expanded form. [6.1]

Simplify each expression. Write all answers with positive exponents only. [6.2]

5. -3^{-2}

6. $(7a^2b^2)^1$

7. $\dfrac{a^{-4}}{a^{-1}}$

8. $\dfrac{(x^{-2})^3(x^{-4})^{-1}}{x}$

9. Write 0.0278×10^3 in scientific notation. [6.2]

Simplify. Write all answers with positive exponents only. [6.3]

10. $\dfrac{35x^3y^3z}{7xyz^3}$

11. $\dfrac{(3a^2b)(12a^2b)}{4ab^2}$

12. $\dfrac{24x^3}{3x^2} - \dfrac{33x^{-1}}{11x^{-2}}$

13. $\dfrac{(3.2 \times 10^{-3})(1.8 \times 10^2)}{3.6 \times 10^{-5}}$

Add or subtract as indicated. [6.4]

14. $8x^2 - 4x + 3x - 12$

15. $(3x^2 - x + 3) - (2x^2 - 8x + 3)$

16. Subtract $5x + 8$ from $-(x + 2)$.

17. Find the value of $2y^2 - y - 4$ when y is 3.

Multiply. [6.5]

18. $2a^2(3a^2 + 5a - 4)$

19. $\left(x - \dfrac{3}{2}\right)\left(x + \dfrac{1}{2}\right)$

20. $(4x + 3)(2x - 2)$

21. $(x + 3)(x - 3)$

Multiply. [6.6]

22. $(x - 3)^2(x + 3)^2$

23. $(2a - b)^2$

24. $(\sqrt{3}x - \sqrt{8}y)(\sqrt{3}x + \sqrt{8}y)$

25. $(a^2 + \sqrt{3})(a^2 - \sqrt{3})$

26. Divide $6x^2 - xy - 2y^2$ by $2x + y$. [4.8]

Divide. [6.8]

27. $\dfrac{3x^2 + 7x + 2}{x + 2}$

28. $\dfrac{6x^2 - 2x - 1}{3x - 1}$

29. **Volume** Find the volume of a cube if the length of a side is 0.5 feet. [6.1]

30. **Volume** Find the volume of a rectangular solid if the length is two thirds the width, and the height is one and a half times the width. [6.3]

The illustration below shows the top grossing Disney animated movies of all time.

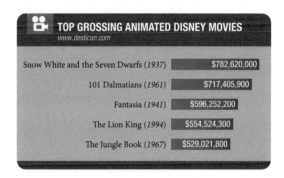

Write the total revenue for the given movies in scientific notation.

31. Snow White and the Seven Dwarfs

32. The Lion King

The illustration below shows the number of stars received by the top rated comic strips of all time at an online rating site.

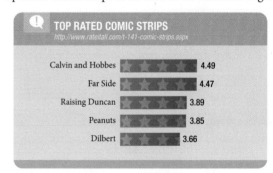

Write the difference in the number of stars received by the following strips in scientific notation.

33. Calvin and Hobbes compared to Far Side

34. Raising Duncan compared to Peanuts

Factoring Polynomials

Rio de Janiero, Brazil
© 2010 Google

7.1 The Greatest Common Factor, Factoring by Grouping

7.2 Factoring Trinomials of the Form $x^2 + bx + c$

7.3 Factoring Trinomials of the Form $ax^2 + bx + c$

7.4 The Difference of Two Squares and Perfect Square Trinomials

7.5 The Sum and Difference of Two Cubes

7.6 Factoring: A General Review

7.7 Solving Quadratic Equations by Factoring

7.8 Applications of Quadratic Equations

Upon its grand opening in 2004, the Millau Viaduct in France was the world's tallest bridge. It has a cable-stayed design, where cables hanging from columns attach to and support the road. The bridge stands at an impressive height of 1,125 feet at one column's summit. Slightly taller than the Eiffel Tower and only a few feet short of the Empire State Building, the viaduct is an architectural marvel. This huge structure crosses the valley of the River Tran and has greatly decongested traffic in the region since it opened.

Suppose you could stand at the top of one of the bridge's columns and toss a rock into the air with an upward velocity of 30 feet per second. As it falls to the River Tran below, the equation that gives the rock's height h at any time t is

$$h = 1125 + 30t - 16t^2$$

Solving the equation to find the height of the rock requires factoring the trinomial in the equation. We will work similar problems in this chapter.

©iStockphoto.com/JoeGough

OBJECTIVES

A Factor the greatest common factor from a polynomial.

B Factor polynomials by grouping.

KEY WORDS

greatest common factor

largest monomial

factoring by grouping

Note As you will see as we progress through the book, factoring is a tool that is used in solving a number of problems. Before seeing how it is used, however, we first must learn how to do it. So, in this section and the sections that follow, we will be developing our factoring skills.

**Video Examples
Section 7.1**

Image © sxc.hu, Kaliyoda, 2009

In flight archery, a person uses a bow to shoot an arrow into the air with a goal of achieving the greatest distance, as opposed to aiming for a specific target. The curved path that the arrow takes from the bow to the ground creates the shape of a parabola, which we will discuss further in Chapter 9. For now, let's say the arrow's path can be represented by the polynomial $-16x^2 + 62x + 8$. To begin factoring this polynomial, we need to find the greatest common factor, which is the primary focus of this section.

Recall the following diagram to illustrate the relationship between multiplication and factoring.

$$\text{Factors} \rightarrow \; 3 \cdot 5 = 15 \; \leftarrow \text{Product}$$

A similar relationship holds for multiplication of polynomials. Reading the following diagram from left to right, we say the product of the binomials $x + 2$ and $x + 3$ is the trinomial $x^2 + 5x + 6$. However, if we read in the other direction, we can say that $x^2 + 5x + 6$ factors into the product of $x + 2$ and $x + 3$.

$$\text{Factors} \; \rightarrow (x + 2)(x + 3) = x^2 + 5x + 6 \leftarrow \text{Product}$$

In this chapter, we develop a systematic method of factoring polynomials.

A Factoring The Greatest Common Factor

In this section, we will apply the distributive property to polynomials to factor from them what is called the greatest common factor.

> **DEFINITION** greatest common factor
>
> The *greatest common factor* for a polynomial is the largest monomial that divides (is a factor of) each term of the polynomial.

We use the term *largest monomial* to mean the monomial with the greatest coefficient and highest power of the variable that is a common factor of each term in the polynomial.

Practice Problems

EXAMPLE 1 Find the greatest common factor for the polynomial $3x^5 + 12x^2$.

Solution The terms of the polynomial are $3x^5$ and $12x^2$. The largest number that divides the coefficients is 3, and the highest power of x that is a factor of x^5 and x^2 is x^2. Therefore, the greatest common factor for $3x^5 + 12x^2$ is $3x^2$; that is, $3x^2$ is the largest monomial that divides each term of $3x^5 + 12x^2$.

1. Find the greatest common factor for $10x^6 + 15x^4$.

EXAMPLE 2 Find the greatest common factor for $8a^3b^2 + 16a^2b^3 + 20a^3b^3$.

Solution The largest number that divides each of the coefficients is 4. The highest power of the variable that is a factor of a^3b^2, a^2b^3, and a^3b^3 is a^2b^2. The greatest common factor for $8a^3b^2 + 16a^2b^3 + 20a^3b^3$ is $4a^2b^2$. It is the largest monomial that is a factor of each term.

2. Find the greatest common factor for $27a^4b^5 + 18a^3b^3 + 9a^3b^2$.

Once we have recognized the greatest common factor of a polynomial, we can apply the distributive property and factor it out of each term. We rewrite the polynomial as the product of its greatest common factor with the polynomial that remains after the greatest common factor has been factored from each term in the original polynomial.

EXAMPLE 3 Factor the greatest common factor from $3x - 15$.

Solution The greatest common factor for the terms $3x$ and 15 is 3. We can rewrite both $3x$ and 15 so that the greatest common factor 3 is showing in each term. It is important to realize that $3x$ means $3 \cdot x$. The 3 and the x are not "stuck" together.

$$3x - 15 = 3 \cdot x - 3 \cdot 5$$

Now, applying the distributive property, we have

$$3 \cdot x - 3 \cdot 5 = 3(x - 5)$$

To check a factoring problem like this, we can multiply 3 and $x - 5$ to get $3x - 15$, which is what we started with. Factoring is simply a procedure by which we change sums and differences into products. In this case, we changed the difference $3x - 15$ into the product $3(x - 5)$. Note, however, that we have not changed the meaning or value of the expression. The expression we end up with is equivalent to the expression we started with.

3. Factor the greatest common factor from $5x + 15$.

EXAMPLE 4 Factor the greatest common factor from $-5x^3 - 15x^2$.

Solution The greatest common factor is $-5x^2$. We rewrite the polynomial as

$$-5x^3 - 15x^2 = -5x^2 \cdot x - (-5x^2) \cdot 3$$

Then we apply the distributive property to get

$$-5x^2 \cdot x - (-5x^2) \cdot 3 = -5x^2(x + 3)$$

To check our work, we simply multiply $-5x^2$ and $(x + 3)$ to get $-5x^3 - 15x^2$, which is our original polynomial.

4. Factor the greatest common factor from $-25x^4 - 35x^3$.

EXAMPLE 5 Factor the greatest common factor from $16x^5 - 20x^4 + 8x^3$.

Solution The greatest common factor is $4x^3$. We rewrite the polynomial so we can see the greatest common factor $4x^3$ in each term; then we apply the distributive property to factor it out.

$$16x^5 - 20x^4 + 8x^3 = 4x^3 \cdot 4x^2 - 4x^3 \cdot 5x + 4x^3 \cdot 2$$
$$= 4x^3(4x^2 - 5x + 2)$$

5. Factor the greatest common factor from $20x^8 - 12x^7 + 16x^6$.

Answers

1. $5x^4$
2. $9a^3b^2$
3. $5(x + 3)$
4. $-5x^3(5x + 7)$
5. $4x^6(5x^2 - 3x + 4)$

6. Factor the greatest common factor from $8xy^3 - 16x^2y^2 + 8x^3y$.

EXAMPLE 6 Factor the greatest common factor from $6x^3y - 18x^2y^2 + 12xy^3$.

Solution The greatest common factor is $6xy$. We rewrite the polynomial in terms of $6xy$ and then apply the distributive property as follows:

$$6x^3y - 18x^2y^2 + 12xy^3 = 6xy \cdot x^2 - 6xy \cdot 3xy + 6xy \cdot 2y^2$$
$$= 6xy(x^2 - 3xy + 2y^2)$$

7. Factor the greatest common factor from $2ab^2 - 6a^2b^2 + 8a^3b^2$.

EXAMPLE 7 Factor the greatest common factor from $3a^2b - 6a^3b^2 + 9a^3b^3$.

Solution The greatest common factor is $3a^2b$.

$$3a^2b - 6a^3b^2 + 9a^3b^3 = 3a^2b(1) - 3a^2b(2ab) + 3a^2b(3ab^2)$$
$$= 3a^2b(1 - 2ab + 3ab^2)$$

B Factoring by Grouping

To develop our next method of factoring, called *factoring by grouping*, we start by examining the polynomial $xc + yc$. The greatest common factor for the two terms is c. Factoring c from each term we have

$$xc + yc = c(x + y)$$

But suppose that c itself was a more complicated expression, such as $a + b$, so that the expression we were trying to factor was $x(a + b) + y(a + b)$, instead of $xc + yc$. The greatest common factor for $x(a + b) + y(a + b)$ is $(a + b)$. Factoring this common factor from each term looks like this:

$$x(a + b) + y(a + b) = (a + b)(x + y)$$

To see how all of this applies to factoring polynomials, consider the polynomial

$$xy + 3x + 2y + 6$$

There is no greatest common factor other than the number 1. However, if we group the terms together two at a time, we can factor an x from the first two terms and a 2 from the last two terms.

$$xy + 3x + 2y + 6 = x(y + 3) + 2(y + 3)$$

The expression on the right can be thought of as having two terms: $x(y + 3)$ and $2(y + 3)$. Each of these expressions contains the common factor $y + 3$, which can be factored out using the distributive property in reverse.

$$x(y + 3) + 2(y + 3) = (y + 3)(x + 2)$$

This last expression is in factored form. The process we used to obtain it is called factoring by grouping. Here are some additional examples:

8. Factor $ax - bx + ay - by$.

EXAMPLE 8 Factor $ax + bx + ay + by$.

Solution We begin by factoring the greatest common factor x from the first two terms and the greatest common factor y from the last two terms.

$$ax + bx + ay + by = x(a + b) + y(a + b)$$
$$= (a + b)(x + y)$$

To convince yourself that this is factored correctly, multiply the two factors $(a + b)$ and $(x + y)$.

EXAMPLE 9 Factor $3ax - 2a + 15x - 10$ by grouping.

Solution First, we factor a from the first two terms and 5 from the last two terms. Then, we factor $3x - 2$ from the remaining two expressions.

$$3ax - 2a + 15x - 10 = a(3x - 2) + 5(3x - 2)$$
$$= (3x - 2)(a + 5)$$

Again, multiplying $(3x - 2)$ and $(a + 5)$ will convince you that these are the correct factors.

EXAMPLE 10 Factor $2x^2 + 5ax - 2xy - 5ay$.

Solution From the first two terms we factor x. From the second two terms we must factor $-y$ so that the binomial that remains after we do so matches the binomial produced by the first two terms.

$$2x^2 + 5ax - 2xy - 5ay = x(2x + 5a) - y(2x + 5a)$$
$$= (2x + 5a)(x - y)$$

Another way to accomplish the same result is to use the commutative property to interchange the middle two terms, and then factor by grouping.

$$2x^2 + 5ax - 2xy - 5ay = 2x^2 - 2xy + 5ax - 5ay \qquad \text{Commutative property}$$
$$= 2x(x - y) + 5a(x - y)$$
$$= (x - y)(2x + 5a)$$

This is the same result we obtained previously.

GETTING READY FOR CLASS

After reading through the preceding section, respond in your own words and in complete sentences.

A. What is the greatest common factor for a polynomial?

B. How would you use the distributive property when factoring a polynomial?

C. When would you try to factor by grouping?

D. After factoring a polynomial, how can you check your result?

9. Factor $5ax - 3ay + 10x - 6y$ by grouping.

10. Factor $3x^2 + 7bx - 3xy - 7by$.

Answers
9. $(5x - 3y)(a + 2)$
10. $(3x + 7b)(x - y)$

Vocabulary Review

Choose the correct words to fill in the blanks below.

 grouping coefficient monomial factors

1. The greatest common factor for a polynomial is the largest _____ that divides each term of the polynomial.

2. The largest monomial is the monomial with the greatest _____ and highest power of the variable.

3. Factoring by _____ is a method of factoring a polynomial that involves grouping terms two at a time to find common _____.

Problems

A Factor the following polynomials by using the greatest common factor.

1. $15x + 25$

2. $14x + 21$

3. $6a + 9$

4. $8a + 10$

5. $4x - 8y$

6. $9x - 12y$

7. $3x^2 - 6x + 9$

8. $2x^2 + 6x + 12$

9. $3a^2 - 3a + 60$

10. $2a^2 - 18a + 32$

11. $24y^2 - 52y + 36$

12. $18y^2 + 48y + 40$

13. $9x^2 - 8x^3$

14. $7x^3 - 4x^2$

15. $13a^2 - 26a^3$

16. $5a^2 - 10a^3$

17. $21x^2y - 28xy^2$

18. $30xy^2 - 25x^2y$

19. $22a^2b^2 - 11ab^2$

20. $15x^3 - 25x^2 + 30x$

21. $7x^3 + 21x^2 + 28x$

22. $16x^4 - 20x^2 - 16x$

23. $121y^4 + 11x^4$

24. $25a^4 + 5b^4$

25. $100x^4 - 50x^3 + 25x^2$

26. $36x^5 + 72x^3 - 81x^2$

27. $8a^2 + 16b^2 + 32c^2$

28. $9a^2 - 18b^2 - 27c^2$

29. $4a^2b - 16ab^2 + 32a^2b^2$

30. $5ab^2 + 10a^2b^2 + 15a^2b$

31. $6a^3b^2 + 6a^3b - 6a^2b^2$

32. $6ab^2 + 3ab^2 - 3b^3$

33. $121a^3b^2 - 22a^2b^3 + 33a^3b^3$ **34.** $20a^4b^3 - 18a^3b^4 + 22a^4b^4$ **35.** $12x^2y^3 - 72x^5y^3 - 36x^4y^4$ **36.** $49xy - 21x^2y^2 + 35x^3y^3$

B Factor the following polynomials by grouping.

37. $xy + 5x + 3y + 15$ **38.** $xy + 2x + 4y + 8$ **39.** $xy + 6x + 2y + 12$ **40.** $xy + 2y + 6x + 12$

41. $ab + 7a - 3b - 21$ **42.** $ab + 3b - 7a - 21$ **43.** $ax - bx + ay - by$ **44.** $ax - ay + bx - by$

45. $2ax + 6x - 5a - 15$ **46.** $3ax + 21x - a - 7$ **47.** $3xb - 4b - 6x + 8$ **48.** $3xb - 4b - 15x + 20$

49. $x^2 + ax + 2x + 2a$ **50.** $x^2 + ax + 3x + 3a$ **51.** $x^2 - ax - bx + ab$ **52.** $x^2 + ax - bx - ab$

Factor the following polynomials by grouping. You can group the terms together two at a time or three at a time. Either way will produce the same result.

53. $ax + ay + bx + by + cx + cy$ **54.** $ax + bx + cx + ay + by + cy$

Factor the following polynomials by grouping the terms together two at a time.

55. $6x^2 + 9x + 4x + 6$ **56.** $6x^2 - 9x - 4x + 6$ **57.** $20x^2 - 2x + 50x - 5$ **58.** $20x^2 + 25x + 4x + 5$

59. $20x^2 + 4x + 25x + 5$ **60.** $20x^2 + 4x - 25x - 5$ **61.** $x^3 + 2x^2 + 3x + 6$ **62.** $x^3 - 5x^2 + 4x - 20$

63. $6x^3 - 4x^2 + 15x - 10$ **64.** $8x^3 - 12x^2 + 14x - 21$ **65.** $3x^3 + 2x^2 - 15x - 10$ **66.** $4x^3 - 3x^2 + 8x - 6$

67. The greatest common factor of the binomial $3x + 6$ is 3, and the greatest common factor of the binomial $2x + 4$ is 2. What is the greatest common factor of their product $(3x + 6)(2x + 4)$ when it has been multiplied out?

68. The greatest common factors of the binomials $4x + 2$ and $5x + 10$ are 2 and 5, respectively. What is the greatest common factor of their product $(4x + 2)(5x + 10)$ when it has been multiplied out?

69. The following factorization is incorrect. Find the mistake, and correct the right-hand side.

$$12x^2 + 6x + 3 = 3(4x^2 + 2x)$$

70. Find the mistake in the following factorization, and then rewrite the right-hand side correctly.

$$10x^2 + 2x + 6 = 2(5x^2 + 3)$$

Applying the Concepts

71. Investing If you invest \$1,000 in an account with an annual interest rate of r compounded annually, the amount of money you have in the account after one year is

$$A = 1,000 + 1,000r$$

Write this formula again with the right side in factored form. Then, find the amount of money in this account at the end of one year if the interest rate is 12%.

72. Investing If you invest P dollars in an account with an annual interest rate of 8% compounded annually, then the amount of money in that account after one year is given by the formula

$$A = P + 0.08P$$

Rewrite this formula with the right side in factored form, and then find the amount of money in the account at the end of one year if \$500 was the initial investment.

73. Biological Growth If 1,000,000 bacteria are placed in a petri dish and the bacteria have a growth rate of r (a percent expressed as a decimal) per hour, then 1 hour later the amount of bacteria will be given by

$$A = 1,000,000 + 1,000,000r$$

a. Factor the right side of the equation.

b. If $r = 30\%$, find the number of bacteria present after one hour.

74. Biological Growth If there are B E. coli bacteria present initially in a petri dish and their growth rate is r (a percent expressed as a decimal) per hour, then after one hour the amount of bacteria present is given by

$$A = B + Br$$

a. Factor the right side of this equation.

b. The following bar graph shows the number of E. coli bacteria present initially and the number of bacteria present hours later. Use the bar chart to find B and A in the preceding equation.

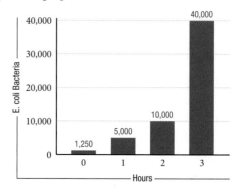

Getting Ready for the Next Section

Multiply.

75. $(x - 7)(x + 2)$

76. $(x - 7)(x - 2)$

77. $(x - 3)(x + 2)$

78. $(x + 3)(x - 2)$

79. $(x + 3)(x^2 - 3x + 9)$

80. $(x - 2)(x^2 + 2x + 4)$

81. $(2x + 1)(x^2 + 4x - 3)$

82. $(3x + 2)(x^2 - 2x - 4)$

83. $3x^4(6x^3 - 4x^2 + 2x)$

84. $2x^4(5x^3 + 4x^2 - 3x)$

85. $\left(x + \dfrac{1}{3}\right)\left(x + \dfrac{2}{3}\right)$

86. $\left(x + \dfrac{1}{4}\right)\left(x + \dfrac{3}{4}\right)$

87. $(6x + 4y)(2x - 3y)$

88. $(8a - 3b)(4a - 5b)$

89. $(9a + 1)(9a - 1)$

90. $(7b + 1)(7b + 1)$

91. $(x - 9)(x - 9)$

92. $(x - 8)(x - 8)$

93. $(x + 2)(x^2 - 2x + 4)$

94. $(x - 3)(x^2 + 3x + 9)$

Find the Mistake

Each sentence below contains a mistake. Circle the mistake and write the correct word(s) or expression(s) on the line provided.

1. The greatest common factor for a polynomial is the monomial with the smallest coefficient and highest power of the variable that divides each term of the polynomial. _____

2. The greatest common factor for the polynomial $12x^4 + 3x^3 - 6x$ is $12x^4$. _____

3. Using the process of factoring by grouping, we can factor the polynomial $8xy + 2x - 12y - 3$ to get $y(8x + 2x - 12 - 3)$.

4. To factor the polynomial $9xy - 27x - 2y + 6$ by grouping, factor the number 3 from the first, second, and third term. _____

Image © sxc.hu, bugdog, 2006

**Video Examples
Section 7.2**

The Peregrine falcon is the fastest-moving animal on Earth, with a diving speed of 200 miles per hour. The falcon has incredible eyesight, 6 times more powerful than a human, and can spot its prey scurrying on the ground from hundreds of feet in the air. The path of the falcon's dive to catch her prey takes the form of the trinomial $x^2 + bx + c$, where the values of the variables depend on the dive's height, length, and speed. In this section, we will factor this type of trinomial in which the coefficient of the squared term is 1.

The more familiar we are with multiplication of binomials the easier factoring trinomials will be. Therefore, recall multiplication of binomials from Chapter 4:

$$(x + 3)(x + 4) = x^2 + 7x + 12$$

$$(x - 5)(x + 2) = x^2 - 3x - 10$$

Using the FOIL method we learned in the last chapter, the first term in the answer is the product of the first terms in each binomial. The last term in the answer is the product of the last terms in each binomial. The middle term in the answer comes from adding the product of the outside terms to the product of the inside terms.

Let's have a and b represent real numbers and look at the product of $(x + a)$ and $(x + b)$.

$$(x + a)(x + b) = x^2 + ax + bx + ab$$

$$= x^2 + (a + b)x + ab$$

The coefficient of the middle term is the sum of a and b. The last term is the product of a and b. Writing this as a factoring problem, we have

$$x^2 + \underset{\underset{\text{Sum}}{\downarrow}}{(a + b)}x + \underset{\underset{\text{Product}}{\downarrow}}{ab} = (x + a)(x + b)$$

A Factoring with a Leading Coefficient of 1

To factor a trinomial in which the coefficient of x^2 is 1, we need only find the numbers a and b whose sum is the coefficient of the middle term and whose product is the constant term (last term).

EXAMPLE 1 Factor $x^2 + 8x + 12$.

Solution The coefficient of x^2 is 1. We need two numbers whose sum is 8 and whose product is 12. The numbers are 6 and 2.

$$x^2 + 8x + 12 = (x + 6)(x + 2)$$

We can easily check our work by multiplying $(x + 6)$ and $(x + 2)$.

$$\text{Check:} \quad (x + 6)(x + 2) = x^2 + 6x + 2x + 12$$
$$= x^2 + 8x + 12$$

EXAMPLE 2 Factor $x^2 - 2x - 15$.

Solution The coefficient of x^2 is again 1. We need to find a pair of numbers whose sum is -2 and whose product is -15. It can be helpful to create a table of products and sums to find our correct pair of numbers. Here are all the possibilities for products that are -15.

Products	Sums
$-1(15) = -15$	$-1 + 15 = 14$
$1(-15) = -15$	$1 + (-15) = -14$
$-5(3) = -15$	$-5 + 3 = -2$
$5(-3) = -15$	$5 + (-3) = 2$

The third line gives us what we want. The factors of $x^2 - 2x - 15$ are $(x - 5)$ and $(x + 3)$.

$$x^2 - 2x - 15 = (x - 5)(x + 3)$$

B Factoring Trinomials Using the Greatest Common Factor

In Example 3, we begin by factoring out the greatest common factor. The first step in factoring any trinomial is to look for the greatest common factor. If the trinomial in question has a greatest common factor other than 1, we factor it out first and then try to factor the trinomial that remains.

EXAMPLE 3 Factor $2x^2 + 10x - 28$.

Solution We begin by factoring out the greatest common factor, which is 2.

$$2x^2 + 10x - 28 = 2(x^2 + 5x - 14)$$

Now, we factor the remaining trinomial by finding a pair of numbers whose sum is 5 and whose product is -14. Here are the possibilities:

Products	Sums
$-1(14) = -14$	$-1 + 14 = 13$
$1(-14) = -14$	$1 + (-14) = -13$
$-7(2) = -14$	$-7 + 2 = -5$
$7(-2) = -14$	$7 + (-2) = 5$

From the last line, we see that the factors of $x^2 + 5x - 14$ are $(x + 7)$ and $(x - 2)$. Here is the complete problem:

$$2x^2 + 10x - 28 = 2(x^2 + 5x - 14)$$
$$= 2(x + 7)(x - 2)$$

Practice Problems

1. Factor $x^2 + 5x + 6$.

2. Factor $x^2 + 2x - 15$.

Note Again, we can check our results by multiplying our factors to see if their product is the original polynomial.

3. Factor $2x^2 - 2x - 24$.

Answers
1. $(x + 2)(x + 3)$
2. $(x + 5)(x - 3)$
3. $2(x - 4)(x + 3)$

4. Factor $-3x^3 - 18x^2 - 15x$.

EXAMPLE 4 Factor $-3x^3 + 3x^2 + 18x$.

Solution We begin by factoring out the greatest common factor, which is $-3x$. Note that when the leading term is negative, we use the negative sign in our greatest common factor. Then we factor the remaining trinomial. Without showing the table of products and sums as we did in Examples 2 and 3, here is the complete problem.

$$-3x^3 + 3x^2 + 18x = -3x(x^2 - x - 6)$$
$$= -3x(x - 3)(x + 2)$$

This approach may also be used to factor trinomials with two variables, as we do in the next example.

5. Factor $x^2 + 7xy + 12y^2$.

EXAMPLE 5 Factor $x^2 + 8xy + 12y^2$.

Solution This time we need two expressions whose product is $12y^2$ and whose sum is $8y$. The two expressions are $6y$ and $2y$ (see Example 1 in this section).

$$x^2 + 8xy + 12y^2 = (x + 6y)(x + 2y)$$

You should convince yourself that these factors are correct by finding their product.

Note Trinomials in which the coefficient of the second-degree term is 1 are the easiest to factor. Success in factoring any type of polynomial is directly related to the amount of time spent working the problems. The more we practice, the more accomplished we become at factoring.

GETTING READY FOR CLASS

After reading through the preceding section, respond in your own words and in complete sentences.

A. When the leading coefficient of a trinomial is 1, what is the relationship between the other two coefficients and the factors of the trinomial?

B. In words, explain how you would factor a trinomial with a leading coefficient other than 1.

C. How can you check to see that you have factored a trinomial correctly?

D. Describe how you would find the factors of $x^2 + 8x + 12$.

Answers

4. $-3x(x + 1)(x + 5)$
5. $(x + 4y)(x + 3y)$

Vocabulary Review

Choose the correct words to fill in the blanks in the paragraph below.

greatest common factor coefficient leading

trinomial sum product

To factor a trinomial in which the _____ of x^2 is 1, find the numbers a and b whose _____ is the coefficient of the middle term and whose _____ is the constant term. If a trinomial has a _____ coefficient other than 1, factor out the _____ first (if possible) and then factor the _____ that remains.

Problems

A Factor the following trinomials.

1. $x^2 + 7x + 12$ **2.** $x^2 + 7x + 10$ **3.** $x^2 + 3x + 2$ **4.** $x^2 + 7x + 6$

5. $a^2 + 10a + 21$ **6.** $a^2 - 7a + 12$ **7.** $x^2 - 7x + 10$ **8.** $x^2 - 3x + 2$

9. $y^2 - 10y + 21$ **10.** $y^2 - 7y + 6$ **11.** $x^2 - x - 12$ **12.** $x^2 - 4x - 5$

13. $y^2 + y - 12$ **14.** $y^2 + 3y - 18$ **15.** $x^2 + 5x - 14$ **16.** $x^2 - 5x - 24$

17. $r^2 - 8r - 9$ **18.** $r^2 - r - 2$ **19.** $x^2 - x - 30$ **20.** $x^2 + 8x + 12$

21. $a^2 + 15a + 56$ **22.** $a^2 - 9a + 20$ **23.** $y^2 - y - 42$ **24.** $y^2 + y - 42$

25. $x^2 + 13x + 42$ **26.** $x^2 - 13x + 42$ **27.** $a^2 - 2a - 63$ **28.** $a^2 - a - 132$

B Factor the following polynomials completely. First factor out the greatest common factor, and then factor the remaining trinomial.

29. $2x^2 + 6x + 4$

30. $3x^2 - 6x - 9$

31. $3a^2 - 3a - 60$

32. $2a^2 - 18a + 28$

33. $100x^2 - 500x + 600$

34. $100x^2 - 900x + 2{,}000$

35. $100p^2 - 1{,}300p + 4{,}000$

36. $100p^2 - 1{,}200p + 3{,}200$

37. $x^4 - x^3 - 12x^2$

38. $x^4 - 11x^3 + 24x^2$

39. $2r^3 + 4r^2 - 30r$

40. $5r^3 + 45r^2 + 100r$

41. $2y^4 - 6y^3 - 8y^2$

42. $3r^3 - 3r^2 - 6r$

43. $x^5 + 4x^4 + 4x^3$

44. $x^5 + 13x^4 + 42x^3$

45. $3y^4 - 12y^3 - 15y^2$

46. $5y^4 - 10y^3 + 5y^2$

47. $4x^4 - 52x^3 + 144x^2$

48. $3x^3 - 3x^2 - 18x$

49. $-2x^3 + 10x^2 - 12x$

50. $-5x^3 + 15x^2 + 20x$

51. $-7x^4 - 21x^3 - 14x^2$

52. $-x^5 + 6x^4 - 8x^3$

Factor the following trinomials.

53. $x^2 + 5xy + 6y^2$

54. $x^2 - 5xy + 6y^2$

55. $x^2 - 9xy + 20y^2$

56. $x^2 + 9xy + 20y^2$

57. $a^2 + 2ab - 8b^2$

58. $a^2 - 2ab - 8b^2$

59. $a^2 - 10ab + 25b^2$

60. $a^2 + 6ab + 9b^2$

61. $a^2 + 10ab + 25b^2$

62. $a^2 - 6ab + 9b^2$

63. $x^2 + 2xa - 48a^2$

64. $x^2 - 3xa - 10a^2$

65. $x^2 - 5xb - 36b^2$

66. $x^2 - 13xb + 36b^2$

67. $2x^3 - 4x^2y - 6xy^2$

68. $3x^3 - 15x^2y + 18xy^2$

69. $x^3y - 7x^2y^2 + 12xy^3$

70. $a^3b - a^2b^2 - 20ab^3$

71. $-2a^3 - 4a^2b + 16ab^2$

72. $-3a^2b - 15ab^2 - 18b^3$

73. $x^4 - 5x^2 + 6$

74. $x^6 - 2x^3 - 15$

75. $x^2 - 80x - 2,000$

76. $x^2 - 190x - 2,000$

77. $x^2 - x + \dfrac{1}{4}$

78. $x^2 - \dfrac{2}{3}x + \dfrac{1}{9}$

79. $x^2 + 0.6x + 0.08$

80. $x^2 + 0.8x + 0.15$

81. If one of the factors of $x^2 + 24x + 128$ is $x + 8$, what is the other factor?

82. If one factor of $x^2 + 260x + 2,500$ is $x + 10$, what is the other factor?

83. What polynomial, when factored, gives $(4x + 3)(x - 1)$?

84. What polynomial factors to $(4x - 3)(x + 1)$?

85. If $x + 3$ is a factor of $x^2 + bx - 15$, what is the value of b?

86. If $x - 4$ is a factor of $x^2 + bx - 20$, what is the value of b?

87. If $x + 5$ is a factor of $x^2 + 2x + c$, what is the value of c?

88. If $x - 4$ is a factor of $x^2 - 12x + c$, what is the value of c?

Maintaining Your Skills

89. iPod Memory The chart shows the memory capacity of each generation of new iPod. Use it to answer the following questions.

a. How much more memory was there in the 2003 generation as compared to the 2001?

b. How much more memory was there in the 2007 generation as compared to the 2003?

c. If the trend from 2007 to 2009 continued, how much capacity was there in the 2011 generation iPod?

90. Mac Users The chart shows the percentage increase in Mac users over a seven-year period.

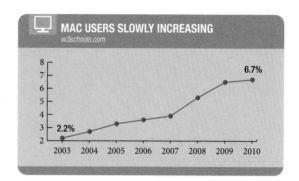

Let x represent the number of years past 2003 and y represent the percentages shown on the graph. Use the first and last points on the graph to find the equation of the line that connects them. Then use the equation to estimate the percentage increase in 2011.

Getting Ready for the Next Section

Multiply using the FOIL method.

91. $(6a + 1)(a + 2)$

92. $(6a - 1)(a - 2)$

93. $(3a + 2)(2a + 1)$

94. $(3a - 2)(2a - 1)$

95. $(6a + 2)(a + 1)$

96. $(3a + 1)(2a + 2)$

97. $2a(a + 4)(2a - 3)$

98. $3a(2a + 1)(3a - 1)$

Find the Mistake

Each sentence below contains a mistake. Circle the mistake and write the correct expression on the line provided.

1. The leading coefficient of the polynomial $x^2 - 2x - 35$ is 2. _____

2. Factoring the polynomial in the above sentence gives us $(x + 7)(x - 5)$. _____

3. For the polynomial $8x^3 - 30x^2 - 108x$, the greatest common factor is $8x^3$. _____

4. Factoring the polynomial in the above sentences gives us $(x - 6)(4x + 9)$. _____

7.3 Factoring Trinomials of the Form $ax^2 + bx + c$

Image © sxc.hu, Chemtec, 2010

OBJECTIVES

A Use trial and error to factor a trinomial whose leading coefficient is other than 1.

B Factor a polynomial by first factoring out the greatest common factor and then factoring the polynomial that remains.

KEY WORDS

trial and error

**Video Examples
Section 7.3**

The sport of fierljeppen originates in the Netherlands and requires a person to jump over a canal using a long pole. The pole is typically between 8 and 13 meters long and has a round plate at the bottom. This plate prevents the pole from sinking into the bottom of the body of water being jumped. The person sprints to the pole that is sticking out of a canal at an angle. Then he grabs the pole and tries to climb to the top to control its swing over the water. If well-trained, the person lands in a sand pit on the opposite side, instead of in the water.

Suppose the path of the jumper, once she grabs onto the pole, is given by the trinomial $-x^2 + 40x - 370$. In this section, we will consider similar trinomials whose greatest common factor is 1 and whose leading coefficient (the coefficient of the squared term) is a number other than 1.

A Factoring $ax^2 + bx + c$ by Trial and Error

Suppose we want to factor the trinomial $2x^2 - 5x - 3$. We know the factors (if they exist) will be a pair of binomials. The product of their first terms is $2x^2$ and the product of their last term is -3. Let us list all the possible factors along with the trinomial that would result if we were to multiply them together. Remember, the middle term comes from the product of the inside terms plus the product of the outside terms.

Binomial Factors	First Term	Middle Term	Last Term
$(2x - 3)(x + 1)$	$2x^2$	$-x$	-3
$(2x + 3)(x - 1)$	$2x^2$	$+x$	-3
$(2x - 1)(x + 3)$	$2x^2$	$+5x$	-3
$(2x + 1)(x - 3)$	$2x^2$	$-5x$	-3

We can see from the last line that the factors of $2x^2 - 5x - 3$ are $(2x + 1)$ and $(x - 3)$. There is no straightforward way, as there was in the previous section, to find the factors, other than by trial and error or by simply listing all the possibilities. We look for possible factors that, when multiplied, will give the correct first and last terms, and then we see if we can adjust them to give the correct middle term.

Practice Problems

1. Factor $2a^2 + 7a + 6$.

EXAMPLE 1 Factor $6a^2 + 7a + 2$.

Solution We list all the possible pairs of factors that, when multiplied together, give a trinomial whose first term is $6a^2$ and whose last term is 2.

Binomial Factors	First Term	Middle Term	Last Term
$(6a + 1)(a + 2)$	$6a^2$	$+13a$	$+2$
$(6a - 1)(a - 2)$	$6a^2$	$-13a$	$+2$
$(3a + 2)(2a + 1)$	$6a^2$	$+7a$	$+2$
$(3a - 2)(2a - 1)$	$6a^2$	$-7a$	$+2$

The factors of $6a^2 + 7a + 2$ are $(3a + 2)$ and $(2a + 1)$ since these give the correct middle term.

Check: $(3a + 2)(2a + 1) = 6a^2 + 7a + 2$

Note Remember, we can always check our results by multiplying the factors we have and comparing that product with our original polynomial.

Notice that in the preceding list we did not include the factors $(6a + 2)$ and $(a + 1)$. We do not need to try these since the first factor has a 2 common to each term and so could be factored again, giving $2(3a + 1)(a + 1)$. Since our original trinomial, $6a^2 + 7a + 2$, did *not* have a greatest common factor of 2, neither of its factors will.

EXAMPLE 2 Factor $4x^2 - x - 3$.

Solution We list all the possible factors that, when multiplied, give a trinomial whose first term is $4x^2$ and whose last term is -3.

Binomial Factors	First Term	Middle Term	Last Term
$(4x + 1)(x - 3)$	$4x^2$	$-11x$	-3
$(4x - 1)(x + 3)$	$4x^2$	$+11x$	-3
$(4x + 3)(x - 1)$	$4x^2$	$-x$	-3
$(4x - 3)(x + 1)$	$4x^2$	$+x$	-3
$(2x + 1)(2x - 3)$	$4x^2$	$-4x$	-3
$(2x - 1)(2x + 3)$	$4x^2$	$+4x$	-3

2. Factor $4x^2 - x - 5$.

The third line shows that the factors are $(4x + 3)$ and $(x - 1)$.

Check: $(4x + 3)(x - 1) = 4x^2 - x - 3$

You will find that the more practice you have at factoring this type of trinomial, the faster you will get the correct factors. You will pick up some shortcuts along the way, or you may come across a system of eliminating some factors as possibilities. Whatever works best for you is the method you should use. Factoring is a very important tool, and you must be good at it.

B Factoring Trinomials Using the Greatest Common Factor

Once again, the first step in any factoring problem is to factor out the greatest common factor if it is other than 1.

Answers

1. $(2a + 3)(a + 2)$
2. $(4x - 5)(x + 1)$

EXAMPLE 3 Factor $12y^3 + 10y^2 - 12y$.

Solution We begin by factoring out the greatest common factor, $2y$.

$$12y^3 + 10y^2 - 12y = 2y(6y^2 + 5y - 6)$$

We now list all possible factors of a trinomial with the first term $6y^2$ and last term -6, along with the associated middle terms.

Possible Factors	Middle Term When Multiplied
$(3y + 2)(2y - 3)$	$-5y$
$(3y - 2)(2y + 3)$	$+5y$
$(6y + 1)(y - 6)$	$-35y$
$(6y - 1)(y + 6)$	$+35y$

The second line gives the correct factors. The complete problem is

$$12y^3 + 10y^2 - 12y = 2y(6y^2 + 5y - 6)$$
$$= 2y(3y - 2)(2y + 3)$$

EXAMPLE 4 Factor $30x^2y - 5xy^2 - 10y^3$.

Solution The greatest common factor is $5y$.

$$30x^2y - 5xy^2 - 10y^3 = 5y(6x^2 - xy - 2y^2)$$
$$= 5y(2x + y)(3x - 2y)$$

EXAMPLE 5 A ball is tossed into the air with an upward velocity of 16 feet per second from the top of a building 32 feet high. The equation that gives the height of the ball above the ground at any time t is

$$h = 32 + 16t - 16t^2$$

Factor the right side of this equation and then find h when t is 2.

Solution We begin by factoring out the greatest common factor, 16. Then, we factor the trinomial that remains.

$$h = 32 + 16t - 16t^2$$
$$h = 16(2 + t - t^2)$$
$$h = 16(2 - t)(1 + t)$$

Letting $t = 2$ in the equation, we have

$$h = 16(0)(3) = 0$$

When t is 2, h is 0.

GETTING READY FOR CLASS

After reading through the preceding section, respond in your own words and in complete sentences.

A. Why would you use the trial and error method of factoring a trinomial?

B. Describe the criteria you would use to set up a table of possible factors of a trinomial.

C. What does it mean if you factor a trinomial and one of your factors has a greatest common factor of 3?

D. Describe how you would look for possible factors of $6a^2 + 7a + 2$.

3. Factor $30y^2 + 5y - 5$.

4. Factor $10x^2y^2 + 5xy^3 - 30y^4$.

5. Assume the ball in Example 5 is tossed from a building that is 48 feet tall and has an upward velocity of 32 feet per second. The equation that gives the height of the ball above the ground at any time t is $h = 48 + 32t - 16t^2$. Factor the right side of this equation and then find h when t is 3.

Answers

3. $5(3y - 1)(2y + 1)$
4. $5y^2(2x - 3y)(x + 2y)$
5. $h = 16(3 - t)(1 + t)$; When t is 3, h is 0.

Vocabulary Review

Choose the correct words to fill in the blanks in the paragraph below.

trial and error factors products polynomial multiply leading coefficient

To factor a polynomial whose _____ is something other than 1, list all the possibilities
in a process called _____. This process involves looking for possible _____
whose products will give the correct first and last terms, and whose sum of the _____ will
give the middle term. To check your results, _____ the factors and compare the product to the
original _____.

Problems

A Factor the following trinomials.

1. $2x^2 + 7x + 3$

2. $2x^2 + 5x + 3$

3. $2a^2 - a - 3$

4. $2a^2 + a - 3$

5. $3x^2 + 2x - 5$

6. $3x^2 - 2x - 5$

7. $3y^2 - 14y - 5$

8. $3y^2 + 14y - 5$

9. $6x^2 + 13x + 6$

10. $6x^2 - 13x + 6$

11. $4x^2 - 12xy + 9y^2$

12. $4x^2 + 12xy + 9y^2$

13. $4y^2 - 11y - 3$

14. $4y^2 + y - 3$

15. $20x^2 - 41x + 20$

16. $20x^2 + 9x - 20$

17. $20a^2 + 48ab - 5b^2$

18. $20a^2 + 29ab + 5b^2$

19. $20x^2 - 21x - 5$

20. $20x^2 - 48x - 5$

21. $12m^2 + 16m - 3$

22. $12m^2 + 20m + 3$

23. $20x^2 + 37x + 15$

24. $20x^2 + 13x - 15$

25. $12a^2 - 25ab + 12b^2$

26. $12a^2 + 7ab - 12b^2$

27. $3x^2 - xy - 14y^2$

28. $3x^2 + 19xy - 14y^2$

29. $14x^2 + 29x - 15$

30. $14x^2 + 11x - 15$

31. $6x^2 - 43x + 55$

32. $6x^2 - 7x - 55$

33. $15t^2 - 67t + 38$

34. $15t^2 - 79t - 34$

35. $28t^2 - 15t - 27$

36. $36t^2 + 33t - 20$

B Factor each of the following polynomials completely. Use the greatest common factor for your first step, then factor the remaining trinomial.

37. $4x^2 + 2x - 6$

38. $6x^2 - 51x + 63$

39. $24a^2 - 50a + 24$

40. $18a^2 + 48a + 32$

41. $10x^3 - 23x^2 + 12x$

42. $10x^4 + 7x^3 - 12x^2$

43. $6x^4 - 11x^3 - 10x^2$

44. $6x^3 + 19x^2 + 10x$

45. $10a^3 - 6a^2 - 4a$

46. $6a^3 + 15a^2 + 9a$

47. $15x^3 - 102x^2 - 21x$

48. $2x^4 - 24x^3 + 64x^2$

49. $35y^3 - 60y^2 - 20y$

50. $14y^4 - 32y^3 + 8y^2$

51. $15a^4 - 2a^3 - a^2$

52. $10a^5 - 17a^4 + 3a^3$

53. $24x^2y - 6xy - 45y$

54. $8x^2y^2 + 26xy^2 + 15y^2$

55. $12x^2y - 34xy^2 + 14y^3$

56. $12x^2y - 46xy^2 + 14y^3$

57. Evaluate the expression $2x^2 + 7x + 3$ and the expression $(2x + 1)(x + 3)$ for $x = 2$.

58. Evaluate the expression $2a^2 - a - 3$ and the expression $(2a - 3)(a + 1)$ for $a = 5$.

59. What polynomial factors to $(2x + 3)(2x - 3)$?

60. What polynomial factors to $(5x + 4)(5x - 4)$?

61. What polynomial factors to $(x + 3)(x - 3)(x^2 + 9)$?

62. What polynomial factors to $(x + 2)(x - 2)(x^2 + 4)$?

63. If $x + 3$ is a factor of $2x^2 + bx - 15$, what is the value of b?

64. If $2x - 5$ is a factor of $6x^2 + bx + 10$, what is the value of b?

65. If $3x + 4$ is a factor of $6x^2 - 7x + c$, what is the value of c?

66. If $5x + 3$ is a factor of $20x^2 - 13x + c$, what is the value of c?

Applying the Concepts

67. Archery Margaret shoots an arrow into the air. The equation for the height (in feet) of the tip of the arrow is

$$h = 8 + 62t - 16t^2$$

Factor the right side of this equation. Then fill in the table for various heights of the arrow, using the factored form of the equation.

Time t (seconds)	0	1	2	3	4
Height h (feet)					

68. Coin Toss At the beginning of every football game, the referee flips a coin to see who will kick off. The equation that gives the height (in feet) of the coin tossed in the air is

$$h = 6 + 29t - 16t^2$$

a. Factor this equation.

b. Use the factored form of the equation to find the height of the quarter after 0 seconds, 1 second, and 2 seconds.

69. Constructing a Box Yesterday I was experimenting with how to cut and fold a certain piece of cardboard to make a box with different volumes. Unfortunately, today I have lost both the cardboard and most of my notes. I remember that I made the box by cutting equal squares from the corners then folding up the side flaps.

I don't remember how big the cardboard was, and I can only find the last page of notes, which says that if x is the length of a side of a small square (in inches), then the volume is $V = 99x - 40x^2 + 4x^3$.

a. Factor the right side of this expression completely.

b. What were the dimensions of the original piece of cardboard?

70. Constructing a Box Repeat Problem 69 if the remaining formula is $V = 15x - 16x^2 + 4x^3$.

...

Getting Ready for the Next Section

Multiply each of the following.

71. $(x + 3)(x - 3)$

72. $(x - 4)(x + 4)$

73. $(x + 5)(x - 5)$

74. $(x - 6)(x + 6)$

75. $(x + 7)(x - 7)$

76. $(x - 8)(x + 8)$

77. $(x + 9)(x - 9)$

78. $(x - 10)(x + 10)$

79. $(2x - 3y)(2x + 3y)$

80. $(5x - 6y)(5x + 6y)$

81. $(x^2 + 4)(x + 2)(x - 2)$

82. $(x^2 + 9)(x + 3)(x - 3)$

83. $(x + 3)^2$

84. $(x - 4)^2$

85. $(x + 5)^2$

86. $(x - 6)^2$

87. $(x + 7)^2$

88. $(x - 8)^2$

89. $(x + 9)^2$

90. $(x - 10)^2$

91. $(2x + 3)^2$

92. $(3x - y)^2$

93. $(4x - 2y)^2$

94. $(5x - 6y)^2$

Find the Mistake

Each sentence below contains a mistake. Circle the mistake and write the correct word(s) or expression(s) on the line provided.

1. $(2x + 2)$ is a factor of the polynomial $6x^2 + 7x + 12$. _____

2. The first step in factoring $8a^3 - 4a^2 - 6a$ is to list all possible factors of a trinomial with the first term $8a^3$ and the last term $-6a$, along with the associated middle terms. _____

3. All of the factors for $8a^3 - 4a^2 - 6a$ are $(4a - 1)(a + 3)$. _____

4. The factors of the polynomial $40x^3z - 16x^2z - 12xz$ are $(5xz + 3)(2xz - 2)$ _____

Landmark Review: Checking Your Progress

1. What is the first thing you should look for when you begin to factor a polynomial?

Factor the following polynomials.

2. $4a^3 + 8a^2 - 10a$

3. $40x^2y + 24xy^2 + 16x^2y^2$

4. $x^2 - 8x - x + 8$

5. $xy + 6x - 2y - 12$

6. $x^3 - 7x^2 - 4x + 28$

7. $x^3 - 8x^2 + 6x - 48$

8. $x^2 + 10x + 21$

9. $x^2 - 3x - 28$

10. $x^2 - 7x + 12$

11. $x^2 + 2x - 24$

12. $6x^2 - 5xy - 6y^2$

13. $3a^2 + 11a + 10$

14. $6y^2 - 7y + 2$

15. $20x^2 - 49xy + 30y^2$

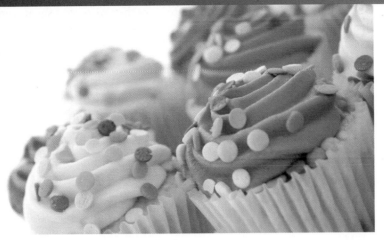

The cupcake craze is in full force with bakeries specializing in the small treats popping up all over the country. Suppose a bakery can bake x hundred cupcakes for a total cost given by the trinomial $100x^2 + 600x + 900$. Upon factoring out the greatest common factor of 100, we find the remaining expression contains a special trinomial called a perfect square trinomial, which we will discuss in this section.

$$100x^2 + 600x + 900 = 100(x^2 + 6x + 9)$$

But first, let's examine how three special products we listed in the previous chapter can also be considered as three special factors.

Recall the following three special products:

$$(a + b)^2 = (a + b)(a + b) = a^2 + 2ab + b^2$$
$$(a - b)^2 = (a - b)(a - b) = a^2 - 2ab + b^2$$
$$(a + b)(a - b) = a^2 - b^2$$

Since factoring is the reverse of multiplication, we can also consider the three special products as three special factorings:

$$a^2 + 2ab + b^2 = (a + b)^2$$
$$a^2 - 2ab + b^2 = (a - b)^2$$
$$a^2 - b^2 = (a + b)(a - b)$$

Any trinomial of the form $a^2 + 2ab + b^2$ or $a^2 - 2ab + b^2$ can be factored by the methods of Section 4.6. The last line is the factoring to obtain the *difference of two squares*. The difference of two squares always factors in this way. Again, these are patterns you must be able to recognize on sight.

A The Difference of Two Squares

EXAMPLE 1 Factor $16x^2 - 25$.

Solution We can see that the first term is a perfect square, and the last term is also. This fact becomes even more obvious if we rewrite the problem as

$$16x^2 - 25 = (4x)^2 - (5)^2$$

The first term is the square of the quantity $4x$, and the last term is the square of 5. The completed problem looks like this:

$$16x^2 - 25 = (4x)^2 - (5)^2$$
$$= (4x + 5)(4x - 5)$$

To check our results, we multiply

$$(4x + 5)(4x - 5) = 16x^2 + 20x - 20x - 25$$
$$= 16x^2 - 25$$

**Video Examples
Section 7.4**

Practice Problems

1. Factor $9x^2 - 25$.

Answer

1. $(3x + 5)(3x - 5)$

2. Factor $49a^2 - 1$.

3. Factor $a^4 - b^4$.

Note If you think the sum of two squares $x^2 + y^2$ factors, you should try it. Write down the factors you think it has, and then multiply them using the FOIL method. You won't get $x^2 + y^2$. This type of polynomial that cannot be factored is considered *prime*.

4. Factor $4x^2 + 12x + 9$.

Note As we have indicated before, perfect square trinomials like the ones in Examples 4 and 5 can be factored by the methods developed in previous sections. Recognizing that they factor to binomial squares simply saves time in factoring.

5. Factor $6x^2 + 24x + 24$.

EXAMPLE 2 Factor $36a^2 - 1$.

Solution We rewrite the two terms to show they are perfect squares and then factor. Remember, 1 is its own square, $1^2 = 1$.

$$36a^2 - 1 = (6a)^2 - (1)^2$$
$$= (6a + 1)(6a - 1)$$

To check our results, we multiply

$$(6a + 1)(6a - 1) = 36a^2 + 6a - 6a - 1$$
$$= 36a^2 - 1$$

EXAMPLE 3 Factor $x^4 - y^4$.

Solution x^4 is the perfect square $(x^2)^2$, and y^4 is $(y^2)^2$.

$$x^4 - y^4 = (x^2)^2 - (y^2)^2$$
$$= (x^2 - y^2)(x^2 + y^2)$$

Notice the factor $(x^2 - y^2)$ is itself the difference of two squares and therefore can be factored again. The factor $(x^2 + y^2)$ is the *sum* of two squares and cannot be factored again. The complete problem is this:

$$x^4 - y^4 = (x^2)^2 - (y^2)^2$$
$$= (x^2 - y^2)(x^2 + y^2)$$
$$= (x + y)(x - y)(x^2 + y^2)$$

B Perfect Square Trinomials

EXAMPLE 4 Factor $25x^2 - 60x + 36$.

Solution Although this trinomial can be factored by the method we used in Section 7.3, we notice that the first and last terms are the perfect squares $(5x)^2$ and $(6)^2$. Before going through the method for factoring trinomials by listing all possible factors, we can check to see if $25x^2 - 60x + 36$ factors to $(5x - 6)^2$. We need only multiply to check.

$$(5x - 6)^2 = (5x - 6)(5x - 6)$$
$$= 25x^2 - 30x - 30x + 36$$
$$= 25x^2 - 60x + 36$$

The trinomial $25x^2 - 60x + 36$ factors to $(5x - 6)(5x - 6) = (5x - 6)^2$ and therefore, it is called a *perfect square trinomial*.

EXAMPLE 5 Factor $5x^2 + 30x + 45$.

Solution We begin by factoring out the greatest common factor, which is 5. Then we notice that the trinomial that remains is a perfect square trinomial.

$$5x^2 + 30x + 45 = 5(x^2 + 6x + 9)$$
$$= 5(x + 3)^2$$

Answers
2. $(7a + 1)(7a - 1)$
3. $(a^2 + b^2)(a + b)(a - b)$
4. $(2x + 3)^2$
5. $6(x + 2)^2$

EXAMPLE 6 Factor $(x - 3)^2 - 25$.

Solution This example has the form $a^2 - b^2$, where a is $x - 3$ and b is 5. We factor it according to the formula for the difference of two squares.

$$(x - 3)^2 - 25 = (x - 3)^2 - 5^2 \qquad \text{Write 25 as } 5^2.$$
$$= [(x - 3) - 5][(x - 3) + 5] \qquad \text{Factor.}$$
$$= (x - 8)(x + 2) \qquad \text{Simplify.}$$

Notice in this example we could have expanded $(x - 3)^2$, subtracted 25, and then factored to obtain the same result.

$$(x - 3)^2 - 25 = x^2 - 6x + 9 - 25 \qquad \text{Expand } (x \cdot 3)^2.$$
$$= x^2 - 6x - 16 \qquad \text{Simplify.}$$
$$= (x - 8)(x + 2) \qquad \text{Factor.}$$

6. Factor $(x - 4)^2 - 36$.

EXAMPLE 7 Factor each of the following.

a. $16x^2 + 56x + 49$ **b.** $16x^2 - 56x + 49$ **c.** $16x^2 - 49$ **d.** $16x^2 + 49$

Solution We use the formulas for perfect square trinomials and the difference of two squares whenever we can.

a. $16x^2 + 56x + 49 = (4x + 7)(4x + 7)$
$$= (4x + 7)^2$$

b. $16x^2 - 56x + 49 = (4x - 7)(4x - 7)$
$$= (4x - 7)^2$$

c. $16x^2 - 49 = (4x)^2 - (7)^2$
$$= (4x - 7)(4x + 7)$$

d. $16x^2 + 49$ represents the sum of two squares and cannot be factored. We say that it is prime.

7. Factor each of the following:
 a. $9x^2 + 30x + 25$
 b. $9x^2 - 30x + 25$
 c. $9x^2 - 25$
 d. $9x^2 + 25$

GETTING READY FOR CLASS

After reading through the preceding section, respond in your own words and in complete sentences.

A. Describe how you factor the difference of two squares.

B. What is a perfect square trinomial?

C. How do you know when you've factored completely?

D. Describe how you would factor $25x^2 - 60x + 36$.

Answers
 6. $(x - 10)(x + 2)$
 7. a. $(3x + 5)^2$ **b.** $(3x - 5)^2$
 c. $(3x + 5)(3x - 5)$ **d.** Prime

Vocabulary Review

Choose the correct words to fill in the blanks below.

 sum square perfect squares difference of two squares trinomial

1. If the first and last terms of a binomial are _____ and the operation symbol that separates the two terms implies subtraction, then the binomial is a special factoring called a _____ .

2. The _____ of two squares $(x^2 + y^2)$ cannot be factored.

3. A perfect square _____ has first and last terms that are perfect squares.

4. If you _____ a binomial, your result will be a perfect square trinomial.

Problems

A Factor the following binomials.

1. $x^2 - 9$

2. $x^2 - 25$

3. $a^2 - 36$

4. $a^2 - 64$

5. $x^2 - 49$

6. $x^2 - 121$

7. $4a^2 - 16$

8. $4a^2 + 16$

9. $9x^2 + 25$

10. $16x^2 - 36$

11. $25x^2 - 169$

12. $x^2 - y^2$

13. $9a^2 - 16b^2$

14. $49a^2 - 25b^2$

15. $9 - m^2$

16. $16 - m^2$

17. $25 - 4x^2$

18. $36 - 49y^2$

19. $2x^2 - 18$

20. $3x^2 - 27$

21. $32a^2 - 128$

22. $3a^3 - 48a$

23. $8x^2y - 18y$

24. $50a^2b - 72b$

25. $a^4 - b^4$ **26.** $a^4 - 16$ **27.** $16m^4 - 81$ **28.** $81 - m^4$

29. $3x^3y - 75xy^3$ **30.** $2xy^3 - 8x^3y$ **31.** $-3x^3 + 48x$ **32.** $-5y^3 + 45y$

B Factor the following trinomials.

33. $x^2 - 2x + 1$ **34.** $x^2 - 6x + 9$ **35.** $x^2 + 2x + 1$ **36.** $x^2 + 6x + 9$

37. $a^2 - 10a + 25$ **38.** $a^2 + 10a + 25$ **39.** $y^2 + 4y + 4$ **40.** $y^2 - 8y + 16$

41. $x^2 - 4x + 4$ **42.** $x^2 + 8x + 16$ **43.** $m^2 - 12m + 36$ **44.** $m^2 + 12m + 36$

45. $4a^2 + 12a + 9$ **46.** $9a^2 - 12a + 4$ **47.** $49x^2 - 14x + 1$ **48.** $64x^2 - 16x + 1$

49. $9y^2 - 30y + 25$ **50.** $25y^2 + 30y + 9$ **51.** $x^2 + 10xy + 25y^2$ **52.** $25x^2 + 10xy + y^2$

53. $9a^2 + 6ab + b^2$ **54.** $9a^2 - 6ab + b^2$ **55.** $4x^2 - 12xy + 9y^2$ **56.** $9x^2 + 24xy + 16y^2$

Factor the following by first factoring out the greatest common factor.

57. $3a^2 + 18a + 27$ **58.** $4a^2 - 16a + 16$ **59.** $2x^2 + 20xy + 50y^2$ **60.** $3x^2 + 30xy + 75y^2$

61. $5x^3 + 30x^2y + 45xy^2$ **62.** $12x^2y - 36xy^2 + 27y^3$ **63.** $-4x^3 + 24x^2 - 36x$ **64.** $-3x^2y + 12xy^2 - 3y^3$

Factor by grouping the first three terms together.

65. $x^2 + 6x + 9 - y^2$ **66.** $x^2 + 10x + 25 - y^2$ **67.** $x^2 + 2xy + y^2 - 9$ **68.** $a^2 + 2ab + b^2 - 25$

69. Find a value for b so that the polynomial $x^2 + bx + 49$ factors to $(x + 7)^2$.

70. Find a value of b so that the polynomial $x^2 + bx + 81$ factors to $(x + 9)^2$.

71. Find the value of b for which the polynomial $x^2 + bx + 25$ factors to $(x - 5)^2$.

72. Find the value of b for which the polynomial $x^2 + bx + 64$ factors to $(x - 8)^2$.

73. Find the value of c for which the polynomial $x^2 + 10x + c$ factors to $(x + 5)^2$.

74. Find the value of a for which the polynomial $ax^2 + 12x + 9$ factors to $(2x + 3)^2$.

75. Find the value of c for which the polynomial $x^2 - 8x + c$ factors to $(x - 4)^2$.

76. Find the value of c for which the polynomial $x^2 - 12x + c$ factors to $(x - 6)^2$.

77. Find the value of a for which the polynomial $ax^2 - 12x + 4$ factors to $(3x - 2)^2$.

78. Find the value of a for which the polynomial $ax^2 - 8x + 1$ factors to $(4x - 1)^2$.

Applying the Concepts

79. Area

 a. What is the area of the following figure?

 b. Factor the answer from part a.

 c. Find a way to cut the figure into two pieces and put them back together to show that the factorization in part b is correct.

80. Area

 a. What is the area of the following figure?

 b. Factor the expression from part a.

 c. Cut and rearrange the figure to show that the factorization is correct.

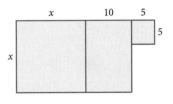

Find the area for the shaded regions; then write your result in factored form.

81.

82.

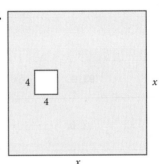

Getting Ready for the Next Section

Multiply each of the following:

83. a. 1^3

 b. 2^3

 c. 3^3

 d. 4^3

 e. 5^3

84. a. $(-1)^3$

 b. $(-2)^3$

 c. $(-3)^3$

 d. $(-4)^3$

 e. $(-5)^3$

85. a. $x(x^2 - x + 1)$

 b. $1(x^2 - x + 1)$

 c. $(x + 1)(x^2 - x + 1)$

86. a. $x(x^2 + x + 1)$

 b. $-1(x^2 + x + 1)$

 c. $(x - 1)(x^2 + x + 1)$

87. a. $x(x^2 - 2x + 4)$

 b. $2(x^2 - 2x + 4)$

 c. $(x + 2)(x^2 - 2x + 4)$

88. a. $x(x^2 + 2x + 4)$

 b. $-2(x^2 + 2x + 4)$

 c. $(x - 2)(x^2 + 2x + 4)$

89. a. $x(x^2 - 3x + 9)$

 b. $3(x^2 - 3x + 9)$

 c. $(x + 3)(x^2 - 3x + 9)$

90. a. $x(x^2 + 3x + 9)$

 b. $-3(x^2 + 3x + 9)$

 c. $(x - 3)(x^2 + 3x + 9)$

91. a. $x(x^2 - 4x + 16)$

 b. $4(x^2 - 4x + 16)$

 c. $(x + 4)(x^2 - 4x + 6)$

92. a. $x(x^2 + 4x + 16)$

 b. $-4(x^2 + 4x + 16)$

 c. $(x - 4)(x^2 + 4x + 16)$

93. a. $x(x^2 - 5x + 25)$

 b. $5(x^2 - 5x + 25)$

 c. $(x + 5)(x^2 - 5x + 25)$

94. a. $x(x^2 + 5x + 25)$

 b. $-5(x^2 + 5x + 25)$

 c. $(x - 5)(x^2 + 5x + 25)$

Find the Mistake

Each sentence below contains a mistake. Circle the mistake and write the correct word(s) or expression on the line provided.

1. The difference of two squares says that $a^2 - b^2 = (a - b)^2$. _____

2. Factoring $81y^6 - 196$ gives us $(9y^3 - 14)(9y^3 - 14)$. _____

3. The polynomial $18x^2 - 84x + 98$ is an example of a difference of two squares. _____

4. The polynomial in the above sentence factors to $3x - 7$. _____

OBJECTIVES

A Factor the sum or difference of two cubes.

KEY WORDS

sum of two cubes

difference of two cubes

A company that sells mountain biking apparel has just received an order for a new helmet and gloves. The helmet is packaged in a box with a volume of a^3 and the gloves are packaged in a box with a volume of b^3. In order to figure out the minimum volume needed for a shipping box to contain both items, we need to set up the following expression:

$$a^3 + b^3$$

This expression can also be called the sum of two cubes, which we were learn how to factor in this section.

Previously, we factored a variety of polynomials. Among the polynomials we factored were polynomials that were the difference of two squares. The formula we used to factor the difference of two squares looks like this:

$$a^2 - b^2 = (a + b)(a - b)$$

If we ran across a binomial that had the form of the difference of two squares, we factored it by applying this formula. For example, to factor $x^2 - 25$, we simply notice that it can be written in the form $x^2 - 5^2$, which looks like the difference of two squares. According to the formula above, this binomial factors into $(x + 5)(x - 5)$.

A Factoring the Sum or Difference of Two Cubes

In this section, we want to use two new formulas that will allow us to factor the sum and difference of two cubes. For example, we want to factor the binomial $x^3 - 8$, which is the difference of two cubes. (To see that it is the difference of two cubes, notice that it can be written $x^3 - 2^3$.) We also want to factor $y^3 + 27$, which is the sum of two cubes. (To see this, notice that $y^3 + 27$ can be written as $y^3 + 3^3$.)

The formulas that allow us to factor the sum of two cubes and the difference of two cubes are not as simple as the formula for factoring the difference of two squares. Here is what they look like:

$$a^3 + b^3 = (a + b)(a^2 - ab + b^2) \qquad \text{Sum of two cubes}$$
$$a^3 - b^3 = (a - b)(a^2 + ab + b^2) \qquad \text{Difference of two cubes}$$

Let's begin our work with these two formulas by showing that they are true. To do so, we multiply out the right side of each formula.

**Video Examples
Section 7.5**

EXAMPLE 1 Verify the formula for a sum of two cubes and the formula for a difference of two cubes.

Solution We verify the formulas by multiplying the right sides and comparing the results with the left sides.

$$
\begin{array}{r}
a^2 - ab + b^2 \\
\underline{a\ \ + b} \\
a^3 - a^2b + ab^2 \\
\underline{a^2b - ab^2 + b^3} \\
a^3 \qquad\qquad + b^3
\end{array}
\qquad\qquad
\begin{array}{r}
a^2 + ab\ + b^2 \\
\underline{a\ \ - b} \\
a^3 + a^2b + ab^2 \\
\underline{-a^2b - ab^2 - b^3} \\
a^3 \qquad\qquad - b^3
\end{array}
$$

The first formula is correct. The second formula is correct.

Here are some examples that use the formulas for factoring the sum and difference of two cubes.

EXAMPLE 2 Factor $x^3 - 8$.

Solution Since the two terms are prefect cubes, we write them as such and apply the formula.

$$x^3 - 8 = x^3 - 2^3$$
$$= (x - 2)(x^2 + 2x + 2^2)$$
$$= (x - 2)(x^2 + 2x + 4)$$

EXAMPLE 3 Factor $y^3 + 27$.

Solution Proceeding as we did in Example 2, we first write 27 as 3^3. Then, we apply the formula for factoring the sum of two cubes, which is $a^3 + b^3 = (a + b)(a^2 - ab + b^2)$.

$$y^3 + 27 = y^3 + 3^3$$
$$= (y + 3)(y^2 - 3y + 3^2)$$
$$= (y + 3)(y^2 - 3y + 9)$$

EXAMPLE 4 Factor $a^3 - \dfrac{1}{8}$.

Solution The first term is the cube of a, the second term is the cube of $\frac{1}{2}$.
$$a^3 - \frac{1}{8} = a^3 - \left(\frac{1}{2}\right)^3$$
$$= \left(a - \frac{1}{2}\right)\left(a^2 + \frac{1}{2}a + \frac{1}{4}\right)$$

EXAMPLE 5 Factor $x^6 - y^6$.

Solution To begin, we have a choice of how we want to write the two terms. We can write the expression as the difference of two squares, $(x^3)^2 - (y^3)^2$, or as the difference of two cubes, $(x^2)^3 - (y^2)^3$. It is easier to use the difference of two squares if we have a choice.

$$x^6 - y^6 = (x^3)^2 - (y^3)^2$$
$$= (x^3 - y^3)(x^3 + y^3)$$
$$= (x - y)(x^2 + xy + y^2)(x + y)(x^2 - xy + y^2)$$

Practice Problems

2. Factor $x^3 - 27$.

3. Factor $y^3 + 64$.

4. Factor $x^3 + \dfrac{1}{125}$.

5. Factor $a^6 + b^6$.

Answers

2. $(x - 3)(x^2 + 3x + 9)$

3. $(y + 4)(y^2 - 4y + 16)$

4. $\left(x + \dfrac{1}{5}\right)\left(x^2 - \dfrac{x}{5} + \dfrac{1}{25}\right)$

5. $(a^2 + b^2)(a^4 - a^2b^2 + b^4)$

Vocabulary Review

Choose the correct words to fill in the blanks below.

sum difference cubes factored

1. The sum of two _____ is given by $a^3 + b^3 = (a + b)(a^2 - ab + b^2)$.

2. The _____ of two cubes is given by $a^3 - b^3 = (a - b)(a^2 + ab + b^2)$.

3. The sum of two cubes can be _____, whereas the sum of two squares cannot.

4. The expression $27 + x^6$ is an example of a _____ of two cubes $3^3 + (x^2)^3$.

Problems

A Factor each of the following polynomials.

1. $x^3 - y^3$

2. $x^3 + y^3$

3. $a^3 + 8$

4. $a^3 - 8$

5. $27 + x^3$

6. $27 - x^3$

7. $y^3 - 1$

8. $y^3 + 1$

9. $y^3 - 64$

10. $y^3 + 64$

11. $125h^3 - t^3$

12. $t^3 + 125h^3$

13. $x^3 - 216$

14. $216 + x^3$

15. $2y^3 - 54$

16. $81 + 3y^3$

17. $2a^3 - 128b^3$

18. $128a^3 + 2b^3$

19. $2x^3 + 432y^3$

20. $432x^3 - 2y^3$

21. $10a^3 - 640b^3$

22. $640a^3 + 10b^3$

23. $10r^3 - 1,250$

24. $10r^3 + 1,250$

25. $64 + 27a^3$

26. $27 - 64a^3$

27. $8x^3 - 27y^3$

28. $27x^3 - 8y^3$

29. $t^3 + \dfrac{1}{27}$

30. $t^3 - \dfrac{1}{27}$

31. $27x^3 - \dfrac{1}{27}$

32. $8x^3 + \dfrac{1}{8}$

33. $64a^3 + 125b^3$

34. $125a^3 - 27b^3$

35. $\dfrac{1}{8}x^3 - \dfrac{1}{27}y^3$

36. $\dfrac{1}{27}x^3 + \dfrac{1}{8}y^3$

37. $a^6 - b^6$

38. $x^6 - 64y^6$

39. $64x^6 - y^6$

40. $x^6 - (3y)^6$

41. $x^6 - (5y)^6$

42. $(4x)^6 - (7y)^6$

43. $x^{27} - 27$

44. $x^{15} + 8$

45. Find a value for c so that $x + c$ is a factor of $x^3 + 8$.

46. Find a value for c so that $x + c$ is a factor of $x^3 - 27$.

47. Find a value for c so that $x + 5$ is a factor of $x^3 + c$.

48. Find a value for c so that $x - 4$ is a factor of $x^3 + c$.

Getting Ready for the Next Section

Multiply each of the following.

49. $2x^3(x + 2)(x - 2)$

50. $3x^2(x + 3)(x - 3)$

51. $3x^2(x - 3)^2$

52. $2x^2(x + 5)^2$

53. $y(y^2 + 25)$

54. $y^3(y^2 + 36)$

55. $(5a - 2)(3a + 1)$

56. $(3a - 4)(2a - 1)$

57. $4x^2(x - 5)(x + 2)$

58. $6x(x - 4)(x + 2)$

59. $2ab^3(b^2 - 4b + 1)$

60. $2a^3b(a^2 + 3a + 1)$

Find the Mistake

Each sentence below contains a mistake. Circle the mistake and write the correct word or expression(s) on the line provided.

1. The difference of two squares says that $a^3 - b^3 = (a - b)(a^2 + ab + b^2)$. _____

2. The expression $a^3 + 8b^3$ is an example of a difference of two cubes. _____

3. The factors of the sum of two cubes in the above sentence are $(a + 8)$ and $(a^2 - ab + 64)$.

4. Factoring $x^{12} - 343$ gives us $(x^4 - 7)(x^8 - 49)$. _____

7.6 Factoring: A General Review

OBJECTIVES

A Factor a variety of polynomials.

©iStockphoto.com/xof711

The Norwegian fjords are a popular location for wingsuit flying. Many of the thrill-seekers opt to leap, flip, or be thrown off the edge of a sheer cliff to begin their flight. A wingsuit flyer wears a special jumpsuit that adds surface area to his body to create lift during a flight. The suit contains extra fabric that stretches from each arm to the flyer's body, as well as between the flyer's legs. At a designated altitude, the flyer may deploy a parachute from a backpack to ensure a safe landing.

Suppose the wingsuit flyer's flight path from the time he jumps to the time he opens the parachute can be given by the following expression:

$$-9x^3 - 12x^2 + 4x + 40$$

In this section, we will review the different methods of factoring that we presented in the previous sections of the chapter. This section is important because it will give you an opportunity to factor a variety of polynomials such as the one above. Prior to this section, the polynomials you worked with were grouped together according to the method used to factor them; that is, in Section 7.4 all the polynomials you factored were either the difference of two squares or perfect square trinomials. What usually happens in a situation like this is that you become proficient at factoring the kind of polynomial you are working with at the time but have trouble when given a variety of polynomials to factor.

**Video Examples
Section 7.6**

A Factoring Polynomials of Any Type

We begin this section with a checklist that can be used in factoring polynomials of any type. When you have finished this section and the problem set that follows, you want to be proficient enough at factoring that the checklist is second nature to you.

> **HOW TO** Factor a Polynomial
>
> **Step 1:** If the polynomial has a greatest common factor other than 1, then factor out the greatest common factor.
>
> **Step 2:** If the polynomial has two terms (a binomial), then see if it is the difference of two squares or the sum or difference of two cubes, then factor accordingly. Remember, if it is the sum of two squares, it will not factor.
>
> **Step 3:** If the polynomial has three terms (a trinomial), then either it is a perfect square trinomial, which will factor into the square of a binomial, or it is not a perfect square trinomial, in which case you can try to use the trial and error method developed in Section 5.3.
>
> **Step 4:** If the polynomial has more than three terms, try to factor it by grouping.
>
> **Step 5:** As a final check, see if any of the factors you have written can be factored further. If you have overlooked a common factor, you can catch it here. Remember, some polynomials cannot be factored, in which case they are considered prime.

Practice Problems

1. Factor $3x^4 - 27x^2$.

Here are some examples illustrating how we use the checklist.

EXAMPLE 1 Factor $2x^5 - 8x^3$.

Solution First, we check to see if the greatest common factor is other than 1. Since the greatest common factor is $2x^3$, we begin by factoring it out. Once we have done so, we notice that the binomial that remains is the difference of two squares.

$$2x^5 - 8x^3 = 2x^3(x^2 - 4) \qquad \textit{Factor out the greatest common factor, } 2x^3.$$
$$= 2x^3(x + 2)(x - 2) \qquad \textit{Factor the difference of two squares.}$$

Note that the greatest common factor $2x^3$ that we factored from each term in the first step of Example 1 remains as part of the answer to the problem, because it is one of the factors of the original binomial. Remember, the expression we end up with when factoring must be equal to the expression we start with. We can't just drop a factor and expect the resulting expression to equal the original expression.

2. Factor $2x^5 + 20x^4 + 50x^3$.

EXAMPLE 2 Factor $3x^4 - 18x^3 + 27x^2$.

Solution Step 1 is to factor out the greatest common factor, $3x^2$. After we have done so, we notice that the trinomial that remains is a perfect square trinomial, which will factor as the square of a binomial.

$$3x^4 - 18x^3 + 27x^2 = 3x^2(x^2 - 6x + 9) \quad \textit{Factor out } 3x^2.$$
$$= 3x^2(x - 3)^2 \qquad \textit{x}^2 - 6x + 9 \textit{ is the square of } x - 3.$$

3. Factor $y^5 + 36y^3$.

EXAMPLE 3 Factor $y^3 + 25y$.

Solution We begin by factoring out the y that is common to both terms. The binomial that remains after we have done so is the sum of two squares, which does not factor, so after the first step we are finished.

$$y^3 + 25y = y(y^2 + 25) \qquad \textit{Factor out the greatest common factor, } y; \textit{ then,}$$
$$\textit{notice that } y^2 + 25 \textit{ cannot factored further.}$$

4. Factor $15a^2 - a - 2$.

EXAMPLE 4 Factor $6a^2 - 11a + 4$.

Solution Here we have a trinomial that does not have a greatest common factor other than 1. Since it is not a perfect square trinomial, we factor it by trial and error; that is, we look for binomial factors of the product whose first terms is $6a^2$ and whose last terms is 4. Then we look for the combination of these types of binomials whose product gives us a middle term of $-11a$. Without showing all the different possibilities, here is the answer:

$$6a^2 - 11a + 4 = (3a - 4)(2a - 1)$$

5. Factor $4x^4 - 12x^3 - 40x^2$.

EXAMPLE 5 Factor $6x^3 - 12x^2 - 48x$.

Solution This trinomial has a greatest common factor of $6x$. The trinomial that remains after the $6x$ has been factored from each term must be factored by trial and error.

$$6x^3 - 12x^2 - 48x = 6x(x^2 - 2x - 8)$$
$$= 6x(x - 4)(x + 2)$$

6. Factor $3ab + 9a + 2b + 6$.

EXAMPLE 6 Factor $xy + 8x + 3y + 24$.

Solution Since our polynomial has four terms, we try factoring by grouping.

$$xy + 8x + 3y + 24 = x(y + 8) + 3(y + 8)$$
$$= (y + 8)(x + 3)$$

Answers

1. $3x^2(x + 3)(x - 3)$
2. $2x^3(x + 5)^2$
3. $y^3(y^2 + 36)$
4. $(5a - 2)(3a + 1)$
5. $4x^2(x - 5)(x + 2)$
6. $(b + 3)(3a + 2)$

Vocabulary Review

The following is a list of steps for factoring. Choose the correct words to fill in the blanks below.

factors cubes greatest common factor trial and error

grouping perfect square trinomial squares

Step 1: If the polynomial has a _____ other than 1, then factor out the greatest common factor.

Step 2: If the polynomial has two terms, then see if it is the difference of two _____ or the sum or difference of two _____, then factor accordingly.

Step 3: If the polynomial has three terms, then either it is a _____, which will factor into the square of a binomial, or it is not a perfect square trinomial, which will factor using _____.

Step 4: If the polynomial has more than three terms, try to factor it by _____.

Step 5: As a final check, see if any _____ can be factored further.

Problems

A Factor each of the following polynomials completely; that is, once you are finished factoring, none of the factors you obtain should be factorable. Also, note that the even-numbered problems are not necessarily similar to the odd-numbered problems that precede them in this exercise set.

1. $x^2 - 81$

2. $x^2 - 18x + 81$

3. $x^2 + 2x - 15$

4. $15x^2 + 11x - 6$

5. $x^2 + 6x + 9$

6. $12x^2 - 11x + 2$

7. $y^2 - 10y + 25$

8. $21y^2 - 25y - 4$

9. $2a^3b + 6a^2b + 2ab$

10. $6a^2 - ab - 15b^2$

11. $x^2 + x + 1$

12. $2x^2 - 4x + 2$

13. $12a^2 - 75$

14. $18a^2 - 50$

15. $9x^2 - 12xy + 4y^2$

16. $x^3 - x^2$

17. $4x^3 + 16xy^2$

18. $16x^2 + 49y^2$

19. $2y^3 + 20y^2 + 50y$

20. $3y^2 - 9y - 30$

21. $a^6 + 4a^4b^2$

22. $5a^2 - 45b^2$

23. $xy + 3x + 4y + 12$

24. $xy + 7x + 6y + 42$

25. $x^4 - 16$

26. $x^4 - 81$

27. $xy - 5x + 2y - 10$

28. $xy - 7x + 3y - 21$

29. $5a^2 + 10ab + 5b^2$

30. $3a^3b^2 + 15a^2b^2 + 3ab^2$

31. $x^2 + 49$

32. $16 - x^4$

33. $3x^2 + 15xy + 18y^2$

34. $3x^2 + 27xy + 54y^2$

35. $2x^2 + 15x - 38$

36. $2x^2 + 7x - 85$

37. $100x^2 - 300x + 200$

38. $100x^2 - 400x + 300$

39. $x^2 - 64$

40. $9x^2 - 4$

41. $x^2 + 3x + ax + 3a$

42. $x^2 + 4x + bx + 4b$

43. $49a^7 - 9a^5$

44. $a^4 - 1$

45. $49x^2 + 9y^2$

46. $12x^4 - 62x^3 + 70x^2$

47. $25a^3 + 20a^2 + 3a$

48. $36a^4 - 100a^2$

49. $xa - xb + ay - by$

50. $xy - bx + ay - ab$

51. $48a^4b - 3a^2b$

52. $18a^4b^2 - 12a^3b^3 + 8a^2b^4$

53. $20x^4 - 45x^2$

54. $16x^3 + 16x^2 + 3x$

55. $3x^2 + 35xy - 82y^2$

56. $3x^2 + 37xy - 86y^2$

57. $16x^5 - 44x^4 + 30x^3$

58. $16x^2 + 16x - 1$

59. $2x^2 + 2ax + 3x + 3a$

60. $2x^2 + 2ax + 5x + 5a$

61. $y^4 - 1$

62. $25y^7 - 16y^5$

63. $12x^4y^2 + 36x^3y^3 + 27x^2y^4$

64. $16x^3y^2 - 4xy^2$

65. $-4x^8 - 32x^5$

66. $-3y^4 + 81y$

The following problems require you to factor out a common binomial factor first.

 For example,

 the expression \rightarrow $(x + 2)^3 - 3(x + 2)^2$

 can be factored as \rightarrow $(x + 2)^2[(x + 2) - 3]$

 which simplifies to \rightarrow $(x + 2)^2(x - 1)$.

Use this technique in problems 67–74.

67. $(x + 4)^3 - 5(x + 4)^2$

68. $(x - 5)^3 + 2(x - 5)^2$

69. $(x + 2)^4 + 5(x + 2)^2$

70. $(x - 3)^4 + 4(x - 3)^2$

71. $(2x + 1)^3 - (2x + 1)$

72. $(3x - 2)^3 - (3x - 2)$

73. $(2x - 3)^3 - 2(2x - 3)$

74. $(3x - 1)^3 - (3x - 1)$

Maintaining Your Skills

75. Facebook Users The chart shows the rise in Facebook users over a six-year period. Give five ordered pairs shown on the chart.

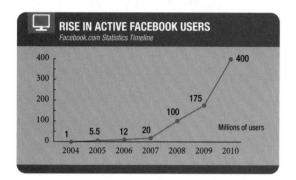

76. Fox Shows The chart lists the top primetime shows on the Fox network. Write a mathematical statement using one of the symbols $<$ or $>$ to compare the following:

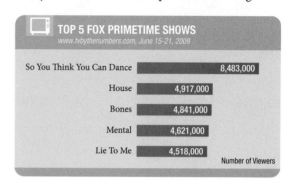

a. *House* to *Lie to Me*

b. *Mental* to *So You Think You Can Dance*

c. *Bones* to *Lie to Me*

Getting Ready for the Next Section

Solve each equation.

77. $3x - 6 = 9$

78. $5x - 1 = 14$

79. $2x + 3 = 0$

80. $4x - 5 = 0$

81. $4x + 3 = 0$

82. $3x - 1 = 0$

Find the Mistake

Each sentence below contains a mistake. Circle the mistake and write the correct word(s) or expression on the line provided.

1. If a binomial is the sum of two squares, it will factor into two consecutive integers. _____

2. To factor $14y^4 - 126y^2$, we first factor out the greatest common factor y^2. _____

3. Once you factored out the greatest common factor from the polynomial in the previous sentence, you are left with a perfect square trinomial. _____

4. To factor the polynomial $9x^2 + 3xy - 4y + 16x^2$, group the first and last terms together and the middle two terms together. _____

OBJECTIVES

A Solve a quadratic equation by writing it in standard form and then factoring.

KEY WORDS

quadratic equation

standard form

quadratic term

linear term

constant term

zero-factor property

In this section, we will use the methods of factoring developed in previous sections, along with a special property of 0, to solve quadratic equations.

Picture a football player throwing a ball straight up in the air. The distance the ball is from the ground can be represented by the quadratic equation $d = -16t^2 + 80t$ where t is the time the ball is in the air. If given a value for d, we can put this equation into standard form, which will enable us to solve for t. Here is a formal definition for a quadratic equation and standard form.

A Solving a Quadratic Equation by Factoring

DEFINITION quadratic equation

Any equation that can be put in the form $ax^2 + bx + c = 0$, where a, b, and c are real numbers ($a \neq 0$), is called a **quadratic equation.** The equation $ax^2 + bx + c = 0$ is called **standard form** for a quadratic equation.

$$\underbrace{a(\text{variable})^2}_{x^2 \text{ term}} + \underbrace{b(\text{variable})}_{x \text{ term}} + \underbrace{c(\text{absence of the variable})}_{\text{Constant term}} = 0$$

The x^2 term is called the *quadratic term* and the x term is called the *linear term*.

The number 0 has a special property. If we multiply two numbers and the product is 0, then one or both of the original two numbers must be 0. In symbols, this property looks like this:

PROPERTY Zero-Factor Property

Let a and b represent real numbers. If $a \cdot b = 0$, then $a = 0$ or $b = 0$.

Suppose we want to solve the quadratic equation $x^2 + 5x + 6 = 0$. We can factor the left side into $(x + 2)(x + 3)$. Then we have

$$x^2 + 5x + 6 = 0$$
$$(x + 2)(x + 3) = 0$$

Now, $(x + 2)$ and $(x + 3)$ both represent real numbers when we replace x with a number. Their product is 0; therefore, either $(x + 3)$ is 0 or $(x + 2)$ is 0. Either way we have a solution to our equation. We use the zero-factor property stated to finish the problem.

$$x^2 + 5x + 6 = 0$$
$$(x + 2)(x + 3) = 0$$
$$x + 2 = 0 \quad \text{or} \quad x + 3 = 0$$
$$x = -2 \qquad \qquad x = -3$$

Note A quadratic equation in standard form may also use a variable other than x, or be missing the linear or constant term, such as the equation about the football in our opener.

**Video Examples
Section 7.7**

Our solution set is $\{-2, -3\}$. Our equation has two solutions. To check our solutions we have to check each one separately to see that they both produce a true statement when used in place of the variable.

$$\text{When} \rightarrow \qquad\qquad\qquad\qquad x = -3$$
$$\text{the equation} \rightarrow \qquad\qquad x^2 + 5x + 6 = 0$$
$$\text{becomes} \rightarrow \qquad (-3)^2 + 5(-3) + 6 \overset{?}{=} 0$$
$$9 + (-15) + 6 = 0$$
$$0 = 0$$

$$\text{When} \rightarrow \qquad\qquad\qquad\qquad x = -2$$
$$\text{the equation} \rightarrow \qquad\qquad x^2 + 5x + 6 = 0$$
$$\text{becomes} \rightarrow \qquad (-2)^2 + 5(-2) + 6 \overset{?}{=} 0$$
$$4 + (-10) + 6 = 0$$
$$0 = 0$$

We have solved a quadratic equation by replacing it with two linear equations in one variable. Based on this process, we have developed the following strategy to solve a quadratic equation by factoring.

HOW TO Strategy for Solving a Quadratic Equation by Factoring

Step 1: Put the equation in standard form; that is, 0 on one side and decreasing powers of the variable on the other.

Step 2: Factor completely.

Step 3: Use the zero-factor property to set each variable factor from Step 2 equal to 0.

Step 4: Solve each equation produced in Step 3.

Step 5: Check each solution, if necessary.

Practice Problems

1. Solve the equation $2x^2 + 5x = 3$.

EXAMPLE 1 Solve the equation $2x^2 - 5x = 12$.

Solution

Step 1: Begin by adding -12 to both sides, so the equation is in standard form.

$$2x^2 - 5x = 12$$
$$2x^2 - 5x - 12 = 0$$

Step 2: Factor the left side completely.

$$(2x + 3)(x - 4) = 0$$

Step 3: Set each factor equal to 0.

$$2x + 3 = 0 \qquad \text{or} \qquad x - 4 = 0$$

Step 4: Solve each of the equations from step 3.

$$2x + 3 = 0 \qquad\qquad x - 4 = 0$$
$$2x = -3 \qquad\qquad\quad x = 4$$
$$x = -\frac{3}{2}$$

Step 5: Substitute each solution into $2x^2 - 5x = 12$ to check.

$$\text{Check: } -\frac{3}{2} \qquad\qquad \text{Check: } 4$$

$$2\left(-\frac{3}{2}\right)^2 - 5\left(-\frac{3}{2}\right) \overset{?}{=} 12 \qquad\qquad 2(4)^2 - 5(4) \overset{?}{=} 12$$

$$2\left(\frac{9}{4}\right) + 5\left(\frac{3}{2}\right) = 12 \qquad\qquad 2(16) - 20 = 12$$

$$\frac{9}{2} + \frac{15}{2} = 12 \qquad\qquad 32 - 20 = 12$$

$$\frac{24}{2} = 12 \qquad\qquad 12 = 12$$

$$12 = 12$$

EXAMPLE 2 Solve the equation $16a^2 - 25 = 0$.

Solution The equation is already in standard form.

$$16a^2 - 25 = 0$$

$(4a - 5)(4a + 5) = 0$	*Factor the left side.*
$4a - 5 = 0 \quad\text{or}\quad 4a + 5 = 0$	*Set each factor equal to 0.*
$4a = 5 \qquad\qquad 4a = -5$	*Solve the resulting equations.*
$a = \dfrac{5}{4} \qquad\qquad a = -\dfrac{5}{4}$	

2. Solve the equation $49a^2 - 16 = 0$.

EXAMPLE 3 Solve the equation $4x^2 = 8x$.

Solution We begin by adding $-8x$ to each side of the equation to put it in standard form. Then we factor the left side of the equation by factoring out the greatest common factor.

$$4x^2 = 8x$$

$4x^2 - 8x = 0$	*Add $-8x$ to each side.*
$4x(x - 2) = 0$	*Factor the left side.*
$4x = 0 \quad\text{or}\quad x - 2 = 0$	*Set each factor equal to 0.*
$x = 0 \qquad\qquad x = 2$	*Solve the resulting equations.*

The solutions are 0 and 2.

3. Solve the equation $5x^2 = -2x$.

EXAMPLE 4 Solve the equation $x(2x + 3) = 44$.

Solution We must multiply out the left side first and then put the equation in standard form.

$$x(2x + 3) = 44$$

$2x^2 + 3x = 44$	*Multiply out the left side.*
$2x^2 + 3x - 44 = 0$	*Add -44 to each side.*
$(2x + 11)(x - 4) = 0$	*Factor the left side.*
$2x + 11 = 0 \quad\text{or}\quad x - 4 = 0$	*Set each factor equal to 0.*
$2x = -11 \qquad\qquad x = 4$	*Solve the resulting equations.*
$x = -\dfrac{11}{2}$	

The two solutions are $-\dfrac{11}{2}$ and 4.

4. Solve the equation $x(13 - x) = 40$.

Answers

2. $-\dfrac{4}{7}, \dfrac{4}{7}$

3. $0, -\dfrac{2}{5}$

4. $5, 8$

5. Solve the equation
$(x + 2)^2 = (x + 1)^2 + x^2$.

EXAMPLE 5 Solve the equation $5^2 = x^2 + (x + 1)^2$.

Solution Before we can put this equation in standard form we must square the binomial. Remember, to square a binomial, we use the formula $(a + b)^2 = a^2 + 2ab + b^2$.

$$5^2 = x^2 + (x + 1)^2$$
$$25 = x^2 + x^2 + 2x + 1 \qquad \text{Expand } 5^2 \text{ and } (x+1)^2.$$
$$25 = 2x^2 + 2x + 1 \qquad \text{Simplify the right side.}$$
$$0 = 2x^2 + 2x - 24 \qquad \text{Add } -25 \text{ to each side.}$$
$$0 = 2(x^2 + x - 12) \qquad \text{Begin factoring.}$$
$$0 = 2(x + 4)(x - 3) \qquad \text{Factor completely.}$$
$$x + 4 = 0 \quad \text{or} \quad x - 3 = 0 \qquad \text{Set each variable factor equal to 0.}$$
$$x = -4 \qquad\qquad x = 3$$

Note, in the second to the last line, that we do not set 2 equal to 0. That is because 2 can never be 0. It is always 2. We only use the zero-factor property to set variable factors equal to 0 because they are the only factors that can possibly be 0.

 Also notice that it makes no difference which side of the equation is 0 when we write the equation in standard form.

Although the equation in the next example is not a quadratic equation, it can be solved by the method shown in the first five examples.

6. Solve the equation
$4x^3 = x^2 + 5x$.

EXAMPLE 6 Solve the equation $24x^3 = -10x^2 + 6x$.

Solution First, we write the equation in standard form.

$$24x^3 + 10x^2 - 6x = 0 \qquad \text{Standard form}$$
$$2x(12x^2 + 5x - 3) = 0 \qquad \text{Factor out } 2x.$$
$$2x(3x - 1)(4x + 3) = 0 \qquad \text{Factor remaining trinomial.}$$
$$2x = 0 \quad \text{or} \quad 3x - 1 = 0 \quad \text{or} \quad 4x + 3 = 0 \qquad \text{Set factors equal to 0 to find solutions.}$$
$$x = 0 \qquad\qquad x = \frac{1}{3} \qquad\qquad x = -\frac{3}{4}$$

GETTING READY FOR CLASS

After reading through the preceding section, respond in your own words and in complete sentences.

A. When is an equation in standard form?

B. What is the first step in solving an equation by factoring?

C. Describe the zero-factor property in your own words.

D. Describe how you would solve $2x^2 - 5x = 12$ for x.

Answers

5. $-1, 3$

6. $0, \frac{5}{4}, -1$

Vocabulary Review

Choose the correct words to fill in the blanks below.

constant standard form zero-factor property quadratic equation

1. A _____ is any equation that can be put in the form $ax^2 + bx + c = 0$, where a, b, and c are real numbers and a does not equal zero.

2. The equation $ax^2 + bx + c = 0$ is called _____ for a quadratic equation.

3. The variable c is considered the _____ term of a quadratic equation.

4. The _____ states that if $a \cdot b = 0$, then $a = 0$ or $b = 0$.

The following is a list of steps for solving a quadratic equation by factoring. Write the correct step number in the blanks.

_____ Use the zero-factor property to set each variable factor equal to zero.

_____ Solve the equation.

_____ Factor completely.

_____ Check each solution.

_____ Put the equation in standard form.

Problems

A The following equations are already in factored form. Use the zero-factor property to set the factors equal to 0 and solve.

1. $(x + 2)(x - 1) = 0$

2. $(x + 3)(x + 2) = 0$

3. $(a - 4)(a - 5) = 0$

4. $(a + 6)(a - 1) = 0$

5. $x(x + 1)(x - 3) = 0$

6. $x(2x + 1)(x - 5) = 0$

7. $(3x + 2)(2x + 3) = 0$

8. $(4x - 5)(x - 6) = 0$

9. $m(3m + 4)(3m - 4) = 0$

10. $m(2m - 5)(3m - 1) = 0$

11. $2y(3y + 1)(5y + 3) = 0$

12. $3y(2y - 3)(3y - 4) = 0$

Solve the following equations.

13. $x^2 + 3x + 2 = 0$

14. $x^2 - x - 6 = 0$

15. $x^2 - 9x + 20 = 0$

16. $x^2 + 2x - 3 = 0$

17. $a^2 - 2a - 24 = 0$

18. $a^2 - 11a + 30 = 0$

19. $100x^2 - 500x + 600 = 0$

20. $100x^2 - 300x + 200 = 0$

21. $x^2 = -6x - 9$

22. $x^2 = 10x - 25$

23. $a^2 - 16 = 0$

24. $a^2 - 36 = 0$

25. $2x^2 + 5x - 12 = 0$

26. $3x^2 + 14x - 5 = 0$

27. $9x^2 + 12x + 4 = 0$

28. $12x^2 - 24x + 9 = 0$

29. $a^2 + 25 = 10a$

30. $a^2 + 16 = 8a$

31. $2x^2 = 3x + 20$

32. $6x^2 = x + 2$

33. $3m^2 = 20 - 7m$

34. $2m^2 = -18 + 15m$

35. $4x^2 - 49 = 0$

36. $16x^2 - 25 = 0$

37. $x^2 + 6x = 0$

38. $x^2 - 8x = 0$

39. $x^2 - 3x = 0$

40. $x^2 + 5x = 0$

41. $2x^2 = 8x$

42. $2x^2 = 10x$

43. $3x^2 = 15x$

44. $5x^2 = 15x$

45. $1,400 = 400 + 700x - 100x^2$

46. $2,700 = 700 + 900x - 100x^2$

47. $6x^2 = -5x + 4$

48. $9x^2 = 12x - 4$

49. $x(2x - 3) = 20$

50. $x(3x - 5) = 12$

51. $t(t + 2) = 80$

52. $t(t + 2) = 99$

53. $4{,}000 = (1{,}300 - 100p)p$

54. $3{,}200 = (1{,}200 - 100p)p$

55. $x(14 - x) = 48$

56. $x(12 - x) = 32$

57. $(x + 5)^2 = 2x + 9$

58. $(x + 7)^2 = 2x + 13$

59. $(y - 6)^2 = y - 4$

60. $(y + 4)^2 = y + 6$

61. $10^2 = (x + 2)^2 + x^2$

62. $15^2 = (x + 3)^2 + x^2$

63. $2x^3 + 11x^2 + 12x = 0$

64. $3x^3 + 17x^2 + 10x = 0$

65. $4y^3 - 2y^2 - 30y = 0$

66. $9y^3 + 6y^2 - 24y = 0$

67. $8x^3 + 16x^2 = 10x$

68. $24x^3 - 22x^2 = -4x$

69. $20a^3 = -18a^2 + 18a$

70. $12a^3 = -2a^2 + 10a$

71. $16t^2 - 32t + 12 = 0$

72. $16t^2 - 64t + 48 = 0$

Simplify each side as much as possible, then solve the equation.

73. $(a - 5)(a + 4) = -2a$

74. $(a + 2)(a - 3) = -2a$

75. $3x(x + 1) - 2x(x - 5) = -42$

76. $4x(x - 2) - 3x(x - 4) = -3$

77. $2x(x + 3) = x(x + 2) - 3$

78. $3x(x - 3) = 2x(x - 4) + 6$

79. $a(a - 3) + 6 = 2a$

80. $a(a - 4) + 8 = 2a$

81. $15(x + 20) + 15x = 2x(x + 20)$

82. $15(x + 8) + 15x = 2x(x + 8)$

83. $15 = a(a + 2)$

84. $6 = a(a - 5)$

Use factoring by grouping to solve the following equations.

85. $x^3 + 3x^2 - 4x - 12 = 0$

86. $x^3 + 5x^2 - 9x - 45 = 0$

87. $x^3 + x^2 - 16x - 16 = 0$

88. $4x^3 + 12x^2 - 9x - 27 = 0$

Getting Ready for the Next Section

Write each sentence as an algebraic equation.

89. The product of two consecutive integers is 72.

90. The product of two consecutive even integers is 80.

91. The product of two consecutive odd integers is 99.

92. The product of two consecutive odd integers is 63.

93. The product of two consecutive even integers is 10 less than 5 times their sum.

94. The product of two consecutive odd integers is 1 less than 4 times their sum.

The following word problems are taken from the book *Academic Algebra*, written by William J. Milne and published by the American Book Company in 1901. Solve each problem.

95. Cost of a Bicycle and a Suit A bicycle and a suit cost $90. How much did each cost, if the bicycle cost 5 times as much as the suit?

96. Cost of a Cow and a Calf A man bought a cow and a calf for $36, paying 8 times as much for the cow as for the calf. What was the cost of each?

97. Cost of a House and a Lot A house and a lot cost $3,000. If the house cost 4 times as much as the lot, what was the cost of each?

98. Daily Wages A plumber and two helpers together earned $7.50 per day. How much did each earn per day, if the plumber earned 4 times as much as each helper?

Find the Mistake

Each sentence below contains a mistake. Circle the mistake and write the correct word(s) or equation(s) on the line provided.

1. Standard form for a linear equation is $a^2 + bx + c = 0$. _____

2. The zero-factor property says that if $(a+8)(b-3) = 0$, then $a + 8 = b - 3$. _____

3. To solve the equation $x^2 = -3x + 108$, we must first factor both sides. _____

4. The factors for the equation $3x^2 + 77 = 40x$ are $x = 7$ and $x = -11$. _____

Photo courtesy of Megan Silva

OBJECTIVES

A Solve number problems by factoring.

B Solve geometry problems by factoring.

C Solve business problems by factoring.

D Use the Pythagorean theorem to solve problems.

KEY WORDS

Pythagorean theorem

hypotenuse

Imagine you went backpacking through Evolution Valley in Kings Canyon National Park, California. You hiked directly east along one leg of a trail for 2 miles, then turned and hiked north for 3 more miles. If you were to draw a straight line from where you began to where you are now, what would be the distance? In order to solve this problem, we need to learn about the Pythagorean theorem which we will introduce in this section.

The Pythagorean theorem is one of the oldest theorems in the history of mathematics. The person whose name we associate with the theorem, Pythagoras (of Samos), was a Greek philosopher and mathematician who lived from about 560 B.C. to 480 B.C. According to the British philosopher Bertrand Russell, Pythagoras was "intellectually one of the most important men that ever lived."

Also in this section, the solutions to the examples show only the essential steps from our Blueprint for Problem Solving. Recall that step 1 is done mentally; we read the problem and mentally list the items that are known and the items that are unknown. This is an essential part of problem solving. However, now that you have had experience with application problems, you should be doing step 1 automatically.

**Video Examples
Section 7.8**

A Number Problems

EXAMPLE 1 The product of two consecutive odd integers is 63. Find the integers.

Solution Let $x =$ the first odd integer; then $x + 2 =$ the second odd integer. An equation that describes the situation is

$$x(x + 2) = 63 \quad \text{Their product is 63.}$$

Then solve the equation.

$$x(x + 2) = 63$$
$$x^2 + 2x = 63$$
$$x^2 + 2x - 63 = 0$$
$$(x - 7)(x + 9) = 0$$
$$x - 7 = 0 \quad \text{or} \quad x + 9 = 0$$
$$x = 7 \quad\quad\quad x = -9$$

If the first odd integer is 7, the next odd integer is $7 + 2 = 9$. If the first odd integer is -9, the next consecutive odd integer is $-9 + 2 = -7$. We have two pairs of consecutive odd integers that are solutions. They are 7 and 9, or -9 and -7.

We check to see that their products are 63.

$$7(9) = 63$$
$$-7(-9) = 63$$

Practice Problems

1. The product of two consecutive even integers is 48. Find the integers.

Answers

1. -8 and -6, or 6 and 8

Suppose we know that the sum of two numbers is 50. We want to find a way to represent each number using only one variable. If we let x represent one of the two numbers, how can we represent the other? Let's suppose for a moment that x turns out to be 30. Then the other number will be 20, because their sum is 50; that is, if two numbers add up to 50 and one of them is 30, then the other must be $50 - 30 = 20$. Generalizing this to any number x, we see that if two numbers have a sum of 50 and one of the numbers is x, then the other must be $50 - x$. The table that follows shows some additional examples.

If two numbers have a sum of	and one of them is	then the other must be
50	x	$50 - x$
100	x	$100 - x$
10	y	$10 - y$
12	n	$12 - n$

Now, let's look at an example that uses this idea.

EXAMPLE 2 The sum of two numbers is 13. Their product is 40. Find the numbers.

Solution If we let x represent one of the numbers, then $13 - x$ must be the other number because their sum is 13. Since their product is 40, we can write

$$x(13 - x) = 40 \qquad \textit{The product of the two numbers is 40.}$$
$$13x - x^2 = 40 \qquad \textit{Multiply the left side.}$$
$$x^2 - 13x = -40 \qquad \textit{Multiply both sides by } -1 \textit{ and reverse the order of the terms on the left side.}$$
$$x^2 - 13x + 40 = 0 \qquad \textit{Add 40 to each side.}$$
$$(x - 8)(x - 5) = 0 \qquad \textit{Factor the left side.}$$
$$x - 8 = 0 \quad \text{or} \quad x - 5 = 0$$
$$x = 8 \qquad\qquad x = 5$$

The two solutions are 8 and 5. If x is 8, then the other number is $13 - x = 13 - 8 = 5$. Likewise, if x is 5, the other number is $13 - x = 13 - 5 = 8$. Therefore, the two numbers we are looking for are 8 and 5. Their sum is 13 and their product is 40.

B Geometry Problems

Many word problems dealing with area can best be described algebraically by quadratic equations.

EXAMPLE 3 The length of a rectangle is 3 more than twice the width. The area is 44 square inches. Find the dimensions (find the length and width).

Solution As shown in Figure 1, let $x =$ the width of the rectangle. Then $2x + 3 =$ the length of the rectangle because the length is three more than twice the width.

FIGURE 1

Since the area is 44 square inches, an equation that describes the situation is

$$x(2x + 3) = 44 \qquad \textit{Length} \cdot \textit{width} = \textit{area}$$

2. The sum of two numbers is 14. Their product is 48. Find the two numbers.

3. The length of a rectangle is 2 more than twice the width. The area is 60 square inches. Find the dimensions.

Answers
2. 6 and 8
3. Width is 5 inches; length is 12 inches.

We now solve the equation.

$$x(2x + 3) = 44$$

$$2x^2 + 3x = 44$$

$$2x^2 + 3x - 44 = 0$$

$$(2x + 11)(x - 4) = 0$$

$$2x + 11 = 0 \quad \text{or} \quad x - 4 = 0$$

$$x = -\frac{11}{2} \qquad x = 4$$

The solution $x = -\frac{11}{2}$ cannot be used since length and width are always given in positive units. The width is 4. The length is 3 more than twice the width or $2(4) + 3 = 11$.

Width = 4 inches

Length = 11 inches

The solutions check in the original problem since $4(11) = 44$.

EXAMPLE 4 The numerical value of the area of a square is twice its perimeter. What is the length of its side?

Solution As shown in Figure 2, let $x =$ the length of its side. Then $x^2 =$ the area of the square and $4x =$ the perimeter of the square.

FIGURE 2

An equation that describes the situation is

$$x^2 = 2(4x) \qquad \textit{The area is 2 times the perimeter.}$$

$$x^2 = 8x$$

$$x^2 - 8x = 0$$

$$x(x - 8) = 0$$

$$x = 0 \quad \text{or} \quad x = 8$$

Since $x = 0$ does not make sense in our original problem, we use $x = 8$. If the side has length 8, then the perimeter is $4(8) = 32$ and the area is $8^2 = 64$. Since 64 is twice 32, our solution is correct.

4. The height of a triangle is 3 inches more than the base. The area is 20 square inches. Find the base.

C Business Problems

5. Using the information in Example 5, find the number of items that should be made if the total cost is to be $700.

EXAMPLE 5 A company can manufacture x hundred items for a total cost of $C = 300 + 500x - 100x^2$. How many items were manufactured if the total cost is $900?

Solution We are looking for x when C is 900. We begin by substituting 900 for C in the cost equation. Then we solve for x.

$$\text{When} \rightarrow \qquad C = 900$$
$$\text{the equation} \rightarrow \qquad C = 300 + 500x - 100x^2$$
$$\text{becomes} \rightarrow \qquad 900 = 300 + 500x - 100x^2$$

We can write this equation in standard form by adding -300, $-500x$, and $100x^2$ to each side. The result looks like this:

$$900 = 300 + 500x - 100x^2$$

$$100x^2 - 500x + 600 = 0 \qquad \textit{Standard form}$$
$$100(x^2 - 5x + 6) = 0 \qquad \textit{Begin factoring.}$$
$$100(x - 2)(x - 3) = 0 \qquad \textit{Factor completely.}$$
$$x - 2 = 0 \quad \text{or} \quad x - 3 = 0 \qquad \textit{Set variable factors equal to 0.}$$
$$x = 2 \qquad\qquad x = 3$$

Our solutions are 2 and 3, which means that the company can manufacture 2 hundred items or 3 hundred items for a total cost of $900.

6. Use the information in Example 6 to find the selling price that should be charged if the weekly revenue is to be $4,200.

EXAMPLE 6 A manufacturer of small portable radios knows that the number of radios she can sell each week is related to the price of the radios by the equation $x = 1,300 - 100p$ (x is the number of radios and p is the price per radio). What price should she charge for the radios to have a weekly revenue of $4,000?

Solution First, we must find the revenue equation. The equation for total revenue is $R = xp$, where x is the number of units sold and p is the price per unit. Since we want R in terms of p, we substitute $1,300 - 100p$ for x in the equation $R = xp$.

$$\text{If} \qquad R = xp$$
$$\text{and} \qquad x = 1,300 - 100p$$
$$\text{then} \qquad R = (1,300 - 100p)p$$

We want to find p when R is 4,000. Substituting 4,000 for R in the equation gives us

$$4,000 = (1,300 - 100p)p$$

If we multiply out the right side, we have

$$4,000 = 1,300p - 100p^2$$

To write this equation in standard form, we add $100p^2$ and $-1,300p$ to each side.

$$100p^2 - 1,300p + 4,000 = 0 \qquad \textit{Add } 100p^2 \textit{ and } -1,300p.$$
$$100(p^2 - 13p + 40) = 0 \qquad \textit{Begin factoring.}$$
$$100(p - 5)(p - 8) = 0 \qquad \textit{Factor completely.}$$
$$p - 5 = 0 \quad \text{or} \quad p - 8 = 0 \qquad \textit{Set variable factors equal to 0.}$$
$$p = 5 \qquad\qquad p = 8$$

If she sells the radios for $5 each or for $8 each, she will have a weekly revenue of $4,000.

D The Pythagorean Theorem

Next, we will work some problems involving the Pythagorean theorem, which uses variables to represent the three sides of a triangle. When calculating the lengths of these sides, the theorem often translates into a quadratic equation we can factor and solve.

FACTS FROM GEOMETRY Pythagorean Theorem

In any right triangle (Figure 3), the square of the longer side (called the hypotenuse) is equal to the sum of the squares of the other two sides (called legs).

$$c^2 = a^2 + b^2$$

FIGURE 3

EXAMPLE 7 The three sides of a right triangle are three consecutive integers. Find the lengths of the three sides.

Solution Let x = the first integer (shortest side)

then $x + 1$ = the next consecutive integer

and $x + 2$ = the last consecutive integer (longest side)

A diagram of the triangle is shown in Figure 4.

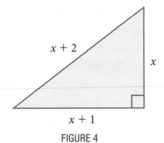

FIGURE 4

The Pythagorean theorem tells us that the square of the longest side $(x + 2)^2$ is equal to the sum of the squares of the two shorter sides, $(x + 1)^2 + x^2$. Here is the equation:

$$(x + 2)^2 = (x + 1)^2 + x^2$$

$x^2 + 4x + 4 = x^2 + 2x + 1 + x^2$ Expand squares.

$x^2 - 2x - 3 = 0$ Standard form

$(x - 3)(x + 1) = 0$ Factor.

$x - 3 = 0$ or $x + 1 = 0$ Set factors equal to 0.

$x = 3$ $x = -1$

Since a triangle cannot have a side with a negative number for its length, we must not use -1 for a solution to our original problem; therefore, the shortest side is 3. The other two sides are the next two consecutive integers, 4 and 5.

7. The longer leg of a right triangle is 2 more than twice the shorter leg. The hypotenuse is 3 more than twice the shorter leg. Find the length of each side. (*Hint*: Let x = the length of the shorter leg. Then the longer leg is $2x + 2$. Now write an expression for the hypotenuse, and you will be on your way.)

8. The hypotenuse of a right triangle is 10 inches, and the lengths of the two sides are given by two consecutive even integers. Find the two sides.

EXAMPLE 8 The hypotenuse of a right triangle is 5 inches, and the lengths of the two legs (the other two sides) are given by two consecutive integers. Find the lengths of the two legs.

Solution If we let $x =$ the length of the shortest side, then the other side must be $x + 1$. A diagram of the triangle is shown in Figure 5.

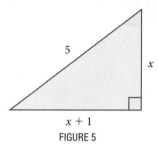

FIGURE 5

The Pythagorean theorem tells us that the square of the longest side, 5^2, is equal to the sum of the squares of the two shorter sides, $x^2 + (x + 1)^2$. Here is the equation:

$$5^2 = x^2 + (x + 1)^2 \qquad \text{Pythagorean theorem}$$
$$25 = x^2 + x^2 + 2x + 1 \qquad \text{Expand } 5^2 \text{ and } (x + 1)^2.$$
$$25 = 2x^2 + 2x + 1 \qquad \text{Simplify the right side.}$$
$$0 = 2x^2 + 2x - 24 \qquad \text{Add } -25 \text{ to each side.}$$
$$0 = 2(x^2 + x - 12) \qquad \text{Begin factoring.}$$
$$0 = 2(x + 4)(x - 3) \qquad \text{Factor completely.}$$
$$x + 4 = 0 \quad \text{or} \quad x - 3 = 0 \qquad \text{Set variable factors equal to 0.}$$
$$x = -4 \qquad\qquad x = 3$$

Since a triangle cannot have a side with a negative number for its length, we cannot use -4; therefore, the shortest side must be 3 inches. The next side is $x + 1 = 3 + 1 = 4$ inches. Since the hypotenuse is 5, we can check our solutions with the Pythagorean theorem as shown in Figure 6.

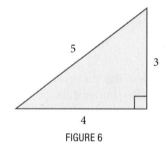

FIGURE 6

GETTING READY FOR CLASS

After reading through the preceding section, respond in your own words and in complete sentences.

A. What are consecutive integers?

B. How would you use factoring to solve a geometry problem?

C. Explain the Pythagorean theorem in words.

D. Write an application problem for which the solution depends on solving the equation $x(x + 1) = 12$.

Answers

8. The lengths of the sides are 6 inches and 8 inches.

Vocabulary Review

Choose the correct words to fill in the blanks below.

hypotenuse standard form Pythagorean theorem consecutive

1. The quadratic equation $20 = x^2 + x$ represents two _____ integers with a product of 20.

2. To write a quadratic equation in _____, have zero on one side of the equation and decreasing powers of the variable on the other.

3. In any right triangle, the square of the _____ is equal to the sum of the squares of the legs.

4. The _____ states that $c^2 = a^2 + b^2$, where c is the hypotenuse and a and b are the legs of a right triangle.

Problems

Solve the following word problems. Be sure to show the equation used.

A Number Problems

1. The product of two consecutive even integers is 80. Find the two integers.

2. The product of two consecutive integers is 72. Find the two integers.

3. The product of two consecutive odd integers is 99. Find the two integers.

4. The product of two consecutive integers is 132. Find the two integers.

5. The product of two consecutive even integers is 10 less than 5 times their sum. Find the two integers.

6. The product of two consecutive odd integers is 1 less than 4 times their sum. Find the two integers.

7. The sum of two numbers is 14. Their product is 48. Find the numbers.

8. The sum of two numbers is 12. Their product is 32. Find the numbers.

9. One number is 2 more than 5 times another. Their product is 24. Find the numbers.

10. One number is 1 more than twice another. Their product is 55. Find the numbers.

11. One number is 4 times another. Their product is 4 times their sum. Find the numbers.

12. One number is 2 more than twice another. Their product is 2 more than twice their sum. Find the numbers.

B Geometry Problems

13. The length of a rectangle is 1 more than the width. The area is 12 square inches. Find the dimensions.

14. The length of a rectangle is 3 more than twice the width. The area is 44 square inches. Find the dimensions.

15. The height of a triangle is twice the base. The area is 9 square inches. Find the base.

16. The height of a triangle is 2 more than twice the base. The area is 20 square feet. Find the base.

17. The hypotenuse of a right triangle is 10 inches. The lengths of the two legs are given by two consecutive even integers. Find the lengths of the two legs.

18. The hypotenuse of a right triangle is 15 inches. One of the legs is 3 inches more than the other. Find the lengths of the two legs.

19. The shorter leg of a right triangle is 5 meters. The hypotenuse is 1 meter longer than the longer leg. Find the length of the longer leg.

20. The shorter leg of a right triangle is 12 yards. If the hypotenuse is 20 yards, how long is the other leg?

c Business Problems

21. A company can manufacture x hundred items for a total cost of $C = 400 + 700x - 100x^2$. How many items are manufactured if the total cost is $1,400?

22. If the total cost C of manufacturing x hundred items is given by the equation $C = 700 + 900x - 100x^2$, how many items are manufactured when C is $2,700?

23. The total cost C of manufacturing x hundred DVDs is given by the equation $C = 600 + 1,000x - 100x^2$. How many DVDs are manufactured if the total cost is $2,200?

24. The total cost C of manufacturing x hundred pen and pencil sets is given by the equation $C = 500 + 800x - 100x^2$. How many pen and pencil sets are manufactured when C is $1,700?

25. A company that manufactures hair ribbons knows that the number of ribbons it can sell each week, x, is related to the price p per ribbon by the equation $x = 1,200 - 100p$. At what price should the company sell the ribbons if it wants the weekly revenue to be $3,200? (*Remember:* The equation for revenue is $R = xp$.)

26. A company manufactures flash drives for home computers. It knows from experience that the number of drives it can sell each day, x, is related to the price p per drive by the equation $x = 800 - 100p$. At what price should the company sell the flash drives if it wants the daily revenue to be $1,200?

27. The relationship between the number of calculators a company sells per week, x, and the price p of each calculator is given by the equation $x = 1,700 - 100p$. At what price should the calculators be sold if the weekly revenue is to be $7,000?

28. The relationship between the number of pencil sharpeners a company can sell each week, x, and the price p of each sharpener is given by the equation $x = 1,800 - 100p$. At what price should the sharpeners be sold if the weekly revenue is to be $7,200?

D Solve the following problems using the Pythagorean theorem.

29. Pythagorean Theorem A 13-foot ladder is placed so that it reaches to a point on the wall that is 2 feet higher than twice the distance from the base of the wall to the base of the ladder.

a. How far from the wall is the base of the ladder?

b. How high does the ladder reach?

30. Pythagorean Theorem A 25-foot ladder is placed so that its height on the wall is 3 feet more than triple the distance from the base to the wall.

a. How far from the wall is the base of the ladder?

b. How high does the ladder reach?

31. Constructing a Box I have a piece of cardboard that is twice as long as it is wide. If I cut a 2-inch by 2-inch square from each corner and fold up the resulting flaps, I get a box with a volume of 32 cubic inches. What are the dimensions of the cardboard?

32. Constructing a Box A cardboard rectangle has a length which is 3 inches more than its width. If 1-inch by 1-inch squares are cut out, and the resulting flaps are folded up, the volume would be 70 cubic inches. What are the dimensions of the cardboard?

33. Projectile Motion A gun fires a bullet almost straight up from the edge of a 100-foot cliff. If the bullet leaves the gun with a speed of 396 feet per second, its height at time t is given by $h(t) = -16t^2 + 396t + 100$, measured from the ground below the cliff.

 a. When will the bullet land on the ground below the cliff? (*Hint:* What is its height when it lands? Remember that we are measuring from the ground below, not from the cliff.)

 b. Make a table showing the bullet's height every five seconds, from the time it is fired ($t = 0$) to the time it lands. (*Note:* It is faster to substitute into the factored form.)

t	h

34. Height of a Projectile If a rocket is fired vertically into the air with a speed of 240 feet per second, its height at time t seconds is given by $h(t) = -16t^2 + 240t$. At what time(s) will the rocket be the following number of feet above the ground?

 a. 704 feet

 b. 896 feet

 c. Why do parts a and b each have two answers?

 d. How long will the rocket be in the air? (*Hint:* How high is it when it hits the ground?)

 e. When the equation for part d is solved, one of the answers is $t = 0$ second. What does this represent?

Maintaining Your Skills

Simplify each expression. (Write all answers with positive exponents only.)

35. $(5x^3)^2(2x^6)^3$

36. 2^{-3}

37. $\dfrac{x^4}{x^{-3}}$

38. $\dfrac{(20x^2y^3)(5x^4y)}{(2xy^5)(10x^2y^3)}$

39. $(2 \times 10^{-4})(4 \times 10^5)$

40. $\dfrac{9 \times 10^{-3}}{3 \times 10^{-2}}$

41. $20ab^2 - 16ab^2 + 6ab^2$

42. Subtract $6x^2 - 5x - 7$ from $9x^2 + 3x - 2$.

Multiply.

43. $2x^2(3x^2 + 3x - 1)$

44. $(2x + 3)(5x - 2)$

45. $(3y - 5)^2$

46. $(a - 4)(a^2 + 4a + 16)$

47. $(2a^2 + 7)(2a^2 - 7)$

48. Divide $15x^{10} - 10x^8 + 25x^6$ by $5x^6$.

Find the Mistake

Each sentence below contains a mistake. Circle the mistake and write the correct word(s) or equation on the line provided.

1. The sum of two consecutive even integers is 34. Their product is 288. To find the two integers, you must solve for x in the equation $x^2 - 34x = 288$. _____

2. The width of a rectangle is $2x$ and its length is three times the width plus 5 more inches. The area of the rectangle is 781 square inches. The longest side of the rectangle measures 11 inches. _____

3. The Pythagorean theorem states that the square of the longest leg is equal to the square of the hypotenuse and the shortest leg. _____

4. The hypotenuse of a triangle is 13 feet. The longest leg of the triangle is twice the shortest leg plus 2 more feet. To find the lengths of the two legs, we must find the square root of the hypotenuse.

EXAMPLES

1. $8x^4 - 10x^3 + 6x^2$
$= 2x^2 \cdot 4x^2 - 2x^2 \cdot 5x + 2x^2 \cdot 3$
$= 2x^2(4x^2 - 5x + 3)$

Greatest Common Factor [7.1]

The largest monomial that divides each term of a polynomial is called the greatest common factor for that polynomial. We begin all factoring by factoring out the greatest common factor.

2. $x^2 + 5x + 6 = (x + 2)(x + 3)$
$x^2 - 5x + 6 = (x - 2)(x - 3)$
$6x^2 - x - 2 = (2x + 1)(3x - 2)$
$6x^2 + 7x + 2 = (2x + 1)(3x + 2)$

Factoring Trinomials [7.2, 7.3]

One method of factoring a trinomial is to list all pairs of binomials whose product of the first terms gives the first term of the trinomial, and whose product of the last terms gives the last term of the trinomial. Then we choose the pair that gives the correct middle term for the original trinomial.

3. $x^2 + 10x + 25 = (x + 5)^2$
$x^2 - 10x + 25 = (x - 5)^2$
$x^2 - 25 = (x + 5)(x - 5)$

Special Factoring [7.4]

$$a^2 + 2ab + b^2 = (a + b)^2$$
$$a^2 - 2ab + b^2 = (a - b)^2$$
$\left.\right\}$ *Perfect square trinomials*

$$a^2 - b^2 = (a + b)(a - b) \qquad \text{\textit{Difference of two squares}}$$

Remember, $a^2 + b^2$ is a prime sum because it does not factor.

4. $x^3 - 27 = (x - 3)(x^2 + 3x + 9)$
$x^3 + 27 = (x + 3)(x^2 - 3x + 9)$

Sum and Difference of Two Cubes [7.5]

$$a^3 - b^3 = (a - b)(a^2 + ab + b^2) \qquad \text{\textit{Difference of two cubes}}$$
$$a^3 + b^3 = (a + b)(a^2 - ab + b^2) \qquad \text{\textit{Sum of two cubes}}$$

5. a. $2x^5 - 8x^3 = 2x^3(x^2 - 4)$
$= 2x^3(x + 2)(x - 2)$

b. $3x^4 - 18x^3 + 27x^2$
$= 3x^2(x^2 - 6x + 9)$
$= 3x^2(x - 3)^2$

c. $6x^3 - 12x^2 - 48x$
$= 6x(x^2 - 2x - 8)$
$= 6x(x - 4)(x + 2)$

d. $x^2 + ax + bx + ab$
$= x(x + a) + b(x + a)$
$= (x + a)(x + b)$

Strategy for Factoring a Polynomial [7.6]

Step 1: If the polynomial has a greatest common factor other than 1, then factor out the greatest common factor.

Step 2: If the polynomial has two terms (a binomial), then see if it is the difference of two squares or the sum or difference of two cubes, and then factor accordingly. Remember, if it is the sum of two squares, it will not factor.

Step 3: If the polynomial has three terms (a trinomial), then it is either a perfect square trinomial that will factor into the square of a binomial, or it is not a perfect square trinomial, in which case you can try to use the trial and error method developed in Section 5.3.

Step 4: If the polynomial has more than three terms, then try to factor it by grouping.

Step 5: As a final check, see if any of the factors you have written can be factored further. If you have overlooked a common factor, you can catch it here. Remember, some polynomials cannot be factored, in which case they are considered prime.

Strategy for Solving a Quadratic Equation [7.7]

Step 1: Write the equation in standard form $ax^2 + bx + c = 0$.

Step 2: Factor completely.

Step 3: Set each variable factor equal to 0.

Step 4: Solve the equations found in step 3.

Step 5: Check solutions, if necessary.

6. Solve $x^2 - 6x = -8$.
$$x^2 - 6x + 8 = 0$$
$$(x - 4)(x - 2) = 0$$
$$x - 4 = 0 \quad \text{or} \quad x - 2 = 0$$
$$x = 4 \qquad\qquad x = 2$$
Both solutions check.

The Pythagorean Theorem [7.8]

In any right triangle, the square of the longest side (called the hypotenuse) is equal to the sum of the squares of the other two sides (called legs).

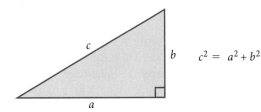

$$c^2 = a^2 + b^2$$

7. The hypotenuse of a right triangle is 5 inches, and the lengths of the two legs (the other two sides) are given by two consecutive integers. Find the lengths of the two legs.

If we let x = the length of the shortest side, then the other side must be $x + 1$. The Pythagorean theorem tells us that

$$5^2 = x^2 + (x + 1)^2$$
$$25 = x^2 + x^2 + 2x + 1$$
$$25 = 2x^2 + 2x + 1$$
$$0 = 2x^2 + 2x - 24$$
$$0 = 2(x^2 + x - 12)$$
$$0 = 2(x + 4)(x - 3)$$
$$x + 4 = 0 \quad \text{or} \quad x - 3 = 0$$
$$x = -4 \qquad\qquad x = 3$$

Since a triangle cannot have a side with a negative number for its length, we cannot use -4. One leg is $x = 3$ and the other leg is $x + 1 = 3 + 1 = 4$.

> **COMMON MISTAKE**
>
> It is a mistake to apply the zero-factor property to numbers other than zero. For example, consider the equation $(x - 3)(x + 4) = 18$. A fairly common mistake is to attempt to solve it with the following steps:
>
> $$(x - 3)(x + 4) = 18$$
> $$x - 3 = 18 \quad \text{or} \quad x + 4 = 18 \leftarrow \textit{Mistake}$$
> $$x = 21 \qquad\qquad x = 14$$
>
> These are obviously not solutions, as a quick check will verify.
>
> $$\text{Check: } x = 21 \qquad\qquad \text{Check: } x = 14$$
> $$(21 - 3)(21 + 4) \overset{?}{=} 18 \qquad (14 - 3)(14 + 4) \overset{?}{=} 18$$
> $$18 \cdot 25 = 18 \qquad\qquad 11 \cdot 18 = 18$$
> $$450 \neq 18 \quad \xleftarrow{\textit{False statements}} \quad 198 \neq 18$$
>
> The mistake is in setting each factor equal to 18. It is not necessarily true that when the product of two numbers is 18, either one of them is itself 18. The correct solution looks like this:
>
> $$(x - 3)(x + 4) = 18$$
> $$x^2 + x - 12 = 18$$
> $$x^2 + x - 30 = 0$$
> $$(x + 6)(x - 5) = 0$$
> $$x + 6 = 0 \quad \text{or} \quad x - 5 = 0$$
> $$x = -6 \qquad\qquad x = 5$$
>
> To avoid this mistake, remember that before you factor a quadratic equation, you must write it in standard form. It is in standard form only when 0 is on one side and decreasing powers of the variable are on the other.

Factor out the greatest common factor. [7.1]

1. $6x + 18$

2. $12a^2b - 24ab + 8ab^2$

Factor by grouping. [7.1]

3. $x^2 + 3ax - 2bx - 6ab$

4. $15y - 5xy - 12 + 4x$

Factor the following completely. [7.2−7.6]

5. $x^2 + x - 12$

6. $x^2 - 4x - 21$

7. $x^2 - 25$

8. $x^4 - 16$

9. $x^2 + 36$

10. $18x^2 + 8xy - 32y^2$

11. $x^3 + 4x^2 - 3x - 12$

12. $x^2 + bx - 3x - 3b$

13. $4x^2 - 6x - 10$

14. $4n^2 + 13n - 12$

15. $12c^2 + c - 6$

16. $12x^3 + 12x^2 - 9x$

17. $x^3 + 125y^3$

18. $54b^3 - 128$

Solve the following equations. [7.7]

19. $x^2 - 2x - 15 = 0$

20. $x^2 - 7x + 12 = 0$

21. $x^2 - 25 = 0$

22. $x^2 = 5x + 14$

23. $x^2 + x = 30$

24. $y^3 = 9y$

25. $2x^2 = -5x + 12$

26. $15x^3 - 65x^2 - 150x = 0$

Solve the following word problems. Be sure to show the equation used. [7.8]

27. Number Problem Two numbers have a sum of 18. Their product is 72. Find the numbers.

28. Consecutive Integers The product of two consecutive even integers is 14 more than their sum. Find the integers.

29. Geometry The length of a rectangle is 1 foot more than 3 times the width. The area is 52 square feet. Find the dimensions.

30. Geometry One leg of a right triangle is 2 feet more than the other. The hypotenuse is 10 feet. Find the lengths of the two legs.

31. Production Cost A company can manufacture x hundred items for a total cost, C, given the equation $C = 100 + 500x - 100x^2$. How many items can be manufactured if the total cost is to be $700?

32. Price and Revenue A manufacturer knows that the number of items he can sell each week, x, is related to the price p of each item by the equation $x = 800 - 100p$. What price should he charge for each item to have a weekly revenue of $1,500? (*Remember: $R = xp$.*)

33. Pythagorean Theorem A 13-foot ladder is placed so that it leans against a wall at a point that is two feet more than twice the distance from the wall to the base of the ladder.

a. How far from the wall is the ladder?

b. How high does the ladder reach?

13 ft

34. Projectile Motion A ball is thrown into the air with an upward velocity of 26 feet per second from a building that is 12 feet high. The equation that gives the height of the ball above the ground at time t is $h = 12 + 26t - 16t^2$

a. At what time(s) will the ball be 22 feet above the ground?

b. When will the ball hit the ground?

h

Rational Expressions, Equations, and Functions

Image © 2010 DigitalGlobe
Perrine Memorial Bridge
Twin Falls, Idaho

8

8.1 Basic Properties and Reducing to Lowest Terms

8.2 Multiplication and Division of Rational Expressions

8.3 Addition and Subtraction of Rational Expressions

8.4 Complex Rational Expressions

8.5 Rational Equations

8.6 Applications

BASE jumping is an extreme sport where participants jump off fixed objects and have mere seconds to deploy their parachute. The hazardous sport is named for the acronym representing the objects enthusiasts jump from: **B**uildings, **A**ntennae, **S**pans, and **E**arth. Despite the danger associated with this sport, there are many popular jumping destinations. Perrine Memorial Bridge in Twin Falls, Idaho is one such location. BASE jumping is allowed here year round without a permit, making it the only man-made structure in the United States where this is true. Standing at 486 feet, more than 10,000 jumps are made from Perrine each year.

Suppose a BASE jumper jumped from Perrine Memorial Bridge and fell 64 feet in the first 2 seconds of her jump. The table here summarizes the distances she will fall.

Input (time in seconds)	Output (distance in feet)
t	$d(t)$
0	0
1	16
2	64
3	144
4	256
5	400

Later in the chapter, you will learn how we arrived at these distances and how to use that formula, which uses a rational expression to determine how long the jumper will fall before she deploys her parachute.

8.1 Basic Properties and Reducing to Lowest Terms

KEY WORDS

rational expression

reduce

lowest terms

rational function

difference quotient

**Video Examples
Section 8.1**

Archaeologists recently discovered the first use of pressurized plumbing in the New World. They unearthed a stone aqueduct in the ancient Mayan city of Palenque, Mexico. Water from a nearby stream would have spilled down the 60-meter-long aqueduct. The downward force of the water would have powered it through a narrow buried conduit for the last 2 meters, thereby building pressure and propelling it up out of the ground. Whether to control flooding or to feed a fountain, the exact purpose for the aqueduct is still unknown.

In order to mathematically understand the concept of pressure, like that used in the aqueduct, we need to explore rational expressions. For pressure P, we have to observe the rational relationship between force F and area A using the following equation:

$$P = \frac{F}{A}$$

Recall from a previous math class that a *rational number* is any number that can be expressed as the ratio of two integers:

$$\text{Rational numbers} = \left\{ \frac{a}{b} \;\middle|\; a \text{ and } b \text{ are integers, } b \neq 0 \right\}$$

A *rational expression* is defined similarly as any expression that can be written as the ratio of two polynomials:

$$\text{Rational expressions} = \left\{ \frac{P}{Q} \;\middle|\; P \text{ and } Q \text{ are polynomials, } Q \neq 0 \right\}$$

Some examples of rational expressions are

$$\frac{2x - 3}{x + 5} \qquad \frac{x^2 - 5x - 6}{x^2 - 1} \qquad \frac{a - b}{b - a}$$

Basic Properties

For rational expressions, multiplying the numerator and denominator by the same nonzero expression may change the form of the rational expression, but it will always produce an expression equivalent to the original one. The same is true when dividing the numerator and denominator by the same nonzero quantity.

PROPERTY Properties of Rational Expressions

If P, Q, and K are polynomials with $Q \neq 0$ and $K \neq 0$, then

$$\frac{P}{Q} = \frac{PK}{QK} \qquad \text{and} \qquad \frac{P}{Q} = \frac{\frac{P}{K}}{\frac{Q}{K}}$$

A Reducing to Lowest Terms

The fraction $\frac{6}{8}$ can be written in lowest terms as $\frac{3}{4}$. The process is shown here:

$$\frac{6}{8} = \frac{3 \cdot 2}{4 \cdot 2} = \frac{3}{4}$$

Reducing $\frac{6}{8}$ to $\frac{3}{4}$ involves dividing the numerator and denominator by 2, the factor they have in common. Before dividing out the common factor 2, we must notice that the common factor *is* 2. (This may not be as obvious with other numbers. We are very familiar with the numbers 6 and 8 and therefore do not have to put much thought into finding what number divides both of them.)

We reduce rational expressions to lowest terms by first factoring the numerator and denominator and then dividing both numerator and denominator by any factors they have in common.

EXAMPLE 1 Reduce $\frac{x^2 - 9}{x - 3}$ to lowest terms.

Solution Factoring, we have

$$\frac{x^2 - 9}{x - 3} = \frac{(x + 3)(x - 3)}{x - 3}$$

The numerator and denominator have the factor $x - 3$ in common. Dividing the numerator and denominator by $x - 3$, we have

$$\frac{(x + 3)\cancel{(x - 3)}}{\cancel{x - 3}} = \frac{x + 3}{1} = x + 3$$

For the problem in Example 1, there is an implied restriction on the variable x: It cannot be 3. If x were 3, the expression $\frac{(x^2 - 9)}{(x - 3)}$ would become $\frac{0}{0}$, an expression that we cannot associate with a real number. For all problems involving rational expressions, we restrict the variable to only those values that result in a nonzero denominator. When we state the relationship

$$\frac{x^2 - 9}{x - 3} = x + 3$$

we are assuming that it is true for all values of x except $x = 3$.

It is also a common mistake, when working a problem like Example 1, to begin by dividing common terms.

$$\frac{x^2 - 9}{x - 3} \neq \frac{x^2}{x} - \frac{9}{3}$$

Factoring the numerator is essential since we can only divide out common factors, not common terms.

Here are some other examples of reducing rational expressions to lowest terms:

EXAMPLE 2 Reduce $\frac{y^2 - 5y - 6}{y^2 - 1}$ to lowest terms.

Solution

$$\frac{y^2 - 5y - 6}{y^2 - 1} = \frac{(y - 6)(y + 1)}{(y - 1)(y + 1)}$$

$$= \frac{y - 6}{y - 1}$$

EXAMPLE 3 Reduce $\frac{2a^3 - 16}{4a^2 - 12a + 8}$ to lowest terms.

Solution

$$\frac{2a^3 - 16}{4a^2 - 12a + 8} = \frac{2(a^3 - 8)}{4(a^2 - 3a + 2)}$$

$$= \frac{2(a - 2)(a^2 + 2a + 4)}{4(a - 2)(a - 1)}$$

$$= \frac{a^2 + 2a + 4}{2(a - 1)}$$

Practice Problems

1. Reduce $\frac{x^2 - 4}{x^2 - 2x - 8}$ to lowest terms.

> *Note* The lines drawn through the $(x - 3)$ in the numerator and denominator indicate that we have divided through by $(x - 3)$. As the problems become more involved, these lines will help keep track of which factors have been divided out and which have not.

2. Reduce $\frac{x^2 + 2x - 8}{x^2 - 4}$ to lowest terms.

3. Reduce $\frac{x^3 - 27}{3x^2 + 3x - 36}$ to lowest terms.

Answers

1. $\frac{x - 2}{x - 4}$

2. $\frac{x + 4}{x + 2}$

3. $\frac{x^2 + 3x + 9}{3(x + 4)}$

4. Reduce $\dfrac{xa + 3x - 4a - 12}{xa - 2a + 3x - 6}$ to lowest terms.

EXAMPLE 4 Reduce $\dfrac{x^2 - 3x + ax - 3a}{x^2 - ax - 3x + 3a}$ to lowest terms.

Solution

$$f(x) = \frac{x^2 - 3x + ax - 3a}{x^2 - ax - 3x + 3a}$$

$$= \frac{x(x - 3) + a(x - 3)}{x(x - a) - 3(x - a)}$$

$$= \frac{(x - 3)(x + a)}{(x - a)(x - 3)}$$

$$= \frac{x + a}{x - a}$$

The answer to Example 4 cannot be reduced further. It is a fairly common mistake to attempt to divide out an x or an a in this last expression. Remember, again we can divide out only the factors common to the numerator and denominator of a rational expression.

The next example involves what we call a trick. The trick is to reverse the order of the terms in a difference by factoring -1 from each term in either the numerator or the denominator. Here is how this is done.

5. Reduce $\dfrac{3 - a}{a - 3}$ to lowest terms.

EXAMPLE 5 Reduce $\dfrac{a - b}{b - a}$ to lowest terms.

Solution The relationship between $a - b$ and $b - a$ is that they are opposites. We can show this fact by factoring -1 from each term in the numerator.

$$\frac{a - b}{b - a} = \frac{-1(-a + b)}{b - a} \qquad \textit{Factor } -1 \textit{ from each term in the numerator.}$$

$$= \frac{-1(b - a)}{b - a} \qquad \textit{Reverse the order of the terms in the numerator.}$$

$$= -1 \qquad \textit{Divide out common factor } b - a.$$

6. Reduce $\dfrac{2x^2 - 8}{2 - x}$ to lowest terms.

EXAMPLE 6 Reduce $\dfrac{x^2 - 25}{5 - x}$ to lowest terms.

Solution We begin by factoring the numerator.

$$\frac{x^2 - 25}{5 - x} = \frac{(x - 5)(x + 5)}{5 - x}$$

The factors $x - 5$ and $5 - x$ are similar but are not exactly the same. We can reverse the order of either by factoring -1 from it. That is: $5 - x = -1(-5 + x) = -1(x - 5)$.

$$\frac{(x - 5)(x + 5)}{5 - x} = \frac{(x - 5)(x + 5)}{-1(x - 5)}$$

$$= \frac{x + 5}{-1}$$

$$= -(x + 5)$$

B Rational Functions

We can extend our knowledge of rational expressions to functions with the following definition:

> **DEFINITION** rational function
>
> A *rational function* is any function that can be written in the form
>
> $$f(x) = \frac{P(x)}{Q(x)}$$
>
> where $P(x)$ and $Q(x)$ are polynomials and $Q(x) \neq 0$.

Answers

4. $\dfrac{x - 4}{x - 2}$

5. -1

6. $-2(x + 2)$

EXAMPLE 7 For the rational function $f(x) = \dfrac{x-4}{x-2}$, find $f(0)$, $f(-4)$, $f(4)$, $f(-2)$, and $f(2)$.

7. If $f(x) = \dfrac{x+5}{x-3}$ find $f(5)$.

Solution To find these function values, we substitute the given value of x into the rational expression, and then simplify if possible.

$$f(0) = \frac{0-4}{0-2} = \frac{-4}{-2} = 2 \qquad f(-2) = \frac{-2-4}{-2-2} = \frac{-6}{-4} = \frac{3}{2}$$

$$f(-4) = \frac{-4-4}{-4-2} = \frac{-8}{-6} = \frac{4}{3} \qquad f(2) = \frac{2-4}{2-2} = \frac{-2}{0} \quad \text{Undefined}$$

$$f(4) = \frac{4-4}{4-2} = \frac{0}{2} = 0$$

Because the rational function in Example 7 is not defined when x is 2, the domain of that function does not include 2. We have more to say about the domain of a rational function next.

The Domain of a Rational Function

If the domain of a rational function is not specified, it is assumed to be all real numbers for which the function is defined. That is, the domain of the rational function

$$f(x) = \frac{P(x)}{Q(x)}$$

is all x for which $Q(x)$ is nonzero.

EXAMPLE 8 Find the domain for each function.

8. Find the domain of $q(x) = \dfrac{x}{2x^2 - 8}$.

a. $f(x) = \dfrac{x-4}{x-2}$ **b.** $g(x) = \dfrac{x^2+5}{x+1}$ **c.** $h(x) = \dfrac{x}{x^2-9}$

Solution

a. The domain for $f(x) = \dfrac{x-4}{x-2}$ is $\{x \mid x \neq 2\}$.

b. The domain for $g(x) = \dfrac{x^2+5}{x+1}$ is $\{x \mid x \neq -1\}$.

c. The domain for $h(x) = \dfrac{x}{x^2-9}$ is $\{x \mid x \neq -3, x \neq 3\}$.

Notice that, for these functions, $f(2)$, $g(-1)$, $h(-3)$, and $h(3)$ are all undefined, and that is why the domains are written as shown.

C Difference Quotients

Figure 1 is an important diagram from calculus. Although it may look complicated, the point of it is simple: The slope of the line passing through the points P and Q is given by the following formula.

$$\text{Slope of line through } PQ = m = \frac{f(x) - f(a)}{x - a}$$

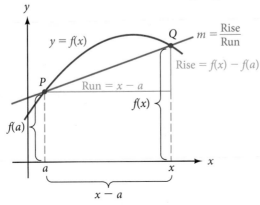

FIGURE 1

9. If $f(x) = 4x + 2$, find $\dfrac{f(x) - f(a)}{x - a}$.

10. If $f(x) = 2x^2 - 5$, find $\dfrac{f(x) - f(a)}{x - a}$
.

The expression $\dfrac{f(x) - f(a)}{x - a}$ is called a *difference quotient*. When $f(x)$ is a polynomial, it will be a rational expression. It is also important to know that a is a constant and x is a variable.

EXAMPLE 9 If $f(x) = 3x - 5$, find $\dfrac{f(x) - f(a)}{x - a}$.

Solution

$$\frac{f(x) - f(a)}{x - a} = \frac{(3x - 5) - (3a - 5)}{x - a}$$

$$= \frac{3x - 3a}{x - a}$$

$$= \frac{3(x - a)}{x - a}$$

$$= 3$$

EXAMPLE 10 If $f(x) = x^2 - 4$, find $\dfrac{f(x) - f(a)}{x - a}$ and simplify.

Solution Because $f(x) = x^2 - 4$ and $f(a) = a^2 - 4$, we have

$$\frac{f(x) - f(a)}{x - a} = \frac{(x^2 - 4) - (a^2 - 4)}{x - a}$$

$$= \frac{x^2 - 4 - a^2 + 4}{x - a}$$

$$= \frac{x^2 - a^2}{x - a}$$

$$= \frac{(x + a)(x - a)}{x - a} \qquad \text{Factor and divide out common factor.}$$

$$= x + a$$

Figure 2 is similar to the one in Figure 1. The main difference is in how we label the points. From Figure 2, we can see another difference quotient that gives us the slope of the line through the points P and Q. Similar to the formula used in Figure 1, h is a constant and x is a variable.

$$\text{Slope of line through } PQ = m = \frac{f(x + h) - f(x)}{h}$$

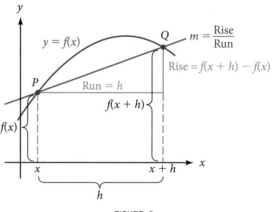

FIGURE 2

Examples 11 and 12 use the same functions used in Examples 9 and 10, but this time the new difference quotient is used.

EXAMPLE 11 If $f(x) = 3x - 5$, find $\dfrac{f(x + h) - f(x)}{h}$.

Solution The expression $f(x + h)$ is given by

$$f(x + h) = 3(x + h) - 5$$
$$= 3x + 3h - 5$$

Using this result gives us

$$\frac{f(x + h) - f(x)}{h} = \frac{(3x + 3h - 5) - (3x - 5)}{h}$$
$$= \frac{3h}{h}$$
$$= 3$$

EXAMPLE 12 If $f(x) = x^2 - 4$, find $\dfrac{f(x + h) - f(x)}{h}$.

Solution The expression $f(x + h)$ is given by

$$f(x + h) = (x + h)^2 - 4$$
$$= x^2 + 2xh + h^2 - 4$$

Using this result gives us

$$\frac{f(x + h) - f(x)}{h} = \frac{(x^2 + 2xh + h^2 - 4) - (x^2 - 4)}{h}$$
$$= \frac{2xh + h^2}{h}$$
$$= \frac{h(2x + h)}{h}$$
$$= 2x + h$$

GETTING READY FOR CLASS

After reading through the preceding section, respond in your own words and in complete sentences.

A. What is a rational expression?

B. Explain how to determine if a rational expression is in "lowest terms."

C. What is a rational function?

D. What is a difference quotient?

11. If $f(x) = 4x + 2$, find $\dfrac{f(x + h) - f(x)}{h}$.

12. If $f(x) = 2x^2 - 5$, find $\dfrac{f(x + h) - f(x)}{h}$.

Answers

11. 4

12. $2(2x + h)$

Vocabulary Review

Choose the correct words to fill in the blanks below.

nonzero rational domain quotient function

1. For _____ expressions, $\dfrac{P}{Q} = \dfrac{PK}{QK}$ and $\dfrac{P}{Q} = \dfrac{\frac{P}{K}}{\frac{Q}{K}}$.

2. Problems involving rational expressions contain a variable with a value that results in a _____ denominator.

3. $P(x)$ and $Q(x)$ in the rational _____ $f(x) = \dfrac{P(x)}{Q(x)}$ are polynomials.

4. The _____ of the rational function $f(x) = \dfrac{P(x)}{Q(x)}$ is all x for which $Q(x)$ is nonzero.

5. The expression $\dfrac{f(x) - f(a)}{(x - a)}$ is called a difference _____.

Problems

A Reduce each rational expression to lowest terms.

1. $\dfrac{x^2 - 16}{6x + 24}$

2. $\dfrac{12x - 9y}{3x^2 + 3xy}$

3. $\dfrac{a^4 - 81}{a - 3}$

4. $\dfrac{a^2 - 4a - 12}{a^2 + 8a + 12}$

5. $\dfrac{20y^2 - 45}{10y^2 - 5y - 15}$

6. $\dfrac{20x^2 - 93x + 34}{4x^2 - 9x - 34}$

7. $\dfrac{12y - 2xy - 2x^2y}{6y - 4xy - 2x^2y}$

8. $\dfrac{250a + 100ax + 10ax^2}{50a - 2ax^2}$

9. $\dfrac{(x - 3)^2(x + 2)}{(x + 2)^2(x - 3)}$

10. $\dfrac{(x - 4)^3(x + 3)}{(x + 3)^2(x - 4)}$

11. $\dfrac{x^3 + 1}{x^2 - 1}$

12. $\dfrac{x^3 - 1}{x^2 - 1}$

13. $\dfrac{4am - 4an}{3n - 3m}$

14. $\dfrac{ad - ad^2}{d - 1}$

15. $\dfrac{ab - a + b - 1}{ab + a + b + 1}$

16. $\dfrac{6cd - 4c - 9d + 6}{6d^2 - 13d + 6}$

17. $\dfrac{21x^2 - 23x + 6}{21x^2 + x - 10}$

18. $\dfrac{36x^2 - 11x - 12}{20x^2 - 39x + 18}$

19. $\dfrac{8x^2 - 6x - 9}{8x^2 - 18x + 9}$

20. $\dfrac{42x^2 + 23x - 10}{14x^2 + 45x - 14}$

21. $\dfrac{4x^2 + 29x + 45}{8x^2 - 10x - 63}$

22. $\dfrac{30x^2 - 61x + 30}{60x^2 + 22x - 60}$

23. $\dfrac{3a^2 + 10ax - 25x^2}{25x^2 - 9a^2}$

24. $\dfrac{6x^2 + xy - 12y^2}{16y^2 - 9x^2}$

25. $\dfrac{a^3 + b^3}{a^2 - b^2}$

26. $\dfrac{a^2 - b^2}{a^3 - b^3}$

27. $\dfrac{8x^4 - 8x}{4x^4 + 4x^3 + 4x^2}$

28. $\dfrac{6x^5 - 48x^3}{12x^3 + 24x^2 + 48x}$

29. $\dfrac{ax + 2x + 3a + 6}{ay + 2y - 4a - 8}$

30. $\dfrac{x^2 - 3ax - 2x + 6a}{x^2 - 3ax + 2x - 6a}$

31. $\dfrac{x^3 + 3x^2 - 4x - 12}{x^2 + x - 6}$

32. $\dfrac{6a^2 - 19ab + 15b^2}{12b^2 + ab - 6a^2}$

33. $\dfrac{9x^2 - 4y^2}{2y^2 + 3xy - 9x^2}$

34. $\dfrac{x^3 + 5x^2 - 4x - 20}{x^2 + 7x + 10}$

35. $\dfrac{x^3 - 8}{x^2 - 4}$

36. $\dfrac{y^2 - 9}{y^3 + 27}$

37. $\dfrac{8x^2 - 27}{4x^2 - 9}$

38. $\dfrac{25y^2 - 4}{125y^3 + 8}$

39. $\dfrac{x + 2}{x^3 + 8}$

40. $\dfrac{x + 3}{x^3 + 27}$

Refer to Examples 5 and 6 in this section, and reduce the following to lowest terms.

41. $\dfrac{x - 4}{4 - x}$

42. $\dfrac{6 - x}{x - 6}$

43. $\dfrac{y^2 - 36}{6 - y}$

44. $\dfrac{1 - y}{y^2 - 1}$

45. $\dfrac{1 - 9a^2}{9a^2 - 6a + 1}$

46. $\dfrac{1 - a^2}{a^2 - 2a + 1}$

Simplify each expression.

47. $\dfrac{(3x - 5) - (3a - 5)}{x - a}$

48. $\dfrac{(2x + 3) - (2a + 3)}{x - a}$

49. $\dfrac{(x^2 - 4) - (a^2 - 4)}{x - a}$

50. $\dfrac{(x^2 - 1) - (a^2 - 1)}{x - a}$

51. $\dfrac{(x^2 + 3x) - (a^2 + 3a)}{x - a}$

52. $\dfrac{(x^2 - 2x) - (a^2 - 2a)}{x - a}$

B State the domain for each rational function.

53. $f(x) = \dfrac{x - 3}{x - 1}$

54. $f(x) = \dfrac{x + 4}{x - 2}$

55. $g(x) = \dfrac{x^2 - 4}{x - 2}$

56. $g(x) = \dfrac{x^2 - 9}{x - 3}$

57. $h(t) = \dfrac{t - 4}{t^2 - 16}$

58. $h(t) = \dfrac{t - 5}{t^2 - 25}$

59. If $g(x) = \dfrac{x + 3}{x - 1}$, find $g(0), g(-3), g(3), g(-1)$, and $g(1)$, if possible.

60. If $g(x) = \dfrac{x - 2}{x - 1}$, find $g(0), g(-2), g(2), g(-1)$, and $g(1)$, if possible.

61. If $h(t) = \dfrac{t - 3}{t + 1}$, find $h(0), h(-3), h(3), h(-1)$, and $h(1)$, if possible.

62. If $h(t) = \dfrac{t - 2}{t + 1}$, find $h(0), h(-2), h(2), h(-1)$, and $h(1)$, if possible.

C For the functions below, evaluate each difference quotient of the form:

a. $\dfrac{f(x) - f(a)}{x - a}$

b. $\dfrac{f(x + h) - f(x)}{h}$

63. $f(x) = 4x$

64. $f(x) = -3x$

65. $f(x) = 5x + 3$

66. $f(x) = 6x - 5$

67. $f(x) = x^2$

68. $f(x) = 3x^2$

69. $f(x) = x^2 + 1$

70. $f(x) = x^2 - 3$

71. $f(x) = x^2 - 3x + 4$

72. $f(x) = x^2 + 4x - 7$

73. $f(x) = 2x^2 + 3x - 4$

74. $f(x) = 5x^2 + 3x - 7$

Applying the Concepts

75. Diet The following rational function is a mathematical model of a person's weekly progress on a weight-loss diet. The quantity $W(x)$ is the weight (in pounds) of the person after x weeks of dieting. Use the function to fill in the table.

$$W(x) = \frac{80(2x + 15)}{x + 6}$$

Weeks	Weight (lb)
x	$W(x)$
0	
1	
4	
12	
24	

76. Drag Racing The following rational function gives the speed $V(x)$, in miles per hour, of a dragster at each second x during a quarter-mile race. Use the function to fill in the table.

$$V(x) = \frac{340x}{x + 3}$$

Time (sec)	Speed (mph)
x	$V(x)$
0	
1	
2	
3	
4	
5	
6	

77. A new refrigerator cost $1,200 and costs approximately $540 per year to operate. Let n represent the total years you own the refrigerator.

$1,200

a. Express the total cost C as a function of the number of years n. (Assume there are no additional costs).

b. Express the average yearly cost \overline{C} as a function of the number of years n.

c. If you operate the refrigerator for 10 years, what is your average yearly cost?

d. If you operate the refrigerator for 15 years, what is your average yearly cost?

e. Is the average yearly cost increasing or decreasing as n increases? Explain your answer.

78. A new washer and dryer cost $2,200 and cost approximately $140 per year to operate. Let n represent the total years you own the washer and dryer.

a. Express the total cost C as a function of the number of years n. (Assume there are no additional costs).

b. Express the average yearly cost \overline{C} as a function of the number of years n.

c. If you operate the washer and dryer for 10 years, what is your average yearly cost?

d. If you operate the washer and dryer for 15 years, what is your average yearly cost?

e. Is the average yearly cost increasing or decreasing as n increases? Explain your answer.

Getting Ready for the Next Section

Multiply or divide, as indicated.

79. $\dfrac{6}{7} \cdot \dfrac{14}{18}$

80. $\dfrac{6}{8} \div \dfrac{3}{5}$

81. $5y^2 \cdot 4x^2$

82. $6x^4 \cdot 12y^5$

Factor.

83. $x^2 - 4$

84. $x^2 - 6x + 9$

85. $x^3 - x^2y$

86. $2y^2 - 2$

One Step Further

The graphs of two rational functions are given in Figures 3 and 4. Use the graphs to find the following.

87. a. $f(2)$

b. $f(-1)$

c. $f(0)$

d. $g(3)$

e. $g(6)$

f. $g(-1)$

g. $f(g(6))$

88. a. $f(4)$

b. $f(2)$

c. $g(-3)$

d. $g(-6)$

e. $g(f(4))$

f. $f(g(-6))$

g. $f(-2) + g(3)$

FIGURE 3

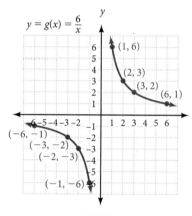

FIGURE 4

Find the Mistake

Each sentence below contains a mistake. Circle the mistake and write the correct word or expression on the line provided.

1. The rational expression $\dfrac{8x^2 + 2x - 1}{16x^2 - 6x - 7}$ reduced to lowest terms is $\dfrac{4x - 1}{4x - 7}$. _____

2. To reduce the rational expression $\dfrac{15x^2 - 16x + 4}{5x^2 - 17x + 6}$ to lowest terms, we must divide the numerator and denominator by the factor $(3x - 2)$. _____

3. The rational function $f(x) = \dfrac{x^3 - 4x^2 + 4x + 16}{x^2 - 6x + 8}$ is undefined when $f(-2)$ and $f(1)$. _____

4. To find the difference quotient for $f(x) = x^2 + x - 7$, we use $f(a) = x + a + 1$. _____

To have a home videotape transferred to DVD, the amount you pay for the service depends on the number of DVDs you have made. In other words, The more DVDs you have made, the lower the charge per copy. The following demand function gives the price (in dollars) per disk $p(x)$ a company charges for making x DVDs. As you can see, it is a rational function.

$$p(x) = \frac{2(x + 60)}{x + 5}$$

The graph in Figure 1 shows this function from $x = 0$ to $x = 100$. As you can see, the more DVDs that are made, the lower the price per copy.

FIGURE 1

If we were interested in finding the revenue function for this situation, we would multiply the number of copies made x by the price per copy $p(x)$. This involves multiplication with a rational expression, which is one of the topics we cover in this section.

In the last section, we found the process of reducing rational expressions to lowest terms to be the same process used in reducing fractions to lowest terms. The similarity also holds for the process of multiplication or division of rational expressions.

Multiplication with fractions is the simplest of the four basic operations. To multiply two fractions, we simply multiply numerators and multiply denominators. That is, if a, b, c, and d are real numbers, with $b \neq 0$ and $d \neq 0$, then

$$\frac{a}{b} \cdot \frac{c}{d} = \frac{ac}{bd}$$

Practice Problems

1. Multiply $\dfrac{3}{4} \cdot \dfrac{10}{21}$.

A Multiplying and Dividing Rational Expressions

EXAMPLE 1 Multiply $\dfrac{6}{7} \cdot \dfrac{14}{18}$.

Solution

$$\dfrac{6}{7} \cdot \dfrac{14}{18} = \dfrac{6(14)}{7(18)} \qquad \text{Multiply numerators and denominators.}$$

$$= \dfrac{2 \cdot 3(2 \cdot 7)}{7(2 \cdot 3 \cdot 3)} \qquad \text{Factor.}$$

$$= \dfrac{2}{3} \qquad \text{Divide out common factors.}$$

Our next example is similar to some of the problems you may have worked previously. We multiply fractions whose numerators and denominators are monomials by multiplying numerators and multiplying denominators and then reducing to lowest terms. Here is how it looks:

2. Multiply $\dfrac{6x^4}{10y^5} \cdot \dfrac{15y^3}{3x^2}$.

EXAMPLE 2 Multiply $\dfrac{8x^3}{27y^8} \cdot \dfrac{9y^3}{12x^2}$.

Solution We multiply numerators and denominators and then divide out common factors.

$$\dfrac{8x^3}{27y^8} \cdot \dfrac{9y^3}{12x^2} = \dfrac{8 \cdot 9x^3y^3}{27 \cdot 12x^2y^8} \qquad \begin{array}{l}\text{Multiply numerators.}\\ \text{Multiply denominators.}\end{array}$$

$$= \dfrac{4 \cdot 2 \cdot 9x^3y^3}{9 \cdot 3 \cdot 4 \cdot 3x^2y^8} \qquad \text{Factor coefficients.}$$

$$= \dfrac{2x}{9y^5} \qquad \text{Divide out common factors.}$$

The product of two rational expressions is the product of their numerators over the product of their denominators.

Once again, we should mention that the little slashes we have drawn through the factors are simply used to denote the factors we have divided out of the numerator and denominator.

3. Let $f(x) = \dfrac{x-5}{x+2}$ and

$g(x) = \dfrac{x^2-4}{3x-15}$.

Find $f(x) \cdot g(x)$.

EXAMPLE 3 Let $f(x) = \dfrac{x-3}{x^2-4}$ and $g(x) = \dfrac{x+2}{x^2-6x+9}$. Find $f(x) \cdot g(x)$.

Solution We begin by multiplying numerators and denominators. We then factor all polynomials and divide out factors common to the numerator and denominator.

$$f(x) \cdot g(x) = \dfrac{x-3}{x^2-4} \cdot \dfrac{x+2}{x^2-6x+9}$$

$$= \dfrac{(x-3)(x+2)}{(x^2-4)(x^2-6x+9)} \qquad \text{Multiply.}$$

$$= \dfrac{(x-3)(x+2)}{(x+2)(x-2)(x-3)(x-3)} \qquad \text{Factor.}$$

$$= \dfrac{1}{(x-2)(x-3)}$$

The first two steps can be combined to save time. In the next example, we'll show how to perform the multiplication and factoring steps together.

Answers

1. $\dfrac{5}{14}$

2. $\dfrac{3x^2}{y^2}$

3. $f(x) \cdot g(x) = \dfrac{x-2}{3}$

EXAMPLE 4 Multiply $\dfrac{2y^2 - 4y}{2y^2 - 2} \cdot \dfrac{y^2 - 2y - 3}{y^2 - 5y + 6}$.

Solution

$$\frac{2y^2 - 4y}{2y^2 - 2} \cdot \frac{y^2 - 2y - 3}{y^2 - 5y + 6} = \frac{2y(y - 2)(y - 3)(y + 1)}{2(y + 1)(y - 1)(y - 3)(y - 2)}$$

$$= \frac{y}{y - 1}$$

Notice in both of the preceding examples that we did not actually multiply the polynomials. It would be senseless to do that because we would then have to factor each of the resulting products to reduce them to lowest terms.

The quotient of two rational expressions is the product of the first and the reciprocal of the second. That is, we find the quotient of two rational expressions the same way we find the quotient of two fractions. Here is an example that reviews division with fractions:

EXAMPLE 5 Divide $\dfrac{6}{8} \div \dfrac{3}{5}$.

Solution

$$\frac{6}{8} \div \frac{3}{5} = \frac{6}{8} \cdot \frac{5}{3} \qquad \textit{Write division in terms of multiplication.}$$

$$= \frac{6(5)}{8(3)} \qquad \textit{Multiply numerators and denominators.}$$

$$= \frac{2 \cdot 3(5)}{2 \cdot 2 \cdot 2(3)} \qquad \textit{Factor.}$$

$$= \frac{5}{4} \qquad \textit{Divide out common factors.}$$

To divide one rational expression by another, we use the definition of division to multiply by the reciprocal of the expression that follows the division symbol.

EXAMPLE 6 Divide $\dfrac{8x^3}{5y^2} \div \dfrac{4x^2}{10y^6}$.

Solution First, we rewrite the problem in terms of multiplication. Then we multiply.

$$\frac{8x^3}{5y^2} \div \frac{4x^2}{10y^6} = \frac{8x^3}{5y^2} \cdot \frac{10y^6}{4x^2}$$

$$= \frac{\overset{2}{8} \cdot \overset{2}{10}x^3 y^6}{4 \cdot 5x^2 y^2}$$

$$= 4xy^4$$

EXAMPLE 7 Divide $\dfrac{x^2 - y^2}{x^2 - 2xy + y^2} \div \dfrac{x^3 + y^3}{x^3 - x^2 y}$.

Solution We begin by writing the problem as the product of the first and the reciprocal of the second and then proceed as in the previous two examples.

$$\frac{x^2 - y^2}{x^2 - 2xy + y^2} \div \frac{x^3 + y^3}{x^3 - x^2 y} \cdot$$

$$= \frac{x^2 - y^2}{x^2 - 2xy + y^2} \cdot \frac{x^3 - x^2 y}{x^3 + y^3} \qquad \begin{array}{l}\textit{Multiply by the reciprocal}\\ \textit{of the divison}\end{array}$$

$$= \frac{(x - y)(x + y)(x^2)(x - y)}{(x - y)(x - y)(x + y)(x^2 - xy + y^2)} \qquad \textit{Factor and multiply.}$$

$$= \frac{x^2}{x^2 - xy + y^2} \qquad \textit{Divide out common factors.}$$

4. Multiply

$$\frac{4x^2 - 16x}{2x^2 - 14x + 24} \cdot \frac{x^2 - 5x + 6}{x^2 + x - 6}.$$

5. Divide $\dfrac{3}{4} \div \dfrac{5}{8}$.

6. Divide $\dfrac{9x^2 y}{10x^3 y^2} \div \dfrac{6xy^3}{5x^5 y^4}$.

Note A common mistake is that students divide the 8 and the 10 in the original problem by 2 before multiplying. It is important to multiply by the reciprocal before dividing out common factors.

7. Divide $\dfrac{4x + 8}{x^2 - x - 6} \div \dfrac{x^2 + 7x + 12}{x^2 - 9}$.

Answers

4. $\dfrac{2x}{x + 3}$

5. $\dfrac{6}{5}$

6. $\dfrac{3x^3}{4}$

7. $\dfrac{4}{x + 4}$

8. Perform the indicated operations.

$$\frac{x^2 - 9x + 20}{x^2 + 5x + 6} \cdot \frac{x + 2}{x^2 - 4x - 5} \div \frac{x - 4}{x^2 + 2x - 3}$$

Here are some more examples of multiplication and division with rational expressions.

EXAMPLE 8 Perform the indicated operations.

$$\frac{a^2 - 8a + 15}{a + 4} \cdot \frac{a + 2}{a^2 - 5a + 6} \div \frac{a^2 - 3a - 10}{a^2 + 2a - 8}$$

Solution First, we rewrite the division as multiplication by the reciprocal. Then we proceed as usual.

$$\frac{a^2 - 8a + 15}{a + 4} \cdot \frac{a + 2}{a^2 - 5a + 6} \div \frac{a^2 - 3a - 10}{a^2 + 2a - 8}$$ *Change division to multiplication by the reciprocal.*

$$= \frac{(a^2 - 8a + 15)(a + 2)(a^2 + 2a - 8)}{(a + 4)(a^2 - 5a + 6)(a^2 - 3a - 10)}$$ *Factor.*

$$= \frac{(a - 5)(a - 3)(a + 2)(a + 4)(a - 2)}{(a + 4)(a - 3)(a - 2)(a - 5)(a + 2)}$$ *Divide out common factors.*

$$= 1$$

Our next example involves factoring by grouping. As you may have noticed, working the problems in this chapter gives you a very detailed review of factoring. You also may have noticed that when multiplying or dividing rational expressions we do not find a common denominator. This process would only complicate our work, and is only necessary when adding or subtracting rational expressions.

9. Multiply.

$$\frac{xy - 2x - 4y + 8}{xy + 3x + 2y + 6} \cdot \frac{xy + 3x - 4y - 12}{xy + 2y - 2x - 4}$$

EXAMPLE 9 Multiply $\dfrac{xa + xb + ya + yb}{xa - xb - ya + yb} \cdot \dfrac{xa + xb - ya - yb}{xa - xb + ya - yb}$.

Solution We will factor each polynomial by grouping, which takes two steps.

$$\frac{xa + xb + ya + yb}{xa - xb - ya + yb} \cdot \frac{xa + xb - ya - yb}{xa - xb + ya - yb}$$

$$= \frac{x(a + b) + y(a + b)}{x(a - b) - y(a - b)} \cdot \frac{x(a + b) - y(a + b)}{x(a - b) + y(a - b)} \quad \left.\right\} \text{ Factor by grouping.}$$

$$= \frac{(a + b)(x + y)(a + b)(x - y)}{(a - b)(x - y)(a - b)(x + y)}$$

$$= \frac{(a + b)^2}{(a - b)^2}$$

10. Multiply $(x^2 - 9) \cdot \dfrac{x + 2}{x + 3}$.

EXAMPLE 10 Multiply $(4x^2 - 36) \cdot \dfrac{12}{4x + 12}$.

Solution We can think of $4x^2 - 36$ as having a denominator of 1. Thinking of it in this way allows us to proceed as we did in the previous examples.

$$(4x^2 - 36) \cdot \frac{12}{4x + 12}$$

$$= \frac{4x^2 - 36}{1} \cdot \frac{12}{4x + 12}$$ *Write $4x^2 - 36$ with denominator 1.*

$$= \frac{4(x - 3)(x + 3)12}{4(x + 3)}$$ *Factor.*

$$= 12(x - 3)$$ *Divide out common factors.*

Answers

8. $\dfrac{x - 1}{x + 1}$

9. $\dfrac{(x - 4)^2}{(x + 2)^2}$

10. $(x - 3)(x + 2)$

EXAMPLE 11 Multiply $3(x - 2)(x - 1) \cdot \dfrac{5}{x^2 - 3x + 2}$.

Solution This problem is very similar to the problem in Example 10. Writing the first rational expression with a denominator of 1, we have

$$\frac{3(x - 2)(x - 1)}{1} \cdot \frac{5}{x^2 - 3x + 2} = \frac{3(x - 2)(x - 1)5}{(x - 2)(x - 1)}$$

$$= 3 \cdot 5$$

$$= 15$$

GETTING READY FOR CLASS

After reading through the preceding section, respond in your own words and in complete sentences.

A. Summarize the steps used to multiply fractions.

B. What is the first step in multiplying two rational expressions?

C. Why is factoring important when multiplying and dividing rational expressions?

D. How is division with rational expressions different than multiplication of rational expressions?

11. Multiply $a(a + 2)(a - 4)\dfrac{a + 5}{a^2 - 4a}$

Vocabulary Review

Choose the correct words to fill in the blanks below.

numerators quotient reciprocal product

1. The _____ of two rational expressions is the product of their _____ over the product of their denominators.

2. The _____ of two rational expressions is the product of the first and the _____ of the second.

Problems

A Perform the indicated operations.

1. $\dfrac{2}{9} \cdot \dfrac{3}{4}$

2. $\dfrac{5}{6} \cdot \dfrac{7}{8}$

3. $\dfrac{3}{4} \div \dfrac{1}{3}$

4. $\dfrac{3}{8} \div \dfrac{5}{4}$

5. $\dfrac{3}{7} \cdot \dfrac{14}{24} \div \dfrac{1}{2}$

6. $\dfrac{6}{5} \cdot \dfrac{10}{36} \div \dfrac{3}{4}$

7. $\dfrac{10x^2}{5y^2} \cdot \dfrac{15y^3}{2x^4}$

8. $\dfrac{8x^3}{7y^4} \cdot \dfrac{14y^6}{16x^2}$

9. $\dfrac{11a^2b}{5ab^2} \div \dfrac{22a^3b^2}{10ab^4}$

10. $\dfrac{8ab^3}{9a^2b} \div \dfrac{16a^2b^2}{18ab^3}$

11. $\dfrac{6x^2}{5y^3} \cdot \dfrac{11z^2}{2x^2} \div \dfrac{33z^5}{10y^8}$

12. $\dfrac{4x^3}{7y^2} \cdot \dfrac{6z^5}{5x^6} \div \dfrac{24z^2}{35x^6}$

Perform the indicated operations. Be sure to write all answers in lowest terms.

13. $\dfrac{x^2 - 9}{x^2 - 4} \cdot \dfrac{x - 2}{x - 3}$

14. $\dfrac{x^2 - 16}{x^2 - 25} \cdot \dfrac{x - 5}{x - 4}$

15. $\dfrac{y^2 - 1}{y + 2} \cdot \dfrac{y^2 + 5y + 6}{y^2 + 2y - 3}$

16. $\dfrac{y - 1}{y^2 - y - 6} \cdot \dfrac{y^2 + 5y + 6}{y^2 - 1}$

17. $\dfrac{3x - 12}{x^2 - 4} \cdot \dfrac{x^2 + 6x + 8}{x - 4}$

18. $\dfrac{x^2 + 5x + 1}{4x - 4} \cdot \dfrac{x - 1}{x^2 + 5x + 1}$

19. $\dfrac{xy}{xy + 1} \div \dfrac{x}{y}$

20. $\dfrac{y}{x} \div \dfrac{xy}{xy - 1}$

21. $\dfrac{1}{x^2 - 9} \div \dfrac{1}{x^2 + 9}$

22. $\dfrac{1}{x^2 - 9} \div \dfrac{1}{(x - 3)^2}$

23. $\dfrac{5x - 10}{6x - 9} \div \dfrac{6 - 3x}{8x - 12}$

24. $\dfrac{10b - 15}{4b + 2} \div \dfrac{6 - 4b}{6b + 3}$

25. $\dfrac{5x + 2y}{25x^2 - 5xy - 6y^2} \cdot \dfrac{20x^2 - 7xy - 3y^2}{4x + y}$

26. $\dfrac{7x + 3y}{42x^2 - 17xy - 15y^2} \cdot \dfrac{12x^2 - 4xy - 5y^2}{2x + y}$

27. $\dfrac{a^2 - 5a + 6}{a^2 - 2a - 3} \div \dfrac{a - 5}{a^2 + 3a + 2}$

28. $\dfrac{a^2 + 7a + 12}{a - 5} \div \dfrac{a^2 + 9a + 18}{a^2 - 7a + 10}$

29. $\dfrac{4t^2 - 1}{6t^2 + t - 2} \div \dfrac{8t^3 + 1}{27t^3 + 8}$

30. $\dfrac{9t^2 - 1}{6t^2 + 7t - 3} \div \dfrac{27t^3 + 1}{8t^3 + 27}$

31. $\dfrac{2x^2 - 5x - 12}{4x^2 + 8x + 3} \div \dfrac{x^2 - 16}{2x^2 + 7x + 3}$

32. $\dfrac{x^2 - 2x + 1}{3x^2 + 7x - 20} \div \dfrac{x^2 + 3x - 4}{3x^2 - 2x - 5}$

33. $\dfrac{2a^2 - 21ab - 36b^2}{a^2 - 11ab - 12b^2} \div \dfrac{10a + 15b}{a^2 - b^2}$

34. $\dfrac{3a^2 + 7ab - 20b^2}{a^2 + 5ab + 4b^2} \div \dfrac{3a^2 - 17ab + 20b^2}{3a - 12b}$

35. $\dfrac{6c^2 - c - 15}{9c^2 - 25} \cdot \dfrac{15c^2 + 22c - 5}{6c^2 + 5c - 6}$

36. $\dfrac{m^2 + 4m - 21}{m^2 - 12m + 27} \cdot \dfrac{m^2 - 7m + 12}{m^2 + 3m - 28}$

37. $\dfrac{6x^2 - 11xy + 3y^2}{2x^2 - 7xy + 6y^2} \cdot \dfrac{3x^2 + 8xy + 4y^2}{2y^2 - 3xy - 9x^2}$

38. $\dfrac{a^2 + ab - 12b^2}{3a^2 + 8ab - 16b^2} \cdot \dfrac{2a^2 + 7ab + 6b^2}{9b^2 + 3ab - 2a^2}$

39. $\dfrac{360x^3 - 490x}{36x^2 + 84x + 49} \cdot \dfrac{30x^2 + 83x + 56}{150x^3 + 65x^2 - 280x}$

40. $\dfrac{490x^2 - 640}{49x^2 - 112x + 64} \cdot \dfrac{28x^2 - 95x + 72}{56x^3 - 62x^2 - 144x}$

41. $\dfrac{x^5 - x^2}{5x^2 - 5x} \cdot \dfrac{10x^4 - 10x^2}{2x^4 + 2x^3 + 2x^2}$

42. $\dfrac{2x^4 - 16x}{3x^6 - 48x^2} \cdot \dfrac{6x^5 + 24x^3}{4x^4 + 8x^3 + 16x^2}$

43. $\dfrac{a^2 - 16b^2}{a^2 - 8ab + 16b^2} \cdot \dfrac{a^2 - 9ab + 20b^2}{a^2 - 7ab + 12b^2} \div \dfrac{a^2 - 25b^2}{a^2 - 6ab + 9b^2}$

44. $\dfrac{a^2 - 6ab + 9b^2}{a^2 - 4b^2} \cdot \dfrac{a^2 - 5ab + 6b^2}{(a - 3b)^2} \div \dfrac{a^2 - 9b^2}{a^2 - ab - 6b^2}$

45. $\dfrac{2y^2 - 7y - 15}{42y^2 - 29y - 5} \cdot \dfrac{12y^2 - 16y + 5}{7y^2 - 36y + 5} \div \dfrac{4y^2 - 9}{49y^2 - 1}$

46. $\dfrac{8y^2 + 18y - 5}{21y^2 - 16y + 3} \cdot \dfrac{35y^2 - 22y + 3}{6y^2 + 17y + 5} \div \dfrac{16y^2 - 1}{9y^2 - 1}$

47. $\dfrac{xy - 2x + 3y - 6}{xy + 2x - 4y - 8} \cdot \dfrac{xy + x - 4y - 4}{xy - x + 3y - 3}$

48. $\dfrac{ax + bx + 2a + 2b}{ax - 3a + bx - 3b} \cdot \dfrac{ax - bx - 3a + 3b}{ax - bx - 2a + 2b}$

49. $\dfrac{xy^2 - y^2 + 4xy - 4y}{xy - 3y + 4x - 12} \div \dfrac{xy^3 + 2xy^2 + y^3 + 2y^2}{xy^2 - 3y^2 + 2xy - 6y}$

50. $\dfrac{4xb - 8b + 12x - 24}{xb^2 + 3b^2 + 3xb + 9b} \div \dfrac{4xb - 8b - 8x + 16}{xb^2 + 3b^2 - 2xb - 6b}$

51. $\dfrac{2x^3 + 10x^2 - 8x - 40}{x^3 + 4x^2 - 9x - 36} \cdot \dfrac{x^2 + x - 12}{2x^2 + 14x + 20}$

52. $\dfrac{x^3 + 2x^2 - 9x - 18}{x^4 + 3x^3 - 4x^2 - 12x} \cdot \dfrac{x^3 + 5x^2 + 6x}{x^2 - x - 6}$

53. $\dfrac{w^3 - w^2x}{wy - w} \div \left(\dfrac{w - x}{y - 1}\right)^2$

54. $\dfrac{a^3 - a^2b}{ac - a} \div \left(\dfrac{a - b}{c - 1}\right)^2$

55. $\dfrac{mx + my + 2x + 2y}{6x^2 - 5xy - 4y^2} \div \dfrac{2mx - 4x + my - 2y}{3mx - 6x - 4my + 8y}$

56. $\dfrac{ax - 2a + 2xy - 4y}{ax + 2a - 2xy - 4y} \div \dfrac{ax + 2a + 2xy + 4y}{ax - 2a - 2xy + 4y}$

57. $\dfrac{1 - 4d^2}{(d - c)^2} \cdot \dfrac{d^2 - c^2}{1 + 2d}$

58. $\dfrac{k^2 - 1}{k^3 + 1} \div \dfrac{k^3 - 1}{k - 1} \cdot (k^2 + k + 1)$

59. $\dfrac{r^2 - s^2}{r^2 + rs + s^2} \cdot \dfrac{r^3 - s^3}{r^2 + s^2} \div \dfrac{r^4 - s^4}{r^2 - s^2}$

60. $\dfrac{r^3 - s^3}{s - r} \cdot \dfrac{(r + s)^2}{r^2 + s^2}$

Use the method shown in Examples 10 and 11 to find the following products.

61. $(3x - 6) \cdot \dfrac{x}{x - 2}$

62. $(4x + 8) \cdot \dfrac{x}{x + 2}$

63. $(x^2 - 25) \cdot \dfrac{2}{x - 5}$

64. $(x^2 - 49) \cdot \dfrac{5}{x + 7}$

65. $(x^2 - 3x + 2) \cdot \dfrac{3}{3x - 3}$

66. $(x^2 - 3x + 2) \cdot \dfrac{-1}{x - 2}$

67. $(y - 3)(y - 4)(y + 3) \cdot \dfrac{-1}{y^2 - 9}$

68. $(y + 1)(y + 4)(y - 1) \cdot \dfrac{3}{y^2 - 1}$

69. $a(a + 5)(a - 5) \cdot \dfrac{a + 1}{a^2 + 5a}$

70. $a(a + 3)(a - 3) \cdot \dfrac{a - 1}{a^2 - 3a}$

The next two problems are intended to give you practice reading, and paying attention to, the instructions that accompany the problems you are working. Working these problems is an excellent way to get ready for a test or a quiz.

71. Work each problem according to the instructions.

 a. Simplify: $\dfrac{16 - 1}{64 - 1}$.

 b. Reduce: $\dfrac{25x^2 - 9}{125x^3 - 27}$.

 c. Multiply: $\dfrac{25x^2 - 9}{125x^3 - 27} \cdot \dfrac{5x - 3}{5x + 3}$.

 d. Divide: $\dfrac{25x^2 - 9}{125x^3 - 27} \div \dfrac{5x - 3}{25x^2 + 15x + 9}$.

72. Work each problem according to the instructions.

 a. Simplify: $\dfrac{64 - 49}{64 + 112 + 49}$.

 b. Reduce: $\dfrac{9x^2 - 49}{9x^2 + 42x + 49}$.

 c. Multiply: $\dfrac{9x^2 - 49}{9x^2 + 42x + 49} \cdot \dfrac{3x + 7}{3x - 7}$.

 d. Divide: $\dfrac{9x^2 - 49}{9x^2 + 42x + 49} \div \dfrac{3x + 7}{3x - 7}$.

The work you did with algebra of functions will help with these.

73. Let $f(x) = \dfrac{x^2 - x - 6}{x - 1}$ and $g(x) = \dfrac{x + 2}{x^2 - 4x + 3}$, find

 a. $f(x) \cdot g(x)$

 b. $f(x) \div g(x)$

 c. $g(x) \div f(x)$

74. Let $f(x) = \dfrac{x^2 - x - 12}{x^2 - 4x + 3}$ and $g(x) = \dfrac{x^2 - x - 12}{x^2 - 5x + 4}$, find

 a. $f(x) \cdot g(x)$

 b. $f(x) \div g(x)$

 c. $g(x) \div f(x)$

Applying the Concepts

75. Demand Equation At the beginning of this section, we introduced the demand equation shown here.

$$p(x) = \frac{2(x + 60)}{x + 5}$$

a. Use the demand equation to fill in the table. Then compare your results with the graph shown in Figure 1 of this section.

Number of Copies	Price per Copy ($)
x	p(x)
1	
10	
20	
50	
100	

b. To find the revenue for selling 50 copies of a tape, we multiply the price per tape by 50. Find the revenue for selling 50 tapes.

c. Find the revenue for selling 100 tapes.

d. Find the revenue equation $R(x)$.

76. Rate Equation The drug concentration t hours after an injection of a certain drug is given by the rational function

$$c(t) = \frac{0.04t + 5}{0.02t^2}$$

a. Complete the following table using the rational function given above.

Time	Concentration
t	c(t)
5	
10	
15	
20	

b. Find the concentration after 1 day and 2 days.

Maintaining Your Skills

77. The chart shows the ratio of a planet's diameter to Earth's diameter. If Earth has a diameter of 7,900 miles, what is the diameter of Venus? Of Saturn?

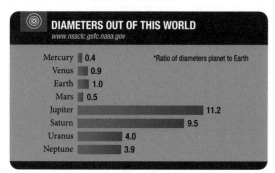

78. The following chart shows the number of dogs registered in the most popular breeds. Use the chart to write the numbers in scientific notation.

a. Labrador Retriever

b. German Shepherd

c. Beagle

Getting Ready for the Next Section

Combine.

79. $\dfrac{4}{9} + \dfrac{2}{9}$

80. $\dfrac{3}{8} + \dfrac{1}{8}$

81. $\dfrac{3}{14} + \dfrac{7}{30}$

82. $\dfrac{3}{10} + \dfrac{11}{42}$

Multiply.

83. $-1(7 - x)$

84. $-1(3 - x)$

85. $-2(2x-5)$

86. $-3(4-7x)$

Factor.

87. $x^2 - 1$

88. $x^2 - 2x - 3$

89. $2x + 10$

90. $x^2 + 4x + 3$

91. $a^3 - b^3$

92. $8y^3 - 27$

93. $4x^2 - 9$

94. $5x^2 - 125$

One Step Further

Perform the following operations. Be sure to write all answers in lowest terms.

95. $\dfrac{x^6 + y^6}{x^4 + 4x^2y^2 + 3y^4} \div \dfrac{x^4 + 3x^2y^2 + 2y^4}{x^4 + 5x^2y^2 + 6y^4}$

96. $\dfrac{x^2 + 9xy + 8y^2}{x^2 + 7xy - 8y^2} \div \dfrac{x^2 - y^2}{x^2 + 5xy - 6y^2}$

97. $\dfrac{a^2(2a + b) + 6a(2a + b) + 5(2a + b)}{3a^2(2a + b) - 2a(2a + b) + (2a + b)} \div \dfrac{a + 1}{a - 1}$

98. $\dfrac{2x^2(x - 3z) - 5x(x - 3z) + 2(x - 3z)}{4x^2(x - 3z) - 11x(x - 3z) + 6(x - 3z)} \div \dfrac{4x - 3}{4x + 1}$

99. $\dfrac{p^3 + q^3}{q - p} \div \dfrac{(p + q)^2}{p^2 - q^2}$

100. $\dfrac{x^3 - y^3}{x^3 + y^3} \cdot \dfrac{x + y}{x - y}$

Find the Mistake

Each sentence below contains a mistake. Circle the mistake and write the correct word(s) or expression on the line provided.

1. Multiplying $\dfrac{x^2 - 4}{x - 2} \cdot \dfrac{x^2 + 2x - 8}{x^2 + 6x + 8}$ we get $x + 2$. _____

2. Multiplying $\dfrac{2x - 6}{4x^2 - 4x - 24} \cdot \dfrac{6x^2 + 2x - 20}{3x - 5}$ we get $\dfrac{2x - 6}{x - 3}$. _____

3. To divide the rational expression $\dfrac{5x^2 + 21x - 20}{15x^2 - 17x + 4}$ by $\dfrac{x^2 - 25}{3x^2 - 4x + 1}$ we multiply the numerators and the denominators.

4. Dividing $\dfrac{4x^2 + 27xy + 18y^2}{x^2 + 7xy + 6y^2} \div \dfrac{16x + 12y}{x^2 - y^2}$, we get $x - y$. _____

A Add and subtract rational
 expressions with the same
 denominator.

B Add and subtract rational
 expressions with different
 denominators.

KEY WORDS

least common denominator

LCD

**Video Examples
Section 8.3**

Practice Problems

1. Add $\dfrac{2}{x} + \dfrac{7}{x}$.

8.3 Addition and Subtraction of Rational Expressions

Have you ever heard of the competitive sport of joggling? Imagine a track runner competing in the 100-meter dash while juggling five balls! Previously, we have used the rate equation $d = rt$ to help us solve some application problems. If we solve this equation for rate r, we get the rational equation $r = \frac{d}{t}$. Suppose a runner can complete the 100-meter dash without juggling in 12 seconds, but he takes 31 seconds to joggle the same distance. After reading this section, we can create two rational expressions for rate r to determine the juggler's difference in speed.

This section is concerned with addition and subtraction of rational expressions. In the first part of this section, we will look at addition of expressions that have the same denominator. In the second part of this section, we will look at addition of expressions that have different denominators.

A Addition and Subtraction with the Same Denominator

To add two expressions that have the same denominator, we simply add numerators and put the sum over the common denominator. Because the process we use to add and subtract rational expressions is the same process used to add and subtract fractions, we will begin with an example involving fractions.

EXAMPLE 1 Add $\dfrac{4}{9} + \dfrac{2}{9}$.

Solution We add fractions with the same denominator by using the distributive property. Here is a detailed look at the steps involved:

$$\frac{4}{9} + \frac{2}{9} = 4\left(\frac{1}{9}\right) + 2\left(\frac{1}{9}\right)$$

$$= (4 + 2)\left(\frac{1}{9}\right) \qquad \textit{Distributive property}$$

$$= 6\left(\frac{1}{9}\right)$$

$$= \frac{6}{9}$$

$$= \frac{2}{3} \qquad \textit{Divide the numerator and denominator by common factor 3.}$$

Note that the important thing about the fractions in this example is that they each have a denominator of 9. If they did not have the same denominator, we could not have written them as two terms with a factor of $\frac{1}{9}$ in common. Without the $\frac{1}{9}$ common to each term, we couldn't apply the distributive property. Without the distributive property, we would not have been able to add the two fractions in this form.

Answer

1. $\dfrac{9}{x}$

In the following examples, we will not show all the steps we showed in Example 1. The steps are shown in Example 1 so you will see why both fractions must have the same denominator before we can add them. In practice, we simply add numerators and place the result over the common denominator.

We add and subtract rational expressions with the same denominator by combining numerators and writing the result over the common denominator. Then we reduce the result to lowest terms, if possible. Example 2 shows this process in detail. If you see the similarities between operations on rational numbers and operations on rational expressions, this chapter will look like an extension of rational numbers rather than a completely new set of topics.

EXAMPLE 2 Add $\dfrac{x}{x^2 - 1} + \dfrac{1}{x^2 - 1}$.

Solution Because the denominators are the same, we simply add numerators.

$$\frac{x}{x^2 - 1} + \frac{1}{x^2 - 1} = \frac{x + 1}{x^2 - 1} \qquad \text{Add numerators.}$$

$$= \frac{\cancel{x + 1}}{(x - 1)(x + 1)} \qquad \text{Factor denominator.}$$

$$= \frac{1}{x - 1} \qquad \text{Divide out common factor } x + 1.$$

Our next example involves subtraction of rational expressions. Pay careful attention to what happens to the signs of the terms in the numerator of the second expression when we subtract it from the first expression.

EXAMPLE 3 Let $f(x) = \dfrac{2x - 5}{x - 2}$ and $g(x) = \dfrac{x - 3}{x - 2}$. Find and simplify $f(x) - g(x)$.

Solution Because each expression has the same denominator, we simply subtract the numerator in the second expression from the numerator in the first expression and write the difference over the common denominator $x - 2$. We must be careful, however, that we subtract both terms in the second numerator. To ensure that we do, we will enclose that numerator in parentheses.

$$f(x) - g(x) = \frac{2x - 5}{x - 2} - \frac{x - 3}{x - 2}$$

$$= \frac{2x - 5 - (x - 3)}{x - 2} \qquad \text{Subtract numerators.}$$

$$= \frac{2x - 5 - x + 3}{x - 2} \qquad \text{Remove parentheses.}$$

$$= \frac{\cancel{x - 2}}{\cancel{x - 2}} \qquad \text{Combine similar terms in the numerator.}$$

$$= 1 \qquad \text{Reduce (or divide).}$$

Note the 3 in the numerator of the second step. It is a common mistake to write this as -3, by forgetting to subtract both terms in the numerator of the second expression. Whenever the expression we are subtracting has two or more terms in its numerator, we have to watch for this mistake.

Next we consider addition and subtraction of fractions and rational expressions that have different denominators.

2. Add $\dfrac{2x}{x^2 - 1} + \dfrac{2}{x^2 - 1}$.

3. Let $f(x) = \dfrac{4x - 7}{x - 4}$ and $g(x) = \dfrac{3x - 3}{x - 4}$.

Find and simplify $f(x) - g(x)$.

Answers

2. $\dfrac{2}{x - 1}$

3. 1

B Addition and Subtraction With Different Denominators

Before we look at an example of addition of fractions with different denominators, we need to review the definition for the *least common denominator (LCD)*.

> **DEFINITION** least common denominator
>
> The *least common denominator* for a set of denominators is the smallest expression that is divisible by each of the denominators. It is also considered the least common multiple of the given set of denominators.

The first step in combining two fractions is to find the LCD. Once we have the common denominator, we rewrite each fraction as an equivalent fraction with the common denominator. After that, we simply add or subtract as we did in our first three examples.

Example 4 is a review of the step-by-step procedure used to add two fractions with different denominators.

4. Add $\dfrac{1}{21} + \dfrac{4}{15}$.

EXAMPLE 4 Add $\dfrac{3}{14} + \dfrac{7}{30}$.

Solution

Step 1: *Find the LCD.*

To do this, we first factor both denominators into prime factors.

Factor 14: $14 = 2 \cdot 7$

Factor 30: $30 = 2 \cdot 3 \cdot 5$

Because the LCD must be divisible by 14, it must have factors of 2 and 7. It must also be divisible by 30 and, therefore, have factors of 2, 3, and 5. We do not need to repeat the 2 that appears in both the factors of 14 and those of 30. Therefore,

$$\text{LCD} = 2 \cdot 3 \cdot 5 \cdot 7 = 210$$

Step 2: *Change to equivalent fractions.*

Because we want each fraction to have a denominator of 210 and at the same time keep its original value, we multiply each by 1 in the appropriate form.

Change $\frac{3}{14}$ to a fraction with denominator 210.

$$\frac{3}{14} \cdot \frac{15}{15} = \frac{45}{210}$$

Change $\frac{7}{30}$ to a fraction with denominator 210.

$$\frac{7}{30} \cdot \frac{7}{7} = \frac{49}{210}$$

Step 3: *Add numerators of equivalent fractions found in step 2.*

$$\frac{45}{210} + \frac{49}{210} = \frac{94}{210}$$

Step 4: *Reduce to lowest terms, if necessary.*

$$\frac{94}{210} = \frac{47}{105}$$

The main idea in adding fractions is to write each fraction again with the LCD for a denominator. In doing so, we must be sure not to change the value of either of the original fractions.

Answer

4. $\dfrac{11}{35}$

EXAMPLE 5 Add $\dfrac{-2}{x^2 - 2x - 3} + \dfrac{3}{x^2 - 9}$.

Solution

Step 1: *Find the LCD.*

$$x^2 - 2x - 3 = (x - 3)(x + 1)$$
$$x^2 - 9 = (x - 3)(x + 3) \qquad LCD = (x - 3)(x + 3)(x + 1)$$

Step 2: *Change to equivalent rational expressions.*

$$\dfrac{-2}{x^2 - 2x - 3} = \dfrac{-2}{(x - 3)(x + 1)} \cdot \dfrac{(x + 3)}{(x + 3)} = \dfrac{-2x - 6}{(x - 3)(x + 3)(x + 1)}$$

$$\dfrac{3}{x^2 - 9} = \dfrac{3}{(x - 3)(x + 3)} \cdot \dfrac{(x + 1)}{(x + 1)} = \dfrac{3x + 3}{(x - 3)(x + 3)(x + 1)}$$

Step 3: *Add numerators.*

$$\dfrac{-2x - 6}{(x - 3)(x + 3)(x + 1)} + \dfrac{3x + 3}{(x - 3)(x + 3)(x + 1)} = \dfrac{x - 3}{(x - 3)(x + 3)(x + 1)}$$

Step 4: *Reduce to lowest terms.*

$$\dfrac{\cancel{x - 3}}{\cancel{(x - 3)}(x + 3)(x + 1)} = \dfrac{1}{(x + 3)(x + 1)}$$

●

EXAMPLE 6 Subtract $\dfrac{x + 4}{2x + 10} - \dfrac{5}{x^2 - 25}$.

Solution We begin by factoring each denominator.

$$\dfrac{x + 4}{2x + 10} - \dfrac{5}{x^2 - 25} = \dfrac{x + 4}{2(x + 5)} - \dfrac{5}{(x + 5)(x - 5)}$$

The LCD is $2(x + 5)(x - 5)$. Completing the problem, we have

$$= \dfrac{x + 4}{2(x + 5)} \cdot \dfrac{(x - 5)}{(x - 5)} - \dfrac{5}{(x + 5)(x - 5)} \cdot \dfrac{2}{2}$$

$$= \dfrac{x^2 - x - 20}{2(x + 5)(x - 5)} - \dfrac{10}{2(x + 5)(x - 5)}$$

$$= \dfrac{x^2 - x - 30}{2(x + 5)(x - 5)}$$

To see if this expression will reduce, we factor the numerator into $(x - 6)(x + 5)$.

$$= \dfrac{(x - 6)\cancel{(x + 5)}}{2\cancel{(x + 5)}(x - 5)}$$

$$= \dfrac{x - 6}{2(x - 5)}$$

●

5. Add $\dfrac{1}{x^2 + 8x + 15} + \dfrac{3}{x^2 - 9}$.

6. Subtract $\dfrac{x + 3}{2x + 8} - \dfrac{4}{x^2 - 16}$.

Answers

5. $\dfrac{4}{(x + 5)(x - 3)}$

6. $\dfrac{x - 5}{2(x - 4)}$

7. Subtract $\dfrac{x-1}{x^2 - x - 2} - \dfrac{x}{x^2 + x - 6}$.

EXAMPLE 7 Subtract $\dfrac{2x-2}{x^2 + 4x + 3} - \dfrac{x-1}{x^2 + 5x + 6}$.

Solution We factor each denominator and build the LCD from those factors.

$$\frac{2x-2}{x^2 + 4x + 3} - \frac{x-1}{x^2 + 5x + 6}$$

$$= \frac{2x-2}{(x+3)(x+1)} - \frac{x-1}{(x+3)(x+2)}$$

$$= \frac{2x-2}{(x+3)(x+1)} \cdot \frac{(x+2)}{(x+2)} - \frac{x-1}{(x+3)(x+2)} \cdot \frac{(x+1)}{(x+1)} \qquad \text{The LCD is } (x+1)(x+2)(x+3).$$

$$= \frac{2x^2 + 2x - 4}{(x+1)(x+2)(x+3)} - \frac{x^2 - 1}{(x+1)(x+2)(x+3)} \qquad \text{Multiply out each numerator.}$$

$$= \frac{(2x^2 + 2x - 4) - (x^2 - 1)}{(x+1)(x+2)(x+3)} \qquad \text{Subtract numerators.}$$

$$= \frac{x^2 + 2x - 3}{(x+1)(x+2)(x+3)} \qquad \text{Factor numerator to see if we can reduce.}$$

$$= \frac{(x+3)(x-1)}{(x+1)(x+2)(x+3)} \qquad \text{Reduce.}$$

$$= \frac{x-1}{(x+1)(x+2)}$$

8. Let $f(x) = \dfrac{x^2}{x-3}$ and $g(x) = \dfrac{12-x}{-x+3}$.

 Find and simplify $f(x) + g(x)$.

EXAMPLE 8 Let $f(x) = \dfrac{x^2}{x-7}$ and $g(x) = \dfrac{6x+7}{7-x}$. Find and simplify $f(x) + g(x)$.

Solution In Section 5.1, we were able to reverse the terms in a factor such as $7 - x$ by factoring -1 from each term. In a problem like this, the same result can be obtained by multiplying the numerator and denominator by -1.

$$f(x) + g(x) = \frac{x^2}{x-7} + \frac{6x+7}{7-x} \cdot \frac{-1}{-1} = \frac{x^2}{x-7} + \frac{-6x-7}{x-7}$$

$$= \frac{x^2 - 6x - 7}{x-7} \qquad \text{Add numerators.}$$

$$= \frac{(x-7)(x+1)}{(x-7)} \qquad \text{Factor numerator.}$$

$$= x + 1 \qquad \text{Divide out } x - 7.$$

For our next example, we will look at a problem in which we combine a whole number and a rational expression.

EXAMPLE 9 Subtract $2 - \dfrac{9}{3x + 1}$.

Solution To subtract these two expressions, we think of 2 as a rational expression with a denominator of 1.

$$2 - \frac{9}{3x + 1} = \frac{2}{1} - \frac{9}{3x + 1}$$

The LCD is $3x + 1$. Multiplying the numerator and denominator of the first expression by $3x + 1$ gives us a rational expression equivalent to 2, but with a denominator of $3x + 1$.

$$\frac{2}{1} \cdot \frac{(3x + 1)}{(3x + 1)} - \frac{9}{3x + 1} = \frac{6x + 2 - 9}{3x + 1}$$

$$= \frac{6x - 7}{3x + 1}$$

The numerator and denominator of this last expression do not have any factors in common other than 1, so the expression is in lowest terms.

EXAMPLE 10 Write an expression for the sum of a number and twice its reciprocal. Then, simplify that expression.

Solution If x is the number, then its reciprocal is $\frac{1}{x}$. Twice its reciprocal is $\frac{2}{x}$. The sum of the number and twice its reciprocal is

$$x + \frac{2}{x}$$

To combine these two expressions, we think of the first term x as a rational expression with a denominator of 1. The LCD is x.

$$x + \frac{2}{x} = \frac{x}{1} + \frac{2}{x}$$

$$= \frac{x}{1} \cdot \frac{x}{x} + \frac{2}{x}$$

$$= \frac{x^2 + 2}{x}$$

> ### GETTING READY FOR CLASS
>
> *After reading through the preceding section, respond in your own words and in complete sentences.*
>
> **A.** How would you use the distributive property to add two rational expressions that have the same denominator?
>
> **B.** What is the definition of the least common denominator?
>
> **C.** Why is factoring important in finding a least common denominator?
>
> **D.** What is the last step in adding or subtracting two rational expressions?

9. Subtract $3 - \dfrac{7}{4x - 7}$.

10. Write an expression for the difference of twice a number and its reciprocal. Then simplify.

Answers

9. $\frac{4(3x - 7)}{4x - 7}$

10. $\frac{2x^2 - 1}{x}$

Vocabulary Review

Choose the correct words to fill in the blanks below.

common	equivalent	numerators
lowest	different	denominator

1. To add rational expressions with the same denominator, add numerators and put the sum over the _____ denominator.

2. LCD is an acronym for least common _____.

3. To add two fractions with _____ denominators, follow these steps:

Step 1: Find the lowest common denominator.

Step 2: Change to _____ fractions.

Step 3: Add _____ of equivalent fractions found in Step 2.

Step 4: Reduce to _____ terms, if necessary.

Problems

A Combine the following rational expressions. Reduce all answers to lowest terms.

1. $\dfrac{x}{x+3} + \dfrac{3}{x+3}$

2. $\dfrac{2}{5x-2} - \dfrac{5x}{5x-2}$

3. $\dfrac{4}{y-4} - \dfrac{y}{y-4}$

4. $\dfrac{8}{y-8} - \dfrac{y}{y-8}$

5. $\dfrac{x}{x^2-y^2} - \dfrac{y}{x^2-y^2}$

6. $\dfrac{x}{x^2-y^2} + \dfrac{y}{x^2-y^2}$

7. $\dfrac{2x-3}{x-2} - \dfrac{x-1}{x-2}$

8. $\dfrac{2x-4}{x+2} - \dfrac{x-6}{x+2}$

9. $\dfrac{1}{a} + \dfrac{2}{a^2} - \dfrac{3}{a^3}$

10. $\dfrac{3}{a} + \dfrac{2}{a^2} - \dfrac{1}{a^3}$

11. $\dfrac{7x-2}{2x+1} - \dfrac{5x-3}{2x+1}$

12. $\dfrac{7x-1}{3x+2} - \dfrac{4x-3}{3x+2}$

B Combine the following fractions.

13. $\dfrac{3}{4} + \dfrac{1}{2}$

14. $\dfrac{5}{6} + \dfrac{1}{3}$

15. $\dfrac{2}{5} - \dfrac{1}{15}$

16. $\dfrac{5}{8} - \dfrac{1}{4}$

17. $\dfrac{5}{6} + \dfrac{7}{8}$

18. $\dfrac{3}{4} + \dfrac{2}{3}$

19. $\dfrac{9}{48} - \dfrac{3}{54}$

20. $\dfrac{6}{28} - \dfrac{5}{42}$

21. $\dfrac{3}{4} - \dfrac{1}{8} + \dfrac{2}{3}$

22. $\dfrac{1}{3} - \dfrac{5}{6} + \dfrac{5}{12}$

23. $\dfrac{1}{3} - \dfrac{1}{2} + \dfrac{5}{6}$

24. $\dfrac{1}{4} - \dfrac{5}{6} - \dfrac{1}{8}$

25. Work each problem according to the instructions.

 a. Multiply: $\dfrac{3}{8} \cdot \dfrac{1}{6}$.

 b. Divide: $\dfrac{3}{8} \div \dfrac{1}{6}$.

 c. Add: $\dfrac{3}{8} + \dfrac{1}{6}$.

 d. Multiply: $\dfrac{x+3}{x-3} \cdot \dfrac{5x+15}{x^2-9}$.

 e. Divide: $\dfrac{x+3}{x-3} \div \dfrac{5x+15}{x^2-9}$.

 f. Subtract: $\dfrac{x+3}{x-3} - \dfrac{5x+15}{x^2-9}$.

26. Work each problem according to the instructions.

 a. Multiply: $\dfrac{16}{49} \cdot \dfrac{1}{28}$.

 b. Divide: $\dfrac{16}{49} \div \dfrac{1}{28}$.

 c. Subtract: $\dfrac{16}{49} - \dfrac{1}{28}$.

 d. Multiply: $\dfrac{3x-2}{3x+2} \cdot \dfrac{15x+6}{9x^2-4}$.

 e. Divide: $\dfrac{3x-2}{3x+2} \div \dfrac{15x+6}{9x^2-4}$.

 f. Subtract: $\dfrac{3x+2}{3x-2} - \dfrac{15x+6}{9x^2-4}$.

B Find and simplify $f(x) - g(x)$ for the following functions.

27. $f(x) = \dfrac{3x+1}{2x-6}, g(x) = \dfrac{x+2}{x-3}$

28. $f(x) = \dfrac{x+1}{x-2}, g(x) = \dfrac{4x+7}{5x-10}$

29. $f(x) = \dfrac{6x+5}{5x-25}, g(x) = \dfrac{x+2}{x-5}$

30. $f(x) = \dfrac{4x+2}{3x+12}, g(x) = \dfrac{x}{x+4}$

Combine the following rational expressions. Reduce all answers to lowest terms.

31. $\dfrac{x+1}{2x-2} - \dfrac{2}{x^2-1}$

32. $\dfrac{x+7}{2x+12} + \dfrac{6}{x^2-36}$

33. $\dfrac{1}{a-b} - \dfrac{3ab}{a^3-b^3}$

34. $\dfrac{1}{a+b} + \dfrac{3ab}{a^3+b^3}$

35. $\dfrac{x-2}{2x+3} - \dfrac{2x-11}{3-4x-4x^2}$

36. $\dfrac{5}{2x-1} + \dfrac{6x+7}{1-4x^2}$

37. $\dfrac{1}{2y-3} - \dfrac{18y}{8y^3-27}$

38. $\dfrac{1}{3y-2} - \dfrac{18y}{27y^3-8}$

39. $\dfrac{x}{x^2-5x+6} - \dfrac{3}{3-x}$

40. $\dfrac{x}{x^2+4x+4} - \dfrac{2}{2+x}$

41. $\dfrac{2}{4t-5} + \dfrac{9}{8t^2-38t+35}$

42. $\dfrac{3}{2t-5} + \dfrac{21}{8t^2-14t-15}$

43. $\dfrac{1}{a^2-5a+6} + \dfrac{3}{a^2-a-2}$

44. $\dfrac{-3}{a^2+a-2} + \dfrac{5}{a^2-a-6}$

45. $\dfrac{1}{8x^3-1} - \dfrac{1}{4x^2-1}$

46. $\dfrac{1}{27x^3-1} - \dfrac{1}{9x^2-1}$

47. $\dfrac{4}{4x^2-9} - \dfrac{6}{8x^2-6x-9}$

48. $\dfrac{9}{9x^2+6x-8} - \dfrac{6}{9x^2-4}$

49. $\dfrac{4a}{a^2+6a+5} - \dfrac{3a}{a^2+5a+4}$

50. $\dfrac{3a}{a^2+7a+10} - \dfrac{2a}{a^2+6a+8}$

51. $\dfrac{2x-1}{x^2+x-6} - \dfrac{x+2}{x^2+5x+6}$

52. $\dfrac{4x+1}{x^2+5x+4} - \dfrac{x+3}{x^2+4x+3}$

53. $\dfrac{2x-8}{3x^2+8x+4} + \dfrac{x+3}{3x^2+5x+2}$

54. $\dfrac{5x+3}{2x^2+5x+3} - \dfrac{3x+9}{2x^2+7x+6}$

55. $\dfrac{2}{x^2 + 5x + 6} - \dfrac{4}{x^2 + 4x + 3} + \dfrac{3}{x^2 + 3x + 2}$

56. $\dfrac{-5}{x^2 + 3x - 4} + \dfrac{5}{x^2 + 2x - 3} + \dfrac{1}{x^2 + 7x + 12}$

57. $\dfrac{2x + 8}{x^2 + 5x + 6} - \dfrac{x + 5}{x^2 + 4x + 3} - \dfrac{x - 1}{x^2 + 3x + 2}$

58. $\dfrac{2x + 11}{x^2 + 9x + 20} - \dfrac{x + 1}{x^2 + 7x + 12} - \dfrac{x + 6}{x^2 + 8x + 15}$

59. $2 + \dfrac{3}{2x + 1}$

60. $3 - \dfrac{2}{2x + 3}$

61. $5 + \dfrac{2}{4 - t}$

62. $7 + \dfrac{3}{5 - t}$

63. $x - \dfrac{4}{2x + 3}$

64. $x - \dfrac{5}{3x + 4} + 1$

65. $\dfrac{x}{x + 2} + \dfrac{1}{2x + 4} - \dfrac{3}{x^2 + 2x}$

66. $\dfrac{x}{x + 3} + \dfrac{7}{3x + 9} - \dfrac{2}{x^2 + 3x}$

67. $\dfrac{1}{x} + \dfrac{x}{2x + 4} - \dfrac{2}{x^2 + 2x}$

68. $\dfrac{1}{x} + \dfrac{x}{3x + 9} - \dfrac{3}{x^2 + 3x}$

69. Let $f(x) = \dfrac{2}{x + 4}$ and $g(x) = \dfrac{x - 1}{x^2 + 3x - 4}$.
Find $f(x) + g(x)$.

70. Let $f(t) = \dfrac{5}{3t - 2}$ and $g(t) = \dfrac{t - 3}{3t^2 + 7t - 6}$.
Find $f(t) - g(t)$.

71. Let $f(x) = \dfrac{2x}{x^2 - x - 2}$ and $g(x) = \dfrac{5}{x^2 + x - 6}$.
Find $f(x) + g(x)$.

72. Let $f(x) = \dfrac{7}{x^2 - x - 12}$ and $g(x) = \dfrac{5}{x^2 + x - 6}$.
Find $f(x) - g(x)$.

Applying the Concepts

73. Optometry The formula

$$P = \frac{1}{a} + \frac{1}{b}$$

is used by optometrists to help determine how strong to make the lenses for a pair of eyeglasses. If a is 10 and b is 0.2, find the corresponding value of P.

74. Quadratic Formula Later in the book we will work with the quadratic formula. The derivation of the formula requires that you can add the fractions below. Add the fractions.

$$\frac{-c}{a} + \frac{b^2}{(2a)^2}$$

75. Painting The formula

$$T = \frac{ab}{a+b}$$

gives the time for two people to paint a room together if their individual times are a and b. If two people take 2 hours and 4 hours to paint a room alone, how long will it take them to paint the room together?

76. Electronics The formula

$$R = \frac{R_1 R_2}{R_1 + R_2}$$

gives the overall resistance of a circuit with individual resistors R_1 and R_2. If two resistors are 4 ohms and 3 ohms, what is the overall resistance of the circuit?

77. Number Problem Write an expression for the sum of a number and 4 times its reciprocal. Then simplify that expression.

78. Number Problem Write an expression for the sum of a number and 3 times its reciprocal. Then simplify that expression.

79. Number Problem Write an expression for the sum of the reciprocals of two consecutive integers. Then simplify that expression.

80. Number Problem Write an expression for the sum of the reciprocals of two consecutive even integers. Then simplify that expression.

Getting Ready for the Next Section

Divide.

81. $\dfrac{3}{4} \div \dfrac{5}{8}$

82. $\dfrac{2}{3} \div \dfrac{5}{6}$

83. $\dfrac{1}{15} \div \dfrac{3}{5}$

84. $\dfrac{4}{9} \div \dfrac{2}{3}$

Multiply.

85. $x\left(1 + \dfrac{2}{x}\right)$

86. $3\left(x + \dfrac{1}{3}\right)$

87. $3x\left(\dfrac{1}{x} - \dfrac{1}{3}\right)$

88. $3x\left(\dfrac{1}{x} + \dfrac{1}{3}\right)$

Factor.

89. $x^2 - 4$

90. $x^2 - x - 6$

91. $x^3 - 27$

92. $x^4 - 16$

One Step Further

Simplify.

93. $\left(1 - \dfrac{1}{x}\right)\left(1 - \dfrac{1}{x+1}\right)\left(1 - \dfrac{1}{x+2}\right)\left(1 - \dfrac{1}{x+3}\right)$

94. $\left(1 + \dfrac{1}{x}\right)\left(1 + \dfrac{1}{x+1}\right)\left(1 + \dfrac{1}{x+2}\right)\left(1 + \dfrac{1}{x+3}\right)$

95. $\left(\dfrac{a^2 - b^2}{u^2 - v^2}\right)\left(\dfrac{av - au}{b - a}\right) + \left(\dfrac{a^2 - av}{u + v}\right)\left(\dfrac{1}{a}\right)$

96. $\left(\dfrac{6r^2}{r^2 - 1}\right)\left(\dfrac{r+1}{3}\right) - \left(\dfrac{2r^2}{r-1}\right)$

97. $\dfrac{18x - 19}{4x^2 + 27x - 7} - \dfrac{12x - 41}{3x^2 + 17x - 28}$

98. $\dfrac{42 - 22y}{3y^2 - 13y - 10} - \dfrac{21 - 13y}{2y^2 - 9y - 5}$

99. $\left(\dfrac{1}{y^2 - 1} \div \dfrac{1}{y^2 + 1}\right)\left(\dfrac{y^3 + 1}{y^4 - 1}\right) + \dfrac{1}{(y+1)^2(y-1)}$

100. $\left(\dfrac{a^3 - 64}{a^2 - 16} \div \dfrac{a^2 - 4a + 16}{a^2 - 4} \div \dfrac{a^2 + 4a + 16}{a^3 + 64}\right) + 4 - a^2$

Find the Mistake

Each sentence below contains a mistake. Circle the mistake and write the correct word or expression on the line provided.

1. If we add the rational expressions $\dfrac{x+5}{3x+9}$ and $\dfrac{4}{x^2-9}$ and reduce to the lowest terms, we get $x - 1$. _____

2. The least common denominator for the rational expressions $\dfrac{2x+3}{3x^2 + 6x - 45}$ and $\dfrac{x+7}{3x^2 + 14x - 5}$ is $3x - 1$. _____

3. If we subtract $\dfrac{5}{2x^2 + x - 3}$ from $\dfrac{2}{x^2 - 1}$, we get $\dfrac{1}{(x+1)(2x+3)}$. _____

4. To subtract $\dfrac{x-4}{x^2 - x - 20}$ from $\dfrac{2x+5}{x^2 - 4x - 5}$, make the expressions equivalent by first using LCD $(x + 4)(2x + 5)$.

**Video Examples
Section 8.4**

Practice Problems

1. Simplify $\dfrac{\dfrac{1}{3}}{\dfrac{3}{4}}$.

Answer

1. $\dfrac{4}{9}$

8.4 Complex Rational Expressions

Every week in Portland, Oregon, bicyclists known as zoobombers convene on a street corner near the local zoo. Bicycles are decorated in streamers, strings of lights, and other ornate details to coincide with the often-costumed riders. Even the bikes themselves are known to cause a stir, falling into a variety of categories, such as swing bikes, mini bikes, tall bikes, and cruisers. Amidst laughter and screams of joy, the bicyclists spend the evening speeding together down the steep hills of the neighborhood.

Imagine you are a spectator on the sidewalk of a street the zoobombers are speeding down. You hear the sound of a single screaming bicyclist zooming by because of the Doppler effect, where the pitch of the sound changes as the bicyclist approaches and then eventually passes you. The formula for the sound frequency h you hear is

$$h = \frac{f}{1 + \dfrac{v}{s}}$$

where f is the actual frequency of the sound being produced, s is the speed of the sound (about 740 miles per hour), and v is the velocity of the bicyclist. This formula uses a *complex rational expression*, a quotient of two rational expressions, which is the focus of this section. You will encounter this exact formula later in the problem set.

A Simplifying Complex Rational Expressions

EXAMPLE 1 Simplify $\dfrac{\dfrac{3}{4}}{\dfrac{5}{8}}$.

Solution There are generally two methods that can be used to simplify complex fractions.

Method 1 We can multiply the numerator and denominator of the complex fractions by the LCD for both of the fractions, which in this case is 8.

$$\frac{\dfrac{3}{4}}{\dfrac{5}{8}} = \frac{\dfrac{3}{4} \cdot 8}{\dfrac{5}{8} \cdot 8} = \frac{6}{5}$$

Method 2 Instead of dividing by $\frac{5}{8}$ we can multiply by $\frac{8}{5}$.

$$\frac{\dfrac{3}{4}}{\dfrac{5}{8}} = \frac{3}{4} \cdot \frac{8}{5} = \frac{24}{20} = \frac{6}{5}$$

Here are some examples of complex fractions involving rational expressions. Most can be solved using either of the two methods shown in Example 1.

EXAMPLE 2 Simplify $\dfrac{\dfrac{1}{x} + \dfrac{1}{y}}{\dfrac{1}{x} - \dfrac{1}{y}}$.

Solution This problem is most easily solved using Method 1. We begin by multiplying both the numerator and denominator by the quantity xy, which is the LCD for all the fractions.

$$\frac{\dfrac{1}{x} + \dfrac{1}{y}}{\dfrac{1}{x} - \dfrac{1}{y}} = \frac{\left(\dfrac{1}{x} + \dfrac{1}{y}\right) \cdot xy}{\left(\dfrac{1}{x} - \dfrac{1}{y}\right) \cdot xy}$$

$$= \frac{\dfrac{1}{x}(xy) + \dfrac{1}{y}(xy)}{\dfrac{1}{x}(xy) - \dfrac{1}{y}(xy)}$$

Apply the distributive property to distribute xy over both terms in the numerator and denominator.

$$= \frac{y + x}{y - x}$$

EXAMPLE 3 Let $f(x) = \dfrac{x - 2}{x^2 - 9}$ and $g(x) = \dfrac{x^2 - 4}{x + 3}$. Simplify $\dfrac{f(x)}{g(x)}$.

Solution Applying Method 2, we have

$$\frac{f(x)}{g(x)} = \frac{\dfrac{x - 2}{x^2 - 9}}{\dfrac{x^2 - 4}{x + 3}} = \frac{x - 2}{x^2 - 9} \cdot \frac{x + 3}{x^2 - 4}$$

$$= \frac{(x - 2)(x + 3)}{(x + 3)(x - 3)(x + 2)(x - 2)}$$

$$= \frac{1}{(x - 3)(x + 2)}$$

EXAMPLE 4 Simplify $\dfrac{1 - \dfrac{4}{x^2}}{1 - \dfrac{1}{x} - \dfrac{6}{x^2}}$.

Solution The easiest way to simplify this complex fraction is to multiply the numerator and denominator by the LCD, x^2.

$$\frac{1 - \dfrac{4}{x^2}}{1 - \dfrac{1}{x} - \dfrac{6}{x^2}} = \frac{x^2\left(1 - \dfrac{4}{x^2}\right)}{x^2\left(1 - \dfrac{1}{x} - \dfrac{6}{x^2}\right)}$$

Multiply numerator and denominator by x^2.

$$= \frac{x^2 \cdot 1 - x^2 \cdot \dfrac{4}{x^2}}{x^2 \cdot 1 - x^2 \cdot \dfrac{1}{x} - x^2 \cdot \dfrac{6}{x^2}}$$

Distributive property

$$= \frac{x^2 - 4}{x^2 - x - 6}$$

Simplify.

$$= \frac{(x - 2)(x + 2)}{(x - 3)(x + 2)}$$

Factor.

$$= \frac{x - 2}{x - 3}$$

Reduce.

2. Simplify $\dfrac{\dfrac{1}{x} - \dfrac{1}{y}}{\dfrac{1}{y} + \dfrac{1}{x}}$.

3. Let $f(x) = \dfrac{x - 1}{x^2 - 16}$ and $g(x) = \dfrac{x^2 - 1}{x + 4}$.

Simplify $\dfrac{f(x)}{g(x)}$.

4. Simplify $\dfrac{1 - \dfrac{9}{x^2}}{1 - \dfrac{5}{x} + \dfrac{6}{x^2}}$.

Answers

2. $\dfrac{y - x}{x + y}$

3. $\dfrac{f(x)}{g(x)} = \dfrac{1}{(x + 4)(x + 1)}$

4. $\dfrac{x + 3}{x - 2}$

5. Simplify $4 - \dfrac{2}{x + \dfrac{1}{2}}$.

EXAMPLE 5 Simplify $2 - \dfrac{3}{x + \dfrac{1}{3}}$.

Solution First, we simplify the expression that follows the subtraction sign.

$$2 - \frac{3}{x + \dfrac{1}{3}} = 2 - \frac{3 \cdot 3}{3\left(x + \dfrac{1}{3}\right)} = 2 - \frac{9}{3x + 1}$$

Now we subtract by rewriting the first term, 2, with the LCD, $3x + 1$.

$$2 - \frac{9}{3x + 1} = \frac{2}{1} \cdot \frac{3x + 1}{3x + 1} - \frac{9}{3x + 1}$$

$$= \frac{6x + 2 - 9}{3x + 1} = \frac{6x - 7}{3x + 1}$$

6. Simplify $\dfrac{1 - 4x^{-2}}{1 + 2x^{-1} - 8x^{-2}}$.

EXAMPLE 6 Simplify $\dfrac{1 - 9x^{-2}}{1 - 5x^{-1} + 6x^{-2}}$.

Solution Let's begin by writing this expression with only positive exponents.

$$\frac{1 - \dfrac{9}{x^2}}{1 - \dfrac{5}{x} + \dfrac{6}{x^2}} \qquad \textit{Negative exponent property}$$

Now we simplify this complex fraction by multiplying the numerator and denominator by x^2.

$$\frac{1 - \dfrac{9}{x^2}}{1 - \dfrac{5}{x} + \dfrac{6}{x^2}} = \frac{x^2\left(1 - \dfrac{9}{x^2}\right)}{x^2\left(1 - \dfrac{5}{x} + \dfrac{6}{x^2}\right)} \qquad \textit{Multiply numerator and denominator by } x^2.$$

$$= \frac{x^2 - 9}{x^2 - 5x + 6} \qquad \textit{Simplify.}$$

$$= \frac{(x - 3)(x + 3)}{(x - 3)(x - 2)} \qquad \textit{Factor.}$$

$$= \frac{x + 3}{x - 2} \qquad \textit{Reduce.}$$

GETTING READY FOR CLASS

After reading through the preceding section, respond in your own words and in complete sentences.

A. What is a complex rational expression?

B. Explain how to simplify a complex fraction using a least common denominator.

C. When is it more efficient to convert a complex fraction to a division problem of rational expressions?

D. Which method of simplifying complex fractions do you prefer? Why?

Answers

5. $\dfrac{8x}{2x + 1}$

6. $\dfrac{x + 2}{x + 4}$

Vocabulary Review

Choose the correct words to fill in the blanks in the paragraph.

numerator reciprocal LCD complex quotient denominator

A _____ rational expression is a _____ of two rational expressions. Method 1 for simplifying a complex rational expression requires multiplying the _____ and denominator of the complex fraction by the _____ for both fractions. Method 2 for simplifying a complex rational expression requires multiplying the numerator of the complex fraction by the _____ of the fraction in the _____ .

Problems

A Simplify each of the following as much as possible.

1. $\dfrac{\frac{3}{4}}{\frac{2}{3}}$

2. $\dfrac{\frac{5}{9}}{\frac{7}{12}}$

3. $\dfrac{\frac{1}{3} - \frac{1}{4}}{\frac{1}{2} + \frac{1}{8}}$

4. $\dfrac{\frac{1}{6} - \frac{1}{3}}{\frac{1}{4} - \frac{1}{8}}$

5. $\dfrac{3 + \frac{2}{5}}{1 - \frac{3}{7}}$

6. $\dfrac{2 + \frac{5}{6}}{1 - \frac{7}{8}}$

7. $\dfrac{\frac{1}{x}}{1 + \frac{1}{x}}$

8. $\dfrac{1 - \frac{1}{x}}{\frac{1}{x}}$

9. $\dfrac{1 + \frac{1}{a}}{1 - \frac{1}{a}}$

10. $\dfrac{1 - \frac{2}{a}}{1 - \frac{3}{a}}$

11. $\dfrac{\frac{1}{x} - \frac{1}{y}}{\frac{1}{x} + \frac{1}{y}}$

12. $\dfrac{\frac{1}{x} + \frac{2}{y}}{\frac{2}{x} + \frac{1}{y}}$

13. $\dfrac{\frac{x - 5}{x^2 - 4}}{\frac{x^2 - 25}{x + 2}}$

14. $\dfrac{\frac{3x + 1}{x^2 - 49}}{\frac{9x^2 - 1}{x - 7}}$

15. $\dfrac{\frac{4a}{2a^3 + 2}}{\frac{8a}{4a + 4}}$

16. $\dfrac{\frac{2a}{3a^3 - 3}}{\frac{4a}{6a - 6}}$

17. $\dfrac{1 - 9x^{-2}}{1 - x^{-1} - 6x^{-2}}$

18. $\dfrac{4 - x^{-2}}{4 + 4x^{-1} + x^{-2}}$

19. $\dfrac{2 + \frac{5}{a} - \frac{3}{a^2}}{2 - \frac{5}{a} + \frac{2}{a^2}}$

20. $\dfrac{3 + \frac{5}{a} - \frac{2}{a^2}}{3 - \frac{10}{a} + \frac{3}{a^2}}$

21. $\dfrac{1 + \dfrac{1}{x + 3}}{1 + \dfrac{7}{x - 3}}$

22. $\dfrac{1 + \dfrac{1}{x - 2}}{1 - \dfrac{3}{x + 2}}$

23. $\dfrac{1 - (a + 1)^{-1}}{1 + (a - 1)^{-1}}$

24. $\dfrac{(a - 1)^{-1} + 1}{(a + 1)^{-1} - 1}$

25. $\dfrac{(x + 3)^{-1} + (x - 3)^{-1}}{(x + 3)^{-1} - (x - 3)^{-1}}$

26. $\dfrac{(x + a)^{-1} + (x - a)^{-1}}{(x + a)^{-1} - (x - a)^{-1}}$

27. $\dfrac{\dfrac{y + 1}{y - 1} + \dfrac{y - 1}{y + 1}}{\dfrac{y + 1}{y - 1} - \dfrac{y - 1}{y + 1}}$

28. $\dfrac{\dfrac{y - 1}{y + 1} - \dfrac{y + 1}{y - 1}}{\dfrac{y - 1}{y + 1} + \dfrac{y + 1}{y - 1}}$

29. $1 - \dfrac{x}{1 - \dfrac{1}{x}}$

30. $x - \dfrac{1}{x - \dfrac{1}{2}}$

31. $1 + \dfrac{1}{1 + \dfrac{1}{1 + 1}}$

32. $1 - \dfrac{1}{1 - \dfrac{1}{1 - \dfrac{1}{2}}}$

33. $\dfrac{1 - \dfrac{1}{x + \dfrac{1}{2}}}{1 + \dfrac{1}{x + \dfrac{1}{2}}}$

34. $\dfrac{2 + \dfrac{1}{x - \dfrac{1}{3}}}{2 - \dfrac{1}{x - \dfrac{1}{3}}}$

35. $\dfrac{\dfrac{1}{x + h} - \dfrac{1}{x}}{h}$

36. $\dfrac{\dfrac{1}{(x + h)^2} - \dfrac{1}{x^2}}{h}$

37. $\dfrac{\dfrac{1}{x + 8} - \dfrac{1}{x}}{8}$

38. $\dfrac{\dfrac{1}{x + 7} - \dfrac{1}{x}}{7}$

39. $\dfrac{1}{(x + 6)^2} - \dfrac{1}{x^2}$

40. $\dfrac{1}{(x + 7)^2} - \dfrac{1}{x^2}$

41. $\dfrac{\dfrac{3}{ab} + \dfrac{4}{bc} - \dfrac{2}{ac}}{\dfrac{5}{abc}}$

42. $\dfrac{\dfrac{x}{yz} - \dfrac{y}{xz} + \dfrac{z}{xy}}{\dfrac{1}{x^2y^2} - \dfrac{1}{x^2z^2} + \dfrac{1}{y^2z^2}}$

43. $\dfrac{\dfrac{t^2 - 2t - 8}{t^2 + 7t + 6}}{\dfrac{t^2 - t - 6}{t^2 + 2t + 1}}$

44. $\dfrac{\dfrac{y^2 - 5y - 14}{y^2 + 3y - 10}}{\dfrac{y^2 - 8y + 7}{y^2 + 6y + 5}}$

45. $\dfrac{5 + \left(\frac{b-1}{4}\right)^{-1}}{\left(\frac{b+5}{7}\right)^{-1} - \left(\frac{b-1}{3}\right)^{-1}}$

46. $\dfrac{(k - 7k + 12)^{-1}}{(k - 3)^{-1} + (k - 4)^{-1}}$

47. $\dfrac{\dfrac{3}{x^2 - x - 6}}{\dfrac{2}{x + 2} - \dfrac{4}{x - 3}}$

48. $\dfrac{\dfrac{9}{a - 7} + \dfrac{8}{2a + 3}}{\dfrac{10}{2a^2 - 11a - 21}}$

49. $\dfrac{\dfrac{1}{m - 4} + \dfrac{1}{m - 5}}{\dfrac{1}{m^2 - 9m + 20}}$

50. $\dfrac{\dfrac{1}{k^2 - 7k + 12}}{\dfrac{1}{k - 3} + \dfrac{1}{k - 4}}$

51. $\dfrac{\dfrac{1}{x + 2} - \dfrac{1}{x - 2}}{\dfrac{1}{x + 2} + \dfrac{1}{x - 2}}$

52. $1 + \dfrac{1}{1 + \dfrac{1}{1 + \dfrac{1}{x}}}$

For the following functions, find $\dfrac{f(x)}{g(x)}$, and simplify.

53. $f(x) = 2 + \dfrac{3}{x} - \dfrac{18}{x^2} - \dfrac{27}{x^3}, g(x) = 2 + \dfrac{9}{x} + \dfrac{9}{x^2}$

54. $f(x) = 3 + \dfrac{5}{x} - \dfrac{12}{x^2} - \dfrac{20}{x^3}, g(x) = 3 + \dfrac{11}{x} + \dfrac{10}{x^2}$

55. $f(x) = 1 + \dfrac{1}{x + 3}, g(x) = 1 - \dfrac{1}{x + 3}$

56. $f(x) = 1 + \dfrac{1}{x - 2}, g(x) = 1 - \dfrac{1}{x - 2}$

Applying the Concepts

To solve the following problems, recall our discussion of difference quotients at the beginning of this chapter.

57. Difference Quotient For each rational function below, find the difference quotient

$$\dfrac{f(x) - f(a)}{x - a}$$

a. $f(x) = \dfrac{4}{x}$

b. $f(x) = \dfrac{1}{x + 1}$

c. $f(x) = \dfrac{1}{x^2}$

58. Difference Quotient For each rational function below, find the difference quotient

$$\dfrac{f(x + h) - f(x)}{h}$$

a. $f(x) = \dfrac{4}{x}$

b. $f(x) = \dfrac{1}{x + 1}$

c. $f(x) = \dfrac{1}{x^2}$

59. Doppler Effect As mentioned in the opening to this section, the change in the pitch of a sound as an object passes is called the Doppler effect, named after C. J. Doppler (1803–1853). A person will *hear* a sound with a frequency, *h*, according to the formula

$$h = \frac{f}{1 + \dfrac{v}{s}}$$

where *f* is the actual frequency of the sound being produced, *s* is the speed of sound (about 740 miles per hour), and *v* is the velocity of the moving object.

a. Examine this fraction, and then explain why *h* and *f* approach the same value as *v* becomes smaller and smaller.

b. Solve this formula for *v*.

60. Work Problem A water storage tank has two drains. It can be shown that the time it takes to empty the tank if both drains are open is given by the formula

$$\frac{1}{\dfrac{1}{a} + \dfrac{1}{b}}$$

where *a* = time it takes for the first drain to empty the tank, and *b* = time for the second drain to empty the tank.

a. Simplify this complex fraction.

b. Find the amount of time needed to empty the tank using both drains if, used alone, the first drain empties the tank in 4 hours and the second drain can empty the tank in 3 hours.

. .

Getting Ready for the Next Section

Multiply.

61. $x(y - 2)$

62. $x(y - 1)$

63. $6\left(\dfrac{x}{2} - 3\right)$

64. $6\left(\dfrac{x}{3} + 1\right)$

65. $xab \cdot \dfrac{1}{x}$

66. $xab\left(\dfrac{1}{b} + \dfrac{1}{a}\right)$

67. $st^2\left(\dfrac{2}{s} - \dfrac{3}{t^2}\right)$

68. $4x^2y\left(\dfrac{1}{2xy} - \dfrac{3}{x^2}\right)$

Factor.

69. $y^2 - 25$

70. $x^2 - 3x + 2$

71. $xa + xb$

72. $xy - y$

Solve.

73. $5x - 4 = 6$

74. $y^2 + y - 20 = 2y$

75. $x^2 - 2x - 12 = 2x$

76. $3x - 4 = 9$

Find the Mistake

Each sentence below contains a mistake. Circle the mistake and write the correct words or expression on the line provided.

1. Simplifying $\dfrac{\frac{2}{m} + \frac{4}{n}}{\frac{2}{m} - \frac{6}{n}}$ gives us $-\dfrac{2}{3}$. _____

2. If $f(x) = \dfrac{x+2}{x^2-16}$ and $g(x) = \dfrac{x^2-x-6}{x+4}$, then $\dfrac{f(x)}{g(x)} = \dfrac{(x-2)^2(x+3)}{(x-4)(x+4)^2}$. _____

3. Simplifying $5 - \dfrac{2}{x + \frac{1}{4}}$ gives us $-\dfrac{3}{4x+1}$. _____

4. The first step to simplify $\dfrac{1 - 36x^{-2}}{1 - 9x^{-1} + 18x^{-2}}$ is to multiply the numerator and denominator by x. _____

Landmark Review: Checking Your Progress

Simplify the following expressions.

1. $\dfrac{3a^2 - 27}{6a - 18}$

2. $\dfrac{y - 3}{y^2 - 6y + 9}$

3. $\dfrac{3x^3 - 6x^2 - 45x}{x^2 + 5x + 6}$

4. $\dfrac{5x^2 + 22x + 8}{25x^2 - 4}$

Perform the indicated operations.

5. $\dfrac{3a^4}{(6a + 18)} \cdot \dfrac{(a + 3)}{a^3}$

6. $\dfrac{(4x^2 - 12x)}{(x^2 - 1)} \cdot \dfrac{(x^2 + x)}{4x}$

7. $\dfrac{(y^2 - 3y)}{(2y^2 - 8)} \div \dfrac{(y^2 - 25)}{(y^2 + 3y - 10)}$

8. $(x^2 + 8x + 15) \cdot \dfrac{2x}{x + 5}$

9. $\dfrac{8}{(4x + 16)} \div \dfrac{2}{(x + 4)}$

10. $\dfrac{4}{y + 3} + \dfrac{24}{y^2 - 9}$

11. $\dfrac{a + 5}{a + 4} + \dfrac{8}{a^2 - 16}$

12. $\dfrac{x^2}{x + 4} - \dfrac{16}{x + 4}$

13. $\dfrac{x + 2}{x - 1} - \dfrac{6}{x^2 - 1}$

14. $\dfrac{x}{(x + 2)(x + 3)} - \dfrac{2}{(x + 2)(x + 3)}$

15. $\dfrac{1 - \frac{3}{a}}{1 - \frac{4}{a}}$

16. $\dfrac{\frac{2}{x} + \frac{1}{y}}{\frac{3}{x} + \frac{1}{y}}$

17. $\dfrac{\frac{4x + 1}{x^2 - 9}}{\frac{16x^2 - 1}{x - 3}}$

18. $\dfrac{\frac{y - 1}{y + 1} - \frac{y + 1}{y - 1}}{\frac{y - 1}{y + 1} + \frac{y + 1}{y - 1}}$

OBJECTIVES

A Solve equations containing rational expressions.

B Solve formulas containing rational expressions.

C Graph rational functions having a vertical asymptote.

KEY WORDS

rational expression

rational equation

extraneous solution

vertical asymptote

rational function

**Video Examples
Section 8.5**

Practice Problems

1. Solve $\dfrac{x}{2} + \dfrac{3}{5} = \dfrac{1}{5}$.

For more than fifty years, the Japanese city of Ito has hosted a wooden wash tub race down the Matsukawa River. Each contestant must sit in a 1-meter-diameter tub and use large wooden rice paddles to steer a 400-meter-long course. We can use the following rational equation to calculate how long it would take one tub racer to complete the course:

$$t = \frac{400}{v + c}$$

where the total time t (in seconds) depends on the speed of the water c (in meters per second) and the speed of the tub v. In this section, we will work more with *rational equations*, such as this one. When working with rational equations, our goal is to find a solution for the variable that makes the equation a true statement. Until now, we have only worked with rational *expressions* where our goal was only to simplify.

A Solving Equations with Rational Expressions

The first step in solving an equation that contains one or more rational expressions is to find the LCD for all denominators in the equation. Then, we multiply both sides of the equation by the LCD to clear the equation of all fractions. That is, after we have multiplied through by the LCD, each term in the resulting equation will have a denominator of 1.

EXAMPLE 1 Solve $\dfrac{x}{2} - 3 = \dfrac{2}{3}$.

Solution The LCD for 2 and 3 is 6. Multiplying both sides by 6, we have

$$6\left(\frac{x}{2} - 3\right) = 6\left(\frac{2}{3}\right)$$

$$6\left(\frac{x}{2}\right) - 6(3) = 6\left(\frac{2}{3}\right)$$

$$3x - 18 = 4$$

$$3x = 22$$

$$x = \frac{22}{3}$$

The solution is $\dfrac{22}{3}$.

Multiplying both sides of an equation by the LCD clears the equation of fractions because the LCD has the property that all the denominators divide it evenly.

Answer

1. $-\dfrac{4}{5}$

EXAMPLE 2 Solve $\dfrac{6}{a-4} = \dfrac{3}{8}$.

2. Solve $\dfrac{4}{x-5} = \dfrac{4}{3}$.

Solution The LCD for $a - 4$ and 8 is $8(a - 4)$. Multiplying both sides by this quantity yields

$$8(a-4) \cdot \frac{6}{a-4} = 8(a-4) \cdot \frac{3}{8}$$

$$48 = (a-4) \cdot 3$$

$$48 = 3a - 12$$

$$60 = 3a$$

$$20 = a$$

The solution is 20, which checks in the original equation.

When we multiply both sides of an equation by an expression containing the variable, we must be sure to check our solutions. The multiplication property of equality does not allow multiplication by 0. If the expression we multiply by contains the variable, then it has the possibility of being 0. In the last example, we multiplied both sides by $8(a - 4)$. This gives a restriction $a \neq 4$ for any solution we come up with.

EXAMPLE 3 Solve $\dfrac{x}{x-2} + \dfrac{2}{3} = \dfrac{2}{x-2}$.

3. Solve $\dfrac{x}{x-5} + \dfrac{5}{2} = \dfrac{5}{x-5}$.

Solution The LCD is $3(x - 2)$. We are assuming $x \neq 2$ when we multiply both sides of the equation by $3(x - 2)$.

$$3(x-2) \cdot \left(\frac{x}{x-2} + \frac{2}{3} \right) = 3(x-2) \cdot \frac{2}{x-2}$$

$$3x + (x-2) \cdot 2 = 3 \cdot 2$$

$$3x + 2x - 4 = 6$$

$$5x - 4 = 6$$

$$5x = 10$$

$$x = 2$$

Note In the process of solving the equation, we multiplied both sides by $3(x - 2)$, solved for x, and got $x = 2$ for our solution. But when x is 2, the quantity $3(x - 2) = 3(2 - 2) = 3(0) = 0$, which means we multiplied both sides of our equation by 0, which is not allowed under the multiplication property of equality.

The only possible solution is $x = 2$. Let's check this value in the original equation.

$$\frac{2}{2-2} + \frac{2}{3} \overset{?}{=} \frac{2}{2-2}$$

$$\frac{2}{0} + \frac{2}{3} \overset{?}{=} \frac{2}{0}$$

The first and last terms are undefined. The proposed solution, $x = 2$, does not check in the original equation. There is no solution to the original equation.

When the proposed solution to an equation is not actually a solution, it is called an *extraneous solution*. In the last example, $x = 2$ is an extraneous solution.

Answers

2. 8

3. No solution

4. Solve $\dfrac{x}{x^2 - 4} - \dfrac{2}{x - 2} = \dfrac{1}{x + 2}$.

EXAMPLE 4 Solve $\dfrac{5}{x^2 - 3x + 2} - \dfrac{1}{x - 2} = \dfrac{1}{3x - 3}$.

Solution Writing the equation again with the denominators in factored form, we have

$$\frac{5}{(x - 2)(x - 1)} - \frac{1}{x - 2} = \frac{1}{3(x - 1)}$$

The LCD is $3(x - 2)(x - 1)$. Multiplying through by the LCD, we have

$$3(x - 2)(x - 1)\frac{5}{(x - 2)(x - 1)} - 3(x - 2)(x - 1) \cdot \frac{1}{(x - 2)}$$

$$= 3(x - 2)(x - 1) \cdot \frac{1}{3(x - 1)}$$

$$3 \cdot 5 - 3(x - 1) \cdot 1 = (x - 2) \cdot 1$$

$$15 - 3x + 3 = x - 2$$

$$-3x + 18 = x - 2$$

$$-4x + 18 = -2$$

$$-4x = -20$$

$$x = 5$$

Checking the proposed solution $x = 5$ in the original equation yields a true statement. Try it and see.

Note We can check the proposed solution in any of the equations obtained before multiplying through by the LCD. We cannot check the proposed solution in an equation obtained after multiplying through by the LCD because, if we have multiplied by 0, the resulting equations will not be equivalent to the original one.

5. Let $f(x) = 2 + \dfrac{9}{x}$ and $g(x) = -\dfrac{10}{x^2}$.

Find all values of x where $f(x) = g(x)$.

EXAMPLE 5 Let $f(x) = 3 + \dfrac{1}{x}$ and $g(x) = \dfrac{10}{x^2}$. Find all values of x where $f(x) = g(x)$.

Solution To clear the equation of denominators, we multiply both sides by x^2.

$$x^2\left(3 + \frac{1}{x}\right) = x^2\left(\frac{10}{x^2}\right)$$

$$3(x^2) + \left(\frac{1}{x}\right)(x^2) = \left(\frac{10}{x^2}\right)(x^2)$$

$$3x^2 + x = 10$$

Rewrite in standard form, and solve.

$$3x^2 + x - 10 = 0$$

$$(3x - 5)(x + 2) = 0$$

$$3x - 5 = 0 \quad \text{or} \quad x + 2 = 0$$

$$x = \frac{5}{3} \qquad\qquad x = -2$$

Both solutions check in the original equation. So $f(x) = g(x)$ when $x = \frac{5}{3}$ or $x = -2$. Remember, we have to check all solutions any time we multiply both sides of the equation by an expression that contains the variable, just to be sure we haven't multiplied by 0.

6. Solve $\dfrac{a + 2}{a^2 + 3a} = \dfrac{-2}{a^2 - 9}$.

EXAMPLE 6 Solve $\dfrac{y - 4}{y^2 - 5y} = \dfrac{2}{y^2 - 25}$.

Solution Factoring each denominator, we find the LCD is $y(y - 5)(y + 5)$. Multiplying each side of the equation by the LCD clears the equation of denominators and leads us to our possible solutions.

$$y(y - 5)(y + 5) \cdot \frac{y - 4}{y(y - 5)} = \frac{2}{(y - 5)(y + 5)} \cdot y(y - 5)(y + 5)$$

$$(y + 5)(y - 4) = 2y$$

$$y^2 + y - 20 = 2y \qquad \text{Multiply out the left side.}$$

$$y^2 - y - 20 = 0 \qquad \text{Add } -2y \text{ to each side.}$$

$$(y - 5)(y + 4) = 0$$

$$y - 5 = 0 \quad \text{or} \quad y + 4 = 0$$

$$y = 5 \qquad\qquad y = -4$$

Answers

4. -1

5. $x = -\frac{5}{2}, -2$

6. 2

The two possible solutions are 5 and -4. If we substitute -4 for y in the original equation, we find that it leads to a true statement. It is therefore a solution. On the other hand, if we substitute 5 for y in the original equation, we find that both sides of the equation are undefined. The only solution to our original equation is $y = -4$. The other possible solution $y = 5$ is extraneous.

B Formulas

EXAMPLE 7 Solve $x = \dfrac{y - 4}{y - 2}$ for y.

Solution To solve for y, we first multiply each side by $y - 2$ to obtain

$$x(y - 2) = y - 4$$

$$xy - 2x = y - 4 \qquad \text{Distributive property}$$

$$xy - y = 2x - 4 \qquad \text{Collect all terms containing } y \text{ on the left side.}$$

$$y(x - 1) = 2x - 4 \qquad \text{Factor } y \text{ from each term on the left side.}$$

$$y = \frac{2x - 4}{x - 1} \qquad \text{Divide each side by } x - 1.$$

EXAMPLE 8 Solve the formula $\dfrac{1}{x} = \dfrac{1}{b} + \dfrac{1}{a}$ for x.

Solution We begin by multiplying both sides by the least common denominator xab. As you can see from our previous examples, multiplying both sides of an equation by the LCD is equivalent to multiplying each term of both sides by the LCD.

$$xab \cdot \frac{1}{x} = \frac{1}{b} \cdot xab + \frac{1}{a} \cdot xab$$

$$ab = xa + xb$$

$$ab = (a + b)x \qquad \text{Factor } x \text{ from the right side.}$$

$$\frac{ab}{a + b} = x$$

We know we are finished because the variable we were solving for is alone on one side of the equation and does not appear on the other side.

C Graphing Rational Functions

> **DEFINITION** rational function
>
> A *rational function* is any function that is a polynomial quotient; that is, a polynomial is divided by another polynomial.

Our next example begins our investigation of the graphs of rational functions.

EXAMPLE 9 Graph the rational function $f(x) = \dfrac{6}{x - 2}$.

Solution To find the y-intercept, we let x equal 0.

$$\text{When } x = 0: \qquad y = \frac{6}{0 - 2} = \frac{6}{-2} = -3 \qquad \text{y-intercept}$$

The graph will not cross the x-axis. If it did, we would have a solution to the equation

$$0 = \frac{6}{x - 2}$$

which has no solution because there is no number to divide 6 by to obtain 0.

7. Solve $x = \dfrac{y - 3}{y + 3}$ for y.

8. Solve $\dfrac{2}{x} = \dfrac{3}{b} + \dfrac{1}{a}$ for x.

9. Graph $f(x) = \dfrac{6}{x - 4}$.

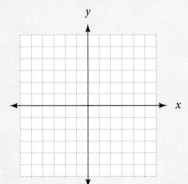

Answers

7. $y = \dfrac{-3(x + 1)}{x - 1}$

8. $x = \dfrac{2ab}{3a + b}$

The graph of our equation is shown in Figure 1 along with a table giving values of x and y that satisfy the equation. Notice that y is undefined when x is 2. This means that the graph will not cross the vertical line $x = 2$. (If it did, there would be a value of y for $x = 2$.) The line $x = 2$ is called a *vertical asymptote* of the graph. The graph will get very close to the vertical asymptote, but will never touch or cross it.

x	y
-4	-1
-1	-2
0	-3
1	-6
2	Undefined
3	6
4	3
5	2

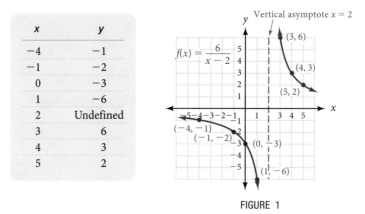

FIGURE 1

If you were to graph $y = \dfrac{6}{x}$ on the coordinate system in Figure 1, you would see that the graph of $y = \dfrac{6}{x-2}$ is the graph of $y = \dfrac{6}{x}$ with all points shifted 2 units to the right.

USING TECHNOLOGY **More About Example 9**

We know the graph of $f(x) = \dfrac{6}{x-2}$ will not cross the vertical asymptote $x = 2$ because replacing x with 2 in the equation gives us an undefined expression, meaning there is no value of y to associate with $x = 2$. We can use a graphing calculator to explore the behavior of this function when x gets closer and closer to 2 by using the table function on the calculator. We want to put our own values for X into the table, so we set the independent variable to Ask. (On a TI-83/84, use the TBLSET key to set up the table.) To see how the function behaves as x gets close to 2, we let X take on values of 1.9, 1.99, and 1.999. Then we move to the other side of 2 and let X become 2.1, 2.01, and 2.001.

The table will look like this:

X	Y_1
1.9	-60
1.99	-600
1.999	-6000
2.1	60
2.01	600
2.001	6000

Again, the calculator asks us for a table increment. Because we are inputting the x values ourselves, the increment value does not matter.

As you can see, the values in the table support the shape of the curve in Figure 1 around the vertical asymptote $x = 2$.

EXAMPLE 10 Graph $g(x) = \dfrac{6}{x + 2}$.

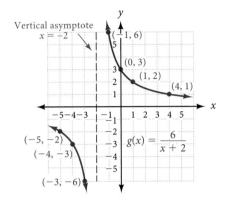

FIGURE 2

10. Graph $f(x) = \dfrac{6}{x + 1}$.

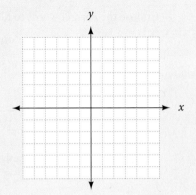

Solution The only difference between this equation and the equation in Example 9 is in the denominator. This graph will have the same shape as the graph in Example 9, but the vertical asymptote will be $x = -2$ instead of $x = 2$. Figure 2 shows the graph.

Notice that the graphs shown in Figures 1 and 2 are both graphs of functions because no vertical line will cross either graph in more than one place. Notice the similarities and differences in our two functions,

$$f(x) = \frac{6}{x - 2} \qquad \text{and} \qquad g(x) = \frac{6}{x + 2}$$

and their graphs. The vertical asymptotes shown in Figures 1 and 2 correspond to the fact that both $f(2)$ and $g(-2)$ are undefined. The domain for the function $f(x)$ is all real numbers except $x = 2$, while the domain for $g(x)$ is all real numbers except $x = -2$. We would write the domain for $f(x)$ as $\{x \mid x \neq 2\}$ and the domain for $g(x)$ as $\{x \mid x \neq -2\}$

GETTING READY FOR CLASS

After reading through the preceding section, respond in your own words and in complete sentences.

A. Explain how a least common denominator can be used to simplify an equation.

B. What is an extraneous solution?

C. How does the location of the vertical asymptote in the graph of a rational function relate to the equation of the function?

D. What is the last step in solving an equation that contains rational expressions?

Vocabulary Review

Choose the correct words to fill in the blanks below.

asymptote rational extraneous function

1. The first step when solving an equation that contains _____ expressions is to find the least common denominator.

2. An _____ solution is a proposed solution to an equation but is not an actual solution.

3. When graphing a rational _____ no vertical line should cross the graph in more than one place.

4. A vertical line that a graph may get very close to but never touch is called a vertical _____ .

Problems

A Solve each of the following equations.

1. $\dfrac{x}{5} + 4 = \dfrac{5}{3}$

2. $\dfrac{x}{5} = \dfrac{x}{2} - 9$

3. $\dfrac{a}{3} + 2 = \dfrac{4}{5}$

4. $\dfrac{a}{4} + \dfrac{1}{2} = \dfrac{2}{3}$

5. $\dfrac{y}{2} + \dfrac{y}{4} + \dfrac{y}{6} = 3$

6. $\dfrac{y}{3} - \dfrac{y}{6} + \dfrac{y}{2} = 1$

7. $\dfrac{5}{2x} = \dfrac{1}{x} + \dfrac{3}{4}$

8. $\dfrac{1}{2a} = \dfrac{2}{a} - \dfrac{3}{8}$

9. $\dfrac{1}{x} = \dfrac{1}{3} - \dfrac{2}{3x}$

10. $\dfrac{5}{2x} = \dfrac{2}{x} - \dfrac{1}{12}$

11. $\dfrac{2x}{x-3} + 2 = \dfrac{2}{x-3}$

12. $\dfrac{2}{x+5} = \dfrac{2}{5} - \dfrac{x}{x+5}$

13. $1 - \dfrac{1}{x} = \dfrac{12}{x^2}$

14. $2 + \dfrac{5}{x} = \dfrac{3}{x^2}$

15. $y - \dfrac{4}{3y} = -\dfrac{1}{3}$

16. $\dfrac{y}{2} - \dfrac{4}{y} = -\dfrac{7}{2}$

17. $\dfrac{x+2}{x+1} = \dfrac{1}{x+1} + 2$

18. $\dfrac{x+6}{x+3} = \dfrac{3}{x+3} + 2$

19. $\dfrac{x + 2}{3x - 1} = \dfrac{x - 12}{4 - 9x - 9x^2}$

20. $\dfrac{2x + 3}{2x + 4} = \dfrac{5x + 4}{16 + 4x - 2x^2}$

21. $6 - \dfrac{5}{x^2} = \dfrac{7}{x}$

22. $10 - \dfrac{3}{x^2} = -\dfrac{1}{x}$

23. $\dfrac{1}{x - 1} - \dfrac{1}{x + 1} = \dfrac{3x}{x^2 - 1}$

24. $\dfrac{5}{x - 1} + \dfrac{2}{x - 1} = \dfrac{4}{x + 1}$

25. $\dfrac{t - 4}{t^2 - 3t} = \dfrac{-2}{t^2 - 9}$

26. $\dfrac{t + 3}{t^2 - 2t} = \dfrac{10}{t^2 - 4}$

27. $\dfrac{3}{y - 4} - \dfrac{2}{y + 1} = \dfrac{5}{y^2 - 3y - 4}$

28. $\dfrac{1}{y + 2} - \dfrac{2}{y - 3} = \dfrac{-2y}{y^2 - y - 6}$

29. $\dfrac{2}{1 + a} = \dfrac{3}{1 - a} + \dfrac{5}{a}$

30. $\dfrac{1}{a + 3} - \dfrac{a}{a^2 - 9} = \dfrac{2}{3 - a}$

31. $\dfrac{3}{2x - 6} - \dfrac{x + 1}{4x - 12} = 4$

32. $\dfrac{2x - 3}{5x + 10} + \dfrac{3x - 2}{4x + 8} = 1$

33. $\dfrac{y + 2}{y^2 - y} - \dfrac{6}{y^2 - 1} = 0$

34. $\dfrac{y + 3}{y^2 - y} - \dfrac{8}{y^2 - 1} = 0$

35. $\dfrac{4}{2x - 6} - \dfrac{12}{4x + 12} = \dfrac{12}{x^2 - 9}$

36. $\dfrac{1}{x + 2} + \dfrac{1}{x - 2} = \dfrac{4}{x^2 - 4}$

37. $\dfrac{2}{y^2 - 7y + 12} - \dfrac{1}{y^2 - 9} = \dfrac{4}{y^2 - y - 12}$

38. $\dfrac{1}{y^2 + 5y + 4} + \dfrac{3}{y^2 - 1} = \dfrac{-1}{y^2 + 3y - 4}$

Given the functions $f(x)$ and $g(x)$, find where $f(x) = g(x)$.

39. $f(x) = \dfrac{2}{x-3} + \dfrac{x}{x^2-9}, g(x) = \dfrac{4}{x+3}$

40. $f(x) = \dfrac{2}{x+5} + \dfrac{3}{x+4}, g(x) = \dfrac{2x}{x^2+9x+20}$

41. $f(x) = \dfrac{3}{2} - \dfrac{1}{x-4}, g(x) = \dfrac{-2}{2x-8}$

42. $f(x) = \dfrac{2}{x} - \dfrac{1}{x+1}, g(x) = \dfrac{-2}{5x+5}$

43. Let $f(x) = \dfrac{1}{x-3}$ and $g(x) = \dfrac{1}{x+3}$. Find x if

 a. $f(x) + g(x) = \dfrac{5}{8}$

 b. $\dfrac{f(x)}{g(x)} = 5$

 c. $f(x) = g(x)$

44. Let $f(x) = \dfrac{4}{x+2}$ and $g(x) = \dfrac{4}{x-2}$. Find x if

 a. $f(x) - g(x) = -\dfrac{4}{3}$

 b. $\dfrac{g(x)}{f(x)} = -7$

 c. $f(x) = -g(x)$

45. Solve each equation.

 a. $6x - 2 = 0$

 b. $\dfrac{6}{x} - 2 = 0$

 c. $\dfrac{x}{6} - 2 = -\dfrac{1}{2}$

 d. $\dfrac{6}{x} - 2 = -\dfrac{1}{2}$

 e. $\dfrac{6}{x^2} + 6 = \dfrac{20}{x}$

46. Solve each equation.

 a. $5x - 2 = 0$

 b. $5 - \dfrac{2}{x} = 0$

 c. $\dfrac{x}{2} - 5 = -\dfrac{3}{4}$

 d. $\dfrac{2}{x} - 5 = -\dfrac{3}{4}$

 e. $-\dfrac{3}{x} + \dfrac{2}{x^2} = 5$

47. Work each problem according to the instructions.

 a. Divide: $\dfrac{6}{x^2-2x-8} \div \dfrac{x+3}{x+2}$.

 b. Add: $\dfrac{6}{x^2-2x-8} + \dfrac{x+3}{x+2}$.

 c. Solve: $\dfrac{6}{x^2-2x-8} + \dfrac{x+3}{x+2} = 2$.

48. Work each problem according to the instructions.

 a. Divide: $\dfrac{-10}{x^2-25} \div \dfrac{x-4}{x-5}$.

 b. Add: $\dfrac{-10}{x^2-25} + \dfrac{x-4}{x-5}$.

 c. Solve: $\dfrac{-10}{x^2-25} + \dfrac{x-4}{x-5} = \dfrac{4}{5}$.

B Solve for the given variable

49. Solve $\dfrac{1}{x} = \dfrac{1}{b} - \dfrac{1}{a}$ for x.

50. Solve $\dfrac{1}{x} = \dfrac{1}{a} - \dfrac{1}{b}$ for x.

Solve for *y*.

51. $x = \dfrac{y - 3}{y - 1}$

52. $x = \dfrac{y - 2}{y - 3}$

53. $x = \dfrac{2y + 1}{3y + 1}$

54. $x = \dfrac{3y + 2}{5y + 1}$

C Graph each function. Show the vertical asymptote.

55. $f(x) = \dfrac{1}{x - 3}$

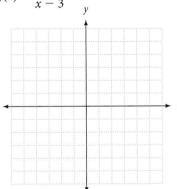

56. $f(x) = \dfrac{1}{x + 3}$

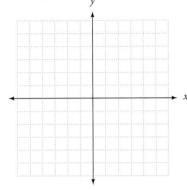

57. $f(x) = \dfrac{4}{x + 2}$

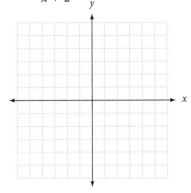

58. $f(x) = \dfrac{4}{x - 2}$

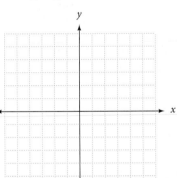

59. $g(x) = \dfrac{2}{x - 4}$

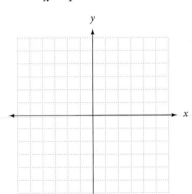

60. $g(x) = \dfrac{2}{x + 4}$

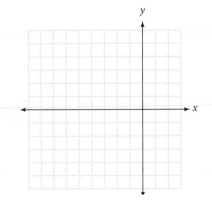

61. $g(x) = \dfrac{6}{x + 1}$

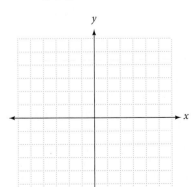

62. $g(x) = \dfrac{6}{x - 1}$

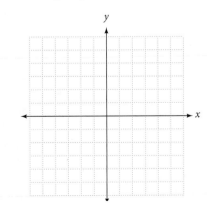

For each function given, find the location of any vertical asymptotes. Also give the domain of the function. (Do not graph.)

63. $f(x) = \dfrac{x}{3x - 5}$

64. $f(x) = \dfrac{2x}{4x + 7}$

65. $\dfrac{3x + 1}{2x - 3}$

66. $g(x) = \dfrac{x^2}{x^2 - 4}$

67. $g(x) = \dfrac{5x}{x^2 - 9}$

68. $g(x) = \dfrac{x + 2}{5x^2 + x}$

69. $h(x) = \dfrac{4 - 3x}{2x^2 + 6x}$

70. $h(x) = \dfrac{x^3}{x^2 + x - 2}$

71. $h(x) = \dfrac{x^2}{x^2 - 3x + 2}$

72. $y(x) = \dfrac{x + 3}{x^2 - 5x + 4}$

73. $y(x) = \dfrac{5}{2x^2 - 5x - 3}$

74. $y(x) = \dfrac{7}{3x^2 + 5x - 2}$

Applying the Concepts

75. Geometry From plane geometry and the principle of similar triangles, the relationship between y_1, y_2, and h shown in Figure 3 can be expressed as

$$\frac{1}{h} = \frac{1}{y_1} + \frac{1}{y_2}$$

Two poles are 12 feet high and 8 feet high. If a cable is attached to the top of each one and stretched to the bottom of the other, what is the height above the ground at which the two wires will meet?

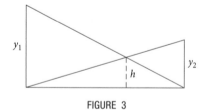

FIGURE 3

76. Kayak Race In a kayak race, the participants must paddle a kayak 450 meters down a river and then return 450 meters up the river to the starting point. Susan has correctly deduced that the total time t (in seconds) depends on the speed of the current c (in meters per second) according to the following expression:

$$t = \frac{450}{v + c} + \frac{450}{v - c}$$

where v is the speed of the kayak (in still water).

a. Fill in the following table.

Time	Speed of Kayak in Still Water	Speed of the Current
t(sec)	v(m/sec)	c(m/sec)
240		1
300		2
	4	3
	3	1
540	3	
	3	3

b. If the kayak race were conducted in the still waters of a lake, do you think that the total time of a given participant would be greater than, equal to, or smaller than the time in the river? Justify your answer.

c. Suppose Peter can paddle his kayak at 4.1 meters per second and that the speed of the current is 4.1 meters per second. What will happen when Peter makes the turn and tries to come back up the river? How does this situation show up in the equation for total time?

Getting Ready for the Next Section

Multiply.

77. $39.3 \cdot 60$

78. $1,100 \cdot 60 \cdot 60$

79. $2x\left(\dfrac{1}{x} + \dfrac{1}{2x}\right)$

80. $3x\left(\dfrac{1}{x} + \dfrac{1}{3x}\right)$

Divide. Round to the nearest tenth, if necessary.

81. $65{,}000 \div 5{,}280$

82. $3{,}960{,}000 \div 5{,}280$

Solve.

83. $12(x + 3) + 12(x - 3) = 3(x^2 - 9)$

84. $40 + 2x = 60 - 3x$

85. $\dfrac{1}{10} - \dfrac{1}{12} = \dfrac{1}{x}$

86. $\dfrac{1}{x} + \dfrac{1}{2x} = 2$

One Step Further

Solve each equation.

87. $\dfrac{12}{x} + \dfrac{8}{x^2} - \dfrac{75}{x^3} - \dfrac{50}{x^4} = 0$

88. $\dfrac{45}{x} + \dfrac{18}{x^2} - \dfrac{80}{x^3} - \dfrac{32}{x^4} = 0$

89. $\dfrac{1}{x^3} - \dfrac{1}{3x^2} - \dfrac{1}{4x} + \dfrac{1}{12} = 0$

90. $\dfrac{1}{x^3} - \dfrac{1}{2x^2} - \dfrac{1}{9x} + \dfrac{1}{18} = 0$

91. Solve $\dfrac{2}{x} + \dfrac{4}{x + a} = \dfrac{-6}{a - x}$ for x.

92. Solve $\dfrac{1}{b - x} - \dfrac{1}{x} = \dfrac{-2}{b + x}$ for x.

93. Solve $\dfrac{s - vt}{t^2} = -16$ for v.

94. Solve $A = P\left(1 + \dfrac{r}{n}\right)$ for r.

95. Solve $\dfrac{1}{p} = \dfrac{1}{f} + \dfrac{1}{g}$ for f.

96. Solve $h = \dfrac{v^2}{2g} + \dfrac{P}{c}$ for p.

Find the Mistake

Each sentence below contains a mistake. Circle the mistake and write the correct word or expression on the line provided.

1. When we solve the rational equation $\dfrac{1}{x - 3} - \dfrac{5}{x + 3} = \dfrac{x}{x^2 - 9}$, we get $x = 3$. _____

2. To solve the rational equation $\dfrac{x + 5}{x^2 - 5x} = \dfrac{20}{x^2 - 25}$, we must use the LCD $x - 5$. _____

3. The graph for $f(x) = \dfrac{3}{x - 1}$ will come very close but not touch the line $x = 3$. _____

4. The graph for $f(x) = \dfrac{2}{x + 2}$ has a horizontal asymptote of $x = -2$. _____

**Video Examples
Section 8.6**

Practice Problems

1. One number is three times another. The sum of their reciprocals is $\frac{4}{9}$. Find the numbers.

Late one night in 2009, an octopus at the Santa Monica Aquarium used her strong arm to open a valve at the top of her tank. Two hundred gallons of water spilled onto the floor and flooded nearby exhibits. Aquarium employees discovered the octopus alive and still in her tank the next morning. Let us imagine for a moment that an employee was tasked with refilling the octopus's tank, but he forgot to close the valve. If water from an inlet pipe could fill the tank in 2 hours, and it took 3 hours to empty the tank through the valve, how long would it take to fill the tank with the valve still open? In this section, we will work with similar application problems, the solutions to which involve equations that contain rational expressions.

As you will see, the solutions to the examples show only the essential steps from our Blueprint for Problem Solving. Recall that Step 1 was done mentally; we read the problem and mentally list the items that are known and the items that are unknown. This is an essential part of problem solving. Now that you have had experience with application problems, however, you should be doing Step 1 automatically.

A Applications

EXAMPLE 1 One number is twice another. The sum of their reciprocals is 2. Find the numbers.

Solution Let $x =$ the smaller number. The larger number is $2x$. Their reciprocals are $\frac{1}{x}$ and $\frac{1}{2x}$. The equation that describes the situation is

$$\frac{1}{x} + \frac{1}{2x} = 2$$

Multiplying both sides by the LCD $2x$, we have

$$2x \cdot \frac{1}{x} + 2x \cdot \frac{1}{2x} = 2x(2)$$
$$2 + 1 = 4x$$
$$3 = 4x$$
$$x = \frac{3}{4}$$

The smaller number is $\frac{3}{4}$. The larger is $\left(2 \cdot \frac{3}{4} \right) = \frac{6}{4} = \frac{3}{2}$. Adding their reciprocals, we have

$$\frac{4}{3} + \frac{2}{3} = \frac{6}{3} = 2$$

The sum of the reciprocals of $\frac{3}{4}$ and $\frac{3}{2}$ is 2.

Answers

1. 3, 9

Chapter 8 Rational Expressions, Equations, and Functions

©iStockphoto.com/FotoW

EXAMPLE 2 Two families from the same neighborhood plan a ski trip together. The first family is driving a newer vehicle and makes the 455-mile trip at a speed 5 miles per hour faster than the second family who is traveling in an older vehicle. The second family takes a half-hour longer to make the trip. What are the speeds of the two families?

Solution The following table will be helpful in finding the equation necessary to solve this problem.

	Distance d (miles)	Rate r (mph)	Time t (hours)
First family			
Second family			

If we let x be the speed of the second family, then the speed of the first family will be $x + 5$. Both families travel the same distance of 455 miles.

Putting this information into the table we have

	d	r	t
First family	455	$x + 5$	
Second family	455	x	

To fill in the last two spaces in the table, we use the relationship $d = r \cdot t$. Since the last column of the table is the time, we solve the equation $d = r \cdot t$ for t and get

$$t = \frac{d}{r}$$

Taking the distance and dividing by the rate (speed) for each family, we complete the table.

	d	r	t
First family	455	$x + 5$	$\frac{455}{x + 5}$
Second family	455	x	$\frac{455}{x}$

Reading the problem again, we find that the time for the second family is longer than the time for the first family by one-half hour. In other words, the time for the second family can be found by adding one-half hour to the time for the first family, or

$$\frac{455}{x + 5} + \frac{1}{2} = \frac{455}{x}$$

Multiplying both sides by the LCD of $2x(x + 5)$ gives

$$2x \cdot (455) + x(x + 5) \cdot 1 = 455 \cdot 2(x + 5)$$
$$910x + x^2 + 5x = 910x + 4550$$
$$x^2 + 5x - 4550 = 0$$
$$(x + 70)(x - 65) = 0$$
$$x = -70 \quad \text{or} \quad x = 65$$

Since we cannot have a negative speed, the only solution is $x = 65$. Then

$$x + 5 = 65 + 5 = 70$$

The speed of the first family is 70 miles per hour, and the speed of the second family is 65 miles per hour.

●

2. If the two families in Example 2 drive 330 miles instead of 455, what are the speeds of the two families?

Answers

2. 55 mph, 60 mph

3. A boat travels 26 miles up a river in the same amount of time it takes to travel 34 miles down the same river. The current is 2 miles per hour, what is the speed of the boat in still water?

EXAMPLE 3 The current of a river is 3 miles per hour. It takes a motorboat a total of 3 hours to travel 12 miles upstream and return 12 miles downstream. What is the speed of the boat in still water?

Solution This time we let $x =$ the speed of the boat in still water. Then, we fill in as much of the table as possible using the information given in the problem. For instance, because we let $x =$ the speed of the boat in still water, the rate upstream (against the current) must be $x - 3$. The rate downstream (with the current) is $x + 3$.

	Distance d (miles)	Rate r (mph)	Time t (hours)
Upstream	12	$x - 3$	
Downstream	12	$x + 3$	

The last two boxes can be filled in using the relationship $t = \dfrac{d}{r}$.

	d	r	t
Upstream	12	$x - 3$	$\dfrac{12}{x - 3}$
Downstream	12	$x + 3$	$\dfrac{12}{x + 3}$

The total time for the trip up and back is 3 hours:

$$\text{Time upstream} + \text{Time downstream} = \text{Total time}$$

$$\frac{12}{x - 3} \quad + \quad \frac{12}{x + 3} \quad = \quad 3$$

Multiplying both sides by $(x - 3)(x + 3)$, we have

$$12(x + 3) + 12(x - 3) = 3(x^2 - 9)$$

$$12x + 36 + 12x - 36 = 3x^2 - 27$$

$$3x^2 - 24x - 27 = 0$$

$$x^2 - 8x - 9 = 0 \qquad \text{Divide both sides by 3.}$$

$$(x - 9)(x + 1) = 0$$

$$x = 9 \quad \text{or} \quad x = -1$$

The speed of the motorboat in still water is 9 miles per hour. (We don't use $x = -1$ because the speed of the motorboat cannot be a negative number.)

EXAMPLE 4 An inlet pipe can fill a pool in 10 hours, while the drain can empty it in 12 hours. If the pool is empty and both the inlet pipe and drain are open, how long will it take to fill the pool?

4. A tub can be filled by the cold water faucet in 8 minutes. The drain empties the tub in 12 minutes. How long will it take to fill the tub if both the faucet and the drain are open?

10 hours to fill pool

12 hours to empty pool

Answers

3. 15 miles per hour
4. 24 minutes

Solution It is helpful to think in terms of how much work is done by each pipe in 1 hour.

Let x = the time it takes to fill the pool with both pipes open.

If the inlet pipe can fill the pool in 10 hours, then in 1 hour it is $\frac{1}{10}$ full. If the outlet pipe empties the pool in 12 hours, then in 1 hour it is $\frac{1}{12}$ empty. If the pool can be filled in x hours with both the inlet pipe and the drain open, then in 1 hour it is $\frac{1}{x}$ full when both pipes are open.

Here is the equation:

$$\begin{bmatrix} \text{Amount filled by} \\ \text{inlet pipe in 1 hour} \end{bmatrix} - \begin{bmatrix} \text{Amount emptied by} \\ \text{the drain in 1 hour} \end{bmatrix} = \begin{bmatrix} \text{Fraction of pool filled} \\ \text{with both pipes in 1 hour} \end{bmatrix}$$

$$\frac{1}{10} - \frac{1}{12} = \frac{1}{x}$$

Multiplying through by $60x$, we have

$$60x \cdot \frac{1}{10} - 60x \cdot \frac{1}{12} = 60x \cdot \frac{1}{x}$$

$$6x - 5x = 60$$

$$x = 60$$

It takes 60 hours to fill the pool if both the inlet pipe and the drain are open.

B More About Graphing Rational Functions

We continue our investigation of the graphs of rational functions by considering the graph of a rational function with binomials in the numerator and denominator.

EXAMPLE 5 Graph the rational function $y = \dfrac{x - 4}{x - 2}$.

Solution In addition to making a table to find some points on the graph, we can analyze the graph as follows:

1. The graph will have a y-intercept of 2, because when $x = 0$, $y = \frac{-4}{-2} = 2$.

2. To find the x-intercept, we let $y = 0$ to get

$$0 = \frac{x - 4}{x - 2}$$

The only way this expression can be 0 is if the numerator is 0, which happens when $x = 4$. (If you want to solve this equation, multiply both sides by $x - 2$. You will get the same solution, $x = 4$.)

3. The graph will have a vertical asymptote at $x = 2$, because $x = 2$ will make the denominator of the function 0, meaning y is undefined when x is 2.

4. The graph will have a *horizontal asymptote* at $y = 1$ because for very large values of x, $\frac{x - 4}{x - 2}$ is very close to 1 but never crosses it. The larger x is, the closer $\frac{x - 4}{x - 2}$ is to 1. The same is true for very small values of x, such as $-1,000$ and $-10,000$.

Putting this information together with the ordered pairs in the table next to the figure, we have the graph shown in Figure 1.

5. Graph the rational function

$$y = \frac{x - 4}{x + 2}.$$

x	y
-1	$\frac{5}{3}$
0	2
1	3
2	Undefined
3	-1
4	0
5	$\frac{1}{3}$

FIGURE 1

USING TECHNOLOGY More About Example 5

In the previous section, we used technology to explore the graph of a rational function around a vertical asymptote. This time, we are going to explore the graph near the horizontal asymptote. In Figure 1, the horizontal asymptote is at $y = 1$. To show that the graph approaches this line as x becomes very large, we use the table function on our graphing calculator, with X taking values of 100, 1,000, and 10,000. To show that the graph approaches the line $y = 1$ on the left side of the coordinate system, we let X become -100, $-1,000$ and $-10,000$.

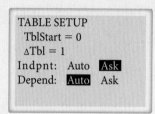

TABLE SETUP
TblStart = 0
ΔTbl = 1
Indpnt: Auto Ask
Depend: Auto Ask

Plot1 Plot2 Plot3
\Y₁ ■ (X − 4)(X − 2)
\Y₂ =
\Y₃ =
\Y₄ =
\Y₅ =
\Y₆ =
\Y₇ =

The table will look like this:

X	Y_1
100	.97959
1000	.998
10000	.9998
−100	1.0196
−1000	1.002
−100000	1.0002

As you can see, as x becomes very large in the positive direction, the graph approaches the line $y = 1$ from below. As x becomes very small in the negative direction, the graph approaches the line $y = 1$ from above.

GETTING READY FOR CLASS

After reading through the preceding section, respond in your own words and in complete sentences.

A. Briefly list the steps in the Blueprint for Problem Solving that you have used previously to solve application problems.

B. Write an application problem for which the solution depends on solving the equation $\frac{1}{2} + \frac{1}{3} = \frac{1}{x}$.

C. How may a table be helpful when solving application problems similar to those in this section?

D. What is a horizontal asymptote?

Vocabulary Review

Choose the correct words to fill in the blanks below.

table asymptote LCD blueprint

1. Using the _____ for problem solving will help you work application problems that involve rational expressions.

2. Multiplying both sides of an equation that contains rational expressions by the _____ will rid the equation of fractions.

3. Putting information into a _____ may help you find the equation necessary for solving an application problem.

4. A horizontal _____ is a horizontal line that a graph of a function may get very close to but never intersects.

Problems

Solve the following word problems. Use techniques illustrated in the examples of this section to set up an equation, then solve it to answer the question.

A Number Problems

1. One number is 3 times another. The sum of their reciprocals is $\frac{20}{3}$. Find the numbers.

2. One number is 3 times another. The sum of their reciprocals is $\frac{4}{5}$. Find the numbers.

3. The sum of a number and its reciprocal is $\frac{10}{3}$. Find the number.

4. The sum of a number and twice its reciprocal is $\frac{27}{5}$. Find the number.

5. The sum of the reciprocals of two consecutive integers is $\frac{7}{12}$. Find the two integers.

6. Find two consecutive even integers, the sum of whose reciprocals is $\frac{3}{4}$.

7. If a certain number is added to the numerator and denominator of $\frac{7}{9}$, the result is $\frac{5}{6}$. Find the number.

8. Find the number you would add to both the numerator and denominator of $\frac{8}{11}$ so that the result would be $\frac{6}{7}$.

Rate Problems

9. The speed of a boat in still water is 5 miles per hour. If the boat travels 3 miles downstream in the same amount of time it takes to travel 1.5 miles upstream, what is the speed of the current?

 a. Let x be the speed of the current. Complete the distance d and rate r columns in the table.

	d (miles)	r (mph)	t (hours)
Upstream			
Downstream			

 b. Now use the distance and rate information to complete the time t column.

 c. What does the problem tell us about the two times? Use this fact to write an equation involving the two expressions for time.

 d. Solve the equation. Write your answer as a complete sentence.

10. A boat, which moves at 18 miles per hour in still water, travels 14 miles downstream in the same amount of time it takes to travel 10 miles upstream. Find the speed of the current.

 a. Let x be the speed of the current. Complete the distance d and rate r columns in the table.

	d (miles)	r (mph)	t (hours)
Upstream			
Downstream			

 b. Now use the distance and rate information to complete the time t column.

 c. What does the problem tell us about the two times? Use this fact to write an equation involving the two expressions for time.

 d. Solve the equation. Write your answer as a complete sentence.

11. The current of a river is 2 miles per hour. A boat travels to a point 8 miles upstream and back again in 3 hours. What is the speed of the boat in still water?

12. A motorboat travels at 4 miles per hour in still water. It goes 12 miles upstream and 12 miles back again in a total of 8 hours. Find the speed of the current of the river.

13. Train A has a speed 15 miles per hour greater than that of train B. If train A travels 150 miles in the same time train B travels 120 miles, what are the speeds of the two trains?

 a. Let x be the speed of the train B. Complete the distance d and rate r columns in the table.

	d (miles)	r (mph)	t (hours)
Train A			
Train B			

 b. Now use the distance and rate information to complete the time column.

 c. What does the problem tell us about the two times? Use this fact to write an equation involving the two expressions for time.

 d. Solve the equation. Write your answer as a complete sentence.

14. A train travels 30 miles per hour faster than a car. If the train covers 120 miles in the same time the car covers 80 miles, what are the speeds of each of them?

 a. Let x be the speed of the car. Complete the distance d and rate r columns in the table.

	d (miles)	r (mph)	t (hours)
Car			
Train			

 b. Now use the distance and rate information to complete the time column.

 c. What does the problem tell us about the two times? Use this fact to write an equation involving the two expressions for time.

 d. Solve the equation. Write your answer as a complete sentence.

15. A small airplane flies 810 miles from Los Angeles to Portland, OR, with an average speed of 270 miles per hour. An hour and a half after the plane leaves, a Boeing 747 leaves Los Angeles for Portland. Both planes arrive in Portland at the same time. What was the average speed of the 747?

16. Lou leaves for a cross-country excursion on a bicycle traveling at 20 miles per hour. His friends are driving the trip and will meet him at several rest stops along the way. The first stop is scheduled 30 miles from the original starting point. If the people driving leave 15 minutes after Lou from the same place, how fast will they have to drive to reach the first rest stop at the same time as Lou?

17. A tour bus leaves Sacramento every Friday evening at 5:00 p.m. for a 270-mile trip to Las Vegas. This week, however, the bus leaves at 5:30 p.m. To arrive in Las Vegas on time, the driver drives 6 miles per hour faster than usual. What is the bus' usual speed?

18. A bakery delivery truck leaves the bakery at 5:00 a.m. each morning on its 140-mile route. One day the driver gets a late start and does not leave the bakery until 5:30 a.m. To finish her route on time the driver drives 5 miles per hour faster than usual. At what speed does she usually drive?

Work Problems

19. A water tank can be filled by an inlet pipe in 8 hours. It takes twice that long for the outlet pipe to empty the tank. How long will it take to fill the tank if both pipes are open?

20. A sink can be filled from the faucet in 5 minutes. It takes only 3 minutes to empty the sink when the drain is open. If the sink is full and both the faucet and the drain are open, how long will it take to empty the sink?

8 hours to fill

Twice as long to empty

5 min to fill

3 min to empty

21. It takes 10 hours to fill a pool with the inlet pipe. It can be emptied in 15 hours with the outlet pipe. If the pool is half full to begin with, how long will it take to fill it from there if both pipes are open?

22. A sink is one-quarter full when both the faucet and the drain are opened. The faucet alone can fill the sink in 6 minutes, while it takes 8 minutes to empty it with the drain. How long will it take to fill the remaining three quarters of the sink?

10 hours to fill pool

15 hours to empty pool

23. A sink has two faucets: one for hot water and one for cold water. The sink can be filled by a cold water faucet in 3.5 minutes. If both faucets are open, the sink is filled in 2.1 minutes. How long does it take to fill the sink with just the hot water faucet open?

24. A water tank is being filled by two inlet pipes. Pipe A can fill the tank in $4\frac{1}{2}$ hours, but both pipes together can fill the tank in 2 hours. How long does it take to fill the tank using only pipe B?

25. A hot water faucet can fill a sink in 3.5 minutes. The cold water faucet can fill the sink in 2.7 minutes. How long does it take to fill the sink if both faucets are turned on?

26. If Amelie can paint a room in 2.3 hours and her roommate can paint it in 4.2 hours, how long will it take to paint the room if they work together?

Miscellaneous Problems

27. Rhind Papyrus Nearly 4,000 years ago, Egyptians worked mathematical exercises involving reciprocals. The *Rhind Papyrus* contains a wealth of such problems, and one of them is as follows:

"A quantity and its two thirds are added together, one third of this is added, then one third of the sum is taken, and the result is 10."

Write an equation and solve this exercise.

28. Photography For clear photographs, a camera must be properly focused. Professional photographers use a mathematical relationship relating the distance from the camera lens to the object being photographed, *a*; the distance from the lens to the sensor creating an image, *b*; and the focal length of the lens, *f*. These quantities, *a*, *b*, and *f*, are related by the equation

$$\frac{1}{a} + \frac{1}{b} = \frac{1}{f}$$

A camera has a focal length of 3 inches. If the lens is 5 inches from the film, how far should the lens be placed from the object being photographed for the camera to be perfectly focused?

The Periodic Table If you take a chemistry class, you will work with the Periodic Table of Elements. Figure 2 shows three of the elements listed in the periodic table. As you can see, the bottom number in each figure is the molecular weight of the element. In chemistry, a mole is the amount of a substance that will give the weight in grams equal to the molecular weight. For example, 1 mole of lead is 207.2 grams.

FIGURE 2

29. Chemistry For the element carbon, 1 mole = 12.01 grams.

 a. To the nearest gram, how many grams of carbon are in 2.5 moles of carbon?

 b. How many moles of carbon are in 39 grams of carbon? Round to the nearest hundredth.

30. Chemistry For the element sulfur, 1 mole = 32.07 grams.

 a. How many grams of sulfur are in 3 moles of sulfur?

 b. How many moles of sulfur are found in 80.2 grams of sulfur?

B Graph each rational function. In each case, show the vertical asymptote, the horizontal asymptote, and any intercepts that exist.

31. $f(x) = \dfrac{x-3}{x-1}$

32. $f(x) = \dfrac{x+4}{x-2}$

33. $f(x) = \dfrac{x+3}{x-1}$

34. $f(x) = \dfrac{x-2}{x-1}$

35. $g(x) = \dfrac{x-3}{x+1}$

36. $g(x) = \dfrac{x-2}{x+1}$

37. $g(x) = \dfrac{x+2}{x+4}$

38. $g(x) = \dfrac{x}{x+2}$

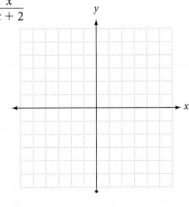

Maintaining Your Skills

39. The chart shows the number of Facebook fans for several popular fan pages. How many more fans does Homer Simpson have than Nutella?

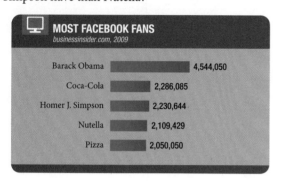

40. The chart shows the number of farmer's markets in the United States. Find the slope of the line that connects the first and last point of the chart. What does this mean?

Getting Ready for the Next Section

Divide.

41. $\dfrac{10x^2}{5x^2}$

42. $\dfrac{-15x^4}{5x^2}$

43. $\dfrac{4x^4y^3}{-2x^2y}$

44. $\dfrac{10a^4b^2}{4a^2b^2}$

45. $4{,}628 \div 25$

46. $7{,}546 \div 35$

47. $\dfrac{(x-3)^2}{x-3}$

48. $\dfrac{(x+4)^2}{(x+4)^4}$

Multiply.

49. $2x^2(2x-4)$

50. $3x^2(x-2)$

51. $(2x-4)(2x^2+4x+5)$

52. $(x-2)(3x^2+6x+15)$

Subtract.

53. $(2x^2-7x+9)-(2x^2-4x)$

54. $(x^2-6xy-7y^2)-(x^2+xy)$

Factor.

55. x^2-a^2

56. x^2-1

57. $x^2-6xy-7y^2$

58. $2x^2-5xy+3y^2$

Find the Mistake

Each sentence below contains a mistake. Circle the mistake and write the correct word or expression on the line provided.

1. One number is three times another. The sum of their reciprocals is 8. The equation to find the numbers is $\dfrac{1}{x}+\dfrac{3}{x}=8$.

2. The rate equation $t=\dfrac{r}{d}$ is very useful for many applications of rational equations.

3. Two hikers leave the same camp for a destination 10 miles away. One is 2 miles per hour faster than the other and arrives $\frac{1}{2}$ an hour sooner. To find the speed of the hikers, you must recognize that their travel time is the same.

4. One worker can mow a park in 3 hours and another can finish in 4 hours. To find how long it takes to finish together, you would use the equation $\dfrac{1}{4}-\dfrac{1}{3}=x$.

Rational Numbers and Expressions [8.1]

A *rational number* is any number that can be expressed as the ratio of two integers.

$$\text{Rational numbers} = \left\{ \frac{a}{b} \,\middle|\, a \text{ and } b \text{ are integers}, b \neq 0 \right\}$$

A *rational expression* is any quantity that can be expressed as the ratio of two polynomials.

$$\text{Rational expressions} = \left\{ \frac{P}{Q} \,\middle|\, P \text{ and } Q \text{ are polynomials}, Q \neq 0 \right\}$$

Properties of Rational Expressions [8.1]

If P, Q, and K are polynomials with $Q \neq 0$ and $K \neq 0$, then

$$\frac{P}{Q} = \frac{PK}{QK} \quad \text{and} \quad \frac{P}{Q} = \frac{\frac{P}{K}}{\frac{Q}{K}}$$

which is to say that multiplying or dividing the numerator and denominator of a rational expression by the same nonzero quantity always produces an equivalent rational expression.

Reducing to Lowest Terms [8.1]

To reduce a rational expression to lowest terms, we first factor the numerator and denominator and then divide the numerator and denominator by any factors they have in common.

Multiplication of Rational Expressions [8.2]

To multiply two rational numbers or rational expressions, multiply numerators and multiply denominators. In symbols,

$$\frac{P}{Q} \cdot \frac{R}{S} = \frac{PR}{QS} \quad (Q \neq 0 \text{ and } S \neq 0)$$

In practice, we don't really multiply, but rather, we factor and then divide out common factors.

Division of Rational Expressions [8.2]

To divide one rational expression by another, we use the definition of division to rewrite our division problem as an equivalent multiplication problem. To divide by a rational expression we multiply by its reciprocal. In symbols,

$$\frac{P}{Q} \div \frac{R}{S} = \frac{P}{Q} \cdot \frac{S}{R} = \frac{PS}{QR} \quad (Q \neq 0, S \neq 0, R \neq 0)$$

EXAMPLES

1. $\frac{3}{4}$ is a rational number. $\frac{x-3}{x^2-9}$ is a rational expression.

2. $\dfrac{x-3}{x^2-9} = \dfrac{\cancel{x-3}}{(\cancel{x-3})(x+3)}$

$$= \dfrac{1}{x+3}$$

3. $\dfrac{x+1}{x^2-4} \cdot \dfrac{x+2}{3x+3}$

$$= \dfrac{\cancel{(x+1)}\cancel{(x+2)}}{(x-2)\cancel{(x+2)}(3)\cancel{(x+1)}}$$

$$= \dfrac{1}{3(x-2)}$$

4. $\dfrac{x^2-y^2}{x^3+y^3} \div \dfrac{x-y}{x^2-xy+y^2}$

$$= \dfrac{x^2-y^2}{x^3+y^3} \cdot \dfrac{x^2-xy+y^2}{x-y}$$

$$= \dfrac{(x+y)\cancel{(x-y)}\cancel{(x^2-xy+y^2)}}{(x+y)\cancel{(x^2-xy+y^2)}\cancel{(x-y)}}$$

$$= 1$$

Least Common Denominator [8.3]

5. The LCD for $\frac{2}{x-3}$ and $\frac{3}{5}$ is $5(x-3)$.

The *least common denominator*, LCD, for a set of denominators is the smallest quantity divisible by each of the denominators.

Addition and Subtraction of Rational Expressions [8.3]

6. $\dfrac{2}{x-3} + \dfrac{3}{5}$

$= \dfrac{2}{x-3} \cdot \dfrac{5}{5} + \dfrac{3}{5} \cdot \dfrac{x-3}{x-3}$

$= \dfrac{3x+1}{5(x-3)}$

If P, Q, and R represent polynomials, $R \neq 0$, then

$$\frac{P}{R} + \frac{Q}{R} = \frac{P+Q}{R} \quad \text{and} \quad \frac{P}{R} - \frac{Q}{R} = \frac{P-Q}{R}$$

When adding or subtracting rational expressions with different denominators, we must find the LCD for all denominators and change each rational expression to an equivalent expression that has the LCD.

Complex Rational Expressions [8.4]

7. $\dfrac{\dfrac{1}{x} + \dfrac{1}{y}}{\dfrac{1}{x} - \dfrac{1}{y}} = \dfrac{xy\left(\dfrac{1}{x} + \dfrac{1}{y}\right)}{xy\left(\dfrac{1}{x} - \dfrac{1}{y}\right)}$

$= \dfrac{y+x}{y-x}$

A rational expression that contains, in its numerator or denominator, other rational expressions is called a *complex fraction*, or complex rational expression. One method of simplifying a complex fraction is to multiply the numerator and denominator by the LCD for all denominators.

Rational Equations [8.5]

8. Solve $\dfrac{x}{2} + 3 = \dfrac{1}{3}$.

$6\left(\dfrac{x}{2}\right) + 6 \cdot 3 = 6 \cdot \dfrac{1}{3}$

$3x + 18 = 2$

$x = -\dfrac{16}{3}$

To solve an equation involving rational expressions, we first find the LCD for all denominators appearing on either side of the equation. We then multiply both sides by the LCD to clear the equation of all fractions and solve as usual.

Reduce to lowest terms. [8.1]

1. $\dfrac{x^2 + 2xy + y^2}{x + y}$

2. $\dfrac{4x^2 - 4x - 3}{2x^2 - x - 1}$

3. Average Speed A person riding a Ferris wheel with a diameter of 65 feet travels once around the wheel in 50 seconds. What is the average speed of the rider in feet per second? Use 3.14 as an approximation for π and round your answer to the nearest tenth. [8.1]

For the functions in Problems 4 and 5, evaluate the given difference quotients. [8.1]

a. $\dfrac{f(x + h) - f(x)}{h}$

b. $\dfrac{f(x) - f(a)}{x - a}$

4. $f(x) = 8x - 5$

5. $f(x) = x^2 + x - 6$

Multiply and divide as indicated. [8.2]

6. $\dfrac{a^2 - 9}{4a + 12} \cdot \dfrac{12(a + 2)^2}{a^2 - a - 6}$

7. $\dfrac{a^4 - 16}{2a^2 + a - 6} \div \dfrac{a^2 + 4}{8a - 12}$

8. $\dfrac{2x^2 + 3x + 1}{2x^2 + 6x + 18} \div \dfrac{2x^2 - 5x - 3}{x^3 - 27}$

Add and subtract as indicated. [8.3]

9. $\dfrac{3}{10} + \dfrac{6}{25}$

10. $\dfrac{1}{2} - \dfrac{5}{32} + \dfrac{3}{16}$

11. $\dfrac{a + 2}{a^2 - 25} - \dfrac{7}{a^2 - 25}$

12. $\dfrac{2}{x + 1} + \dfrac{4}{2x - 1}$

13. $\dfrac{5x}{x^2 + 3x - 4} - \dfrac{4x}{x^2 + 2x - 3}$

14. $\dfrac{2x + 4}{x^2 - 10x - 39} - \dfrac{x + 2}{x^2 - 18x + 65}$

Simplify each complex rational expression. [8.4]

15. $\dfrac{2 - \dfrac{1}{a + 6}}{2 + \dfrac{1}{a + 6}}$

16. $\dfrac{1 - \dfrac{25}{x^2}}{1 + \dfrac{7}{x} + \dfrac{10}{x^2}}$

Solve each of the following equations. [8.5]

17. $5 + \dfrac{2}{x} = \dfrac{6}{5}$

18. $\dfrac{2x}{x + 4} - 3 = \dfrac{-8}{x + 4}$

19. $\dfrac{9}{y + 3} + \dfrac{y + 4}{7y} = \dfrac{1}{7}$

20. $1 - \dfrac{10}{x} = \dfrac{-21}{x^2}$

Solve the following applications. Be sure to show the equation in each case. [8.6]

21. Number Problem What number must be added to the denominator of $\dfrac{13}{54}$ to make the result $\dfrac{1}{4}$?

22. Speed of a Boat The current of a river is 5 miles per hour. It takes a motorboat a total of 14 hours to travel 24 miles upstream and return 24 miles downstream. What is the speed of the boat in still water?

23. Filling a Pool An inlet pipe can fill a pool in 6 hours, and the drain can empty it in 10 hours. If the pool is one-third full and both the inlet pipe and the drain are left open, how long will it take to fill the pool the rest of the way?

Graphing Linear Equations and Inequalities in Two Variables

9

9.1 The Rectangular Coordinate System

9.2 Solutions and Graphs of Linear Equations

9.3 Graphing Linear Equations Using Intercepts

9.4 Graphing Linear Equations Using Slope

9.5 Finding the Equation of a Line

9.6 Graphing Linear Inequalities in Two Variables

Ligurian Sea, Italy
Image © 2010 European Space Imaging

Italy's coast of the Ligurian Sea is known for its rugged cliffs and historic settlements. If you are lucky enough to visit, you will find five fishing villages connected by a walking trail, known as Sentiero Azzurro ("Light Blue Trail"). Constructed over the centuries as a connection between these villages, the trail has become a must-see for visitors. The walk varies in difficulty and takes approximately 5 hours. Some sections are easy strolls, others are very challenging hikes. One leg of the trail begins with an easy walk and finishes with a climb of 368 stairs. It winds through olive orchards and vineyards and can be rough in places, but offers spectacular views of the bay and the unique coastline. The difference in height for the Sentiero Azzurro from bottom to top is more than 600 meters and the length is more than 10 km.

©iStockphoto.com/danbreckwoldt

From the above information, we know that the steepness of the trail varies along its length. However, we can use the knowledge of the overall change in height from the bottom of the trail to the top to potentially determine an average steepness, or slope. In mathematics, slope is the difference between the rise of a path compared to the distance over which that rise takes place. Calculation of slope is one of the topics we will explore in this chapter.

603

A Create a scatter diagram or line graph from a table of data.

B Graph ordered pairs on a rectangular coordinate system.

9.1 The Rectangular Coordinate System

Image © Katherine Heistand Shields, 2010

KEY WORDS

scatter diagram

line graph

x-coordinate

y-coordinate

axis

quadrant

**Video Examples
Section 9.1**

In New Jersey, a fast food restaurant called Cluck-U Chicken markets a wide range of hot sauces for its famous buffalo wings. Their 911 hot sauce is so spicy, the restaurant requires all tasters to sign a waiver upon order. According to Cluck-U's corporate office, the Scoville rating (a scale to measure the spicy heat of the chili pepper) of the 911 sauce is 450,000 units. Look at Table 1 below, which compares other peppers with the 911 hot sauce. Figure 1 is a bar graph, which is a visual representation of the same information. The ratings in the table and graph are approximate because each entry on the Scoville scale typically appears as a range.

Table 1

Spicy Peppers	
Pepper or Pepper-Infused Item	**Scoville Units**
Green pepper	0
Carolina cayenne pepper	100,000
Bird's eye pepper	200,000
Habanero pepper	300,000
Cluck-U's 911 sauce	450,000

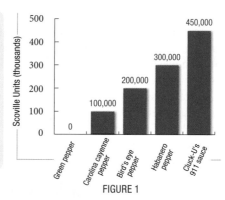

FIGURE 1

A Scatter Diagrams and Line Graphs

The information in the table in the introduction can be visualized with a *scatter diagram*, also called a *scatter plot* and *line graph* as well. Figure 2 is a scatter diagram of the information in Table 1. We use dots instead of the bars shown in Figure 1 to show the Scoville rating of each pepper item. Figure 3 is called a *line graph*. It is constructed by taking the dots in Figure 2 and connecting each one to the next with a straight line. Notice that we have labeled the axes in these two figures a little differently than we did with the bar chart by making the axes intersect at the number 0.

FIGURE 2

FIGURE 3

EXAMPLE 1 The tables below give the first five odd numbers and the first five counting numbers squared as paired data. In each case construct a scatter diagram.

Odd Numbers	
Position	Value
1	1
2	3
3	5
4	7
5	9

Squares	
Position	Value
1	1
2	4
3	9
4	16
5	25

Solution The two scatter diagrams are based on the data from these tables shown here. Notice how the dots in Figure 4 seem to line up in a straight line, whereas the dots in Figure 5 give the impression of a curve. We say the points in Figure 4 suggest a linear relationship between the two sets of data, whereas the points in Figure 5 suggest a nonlinear relationship.

FIGURE 4

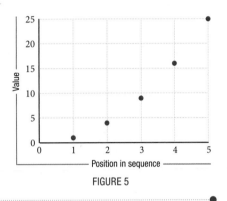

FIGURE 5

B Graphing Ordered Pairs

As you know, each dot in Figures 4 and 5 corresponds to a pair of numbers, one of which is associated with the horizontal axis and the other with the vertical axis. Paired data play a very important role in the equations we will solve in the next section. To prepare ourselves for those equations, we need to expand the concept of paired data to include negative numbers. At the same time, we want to standardize the position of the axes in the diagrams that we use to visualize paired data.

Our ability to graph paired data is due to the invention of the rectangular coordinate system also caled the *Cartesian coordinate system* after the French philosopher René Descartes (1595–1650). Descartes is the person usually credited with the invention of the rectangular coordinate system. As a philosopher, Descartes is responsible for the statement "I think, therefore I am." Until Descartes invented his coordinate system in 1637, algebra and geometry were treated as separate subjects. The rectangular coordinate system allows us to connect algebra and geometry by associating geometric shapes with algebraic equations.

Here is the formal definition for a graph's coordinates:

> **DEFINITION** *x*-coordinate, *y*-coordinate
>
> A pair of numbers enclosed in parentheses and separated by a comma, such as $(-2, 1)$, is called an ordered pair of numbers. The first number in the pair is called the *x*-coordinate of the ordered pair; the second number is called the *y*-coordinate. For the ordered pair $(-2, 1)$, the *x*-coordinate is -2 and the *y*-coordinate is 1.

Practice Problems

1. The table below gives the position and value of the first five counting numbers cubed. Construct a scatter diagram using these data.

Cubes	
Position	Value
1	1
2	8
3	27
4	64
5	125

Ordered pairs of numbers are important in the study of mathematics because they give us a way to visualize solutions to equations. To see the visual component of ordered pairs, we need the diagram shown in Figure 6. It is called the *rectangular coordinate system*.

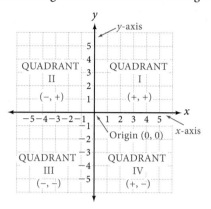

FIGURE 6

The rectangular coordinate system is built from two number lines oriented perpendicular to each other. The horizontal number line is exactly the same as our real number line and is called the *x-axis*. The vertical number line is also the same as our real number line with the positive direction up and the negative direction down. It is called the *y-axis*. The point where the two axes intersect is called the *origin*. As you can see from Figure 6, the axes divide the plane into four *quadrants*, which are numbered I through IV in a counterclockwise direction.

To graph the ordered pair (a, b), we start at the origin and move a units forward or back (forward if a is positive and back if a is negative). Then we move b units up or down (up if b is positive, down if b is negative). The point where we end up is the graph of the ordered pair (a, b). To graph the ordered pair $(5, 2)$, we start at the origin and move 5 units to the right. Then, from that position, we move 2 units up.

Note It is very important that you graph ordered pairs quickly and accurately. Remember, the first coordinate goes with the horizontal axis and the second coordinate goes with the vertical axis.

2. Graph the ordered pairs $(2, 3)$, $(2, -3)$, $(-2, 3)$, and $(-2, -3)$.

EXAMPLE 2 Graph the ordered pairs $(3, 4)$, $(3, -4)$, $(-3, 4)$, and $(-3, -4)$.

Solution

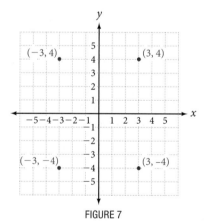

FIGURE 7

We can see in Figure 7 that when we graph ordered pairs, the *x*-coordinate corresponds to movement parallel to the *x*-axis (horizontal) and the *y*-coordinate corresponds to movement parallel to the *y*-axis (vertical).

EXAMPLE 3 Graph the ordered pairs $(-1, 3)$, $(2, 5)$, $(0, 0)$, $(0, -3)$, and $(4, 0)$.

Solution

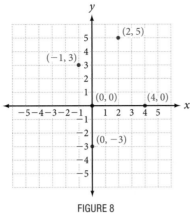

FIGURE 8

3. Graph the ordered pairs $(-2, 1)$, $(3, 5)$, $(0, 2)$, $(-5, 0)$, and $(-3, -3)$.

Note If we do not label the axes of a coordinate system, we assume that each square is one unit long and one unit wide.

Review the three points in Example 3 that contain a zero as either an *x*- or a *y*-coordinate. It is important to note that ordered pairs that sit on an axis or at the origin are not associated with any quadrant.

GETTING READY FOR CLASS

After reading through the preceding section, respond in your own words and in complete sentences.

A. What is an ordered pair of numbers?

B. Explain in words how you would graph the ordered pair $(3, 4)$.

C. How does a scatter diagram differ from a line graph?

D. Where is the origin on a rectangular coordinate system and why is it important?

Vocabulary Review

Choose the correct words to fill in the blanks below.

x-axis	line graph	ordered pair	scatter diagram/plot
y-axis	origin	quadrants	

1. A _____ uses dots on a graph to identify data.

2. Connecting the dots of a scatter diagram gives a _____.

3. A pair of numbers enclosed in parentheses and separated by a comma is called an _____.

4. A rectangular coordinate system has a horizontal number line called the _____ and a vertical number line called the _____.

5. The *x*-axis and *y*-axis divide the rectangular coordinate system into four _____.

6. The point where the *x*-and *y*-axis intersect is called the _____.

Problems

A 1. Non-Camera Phone Sales The table here shows what are the estimated sales of non-camera phones for the years 2006–2010. Use the information from the table and chart to construct a line graph.

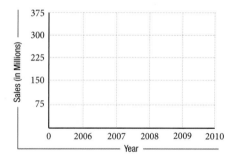

Year	Sales (in millions)
2006	300
2007	250
2008	175
2009	150
2010	125

2. Camera Phone Sales The table here shows the estimated sales of camera phones from 2006 to 2010. Use the information from the table to construct a line graph.

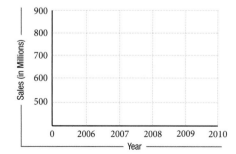

Year	Sales (in millions)
2006	500
2007	650
2008	750
2009	875
2010	900

B Graph the following ordered pairs.

3. $(3, 2)$ **4.** $(3, -2)$ **5.** $(-3, 2)$ **6.** $(-3, -2)$ **7.** $(5, 1)$ **8.** $(5, -1)$

9. $(1, 5)$ **10.** $(1, -5)$ **11.** $(-1, 5)$ **12.** $(-1, -5)$ **13.** $\left(2, \frac{1}{2}\right)$ **14.** $\left(3, \frac{3}{2}\right)$

15. $\left(-4, -\frac{5}{2}\right)$ **16.** $\left(-5, -\frac{3}{2}\right)$ **17.** $(3, 0)$ **18.** $(-2, 0)$ **19.** $(0, 5)$ **20.** $(0, 0)$

ODD-NUMBERED PROBLEMS

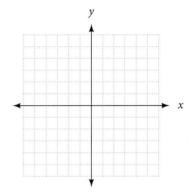

EVEN-NUMBERED PROBLEMS

Give the coordinates of each numbered point in the figure and list which quadrant each point is in.

21. **22.**

23. **24.**

25. **26.**

27. **28.**

29. **30.**

Graph the points $(4, 3)$ and $(-4, -1)$, and draw a straight line that passes through both of them. Then answer the following questions.

31. Does the graph of $(2, 2)$ lie on the line?

32. Does the graph of $(-2, 0)$ lie on the line?

33. Does the graph of $(0, -2)$ lie on the line?

34. Does the graph of $(-6, 2)$ lie on the line?

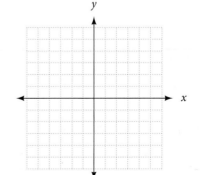

Graph the points $(-2, 4)$ and $(2, -4)$, and draw a straight line that passes through both of them. Then answer the following questions.

35. Does the graph of $(0, 0)$ lie on the line?

36. Does the graph of $(-1, 2)$ lie on the line?

37. Does the graph of $(2, -1)$ lie on the line?

38. Does the graph of $(1, -2)$ lie on the line?

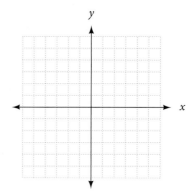

Draw a straight line that passes through the points $(3, 4)$ and $(3, -4)$. Then answer the following questions.

39. Is the graph of $(3, 0)$ on this line?

40. Is the graph of $(0, 3)$ on this line?

41. Is there any point on this line with an x-coordinate other than 3?

42. If you extended the line, would it pass through a point with a y-coordinate of 10?

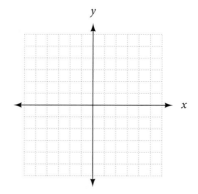

Draw a straight line that passes through the points $(3, 4)$ and $(-3, 4)$. Then answer the following questions.

43. Is the graph of $(4, 0)$ on this line?

44. Is the graph of $(0, 4)$ on this line?

45. Is there any point on this line with a y-coordinate other than 4?

46. If you extended the line, would it pass through a point with an x-coordinate of 10?

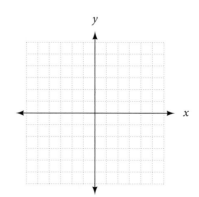

Applying the Concepts

47. Hourly Wages Jane takes a job at the local Marcy's department store. Her job pays $8.00 per hour. The graph shows how much Jane earns for working from 0 to 40 hours in a week.

a. List three ordered pairs that lie on the line graph.

b. How much will she earn for working 40 hours?

c. If her check for one week is $240, how many hours did she work?

d. She works 35 hours one week, but her paycheck before deductions are subtracted out is for $260. Is this correct? Explain.

48. Hourly Wages Judy takes a job at Gigi's boutique. Her job pays $6.00 per hour plus $50 per week in commission. The graph shows how much Judy earns for working from 0 to 40 hours in a week.

a. List three ordered pairs that lie on the line graph.

b. How much will she earn for working 40 hours?

c. If her check for one week is $230, how many hours did she work?

d. She works 35 hours one week, but her paycheck before deductions are subtracted out is for $260. Is this correct? Explain.

49. Baseball Attendance The graph gives the attendance at Major League Baseball games over a ten-year period. If x represents the year in question and y represents attendance in millions, write five ordered pairs that describe the information in the table.

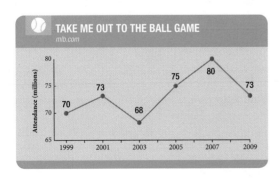

50. FIFA World Cup Goals The graph shows the number of goals scored by the US Men's National Team in FIFA World Cup competition. Write five ordered pairs that lie on the graph.

51. Right triangle ABC has legs of length 5. Point C is the ordered pair $(6, 2)$. Find the coordinates of A and B.

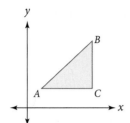

52. Right triangle ABC has legs of length 7. Point C is the ordered pair $(-8, -3)$. Find the coordinates of A and B.

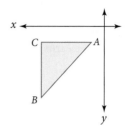

53. Rectangle $ABCD$ has a length of 5 and a width of 3. Point D is the ordered pair $(7, 2)$. Find points A, B, and C.

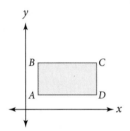

54. Rectangle $ABCD$ has a length of 5 and a width of 3. Point D is the ordered pair $(-1, 1)$. Find points A, B, and C.

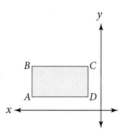

Getting Ready for the Next Section

55. Let $2x + 3y = 6$.
 a. Find x if $y = 4$.

 b. Find x if $y = -2$.

 c. Find y if $x = 3$.

 d. Find y if $x = 9$.

56. Let $2x - 5y = 20$.
 a. Find x if $y = 0$.

 b. Find x if $y = -6$.

 c. Find y if $x = 0$.

 d. Find y if $x = 5$.

57. Let $y = 2x - 1$.
 a. Find x if $y = 7$.

 b. Find x if $y = 3$.

 c. Find y if $x = 0$.

 d. Find y if $x = 5$.

58. Let $y = 3x - 2$.
 a. Find x if $y = 4$.

 b. Find x if $y = 3$.

 c. Find y if $x = 2$.

 d. Find y if $x = -3$.

59. Find y when x is 4 in the formula $3x + 2y = 6$.

60. Find y when x is 0 in the formula $3x + 2y = 6$.

61. Find y when x is 0 in $y = -\dfrac{1}{3}x + 2$.

62. Find y when x is 3 in $y = -\dfrac{1}{3}x + 2$.

63. Find y when x is 2 in $y = \dfrac{3}{2}x - 3$.

64. Find y when x is 4 in $y = \dfrac{3}{2}x - 3$.

65. Solve $5x + y = 4$ for y.

66. Solve $-3x + y = 5$ for y.

67. Solve $3x - 2y = 6$ for y.

68. Solve $2x - 3y = 6$ for y.

Find the Mistake

Each sentence below contains a mistake. Circle the mistake and write the correct word or number on the line provided.

1. A scatter diagram uses bars to show data in the form of order pairs. _____

2. In the ordered pair $(-3, 6)$, the x–coordinate is 6. _____

3. To graph the ordered pair $(7, 3)$, start at the origin, move 7 units to the left, and 3 units up. _____

4. The ordered pair $(-2, 5)$ appears in the first quadrant. _____

OBJECTIVES

A Find solutions to linear equations in two variables.

B Determine whether an ordered pair is a solution to a linear equation in two variables.

C Graph a linear equation in two variables.

D Graph horizontal lines, vertical lines, and lines through the origin.

KEY WORDS

ordered pair

graph

linear equation in two variables

horizontal

vertical

origin

**Video Examples
Section 9.2**

Image © Katherine Heistand Shields, 2010

A new robot named EMILY (Emergency Integrated Lifesaving Lanyard) patrols dangerous ocean waters off the coast of Malibu, California. Currently, a lifeguard controls EMILY by remote. However, developers will soon equip the robot with sonar and an autonomous system able to detect swimmers in distress without the help of a lifeguard on shore. EMILY's electric-powered impeller drives the robot through even the roughest surf at a speed 6 times faster than that of a human. A swimmer in distress can use EMILY as a flotation device until further help arrives, or EMILY can tow the swimmer back to shore.

Using the information about EMILY's speed, we can write the following equation, which uses two variables:

$$y = 6x$$

where x is a human's swimming speed and y is EMILY's speed. In this section, we will begin to investigate equations in two variables, such as the one above. As you will see, equations in two variables have pairs of numbers for solutions. Because we know how to use paired data to construct tables, histograms, and other charts, we can take our work with paired data further by using equations in two variables to construct tables of paired data.

A Solving Linear Equations

Let's begin this section by reviewing the relationship between equations in one variable and their solutions.

If we solve the equation $3x - 2 = 10$, the solution is $x = 4$. If we graph this solution, we simply draw the real number line and place a dot at the point whose coordinate is 4. The relationship between linear equations in one variable, their solutions, and the graphs of those solutions look like this:

Equation	*Solution*	*Graph of Solution Set*
$3x - 2 = 10$	$x = 4$	• at 4 (0 and 4 marked)
$x + 5 = 7$	$x = 2$	• at 2 (0 and 2 marked)
$2x = -6$	$x = -3$	• at −3 (−3 and 0 marked)

When the equation has one variable, the solution is a single number whose graph is a point on a line.

Now, consider the equation $2x + y = 3$. The first thing we notice is that there are two variables instead of one. Therefore, a solution to the equation $2x + y = 3$ will be not a single number but a pair of numbers, one for x and one for y, that makes the equation

a true statement. One pair of numbers that works is $x = 2$, $y = -1$ because when we substitute them for x and y in the equation, we get a true statement.

$$2(2) + (-1) \stackrel{?}{=} 3$$

$$4 - 1 = 3$$

$$3 = 3 \qquad \text{A true statement}$$

The pair of numbers $x = 2$, $y = -1$ is written as $(2, -1)$. As you know from Section 3.1, $(2, -1)$ is called an *ordered pair* because it is a pair of numbers written in a specific order. The first number is always associated with the variable x, and the second number is always associated with the variable y. We call the first number in the ordered pair the *x-coordinate* and the second number the *y-coordinate* of the ordered pair.

Let's look back to the equation $2x + y = 3$. The ordered pair $(2, -1)$ is not the only solution. Another solution is $(0, 3)$ because when we substitute 0 for x and 3 for y we get

$$2(0) + 3 \stackrel{?}{=} 3$$

$$0 + 3 = 3$$

$$3 = 3 \qquad \text{A true statement}$$

Still another solution is the ordered pair $(5, -7)$ because

$$2(5) + (-7) \stackrel{?}{=} 3$$

$$10 - 7 = 3$$

$$3 = 3 \qquad \text{A true statement}$$

As a matter of fact, for any number we want to use for x, there is another number we can use for y that will make the equation a true statement. There is an infinite number of ordered pairs that satisfy (are solutions to) the equation $2x + y = 3$; we have listed just a few of them.

EXAMPLE 1 Given the equation $2x + 3y = 6$, complete the following ordered pairs so they will be solutions to the equation: $(0,\)$, $(\ , 1)$, $(3,\)$.

Solution To complete the ordered pair $(0,\)$, we substitute 0 for x in the equation and then solve for y.

$$2(0) + 3y = 6$$

$$3y = 6$$

$$y = 2$$

The ordered pair is $(0, 2)$.

To complete the ordered pair $(\ , 1)$, we substitute 1 for y in the equation and solve for x.

$$2x + 3(1) = 6$$

$$2x + 3 = 6$$

$$2x = 3$$

$$x = \frac{3}{2}$$

The ordered pair is $\left(\frac{3}{2}, 1\right)$.

To complete the ordered pair $(3,\)$, we substitute 3 for x in the equation and solve for y.

$$2(3) + 3y = 6$$

$$6 + 3y = 6$$

$$3y = 0$$

$$y = 0$$

The ordered pair is $(3, 0)$.

Note If this discussion seems a little long and confusing, you may want to look over some of the examples first and then come back and read this. Remember, it isn't always easy to read material in mathematics. What is important is that you understand what you are doing when you work problems. The reading is intended to assist you in understanding what you are doing. It is important to read everything in the book, but you don't always have to read it in the order it is written.

Practice Problems

1. For the equation $2x + 5y = 10$, complete the ordered pairs $(0,\)$, $(\ , 1)$, and $(5,\)$.

Answers

1. $(0, 2), \left(\frac{5}{2}, 1\right), (5, 0)$

Notice in each case that once we have used a number in place of one of the variables, the equation becomes a linear equation in one variable. We then use the method explained in Chapter 2 to solve for that variable.

2. Complete the table for $3x - 2y = 12$.

x	y
0	
	3
	0
-3	

EXAMPLE 2 Complete the following table for the equation $2x - 5y = 20$.

x	y
0	
	2
	0
-5	

Solution Filling in the table is equivalent to completing the following ordered pairs: $(0,\), (\ , 2), (\ , 0), (-5,\)$. So we proceed as in Example 1.

When $x = 0$, we have

$$2(0) - 5y = 20$$
$$0 - 5y = 20$$
$$-5y = 20$$
$$y = -4$$

When $y = 2$, we have

$$2x - 5(2) = 20$$
$$2x - 10 = 20$$
$$2x = 30$$
$$x = 15$$

When $y = 0$, we have

$$2x - 5(0) = 20$$
$$2x - 0 = 20$$
$$2x = 20$$
$$x = 10$$

When $x = -5$, we have

$$2(-5) - 5y = 20$$
$$-10 - 5y = 20$$
$$-5y = 30$$
$$y = -6$$

The completed table looks like this:

x	y
0	-4
15	2
10	0
-5	-6

The table above is equivalent to the ordered pairs $(0, -4)$, $(15, 2)$, $(10, 0)$, and $(-5, -6)$.

3. Complete the table for $y = 3x - 2$.

x	y
0	
2	
	7
	3

EXAMPLE 3 Complete the following table for the equation $y = 2x - 1$.

x	y
0	
5	
	7
	3

Solution When $x = 0$, we have

$$y = 2(0) - 1$$
$$y = 0 - 1$$
$$y = -1$$

When $x = 5$, we have

$$y = 2(5) - 1$$
$$y = 10 - 1$$
$$y = 9$$

When $y = 7$, we have When $y = 3$, we have

$$7 = 2x - 1$$ $$3 = 2x - 1$$

$$8 = 2x$$ $$4 = 2x$$

$$4 = x$$ $$2 = x$$

The completed table is

x	y
0	−1
5	9
4	7
2	3

which means the ordered pairs $(0, -1)$, $(5, 9)$, $(4, 7)$, and $(2, 3)$ are among the solutions to the equation $y = 2x - 1$.

B Determining if an Ordered Pair is a Solution

EXAMPLE 4 Which of the ordered pairs $(2, 3)$, $(1, 5)$, and $(-2, -4)$ are solutions to the equation $y = 3x + 2$?

Solution If an ordered pair is a solution to the equation, then it must satisfy the equation; that is, when the coordinates are used in place of the variables in the equation, the equation becomes a true statement.

Try $(2, 3)$ in $y = 3x + 2$:

$$3 \overset{?}{=} 3(2) + 2$$

$$3 = 6 + 2$$

$$3 = 8 \qquad \text{A false statement}$$

Try $(1, 5)$ in $y = 3x + 2$:

$$5 \overset{?}{=} 3(1) + 2$$

$$5 = 3 + 2$$

$$5 = 5 \qquad \text{A true statement}$$

Try $(-2, -4)$ in $y = 3x + 2$:

$$-4 \overset{?}{=} 3(-2) + 2$$

$$-4 = -6 + 2$$

$$-4 = -4 \qquad \text{A true statement}$$

The ordered pairs $(1, 5)$ and $(-2, -4)$ are solutions to the equation $y = 3x + 2$, and $(2, 3)$ is not.

C Graphing Linear Equations

At the end of the previous section we used a line graph to obtain a visual picture of *some* of the solutions to the equations $y = 1.5x + 15$. In this section we will use the rectangular coordinate system introduced in Section 3.1 to obtain a visual picture of *all* solutions to a linear equation in two variables. The process we use to obtain a visual picture of all solutions to an equation is called *graphing*. The picture itself is called the *graph* of the equation.

4. Which of the following ordered pairs is a solution to $y = 4x + 1$?
$(0, 1)$, $(3, 11)$, $(2, 9)$

Answers
4. $(0, 1)$ and $(2, 9)$

5. Graph the equation $x + y = 3$.

EXAMPLE 5 Graph the solution set for $x + y = 5$.

Solution We know from the previous section that an infinite number of ordered pairs are solutions to the equation $x + y = 5$. We can't possibly list them all. What we can do is list a few of them and see if there is any pattern to their graphs.

Some ordered pairs that are solutions to $x + y = 5$ are $(0, 5)$, $(2, 3)$, $(3, 2)$, $(5, 0)$. The graph of each is shown in Figure 1.

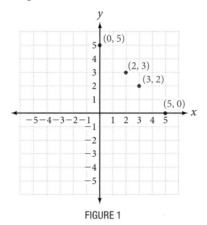

FIGURE 1

Now, by passing a straight line through these points we can graph the solution set for the equation $x + y = 5$. Linear equations in two variables always have graphs that are straight lines. The graph of the solution set for $x + y = 5$ is shown in Figure 2.

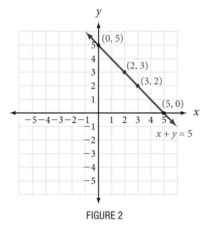

FIGURE 2

Every ordered pair that satisfies $x + y = 5$ has its graph on the line, and any point on the line has coordinates that satisfy the equation. So there is a one-to-one correspondence between points on the line and solutions to the equation. To summarize, the set of all points of a graph that satisfy an equation is the equation's solution set.

Here is the precise definition for a linear equation in two variables.

> **DEFINITION** linear equation in two variables
>
> Any equation that can be put in the form $Ax + By = C$, where A, B, and C are real numbers and A and B are not both 0, is called a ***linear equation in two variables***. The graph of any equation of this form is a straight line (that is why these equations are called "linear"). The form $Ax + By = C$ is called ***standard form***.

To graph a linear equation in two variables, as we did in Example 5, we simply graph its solution set; that is, we draw a line through all the points whose coordinates satisfy the equation. Here are the steps to follow:

HOW TO To Graph a Linear Equation in Two Variables

Step 1: Find any three ordered pairs that satisfy the equation. This can be done by using a convenient number for one variable and solving for the other variable.

Step 2: Graph the three ordered pairs found in step 1. Actually, we need only two points to graph a straight line. The third point serves as a check. If all three points do not line up, there is a mistake in our work.

Step 3: Draw a straight line through the three points graphed in step 2.

EXAMPLE 6 Graph the equation $y = 3x - 1$.

Solution Because $y = 3x - 1$ can be put in the form $ax + by = c$, it is a linear equation in two variables. Hence, the graph of its solution set is a straight line. We can find some specific solutions by substituting numbers for x and then solving for the corresponding values of y. We are free to choose any numbers for x, so let's use 0, 2, and -1.

Let $x = 0$:
$$y = 3(0) - 1$$
$$y = 0 - 1$$
$$y = -1$$

The ordered pair $(0, -1)$ is one solution.

Let $x = 2$:
$$y = 3(2) - 1$$
$$y = 6 - 1$$
$$y = 5$$

The ordered pair $(2, 5)$ is a second solution.

Let $x = -1$:
$$y = 3(-1) - 1$$
$$y = -3 - 1$$
$$y = -4$$

The ordered pair $(-1, -4)$ is a third solution.

In table form

x	y
0	-1
2	5
-1	-4

Next, we graph the ordered pairs $(0, -1)$, $(2, 5)$, $(-1, -4)$ and draw a straight line through them. The line we have drawn in Figure 3 is the graph of $y = 3x - 1$.

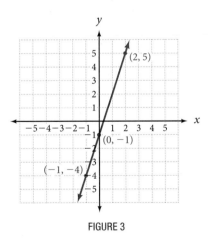

FIGURE 3

Example 6 illustrates the connection between algebra and geometry that we mentioned previously. Descartes' rectangular coordinate system allows us to associate the equation $y = 3x - 1$ (an algebraic concept) with a specific straight line (a geometric concept). The study of the relationship between equations in algebra and their associated geometric figures is called *analytic geometry*.

Note The meaning of the convenient numbers referred to in step 1 will become clear as you read the next two examples.

6. Graph the equation $y = 2x + 3$.

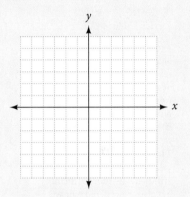

Note It may seem that we have simply picked the numbers 0, 2, and -1 out of the air and used them for x. In fact we have done just that. Could we have used numbers other than these? The answer is yes, we can substitute any number for x; for a linear equation in two variables, there will always be a value of y to go with it.

7. Graph $y = \frac{3}{2}x - 3$.

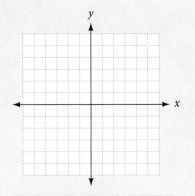

Note In Example 7 the values of x we used, -3, 0, and 3, are referred to as convenient values of x because they are easier to work with than some other numbers. For instance, if we let $x = 2$ in our original equation, we would have to add $-\frac{2}{3}$ and 2 to find the corresponding value of y. Not only would the arithmetic be more difficult but also the ordered pair we obtained would have a fraction for its y-coordinate, making it more difficult to graph accurately.

EXAMPLE 7 Graph the equation $y = -\frac{1}{3}x + 2$.

Solution We need to find three ordered pairs that satisfy the equation. To do so, we can let x equal any numbers we choose and find corresponding values of y. But, every value of x we substitute into the equation is going to be multiplied by $-\frac{1}{3}$. For our convenience, let's use numbers for x that are divisible by 3, like -3, 0, and 3. That way, when we multiply them by $-\frac{1}{3}$, the result will be an integer.

Let $x = -3$: $y = -\frac{1}{3}(-3) + 2$

$$y = 1 + 2$$
$$y = 3$$

The ordered pair $(-3, 3)$ is one solution.

Let $x = 0$: $y = -\frac{1}{3}(0) + 2$

$$y = 0 + 2$$
$$y = 2$$

The ordered pair $(0, 2)$ is a second solution.

Let $x = 3$: $y = -\frac{1}{3}(3) + 2$

$$y = -1 + 2$$
$$y = 1$$

The ordered pair $(3, 1)$ is a third solution.

In table form

x	y
-3	3
0	2
3	1

Graphing the ordered pairs $(-3, 3)$, $(0, 2)$, and $(3, 1)$ and drawing a straight line through their graphs, we have the graph of the equation $y = -\frac{1}{3}x + 2$, as shown in Figure 4.

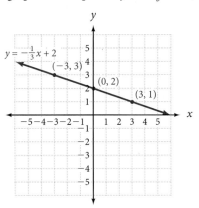

FIGURE 4

8. Graph the solution set for $2x - 4y = 8$.

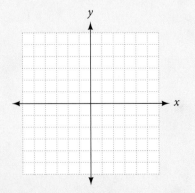

EXAMPLE 8 Graph the solution set for $3x - 2y = 6$.

Solution It will be easier to find convenient values of x to use in the equation if we first solve the equation for y. To do so, we add $-3x$ to each side, and then we multiply each side by $-\frac{1}{2}$.

$3x - 2y = 6$	*Original equation*
$-2y = -3x + 6$	*Add $-3x$ to each side.*
$-\frac{1}{2}(-2y) = -\frac{1}{2}(-3x + 6)$	*Multiply each side by $-\frac{1}{2}$.*
$y = \frac{3}{2}x - 3$	*Simplify each side.*

Now, because each value of x will be multiplied by $\frac{3}{2}$, it will be to our advantage to choose values of x that are divisible by 2. That way, we will obtain values of y that do not contain fractions. This time, let's use 0, 2, and 4 for x.

When $x = 0$: $\qquad y = \frac{3}{2}(0) - 3$

$\qquad\qquad\qquad y = 0 - 3$

$\qquad\qquad\qquad y = -3$

The ordered pair $(0, -3)$ is one solution.

When $x = 2$: $\qquad y = \frac{3}{2}(2) - 3$

$\qquad\qquad\qquad y = 3 - 3$

$\qquad\qquad\qquad y = 0$

The ordered pair $(2, 0)$ is a second solution.

When $x = 4$: $\qquad y = \frac{3}{2}(4) - 3$

$\qquad\qquad\qquad y = 6 - 3$

$\qquad\qquad\qquad y = 3$

The ordered pair $(4, 3)$ is a third solution.

Graphing the ordered pairs $(0, -3)$, $(2, 0)$, and $(4, 3)$ and drawing a line through them, we have the graph shown in Figure 5.

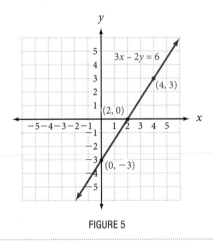

FIGURE 5

Note After reading through Example 8, many students ask why we didn't use -2 for x when we were finding ordered pairs that were solutions to the original equation. The answer is, we could have. If we were to let $x = -2$, the corresponding value of y would have been -6. As you can see by looking at the graph in Figure 5, the ordered pair $(-2, -6)$ is on the graph.

D Horizontal and Vertical Lines; Lines Through The Origin

EXAMPLE 9 Graph each of the following lines.

a. $y = \frac{1}{2}x$ \qquad **b.** $x = 3$ \qquad **c.** $y = -2$

Solution

a. The line $y = \frac{1}{2}x$ passes through the origin because $(0, 0)$ satisfies the equation. To sketch the graph we need at least one more point on the line. When x is 2, we obtain the point $(2, 1)$, and when x is -4, we obtain the point $(-4, -2)$. The graph of $y = \frac{1}{2}x$ is shown in Figure 6A.

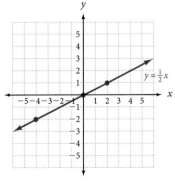

FIGURE 6A

9. Graph each of the following lines.

a. $y = \frac{2}{3}x$

b. $x = 3$

c. $y = -2$

Note The equations in parts b and c are linear equations in two variables where the second variable is not explicit; in other words, $x = 3$ is the same as

$$0y + x = 3$$

and $y = -2$ is the same as

$$0x + y = -2$$

b. The line $x = 3$ is the set of all points whose x-coordinate is 3. The variable y does not appear in the equation, so the y-coordinate can be any number. Note that we can write our equation as a linear equation in two variables by writing it as $x + 0y = 3$. Because the product of 0 and y will always be 0, y can be any number. The graph of $x = 3$ is the vertical line shown in Figure 6B.

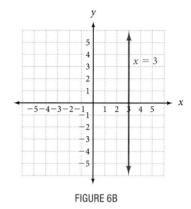

FIGURE 6B

c. The line $y = -2$ is the set of all points whose y-coordinate is -2. The variable x does not appear in the equation, so the x-coordinate can be any number. Again, we can write our equation as a linear equation in two variables by writing it as $0x + y = -2$. Because the product of 0 and x will always be 0, x can be any number. The graph of $y = -2$ is the horizontal line shown in Figure 6C.

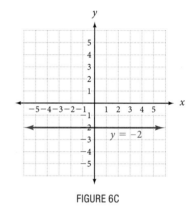

FIGURE 6C

FACTS FROM GEOMETRY Special Equations and Their Graphs

For the equations below, m, a, and b are real numbers.

Through the Origin Vertical Line Horizontal Line

FIGURE 7A *Any equation of the form $y = mx$ has a graph that passes through the origin.*

FIGURE 7B *Any equation of the form $x = a$ has a vertical line for its graph.*

FIGURE 7C *Any equation of the form $y = b$ has a horizontal line for its graph.*

GETTING READY FOR CLASS

After reading through the preceding section, respond in your own words and in complete sentences.

A. How can you tell if an ordered pair is a solution to an equation?

B. How would you find a solution to $y = 3x - 5$?

C. How many solutions are there to an equation that contains two variables?

D. What type of equation has a line that passes through the origin?

Vocabulary Review

Choose the correct words to fill in the blanks below.

solution	graphing	graph	first	vertical
linear	ordered pairs	straight line	second	horizontal

1. The _____ number of an ordered pair is called the x-coordinate.

2. The _____ number of an ordered pair is called the y-coordinate.

3. An ordered pair is a _____ to an equation if it satisfies the equation.

4. The process of obtaining a visual picture of all solutions to an equation is called _____.

5. A _____ equation in two variables is any equation that can be put into the form $ax + bx = c$.

6. The first step to graph a linear equation in two variables is to find any three _____ that satisfy the equation.

7. The second step to graph a linear equation in two variables is to _____ the three ordered pairs that satisfy the equation.

8. The third step to graph a linear equation in two variables is to draw a _____ through the three ordered pairs that satisfy the equation.

9. Any equation of the form $y = b$ has a _____ line for it's graph, whereas any equation of the form $x = a$ has a _____ line for it's graph.

Problems

A For each equation, complete the given ordered pairs.

1. $2x + y = 6$ \quad (0,), (, 0), (, −6)

2. $3x − y = 5$ \quad (0,), (1,), (, 5)

3. $3x + 4y = 12$ \quad (0,), (, 0), (−4,)

4. $5x − 5y = 20$ \quad (0,), (, −2), (1,)

5. $y = 4x − 3$ \quad (1,), (, 0), (5,)

6. $y = 3x − 5$ \quad (, 13), (0,), (−2,)

7. $y = 7x − 1$ \quad (2,), (, 6), (0,)

8. $y = 8x + 2$ \quad (3,), (, 0), (, −6)

9. $x = −5$ \quad (, 4), (, −3), (, 0)

10. $y = 2$ \quad (5,), (−8,), $\left(\dfrac{1}{2}, \ \right)$

For each of the following equations, complete the given table.

11. $y = 3x$

x	y
1	3
−3	
	12
	18

12. $y = -2x$

x	y
−4	
0	
	10
	12

13. $y = 4x$

x	y
0	
	−2
−3	
	12

14. $y = -5x$

x	y
3	
	0
−2	
	−20

15. $x + y = 5$

x	y
2	
3	
	0
	−4

16. $x - y = 8$

x	y
0	
4	
	−3
	−2

17. $2x - y = 4$

x	y
	0
	2
1	
−3	

18. $3x - y = 9$

x	y
	0
	−9
5	
−4	

19. $y = 6x - 1$

x	y
0	
	−7
−3	
	8

20. $y = 5x + 7$

x	y
0	
	−2
	−4
	−8

21. $y = -2x + 3$

x	y
0	
	−2
	2
	−7

22. $y = -3x + 1$

x	y
0	
	7
	−5
−3	

B For the following equations, tell which of the given ordered pairs are solutions.

23. $2x - 5y = 10$ $(2, 3), (0, -2), \left(\dfrac{5}{2}, 1\right)$

24. $3x + 7y = 21$ $(0, 3), (7, 0), (1, 2)$

25. $y = 7x - 2$ $(1, 5), (0, -2), (-2, -16)$

26. $y = 8x - 3$ $(0, 3), (5, 16), (1, 5)$

27. $y = 6x$ $(1, 6), (-2, 12), (0, 0)$

28. $y = -4x$ $(0, 0), (2, 4), (-3, 12)$

29. $x + y = 0$ $(1, 1), (2, -2), (3, 3)$

30. $x - y = 1$ $(0, 1), (0, -1), (1, 2)$

31. $x = 3$ $(3, 0), (3, -3), (5, 3)$

32. $y = -4$ $(3, -4), (-4, 4), (0, -4)$

C D For the following equations, complete the given ordered pairs, and use the results to graph the solution set for the equation.

33. $x + y = 4$ $(0, \), (2, \), (\ , 0)$

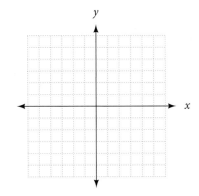

34. $x - y = 3$ $(0, \), (2, \), (\ , 0)$

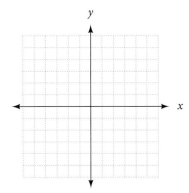

35. $x + y = 3$ $(0, \), (2, \), (\ , -1)$

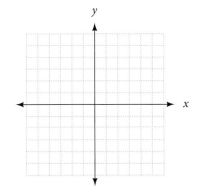

36. $x - y = 4$ $(1, \), (-1, \), (\ , 0)$

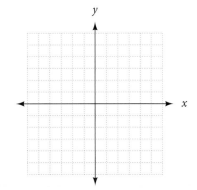

37. $y = 2x$ $(0, \), (-2, \), (2, \)$

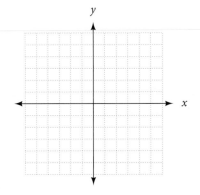

38. $y = \dfrac{1}{2}x$ $(0, \), (-2, \), (2, \)$

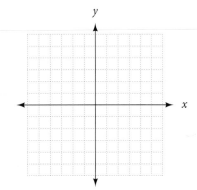

39. $y = \dfrac{1}{3}x$ $(-3, \), (0, \), (3, \)$

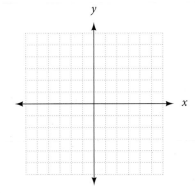

40. $y = 3x$ $(-2, \), (0, \), (2, \)$

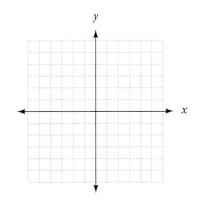

41. $y = 2x + 1$ $(0,\), (-1,\), (1,\)$

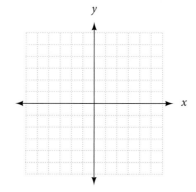

42. $y = -2x + 1$ $(0,\), (-1,\), (1,\)$

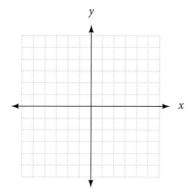

43. $y = 4$ $(0,\), (-1,\), (2,\)$

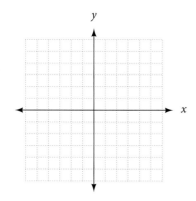

44. $x = 3$ $(\ , -2), (\ , 0), (\ , 5)$

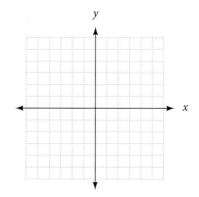

45. $y = \dfrac{1}{2}x + 3$ $(-2,\), (0,\), (2,\)$

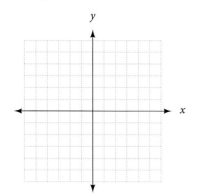

46. $y = \dfrac{1}{2}x - 3$ $(-2,\), (0,\), (2,\)$

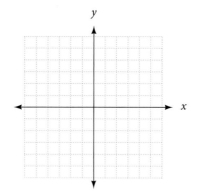

47. $y = -\dfrac{2}{3}x + 1$ $(-3,\), (0,\), (3,\)$

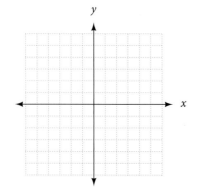

48. $y = -\dfrac{2}{3}x - 1$ $(-3,\), (0,\), (3,\)$

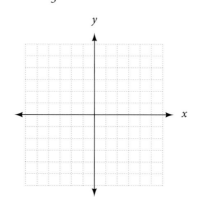

C Solve each equation for y. Then, complete the given ordered pairs, and use them to draw the graph.

49. $2x + y = 3$ $(-1, \), (0, \), (1, \)$

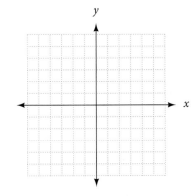

50. $3x + y = 2$ $(-1, \), (0, \), (1, \)$

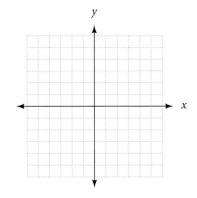

51. $3x + 2y = 6$ $(0, \), (2, \), (4, \)$

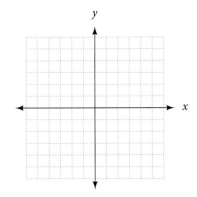

52. $2x + 3y = 6$ $(0, \), (3, \), (6, \)$

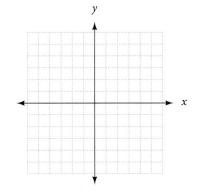

53. $-x + 2y = 6$ $(-2, \), (0, \), (2, \)$

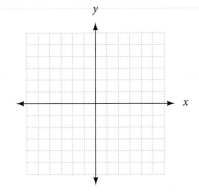

54. $-x + 3y = 6$ $(-3, \), (0, \), (3, \)$

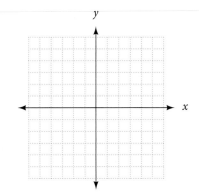

Find three solutions to each of the following equations, and then graph the solution set.

55. $y = -\dfrac{1}{2}x$

56. $y = -2x$

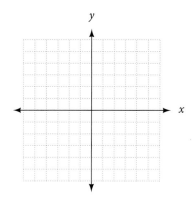

57. $y = 3x - 1$

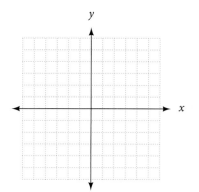

58. $y = -3x - 1$

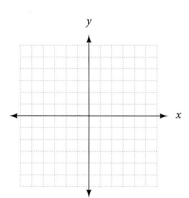

59. $-2x + y = 1$

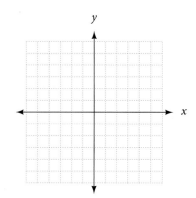

60. $-3x + y = 1$

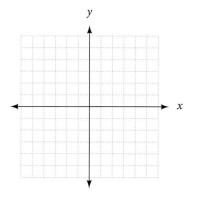

61. $3x + 4y = 8$

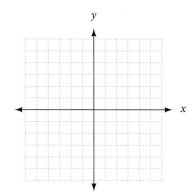

62. $3x - 4y = 8$

63. $x = -2$

64. $y = 3$

65. $y = 2$

66. $x = -3$

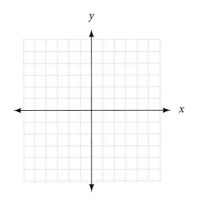

Graph each equation.

67. $y = \dfrac{3}{4}x + 1$

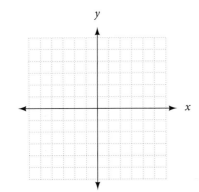

68. $y = \dfrac{2}{3}x + 1$

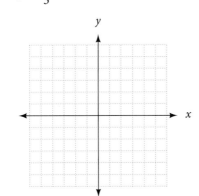

69. $y = \dfrac{1}{3}x + \dfrac{2}{3}$

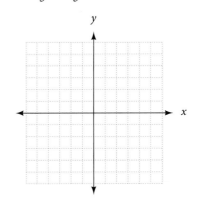

70. $y = \dfrac{1}{2}x + \dfrac{1}{2}$

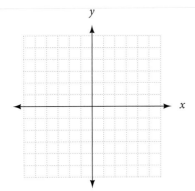

71. $y = \dfrac{2}{3}x + \dfrac{2}{3}$

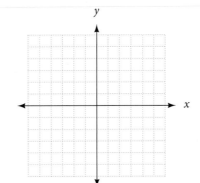

72. $y = -\dfrac{3}{4}x + \dfrac{3}{2}$

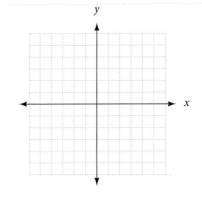

D For each equation in each table below, indicate whether the graph is horizontal (H), or vertical (V), or whether it passes through the origin (O).

73.

Equation	H, V, and/or O
$x = 3$	
$y = 3$	
$y = 3x$	
$y = 0$	

74.

Equation	H, V, and/or O
$x = \dfrac{1}{2}$	
$y = \dfrac{1}{2}$	
$y = \dfrac{1}{2}x$	
$x = 0$	

75.

Equation	H, V, and/or O
$x = -\dfrac{3}{5}$	
$y = -\dfrac{3}{5}$	
$y = -\dfrac{3}{5}x$	
$x = 0$	

76.

Equation	H, V, and/or O
$x = -4$	
$y = -4$	
$y = -4x$	
$y = 0$	

77. Use the graph to complete the table.

x	y
	−3
−2	
0	
	0
6	

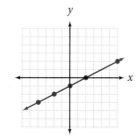

78. Use the graph at the right to complete the table. (*Hint:* Some parts have two answers.)

x	y
−3	6
	4
0	3
	1
6	

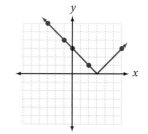

Applying the Concepts

79. Perimeter If the perimeter of a rectangle is 30 inches, then the relationship between the length l and the width w is given by the equation

$$2l + 2w = 30$$

What is the length when the width is 3 inches?

80. Perimeter The relationship between the perimeter P of a square and the length of its side s is given by the formula

$$P = 4s$$

If each side of a square is 5 inches, what is the perimeter? If the perimeter of a square is 28 inches, how long is a side?

81. Janai earns $12 per hour working as a math tutor. We can express the amount she earns each week, y, for working x hours with the equation $y = 12x$. Indicate with a yes or no, which of the following could be one of Janai's paychecks before deductions are taken out. If you answer no, explain your answer.

 a. $60 for working five hours

 b. $100 for working nine hours

 c. $80 for working seven hours

 d. $168 for working 14 hours

82. Erin earns $15 per hour working as a graphic designer. We can express the amount she earns each week, y, for working x hours with the equation $y = 15x$. Indicate with a yes or no which of the following could be one of Erin's paychecks before deductions are taken out. If you answer no, explain your answer.

 a. $75 for working five hours

 b. $125 for working nine hours

 c. $90 for working six hours

 d. $500 for working 35 hours

83. The equation $V = -45{,}000t + 600{,}000$, can be used to find the value, V, of a small crane at the end of t years.

 a. What is the value of the crane at the end of five years?

 b. When is the crane worth $330,000?

 c. Is it true that the crane with be worth $150,000 after nine years?

 d. How much did the crane cost?

84. The equation $V = -400t + 2{,}500$, can be used to find the value, V, of a notebook computer at the end of t years.

 a. What is the value of the notebook computer at the end of four years?

 b. When is the notebook computer worth $1,700?

 c. Is it true that the notebook computer with be worth $100 after five years?

 d. How much did the notebook computer cost?

Getting Ready for the Next Section

85. Let $3x + 2y = 6$.

 a. Find x when $y = 0$.

 b. Find y when $x = 0$.

86. Let $2x - 5y = 10$.

 a. Find x when $y = 0$.

 b. Find y when $x = 0$.

87. Let $-x + 2y = 4$.

 a. Find x when $y = 0$.

 b. Find y when $x = 0$.

88. Let $3x - y = 6$.

 a. Find x when $y = 0$.

 b. Find y when $x = 0$.

89. Let $y = -\dfrac{1}{3}x + 2$.

 a. Find x when $y = 0$.

 b. Find y when $x = 0$.

90. Let $y = \dfrac{3}{2}x - 3$.

 a. Find x when $y = 0$.

 b. Find y when $x = 0$.

Find the Mistake

Each sentence below contains a mistake. Circle the mistake and write the correct word(s) or number on the line provided.

 1. A solution to a linear equation in two variables will be a single number that makes the equation a true statement.

 2. A solution to the equation $3y = 6x + 9$ is $(2, 6)$. _____

 3. To graph a linear equation in two variables, you must find only one solution. _____

 4. The line $y = \dfrac{1}{2}x$ crosses the x–axis but not the y–axis. _____

A Find the intercepts of a line from the equation of the line.

B Use intercepts to graph a line.

9.3 Graphing Linear Equations Using Intercepts

©iStockphoto.com/dlanier

KEY WORDS

x-intercept

y-intercept

**Video Examples
Section 9.3**

Some talented water skiers ski on their bare feet while being towed behind a motor boat. During training, the barefoot skiers may use foot skis, which are small thin skis just a few inches longer and a little wider than the skier's foot. With foot skis, the skier can learn maneuvers faster and with fewer wipeouts than without. Normal water skiing speeds reach 30-45 miles per hour. An appropriate speed for barefoot water skiing needs to be faster and depends on the skier's body weight using the following linear equation:

$$y = \frac{x}{10} + 20$$

where *x* is the skier's weight in pounds. In this section, we will practice finding *x*- and *y*-intercepts for equations similar to the barefoot skier's equation.

A Intercepts

In this section we continue our work with graphing lines by finding the points where a line crosses the axes of our coordinate system. To do so, we use the fact that any point on the *x*-axis has a *y*-coordinate of 0 and any point on the *y*-axis has an *x*-coordinate of 0. We begin with the following definition:

> **DEFINITION** *x*-intercept, *y*-intercept
>
> The *x-intercept* of a straight line is the *x*-coordinate of the point where the graph intersects the *x*-axis and the *y*-coordinate is zero. The *y-intercept* is defined similarly. It is the *y*-coordinate of the point where the graph intersects the *y*-axis and the *x*-coordinate is zero.

B Using Intercepts to Graph Lines

If the *x*-intercept is *a*, then the point (*a*, 0) lies on the graph. (This is true because any point on the *x*-axis has a *y*-coordinate of 0.)

If the *y*-intercept is *b*, then the point (0, *b*) lies on the graph. (This is true because any point on the *y*-axis has an *x*-coordinate of 0.)

Graphically, the relationship is shown in Figure 1.

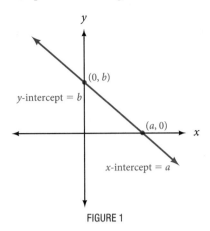

FIGURE 1

EXAMPLE 1 Find the x- and y-intercepts for $3x - 2y = 6$, and then use them to draw the graph.

Solution To find where the graph crosses the x-axis, we let $y = 0$. (The y-coordinate of any point on the x-axis is 0.)

x-intercept:

$$\begin{aligned}
\text{When} \rightarrow \qquad & y = 0 \\
\text{the equation} \rightarrow \qquad & 3x - 2y = 6 \\
\text{becomes} \rightarrow \qquad & 3x - 2(0) = 6 \\
& 3x - 0 = 6 \\
& x = 2 \qquad \text{Multiply each side by } \tfrac{1}{3}.
\end{aligned}$$

The graph crosses the x-axis at $(2, 0)$, which means the x-intercept is 2.

y-intercept:

$$\begin{aligned}
\text{When} \rightarrow \qquad & x = 0 \\
\text{the equation} \rightarrow \qquad & 3x - 2y = 6 \\
\text{becomes} \rightarrow \qquad & 3(0) - 2y = 6 \\
& 0 - 2y = 6 \\
& -2y = 6 \\
& y = -3 \qquad \text{Multiply each side by } -\tfrac{1}{2}.
\end{aligned}$$

The graph crosses the y-axis at $(0, -3)$, which means the y-intercept is -3.

Plotting the x- and y-intercepts and then drawing a line through them, we have the graph of $3x - 2y = 6$, as shown in Figure 2.

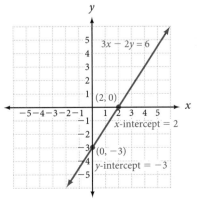

FIGURE 2

Practice Problems

1. Find the x- and y-intercepts for $2x - 5y = 10$, and use them to draw the graph.

2. Graph $3x - y = 6$ by first finding the intercepts.

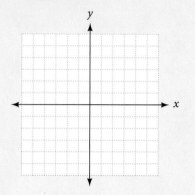

EXAMPLE 2 Graph $-x + 2y = 4$ by finding the intercepts and using them to draw the graph.

Solution Again, we find the x-intercept by letting $y = 0$ in the equation and solving for x. Similarly, we find the y-intercept by letting $x = 0$ and solving for y.

x-intercept:

$$\text{When} \rightarrow \qquad\qquad y = 0$$
$$\text{the equation} \rightarrow \quad -x + 2y = 4$$
$$\text{becomes} \rightarrow \quad -x + 2(0) = 4$$
$$-x + 0 = 4$$
$$-x = 4$$
$$x = -4 \qquad \textit{Multiply each side by } -1.$$

The x-intercept is -4, indicating that the point $(-4, 0)$, is on the graph of $-x + 2y = 4$.

y-intercept:

$$\text{When} \rightarrow \qquad\qquad x = 0$$
$$\text{the equation} \rightarrow \quad -x + 2y = 4$$
$$\text{becomes} \rightarrow \quad -0 + 2y = 4$$
$$2y = 4$$
$$y = 2 \qquad \textit{Multiply each side by } \tfrac{1}{2}.$$

The y-intercept is 2, indicating that the point $(0, 2)$ is on the graph of $-x + 2y = 4$.

Plotting the intercepts and drawing a line through them, we have the graph of $-x + 2y = 4$, as shown in Figure 3.

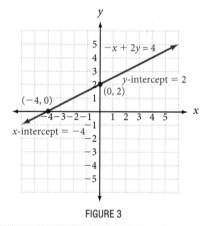

FIGURE 3

Graphing a line by finding the intercepts, as we have done in Examples 1 and 2, is an easy method of graphing if the equation has the form $ax + by = c$ and both the numbers a and b divide the number c evenly.

In our next example, we use the intercepts to graph a line in which y is given in terms of x.

EXAMPLE 3 Use the intercepts for $y = -\frac{1}{3}x + 2$ to draw its graph.

Solution We graphed this line previously in Example 7 of Section 3.2 by substituting three different values of x into the equation and solving for y. This time we will graph the line by finding the intercepts.

x-intercept:

$$\text{When} \rightarrow \qquad y = 0$$

$$\text{the equation} \rightarrow \qquad y = -\frac{1}{3}x + 2$$

$$\text{becomes} \rightarrow \qquad 0 = -\frac{1}{3}x + 2$$

$$-2 = -\frac{1}{3}x \qquad \text{Add } -2 \text{ to each side.}$$

$$6 = x \qquad \text{Multiply each side by } -3.$$

The x-intercept is 6, which means the graph passes through the point $(6, 0)$.

y-intercept:

$$\text{When} \rightarrow \qquad x = 0$$

$$\text{the equation} \rightarrow \qquad y = -\frac{1}{3}x + 2$$

$$\text{becomes} \rightarrow \qquad y = -\frac{1}{3}(0) + 2$$

$$y = 2$$

The y-intercept is 2, which means the graph passes through the point $(0, 2)$. The graph of $y = -\frac{1}{3}x + 2$ is shown in Figure 4. Compare this graph, and the method used to obtain it, with Example 7 in Section 3.2.

FIGURE 4

3. Use the intercepts to graph
$y = \frac{3}{2}x - 3$.

GETTING READY FOR CLASS

After reading through the preceding section, respond in your own words and in complete sentences.

A. What is the x-intercept for a graph?

B. What is the y-intercept for a graph?

C. How do you find the y-intercept for a line from the equation $3x - y = 6$?

D. How do you graph a line for $y = -\frac{1}{2}x + 3$ using its intercepts?

Vocabulary Review

Choose the correct words to fill in the blanks below.

x-axis y-axis x-intercept y-intercept

1. The _____ of a straight line is the x-coordinate of the point where the graph crosses the x-axis.
2. The _____ of a straight line is the y-coordinate of the point where the graph crosses the y-axis.
3. Any point on the _____ has a y-coordinate of 0.
4. Any point on the _____ has a x-coordinate of 0.

Problems

A Complete each table.

1.

Equation	x-intercept	y-intercept
$3x + 4y = 12$		
$3x + 4y = 4$		
$3x + 4y = 3$		
$3x + 4y = 2$		

2.

Equation	x-intercept	y-intercept
$-2x + 3y = 6$		
$-2x + 3y = 3$		
$-2x + 3y = 2$		
$-2x + 3y = 1$		

3.

Equation	x-intercept	y-intercept
$x - 3y = 2$		
$y = \frac{1}{3}x - \frac{2}{3}$		
$x - 3y = 0$		
$y = \frac{1}{3}x$		

4.

Equation	x-intercept	y-intercept
$x - 2y = 1$		
$y = \frac{1}{2}x - \frac{1}{2}$		
$x - 2y = 0$		
$y = \frac{1}{2}x$		

B Find the x- and y-intercepts for the following equations. Then use the intercepts to graph each equation.

5. $2x + y = 4$

6. $2x + y = 2$

7. $-x + y = 3$

8. $-x + y = 4$

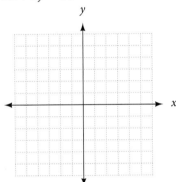

9. $-x + 2y = 2$

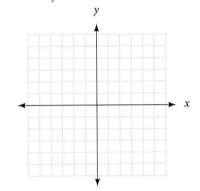

10. $-x + 2y = 4$

11. $5x + 2y = 10$

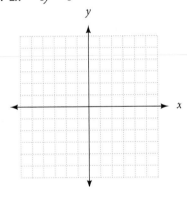

12. $2x + 5y = 10$

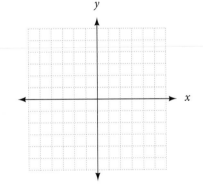

13. $4x - 2y = 8$

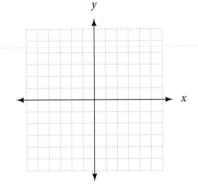

14. $2x - 4y = 8$

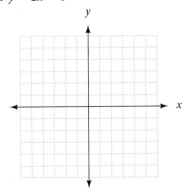

15. $-4x + 5y = 20$

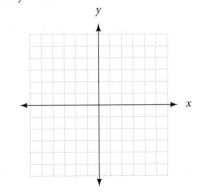

16. $-5x + 4y = 20$

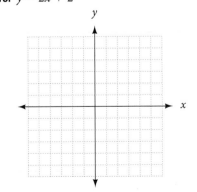

17. $y = 2x - 6$

18. $y = 2x + 6$

19. $y = 2x + 2$

20. $y = -2x + 2$

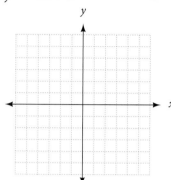

21. $y = 2x - 1$

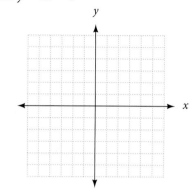

22. $y = -2x - 1$

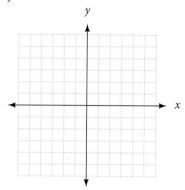

23. $y = \frac{1}{2}x + 3$

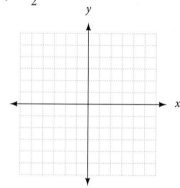

24. $y = \frac{1}{2}x - 3$

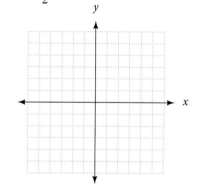

25. $y = -\frac{1}{3}x - 2$

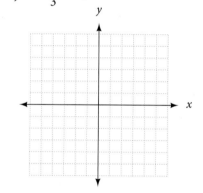

26. $y = -\frac{1}{3}x + 2$

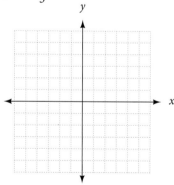

27. $y = -\frac{2}{3}x + 4$

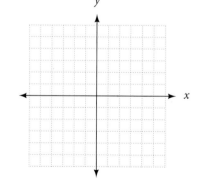

28. $y = -\frac{1}{2}x + 2$

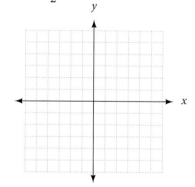

For each of the following lines the x-intercept and the y-intercept are both 0, which means the graph of each will go through the origin, $(0, 0)$. Graph each line by finding a point on each, other than the origin, and then drawing a line through that point and the origin.

29. $y = -2x$

30. $y = \frac{1}{2}x$

31. $y = -\frac{1}{3}x$

32. $y = -3x$

33. $y = \frac{2}{3}x$

34. $y = \frac{3}{2}x$

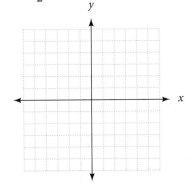

The next two problems are intended to give you practice reading, and paying attention to, the instructions that accompany the problems you are working. Working these problems is an excellent way to get ready for a test or a quiz.

35. Work each problem according to the instructions.

 a. Solve: $2x - 3 = -3$.

 b. Find the x-intercept: $2x - 3y = -3$.

 c. Find y when x is 0: $2x - 3y = -3$.

 d. Graph: $2x - 3y = -3$.

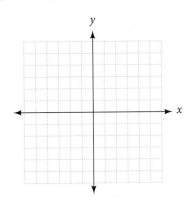

 e. Solve for y: $2x - 3y = -3$.

36. Work each problem according to the instructions.

 a. Solve: $3x - 4 = -4$.

 b. Find the y-intercept: $3x - 4y = -4$.

 c. Find x when y is 0: $3x - 4y = -4$.

 d. Graph: $3x - 4y = -4$.

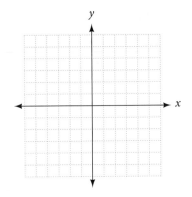

 e. Solve for y: $3x - 4y = -4$.

From the graphs below, find the *x*- and *y*-intercepts for each line.

37.

38.

39.

40.

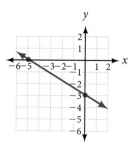

41. Graph the line that passes through the point $(-4, 4)$ and has an *x*-intercept of -2. What is the *y*-intercept of this line?

42. Graph the line that passes through the point $(-3, 4)$ and has a *y*-intercept of 1. What is the *x*-intercept of this line?

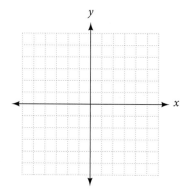

43. A line passes through the point $(1, 4)$ and has a *y*-intercept of 3. Graph the line and name its *x*-intercept.

44. A line passes through the point $(3, 4)$ and has an *x*-intercept of 1. Graph the line and name its *y*-intercept.

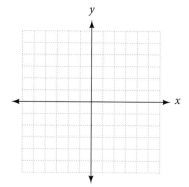

45. Graph the line that passes through the points $(-2, 5)$ and $(5, -2)$. What are the *x*- and *y*-intercepts for this line?

46. Graph the line that passes through the points $(5, 3)$ and $(-3, -5)$. What are the *x*- and *y*-intercepts for this line?

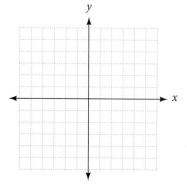

47. The vertical line $x = 3$ has only one intercept. Graph $x = 3$, and name its intercept. [Remember, ordered pairs (x, y) that are solutions to the equation $x = 3$ are ordered pairs with an x-coordinate of 3 and any y-coordinate.]

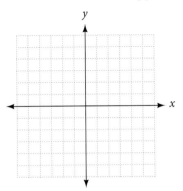

48. Graph the vertical line $x = -2$. Then name its intercept.

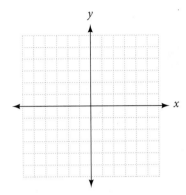

49. The horizontal line $y = 4$ has only one intercept. Graph $y = 4$, and name its intercept. [Ordered pairs (x, y) that are solutions to the equation $y = 4$ are ordered pairs with a y-coordinate of 4 and any x-coordinate.]

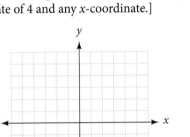

50. Graph the horizontal line $y = -3$. Then name its intercept.

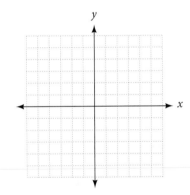

Applying the Concepts

51. Seating Capacity A theater has a capacity of 500 people. Because a different price is charged for students than for general admission, the owners would like to know what combinations of students and general tickets can be sold when the show is sold out. The relationship between student and general tickets can be written as

$$x + y = 500$$

Graph this equation where the x-axis is student tickets and the y-axis is general tickets. Find the intercepts first, and limit your graph to the first quadrant.

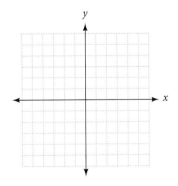

52. Ladder Angles A 13-foot ladder is placed against a barn. As the top of the ladder moves up and down the wall, the angles made by the ladder and the wall, α, and the ladder and the ground, β, can be written as

$$\alpha + \beta = 90.$$

Graph this equation on a coordinate system where α is the y-axis and β is the x-axis. Find the intercepts first, and limit your graph to the first quadrant only.

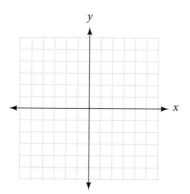

53. Working Two Jobs Maggie has a job working in an office for $10 an hour and another job driving a tractor pays $12 an hour. Maggie works at both jobs and earns $480 in one week.

a. Write an equation that gives the number of hours x, she worked for $10 per hour and the number of hours, y, she worked for $12 per hour.

b. Find the x- and y-intercepts for this equation.

c. Graph this equation from the intercepts, using only the first quadrant.

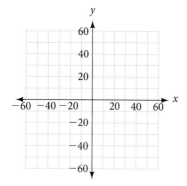

d. From the graph, find how many hours she worked at $12, if she worked 36 hours at $10 per hour?

e. From the graph, find how many hours she worked at $10, if she worked 25 hours at $12 per hour?

54. Ticket prices Devin is selling tickets to a BBQ. Adult tickets cost $6.00 and children's tickets cost $4.00. The total amount of money he collects is $240.

a. Write an equation that gives the number of adult tickets x, he sold for $6 and the children's tickets, y, he sold for $4

b. Find the x- and y-intercepts for this equation.

c. Graph this equation from the intercepts, using only the first quadrant.

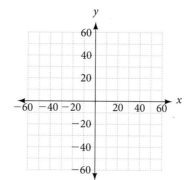

d. From the graph, find how many children's tickets were sold if 30 adult tickets sold?

e. From the graph, find how many adult tickets were sold if 30 children's tickets sold?

55. Brian drives a certain amount of time at 50 mph and another amount at 60 mph. Together he drives 450 miles.

 a. Write an equation involving x (time driving at 50 mph) and y (time driving at 60 mph).

 b. Find the x- and y- intercepts for the equation in part a.

 c. Graph this equation in the first quadrant only.

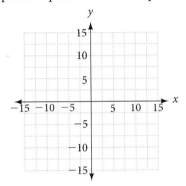

 d. From the graph, find how many hours Brian drove at 60 mph if he drove a total of 8 hours?

 e. If Brian drove at 50 mph for 4 hours, how long did he drive at 60 mph?

56. Juan tutors in the math lab for $8 per hour and privately for $12 per hour. He earns $240 in one week.

 a. Write and equation involving x (number of math lab hours) and y (number of private hours).

 b. Find the x- and y- intercepts for this equation.

 c. Graph this equation in the first quadrant only.

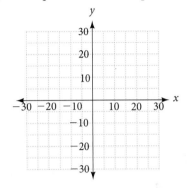

 d. From the graph, find how many hours did Juan tutor in the math lab if he tutored privately for 10 hours?

 e. If Juan is limited to tutoring 19 hours in the math lab, how many hours must he tutor privately?

Getting Ready for the Next Section

57. Evaluate

 a. $\dfrac{5-2}{3-1}$

 b. $\dfrac{2-5}{1-3}$

58. Evaluate

 a. $\dfrac{-4-1}{5-(-2)}$

 b. $\dfrac{1+4}{-2-5}$

59. Evaluate the following expressions when $x = 3$, and $y = 5$.

 a. $\dfrac{y-2}{x-1}$

 b. $\dfrac{2-y}{1-x}$

60. Evaluate the following expressions when $x = 4$, and $y = -1$.

 a. $\dfrac{-4-y}{5-x}$

 b. $\dfrac{y+4}{x-5}$

Find the Mistake

Each sentence below contains a mistake. Circle the mistake and write the correct word or number on the line provided.

1. The y–intercept is where the graph of a line crosses the x–axis at the point $(0, y)$. _____

2. The x–intercept for the linear equation $9x = \frac{1}{3}y + 5$ is $(-15, 0)$. _____

3. The y–intercept for the linear equation $2x + 7y = 3$ is $\left(0, \frac{3}{2}\right)$. _____

4. The intercepts for the linear equation $15 = 3y - 5x$ are $(5, 0)$ and $(0, -3)$. _____

OBJECTIVES

A Find the slope of a line from two points on the line.

B Graph a line given the slope and *y*-intercept.

KEY WORDS

slope

rise

run

**Video Examples
Section 9.4**

Heli-skiing is a sport where a helicopter transports a skier to the top of the mountain he will ski down. The skier leaps out of the helicopter's cabin and onto an untracked slope, often inaccessible from the ground.

Let's suppose for every 30 feet of horizontal distance the heli-skier covers, his elevation drops 10 feet. In this section, we will learn how to use these numbers to assign a numerical value to the slope of the hill.

In defining the slope of a straight line, we are looking for a number to associate with a straight line that does two things. First of all, we want the slope of a line to measure the "steepness" of the line; that is, in comparing two lines, the slope of the steeper line should have the larger numerical value. Second, we want a line that *rises* going from left to right to have a *positive* slope. We want a line that *falls* going from left to right to have a *negative* slope. (A horizontal line that neither rises nor falls going from left to right must, therefore, have 0 slope; whereas, a vertical line has an undefined slope.) These are illustrated in Figure 1.

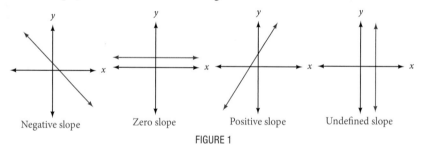

FIGURE 1

Suppose we know the coordinates of two points on a line. Because we are trying to develop a general formula for the slope of a line, we will use general points—let's call the two points $P_1(x_1, y_1)$ and $P_2(x_2, y_2)$. They represent the coordinates of any two different points on our line. We define the *slope* of our line to be the ratio of the vertical change of *y*-coordinates to the horizontal change of *x*-coordinates as we move from point (x_1, y_1) to point (x_2, y_2) on the line. (See Figure 2 on next page.)

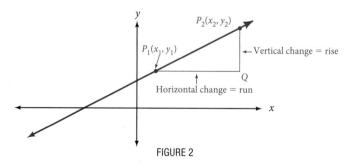

FIGURE 2

We call the vertical change the *rise* in the graph and the horizontal change the *run* in the graph. The slope, then, is

$$\text{Slope} = \frac{\text{vertical change}}{\text{horizontal change}} = \frac{\text{rise}}{\text{run}}$$

We would like to have a numerical value to associate with the rise in the graph and a numerical value to associate with the run in the graph. A quick study of Figure 2 shows that the coordinates of point Q must be (x_2, y_1), because Q is directly below point P_2 and right across from point P_1. We can draw our diagram again in the manner shown in Figure 3. It is apparent from this graph that the rise can be expressed as $(y_2 - y_1)$ and the run as $(x_2 - x_1)$. We usually denote the slope of a line by the letter m. The complete definition of slope follows along with a diagram (Figure 3) that illustrates the definition.

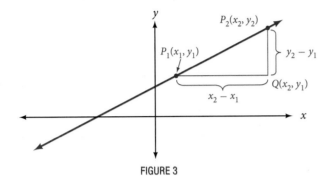

FIGURE 3

DEFINITION slope

If points (x_1, y_1) and (x_2, y_2) are any two different points and $x_1 \neq x_2$, then the *slope* of the line on which they lie is, denoted by m,

$$\text{Slope} = m = \frac{\text{rise}}{\text{run}} = \frac{y_2 - y_1}{x_2 - x_1}$$

This definition of the *slope* of a line does just what we want it to do. If the line rises going from left to right, the slope will be positive. If the line falls from left to right, the slope will be negative. Also, the steeper the line, the larger numerical value the slope will have.

A Finding the Slope Given Two Points on a Line

EXAMPLE 1 Find the slope of the line between the points $(1, 2)$ and $(3, 5)$.

Solution We can let

$$(x_1, y_1) = (1, 2)$$

and

$$(x_2, y_2) = (3, 5)$$

then

$$m = \frac{y_2 - y_1}{x_2 - x_1} = \frac{5 - 2}{3 - 1} = \frac{3}{2}$$

The slope is $\frac{3}{2}$. For every vertical change of 3 units, there will be a corresponding horizontal change of 2 units. (See Figure 4.)

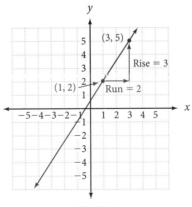

FIGURE 4

Practice Problems

1. Find the slope of the line between the points $(4, -3)$ and $(-2, 4)$.

Answers

1. $-\frac{7}{6}$

2. Find the slope of the line through (3, 4) and (1, −2).

EXAMPLE 2 Find the slope of the line through (−2, 1) and (5, −4).

Solution It makes no difference which ordered pair we call (x_1, y_1) and which we call (x_2, y_2).

$$\text{Slope} = m = \frac{y_2 - y_1}{x_2 - x_1} = \frac{-4 - 1}{5 - (-2)} = -\frac{5}{7}$$

The slope is $-\frac{5}{7}$. Every vertical change of −5 units (down 5 units) is accompanied by a horizontal change of 7 units (to the right 7 units). (See Figure 5.)

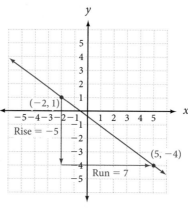

FIGURE 5

B Graphing a Line from Its Slope

3. Graph the line with slope $\frac{1}{4}$ and y-intercept −3.

EXAMPLE 3 Graph the line with slope $\frac{3}{2}$ and y-intercept 1.

Solution Because the y-intercept is 1, we know that one point on the line is (0, 1). So, we begin by plotting the point (0, 1), as shown in Figure 6.

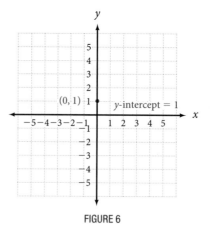

FIGURE 6

There are many lines that pass through the point shown in Figure 6, but only one of those lines has a slope of $\frac{3}{2}$. The slope, $\frac{3}{2}$, can be thought of as the rise in the graph divided by the run in the graph. Therefore, if we start at the point (0, 1) and move 3 units up (that's a rise of 3) and then 2 units to the right (a run of 2), we will be at another point on the graph. Figure 7 shows that the point we reach by doing so is the point (2, 4).

Answers

2. 3

$$\text{Slope} = m = \frac{\text{rise}}{\text{run}} = \frac{3}{2}$$

FIGURE 7

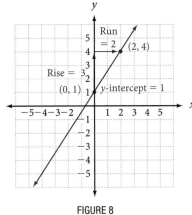

FIGURE 8

To graph the line with slope $\frac{3}{2}$ and y-intercept 1, we simply draw a line through the two points in Figure 7 to obtain the graph shown in Figure 8.

●

EXAMPLE 4 Find the slope of the line containing $(3, -1)$ and $(3, 4)$.

Solution Using the definition for slope, we have

$$m = \frac{y_2 - y_1}{x_2 - x_1} = \frac{4 - (-1)}{3 - 3} = \frac{5}{0}$$

The expression $\frac{5}{0}$ is undefined; that is, there is no real number to associate with it. In this case, we say the line *has an undefined slope.*

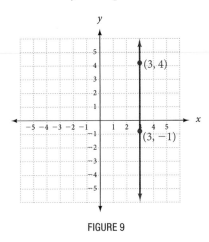

FIGURE 9

The graph of our line is shown in Figure 9. All vertical lines have an undefined slope. (And all horizontal lines, as we mentioned earlier, have a slope of zero.)

●

As a final note, the following summary reminds us that all horizontal lines have equations of the form $y = b$ and slopes of 0. Because they cross the y-axis at b, the y-intercept is b; there is no x-intercept. Vertical lines have an undefined slope and equations of the form $x = a$. Each will have an x-intercept at a and no y-intercept. Finally, equations of the form $y = mx$ have graphs that pass through the origin. The slope is always m and both the x-intercept and the y-intercept are 0.

4. Find the slope of the line through $(2, -3)$ and $(-1, -3)$.

Note Exceptions to this discussion are the lines $y = 0$ and $x = 0$. For $y = 0$, the x-intercepts are infinite; whereas $x = 0$ has infinite y-intercepts.

Answer
4. 0

FACTS FROM GEOMETRY Special Equations and Their Graphs, Slopes, and Intercepts

For the equations below, m, a, and b are real numbers.

Through the Origin	Vertical Line	Horizontal Line
FIGURE 10A	FIGURE 10B	FIGURE 10C

Equation: $y = mx$ Equation: $x = a$ Equation: $y = b$
Slope $= m$ Undefined Slope $= 0$
x-intercept $= 0$ x-intercept $= a$ No x-intercept
y-intercept $= 0$ No y-intercept y-intercept $= b$

GETTING READY FOR CLASS

After reading through the preceding section, respond in your own words and in complete sentences.

A. Using x- and y-coordinates, how do you find the slope of a line?

B. Would you rather climb a hill with a slope of 1 or a slope of 3? Explain why.

C. Describe how you would graph a line from its slope and y-intercept.

D. Describe how to obtain the slope of a line if you know the coordinates of two points on the line.

Vocabulary Review

Choose the correct words to fill in the blanks below.

vertical undefined ratio zero horizontal slope

1. The slope of a line is the _____ of the vertical change of y-coordinates to the horizontal change of x-coordinates as we move from point (x_1, y_1) to point (x_2, y_2) on the line.

2. The _____ change between one point on a straight line to another point is called the rise.

3. The _____ change between one point on a straight line to another point is called the run.

4. The _____ of a line is given by the equation $m = \dfrac{y_2 - y_1}{x_2 - x_1}$.

5. A horizontal line has a(n) _____ slope.

6. A vertical line has a(n) _____ slope.

Problems

A Find the slope of the line through the following pairs of points. Then plot each pair of points, draw a line through them, and indicate the rise and run in the graph in the same manner shown in Examples 1 and 2.

1. $(2, 1), (4, 4)$

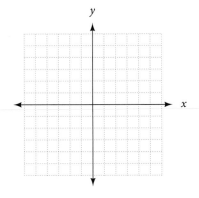

2. $(3, 1), (5, 4)$

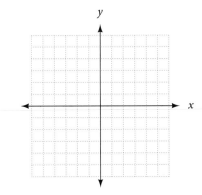

3. $(1, 4), (5, 2)$

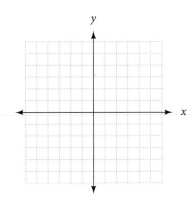

4. $(1, 3), (5, 2)$

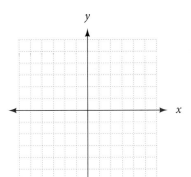

5. $(1, -3), (4, 2)$

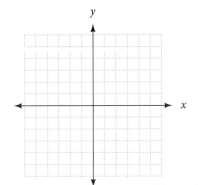

6. $(2, -3), (5, 2)$

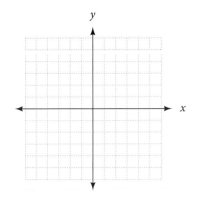

7. $(-3, -2), (1, 3)$

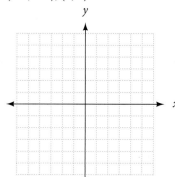

8. $(-3, -1), (1, 4)$

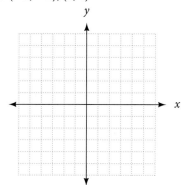

9. $(-3, 2), (3, -2)$

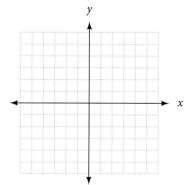

10. $(-3, 3), (3, -1)$

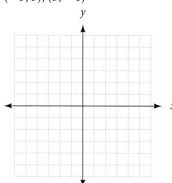

11. $(2, -5), (3, -2)$

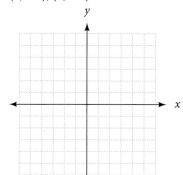

12. $(2, -4), (3, -1)$

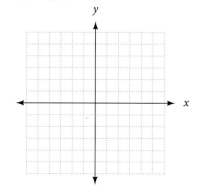

13. $(2, 4), (2, -3)$

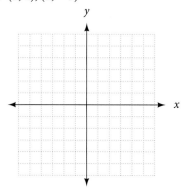

14. $(2, 3), (-1, 3)$

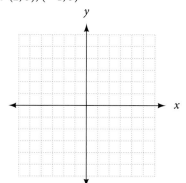

B In each of the following problems, graph the line with the given slope and y-intercept b.

15. $m = \dfrac{2}{3}, b = 1$

16. $m = \dfrac{3}{4}, b = -2$

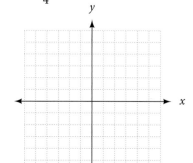

17. $m = \dfrac{3}{2}, b = -3$

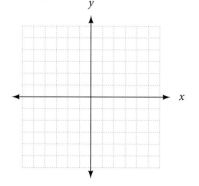

18. $m = \dfrac{4}{3}, b = 2$

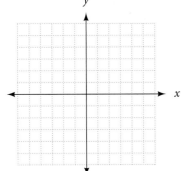

19. $m = -\dfrac{4}{3}, b = 5$

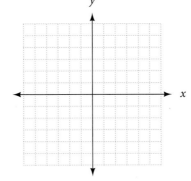

20. $m = -\dfrac{3}{5}, b = 4$

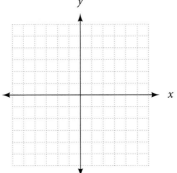

21. $m = 2, b = 1$

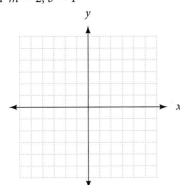

22. $m = -2, b = 4$

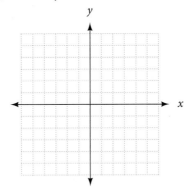

23. $m = 3, b = -1$

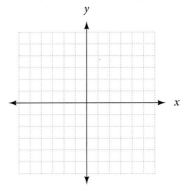

24. $m = 3, b = -2$

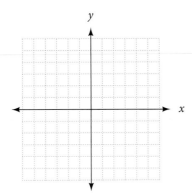

25. $m = 1, b = 0$

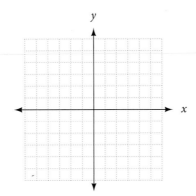

26. $m = -1, b = 0$

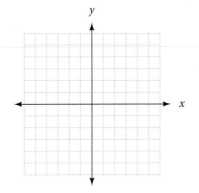

Find the slope and y-intercept for each line.

27.

28.

29.

30.

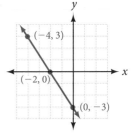

31. Graph the line that has an *x*-intercept of 3 and a *y*-intercept of −2. What is the slope of this line?

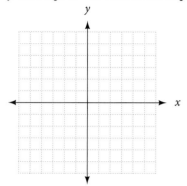

32. Graph the line that has an *x*-intercept of 2 and a *y*-intercept of −3. What is the slope of this line?

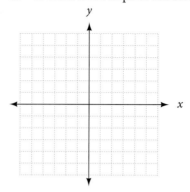

33. Graph the line with *x*-intercept 4 and *y*-intercept 2. What is the slope of this line?

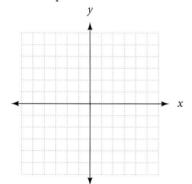

34. Graph the line with *x*-intercept −4 and *y*-intercept −2. What is the slope of this line?

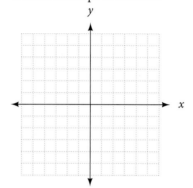

35. Graph the line $y = 2x - 3$, then name the slope and *y*-intercept by looking at the graph.

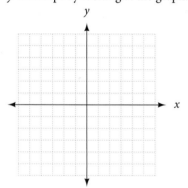

36. Graph the line $y = -2x + 3$, then name the slope and *y*-intercept by looking at the graph.

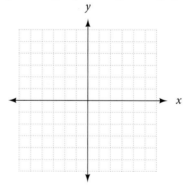

37. Graph the line $y = \frac{1}{2}x + 1$, then name the slope and *y*-intercept by looking at the graph.

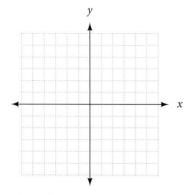

38. Graph the line $y = -\frac{1}{2}x - 2$, then name the slope and *y*-intercept by looking at the graph.

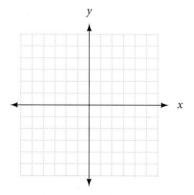

39. Find y if the line through $(4, 2)$ and $(6, y)$ has a slope of 2.

40. Find y if the line through $(1, y)$ and $(7, 3)$ has a slope of 6.

41. Find y if the line through $(3, 2)$ and $(3, y)$ has an undefined slope.

42. Find y if the line through $(2, 3)$ and $(5, y)$ has a slope of 0.

For each equation in each table, give the slope of the graph.

43.

Equation	Slope
$x = 3$	
$y = 3$	
$y = 3x$	

44.

Equation	Slope
$y = \frac{3}{2}$	
$x = \frac{3}{2}$	
$y = \frac{3}{2}x$	

45.

Equation	Slope
$y = -\frac{2}{3}$	
$x = -\frac{2}{3}$	
$y = -\frac{2}{3}x$	

46.

Equation	Slope
$x = -2$	
$y = -2$	
$y = -2x$	

Applying the Concepts

47. Garbage Production The table and completed line graph gives the annual production of garbage in the United States for some specific years. Find the slope of each of the four line segments, A, B, C, and D.

48. Grass Height The table and completed line graph gives the growth of a certain grass species over time. Find the slopes of the line segments labeled A, B, and C.

Year	Garbage (millions of tons)
1960	88
1970	121
1980	152
1990	205
2000	224
2007	254

Day	Plant Height
0	0
2	1
4	3
6	6
8	13
10	23

49. Non-Camera Phone Sales The table and line graph here each show the estimated non-camera phone sales each year through 2010. Find the slope of each of the three line segments, *A*, *B*, and *C*.

Year	Sales (in millions)
2006	300
2007	250
2008	175
2009	150
2010	125

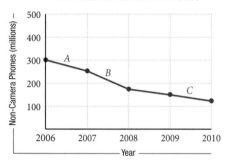

Estimated Non-Camera Phone Sales

50. Camera Phone Sales The table and line graph here each show the estimated sales of camera phones from 2006 to 2010. Find the slopes of line segments *A*, *B*, and *C*.

Year	Sales (in millions)
2006	500
2007	650
2008	750
2009	875
2010	900

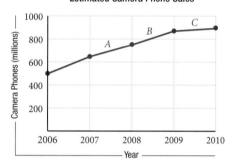

Estimated Camera Phone Sales

51. The federal budget deficit has become a very common political topic. The following data lists the total national outstanding debt over some selected years. Use this data to answer the questions that follow.

Year	Debt (trillions of $)
1980	0.9
1990	3.2
2000	5.7
2010	13.6

a. Using the table, sketch a line graph of *A*, *B*, and *C*. Add a graph.

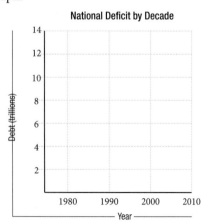

National Deficit by Decade

b. Calculate the slopes of the line segments.

c. Based on these slopes, is the rate of change of the debt increasing or decreasing?

Nike Women's Marathon The following chart shows the number of finishers at the Nike Women's Marathon in San Francisco, CA from 2005 to 2010.

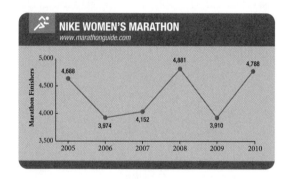

52. Find the slope of the line from 2005 to 2010. Explain in words what the slope represents.

53. Find the slope of the line from 2009 to 2010. Explain in words what the slope represents.

54. Heating a Block of Ice A block of ice with an initial temperature of −20°C is heated at a steady rate. The graph shows how the temperature changes as the ice melts to become water and the water boils to become steam and water.

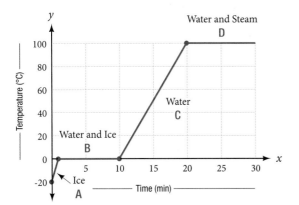

a. How long does it take all the ice to melt?

b. From the time the heat is applied to the block of ice, how long is it before the water boils?

c. Find the slope of the line segment labeled A. What units would you attach to this number?

d. Find the slope of the line segment labeled C. Be sure to attach units to your answer.

e. Is the temperature changing faster during the 1st minute or the 16th minute?

55. Slope of a Highway A sign at the top of the Cuesta Grade, outside of San Luis Obispo, reads "7% downgrade next 3 miles." The following diagram is a model of the Cuesta Grade that illustrates the information on that sign.

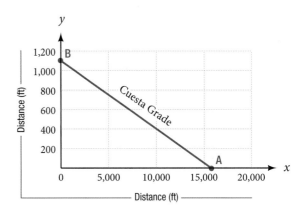

a. At point B, the graph crosses the y-axis at 1,106 feet. How far is it from the origin to point A?

b. What is the slope of the Cuesta Grade?

56. Facebook Users The graph below shows the number of Facebook users over a six year period. Using the graph, find the slope of the line connecting the first (2004, 1) and last (2010, 400) points on the line. Explain in words what the slope represents.

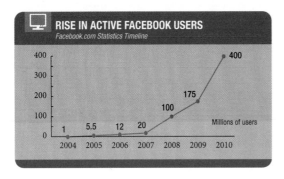

57. Mac Users The graph shows the average increase in the percentage of computer users who use Macs since 2003. Find the slope of the line connecting the first (2003, 2.2) and last (2010, 6.7) points on the line. Explain in words what the slope represents.

Getting Ready for the Next Section

Solve each equation for y.

58. $-2x + y = 4$

59. $-4x + y = -2$

60. $2x + y = 3$

61. $3x + 2y = 6$

62. $x + 2y = 4$

63. $x - 3y = 6$

64. $2x - 3y = 12$

65. $3x + 4y = 12$

66. $4x - 5y = 20$

67. $-2x - 5y = 10$

68. $-y - 3 = -2(x + 4)$

69. $-y + 5 = 2(x + 2)$

70. $-y - 3 = -\dfrac{2}{3}(x + 3)$

71. $-y - 1 = -\dfrac{1}{2}(x + 4)$

72. $-\dfrac{y - 1}{x} = \dfrac{3}{2}$

73. $-\dfrac{y + 1}{x} = \dfrac{3}{2}$

Find the Mistake

Each sentence below contains a mistake. Circle the mistake and write the correct word or number on the line provided.

1. A horizontal line has a positive slope. _____

2. A line that rises from the left to the right has a negative slope. _____

3. The slope of the line that passes through the points $(-1, 6)$ and $(4, -9)$ is $-\dfrac{1}{3}$. _____

4. Use the point $(3, -2)$ to graph a line with a y–intercept -6 and a slope $\dfrac{5}{2}$. _____

Landmark Review: Checking Your Progress

1. What is the name of the point with the coordinates $(0, 0)$?

2. Name the x-coordinate of any point that is on the line passing through the points $(4, 3)$ and $(4, 8)$.

3. Is the point $(5, 3)$ on the line that passes through $(2, 4)$ and $(4, 4)$?

For each equation, complete the ordered pairs.

4. $4x - y = 3$ $(0,\ \), (\ \ , 0), (3,\ \)$

5. $2y - 3x = 15$ $(\ \ , 3), (1,\ \), (6,\ \)$

Find the x- and y-intercepts for the following equations.

6. $x + y = 1$

7. $x - 2y = 6$

Find the slope of the line through the following pairs of points.

8. $(1, 3), (5, 4)$

9. $(4, 2), (3, 4)$

10. $(3, 4), (3, 2)$

11. Would you expect a line with a positive slope to rise or fall moving from left to right?

12. If a line has a slope of 0 and doesn't pass through the origin, what do you know about the x-intercept?

9.5 Finding the Equation of a Line

OBJECTIVES

A Find the equation of a line given the slope and *y*-intercept of the line.

B Find the slope and *y*-intercept of a line given the equation of the line.

C Find the equation of a line given a point on the line and the slope of the line.

D Find the equation of a line given two points on the line.

E Find the equations of parallel and perpendicular lines.

KEY WORDS

slope-intercept form

point-slope form

parallel

perpendicular

Every winter, the Angel Fire Resort in New Mexico hosts the World Shovel Racing Championships. At the top of a snow-covered slope, contestants sit straddling the handle of a stock grain scoop shovel. Then they speed down the 1,000-foot course using the handle to steer. Helmets are required because wipeouts are common. Let's say the fastest racer finished the course in 17 seconds and the runner up finished in 18 seconds. That gives us the two points (17, 1000) and (18, 1000) that could help us draw the graph for the race. In this section, we will learn ways to find the equation of a line using slope, intercepts, and given points.

To this point in the chapter, most of the problems we have worked have used the equation of a line to find different types of information about the line. For instance, given the equation of a line, we can find points on the line, the graph of the line, the intercepts, and the slope of the line. In this section, we reverse things somewhat and use information about a line, such as its slope and *y*-intercept, to find the line's equation.

There are three main types of problems to solve in this section.

1. Find the equation of a line from the slope and the *y*-intercept.

2. Find the equation of a line given one point on the line and the slope of the line.

3. Find the equation of a line given two points on the line.

Examples 1 and 2 illustrate the first type of problem. Example 5 solves the second type of problem. We solve the third type of problem in Example 6.

A The Slope-Intercept Form of an Equation of a Straight Line

EXAMPLE 1 Find the equation of the line with slope $\frac{3}{2}$ and *y*-intercept 1.

Solution We graphed the line with slope $\frac{3}{2}$ and *y*-intercept 1 in Example 3 of the previous section. Figure 1 shows that graph.

FIGURE 1

**Video Examples
Section 9.5**

Practice Problems

1. Find the equation of the line with slope $\frac{4}{3}$ and *y*-intercept 2.

Answer

1. $y = \frac{4}{3}x + 2$

9.5 Finding the Equation of a Line

657

What we want to do now is find the equation of the line shown in Figure 1. To do so, we take any other point (x, y) on the line and apply our slope formula to that point and the point $(0, 1)$. We set that result equal to $\frac{3}{2}$, because $\frac{3}{2}$ is the given slope of our line.

$$m = \frac{\text{rise}}{\text{run}} = \frac{y - 1}{x - 0} = \frac{3}{2} \qquad \text{Slope} = \frac{\text{vertical change}}{\text{horizontal change}}.$$

$$\frac{y - 1}{x} = \frac{3}{2} \qquad x - 0 = x.$$

$$y - 1 = \frac{3}{2}x \qquad \text{Multiply each side by } x.$$

$$y = \frac{3}{2}x + 1 \qquad \text{Add 1 to each side.}$$

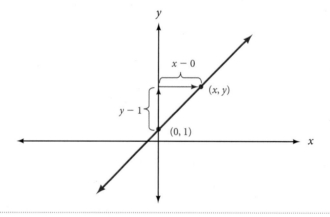

What is interesting and useful about the equation we have just found is that the number in front of x is the slope of the line and the constant term is the y-intercept. It is no coincidence that it turned out this way. Whenever an equation has the form $y = mx + b$, the graph is always a straight line with slope m and y-intercept b. To see that this is true in general, suppose we want the equation of a line with slope m and y-intercept b. Because the y-intercept is b, then the point $(0, b)$ is on the line. If (x, y) is any other point on the line, then we apply our slope formula to get

$$\frac{y - b}{x - 0} = m \qquad \text{Slope} = \frac{\text{vertical change}}{\text{horizontal change}}.$$

$$\frac{y - b}{x} = m \qquad x - 0 = x.$$

$$y - b = mx \qquad \text{Multiply each side by } x.$$

$$y = mx + b \qquad \text{Add } b \text{ to each side.}$$

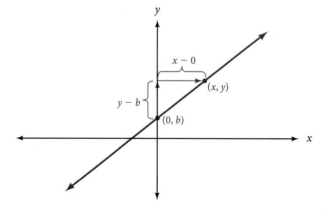

Here is a summary of what we have just found:

> **Slope-Intercept Form of the Equation of a Line**
>
> The equation of the non-vertical line with slope m and y-intercept b is always given by
> $$y = mx + b$$

EXAMPLE 2 Find the equation of the line with slope $-\frac{4}{3}$ and y-intercept 5. Then, graph the line.

Solution Substituting $m = -\frac{4}{3}$ and $b = 5$ into the equation $y = mx + b$, we have

$$y = -\frac{4}{3}x + 5$$

Finding the equation from the slope and y-intercept is just that easy. If the slope is m and the y-intercept is b, then the equation is always $y = mx + b$.

Because the y-intercept is 5, the graph goes through the point $(0, 5)$. To find a second point on the graph, we start at $(0, 5)$ and move 4 units down (that's a rise of -4) and 3 units to the right (a run of 3). The point we reach is $(3, 1)$. Drawing a line that passes through $(0, 5)$ and $(3, 1)$, we have the graph of our equation. (Note that we could also let the rise = 4 and the run = -3 and obtain the same graph.) The graph is shown in Figure 2.

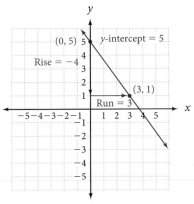

FIGURE 2

B Finding the Slope and *y*-Intercept

EXAMPLE 3 Find the slope and y-intercept for $-2x + y = -4$. Then, use them to draw the graph.

Solution To identify the slope and y-intercept from the equation, the equation must be in the form $y = mx + b$ (slope-intercept form). To write our equation in this form, we must solve the equation for y. To do so, we simply add $2x$ to each side of the equation.

$$-2x + y = -4 \qquad \textit{Original equation}$$
$$y = 2x - 4 \qquad \textit{Add 2x to each side.}$$

The equation is now in slope-intercept form, so the slope must be 2 and the y-intercept must be -4. The graph, therefore, crosses the y-axis at $(0, -4)$. Because the slope is 2, we can let the rise = 2 and the run = 1 because $y = 2x - 4$ can also be written as $y = \frac{2}{1}x - 4$. We can then find a second point on the graph, which is $(1, -2)$. The graph is shown in Figure 3.

2. Find the equation of the line with slope $\frac{2}{3}$ and y-intercept 1. Then graph the line.

3. Find the slope and y-intercept for $3x - 2y = -6$. Then use them to draw the graph.

Answers

2. $y = \frac{2}{3}x + 1$

3. Slope is $\frac{3}{2}$; y-intercept is 3.

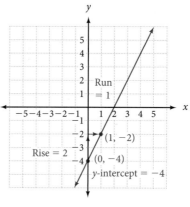

FIGURE 3

4. Find the slope and y-intercept for $4x - 2y = 12$.

EXAMPLE 4 Find the slope and y-intercept for $3x - 2y = 6$.

Solution To find the slope and y-intercept from the equation, we must write the equation in the form $y = mx + b$. This means we must solve the equation $3x - 2y = 6$ for y.

$$3x - 2y = 6 \qquad \text{Original equation}$$

$$-2y = -3x + 6 \qquad \text{Add } -3x \text{ to each side.}$$

$$-\frac{1}{2}(-2y) = -\frac{1}{2}(-3x + 6) \qquad \text{Multiply each side by } -\frac{1}{2}.$$

$$y = \frac{3}{2}x - 3 \qquad \text{Simplify each side.}$$

Now that the equation is written in slope-intercept form, we can identify the slope as $\frac{3}{2}$ and the y-intercept as -3. We use the same process as in Example 3 to graph the line. The graph is shown in Figure 4.

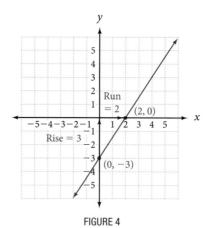

FIGURE 4

C The Point-Slope Form of an Equation of a Straight Line

A second useful form of the equation of a non-vertical straight line is the point-slope form.

Let line l contain the point (x_1, y_1) and have slope m. If (x, y) is any other point on l, then by the definition of slope we have

$$\frac{y - y_1}{x - x_1} = m$$

Multiplying both sides by $(x - x_1)$ gives us

$$(x - x_1) \cdot \frac{y - y_1}{x - x_1} = m(x - x_1)$$

$$y - y_1 = m(x - x_1)$$

This last equation is known as the *point-slope form* of the equation of a straight line.

> **Point-Slope Form of the Equation of a Line**
>
> The point-slope form of the equation of the non-vertical line through (x_1, y_1) with slope m is given by
>
> $$y - y_1 = m(x - x_1)$$

This form is used to find the equation of a line, either given one point on the line and the slope, or given two points on the line.

EXAMPLE 5 Using the point-slope form, find the equation of the line with slope -2 that contains the point $(-4, 3)$. Write the answer in slope-intercept form.

Solution Using $(x_1, y_1) = (-4, 3)$ and $m = -2$

in $\qquad y - y_1 = m(x - x_1) \qquad$ *Point-slope form*

gives us $\qquad y - 3 = -2(x + 4) \qquad$ *Note:* $x - (-4) = x + 4.$

$\qquad\qquad y - 3 = -2x - 8 \qquad$ *Multiply out right side.*

$\qquad\qquad\quad y = -2x - 5 \qquad$ *Add 3 to each side.*

Figure 5 is the graph of the line that contains $(-4, 3)$ and has a slope of -2. Notice that the y-intercept on the graph matches that of the equation we found, which we wrote in slope-intercept form.

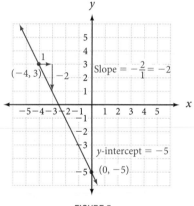

FIGURE 5

D Finding an Equation Given Two Points

EXAMPLE 6 Find the equation of the line that passes through the points $(-3, 3)$ and $(3, -1)$.

Solution We begin by finding the slope of the line:

$$m = \frac{3 - (-1)}{-3 - 3} = \frac{4}{-6} = -\frac{2}{3}$$

Using $(x_1, y_1) = (3, -1)$ and $m = -\frac{2}{3}$ in $y - y_1 = m(x - x_1)$ yields

$$y + 1 = -\frac{2}{3}(x - 3)$$

$$y + 1 = -\frac{2}{3}x + 2 \qquad \text{Multiply out right side.}$$

$$y = -\frac{2}{3}x + 1 \qquad \text{Add } -1 \text{ to each side.}$$

5. Find the equation of the line with slope 3 that contains the point $(-1, 2)$.

6. Find the equation of the line that passes through $(2, 5)$, and $(6, -3)$.

Answers

5. $y = 3x + 5$

6. $y = -2x + 9$

Note In Example 6 we could have used the point $(-3, 3)$ instead of $(3, -1)$ and obtained the same equation; that is, using $(x_1, y_1) = (-3, 3)$ and $m = -\frac{2}{3}$ in $y - y_1 = m(x - x_1)$ gives us

$$y - 3 = -\frac{2}{3}(x + 3)$$

$$y - 3 = -\frac{2}{3}x - 2$$

$$y = -\frac{2}{3}x + 1$$

which is the same result we obtained using $(3, -1)$.

Figure 6 shows the graph of the line that passes through the points $(-3, 3)$ and $(3, -1)$. As you can see, the slope and y-intercept are $-\frac{2}{3}$ and 1, respectively.

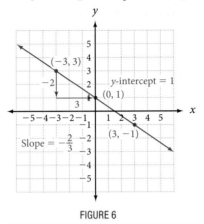

FIGURE 6

Methods of Graphing Lines

Here is a list of the graphing methods we have discovered in the last few sections.

1. Substitute convenient values of x into the equation, and find the corresponding values of y. We used this method first for equations like $y = 2x - 3$. To use this method for equations that looked like $2x - 3y = 6$, we first solved them for y.

2. Find the x- and y-intercepts. This method works best for equations of the form $3x + 2y = 6$ where the numbers in front of x and y divide the constant term evenly.

3. Find the slope and y-intercept. This method works best when the equation has the form $y = mx + b$ and b is an integer.

E Parallel and Perpendicular Lines

Now we can use what we have learned about linear equations to find the equations for parallel and perpendicular lines. Parallel lines are two lines that are nonintersecting and rise or fall at the same rate. In other words, two lines are parallel if and only if they have the same slope.

Here is a formal definition for parallel lines:

Note It is important to note that two different vertical lines are parallel; however, they both have the same undefined slope.

> **DEFINITION** parallel
>
> Two straight lines are *parallel* if they have the same slope. In other words, for $y = mx + b$, all equations of the form $y = mx + c$ are parallel if $b \neq c$.

7. Find the slope of a line parallel to the line that passes through $(-1, 3)$ and has a y-intercept of 1.

EXAMPLE 7 Find the slope of a line parallel to the line that passes through $(-3, 1)$ and has a y-intercept of 3.

Solution Since the y-intercept of the line is 3, we can determine that another point on the line is $(0, 3)$. We begin by finding the slope of the line.

$$m = \frac{1 - 3}{-3 - 0} = \frac{-2}{-3} = \frac{2}{3}$$

Since we know the y-intercept is $(0, 3)$ and using the slope $m = \frac{2}{3}$, we can write an equation for our line in slope-intercept form.

$$y = \frac{2}{3}x + 3$$

Two lines are parallel if they have the same slope, therefore, any line with the slope $m = \frac{2}{3}$ will be parallel to our original line.

Answer

7. -2

Although it is not as obvious, it is also true that two nonvertical lines are perpendicular if and only if the product of their slopes is −1. This is the same as saying their slopes are negative reciprocals. The following are graphs of perpendicular lines. Notice the slopes of the corresponding equations; the slope of one is the negative reciprocal of the slope of the other.

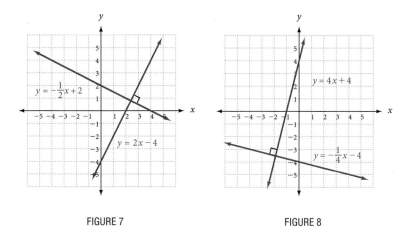

FIGURE 7 FIGURE 8

The following definition can be assumed for perpendicular lines:

> **DEFINITION** perpendicular
>
> Two non-vertical lines are *perpendicular* if the product of their slopes is −1. That is, $y = mx + b$ is perpendicular to $y = \left(-\frac{1}{m}\right)x + c$, if m is any real number except 0.

EXAMPLE 8 Find the slope of a line perpendicular to the line $y = 3x - 2$.

Solution Since the equation is already written in slope-intercept form, we know that the slope of the first line is 3. Any line with a slope that, when multiplied by 3, results in a product of −1 is perpendicular to this line. For this to be true, the slope must be the negative reciprocal of 3, which is $-\frac{1}{3}$.

 Furthermore, any line of the form $y = -\frac{1}{3}x + a$, where a is any real number, will provide a graph for our solution. ●

> *Note* Also remember that a vertical line and a horizontal line are perpendicular.

8. Find the slope of a line perpendicular to the line $y = \frac{1}{3}x - 2$.

> **GETTING READY FOR CLASS**
>
> *After reading through the preceding section, respond in your own words and in complete sentences.*
>
> **A.** What are m and b in the equation $y = mx + b$? What form is this equation?
>
> **B.** How would you find the slope and y-intercept for the line $4x - 3y = 9$?
>
> **C.** What is the point-slope form of the equation of a line?
>
> **D.** Find the equation of the line that passes through the points (8, 1) and (4, −2).

Answer

8. −3

Vocabulary Review

Choose the correct words to fill in the blanks below.

substitute	slope	slope-intercept	y-intercept
parallel	perpendicular	point-slope	

1. The equation of a line with a slope m and y-intercept b given in _____ form is $y = mx + b$.

2. The equation of a line through (x_1, y_1) with slope m given in _____ form is $y - y_1 = m(x - x_1)$.

3. The first method used to graph lines is to _____ convenient values of x into the equation, and find the corresponding values of y.

4. The second method used to graph lines is to find the x-intercept and then _____.

5. The third method used to graph lines is to find the _____ and the y-intercept.

6. Two straight lines are _____ if they have the same slope.

7. Two straight non-vertical lines are _____ if the product of their slopes is -1.

Problems

A In each of the following problems, give the equation of the line with the given slope and y-intercept.

1. $m = \dfrac{2}{3}, b = 1$ **2.** $m = \dfrac{3}{4}, b = -2$ **3.** $m = \dfrac{3}{2}, b = -1$ **4.** $m = \dfrac{4}{3}, b = 2$

5. $m = -\dfrac{2}{3}, b = 3$ **6.** $m = -\dfrac{3}{5}, b = 4$ **7.** $m = 2, b = -4$ **8.** $m = -2, b = 4$

B Find the slope and y-intercept for each of the following equations by writing them in the form $y = mx + b$. Then, graph each equation.

9. $-2x + y = 4$ **10.** $-2x + y = 2$ **11.** $3x + y = 3$

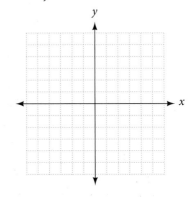

12. $3x + y = 6$

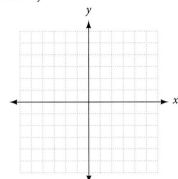

13. $3x + 2y = 6$

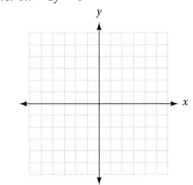

14. $2x + 3y = 6$

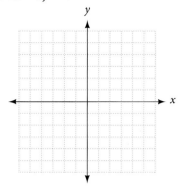

15. $4x - 5y = 20$

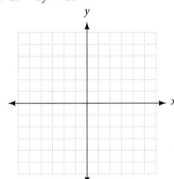

16. $2x - 5y = 10$

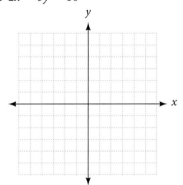

17. $-2x - 5y = 10$

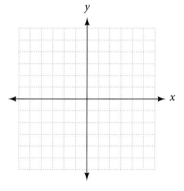

18. $-4x + 5y = 20$

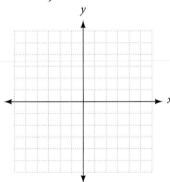

19. $-3x + 2y = -6$

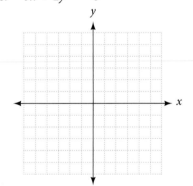

20. $2x - 3y = -6$

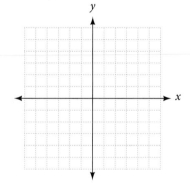

C For each of the following problems, the slope and one point on a line are given. In each case use the point-slope form to find the equation of that line. Write your answers in slope-intercept form.

21. $(-2, -5), m = 2$

22. $(-1, -5), m = 2$

23. $(-4, 1), m = -\dfrac{1}{2}$

24. $(-2, 1), m = -\dfrac{1}{2}$

25. $(2, -3)$, $m = \dfrac{3}{2}$　　　　**26.** $(3, -4)$, $m = \dfrac{4}{3}$　　　　**27.** $(-1, 4)$, $m = -3$　　　　**28.** $(-2, 5)$, $m = -3$

29. $(3, -4)$, $m = 0$　　　　**30.** $(-5, 2)$, $m = 0$　　　　**31.** $(-4, -3)$, m is undefined　　**32.** $(-5, 1)$, m is undefined

D Find the equation of the line that passes through each pair of points. Write your answers in slope-intercept form, if possible.

33. $(-2, -4)$, $(1, -1)$　　　**34.** $(2, 4)$, $(-3, -1)$　　　**35.** $(-1, -5)$, $(2, 1)$　　　**36.** $(-1, 6)$, $(1, 2)$

37. $(-3, -2)$, $(3, 6)$　　　**38.** $(-3, 6)$, $(3, -2)$　　　**39.** $(-3, -1)$, $(3, -5)$　　　**40.** $(-3, -5)$, $(3, 1)$

41. $(-4, 2)$, $(3, 2)$　　　**42.** $(-1, 7)$, $(1, 7)$　　　**43.** $(-5, 1)$, $(-5, 6)$　　　**44.** $(2, -3)$, $(2, 1)$

Find the slope and y-intercept for each line. Then write the equation of each line in slope-intercept form.

45.

46.

47.

48.

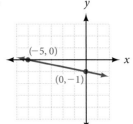

The next two problems are intended to give you practice reading, and paying attention to, the instructions that accompany the problems you are working. Working these problems is an excellent way to get ready for a test or a quiz.

49. Work each problem according to the instructions given.
 a. Solve: $-2x + 1 = 6$.

 b. Write in slope-intercept form: $-2x + y = 6$.

 c. Find the y-intercept: $-2x + y = 6$.

 d. Find the slope: $-2x + y = 6$.

 e. Graph: $-2x + y = 6$.

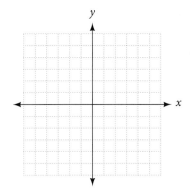

50. Work each problem according to the instructions given.
 a. Solve: $x + 3 = -6$.

 b. Write in slope-intercept form: $x + 3y = -6$.

 c. Find the y-intercept: $x + 3y = -6$.

 d. Find the slope: $x + 3y = -6$.

 e. Graph: $x + 3y = -6$.

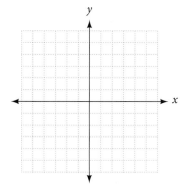

51. Find the equation of the line with x-intercept 3 and y-intercept 2.

52. Find the equation of the line with x-intercept 2 and y-intercept 3.

53. Find the equation of the line with x-intercept -2 and y-intercept -5.

54. Find the equation of the line with x-intercept -3 and y-intercept -5.

55. The equation of the vertical line that passes through the points $(3, -2)$ and $(3, 4)$ is either $x = 3$ or $y = 3$. Which one is it?

56. The equation of the horizontal line that passes through the points $(2, 3)$ and $(-1, 3)$ is either $x = 3$ or $y = 3$. Which one is it?

E **57. Parallel Lines** Provide an equation for a line parallel to the line given by $y = 4x - 2$.

58. Parallel Lines Find the slope of a line parallel to the line that crosses the point $(3, 1)$ and has a y-intercept of 2.

59. Parallel Lines Find the slope of a line parallel to the line that passes through (2, 3) and (4, 1).

60. Perpendicular Lines Find the equation of a line perpendicular to the line given by $y = \frac{1}{3}x - 2$.

61. Perpendicular Lines If a line contains the x-intercept 4 and the y-intercept –2, find the slope of any perpendicular line.

62. Perpendicular Lines One line passes through the points (3, 4) and (–3, 1). Find the slope of a second line that is perpendicular to the first.

Applying the Concepts

63. Value of a Copy Machine Cassandra buys a new color copier for her small business. It will cost $21,000 and will decrease in value each year. The graph below shows the value of the copier after the first 5 years of ownership.

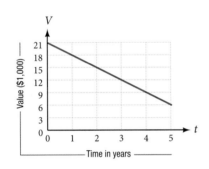

a. How much is the copier worth after 5 years?

b. After how many years is the copier worth $12,000?

c. Find the slope of this line.

d. By how many dollars per year is the copier decreasing in value?

e. Find the equation of this line where V is the value after t years.

64. Value of a Forklift Elliot buys a new forklift for his business. It will cost $140,000 and will decrease in value each year. The graph below shows the value of the forklift after the first 6 years of ownership.

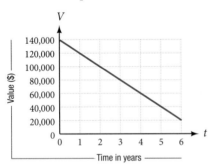

a. How much is the forklift worth after 6 years?

b. After how many years is the forklift worth $80,000?

c. Find the slope of this line.

d. By how many dollars per year is the forklift decreasing in value?

e. Find the equation of this line where V is the value after t years.

65. Salesperson's Income Kevin starts a new job in sales next month. He will earn $1,000 per month plus a certain amount for each shirt he sells. The graph below shows the amount Kevin will earn per month based on how many shirts he sells.

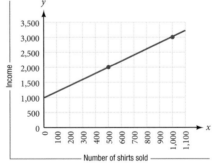

a. How much will he earn for selling 1,000 shirts?

b. How many shirts must he sell to earn $2,000 for a month?

c. Find the slope of this line.

d. How much money does Kevin earn for each shirt he sells?

e. Find the equation of this line where y is the amount he earns for selling x number of shirts.

66. Movie Prices The chart shows the price of movie tickets over a nine-year period. Let x represent the number of years past 2000 and y represent the prices shown on the graph. Use the first and last points on the graph to find the equation of the line that connects them. Then use the equation to estimate the price in 2011.

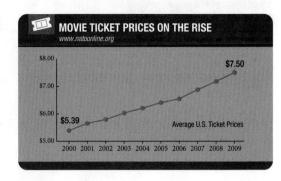

Find the Mistake

Each sentence below contains a mistake. Circle the mistake and write the correct word(s) on the line provided.

1. The equation $y = -\frac{4}{7}x + 2$ is in point-slope form. _____

2. For the equation $-8x + 2y = -4$, the slope is –2 and the y–intercept is 4. _____

3. The line represented by the equation $(y - 2) = 6(x - 4)$ has a y–intercept of 6. _____

4. The lines represented by the equations $y = \frac{7}{2}x + \frac{1}{3}$ and $y = -\frac{2}{7}x - 5$ are parallel. _____

©iStockphoto.com/tarantas

**Video Examples
Section 9.6**

At the U.S. National Handcar Races in Truckee, California, five-person teams face off on a railroad track. One team member gives the 1000-pound handcar a push while the other four members, standing on its flatbed, use the hand pump to propel the handcar down a 1000-foot stretch of track. The team with the fastest time wins the race. The races are held each year in honor of California's great railroad history, which aside from a few local lines began in 1869 with the completion of the First Continental Railroad. The handcar was once used for inspection and maintenance of the tracks, before an engine-powered version replaced it.

Let's assume that in 2009 the fastest handcar race time was 34 seconds, which was 7 seconds slower than the fastest time in 2010. Using this information, we can create an inequality that includes all the race times for both years. In this section, we will work with graphs of linear inequalities, such as this one.

A Graphing Linear Inequalities in Two Variables

A linear inequality in two variables is any expression that can be put in the form

$$Ax + By < C$$

where A, B, and C are real numbers (A and B not both 0). The inequality symbol can be any of the following four: $<$, \leq, $>$, \geq.

Some examples of linear inequalities are

$$2x + 3y < 6 \qquad y \geq 2x + 1 \qquad x - y \leq 0$$

Although not all of these inequalities have the form $Ax + By < C$, each one can be put in that form with the appropriate symbol.

In an earlier chapter, we graphed the solution set for a linear inequality in one variable on a number line. The solution set for a linear inequality in two variables is graphed as a section of the coordinate plane. The boundary for the section is found by replacing the inequality symbol with an equals sign and graphing the resulting equation using the methods summarized in the previous section. The boundary is included in the solution set (and represented with a solid line) if the inequality symbol used originally is \leq or \geq. The boundary is not included (and is represented with a broken line) if the original symbol is $<$ or $>$.

EXAMPLE 1 Graph the linear inequality $x + y \leq 4$.

Solution The boundary for the graph is the graph of $x + y = 4$. The boundary (a solid line) is included in the solution set because the inequality symbol is \leq.

The graph of the boundary is shown in Figure 1.

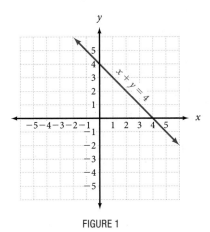

FIGURE 1

The boundary separates the coordinate plane into two sections, or regions: the region above the boundary and the region below the boundary. The solution set for $x + y \leq 4$ is one of these two regions along with the boundary. To find the correct region, we simply choose any convenient point that is *not* on the boundary. We then substitute the coordinates of the point into the original inequality $x + y \leq 4$. If the point we choose satisfies the inequality, then it is a member of the solution set, and we can assume that all points on the same side of the boundary as the chosen point are also in the solution set. If the coordinates of our point do not satisfy the original inequality, then the solution set lies on the other side of the boundary.

In this example, a convenient point not on the boundary is the origin. Substituting $(0, 0)$ into $x + y \leq 4$ gives us

$$0 + 0 \overset{?}{\leq} 4$$

$$0 \leq 4 \qquad \text{A true statement}$$

Because the origin is a solution to the inequality $x + y \leq 4$, and the origin is below the boundary, all other points below the boundary are also solutions.

The graph of $x + y \leq 4$ is shown in Figure 2.

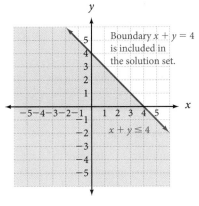

FIGURE 2

The region above the boundary is described by the inequality $x + y > 4$.

Practice Problems

1. Graph the linear inequality $x + y \geq 3$.

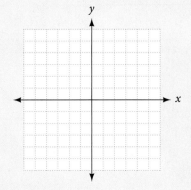

Here is a list of steps to follow when graphing the solution set for linear inequalities in two variables:

> **HOW TO** Graph the Solution Set for Linear Inequalities in Two Variables
>
> **Step 1:** Replace the inequality symbol with an equals sign. The resulting equation represents the boundary for the solution set.
>
> **Step 2:** Graph the boundary found in step 1 using a *solid line* if the boundary is included in the solution set (that is, if the original inequality symbol was either \leq or \geq). Use a *broken line* to graph the boundary if it is *not* included in the solution set; that is, if the original inequality was either $<$ or $>$.
>
> **Step 3:** Choose any convenient point not on the boundary and substitute the coordinates into the *original* inequality. If the resulting statement is *true*, the graph lies on the *same* side of the boundary as the chosen point. If the resulting statement is *false*, the solution set lies on the *opposite* side of the boundary.

2. Graph the linear inequality for $y < \frac{1}{2}x + 3$.

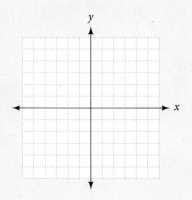

EXAMPLE 2 Graph the linear inequality $y < 2x - 3$.

Solution The boundary is the graph of $y = 2x - 3$. The boundary is not included because the original inequality symbol is $<$. We therefore use a broken line to represent the boundary, as shown in Figure 3.

A convenient test point is again the origin. Using $(0, 0)$ in $y < 2x - 3$, we have

$$0 \overset{?}{<} 2(0) - 3$$

$$0 < -3 \qquad \text{A false statement}$$

FIGURE 3

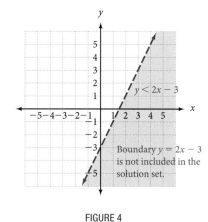

FIGURE 4

Because our test point gives us a false statement and it lies above the boundary, the solution set must lie on the other side of the boundary, as shown in Figure 4.

3. Graph the linear inequality $3x - 2y \leq 6$.

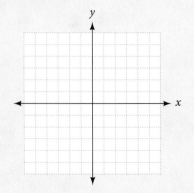

EXAMPLE 3 Graph the linear inequality $2x + 3y \leq 6$.

Solution We begin by graphing the boundary $2x + 3y = 6$. The boundary is included in the solution because the inequality symbol is \leq.

If we use $(0, 0)$ as our test point, we see that it yields a true statement when its coordinates are substituted into $2x + 3y \leq 6$. The graph, therefore, lies below the boundary, as shown in Figure 5.

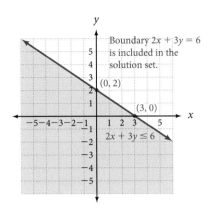

FIGURE 5

The ordered pair $(0, 0)$ is a solution to $2x + 3y \leq 6$; all points on the same side of the boundary as $(0, 0)$ also must be solutions to the inequality $2x + 3y \leq 6$.

EXAMPLE 4 Graph the linear inequality $x \leq 5$.

Solution The boundary is $x = 5$, which is a vertical line. All points to the left have x-coordinates less than 5, and all points to the right have x-coordinates greater than 5, as shown in Figure 6.

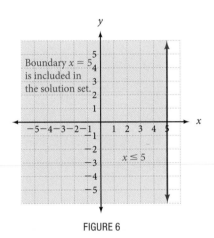

FIGURE 6

4. Graph the linear inequality $y > -2$.

GETTING READY FOR CLASS

After reading through the preceding section, respond in your own words and in complete sentences.

A. When graphing a linear inequality in two variables, how do you find the equation of the boundary line?

B. What is the significance of a broken line in the graph of an inequality?

C. When graphing a linear inequality in two variables, how do you know which side of the boundary line to shade?

D. Why is the coordinate $(0, 0)$ a convenient test point?

EXERCISE SET 9.6

Vocabulary Review

Choose the correct words to fill in the blanks below.

boundary solution set broken linear inequality solid

1. A _____ in two variables is any expression that can be put in the form $Ax + Bx < C$, where the inequality symbol can be $<$, $>$, \geq, or \leq.

2. The _____ for a linear inequality is a section of the coordinate plane.

3. The _____ for the solution set of a linear inequality can be found by replacing the inequality symbol with an equals sign and graphing the resulting equation.

4. Use a _____ line for the boundary of a linear inequality if the original inequality symbol was \geq or \leq.

5. Use a _____ line for the boundary of a linear inequality if the original inequality symbol was $>$ or $<$.

Problems

A Graph the following linear inequalities.

1. $2x - 3y < 6$

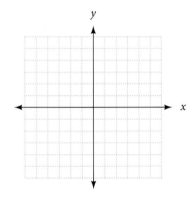

2. $3x + 2y \geq 6$

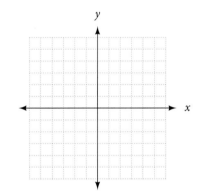

3. $x - 2y \leq 4$

4. $2x + y > 4$

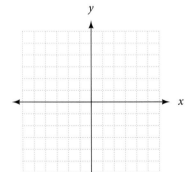

5. $x - y \leq 2$

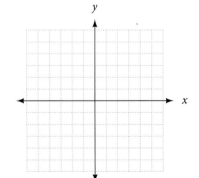

6. $x - y \leq 1$

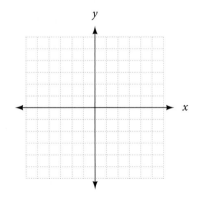

7. $3x - 4y \geq 12$

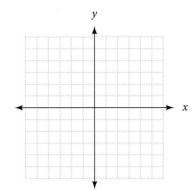

8. $4x + 3y < 12$

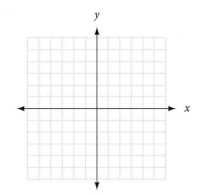

9. $5x - y \leq 5$

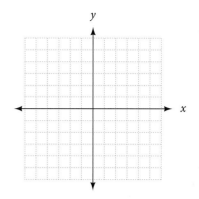

10. $4x + y > 4$

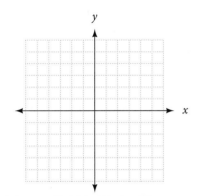

11. $2x + 6y \leq 12$

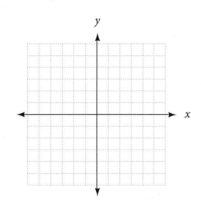

12. $x - 5y > 5$

13. $x \geq 1$

14. $x < 5$

15. $x \geq -3$

16. $y \leq -4$

17. $y < 2$

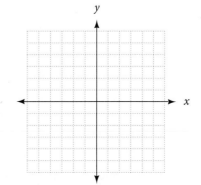

18. $3x - y > 1$

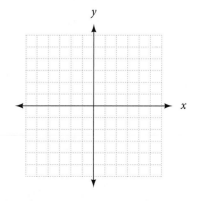

19. $2x + y > 3$

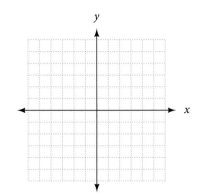

20. $5x + 2y < 2$

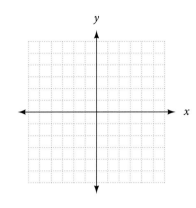

21. $y \leq 3x - 1$

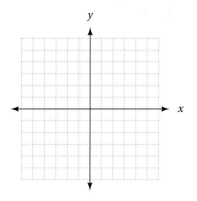

22. $y \geq 3x + 2$

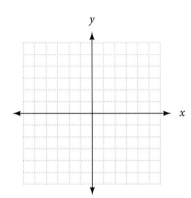

23. $y \leq -\dfrac{1}{2}x + 2$

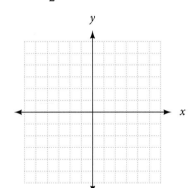

24. $y < \dfrac{1}{3}x + 3$

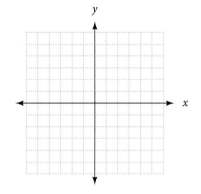

The next two problems are intended to give you practice reading, and paying attention to, the instructions that accompany the problems you are working.

25. Work each problem according to the instructions.

 a. Solve: $4 + 3y < 12$.

 b. Solve: $4 - 3y < 12$.

 c. Solve for y: $4x + 3y = 12$.

 d. Graph: $y < -\dfrac{4}{3}x + 4$.

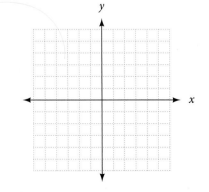

26. Work each problem according to the instructions.

 a. Solve: $3x + 2 \geq 6$.

 b. Solve: $-3x + 2 \geq 6$.

 c. Solve for y: $3x + 2y = 6$.

 d. Graph: $y \geq -\dfrac{3}{2}x + 3$.

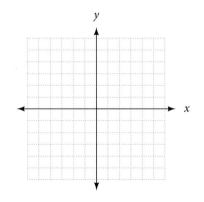

27. Find the equation of the line in part a, then use this information to find the inequalities for the graphs on parts b and c.

a.

b.

c.

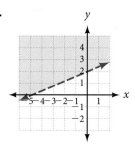

28. Find the equation of the line in part a, then use this information to find the inequalities for the graphs on parts b and c.

a.

b.

c.

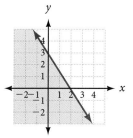

Review Problems

29. Simplify the expression $7 - 3(2x - 4) - 8$.

30. Find the value of $x^2 - 2xy + y^2$ when $x = 3$ and $y = -4$.

Solve each equation.

31. $-\dfrac{3}{2}x = 12$

32. $2x - 4 = 5x + 2$

33. $8 - 2(x + 7) = 2$

34. $3(2x - 5) - (2x - 4) = 6 - (4x + 5)$

35. Solve the formula $P = 2l + 2w$ for w.

36. Solve the formula $C = 2\pi r$ for r.

Solve each inequality, and graph the solution on a number line.

37. $-4x < 20$

38. $3 - 2x > 5$

39. $3 - 4(x - 2) \geq -5x + 6$

 0

40. $4 - 6(x + 1) < -2x + 3$

 0

41. Solve the inequality $3x - 2y \leq 12$ for y.

42. What number is 12% of 2,000?

43. Geometry The length of a rectangle is 5 inches more than 3 times the width. If the perimeter is 26 inches, find the length and width.

44. The length of a rectangle is 1 inch less than twice the width. If the perimeter is 28 inches, find the length and width.

Find the Mistake

Each sentence below contains a mistake. Circle the mistake and write the correct word(s) on the line provided.

1. The solution set for the linear inequality $y < 6x - \frac{3}{5}$ contains the points through which the boundary line passes.

2. The boundary for the linear inequality $2 - 8y \geq x + 6$ is represented by a broken line. _____

3. If a test point substituted into an original inequality results in a false statement, the solution set for the inequality lies on the same side as the chosen point. _____

4. The solution set for the linear inequality $x < -8$, includes all points to the right of the line. _____

Linear Equation in Two Variables [9.2]

A linear equation in two variables is any equation that can be put in the form $Ax + By = C$. The graph of every linear equation is a straight line.

Strategy for Graphing Linear Equations in Two Variables By Plotting Points [9.2]

Step 1: Find any three ordered pairs that satisfy the equation. This can be done by using a convenient number for one variable and solving for the other variable.

Step 2: Graph the three ordered pairs found in step 1. Actually, we need only two points to graph a straight line. The third point serves as a check. If all three points do not line up, there is a mistake in our work.

Step 3: Draw a straight line through the three points graphed in step 2.

Intercepts [9.3]

The *x-intercept* of an equation is the *x-coordinate* of the point where the graph intersects the *x-axis*. The *y-intercept* is the *y-coordinate* of the point where the graph intersects the *y-axis*. We find the *y-intercept* by substituting $x = 0$ into the equation and solving for *y*. The *x-intercept* is found by letting $y = 0$ and solving for *x*.

Slope of a Line [9.4]

The *slope* of the line containing the points (x_1, y_1) and (x_2, y_2) is given by

$$\text{Slope} = m = \frac{y_2 - y_1}{x_2 - x_1} = \frac{\text{rise}}{\text{run}}$$

Slope-Intercept Form of a Straight Line [9.5]

The equation of the non-vertical line with a slope of *m* and a *y-intercept* of *b* is

$$y = mx + b$$

EXAMPLES

1. The equation $3x + 2y = 6$ is an example of a linear equation in two variables.

2. The graph of $y = -\frac{2}{3}x - 1$ is shown below.

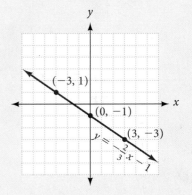

3. To find the *x-intercept* for $3x + 2y = 6$, we let $y = 0$ and get

$$3x = 6$$
$$x = 2$$

In this case the *x-intercept* is 2, and the graph crosses the *x-axis* at $(2, 0)$.

4. The slope of the line through $(3, -5)$ and $(-2, 1)$ is

$$m = \frac{-5 - 1}{3 - (-2)} = \frac{-6}{5} = -\frac{6}{5}$$

5. The equation of the line with a slope of 2 and a *y-intercept* of 5 is

$$y = 2x + 5$$

Point-Slope Form of a Straight Line [9.5]

6. The equation of the line through (1, 2) with a slope of 3 is
$$y - 2 = 3(x - 1)$$
$$y - 2 = 3x - 3$$
$$y = 3x - 1$$

If a non-vertical line has a slope of m and contains the point (x_1, y_1), the equation can be written as

$$y - y_1 = m(x - x_1)$$

To Graph a Linear Inequality in Two Variables [9.6]

7. Graph $x - y \geq 3$.

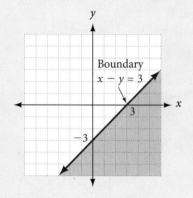

Step 1: Replace the inequality symbol with an equals sign. The resulting equation represents the boundary for the solution set.

Step 2: Graph the boundary found in step 1, using a *solid line* if the original inequality symbol was either \leq, or \geq. Use a *broken line* otherwise.

Step 3: Choose any convenient point not on the boundary and substitute the coordinates into the *original* inequality. If the resulting statement is *true*, the graph lies on the *same* side of the boundary as the chosen point. If the resulting statement is *false*, the solution set lies on the *opposite* side of the boundary.

Graph the ordered pairs. [9.1]

1. $(2, -1)$ **2.** $(-4, 3)$

3. $(-3, -2)$ **4.** $(0, -4)$

5. Fill in the following ordered pairs for the equation $3x - 2y = 6$. [9.2]

$$(0, \quad) \; (\quad, 0) \; (4, \quad) \; (\quad, -6)$$

6. Which of the following ordered pairs are solutions to $y = -3x + 7$? [9.2]

$$(0, 7) \; (2, -1) \; (4, -5) \; (-5, -3)$$

Graph each line. [9.3]

7. $y = -\dfrac{1}{2}x + 4$ **8.** $x = -3$

Find the x- and y-intercepts. [9.3]

9. $8x - 4y = 16$ **10.** $y = \dfrac{3}{2}x + 6$

11.

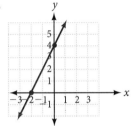

Find the slope of the line through each pair of points. [9.4]

12. $(3, 2), (-5, 6)$ **13.** $(0, 9), (7, 1)$

Find the slope of each line. [3.4]

14.

15.

16.

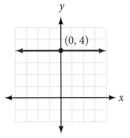

17. Find the equation of the line through $(4, 1)$ with a slope of $-\dfrac{1}{2}$. [9.5]

18. Find the equation of the line with a slope of 3 and y-intercept -5. [9.5]

19. Find the equation of the line passing through the points $(3, -4)$ and $(-6, 2)$. [9.5]

20. A straight line has an x-intercept -2 and passes through the point $(-4, 6)$. Find its equation. [9.5]

21. Write the equation, in slope-intercept form, of the line whose graph is shown below. [9.5]

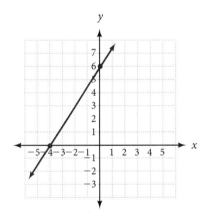

Find the equation for each line. [9.5]

22. The line through $(-6, 1)$ that has slope $m = -3$

23. The line through $(6, 8)$ and $(-3, 2)$

24. The line which contains the point $(-4, 1)$ and is perpendicular to the line $y = -\dfrac{2}{3}x + 4$

25. The line which contains the point $(0, -6)$ and is parallel to the line $4x - 2y = 3$

Graph each linear inequality in two variables. [9.6]

26. $y > x - 6$ **27.** $6x - 9y \le 18$

Systems of Linear Equations

10.1 Solving Linear Systems by Graphing

10.2 The Substitution Method

10.3 The Elimination Method

10.4 Systems of Linear Equations in Three Variables

10.5 Systems of Linear Inequalites and Applications

Gray Buildings © 2008 Sanborn
Image © 2010 Digital Globe
Las Vegas, Nevada

Google

If you have ever been to Las Vegas, Nevada, you have most likely seen the Stratosphere Hotel and Casino, which contains the tallest observation deck in the United States. The observation deck stands at 869 feet above the ground, offers 360-degree views, and provides visitors access to some incredible thrill rides. The Big Shot is a roller coaster circling the tower at nearly 45 miles per hour, and is the highest coaster in the United States. The second highest thrill ride in the world, Insanity, dangles riders out over The Strip on a 64-foot arm as they spin around at three times regular g-forces. The X-Scream is a seesaw that holds patrons out over the edge of the tower more than 80 stories from the ground. The newest Stratosphere attraction, the SkyJump, allows jumpers to descend over 100 stories down to street level wearing a special suit and guided by wires. The controlled descent allows visitors the experience of a sky dive without the parachute or the airplane.

©iStockphoto.com/Kasia75

Suppose a ticket to ride Insanity costs $12, a ticket for X-Scream costs $14, a ticket for Big Shot costs $13, and a ticket for SkyJump costs $110. Let's also suppose that in one day, the thrill rides were ridden 105 times bringing in a total of $2,810. How many tickets for each ride were sold? In order to answer this question, you will need to know how to work with systems of equations, which is the topic of this chapter.

A Solve a system of linear equations in two variables by graphing.

Image © sxc.hu, AmyJacobs, 2007

KEY WORDS

- system of linear equations
- point of intersection
- consistent system
- inconsistent system
- dependent equations
- independent equations

Each year, spectators at the QuikTrip Air and Rocket Racing Show see a variety of planes, such as rocket racers, combat aircraft, biplanes, and gliders take to the skies above an airfield in Tulsa, Oklahoma. The wide-eyed ticket holders, packed tightly into the grandstands, watch planes perform dramatic acrobatic stunts. During some acts, two or more planes fly in the air at once. It it extremely important that the planes' routines are well-planned to avoid any collisions. Suppose we were to assign a linear equation, in the form of $Ax + By = C$, to the path of each plane of a two-plane act. When we consider these two equations together, we are working with a *system of linear equations*, which is the focus of this section. Then we would be able to determine if the planes were flying directly parallel to each other or if their paths will intersect.

A Solving a System of Linear Equations by Graphing

In a system of linear equations, both equations contain two variables and have graphs that are straight lines. The following are systems of linear equations:

$$x + y = 3 \qquad y = 2x + 1 \qquad 2x - y = 1$$
$$3x + 4y = 2 \qquad y = 3x + 2 \qquad 3x - 2y = 6$$

The solution set for a system of linear equations is all ordered pairs that are solutions to both equations. The intersection of the graphs are a point whose coordinates are solutions to a system; that is, if we graph both equations on a coordinate system, we can read the coordinates of the point of intersection as a solution to our system. Here is an example.

EXAMPLE 1 Solve the following system by graphing.

$$x + y = 4$$
$$x - y = -2$$

Solution Graph each equation separately. Figure 1 shows both graphs, without showing the work necessary to get them. We can see the graphs intersect at the point (1, 3). Therefore, the point (1, 3) must be a solution to our system because it is the only ordered pair that lies on both lines. Its coordinates satisfy both equations.

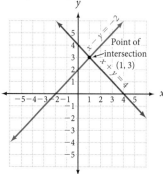

Point of intersection (1, 3)

FIGURE 1

Practice Problems

1. Solve the system by graphing.

$$x + y = 3$$
$$x - y = 5$$

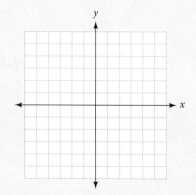

Answer

1. $(4, -1)$

We can check our results by substituting the coordinates $x = 1$, $y = 3$ into both equations to see if they work.

When →	$x = 1$	When →	$x = 1$
and →	$y = 3$	and →	$y = 3$
the equation →	$x + y = 4$	the equation →	$x - y = -2$
becomes →	$1 + 3 \stackrel{?}{=} 4$	becomes →	$1 - 3 \stackrel{?}{=} -2$
	$4 = 4$		$-2 = -2$

The point $(1, 3)$ satisfies both equations.

Here are some steps to follow when solving linear systems by graphing.

> **HOW TO** Solving a Linear System by Graphing
>
> **Step 1:** Graph the first equation by the methods described in Chapter 3.
>
> **Step 2:** Graph the second equation on the same set of axes used for the first equation.
>
> **Step 3:** Read the coordinates of the point of intersection of the two graphs.
>
> **Step 4:** Check the solution in both equations.

EXAMPLE 2 Solve the following system by graphing.

$$x + 2y = 8$$
$$2x - 3y = 2$$

Solution Graphing each equation on the same coordinate system, we have the lines shown in Figure 2.

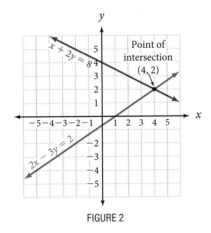

FIGURE 2

From Figure 2, we can see the solution for our system is $(4, 2)$. We check this solution as follows:

When →	$x = 4$	When →	$x = 4$
and →	$y = 2$	and →	$y = 2$
the equation →	$x + 2y = 8$	the equation →	$2x - 3y = 2$
becomes →	$4 + 2(2) \stackrel{?}{=} 8$	becomes →	$2(4) - 3(2) \stackrel{?}{=} 2$
	$4 + 4 = 8$		$8 - 6 = 2$
	$8 = 8$		$2 = 2$

The point $(4, 2)$ satisfies both equations and, therefore, must be the solution to our system.

**Video Examples
Section 10.1**

2. Solve the system by graphing.

$$x + 2y = 6$$
$$3x - y = 4$$

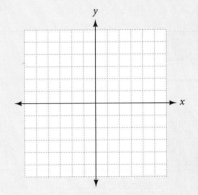

Answer

2. $(2, 2)$

3. Solve by graphing.

$$y = 3x - 1$$
$$x = 2$$

4. Solve by graphing.

$$y = x + 3$$
$$y = x - 2$$

5. Graph the system.

$$6x + 2y = 12$$
$$9x + 3y = 18$$

EXAMPLE 3 Solve this system by graphing.

$$y = 2x - 3$$
$$x = 3$$

Solution Graphing both equations on the same set of axes, we have Figure 3.

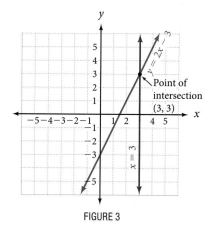

FIGURE 3

The solution to the system is the point $(3, 3)$.

EXAMPLE 4 Solve by graphing.

$$y = x - 2$$
$$y = x + 1$$

Solution Graphing both equations produces the lines shown in Figure 4. We can see in Figure 4 that the lines are parallel and therefore do not intersect. Our system has no ordered pair as a solution because there is no ordered pair that satisfies both equations. There is no solution to the system.

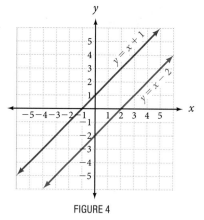

FIGURE 4

EXAMPLE 5 Graph the system.

$$2x + y = 4$$
$$4x + 2y = 8$$

Solution Both graphs are shown in Figure 5. The two graphs coincide. The reason becomes apparent when we multiply both sides of the first equation by 2.

$$2x + y = 4$$
$$2(2x + y) = 2(4) \qquad \text{Multiply both sides by 2.}$$
$$4x + 2y = 8$$

The equations have the same solution set. We say the lines coincide. Any ordered pair that is a solution to one is a solution to the system. The system has an infinite number of solutions. (Any point on the line is a solution to the system.)

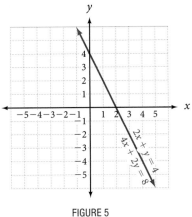

FIGURE 5

We sometimes use special vocabulary to describe the special cases shown in Examples 4 and 5. When a system of equations has no solution because the lines are parallel (as in Example 4), we say the system is *inconsistent*. When the lines coincide (as in Example 5), we say the equations are *dependent*.

The special cases illustrated in the previous two examples do not happen often. Usually, a system has a single ordered pair as a solution in which case we call the system *consistent* and the equations *independent*. Solving a system of linear equations by graphing is useful only when the ordered pair in the solution set has integers for coordinates. Two other solution methods work well in all cases. We will develop the other two methods in the next two sections.

Here is a summary of three possible types of solutions to a system of equations in two variables:

One Solution	**No Solution**	**Infinite Solutions**
Lines intersect at a single point	Lines are parallel and never cross	Lines coincide

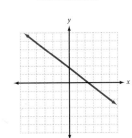

System is consistent.	System is inconsistent.	Equations are dependent.

GETTING READY FOR CLASS

After reading through the preceding section, respond in your own words and in complete sentences.

A. What is a system of linear equations?

B. What is a solution to a system of linear equations?

C. How do you solve a system of linear equations by graphing?

D. Under what conditions will a system of linear equations not have a solution?

EXERCISE SET 10.1

Vocabulary Review

Choose the correct words to fill in the blanks below.

> intersection axes ordered pairs system

1. A _____ of linear equations is two linear equations considered together.

2. The solution set for a system of linear equations is all _____ that are solutions to both equations.

3. To solve a system of linear equations, first graph the two equations on the same _____.

4. Once you have graphed a system of linear equations, check the point of _____ as a solution to both equations.

Problems

A Solve the following systems of linear equations by graphing. If the system is inconsistent or dependent, state so.

1. $x + y = 3$
$x - y = 1$

2. $x + y = 2$
$x - y = 4$

3. $x + y = 1$
$-x + y = 3$

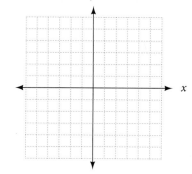

4. $x + y = 1$
$x - y = -5$

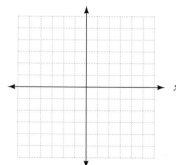

5. $x + y = 8$
$-x + y = 2$

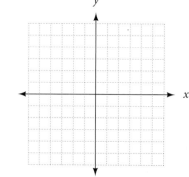

6. $x + y = 6$
$-x + y = -2$

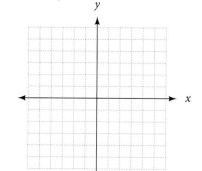

7. $3x - 2y = 6$
$x - y = 1$

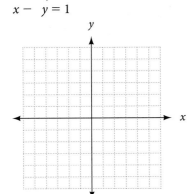

8. $5x - 2y = 10$
$x - y = -1$

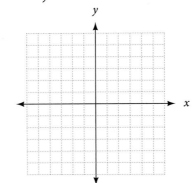

9. $6x - 2y = 12$
$3x + y = -6$

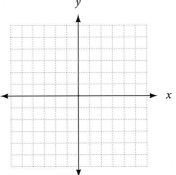

10. $4x - 2y = 8$
$2x + y = -4$

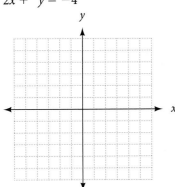

11. $4x + y = 4$
$3x - y = 3$

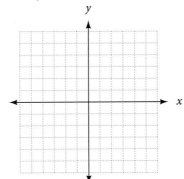

12. $5x - y = 10$
$2x + y = 4$

13. $x + 2y = 0$
$2x - y = 0$

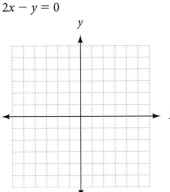

14. $3x + y = 0$
$5x - y = 0$

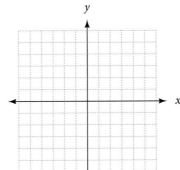

15. $3x - 5y = 15$
$-2x + y = 4$

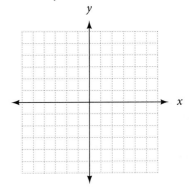

16. $2x - 4y = 8$
$2x - y = -1$

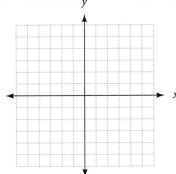

17. $y = 2x + 1$
$y = -2x - 3$

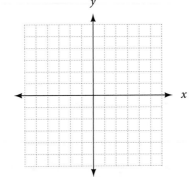

18. $y = 3x - 4$
$y = -2x + 1$

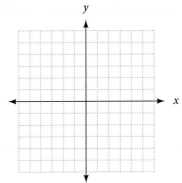

19. $x + 3y = 3$
$y = x + 5$

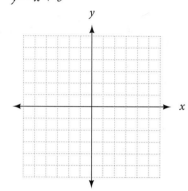

20. $2x + y = -2$
$y = x + 4$

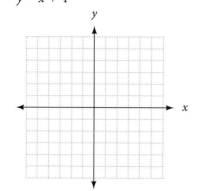

21. $x + y = 2$
$x = -3$

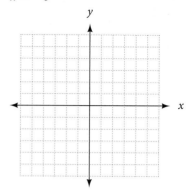

22. $x + y = 6$
$y = 2$

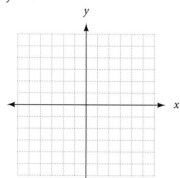

23. $x = -4$
$y = 6$

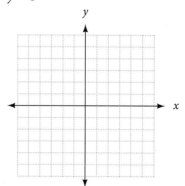

24. $x = 5$
$y = -1$

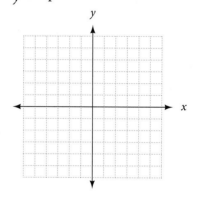

25. $x + y = 4$
$2x + 2y = -6$

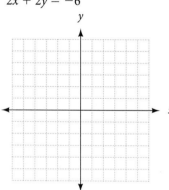

26. $x - y = 3$
$2x - 2y = 6$

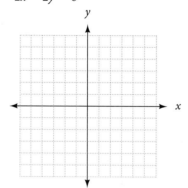

27. $4x - 2y = 8$
$2x - y = 4$

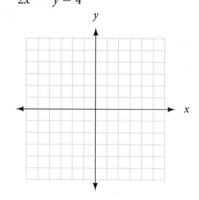

28. $3x - 6y = 6$
$x - 2y = 4$

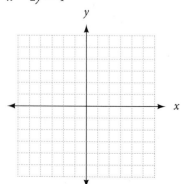

29. $y = 3x + 1$
$y = -2x + 1$

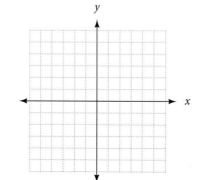

30. $y = -4x$
$y = \dfrac{1}{2}x$

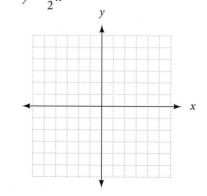

31. As you probably have guessed by now, it can be difficult to solve a system of equations by graphing if the solution to the system contains a fraction. The solution to the following system is $\left(\frac{1}{2}, 1\right)$. Solve the system by graphing.

$y = -2x + 2$
$y = 4x - 1$

32. The solution to the following system is $\left(\frac{1}{3}, -2\right)$. Solve the system by graphing.

$y = 3x - 3$
$y = -3x - 1$

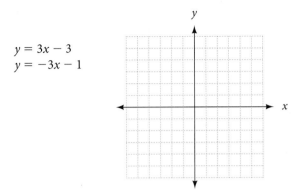

33. A second difficulty can arise in solving a system of equations by graphing if one or both of the equations is difficult to graph. The solution to the following system is $(2, 1)$. Solve the system by graphing.

$3x - 8y = -2$
$x - y = 1$

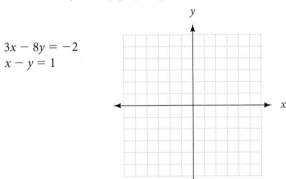

34. The solution to the following system is $(-3, 2)$. Solve the system by graphing.

$2x + 5y = 4$
$x - y = -5$

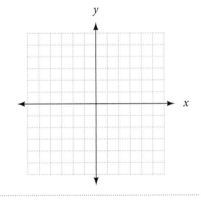

Applying the Concepts

35. Suppose you have 45 poker chips with $5 and $25 values. The total value of chips is $465. If x is the number of $5 chips and y is the number of $25 chips, then the following system of equations represents this situation:

$$x + y = 45$$
$$5x + 25y = 465$$

Solve this system by graphing each equation.

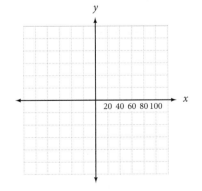

36. The stratosphere's Big Shot holds 16 passengers. Tickets can be purchased for 1 ride ($13), or for all 3 rides ($31). The total value of tickets purchased is $388. If x is the number of one-ride tickets and y is the number of three-ride tickets, then the following system of equations represents this situation:

$$x + y = 16$$
$$13x + 31y = 388$$

Solve this system by graphing each equation.

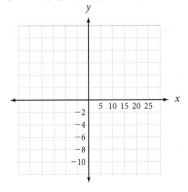

37. Job Comparison Jane is deciding between two sales positions. She can work for Marcy's and receive $8.00 per hour, or she can work for Gigi's, where she earns $6.00 per hour but also receives a $50 commission per week. The two lines in the following figure represent the money Jane will make for working at each of the jobs.

a. From the figure, how many hours would Jane have to work to earn the same amount at each of the positions?

b. If Jane expects to work less than 20 hours a week, which job should she choose?

c. If Jane expects to work more than 30 hours a week, which job should she choose?

38. Truck Rental You need to rent a moving truck for two days. Rider Moving Trucks charges $50 per day and $0.50 per mile. UMove Trucks charges $45 per day and $0.75 per mile. The following figure represents the cost of renting each of the trucks for two days.

a. From the figure, after how many miles would the trucks cost the same?

b. Which company will give you a better deal if you drive less than 30 miles?

c. Which company will give you a better deal if you drive more than 60 miles?

Getting Ready for the Next Section

Solve each equation.

39. $x + (2x - 1) = 2$

40. $2x - 3(2x - 8) = 12$

41. $2(3y - 1) - 3y = 4$

42. $-2x + 4(3x + 6) = 14$

43. $4x + 2(-2x + 4) = 8$

44. $1.5x + 15 = 0.75x + 24.95$

Solve each equation for the indicated variable.

45. $x - 3y = -1$ for x

46. $-3x + y = 6$ for y

47. Let $y = 2x - 1$. If $x = 1$, find y.

48. Let $y = 2x - 8$. If $x = 5$, find y.

49. Let $x = 3y - 1$. If $y = 2$, find x.

50. Let $y = 3x + 6$. If $y = -6$, find x.

Let $y = 1.5x + 15$ for the next four problems.

51. If $x = 13$, find y. **52.** If $x = 14$, find y. **53.** If $y = 21$, find x. **54.** If $y = 42$, find x.

Let $y = 0.75x + 24.95$ for the next four problems.

55. If $x = 12$, find y. **56.** If $x = 16$, find y. **57.** If $y = 39.95$, find x. **58.** If $y = 41.45$, find x.

Find the Mistake

Each sentence below contains a mistake. Circle the mistake and write the correct word(s) or number(s) on the line provided.

1. The point of intersection for two straight line graphs is a solution to only one of the equations. _____

2. The point of intersection for the equations $y = 6$ and $y = 4x + 2$ is (2, 6). _____

3. The point of intersection for the equations $x = 5$ and $x = 9$ is (5, 9). _____

4. When lines coincide, we say that the equations are inconsistent. _____

KEY WORDS

substitution

Video Examples
Section 10.2

Practice Problems

1. Solve the system by substitution.

$$x + y = 3$$
$$y = x + 5$$

Note Sometimes this method of solving systems of equations is confusing the first time you see it. If you are confused, you may want to read through this first example more than once and try it on your own before attempting the previous practice problem.

Answer

1. $(-1, 4)$

694

10.2 The Substitution Method

© iStockphoto.com/cstewart

Art and physical endurance enthusiasts race in an annual event called the Kinetic Sculpture Grand Championship. Teams have each engineered an amphibious work of art that can travel over any terrain. The vehicle must be able to hold its pilots, who power it over sand dunes, across a harbor, up and down steep city streets, and through a swamp. The vehicles are built and decorated as if they were elaborate parade floats, many with animated parts and their pilots dressed in costume.

Let's suppose a kinetic sculpture team races its vehicle on a river. It takes 5 minutes for the vehicle to race down the river going with a current of 3 miles per hour, and 12 minutes to travel the same distance back up the river against the current. After working through this section, you will be able to set up a linear system of equations for this problem and use an algebraic method, called the substitution method, to solve the system.

A Substitution Method

EXAMPLE 1 Solve the system by substitution.

$$x + y = 2$$
$$y = 2x - 1$$

Solution If we were to solve this system by the methods used in the previous section, we would have to graph both lines and see where they intersect. There is no need to do this, however, because the second equation tells us that y is $2x - 1$. We can replace the y variable in the first equation with the expression $2x - 1$ from the second equation; that is, we *substitute* $2x - 1$ from the second equation for y in the first equation. Here is what it looks like:

$$x + (2x - 1) = 2$$

The equation we end up with contains only the variable x. The y variable has been eliminated by substitution.

Solving the resulting equation, we have

$$x + (2x - 1) = 2$$
$$3x - 1 = 2$$
$$3x = 3$$
$$x = 1$$

This is the x-coordinate of the solution to our system. To find the y-coordinate, we substitute $x = 1$ into the second equation of our system. (We could substitute $x = 1$ into the first equation also and have the same result.)

$$y = 2(1) - 1$$
$$y = 2 - 1$$
$$y = 1$$

Chapter 10 Systems of Linear Equations

The solution to our system is the ordered pair (1, 1). It satisfies both of the original equations. Figure 1 provides visual evidence that the substitution method yields the correct solution.

EXAMPLE 2 Solve the system by substitution.

$$2x - 3y = 12$$
$$y = 2x - 8$$

Solution Again, the second equation says y is $2x - 8$. Because we are looking for the ordered pair that satisfies both equations, the y in the first equation must also be $2x - 8$. Substituting $2x - 8$ from the second equation for y in the first equation, we have

$$2x - 3(2x - 8) = 12$$

This equation can still be read as $2x - 3y = 12$ because $2x - 8$ is the same as y. Solving the equation, we have

$$2x - 3(2x - 8) = 12$$
$$2x - 6x + 24 = 12$$
$$-4x + 24 = 12$$
$$-4x = -12$$
$$x = 3$$

To find the y-coordinate of our solution, we substitute $x = 3$ into the second equation in the original system.

$$\text{When} \rightarrow \qquad x = 3$$
$$\text{the equation} \rightarrow \quad y = 2x - 8$$
$$\text{becomes} \rightarrow \qquad y = 2(3) - 8$$
$$y = 6 - 8 = -2$$

The solution to our system is $(3, -2)$.

EXAMPLE 3 Solve the following system by solving the first equation for x and then using the substitution method.

$$x - 3y = -1$$
$$2x - 3y = 4$$

Solution We solve the first equation for x by adding $3y$ to both sides to get

$$x = 3y - 1$$

Using this value of x in the second equation, we have

$$2(3y - 1) - 3y = 4$$
$$6y - 2 - 3y = 4$$
$$3y - 2 = 4$$
$$3y = 6$$
$$y = 2$$

Next, we find x.

$$\text{When} \rightarrow \qquad y = 2$$
$$\text{the equation} \rightarrow \quad x = 3y - 1$$
$$\text{becomes} \rightarrow \qquad x = 3(2) - 1$$
$$x = 6 - 1$$
$$x = 5$$

The solution to our system is $(5, 2)$

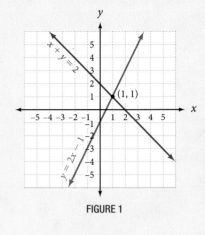

FIGURE 1

2. Solve the system by substitution.
$$5x - 4y = -2$$
$$y = 2x + 2$$

3. Solve the first equation for x, and then substitute the result into the second equation to solve the system by substitution.
$$x - 4y = -5$$
$$3x - 2y = 5$$

Answers
2. $(-2, -2)$
3. $(3, 2)$

Here are the steps to use in solving a system of equations by the substitution method.

> **HOW TO** **Strategy for Solving a System of Linear Equations by the Substitution Method**
>
> **Step 1:** Solve either one of the equations for x or y. (This step is not necessary if one of the equations is already in the correct form, as in Examples 1 and 2.)
>
> **Step 2:** Substitute the expression for the variable obtained in step 1 into the other equation and solve it.
>
> **Step 3:** Substitute the solution from step 2 into any equation in the system that contains both variables and solve it.
>
> **Step 4:** Check your results, if necessary.

4. Solve the system by substitution.

$$5x - y = 1$$
$$-2x + 3y = 10$$

EXAMPLE 4 Solve the system by substitution.

$$-2x + 4y = 14$$
$$-3x + y = 6$$

Solution We can solve either equation for either variable. If we look at the system closely, it becomes apparent that solving the second equation for y is the easiest way to go. If we add $3x$ to both sides of the second equation, we have

$$y = 3x + 6$$

Substituting the expression $3x + 6$ back into the first equation in place of y yields the following result.

$$-2x + 4(3x + 6) = 14$$
$$-2x + 12x + 24 = 14$$
$$10x + 24 = 14$$
$$10x = -10$$
$$x = -1$$

Substituting $x = -1$ into the equation $y = 3x + 6$ leaves us with

$$y = 3(-1) + 6$$
$$y = -3 + 6$$
$$y = 3$$

The solution to our system is $(-1, 3)$.

5. Solve the system by substitution.

$$6x + 3y = 1$$
$$y = -2x - 5$$

EXAMPLE 5 Solve the system by substitution.

$$4x + 2y = 8$$
$$y = -2x + 4$$

Solution Substituting the expression $-2x + 4$ for y from the second equation into the first equation, we have

$$4x + 2(-2x + 4) = 8$$
$$4x - 4x + 8 = 8$$
$$8 = 8 \qquad \text{A true statement}$$

Both variables have been eliminated, and we are left with a true statement. A true statement in this situation tells us the lines coincide; that is, the equations $4x + 2y = 8$ and $y = -2x + 4$ have exactly the same graph. Any point on that graph has coordinates that satisfy both equations and is a solution to the system. This system is dependent. If the result had been a false statement, this would tell us that the system is inconsistent, that is, the lines are parallel. If the resulting statement is false, there is no solution to the system.

Answers

4. $(1, 4)$

5. No solution. The system is inconsistent.

EXAMPLE 6 The following table shows two contract rates charged by a cellular phone company. How many text messages must a person send so that the cost will be the same regardless of which plan is chosen?

	Flat Monthly Rate	Per Text Charge
Plan 1	$40	$0.15
Plan 2	$70	$0.05

Solution If we let y = the monthly charge for x number of text messages, then the equations for each plan are

Plan 1: $y = 0.15x + 40$

Plan 2: $y = 0.05x + 70$

We can solve this system using the substitution method by replacing the variable y in Plan 2 with the expression $0.15x + 40$ from Plan 1. If we do so, we have

$0.15x + 40 = 0.05x + 70$

$0.10x + 40 = 70$

$0.10x = 30$

$x = 300$

The monthly bill is based on the number of text messages you send. We can use this value of x to gather more information by plugging it into either plan equation to find y.

Plan 1: $y = 0.15x + 40$ or Plan 2: $y = 0.05x + 70$

$y = 0.15(300) + 40$ $y = 0.05(300) + 70$

$y = 45 + 40$ $y = 15 + 70$

$y = 85$ $y = 85$

Therefore, when you send 300 text messages in a month, the total cost for that month will be $85 regardless of which plan you used.

GETTING READY FOR CLASS

After reading through the preceding section, respond in your own words and in complete sentences.

A. What is the first step in solving a system of linear equations by substitution?

B. When would substitution be more efficient than graphing in solving two linear equations?

C. What does it mean when you solve a system of linear equations by the substitution method and you end up with the statement $8 = 8$?

D. How would you begin solving the following system using the substitution method?

$$x + y = 2$$
$$y = 2x - 1$$

6. The rates for two garbage companies are given in the following table. How many bags of trash must be picked up in one month for the two companies to charge the same amount?

	Flat Rate	Per Bag Charge
Co. 1	$13.00	$1.50
Co. 2	$20.00	$1.00

Answer

6. 14 bags

Vocabulary Review

The following is a list of steps for solving a system of linear equations by the substitution method. Choose the correct words to fill in the blanks below.

substitute variables check equations

Step 1: Solve either one of the _____ for x or y.

Step 2: _____ the expression for one variable into the other equation and solve it.

Step 3: Substitute the solution from Step 2 into any equation in the system that contains both _____ and solve it.

Step 4: _____ your results.

Problems

A Solve the following systems by substitution. Substitute the expression in the second equation into the first equation and solve.

1. $x + y = 11$
$y = 2x - 1$

2. $x - y = -3$
$y = 3x + 5$

3. $x + y = 20$
$y = 5x + 2$

4. $3x - y = -1$
$x = 2y - 7$

5. $-2x + y = -1$
$y = -4x + 8$

6. $4x - y = 5$
$y = -4x + 1$

7. $3x - 2y = -2$
$x = -y + 6$

8. $2x - 3y = 17$
$x = -y + 6$

9. $5x - 4y = -16$
$y = 4$

10. $6x + 2y = 18$
$x = 3$

11. $5x + 4y = 7$
$y = -3x$

12. $10x + 2y = -6$
$y = -5x$

Solve the following systems by solving one of the equations for x or y and then using the substitution method.

13. $x + 3y = 4$
$x - 2y = -1$

14. $x - y = 5$
$x + 2y = -1$

15. $2x + y = 1$
$x - 5y = 17$

16. $2x - 2y = 2$
$x - 3y = -7$

17. $3x + 5y = -3$
$x - 5y = -5$

18. $2x - 4y = -4$
$x + 2y = 8$

19. $5x + 3y = 0$
$x - 3y = -18$

20. $x - 3y = -5$
$x - 2y = 0$

21. $-3x - 9y = 7$
$x + 3y = 12$

22. $2x + 6y = -18$
$x + 3y = -9$

23. $3x + 2y = -4$
$6x + 4y = -8$

24. $y = 4x + 7$
$y = 4x - 3$

Solve the following systems using the substitution method.

25. $5x - 8y = 7$
$\quad y = 2x - 5$

26. $3x + 4y = 10$
$\quad y = 8x - 15$

27. $7x - 6y = -1$
$\quad x = 2y - 1$

28. $4x + 2y = 3$
$\quad x = 4y - 3$

29. $-3x + 2y = 6$
$\quad y = 3x$

30. $-2x - y = -3$
$\quad y = -3x$

31. $5x - 6y = -4$
$\quad x = y$

32. $2x - 4y = 0$
$\quad y = x$

33. $3x + 3y = 9$
$\quad y = 2x - 12$

34. $7x + 6y = -9$
$\quad y = -2x + 1$

35. $7x - 11y = 16$
$\quad y = 10$

36. $9x - 7y = -14$
$\quad x = 7$

37. $-4x + 4y = -8$
$\quad y = x - 2$

38. $-4x + 2y = -10$
$\quad y = 2x - 5$

39. $-6x + 3y = 4$
$\quad y = 2x + 1$

40. $5x - 10y = 4$
$\quad y = \dfrac{1}{2}x + \dfrac{2}{5}$

Solve each system by substitution. You can eliminate the decimals if you like, but you don't have to. The solution will be the same in either case.

41. $0.05x + 0.10y = 1.70$
$\quad y = 22 - x$

42. $0.20x + 0.50y = 3.60$
$\quad y = 12 - x$

Applying the Concepts

43. In the introduction to the section, it took 5 minutes to race down river with a current of 3 mph, and it took 12 minutes to travel back upriver against the current. Use the following system of equations to find the rate r of the kinetic sculpture team.

$$5(r + 3) = d$$
$$12(r - 3) = d$$

44. San Francisco Bay has a current of approximately 4 mph. It takes 20 minutes to power a boat with the current, but 35 minutes against the current. Use the following system of equations to find the rate r of the boat.

4 mph

$$20(r + 4) = d$$
$$35(r - 4) = d$$

45. Gas Mileage Daniel is trying to decide whether to buy a car or a truck. The truck he is considering will cost him $150 a month in loan payments, and it gets 20 miles per gallon in gas mileage. The car will cost $180 a month in loan payments, but it gets 35 miles per gallon in gas mileage. Daniel estimates that he will pay $2.50 per gallon for gas. This means that the monthly cost to drive the truck x miles will be $y = \frac{2.50}{20}x + 150$. The total monthly cost to drive the car x miles will be $y = \frac{2.50}{35}x + 180$. The following figure shows the graph of each equation.

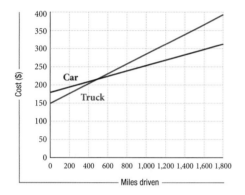

a. At how many miles do the car and the truck cost the same to operate?

b. If Daniel drives more than 1,200 miles, which will be cheaper?

c. If Daniel drives fewer than 300 miles, which will be cheaper?

d. Why do the graphs appear in the first quadrant only?

46. Video Production Pat runs a small company that duplicates DVDs. The daily cost and daily revenue for a company duplicating DVDs are shown in the following figure. The daily cost for duplicating x DVDs is $y = \frac{6}{5}x + 20$; the daily revenue (the amount of money he brings in each day) for duplicating x DVDs is $y = 1.7x$. The graphs of the two lines are shown in the following figure.

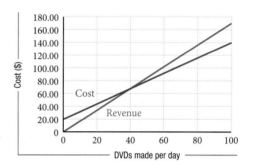

a. Pat will "break even" when his cost and his revenue are equal. How many DVDs does he need to duplicate to break even?

b. Pat will incur a loss when his revenue is less than his cost. If he duplicates 30 DVDs in one day, will he incur a loss?

c. Pat will make a profit when his revenue is larger than his costs. For what values of x will Pat make a profit?

d. Why does the graph appear in the first quadrant only?

Getting Ready For the Next Section

Simplify each of the following.

47. $(x + y) + (x - y)$

48. $(x + 2y) + (-x + y)$

49. $3(2x - y) + (x + 3y)$

50. $3(2x + 3y) - 2(3x + 5y)$

51. $-4(3x + 5y) + 5(5x + 4y)$

52. $(3x + 8y) - (3x - 2y)$

53. $6\left(\dfrac{1}{2}x - \dfrac{1}{3}y\right)$

54. $12\left(\dfrac{1}{4}x + \dfrac{2}{3}y\right)$

55. $-4(3x - 2y)$

56. $-8(2x - 3y)$

57. $4(3x - 2y) + 2(5x + 4y)$

58. $-3(4x - 3y) + 4(3x - 5y)$

59. Let $x + y = 4$. If $x = 3$, find y.

60. Let $x + 2y = 4$. If $x = 3$, find y.

61. Let $x + 3y = 3$. If $x = 3$, find y.

62. Let $2x + 3y = -1$. If $y = -1$, find x.

63. Let $3x + 5y = -7$. If $x = 6$, find y.

64. Let $3x - 2y = 12$. If $y = 6$, find x.

Find the Mistake

The table below shows systems and the first step taken to solve each system by the substitution method. Circle the mistake(s) in the first step and write the correct way to begin on the line provided.

System	First Step
1. $x = 6y - 5$ $2x + 4y = 1$	Substitute the value for x from the first equation for y in the second equation. _____
2. $x - 3y = 9$ $7y - 3x = -2$	Subtract x from both sides of the first equation to isolate y. _____
3. $y = 8x + 4$ $3x = 7y$	Divide both sides of the second equation by 3 to isolate x. _____
4. $x = \dfrac{1}{2}y + \dfrac{3}{4}$ $4y - 3x = 12$	Multiply all terms in the first equation by the LCD 4. _____

©iStockphoto.com/Vipre77

Video Examples
Section 10.3

Remember the exciting QuikTrip Air and Rocket Racing Show we discussed earlier in the chapter. Let's suppose 1720 general admission tickets were sold at the main gate for a total of $24,590. If it cost $15 for adults and $10 for children, how many of each kind of ticket were sold? In this section, we will learn another method of solving systems of linear equations in two variables, such as the one needed to answer the above question. This method is called the elimination method and makes use of the addition property.

The addition property states that if equal quantities are added to both sides of an equation, the solution set is unchanged. In the past, we have used this property to help solve equations in one variable. We will now use it to solve systems of linear equations. Here is another way to state the addition property of equality.

Let A, B, C, and D represent algebraic expressions.

$$\begin{aligned} \text{If} \rightarrow \quad & A = B \\ \text{and} \rightarrow \quad & C = D \\ \text{then} \rightarrow \quad & A + C = B + D \end{aligned}$$

Because C and D are equal (that is, they represent the same number), what we have done is added the same amount to both sides of the equation $A = B$. Let's see how we can use this form of the addition property of equality to solve a system of linear equations by way of the elimination, or addition, method.

A The Elimination Method

Practice Problems

1. Solve the following system using the elimination method.

$$x + y = 3$$
$$x - y = 5$$

Note The graphs with our first three examples are not part of their solutions. The graphs are there simply to show you that the results we obtain by the elimination method are consistent with the results we would obtain by graphing.

EXAMPLE 1 Solve the following system using the elimination method.

$$x + y = 4$$
$$x - y = 2$$

Solution The system is written in the form of the addition property of equality as shown above. It looks like this:

$$A = B$$
$$C = D$$

where A is $x + y$, B is 4, C is $x - y$, and D is 2.

We use the addition property of equality to add the left sides together and the right sides together.

$$\begin{array}{r} x + y = 4 \\ \underline{x - y = 2} \\ 2x + 0 = 6 \end{array}$$

Answer

1. $(4, -1)$

We now solve the resulting equation for x.

$$2x + 0 = 6$$

$$2x = 6$$

$$x = 3$$

The value we get for x is the value of the x-coordinate of the point of intersection of the two lines $x + y = 4$ and $x - y = 2$. To find the y-coordinate, we simply substitute $x = 3$ into either of the two original equations. Using the first equation, we get

$$3 + y = 4$$

$$y = 1$$

The solution to our system is the ordered pair $(3, 1)$. It satisfies both equations.

When →	$x = 3$	When →	$x = 3$
and →	$y = 1$	and →	$y = 1$
the equation →	$x + y = 4$	the equation →	$x - y = 2$
becomes →	$3 + 1 \stackrel{?}{=} 4$	becomes →	$3 - 1 \stackrel{?}{=} 2$
	$4 = 4$		$2 = 2$

Figure 1 is visual evidence that the solution to our system is $(3, 1)$.

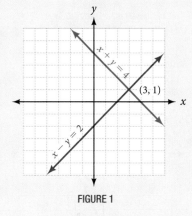

FIGURE 1

The most important part of this method of solving linear systems is eliminating one of the variables when we add the left and right sides together. In our first example, the equations were written so that the y variable was eliminated when we added the left and right sides together. If the equations are not set up this way to begin with, we have to work on one or both of them separately before we can add them together to eliminate one variable.

EXAMPLE 2 Solve the following system using the elimination method.

$$x + 2y = 4$$
$$x - y = -5$$

Solution Notice that if we were to add the equations together as they are, the resulting equation would have terms in both x and y. Let's eliminate the variable x by multiplying both sides of the second equation by -1 before we add the equations together. (As you will see, we can choose to eliminate either the x or the y variable.) Multiplying both sides of the second equation by -1 will not change its solution, so we do not need to be concerned that we have altered the system.

$$
\begin{array}{ll}
x + 2y = 4 & \xrightarrow{\text{No change}} \quad x + 2y = 4 \\
x - y = -5 & \xrightarrow[\text{Multiply by } -1.]{} \quad -x + y = 5
\end{array}
$$

$$0 + 3y = 9 \qquad \text{Add left and right sides.}$$

$$3y = 9$$

$$y = 3 \qquad \left\{ \begin{array}{l} y\text{-coordinate of the} \\ \text{point of intersection} \end{array} \right.$$

Substituting $y = 3$ into either of the two original equations, we get $x = -2$. The solution to the system is $(-2, 3)$. It satisfies both equations. Figure 2 shows the solution to the system as the point where the two lines cross.

2. Solve the following system using the elimination method.

$$x + 3y = 5$$
$$x - y = 1$$

FIGURE 2

Answer

2. $(2, 1)$

3. Solve the following system using the elimination method.

$$3x - y = 7$$
$$x + 2y = 7$$

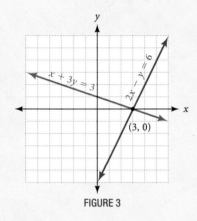

FIGURE 3

4. Solve the system using the elimination method.

$$3x + 2y = 3$$
$$2x + 5y = 13$$

> *Note* If you are having trouble understanding this method of solution, it is probably because you can't see why we chose to multiply by 3 and −2 in the first step of Example 4. Look at the result of doing so: the $6x$ and $-6x$ will add to 0. We chose to multiply by 3 and −2 because they produce $6x$ and $-6x$, which will add to 0.

EXAMPLE 3 Solve the following system using the elimination method.

$$2x - y = 6$$
$$x + 3y = 3$$

Solution Let's eliminate the y variable from the two equations. We can do this by multiplying the first equation by 3 and leaving the second equation unchanged.

$$2x - y = 6 \xrightarrow{\text{Multiply by 3.}} 6x - 3y = 18$$
$$x + 3y = 3 \xrightarrow{\text{No change}} x + 3y = 3$$

The important thing about our system now is that the *coefficients* (the numbers in front) of the y variables are opposites. When we add the terms on each side of the equal sign, then the terms in y will add to zero and be eliminated.

$$\begin{array}{r} 6x - 3y = 18 \\ x + 3y = 3 \\ \hline 7x = 21 \end{array} \quad \text{Add corresponding terms.}$$

This gives us $x = 3$. Using this value of x in the second equation of our original system, we have

$$3 + 3y = 3$$
$$3y = 0$$
$$y = 0$$

We could substitute $x = 3$ into any of the equations with both x and y variables and also get $y = 0$. The solution to our system is the ordered pair $(3, 0)$. Figure 3 is a picture of the system of equations showing the solution $(3, 0)$.

EXAMPLE 4 Solve the system using the elimination method.

$$2x + 3y = -1$$
$$3x + 5y = -2$$

Solution Let's eliminate x from the two equations. If we multiply the first equation by 3 and the second by −2, the coefficients of x will be 6 and −6, respectively. The x terms in the two equations will then add to zero.

$$2x + 3y = -1 \xrightarrow{\text{Multiply by 3.}} 6x + 9y = -3$$
$$3x + 5y = -2 \xrightarrow{\text{Multiply by } -2.} -6x - 10y = 4$$

We now add the left and right sides of our new system together.

$$\begin{array}{r} 6x + 9y = -3 \\ -6x - 10y = 4 \\ \hline -y = 1 \\ y = -1 \end{array}$$

Substituting $y = -1$ into the first equation in our original system, we have

$$2x + 3(-1) = -1$$
$$2x - 3 = -1$$
$$2x = 2$$
$$x = 1$$

The solution to our system is $(1, -1)$. It is the only ordered pair that satisfies both equations.

EXAMPLE 5 Solve the system using the elimination method.

$$3x + 5y = -7$$
$$5x + 4y = 10$$

Solution Let's eliminate y by multiplying the first equation by -4 and the second equation by 5.

$$3x + 5y = -7 \xrightarrow{\text{Multiply by } -4.} -12x - 20y = 28$$
$$5x + 4y = 10 \xrightarrow[\text{Multiply by 5.}]{} \underline{25x + 20y = 50}$$
$$13x \qquad\;\; = 78$$
$$x = 6$$

Substitute $x = 6$ into either equation in our original system, and the result will be $y = -5$. Therefore, the solution is $(6, -5)$.

EXAMPLE 6 Solve the system using the elimination method.

$$\frac{1}{2}x - \frac{1}{3}y = 2$$
$$\frac{1}{4}x + \frac{2}{3}y = 6$$

Solution Although we could solve this system without clearing the equations of fractions, there is probably less chance for error if we have only integer coefficients to work with. So let's begin by multiplying both sides of the top equation by 6 and both sides of the bottom equation by 12, to clear each equation of fractions.

$$\frac{1}{2}x - \frac{1}{3}y = 2 \xrightarrow{\text{Multiply by 6.}} 3x - 2y = 12$$
$$\frac{1}{4}x + \frac{2}{3}y = 6 \xrightarrow[\text{Multiply by 12.}]{} 3x + 8y = 72$$

Now we can eliminate x by multiplying the top equation by -1 and leaving the bottom equation unchanged.

$$3x - 2y = 12 \xrightarrow{\text{Multiply by } -1.} -3x + 2y = -12$$
$$3x + 8y = 72 \xrightarrow[\text{No change}]{} \underline{3x + 8y = \;\;72}$$
$$10y = \;\;60$$
$$y = \;\;\;6$$

We can substitute $y = 6$ into any equation that contains both x and y. Let's use $3x - 2y = 12$.

$$3x - 2(6) = 12$$
$$3x - 12 = 12$$
$$3x = 24$$
$$x = 8$$

The solution to the system is $(8, 6)$.

Our next two examples will remind us what happens when we try to solve a system of equations consisting of parallel lines and a system in which the lines coincide.

5. Solve the system using the elimination method.

$$5x + 4y = -6$$
$$2x + 3y = -8$$

6. Solve the system using the elimination method.

$$\frac{1}{3}x + \frac{1}{2}y = 1$$
$$x + \frac{3}{4}y = 0$$

7. Solve the system using the
elimination method.

$$x - 3y = 2$$
$$-3x + 9y = 2$$

EXAMPLE 7 Solve the system using the elimination method.

$$2x - y = 2$$
$$4x - 2y = 12$$

Solution Let us choose to eliminate y from the system. We can do this by multiplying the first equation by -2 and leaving the second equation unchanged.

$$2x - \ y = \ 2 \xrightarrow{\text{Multiply by } -2.} -4x + 2y = -4$$

$$4x - 2y = 12 \xrightarrow{\quad\text{No change}\quad} 4x - 2y = \ 12$$

If we add both sides of the resulting system, we have

$$\begin{array}{r} -4x + 2y = -4 \\ \underline{4x - 2y = \ 12} \\ 0 + 0 = \ \ 8 \end{array}$$

$$0 = \ \ 8 \qquad \text{A false statement.}$$

Both variables have been eliminated and we end up with the false statement $0 = 8$. We have tried to solve a system that consists of two parallel lines. There is no solution and the system is inconsistent. Figure 4 is a visual representation of the situation and is conclusive evidence that there is no solution to our system.

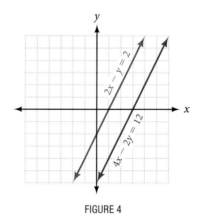

FIGURE 4

EXAMPLE 8 Solve the system using the elimination method.

$$4x - 3y = 2$$
$$8x - 6y = 4$$

8. Solve the system using the
elimination method.

$$5x - y = 1$$
$$10x - 2y = 2$$

Solution Multiplying the top equation by -2 and adding, we can eliminate the variable x.

$$4x - 3y = 2 \xrightarrow{\text{Multiply by } -2.} -8x + 6y = -4$$

$$8x - 6y = 4 \xrightarrow{\quad\text{No change}\quad} \underline{8x - 6y = \ \ 4}$$

$$0 = \ \ 0$$

Both variables have been eliminated, and the resulting statement $0 = 0$ is true. In this case the lines coincide because the equations are equivalent. The solution set consists of all ordered pairs that satisfy either equation, and the system is dependent.

The preceding two examples remind us of the the two special cases in which the graphs of the equations in the system either coincide or are parallel.

Here is a summary:

Both variables are eliminated and the resulting statement is false.	↔	The lines are parallel and there is no solution to the system.
Both variables are eliminated and the resulting statement is true.	↔	The lines coincide and there is an infinite number of solutions to the system.

The main idea in solving a system of linear equations by the elimination method is to use the multiplication property of equality on one or both of the original equations, if necessary, to make the coefficients of either variable opposites. The following box shows some steps to follow when solving a system of linear equations by the elimination method.

HOW TO Solving a System of Linear Equations by the Elimination Method

Step 1: Decide which variable to eliminate. (In some cases one variable will be easier to eliminate than the other. With some practice you will notice which one it is.)

Step 2: Use the multiplication property of equality on each equation separately to make the coefficients of the variable that is to be eliminated opposites.

Step 3: Add the respective left and right sides of the system together.

Step 4: Solve for the variable remaining.

Step 5: Substitute the value of the variable from step 4 into an equation containing both variables and solve for the other variable.

Step 6: Check your solution in both equations, if necessary.

GETTING READY FOR CLASS

After reading through the preceding section, respond in your own words and in complete sentences.

A. How do you use the addition property of equality in the elimination method of solving a system of linear equations?

B. What happens when you use the elimination method to solve a system of linear equations consisting of two parallel lines?

C. How would you use the multiplication property of equality to solve a system of linear equations?

D. What is the first step in solving a system of linear equations that contains fractions?

Vocabulary Review

Choose the correct words to fill in the blanks below.

substitute solution add variable solve multiplication property of equality

Solving a System of Linear Equations by the Elimination Method

Step 1: Decide which _____ to eliminate.

Step 2: Use the _____ on each equation separately to make the coefficients of the variable that is to be eliminated opposites.

Step 3: _____ the respective left and right sides of the system together.

Step 4: _____ for the variable remaining.

Step 5: _____ the value of the variable from step 4 into an equation containing both variable and solve for the other variable.

Step 6: Check your _____ in both equations.

Problems

A Solve the following systems of linear equations by elimination.

1. $x + y = 3$
$\quad x - y = 1$

2. $x + y = -2$
$\quad x - y = 6$

3. $x + y = 10$
$\quad -x + y = 4$

4. $x - y = 1$
$\quad -x - y = -7$

5. $x - y = 7$
$\quad -x - y = 3$

6. $x - y = 4$
$\quad 2x + y = 8$

7. $x + y = -1$
$\quad 3x - y = -3$

8. $2x - y = -2$
$\quad -2x - y = 2$

9. $\quad 3x + 2y = 1$
$\quad -3x - 2y = -1$

10. $-2x - 4y = 1$
$\quad 2x + 4y = -1$

11. $3x - 2y = 4$
$\quad -3x + 2y = 6$

12. $x - 2y = 5$
$\quad -x + 2y = 5$

Solve each of the following systems by eliminating the y variable.

13. $3x - y = 4$
$2x + 2y = 24$

14. $2x + y = 3$
$3x + 2y = 1$

15. $5x - 3y = -2$
$10x - y = 1$

16. $4x - y = -1$
$2x + 4y = 13$

17. $11x - 4y = 11$
$5x + y = 5$

18. $3x - y = 7$
$10x - 5y = 25$

19. $2x - y = 6$
$x + 4y = 3$

20. $2x + y = 5$
$-2x + 2y = 4$

Solve each of the following systems by eliminating the x variable.

21. $3x - 5y = 7$
$-x + y = -1$

22. $4x + 2y = 32$
$x + y = -2$

23. $-x - 8y = -1$
$-2x + 4y = 13$

24. $-x + 10y = 1$
$-5x + 15y = -9$

25. $-3x - y = 7$
$6x + 7y = 11$

26. $-5x + 2y = -6$
$10x + 7y = 34$

27. $-3x + 2y = 6$
$6x - y = -9$

28. $-4x + 5y = 20$
$8x - 6y = -12$

Solve each of the following systems of linear equations by the elimination method.

29. $6x - y = -8$
$2x + y = -16$

30. $5x - 3y = -3$
$3x + 3y = -21$

31. $x + 3y = 9$
$2x - y = 4$

32. $x + 2y = 0$
$2x - y = 0$

33. $x - 6y = 3$
$4x + 3y = 21$

34. $8x + y = -1$
$4x - 5y = 16$

35. $2x + 9y = 2$
$5x + 3y = -8$

36. $5x + 2y = 11$
$7x + 8y = 7$

37. $\dfrac{1}{3}x + \dfrac{1}{4}y = \dfrac{7}{6}$

$\dfrac{3}{2}x - \dfrac{1}{3}y = \dfrac{7}{3}$

38. $\dfrac{7}{12}x - \dfrac{1}{2}y = \dfrac{1}{6}$

$\dfrac{2}{5}x - \dfrac{1}{3}y = \dfrac{11}{15}$

39. $3x + 2y = -1$

$6x + 4y = 0$

40. $8x - 2y = 2$

$4x - y = 2$

41. $11x + 6y = 17$

$5x - 4y = 1$

42. $3x - 8y = 7$

$10x - 5y = 45$

43. $\dfrac{1}{2}x + \dfrac{1}{6}y = \dfrac{1}{3}$

$-x - \dfrac{1}{3}y = -\dfrac{1}{6}$

44. $-\dfrac{1}{3}x - \dfrac{1}{2}y = -\dfrac{2}{3}$

$-\dfrac{2}{3}x - y = -\dfrac{4}{3}$

45. Multiply both sides of the second equation in the following system by 100, and then solve as usual.

$$x + y = 22$$
$$0.05x + 0.10y = 1.70$$

46. Multiply both sides of the second equation in the following system by 100, and then solve as usual.

$$x + y = 15{,}000$$
$$0.06x + 0.07y = 980$$

Applying the Concepts

47. Recall at the beginning of this section that adult tickets x for the QuikTrip Air and Rocket Racing Show cost $15 and children's tickets y cost $10 Suppose 1,720 tickets are sold at the gate for a total of $19,750.

$$x + y = 1{,}720$$
$$15x - 10y = 19{,}750$$

Solve the above system of equations to find the number of adult and children tickets sold.

48. A local community theater sells adult tickets x for $10 and children's tickets y for $5. For one evening performance of a popular musical, the theater sells out 375 seats and collects a revenue of $2,875. The system of equations for this situation is

$$x + y = 375$$
$$10x + 5y = 2{,}875$$

Solve this system to find the number of adult and children tickets sold.

Getting Ready for the Next Section

49. One number is eight more than five times another; their sum is 26. Find the numbers.

50. One number is three less than four times another; their sum is 27. Find the numbers.

51. The difference of two positive numbers is nine. The larger number is six less than twice the smaller number. Find the numbers.

52. The difference of two positive numbers is 17. The larger number is one more than twice the smaller number. Find the numbers.

53. The length of a rectangle is five inches more than three times the width. The perimeter is 58 inches. Find the length and width.

54. The length of a rectangle is three inches less than twice the width. The perimeter is 36 inches. Find the length and width.

55. John has $1.70 in nickels and dimes in his pocket. He has four more nickels than he does dimes. How many of each does he have?

56. Jamie has $2.65 in dimes and quarters in her pocket. She has two more dimes than she does quarters. How many of each does she have?

Find the Mistake

The table below shows systems and the first step taken to solve each system by the elimination method. Circle the mistake(s) in the first step and write the correct way to begin on the line provided.

System	First Step
1. $4x - y = -8$ $-4x - y = 8$	Multiply both sides of the equation by –1 to eliminate the variable y.
2. $x - 7y = 11$ $3x + 2y = 15$	Multiply both sides of the first equation by 3 to eliminate x.
3. $5x + y = 9$ $-5x + 3y = 21$	Multiply both sides of the first equation by 3 to eliminate y.

Circle the mistake in the sentence below and write the correct word or phrase on the line provided.

4. When solving a system by the elimination method, if both variables are eliminated and the resulting statement is true, then the lines are parallel. _____

Landmark Review: Checking Your Progress

Solve the following systems.

1. $y = 2x + 5$
 $y = -3x - 5$

2. $2x - y = -2$
 $2x - 3y = 6$

3. $2x - y = 1$
 $y = -2x + 5$

4. $2x - 5y = 0$
 $x - 5y = -9$

5. $-3x + y = 8$
 $x = -y + 4$

6. $-6x + 2y = -9$
 $y = 3x + 4$

7. $x - y = 11$
 $x + y = -1$

8. $-3x + y = -15$
 $-2x + 3y = -17$

9. $2x + 5y = 17$
 $x - 4y = -24$

10. $x + 4y = 5$
 $3x + 8y = 14$

10.4 Systems of Linear Equations in Three Variables

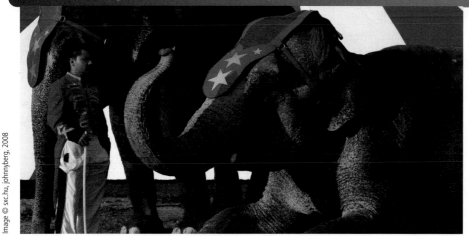

Image © sxc.hu, johnnyberg, 2008

OBJECTIVES

A Solve systems of linear equations in three variables.

KEY WORDS

system in three variables

solution set

planes

If a human faced off against an elephant in a hot dog bun eating contest, who would win? Each July, three brave humans sign up to find out firsthand. Three elephants, Susie, Minnie, and Bunny, from Ringling Bros. and Barnum & Bailey Circus stand behind a table on which a tall pile of hot dog buns sits. A few feet away, three competitive eating contestants stand in front of their own table outfitted with stacks of buns and large cups of water in which they can dunk the buns to help their cause. When the announcer yells "Go!", the two teams begin devouring the buns. They have 6 minutes to eat as many as possible. The 2010 final tally: 40 dozen buns for the elephants, 15 dozen for the humans.

We can use this scenario to help us introduce the topic of this section: systems of linear equations in three variables. Up until now, we have worked with systems in two variables. Now we will work with those with three unknowns. For instance, let's assume the individual totals for the amount of buns Susie, Minnie, and Bunny ate are x, y, and z, respectively and the total number of buns consumed by the elephants was 401 dozen. Suppose Minnie ate twice as many buns as Susie plus 10 more buns, and Bunny ate as much as Minnie plus one third the amount the humans ate. After reading through this section, you will be able to assign three linear equations and solve for the three variables in this problem.

**Video Examples
Section 10.4**

A Solving Systems in Three Variables

A solution to an equation in three variables such as

$$2x + y - 3z = 6$$

is an ordered triple of numbers (x, y, z). For example, the ordered triples $(0, 0, -2)$, $(2, 2, 0)$, and $(0, 9, 1)$ are solutions to the equation $2x + y - 3z = 6$, because they produce a true statement when their coordinates are substituted for x, y, and z in the equation.

> **DEFINITION** solution set
>
> The **solution set** for a system of three linear equations in three variables is the set of ordered triples that satisfies all three equations.

EXAMPLE 1 Solve the system.

$$x + y + z = 6 \qquad \text{Equation (1)}$$
$$2x - y + z = 3 \qquad \text{Equation (2)}$$
$$x + 2y - 3z = -4 \qquad \text{Equation (3)}$$

Practice Problems

1. Solve the system.

$$x + 2y + z = 2$$
$$x + y - z = 6$$
$$x - y + 2z = -7$$

Answer

1. $(1, 2, -3)$

Note Solving a system in three variables can be overwhelming at first. Don't get discouraged. Remember to read the solution's explanation slowly, and you may need to read it more than once to fully grasp each step.

Solution We want to find the ordered triple (x, y, z) that satisfies all three equations. We have numbered the equations so it will be easier to keep track of where they are and what we are doing.

There are many ways to proceed. The main idea is to take two different pairs of equations and eliminate the same variable from each pair. We begin by adding equations (1) and (2) to eliminate the y-variable. The resulting equation is numbered (4).

$$
\begin{array}{ll}
x + y + z = 6 & (1) \\
2x - y + z = 3 & (2) \\
\hline
3x \quad\ + 2z = 9 & \text{Equation (4)}
\end{array}
$$

Adding twice equation (2) to equation (3) will also eliminate the variable y. The resulting equation is numbered (5):

$$
\begin{array}{ll}
4x - 2y + 2z = \ \ 6 & \text{Twice (2)} \\
x + 2y - 3z = -4 & (3) \\
\hline
5x \qquad - z = \ \ 2 & \text{Equation (5)}
\end{array}
$$

Equations (4) and (5) form a linear system in two variables. By multiplying equation (5) by 2 and adding the result to equation (4), we succeed in eliminating the variable z from the new pair of equations.

$$
\begin{array}{ll}
3x + 2z = 9 & (4) \\
10x - 2z = 4 & \text{Twice (5)} \\
\hline
13x \qquad = 13 & \\
x \qquad = 1 &
\end{array}
$$

Substituting $x = 1$ into equation (4), we have

$$
\begin{aligned}
3(1) + 2z &= 9 \\
2z &= 6 \\
z &= 3
\end{aligned}
$$

Using $x = 1$ and $z = 3$ in equation (1) gives us

$$
\begin{aligned}
1 + y + 3 &= 6 \\
y + 4 &= 6 \\
y &= 2
\end{aligned}
$$

The solution is the ordered triple $(1, 2, 3)$.

2. Solve the system.

$$
\begin{aligned}
3x - 2y + z &= 2 \\
3x + y + 3z &= 7 \\
x + 4y - z &= 4
\end{aligned}
$$

EXAMPLE 2 Solve the system.

$$
\begin{array}{ll}
2x + y - z = 3 & (1) \\
3x + 4y + z = 6 & (2) \\
2x - 3y + z = 1 & (3)
\end{array}
$$

Solution It is easiest to eliminate z from the equations using the elimination method. The equation produced by adding (1) and (2) is

$$
5x + 5y = 9 \qquad (4)
$$

The equation that results from adding (1) and (3) is

$$
4x - 2y = 4 \qquad (5)
$$

Answer

2. $(1, 1, 1)$

Equations (4) and (5) form a linear system in two variables. We can eliminate the variable y from this system as follows:

$$5x + 5y = 9 \xrightarrow{\text{Multiply by 2.}} 10x + 10y = 18$$

$$4x - 2y = 4 \xrightarrow[\text{Multiply by 5.}]{} 20x - 10y = 20$$

$$30x \qquad = 38$$

$$x = \frac{38}{30}$$

$$= \frac{19}{15}$$

Substituting $x = \frac{19}{15}$ into equation (5) or equation (4) and solving for y gives

$$y = \frac{8}{15}$$

Using $x = \frac{19}{15}$ and $y = \frac{8}{15}$ in equation (1), (2), or (3) and solving for z results in

$$z = \frac{1}{15}$$

The ordered triple that satisfies all three equations is $\left(\frac{19}{15}, \frac{8}{15}, \frac{1}{15} \right)$.

EXAMPLE 3 Solve the system.

$$2x + 3y - z = 5 \qquad (1)$$

$$4x + 6y - 2z = 10 \qquad (2)$$

$$x - 4y + 3z = 5 \qquad (3)$$

Solution Multiplying equation (1) by -2 and adding the result to equation (2) looks like this:

$$-4x - 6y + 2z = -10 \qquad -2 \text{ times } (1)$$

$$4x + 6y - 2z = 10 \qquad (2)$$

$$0 = 0$$

All three variables have been eliminated, and we are left with a true statement. This implies that the two equations are dependent. With a system of three equations in three variables, however, a dependent system can have no solution or an infinite number of solutions. After we have concluded the examples in this section, we will discuss the geometry behind these systems. Doing so will give you some additional insight into dependent systems.

EXAMPLE 4 Solve the system.

$$x - 5y + 4z = 8 \qquad (1)$$

$$3x + y - 2z = 7 \qquad (2)$$

$$-9x - 3y + 6z = 5 \qquad (3)$$

Solution Multiplying equation (2) by 3 and adding the result to equation (3) produces

$$9x + 3y - 6z = 21 \qquad 3 \text{ times } (2)$$

$$-9x - 3y + 6z = 5 \qquad (3)$$

$$0 = 26$$

In this case, all three variables have been eliminated, and we are left with a false statement. The two equations are inconsistent; there are no ordered triples that satisfy both equations. There is no solution to the system. If equations (2) and (3) have no ordered triples in common, then certainly (1), (2), and (3) do not either.

3. Solve the system.

$$3x + 5y - 2z = 1$$
$$4x - 3y - z = 5$$
$$x - 8y + z = 4$$

4. Solve the system.

$$3x - y + 2z = 4$$
$$-2x + 2y - 5z = 1$$
$$5x - 3y + 7z = 5$$

Answers

3. Dependent (no unique solution)
4. Inconsistent (no solution)

5. Solve the system.
$$x + 2y = 0$$
$$3y + z = -3$$
$$2x - z = 5$$

EXAMPLE 5 Solve the system.

$$x + 3y = 5 \quad (1)$$
$$6y + z = 12 \quad (2)$$
$$x - 2z = -10 \quad (3)$$

Solution It may be helpful to rewrite the system as

$$x + 3y \quad\quad = 5 \quad (1)$$
$$6y + z = 12 \quad (2)$$
$$x \quad\quad - 2z = -10 \quad (3)$$

Equation (2) does not contain the variable x. If we multiply equation (3) by -1 and add the result to equation (1), we will be left with another equation that does not contain the variable x.

$$x + 3y \quad\quad = 5 \quad\quad (1)$$
$$-x \quad\quad + 2z = 10 \quad\quad -1 \text{ times (3)}$$
$$\overline{\quad\quad 3y + 2z = 15 \quad\quad (4)}$$

Equations (2) and (4) form a linear system in two variables. Multiplying equation (2) by -2 and adding the result to equation (4) eliminates the variable z.

$$6y + z = 12 \quad \xrightarrow{\text{Multiply by } -2.} \quad -12y - 2z = -24$$
$$3y + 2z = 15 \quad \xrightarrow[\text{No Change}]{} \quad \underline{\quad 3y + 2z = 15 \quad}$$
$$-9y + \quad = -9$$
$$y = 1$$

Using $y = 1$ in equation (4) and solving for z, we have

$$z = 6$$

Substituting $y = 1$ into equation (1) gives

$$x = 2$$

The ordered triple that satisfies all three equations is $(2, 1, 6)$.

The Geometry Behind Linear Equations in Three Variables

We can graph an ordered triple on a coordinate system with three axes. The graph will be a point in space. The coordinate system is drawn in perspective; you have to imagine that the x-axis comes out of the paper and is perpendicular to both the y-axis and the z-axis. To graph the point $(3, 4, 5)$, we move 3 units in the x-direction, 4 units in the y-direction, and then 5 units in the z-direction, as shown in Figure 1.

FIGURE 1

 Although in actual practice it is sometimes difficult to graph equations in three variables, if we were to graph a linear equation in three variables, we would find that the graph was a plane in space. A system of three equations in three variables is represented by three planes in space.

Answer

5. $(4, -2, 3)$

There are a number of possible ways in which these three planes can intersect, some of which are shown below. And there are still other possibilities that are not among these shown.

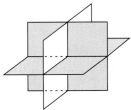

CASE 1: ONE SOLUTION
The three planes have exactly one point in common. In this case we get one solution to our system, as in Examples 1, 2, and 5.

CASE 2: INCONSISTENT SYSTEM, THREE PARALLEL PLANES
The three planes have no points in common because they are all parallel to one another. The system they represent is an inconsistent system.

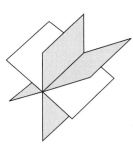

CASE 3: DEPENDENT SYSTEM
The three planes intersect in a line. Any point on the line is a solution to the system of equations represented by the planes, so there is an infinite number of solutions to the system. This is an example of a dependent system.

CASE 4: INCONSISTENT SYSTEM, TWO PARALLEL LINES
Two of the planes are parallel; the third plane intersects each of the parallel planes. In this case, the three planes have no points in common. There is no solution to the system; it is an inconsistent system.

In Example 3, we found that equations (1) and (2) were dependent equations. They represent the same plane. That is, they have all their points in common. But the system of equations that they came from has either no solution or an infinite number of solutions. It all depends on the third plane. If the third plane coincides with the first two, then the solution to the system is a plane. If the third plane is parallel to the first two, then there is no solution to the system. Finally, if the third plane intersects the first two but does not coincide with them, then the solution to the system is that line of intersection.

In Example 4, we found that trying to eliminate a variable from the second and third equations resulted in a false statement. This means that the two planes represented by these equations are parallel. It makes no difference where the third plane is; there is no solution to the system in Example 4. (If we were to graph the three planes from Example 4, we would obtain a diagram similar to Case 2 or Case 4 above.)

If, in the process of solving a system of linear equations in three variables, we eliminate all the variables from a pair of equations and are left with a false statement, we will say the system is inconsistent. If we eliminate all the variables and are left with a true statement, then we will say the system is a dependent one.

GETTING READY FOR CLASS

After reading through the preceding section, respond in your own words and in complete sentences.

A. What is an ordered triple of numbers?

B. Explain what it means for (1, 2, 3) to be a solution to a system of liner equations in three variables.

C. Explain in a general way the procedure you would use to solve a system of three linear equations in three variables.

D. How do you know when a system of linear equations in three variables has no solution?

Vocabulary Review

Choose the correct words to fill in the blanks below.

 dependent triples infinite inconsistent

1. The solution set for a system of three linear equations in three variables is the set of ordered _____ that satisfies all three equations.

2. If three variables are eliminated and we are left with a true statement, we consider the system to be _____.

3. With a system of three equations in three variables, a dependent system can have no solution or an _____ number of solutions.

4. If three variables are eliminated and we are left with a false statement, we consider the system to be _____ with no solutions.

Problems

A Solve the following systems.

1. $x + y + z = 4$
$x - y + 2z = 1$
$x - y - 3z = -4$

2. $x - y - 2z = -1$
$x + y + z = 6$
$x + y - z = 4$

3. $x + y + z = 6$
$x - y + 2z = 7$
$2x - y - 4z = -9$

4. $x + y + z = \ \ \ 0$
$x + y - z = \ \ \ 6$
$x - y + 2z = -7$

5. $x + 2y + z = 3$
$2x - y + 2z = 6$
$3x + y - z = 5$

6. $2x + y - 3z = -14$
$x - 3y + 4z = 22$
$3x + 2y + z = 0$

7. $2x + 3y - 2z = 4$
$x + 3y - 3z = 4$
$3x - 6y + z = -3$

8. $4x + y - 2z = 0$
$2x - 3y + 3z = 9$
$-6x - 2y + z = 0$

9. $-x + 4y - 3z = 2$
$2x - 8y + 6z = 1$
$3x - \ \ y + z = 0$

10. $4x + 6y - 8z = 1$
$-6x - 9y + 12z = 0$
$x - 2y - 2z = 3$

11. $\frac{1}{2}x - y + z = 0$
$2x + \frac{1}{3}y + z = \ \ \ 2$
$x + y + z = -4$

12. $\frac{1}{3}x + \frac{1}{2}y + z = -1$
$x - y + \frac{1}{5}z = -1$
$x + y + z = -5$

13. $2x - y - 3z = 1$
$x + 2y + 4z = 3$
$4x - 2y - 6z = 2$

14. $3x + 2y + z = 3$
$x - 3y + z = 4$
$-6x - 4y - 2z = 1$

15. $2x - y + 3z = 4$
$x + 2y - z = -3$
$4x + 3y + 2z = -5$

16. $6x - 2y + z = 5$
$3x + y + 3z = 7$
$x + 4y - z = 4$

17. $x + y = 9$
$y + z = 7$
$x - z = 2$

18. $x - y = -3$
$x + z = 2$
$y - z = 7$

19. $2x + y = 2$
$y + z = 3$
$4x - z = 0$

20. $2x + y = 6$
$3y - 2z = -8$
$x + z = 5$

21. $2x - 3y = 0$
$6y - 4z = 1$
$x + 2z = 1$

22. $3x + 2y = 3$
$y + 2z = 2$
$6x - 4z = 1$

23. $x + y - z = 2$
$2x + y + 3z = 4$
$x - 2y + 2z = 6$

24. $x + 2y - 2z = 4$
$3x + 4y - z = -2$
$2x + 3y - 3z = -5$

25. $2x + 3y = -\dfrac{1}{2}$
$4x + 8z = 2$
$3y + 2z = -\dfrac{3}{4}$

26. $3x - 5y = 2$
$4x + 6z = \dfrac{1}{3}$
$5y - 7z = \dfrac{1}{6}$

27. $\dfrac{1}{3}x + \dfrac{1}{2}y - \dfrac{1}{6}z = 4$
$\dfrac{1}{4}x - \dfrac{3}{4}y + \dfrac{1}{2}z = \dfrac{3}{2}$
$\dfrac{1}{2}x - \dfrac{2}{3}y - \dfrac{1}{4}z = -\dfrac{16}{3}$

28. $-\dfrac{1}{4}x + \dfrac{3}{8}y + \dfrac{1}{2}z = -1$
$\dfrac{2}{3}x - \dfrac{1}{6}y - \dfrac{1}{2}z = 2$
$\dfrac{3}{4}x - \dfrac{1}{2}y - \dfrac{1}{8}z = 1$

29. $x - \dfrac{1}{2}y - \dfrac{1}{3}z = -\dfrac{4}{3}$
$\dfrac{1}{3}x - \dfrac{1}{2}z = 5$
$-\dfrac{1}{4}x + \dfrac{2}{3}y - z = -\dfrac{3}{4}$

30. $x + \dfrac{1}{3}y - \dfrac{1}{2}z = -\dfrac{3}{2}$
$\dfrac{1}{2}x - y + \dfrac{1}{3}z = 8$
$\dfrac{1}{3}x - \dfrac{1}{4}y - z = -\dfrac{5}{6}$

Applying the Concepts

31. Electric Current In the following diagram of an electrical circuit, x, y, and z represent the amount of current (in amperes) flowing across the 5-ohm, 20-ohm, and 10-ohm resistors, respectively. (In circuit diagrams, resistors are represented by ⧦ and cells that provide electrical energy are ⊣⊢.)

The system of equations used to find the three currents x, y, and z is

$$x - y - z = 0$$
$$5x + 20y = 80$$
$$20y - 10z = 50$$

Solve the system for all variables.

32. Cost of a Rental Car If a car rental company charges $10 a day and 8¢ a mile to rent one of its cars, then the cost z, in dollars, to rent a car for x, days and drive y miles can be found from the equation

$$z = 10x + 0.08y$$

a. How much does it cost to rent a car for 2 days and drive it 200 miles under these conditions?

b. A second company charges $12 a day and 6¢ a mile for the same car. Write an equation that gives the cost z, in dollars, to rent a car from this company for x days and drive it y miles.

c. A car is rented from each of the companies mentioned in a. and b. for 2 days. To find the mileage at which the cost of renting the cars from each of the two companies will be equal, solve the following system for y:

$$z = 10x + 0.08y$$
$$z = 12x + 0.06y$$
$$x = 2$$

33. Hot Dogs Refer to the hot dog eating contest at the beginning of this section. Set up a system of three equations in three unknowns, then solve it, to find the number of hot dog buns Susie, Minnie, and Bunny each ate.

34. Hot Dogs If the organizers of the hot dog eating contest paid $1.20 for each dozen hot dog buns, how much did they spend on the buns consumed by the elephants and the humans?

Getting Ready for the Next Section

Translate into symbols.

35. Two more than 3 times a number

36. One less than twice a number

Apply the distributive property, then simplify.

37. $10(0.2x + 0.5y)$

38. $100(0.09x + 0.08y)$

Solve.

39. $x + (3x + 2) = 26$

40. $5x = 2,500$

Solve each system.

41. $-2y - 4z = -18$
$\quad -7y + 4z = 27$

42. $\quad -x + 2y = 200$
$\quad\quad 4x - 2y = 1{,}300$

43. The sum of three numbers is 20. The largest number equals the sum of the two smaller numbers, and the middle number is one more than twice the smaller number. Find these numbers.

44. The sum of three numbers is 34. The largest number is six more than the sum of the two smaller numbers, and the middle number is one less than twice the smaller number. Find the three numbers.

45. A 20 foot board is cut into three pieces. The longest piece is 2 ft. more than the sum of the two smaller pieces, and the middle piece is twice the length of the shortest piece. Find the length of each piece.

46. A 10 foot board is cut into three pieces. The longest piece is the sum of the lengths of the two smaller pieces, and the middle piece is one foot longer than the shortest piece. Find the length of each piece.

Find the Mistake

Each sentence below contains a mistake. Circle the mistake and write the correct word or phrase on the line provided.

1. The solution to the following system is (3, 3, 2). _____
$\quad x + y + z = -2$
$\quad 2x + 3y + z = -3$
$\quad 2x - y - z = 11$

2. It is easiest to solve the following system by eliminating y from each equation. _____
$\quad x + 2y + z = -7$
$\quad 2x + y - z = 4$
$\quad 2x - 2y + z = 2$

3. The following system has no solution because the equations represent parallel lines. _____
$\quad x + 2y + 2 = -8$
$\quad 2x + y + 2z = -7$
$\quad 4x + 2y + 4z = -14$

4. The following system has an infinite number of solutions. _____
$\quad x + 3y + z = 14$
$\quad 2x - y - 3z = 2$
$\quad 2x + 2y + 2z = 16$

OBJECTIVES

A Graph the solution to a system of linear inequalities in two variables.

B Solve application problems using systems of inequalities, including break-even analysis and supply-demand equilibrium equations.

KEY WORDS

system of linear inequalities

fixed cost

variable cost

revenue

profit

break even point

equilibrium point

supply

demand

**Video Examples
Section 10.5**

Imagine you work at a local coffee house and your boss wants you to create a new house blend using the three most popular coffees: the dark roast, the medium roast, and the flavored blend. The dark roast coffee costs $20.00 per pound, the medium roast costs $14.50 per pound, and the flavored blend costs $16.00 per pound. Your boss wants the cost of the new blend to be at most $17.00 per pound and is requesting to see possible combinations to serve customers. As we have discussed, words like "at most" imply the use of inequalities. Working through this section will help us visualize solutions to systems of linear inequalities in two and three variables, such as the coffee house example.

A Graphing Solutions to Linear Inequalities

In Section 2.7, we graphed linear inequalities in two variables. To review, we graph the boundary line, using a solid line if the boundary is part of the solution set and a broken line if the boundary is not part of the solution set. Then we test any point that is not on the boundary line in the original inequality. A true statement tells us that the point lies in the solution set, a false statement tells us the solution set is the other region.

Figure 1 shows the graph of the inequality $x + y < 4$. Note that the boundary is not included in the solution set, and is therefore drawn with a broken line. Figure 2 shows the graph of $-x + y \le 3$. Note that the boundary is drawn with a solid line, because it is part of the solution set.

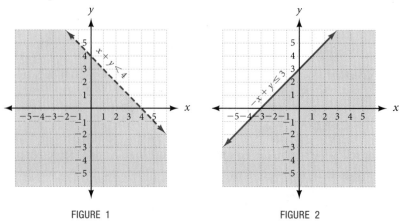

FIGURE 1 FIGURE 2

If we form a system of inequalities with the two inequalities, the solution set will be all the points common to both solution sets shown in the two figures above. It is the intersection of the two solution sets.

Therefore, the solution set for the system of inequalities from Figure 1 and 2 is all the ordered pairs that satisfy both inequalities. It is the set of points that are below the line $x + y = 4$, and also below (and including) the line $-x + y = 3$. The graph of the solution set to this system is shown in Figure 3. We have written the system in Figure 3 with the word *and* just to remind you that the solution set to a system of equations or inequalities is all the points that satisfy both equations or inequalities.

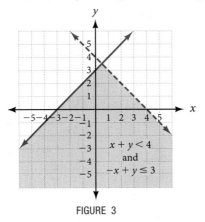

$$x + y < 4$$
$$\text{and}$$
$$-x + y \le 3$$

FIGURE 3

EXAMPLE 1 Graph the solution to the system of linear inequalities.

$$y < \frac{1}{2}x + 3$$

$$y \ge \frac{1}{2}x - 2$$

Solution Figures 4 and 5 show the solution set for each of the inequalities separately.

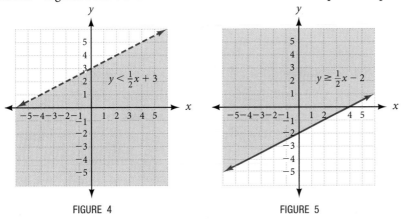

FIGURE 4 FIGURE 5

Figure 6 is the solution set to the system of inequalities. It is the region consisting of points whose coordinates satisfy both inequalities.

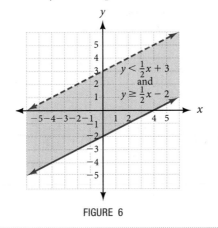

$$y < \frac{1}{2}x + 3$$
$$\text{and}$$
$$y \ge \frac{1}{2}x - 2$$

FIGURE 6

Practice Problems

1. Graph the solution set to the system.

$$y \le \frac{1}{3}x + 1$$

$$y > \frac{1}{3}x - 3$$

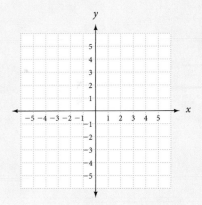

2. Graph the solution set to the system.

$$x + y < 3$$
$$x \geq 0$$
$$y \geq 0$$

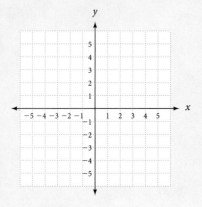

EXAMPLE 2 Graph the solution to the system of linear inequalities.

$$x + y < 4$$
$$x \geq 0$$
$$y \geq 0$$

Solution We graphed the first inequality, $x + y < 4$, in Figure 1 at the beginning of this section. The solution set to the inequality $x \geq 0$, shown in Figure 7, is all the points to the right of the y-axis; that is, all the points with x-coordinates that are greater than or equal to 0. Figure 8 shows the graph of $y \geq 0$. It consists of all points with y-coordinates greater than or equal to 0; that is, all points from the x-axis up.

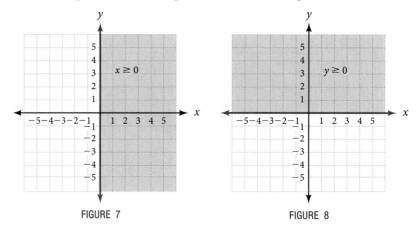

FIGURE 7 FIGURE 8

The regions shown in Figures 7 and 8 overlap in the first quadrant. Therefore, putting all three regions together we have the points in the first quadrant that are below the line $x + y = 4$. This region is shown in Figure 9, and it is the solution to our system of inequalities.

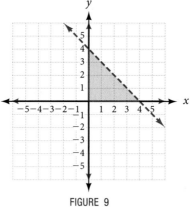

FIGURE 9

Extending the discussion in Example 2 we can name the points in each of the four quadrants using systems of inequalities.

FIGURE 10 FIGURE 11

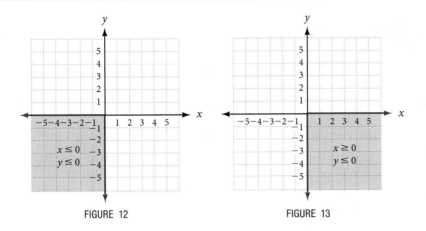

FIGURE 12 FIGURE 13

EXAMPLE 3 Graph the solution to the system of linear inequalities.

$$x \leq 4$$
$$y \geq -3$$

Solution The solution to this system will consist of all points to the left of and including the vertical line $x = 4$ that intersect with all points above and including the horizontal line $y = -3$. The solution set is shown in Figure 14.

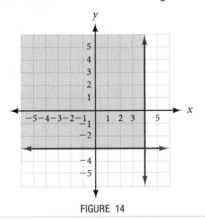

FIGURE 14

EXAMPLE 4 Graph the solution set for the following system.

$$x - 2y \leq 4$$
$$x + y \leq 4$$
$$x \geq -1$$

Solution We have three linear inequalities, representing three sections of the coordinate plane. The graph of the solution set for this system will be the intersection of these three sections. The graph of $x - 2y \leq 4$ is the section above and including the boundary $x - 2y = 4$. The graph of $x + y \leq 4$ is the section below and including the boundary line $x + y = 4$. The graph of $x \geq -1$ is all the points to the right of, and including, the vertical line $x = -1$. The intersection of these three graphs is shown in Figure 15.

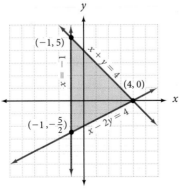

FIGURE 15

3. Graph the solution to the system
$$x > -3$$
$$y < 4$$

4. Graph the solution set for the following system.
$$x - y < 5$$
$$x + y < 5$$
$$x > 1$$

5. How would the graph for Example 5 change if they had reserved only 300 tickets at the $15 rate?

B Application Problems

EXAMPLE 5 A college basketball arena plans on charging $20 for certain seats and $15 for others. They want to bring in more than $18,000 from all ticket sales and have reserved at least 500 tickets at the $15 rate. Find a system of inequalities describing all possibilities and sketch the graph. If 620 tickets are sold for $15, at least how many tickets are sold for $20?

Solution Let x = the number of $20 tickets and y = the number of $15 tickets. We need to write a list of inequalities that describe this situation. That list will form our system of inequalities. First of all, we note that we cannot use negative numbers for either x or y. So, we have our first inequalities:

$$x \geq 0$$

$$y \geq 0$$

Next, we note that they are selling at least 500 tickets for $15, so we can replace our second inequality with $y \geq 500$. Now our system is

$$x \geq 0$$

$$y \geq 500$$

Now the amount of money brought in by selling $20 tickets is $20x$, and the amount of money brought in by selling $15 tickets is $15y$. It the total income from ticket sales is to be more than $18,000, then $20x + 15y$ must be greater than 18,000. This gives us our last inequality and completes our system.

$$20x + 15y > 18,000$$

$$x \geq 0$$

$$y \geq 500$$

We have used all the information in the problem to arrive at this system of inequalities. The solution set contains all the values of x and y that satisfy all the conditions given in the problem. Figure 16 is the graph of the solution set.

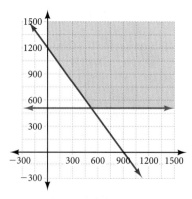

FIGURE 16

If 620 tickets are sold for $15, then we substitute 620 for y in our first inequality to obtain

$$20x + 15(620) > 18000 \qquad \text{Substitute 620 for y.}$$

$$20x + 9300 > 18000 \qquad \text{Multiply.}$$

$$20x > 8700 \qquad \text{Add } -9300 \text{ to each side.}$$

$$x > 435 \qquad \text{Divide each side by 20.}$$

If they sell 620 tickets for $15 each, then they need to sell more than 435 tickets at $20 each to bring in more than $18,000.

Answer

5. The horizontal line would move down to $y = 300$ from $y = 500$.

EXAMPLE 6 Make-A-Sale is a new company designed to assist clients selling household items over the internet. Start-up costs for the company are $40,000, and they expect variable costs to be $6.00 for each item sold. They plan to charge $14 per item.

a. Find the total cost, revenue, and profit for selling x items.

b. If they sell 3,000 items, find their cost, revenue, and profit.

c. If they sell 12,000 items, find their cost, revenue, and profit.

d. Find the break-even point (when revenue equals cost). Include a graph of the inequalities $R(x) \leq C(x)$ and $R(x) \geq C(x)$ to show intervals of profit and loss.

Solution

a. The total cost includes the fixed cost ($40,000) plus the variable cost ($6 per item). So the total cost is given by $C(x) = 40,000 + 6x$.

The revenue is the price times the number sold, so the revenue is $R(x) = 14x$.

The profit is the revenue minus the cost, so the profit is given by

$$P(x) = R(x) - C(x)$$
$$= (14x) - (40,000 + 6x)$$
$$= 8x - 40,000$$

b. Substituting $x = 3,000$ into the cost, revenue, and profit functions:

$$C(3,000) = 40,000 + 6(3,000) = \$58,000$$
$$R(3,000) = 14(3,000) = \$42,000$$
$$P(3,000) = 8(3,000) - 40,000 = -\$16,000$$

Note that $P(3,000) = R(3,000) - C(3,000)$

c. Substituting $x = 12,000$ into the cost, revenue, and profit functions:

$$C(12,000) = 40,000 + 6(12,000) = \$112,000$$
$$R(12,000) = 14(12,000) = \$168,000$$
$$P(12,000) = 8(12,000) - 40,000 = \$56,000$$

Again, note that $P(12,000) = R(12,000) - C(12,000)$

d. Set the revenue equal to the cost to find the break-even point.

$$R(x) = C(x)$$
$$14x = 40,000 + 6x$$
$$8x = 40,000$$
$$x = 5,000$$

Figure 17 indicates areas of profit ($x > 5,000$) and areas of loss ($x < 5,000$).

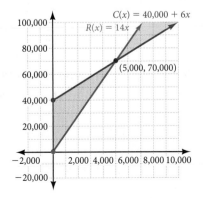

FIGURE 17

6. Suppose a company's fixed costs are $70,000 and variable costs are $8 per item. They plan to charge $15 per item.

a. Find the total cost, revenue, and profit for selling x items.

b. If they sell 7,500 items, find their cost, revenue, and profit.

c. If they sell 20,000 items, find their cost, revenue, and profit.

d. Find the break-even point and sketch the areas of profit and loss.

Answers

6 a. $C(x) = 70,000 + 8x$
$R(x) = 15x$
$P(x) = 7x - 70,000$

b $C(7,500) = \$130,000$
$R(7,500) = \$112,500$
$P(7,500) = \$17,500$

c. $C(20,000) = \$230,000$
$R(20,000) = \$300,000$
$P(20,000) = -\$70,000$

d. 10,000 items

As the price (p) of an item increases, the supply (s) will increase since producers will want to sell at the higher price. However, the demand (d) will decrease since consumers will not want to purchase the item at a higher price. The equilibrium point represents the price and quantity that the consumer and the producer agree.

FIGURE 18

7. Find the equilibrium price and quantity for the following supply and demand equations.

$$d(p) = 7,000 - 12p$$
$$s(p) = 1,000 + 18p$$

EXAMPLE 7 Find the equilibrium price and quantity for the following supply and demand equations:

$$d(p) = 5,000 - 4p$$
$$s(p) = 1,000 + 6p$$

Solution The equilibrium point occurs when $d(p) = s(p)$

$$5,000 - 4p = 1,000 + 6p$$
$$5,000 - 10p = 1,000$$
$$-10p = -4,000$$
$$p = 400$$

The equilibrium price is $400 per unit. Substituting this price into either the supply or demand function, we get

$$d(400) = 5,000 - 4(400) = 3,400$$
$$p(400) = 1,000 + 6(400) = 3,400$$

The equilibrium quantity is 3,400.

GETTING READY FOR CLASS

After reading through the preceding section, respond in your own words and in complete sentences.

A. What is the solution set for a system of inequalities?

B. If an ordered pair satisfied only one inequality in a system of inequalities, is the ordered pair part of the solution set for that system? Explain.

C. Explain how you would graph the solution set for a system of three linear inequalities.

D. Give an example of an application problem that involves a system of linear inequalities.

Vocabulary Review

Choose the correct words to fill in the blanks below.

shaded first second

third fourth common

1. The solution set to a system of inequalities will be all the points _____ to both solution sets.

2. When graphing a system of inequalities, the solution set is represented by a _____ portion of the graph.

3. For the system $x \leq 0$ and $y \geq 0$, the solution set consists of points in the _____ quadrant.

4. For the system $x \geq 0$ and $y \geq 0$, the solution set consists of points in the _____ quadrant.

5. For the system $x \geq 0$ and $y \leq 0$, the solution set consists of points in the _____ quadrant.

6. For the system $x \leq 0$ and $y \leq 0$, the solution set consists of points in the _____ quadrant.

Problems

A Graph the solution set for each system of linear inequalities.

1. $x + y < 5$
 $2x - y > 4$

2. $x + y < 5$
 $2x - y < 4$

3. $y < \dfrac{1}{3}x + 4$

 $y \geq \dfrac{1}{3}x - 3$

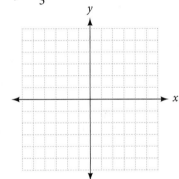

4. $y < 2x + 4$
 $y \geq 2x - 3$

5. $x \geq -3$
 $y < -2$

6. $x \leq 4$
 $y \geq -2$

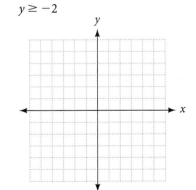

7. $1 \leq x \leq 3$
$2 \leq y \leq 4$

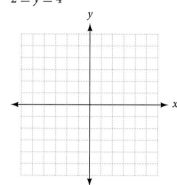

8. $-4 \leq x \leq -2$
$1 \leq y \leq 3$

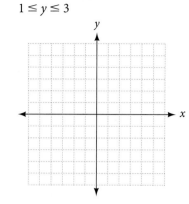

9. $x + y \leq 4$
$x \geq 0$
$y \geq 0$

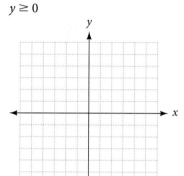

10. $x - y \leq 2$
$x \geq 0$
$y \leq 0$

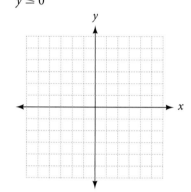

11. $x + y \leq 3$
$x - 3y \leq 3$
$x \geq -2$

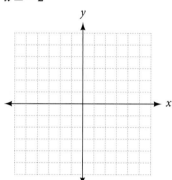

12. $x - y \leq 4$
$x + 2y \leq 4$
$x \geq -1$

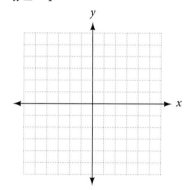

13. $x + y \leq 2$
$-x + y \leq 2$
$y \geq -2$

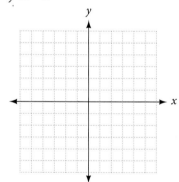

14. $x - y \leq 3$
$-x - y \leq 3$
$y \leq -1$

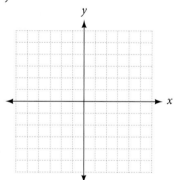

15. $x + y < 5$
$y > x$
$y \geq 0$

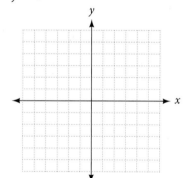

16. $x + y < 5$
$y > x$
$x \geq 0$

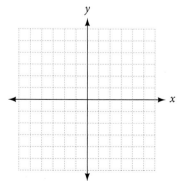

17. $2x + 3y \leq 6$
$x \geq 0$
$y \geq 0$

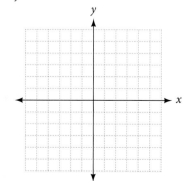

18. $x + 2y \leq 10$
$3x + 2y \leq 12$
$x \geq 0$
$y \geq 0$

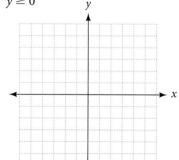

19. $x + 4y > 12$
$2x + y > 6$
$x > 0$
$y > 0$

20. $3x + y > 6$
$2x + 3y > 12$
$x > 0$
$y > 0$

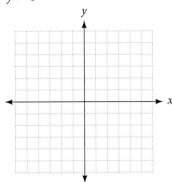

For each figure below, find a system of inequalities that describes the shaded region.

21.

FIGURE 19

22.

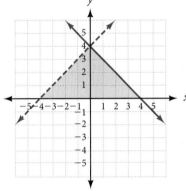

FIGURE 20

For each figure below, find a system of inequalities that describes the shaded region.

23.

FIGURE 21

24.

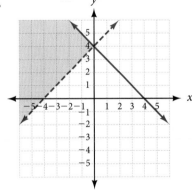

FIGURE 22

Applying the Concepts

25. Office Supplies An office worker wants to purchase some $0.55 postage stamps and also some $0.65 postage stamps totaling no more than $40. She also wants to have at least twice as many $0.55 stamps and more than 15 $0.55 stamps.

a. Find a system of inequalities describing all the possibilities and sketch the graph.

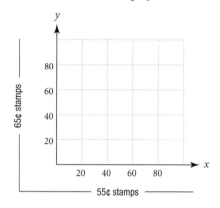

b. If she purchases 20 $0.55 stamps, what is the maximum number of $0.65 stamps she can purchase?

26. Inventory A store sells two brands of DVD players. Customer demand indicates that it is necessary to stock at least twice as many DVD players of brand A as of brand B. At least 30 of brand A and 15 of brand B must be on hand. In the store, there is room for not more than 100 DVD players.

a. Find a system of inequalities describing all possibilities, then sketch the graph.

b. If there are 35 DVD players of brand A, what is the most number of brand B DVD players on hand?

Find the break-even point for the following fixed costs, variable costs, and price. Include a graph showing areas of profit and loss.

27. Fixed cost = $24,000
Variable cost = $14
Price = $22

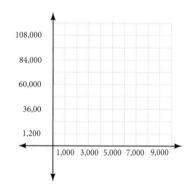

28. Fixed cost = $50,000
Variable cost = $24
Price = $44

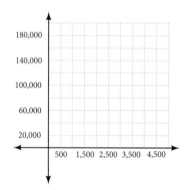

29. Fixed cost = $120,000
Variable cost = $175
Price = $225

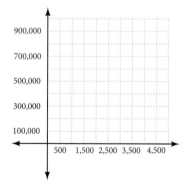

30. Fixed cost = $175,000
Variable cost = $215
Price = $315

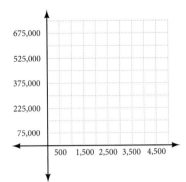

31. Fixed cost = $19,500
Variable cost = $6.50
Price = $9.50

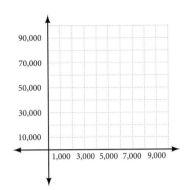

32. Fixed cost = $21,300
Variable cost = $9.75
Price = $12.25

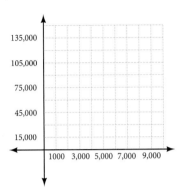

33. Fixed cost = $1,200
Variable cost = $8.25
Price = $9.75

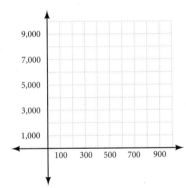

34. Fixed cost = $6,000
Variable cost = $11.80
Price = $14.30

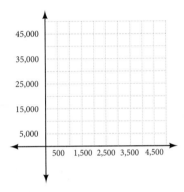

Find the equilibrium price and quantity for the following supply and demand equations.

35. $d(p) = 4,200 - 6p$
$s(p) = 4p$

36. $d(p) = 6,700 - 9p$
$s(p) = 2,200 + 6p$

37. $d(p) = 10,000 - 11p$
$s(p) = 2,000 + 5p$

38. $d(p) = 12,500 - 17p$
$s(p) = 2,500 + 3p$

39. $d(p) = 24,000 - 24p$
$s(p) = 6,000 + 6p$

40. $d(p) = 300,000 - 47p$
$s(p) = 3p$

41. Websites View-Me is a company which assists clients in creating and maintaining website addresses. Start-up costs for the company are $75,000, and they expect monthly costs to maintain each website to be $24. They plan to charge $40 per month for their service.

 a. Find the total cost, revenue, and profit for maintaining *x* websites.

 b. If they maintain 2,000 websites, find their cost, revenue, and profit.

 c. If they maintain 8,000 websites, find their cost, revenue, and profit.

 d. Find the break-even point. Include a graph to show intervals of profit and loss.

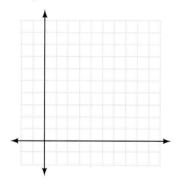

42. Tax Returns Give-It-Back is a company which prepares tax returns for clients. Start-up costs for the company are $24,000, and each return costs the company $48 to process and file. They charge $80 to prepare an individual tax return.

 a. Find the total cost, revenue, and profit for preparing *x* tax returns.

 b. If they prepare 500 tax returns, find their cost, revenue, and profit.

 c. Find the break-even point. Include a graph to show intervals of profit and loss.

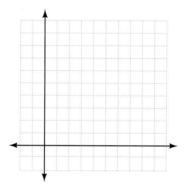

 d. If the company wants to make a net profit of $60,000, how many tax returns would they need to prepare?

Maintaining Your Skills

For each of the following straight lines, identify the *x*-intercept, *y*-intercept, and slope, and sketch the graph.

43. $2x + y = 6$

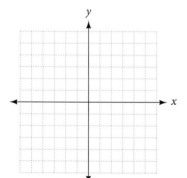

44. $y = \dfrac{3}{2}x + 4$

45. $x = -2$

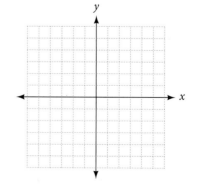

Find the equation for each line.

46. Give the equation of the line through $(-1, 3)$ that has slope $m = 2$.

47. Give the equation of the line through $(-3, 2)$ and $(4, -1)$.

48. Line l contains the point $(5, -3)$ and has a graph parallel to the graph of $2x - 5y = 10$. Find the equation for l.

49. Give the equation of the vertical line through $(4, -7)$.

State the domain and range for the following relations, and indicate which relations are also functions.

50. $\{(-2, 0), (-3, 0), (-2, 1)\}$

51. $y = x^2 - 9$

Let $f(x) = x - 2$, $g(x) = 3x + 4$ and $h(x) = 3x^2 - 2x - 8$, and find the following.

52. $f(3) + g(2)$

53. $h(0) + g(0)$

54. $f(g(2))$

55. $g(f(2))$

56. $h(f(x))$

57. $f(g(x))$

58. $\dfrac{h(x)}{f(x)}$

59. $\dfrac{g(x)}{h(x)}$

Find the Mistake

Each sentence below contains a mistake. Circle the mistake and write the correct word or phrase on the line provided.

1. The solution set for the system of inequalities $x \leq 0$ and $y \geq 0$ is all the points in the first quadrant only.

2. The solution set for the system of inequalities $x \leq 0$ and $y \leq 0$ is all the points in the second quadrant.

Write true or false for each statement below. Graph each system to support your answer.

3. True or False: The solution for the following system of inequalities contains the point $(-1, 2)$.

$$x - y \leq 2$$
$$x + 3y \leq -2$$
$$x \geq -2$$

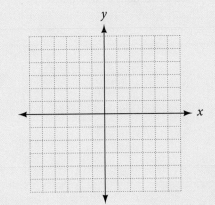

4. True or False: The solution for the following system of inequalities contains the point $(3, 3)$.

$$3x + 2y \geq 12$$
$$3x - y \geq -6$$
$$y \geq 0$$

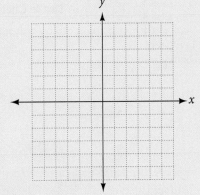

Definitions [10.1]

1. A *system of linear equations*, as the term is used in this book, is two linear equations that each contain the same two variables.

2. The *solution set* for a system of equations is the set of all ordered pairs that satisfy *both* equations. The solution set to a system of linear equations will contain:

 Case I One ordered pair when the graphs of the two equations intersect at only one point (this is the most common situation)

 Case II No ordered pairs when the graphs of the two equations are parallel lines

 Case III An infinite number of ordered pairs when the graphs of the two equations coincide (are the same line)

EXAMPLES

1. The solution to the system
$$x + 2y = 4$$
$$x - y = 1$$
is the ordered pair (2, 1). It is the only ordered pair that satisfies both equations.

Strategy for Solving a System by Graphing [10.1]

Step 1: Graph the first equation.

Step 2: Graph the second equation on the same set of axes.

Step 3: Read the coordinates of the point where the graphs cross each other (the coordinates of the point of intersection.)

Step 4: Check the solution to see that it satisfies *both* equations.

2. Solving the system in Example 1 by graphing looks like

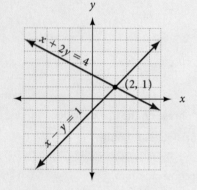

Strategy for Solving a System by the Substitution Method [10.2]

Step 1: Solve either of the equations for one of the variables (this step is not necessary if one of the equations has the correct form already.)

Step 2: Substitute the results of step 1 into the other equation, and solve.

Step 3: Substitute the results of step 2 into an equation with both x and y variables, and solve. (The equation produced in step 1 is usually a good one to use.)

Step 4: Check your solution, if necessary.

3. We can apply the substitution method to the system in Example 1 by first solving the second equation for x to get $x = y + 1$. Substituting this expression for x into the first equation, we have
$$(y + 1) + 2y = 4$$
$$3y + 1 = 4$$
$$3y = 3$$
$$y = 1$$
Using $y = 1$ in either of the original equations gives $x = 2$.

Strategy for Solving a System by the Elimination Method [10.3]

Step 1: Look the system over to decide which variable will be easier to eliminate.

Step 2: Use the multiplication property of equality on each equation separately to ensure that the coefficients of the variable to be eliminated are opposites.

Step 3: Add the left and right sides of the system produced in step 2, and solve the resulting equation.

Step 4: Substitute the solution from step 3 back into any equation with both x and y variables, and solve.

Step 5: Check your solution in both equations, if necessary.

4. We can eliminate the y variable from the system in Example 1 by multiplying both sides of the second equation by 2 and adding the result to the first equation

$$
\begin{array}{ll}
x + 2y = 4 \xrightarrow{\text{No Change}} & x + 2y = 4 \\
x - y = 1 \xrightarrow{\text{Multiply by 2}} & 2x - 2y = 2 \\
\hline
& 3x = 6 \\
& x = 2
\end{array}
$$

Substituting $x = 2$ into either of the original two equations gives $y = 1$. The solution is $(2, 1)$.

Special Cases [10.1, 10.2, 10.3]

In some cases, using the elimination or substitution method eliminates both variables. The situation is interpreted as follows.

1. If the resulting statement is *false*, then the lines are parallel and there is no solution to the system. This is called an inconsistent system.

2. If the resulting statement is *true*, then the equations represent the same line (the lines coincide). In this case any ordered pair that satisfies either equation is a solution to the system. This is called a dependent system.

Systems of Linear Equations [10.4]

A system of linear equations consists of two or more linear equations considered simultaneously. The solution set to a linear system in two variables is the set of ordered pairs that satisfy both equations. The solution set to a linear system in three variables consists of all the ordered triples that satisfy each equation in the system.

5. The solution to the system

$$
\begin{array}{l}
x + 2y = 4 \\
x - y = 1
\end{array}
$$

is the ordered pair $(2, 1)$. It is the only ordered pair that satisfies both equations.

Inconsistent and Dependent Equations [10.4]

A system of two linear equations that have no solutions in common is said to be an *inconsistent* system, whereas two linear equations that have all their solutions in common are said to be *dependent* equations.

6. If the two lines are parallel, then the system will be inconsistent and the solution is \varnothing. If the two lines coincide, then the equations are dependent.

Systems of Linear Inequalities [10.5]

A system of linear inequalities is two or more linear inequalities considered at the same time. To find the solution set to the system, we graph each of the inequalities on the same coordinate system. The solution set is the region that is common to all the regions graphed.

7. The solution set for the system

$$
\begin{array}{l}
x + y < 4 \\
-x + y \le 3
\end{array}
$$

is shown below.

CHAPTER 10 Test

1. Write the solution to the system which is graphed below.
 [10.1]

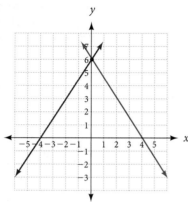

Solve each system by graphing. [10.1]

2. $4x - 2y = 8$
 $y = \frac{2}{3}x$

3. $3x - 2y = 13$
 $y = 4$

4. $2x - 2y = -12$
 $-3x - y = 2$

Solve each system by the substitution method. [10.2]

5. $3x - y = 12$
 $y = 2x - 8$

6. $3x + 6y = 3$
 $x = 4y - 17$

7. $2x - 3y = -18$
 $3x + y = -5$

8. $2x - 3y = 13$
 $x - 4y = -1$

Solve each system by the elimination method. [10.3]

9. $x - y = -9$
 $2x + 3y = 7$

10. $3x - y = 1$
 $5x - y = 3$

11. $2x + 3y = -3$
 $x + 6y = 12$

12. $2x + 3y = 4$
 $4x + 6y = 8$

13. Solve the system using the elimination method. [10.4]
 $x + 4y + 2z = 5$
 $x - 2y - 4z = -3$
 $4x - y - 2z = 2$

Graph the solution set for each system of linear inequalities.
[10.5]

14. $x + 2y < 6$
 $6x + 3y \geq 6$

15. $y \leq \frac{1}{4}x + 2$
 $x \geq 0$
 $y \geq 0$

Find a system of inequalities that describes the shaded region.
[10.6]

16.

17.

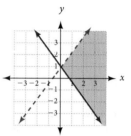

Inequalities, Graphs, and Functions

11

11.1 Interval Notation and Linear Inequalities

11.2 Sets and Compound Inequalities

11.3 Absolute Value Equations

11.4 Absolute Value Inequalities

11.5 Graphs of Equations

11.6 Introduction to Functions

11.7 Function Notation

Google © Google 2010 Altamont Pass, California

Wind power generation has been extensively studied as a replacement for traditional power produced from fossil fuels. Offshore wind has been viewed as a potentially limitless source of power since the ocean has stronger and more constant winds compared to land. Recently scientists analyzed five years of wind data from eleven meteorological stations off the Atlantic coast of the United States from Florida to Maine. They found that combining power from all stations with a transmission cable could prevent large power fluctuations. In a simulation, underwater transmission cable stretched more than 1,550 miles and connected all eleven stations. Although individual sites showed erratic patterns, the total power output changed very little. For example,

the power output of an individual station would regularly drop to zero and could fluctuate by more than 50 percent in an hour. The output of the entire grid, however, did not change more than 10 percent in any given hour. Grid power never dropped to zero during the entire five-year period.

If you consider each individual wind station, the power generated by that station would be its output of electricity as a function of the input of the wind. If you consider all the stations together, the composition of those functions will give you the output of the entire grid. Functions and combinations of functions will be one of the topics we will cover in this chapter.

OBJECTIVES

A Solve a linear inequality in one variable and graph the solution set.

B Write solutions to inequalities using interval notation.

C Solve application problems using inequalities.

KEY WORDS

linear inequality

addition property for inequalities

multiplication property for inequalities

set notation

interval notation

line graph

**Video Examples
Section 11.1**

Note Because subtraction is defined as addition of the opposite, our new property holds for subtraction as well as addition. That is, we can subtract the same quantity from each side of an inequality and always be sure that we have not changed the solution.

The 'Apapane, a native Hawaiian bird, feeds mainly on the nectar of the 'ohi'a lehua blossom. The 'Apapane lives in high altitude regions where the 'ohi'a blossoms are found and where the birds are protected from mosquitoes which transmit avian malaria and avian pox. Predators of the 'Apapane include the rat, feral cat, mongoose, and owl. According to the *U.S. Geological Survey*:

Annual survival rates based on 1,584 recaptures of 429 banded individuals: 0.72 ± 0.11 for adults and 0.13 ± 0.07 for juveniles.

The number following the plus or minus sign represents the margin of error for the data collection. Using this margin of error, we can write the survival rate for the adults as an inequality:

$$0.61 \leq r \leq 0.83$$

Inequalities are what we will study in this section.

> **DEFINITION** linear inequality
>
> A *linear inequality* in one variable is any inequality that can be put in the form
>
> $$ax + b < c \qquad (a, b, \text{ and } c \text{ constants}, a \neq 0)$$
>
> where the inequality symbol ($<$) can be replaced with any of the other three inequality symbols (\leq, $>$, or \geq).

Some examples of *linear inequalities* are

$$3x - 2 \geq 7 \qquad -5y < 25 \qquad 3(x - 4) > 2x$$

A Solving Linear Inequalities

Our first property for inequalities is similar to the addition property we used when solving equations.

> **PROPERTY** Addition Property for Inequalities
>
> For any algebraic expressions, A, B, and C,
>
> $$\text{if} \qquad A < B$$
>
> $$\text{then} \qquad A + C < B + C$$
>
> *In words:* Adding the same quantity to both sides of an inequality will not change the solution set.

EXAMPLE 1 Solve $3x + 3 < 2x - 1$ and graph the solution.

Solution We use the addition property for inequalities to write all the variable terms on one side and all constant terms on the other side.

$$3x + 3 < 2x - 1$$
$$3x + (-2x) + 3 < 2x + (-2x) - 1 \qquad \text{Add } -2x \text{ to each side.}$$
$$x + 3 < -1$$
$$x + 3 + (-3) < -1 + (-3) \qquad \text{Add } -3 \text{ to each side.}$$
$$x < -4$$

The solution set is all real numbers that are less than -4. To show this we can use *set notation* and write

$$\{x \mid x < -4\}$$

We can graph the solution set on the number line using an open circle at -4 to show that -4 is not part of the solution set. This is the format you may have used when graphing inequalities in beginning algebra.

Here is an equivalent graph that uses a parenthesis opening left, instead of an open circle, to represent the end point of the graph.

This graph gives rise to the following notation, called *interval notation*, that is an alternative way to write the solution set.

$$(-\infty, -4)$$

The preceding expression indicates that the solution set is all real numbers from negative infinity up to, but not including, -4.

We have three equivalent representations for the solution set to our original inequality. Here are all three together:

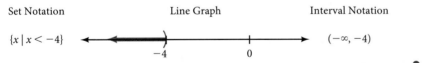

Set Notation Line Graph Interval Notation

$\{x \mid x < -4\}$ $(-\infty, -4)$

B Interval Notation and Graphing

The following table shows the connection between set notation, interval notation, and number line graphs. We have included the graphs with open and closed circles that we have used previously. For the remainder of this book, we will show our graphs using the parentheses/brackets method.

Practice Problems

1. Solve $4x - 2 < 3x + 4$ and graph the solution.

Note The English mathematician John Wallis (1616–1703) was the first person to use the ∞ symbol to represent infinity. When we encounter the interval $(3, \infty)$, we read it as "the interval from 3 to infinity," and we mean the set of real numbers that are greater than three. Likewise, the interval $(-\infty, -4)$ is read "the interval from negative infinity to -4," which is all real numbers less than -4.

Inequality notation	Interval notation	Graph using parentheses/brackets	Graph using open and closed circles
$x < 2$	$(-\infty, 2)$		
$x \leq 2$	$(-\infty, 2]$		
$x \geq -3$	$[-3, \infty)$		
$x > -3$	$(-3, \infty)$		

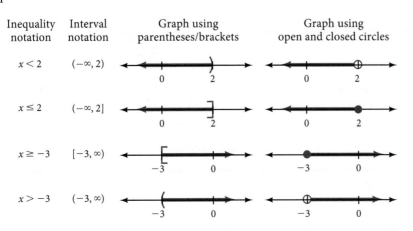

Before we state the multiplication property for inequalities, we will take a look at what happens to an inequality statement when we multiply both sides by a positive number and what happens when we multiply by a negative number.

We begin by writing three true inequality statements:

$$3 < 5 \qquad -3 < 5 \qquad -5 < -3$$

We multiply both sides of each inequality by a positive number, say, 4:

$$4(3) < 4(5) \qquad 4(-3) < 4(5) \qquad 4(-5) < 4(-3)$$
$$12 < 20 \qquad\quad -12 < 20 \qquad\quad -20 < -12$$

Notice in each case that the resulting inequality symbol points in the same direction as the original inequality symbol. Multiplying both sides of an inequality by a positive number preserves the *sense* of the inequality.

Let's take the same three original inequalities and multiply both sides by -4:

$$3 < 5 \qquad\qquad -3 < 5 \qquad\qquad -5 < -3$$
$$-4(3) > -4(5) \qquad -4(-3) > -4(5) \qquad -4(-5) > -4(-3)$$
$$-12 > -20 \qquad\quad 12 > -20 \qquad\qquad 20 > 12$$

Notice in this case that the resulting inequality symbol always points in the opposite direction from the original one. Multiplying both sides of an inequality by a negative number *reverses* the sense of the inequality. Keeping this in mind, we will now state the multiplication property for inequalities.

PROPERTY Multiplication Property for Inequalities

Let A, B, and C represent algebraic expressions.

$$\begin{aligned} &\text{If} && A < B \\ &\text{then} && AC < BC && \text{if} && C \text{ is positive } (C > 0) \\ &\text{or} && AC > BC && \text{if} && C \text{ is negative } (C < 0) \end{aligned}$$

In words: Multiplying both sides of an inequality by a positive number always produces an equivalent inequality. Multiplying both sides of an inequality by a negative number reverses the sense of the inequality.

The multiplication property for inequalities does not limit what we can do with inequalities. We are still free to multiply both sides of an inequality by any nonzero number we choose. If the number we multiply by happens to be *negative*, then we *must also reverse* the direction of the inequality.

EXAMPLE 2 Find the solution set for $-2y - 3 \le 7$.

Solution We begin by adding 3 to each side of the inequality:

$$-2y - 3 \le 7$$
$$-2y \le 10 \qquad\qquad \text{Add 3 to both sides.}$$
$$-\frac{1}{2}(-2y) \ge -\frac{1}{2}(10) \qquad \text{Multiply by } -\frac{1}{2} \text{ and reverse the direction of}$$
$$\text{the inequality symbol.}$$
$$y \ge -5$$

The solution set is all real numbers that are greater than or equal to -5. The following are three equivalent ways to represent this solution set.

Set Notation	Line Graph	Interval Notation
$\{y \mid y \ge -5\}$		$[-5, \infty)$

Notice how a bracket is used with interval notation to show that -5 is part of the solution set.

Note Because division is defined as multiplication by the reciprocal, we can apply our new property to division as well as to multiplication. We can divide both sides of an inequality by any nonzero number as long as we reverse the direction of the inequality when the number we are dividing by is negative.

2. Find the solution set for $-3y - 2 < 7$.

Answer

2. $y > -3$

When our inequalities become more complicated, we use the same basic steps we used when we were solving equations. That is, we simplify each side of the inequality before we apply the addition property or multiplication property. When we have solved the inequality, we graph the solution on a number line.

EXAMPLE 3 Solve $3(2x - 4) - 7x \leq -3x$.

Solution We begin by using the distributive property to separate terms. Next, simplify both sides.

$$3(2x - 4) - 7x \leq -3x \qquad \text{Original inequality}$$
$$6x - 12 - 7x \leq -3x \qquad \text{Distributive property}$$
$$-x - 12 \leq -3x \qquad 6x - 7x = (6 - 7)x = -x$$
$$-12 \leq -2x \qquad \text{Add } x \text{ to both sides.}$$
$$-\frac{1}{2}(-12) \geq -\frac{1}{2}(-2x) \qquad \text{Multiply both sides by } -\frac{1}{2} \text{ and reverse the direction of the inequality symbol.}$$
$$6 \geq x$$

This last line is equivalent to $x \leq 6$. The solution set can be represented with any of the three following items.

Set Notation	Line Graph	Interval Notation
$\{x \mid x \leq 6\}$		$(-\infty, 6]$

EXAMPLE 4 Solve and graph $2(1 - 3x) + 4 < 4x - 14$.

Solution
$$2 - 6x + 4 < 4x - 14 \qquad \text{Distributive property}$$
$$-6x + 6 < 4x - 14 \qquad \text{Simplify.}$$
$$-6x + 6 + (-6) < 4x - 14 + (-6) \qquad \text{Add } -6 \text{ to both sides.}$$
$$-6x < 4x - 20$$
$$-6x + (-4x) < 4x + (-4x) - 20 \qquad \text{Add } -4x \text{ to both sides.}$$
$$-10x < -20$$
$$\left(-\frac{1}{10}\right)(-10x) > \left(-\frac{1}{10}\right)(-20) \qquad \text{Multiply by } -\frac{1}{10}, \text{ reverse the direction of the inequality.}$$
$$x > 2$$

EXAMPLE 5 Solve $2x - 3y < 6$ for y.

Solution We can solve this formula for y by first adding $-2x$ to each side and then multiplying each side by $-\frac{1}{3}$. When we multiply by $-\frac{1}{3}$ we must reverse the direction of the inequality symbol. Because this is a formula, we will not graph the solution.

$$2x - 3y < 6 \qquad \text{Original formula}$$
$$2x + (-2x) - 3y < (-2x) + 6 \qquad \text{Add } -2x \text{ to each side.}$$
$$-3y < -2x + 6$$
$$-\frac{1}{3}(-3y) > -\frac{1}{3}(-2x + 6) \qquad \text{Multiply each side by } -\frac{1}{3}.$$
$$y > \frac{2}{3}x - 2 \qquad \text{Distributive property}$$

3. Solve $4(3x - 2) - 2x \geq 4x$.

> *Note* In Examples 2 and 3, notice that each time we multiplied both sides of the inequality by a negative number we also reversed the direction of the inequality symbol. Failure to do so would cause our graph to lie on the wrong side of the endpoint.

4. Solve and graph $3(1 - 2x) + 5 < 3x - 1$.

5. Solve $2x - 3y < 6$ for x.

Answers

3. $x \geq \frac{4}{3}$

4. $x > 1$

5. $x < \frac{3}{2}y + 3$

C Applications

EXAMPLE 6 A company that manufactures ink cartridges for printers finds that they can sell x cartridges each week at a price of p dollars each, according to the formula $x = 1,300 - 100p$. What price should they charge for each cartridge if they want to sell at least 300 cartridges a week?

6. In Example 6, what price should they charge per cartridge if they want to sell at least 400 cartridges a week?

Solution Because x is the number of cartridges they sell each week, an inequality that corresponds to selling at least 300 cartridges a week is

$$x \geq 300$$

Substituting $1,300 - 100p$ for x gives us an inequality in the variable p.

$$1,300 - 100p \geq 300$$
$$-100p \geq -1,000 \qquad \text{Add } -1,300 \text{ to each side.}$$
$$p \leq 10 \qquad \text{Divide each side by } -100, \text{ and reverse the direction of the inequality symbol.}$$

To sell at least 300 cartridges each week, the price per cartridge should be no more than $10. That is, selling the cartridges for $10 or less will produce weekly sales of 300 or more cartridges.

EXAMPLE 7 The formula $F = \frac{9}{5}C + 32$ gives the relationship between the Celsius and Fahrenheit temperature scales. If the temperature on a certain day is below 104° Fahrenheit, what is the temperature in degrees Celsius?

7. If the temperature range is below 77° Fahrenheit, what is the temperature range in Celsius?

Solution From the given information, we can write $F \leq 104$. However, because F is equal to $\frac{9}{5}C + 32$, we can also write

$$\frac{9}{5}C + 32 \leq 104$$
$$\frac{9}{5}C \leq 72 \qquad \text{Add } -32 \text{ to each number.}$$
$$\frac{5}{9}\left(\frac{9}{5}C\right) \leq \frac{5}{9}(72) \qquad \text{Multiply each number by } \frac{5}{9}.$$
$$C \leq 40$$

A temperature below 104° Fahrenheit corresponds to a temperature below 40° Celsius.

GETTING READY FOR CLASS

After reading through the preceding section, respond in your own words and in complete sentences.

A. What is a linear inequality in one variable?

B. What is the addition property for inequalities?

C. When we use interval notation to denote a section of the real number line, when do we use parentheses () and when do we use brackets []?

D. Explain the difference between the multiplication property of equality and the multiplication property for inequalities.

Answers

6. $p \leq 9$
7. $C \leq 25$

Vocabulary Review

Choose the correct words to fill in the blanks below.

 linear inequality in one variable set interval

 multiplication property for inequalities addition property for inequalities

1. A _____ can be put in the form $ax + b < c$.

2. The _____ says that adding the same quantity to both sides of an inequality will not change the solution set.

3. Using _____ notation, we can write the solution set to the inequality $3x + 3 < 2x - 1$ as $\{x \mid x < -4\}$.

4. The solution set $[-5, \infty]$ for the inequality $-2y + 3 < 7$ is written using _____ notation.

5. The _____ says that multiplying both sides of an inequality by a negative number reverses the sense of the inequality.

Problems

A Solve each of the following inequalities, and graph each solution.

1. $2x \le 3$

2. $5x \ge -115$

3. $\dfrac{1}{2}x > 2$

4. $\dfrac{1}{3}x > 4$

5. $-5x \le 25$

6. $-7x \ge 35$

7. $-\dfrac{3}{2}x > -6$

8. $-\dfrac{2}{3}x < -8$

9. $-12 \le 2x$

10. $-20 \ge 4x$

11. $-1 \ge -\dfrac{1}{4}x$

12. $-1 \le -\dfrac{1}{5}x$

Solve each of the following inequalities, and graph each solution.

13. $-3x + 1 > 10$

14. $-2x - 5 \le 15$

15. $\dfrac{1}{2} - \dfrac{m}{12} \le \dfrac{7}{12}$

16. $\dfrac{1}{2} - \dfrac{m}{10} > -\dfrac{1}{5}$

17. $\dfrac{1}{2} \geq -\dfrac{1}{6} - \dfrac{2}{9}x$

18. $\dfrac{9}{5} > -\dfrac{1}{5} - \dfrac{1}{2}x$

19. $-40 \leq 30 - 20y$

20. $-20 > 50 - 30y$

21. $\dfrac{2}{3}x - 3 < 1$

22. $\dfrac{3}{4}x - 2 > 7$

23. $10 - \dfrac{1}{2}y \leq 36$

24. $8 - \dfrac{1}{3}y \geq 20$

25. $4 - \dfrac{1}{2}x < \dfrac{2}{3}x - 5$

26. $5 - \dfrac{1}{3}x > \dfrac{1}{4}x + 2$

27. $0.03x - 0.4 \leq 0.08x + 1.2$

28. $2.0 - 0.7x < 1.3 - 0.3x$

29. $3 - \dfrac{x}{5} < 5 - \dfrac{x}{4}$

30. $-2 + \dfrac{x}{3} \geq \dfrac{x}{2} - 5$

A B Simplify each side first, then solve the following inequalities. Write your answers with interval notation.

31. $2(3y + 1) \leq -10$

32. $3(2y - 4) > 0$

33. $-(a + 1) - 4a \leq 2a - 8$

34. $-(a - 2) - 5a \le 3a + 7$

35. $\frac{1}{3}t - \frac{1}{2}(5 - t) < 0$

36. $\frac{1}{4}t - \frac{1}{3}(2t - 5) < 0$

37. $-2 \le 5 - 7(2a + 3)$

38. $1 < 3 - 4(3a - 1)$

39. $-\frac{1}{3}(x + 5) \le -\frac{2}{9}(x - 1)$

40. $-\frac{1}{2}(2x + 1) \le -\frac{3}{8}(x + 2)$

41. $5(x - 2) - 7(x + 1) \le -4x + 3$

42. $-3(1 - 2x) - 3(x - 4) < -3 - 4x$

43. $\frac{2}{3}x - \frac{1}{3}(4x - 5) < 1$

44. $\frac{1}{4}x - \frac{1}{2}(3x + 1) \ge 2$

45. $20x + 9{,}300 > 18{,}000$

46. $20x + 4{,}800 > 18{,}000$

47. $0.04x + 0.06(1{,}200 - x) \ge 54$

48. $0.08x + 0.05\,(900 - x) \ge 60$

The next two problems are intended to give you practice reading, and paying attention to, the instructions that accompany the problems you are working.

49. Work each problem according to the instructions.

 a. Evaluate when $x = 0$: $-\frac{1}{2}x + 1$

 b. Solve: $-\frac{1}{2}x + 1 = -7$

 c. Is 0 a solution to $-\frac{1}{2}x + 1 < -7$?

 d. Solve: $-\frac{1}{2}x + 1 < -7$

50. Work each problem according to the instructions.

 a. Evaluate when $x = 0$: $-\frac{2}{3}x - 5$.

 b. Solve: $-\frac{2}{3}x - 5 = 1$.

 c. Is 0 a solution to $-\frac{2}{3}x - 5 > 1$?

 d. Solve: $-\frac{2}{3}x - 5 > 1$.

Applying the Concepts

51. **Geometry Problems** The length of a rectangle is 3 times the width. If the perimeter is to be at least 48 meters, what are the possible values for the width? (If the perimeter is at least 48 meters, then it is greater than or equal to 48 meters.)

52. **Geometry Problems** The length of a rectangle is 3 more than twice the width. If the perimeter is to be at least 51 meters, what are the possible values for the width? (If the perimeter is at least 51 meters, then it is greater than or equal to 51 meters.)

53. **Geometry Problems** The numerical values of the three sides of a triangle are given by three consecutive even integers. If the perimeter is greater than 24 inches, what are the possibilities for the shortest side?

54. **Geometry Problems** The numerical values of the three sides of a triangle are given by three consecutive odd integers. If the perimeter is greater than 27 inches, what are the possibilities for the shortest side?

55. **Car Heaters** If you have ever sat in a cold car early in the morning, you know that the heater does not work until the engine warms up. This is because the heater relies on the heat coming off the engine. Write an equation using an inequality sign to express when the heater will work if the heater works only after the engine is 100°F.

56. **Exercise** When Kate exercises, she either swims or runs. She wants to spend a minimum of 8 hours a week exercising, and she wants to swim 3 times the amount she runs. What is the minimum amount of time she must spend doing each exercise?

57. **Profit and Loss** Movie theaters pay a certain price for the movies they show. Suppose a theater pays $1,500 for each showing of a popular movie. If they charge $7.50 for each ticket they sell, then they will lose money if ticket sales are less than $1,500. However, they will make a profit if ticket sales are greater than $1,500. What is the range of tickets they can sell and still lose money? What is the range of tickets they can sell and make a profit?

58. **Stock Sales** Suppose you purchase x shares of a stock at $12 per share. After 6 months you decide to sell all your shares at $20 per share. Your broker charges you $15 for the trade. If your profit is at least $3,985, how many shares did you purchase in the first place?

59. Art Supplies A store selling art supplies finds that they can sell x sketch pads each week at a price of p dollars each, according to the formula $x = 900 - 300p$. What price should they charge if they want to sell

a. at least 300 pads each week?

b. more than 600 pads each week?

c. less than 525 pads each week?

d. at most 375 pads each week?

60. Temperature Range Each of the following temperature ranges is in degrees Fahrenheit. Use the formula $F = \frac{9}{5}C + 32$ to find the corresponding temperature range in degrees Celsius.

a. below 113°

b. above 86°

c. below 14°

d. above −4°

61. Fuel Efficiency The fuel efficiency (mpg rating) for cars has been increasing steadily since 1980. The formula for a car's fuel efficiency for a given year between 1980 and 1996 is

$$E = 0.36x + 15.9$$

where E is miles per gallon and x is the number of years after 1980 (U.S. Federal Highway Administration, *Highway Statistics*, annual).

a. In what years was the average fuel efficiency for cars less than 17 mpg?

b. In what years was the average fuel efficiency for cars more than 20 mpg?

62. Student Loan When considering how much debt to incur in student loans, you learn that it is wise to keep your student loan payment to 8% or less of your starting monthly income. Suppose you anticipate a starting annual salary of $36,000. Set up and solve an inequality that represents the amount of monthly debt for student loans that would be considered manageable.

63. Wage The annual pay for a mathematics major graduating from college is given by the formula $P = 46{,}400 + 2{,}800t$, where t represents years worked after college. How many years would a mathematics major expect to work to be earning at least $90,000 per year?

64. Wage The pay per year for a chemical engineering major graduating from college is given by the formula $P = 64{,}800 + 2{,}880t$, where t represents years worked after college. How many years would a chemical engineering major expect to work to be earning at least $108,000 per year?

65. Wage The pay for a government major graduating from college is given by the formula $P = 41{,}500 + 3{,}050t$, where t represents years worked after college. How many years would a government major expect to work to be earning more than the mathematics major from problem 63?

66. Wage Is it reasonable to expect that the government major from problem 65 will exceed the chemical engineering majors salary from problem 64? Solve an inequality to answer this question.

Getting Ready for the Next Section

Solve each inequality. Do not graph.

67. $2x - 1 \geq 3$

68. $3x + 1 \geq 7$

69. $-2x > -8$

70. $-3x > -12$

71. $-3 > 4x + 1$

72. $4x + 1 \leq 9$

73. $-4x + 3 < 15$

74. $-5x - 1 > -16$

One Step Further

Assume that a, b, and c are positive, and solve each formula for x.

75. $ax + b < c$

76. $ax - b > -c$

77. $\dfrac{x}{a} + \dfrac{y}{b} < 1$

78. $\dfrac{x}{a} - \dfrac{y}{b} \geq 1$

Assume that $0 < a < b < c < d$ and solve each formula for x.

79. $ax + b < cx + d$

80. $cx + b < ax + d$

81. $\dfrac{x}{a} + \dfrac{1}{b} < \dfrac{x}{c} + \dfrac{1}{d}$

82. $\dfrac{3}{a} - \dfrac{x}{b} \leq \dfrac{5}{c} - \dfrac{x}{d}$

Find the Mistake

Write true or false for each statement below. If false, circle what is false and write the correct answer on the line provided.

1. True or False: The graph of the solution for the inequality $-16 \leq 4x$ will use a parenthesis to show that -4 is included in the solution. _____

2. True or False: The inequality $-10 > 30 - 5y$ has a solution set that does not include the number 7. _____

3. True or False: 3 is a solution to $-\dfrac{2}{3}x - 5 < -7$. _____

4. True or False: The interval notation for the solution set to $5 > \dfrac{1}{2} + 3(4x - 1)$ begins with a parentheses and ends with a bracket. _____

OBJECTIVES

A Identify elements of a set.

B Find the union and intersection of two sets.

C Graph compound inequalities.

D Solve compound inequalities.

KEY WORDS

set

elements

subset

empty set

union

intersection

set builder notation

compound inequality

open interval

closed interval

continued inequality

half-open interval

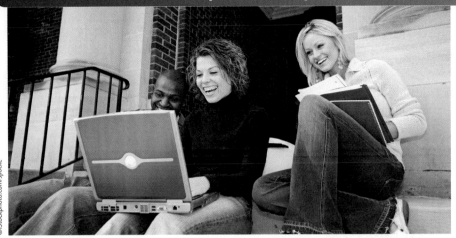

Suppose a recent survey of web users asked to name a favorite search engine with the following four answer choices:

 Ⓐ Google Ⓑ Bing Ⓒ Both Ⓓ Neither

The results of this survey are displayed in an illustration called a Venn diagram, shown here:

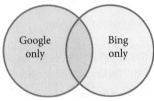

If we let $A = \{$all Google users$\}$ and $B = \{$all Bing users$\}$, then the notation

 $A \cap B$ represents those who use both Google **and** Bing. This is the *intersection* of the two sets.

 $A \cup B$ represents those who use Google **or** Bing or both. This is the *union* of the two sets.

We can use sets to represent everyday information like this survey. In this section, we will work with sets and use Venn diagrams to give visual representations of the relationships between different sets.

A Sets and Subsets

The concept of a set can be considered the starting point for all the branches of mathematics.

> **DEFINITION** set
>
> A *set* is a collection of objects or things. The objects in the set are called *elements*, or *members*, of the set.

Sets are usually denoted by capital letters, and elements of sets are denoted by lowercase letters. We use braces, { }, to enclose the elements of a set.

To show that an element is contained in a set, we use the symbol \in. That is,

 $x \in A$ is read "x is an element (member) of set A"

For example, if A is the set $\{1, 2, 3\}$, then $2 \in A$. However, $5 \notin A$ means 5 is not an element of set A.

Video Examples
Section 11.2

> **Note** The symbol \in means "is a member of a set."

DEFINITION subset

Set A is a *subset* of set B, written $A \subset B$, if every element in A is also an element of B. That is

$$A \subset B \qquad \text{if and only if} \qquad A \text{ is contained in } B$$

EXAMPLE 1 The set of numbers used to count things is $\{1, 2, 3, . . . \}$. The dots mean the set continues indefinitely in the same manner. This is an example of an *infinite set*.

EXAMPLE 2 The set of all numbers represented by the dots on the faces of a regular die is $\{1, 2, 3, 4, 5, 6\}$. This set is a subset of the set in Example 1. It is an example of a *finite set* because it has a limited number of elements.

DEFINITION empty set

The set with no members is called the *empty,* or *null set.* It is denoted by the symbol \varnothing. The empty set is considered a subset of every set.

B Operations with Sets

The two basic operations used to combine sets are *union* and *intersection*.

DEFINITION union

The *union* of two sets A and B, written $A \cup B$, is the set of all elements that are either in A or in B, or in both A and B. The key word here is *or.* For an element to be in $A \cup B$, it must be in A or B. These sets are also called *disjunctions.* In symbols, the definition looks like this:

$$x \in A \cup B \qquad \text{if and only if} \qquad x \in A \text{ or } x \in B$$

DEFINITION intersection

The *intersection* of two sets A and B, written $A \cap B$, is the set of elements in both A and B. The key word in this definition is the word *and.* These sets are also called *conjunctions.* For an element to be in $A \cap B$, it must be in both A and B. In symbols:

$$x \in A \cap B \qquad \text{if and only if} \qquad x \in A \text{ and } x \in B$$

EXAMPLE 3 Let $A = \{1, 3, 5\}$, $B = \{0, 2, 4\}$, and $C = \{1, 2, 3, . . . \}$.

a. $A \cup B = \{0, 1, 2, 3, 4, 5\}$

b. $A \cap B = \varnothing$ (A and B have no elements in common.)

c. $A \cap C = \{1, 3, 5\} = A$

d. $B \cup C = \{0, 1, 2, 3...\}$

Another notation we can use to describe sets is called *set-builder notation.* Here is how we write our definition for the union of two sets A and B using set-builder notation:

$$A \cup B = \{x \mid x \in A \text{ or } x \in B\}$$

The right side of this statement is read "the set of all x such that x is a member of A or x is a member of B." As you can see, the vertical line after the first x is read "such that."

EXAMPLE 4 If $A = \{1, 2, 3, 4, 5, 6\}$, find $C = \{x \mid x \in A \text{ and } x \geq 4\}$.

Solution We are looking for all the elements of A that are also greater than or equal to 4. They are 4, 5, and 6. Using set notation, we have

$$C = \{4, 5, 6\}$$

4. Let $A = \{2, 4, 6, 8, 10\}$ and find $B = \{x \mid x \in A \text{ and } x < 8\}$.

EXAMPLE 5 Use the information in the Venn diagram to find the following sets.

a. A

b. B

c. $A \cup B$

d. $A \cap B$

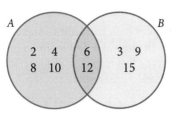

5. Use the Venn Diagram below to find the following sets.

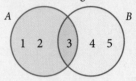

a. A
b. B
c. $A \cup B$
d. $A \cap B$

Solution

a. Set A includes the numbers in set A, as well as those common to both sets.
$A = \{2, 4, 6, 8, 10, 12\}$

b. Set B includes the numbers in set B, as well as those common to both sets.
$B = \{3, 6, 9, 12, 15\}$

c. The union of A and B includes all the numbers in either or both sets.
$A \cup B = \{2, 3, 4, 6, 8, 9, 10, 12, 15\}$

d. A intersect B includes only numbers common to both sets.
$A \cap B = \{6, 12\}$

C Graphing Compound Inequalities

Before we solve more inequalities, let's review some of the details involved with graphing more complicated inequalities.

Previously we defined the union of two sets A and B to be the set of all elements that are in either A or B. The word *or* is the key word in the definition. The intersection of two sets A and B is the set of all elements contained in both A and B, the key word here being *and*. We can use the words *and* and *or*, together with our methods of graphing inequalities, to graph some *compound inequalities*.

EXAMPLE 6 Graph: $\{x \mid x \leq -2 \text{ or } x > 3\}$.

6. Graph $\{x \mid x < -3 \text{ or } x \geq 2\}$.

Solution The two inequalities connected by the word *or* are referred to as a *compound inequality*. We begin by graphing each inequality separately.

Because the two inequalities are connected by the word *or*, we graph their union. That is, we graph all points on either graph.

To represent this set of numbers with interval notation we use two intervals connected with the symbol for the union of two sets. Here is the equivalent set of numbers described with interval notation:

$$(-\infty, -2] \cup (3, \infty)$$

Note The square bracket indicates -2 is included in the solution set, and the parenthesis indicates 3 is not included.

Answers
4. $\{2, 4, 6\}$
5. a. $\{1, 2, 3\}$ **b.** $\{3, 4, 5\}$
c. $\{1, 2, 3, 4, 5\}$ **d.** $\{3\}$

7. Graph $\{x \mid x \leq 1 \text{ or } x \geq -4\}$.

EXAMPLE 7 Graph: $\{x \mid x \geq -5 \text{ or } x \leq 2\}$.

Solution We begin by graphing each inequality separately.

Since the third graph includes both inequalities and is completely shaded, the solution can be written as $(-\infty, \infty)$ in interval notation. In set builder notation, the solution uses the symbol \mathbb{R} to represent the real numbers and is written $\{x \mid x = \mathbb{R}\}$.

8. Graph $\{x \mid x < 3 \text{ and } x \geq -2\}$.

EXAMPLE 8 Graph: $\{x \mid x > -1 \text{ and } x < 2\}$.

Solution We first graph each inequality separately.

Because the two inequalities are connected by the word *and*, we graph their intersection — the part they have in common.

This graph corresponds to the interval $(-1, 2)$, which is called an *open interval* because neither endpoint is included in the interval.

9. Graph $\{x \mid x \geq 3 \text{ and } x \geq -2\}$.

EXAMPLE 9 Graph: $\{x \mid x > 6 \text{ and } x > -3\}$.

Solution We begin by graphing each inequality separately.

Since all the numbers greater than -3 are also greater than 6, the solution set it only those numbers greater than 6. We write the solution in set builder notation as $\{x \mid x > 6\}$.

10. Graph $\{x \mid x \leq -3 \text{ and } x \geq -1\}$.

EXAMPLE 10 Graph: $\{x \mid x \leq 2 \text{ and } x \geq 4\}$.

Solution We begin by graphing each inequality separately.

There are no numbers less than 2 and greater than 4.

$\{x \mid x \leq 2\} \cap \{x \geq 4\} = \varnothing$

There is no solution to this inequality. The blank number line represent the null set, \varnothing.

Notation Sometimes compound inequalities that use the word *and* as the connecting word can be written in a shorter form. For example, the compound inequality $-3 \le x$ and $x \le 4$ can be written $-3 \le x \le 4$. Inequalities of the form $-3 \le x \le 4$ are called *continued inequalities* or *double inequalities*. The graph of $-3 \le x \le 4$ is

The corresponding interval is $[-3, 4]$, which is called a *closed interval* because both endpoints are included in the interval.

D Solving Compound Inequalities

EXAMPLE 11 Solve and graph $-3 \le 2x - 5 \le 3$.

Solution We can extend our properties for addition and multiplication to cover this situation. If we add a number to the middle expression, we must add the same number to the outside expressions. We do the same for multiplication, remembering to reverse the direction of the inequality symbols if we multiply by a negative number.

$$-3 \le 2x - 5 \le 3$$

$$2 \le \ \ 2x \ \ \le 8 \qquad \text{Add 5 to all three members.}$$

$$1 \le \ \ \ x \ \ \ \le 4 \qquad \text{Multiply through by } \tfrac{1}{2}.$$

Here are three ways to write this solution set:

Set Notation	Line Graph	Interval Notation
$\{x \mid 1 \le x \le 4\}$		$[1, 4]$

EXAMPLE 12 Solve the compound inequality.

$$3t + 7 \le -4 \qquad \text{or} \qquad 3t + 7 \ge 4$$

Solution We solve each half of the compound inequality separately, then we graph the solution set.

$$3t + 7 \le -4 \qquad \text{or} \quad 3t + 7 \ge 4$$

$$3t \le -11 \qquad \text{or} \qquad 3t \ge -3 \qquad \text{Add } -7.$$

$$t \le -\frac{11}{3} \qquad \text{or} \qquad t \ge -1 \qquad \text{Multiply by } \tfrac{1}{3}.$$

The solution set can be written in any of the following ways:

Set Notation	Line Graph	Interval Notation
$\{t \mid t \le -\frac{11}{3}$ or $t \ge -1\}$		$\left(-\infty, -\frac{11}{3}\right] \cup \left[-1, \infty\right)$

GETTING READY FOR CLASS

After reading through the preceding section, respond in your own words and in complete sentences.

A. What is a compound inequality?

B. Explain the shorthand notation that can be used to write two inequalities connected by the word *and*.

C. Write two inequalities connected by the word *and* that together are equivalent to $-1 < x < 2$.

D. Explain how you would graph the compound inequality $x < 2$ or $x > -3$.

11. Solve and graph
$-7 \le 2x + 1 \le 7$.

12. Graph the solution for the compound inequality
$3t - 6 \le -3$ or $3t - 6 \ge 3$.

Note As we have seen, an interval can appear open on one side and closed on the other side, such as $[-1, \infty)$. This interval is called a *half-open interval*.

Vocabulary Review

Choose the correct words to fill in the blanks below.

set	union	subset	continued	compound	closed
empty set	set-builder	intersection	elements	open	

1. A _____ is a collection of objects or things, also called _____.

2. If every element in A is also an element of B, then set A is a _____ of set B.

3. The set with no members is called the _____.

4. The _____ of two sets A and B with elements in either in A or B is written $A \cup B$.

5. The _____ of two sets A and B with elements in both A or B is written $A \cap B$.

6. Using _____ notation, we can write the union of two sets A and B as $A \cup B = \{x \mid x \in A \text{ or } x \in B\}$.

7. Two inequalities connected by the word *or* is called a _____ inequality.

8. A compound inequality written with the word *and* implied is called a _____ inequality.

9. A(n) _____ interval includes both endpoints, whereas a(n) _____ interval includes neither.

Problems

A Identify each statement as true or false.

1. The set of prime numbers is an example of an infinite set.

2. The set of odd numbers is an example of an infinite set.

3. The set of whole numbers less than 10 is an example of an infinite set.

4. The set of integers less than -20 is an example of an infinite set.

5. The set $\{1, 3, 5\}$ is a subset of $\{1, 3, 5 \ldots\}$.

6. The set $\{-2, 4, 6\}$ is a subset of $\{2, 4, 6 \ldots\}$.

7. The set of whole numbers is a subset of the set of real numbers.

8. The set of real numbers is a subset of the set of whole numbers.

B For the following problems, let $A = \{0, 2, 4, 6\}$, $B = \{1, 2, 3, 4, 5\}$, and $C = \{1, 3, 5, 7\}$.

9. $A \cup B$

10. $A \cup C$

11. $A \cap B$

12. $A \cap C$

13. $A \cup (B \cap C)$

14. $C \cup (A \cap B)$

15. $\{x \mid x \in A \text{ and } x < 4\}$

16. $\{x \mid x \in B \text{ and } x > 3\}$

17. $\{x \mid x \in A \text{ or } x \in C\}$

18. $\{x \mid x \in A \text{ or } x \in B\}$

19. $\{x \mid x \in B \text{ and } x \neq 3\}$

20. $\{x \mid x \in C \text{ and } x \neq 5\}$

Use the following Venn diagrams to find each set.

21. a. A
 b. B
 c. $A \cup B$
 d. $A \cap B$

22. a. A
 b. B
 c. $A \cup B$
 d. $A \cap B$

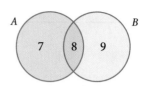

23. a. A
 b. B
 c. $A \cup B$
 d. $A \cap B$

24. a. A
 b. B
 c. $A \cup B$
 d. $A \cap B$

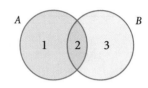

Construct a Venn diagram to represent the following sets.

25. $A = \{1, 3, 5, 7\}$
 $B = \{3, 4, 5\}$

26. $A = \{2, 3, 4\}$
 $B = \{4, 5, 6\}$

Use interval notation to describe each of the graphs.

27.

28.

29.

30.

31.

32.

C Graph the following compound inequalities and write each solution set using interval notation.

33. $x < -1$ or $x > 5$

34. $x \leq -2$ or $x \geq -1$

35. $x < 5$ and $x > 1$

36. $x \leq 6$ and $x > -1$

37. $x \leq 7$ and $x > 0$

38. $x > 2$ and $x < 4$

39. $x < 2$ or $x > 4$

40. $x \leq 2$ or $x \geq 4$

41. $-1 < x < 3$

42. $-1 \leq x \leq 3$

43. $-3 < x \leq -2$

44. $-5 \leq x \leq 0$

45. $-3 \leq x < 2$

46. $-5 < x \leq -1$

D Solve the following continued inequalities. Use both a line graph and interval notation to write each solution set.

47. $-2 \leq m - 5 \leq 2$

48. $-3 \leq m + 1 \leq 3$

49. $-40 < 20a + 20 < 20$

50. $-60 < 50a - 40 < 60$

51. $0.5 \leq 0.3a - 0.7 \leq 1.1$

52. $0.1 \leq 0.4a + 0.1 \leq 0.3$

53. $3 < \frac{1}{2}x + 5 < 6$

54. $5 < \frac{1}{4}x + 1 < 9$

55. $4 < 6 + \frac{2}{3}x < 8$

56. $3 < 7 + \frac{4}{5}x < 15$

57. $-2 < -\frac{1}{2}x + 1 \leq 1$

58. $-3 \leq -\frac{1}{3}x - 1 < 2$

59. $-\frac{1}{2} \leq \frac{3x + 1}{2} \leq \frac{1}{2}$

60. $-\frac{5}{6} \leq \frac{2x + 5}{3} \leq \frac{5}{6}$

61. $-1.5 \leq \frac{2x - 3}{4} \leq 3.5$

62. $-1.25 \leq \frac{2x + 3}{4} \leq .75$

63. $-\frac{3}{4} \leq \frac{4x - 3}{2} \leq 1.5$

64. $-\frac{5}{6} \leq \frac{5x + 2}{3} \leq \frac{5}{6}$

Graph the solution sets for the following compound inequalities. Then write each solution set using interval notation.

65. $x + 5 \le -2$ or $x + 5 \ge 2$

66. $3x + 2 < -3$ or $3x + 2 > 3$

67. $5y + 1 \le -4$ or $5y + 1 \ge 4$

68. $7y - 5 \le -2$ or $7y - 5 \ge 2$

69. $2x + 5 < 3x - 1$ or $x - 4 > 2x + 6$

70. $3x - 1 > 2x + 4$ or $5x - 2 < 3x + 4$

71. $3x + 1 < -8$ or $-2x + 1 \le -3$

72. $2x - 5 \le -1$ or $-3x - 6 < -15$

Translate each of the following phrases into an equivalent inequality statement.

73. x is greater than -2 and at most 4

74. x is less than 9 and at least -3

75. x is less than -4 or at least 1

76. x is at most 1 or more than 6

Applying the Concepts

77. Temperature Range Each of the following temperature ranges is in degrees Fahrenheit. Use the formula

$$F = \frac{9}{5}C + 32$$

to find the corresponding temperature range in degrees Celsius.

a. 95° to 113°

b. 68° to 86°

c. −13° to 14°

d. −4° to 23°

78. Calories Burned The chart shows the range of calories burned during the different events of an ironman triathalon.

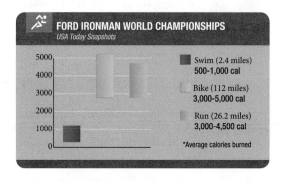

FORD IRONMAN WORLD CHAMPIONSHIPS
USA Today Snapshots

Swim (2.4 miles)
500-1,000 cal

Bike (112 miles)
3,000-5,000 cal

Run (26.2 miles)
3,000-4,500 cal

*Average calories burned

a. Write an inequality that describes the number of calories burned during the entire race.

b. If an athlete burned 750 calories during the swim, write an inequality that describes the number of calories burned during the entire race.

79. Here is what the United States Geological Survey has to say about the survival rates of the Apapane, one of the endemic birds of Hawaii.

> *Annual survival rates based on 1,584 recaptures of 429 banded individuals 0.72 ± 0.11 for adults and 0.13 ± 0.07 for juveniles.*

Write the survival rates using inequalities. Then give the survival rates in terms of percent.

80. Survival Rates for Sea Gulls Here is part of a report concerning the survival rates of Western Gulls that appeared on the web site of Cornell University.

> *Survival of eggs to hatching is 70%–80%; of hatched chicks to fledglings 50%–70%; of fledglings to age of first breeding <50%.*

Write the survival rates using inequalities without percent.

Getting Ready for the Next Section

To understand all of the explanations and examples in the next section you must be able to work the problems below.

Solve each equation.

81. $2a - 1 = -7$

82. $3x - 6 = 9$

83. $\dfrac{2}{3}x - 3 = 7$

84. $\dfrac{2}{3}x - 3 = -7$

85. $x - 5 = x - 7$

86. $x + 3 = x + 8$

87. $x - 5 = -x - 7$

88. $x + 3 = -x + 8$

One Step Further

Assume that a, b, and c are positive, and solve each formula for x.

89. $-c < ax + b < c$

90. $-1 < \dfrac{ax + b}{c} < 1$

Assume that $0 < a < b < c < d$ and solve each formula for x.

91. $-d < \dfrac{ax + b}{c} < d$

92. $\dfrac{x}{a} + \dfrac{1}{b} < \dfrac{x}{c}$

Find the Mistake

Each sentence below contains a mistake. Circle the mistake and write the correct word or phrase on the line provided.

1. The notation $x \in A$ is read "x is a set of element A." _____

2. The intersection of two sets A and B is written $A \cup B$. _____

3. The interval $[-6, 8]$ for the continued inequality $-6 \le x \le 8$ is called an open interval. _____

4. To solve the compound inequality $-2 \le 4x - 3 \le 9$, subtract 9 from the first two parts of the inequality. _____

11.3 Absolute Value Equations

OBJECTIVES

A Solve equations with absolute value symbols.

©iStockphoto.com/tarantas

At the U.S. National Handcar Races in Truckee, California, five-person teams face off on a railroad track. One team member gives the 1000-pound handcar a push while the other four members, standing on its flatbed, use the hand pump to propel the handcar down a 1000-foot stretch of track. The team with the fastest time wins the race. The races are held each year in honor of California's great railroad history, which aside from a few local lines began in 1869 with the completion of the First Continental Railroad.

Today people traveling across the country by train use Amtrak, the government-established long-distance passenger train for inter-city travel in the United States. Amtrak's annual passenger revenue for the years 1985–1995 is modeled approximately by the formula

$$R = -60\,|x - 11| + 962$$

where R is the annual revenue in millions of dollars and x is the number of years after 1980 (Association of American Railroads, Washington, DC, *Railroad Facts, Statistics of Railroads of Class 1*, annual). In what year was the passenger revenue $722 million?

**Video Examples
Section 11.3**

A Solve Equations with Absolute Values

Earlier, we defined the *absolute value* of x, $|x|$, to be the distance between x and 0 on the number line. The absolute value of a number measures its distance from 0.

EXAMPLE 1 Solve $|x| = 5$ for x.

Solution Using the definition of absolute value, we can read the equation as, "The distance between x and 0 on the number line is 5." If x is 5 units from 0, then x can be 5 or -5.

$$\text{If } |x| = 5 \quad \text{then } x = 5 \quad \text{or} \quad x = -5$$

Practice Problems

1. Solve $|x| = 3$ for x.

In general, we can see that any equation of the form $|a| = b$ is equivalent to the equations $a = b$ or $a = -b$, as long as $b > 0$. Furthermore, $|a| = 0$ is equivalent to $a = 0$.

Answer

1. $-3, 3$

2. Solve $|3x - 6| = 9$.

EXAMPLE 2 Solve $|2a - 1| = 7$.

Solution We can read this question as "$2a - 1$ is 7 units from 0 on the number line." The quantity $2a - 1$ must be equal to 7 or -7.

$$|2a - 1| = 7$$
$$2a - 1 = 7 \quad \text{or} \quad 2a - 1 = -7$$

We have transformed our absolute value equation into two equations that do not involve absolute value. We can solve each equation using the method in Section 1.1.

$$2a - 1 = 7 \quad \text{or} \quad 2a - 1 = -7$$
$$2a = 8 \qquad\qquad 2a = -6 \qquad \text{Add 1 to both sides.}$$
$$a = 4 \qquad\qquad a = -3 \qquad \text{Multiply by } \tfrac{1}{2}.$$

The solutions are -3 and 4.

To check our solutions, we put them into the original absolute value equation.

When \rightarrow $\quad a = 4$ \qquad When \rightarrow $\quad a = -3$

the equation $\rightarrow |2a - 1| = 7 \qquad$ the equation $\rightarrow |2a - 1| = 7$

becomes $\rightarrow \quad |2(4) - 1| = 7 \qquad$ becomes $\rightarrow |2(-3) - 1| = 7$

$$|7| = 7 \qquad\qquad\qquad |-7| = 7$$
$$7 = 7 \qquad\qquad\qquad 7 = 7$$

3. Solve $|4x - 3| + 2 = 3$.

EXAMPLE 3 Solve $\left|\dfrac{2}{3}x - 3\right| + 5 = 12$.

Solution To use the definition of absolute value to solve this equation, we must isolate the absolute value on the left side of the equal sign. To do so, we add -5 to both sides of the equation to obtain

$$\left|\frac{2}{3}x - 3\right| = 7$$

Now that the equation is in the correct form, we can write

$$\frac{2}{3}x - 3 = 7 \quad \text{or} \quad \frac{2}{3}x - 3 = -7$$
$$\frac{2}{3}x = 10 \qquad\qquad \frac{2}{3}x = -4 \qquad \text{Add 3 to both sides.}$$
$$x = 15 \qquad\qquad x = -6 \qquad \text{Multiply by } \tfrac{3}{2}.$$

The solutions are -6 and 15.

4. Solve $|7a - 1| = -2$.

EXAMPLE 4 Solve $|3a - 6| = -4$.

Solution The solution set is \varnothing because the right side is negative but the left side cannot be negative. No matter what we try to substitute for the variable a, the quantity $|3a - 6|$ will always be positive or zero. It can never be -4.

Note The symbol \Leftrightarrow means "if and only if" and "is equivalent to."

Consider the statement $|a| = |b|$. What can we say about a and b? We know they are equal in absolute value. By the definition of absolute value, they are the same distance from 0 on the number line. They must be equal to each other or opposites of each other. In symbols, we write

$$|a| = |b| \quad \Leftrightarrow \quad a = b \quad \text{or} \quad a = -b$$

$\qquad\qquad\uparrow \qquad\qquad\qquad \uparrow \qquad\qquad\qquad \uparrow$

\qquad Equal in $\qquad\qquad$ Equals \quad or \quad Opposites

\qquad absolute value

Answers

2. $-1, 5$

3. $1, \dfrac{1}{2}$

4. No solution

EXAMPLE 5 Solve $|3a + 2| = |2a + 3|$.

Solution The quantities $3a + 2$ and $2a + 3$ have equal absolute values. They are, therefore, the same distance from 0 on the number line. They must be equals or opposites.

$$|3a + 2| = |2a + 3|$$

Equals		Opposites
$3a + 2 = 2a + 3$	or	$3a + 2 = -(2a + 3)$
$a + 2 = 3$		$3a + 2 = -2a - 3$
$a = 1$		$5a + 2 = -3$
		$5a = -5$
		$a = -1$

The solutions are -1 and 1.

 It makes no difference in the outcome of the problem if we take the opposite of the first or second expression. It is very important, once we have decided which one to take the opposite of, that we take the opposite of both its terms and not just the first term. That is, the opposite of $2a + 3$ is $-(2a + 3)$, which we can think of as $-1(2a + 3)$. Distributing the -1 across *both* terms, we have

$$-1(2a + 3) = -2a - 3$$

EXAMPLE 6 Solve $|x - 5| = |x - 7|$.

Solution As was the case in Example 5, the quantities $x - 5$ and $x - 7$ must be equal or they must be opposites, because their absolute values are equal.

Equals		Opposites
$x - 5 = x - 7$	or	$x - 5 = -(x - 7)$
$-5 = -7$		$x - 5 = -x + 7$
No solution here		$2x - 5 = 7$
		$2x = 12$
		$x = 6$

Because the first equation leads to a false statement, it will not give us a solution. (If either of the two equations were to reduce to a true statement, it would mean all real numbers would satisfy the original equation.) In this case, our only solution is 6.

GETTING READY FOR CLASS

After reading through the preceding section, respond in your own words and in complete sentences.

A. Why do some equations that involve absolute value have two solutions instead of one?

B. Translate $|x| = 6$ into words using the definition of absolute value.

C. Explain in words what the equation $|x - 3| = 4$ means with respect to distance on the number line.

D. When is the statement $|x| = x$ true?

5. Solve $|4x - 5| = |3x + 9|$.

6. Solve $|x + 3| = |x + 8|$.

EXERCISE SET 11.3

Problems

A Use the definition of absolute value to solve each of the following equations.

1. $|x| = 4$

2. $|x| = 7$

3. $2 = |a|$

4. $5 = |a|$

5. $|x| = -3$

6. $|x| = -4$

7. $|a| + 2 = 3$

8. $|a| - 5 = 2$

9. $|y| + 4 = 3$

10. $|y| + 3 = 1$

11. $|a - 4| = \dfrac{5}{3}$

12. $|a + 2| = \dfrac{7}{5}$

13. $\left|\dfrac{3}{5}a + \dfrac{1}{2}\right| = 1$

14. $\left|\dfrac{2}{7}a + \dfrac{3}{4}\right| = 1$

15. $60 = |20x - 40|$

16. $800 = |400x - 200|$

17. $|2x + 1| = -3$

18. $|2x - 5| = -7$

19. $\left|\dfrac{3}{4}x - 6\right| = 9$

20. $\left|\dfrac{4}{5}x - 5\right| = 15$

21. $\left|1 - \dfrac{1}{2}a\right| = 3$

22. $\left|2 - \dfrac{1}{3}a\right| = 10$

23. $|2x - 5| = 3$

24. $|3x + 1| = 4$

25. $|4 - 7x| = 5$

26. $|9 - 4x| = 1$

27. $\left|3 - \dfrac{2}{3}y\right| = 5$

28. $\left|-2 - \dfrac{3}{4}y\right| = 6$

Solve each equation.

29. $|3x + 4| + 1 = 7$

30. $|5x - 3| - 4 = 3$

31. $|3 - 2y| + 4 = 3$

32. $|8 - 7y| + 9 = 1$

33. $3 + |4t - 1| = 8$

34. $2 + |2t - 6| = 10$

35. $\left|9 - \dfrac{3}{5}x\right| + 6 = 12$

36. $\left|4 - \dfrac{2}{7}x\right| + 2 = 14$

37. $5 = \left|\dfrac{2}{7}x + \dfrac{4}{7}\right| - 3$

38. $7 = \left|\dfrac{3}{5}x + \dfrac{1}{5}\right| + 2$

39. $2 = -8 + \left|4 - \dfrac{1}{2}y\right|$

40. $1 = -3 + \left|2 - \dfrac{1}{4}y\right|$

41. $|3(x + 1)| - 4 = -1$

42. $|2(2x + 3)| - 5 = -1$

43. $|1 + 3(2x - 1)| = 5$

44. $|3 + 4(3x + 1)| = 7$

45. $3 = -2 + \left|5 - \dfrac{2}{3}a\right|$

46. $4 = -1 + \left|6 - \dfrac{4}{5}a\right|$

47. $6 = |7(k + 3) - 4|$

48. $5 = |6(k - 2) + 1|$

Solve the following equations.

49. $|3a + 1| = |2a - 4|$

50. $|5a + 2| = |4a + 7|$

51. $\left|x - \dfrac{1}{3}\right| = \left|\dfrac{1}{2}x + \dfrac{1}{6}\right|$

52. $\left|\dfrac{1}{10}x - \dfrac{1}{2}\right| = \left|\dfrac{1}{5}x + \dfrac{1}{10}\right|$

53. $|y - 2| = |y + 3|$

54. $|y - 5| = |y - 4|$

55. $|3x - 1| = |3x + 1|$

56. $|5x - 8| = |5x + 8|$

57. $|0.03 - 0.01x| = |0.04 + 0.05x|$

58. $|0.07 - 0.01x| = |0.08 - 0.02x|$

59. $|x - 2| = |2 - x|$

60. $|x - 4| = |4 - x|$

61. $\left|\dfrac{x}{5} - 1\right| = \left|1 - \dfrac{x}{5}\right|$

62. $\left|\dfrac{x}{3} - 1\right| = \left|1 - \dfrac{x}{3}\right|$

63. $\left|\dfrac{2}{3}b - \dfrac{1}{4}\right| = \left|\dfrac{1}{6}b + \dfrac{1}{2}\right|$

64. $\left|-\dfrac{1}{4}x + 1\right| = \left|\dfrac{1}{2}x - \dfrac{1}{3}\right|$

65. $|0.1a - 0.04| = |0.3a + 0.08|$

66. $|-0.4a + 0.6| = |1.3 - 0.2a|$

67. Work each problem.

 a. Solve: $4x - 5 = 0$.

 b. Solve: $|4x - 5| = 0$.

 c. Solve: $4x - 5 = 3$.

 d. Solve: $|4x - 5| = 3$.

 e. Solve: $|4x - 5| = |2x + 3|$.

68. Work each problem.

 a. Solve: $3x + 6 = 0$.

 b. Solve: $|3x + 6| = 0$.

 c. Solve: $3x + 6 = 4$.

 d. Solve: $|3x + 6| = 4$.

 e. Solve: $|3x + 6| = |7x + 4|$.

Applying the Concepts

69. Amtrak Amtrak's annual passenger revenue for the years 1985–1995 is modeled approximately by the formula

$$R = -60|x - 11| + 962$$

where R is the annual revenue in millions of dollars and x is the number of years after 1980 (Association of American Railroads, Washington, DC, *Railroad Facts, Statistics of Railroads of Class 1*, annual). In what year was the passenger revenue $722 million?

70. Corporate Profits The corporate profits for various U.S. industries vary from year to year. An approximate model for profits of U.S. communications companies during a given year between 1990 and 1997 is given by

$$P = -3400|x - 5.5| + 36000$$

where P is the annual profits (in millions of dollars) and x is the number of years after 1990 (U.S. Bureau of Economic Analysis, Income and Product Accounts of the U.S. (1929 – 1994), *Survey of Current Business*, September 1998). Use the model to determine the years in which profits of communication companies were $31.5 billion ($31,500 million).

Getting Ready for the Next Section

To understand all of the explanations and examples in the next section you must be able to work the problems below.

Solve each inequality. Do not graph the solution set.

71. $2x - 5 < 3$

72. $-3 < 2x - 5$

73. $-4 \leq 3a + 7$

74. $3a + 2 \leq 4$

75. $4t - 3 \leq -9$

76. $4t - 3 \geq 9$

One Step Further

Solve each formula for x. (Assume a, b, and c are positive.)

77. $|x - a| = b$

78. $|x + a| - b = 0$

79. $|ax + b| = c$

80. $|ax - b| - c = 0$

81. $\left| \dfrac{x}{a} + \dfrac{y}{b} \right| = 1$

82. $\left| \dfrac{x}{a} + \dfrac{y}{b} \right| = c$

Find the Mistake

Each sentence below contains a mistake. Circle the mistake and write the correct word or phrase on the line provided.

1. All absolute value equations have two solutions. _____

2. There are two solutions to the absolute value equation $|2x - 3| = |2x + 4|$. _____

3. There is no solution for the absolute equation $|y - 7| = |7 - y|$. _____

4. The absolute value equation $|6a - 4| = -9$ has a negative number for a solution. _____

11.4 Absolute Value Inequalities

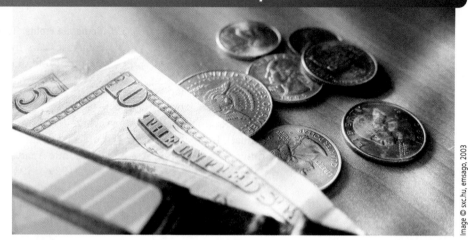

Image © sxc.hu, emsago, 2003

**Video Examples
Section 11.4**

A student survey conducted by the University of Minnesota, found that 30% of students were solely responsible for their finances. The survey was reported to have a margin of error plus or minus 3.74%. This means that the difference between the sample estimate of 30% and the actual percent of students who are responsible for their own finances is most likely less than 3.74%. We can write this as an inequality:

$$|x - 0.30| \le 0.0374$$

where x represents the true percent of students who are responsible for their own finances.

A Solving Inequalities with Absolute Values

In this section, we will apply the definition of absolute value to solve inequalities involving absolute value. Again, the absolute value of x, which is denoted $|x|$, represents the distance that x is from 0 on the number line. We will begin by considering three absolute value expressions and their verbal translations:

Expression	In Words
$\|x\| = 7$	x is exactly 7 units from 0 on the number line.
$\|a\| < 5$	a is less than 5 units from 0 on the number line.
$\|y\| \ge 4$	y is greater than or equal to 4 units from 0 on the number line.

Once we have translated the expression into words, we can use the translation to graph the original equation or inequality. The graph is then used to write a final equation or inequality that does not involve absolute value.

Original Expression	Graph	Final Expression
$\|x\| = 7$		$x = -7$ or $x = 7$
$\|a\| < 5$		$-5 < a < 5$
$\|y\| \ge 4$		$y \le -4$ or $y \ge 4$

Although we will not always write out the verbal translation of an absolute value inequality, it is important that we understand the translation. Our second expression, $|a| < 5$, means a is within 5 units of 0 on the number line. The graph of this relationship is

which can be written with the following continued inequality:

$$-5 < a < 5$$

We can follow this same kind of reasoning to solve more complicated absolute value inequalities.

EXAMPLE 1 Solve and graph $|2x - 5| < 3$.

Solution The absolute value of $2x - 5$ is the distance that $2x - 5$ is from 0 on the number line. We can translate the inequality as, "$2x - 5$ is less than 3 units from 0 on the number line." That is, $2x - 5$ must appear between -3 and 3 on the number line.

A picture of this relationship is

Using the picture, we can write an inequality without absolute value that describes the situation:

$$-3 < 2x - 5 < 3$$

Next, we solve the continued inequality by first adding 5 to all three members and then multiplying all three by $\frac{1}{2}$.

$$-3 < 2x - 5 < 3$$
$$2 < \quad 2x \quad < 8 \qquad \text{Add 5 to all three expressions.}$$
$$1 < \quad x \quad < 4 \qquad \text{Multiply each expression by } \frac{1}{2}.$$

The graph of the solution set is

We can see from the solution that for the absolute value of $2x - 5$ to be within 3 units of 0 on the number line, x must be between 1 and 4.

EXAMPLE 2 Solve and graph $|3a + 7| \leq 4$.

Solution We can read the inequality as, "The distance between $3a + 7$ and 0 is less than or equal to 4." Or, "$3a + 7$ is within 4 units of 0 on the number line." This relationship can be written without absolute value as

$$-4 \leq 3a + 7 \leq 4$$

Solving as usual, we have

$$-4 \leq 3a + 7 \leq 4$$
$$-11 \leq \quad 3a \quad \leq -3 \qquad \text{Add } -7 \text{ to all three members.}$$
$$-\frac{11}{3} \leq \quad a \quad \leq -1 \qquad \text{Multiply each expression by } \frac{1}{3}.$$

We can see from Examples 1 and 2 that to solve an inequality involving absolute value, we must be able to write an equivalent expression that does not involve absolute value.

Practice Problems

1. Solve and graph $|2x + 1| \leq 7$.

2. Solve and graph $|4a - 3| < 5$.

3. Solve and graph $|x + 2| > 7$.

EXAMPLE 3 Solve and graph $|x - 3| > 5$.

Solution We interpret the absolute value inequality to mean that $x - 3$ is more than 5 units from 0 on the number line. The quantity $x - 3$ must be either above 5 or below -5. Here is a picture of the relationship:

An inequality without absolute value that also describes this situation is

$$x - 3 < -5 \quad \text{or} \quad x - 3 > 5$$

Adding 3 to both sides of each inequality we have

$$x < -2 \quad \text{or} \quad x > 8$$

Here are three ways to write our result:

Set Notation $\{x \mid x < -2 \text{ or } x > 8\}$ Interval Notation $(-\infty, -2) \cup (8, \infty)$

We can use the results of our first few examples and the material in the previous section to summarize the information we have related to absolute value equations and inequalities in the table below. If c is a positive real number, then each of the following statements on the left is equivalent to the corresponding statement on the right.

Note The inequalities in the table may also be written using the symbols \leq and \geq. Corresponding line graphs would then use a bracket at each value instead of a parenthesis.

With Absolute Value	Without Absolute Value	Line Graph
$\|x\| = c$ $\|ax + b\| = c$	$x = -c$ or $x = c$ $ax + b = -c$ or $ax + b = c$	
$\|x\| < c$ $\|ax + b\| < c$	$-c < x < c$ $-c < ax + b < c$	
$\|x\| > c$ $\|ax + b\| > c$	$x < -c$ or $x > c$ $ax + b < -c$ or $ax + b > c$	

4. Solve and graph $|2x + 5| - 2 < 9$.

EXAMPLE 4 Solve and graph $|2x + 3| + 4 < 9$.

Solution Before we can apply the method of solution we used in the previous examples, we must isolate the absolute value on one side of the inequality. To do so, we add -4 to each side.

$$|2x + 3| + 4 < 9$$
$$|2x + 3| + 4 + (-4) < 9 + (-4)$$
$$|2x + 3| < 5$$

From this last line, we know that $2x + 3$ must be between -5 and 5.

$$-5 < 2x + 3 < 5$$
$$-8 < \quad 2x \quad < 2 \qquad \text{Add } -3 \text{ to each expression.}$$
$$-4 < \quad x \quad < 1 \qquad \text{Multiply each expression by } \tfrac{1}{2}.$$

Here are three equivalent ways to write our solution

Set Notation $\{x \mid -4 < x < 1\}$ Interval Notation $(-4, 1)$

EXAMPLE 5 Solve and graph $|4 - 2t| > 2$.

Solution The inequality indicates that $4 - 2t$ is less than -2 or greater than 2. Writing this without absolute value symbols, we have

$$4 - 2t < -2 \qquad \text{or} \qquad 4 - 2t > 2$$

To solve these inequalities we begin by adding -4 to each side.

$$4 + (-4) - 2t < -2 + (-4) \quad \text{or} \quad 4 + (-4) - 2t > 2 + (-4)$$

$$-2t < -6 \qquad\qquad\qquad -2t > -2$$

$$-\frac{1}{2}(-2t) > -\frac{1}{2}(-6) \qquad\qquad -\frac{1}{2}(-2t) < -\frac{1}{2}(-2)$$

$$t > 3 \qquad\qquad\qquad\qquad t < 1$$

Although in situations like this we are used to seeing the "less than" symbol written first, the meaning of the solution is clear. We want to graph all real numbers that are either greater than 3 or less than 1. Here is the graph:

Because absolute value always results in a nonnegative quantity, we sometimes come across special solution sets when a negative number appears on the right side of an absolute value inequality.

EXAMPLE 6 Solve $|7y - 1| < -2$.

Solution The *left* side is never negative because it is an absolute value. The *right* side is negative. We have a positive quantity (or zero) less than a negative quantity, which is impossible. The solution set is the empty set, \varnothing. There is no real number to substitute for y to make this inequality a true statement.

EXAMPLE 7 Solve $|6x + 2| > -5$.

Solution This is the opposite case from that in Example 7. No matter what real number we use for x on the *left* side, the result will always be positive, or zero. The *right* side is negative. We have a positive quantity (or zero) greater than a negative quantity. Every real number we choose for x gives us a true statement. The solution set is the set of all real numbers.

GETTING READY FOR CLASS

After reading through the preceding section, respond in your own words and in complete sentences.

A. Write an inequality containing absolute value, the solution to which is all the numbers between -5 and 5 on the number line.

B. Translate $x \geq |3|$ into words using the definition of absolute value.

C. Explain in words what the inequality $|x - 5| < 2$ means with respect to distance on the number line.

D. Why is there no solution to the inequality $|2x - 3| < 0$?

5. Solve and graph $|5 - 2t| > 3$.

6. Solve $|8y + 3| \leq -5$.

7. Solve $|2x - 9| \geq -7$.

EXERCISE SET 11.4

Vocabulary Review

Match the following absolute value equations on the left to their equivalents without absolute values on the right.

1. $|x| = c$ a. $ax + b < -c$ or $ax + b < c$
2. $|x| < c$ b. $x < -c$ or $x > c$
3. $|x| > c$ c. $-c < ax + b < c$
4. $|ax + b| = c$ d. $ax + b = -c$ or $ax + b = c$
5. $|ax + b| < c$ e. $x = -c$ or $x = c$
6. $|ax + b| > c$ f. $-c < x < c$

Problems

A Solve each of the following inequalities using the definition of absolute value. Graph the solution set in each case and write the solution set in interval notation.

1. $|x| < 3$

2. $|x| \leq 7$

3. $|x| \geq 2$

4. $|x| > 4$

5. $|x| + 2 < 5$

6. $|x| - 3 < -1$

7. $|t| - 3 > 4$

8. $|t| + 5 > 8$

9. $|y| < -5$

10. $|y| > -3$

11. $|x| \geq -2$

12. $|x| \leq -4$

13. $|x - 3| < 7$

14. $|x + 4| < 2$

15. $|a + 5| \geq 4$

16. $|a - 6| \geq 3$

17. $|x - 5| < 3$

18. $|x - 3| \leq 5$

Solve each inequality and graph the solution set and write each solution set using interval notation.

19. $|a - 1| < -3$

20. $|a + 2| \geq -5$

21. $|2x - 4| < 6$

22. $|2x + 6| < 2$

23. $|3y + 9| \geq 6$

24. $|5y - 1| \geq 4$

25. $|2k + 3| \geq 7$

26. $|2k - 5| \geq 3$

27. $|x - 3| + 2 < 6$

28. $|x + 4| - 3 < -1$

29. $|2a + 1| + 4 \geq 7$

30. $|2a - 6| - 1 \geq 2$

31. $|3x + 5| - 8 < 5$

32. $|6x - 1| - 4 \leq 2$

Solve each inequality and write your answer using interval notation. Keep in mind that if you multiply or divide both sides of an inequality by a negative number you must reverse the sense of the inequality.

33. $|x - 3| \leq 5$

34. $|a + 4| < 6$

35. $|3y + 1| < 5$

36. $|2x - 5| \leq 3$

37. $|a + 4| \geq 1$

38. $|y - 3| > 6$

39. $|2x + 5| > 2$

40. $|-3x + 1| \geq 7$

41. $|-5x + 3| \leq 8$

42. $|-3x + 4| \leq 7$

43. $|-3x + 7| < 2$

44. $|-4x + 2| < 6$

Solve each inequality and graph the solution set.

45. $|5 - x| > 3$

46. $|7 - x| > 2$

47. $\left|3 - \frac{2}{3}x\right| \geq 5$

48. $\left|3 - \frac{3}{4}x\right| \geq 9$

49. $\left|2 - \frac{1}{2}x\right| > 1$

50. $\left|3 - \frac{1}{3}x\right| > 1$

51. $\left|\frac{1}{3}x - 2\right| > 4$

52. $\left|\frac{2}{5}x + 4\right| \geq 3$

Solve each inequality.

53. $|x - 1| < 0.01$

54. $|x + 1| < 0.01$

55. $|2x + 1| \geq \frac{1}{5}$

56. $|2x - 1| \geq \frac{1}{8}$

57. $|3x - 2| \leq \frac{1}{3}$

58. $|2x + 5| < \frac{1}{2}$

59. $\left|\frac{3x + 1}{2}\right| > \frac{1}{2}$

60. $\left|\frac{2x - 5}{3}\right| \geq \frac{1}{6}$

61. $\left|\frac{4 - 3x}{2}\right| \geq 1$

62. $\left|\frac{2x - 3}{4}\right| < 0.35$

63. $\left|\frac{3x - 2}{5}\right| \leq \frac{1}{2}$

64. $\left|\frac{4x - 3}{2}\right| \leq \frac{1}{3}$

65. $\left|2x - \frac{1}{5}\right| < 0.3$

66. $\left|3x - \frac{3}{5}\right| < 0.2$

67. $\left|-2x + \frac{1}{2}\right| < \frac{1}{3}$

68. $\left|-4x - \frac{1}{3}\right| \leq \frac{1}{6}$

69. Write the continued inequality $-4 \leq x \leq 4$ as a single inequality involving absolute value.

70. Write the continued inequality $-8 \leq x \leq 8$ as a single inequality involving absolute value.

71. Write $-1 \leq x - 5 \leq 1$ as a single inequality involving absolute value.

72. Write $-3 \leq x + 2 \leq 3$ as a single inequality involving absolute value.

73. Work each problem according to the instructions given.

 a. Evaluate $|5x + 3|$ when $x = 0$.

 b. Solve: $|5x + 3| = 7$.

 c. Is 0 a solution to $|5x + 3| > 7$?

 d. Solve: $|5x + 3| > 7$.

74. Work each problem according to the instructions given.

 a. Evaluate $|-2x - 5|$ when $x = 0$.

 b. Solve: $|-2x - 5| = 1$.

 c. Is 0 a solution to $|-2x - 5| > 1$?

 d. Solve: $|-2x - 5| > 1$.

75. Wavelengths of Light When white light from the sun passes through a prism, it is broken down into bands of light that form colors. The wavelength v (in nanometers) of some common colors are

Blue:	$424 < v < 491$
Green:	$491 < v < 575$
Yellow:	$575 < v < 585$
Orange:	$585 < v < 647$
Red:	$647 < v < 700$

When a fireworks display made of copper is burned, it lets out light with wavelengths v that satisfy the relationship $|v - 455| < 23$. Write this inequality without absolute values, find the range of possible values for v, and then using the preceding list of wavelengths, determine the color of that copper fireworks display.

76. Speed Limits One of the highest interstate speed limits for cars in the United States is 75 miles per hour in Nebraska, Nevada, New Mexico, Oklahoma, South Dakota, Utah, and Wyoming. To discourage passing, minimum speeds are also posted, so that the difference between the fastest and slowest moving traffic is no more than 20 miles per hour. Write an absolute value inequality that describes the relationship between the minimum allowable speed and a maximum speed of 75 miles per hour.

One Step Further

Assume a, b, and c, are positive, and solve each inequality for x.

77. $x - a < b$

78. $x - a > b$

79. $ax - b > c$

80. $ax - b < c$

81. $ax + b \leq c$

82. $ax + b \geq c$

83. $\left| \dfrac{x}{a} + \dfrac{1}{b} \right| > c$

84. $\left| \dfrac{x}{a} + \dfrac{1}{b} \right| < c$

85. $|(x + a)^2 - x^2| < 3a^2$

86. $|(x - a)^2 - x^2| < 2a^2$

87. $|x - a^2| \leq 2a^2$

88. $|x - 2b^2| < 5b^2$

Find the Mistake

Write true or false for each statement below. If false, circle what is false and write the correct answer on the line provided.

1. True or False: The expression $|b| > 9$ is read as "b is less than or equal to 9 units from zero on the number line."

2. True or False: When solving $-3a < -9$ for a, we must reverse the direction of the inequality symbol.

3. True or False: When an absolute value expression is greater than a negative value, the solution set will be \varnothing.

4. True or False: When an absolute value expression is less than a negative value, the solution set will be \varnothing.

Landmark Review: Checking Your Progress

Solve and graph each of the following inequalities.

1. $-24 \leq 3x$ ⟵—————————⟶

2. $-18 < 3x$ ⟵—————————⟶

3. $\frac{4}{7}x - 5 \leq 11$ ⟵—————————⟶

4. $3 - \frac{2}{5}x \geq -1$ ⟵—————————⟶

Let $A = \{0, 2, 4, 6\}$, $B = \{1, 2, 3, 4, 5\}$, and $C = \{1, 3, 5, 7\}$.

5. Find $A \cup B$ **6.** Find $A \cap C$ **7.** Find $A \cap B$ **8.** Find $A \cup C$

Solve the following continued inequalities. Use both a line graph and interval notation to write each solution set.

9. $3 \leq \frac{1}{2}t - 5 \leq 6$ ⟵—————————⟶

10. $1 < \frac{1}{2}y - 4 < 4$ ⟵—————————⟶

Solve each equation.

11. $\left|2 - \frac{3}{7}y\right| = 4$

12. $|4(t + 5)| + 1 = 5$

Solve each inequality and graph the solution set.

13. $|4x + 4| \geq 8$ ⟵—————————⟶

14. $|-6x + 6| < 6$ ⟵—————————⟶

11.5 Graphs of Equations

OBJECTIVES

A Graph ordered pairs on a rectangular coordinate system.

B Graph equations by making a table of values.

C Find intercepts to graph equations.

D Graph horizontal and vertical lines.

KEY WORDS

rectangular coordinate system

x-axis

y-axis

bar chart

line graph

ordered pair

x-coordinate

y-coordinate

origin

intercepts

In this section, we place our work with charts and graphs in a more formal setting. Our foundation will be the *rectangular coordinate system*, because it gives us a link between algebra and geometry. With it we notice relationships between certain equations and different lines and curves.

Table 1 gives the wholesale price of a popular intermediate algebra text at the beginning of each year in which a new edition was published. (The wholesale price is the price the bookstore pays for the book, not the price you pay for it.)

Table 1	Price of a Textbook	
Edition	**Year Published**	**Wholesale Price ($)**
First	1991	30.50
Second	1995	39.25
Third	1999	47.50
Fourth	2003	55.00
Fifth	2007	65.75

The information in Table 1 is represented visually in Figures 1 and 2. The diagram in Figure 1 is called a *bar chart*. The diagram in Figure 2 is called a *line graph*. The data in Table 1 is called *paired data* because each number in the year column is paired with a specific number in the price column.

FIGURE 1

FIGURE 2

Video Examples
Section 11.5

A Ordered Pairs

Paired data play an important role in equations that contain two variables. Working with these equations is easier if we standardize the terminology and notation associated with paired data. So here is a definition that will do just that:

> **DEFINITION** ordered pair, *x*-coordinate, *y*-coordinate
>
> A pair of numbers enclosed in parentheses and separated by a comma, such as $(-2, 1)$, is called an **ordered pair** of numbers. The first number in the pair is called the **x-coordinate** of the ordered pair; the second number is called the **y-coordinate**. For the ordered pair $(-2, 1)$, the *x*-coordinate is -2 and the *y*-coordinate is 1.

Rectangular Coordinate System

A rectangular coordinate system allows us to connect algebra and geometry by associating geometric shapes (the lines and curves shown in the diagrams) with algebraic equations. The French philosopher and mathematician René Descartes (1596–1650) is usually credited with the invention of the rectangular coordinate system, which is often referred to as the Cartesian coordinate system in his honor. As a philosopher, Descartes is responsible for the statement, "I think, therefore, I am." Until Descartes invented his coordinate system in 1637, algebra and geometry were treated as separate subjects.

A *rectangular coordinate system* is made by drawing two real number lines at right angles to each other. The two number lines, called *axes*, cross each other at 0. This point is called the *origin*. Positive directions are to the right and up. Negative directions are to the left and down. The rectangular coordinate system is shown in Figure 3.

> *Note* Any point in quadrant I has positive $(+, +)$ *x*- and *y*-coordinates. Points in quadrant II have negative *x*-coordinates and positive *y*-coordinates $(-, +)$. In quadrant III, both coordinates are negative $(-, -)$. In quadrant IV, the form is $(+, -)$. If a point lies on an axis or at the origin, then it is not considered in any quadrant.

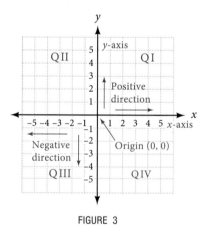

FIGURE 3

The horizontal number line is called the *x-axis*, and the vertical number line is called the *y-axis*. The two number lines divide the coordinate system into four quadrants, which we number, Q I through Q IV, in a counterclockwise direction. Points on the axes are not considered as being in any quadrant.

Graphing Ordered Pairs

To graph the ordered pair (a, b) on a rectangular coordinate system, we start at the origin and move *a* units right or left (right if *a* is positive, left if *a* is negative). Then we move *b* units up or down (up if *b* is positive, down if *b* is negative). The point where we end up is the graph of the ordered pair (a, b).

EXAMPLE 1 Plot (graph) the ordered pairs $(2, 5)$, $(-2, 5)$, $(-2, -5)$, and $(2, -5)$.

Solution To graph the ordered pair $(2, 5)$, we start at the origin and move 2 units to the right, then 5 units up. We are now at the point whose coordinates are $(2, 5)$. We graph the other three ordered pairs in a similar manner (see Figure 4).

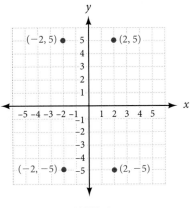

FIGURE 4

EXAMPLE 2 Graph the ordered pairs $(1, -3)$, $\left(\frac{1}{2}, 2\right)$, $(3, 0)$, $(0, -2)$, $(-1, 0)$, and $(0, 5)$.

Solution

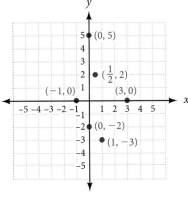

FIGURE 5

From Figure 5, we see that any point on the x-axis has a y-coordinate of 0 (it has no vertical displacement), and any point on the y-axis has an x-coordinate of 0 (no horizontal displacement).

B Graphing Equations

We can plot a single point from an ordered pair, but to draw a line or a curve, we need more points.

To graph an equation in two variables, we simply graph its solution set. That is, we draw a line or smooth curve through all the points whose coordinates satisfy the equation.

EXAMPLE 3 Graph the equation $y = -\frac{1}{3}x$.

Solution The graph of this equation will be a straight line. We need to find three ordered pairs that satisfy the equation. To do so, we can let x equal any numbers we choose and find corresponding values of y. However, because every value of x we substitute into the equation is going to be multiplied by $-\frac{1}{3}$, let's use numbers for x that are divisible by 3, like -3, 0, and 3. That way, when we multiply them by $-\frac{1}{3}$, the result will be an integer.

Practice Problems

1. Graph the ordered pairs $(2, 3)$, $(2, -3)$, $(-2, 3)$, and $(-2, -3)$.

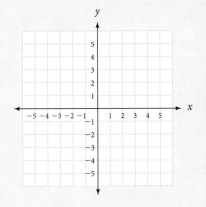

2. Graph the ordered pairs $(-2, 1)$, $(3, 5)$, $(0, 2)$, and $(-3, -3)$.

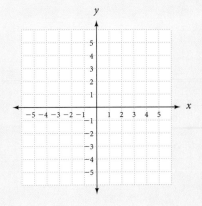

3. Graph the equation $y = \frac{3}{2}x$.

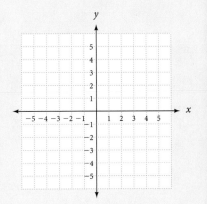

$$\text{Let } x = -3; \quad y = -\frac{1}{3}(-3) = 1$$

The ordered pair $(-3, 1)$ is one solution.

$$\text{Let } x = 0; \quad y = -\frac{1}{3}(0) = 0$$

The ordered pair $(0, 0)$ is a second solution.

$$\text{Let } x = 3; \quad y = -\frac{1}{3}(3) = -1$$

The ordered pair $(3, -1)$ is a third solution.

In table form

x	y
-3	1
0	0
3	-1

Plotting the ordered pairs $(-3, 1)$, $(0, 0)$, and $(3, -1)$ and drawing a straight line through their graphs, we have the graph of the equation $y = -\frac{1}{3}x$ as shown in Figure 6.

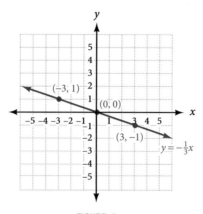

FIGURE 6

Example 3 illustrates again the connection between algebra and geometry that we mentioned earlier in this section. Descartes's rectangular coordinate system allows us to associate the equation $y = -\frac{1}{3}x$ (an algebraic concept) with a specific straight line (a geometric concept). The study of the relationship between equations in algebra and their associated geometric figures is called *analytic geometry*.

Lines Through the Origin

As you can see from Figure 6, the graph of the equation $y = -\frac{1}{3}x$ is a straight line that passes through the origin. The same will be true of the graph of any equation that has the same form as $y = -\frac{1}{3}x$. Here are three more equations that have that form, along with their graphs.

Graph of $y = 2x$ **Graph of $y = \frac{1}{2}x$**

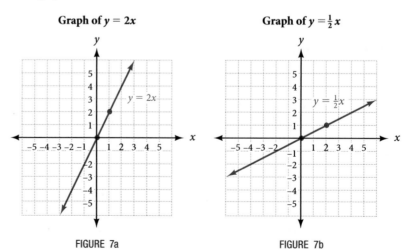

FIGURE 7a FIGURE 7b

Graph of $y = x$

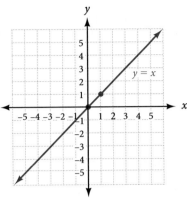

FIGURE 7c

Here is a summary of this discussion:

> **RULE** Lines Through the Origin
>
> The graph of any equation of the form
>
> $$y = mx$$
>
> where m is a real number, will be a straight line through the origin.

EXAMPLE 4 Graph the equation $y = -\dfrac{1}{3}x + 2$.

Solution Again, we need ordered pairs that are solutions to our equation. Noticing the similarity of this equation to our previous equation, we choose the same values of x for our inputs.

Input x	Calculate using the equation	Output y	Form ordered pairs
-3	$y = -\frac{1}{3}(-3) + 2 = 1 + 2 =$	3	$(-3, 3)$
0	$y = -\frac{1}{3}(0) + 2 = 0 + 2 =$	2	$(0, 2)$
3	$y = -\frac{1}{3}(3) + 2 = -1 + 2 =$	1	$(3, 1)$

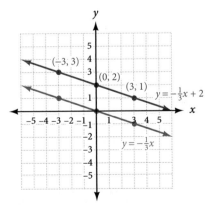

FIGURE 8

Notice that we have included the graph of $y = -\frac{1}{3}x$ along with the graph of $y = -\frac{1}{3}x + 2$. We can see that the graph of $y = -\frac{1}{3}x + 2$ looks just like the graph of $y = -\frac{1}{3}x$, but all points are moved up vertically 2 units.

4. Graph the equation $y = \dfrac{3}{2}x - 3$.

5. Graph the equation $y = \frac{3}{2}x + 4$.

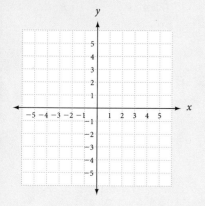

EXAMPLE 5 Graph the equation $y = -\frac{1}{3}x - 4$.

Solution We create a table from ordered pairs, then we graph the information in the table. However, we are expecting the graph of $y = -\frac{1}{3}x - 4$ to be 4 units below the graph of $y = -\frac{1}{3}x$.

Input x	Calculate using the equation	Output y	Form ordered pairs
-3	$y = -\frac{1}{3}(-3) - 4 = 1 - 4 =$	-3	$(-3, -3)$
0	$y = -\frac{1}{3}(0) - 4 = 0 - 4 =$	-4	$(0, -4)$
3	$y = -\frac{1}{3}(3) - 4 = -1 - 4 =$	-5	$(3, -5)$

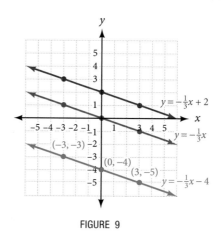

FIGURE 9

Vertical Translations

We know that the graph of $y = mx$ is a line that passes through the origin. From our previous two examples we can generalize as follows:

If K is a positive number, then:	
The graph of	Is the graph of $y = mx$ translated
$y = mx + K$	K units up
$y = mx - K$	K units down

EXAMPLE 6 Find an equation for the blue line.

6. Find the equation of the blue line.

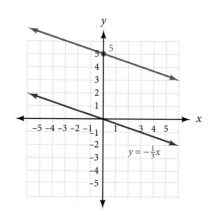

FIGURE 10

Solution The blue line is parallel to the line $y = -\frac{1}{3}x$, but translated up 5 units from $y = -\frac{1}{3}x$. According to what we have done up to this point, the equation for the blue line is

$$y = -\frac{1}{3}x + 5$$

EXAMPLE 7 Graph the equation $y = x^2$.

Solution We input values of x then calculate the values of y using the equation. The result is a set of ordered pairs that we plot and then connect with a smooth curve.

Input x	Calculate using the equation	Output y	Form ordered pairs
-3	$y = (-3)^2 =$	9	$(-3, 9)$
-2	$y = (-2)^2 =$	4	$(-2, 4)$
-1	$y = (-1)^2 =$	1	$(-1, 1)$
0	$y = (0)^2 =$	0	$(0, 0)$
1	$y = (1)^2 =$	1	$(1, 1)$
2	$y = (2)^2 =$	4	$(2, 4)$
3	$y = (3)^2 =$	9	$(3, 9)$

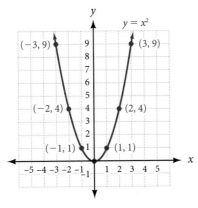

FIGURE 11

EXAMPLE 8 Graph $y = x^2 - 4$.

Solution We make a table as we did in the previous example. If the vertical translation idea works for this type of equation as it did with our straight lines, we expect this graph to be the graph in Example 7 translated down 4 units.

Input x	Calculate using the equation	Output y	Form ordered pairs
-3	$y = (-3)^2 - 4 =$	5	$(-3, 5)$
-2	$y = (-2)^2 - 4 =$	0	$(-2, 0)$
-1	$y = (-1)^2 - 4 =$	-3	$(-1, -3)$
0	$y = (0)^2 - 4 =$	-4	$(0, -4)$
1	$y = (1)^2 - 4 =$	-3	$(1, -3)$
2	$y = (2)^2 - 4 =$	0	$(2, 0)$
3	$y = (3)^2 - 4 =$	5	$(3, 5)$

7. Graph the equation $y = 2x^2$.

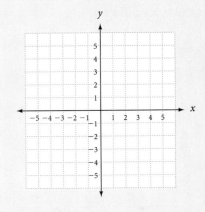

Note Recall our discussion earlier in the book about raising a negative number to a specific power. It is very important to enclose the negative number in a set of parentheses before working the exponent.

8. Graph $y = 2x^2 - 2$.

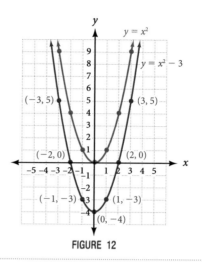

FIGURE 12

As you can see, the graph of $y = x^2 - 4$ is the graph of $y = x^2$ translated down 4 units. We generalize this as follows:

If K is a positive number, then:	
The graph of	**is the graph of $y = x^2$ translated**
$y = x^2 + K$	K units up
$y = x^2 - K$	K units down

9. Graph $y = |x| + 2$.

EXAMPLE 9 Graph $y = |x|$ and $y = |x| - 4$.

Solution We let x take on values of -4, -3, -2, -1, 0, 1, 2, 3, and 4. The corresponding values of y are shown in the tables.

| $y = |x|$ | |
| --- | --- |
| **Input** x | **Output** y |
| -4 | 4 |
| -3 | 3 |
| -2 | 2 |
| -1 | 1 |
| 0 | 0 |
| 1 | 1 |
| 2 | 2 |
| 3 | 3 |
| 4 | 4 |

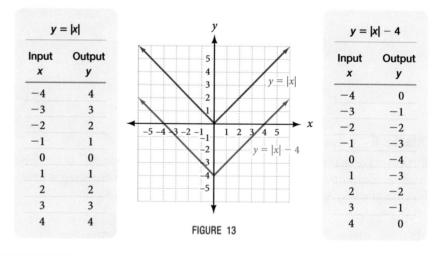

FIGURE 13

| $y = |x| - 4$ | |
| --- | --- |
| **Input** x | **Output** y |
| -4 | 0 |
| -3 | -1 |
| -2 | -2 |
| -1 | -3 |
| 0 | -4 |
| 1 | -3 |
| 2 | -2 |
| 3 | -1 |
| 4 | 0 |

As you can see, the graph of $y = x - 4$ is the graph of $y = x$ translated down 4 units. We will do more with these vertical translations later in the chapter. For now, we simply want to notice how the relationship between the equations can be used to predict how one graph will look when the other graph is given.

C Intercepts and Graphs

DEFINITION intercepts

An *x-intercept* of the graph of an equation is the coordinates of a point where the graph intersects the *x*-axis. The *y-intercept* of the graph of an equation is the coordinates of a point where the graph intersects the *y*-axis.

Because any point on the *x*-axis has a *y*-coordinate of 0, we can find the *x*-intercept by letting $y = 0$ and solving the equation for *x*. We find the *y*-intercept by letting $x = 0$ and solving for *y*.

EXAMPLE 10 Find the *x*- and *y*-intercepts for $2x + 3y = 6$; then graph the equation.

Solution To find the *y*-intercept, we let $x = 0$.

When → $\qquad x = 0$

we have → $\quad 2(0) + 3y = 6$

$$3y = 6$$
$$y = 2$$

y-intercept $= (0, 2)$. To find the *x*-intercept, we let $y = 0$.

When → $\qquad y = 0$

we have → $\quad 2x + 3(0) = 6$

$$2x = 6$$
$$x = 3$$

x-intercept $= (3, 0)$. We use these results to graph the solution set for $2x + 3y = 6$. The graph is shown in Figure 14.

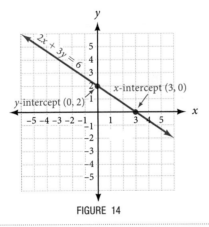

FIGURE 14

EXAMPLE 11 Find the intercepts for $y = x^2 - 4$.

Solution We drew the graph of this equation in Example 8. Looking back to that graph we see that the *x*-intercepts are $(-2, 0)$ and $(2, 0)$, and the *y*-intercept is $(0, -4)$. Let's see if we obtain the same results using the algebraic method shown in Example 10.

x-intercept

When → $\quad y = 0$,

we have → $\quad 0 = x^2 - 4$

$$0 = (x + 2)(x - 2)$$
$$x + 2 = 0 \quad \text{or} \quad x - 2 = 0$$
$$x = -2 \qquad\qquad x = 2$$

The *x*-intercepts are $(-2, 0)$ and $(2, 0)$.

y-intercept

When → $\quad x = 0$,

we have → $\quad y = 0^2 - 4 = -4$

The *y*-intercept is $(0, -4)$.

10. Graph the solution set for $2x - 4y = 8$.

Note Graphing straight lines by finding the intercepts works best when the coefficients of *x* and *y* are factors of the constant term.

11. Find the intercepts of $y = -2x^2 + 8$.

Answers

11. *y*-intercept: $(0, 8)$,
\quad *x*-intercepts: $(-2, 0)$, $(2, 0)$

12. Graph.

a. $x = -4$

b. $y = 3$

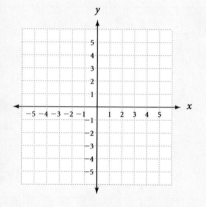

D Horizontal and Vertical Lines

EXAMPLE 12 Graph each of the following lines.

a. $x = 3$ **b.** $y = -2$

Solution

a. The line $x = 3$ is the set of all points whose x-coordinate is 3. The variable y does not appear in the equation, so the y-coordinate can be any number. Note that we can write our equation as a linear equation in two variables by writing it as $x + 0y = 3$. Because the product of 0 and y will always be 0, y can be any number. The graph of $x = 3$ is the vertical line shown in Figure 15a.

b. The line $y = -2$ is the set of all points whose y-coordinate is -2. The variable x does not appear in the equation, so the x-coordinate can be any number. Again, we can write our equation as a linear equation in two variables by writing it as $0x + y = -2$. Because the product of 0 and x will always be 0, x can be any number. The graph of $y = -2$ is the horizontal line shown in Figure 15b.

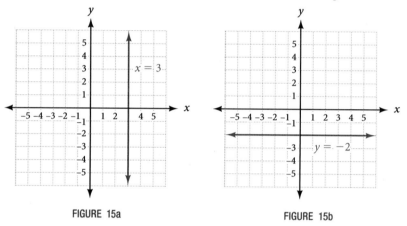

FIGURE 15a FIGURE 15b

FACTS FROM GEOMETRY Special Equations and Their Graphs

For the equations below, m, a, and b are real numbers.

Through the Origin

FIGURE 16a *Any equation of the form $y = mx$ has a graph that passes through the origin.*

FIGURE 16b *Any equation of the form $x = a$ has a vertical line for its graph.*

FIGURE 16c *Any equation of the form $y = b$ has a horizontal line for its graph.*

USING TECHNOLOGY Graphing Calculators

Graphing With Trace and Zoom

All graphing calculators have the ability to graph a function and then trace over the points on the graph, giving their coordinates. Furthermore, all graphing calculators can zoom in and out on a graph that has been drawn.

To graph an equation on a graphing calculator, we first set the graph window. The counterpart values of y are Ymin and Ymax. We will use the notation

$$\text{Window:} \quad -5 \le x \le 4 \text{ and } -3 \le y \le 2$$

to stand for a window in which Xmin $= -5$, Xmax $= 4$, Ymin $= -3$, and Ymax $= 2$

Set your calculator to the following window:

$$\text{Window:} \quad -10 \le x \le 10 \text{ and } -10 \le y \le 10$$

Graph the equation $Y = -X + 8$ and compare your results with this graph:

FIGURE 17

Use the Trace feature of your calculator to name three points on the graph. Next, use the Zoom feature of your calculator to zoom out so your window is twice as large.

Solving for *y* First

To graph the equation from Example 10, $2x + 3y = 6$, on a graphing calculator, you must first solve it for y. When you do so, you will get $y = -\frac{2}{3}x + 2$, which results with the graph in Figure 14. Use this window:

$$\text{Window:} \quad -6 \le x \le 6 \text{ and } -6 \le y \le 6$$

Hint on Tracing

If you are going to use the Trace feature and you want the x-coordinates to be exact numbers, set your window so the range of X inputs is a multiple of the number of horizontal pixels on your calculator screen. On the TI-83/84, the screen is 94 pixels wide. Here are a few convenient trace windows:

To trace with x to the nearest tenth use $-4.7 \le x \le 4.7$ or $0 \le x \le 9.4$

To trace with x to the nearest integer use $-47 \le x \le 47$ or $0 \le x \le 94$

Graph each equation using the indicated window.

1. $y = \frac{1}{2}x - 3$ $-10 \le x \le 10$ and $-10 \le y \le 10$

2. $y = \frac{1}{2}x^2 - 3$ $-10 \le x \le 10$ and $-10 \le y \le 10$

3. $y = \frac{1}{2}x^2 - 3$ $-4.7 \le x \le 4.7$ and $-10 \le y \le 10$

4. $y = x^3$ $-10 \le x \le 10$ and $-10 \le y \le 10$

5. $y = x^3 - 5$ $-4.7 \le x \le 4.7$ and $-10 \le y \le 10$

GETTING READY FOR CLASS

After reading through the preceding section, respond in your own words and in complete sentences.

A. Explain how you would construct a rectangular coordinate system from two real number lines.

B. Explain in words how you would graph the ordered pair $(2, -3)$.

C. How can you tell if an ordered pair is a solution to the equation $y = \frac{1}{3}x + 5$?

D. Explain the difference between the x- and y-intercepts of the graph $y = x^2 - 3$.

EXERCISE SET 11.5

Vocabulary Review

Choose the correct words to fill in the blanks below.

y-intercept	horizontal	x-coordinate	y-axis	vertical
x-intercept	y-coordinate	origin	x-axis	

1. The horizontal number line on a rectangular coordinate system is called the _____, whereas the vertical number line is called the _____.

2. For the ordered pair $(-9, 6)$, the _____ is -9 and the _____ is 6.

3. The graph of an equation $y = mx$ will be a straight line through the _____.

4. The _____ is the pair of coordinates of a point where a graph crosses the x-axis.

5. The _____ is the pair of coordinates of a point where a graph crosses the y-axis.

6. Any equation of the form $x = a$ has a _____ line for its graph.

7. Any equation of the form $y = b$ has a _____ line for its graph.

Problems

A Graph each of the following ordered pairs on a rectangular coordinate system.

1.
 a. $(-1, 2)$
 b. $(-1, -2)$
 c. $(5, 0)$
 d. $(0, 2)$
 e. $(-5, -5)$
 f. $\left(\dfrac{1}{2}, 2\right)$

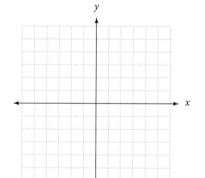

2.
 a. $(1, 2)$
 b. $(1, -2)$
 c. $(0, -3)$
 d. $(4, 0)$
 e. $(-4, -1)$
 f. $\left(3, \dfrac{1}{4}\right)$

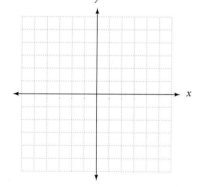

Give the coordinates of each point.

3.

4.

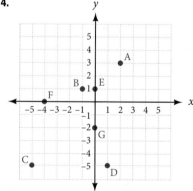

5. Which of the following tables could be produced from the equation $y = 2x - 6$?

a.
x	y
0	6
1	4
2	2
3	0

b.
x	y
0	-6
1	-4
2	-2
3	0

c.
x	y
0	-6
1	-5
2	-4
3	-3

6. Which of the following tables could be produced from the equation $3x - 5y = 15$?

d.
x	y
0	5
-3	0
10	3

e.
x	y
0	-3
5	0
10	3

f.
x	y
0	-3
-5	0
10	-3

7. The graph shown here is the graph of which of the following equations?

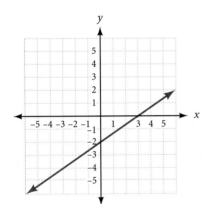

a. $y = \dfrac{3}{2}x - 3$

b. $y = \dfrac{2}{3}x - 2$

c. $y = -\dfrac{2}{3}x + 2$

8. The graph shown here is the graph of which of the following equations?

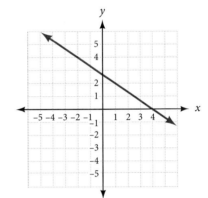

a. $3x - 2y = 8$

b. $2x - 3y = 8$

c. $2x + 3y = 8$

For each problem below, the equation of the red graph is given. Find the equation for the blue graph.

9.

$y = x$

10.

$y = x^2$

11.

$y = |x|$

12.

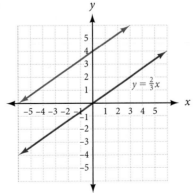

$y = \dfrac{2}{3}x$

Graph each of the following. Use one coordinate system for each problem.

13. a. $y = 2x$ **b.** $y = 2x + 3$ **c.** $y = 2x - 5$

 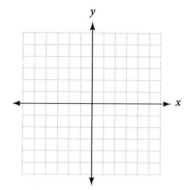

14. a. $y = \dfrac{1}{3}x$ **b.** $y = \dfrac{1}{3}x + 1$ **c.** $y = \dfrac{1}{3}x - 3$

 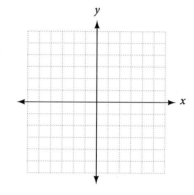

15. a. $y = \dfrac{1}{2}x^2$ **b.** $y = \dfrac{1}{2}x^2 - 2$ **c.** $y = \dfrac{1}{2}x^2 + 2$

 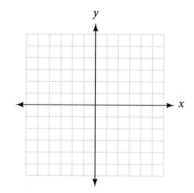

16. a. $y = 2x^2$ **b.** $y = 2x^2 - 5$ **c.** $y = 2x^2 + 1$

 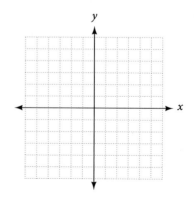

17. Graph the straight line $0.02x + 0.03y = 0.06$.

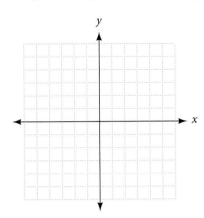

18. Graph the straight line $0.05x - 0.03y = 0.15$.

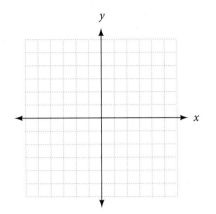

B D Graph each of the following lines.

19. a. $y = 2x$

b. $x = -3$

c. $y = 2$

20. a. $y = 3x$

b. $x = -2$

c. $y = 4$

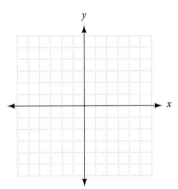

21. a. $y = -\dfrac{1}{2}x$

b. $x = 4$

c. $y = -3$

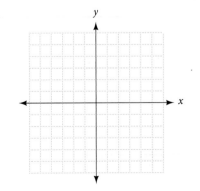

22. a. $y = -\dfrac{1}{3}x$

b. $x = 1$

c. $y = -5$

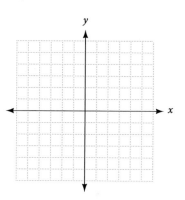

C Find the intercepts for each graph. Then use them to help sketch the graph.

23. $y = x^2 - 9$

24. $y = x^2$

25. $y = 2x - 4$

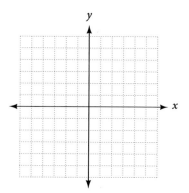

26. $y = 4x - 2$

27. $y = \dfrac{1}{2}x + 1$

28. $y = -\dfrac{1}{2}x + 1$

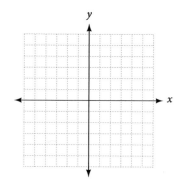

29. $y = 3x$

30. $y = -\dfrac{1}{3}x$

31. $y = x^2 - x$

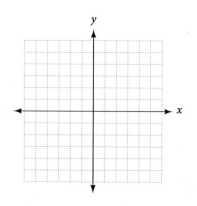

32. $y = x^2 + 3$

33. $y = x - 3$

34. $y = x + 2$

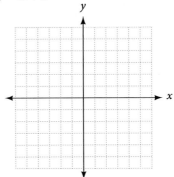

35. a. Solve: $4x + 12 = -16$.
 b. Find x when y is 0: $4x + 12y = -16$.
 c. Find y when x is 0: $4x + 12y = -16$.
 d. Graph: $4x + 12y = -16$.

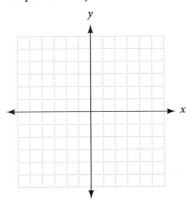

 e. Solve for y: $4x + 12y = -16$.

36. a. Solve: $3x - 8 = -12$.
 b. Find x when y is 0: $3x - 8y = -12$.
 c. Find y when x is 0: $3x - 8y = -12$.
 d. Graph: $3x - 8y = -12$.

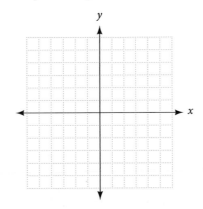

 e. Solve for y: $3x - 8y = -12$.

Applying the Concepts

37. Movie Tickets The graph shows the rise in movie ticket prices from 2000 to 2009. Use the chart to answer the following questions.

 a. Could the graph contain the point (2002, 5.75)?

 b. Could the graph contain the point (2005, 5.5)?

 c. Could the graph contain the point (2007, 6.75)?

38. Gym Memberships The chart shows the rise in health club memberships from 2000 to 2005. Using years as x and millions of members as y, write 5 ordered pairs that describe the information in the chart.

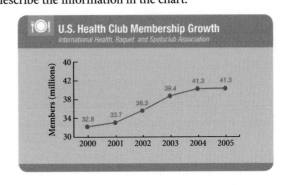

39. Hourly Wages Suppose you have a job that pays $9.50 per hour, and you work anywhere from 0 to 40 hours per week. Table 2 gives the amount of money you will earn in 1 week for working various hours. Construct a line graph from the information in Table 2.

Table 2

Weekly Wages

Hours worked	Pay ($)
0	0
10	95
20	190
30	285
40	380

Weekly Wages

40. Softball Toss Chaudra is tossing a softball into the air with an underhand motion. It takes exactly 2 seconds for the ball to come back to her. Table 3 shows the distance the ball is above her hand at quarter-second intervals. Construct a line graph from the information in the table.

Table 3

Tossing a softball into the air

Time (sec)	Distance (ft)
0	0
0.25	7
0.5	12
0.75	15
1	16
1.25	15
1.5	12
1.75	7
2	0

Tossing a Softball Into the Air

41. Intensity of Light Table 4 gives the intensity of light that falls on a surface at various distances from a 100-watt light bulb. Construct a bar chart from the information in Table 4.

Table 4	Light intensity from a 100-watt light bulb
Distance above surface (ft)	Intensity (lumens/sq ft)
1	120.0
2	30.0
3	13.3
4	7.5
5	4.8
6	3.3

Light Intensity from a 100-watt Light Bulb

42. Value of a Painting Presley purchased a piece of abstract art in 1990 for $125. Table 5 shows the value of the painting at various times, assuming that it doubles in value every 5 years. Construct a bar chart from the information in the table.

Table 5	Value of a Painting
Year	Value ($)
1990	125
1995	250
2000	500
2005	1,000
2010	2,000

Value of a Painting

43. Wind Power Table 6 shows the power output generated by different sized wind turbines. The data is based on power output generated at 33 mph winds, which is considered the ideal power generation speed. Construct a line graph from the information in the table.

Table 6	Power Output from a Wind Turbine
Blade diameter (m)	**Power output(kw)**
10	25
17	100
27	225
33	300
40	500

Source: Danish Wind Industry Association, American Wind Energy Association

Power Output from a Wind Turbine

44. Solar Power Table 7 shows the voltage and current from various power resistors for a typical PV Solar Panel.

Table 7	Power Output from a PV Solar Panel	
Current (Amps)	**Voltage (Volts)**	**Power**
0.00	30.9	
0.29	28.8	
0.54	27.0	
0.90	22.5	
1.03	3.1	
1.10	0.0	

Source: Danish Wind Industry Association, American Wind Energy Association

a. Construct a line graph from the information in the table for current and voltage.

Power Output

b. The power output is the current multiplied by the volts. Compute the power output for each resistance.

c. Identify the point on your line graph with the longest power output. This represents the peak power generated from the solar panel.

45. Reading Graphs The graph shows the number of people in line at a theater box office to buy tickets for a movie that starts at 7:30. The box office opens at 6:45.

a. How many people are in line at 6:30?

b. How many people are in line when the box office opens?

c. How many people are in line when the show starts?

d. At what times are there 60 people in line?

e. How long after the show starts is there no one left in line?

46. Facebook Users The graph shows the number of active Facebook users from 2004 to 2010. If x represents the year and y represents million of users, write five ordered pairs that describe the information in the graph.

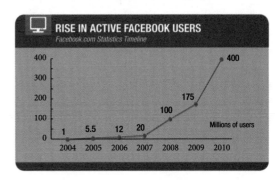

RISE IN ACTIVE FACEBOOK USERS
Facebook.com Statistics Timeline

Getting Ready for the Next Section

Complete each table using the given equation.

47. $y = 7.5x$

x	y
0	
10	
20	

48. $h = 32t - 16t^2$

t	h
0	
1	
-1	

49. $y = 7.5x$

x	y
0	
$\frac{1}{2}$	
1	

50. $h = 32t - 16t^2$

t	h
-3	
0	
3	

Find the Mistake

Write true or false for each statement below. If false, circle what is false and write the correct answer on the line provided.

1. True or False: On a rectangular coordinate system, the vertical axis is called the x-axis. _____

2. True or False: Any point on the x-axis has a y-coordinate of 0. _____

3. True or False: The graph of a line that passes through the origin will have the ordered pair $(0, 0)$ as a solution. _____

4. True or False: The x-intercept for the graph of the equation $4x + 7y = 12$ is 3. _____

11.6 Introduction to Functions

OBJECTIVES

A Construct a table or a graph from a function rule.

B Identify the domain and range of a function.

C Determine whether a relation is also a function.

D Use the vertical line test to determine whether a graph is a function.

KEY WORDS

relation

function

domain

range

function map

vertical line test

Before the harvest each fall, a Nebraskan farmer climbs onto his riding lawn mower and rides through acres of corn. He uses the mower to carve an intricate maze in his cornfield. One of his most impressive designs took the shape of a Star Wars starship. He is known to fly in a plane over his freshly cut maze to ensure its perfection. Thousands flock to Grandpa John's Amazing Maze; each visitor paying $5.00 to walk through it.

In this section, we will introduce you to functions. Using Grandpa John's Amazing Maze as an example, the amount of money the maze earns for the farmer each season depends on the number of people who visit it. In mathematics, we say that his seasonal earnings are a *function* of how many people visit the maze.

Relations

Before we begin our discussion of functions, let's examine what happens when a coordinate from a set of inputs is paired with one or more elements from a set of outputs.

Table 1 shows the prices of used Ford Mustangs that were listed in the local newspaper. The diagram in Figure 1 is called a *scatter diagram*. It gives a visual representation of the data in Table 1.

Table 1

Used Mustang Prices

Year x	Price ($) y
2009	17,995
2009	16,945
2009	25,985
2008	15,995
2008	14,867
2007	16,960
2006	14,999
2006	19,150
2005	11,352

FIGURE 1

Ordered Pairs

(2009, 17,995)
(2009, 16,945)
(2009, 25,985)
(2008, 15,995)
(2008, 14,867)
(2007, 16,960)
(2006, 14,999)
(2006, 19,150)
(2005, 11,352)

**Video Examples
Section 11.6**

Looking at the graph, you can see that the year 2009 is paired with three different prices: $17,995, $16,945, and $25,985. This data is considered a *relation*. Furthermore, all sets of paired data are considered relations where each number in the domain is paired with one or more numbers in the range. The following are two formal definitions of a relation.

A Constructing Tables and Graphs

A special case occurs when each number in a domain is paired with exactly one number in a range. This is called a *function.*

Suppose you have a job that pays $9.50 per hour and that requires you to work anywhere from 0 to 40 hours per week. If we let the variable x represent hours and the variable y represent the money you make, then the relationship between x and y can be written as

$$y = 9.5x \qquad \text{for} \qquad 0 \le x \le 40$$

EXAMPLE 1 Construct a table and graph for the function

$$y = 9.5x \qquad \text{for} \qquad 0 \le x \le 40$$

Solution Table 2 gives some of the paired data that satisfy the equation $y = 9.5x$. Figure 2 is the graph of the equation with the restriction $0 \le x \le 40$.

Table 2 Weekly Wages		
Hours Worked	**Rule**	**Pay**
x	$y = 9.5x$	y
0	$y = 9.5(0)$	0
10	$y = 9.5(10)$	95
20	$y = 9.5(20)$	190
30	$y = 9.5(30)$	285
40	$y = 9.5(40)$	380

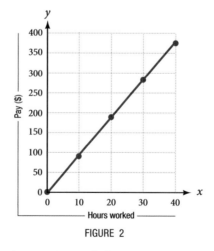

FIGURE 2

Ordered Pairs

(0, 0)
(10, 95)
(20, 190)
(30, 285)
(40, 380)

The equation $y = 9.5x$ with the restriction $0 \le x \le 40$, Table 2, and Figure 2 are three ways to describe the same relationship between the number of hours you work in one week and your gross pay for that week. In all three, we *input* values of x, and then use the function rule to *output* values of y.

Practice Problems

1. Construct a table and graph for $y = 8x, 0 \le x \le 40.$

B Domain and Range of a Function

We began this discussion by saying that the number of hours worked during the week was from 0 to 40, so these are the values that x can assume. From the line graph in Figure 2, we see that the values of y range from 0 to 380. We call the complete set of values that x can assume the *domain* of the function. The values that are assigned to y are called the *range* of the function.

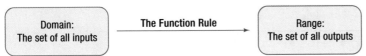

Domain:
The set of all inputs → The Function Rule → Range:
The set of all outputs

EXAMPLE 2 State the domain and range for the function

$$y = 9.5x, \quad 0 \le x \le 40$$

Solution From the previous discussion, we have

Domain $= \{x \,|\, 0 \le x \le 40\} = [0, 40]$

Range $= \{y \,|\, 0 \le y \le 380\} = [0, 380]$

We are familiar with writing domain and range using set-builder notation and interval notation, shown in Example 2. We can also write the domain and range using *roster notation*. This notation lists all the elements in a set one by one. In the case of Example 1, our set consists of the coordinates

$$(0, 0), (10, 95), (20, 190), (30, 285), (40, 380)$$

Therefore, using roster notation, we give the domain as {0, 10, 20, 30, 40}, and the range as {0, 95, 190, 285, 380}. Here is a summary of the three types of notation using the values from our first example:

	Set-Builder Notation	Roster Notation	Interval Notation	
Domain	$\{x \,	\, 0 \le x \le 40\}$	{0, 10, 20, 30, 40}	[0, 40]
Range	$\{y \,	\, 0 \le y \le 380\}$	{0, 95, 190, 285, 380}	[0, 380]

Function Maps

Another way to visualize the relationship between x and y is with the diagram in Figure 2, which we call a *function map*.

Domain Rule: Multiply by 9.5. Range
10 → 95
20 → 190
x → $9.5x$

FIGURE 3

Although the diagram in Figure 3 does not show all the values that x and y can assume, it does give us a visual description of how x and y are related. It shows that values of y in the range come from values of x in the domain according to a specific rule (multiply by 9.5 each time).

We are now ready for the formal definition of a *function*.

DEFINITION function

A *function* is a rule that pairs each element in one set, called the *domain*, with exactly one element from a second set, called the *range*. In other words, a function is a rule for which each input is paired with exactly one output.

2. State the domain and range for $y = 8x, 0 \le x \le 40$.

3. Determine the domain and range for the following relations.

a. $\{(3, 3), (1, -1), (0, -3), (a, b)\}$

b.

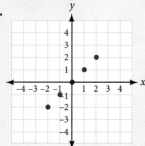

c.

Input	Output
5	0 → 2
7	1 5
3	4 −3
−2	

EXAMPLE 3 Determine the domain and range of the following relations.

a. $\{(5, 7), (0, -3), (2, 1), (x, y)\}$

b.

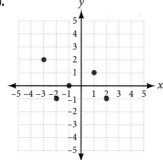

c.

Input	Rule: Population in millions	Output
Ohio		25
Florida		19
Michigan		11
Texas		10

Solution

a. For the set of ordered pairs, the first coordinate is the domain and the second coordinate is the range.

Domain $= \{5, 0, 2, x\}$

Range $= \{7, -3, 1, y\}$

b. The points on the graph represent ordered pairs. The set of x-coordinates are the domain and the set of y-coordinates are the range.

The relation is $\{(-3, 2), (-2, -1), (-1, 0), (1, 1), (2, -1)\}$.

Domain $= \{-3, -2, -1, 1, 2\}$

Range $= \{2, -1, 0, 1\}$

c. The domain is the set of states, and the range is the set of numbers that correspond to each state.

Domain $= \{$Ohio, Florida, Michigan, Texas$\}$

Range $= \{10, 11, 19, 25\}$

EXAMPLE 4 Kendra tosses a softball into the air with an underhand motion. The distance of the ball above her hand is given by the function

$$h = 32t - 16t^2 \qquad \text{for} \qquad 0 \le t \le 2$$

where h is the height of the ball in feet and t is the time in seconds. Construct a table that gives the height of the ball at quarter-second intervals, starting with $t = 0$ and ending with $t = 2$, then graph the function.

Solution We construct Table 3 using the following values of t: $0, \frac{1}{4}, \frac{1}{2}, \frac{3}{4}, 1, \frac{5}{4}, \frac{3}{2}, \frac{7}{4}, 2$. Then we construct the graph in Figure 4 from the table. The graph appears only in the first quadrant because neither t nor h can be negative.

Answers

3. a. Domain $= \{3, 1, 0, a\}$
Range $= \{-3, -1, 3, b\}$
b. Domain $= \{-2, -1, 0, 1, 2\}$
Range $= \{-2, -1, 0, 1, 2\}$
c. Domain $= \{-2, 3, 5, 7\}$
Range $= \{0, 1, 2, 4\}$

Table 3 Tossing a Softball into the Air

Input		Output
Time (sec) t	Function Rule $h = 32t - 16t^2$	Distance (ft) h
0	$h = 32(0) - 16(0)^2 = 0 - 0 = 0$	0
$\frac{1}{4}$	$h = 32\left(\frac{1}{4}\right) - 16\left(\frac{1}{4}\right)^2 = 8 - 1 = 7$	7
$\frac{1}{2}$	$h = 32\left(\frac{1}{2}\right) - 16\left(\frac{1}{2}\right)^2 = 16 - 4 = 12$	12
$\frac{3}{4}$	$h = 32\left(\frac{3}{4}\right) - 16\left(\frac{3}{4}\right)^2 = 24 - 9 = 15$	15
1	$h = 32(1) - 16(1)^2 = 32 - 16 = 16$	16
$\frac{5}{4}$	$h = 32\left(\frac{5}{4}\right) - 16\left(\frac{5}{4}\right)^2 = 40 - 25 = 15$	15
$\frac{3}{2}$	$h = 32\left(\frac{3}{2}\right) - 16\left(\frac{3}{2}\right)^2 = 48 - 36 = 12$	12
$\frac{7}{4}$	$h = 32\left(\frac{7}{4}\right) - 16\left(\frac{7}{4}\right)^2 = 56 - 49 = 7$	7
2	$h = 32(2) - 16(2)^2 = 64 - 64 = 0$	0

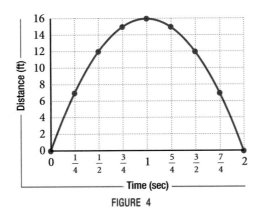

FIGURE 4

4. Kendra is tossing a softball into the air so that the distance h the ball is above her hand t seconds after she begins the toss is given by $h = 48t - 16t^2$, $0 \le t \le 3$. Construct a table and line graph for this function.

t	h
0	
$\frac{1}{2}$	
1	
$\frac{3}{2}$	
2	
$\frac{5}{2}$	
3	

Here is a summary of what we know about functions as it applies to this example: We input values of t and output values of h according to the function rule

$$h = 32t - 16t^2 \quad \text{for} \quad 0 \le t \le 2$$

The domain is given by the inequality that follows the equation; it is

$$\text{Domain} = \{t \mid 0 \le t \le 2\} = [0, 2]$$

The range is the set of all outputs that are possible by substituting the values of t from the domain into the equation. From our table and graph, it seems that the range is

$$\text{Range} = \{h \mid 0 \le h \le 16\} = [0, 16]$$

As you can see from the examples we have done to this point, the function rule produces ordered pairs of numbers. We use this result to write an alternative definition for a function.

ALTERNATE DEFINITION function

A *function* is a set of ordered pairs in which no two different ordered pairs have the same first coordinate. The set of all first coordinates is called the *domain* of the function. The set of all second coordinates is called the *range* of the function. The restriction on first coordinates keeps us from assigning a number in the domain to more than one number in the range.

C Determining Relations and Functions

Here are some facts that will help clarify the distinction between relations and functions:

1. Any rule that assigns numbers from one set to numbers in another set is a relation. If that rule makes the assignment so no input has more than one output, then it is also a function.
2. Any set of ordered pairs is a relation. If none of the first coordinates of those ordered pairs is repeated, the set of ordered pairs is also a function.
3. Every function is a relation.
4. Not every relation is a function.

EXAMPLE 5 Sketch the graph of $x = y^2$.

Solution Without going into much detail, we graph the equation $x = y^2$ by finding a number of ordered pairs that satisfy the equation, plotting these points, then drawing a smooth curve that connects them. A table of values for x and y that satisfy the equation follows, along with the graph of $x = y^2$ shown in Figure 5.

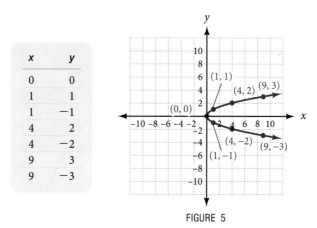

x	y
0	0
1	1
1	−1
4	2
4	−2
9	3
9	−3

FIGURE 5

As you can see from looking at the table and the graph in Figure 5, several ordered pairs whose graphs lie on the curve have repeated first coordinates, for instance $(1, 1)$ and $(1, −1)$, $(4, 2)$ and $(4, −2)$, as well as $(9, 3)$ and $(9, −3)$. Therefore, the graph is not the graph of a function.

D Vertical Line Test

Look back at the scatter diagram for used Mustang prices shown in Figure 1. Notice that some of the points on the diagram lie above and below each other along vertical lines. This is an indication that the data do not constitute a function. Two data points that lie on the same vertical line must have come from two ordered pairs with the same first coordinates.

Now, look at the graph shown in Figure 5. The reason this graph is the graph of a relation, but not of a function, is that some points on the graph have the same first coordinates, for example, the points $(4, 2)$ and $(4, −2)$. Furthermore, any time two points on a graph have the same first coordinates, those points must lie on a vertical line. [To convince yourself, connect the points $(4, 2)$ and $(4, −2)$ with a straight line. You will see that it must be a vertical line.] This allows us to write the following test that uses the graph to determine whether a relation is also a function.

> **RULE** Vertical Line Test
>
> If a vertical line crosses the graph of a relation in more than one place, the relation cannot be a function. If no vertical line can be found that crosses a graph in more than one place, then the graph represents a function.

5. Use the equation $x = y^2 − 4$ to fill in the following table. Then use the table to sketch the graph.

x	y
	−3
	−2
	−1
	0
	1
	2
	3

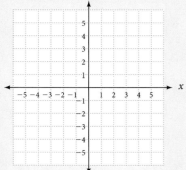

If we look back to the graph of $h = 32t - 16t^2$ as shown in Figure 4, we see that no vertical line can be found that crosses this graph in more than one place. The graph shown in Figure 4 is therefore the graph of a function.

EXAMPLE 6 Match each relation with its graph, then indicate which relations are functions.

a. $y = |x| - 4$ **b.** $y = x^2 - 4$ **c.** $y = 2x + 2$

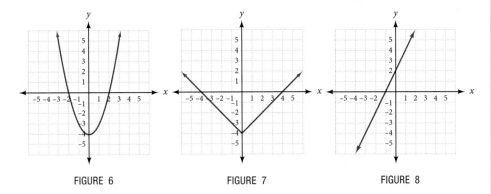

FIGURE 6 FIGURE 7 FIGURE 8

Solution Using the basic graphs for a guide along with our knowledge of translations, we have the following:

a. Figure 7 **b.** Figure 6 **c.** Figure 8

And, since all graphs pass the vertical line test, all are functions.

6. Graph $y = |x| - 3$.

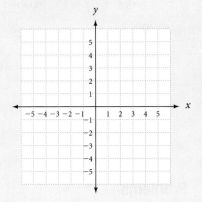

GETTING READY FOR CLASS

After reading through the preceding section, respond in your own words and in complete sentences.

A. What is a relation?

B. What is a function?

C. Which variable is usually associated with the domain of a function? Which variable is associated with the range?

D. What is the vertical line test?

Vocabulary Review

Choose the correct words to fill in the blanks below.

function	domain	right	left
range	relation	vertical line test	

1. The _____ of a function is the complete set of inputs, whereas the _____ is the set of all outputs.

2. On a function map, the values for the domain of the function appear on the _____ and the values for the range appear on the _____.

3. A _____ is a rule that pairs each element in one set with exactly one element from a second set.

4. A _____ is a rule that pairs each element in one set with one or more elements from a second set.

5. The _____ can be used to determine whether the graph of a relation is a function.

Problems

A For each of the following functions, construct a table and graph.

1. $y = 4x$ for $-2 \leq x \leq 5$

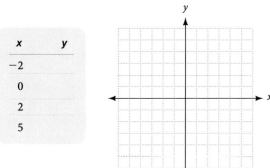

x	y
-2	
0	
2	
5	

2. $y = -3x$ for $-3 \leq x \leq 2$

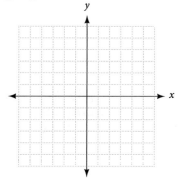

x	y
-3	
-1	
0	
1	
2	

3. $y = 100 - 4x$ for $0 \leq x \leq 25$

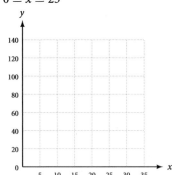

x	y
0	
5	
10	
15	
20	
25	

4. $y = 250 - 10x$ for $0 \leq x \leq 25$

x	y
0	
5	
10	
15	
20	
25	

5. $y = 1600 - 10x$ for $0 \le x \le 160$

x	y
0	
40	
80	
120	
160	

6. $y = 1200 - 40x$ for $0 \le x \le 30$

x	y
0	
10	
20	
30	

7. $y = 40 + 2.5x$ for $0 \le x \le 100$

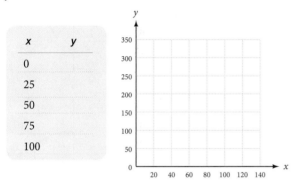

x	y
0	
25	
50	
75	
100	

8. $y = 100 + 0.5x$ for $0 \le x \le 40$

x	y
0	
10	
20	
30	
40	

B C For each of the following relations, give the domain and range using roster notation, and indicate which are also functions.

9. $(1, 2), (3, 4), (5, 6), (7, 8)$

10. $(2, 1), (4, 3), (6, 5), (8, 7)$

11. $(2, 5), (3, 4), (1, 4), (0, 6)$

12. $(0, 4), (1, 6), (2, 4), (1, 5)$

13. $(a, 3), (b, 4), (c, 3), (d, 5)$

14. $(a, 5), (b, 5), (c, 4), (d, 5)$

15. $(a, 1), (a, 2), (a, 3), (a, 4)$

16. $(a, 1), (b, 1), (c, 1), (d, 1)$

Give the domain and range for the following function maps.

17.

18.

19.

20.
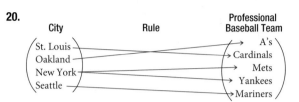

C D State whether each of the following graphs represents the graph of a function.

21.

22.

23.

24.

25.

26.

27.

28.

29.

30.

31.

32.

Determine the domain and range of the following functions. Assume the *entire* function is shown. Write your answers in set-builder notation and interval notation.

33.

34.

35.

36.

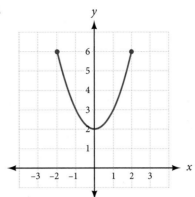

Graph each of the following relations. In each case, use the graph to find the domain and range, and indicate whether the graph is the graph of a function.

37. $y = x^2 - 1$

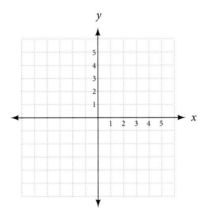

38. $y = x^2 + 1$

39. $y = x^2 + 4$

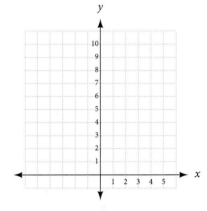

40. $y = x^2 - 9$

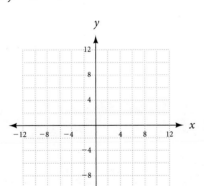

41. $x = y^2 - 1$

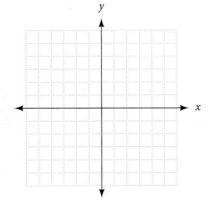

42. $x = y^2 + 1$

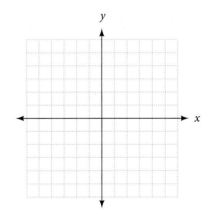

43. $y = (x + 2)^2$

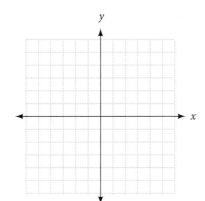

44. $y = (x - 3)^2$

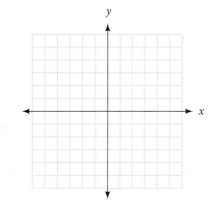

45. $x = (y + 1)^2$

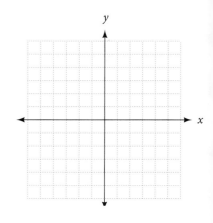

46. $x = 3 - y^2$

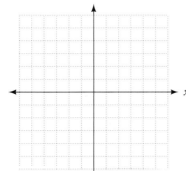

47. $y = |x| + 2$

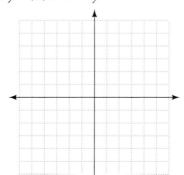

48. $x = |y| - 3$

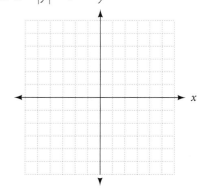

Applying the Concepts

49. iPod Memory The chart shows the memory capacity for different generations of iPods from 2001 to 2009. Using the chart, list all the values in the domain and range for the memory capacity.

50. Red Box Kiosks The chart shows the number of Redbox DVD rental kiosks installed from 2005 to 2010. Use the chart to state the domain and range of the function for Redbox kiosk installation.

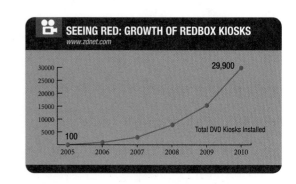

51. Weekly Wages Suppose you have a job that pays $8.50 per hour and you work anywhere from 10 to 40 hours per week.

a. Write an equation, with a restriction on the variable *x*, that gives the amount of money, *y*, you will earn for working *x* hours in one week.

b. Use the function rule you have written in part **a.** to complete Table 4.

Table 4 Weekly Wages		
Hours Worked	Function Rule	Gross Pay ($)
x		*y*
10		
20		
30		
40		

c. Construct a line graph from the information in Table 4.

Weekly Wages

d. State the domain and range of this function.

e. What is the minimum amount you can earn in a week with this job? What is the maximum amount?

52. Weekly Wages The ad shown here was in the local newspaper. Suppose you are hired for the job described in the ad.

a. If *x* is the number of hours you work per week and *y* is your weekly gross pay, write the equation for *y*. (Be sure to include any restrictions on the variable *x* that are given in the ad.)

b. Use the function rule you have written in part a to complete Table 5.

Table 5 Weekly Wages		
Hours Worked	Function Rule	Gross Pay ($)
x		*y*
15		
20		
25		
30		

c. Construct a line graph from the information in Table 5.

Weekly Wages

d. State the domain and range of this function.

e. What is the minimum amount you can earn in a week with this job? What is the maximum amount?

53. Global Solar Energy Demand The chart shows the global solar power demand from 2008 to 2012. Using the chart:

a. List the values for the domain and for the range.

b. Did demand increase more in 2011 or 2012 (projected)?

c. In what year did demand decrease?

PV Module Demand (MW)

Source: wikinvest.com/industry/solar_power

54. Global Wind Power Capacity The chart shows the global wind power capacity from 1996 to 2010:

Use the information from the chart to answer the following questions.

a. List the values for the domain and for the range.

b. Did capacity increase more in 2006 or 2007?

c. In which year did capacity increase the most?

55. Profits Match each of the following statements to the appropriate graph indicated by labels I-IV.

a. Sarah works 25 hours to earn $250.

b. Justin works 35 hours to earn $560.

c. Rosemary works 30 hours to earn $360.

d. Marcus works 40 hours to earn $320.

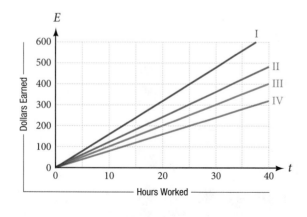

FIGURE 9

56. Find an equation for each of the functions shown in the Figure 9. Show dollars earned, E, as a function of hours worked, t. Then indicate the domain and range of each function.

a. Graph I: $E =$

Domain $= \{ t \mid \qquad \}$

Range $= \{ E \mid \qquad \}$

b. Graph II: $E =$

Domain $= \{ t \mid \qquad \}$

Range $= \{ E \mid \qquad \}$

c. Graph III: $E =$

Domain $= \{ t \mid \qquad \}$

Range $= \{ E \mid \qquad \}$

d. Graph IV: $E =$

Domain $= \{ t \mid \qquad \}$

Range $= \{ E \mid \qquad \}$

Getting Ready for the Next Section

57. If $s = \dfrac{60}{t}$, find s when

 a. $t = 10$

 b. $t = 8$

58. If $y = 3x^2 + 2x - 1$, find y when

 a. $x = 0$

 b. $x = -2$

59. Find the value of $x^2 + 2$ for

 a. $x = 5$

 b. $x = -2$

60. Find the value of $125 \cdot 2^t$ for

 a. $t = 0$

 b. $t = 1$

For the equation $y = x^2 - 3$:

61. Find y if x is 2.

62. Find y if x is -2.

63. Find y if x is 0.

64. Find y if x is -4.

65. For which value of x is $y = -3$?

66. For which values of x is $y = 6$? (There are two answers)

The problems that follow review some of the more important skills you have learned in previous sections and chapters.

67. If $x - 2y = 4$, and $x = \dfrac{8}{5}$ find y.

68. If $\dfrac{x^2}{25} + \dfrac{y^2}{9} = 1$, find y when x is -4.

69. Let $x = 0$ and $y = 0$ in $y = a(x - 8)^2 + 70$ and solve for a.

70. Find R if $p = 2.5$ and $R = (900 - 300p)p$.

Find the Mistake

Each sentence below contains a mistake. Circle the mistake and write the correct word or phrase on the line provided.

1. The set of all inputs for a function is called the range. _____

2. For the relation (1, 2), (3, 4), (5, 6), (7, 8), the domain is {2, 4, 6, 8}. _____

3. Graphing the relation $y = x^2 - 3$ will show that the domain is $\{\, x \mid x \geq -3 \,\}$. _____

4. If a horizontal line crosses the graph of a relation in more than one place, the relation cannot be a function.

A Use function notation to find the value of a function for a given value of the variable.

B Use graphs to visualize the relationship between a function and a variable.

C Use function notation in formulas.

11.7 Function Notation

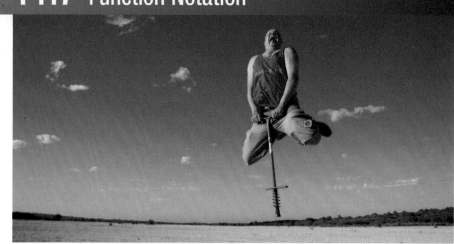

©iStockphoto.com/tap10

KEY WORDS

dependent variable

independent variable

function notation

inputs

outputs

A Evaluate a Function at a Point

Pogopalooza is an annual world championship for stunt pogo stick athletes. Pogo jumpers compete in various events, such as the most or the least jumps in a minute, the highest jump, and numerous exhibitions of acrobatic stunts. Imagine a group of local stunt pogo enthusiasts are holding a pogo stick competition in your city. They have gathered a variety a sponsors, and their budget allows them to hire you to market and promote the event. If the job pays $8.50 per hour for working from 0 to 40 hours a week, then the amount of money y earned in one week is a function of the number of hours worked x. The exact relationship between x and y is written

$$y = 8.5x \quad \text{for} \quad 0 \le x \le 40$$

Because the amount of money earned y depends on the number of hours worked x, we call y the *dependent variable* and x the *independent variable*. Furthermore, if we let f represent all the ordered pairs produced by the equation, then we can write

$$f = \{(x, y) \mid y = 8.5x \quad \text{and} \quad 0 \le x \le 40\}$$

Once we have named a function with a letter, we can use an alternative notation to represent the dependent variable y. The alternative notation for y is $f(x)$. It is read "f of x" and can be used instead of the variable y when working with functions. The notation y and the notation $f(x)$ are equivalent. That is,

$$y = 8.5x \Leftrightarrow f(x) = 8.5x$$

When we use the notation $f(x)$ we are using *function notation*. The benefit of using function notation is that we can write more information with fewer symbols than we can by using just the variable y. For example, asking how much money a person will make for working 20 hours is simply a matter of asking for $f(20)$. Without function notation, we would have to say, "Find the value of y that corresponds to a value of $x = 20$." To illustrate further, using the variable y, we can say "y is 170 when x is 20." Using the notation $f(x)$, we simply say "$f(20) = 170$." Each expression indicates that you will earn $170 for working 20 hours.

Note An important thing to remember is that f is the function name, it is not a variable and cannot be replaced with a number. Furthermore, $f(x)$ is read "f of x" but does not imply multiplication.

Practice Problems

1. If $f(x) = 8x$, find
 a. $f(0)$
 b. $f(5)$
 c. $f(10.5)$

EXAMPLE 1 If $f(x) = 7.5x$, find $f(0)$, $f(10)$, and $f(20)$.

Solution To find $f(0)$, we substitute 0 for x in the expression $7.5x$ and simplify. We find $f(10)$ and $f(20)$ in a similar manner — by substitution.

$$\text{If} \qquad f(x) = 7.5x$$
$$\text{then} \qquad f(0) = 7.5(0) = 0$$
$$f(10) = 7.5(10) = 75$$
$$f(20) = 7.5(20) = 150$$

Answers

1. **a.** 0 **b.** 40 **c.** 84

If we changed the example in the discussion that opened this section so the hourly wage was $6.50 per hour, we would have a new equation to work with, namely,

$$y = 6.5x \quad \text{for} \quad 0 \le x \le 40$$

Suppose we name this new function with the letter g. Then

$$g = \{(x, y) \mid y = 6.5x \quad \text{and} \quad 0 \le x \le 40\}$$

and

$$g(x) = 6.5x$$

If we want to talk about both functions in the same discussion, having two different letters, f and g, makes it easy to distinguish between them. For example, since $f(x) = 7.5x$ and $g(x) = 6.5x$, asking how much money a person makes for working 20 hours is simply a matter of asking for $f(20)$ or $g(20)$, avoiding any confusion over which hourly wage we are talking about.

The diagrams shown in Figure 1 further illustrate the similarities and differences between the two functions we have been discussing.

FIGURE 1

B Function Notation and Graphs

We can visualize the relationship between x and $f(x)$ on the graph of the function. Figure 2 shows the graph of $f(x) = 7.5x$ along with two additional line segments. The horizontal line segment corresponds to $x = 20$, and the vertical line segment corresponds to $f(20)$. (Note that the domain is restricted to $0 \le x \le 40$.)

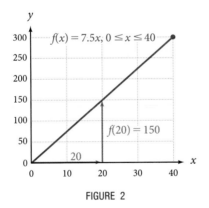

FIGURE 2

We can use functions and function notation to talk about numbers in the following chart showing the number of millionaire households in the U.S. Let's let x represent one of the years in the chart.

**Video Examples
Section 11.7**

If the function *f* pairs each year in the chart with the number of millionaire households for that year, then each statement below is true:

$$f(2006) = 9.0$$

The domain of $f = \{2000, 2002, 2004, 2006, 2008, 2010\}$

In general, when we refer to the function *f* we are referring to the domain, the range, and the rule that takes elements in the domain and outputs elements in the range. When we talk about $f(x)$ we are talking about the rule itself, or an element in the range, or the variable *y*.

The function f

Domain of *f*	$y = f(x)$	Range of *f*
Inputs	Rule	Outputs

C Using Function Notation

The remaining examples in this section show a variety of ways to use and interpret function notation.

EXAMPLE 2 If Lorena takes *t* minutes to run a mile, then her average speed *s*, in miles per hour, is given by the formula

$$s(t) = \frac{60}{t} \qquad \text{for} \qquad t > 0$$

Find $s(10)$ and $s(8)$, and then explain what they mean.

Solution To find $s(10)$, we substitute 10 for *t* in the equation and simplify.

$$s(10) = \frac{60}{10} = 6$$

In words: When Lorena runs a mile in 10 minutes, her average speed is 6 miles per hour. We calculate $s(8)$ by substituting 8 for *t* in the equation. Doing so gives us

$$s(8) = \frac{60}{8} = 7.5$$

In words: Running a mile in 8 minutes is running at a rate of 7.5 miles per hour.

EXAMPLE 3 A painting is purchased as an invest-ment for $125. If its value increases continuously so that it doubles every 5 years, then its value is given by the function

$$V(t) = 125 \cdot 2^{t/5} \qquad \text{for} \qquad t \geq 0$$

where *t* is the number of years since the painting was purchased, and *V* is its value (in dollars) at time *t*. Find $V(5)$ and $V(10)$, and explain what they mean.

Solution The expression $V(5)$ is the value of the painting when $t = 5$ (5 years after it is purchased). We calculate $V(5)$ by substituting 5 for *t* in the equation $V(t) = 125 \cdot 2^{t/5}$. Here is our work:

$$V(5) = 125 \cdot 2^{5/5} = 125 \cdot 2^1 = 125 \cdot 2 = 250$$

In words: After 5 years, the painting is worth $250.

2. When Lorena runs a mile in *t* minutes, then her average speed in feet per second is given by

$$s(t) = \frac{88}{t}, t > 0$$

a. Find $s(8)$ and explain what it means.

b. Find $s(11)$ and explain what it means.

3. A medication has a half-life of 5 days. If the concentration of the medication in a patient's system is 80 ng/mL, and the patient stops taking it, then *t* days later the concentration will be

$$C(t) = 80\frac{1}{2}^{t/5}$$

Find each of the following, and explain what they mean.

a. $C(5)$

b. $C(10)$

Note Remember, as stated in the order of operations, work the exponents before any multiplication the problem requires.

Answers

2. a. $s(8) = 11$; runs a mile in 8 minutes, average speed is 11 feet per second.

b. $s(11) = 8$; runs a mile in 11 minutes, average speed is 8 feet per second.

3. a. $C(5) = 40$ ng/mL; after 5 days the concentration is 40 ng per mL.

b. $C(10) = 20$ ng/mL; after 10 days the concentration is 20 ng per mL.

©Marissa McKeague, 2011

The expression $V(10)$ is the value of the painting after 10 years. To find this number, we substitute 10 for t in the equation:

$$V(10) = 125 \cdot 2^{10/5} = 125 \cdot 2^2 = 125 \cdot 4 = 500$$

In words: The value of the painting 10 years after it is purchased is $500.

EXAMPLE 4 A balloon has the shape of a sphere with a radius of 3 inches. Use the following formulas to find the volume and surface area of the balloon.

$$V(r) = \frac{4}{3}\pi r^3 \qquad S(r) = 4\pi r^2$$

Solution As you can see, we have used function notation to write the formulas for volume and surface area, because each quantity is a function of the radius. To find these quantities when the radius is 3 inches, we evaluate $V(3)$ and $S(3)$.

$$V(3) = \frac{4}{3}\pi \cdot 3^3 = \frac{4}{3}\pi \cdot 27$$

$$= 36\pi \text{ cubic inches, or } 113 \text{ cubic inches}$$
(to the nearest whole number)

$$S(3) = 4\pi \cdot 3^2$$

$$= 36\pi \text{ square inches, or } 113 \text{ square inches}$$
(to the nearest whole number)

The fact that $V(3) = 36\pi$ means that the ordered pair $(3, 36\pi)$ belongs to the function V. Likewise, the fact that $S(3) = 36\pi$ tells us that the ordered pair $(3, 36\pi)$ is a member of function S.

We can generalize the discussion at the end of Example 4 this way:

$$(a, b) \in f \qquad \text{if and only if} \qquad f(a) = b$$

USING TECHNOLOGY More About Example 4

If we look at Example 4, we see that when the radius of a sphere is 3, the numerical values of the volume and surface area are equal. How unusual is this? Are there other values of r for which $V(r)$ and $S(r)$ are equal? We can answer this question by looking at the graphs of both V and S.
 To graph the function $V(r) = \frac{4}{3}\pi r^3$, set $Y_1 = 4\pi X^3/3$. To graph $S(r) = 4\pi r^2$, set $Y_2 = 4\pi X^2$. Graph the two functions in each of the following windows:

Window 1: X from -4 to 4, Y from -2 to 10

Window 2: X from 0 to 4, Y from 0 to 50

Window 3: X from 0 to 4, Y from 0 to 150

Then use the Trace and Zoom features of your calculator to locate the point in the first quadrant where the two graphs intersect. How do the coordinates of this point compare with the results in Example 4?

EXAMPLE 5 If $f(x) = 3x^2 + 2x - 1$, find $f(0)$, $f(3)$, and $f(-2)$.
Solution Since $f(x) = 3x^2 + 2x - 1$, we have

$$f(0) = 3(0)^2 + 2(0) - 1 = 0 + 0 - 1 = -1$$

$$f(3) = 3(3)^2 + 2(3) - 1 = 27 + 6 - 1 = 32$$

$$f(-2) = 3(-2)^2 + 2(-2) - 1 = 12 - 4 - 1 = 7$$

4. The following formulas give the circumference and area of a circle with radius r. Use the formulas to find the circumference and area of a circular plate if the radius is 5 inches.
 a. $C(r) = 2\pi r$
 b. $A(r) = \pi r^2$

5. If $f(x) = 4x^2 - 3$, find
 a. $f(0)$
 b. $f(3)$
 c. $f(-2)$

Answers
4. $C(5) = 10\pi$
 $A(5) = 25\pi$
5. **a.** -3 **b.** 33 **c.** 13

In Example 5, the function f is defined by the equation $f(x) = 3x^2 + 2x - 1$. We could just as easily have said $y = 3x^2 + 2x - 1$. That is, $y = f(x)$. Saying $f(-2) = 7$ is exactly the same as saying y is 7 when x is -2.

EXAMPLE 6 If $f(x) = 4x - 1$ and $g(x) = x^2 + 2$, find the value of each for $x = 5, -2, 0, z, a,$ and $a + 3$.

Solution

$$f(5) = 4(5) - 1 = 19 \quad \text{and} \quad g(5) = 5^2 + 2 = 27$$

$$f(-2) = 4(-2) - 1 = -9 \quad \text{and} \quad g(-2) = (-2)^2 + 2 = 6$$

$$f(0) = 4(0) - 1 = -1 \quad \text{and} \quad g(0) = 0^2 + 2 = 2$$

$$f(z) = 4z - 1 \quad \text{and} \quad g(z) = z^2 + 2$$

$$f(a) = 4a - 1 \quad \text{and} \quad g(a) = a^2 + 2$$

$$f(a + 3) = 4(a + 3) - 1 \qquad g(a + 3) = (a + 3)^2 + 2$$

$$= 4a + 12 - 1 \qquad\qquad = (a^2 + 6a + 9) + 2$$

$$= 4a + 11 \qquad\qquad\qquad = a^2 + 6a + 11$$

USING TECHNOLOGY More About Example 6

Most graphing calculators can use tables to evaluate functions. To work Example 6 using a graphing calculator table, set Y_1 equal to $4X - 1$ and Y_2 equal to $X^2 + 2$. Then set the independent variable in the table to Ask instead of Auto. Go to your table and input 5, -2, and 0. Under Y_1 in the table, you will find $f(5)$, $f(-2)$, and $f(0)$. Under Y_2, you will find $g(5)$, $g(-2)$, and $g(0)$.

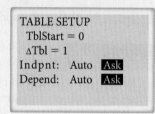

The table will look like this:

X	Y_1	Y_2
5	19	27
-2	-9	6
0	-1	2

Although the calculator asks us for a table increment, the increment doesn't matter because we are inputting the X values ourselves.

6. If $f(x) = 2x + 1$ and $g(x) = x^2 - 3$, find

 a. $f(5)$

 b. $g(5)$

 c. $f(-2)$

 d. $g(-2)$

 e. $f(a)$

 f. $g(a)$

Answers

6. a. 11 **b.** 22 **c.** -3

 d. 1 **e.** $2a + 1$ **f.** $a^2 - 3$

EXAMPLE 7 If the function f is given by $f = \{(-2, 0), (3, -1), (2, 4), (7, 5)\}$, find $f(-2), f(3), f(2)$, and $f(7)$

Solution $f(-2) = 0, f(3) = -1, f(2) = 4$, and $f(7) = 5$.

EXAMPLE 8 If $f(x) = 2x^2$ and $g(x) = 3x - 1$, find
a. $f(g(2))$ **b.** $g(f(2))$

Solution The expression $f(g(2))$ is read "f of g of 2."

a. Because $g(2) = 3(2) - 1 = 5$,
$$f(g(2)) = f(5) = 2(5)^2 = 50$$

b. Because $f(2) = 2(2)^2 = 8$,
$$g(f(2)) = g(8) = 3(8) - 1 = 23$$

EXAMPLE 9 Let $f(x) = x^2 + 2x$ and $g(x) = 3x - 5$. Evaluate the following:
a. $f(3) + g(-3)$
b. $2 f(x) + 4 g(x)$
c. $g(x^2)$
d. $[f(-4)]^2$

Solution

a. Because $f(3) = (3)^2 + 2(3)$ and $g(-3) = 3(-3) - 5$ then
$$f(3) + g(-3) = (3)^2 + 2(3) + 3(-3) - 5 = 9 + 6 - 9 - 5 = 1$$

b. Because $2 f(x) = 2(x^2 + 2x)$ and $4 g(x) = 4(3x - 5)$ then
$$2 f(x) + 4 g(x) = 2(x^2 + 2x) + 4(3x - 5)$$
$$= 2x^2 + 4x + 12x - 20$$
$$= 2x^2 + 16x - 20$$

c. $g(x^2) = 3x^2 - 5$
d. $[f(-4)]^2 = [(-4)^2 + 2(-4)]^2 = (16 - 8)^2 = 64$

GETTING READY FOR CLASS

After reading through the preceding section, respond in your own words and in complete sentences.

A. Explain what you are calculating when you find $f(2)$ for a given function f.

B. If $s(t) = \frac{60}{t}$ how do you find $s(10)$?

C. If $f(2) = 3$ for a function f, what is the relationship between the numbers 2 and 3 and the graph of f?

D. If $f(6) = 0$ for a particular function f, then you can immediately graph one of the intercepts. Explain.

7. If $f = \{(-4, 1), (2, -3), (7, 9)\}$, find
a. $f(-4)$
b. $f(2)$
c. $f(7)$

8. If $f(x) = 3x^2$ and $g(x) = 4x + 1$, find
a. $f(g(2))$
b. $g(f(2))$

9. Let $f(x) = 2x^2 + 3x$ and $g(x) = 5x - 6$. Evaluate the following:
a. $f(-2) - g(1)$
b. $3 f(x) - 2 g(x)$
c. $f(x^2)$
d. $[g(-3)]^2$

Answers
7. a. 1 **b.** -3 **c.** 9
8. a. 243 **b.** 49
9. a. 3 **b.** $6x^2 - x + 12$
 c. $2x^4 + 3x^2$ **d.** 441

EXERCISE SET 11.7

Vocabulary Review

Choose the correct words to fill in the blanks below.

dependent independent f(x) function f

1. The variable that represents the input of a function is called the _____ variable.
2. The variable that represents the output of a function is called the _____ variable.
3. When we use the notation _____ , we are using function notation.
4. The _____ refers to the domain, the range, and the rule that takes elements in the domain and outputs elements in the range.

Problems

A Let $f(x) = 2x - 5$ and $g(x) = x^2 + 3x + 4$. Evaluate the following.

1. $f(2)$

2. $f(3)$

3. $f(-3)$

4. $g(-2)$

5. $g(-1)$

6. $f(-4)$

7. $g(-3)$

8. $g(2)$

9. $g(a)$

10. $f(a)$

11. $f(a + 6)$

12. $g(a + 6)$

Let $f(x) = 3x^2 - 4x + 1$ and $g(x) = 2x - 1$. Evaluate the following.

13. $f(0)$

14. $g(0)$

15. $g(-4)$

16. $f(1)$

17. $f(-1)$

18. $g(-1)$

19. $g\left(\dfrac{1}{2}\right)$

20. $g\left(\dfrac{1}{4}\right)$

21. $f(a)$

22. $g(a)$

23. $f(a + 2)$

24. $g(a + 2)$

If $f = \{(1, 4), (-2, 0), \left(3, \frac{1}{2}\right), (\pi, 0)\}$ and $g = \{(1, 1),(-2, 2), \left(\frac{1}{2}, 0\right)\}$, find each of the following values of f and g.

25. $f(1)$

26. $g(1)$

27. $g\left(\frac{1}{2}\right)$

28. $f(3)$

29. $g(-2)$

30. $f(\pi)$

31. $f(g(1))$

32. $f(-2)-g(-2)$

33. For which value of x is $g(x) = f(\pi)$?

34. Is $f(1) = g(1)$?

Let $f(x) = x^2 - 2x$ and $g(x) = 5x - 4$. Evaluate the following.

35. $f(-4)$

36. $g(-3)$

37. $f(-2) + g(-1)$

38. $f(-1) + g(-2)$

39. $2f(x) - 3g(x)$

40. $f(x) - g(x^2)$

41. $f(g(3))$

42. $g(f(3))$

43. $f(g(2)) - g(f(2))$

44. $g(f(1)) - f(g(1))$

45. $[f(-3)]^2$

46. $[g(3)]^2$

Let $f(x) = \dfrac{1}{x + 3}$ and $g(x) = \dfrac{1}{x} + 1$. Evaluate the following.

47. $f\left(\frac{1}{3}\right)$

48. $g\left(\frac{1}{3}\right)$

49. $f\left(-\frac{1}{2}\right)$

50. $g\left(-\frac{1}{2}\right)$

51. $f(-3)$

52. $g(0)$

53. $g(-2)$

54. $f(-1)$

55. For the function $f(x) = x^2 - 4$, evaluate each of the following expressions.

 a. $f(a) - 3$

 b. $f(a - 3)$

 c. $f(x) + 2$

 d. $f(x + 2)$

 e. $f(a + b)$

 f. $f(x + h)$

56. For the function $f(x) = 3x^2$, evaluate each of the following expressions.

 a. $f(a) - 2$

 b. $f(a - 2)$

 c. $f(x) + 5$

 d. $f(x + 5)$

 e. $f(a + b)$

 f. $f(x + h)$

B

57. Graph the function $f(x) = \frac{1}{2}x + 2$. Then draw and label the line segments that represent $x = 4$ and $f(4)$.

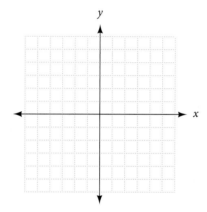

58. Graph the function $f(x) = -\frac{1}{2}x + 6$. Then draw and label the line segments that represent $x = 4$ and $f(4)$.

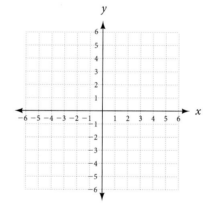

59. For the function $f(x) = \frac{1}{2}x + 2$, find the value of x for which $f(x) = x$.

60. For the function $f(x) = -\frac{1}{2}x + 6$, find the value of x for which $f(x) = x$.

61. Graph the function $f(x) = x^2$. Then draw and label the line segments that represent $x = 1$ and $f(1)$, $x = 2$ and $f(2)$ and, finally, $x = 3$ and $f(3)$.

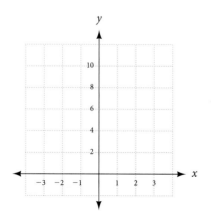

62. Graph the function $f(x) = x^2 - 2$. Then draw and label the line segments that represent $x = 2$ and $f(2)$ and the line segments corresponding to $x = 3$ and $f(3)$.

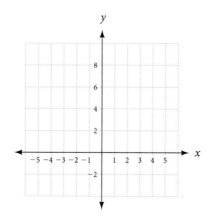

Applying the Concepts

63. Investing in Art Charlie purchased a painting as an investment for $150. If its value increases continuously so that it doubles every 3 years, then its value is given by the function

$$V(t) = 150 \cdot 2^{t/3} \qquad \text{for} \qquad t \geq 0$$

where t is the number of years since the painting was purchased, and $V(t)$ is its value (in dollars) at time t. Find $V(3)$ and $V(6)$, and then explain what they mean.

64. Average Speed If Minke takes t minutes to run a mile, then her average speed $s(t)$, in miles per hour, is given by the formula

$$s(t) = \frac{60}{t} \qquad \text{for} \qquad t > 0$$

Find $s(4)$ and $s(5)$, and then explain what they mean.

65. iPhone Sales The chart below shows the percentage of respondents who purchased the iPhone on each of its first day of sales, shown by their current carrier. Suppose x represents one of the carriers in the chart. We have three functions f, g, and h that do the following:

f pairs each year with the percentage of respondents who were AT&T customers for that year.

g pairs each year with the percentage of respondents who were Verizon customers for that year.

h pairs each year with the percentage of respondents who were T-Mobile customers for that year.

For each statement below, indicate whether the statement is true or false.

a. The domain of g is {2007, 2008, 2009, 2010}

b. The domain of f is $\{x \,|\, 2008 \leq x \leq 2009\}$

c. $f(2008) > g(2008)$

d. $h(2009) > 30\%$

e. $h(2008) < g(2008) < f(2010)$

66. Newspaper Sales Suppose x represents one of the publications in the chart. We have three functions f, g, and h that do the following:

f pairs each publication with the number of paper subscribers for that publication.

g pairs each publication with the number of online viewers for that publication.

h is such that $h(x) = f(x) + g(x)$.

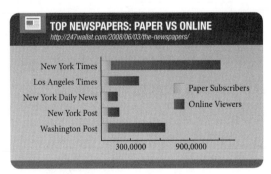

For each statement below, indicate whether the statement is true or false.

a. The domain of f is {New York Times, Los Angeles Times, New York Daily News, New York Washington Post}

b. $h(\text{Los Angeles Times}) = 600{,}000$

c. $g(\text{New York Times}) > f(\text{New York Times})$

d. $g(\text{New York Daily News}) < g(\text{Washington Post})$

e. $h(\text{New York Times}) > h(\text{Los Angeles Times}) > h(\text{New York Post})$

Straight-Line Depreciation Straight-line depreciation is an accounting method used to help spread the cost of new equipment over a number of years. It takes into account both the cost when new and the salvage value, which is the value of the equipment at the time it gets replaced.

67. Value of a Copy Machine The function
$V(t) = -3,300t + 18,000$, where V is value and t is time in years, can be used to find the value of a large copy machine during the first 5 years of use.

a. What is the value of the copier after 3 years and 9 months?

b. What is the salvage value of this copier if it is replaced after 5 years?

c. State the domain of this function.

d. Sketch the graph of this function.

Value of a Copy Machine

e. What is the range of this function?

f. After how many years will the copier be worth only $10,000?

68. Step Function Figure 3 shows the graph of the step function C that was used to calculate the first-class postage on a large envelope weighing x ounces in 2010. Use this graph to answer the following questions.

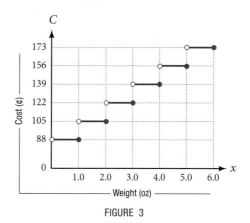
FIGURE 3

a. Fill in the following table:

Weight (ounces)	0.6	1.0	1.1	2.5	3.0	4.8	5.0	5.3
Cost (cents)								

b. If a letter cost 122 cents to mail, how much does it weigh? State your answer in words and as an inequality.

c. If the entire function is shown in Figure 3, state the domain.

d. State the range of the function shown in Figure 3.

69. Wind Turbine The annual output from a wind turbine with an average wind velocity of 10mph is given by the function $f(d) = 13.28d^2$, where d is the diameter (in feet) of the blade and f is kWh/year.

a. Complete the table for various diameter blades.

d (feet)	f(d) (kWh/year)
0	
5	
10	
15	
20	

Source: builditsolar.com/Projects/Wind/Wind.htm

b. What is the range of this function for $0 \le d \le 20$?

70. Wind Turbine The annual output from a wind turbine with a 10–foot–diameter blade is given by the function $f(v) = 1.328v^3$, where v is the average velocity of the wind (in mph) and f is kWh/year.

a. Complete the table for various diameter blades.

v (mph)	f(v) (kWh/year)
0	
10	
20	
30	
40	

Source: builditsolar.com/Projects/Wind/Wind.htm

b. What is the range of this function for $0 \le v \le 40$?

Getting Ready for the Next Section

Multiply.

71. $x(35 - 0.1x)$

72. $0.6(M - 70)$

73. $(4x - 3)(x - 1)$

74. $(4x - 3)(4x^2 - 7x + 3)$

Simplify.

75. $(35x - 0.1x^2) - (8x + 500)$

76. $(4x - 3) + (4x^2 - 7x + 3)$

77. $(4x^2 + 3x + 2) - (2x^2 - 5x - 6)$

78. $(4x^2 + 3x + 2) + (2x^2 - 5x - 6)$

79. $4(2)^2 - 3(2)$

80. $4(-1)^2 - 7(-1)$

Simplify.

81. $2(x - 3)^2 - 5(x - 3)$

82. $5(3x - 7) - 4$

83. $-8(-3x + 1) + 2$

84. $4(x + 1)^2 - 6(x + 1)$

Find the Mistake

Each sentence below contains a mistake. Circle the mistake and write the correct word or phrase on the line provided.

1. To evaluate the function $f(x) = 10x + 3$ for $f(1)$ we begin by setting $10x + 3$ equal to 1. _____

2. For the average speed function $s(t) = \dfrac{60}{t}$, finding $s(6)$ shows an average speed of 6 miles per hour. _____

3. If the volume of a ball is given by $V(r) = \dfrac{4}{3}\pi r^2$ where r is the radius, we see that the volume decreases as the radius of the ball increases. _____

4. If $f(x) = x^2$ and $g(x) = 2x + 4$, evaluating $f(g(2))$ is the same as multiplying $f(2) \cdot g(2)$. _____

Chapter **11** Summary

The numbers in brackets refer to the section(s) in which the topic can be found.

Addition Property for Inequalities [11.1]

1. Adding 5 to both sides of the inequality $x - 5 < -2$ gives

$$x - 5 + 5 < -2 + 5$$

$$x < 3$$

For expressions A, B, and C,

$$\text{if} \qquad A < B$$

$$\text{then} \qquad A + C < B + C$$

Adding the same quantity to both sides of an inequality never changes the solution set.

Multiplication Property for Inequalities [11.1]

2. Multiplying both sides of $-2x \geq 6$ by $-\frac{1}{2}$ gives

$$-2x \geq 6$$

$$-\frac{1}{2}(-2x) \leq -\frac{1}{2}(6)$$

$$x \leq -3$$

For expressions A, B, and C,

$$\text{if} \qquad A < B$$

$$\text{then} \qquad AC < BC \qquad \text{if} \qquad C > 0 \ (C \text{ is positive})$$

$$\text{or} \qquad AC > BC \qquad \text{if} \qquad C < 0 \ (C \text{ is negative})$$

We can multiply both sides of an inequality by the same nonzero number without changing the solution set as long as each time we multiply by a negative number we also reverse the direction of the inequality symbol.

Compound Inequalities [11.2]

3. Graph $x < -3$ or $x > 1$

Graph $-2 \leq x \leq 3$

Two inequalities connected by the word *and* or *or* form a compound inequality. If the connecting word is *or*, we graph all points that are on either graph. If the connecting word is *and*, we graph only those points that are common to both graphs. The inequality $-2 \leq x \leq 3$ is equivalent to the compound inequality $-2 \leq x$ and $x \leq 3$.

Absolute Value Equations [11.3]

4. To solve

$$|2x - 1| + 2 = 7$$

we first isolate the absolute value on the left side by adding -2 to each side to obtain

$$|2x - 1| = 5$$

$$\begin{array}{ll} 2x - 1 = 5 \quad \text{or} & 2x - 1 = -5 \\ 2x = 6 & 2x = -4 \\ x = 3 & x = -2 \end{array}$$

To solve an equation that involves absolute value, we isolate the absolute value on one side of the equation and then rewrite the absolute value equation as two separate equations that do not involve absolute value. In general, if b is a positive real number, then

$$|a| = b \quad \text{is equivalent to} \quad a = b \quad \text{or} \quad a = -b$$

Absolute Value Inequalities [11.4]

To solve an inequality that involves absolute value, we first isolate the absolute value on the left side of the inequality symbol. Then we rewrite the absolute value inequality as an equivalent continued or compound inequality that does not contain absolute value symbols. In general, if b is a positive real number, then

$$|a| < b \text{ is equivalent to } -b < a < b$$

and

$$|a| > b \text{ is equivalent to } a < -b \text{ or } a > b$$

COMMON MISTAKE

A very common mistake in solving inequalities is to forget to reverse the direction of the inequality symbol when multiplying both sides by a negative number. When this mistake occurs, the graph of the solution set is always drawn on the wrong side of the endpoint.

5. To solve

$$|x - 3| + 2 < 6$$

we first add -2 to both sides to obtain

$$|x - 3| < 4$$

which is equivalent to

$$-4 < x - 3 < 4$$
$$-1 < x < 7$$

Intercepts [11.5]

The *x-intercept* of an equation is the *x-coordinate* of the point where the graph crosses the *x*-axis. The *y-intercept* is the *y-coordinate* of the point where the graph crosses the *y-axis*. We find the *y*-intercept by substituting $x = 0$ into the equation and solving for *y*. The *x*-intercept is found by letting $y = 0$ and solving for *x*.

6. To find the *x*-intercept for $3x + 2y = 6$, we let $y = 0$ and get

$$3x = 6$$
$$x = 2$$

In this case the *x*-intercept is 2, and the graph crosses the *x*-axis at $(2, 0)$.

Relations and Functions [11.6]

A *relation* is any set of ordered pairs. The set of all first coordinates is called the *domain* of the relation, and the set of all second coordinates is the *range* of the relation. A *function* is a rule that pairs each element in one set, called the *domain,* with exactly one element from a second set, called the *range.* A function is a relation in which no two different ordered pairs have the same first coordinates.

7. The relation

$$\{(8, 1), (6, 1), (-3, 0)\}$$

is also a function because no ordered pairs have the same first coordinates. The domain is $\{8, 6, -3\}$ and the range is $\{1, 0\}$.

Vertical Line Test [11.6]

If a vertical line crosses the graph of a relation in more than one place, the relation cannot be a function. If no vertical line can be found that crosses the graph in more than one place, the relation must be a function.

8. The graph of $x = y^2$ shown in Figure 5 in Section 2.2 fails the vertical line test. It is not the graph of a function.

Function Notation [11.7]

The alternative notation for y is $f(x)$. It is read "f of x" and can be used instead of the variable y when working with functions. The notation y and the notation $f(x)$ are equivalent; that is, $y = f(x)$.

9. If $f(x) = 5x - 3$, then
$$f(0) = 5(0) - 3$$
$$= -3$$
$$f(1) = 5(1) - 3$$
$$= 2$$
$$f(-2) = 5(-2) - 3$$
$$= -13$$
$$f(a) = 5a - 3$$

Solve the following inequalities. Write the solution set using interval notation, then graph the solution set. [11.1]

1. $-4t \le 16$

2. $-7 - \dfrac{3}{4}x \le 2$

3. $1.2x + 1.5 < 0.6x + 3.9$

4. $2(3y - 2) > 5(y + 3)$

Solve the following equations. [11.3]

5. $\left| \dfrac{1}{6}x - 2 \right| = \dfrac{1}{3}$

6. $\left| \dfrac{4}{7}a + 8 \right| = 4$

7. $|3y + 9| - 2 = 7$

8. $\left| \dfrac{3}{4} - \dfrac{1}{17}x \right| + 4 = 3$

Solve the following inequalities and graph the solutions. [11.4]

9. $|4y + 7| < 3$

10. $|8x + 4| - 3 \ge 9$

11. $|11x - 8| \ge -6$

12. $|2 - 9t| < -7$

For each of the following straight lines, identify the x-intercept, y-intercept, and slope, and sketch the graph. [11.5]

13. $3rd - y = 9$

14. $y = -\dfrac{1}{2}x + 4$

15. $y = \dfrac{3}{5}x - 6$

16. $x = -4$

State the domain and range for the following relations, and indicate which relations are also functions. [11.6]

17. $\{(0, 3), (2, 5), (6, -4)\}$

18. $y = 3x^2 + 2$

19. $\{(0, 5), (-6, 5), (0, 7)\}$

20. $y = x - 5$

Determine the domain and range of the following functions. Assume the *entire* function is shown. [11.6]

21.

22.
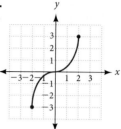

Let $f(x) = x - 5$, $g(x) = 2x - 5$ and $h(x) = x^2 - 5x + 6$, and find the following. [11.7]

23. $f(0) + g(-3)$

24. $h(0) - g(0)$

25. $f(g(-1))$

26. $g(f(3))$

Rational Exponents and Radicals

12

12.1 Rational Exponents

12.2 Simplifying Radicals

12.3 Addition and Subtraction of Radical Expressions

12.4 Multiplication and Division of Radical Expressions

12.5 Radical Equations and Functions

12.6 Complex Numbers

Image © 2010 GeoEye
Chichen Itza, Mexico

Google

On Mexico's Yucatan Peninsula lies one of the world's great mysteries: the ruins of the Mayan city of Chichen Itza. The Temple of Kukulkan is the largest and most recognizable structure at Chichen Itza. This 90-foot-tall pyramid is a 4-sided structure, each side containing a stairway with 91 steps. One of the most puzzling facts about these stairs can be observed from inside of the Temple. If you were to clap your hands inside the pyramid, the echoes from each of the stairs in the staircases sound similar to the primary call of the Mayan sacred bird, the Quetzal. Experts believe that the Mayan people intentionally coded the echoes within their temple to sound like their sacred bird. If this is true, this ancient people's knowledge of sound and acoustics remains astounding.

The time delay of an echo is the time it takes for the sound to travel to the reflection surface and back to the origin of the sound. In other words, to begin calculating the echo's time delay, you must first double the distance to the reflection surface. Therefore, if you were to clap your hands inside the Temple of Kukulkan at a distance of $15\sqrt{5}$ feet from the base of the stairs, you would have to know how to double $15\sqrt{5}$. The formula to calculate the time delay of an echo, as well as the techniques needed to double expressions involving square roots, will be covered as you work through this chapter.

OBJECTIVES

A Simplify radical expressions using the definition for roots.

B Simplify expressions with rational exponents.

KEY WORDS

positive square root

root

radical expression

index

radical sign

radicand

rational exponent

golden rectangle

golden ratio

**Video Examples
Section 12.1**

Kite fighting is a sport where two competitors each fly a kite and attempt to cut down the other's kite string with their own. In some contests, a competitor uses a small nylon kite tethered by a string coated in a paste made from glue and powdered glass. This paste helps to strengthen the line and gives it the sharpness it needs to slice through the opponent's string.

Suppose we are competing in a kite fighting competition. Our opponent's kite is already flying in the air at 100 feet high. Using the Pythagorean theorem, can you set up an equation that would give us the length of the string that is tethering the kite? In this section, we will work with properties of exponents, such as those needed to solve problems that include the Pythagorean theorem. Then we can return to this question to find out how much line we will need to let out for our kite to reach the same height as our opponent's.

FIGURE 1

Figure 2 shows a square in which each of the four sides is 1 inch long. To find the square of the length of the diagonal c, we apply the Pythagorean theorem:

$$c^2 = 1^2 + 1^2$$
$$c^2 = 2$$

FIGURE 2

Because we know that c is positive and that its square is 2, we call c the *positive square root* of 2, and we write $c = \sqrt{2}$. Associating numbers, such as $\sqrt{2}$, with the diagonal of a square or rectangle allows us to analyze some interesting items from geometry. One particularly interesting geometric object that we will study in this section is called the *golden rectangle*, shown in Figure 3. It is constructed from a right triangle, and the length of the diagonal is found from the Pythagorean theorem. We will come back to this figure at the end of this section.

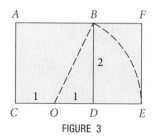

FIGURE 3

A Radical Expressions

In Chapter 0, we developed notation (exponents) to give us the square, cube, or any other power of a number. For instance, if we wanted the square of 3, we wrote $3^2 = 9$. If we wanted the cube of 3, we wrote $3^3 = 27$. In this section, we will develop notation that will take us in the reverse direction; that is, from the square of a number, say 25, back to the original number, 5.

> **DEFINITION** positive square root
>
> If x is a nonnegative real number, then the expression \sqrt{x} is called the **positive square root** of x and is such that
>
> $$(\sqrt{x})^2 = x$$
>
> *In words:* \sqrt{x} is the positive number we square to get x.

The negative square root of x, $-\sqrt{x}$, is defined in a similar manner.

EXAMPLE 1 The positive square root of 64 is 8 because 8 is the positive number with the property $8^2 = 64$. The negative square root of 64 is -8 because -8 is the negative number whose square is 64. We can summarize both these facts by saying

$$\sqrt{64} = 8 \qquad \text{and} \qquad -\sqrt{64} = -8$$

The higher roots, cube roots, fourth roots, and so on, are defined by definitions similar to that of square roots.

> **DEFINITION** roots
>
> If x is a real number and n is a positive integer, then
>
> Positive square root of x, \sqrt{x}, is such that $(\sqrt{x})^2 = x$ $x \geq 0$
>
> Cube root of x, $\sqrt[3]{x}$, is such that $(\sqrt[3]{x})^3 = x$
>
> Positive fourth root of x, $\sqrt[4]{x}$, is such that $(\sqrt[4]{x})^4 = x$ $x \geq 0$
>
> Fifth root of x, $\sqrt[5]{x}$, is such that $(\sqrt[5]{x})^5 = x$
>
> \vdots \vdots \vdots
> \vdots \vdots \vdots
> \vdots \vdots \vdots
>
> The nth root of x, $\sqrt[n]{x}$, is such that $(\sqrt[n]{x})^n = x$ $x \geq 0$ if n is even.

The following is a table of the most common roots used in this book. Any of the roots that are unfamiliar should be memorized.

Square Roots		Cube Roots	Fourth Roots
$\sqrt{0} = 0$	$\sqrt{49} = 7$	$\sqrt[3]{0} = 0$	$\sqrt[4]{0} = 0$
$\sqrt{1} = 1$	$\sqrt{64} = 8$	$\sqrt[3]{1} = 1$	$\sqrt[4]{1} = 1$
$\sqrt{4} = 2$	$\sqrt{81} = 9$	$\sqrt[3]{8} = 2$	$\sqrt[4]{16} = 2$
$\sqrt{9} = 3$	$\sqrt{100} = 10$	$\sqrt[3]{27} = 3$	$\sqrt[4]{81} = 3$
$\sqrt{16} = 4$	$\sqrt{121} = 11$	$\sqrt[3]{64} = 4$	
$\sqrt{25} = 5$	$\sqrt{144} = 12$	$\sqrt[3]{125} = 5$	
$\sqrt{36} = 6$	$\sqrt{169} = 13$		

Note It is a common mistake to assume that an expression like $\sqrt{25}$ indicates both square roots, 5 and -5. The expression $\sqrt{25}$ indicates only the positive square root of 25, which is 5. If we want the negative square root, we must use a negative sign: $-\sqrt{25} = -5$.

Practice Problems

1. Give the two square roots of 36.

Answers

1. $6, -6$

Notation An expression like $\sqrt[3]{8}$ that involves a root is called a *radical expression*. In the expression $\sqrt[3]{8}$, the 3 is called the *index*, the $\sqrt{}$ is the *radical sign*, and 8 is called the *radicand*. The index of a radical must be a positive integer greater than 1. If no index is written, it is assumed to be 2.

Roots and Negative Numbers

When dealing with negative numbers and radicals, the only restriction concerns negative numbers under even roots. We can have negative signs in front of radicals and negative numbers under odd roots and still obtain real numbers. Here are some examples to help clarify this. In the last section of this chapter, we will see how to deal with even roots of negative numbers.

2. Simplify, if possible.

 a. $\sqrt[3]{-64}$

 b. $\sqrt{-25}$

 c. $-\sqrt{4}$

 d. $\sqrt[5]{-1}$

 e. $\sqrt[4]{-16}$

EXAMPLE 2 Simplify each expression, if possible.

 a. $\sqrt[3]{-8} = -2$ because $(-2)^3 = -8$.

 b. $\sqrt{-4}$ is not a real number because there is no real number whose square is -4.

 c. $-\sqrt{25} = -5$, because -5 is the negative square root of 25.

 d. $\sqrt[5]{-32} = -2$ because $(-2)^5 = -32$.

 e. $\sqrt[4]{-81}$ is not a real number because there is no real number we can raise to the fourth power and obtain -81.

Variables Under a Radical

From the preceding examples, it is clear that we must be careful that we do not try to take an even root of a negative number. For the purpose of this section, we will assume that all variables appearing under a radical sign represent nonnegative numbers. We will address even roots of negative numbers later in this book.

3. Simplify.

 a. $\sqrt{81a^4b^8}$

 b. $\sqrt[3]{8x^3y^9}$

 c. $\sqrt[4]{81a^4b^8}$

EXAMPLE 3 Assume all variables represent nonnegative numbers, and simplify each expression as much as possible.

 a. $\sqrt{25a^4b^6} = 5a^2b^3$ because $(5a^2b^3)^2 = 25a^4b^6$.

 b. $\sqrt[3]{x^6y^{12}} = x^2y^4$ because $(x^2y^4)^3 = x^6y^{12}$.

 c. $\sqrt[4]{81r^8s^{20}} = 3r^2s^5$ because $(3r^2s^5)^4 = 81r^8s^{20}$.

B Rational Numbers as Exponents

We will now develop a second kind of notation involving exponents that will allow us to designate square roots, cube roots, and so on in another way.

Consider the equation $x = 8^{1/3}$. The exponent $\frac{1}{3}$ is called a rational exponent. Although we have not encountered fractions as exponents before, let's assume that all the properties of exponents hold in this case. Cubing both sides of the equation, we have

$$x^3 = (8^{1/3})^3$$

$$x^3 = 8^{(1/3)(3)}$$

$$x^3 = 8^1$$

$$x^3 = 8$$

The last line tells us that x is the number whose cube is 8. It must be true, then, that x is the cube root of 8, $x = \sqrt[3]{8}$. Because we started with $x = 8^{1/3}$, it follows that

$$8^{1/3} = \sqrt[3]{8}$$

Answers

2. a. -4 **b.** Not a real number

 c. -2 **d.** -1

 e. Not a real number

3. a. $9a^2b^4$ **b.** $2xy^3$ **c.** $3ab^2$

It seems reasonable, then, to define rational exponents as indicating roots. Here is the formal property:

> **PROPERTY** **Rational Exponent Property**
>
> If x is a real number and n is a positive integer greater than 1, then
>
> $$x^{1/n} = \sqrt[n]{x} \qquad (x \geq 0 \text{ when } n \text{ is even})$$
>
> *In words:* The quantity $x^{1/n}$ is the nth root of x.

With this property, we have a way of representing roots with exponents. Here are some examples:

EXAMPLE 4 Write each expression as a root and then simplify, if possible.

a. $8^{1/3} = \sqrt[3]{8} = 2$

b. $36^{1/2} = \sqrt{36} = 6$

c. $-25^{1/2} = -\sqrt{25} = -5$

d. $(-25)^{1/2} = \sqrt{-25}$, which is not a real number

e. $\left(\dfrac{4}{9}\right)^{1/2} = \sqrt{\dfrac{4}{9}} = \dfrac{2}{3}$

The properties of exponents developed in Chapter 0 were applied to integer exponents only. We will now extend these properties to include rational exponents also. We do so without proof.

> **PROPERTY** **Properties of Exponents**
>
> If a and b are real numbers and r and s are rational numbers, and a and b are nonnegative whenever r and s indicate even roots, then
>
> **1.** $a^r \cdot a^s = a^{r+s}$ Product Property for Exponents
>
> **2.** $(a^r)^s = a^{rs}$ Power Property for Exponents
>
> **3.** $(ab)^r = a^r b^r$ Distributive Property for Exponents
>
> **4.** $a^{-r} = \dfrac{1}{a^r}$ $(a \neq 0)$ Negative Exponent Property
>
> **5.** $\left(\dfrac{a}{b}\right)^r = \dfrac{a^r}{b^r}$ $(b \neq 0)$ Expanded Distributive Property for Exponents
>
> **6.** $\dfrac{a^r}{a^s} = a^{r-s}$ $(a \neq 0)$ Quotient Property for Exponents
>
> **7.** $a^{1/r} = \sqrt[r]{a}$ $(a \neq 0)$ Rational Exponent Property

Sometimes rational exponents can simplify our work with radicals. Here are Examples 3b and 3c again, but this time we will work them using rational exponents.

EXAMPLE 5 Write each radical with a rational exponent, then simplify.

a. $\sqrt[3]{x^6 y^{12}} = (x^6 y^{12})^{1/3}$

 $= (x^6)^{1/3}(y^{12})^{1/3}$

 $= x^2 y^4$

b. $\sqrt[4]{81 r^8 s^{20}} = (81 r^8 s^{20})^{1/4}$

 $= 81^{1/4}(r^8)^{1/4}(s^{20})^{1/4}$

 $= 3 r^2 s^5$

4. Simplify.

a. $9^{1/2}$

b. $27^{1/3}$

c. $-49^{1/2}$

d. $(-49)^{1/2}$

e. $\left(\dfrac{16}{25}\right)^{1/2}$

Note In part d of Example 4, we use parentheses to notate that we are asking for the square root of -25. If the parentheses are absent, as is part c, we consider the negative sign as a -1 multiplier and work the exponent first.

$$-25^{1/2} = -(25^{1/2})$$
$$= -\sqrt{25}$$
$$= -5$$

5. Simplify.

a. $\sqrt[3]{8 x^3 y^9}$

b. $\sqrt[4]{81 a^4 b^8}$

Answers

4. a. 3 **b.** 3 **c.** -7

 d. Not a real number **e.** $\dfrac{4}{5}$

5. a. $2xy^3$ **b.** $3ab^2$

So far, the numerators of all the rational exponents we have encountered have been 1. The next theorem extends the work we can do with rational exponents to rational exponents with numerators other than 1.

We can extend our properties of exponents with the following theorem.

> **THEOREM** Rational Exponent Theorem
>
> If a is a nonnegative real number, m is an integer, and n is a positive integer, then
> $$a^{m/n} = (a^{1/n})^m = (a^m)^{1/n}$$

Proof We can prove the rational exponent theorem using the properties of exponents. Because $m/n = m(1/n)$, we have

$$a^{m/n} = a^{m(1/n)} \qquad\qquad a^{m/n} = a^{(1/n)(m)}$$
$$= (a^m)^{1/n} \qquad\qquad = (a^{1/n})^m$$

Here are some examples that illustrate how we use this theorem:

EXAMPLE 6 Simplify as much as possible.

6. Simplify.

a. $9^{3/2}$

b. $16^{3/4}$

c. $8^{-2/3}$

d. $\left(\dfrac{16}{81}\right)^{-3/4}$

a. $8^{2/3} = (8^{1/3})^2$ *Rational exponent theorem*

$\phantom{8^{2/3}} = 2^2$ *Rational exponent property*

$\phantom{8^{2/3}} = 4$ *The square of 2 is 4.*

b. $25^{3/2} = (25^{1/2})^3$ *Rational exponent theorem*

$\phantom{25^{3/2}} = 5^3$ *Rational exponent property*

$\phantom{25^{3/2}} = 125$ *The cube of 5 is 125.*

c. $9^{-3/2} = (9^{1/2})^{-3}$ *Rational exponent theorem*

$\phantom{9^{-3/2}} = 3^{-3}$ *Rational exponent property*

$\phantom{9^{-3/2}} = \dfrac{1}{3^3}$ *Negative exponent property*

$\phantom{9^{-3/2}} = \dfrac{1}{27}$ *The cube of 3 is 27.*

d. $\left(\dfrac{27}{8}\right)^{-4/3} = \left[\left(\dfrac{27}{8}\right)^{1/3}\right]^{-4}$ *Rational exponent theorem*

$\phantom{\left(\dfrac{27}{8}\right)^{-4/3}} = \left(\dfrac{3}{2}\right)^{-4}$ *Rational exponent property*

$\phantom{\left(\dfrac{27}{8}\right)^{-4/3}} = \left(\dfrac{2}{3}\right)^{4}$ *Negative exponent property*

$\phantom{\left(\dfrac{27}{8}\right)^{-4/3}} = \dfrac{16}{81}$ *The fourth power of $\frac{2}{3}$ is $\frac{16}{81}$.*

In each part of Example 6, you may have noticed that we first worked the $a^{1/n}$ part of each expression before raising to the power m. This approach typically results in easier calculations than raising a to the power m and then working the rational exponent. In the end, either approach will yield the same correct answer.

The following examples show the application of the properties of exponents to rational exponents.

Answers

6. a. 27 **b.** 8 **c.** $\frac{1}{4}$ **d.** $\frac{27}{8}$

EXAMPLE 7 Assume all variables represent positive quantities, and simplify as much as possible.

a. $x^{1/3} \cdot x^{5/6} = x^{1/3 + 5/6}$ Product property for exponents

$\qquad\qquad = x^{2/6 + 5/6}$ LCD is 6.

$\qquad\qquad = x^{7/6}$ Add fractions.

b. $(y^{2/3})^{3/4} = y^{(2/3)(3/4)}$ Power property or exponents

$\qquad\qquad = y^{1/2}$ Multiply fractions: $\dfrac{2}{3} \cdot \dfrac{3}{4} = \dfrac{6}{12} = \dfrac{1}{2}$.

c. $\dfrac{z^{1/3}}{z^{1/4}} = z^{1/3 - 1/4}$ Quotient property for exponents

$\qquad\qquad = z^{4/12 - 3/12}$ LCD is 12.

$\qquad\qquad = z^{1/12}$ Subtract fractions.

d. $\left(\dfrac{a^{-1/3}}{b^{1/2}}\right)^6 = \dfrac{(a^{-1/3})^6}{(b^{1/2})^6}$ Expanded distributive property for exponents

$\qquad\qquad = \dfrac{a^{-2}}{b^3}$ Power property for exponents

$\qquad\qquad = \dfrac{1}{a^2 b^3}$ Negative exponent property

e. $\dfrac{(x^{-3} y^{1/2})^4}{x^{10} y^{3/2}} = \dfrac{(x^{-3})^4 (y^{1/2})^4}{x^{10} y^{3/2}}$ Distributive property for exponents

$\qquad\qquad = \dfrac{x^{-12} y^2}{x^{10} y^{3/2}}$ Power property for exponents

$\qquad\qquad = x^{-22} y^{1/2}$ Quotient property for exponents

$\qquad\qquad = \dfrac{y^{1/2}}{x^{22}}$ Negative exponent property

FACTS FROM GEOMETRY **More about the Pythagorean Theorem**

Now that we have had some experience working with square roots, we can rewrite the Pythagorean theorem using a square root. If triangle ABC is a right triangle with $C = 90°$, then the length of the longest side is the *positive square root* of the sum of the squares of the other two sides (see Figure 4).

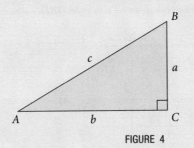

$$c = \sqrt{a^2 + b^2}$$

FIGURE 4

Constructing a Golden Rectangle From a Square of Side 2

Recall the golden rectangle. Its origins can be traced back over 2,000 years to the Greek civilization that produced Pythagoras, Socrates, Plato, Aristotle, and Euclid. The most important mathematical work to come from that Greek civilization was Euclid's *Elements,* an elegantly written summary of all that was known about geometry at that time in history. Euclid's *Elements,* according to Howard Eves, an authority on the history of mathematics, exercised a greater influence on scientific thinking than any other work. Here is how we construct a golden rectangle from a square of side 2, using the same method that Euclid used in his *Elements.*

7. Simplify.

a. $x^{1/2} \cdot x^{1/4}$

b. $(y^{3/5})^{5/6}$

c. $\dfrac{z^{3/4}}{z^{2/3}}$

d. $\left(\dfrac{a^{3/4}}{b^{-1/2}}\right)^4$

e. $\dfrac{(x^{1/3} y^{-3})^6}{x^4 y^{10}}$

Answers

7. a. $x^{3/4}$ **b.** $y^{1/2}$ **c.** $z^{1/12}$
 d. $a^3 b^2$ **e.** $\dfrac{1}{x^2 y^{28}}$

Step 1: Draw a square with a side of length 2. Connect the midpoint of side *CD* to corner *B*. (Note that we have labeled the midpoint of segment *CD* with the letter *O*.)

Step 2: Drop the diagonal from step 1 down so it aligns with side *CD*.

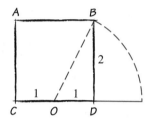

Step 3: Form rectangle *ACEF*. This is a golden rectangle.

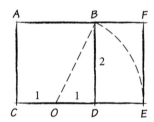

All golden rectangles are constructed from squares. Every golden rectangle, no matter how large or small it is, will have the same shape. To associate a number with the shape of the golden rectangle, we use the ratio of its length to its width. This ratio is called the *golden ratio*. To calculate the golden ratio, we must first find the length of the diagonal we used to construct the golden rectangle. Figure 5 shows the golden rectangle we constructed from a square of side 2. The length of the diagonal *OB* is found by applying the Pythagorean theorem to triangle *OBD*.

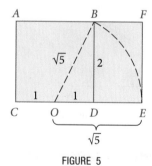

FIGURE 5

The length of segment *OE* is equal to the length of diagonal *OB*; both are $\sqrt{5}$. Because the distance from *C* to *O* is 1, the length *CE* of the golden rectangle is $1 + \sqrt{5}$. Now we can find the golden ratio:

$$\text{Golden ratio} = \frac{\text{Length}}{\text{Width}} = \frac{CE}{EF} = \frac{1 + \sqrt{5}}{2}$$

USING TECHNOLOGY Graphing Calculators — A Word of Caution

Some graphing calculators give surprising results when evaluating expressions such as $(-8)^{2/3}$. As you know from reading this section, the expression $(-8)^{2/3}$ simplifies to 4, either by taking the cube root first and then squaring the result, or by squaring the base first and then taking the cube root of the result. Here are three different ways to evaluate this expression on your calculator:

1. $(-8)\wedge(2/3)$ To evaluate $(-8)^{2/3}$
2. $((-8)\wedge2)\wedge(1/3)$ To evaluate $((-8)^2)^{1/3}$
3. $((-8)\wedge(1/3))\wedge2$ To evaluate $((-8)^{1/3})^2$

Note any differences in the results.

Next, graph each of the following functions, one at a time.

1. $Y_1 = X^{2/3}$ 2. $Y_2 = (X^2)^{1/3}$ 3. $Y_3 = (X^{1/3})^2$

The correct graph is shown in Figure 6. Note which of your graphs match the correct graph.

FIGURE 6

Different calculators evaluate exponential expressions in different ways. You should use the method (or methods) that gave you the correct graph.

GETTING READY FOR CLASS

After reading through the preceding section, respond in your own words and in complete sentences.

A. Define positive square root.

B. Explain why a square root of -4 is not a real number.

C. For the expression $a^{m/n}$, explain the significance of the numerator m and the significance of the denominator n in the exponent.

D. Briefly explain the concept behind the golden rectangle.

EXERCISE SET 12.1

Vocabulary Review

Choose the correct words to fill in the blanks below.

ratio	cube	positive	index	root
rational	radical	radicand	exponent	

1. If x is a nonnegative real number, then the expression \sqrt{x} is called the _____ square root of x where x is greater than zero.

2. The _____ root of x, $\sqrt[3]{x}$ is such that $(\sqrt[3]{x})^3 = x$.

3. An expression like $\sqrt[3]{64}$ that involves a root is called a _____ expression.

4. In the expression $\sqrt[3]{64}$, the 3 is called the index, the $\sqrt{}$ is called the radical sign, and 64 is called the _____.

5. If no _____ of a radical is written, it is assumed to be 2.

6. The quantity x with a fractional exponent $\frac{1}{n}$, such that $x^{1/n}$, is considered the nth _____ of x.

7. The properties of exponents can be extended from integer exponents to also include _____ exponents.

8. The rational _____ theorem states that if a is a nonnegative real number, m is an integer, and n is a positive integer, then $a^{m/n} = (a^{1/n})^m = (a^m)^{1/n}$.

9. The ratio of length to width, known as the golden _____, is used to associate a number with the shape of a golden rectangle.

Problems

A Find each of the following roots, if possible.

1. $\sqrt{144}$

2. $-\sqrt{144}$

3. $\sqrt{-144}$

4. $\sqrt{-49}$

5. $-\sqrt{49}$

6. $\sqrt{49}$

7. $\sqrt[3]{-27}$

8. $-\sqrt[3]{27}$

9. $\sqrt[4]{16}$

10. $-\sqrt[4]{16}$

11. $\sqrt[4]{-16}$

12. $-\sqrt[4]{-16}$

13. $\sqrt{0.04}$

14. $\sqrt{0.81}$

15. $\sqrt[3]{0.008}$

16. $\sqrt[3]{0.125}$

Simplify each expression. Assume all variables represent nonnegative numbers.

17. $\sqrt{36a^8}$

18. $\sqrt{49a^{10}}$

19. $\sqrt[3]{27a^{12}}$

20. $\sqrt[3]{8a^{15}}$

21. $\sqrt[3]{x^3y^6}$

22. $\sqrt[3]{x^6y^3}$

23. $\sqrt[5]{32x^{10}y^5}$

24. $\sqrt[5]{32x^5y^{10}}$

25. $\sqrt[4]{16a^{12}b^{20}}$

26. $\sqrt[4]{81a^{24}b^8}$

27. $\sqrt[3]{-27x^9y12}$

28. $\sqrt[3]{-8x^{15}y^{21}}$

B Convert the following radicals to expressions with rational exponents and simplify where possible. Assume all variables represent positive numbers.

29. $\sqrt[5]{x^4}$

30. $\sqrt[8]{y^4}$

31. $\sqrt[3]{n^2}$

32. \sqrt{a}

33. $\sqrt[6]{a^3}$

34. $\sqrt[10]{y^2}$

35. $\sqrt[8]{m^5}$

36. $\sqrt[12]{b^3}$

Use the rational exponent property to write each of the following with the appropriate root. Then simplify.

37. $36^{1/2}$

38. $49^{1/2}$

39. $-9^{1/2}$

40. $-16^{1/2}$

41. $8^{1/3}$

42. $-8^{1/3}$

43. $(-8)^{1/3}$

44. $-27^{1/3}$

45. $32^{1/5}$

46. $81^{1/4}$

47. $\left(\dfrac{81}{25}\right)^{1/2}$

48. $\left(\dfrac{9}{16}\right)^{1/2}$

49. $\left(\dfrac{64}{125}\right)^{1/3}$

50. $\left(\dfrac{8}{27}\right)^{1/3}$

51. $\left(\dfrac{16}{81}\right)^{1/4}$

52. $\left(-\dfrac{27}{8}\right)^{1/3}$

Use the rational exponent theorem to simplify each of the following as much as possible.

53. $27^{2/3}$

54. $8^{4/3}$

55. $25^{3/2}$

56. $9^{3/2}$

57. $16^{3/4}$

58. $81^{3/4}$

59. $(-27)^{2/3}$

60. $(-32)^{3/5}$

Simplify each expression. Remember, negative exponents give reciprocals.

61. $27^{-1/3}$ **62.** $9^{-1/2}$ **63.** $81^{-3/4}$ **64.** $4^{-3/2}$

65. $\left(\dfrac{25}{36}\right)^{-1/2}$ **66.** $\left(\dfrac{16}{49}\right)^{-1/2}$ **67.** $\left(\dfrac{81}{16}\right)^{-3/4}$ **68.** $\left(\dfrac{27}{8}\right)^{-2/3}$

69. $16^{1/2} + 27^{1/3}$ **70.** $25^{1/2} + 100^{1/2}$ **71.** $8^{-2/3} + 4^{-1/2}$ **72.** $49^{-1/2} + 25^{-1/2}$

Use the properties of exponents to simplify each of the following as much as possible. Assume all bases are positive.

73. $x^{3/5} \cdot x^{1/5}$ **74.** $x^{3/4} \cdot x^{5/4}$ **75.** $(a^{3/4})^{4/3}$ **76.** $(a^{2/3})^{3/4}$

77. $\dfrac{x^{1/5}}{x^{3/5}}$ **78.** $\dfrac{x^{2/7}}{x^{5/7}}$ **79.** $\dfrac{x^{5/6}}{x^{2/3}}$ **80.** $\dfrac{x^{7/8}}{x^{8/7}}$

81. $(a^{2/3}b^{1/2}c^{1/4})^{1/3}$ **82.** $(a^{5/8}b^{1/2}c^{3/4})^{2/5}$ **83.** $(x^{3/5}y^{5/6}z^{1/3})^{3/5}$ **84.** $(x^{3/4}y^{1/8}z^{5/6})^{4/5}$

85. $\dfrac{a^{3/4}b^2}{a^{7/8}b^{1/4}}$ **86.** $\dfrac{a^{1/3}b^4}{a^{3/5}b^{1/3}}$ **87.** $\dfrac{(y^{2/3})^{3/4}}{(y^{1/3})^{3/5}}$ **88.** $\dfrac{(y^{5/4})^{2/5}}{(y^{1/4})^{4/3}}$

89. $\left(\dfrac{x^{-1/2}}{y^{1/4}}\right)^4$ **90.** $\left(\dfrac{a^{-3/4}}{b^{2/3}}\right)^{12}$ **91.** $\left(\dfrac{a^{-1/4}}{b^{1/2}}\right)^8$ **92.** $\left(\dfrac{a^{-1/5}}{b^{1/3}}\right)^{15}$

93. $\dfrac{(r^{-2}s^{1/3})^6}{r^8s^{3/2}}$ **94.** $\dfrac{(r^{-5}s^{1/2})^4}{r^{12}s^{5/2}}$ **95.** $\dfrac{(25a^6b^4)^{1/2}}{(8a^{-9}b^3)^{-1/3}}$ **96.** $\dfrac{(27a^3b^6)^{1/3}}{(81a^8b^{-4})^{1/4}}$

97. Show that the expression $(a^{1/2} + b^{1/2})^2$ is not equal to $a + b$ by replacing a with 9 and b with 4 in both expressions and then simplifying each.

98. Show that the statement $(a^2 + b^2)^{1/2} = a + b$ is not, in general, true by replacing a with 3 and b with 4 and then simplifying both sides.

99. You may have noticed, if you have been using a calculator to find roots, that you can find the fourth root of a number by pressing the square root button twice. Written in symbols, this fact looks like this:

$$\sqrt{\sqrt{a}} = \sqrt[4]{a} \qquad (a \geq 0)$$

Show that this statement is true by rewriting each side with exponents instead of radical notation and then simplifying the left side.

100. Show that the statement is true by rewriting each side with exponents instead of radical notation and then simplifying the left side.

$$\sqrt[3]{\sqrt{a}} = \sqrt[6]{a} \qquad (a \geq 0)$$

Applying the Concepts

101. Maximum Speed The maximum speed v that an automobile can travel around a curve of radius r without skidding is given by the equation

$$v = \left(\frac{5r}{2}\right)^{1/2}$$

where v is in miles per hour and r is measured in feet. What is the maximum speed a car can travel around a curve with a radius of 250 feet without skidding?

102. Relativity The equation

$$L = \left(1 - \frac{v^2}{c^2}\right)^{1/2}$$

gives the relativistic length of a 1-foot ruler traveling with velocity v. Find L if

$$\frac{v}{c} = \frac{3}{5}$$

103. Golden Ratio The golden ratio is the ratio of the length to the width in any golden rectangle. The exact value of this number is $\frac{1 + \sqrt{5}}{2}$. Use a calculator to find a decimal approximation of this number and round it to the nearest thousandth.

104. Golden Ratio The reciprocal of the golden ratio is $\frac{2}{1 + \sqrt{5}}$. Find a decimal approximation to this number that is accurate of the nearest thousandth.

105. Chemistry Figure 7 shows part of a model of a magnesium oxide (MgO) crystal. Each corner of the square is at the center of one oxygen ion (O^{2-}), and the center of the middle ion is at the center of the square. The radius for each oxygen ion is 60 picometers (pm), and the radius for each magnesium ion (Mg^{2+}) is 150 picometers.

a. Find the length of the side of the square. Write your answer in picometers.

Mg^{2+}

O^{2-}

FIGURE 7

b. Find the length of the diagonal of the square. Write your answer in picometers.

c. If 1 meter is 10^{12} picometers, give the length of the diagonal of the square in meters.

106. Geometry The length of each side of the cube shown in Figure 8 is 1 inch.

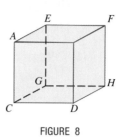

FIGURE 8

a. Find the length of the diagonal CH.

b. Find the length of the diagonal CF.

Below is a list of some well known pyramids in the world.

Pyramid Name	Location Culture	b	h	Slant height (ft)	Volume (ft³)
Temple of Kukulkan	Mexico/Maya	181 ft	98 ft		
Temple IV at Tecal	Guatemala/Maya	118 ft	187 ft		
Pyramid of the Sun	Mexico/Aztec	733 ft	246 ft		
Great Pyramid of Khufu	Gizal/Egypt	755 ft	455 ft		
Luxor Hotel	Las Vegas/Modern	600 ft	350 ft		
Transamerica Pyramid	San Francisco/Modern	175 ft	850 ft		

Source: skyscrapercity.com

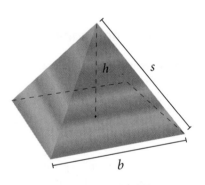

107. The slant height s of a pyramid with a square base with side length b and height h can be found using the formula

$$s = \sqrt{\frac{b^2}{4} + h^2}.$$

Use this formula to find the slant height of each pyramid in the table.

108. The volume v of a pyramid with a square base with side length b and height h is given by the formula

$$v = \frac{1}{3}b^2h.$$

Find the volume of each pyramid. (Round to the nearest thousand cubic feet.)

109. If you were to climb to the top of each pyramid along its slant height, which pyramid is the largest?

110. Which pyramid has the largest volume?

Getting Ready for the Next Section

Simplify. Assume all variable are positive real numbers.

111. $\sqrt{25}$

112. $\sqrt{4}$

113. $\sqrt{6^2}$

114. $\sqrt{3^2}$

115. $\sqrt{16x^4y^2}$

116. $\sqrt{4x^6y^8}$

117. $\sqrt{(5y)^2}$

118. $\sqrt{(8x^3)^2}$

119. $\sqrt[3]{27}$

120. $\sqrt[3]{-8}$

121. $\sqrt[3]{2^3}$

122. $\sqrt[3]{(-5)^3}$

123. $\sqrt[3]{8a^3b^3}$

124. $\sqrt[3]{64a^6b^3}$

125. $\sqrt[4]{625x^8y^4}$

126. $\sqrt[5]{(-x)^5}$

Fill in the blank.

127. $50 = \underline{\quad} \cdot 2$

128. $12 = \underline{\quad} \cdot 3$

129. $48x^4y^3 = \underline{\quad} \cdot y$

130. $40a^5b^4 = \underline{\quad} \cdot 5a^2b$

131. $12x^7y^6 = \underline{\quad} \cdot 3x$

132. $54a^6b^2c^4 = \underline{\quad} \cdot 2b^2c$

133. $10x^2y^5 = \underline{\quad} \cdot 5xy$

134. $-40a^4b^7 = \underline{\quad} \cdot 5a^3b^2$

Find the Mistake

Each sentence below contains a mistake. Circle the mistake and write the correct word or expression on the line provided.

1. \sqrt{x} is the positive square number we add to get x. _____

2. If we raise the fifth root of x to the fifth power, we get x^2. _____

3. For the radical expression $\sqrt[4]{16}$, the 4 is the index and the 16 is the radical. _____

4. Using the rational exponent theorem, we can say that $16^{-3/2} = (16^{1/2})^{-3} = 64$. _____

12.2 Simplifying Radicals

OBJECTIVES

A Write radical expressions in simplified form.

B Rationalize a denominator that contains only one term.

KEY WORDS

product property for radicals

quotient property for radicals

simplified form

rationalize the denominator

perfect square

absolute value property for radicals

spiral of roots

recursive

**Video Examples
Section 12.2**

Note There is not a property for radicals that says the *n*th root of a sum is the sum of the *n*th roots. That is,

$$\sqrt[n]{a + b} \neq \sqrt[n]{a} + \sqrt[n]{b}$$

Nikola Tesla was an inventor and engineer in the late 19th and early 20th centuries. Tesla acquired the first patent for a remote-controlled device. He demonstrated this device, a small boat with an iron hull, on an indoor pond. Onlookers were in awe of this new robotic contraption, having never seen anything like it before. Tesla prompted the audience to ask the boat a question, such as "What is the cube root of 64?" Then, as the boat continued to move over the surface of the pond, Tesla used the remote control to make the lights on the boat flash four times. In this section, we will continue our work with radicals, including cube roots.

Earlier in this chapter, we showed how the Pythagorean theorem can be used to construct a golden rectangle. In a similar manner, the Pythagorean theorem can be used to construct the attractive spiral shown here.

This spiral is called the Spiral of Roots because each of the diagonals is the positive square root of one of the positive integers. At the end of this section, we will use the Pythagorean theorem and some of the material in this section to construct this spiral.

In this section, we will use radical notation instead of rational exponents. We will begin by stating the first two properties of radicals. Following this, we will give a definition for simplified form for radical expressions. The examples in this section show how we use the properties of radicals to write radical expressions in simplified form.

The Spiral of Roots

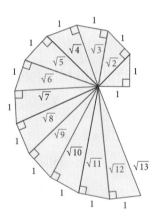

A Simplified Form

Here are the first two properties of radicals. For these two properties, we will assume *a* and *b* are nonnegative real numbers whenever *n* is an even number.

PROPERTY Product Property for Radicals

$$\sqrt[n]{ab} = \sqrt[n]{a} \cdot \sqrt[n]{b}$$

In words: The *n*th root of a product is the product of the *n*th roots.

Proof of the Product Property for Radicals

$$\sqrt[n]{ab} = (ab)^{1/n} \qquad \text{Definition of fractional exponents}$$

$$= a^{1/n}b^{1/n} \qquad \text{Exponents distribute over products.}$$

$$= \sqrt[n]{a} \cdot \sqrt[n]{b} \qquad \text{Definition of fractional exponents}$$

> **PROPERTY** Quotient Property for Radicals
>
> $$\sqrt[n]{\frac{a}{b}} = \frac{\sqrt[n]{a}}{\sqrt[n]{b}} = \quad (b \neq 0)$$
>
> *In words:* The *n*th root of a quotient is the quotient of the *n*th roots.

The proof of the quotient property for radicals is similar to the proof of the product property for radicals.

These two properties of radicals allow us to change the form of and simplify radical expressions without changing their value.

> **RULE** Simplified Form for Radical Expressions
>
> A radical expression is in *simplified form* if
>
> 1. None of the factors of the radicand (the quantity under the radical sign) can be written as powers greater than or equal to the index—that is, no perfect squares can be factors of the quantity under a square root sign, no perfect cubes can be factors of what is under a cube root sign, and so forth.
>
> 2. There are no fractions under the radical sign.
>
> 3. There are no radicals in the denominator.

Satisfying the first condition for simplified form actually amounts to taking as much out from under the radical sign as possible. The following examples illustrate the first condition for simplified form.

EXAMPLE 1 Write $\sqrt{50}$ in simplified form.

Solution The largest perfect square that divides 50 is 25. We write 50 as $25 \cdot 2$ and apply the product property for radicals.

$$\sqrt{50} = \sqrt{25 \cdot 2} \qquad 50 = 25 \cdot 2$$
$$= \sqrt{25}\sqrt{2} \qquad \text{Product property for radicals}$$
$$= 5\sqrt{2} \qquad \sqrt{25} = 5$$

We have taken as much as possible out from under the radical sign — in this case, factoring 25 from 50 and then writing $\sqrt{25}$ as 5.

EXAMPLE 2 Write $\sqrt{48x^4y^3}$, where $x, y \geq 0$ in simplified form.

Solution The largest perfect square that is a factor of the radicand is $16x^4y^2$. Applying the product property for radicals again, we have

$$\sqrt{48x^4y^3} = \sqrt{16x^4y^2 \cdot 3y}$$
$$= \sqrt{16x^4y^2}\sqrt{3y}$$
$$= 4x^2y\sqrt{3y}$$

Practice Problems

1. Write $\sqrt{18}$ in simplified form.

2. Write $\sqrt{50x^2y^3}$ in simplified form. Assume $x, y \geq 0$.

Answers

1. $3\sqrt{2}$
2. $5xy\sqrt{2y}$

3. Write $\sqrt[3]{54a^4b^3}$ in simplified form.

EXAMPLE 3 Write $\sqrt[3]{40a^5b^4}$ in simplified form.

Solution We now want to factor the largest perfect cube from the radicand. We write $40a^5b^4$ as $8a^3b^3 \cdot 5a^2b$ and proceed as we did in Examples 1 and 2.

$$\sqrt[3]{40a^5b^4} = \sqrt[3]{8a^3b^3 \cdot 5a^2b}$$
$$= \sqrt[3]{8a^3b^3}\sqrt[3]{5a^2b}$$
$$= 2ab\sqrt[3]{5a^2b}$$

Here are some further examples concerning the first condition for simplified form.

4. Write each expression in simplified form.

 a. $\sqrt{75x^5y^8}$

 b. $\sqrt[4]{48a^8b^5c^4}$

EXAMPLE 4 Write each expression in simplified form:

a. $\sqrt{12x^7y^6} = \sqrt{4x^6y^6 \cdot 3x}$
$$= \sqrt{4x^6y^6}\sqrt{3x}$$
$$= 2x^3y^3\sqrt{3x}$$

b. $\sqrt[3]{54a^6b^2c^4} = \sqrt[3]{27a^6c^3 \cdot 2b^2c}$
$$= \sqrt[3]{27a^6c^3}\sqrt[3]{2b^2c}$$
$$= 3a^2c\sqrt[3]{2b^2c}$$

B Rationalizing the Denominator

The quotient property of radicals is used to simplify a radical that contains a fraction.

5. Simplify $\sqrt{\dfrac{5}{9}}$.

EXAMPLE 5 Simplify $\sqrt{\dfrac{3}{4}}$.

Solution Applying quotient property for radicals, we have

$$\sqrt{\frac{3}{4}} = \frac{\sqrt{3}}{\sqrt{4}} \qquad \textit{Quotient property for radicals}$$

$$= \frac{\sqrt{3}}{2} \qquad \sqrt{4} = 2$$

The last expression is in simplified form because it satisfies all three conditions for simplified form.

6. Write $\sqrt{\dfrac{2}{3}}$ in simplified form.

EXAMPLE 6 Write $\sqrt{\dfrac{5}{6}}$ in simplified form.

Solution Proceeding as in Example 5, we have

$$\sqrt{\frac{5}{6}} = \frac{\sqrt{5}}{\sqrt{6}}$$

The resulting expression satisfies the second condition for simplified form because neither radical contains a fraction. It does, however, violate condition 3 because it has a radical in the denominator. Getting rid of the radical in the denominator is called *rationalizing the denominator* and is accomplished, in this case, by multiplying the numerator and denominator by $\sqrt{6}$.

$$\frac{\sqrt{5}}{\sqrt{6}} = \frac{\sqrt{5}}{\sqrt{6}} \cdot \frac{\sqrt{6}}{\sqrt{6}}$$

$$= \frac{\sqrt{30}}{(\sqrt{6})^2}$$

$$= \frac{\sqrt{30}}{6}$$

Answers

3. $3ab\sqrt[3]{2a}$

4. a. $5x^2y^4\sqrt{3x}$ **b.** $2a^2bc\sqrt[4]{3b}$

5. $\dfrac{\sqrt{5}}{3}$

6. $\dfrac{\sqrt{6}}{3}$

EXAMPLE 7 Rationalize the denominator.

a. $\dfrac{4}{\sqrt{3}} = \dfrac{4}{\sqrt{3}} \cdot \dfrac{\sqrt{3}}{\sqrt{3}}$

$= \dfrac{4\sqrt{3}}{(\sqrt{3})^2}$

$= \dfrac{4\sqrt{3}}{3}$

b. $\dfrac{2\sqrt{3x}}{\sqrt{5y}} = \dfrac{2\sqrt{3x}}{\sqrt{5y}} \cdot \dfrac{\sqrt{5y}}{\sqrt{5y}}$

$= \dfrac{2\sqrt{15xy}}{(\sqrt{5y})^2}$

$= \dfrac{2\sqrt{15xy}}{5y}$

7. Rationalize the denominator.

a. $\dfrac{5}{\sqrt{2}}$

b. $\dfrac{3\sqrt{5x}}{\sqrt{2y}}$

When the denominator involves a cube root, we must multiply by a radical that will produce a perfect cube under the cube root sign in the denominator, as Example 8 illustrates.

EXAMPLE 8 Rationalize the denominator in $\dfrac{7}{\sqrt[3]{4}}$.

Solution Because $4 = 2^2$, we can multiply both numerator and denominator by $\sqrt[3]{2}$ and obtain $\sqrt[3]{2^3}$ in the denominator.

$\dfrac{7}{\sqrt[3]{4}} = \dfrac{7}{\sqrt[3]{2^2}}$

$= \dfrac{7}{\sqrt[3]{2^2}} \cdot \dfrac{\sqrt[3]{2}}{\sqrt[3]{2}}$

$= \dfrac{7\sqrt[3]{2}}{\sqrt[3]{2^3}}$

$= \dfrac{7\sqrt[3]{2}}{2}$

8. Rationalize the denominator.

$\dfrac{5}{\sqrt[3]{9}}$

EXAMPLE 9 Simplify $\sqrt{\dfrac{12x^5y^3}{5z}}$.

9. Simplify $\sqrt{\dfrac{48x^3y^4}{7z}}$.

Solution We use the quotient property for radicals to write the numerator and denominator as two separate radicals.

$\sqrt{\dfrac{12x^5y^3}{5z}} = \dfrac{\sqrt{12x^5y^3}}{\sqrt{5z}}$

Simplifying the numerator, we have

$\dfrac{\sqrt{12x^5y^3}}{\sqrt{5z}} = \dfrac{\sqrt{4x^4y^2}\sqrt{3xy}}{\sqrt{5z}}$

$= \dfrac{2x^2y\sqrt{3xy}}{\sqrt{5z}}$

To rationalize the denominator, we multiply the numerator and denominator by $\sqrt{5z}$.

$\dfrac{2x^2y\sqrt{3xy}}{\sqrt{5z}} \cdot \dfrac{\sqrt{5z}}{\sqrt{5z}} = \dfrac{2x^2y\sqrt{15xyz}}{(\sqrt{5z})^2}$

$= \dfrac{2x^2y\sqrt{15xyz}}{5z}$

Answers

7. **a.** $\dfrac{5\sqrt{2}}{2}$ **b.** $\dfrac{3\sqrt{10xy}}{2y}$

8. $\dfrac{5\sqrt[3]{3}}{3}$

9. $\dfrac{4xy^2\sqrt{21xz}}{7z}$

Square Root of a Perfect Square

So far in this chapter, we have assumed that all our variables are nonnegative when they appear under a square root symbol. There are times, however, when this is not the case.

Consider the following two statements:

$$\sqrt{3^2} = \sqrt{9} = 3 \qquad \text{and} \qquad \sqrt{(-3)^2} = \sqrt{9} = 3$$

Whether we operate on 3 or -3, the result is the same: both expressions simplify to 3. The other operation we have worked with in the past that produces the same result is absolute value. That is,

$$|3| = 3 \qquad \text{and} \qquad |-3| = 3$$

This leads us to the next property of radicals.

> **PROPERTY** Absolute Value Property for Radicals
>
> If a is a real number, then $\sqrt{a^2} = |a|$.

The result of this discussion and the absolute value property for radicals is simply this:

If we know a is positive, then $\sqrt{a^2} = a$.

If we know a is negative, then $\sqrt{a^2} = |a|$.

If we don't know if a is positive or negative, then $\sqrt{a^2} = |a|$.

10. Simplify each expression. Do not assume the variables represent nonnegative numbers.

 a. $\sqrt{16x^2}$

 b. $\sqrt{25x^3}$

 c. $\sqrt{x^2 + 10x + 25}$

 d. $\sqrt{2x^3 + 7x^2}$

EXAMPLE 10 Simplify each expression. Do *not* assume the variables represent positive numbers.

a. $\sqrt{9x^2} = 3|x|$

b. $\sqrt{x^3} = |x|\sqrt{x}$

c. $\sqrt{x^2 - 6x + 9} = \sqrt{(x-3)^2} = |x-3|$

d. $\sqrt{x^3 - 5x^2} = \sqrt{x^2(x-5)} = |x|\sqrt{x-5}$

As you can see, the square root of a sum or difference does not equal the sum or difference of the square roots. This is a common mistake. We must use absolute value symbols when we take a square root of a perfect square, unless we know the base of the perfect square is a positive number. The same idea holds for higher even roots, but not for odd roots. With odd roots, no absolute value symbols are necessary.

EXAMPLE 11 Simplify each expression.

a. $\sqrt[3]{(-2)^3} = \sqrt[3]{-8} = -2$

b. $\sqrt[3]{(-5)^3} = \sqrt[3]{-125} = -5$

We can extend this discussion to all roots as follows:

11. Simplify each expression.

 a. $\sqrt[3]{(-3)^3}$

 b. $\sqrt[3]{(-1)^3}$

> **PROPERTY** Extending the Absolute Value Property for Radicals
>
> If a is a real number, then
>
> $$\sqrt[n]{a^n} = |a| \qquad \text{if} \qquad n \text{ is even}$$
> $$\sqrt[n]{a^n} = a \qquad \text{if} \qquad n \text{ is odd}$$

Answers

10. a. $4|x|$ **b.** $5|x|\sqrt{x}$ **c.** $|x + 5|$

 d. $|x|\sqrt{2x + 7}$

11. a. -3 **b.** -1

The Spiral of Roots

To visualize the square roots of the positive integers, we can construct a spiral of roots. To begin, we draw two line segments, each of length 1, at right angles to each other. Then we use the Pythagorean theorem to find the length of the diagonal. Figure 1 illustrates this procedure.

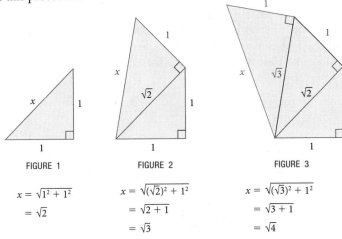

FIGURE 1 FIGURE 2 FIGURE 3

$$x = \sqrt{1^2 + 1^2}$$
$$= \sqrt{2}$$

$$x = \sqrt{(\sqrt{2})^2 + 1^2}$$
$$= \sqrt{2 + 1}$$
$$= \sqrt{3}$$

$$x = \sqrt{(\sqrt{3})^2 + 1^2}$$
$$= \sqrt{3 + 1}$$
$$= \sqrt{4}$$

Next, we construct a second triangle by connecting a line segment of length 1 to the end of the first diagonal so that the angle formed is a right angle. We find the length of the second diagonal using the Pythagorean theorem. Figure 2 illustrates this procedure. Continuing to draw new triangles by connecting line segments of length 1 to the end of each new diagonal, so that the angle formed is a right angle, the spiral of roots begins to appear (Figure 3).

The Spiral of Roots and Function Notation

Looking over the diagrams and calculations in the preceding discussion, we see that each diagonal in the spiral of roots is found by using the length of the previous diagonal.

First diagonal: $\sqrt{1^2 + 1^2} = \sqrt{2}$

Second diagonal: $\sqrt{(\sqrt{2})^2 + 1^2} = \sqrt{3}$

Third diagonal: $\sqrt{(\sqrt{3})^2 + 1^2} = \sqrt{4}$

Fourth diagonal: $\sqrt{(\sqrt{4})^2 + 1^2} = \sqrt{5}$

A process like this one, in which the answer to one calculation is used to find the answer to the next calculation, is called a *recursive* process. In this particular case, we can use function notation to model the process. If we let x represent the length of any diagonal, then the length of the next diagonal is given by

$$f(x) = \sqrt{x^2 + 1}$$

To begin the process of finding the diagonals, we let $x = 1$.

$$f(1) = \sqrt{1^2 + 1} = \sqrt{2}$$

To find the next diagonal, we substitute $\sqrt{2}$ for x to obtain

$$f(f(1)) = f(\sqrt{2}) = \sqrt{(\sqrt{2})^2 + 1} = \sqrt{3}$$

$$f(f(f(1))) = f(\sqrt{3}) = \sqrt{(\sqrt{3})^2 + 1} = \sqrt{4}$$

We can concisely describe the process of finding the diagonals of the spiral of roots this way:

$$f(1), f(f(1)), f(f(f(1))), \ldots \qquad \text{where } f(x) = \sqrt{x^2 + 1}$$

First input
x

function machine

Output $f(x)$

USING TECHNOLOGY

As our preceding discussion indicates, the length of each diagonal in the spiral of roots is used to calculate the length of the next diagonal. The $\boxed{\text{ANS}}$ key on a graphing calculator can be used effectively in a situation like this. To begin, we store the number 1 in the variable $\boxed{\text{ANS}}$. Next, we key in the formula used to produce each diagonal using $\boxed{\text{ANS}}$ for the variable. After that, it is simply a matter of pressing $\boxed{\text{ENTER}}$, as many times as we like, to produce the lengths of as many diagonals as we like. Here is a summary of what we do:

Enter This	Display Shows
1 $\boxed{\text{ENTER}}$	1.000
$\sqrt{}$ (ANS2 + 1) $\boxed{\text{ENTER}}$	1.414
$\boxed{\text{ENTER}}$	1.732
$\boxed{\text{ENTER}}$	2.000
$\boxed{\text{ENTER}}$	2.236

If you continue to press the $\boxed{\text{ENTER}}$ key, you will produce decimal approximations for as many of the diagonals in the spiral of roots as you like.

GETTING READY FOR CLASS

After reading through the preceding section, respond in your own words and in complete sentences.

A. Explain why this statement is false: "The square root of a sum is the sum of the square roots."

B. What is simplified form for an expression that contains a square root?

C. Why is it not necessarily true that $\sqrt{a^2} = a$?

D. What does it mean to rationalize the denominator in an expression?

Vocabulary Review

Choose the correct words to fill in the blanks below.

product absolute value spiral of roots quotient simplified form

1. The _____ property for radicals states that the nth root of a product is the product of the nth roots.

2. The _____ property for radicals is used to rationalize the denominator of a radical that contains a fraction.

3. A radical expression is in _____ if none of the factors of the radicand can be written as powers greater than or equal to the index, there are no fractions under the radical sign, and there are no radicals in the denominator.

4. The _____ property for radicals states that if a is a real number, then $\sqrt{a^2} = |a|$.

5. We use the Pythagorean theorem to calculate the hypotenuse of each triangle that makes up a _____.

Problems

A Use the product property for radicals to write each of the following expressions in simplified form. (Assume all variables are nonnegative through Problem 74.)

1. $\sqrt{8}$

2. $\sqrt{32}$

3. $\sqrt{98}$

4. $\sqrt{75}$

5. $\sqrt{288}$

6. $\sqrt{128}$

7. $\sqrt{80}$

8. $\sqrt{200}$

9. $\sqrt{48}$

10. $\sqrt{27}$

11. $\sqrt{675}$

12. $\sqrt{972}$

13. $\sqrt[3]{54}$

14. $\sqrt[3]{24}$

15. $\sqrt[3]{128}$

16. $\sqrt[3]{162}$

17. $\sqrt[3]{432}$

18. $\sqrt[3]{1,536}$

19. $\sqrt[5]{64}$

20. $\sqrt[4]{48}$

21. $\sqrt{18x^3}$

22. $\sqrt{27x^5}$

23. $\sqrt[4]{32y^7}$

24. $\sqrt[5]{32y^7}$

25. $\sqrt[3]{40x^4y^7}$

26. $\sqrt[3]{128x^6y^2}$

27. $\sqrt{48a^2b^3c^4}$

28. $\sqrt{72a^4b^3c^2}$

29. $\sqrt[3]{48a^2b^3c^4}$

30. $\sqrt[3]{72a^4b^3c^2}$

31. $\sqrt[5]{64x^8y^{12}}$

32. $\sqrt[4]{32x^9y^{10}}$

33. $\sqrt[5]{243x^7y^{10}z^5}$

34. $\sqrt[5]{64x^8y^4z^{11}}$

35. $\sqrt[3]{250x^5y^{10}}$

36. $\sqrt[3]{-128x^{15}y^{20}}$

37. $\sqrt[3]{-27x^{10}y^{13}}$

38. $\sqrt[4]{48x^{13}y^{23}}$

Substitute the given numbers into the expression $\sqrt{b^2 - 4ac}$, and then simplify.

39. $a = 2, b = -6, c = 3$

40. $a = 6, b = 7, c = -5$

41. $a = 1, b = 2, c = 6$

42. $a = 2, b = 5, c = 3$

43. $a = \dfrac{1}{2}, b = -\dfrac{1}{2}, c = -\dfrac{5}{4}$

44. $a = \dfrac{7}{4}, b = -\dfrac{3}{4}, c = -2$

B Rationalize the denominator in each of the following expressions.

45. $\dfrac{2}{\sqrt{3}}$

46. $\dfrac{3}{\sqrt{2}}$

47. $\dfrac{5}{\sqrt{6}}$

48. $\dfrac{7}{\sqrt{5}}$

49. $\sqrt{\dfrac{1}{2}}$

50. $\sqrt{\dfrac{1}{3}}$

51. $\sqrt{\dfrac{1}{5}}$

52. $\sqrt{\dfrac{1}{6}}$

53. $\dfrac{4}{\sqrt[3]{2}}$

54. $\dfrac{5}{\sqrt[3]{3}}$

55. $\dfrac{2}{\sqrt[3]{9}}$

56. $\dfrac{3}{\sqrt[3]{4}}$

57. $\sqrt[4]{\dfrac{3}{2x^2}}$

58. $\sqrt[4]{\dfrac{5}{3x^2}}$

59. $\sqrt[4]{\dfrac{8}{y}}$

60. $\sqrt[4]{\dfrac{27}{y}}$

61. $\sqrt[3]{\dfrac{4x}{3y}}$

62. $\sqrt[3]{\dfrac{7x}{6y}}$

63. $\sqrt[3]{\dfrac{2x}{9y}}$

64. $\sqrt[3]{\dfrac{5x}{4y}}$

65. $\sqrt[4]{\dfrac{1}{8x^3}}$

66. $\sqrt[4]{\dfrac{8}{9x^3}}$

67. $\sqrt[4]{\dfrac{9}{8x^2}}$

68. $\sqrt[4]{\dfrac{5}{27y}}$

Write each of the following in simplified form.

69. $\sqrt{\dfrac{27x^3}{5y}}$

70. $\sqrt{\dfrac{12x^5}{7y}}$

71. $\sqrt{\dfrac{75x^3y^2}{2z}}$

72. $\sqrt{\dfrac{50x^2y^3}{3z}}$

73. $\sqrt[3]{\dfrac{16a^4b^3}{9c}}$

74. $\sqrt[3]{\dfrac{54a^5b^4}{25c^2}}$

75. $\sqrt[3]{\dfrac{8x^3y^6}{9z}}$

76. $\sqrt[3]{\dfrac{27x^6y^3}{2z^2}}$

Simplify each expression. Do *not* assume the variables represent positive numbers.

77. $\sqrt{25x^2}$

78. $\sqrt{49x^2}$

79. $\sqrt{27x^3y^2}$

80. $\sqrt{40x^3y^2}$

81. $\sqrt{x^2 - 10x + 25}$

82. $\sqrt{x^2 - 16x + 64}$

83. $\sqrt{4x^2 + 12x + 9}$

84. $\sqrt{16x^2 + 40x + 25}$

85. $\sqrt{4a^4 + 16a^3 + 16a^2}$ **86.** $\sqrt{9a^4 + 18a^3 + 9a^2}$ **87.** $\sqrt{4x^3 - 8x^2}$ **88.** $\sqrt{18x^3 - 9x^2}$

89. Show that the statement $\sqrt{a + b} = \sqrt{a} + \sqrt{b}$ is not true by replacing a with 9 and b with 16 and simplifying both sides.

90. Find a pair of values for a and b that will make the statement $\sqrt{a + b} = \sqrt{a} + \sqrt{b}$ true.

Applying the Concepts

91. Diagonal Distance The distance d between opposite corners of a rectangular room with length l and width w is given by
$$d = \sqrt{l^2 + w^2}$$
How far is it between opposite corners of a living room that measures 10 by 15 feet?

92. Radius of a Sphere The radius r of a sphere with volume V can be found by using the formula
$$r = \sqrt[3]{\frac{3V}{4\pi}}$$
Find the radius of a sphere with volume 9 cubic feet. Write your answer in simplified form.
(Use $\frac{22}{7}$ for π.)

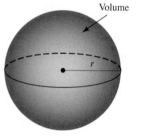
Volume

93. Distance to the Horizon If you are at a point k miles above the surface of the earth, the distance you can see towards the horizon, in miles, is approximated by the equation
$$d = \sqrt{8000k + k^2}$$

a. How far towards the horizon can you see from a point that is 1 mile above the surface of the earth?

b. How far can you see from a point that is 2 miles above the surface of the earth?

c. How far can you see from a point that is 3 miles above the surface of the earth?

94. Investing If you invest P dollars and you want the investment to grow to A dollars in t years, the interest rate that must be earned if interest is compounded annually is given by the formula
$$r = \sqrt[t]{\frac{A}{P}} - 1$$
If you invest $4,000 and want to have $7,000 in 8 years, what interest rate must be earned?

95. Spiral of Roots Construct your own spiral of roots by using a ruler. Draw the first triangle by using two 1-inch lines. The first diagonal will have a length of $\sqrt{2}$ inches. Each new triangle will be formed by drawing a 1-inch line segment at the end of the previous diagonal so the angle formed is 90°.

96. Spiral of Roots Construct a spiral of roots by using line segments of length 2 cm. The length of the first diagonal will be $2\sqrt{2}$ cm. The length of the second diagonal will be $2\sqrt{3}$ cm.

97. Spiral of Roots If $f(x) = \sqrt{x^2 + 1}$, find the first six terms in the following sequence. Use your results to predict the value of the 10th term and the 100th term.

$$f(1), f(f(1)), f(f(f(1))), \ldots$$

98. Spiral of Roots If $f(x) = \sqrt{x^2 + 4}$, find the first six terms in the following sequence. Use your results to predict the value of the 10th term and the 100th term. (The numbers in this sequence are the lengths of the diagonals of the spiral you drew in Problem 94.)

$$f(2), f(f(2)), f(f(f(2))), \ldots$$

Getting Ready for the Next Section

Simplify the following.

99. $5x - 4x + 6x$

100. $12x + 8x - 7x$

101. $35xy^2 - 8xy^2$

102. $20a^2b + 33a^2b$

103. $\dfrac{1}{2}x + \dfrac{1}{3}x$

104. $\dfrac{2}{3}x + \dfrac{5}{8}x$

105. $5(x + 1)^2 - 12(x + 1)^2$

106. $-7(a + 2)^3 + 4(a + 2)^3$

Write in simplified form for radicals.

107. $\sqrt{18}$

108. $\sqrt{8}$

109. $\sqrt{75xy^3}$

110. $\sqrt{12xy}$

111. $\sqrt[3]{8a^4b^2}$

112. $\sqrt[3]{27ab^2}$

113. $\sqrt[4]{32x^8y^5}$

114. $\sqrt[4]{81a^6b^{11}}$

Find the Mistake

Write true or false for each statement below. If false, circle what is false and write the correct answer on the line provided.

1. True or False: To write the radical expression $\sqrt{4x^7y^2}$ in simplified form, we apply the quotient property for radicals. _____

2. True or False: The radical expression in statement 1 above written in simplified form is $2x^3y\sqrt{xy}$.

3. True or False: To rationalize the denominator of the radical expression $\sqrt{\dfrac{8x^{4y}}{3x}}$, we use the quotient property for radicals and then multiply the numerator and denominator by $\sqrt{8x^{4y}}$. _____

4. True of False: For the spiral of roots, each new line segment drawn to the end of a diagonal is at a complementary angle to the previous line segment. _____

OBJECTIVES

A Add and subtract radicals.

B Construct golden rectangles from squares.

KEY WORDS

similar radicals

like radicals

Image © Luca Giarelli

As the sun sets on a high mountain valley called Val Camonica in Northern Italy, shadows bring out ancient rock carvings on the surrounding cliffs. Archaeologists believe the Camuni tribe may have lit the carvings with candlelight, thus making them "move" as the candle flame flickered. The location of the nearby rock walls would also make a storyteller's voice echo. This recurring echo may have provided a cinematic experience for the tribe.

The echo's time delay is the time it takes for the sound of the storyteller's voice to travel to the nearby rock wall, bounce off it, and be audible to the storyteller and audience. To determine this time delay, we need to know the distance of the storyteller to the nearby wall. Suppose the storyteller was standing $12\sqrt{3}$ meters from the opposing rock wall. We can use the following formula for the echo's time delay t:

$$t = \frac{d}{s}$$

where d is the doubled distance (distance of the storyteller to the wall and back), and s is the speed of sound. Since the formula calls for us to double the distance, we need to know how to add $12\sqrt{3}$ and $12\sqrt{3}$ together. In this section, we will learn how to apply our knowledge of similar terms to the addition and subtraction of radical expressions, such as these.

A Combining Similar Terms

> **DEFINITION** similar, or like, radicals
>
> Two radicals are said to be *similar,* or *like, radicals* if they have the same index and the same radicand.

The expressions $5\sqrt[3]{7}$ and $-8\sqrt[3]{7}$ are similar since the index is 3 in both cases and the radicands are 7. The expressions $3\sqrt[4]{5}$ and $7\sqrt[3]{5}$ are not similar because they have different indices, and the expressions $2\sqrt[5]{8}$ and $3\sqrt[5]{9}$ are not similar because the radicands are not the same.

We add and subtract radical expressions in the same way we add and subtract polynomials—by combining similar terms under the distributive property.

EXAMPLE 1 Combine $5\sqrt{3} - 4\sqrt{3} + 6\sqrt{3}$.

Solution All three radicals are similar. We apply the distributive property to get

$$5\sqrt{3} - 4\sqrt{3} + 6\sqrt{3} = (5 - 4 + 6)\sqrt{3}$$
$$= 7\sqrt{3}$$

**Video Examples
Section 12.3**

Practice Problems

1. Combine $3\sqrt{5} - 2\sqrt{5} + 4\sqrt{5}$.

Answer

1. $5\sqrt{5}$

EXAMPLE 2 Combine $3\sqrt{8} + 5\sqrt{18}$.

Solution The two radicals do not seem to be similar. We must write each in simplified form before applying the distributive property.

$$3\sqrt{8} + 5\sqrt{18} = 3\sqrt{4 \cdot 2} + 5\sqrt{9 \cdot 2}$$
$$= 3\sqrt{4} \cdot \sqrt{2} + 5\sqrt{9} \cdot \sqrt{2}$$
$$= 3 \cdot 2\sqrt{2} + 5 \cdot 3\sqrt{2}$$
$$= 6\sqrt{2} + 15\sqrt{2}$$
$$= (6 + 15)\sqrt{2}$$
$$= 21\sqrt{2}$$

The result of Example 2 can be generalized to the following rule for sums and differences of radical expressions.

> **RULE** Adding or Subtracting Radical Expressions
>
> To add or subtract radical expressions, put each in simplified form and apply the distributive property, if possible. We can add only similar radicals. We must write each expression in simplified form for radicals before we can tell if the radicals are similar.

EXAMPLE 3 Combine $7\sqrt{75xy^3} - 4y\sqrt{12xy}$, where $x, y \geq 0$.

Solution We write each expression in simplified form and combine similar radicals.

$$7\sqrt{75xy^3} - 4y\sqrt{12xy} = 7\sqrt{25y^2} \cdot \sqrt{3xy} - 4y\sqrt{4} \cdot \sqrt{3xy}$$
$$= 35y\sqrt{3xy} - 8y\sqrt{3xy}$$
$$= (35y - 8y)\sqrt{3xy}$$
$$= 27y\sqrt{3xy}$$

EXAMPLE 4 Combine $10\sqrt[3]{8a^4b^2} + 11a\sqrt[3]{27ab^2}$.

Solution Writing each radical in simplified form and combining similar terms, we have

$$10\sqrt[3]{8a^4b^2} + 11a\sqrt[3]{27ab^2} = 10\sqrt[3]{8a^3} \cdot \sqrt[3]{ab^2} + 11a\sqrt[3]{27} \cdot \sqrt[3]{ab^2}$$
$$= 20a\sqrt[3]{ab^2} + 33a\sqrt[3]{ab^2}$$
$$= 53a\sqrt[3]{ab^2}$$

EXAMPLE 5 Combine $\dfrac{\sqrt{3}}{2} + \dfrac{1}{\sqrt{3}}$.

Solution We begin by writing the second term in simplified form.

$$\frac{\sqrt{3}}{2} + \frac{1}{\sqrt{3}} = \frac{\sqrt{3}}{2} + \frac{1}{\sqrt{3}} \cdot \frac{\sqrt{3}}{\sqrt{3}}$$
$$= \frac{\sqrt{3}}{2} + \frac{\sqrt{3}}{3}$$
$$= \frac{1}{2}\sqrt{3} + \frac{1}{3}\sqrt{3}$$
$$= \left(\frac{1}{2} + \frac{1}{3}\right)\sqrt{3}$$
$$= \frac{5}{6}\sqrt{3}$$
$$= \frac{5\sqrt{3}}{6}$$

2. Combine $4\sqrt{50} + 3\sqrt{8}$.

3. Assume $x, y \geq 0$ and combine.
$$4\sqrt{18x^2y} - 3x\sqrt{50y}$$

4. Combine
$$2\sqrt[3]{27a^2b^4} + 3b\sqrt[3]{125a^2b}.$$

5. Combine $\dfrac{\sqrt{5}}{3} + \dfrac{1}{\sqrt{5}}$.

Answers

2. $26\sqrt{2}$

3. $-3x\sqrt{2y}$

4. $21b\sqrt[3]{a^2b}$

5. $\dfrac{8\sqrt{5}}{15}$

6. Construct a golden rectangle from a square of side 6. Then show that the ratio of the length to the width is the golden ratio.

B The Golden Rectangle

Recall our discussion of the golden rectangle from the first section of this chapter. Now that we have practiced adding and subtracting radical expressions, let's put our new skills to use and find a golden rectangle of a different size.

EXAMPLE 6 Construct a golden rectangle from a square of side 4. Then show that the ratio of the length to the width is the golden ratio $\frac{1+\sqrt{5}}{2}$.

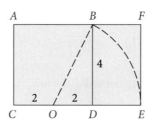

FIGURE 1

Solution Figure 1 shows the golden rectangle constructed from a square of side 4. The length of the diagonal OB is found from the Pythagorean theorem.

$$OB = \sqrt{2^2 + 4^2} = \sqrt{4 + 16} = \sqrt{20} = 2\sqrt{5}$$

The ratio of the length to the width for the rectangle is the golden ratio.

$$\text{Golden ratio} = \frac{CE}{EF} = \frac{2 + 2\sqrt{5}}{4} = \frac{2(1 + \sqrt{5})}{2 \cdot 2} = \frac{1 + \sqrt{5}}{2}$$

As you can see, showing that the ratio of length to width in this rectangle is the golden ratio depends on our ability to write $\sqrt{20}$ as $2\sqrt{5}$ and our ability to reduce to lowest terms by factoring and then dividing out the common factor 2 from the numerator and denominator.

GETTING READY FOR CLASS

After reading through the preceding section, respond in your own words and in complete sentences.

A. What are similar radicals?

B. When can we add two radical expressions?

C. What is the first step when adding or subtracting expressions containing radicals?

D. How would you construct a golden rectangle from a square of side 8?

Vocabulary Review

Choose the correct words to fill in the blanks in the paragraph below.

radicand distributive similar simplified form radicals

To add or subtract radical expressions, put each in _____ and apply the _____ property, if possible. We can add only _____ radicals, which are radicals with the same index and the same _____. We must write each expression in simplified form for radicals before we can tell if the _____ are similar.

Problems

A Combine the following expressions. (Assume any variables under an even root are nonnegative.)

1. $3\sqrt{5} + 4\sqrt{5}$

2. $6\sqrt{3} - 5\sqrt{3}$

3. $3x\sqrt{7} - 4x\sqrt{7}$

4. $6y\sqrt{a} + 7y\sqrt{a}$

5. $5\sqrt[3]{10} - 4\sqrt[3]{10}$

6. $6\sqrt[4]{2} + 9\sqrt[4]{2}$

7. $8\sqrt[5]{6} - 2\sqrt[5]{6} + 3\sqrt[5]{6}$

8. $7\sqrt[6]{7} - \sqrt[6]{7} + 4\sqrt[6]{7}$

9. $3x\sqrt{2} - 4x\sqrt{2} + x\sqrt{2}$

10. $5x\sqrt{6} - 3x\sqrt{6} - 2x\sqrt{6}$

11. $\sqrt{20} - \sqrt{80} + \sqrt{45}$

12. $\sqrt{8} - \sqrt{32} - \sqrt{18}$

13. $4\sqrt{8} - 2\sqrt{50} - 5\sqrt{72}$

14. $\sqrt{48} - 3\sqrt{27} + 2\sqrt{75}$

15. $5x\sqrt{8} + 3\sqrt{32x^2} - 5\sqrt{50x^2}$

16. $2\sqrt{50x^2} - 8x\sqrt{18} - 3\sqrt{72x^2}$

17. $5\sqrt{5y} + \sqrt{18y}$

18. $\sqrt{250x} - 2\sqrt{40x} + 5\sqrt{10x^3}$

19. $5\sqrt[3]{16} - 4\sqrt[3]{54}$

20. $\sqrt[3]{81} + 3\sqrt[3]{24}$

21. $\sqrt[3]{x^4y^2} + 7x\sqrt[3]{xy^2}$

22. $2\sqrt[3]{x^8y^6} - 3y^2\sqrt[3]{8x^8}$

23. $5a^2\sqrt{27ab^3} - 6b\sqrt{12a^5b}$

24. $9a\sqrt{20a^3b^2} + 7b\sqrt{45a^5}$

25. $\sqrt[3]{24a^5b^4} + \sqrt[3]{81a^8b^4}$

26. $\sqrt[4]{32x^7y^5} - \sqrt[4]{162x^7y^4}$

27. $b\sqrt[3]{24a^5b} + 3a\sqrt[3]{81a^2b^4}$

28. $7\sqrt[3]{a^4b^3c^2} - 6ab\sqrt[3]{ac^2}$

29. $5x\sqrt[4]{3y^5} + y\sqrt[4]{243x^4y} + \sqrt[4]{48x^4y^5}$

30. $x\sqrt[4]{5xy^8} + y\sqrt[4]{405x^5y^4} + y^2\sqrt[4]{80x^5}$

31. $\dfrac{\sqrt{2}}{2} + \dfrac{1}{\sqrt{2}}$

32. $\dfrac{\sqrt{3}}{3} + \dfrac{1}{\sqrt{3}}$

33. $\dfrac{\sqrt{5}}{3} + \dfrac{1}{\sqrt{5}}$

34. $\dfrac{\sqrt{6}}{2} + \dfrac{1}{\sqrt{6}}$

35. $\sqrt{x} - \dfrac{1}{\sqrt{x}}$

36. $\sqrt{x} + \dfrac{1}{\sqrt{x}}$

37. $\dfrac{\sqrt{18}}{6} + \sqrt{\dfrac{1}{2}} + \dfrac{\sqrt{2}}{2}$

38. $\dfrac{\sqrt{12}}{6} + \sqrt{\dfrac{1}{3}} + \dfrac{\sqrt{3}}{3}$

39. $\sqrt{6} - \sqrt{\dfrac{2}{3}} + \sqrt{\dfrac{1}{6}}$

40. $\sqrt{15} - \sqrt{\dfrac{3}{5}} + \sqrt{\dfrac{5}{3}}$

41. $\sqrt[3]{25} + \dfrac{3}{\sqrt[3]{5}}$

42. $\sqrt[4]{8} + \dfrac{1}{\sqrt[4]{2}}$

43. $\dfrac{\sqrt{45}}{10} + \sqrt{\dfrac{1}{5}} + \dfrac{\sqrt{5}}{5}$

44. $\sqrt[3]{49} + \dfrac{4}{\sqrt[3]{7}}$

45. Use a calculator to find a decimal approximation for $\sqrt{12}$ and for $2\sqrt{3}$.

46. Use a calculator to find decimal approximations for $\sqrt{50}$ and $5\sqrt{2}$.

47. Use a calculator to find a decimal approximation for $\sqrt{8} + \sqrt{18}$. Is it equal to the decimal approximation for $\sqrt{26}$ or $\sqrt{50}$?

48. Use a calculator to find a decimal approximation for $\sqrt{3} + \sqrt{12}$. Is it equal to the decimal approximation for $\sqrt{15}$ or $\sqrt{27}$?

Each of the following statements is false. Correct the right side of each one to make the statement true.

49. $3\sqrt{2x} + 5\sqrt{2x} = 8\sqrt{4x}$

50. $5\sqrt{3} - 7\sqrt{3} = -2\sqrt{9}$

51. $\sqrt{9 + 16} = 3 + 4$

52. $\sqrt{36 + 64} = 6 + 8$

Applying the Concepts

53. Golden Rectangle Construct a golden rectangle from a square of side 8. Then show that the ratio of the length to the width is the golden ratio $\frac{1+\sqrt{5}}{2}$.

54. Golden Rectangle Construct a golden rectangle from a square of side 10. Then show that the ratio of the length to the width is the golden ratio $\frac{1+\sqrt{5}}{2}$.

55. Golden Rectangle Use a ruler to construct a golden rectangle from a square of side 1 inch. Then show that the ratio of the length to the width is the golden ratio.

56. Golden Rectangle Use a ruler to construct a golden rectangle from a square of side $\frac{2}{3}$ inch. Then show that the ratio of the length to the width is the golden ratio.

57. Golden Rectangle To show that all golden rectangles have the same ratio of length to width, construct a golden rectangle from a square of side $2x$. Then show that the ratio of the length to the width is the golden ratio.

58. Golden Rectangle To show that all golden rectangles have the same ratio of length to width, construct a golden rectangle from a square of side x. Then show that the ratio of the length to the width is the golden ratio.

59. Isosceles Right Triangles A triangle is isosceles if it has two equal sides, and a triangle is a right triangle if it has a right angle in it. Sketch an isosceles right triangle, and find the ratio of the hypotenuse to a leg.

60. Equilateral Triangles A triangle is equilateral if it has three equal sides. The triangle in the figure is equilateral with each side of length $2x$. Find the ratio of the height to a side.

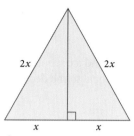

61. Pyramids The following solid is called a regular square pyramid because its base is a square and all eight edges are the same length, 5. It is also true that the vertex, V, is directly above the center of the base.

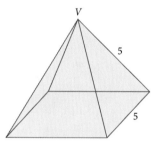

a. Find the ratio of a diagonal of the base to the length of a side.

b. Find the ratio of the area of the base to the diagonal of the base.

c. Find the ratio of the area of the base to the perimeter of the base.

62. Pyramids Refer to the diagram of a square pyramid below. Find the ratio of the height h of the pyramid to the altitude a.

Getting Ready for the Next Section

Simplify the following.

63. $3 \cdot 2$

64. $5 \cdot 7$

65. $(x + y)(4x - y)$

66. $(2x + y)(x - y)$

67. $(x + 3)^2$

68. $(3x - 2y)^2$

69. $(x - 2)(x + 2)$

70. $(2x + 5)(2x - 5)$

Simplify the following expressions.

71. $2\sqrt{18}$

72. $5\sqrt{36}$

73. $(\sqrt{6})^2$

74. $(\sqrt{2})^2$

75. $(3\sqrt{x})^2$

76. $(2\sqrt{y})^2$

77. $(4\sqrt{3x})^2$

78. $(-5\sqrt{2x})^2$

Rationalize the denominator.

79. $\dfrac{\sqrt{3}}{\sqrt{2}}$

80. $\dfrac{\sqrt{5}}{\sqrt{6}}$

81. $\dfrac{\sqrt{6}}{\sqrt{8}}$

82. $\dfrac{\sqrt{10}}{\sqrt{12}}$

Find the Mistake

Write true or false for each statement below. If false, circle what is false and write the correct answer on the line provided.

1. True or False: When adding or subtracting radical expressions, put each in simplified form and apply the product property for radicals. _____

2. True or False: We can only tell if radicals are similar if they are in simplified form. _____

3. True or False: When combining $3\sqrt{5} + 5\sqrt{3} - 2\sqrt{5}$, all three radicals are similar and can be combined to get $5\sqrt{5} - \sqrt{3}$. _____

4. True or False: Combining $4\sqrt{8} + 3\sqrt{32}$, we get $13\sqrt{2}$. _____

**Video Examples
Section 12.4**

Image © sxc.hu, cempey, 2006

The insects of a London park now have a new hotel in which to stay. The city held a unique competition that requested participants to build a structure that could house multiple varieties of bugs at once. A 1.5 meter tall box with a façade mimicking the multi-polygon design of a dragonfly's wing, also called voronoi, won first prize. The designers stacked 25 layers of 20mm thick birch plywood front to back, with each layer having the voronoi design. Each polygonal hole caters to a specific type of insect, and is loosely stuffed with recycled or organic materials. Beetles, bees, butterflies, spiders, and more are all encouraged to make their homes in this new Bug Hotel.

A pentagon is one of the shapes present in the Bug Hotel's façade. We can find the area A of a regular pentagon with a side length t by using the formula $A = \frac{t^2\sqrt{25 + 10\sqrt{5}}}{4}$. This formula uses a radical expression to find the pentagon's area.

In this section, we will look at multiplication and division of expressions like this one that contain radicals. As you will see, multiplication of expressions that contain radicals is very similar to multiplication of polynomials. The division problems in this section are just an extension of the work we did previously when we rationalized denominators.

A Multiplication of Radical Expressions

EXAMPLE 1 Multiply $(3\sqrt{5})(2\sqrt{7})$.

Solution We can rearrange the order and grouping of the numbers in this product by applying the commutative and associative properties. Following this, we apply the product property for radicals and multiply.

$$(3\sqrt{5})(2\sqrt{7}) = (3 \cdot 2)(\sqrt{5}\sqrt{7}) \qquad \text{Communicative and associative properties}$$
$$= (3 \cdot 2)(\sqrt{5 \cdot 7}) \qquad \text{Product property for radicals}$$
$$= 6\sqrt{35} \qquad \text{Multiply.}$$

In practice, it is not necessary to show the first two steps.

EXAMPLE 2 Multiply $\sqrt{3}(2\sqrt{6} - 5\sqrt{12})$.

Solution Applying the distributive property, we have
$$\sqrt{3}(2\sqrt{6} - 5\sqrt{12}) = \sqrt{3} \cdot 2\sqrt{6} - \sqrt{3} \cdot 5\sqrt{12}$$
$$= 2\sqrt{18} - 5\sqrt{36}$$

Writing each radical in simplified form gives
$$2\sqrt{18} - 5\sqrt{36} = 2\sqrt{9}\sqrt{2} - 5\sqrt{36}$$
$$= 6\sqrt{2} - 30$$

Practice Problems

1. Multiply $(7\sqrt{3})(5\sqrt{11})$.

2. Multiply $\sqrt{2}(3\sqrt{5} - 4\sqrt{2})$.

Answers

1. $35\sqrt{33}$
2. $3\sqrt{10} - 8$

EXAMPLE 3 Multiply $(\sqrt{3} + \sqrt{5})(4\sqrt{3} - \sqrt{5})$.

Solution The same principle that applies when multiplying two binomials applies to this product. We must multiply each term in the first expression by each term in the second one. Any convenient method can be used. Let's use the FOIL method.

$$\overset{}{(\sqrt{3} + \sqrt{5})(4\sqrt{3} - \sqrt{5})} = \overset{F}{\sqrt{3} \cdot 4\sqrt{3}} - \overset{O}{\sqrt{3} \cdot \sqrt{5}} + \overset{I}{\sqrt{5} \cdot 4\sqrt{3}} - \overset{L}{\sqrt{5} \cdot \sqrt{5}}$$

$$= 4 \cdot 3 - \sqrt{15} + 4\sqrt{15} - 5$$

$$= 12 + 3\sqrt{15} - 5$$

$$= 7 + 3\sqrt{15}$$

3. Multiply $(\sqrt{2} + \sqrt{7})(\sqrt{2} - 3\sqrt{7})$.

EXAMPLE 4 Expand and simplify $(\sqrt{x} + 3)^2$.

Solution 1 We can write this problem as a multiplication problem and proceed as we did in Example 3.

$$(\sqrt{x} + 3)^2 = (\sqrt{x} + 3)(\sqrt{x} + 3)$$

$$= \overset{F}{\sqrt{x} \cdot \sqrt{x}} + \overset{O}{3\sqrt{x}} + \overset{I}{3\sqrt{x}} + \overset{L}{3 \cdot 3}$$

$$= x + 3\sqrt{x} + 3\sqrt{x} + 9$$

$$= x + 6\sqrt{x} + 9$$

Solution 2 We can obtain the same result by applying the formula for the square of a sum: $(a + b)^2 = a^2 + 2ab + b^2$.

$$(\sqrt{x} + 3)^2 = (\sqrt{x})^2 + 2(\sqrt{x})(3) + 3^2$$

$$= x + 6\sqrt{x} + 9$$

4. Expand and simplify $(\sqrt{x} + 5)^2$.

EXAMPLE 5 Expand $(3\sqrt{x} - 2\sqrt{y})^2$ and simplify the result.

Solution Let's apply the formula for the square of a difference, $(a - b)^2 = a^2 - 2ab + b^2$.

$$(3\sqrt{x} - 2\sqrt{y})^2 = (3\sqrt{x})^2 - 2(3\sqrt{x})(2\sqrt{y}) + (2\sqrt{y})^2$$

$$= 9x - 12\sqrt{xy} + 4y$$

5. Expand $(5\sqrt{a} - 3\sqrt{b})^2$ and simplify the result.

EXAMPLE 6 Expand and simplify $(\sqrt{x + 2} - 1)^2$.

Solution Applying the formula $(a - b)^2 = a^2 - 2ab + b^2$, we have

$$(\sqrt{x + 2} - 1)^2 = (\sqrt{x + 2})^2 - 2\sqrt{x + 2}(1) + 1^2$$

$$= x + 2 - 2\sqrt{x + 2} + 1$$

$$= x + 3 - 2\sqrt{x + 2}$$

6. Expand and simplify $(\sqrt{x + 3} - 1)^2$.

EXAMPLE 7 Multiply $(\sqrt{6} + \sqrt{2})(\sqrt{6} - \sqrt{2})$.

Solution We notice the product is of the form $(a + b)(a - b)$, which always gives the difference of two squares, $a^2 - b^2$.

$$(\sqrt{6} + \sqrt{2})(\sqrt{6} - \sqrt{2}) = (\sqrt{6})^2 - (\sqrt{2})^2$$

$$= 6 - 2$$

$$= 4$$

7. Multiply $(\sqrt{5} + \sqrt{3})(\sqrt{5} - \sqrt{3})$.

Answers

3. $-19 - 2\sqrt{14}$
4. $x + 10\sqrt{x} + 25$
5. $25a - 30\sqrt{ab} + 9b$
6. $x + 4 - 2\sqrt{x + 3}$
7. 2

In Example 7, the two expressions $(\sqrt{6} + \sqrt{2})$ and $(\sqrt{6} - \sqrt{2})$ are called *conjugates*. In general, the conjugate of $\sqrt{a} + \sqrt{b}$ is $\sqrt{a} - \sqrt{b}$. If a and b are integers, multiplying conjugates of this form always produces a rational number. That is, if a and b are positive integers, then

$$(\sqrt{a} + \sqrt{b})(\sqrt{a} - \sqrt{b}) = \sqrt{a}\sqrt{a} - \sqrt{a}\sqrt{b} + \sqrt{a}\sqrt{b} - \sqrt{b}\sqrt{b}$$
$$= a - \sqrt{ab} + \sqrt{ab} - b$$
$$= a - b$$

which is rational if a and b are rational.

B Division of Radical Expressions

Division with radical expressions is the same as rationalizing the denominator. In Section 6.2, we were able to divide $\sqrt{3}$ by $\sqrt{2}$ by rationalizing the denominator:

$$\frac{\sqrt{3}}{\sqrt{2}} = \frac{\sqrt{3}}{\sqrt{2}} \cdot \frac{\sqrt{2}}{\sqrt{2}} = \frac{\sqrt{6}}{2}$$

We can accomplish the same result with expressions such as

$$\frac{6}{\sqrt{5} - \sqrt{3}}$$

by multiplying the numerator and denominator by the conjugate of the denominator.

8. Divide $\dfrac{3}{\sqrt{7} - \sqrt{3}}$.

EXAMPLE 8 Divide $\dfrac{6}{\sqrt{5} - \sqrt{3}}$. (Hint: rationalize the denominator.)

Solution Because the product of two conjugates is a rational number, we multiply the numerator and denominator by the conjugate of the denominator.

$$\frac{6}{\sqrt{5} - \sqrt{3}} = \frac{6}{\sqrt{5} - \sqrt{3}} \cdot \frac{(\sqrt{5} + \sqrt{3})}{(\sqrt{5} + \sqrt{3})}$$

$$= \frac{6\sqrt{5} + 6\sqrt{3}}{(\sqrt{5})^2 - (\sqrt{3})^2}$$

$$= \frac{6\sqrt{5} + 6\sqrt{3}}{5 - 3}$$

$$= \frac{6\sqrt{5} + 6\sqrt{3}}{2}$$

$$= \frac{2(3\sqrt{5} + 3\sqrt{3})}{2} \qquad \textit{Divide out the common factor, 2.}$$

$$= 3\sqrt{5} + 3\sqrt{3}$$

9. Rationalize the denominator.

$$\frac{\sqrt{10} - 3}{\sqrt{10} + 3}$$

EXAMPLE 9 Rationalize the denominator $\dfrac{\sqrt{5} - 2}{\sqrt{5} + 2}$.

Solution To rationalize the denominator, we multiply the numerator and denominator by the conjugate of the denominator.

$$\frac{\sqrt{5} - 2}{\sqrt{5} + 2} = \frac{\sqrt{5} - 2}{\sqrt{5} + 2} \cdot \frac{(\sqrt{5} - 2)}{(\sqrt{5} - 2)}$$

$$= \frac{5 - 2\sqrt{5} - 2\sqrt{5} + 4}{(\sqrt{5})^2 - 2^2}$$

$$= \frac{9 - 4\sqrt{5}}{5 - 4}$$

$$= \frac{9 - 4\sqrt{5}}{1}$$

$$= 9 - 4\sqrt{5}$$

Answers

8. $\frac{3\sqrt{7} + 3\sqrt{3}}{4}$

9. $19 - 6\sqrt{10}$

We have worked with the golden rectangle more than once in this chapter. Remember the following diagram, the general representation of a golden rectangle:

By now you know that, in any golden rectangle constructed from a square (of any size), the ratio of the length to the width will be

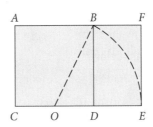

$$\frac{1 + \sqrt{5}}{2}$$

which we call the golden ratio. What is interesting is that the smaller rectangle on the right, *BFED*, is also a golden rectangle. We will now use the mathematics developed in this section to confirm this fact.

EXAMPLE 10 A golden rectangle constructed from a square of side 2 is shown in Figure 1. Show that the smaller rectangle *BDEF* is also a golden rectangle by finding the ratio of its length to its width.

Solution First, find expressions for the length and width of the smaller rectangle.

$$\text{Length} = EF = 2$$

$$\text{Width} = DE = \sqrt{5} - 1$$

Next, we find the ratio of length to width.

$$\text{Ratio of length to width} = \frac{EF}{DE} = \frac{2}{\sqrt{5} - 1}$$

FIGURE 1

To show that the small rectangle is a golden rectangle, we must show that the ratio of length to width is the golden ratio. We do so by rationalizing the denominator.

$$\frac{2}{\sqrt{5} - 1} = \frac{2}{\sqrt{5} - 1} \cdot \frac{\sqrt{5} + 1}{\sqrt{5} + 1}$$

$$= \frac{2(\sqrt{5} + 1)}{5 - 1}$$

$$= \frac{2(\sqrt{5} + 1)}{4}$$

$$= \frac{\sqrt{5} + 1}{2} \qquad \textit{Divide out common factor 2.}$$

Because addition is commutative, this last expression is the golden ratio. Therefore, the small rectangle in Figure 1 is a golden rectangle.

●

GETTING READY FOR CLASS

After reading through the preceding section, respond in your own words and in complete sentences.

A. Explain why $(\sqrt{5} + \sqrt{2})^2 \neq 5 + 2$.

B. Explain in words how you would rationalize the denominator in the expression $\frac{\sqrt{3}}{\sqrt{5} - \sqrt{2}}$.

C. What are conjugates?

D. What result is guaranteed when multiplying radical expressions that are conjugates?

10. If side *AC* in Figure 1 were 4 instead of 2, show that the smaller rectangle *BDEF* is also a golden rectangle by finding the ratio of its length to its width.

Answer

10. $\dfrac{4}{2\sqrt{5} - 2} = \dfrac{4}{2\sqrt{5} - 2} \cdot \dfrac{2\sqrt{5} + 2}{2\sqrt{5} + 2}$

$\qquad = \dfrac{8\sqrt{5} + 8}{4(5) - 4}$

$\qquad = \dfrac{8(\sqrt{5} + 1)}{16}$

$\qquad = \dfrac{\sqrt{5} + 1}{2}$

Vocabulary Review

Choose the correct words to fill in the blanks below.

denominator polynomials radical conjugate

1. Multiplication of radical expressions is very similar to multiplication of _____.

2. The expression $(\sqrt{a} - \sqrt{b})$ is the _____ of the expression $(\sqrt{a} + \sqrt{b})$.

3. Division with _____ expressions is the same as rationalizing the denominator.

4. We can rationalize a denominator that contains a radical expression by multiplying the numerator and denominator by the conjugate of the _____.

Problems

A Multiply. (Assume all expressions appearing under a square root symbol represent nonnegative numbers throughout this exercise set.)

1. $\sqrt{6} \cdot \sqrt{3}$

2. $\sqrt{6} \cdot \sqrt{2}$

3. $(2\sqrt{3})(5\sqrt{7})$

4. $(3\sqrt{5})(2\sqrt{7})$

5. $(4\sqrt{6})(2\sqrt{15})(3\sqrt{10})$

6. $(4\sqrt{35})(2\sqrt{21})(5\sqrt{15})$

7. $(3\sqrt[3]{3})(6\sqrt[3]{9})$

8. $(2\sqrt[3]{2})(6\sqrt[3]{4})$

9. $\sqrt{3}(\sqrt{2} - 3\sqrt{3})$

10. $\sqrt{2}(5\sqrt{3} + 4\sqrt{2})$

11. $6\sqrt[3]{4}(2\sqrt[3]{2} + 1)$

12. $7\sqrt[3]{5}(3\sqrt[3]{25} - 2)$

13. $(\sqrt{3} + \sqrt{2})(3\sqrt{3} - \sqrt{2})$

14. $(\sqrt{5} - \sqrt{2})(3\sqrt{5} + 2\sqrt{2})$

15. $(\sqrt{x} + 5)(\sqrt{x} - 3)$

16. $(\sqrt{x} + 4)(\sqrt{x} + 2)$

17. $(3\sqrt{6} + 4\sqrt{2})(\sqrt{6} + 2\sqrt{2})$

18. $(\sqrt{7} - 3\sqrt{3})(2\sqrt{7} - 4\sqrt{3})$

19. $(\sqrt{3} + 4)^2$

20. $(\sqrt{5} - 2)^2$

21. $(\sqrt{x} - 3)^2$

22. $(\sqrt{x} + 4)^2$

23. $(2\sqrt{a} - 3\sqrt{b})^2$

24. $(5\sqrt{a} - 2\sqrt{b})^2$

25. $(\sqrt{x-4} + 2)^2$

26. $(\sqrt{x-3} + 2)^2$

27. $(\sqrt{x-5} - 3)^2$

28. $(\sqrt{x-3} - 4)^2$

29. $(\sqrt{3} - \sqrt{2})(\sqrt{3} + \sqrt{2})$

30. $(\sqrt{5} - \sqrt{2})(\sqrt{5} + \sqrt{2})$

31. $(\sqrt{a} + 7)(\sqrt{a} - 7)$

32. $(\sqrt{a} + 5)(\sqrt{a} - 5)$

33. $(5 - \sqrt{x})(5 + \sqrt{x})$

34. $(3 - \sqrt{x})(3 + \sqrt{x})$

35. $(\sqrt{x-4} + 2)(\sqrt{x-4} - 2)$

36. $(\sqrt{x+3} + 5)(\sqrt{x+3} - 5)$

37. $(\sqrt{3} + 1)^3$

38. $(\sqrt{5} - 2)^3$

39. $(\sqrt{3} - 3)^3$

40. $(\sqrt{2} - 3)^3$

B Rationalize the denominator in each of the following.

41. $\dfrac{\sqrt{2}}{\sqrt{6} - \sqrt{2}}$

42. $\dfrac{\sqrt{5}}{\sqrt{5} + \sqrt{3}}$

43. $\dfrac{\sqrt{5}}{\sqrt{5} + 1}$

44. $\dfrac{\sqrt{7}}{\sqrt{7} - 1}$

45. $\dfrac{\sqrt{x}}{\sqrt{x} - 3}$

46. $\dfrac{\sqrt{x}}{\sqrt{x} + 2}$

47. $\dfrac{\sqrt{5}}{2\sqrt{5} - 3}$

48. $\dfrac{\sqrt{7}}{3\sqrt{7} - 2}$

49. $\dfrac{3}{\sqrt{x} - \sqrt{y}}$

50. $\dfrac{2}{\sqrt{x} + \sqrt{y}}$

51. $\dfrac{\sqrt{6} + \sqrt{2}}{\sqrt{6} - \sqrt{2}}$

52. $\dfrac{\sqrt{5} - \sqrt{3}}{\sqrt{5} + \sqrt{3}}$

53. $\dfrac{\sqrt{7} - 2}{\sqrt{7} + 2}$

54. $\dfrac{\sqrt{11} + 3}{\sqrt{11} - 3}$

55. $\dfrac{\sqrt{a} + \sqrt{b}}{\sqrt{a} - \sqrt{b}}$

56. $\dfrac{\sqrt{a} - \sqrt{b}}{\sqrt{a} + \sqrt{b}}$

57. $\dfrac{\sqrt{x} + 2}{\sqrt{x} - 2}$

58. $\dfrac{\sqrt{x} - 3}{\sqrt{x} + 3}$

59. $\dfrac{2\sqrt{3} - \sqrt{7}}{3\sqrt{3} + \sqrt{7}}$

60. $\dfrac{5\sqrt{6} + 2\sqrt{2}}{\sqrt{6} - \sqrt{2}}$

61. $\dfrac{3\sqrt{x} + 2}{1 + \sqrt{x}}$

62. $\dfrac{5\sqrt{x} - 1}{2 + \sqrt{x}}$

63. $\dfrac{2\sqrt{x} - 1}{1 - 3\sqrt{x}}$

64. $\dfrac{4\sqrt{x} - 3}{2\sqrt{x} + 3}$

65. Show that the product below is 5.

$$(\sqrt[3]{2} + \sqrt[3]{3})(\sqrt[3]{4} - \sqrt[3]{6} + \sqrt[3]{9})$$

66. Show that the product below is $x + 8$.

$$(\sqrt[3]{x} + 2)(\sqrt[3]{x^2} - 2\sqrt[3]{x} + 4)$$

Each of the following statements below is false. Correct the right side of each one to make it true.

67. $5(2\sqrt{3}) = 10\sqrt{15}$

68. $3(2\sqrt{x}) = 6\sqrt{3x}$

69. $(\sqrt{x} + 3)^2 = x + 9$

70. $(\sqrt{x} - 7)^2 = x - 49$

71. $(5\sqrt{3})^2 = 15$

72. $(3\sqrt{5})^2 = 15$

73. $(x + \sqrt{3})^2 = x^2 + 3$

74. $(\sqrt{x} - \sqrt{3})^2 = x - 3$

Applying the Concepts

75. Gravity If an object is dropped from the top of a 100-foot building, the amount of time t (in seconds) that it takes for the object to be h feet from the ground is given by the formula

$$t = \frac{\sqrt{100 - h}}{4}$$

How long does it take before the object is 50 feet from the ground? How long does it take to reach the ground? (When it is on the ground, h is 0.)

76. Gravity Use the formula given in Problem 73 to determine h if t is 1.25 seconds.

77. Golden Rectangle Rectangle $ACEF$ in Figure 2 is a golden rectangle. If side AC is 6 inches, show that the smaller rectangle $BDEF$ is also a golden rectangle.

78. Golden Rectangle Rectangle $ACEF$ in Figure 2 is a golden rectangle. If side AC is 1 inch, show that the smaller rectangle $BDEF$ is also a golden rectangle.

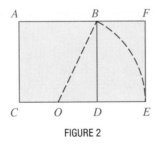

FIGURE 2

79. Golden Rectangle If side AC in Figure 2 is $2x$, show that rectangle $BDEF$ is a golden rectangle.

80. Golden Rectangle If side AC in Figure 2 is x, show that rectangle $BDEF$ is a golden rectangle.

Getting Ready for the Next Section

Simplify.

81. $(t + 5)^2$ **82.** $(x - 4)^2$ **83.** $\sqrt{x} \cdot \sqrt{x}$ **84.** $\sqrt{3x} \cdot \sqrt{3x}$

Solve.

85. $3x + 4 = 5^2$ **86.** $4x - 7 = 3^2$

87. $t^2 + 7t + 12 = 0$ **88.** $x^2 - 3x - 10 = 0$

89. $t^2 + 10t + 25 = t + 7$ **90.** $x^2 - 4x + 4 = x - 2$

91. $(x + 4)^2 = x + 6$ **92.** $(x - 6)^2 = x - 4$

93. Is 7 a solution to $\sqrt{3x + 4} = 5$? **94.** Is 4 a solution to $\sqrt{4x - 7} = -3$?

95. Is -6 a solution to $t + 5 = \sqrt{t + 7}$? **96.** Is -3 a solution to $t + 5 = \sqrt{t + 7}$?

Maintaining Your Skills

97. The chart shows where the most iPhones are being sold. If Apple has sold 50 million iPhones, how many iPhones are in North America? In Asia? In Africa?

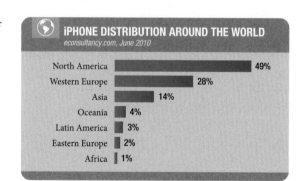

Find the Mistake

Each sentence below contains a mistake. Circle the mistake and write the correct word or expression on the line provided.

1. To multiply $\sqrt{5}(\sqrt{7} + 6\sqrt{5})$, we use the communicative property to get our answer $30 + \sqrt{35}$.

2. To multiply $(2\sqrt{2} + \sqrt{7})(4\sqrt{2} - 3\sqrt{7})$, we apply the associative property and regroup the terms to get our answer $(6\sqrt{2}) - (4\sqrt{7})$. _____

3. To divide 4 by $(2\sqrt{5} - 2)$, we rationalize the denominator by multiplying the numerator and the denominator by 4.

4. Dividing $\frac{(2\sqrt{3} - \sqrt{2})}{(4\sqrt{3} + \sqrt{2})}$ we get $\frac{(24 - 3\sqrt{6})}{44}$. _____

Landmark Review: Checking Your Progress

Find each of the following roots if possible.

1. $\sqrt{81}$

2. $-\sqrt{81}$

3. $\sqrt{-81}$

4. $\sqrt[3]{-27}$

Simplify each expression. Assume all variables represent nonnegative numbers.

5. $\sqrt[3]{x^6 y^{12}}$

6. $64^{1/2}$

7. $\left(\frac{16}{25}\right)^{-1/2}$

8. $81^{1/2} + 100^{1/2}$

Write each of the following expressions in simplified form.

9. $\sqrt{164}$

10. $\sqrt{18x^5 y^6}$

11. $\sqrt[5]{486x^{12} y^{10} z^{18}}$

12. $\frac{4}{\sqrt{5}}$

Perform the indicated operations.

13. $6\sqrt{150} - 4\sqrt{150} + 3\sqrt{216}$

14. $\frac{\sqrt{2}}{3} + \frac{1}{\sqrt{2}}$

15. $\sqrt{6} \cdot \sqrt{2}$

16. $(3\sqrt{5})(2\sqrt{10})(4\sqrt{8})$

17. $(2\sqrt{3} + 3\sqrt{2})(\sqrt{3} + 4\sqrt{2})$

18. $(\sqrt{5} - 1)^2$

19. $\frac{\sqrt{2}}{\sqrt{6} - \sqrt{2}}$

20. $\frac{\sqrt{5} + 3}{\sqrt{5} - 3}$

OBJECTIVES

A Solve equations containing radicals.

B Graph simple square root and cube root functions in two variables.

KEY WORDS

squaring property of equality

radical function

extraneous solution

radical equation

**Video Examples
Section 12.5**

Image © bewinca, 2007

The Three Gorges Dam stretches across the Yangtze River in China. Every year, the Chinese government spends more than a million dollars to remove tons of domestic garbage from the river. The garbage has the potential to block and jam the shipping locks in the dam. It may also damage boats and pollute the water. Approximately 150 million people live upstream from the dam. Unfortunately, it is common practice to allow household waste and agricultural pollutants to run off into the river. In this section, we will work more with radicals and solving equations that contain them. For instance, if a pollutant enters the river at a constant rate, its plume disperses down the river and may be represented by the equation $y = \sqrt{x}$. We will learn how to solve for the variables in this equation. The first step is to eliminate the radical. To do so, we need an additional property.

A Solving Equations with Radicals

> **PROPERTY** Squaring Property of Equality
>
> If both sides of an equation are squared, the solutions to the original equation are solutions to the resulting equation.

We will never lose solutions to our equations by squaring both sides. We may, however, introduce *extraneous solutions*. Extraneous solutions satisfy the equation obtained by squaring both sides of the original equation, but do not satisfy the original equation.

We know that if two real numbers a and b are equal, then so are their squares:

$$\text{If} \quad a = b$$
$$\text{then} \quad a^2 = b^2$$

On the other hand, extraneous solutions are introduced when we square opposites. That is, even though opposites are not equal, their squares are. For example,

$$5 = -5 \qquad \text{A false statement}$$
$$(5)^2 = (-5)^2 \qquad \text{Square both sides.}$$
$$25 = 25 \qquad \text{A true statement}$$

We are free to square both sides of an equation any time it is convenient. We must be aware, however, that doing so may introduce extraneous solutions. We must, therefore, check all our solutions in the original equation if at any time we square both sides of the original equation.

EXAMPLE 1 Solve $\sqrt{3x+4} = 5$.

Solution We square both sides and proceed as usual.

$$\sqrt{3x+4} = 5$$
$$(\sqrt{3x+4})^2 = 5^2$$
$$3x + 4 = 25$$
$$3x = 21$$
$$x = 7$$

Checking $x = 7$ in the original equation, we have

$$\sqrt{3(7)+4} \stackrel{?}{=} 5$$
$$\sqrt{21+4} \stackrel{?}{=} 5$$
$$\sqrt{25} \stackrel{?}{=} 5$$
$$5 = 5$$

The solution 7 satisfies the original equation.

1. Solve $\sqrt{2x+4} = 4$.

EXAMPLE 2 Solve $\sqrt{4x-7} = -3$.

Solution Squaring both sides, we have

$$\sqrt{4x-7} = -3$$
$$(\sqrt{4x-7})^2 = (-3)^2$$
$$4x - 7 = 9$$
$$4x = 16$$
$$x = 4$$

Checking $x = 4$ in the original equation gives

$$\sqrt{4(4)-7} \stackrel{?}{=} -3$$
$$\sqrt{16-7} \stackrel{?}{=} -3$$
$$\sqrt{9} \stackrel{?}{=} -3$$
$$3 = -3$$

The value 4 produces a false statement when checked in the original equation. Because $x = 4$ was the only possible solution, there is no solution to the original equation. The possible solution $x = 4$ is an extraneous solution. It satisfies the equation obtained by squaring both sides of the original equation, but does not satisfy the original equation.

2. Solve $\sqrt{7x-3} = -5$.

Note The fact that there is no solution to the equation in Example 2 was obvious to begin with. Notice that the left side of the equation is the positive square root of $4x - 7$, which must be a positive number or 0. The right side of the equation is -3. Because we cannot have a number that is either positive or zero equal to a negative number, there is no solution to the equation. It is always important to check our work just in case any extraneous solutions exist.

EXAMPLE 3 Solve $\sqrt{5x-1} + 3 = 7$.

Solution We must isolate the radical on the left side of the equation. If we attempt to square both sides without doing so, the resulting equation will also contain a radical. Adding -3 to both sides, we have

$$\sqrt{5x-1} + 3 = 7$$
$$\sqrt{5x-1} = 4$$

We can now square both sides and proceed as usual.

$$(\sqrt{5x-1})^2 = 4^2$$
$$5x - 1 = 16$$
$$5x = 17$$
$$x = \frac{17}{5}$$

3. Solve $\sqrt{4x+5} + 2 = 7$.

Answers
1. 6
2. No solution
3. 5

Checking $x = \dfrac{17}{5}$, we have

$$\sqrt{5\left(\dfrac{17}{5}\right) - 1} + 3 \overset{?}{=} 7$$

$$\sqrt{17 - 1} + 3 \overset{?}{=} 7$$

$$\sqrt{16} + 3 \overset{?}{=} 7$$

$$4 + 3 \overset{?}{=} 7$$

$$7 = 7$$

The solution is $\dfrac{17}{5}$.

4. Solve $t - 6 = \sqrt{t - 4}$.

EXAMPLE 4 Solve $t + 5 = \sqrt{t + 7}$.

Solution This time, squaring both sides of the equation results in a quadratic equation.

$$(t + 5)^2 = (\sqrt{t + 7})^2 \qquad \textit{Square both sides.}$$

$$t^2 + 10t + 25 = t + 7$$

$$t^2 + 9t + 18 = 0 \qquad \textit{Standard form}$$

$$(t + 3)(t + 6) = 0 \qquad \textit{Factor the left side.}$$

$$t + 3 = 0 \quad \text{or} \quad t + 6 = 0 \qquad \textit{Set factors equal to 0.}$$

$$t = -3 \qquad\qquad t = -6$$

We must check each solution in the original equation.

Check $t = -3$	Check $t = -6$
$-3 + 5 \overset{?}{=} \sqrt{-3 + 7}$	$-6 + 5 \overset{?}{=} \sqrt{-6 + 7}$
$2 \overset{?}{=} \sqrt{4}$	$-1 \overset{?}{=} \sqrt{1}$
$2 = 2$ *A true statement*	$-1 = 1$ *A false statement*

Because $t = -6$ does not check, our only solution is -3.

5. Solve $\sqrt{x - 9} = \sqrt{x} - 3$.

EXAMPLE 5 Solve $\sqrt{x - 3} = \sqrt{x} - 3$.

Solution We begin by squaring both sides. Note carefully what happens when we square the right side of the equation, and compare the square of the right side with the square of the left side. You must convince yourself that these results are correct. (The note in the margin will help if you are having trouble convincing yourself that what is written below is true.)

$$(\sqrt{x - 3})^2 = (\sqrt{x} - 3)^2$$

$$x - 3 = x - 6\sqrt{x} + 9$$

Now we still have a radical in our equation, so we will have to square both sides again. Before we do, though, let's isolate the remaining radical.

$$x - 3 = x - 6\sqrt{x} + 9$$

$$-3 = -6\sqrt{x} + 9 \qquad \textit{Add } -x \textit{ to each side.}$$

$$-12 = -6\sqrt{x} \qquad \textit{Add } -9 \textit{ to each side.}$$

$$2 = \sqrt{x} \qquad \textit{Divide each side by } -6.$$

$$4 = x \qquad \textit{Square each side.}$$

Note It is very important that you realize that the square of $(\sqrt{x} - 3)$ is not $x + 9$. Remember, when we square a difference with two terms, we use the formula

$$(a - b)^2 = a^2 - 2ab + b^2$$

Applying this formula to $(\sqrt{x} - 3)^2$, we have

$$(\sqrt{x} - 3)^2 =$$
$$(\sqrt{x})^2 - 2(\sqrt{x})(3) + 3^2$$
$$= x - 6\sqrt{x} + 9$$

Answers
4. 8
5. 9

Our only possible solution is $x = 4$, which we check in our original equation as follows:

$$\sqrt{4 - 3} \stackrel{?}{=} \sqrt{4} - 3$$
$$\sqrt{1} \stackrel{?}{=} 2 - 3$$
$$1 = -1 \qquad \text{A false statement}$$

Substituting 4 for x in the original equation yields a false statement. Because 4 was our only possible solution, there is no solution to our equation.

Here is another example of an equation for which we must apply our squaring property twice before all radicals are eliminated.

EXAMPLE 6 Solve $\sqrt{x + 1} = 1 - \sqrt{2x}$.

6. Solve $\sqrt{x + 4} = 2 - \sqrt{3x}$.

Solution This equation has two separate terms involving radical signs.
 Squaring both sides gives

$$x + 1 = 1 - 2\sqrt{2x} + 2x$$
$$-x = -2\sqrt{2x} \qquad \text{Add } -2x \text{ and } -1 \text{ to both sides.}$$
$$x^2 = 4(2x) \qquad \text{Square both sides.}$$
$$x^2 - 8x = 0 \qquad \text{Standard form}$$

Our equation is a quadratic equation in standard form. To solve for x, we factor the left side and set each factor equal to 0.

$$x(x - 8) = 0 \qquad \text{Factor left side.}$$
$$x = 0 \quad \text{or} \quad x - 8 = 0 \qquad \text{Set factors equal to 0.}$$
$$x = 8$$

Because we squared both sides of our equation, we have the possibility that one or both of the solutions are extraneous. We must check each one in the original equation.

Check $x = 8$ Check $x = 0$

$$\sqrt{8 + 1} \stackrel{?}{=} 1 - \sqrt{2 \cdot 8} \qquad\qquad \sqrt{0 + 1} \stackrel{?}{=} 1 - \sqrt{2 \cdot 0}$$
$$\sqrt{9} \stackrel{?}{=} 1 - \sqrt{16} \qquad\qquad \sqrt{1} \stackrel{?}{=} 1 - \sqrt{0}$$
$$3 \stackrel{?}{=} 1 - 4 \qquad\qquad 1 \stackrel{?}{=} 1 - 0$$
$$3 = -3 \quad \text{A false statement} \qquad 1 = 1 \quad \text{A true statement}$$

Because 8 does not check, it is an extraneous solution. Our only solution is 0.

EXAMPLE 7 Solve $\sqrt{x + 1} = \sqrt{x + 2} - 1$.

7. Solve $\sqrt{x + 2} = \sqrt{x + 3} - 1$.

Solution Squaring both sides we have

$$(\sqrt{x + 1})^2 = (\sqrt{x + 2} - 1)^2$$
$$x + 1 = x + 2 - 2\sqrt{x + 2} + 1$$

Once again, we are left with a radical in our equation. Before we square each side again, we must isolate the radical on the right side of the equation.

$$x + 1 = x + 3 - 2\sqrt{x + 2} \qquad \text{Simplify the right side.}$$
$$1 = 3 - 2\sqrt{x + 2} \qquad \text{Add } -x \text{ to each side.}$$
$$-2 = -2\sqrt{x + 2} \qquad \text{Add } -3 \text{ to each side.}$$
$$1 = \sqrt{x + 2} \qquad \text{Divide each side by } -2.$$
$$1 = x + 2 \qquad \text{Square both sides.}$$
$$-1 = x \qquad \text{Add } -2 \text{ to each side.}$$

Answers
6. 0
7. -2

Checking our only possible solution, -1, in our original equation, we have

$$\sqrt{-1 + 1} \stackrel{?}{=} \sqrt{-1 + 2} - 1$$
$$\sqrt{0} \stackrel{?}{=} \sqrt{1} - 1$$
$$0 \stackrel{?}{=} 1 - 1$$
$$0 = 0 \qquad \text{A true statement}$$

Our solution checks.

8. Solve
$\sqrt{5x - 4} - \sqrt{x + 5} = \sqrt{x - 3}$.

EXAMPLE 8 Solve $\sqrt{2x - 10} = \sqrt{x - 4} + \sqrt{x - 12}$.

Solution We start by squaring both sides of the equation.

$$(\sqrt{2x - 10})^2 = (\sqrt{x - 4} + \sqrt{x - 12})^2$$
$$2x - 10 = x - 4 + 2\sqrt{x^2 - 16x + 48} + x - 12 \quad (a + b)^2 = a^2 + 2ab + b^2$$
$$2x - 10 = 2x - 16 + 2\sqrt{x^2 - 16x + 48} \qquad \text{Combine like terms.}$$

Then we need to square both sides again and put the equation in standard form.

$$6 = 2\sqrt{x^2 - 16x + 48} \qquad \text{Add } -2x \text{ and 16 to both sides.}$$
$$3 = \sqrt{x^2 - 16x + 48} \qquad \text{Divide both sides by 2.}$$
$$3^2 = (\sqrt{x^2 - 16x + 48})^2 \qquad \text{Square both sides.}$$
$$9 = x^2 - 16x + 48$$
$$x^2 - 16x + 39 = 0 \qquad \text{Put the equation in standard form.}$$
$$(x - 3)(x - 13) = 0 \qquad \text{Factor.}$$
$$x - 3 = 0 \quad \text{or} \quad x - 13 = 0 \qquad \text{Zero factor property}$$
$$x = 3 \quad \text{or} \quad x = 13$$

Now we must check our answers in the original equation.

$$\sqrt{2(3) - 10} \stackrel{?}{=} \sqrt{3 - 4} + \sqrt{3 - 12} \qquad \text{Check } x = 3.$$
$$\sqrt{6 - 10} \stackrel{?}{=} \sqrt{-1} + \sqrt{-9}$$

Since the square root of a negative number is not a real number, $x = 3$ is not a solution.

$$\sqrt{2(13) - 10} \stackrel{?}{=} \sqrt{13 - 4} + \sqrt{13 - 12} \qquad \text{Check } x = 13.$$
$$\sqrt{16} \stackrel{?}{=} \sqrt{9} + \sqrt{1}$$
$$4 = 3 + 1$$

So $x = 13$ is the only real solution.

It is also possible to raise both sides of an equation to powers greater than 2. We only need to check for extraneous solutions when we raise both sides of an equation to an even power. Raising both sides of an equation to an odd power will not produce extraneous solutions.

9. Solve $\sqrt[3]{3x - 7} = 2$.

EXAMPLE 9 Solve $\sqrt[3]{4x + 5} = 3$.

Solution Cubing both sides, we have

$$(\sqrt[3]{4x + 5})^3 = 3^3$$
$$4x + 5 = 27$$
$$4x = 22$$
$$x = \frac{22}{4}$$
$$x = \frac{11}{2}$$

We do not need to check $x = \frac{11}{2}$ because we raised both sides to an odd power.

Answers

8. 4
9. 5

B Graphing Radical Functions

We end this section by looking at graphs of some functions that contain radicals.

EXAMPLE 10 Graph $y = \sqrt{x}$ and $y = \sqrt[3]{x}$.

Solution The graphs are shown in Figures 1 and 2. Notice that aside from beginning at the origin, the graph of $y = \sqrt{x}$ appears in the first quadrant because in the equation $y = \sqrt{x}$, x and y cannot be negative.

The graph of $y = \sqrt[3]{x}$ appears in Quadrants 1 and 3 because the cube root of a positive number is also a positive number, and the cube root of a negative number is a negative number. That is, when x is positive, y will be positive, and when x is negative, y will be negative.

The graphs of both equations will contain the origin, because $y = 0$ when $x = 0$ in both equations.

x	y
−4	Undefined
−1	Undefined
0	0
1	1
4	2
9	3
16	4

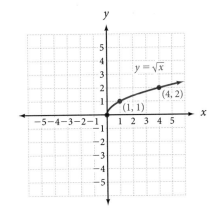

FIGURE 1

x	y
−27	−3
−8	−2
−1	−1
0	0
1	1
8	2
27	3

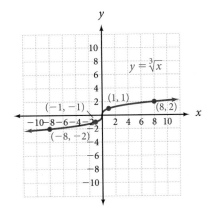

FIGURE 2

10. Graph $y = \sqrt{x} + 3$ and $y = \sqrt{x + 3}$.

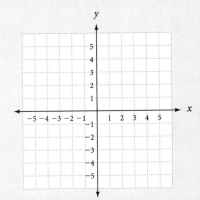

GETTING READY FOR CLASS

After reading through the preceding section, respond in your own words and in complete sentences.

A. What is the squaring property of equality?

B. Under what conditions do we obtain extraneous solutions to equations that contain radical expressions?

C. If we have raised both sides of an equation to a power, when is it not necessary to check for extraneous solutions?

D. Why does the graph of $y = \sqrt{x}$ appear in the first quadrant only?

Vocabulary Review

Choose the correct words to fill in the blanks below.

negative solutions original extraneous radical

1. The first step to solving a radical equation is to eliminate the _____.

2. The squaring property of equality states that if both sides of an equation are squared, the solutions to the original equation are _____ to the resulting equation.

3. _____ solutions satisfy the equation obtained by squaring both sides of the original equation, but do not satisfy the original equation.

4. It is very important to check any solutions in the _____ equation to avoid extraneous solutions.

5. The graph of $y = \sqrt{x}$ appears in the first quadrant only because x and y cannot be _____.

Problems

A Solve each of the following equations.

1. $\sqrt{2x + 1} = 3$

2. $\sqrt{3x + 1} = 4$

3. $\sqrt{4x + 1} = -5$

4. $\sqrt{6x + 1} = -5$

5. $\sqrt{2y - 1} = 3$

6. $\sqrt{3y - 1} = 2$

7. $\sqrt{5x - 7} = -1$

8. $\sqrt{8x + 3} = -6$

9. $\sqrt{2x - 3} - 2 = 4$

10. $\sqrt{3x + 1} - 4 = 1$

11. $\sqrt{4a + 1} + 3 = 2$

12. $\sqrt{5a - 3} + 6 = 2$

13. $\sqrt[4]{3x + 1} = 2$

14. $\sqrt[4]{4x + 1} = 3$

15. $\sqrt[3]{2x - 5} = 1$

16. $\sqrt[3]{5x + 7} = 2$

17. $\sqrt[3]{3a + 5} = -3$

18. $\sqrt[3]{2a + 7} = -2$

19. $\sqrt{y-3}=y-3$

20. $\sqrt{y+3}=y-3$

21. $\sqrt{a+2}=a+2$

22. $\sqrt{a+10}=a-2$

23. $\sqrt{2x+4}=\sqrt{1-x}$

24. $\sqrt{3x+4}=-\sqrt{2x+3}$

25. $\sqrt{4a+7}=-\sqrt{a+2}$

26. $\sqrt{7a-1}=\sqrt{2a+4}$

27. $\sqrt[4]{5x-8}=\sqrt[4]{4x-1}$

28. $\sqrt[4]{6x+7}=\sqrt[4]{x+2}$

29. $x+1=\sqrt{5x+1}$

30. $x-1=\sqrt{6x+1}$

31. $t+5=\sqrt{2t+9}$

32. $t+7=\sqrt{2t+13}$

33. $\sqrt{y-8}=\sqrt{8-y}$

34. $\sqrt{2y+5}=\sqrt{5y+2}$

35. $\sqrt[3]{3x+5}=\sqrt[3]{5-2x}$

36. $\sqrt[3]{4x+9}=\sqrt[3]{3-2x}$

The following equations will require that you square both sides twice before all the radicals are eliminated. Solve each equation using the methods shown in Examples 5, 6, 7, and 8.

37. $\sqrt{x-8}=\sqrt{x}-2$

38. $\sqrt{x+3}=\sqrt{x}-3$

39. $\sqrt{x+1}=\sqrt{x}+1$

40. $\sqrt{x-1}=\sqrt{x}-1$

41. $\sqrt{x+8}=\sqrt{x-4}+2$

42. $\sqrt{x+5}=\sqrt{x-3}+2$

43. $\sqrt{x-5} - 3 = \sqrt{x-8}$

44. $\sqrt{x-3} - 4 = \sqrt{x-3}$

45. $\sqrt{x+4} = 2 - \sqrt{2x}$

46. $\sqrt{5x+1} = 1 + \sqrt{5x}$

47. $\sqrt{2x+4} = \sqrt{x+3} + 1$

48. $\sqrt{2x-1} = \sqrt{x-4} + 2$

49. $\sqrt{2x-5} + \sqrt{3x-5} = \sqrt{2x+3}$

50. $\sqrt{3x+5} + \sqrt{3x-3} = \sqrt{4x+4}$

51. $\sqrt{2a-2} + \sqrt{4a+3} = \sqrt{2a+5}$

52. $\sqrt{2a-3} + \sqrt{a+4} = \sqrt{4a+5}$

53. $\sqrt{4x-5} - \sqrt{x-5} = \sqrt{3x-4}$

54. $\sqrt{4x+1} - \sqrt{2x+4} = \sqrt{x-5}$

Applying the Concepts

55. Solving a Formula Solve the following formula for h:

$$t = \frac{\sqrt{100-h}}{4}$$

56. Solving a Formula Solve the following formula for h:

$$t = \sqrt{\frac{2h - 40t}{g}}$$

57. Pendulum Clock The length of time T in seconds it takes the pendulum of a clock to swing through one complete cycle is given by the formula

$$T = 2\pi\sqrt{\frac{L}{32}}$$

where L is the length, in feet, of the pendulum, and π is approximately $\frac{22}{7}$. How long must the pendulum be if one complete cycle takes 2 seconds?

58. Pendulum Clock Solve the formula in Problem 51 for L.

59. Similar Rectangles Two rectangles are similar if their vertices lie along the same diagonal, as shown in the following diagram.

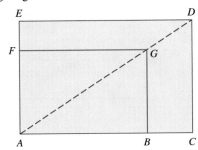

If two rectangles are similar, then corresponding sides are in proportion, which means that $\frac{ED}{DC} = \frac{FG}{GB}$. Use these facts in the following diagram to express the length of the larger rectangle l in terms of x.

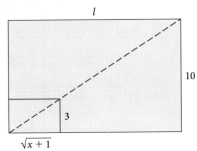

60. Volume Recall that the volume of a box V can be found from the formula $V = (\text{length})(\text{width})(\text{height})$. Find the volume of the following box in terms of x.

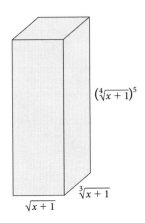

Pollution A long straight river, 100 meters wide, is flowing at 1 meter per second. A pollutant is entering the river at a constant rate from one of its banks. As the pollutant disperses in the water, it forms a plume that is modeled by the equation $y = \sqrt{x}$. Use this information to answer the following questions.

Pollution plume

61. How wide is the plume 25 meters down river from the source of the pollution?

62. How wide is the plume 100 meters down river from the source of the pollution?

63. How far down river from the source of the pollution does the plume reach halfway across the river?

64. How far down the river from the source of the pollution does the plume reach the other side of the river?

65. For the situation described in the instructions and modeled by the equation $y = \sqrt{x}$, what is the range of values that y can assume?

66. If the river was moving at 2 meters per second, would the plume be larger or smaller 100 meters downstream from the source?

B Graph each equation.

67. $y = 2\sqrt{x}$

68. $y = -2\sqrt{x}$

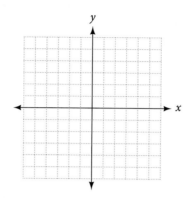

69. $y = \sqrt{x} - 2$

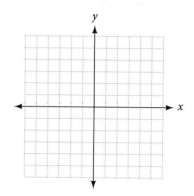

70. $y = \sqrt{x} + 2$

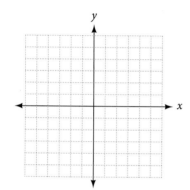

71. $y = \sqrt{x - 2}$

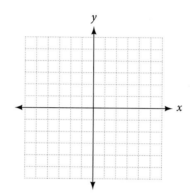

72. $y = \sqrt{x + 2}$

73. $y = 3\sqrt[3]{x}$

74. $y = -3\sqrt[3]{x}$

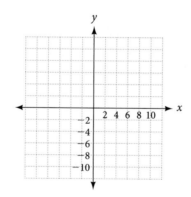

75. $y = \sqrt[3]{x} + 3$

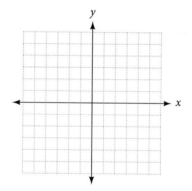

76. $y = \sqrt[3]{x} - 3$

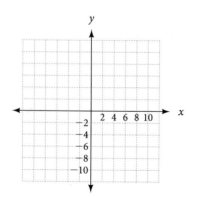

77. $y = \sqrt[3]{x + 3}$

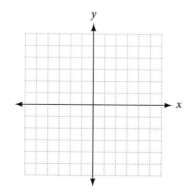

78. $y = \sqrt[3]{x - 3}$

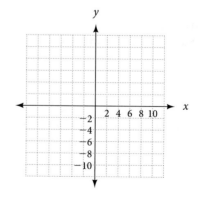

Getting Ready for the Next Section

Simplify.

79. $\sqrt{25}$ **80.** $\sqrt{49}$ **81.** $\sqrt{12}$ **82.** $\sqrt{50}$

83. $(-1)^{15}$ **84.** $(-1)^{20}$ **85.** $(-1)^{50}$ **86.** $(-1)^5$

Solve each equation.

87. $3x = 12$ **88.** $4 = 8y$ **89.** $4x - 3 = 5$ **90.** $7 = 2y - 1$

Perform the indicated operation.

91. $(3 + 4x) + (7 - 6x)$ **92.** $(2 - 5x) + (-1 + 7x)$ **93.** $(7 + 3x) - (5 + 6x)$ **94.** $(5 - 2x) - (9 - 4x)$

95. $(3 - 4x)(2 + 5x)$ **96.** $(8 + x)(7 - 3x)$ **97.** $2x(4 - 6x)$ **98.** $3x(7 + 2x)$

99. $(2 + 3x)^2$ **100.** $(3 + 5x)^2$ **101.** $(2 - 3x)(2 + 3x)$ **102.** $(4 - 5x)(4 + 5x)$

Find the Mistake

Each sentence below contains a mistake. Circle the mistake and write the correct word or phrase on the line provided.

1. Solving $\sqrt{6x + 1} = -5$ gives us a solution of $x = 4$. _____

2. Solving $\sqrt[4]{3x - 8} = 2$ gives us an extraneous solution. _____

3. Solving $\sqrt{x + 8} = x + 6$ gives a solution of $x = -7$. _____

4. The graphs of $y = \sqrt{x - 1}$ and $y = \sqrt{2x - 6}$ will intersect at $(-5, 2)$. _____

12.6 Complex Numbers

©iStockphoto.com/phuxy

In 2010, a magnitude 7.2 earthquake struck Baja California. Residents in the capital city of Mexicali reportedly felt intense shaking that lasted for nearly a minute and a half. Widespread soil liquefaction damaged much of the city's buildings and infrastructure. Earthquake liquefaction occurs when shaking from a seismic event turns the once solid ground to a consistency similar to quicksand. Buildings and bridges can literally sink into the ground! Some civil engineers study soil liquefaction using a special group of numbers called complex numbers, which is the focus of this section.

The equation $x^2 = -9$ has no real number solutions because the square of a real number is always positive. We have been unable to work with square roots of negative numbers like $\sqrt{-25}$ and $\sqrt{-16}$ for the same reason. Complex numbers allow us to expand our work with radicals to include square roots of negative numbers and to solve equations like $x^2 = -9$ and $x^2 = -64$. Our work with complex numbers is based on the following definition.

A Square Roots of Negative Numbers

> **DEFINITION** the number *i*
>
> The **number i,** called the *imaginary unit*, is such that $i = \sqrt{-1}$ (which is the same as saying $i^2 = -1$).

The number *i*, as we have defined it here, is not a real number. Because of the way we have defined *i*, we can use it to simplify square roots of negative numbers.

> **PROPERTY** Square Roots of Negative Numbers
>
> If *a* is a positive number, then $\sqrt{-a}$ can always be written as $i\sqrt{a}$. That is,
>
> $$\sqrt{-a} = i\sqrt{a} \qquad \text{if } a \text{ is a positive number}$$

To justify our rule, we simply square the quantity $i\sqrt{a}$ to obtain $-a$. Here is what it looks like when we do so:

$$(i\sqrt{a})^2 = i^2 \cdot (\sqrt{a})^2$$
$$= -1 \cdot a$$
$$= -a$$

Here are some examples that illustrate the use of our new rule:

EXAMPLE 1 Write each square root in terms of the number i.

a. $\sqrt{-25} = i\sqrt{25} = i \cdot 5 = 5i$

b. $\sqrt{-49} = i\sqrt{49} = i \cdot 7 = 7i$

c. $\sqrt{-12} = i\sqrt{12} = i \cdot 2\sqrt{3} = 2i\sqrt{3}$

d. $\sqrt{-17} = i\sqrt{17}$

B Powers of i

If we assume all the properties of exponents hold when the base is i, we can write any power of i as i, -1, $-i$, or 1. Using the fact that $i^2 = -1$, we have

$$i^1 = i$$

$$i^2 = -1$$

$$i^3 = i^2 \cdot i = -1(i) = -i$$

$$i^4 = i^2 \cdot i^2 = -1(-1) = 1$$

Because $i^4 = 1$, i^5 will simplify to i, and we will begin repeating the sequence i, -1, $-i$, 1 as we simplify higher powers of i: Any power of i simplifies to i, -1, $-i$, or 1. The easiest way to simplify higher powers of i is to write them in terms of i^2. For instance, to simplify i^{21}, we would write it as

$$(i^2)^{10} \cdot i \quad\quad \text{because} \quad\quad 2 \cdot 10 + 1 = 21$$

Then, because $i^2 = -1$, we have

$$(-1)^{10} \cdot i = 1 \cdot i = i$$

EXAMPLE 2 Simplify as much as possible.

a. $i^{30} = (i^2)^{15} = (-1)^{15} = -1$

b. $i^{11} = (i^2)^5 \cdot i = (-1)^5 \cdot i = (-1)i = -i$

c. $i^{40} = (i^2)^{20} = (-1)^{20} = 1$

DEFINITION complex number

A *complex number* is any number that can be put in the form

$$a + bi$$

where a and b are real numbers and $i = \sqrt{-1}$. The form $a + bi$ is called *standard form* for complex numbers. The number a is called the *real part* of the complex number. The number b is called the *imaginary part* of the complex number.

Every real number is a complex number. For example, 8 can be written as $8 + 0i$. Likewise, $-\frac{1}{2}$, π, $\sqrt{3}$, and 29 are complex numbers because they can all be written in the form $a + bi$.

$$-\frac{1}{2} = -\frac{1}{2} + 0i \quad\quad \pi = \pi + 0i \quad\quad \sqrt{3} = \sqrt{3} + 0i \quad\quad -9 = -9 + 0i$$

The real numbers occur when $b = 0$. When $b \neq 0$, we have complex numbers that contain i, such as $2 + 5i$, $6 - i$, $4i$, and $\frac{1}{2}i$. These numbers are called *imaginary numbers*. The following diagram explains this further.

Practice Problems

1. Write each square root in terms of the number i.

 a. $\sqrt{-36}$

 b. $-\sqrt{-64}$

 c. $\sqrt{-18}$

 d. $-\sqrt{-19}$

Note In Example 1 parts c and d, we wrote i before the radical simply to avoid confusion. If we were to write the answer to 3 as $2\sqrt{3}i$, some people would think the i was under the radical sign, but it is not.

2. Simplify.

 a. i^{20}

 b. i^{23}

 c. i^{50}

Answers

1. a. $6i$ **b.** $-8i$ **c.** $3i\sqrt{2}$

 d. $i\sqrt{19}$

2. a. 1 **b.** $-i$ **c.** -1

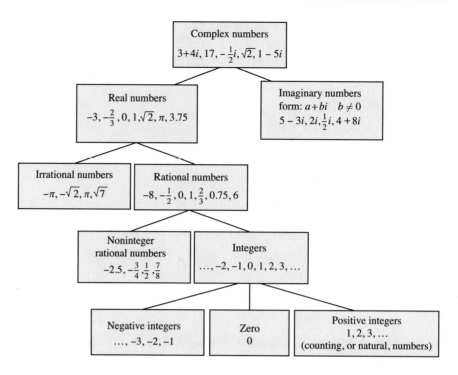

C Equality for Complex Numbers

Two complex numbers are equal if and only if their real parts are equal and their imaginary parts are equal. That is, for real numbers a, b, c, and d,

$$a + bi = c + di \quad \text{if and only if} \quad a = c \quad \text{and} \quad b = d$$

3. Find x and y if
$$4x + 7i = 8 - 14yi.$$

EXAMPLE 3 Find x and y if $3x + 4i = 12 - 8yi$.

Solution Because the two complex numbers are equal, their real parts are equal and their imaginary parts are equal.

$$3x = 12 \quad \text{and} \quad 4 = -8y$$
$$x = 4 \qquad\qquad y = -\frac{1}{2}$$

4. Find x and y if
$$(2x - 1) + 9i = 5 + (4y + 1)i.$$

EXAMPLE 4 Find x and y if $(4x - 3) + 7i = 5 + (2y - 1)i$.

Solution The real parts are $4x - 3$ and 5. The imaginary parts are 7 and $2y - 1$.

$$4x - 3 = 5 \quad \text{and} \quad 7 = 2y - 1$$
$$4x = 8 \qquad\qquad 8 = 2y$$
$$x = 2 \qquad\qquad y = 4$$

D Addition and Subtraction of Complex Numbers

To add two complex numbers, add their real parts and their imaginary parts. That is, if a, b, c, and d are real numbers, then

$$(a + bi) + (c + di) = (a + c) + (b + d)i$$

If we assume that the commutative, associative, and distributive properties hold for the number i, then the definition of addition is simply an extension of these properties.

We define subtraction in a similar manner. If a, b, c, and d are real numbers, then

$$(a + bi) - (c + di) = (a - c) + (b - d)i$$

Answers

3. $x = 2$, $y = -\frac{1}{2}$

4. $x = 3$, $y = 2$

EXAMPLE 5 Add or subtract as indicated.

a. $(3 + 4i) + (7 - 6i) = (3 + 7) + (4 - 6)i = 10 - 2i$

b. $(7 + 3i) - (5 + 6i) = (7 - 5) + (3 - 6)i = 2 - 3i$

c. $(5 - 2i) - (9 - 4i) = (5 - 9) + (-2 + 4)i = -4 + 2i$

E Multiplication of Complex Numbers

Because complex numbers have the same form as binomials, we find the product of two complex numbers the same way we find the product of two binomials.

EXAMPLE 6 Multiply $(3 - 4i)(2 + 5i)$.

Solution Multiplying each term in the second complex number by each term in the first, we have

$$
\begin{array}{cccc}
\mathsf{F} & \mathsf{O} & \mathsf{I} & \mathsf{L}
\end{array}
$$
$$(3 - 4i)(2 + 5i) = 3 \cdot 2 + 3 \cdot 5i - 2 \cdot 4i - 4i(5i)$$
$$= 6 + 15i - 8i - 20i^2$$

Combining similar terms and using the fact that $i^2 = -1$, we can simplify as follows:

$$6 + 15i - 8i - 20i^2 = 6 + 7i - 20(-1)$$
$$= 6 + 7i + 20$$
$$= 26 + 7i$$

The product of the complex numbers $3 - 4i$ and $2 + 5i$ is the complex number $26 + 7i$.

EXAMPLE 7 Multiply $2i(4 - 6i)$.

Solution Applying the distributive property gives us

$$2i(4 - 6i) = 2i \cdot 4 - 2i \cdot 6i$$
$$= 8i - 12i^2$$
$$= 12 + 8i$$

EXAMPLE 8 Expand $(3 + 5i)^2$.

Solution We treat this like the square of a binomial. Remember, $(a + b)^2 = a^2 + 2ab + b^2$.

$$(3 + 5i)^2 = 3^2 + 2(3)(5i) + (5i)^2$$
$$= 9 + 30i + 25i^2$$
$$= 9 + 30i - 25$$
$$= -16 + 30i$$

EXAMPLE 9 Multiply $(2 - 3i)(2 + 3i)$.

Solution This product has the form $(a - b)(a + b)$, which we know results in the difference of two squares, $a^2 - b^2$.

$$(2 - 3i)(2 + 3i) = 2^2 - (3i)^2$$
$$= 4 - 9i^2$$
$$= 4 + 9$$
$$= 13$$

The product of the two complex numbers $2 - 3i$ and $2 + 3i$ is the real number 13. The two complex numbers $2 - 3i$ and $2 + 3i$ are called complex conjugates. The fact that their product is a real number is very useful.

5. Add or subtract as indicated.
 a. $(2 + 6i) + (3 - 4i)$
 b. $(6 + 5i) - (4 + 3i)$
 c. $(7 - i) - (8 - 2i)$

6. Multiply $(2 + 3i)(1 - 4i)$.

7. Multiply $-3i(2 + 3i)$.

8. Expand $(2 + 4i)^2$.

9. Multiply $(3 - 5i)(3 + 5i)$.

Answers

5. **a.** $5 + 2i$ **b.** $2 + 2i$ **c.** $-1 + i$
6. $14 - 5i$
7. $9 - 6i$
8. $-12 + 16i$
9. 34

10. Divide $\dfrac{3 + 2i}{2 - 5i}$.

DEFINITION complex conjugates

The complex numbers $a + bi$ and $a - bi$ are called *complex conjugates*. One important property they have is that their product is the real number $a^2 + b^2$. Here's why:

$$(a + bi)(a - bi) = a^2 - (bi)^2$$
$$= a^2 - b^2 i^2$$
$$= a^2 - b^2(-1)$$
$$= a^2 + b^2$$

F Division With Complex Numbers

The fact that the product of two complex conjugates is a real number is the key to division with complex numbers.

EXAMPLE 10 Divide $\dfrac{2 + i}{3 - 2i}$.

Solution We want a complex number in standard form that is equivalent to the quotient $\frac{2+i}{3-2i}$. We need to eliminate i from the denominator. Multiplying the numerator and denominator by $3 + 2i$ will give us what we want.

$$\frac{2 + i}{3 - 2i} = \frac{2 + i}{3 - 2i} \cdot \frac{(3 + 2i)}{(3 + 2i)}$$
$$= \frac{6 + 4i + 3i + 2i^2}{9 - 4i^2}$$
$$= \frac{6 + 7i - 2}{9 + 4}$$
$$= \frac{4 + 7i}{13}$$
$$= \frac{4}{13} + \frac{7}{13}i$$

Dividing the complex number $2 + i$ by $3 - 2i$ gives the complex number $\frac{4}{13} + \frac{7}{13}i$.

11. Divide $\dfrac{3 + 2i}{i}$.

EXAMPLE 11 Divide $\dfrac{7 - 4i}{i}$.

Solution The conjugate of the denominator is $-i$. Multiplying numerator and denominator by this number, we have

$$\frac{7 - 4i}{i} = \frac{7 - 4i}{i} \cdot \frac{-i}{-i}$$
$$= \frac{-7i + 4i^2}{-i^2}$$
$$= \frac{-7i + 4(-1)}{-(-1)}$$
$$= -4 - 7i$$

GETTING READY FOR CLASS

After reading through the preceding section, respond in your own words and in complete sentences.

A. What is the number i?

B. Explain why every real number is a complex number.

C. What kind of number will always result when we multiply complex conjugates?

D. Explain how to divide complex numbers.

Answers

10. $-\dfrac{4}{29} + \dfrac{19}{29}i$

11. $2 - 3i$

Vocabulary Review

Choose the correct words to fill in the blanks below.

zero equal complex number number i conjugates imaginary

1. The _____ is such that $i = \sqrt{-1}$, which is not a real number.

2. A _____ is any number than can be put in the form $a + bi$.

3. Every real number is a complex number because the variable b in the standard form for complex numbers
 is _____ .

4. Two complex numbers are equal if and only if their real parts are _____ and their imaginary parts
 are equal.

5. To add two complex numbers, add their real parts and their _____ parts, and assume that the
 commutative, associative, and distributive properties hold true for the number i.

6. The complex numbers $a + bi$ and $a - bi$ are complex _____ , and have a product that is the real
 number $a^2 + b^2$.

Problems

A Write the following in terms of i, and simplify as much as possible.

1. $\sqrt{-36}$

2. $\sqrt{-49}$

3. $-\sqrt{-25}$

4. $-\sqrt{-81}$

5. $\sqrt{-72}$

6. $\sqrt{-48}$

7. $-\sqrt{-12}$

8. $-\sqrt{-75}$

B Write each of the following as i, -1, $-i$, or 1.

9. i^{28}

10. i^{31}

11. i^{26}

12. i^{37}

13. i^{75}

14. i^{42}

15. $(-i)^{12}$

16. $(-i)^{22}$

C Find x and y so each of the following equations is true.

17. $2x + 3yi = 6 - 3i$

18. $4x - 2yi = 4 + 8i$

19. $2 - 5i = -x + 10yi$

20. $4 + 7i = 6x - 14yi$

21. $2x + 10i = -16 - 2yi$

22. $4x - 5i = -2 + 3yi$

23. $(2x - 4) - 3i = 10 - 6yi$

24. $(4x - 3) - 2i = 8 + yi$

25. $(7x - 1) + 4i = 2 + (5y + 2)i$

26. $(5x + 2) - 7i = 4 + (2y + 1)i$

D Combine the following complex numbers.

27. $(2 + 3i) + (3 + 6i)$

28. $(4 + i) + (3 + 2i)$

29. $(3 - 5i) + (2 + 4i)$

30. $(7 + 2i) + (3 - 4i)$

31. $(5 + 2i) - (3 + 6i)$

32. $(6 + 7i) - (4 + i)$

33. $(3 - 5i) - (2 + i)$

34. $(7 - 3i) - (4 + 10i)$

35. $[(3 + 2i) - (6 + i)] + (5 + i)$

36. $[(4 - 5i) - (2 + i)] + (2 + 5i)$

37. $[(7 - i) - (2 + 4i)] - (6 + 2i)$

38. $[(3 - i) - (4 + 7i)] - (3 - 4i)$

39. $(3 + 2i) - [(3 - 4i) - (6 + 2i)]$

40. $(7 - 4i) - [(-2 + i) - (3 + 7i)]$

41. $(4 - 9i) + [(2 - 7i) - (4 + 8i)]$

42. $(10 - 2i) - [(2 + i) - (3 - i)]$

E Find the following products.

43. $3i(4 + 5i)$

44. $2i(3 + 4i)$

45. $6i(4 - 3i)$

46. $11i(2 - i)$

47. $(3 + 2i)(4 + i)$

48. $(2 - 4i)(3 + i)$

49. $(4 + 9i)(3 - i)$

50. $(5 - 2i)(1 + i)$

51. $(1 + i)^3$

52. $(1 - i)^3$

53. $(2 - i)^3$

54. $(2 + i)^3$

55. $(2 + 5i)^2$

56. $(3 + 2i)^2$

57. $(1 - i)^2$

58. $(1 + i)^2$

59. $(3 - 4i)^2$

60. $(6 - 5i)^2$

61. $(2 + i)(2 - i)$

62. $(3 + i)(3 - i)$

63. $(6 - 2i)(6 + 2i)$

64. $(5 + 4i)(5 - 4i)$

65. $(2 + 3i)(2 - 3i)$

66. $(2 - 7i)(2 + 7i)$

67. $(10 + 8i)(10 - 8i)$

68. $(11 - 7i)(11 + 7i)$

69. $(\sqrt{3} + 2i)(\sqrt{3} - 2i)$

70. $(\sqrt{7} + \sqrt{5}i)(\sqrt{7} - \sqrt{5}i)$

F Find the following quotients. Write all answers in standard form for complex numbers.

71. $\dfrac{2 - 3i}{i}$

72. $\dfrac{3 + 4i}{i}$

73. $\dfrac{5 + 2i}{-i}$

74. $\dfrac{4 - 3i}{-i}$

75. $\dfrac{4}{2 - 3i}$

76. $\dfrac{3}{4 - 5i}$

77. $\dfrac{6}{-3 + 2i}$

78. $\dfrac{-1}{-2 - 5i}$

79. $\dfrac{2 + 3i}{2 - 3i}$

80. $\dfrac{4 - 7i}{4 + 7i}$

81. $\dfrac{5 + 4i}{3 + 6i}$

82. $\dfrac{2 + i}{5 - 6i}$

Applying the Concepts

83. Electric Circuits Complex numbers may be applied to electrical circuits. Electrical engineers use the fact that resistance R, electrical current I and voltage V are related by the formula $V = RI$. (Voltage is measured in volts, resistance in ohms, and current in amperes.) Find the resistance to electrical flow in a circuit that has a voltage $V = (80 + 20i)$ volts and current $I = (-6 + 2i)$ amps.

84. Electric Circuits Refer to the information about electrical circuits in Problem 83, and find the current in a circuit that has a resistance of $(4 + 10i)$ ohms and a voltage of $(5 - 7i)$ volts.

Maintaining Your Skills.

Solve each equation.

85. $\dfrac{t}{3} - \dfrac{1}{2} = -1$

86. $\dfrac{x}{x-2} + \dfrac{2}{3} = \dfrac{2}{x-2}$

87. $2 + \dfrac{5}{y} = \dfrac{3}{y^2}$

88. $1 - \dfrac{1}{y} = \dfrac{12}{y^2}$

Solve each application problem.

89. The sum of a number and its reciprocal is $\dfrac{41}{20}$. Find the number.

90. It takes an inlet pipe 8 hours to fill a tank. The drain can empty the tank in 6 hours. If the tank is full and both the inlet pipe and drain are open, how long will it take to drain the tank?

91. The chart shows the average movie ticket price from 2000 to 2009. Find the slope of the line that connects the first and last point. Then, use it to predict the cost of a movie ticket in 2015. Round to the nearest cent.

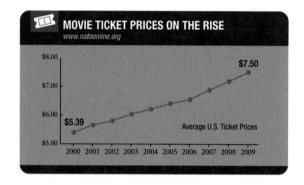

Find the Mistake

Each sentence below contains a mistake. Circle the mistake and write the correct word or phrase on the line provided.

1. The number i is such that $i = \sqrt{-1}$ is the same as saying $i^2 = \dfrac{1}{i}$. _____

2. The form $a + bi$ is called standard form for complex numbers and the i is the imaginary part of the number.

3. The product of the complex numbers $(5 - 3i)$ and $(2 + 4i)$ is $48i$. _____

4. The product of two complex conjugates $6 + 4i$ and $6 - 4i$ is $6^2 - 4^2$._____

Chapter **12** Summary

The numbers in brackets refer to the section(s) in which the topic can be found.

Square Roots [12.1]

Every positive real number x has two square roots. The *positive square root* of x is written \sqrt{x}, and the *negative square root* of x is written $-\sqrt{x}$. Both the positive and the negative square roots of x are numbers we square to get x; that is,

$$\left.\begin{array}{l} (\sqrt{x})^2 = x \\ \text{and} \quad (-\sqrt{x})^2 = x \end{array}\right\} \text{ for } x \geq 0$$

EXAMPLES

1. The number 49 has two square roots, 7 and -7. They are written like this:

$$\sqrt{49} = 7 \qquad -\sqrt{49} = -7$$

General Roots [12.1]

In the expression $\sqrt[n]{a}$, n is the *index*, a is the *radicand*, and $\sqrt{}$ is the *radical sign*. The expression $\sqrt[n]{a}$ is such that

$$(\sqrt[n]{a})^n = a \qquad a \geq 0 \text{ when } n \text{ is even}$$

2. $\sqrt[3]{8} = 2$
$\sqrt[3]{-27} = -3$

Rational Exponents [12.1]

Rational exponents are used to indicate roots. The relationship between rational exponents and roots is as follows:

$$a^{1/n} = \sqrt[n]{a} \qquad \text{and} \qquad a^{m/n} = (a^{1/n})^m = (a^m)^{1/n}$$

$$a \geq 0 \text{ when } n \text{ is even}$$

3. $25^{1/2} = \sqrt{25} = 5$
$8^{2/3} = (\sqrt[3]{8})^2 = 2^2 = 4$
$9^{3/2} = (\sqrt{9})^3 = 3^3 = 27$

Properties of Radicals [12.2]

If a and b are nonnegative real numbers whenever n is even, then

1. $\sqrt[n]{ab} = \sqrt[n]{a}\,\sqrt[n]{b}$ *Product property for radicals*

2. $\sqrt[n]{\dfrac{a}{b}} = \dfrac{\sqrt[n]{a}}{\sqrt[n]{b}}$ $(b \neq 0)$ *Quotient property for radicals*

4. $\sqrt{4 \cdot 5} = \sqrt{4} \cdot \sqrt{5} = 2\sqrt{5}$

$\sqrt{\dfrac{7}{9}} = \dfrac{\sqrt{7}}{\sqrt{9}} = \dfrac{\sqrt{7}}{3}$

Simplifying Radicals [12.2]

A radical expression is said to be in *simplified form* if

1. there is no factor of the radicand that can be written as a power greater than or equal to the index;

2. there are no fractions under the radical sign; and

3. there are no radicals in the denominator.

5. $\sqrt{\dfrac{4}{5}} = \dfrac{\sqrt{4}}{\sqrt{5}}$

$= \dfrac{2}{\sqrt{5}} \cdot \dfrac{\sqrt{5}}{\sqrt{5}}$

$= \dfrac{2\sqrt{5}}{5}$

6. $5\sqrt{3} - 7\sqrt{3} = (5 - 7)\sqrt{3}$
$= -2\sqrt{3}$

$\sqrt{20} + \sqrt{45} = 2\sqrt{5} + 3\sqrt{5}$
$= (2 + 3)\sqrt{5}$
$= 5\sqrt{5}$

Addition and Subtraction of Radical Expressions [12.3]

We add and subtract radical expressions by using the distributive property to combine similar radicals. Similar radicals are radicals with the same index and the same radicand.

Multiplication of Radical Expressions [12.4]

7. $(\sqrt{x} + 2)(\sqrt{x} + 3)$
$= \sqrt{x} \cdot \sqrt{x} + 3\sqrt{x} + 2\sqrt{x} + 2 \cdot 3$
$= x + 5\sqrt{x} + 6$

We multiply radical expressions in the same way that we multiply polynomials. We can use the distributive property and the FOIL method.

Rationalizing the Denominator [12.2, 12.4]

8. $\dfrac{3}{\sqrt{2}} = \dfrac{3}{\sqrt{2}} \cdot \dfrac{\sqrt{2}}{\sqrt{2}} = \dfrac{3\sqrt{2}}{2}$

$\dfrac{3}{\sqrt{5} - \sqrt{3}} = \dfrac{3}{\sqrt{5} - \sqrt{3}} \cdot \dfrac{\sqrt{5} + \sqrt{3}}{\sqrt{5} + \sqrt{3}}$

$= \dfrac{3\sqrt{5} + 3\sqrt{3}}{5 - 3}$

$= \dfrac{3\sqrt{5} + 3\sqrt{3}}{2}$

When a fraction contains a square root in the denominator, we rationalize the denominator by multiplying the numerator and denominator by

1. The square root itself if there is only one term in the denominator, or

2. The conjugate of the denominator if there are two terms in the denominator.

Rationalizing the denominator is also called division of radical expressions.

Squaring Property of Equality [12.5]

9. $\sqrt{2x + 1} = 3$
$(\sqrt{2x + 1})^2 = 3^2$
$2x + 1 = 9$
$x = 4$

We may square both sides of an equation any time it is convenient to do so, as long as we check all resulting solutions in the original equation.

Complex Numbers [12.6]

10. $3 + 4i$ is a complex number.

Addition
$(3 + 4i) + (2 - 5i) = 5 - i$

Multiplication
$(3 + 4i)(2 - 5i)$
$= 6 - 15i + 8i - 20i^2$
$= 6 - 7i + 20$
$= 26 - 7i$

Division
$\dfrac{2}{3 + 4i} = \dfrac{2}{3 + 4i} \cdot \dfrac{3 - 4i}{3 - 4i}$

$= \dfrac{6 - 8i}{9 + 16}$

$= \dfrac{6}{25} - \dfrac{8}{25}i$

A *complex number* is any number that can be put in the form

$$a + bi$$

where a and b are real numbers and $i = \sqrt{-1}$. The *real part* of the complex number is a, and b is the *imaginary part*.

 If a, b, c, and d are real numbers, then we have the following definitions associated with complex numbers:

1. *Equality*

$$a + bi = c + di \quad \text{if and only if} \quad a = c \text{ and } b = d$$

2. *Addition and subtraction*

$$(a + bi) + (c + di) = (a + c) + (b + d)i$$
$$(a + bi) - (c + di) = (a - c) + (b - d)i$$

3. *Multiplication*

$$(a + bi)(c + di) = (ac - bd) + (ad + bc)i$$

4. *Division is similar to rationalizing the denominator.*

Simplify each of the following. (Assume all variable bases are positive integers and all variable exponents are positive real numbers throughout this test.) [12.1]

1. $16^{-3/4}$

2. $\left(\dfrac{81}{64}\right)^{-1/2}$

3. $x^{1/7} \cdot x^{-2/3}$

4. $\dfrac{(a^{1/3}b^2)^{-1}}{(a^{2/5}b^{-5})^{-2}}$

5. $\sqrt{25x^6y^{20}}$

6. $\sqrt[4]{81x^4y^{20}}$

7. $\dfrac{(8a^{12}b^3)^{1/3}}{(49a^{12}b^8)^{1/2}}$

8. $\dfrac{(x^{1/n}y^n)^{n^2}}{(x^ny^{1/n})^{n^2}}$

Write in simplified form. [12.2]

9. $\sqrt{27x^5y^3}$

10. $\sqrt[3]{128x^2y^7}$

11. $\sqrt{\dfrac{3}{5}}$

12. $\sqrt{\dfrac{8a^2b^7}{7c}}$

Rationalize the denominator. [12.2]

13. $\dfrac{7}{\sqrt{2}+1}$

14. $\dfrac{\sqrt{x}+\sqrt{3}}{\sqrt{x}-\sqrt{3}}$

Combine. [12.3]

15. $\dfrac{5}{x^{1/2}} - x^{1/2}$

16. $\dfrac{x^3}{(x^3-17)^{1/2}} - (x^3-17)^{1/2}$

17. $3\sqrt{8} - 2\sqrt{18}$

18. $4\sqrt[3]{32a^6b^3} - 5a^2\sqrt[3]{4b^3}$

Multiply. [12.4]

19. $4a^{3/2}(a^{5/2} - 3a^{1/2})$

20. $(2a^{5/2} - 3)^2$

21. $(\sqrt{x} - 1)(\sqrt{x} + 5)$

22. $(2\sqrt{3} - \sqrt{2})^2$

Solve for x. [12.5]

23. $\sqrt{2x-3} = x - 3$

24. $\sqrt[3]{4x-2} = -2$

25. $\sqrt{x+6} = \sqrt{x-9} + 3$

Graph. [12.5]

26. $y = \sqrt{x+1}$

27. $y = \sqrt[3]{x} - 2$

Match each equation with its graph. [12.5]

28. $y = \sqrt{x}$

29. $y = \sqrt{x-1}$

30. $y = \sqrt{x} + 1$

31. $y = \sqrt[3]{x} + 1$

32. $y = \sqrt[3]{x-1}$

33. $y = -\sqrt[3]{x+1}$

A. **B.**

C. **D.**

E. **F.**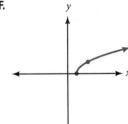

34. Find x and y so that the following equation is true. [12.6]

$$8 - (x-2)i = (7y+1) + 4i$$

Perform the indicated operations. [12.6]

35. $(2+i) - [(5-2i) - (3+4i)]$

36. $(5+3i)(2-i)$

37. $(3-6i)^2$

38. $\dfrac{3i+2}{3i-2}$

39. Show that i^{36} can be written as 1. [12.6]

Quadratic Equations and Functions

13

13.1 Completing the Square

13.2 The Quadratic Formula

K nown as the "Gateway to the West," the St. Louis Gateway Arch is one of the most recognizable structures in the United States. As the tallest monument in the country, it stands 630 feet tall and spans 630 feet across its base. The cross-sections of its legs are isosceles triangles and are hollow to accommodate a unique tram system. Visitors ride these trams up the inside of the arch's legs to an enclosed observation deck at the top. Nearly a million visitors ride the trams each year. The trams have been in operation for over 30 years, traveling a total of 250,000 miles and carrying over 25 million passengers.

While the architect used what is called a catenary curve in the arch's design, the shape of the arch can be closely modeled by the quadratic equation $y = -\frac{1}{150}x^2 + \frac{21}{5}x$. The equation can be graphed on a rectangular coordinate system, and the result is called a parabola, shown below.

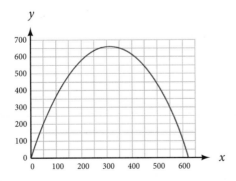

The parabola is one of the topics covered in this chapter. You will also be asked to sketch this graph in one of the problem sets.

OBJECTIVES

A Solve quadratic equations by taking the square root of both sides.

B Solve quadratic equations by completing the square.

C Use quadratic equations to solve for missing parts of right triangles.

KEY WORDS

square root property

completing the square

Video Examples Section 13.1

Table 1 is taken from the trail map given to skiers at the Northstar at Tahoe Ski Resort in Lake Tahoe, California. The table gives the length of each chair lift at Northstar, along with the change in elevation from the beginning of the lift to the end of the lift.

Table 1	Lift Information from the Trail Map for the Northstar at Tahoe Ski Resort	
Lift	Vertical Rise (feet)	Length (feet)
Big Springs Gondola	480	4,100
Bear Paw Double	120	790
Echo Triple	710	4,890
Aspen Express Quad	900	5,100
Forest Double	1,170	5,750
Lookout Double	960	4,330
Comstock Express Quad	1,250	5,900
Rendezvous Triple	650	2,900
Schaffer Camp Triple	1,860	6,150
Chipmunk Tow Lift	28	280
Bear Cub Tow Lift	120	750

Right triangles are good mathematical models for chair lifts. Can you picture why? In this section, we will use our knowledge of right triangles, along with the new material developed in the section, to solve problems involving chair lifts and a variety of other examples.

A The Square Root Method

In this section, we will develop the first of our new methods of solving quadratic equations. Remember, a quadratic equation is an equation in the form $ax^2 + bx + c = 0$, where a, b, c are constants and a is not 0. The new method is called *completing the square*. Completing the square on a quadratic equation allows us to obtain solutions, regardless of whether the equation can be factored. Before we solve equations by completing the square, we need to learn how to solve equations by taking square roots of both sides.

Consider the following equation:

$$x^2 = 16$$

We could solve the equation $x^2 = 16$ by writing it in standard form and factoring the left side. We can shorten our work considerably, however, if we simply notice that x must be either the positive square root of 16 or the negative square root of 16. That is,

$$\text{If} \qquad x^2 = 16$$

$$\text{then} \qquad x = \sqrt{16} \quad \text{or} \quad x = -\sqrt{16}$$

$$x = 4 \qquad\qquad x = -4$$

We can generalize this result as follows:

> **PROPERTY** **Square Root Property for Equations**
> If $a^2 = b$, where b is a real number, then $a = \sqrt{b}$ or $a = -\sqrt{b}$.

Notation The expression $a = b$ or $a = -b$ can be written in shorthand form as $a = \pm b$. The symbol \pm is read "plus or minus."

We can apply the square root property for equations to some fairly complicated quadratic equations.

EXAMPLE 1 Solve $(2x - 3)^2 = 25$.

Solution

$$(2x - 3)^2 = 25$$

$$2x - 3 = \pm\sqrt{25} \qquad \text{Square root property for equations}$$

$$2x - 3 = \pm 5 \qquad \sqrt{25} = 5$$

$$2x = 3 \pm 5 \qquad \text{Add 3 to both sides.}$$

$$x = \frac{3 \pm 5}{2} \qquad \text{Divide both sides by 2.}$$

The last equation can be written as two separate statements:

$$x = \frac{3 + 5}{2} \quad \text{or} \quad x = \frac{3 - 5}{2}$$

$$= \frac{8}{2} \qquad\qquad = \frac{-2}{2}$$

$$= 4 \qquad\qquad = -1$$

The solutions are $4, -1$.

Notice that we could have solved the equation in Example 1 by expanding the left side, writing the resulting equation in standard form, and then factoring. The problem would look like this:

$$(2x - 3)^2 = 25 \qquad \text{Original equation}$$

$$4x^2 - 12x + 9 = 25 \qquad \text{Expand the left side.}$$

$$4x^2 - 12x - 16 = 0 \qquad \text{Add } -25 \text{ to each side.}$$

$$4(x^2 - 3x - 4) = 0 \qquad \text{Begin factoring.}$$

$$4(x - 4)(x + 1) = 0 \qquad \text{Factor completely.}$$

$$x - 4 = 0 \quad \text{or} \quad x + 1 = 0 \qquad \text{Set variable factors equal to 0.}$$

$$x = 4 \qquad\qquad x = -1$$

Notice that solving the equation by factoring leads to the same two solutions.

Practice Problems

1. Solve $(3x + 2)^2 = 16$.

Answers

1. $\frac{2}{3}, -2$

2. Solve $(4x - 3)^2 = -50$.

> **Note** We cannot solve the equation in Example 2 by factoring. If we expand the left side and write the resulting equation in standard form, we are left with a quadratic equation that does not factor.
>
> $$(3x - 1)^2 = -12$$
>
> Equation from Example 2
>
> $$9x^2 - 6x + 1 = -12$$
>
> Expand the left side.
>
> $$9x^2 - 6x + 13 = 0$$
>
> Standard form, but not factorable.

3. Solve $x^2 + 10x + 25 = 20$.

EXAMPLE 2 Solve $(3x - 1)^2 = -12$ for x.

Solution

$$(3x - 1)^2 = -12$$
$$3x - 1 = \pm\sqrt{-12} \qquad \text{Square root property}$$
$$3x - 1 = \pm 2i\sqrt{3} \qquad \sqrt{-12} = 2i\sqrt{3}$$
$$3x = 1 \pm 2i\sqrt{3} \qquad \text{Add 1 to both sides.}$$
$$x = \frac{1 \pm 2i\sqrt{3}}{3} \qquad \text{Divide both sides by 3.}$$

The solutions are $\dfrac{1 + 2i\sqrt{3}}{3}$, $\dfrac{1 - 2i\sqrt{3}}{3}$. We can combine these as $\dfrac{1 \pm 2i\sqrt{3}}{3}$.

Both solutions are complex. Here is a check of the first solution:

$$\text{When} \rightarrow \qquad\qquad\qquad x = \frac{1 + 2i\sqrt{3}}{3}$$
$$\text{the equation} \rightarrow \qquad\qquad (3x - 1)^2 = -12$$
$$\text{becomes} \rightarrow \qquad \left(3 \cdot \frac{1 + 2i\sqrt{3}}{3} - 1\right)^2 \overset{?}{=} -12$$
$$(1 + 2i\sqrt{3} - 1)^2 \overset{?}{=} -12$$
$$(2i\sqrt{3})^2 \overset{?}{=} -12$$
$$4 \cdot i^2 \cdot 3 \overset{?}{=} -12$$
$$12(-1) \overset{?}{=} -12$$
$$-12 = -12$$

EXAMPLE 3 Solve $x^2 + 6x + 9 = 12$.

Solution We can solve this equation as we have the equations in Examples 1 and 2 if we first write the left side as $(x + 3)^2$.

$$x^2 + 6x + 9 = 12 \qquad \text{Original equation}$$
$$(x + 3)^2 = 12 \qquad \text{Write } x^2 + 6x + 9 \text{ as } (x + 3)^2.$$
$$x + 3 = \pm 2\sqrt{3} \qquad \text{Square root property}$$
$$x = -3 \pm 2\sqrt{3} \qquad \text{Add } -3 \text{ to each side.}$$

We have two irrational solutions: $-3 + 2\sqrt{3}$ and $-3 - 2\sqrt{3}$. What is important about this problem, however, is the fact that the equation was easy to solve because the left side was a perfect square trinomial.

B Completing the Square

The method of completing the square is simply a way of transforming any quadratic equation into an equation of the form found in the preceding three examples. The key to understanding the method of completing the square lies in recognizing the relationship between the last two terms of any perfect square trinomial whose leading coefficient is 1.

Consider the following list of perfect square trinomials and their corresponding binomial squares:

$$x^2 - 6x + 9 = (x - 3)^2$$
$$x^2 + 8x + 16 = (x + 4)^2$$
$$x^2 - 10x + 25 = (x - 5)^2$$
$$x^2 + 12x + 36 = (x + 6)^2$$

Answers

2. $\dfrac{3 \pm 5i\sqrt{2}}{4}$

3. $-5 \pm 2\sqrt{5}$

In each case, the leading coefficient is 1. A more important observation comes from noticing the relationship between the linear and constant terms (middle and last terms) in each trinomial. Observe that the constant term in each case is the square of half the coefficient of x in the middle term. For example, in the last expression, the constant term 36 is the square of half of 12, where 12 is the coefficient of x in the middle term. (Notice also that the second terms in all the binomials on the right side are half the coefficients of the middle terms of the trinomials on the left side.) We can use these observations to build our own perfect square trinomials and, in doing so, solve some quadratic equations.

EXAMPLE 4 Solve $x^2 - 6x + 5 = 0$ by completing the square.

Solution We begin by adding -5 to both sides of the equation. We want just $x^2 - 6x$ on the left side so that we can add on our own final term to get a perfect square trinomial.

$$x^2 - 6x + 5 = 0$$

$$x^2 - 6x \quad\quad = -5 \qquad \text{Add } -5 \text{ to both sides.}$$

Now we can add 9 to both sides and the left side will be a perfect square.

$$x^2 - 6x + 9 = -5 + 9$$

$$(x - 3)^2 = 4$$

The final line is in the form of the equations we solved previously.

$$x - 3 = \pm 2$$

$$x = 3 \pm 2 \qquad \text{Add 3 to both sides.}$$

$$x = 3 + 2 \quad \text{or} \quad x = 3 - 2$$

$$x = 5 \qquad\qquad x = 1$$

The two solutions are 5 and 1.

4. Solve $x^2 + 3x - 4 = 0$ by completing the square.

Note The equation in Example 4 can be solved quickly by factoring.

$$x^2 - 6x + 5 = 0$$
$$(x - 5)(x - 1) = 0$$
$$x - 5 = 0 \quad \text{or} \quad x - 1 = 0$$
$$x = 5 \qquad\qquad x = 1$$

The reason we didn't solve it by factoring is we want to practice completing the square on some simple equations.

EXAMPLE 5 Solve $x^2 + 5x - 2 = 0$ by completing the square.

Solution We must begin by adding 2 to both sides. (The left side of the equation, as it is, is not a perfect square, because it does not have the correct constant term. We will simply "move" that term to the other side and use our own constant term.)

$$x^2 + 5x = 2 \qquad \text{Add 2 to each side.}$$

We complete the square by adding the square of half the coefficient of the linear term to both sides.

$$x^2 + 5x + \frac{25}{4} = 2 + \frac{25}{4} \qquad \text{Half of 5 is } \frac{5}{2}, \text{ the square of which}$$
$$\text{is } \left(\frac{5}{2}\right)^2 = \frac{25}{4}.$$

$$\left(x + \frac{5}{2}\right)^2 = \frac{33}{4} \qquad 2 + \frac{25}{4} = \frac{8}{4} + \frac{25}{4} = \frac{33}{4}$$

$$x + \frac{5}{2} = \pm\sqrt{\frac{33}{4}} \qquad \text{Square root property}$$

$$x + \frac{5}{2} = \pm\frac{\sqrt{33}}{2} \qquad \text{Simplify the radical.}$$

$$x = -\frac{5}{2} \pm \frac{\sqrt{33}}{2} \qquad \text{Add } -\frac{5}{2} \text{ to both sides.}$$

$$x = \frac{-5 \pm \sqrt{33}}{2}$$

The solutions are $\dfrac{-5 \pm \sqrt{33}}{2}$.

5. Solve $x^2 - 5x + 2 = 0$ by completing the square.

Answers

4. $1, -4$

5. $\dfrac{5 \pm \sqrt{17}}{2}$

We can use a calculator to get decimal approximations to these solutions. If $\sqrt{33} \approx 5.74$, then

$$\frac{-5 + 5.74}{2} = 0.37$$

$$\frac{-5 - 5.74}{2} = -5.37$$

6. Solve $5x^2 - 3x + 2 = 0$.

EXAMPLE 6 Solve $3x^2 - 8x + 7 = 0$.

Solution

$$3x^2 - 8x + 7 = 0$$

$$3x^2 - 8x = -7 \qquad \textit{Add } -7 \textit{ to both sides.}$$

We cannot complete the square on the left side because the leading coefficient is not 1. We take an extra step and divide both sides by 3.

$$\frac{3x^2}{3} - \frac{8x}{3} = -\frac{7}{3}$$

$$x^2 - \frac{8}{3}x = -\frac{7}{3}$$

Half of $\frac{8}{3}$ is $\frac{4}{3}$, the square of which is $\left(\frac{4}{3}\right)^2 = \frac{16}{9}$.

$$x^2 - \frac{8}{3}x + \frac{16}{9} = -\frac{7}{3} + \frac{16}{9} \qquad \textit{Add } -\frac{6}{9} \textit{ to both sides.}$$

$$\left(x - \frac{4}{3}\right)^2 = -\frac{5}{9} \qquad \textit{Simplify right side.}$$

$$x - \frac{4}{3} = \pm\sqrt{-\frac{5}{9}} \qquad \textit{Square root property}$$

$$x - \frac{4}{3} = \pm\frac{i\sqrt{5}}{3} \qquad \sqrt{-\frac{5}{9}} = \frac{\sqrt{-5}}{3} = \frac{i\sqrt{5}}{3}$$

$$x = \frac{4}{3} \pm \frac{i\sqrt{5}}{3} \qquad \textit{Add } \frac{4}{3} \textit{ to both sides.}$$

$$x = \frac{4 \pm i\sqrt{5}}{3}$$

The solutions are $\dfrac{4 \pm i\sqrt{5}}{3}$.

7. Find the x-intercepts of the function $f(x) = x^2 - 4x + 1$.

EXAMPLE 7 Find the x-intercepts of the function $f(x) = x^2 + 6x + 7$.

Solution Recall that the x-intercepts occur where $f(x) = 0$.

$$x^2 + 6x + 7 = 0$$

$$x^2 + 6x = -7 \qquad \textit{Add } -7 \textit{ to both sides.}$$

Half of 6 is 3, the square of which is $3^2 = 9$.

$$x^2 + 6x + 9 = -7 + 9 \qquad \textit{Add 9 to both sides.}$$

$$(x + 3)^2 = 2 \qquad \textit{Simplify right side.}$$

$$x + 3 = \pm\sqrt{2} \qquad \textit{Square root property}$$

$$x = -3 \pm \sqrt{2} \qquad \textit{Add } -3 \textit{ to both sides.}$$

The x-intercepts are $-3 + \sqrt{2}$ and $-3 - \sqrt{2}$.

Note The x-intercepts cannot be complex, since they represent points on the x-axis.

Answers

6. $\dfrac{3 \pm i\sqrt{31}}{10}$

7. $2 \pm \sqrt{3}$

HOW TO Solve a Quadratic Equation by Completing the Square

To summarize the method used in the preceding two examples, we list the following steps:

Step 1: Write the equation in the form $ax^2 + bx = c$.

Step 2: If the leading coefficient is not 1, divide both sides by the coefficient so that the resulting equation has a leading coefficient of 1. That is, if $a \neq 1$, then divide both sides by a.

Step 3: Add the square of half the coefficient of the linear term to both sides of the equation.

Step 4: Write the left side of the equation as the square of a binomial, and simplify the right side if possible.

Step 5: Apply the square root property for equations, and solve as usual.

C Quadratic Equations and Right Triangles

FACTS FROM GEOMETRY More Special Triangles

The triangles shown in Figures 1 and 2 occur frequently in mathematics.

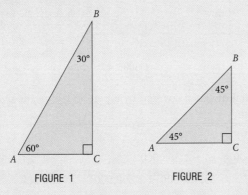

FIGURE 1 FIGURE 2

Note that both of the triangles are right triangles. We refer to the triangle in Figure 1 as a 30°–60°–90° triangle, and the triangle in Figure 2 as a 45°–45°–90° triangle.

EXAMPLE 8 If the shortest side in a 30°–60°–90° triangle is 1 inch, find the lengths of the other two sides.

Solution In Figure 3, triangle ABC is a 30°–60°–90° triangle in which the shortest side AC is 1 inch long. Triangle DBC is also a 30°–60°–90° triangle in which the shortest side DC is 1 inch long.

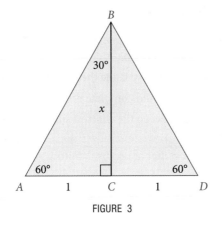

FIGURE 3

8. If the shortest side in a 30°–60°–90° triangle is 2 inches long, find the lengths of the other two sides.

Notice that the large triangle ABD is an equilateral triangle because each of its interior angles is 60°. Each side of triangle ABD is 2 inches long. Side AB in triangle ABC is therefore 2 inches. To find the length of side BC, we use the Pythagorean theorem.

$$BC^2 + AC^2 = AB^2$$
$$x^2 + 1^2 = 2^2$$
$$x^2 + 1 = 4$$
$$x^2 = 3$$
$$x = \sqrt{3} \text{ inches}$$

Note that we write only the positive square root because x is the length of a side in a triangle and is therefore a positive number.

9. Table 1 in the introduction to this section gives the vertical rise of the Lookout Double chair lift as 960 feet and the length of the chair lift as 4,330 feet. To the nearest foot, find the horizontal distance covered by a person riding the lift.

EXAMPLE 9 Table 1 in the introduction to this section gives the vertical rise of the Forest Double chair lift as 1,170 feet and the length of the chair lift as 5,750 feet. To the nearest foot, find the horizontal distance covered by a person riding this lift.

Solution Figure 4 is a model of the Forest Double chair lift. A rider gets on the lift at point A and exits at point B. The length of the lift is AB.

FIGURE 4

To find the horizontal distance covered by a person riding the chair lift, we use the Pythagorean theorem.

$5,750^2 = x^2 + 1,170^2$	Pythagorean theorem
$33,062,500 = x^2 + 1,368,900$	Simplify squares.
$x^2 = 33,062,500 - 1,368,900$	Solve for x^2.
$x^2 = 31,693,600$	Simplify the right side.
$x = \sqrt{31,693,600}$	Square root property
$\approx 5,630 \text{ feet}$	Round to the nearest foot.

A rider getting on the lift at point A and riding to point B will cover a horizontal distance of approximately 5,630 feet.

GETTING READY FOR CLASS

After reading through the preceding section, respond in your own words and in complete sentences.

A. What kind of equation do we solve using the method of completing the square?

B. Explain in words how you would complete the square on $x^2 - 16x = 4$.

C. What is the relationship between the shortest side and the longest side in a 30°–60°–90° triangle?

D. What two expressions together are equivalent to $x = \pm 4$?

Answer

9. 4,222 feet

Vocabulary Review

The following is a list of steps for solving a quadratic equation by completing the square. Write the correct step number in the blanks.

_____ If the leading coefficient is not 1, divide both sides by the coefficient so that the resulting equation has a leading coefficient of 1.

_____ Write the equation in the form $ax^2 + bx = c$.

_____ Write the left side of the equation as the square of a binomial, and simplify the right side if possible.

_____ Add the square of half the coefficient of the linear term to both sides of the equation.

_____ Apply the square root property for equations, and solve as usual.

Problems

A Solve the following equations.

1. $x^2 = 25$ **2.** $x^2 = 16$ **3.** $a^2 = -9$ **4.** $a^2 = -49$

5. $y^2 = \dfrac{3}{4}$ **6.** $y^2 = \dfrac{5}{9}$ **7.** $x^2 + 12 = 0$ **8.** $x^2 + 8 = 0$

9. $4a^2 - 45 = 0$ **10.** $9a^2 - 20 = 0$ **11.** $(2y - 1)^2 = 25$ **12.** $(3y + 7)^2 = 1$

13. $(2a + 3)^2 = -9$ **14.** $(3a - 5)^2 = -49$ **15.** $(5x + 2)^2 = -8$ **16.** $(6x - 7)^2 = -75$

17. $x^2 + 8x + 16 = -27$ **18.** $x^2 - 12x + 36 = -8$ **19.** $4a^2 - 12a + 9 = -4$ **20.** $9a^2 - 12a + 4 = -9$

B Copy each of the following, and fill in the blanks so the left side of each is a perfect square trinomial. That is, complete the square.

21. $x^2 + 12x +$ ___ $= (x +$ ___ $)^2$ **22.** $x^2 + 6x +$ ___ $= (x +$ ___ $)^2$

23. $x^2 - 4x +$ ___ $= (x -$ ___ $)^2$ **24.** $x^2 - 2x +$ ___ $= (x -$ ___ $)^2$

25. $a^2 - 10a +$ ___ $= (a -$ ___ $)^2$ **26.** $a^2 - 8a +$ ___ $= (a -$ ___ $)^2$

27. $x^2 + 5x +$ ___ $= (x +$ ___$)^2$

28. $x^2 + 3x +$ ___ $= (x +$ ___$)^2$

29. $y^2 - 7y +$ ___ $= (y -$ ___$)^2$

30. $y^2 - y +$ ___ $= (y -$ ___$)^2$

31. $x^2 + \dfrac{1}{2}x +$ ___ $= (x +$ ___$)^2$

32. $x^2 - \dfrac{3}{4}x +$ ___ $= (x -$ ___$)^2$

33. $x^2 + \dfrac{2}{3}x +$ ___ $= (x +$ ___$)^2$

34. $x^2 - \dfrac{4}{5}x +$ ___ $= (x -$ ___$)^2$

Solve each of the following quadratic equations by completing the square.

35. $x^2 + 4x = 12$

36. $x^2 - 2x = 8$

37. $x^2 + 12x = -27$

38. $x^2 - 6x = 16$

39. $a^2 - 2a + 5 = 0$

40. $a^2 + 10a + 22 = 0$

41. $y^2 - 8y + 1 = 0$

42. $y^2 + 6y - 1 = 0$

43. $x^2 - 5x - 3 = 0$

44. $x^2 - 5x - 2 = 0$

45. $2x^2 - 4x - 8 = 0$

46. $3x^2 - 9x - 12 = 0$

47. $3t^2 - 8t + 1 = 0$

48. $5t^2 + 12t - 1 = 0$

49. $4x^2 - 3x + 5 = 0$

50. $7x^2 - 5x + 2 = 0$

51. $3x^2 + 4x - 1 = 0$

52. $2x^2 + 6x - 1 = 0$

53. $2x^2 - 10x = 11$

54. $25x^2 - 20x = 1$

55. $4x^2 - 10x + 11 = 0$

56. $4x^2 - 6x + 1 = 0$

57. $27x^2 - 90x + 71 = 0$

58. $18x^2 + 12x - 23 = 0$

59. Consider the equation $x^2 = -9$.
 a. Can it be solved by factoring?

 b. Solve it.

60. Consider the equation $x^2 - 10x + 18 = 0$.
 a. Can it be solved by factoring?

 b. Solve it.

61. Solve the equation $x^2 - 6x = 0$
 a. by factoring.

 b. by completing the square.

62. Solve the equation $x^2 + ax = 0$
 a. by factoring.

 b. by completing the square.

63. Solve the equation $x^2 + 2x = 35$
 a. by factoring.

 b. by completing the square.

64. Solve the equation $8x^2 - 10x - 25 = 0$
 a. by factoring.

 b. by completing the square.

65. Is $x = -3 + \sqrt{2}$ a solution to $x^2 - 6x = 7$?

66. Is $x = 2 - \sqrt{5}$ a solution to $x^2 - 4x = 1$?

67. Solve each equation.
 a. $5x - 7 = 0$

 b. $5x - 7 = 8$

 c. $(5x - 7)^2 = 8$

 d. $\sqrt{5x - 7} = 8$

 e. $\dfrac{5}{2} - \dfrac{7}{2x} = \dfrac{4}{x}$

68. Solve each equation.
 a. $5x + 11 = 0$

 b. $5x + 11 = 9$

 c. $(5x + 11)^2 = 9$

 d. $\sqrt{5x + 11} = 9$

 e. $\dfrac{5}{3} - \dfrac{11}{3x} = \dfrac{3}{x}$

Find the x-intercepts of the given function.

69. $f(x) = x^2 - 4x - 5$ **70.** $f(x) = x^2 + 5x - 6$ **71.** $f(x) = x^2 + 3x - 5$ **72.** $f(x) = x^2 - 5x + 1$

73. $f(x) = x^2 - 7x - 9$ **74.** $f(x) = x^2 + 3x + 1$ **75.** $f(x) = x^2 + 4x + 9$ **76.** $f(x) = x^2 - 6x + 12$

77. $f(x) = 2x^2 + 3x - 4$ **78.** $f(x) = 3x^2 - 2x - 4$ **79.** $f(x) = 4x^2 + 4x - 11$ **80.** $f(x) = 9x^2 - 12x - 28$

C Applying the Concepts

81. Geometry If the shortest side in a 30°–60°–90° triangle is a $\frac{1}{2}$ meter long, find the lengths of the other two sides.

82. Geometry If the length of the longest side of a 30°–60°–90° triangle is x, find the lengths of the other two sides in terms of x.

83. Geometry If the length of the shorter sides of a 45°–45°–90° triangle is 1 inch, find the length of the hypotenuse.

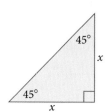

84. Geometry If the length of the shorter sides of a 45°–45°–90° triangle is x, find the length of the hypotenuse, in terms of x.

85. Chair Lift Use Table 1 from the introduction to this section to find the horizontal distance covered by a person riding the Bear Paw Double chair lift. Round your answer to the nearest foot.

86. Fermat's Last Theorem Fermat's last theorem states that if n is an integer greater than 2, then there are no positive integers x, y, and z that will make the formula $x^n + y^n = z^n$ true. Use the formula $x^n + y^n = z^n$ to find the following.

 a. Find z if $n = 2$, $x = 6$, and $y = 8$.

 b. Find y if $n = 2$, $x = 5$, and $z = 13$.

87. Interest Rate Suppose a loan of $3,000 that charges an annual interest rate r (compounded yearly) increases to $3,456 after 2 years. Using the formula $A = P(1 + r)^t$, we have

$$3,456 = 3,000(1 + r)^2$$

Solve for r to find the annual interest rate.

88. Special Triangles In Figure 5, triangle ABC has angles 45° and 30°, and height x. Find the lengths of sides AB, BC, and AC, in terms of x.

FIGURE 5

89. Length of an Escalator An escalator in a department store is made to carry people a vertical distance of 20 feet between floors. How long is the escalator if it makes an angle of 45° with the ground? (See Figure 6.)

FIGURE 6

90. Dimensions of a Tent A two-person tent is to be made so the height at the center is 4 feet. If the sides of the tent are to meet the ground at an angle of 60° and the tent is to be 6 feet in length, how many square feet of material will be needed to make the tent? (Figure 7; assume that the tent has a floor and is closed at both ends.) Give your answer to the nearest tenth of a square foot.

FIGURE 7

Getting Ready for the Next Section

Simplify.

91. $49 - 4(6)(-5)$

92. $49 - 4(6)(2)$

93. $(-27)^2 - 4(0.1)(1,700)$

94. $25 - 4(4)(-10)$

95. $-7 + \dfrac{169}{12}$

96. $-7 - \dfrac{169}{12}$

97. $\dfrac{-4 + \sqrt{36}}{2}$

98. $\dfrac{-5 - \sqrt{81}}{4}$

Factor.

99. $27t^3 - 8$

100. $125t^3 + 1$

101. $2x^3 + 54$

102. $8y^3 - 27$

Maintaining Your Skills

Volunteering The chart shows how many hours per week volunteers spend volunteering. Use the chart to answer the following questions.

103. If 2,000 volunteers were surveyed, how many say

 a. they volunteer less than 1 hour per week?

 b. they volunteer between 2 and 5 hours per week?

104. If 785 people said they volunteer between 1 and 2 hours per week, how many volunteers were surveyed?

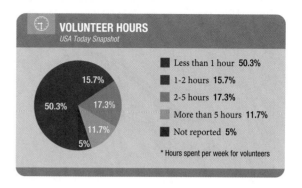

VOLUNTEER HOURS
USA Today Snapshot

15.7% 50.3% 17.3% 11.7% 5%

■ Less than 1 hour **50.3%**
■ 1-2 hours **15.7%**
□ 2-5 hours **17.3%**
□ More than 5 hours **11.7%**
■ Not reported **5%**

* Hours spent per week for volunteers

Find the Mistake

Each sentence below contains a mistake. Circle the mistake and write the correct word or expression on the line provided.

1. The square root property for equations says that if $a^2 = b$, then $a = \sqrt{b}$. _____

2. Solving $(4x + 1)^2 = 16$ gives us only one solution. _____

3. The first step to solving a quadratic equation by completing the square is to put the equation in standard form. _____

4. The method of completing the square says that adding the square of half the coefficient of the constant term to both sides will give a perfect square trinomial. _____

13.2 The Quadratic Formula

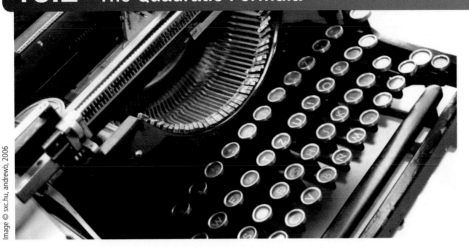

OBJECTIVES

A Solve quadratic equations using the quadratic formula.

B Solve application problems using quadratic equations.

KEY WORDS

quadratic formula

On April 22nd of each year, the people of Springfield, Missouri celebrate Secretaries Day by hurling typewriters off a fifty-foot-high platform. Each participating secretary takes her turn tossing a typewriter off the platform toward a large bull's-eye chalked on the ground below. The machine shatters as it hits the pavement. Then a representative from the sponsoring radio station measures the distance of the largest remnant from the center of the bull's eye. The participant with the closest measurement is the winner.

Let's suppose one of the typewriters is thrown downward with an initial velocity of 15 feet per second. The distance $s(t)$, in feet, the typewriter travels in t seconds is given by the function $s(t) = 15t + 16t^2$. After working similar problems in this section, you will be able to return to this one to determine how long it takes for the typewriter to reach the ground.

In this section, we will use the method of completing the square from the preceding section to derive the quadratic formula. The *quadratic formula* is a very useful tool in mathematics. It allows us to solve all types of quadratic equations.

A The Quadratic Formula

> **THEOREM** The Quadratic Formula
>
> For any quadratic equation in the form $ax^2 + bx + c = 0$, $a \neq 0$, the two solutions are
>
> $$x = \frac{-b + \sqrt{b^2 - 4ac}}{2a} \quad \text{and} \quad x = \frac{-b - \sqrt{b^2 - 4ac}}{2a}$$

**Video Examples
Section 13.2**

Proof We will prove the quadratic formula by completing the square on $ax^2 + bx + c = 0$.

$$ax^2 + bx + c = 0$$

$$ax^2 + bx = -c \qquad \text{Add } -c \text{ to both sides.}$$

$$x^2 + \frac{b}{a}x = -\frac{c}{a} \qquad \text{Divide both sides by } a.$$

To complete the square on the left side, we add the square of $\frac{1}{2}$ of $\frac{b}{a}$ to both sides $\left(\frac{1}{2} \text{ of } \frac{b}{a} \text{ is } \frac{b}{2a}\right)$.

$$x^2 + \frac{b}{a}x + \left(\frac{b}{2a}\right)^2 = -\frac{c}{a} + \left(\frac{b}{2a}\right)^2$$

We now simplify the right side as a separate step. We combine the two terms by writing each with the least common denominator $4a^2$.

$$-\frac{c}{a} + \left(\frac{b}{2a}\right)^2 = -\frac{c}{a} + \frac{b^2}{4a^2} = \frac{4a}{4a}\left(\frac{-c}{a}\right) + \frac{b^2}{4a^2} = \frac{-4ac + b^2}{4a^2}$$

It is convenient to write this last expression as

$$\frac{b^2 - 4ac}{4a^2}$$

Continuing with the proof, we have

$$x^2 + \frac{b}{a}x + \left(\frac{b}{2a}\right)^2 = \frac{b^2 - 4ac}{4a^2}$$

$$\left(x + \frac{b}{2a}\right)^2 = \frac{b^2 - 4ac}{4a^2} \qquad \textit{Write left side as a binomial square.}$$

$$x + \frac{b}{2a} = \pm\frac{\sqrt{b^2 - 4ac}}{2a} \qquad \textit{Square root property for equations}$$

$$x = -\frac{b}{2a} \pm \frac{\sqrt{b^2 - 4ac}}{2a} \qquad \textit{Add } -\frac{b}{2a} \textit{ to both sides.}$$

$$= \frac{-b \pm \sqrt{b^2 - 4ac}}{2a}$$

Our proof is now complete. What we have is this: If our equation is in the form $ax^2 + bx + c = 0$ (standard form), where $a \neq 0$, the two solutions are always given by the formula

$$x = \frac{-b \pm \sqrt{b^2 - 4ac}}{2a}$$

This formula is known as the *quadratic formula*. If we substitute the coefficients a, b, and c of any quadratic equation in standard form into the formula, we need only perform some basic arithmetic to arrive at the solution set.

EXAMPLE 1 Solve $x^2 - 5x - 6 = 0$ using the quadratic formula.

Solution To use the quadratic formula, we must make sure the equation is in standard form; identify a, b, and c; substitute them into the formula; and work out the arithmetic.

For the equation $x^2 - 5x - 6 = 0$, $a = 1$, $b = -5$, and $c = -6$.

$$x = \frac{-b \pm \sqrt{b^2 - 4ac}}{2a}$$

$$= \frac{-(-5) \pm \sqrt{(-5)^2 - 4(1)(-6)}}{2(1)}$$

$$= \frac{5 \pm \sqrt{49}}{2}$$

$$= \frac{5 \pm 7}{2}$$

$$x = \frac{5 + 7}{2} \quad \text{or} \quad x = \frac{5 - 7}{2}$$

$$x = \frac{12}{2} \qquad\qquad x = -\frac{2}{2}$$

$$x = 6 \qquad\qquad x = -1$$

The two solutions are 6 and -1.

EXAMPLE 2 Solve $2x^2 = -4x + 3$ using the quadratic formula.

Solution Before we can identify a, b, and c, we must write the equation in standard form. To do so, we add $4x$ and -3 to each side of the equation.

$$2x^2 = -4x + 3$$

$$2x^2 + 4x - 3 = 0 \qquad \textit{Add } 4x \textit{ and } -3 \textit{ to each side.}$$

Practice Problems

1. Solve $x^2 - 9x + 18 = 0$ using the quadratic formula.

Note Whenever the solutions to our quadratic equations turn out to be rational numbers, as in Example 1, it means the original equation could have been solved by factoring. (We didn't solve the equation in Example 1 by factoring because we were trying to get some practice with the quadratic formula.)

2. Solve $6x^2 + 2 = -7x$ using the quadratic formula.

Answers

1. 3, 6

2. $-\frac{1}{2}, -\frac{2}{3}$

Now that the equation is in standard form, we see that $a = 2$, $b = 4$, and $c = -3$. Using the quadratic formula we have

$$x = \frac{-b \pm \sqrt{b^2 - 4ac}}{2a}$$

$$= \frac{-4 \pm \sqrt{4^2 - 4(2)(-3)}}{2(2)}$$

$$= \frac{-4 \pm \sqrt{40}}{4}$$

$$= \frac{-4 \pm 2\sqrt{10}}{4}$$

We can reduce the final expression in the preceding equation to lowest terms by factoring 2 from the numerator and denominator and then dividing it out.

$$x = \frac{2(-2 \pm \sqrt{10})}{2 \cdot 2}$$

$$= \frac{-2 \pm \sqrt{10}}{2}$$

Our two solutions are $\dfrac{-2 + \sqrt{10}}{2}$ and $\dfrac{-2 - \sqrt{10}}{2}$.

EXAMPLE 3 Solve $x^2 - 6x = -7$ using the quadratic formula.

Solution We begin by writing the equation in standard form.

$$x^2 - 6x = -7$$

$$x^2 - 6x + 7 = 0 \qquad \text{Add 7 to each side.}$$

Using $a = 1$, $b = -6$, and $c = 7$ in the quadratic formula

$$x = \frac{-b \pm \sqrt{b^2 - 4ac}}{2a}$$

we have

$$x = \frac{-(-6) \pm \sqrt{(-6)^2 - 4(1)(7)}}{2(1)}$$

$$= \frac{6 \pm \sqrt{36 - 28}}{2}$$

$$= \frac{6 \pm \sqrt{8}}{2}$$

$$= \frac{6 \pm 2\sqrt{2}}{2}$$

The two terms in the numerator have a 2 in common. We reduce to lowest terms by factoring the 2 from the numerator and then dividing numerator and denominator by 2.

$$= \frac{2(3 \pm \sqrt{2})}{2}$$

$$= 3 \pm \sqrt{2}$$

The two solutions are $3 + \sqrt{2}$ and $3 - \sqrt{2}$. For practice, let's check our solutions in the original equation $x^2 - 6x = -7$.

Checking $x = 3 + \sqrt{2}$, we have

$$(3 + \sqrt{2})^2 - 6(3 + \sqrt{2}) \stackrel{?}{=} -7$$

$$9 + 6\sqrt{2} + 2 - 18 - 6\sqrt{2} \stackrel{?}{=} -7 \qquad \text{Multiply.}$$

$$11 - 18 + 6\sqrt{2} - 6\sqrt{2} \stackrel{?}{=} -7 \qquad \text{Add 9 and 2.}$$

$$-7 + 0 \stackrel{?}{=} -7 \qquad \text{Subtract.}$$

$$-7 = -7 \qquad \text{A true statement}$$

3. Solve $x^2 - 5x = 4$ using the quadratic formula.

Answer

3. $\dfrac{5 \pm \sqrt{41}}{2}$

Checking $x = 3 - \sqrt{2}$, we have

$$(3 - \sqrt{2})^2 - 6(3 - \sqrt{2}) \stackrel{?}{=} -7$$

$$9 - 6\sqrt{2} + 2 - 18 + 6\sqrt{2} \stackrel{?}{=} -7 \qquad \text{Multiply.}$$

$$11 - 18 - 6\sqrt{2} + 6\sqrt{2} \stackrel{?}{=} -7 \qquad \text{Add 9 and 2.}$$

$$-7 + 0 \stackrel{?}{=} -7 \qquad \text{Subtract.}$$

$$-7 = -7 \qquad \text{A true statement}$$

As you can see, both solutions yield true statements when used in place of the variable in the original equation.

4. Solve $\dfrac{x^2}{2} + x = \dfrac{1}{3}$. using the quadratic formula.

EXAMPLE 4 Solve $\dfrac{1}{10}x^2 - \dfrac{1}{5}x = -\dfrac{1}{2}$ using the quadratic formula.

Solution It will be easier to apply the quadratic formula if we clear the equation of fractions. Multiplying both sides of the equation by the LCD 10 gives us

$$x^2 - 2x = -5$$

Next, we add 5 to both sides to put the equation into standard form.

$$x^2 - 2x + 5 = 0 \qquad \text{Add 5 to both sides.}$$

Applying the quadratic formula with $a = 1$, $b = -2$, and $c = 5$, we have

$$x = \frac{-(-2) \pm \sqrt{(-2)^2 - 4(1)(5)}}{2(1)} = \frac{2 \pm \sqrt{-16}}{2} = \frac{2 \pm 4i}{2}$$

Dividing the numerator and denominator by 2, we have the two solutions.

$$x = 1 \pm 2i$$

The two solutions are $1 + 2i$ and $1 - 2i$.

5. Solve $(3x - 2)(x - 1) = -3$ using the quadratic formula.

EXAMPLE 5 Solve $(2x - 3)(2x - 1) = -4$ using the quadratic formula.

Solution We multiply the binomials on the left side and then add 4 to each side to write the equation in standard form. From there we identify a, b, and c and apply the quadratic formula.

$$(2x - 3)(2x - 1) = -4$$

$$4x^2 - 8x + 3 = -4 \qquad \text{Multiply binomials on left side.}$$

$$4x^2 - 8x + 7 = 0 \qquad \text{Add 4 to each side.}$$

Placing $a = 4$, $b = -8$, and $c = 7$ in the quadratic formula we have

$$x = \frac{-(-8) \pm \sqrt{(-8)^2 - 4(4)(7)}}{2(4)}$$

$$= \frac{8 \pm \sqrt{64 - 112}}{8}$$

$$= \frac{8 \pm \sqrt{-48}}{8}$$

$$= \frac{8 \pm 4i\sqrt{3}}{8} \qquad \sqrt{-48} = i\sqrt{48} = i\sqrt{16}\sqrt{3} = 4i\sqrt{3}$$

Note It would be a mistake to try to reduce this final expression further. Sometimes students will try to divide the 2 in the denominator into the 2 in the numerator, which is a mistake. Remember, when we reduce to lowest terms, we do so by dividing the numerator and denominator by any factors they have in common. In this case 2 is not a factor of the numerator. This expression is in lowest terms.

To reduce this final expression to lowest terms, we factor a 4 from the numerator and then divide the numerator and denominator by 4.

$$= \frac{4(2 \pm i\sqrt{3})}{4 \cdot 2}$$

$$= \frac{2 \pm i\sqrt{3}}{2}$$

Although the equation in our next example is not a quadratic equation, we solve it by using both factoring and the quadratic formula.

Answers

4. $\dfrac{-3 \pm \sqrt{15}}{3}$

5. $\dfrac{5 \pm i\sqrt{35}}{6}$

EXAMPLE 6 Solve $27t^3 - 8 = 0$.

Solution It would be a mistake to add 8 to each side of this equation and then take the cube root of each side because we would lose two of our solutions. Instead, we factor the left side, and then set the factors equal to 0.

$$27t^3 - 8 = 0 \qquad \text{Equation in standard form}$$

$$(3t - 2)(9t^2 + 6t + 4) = 0 \qquad \text{Factor as the difference of two cubes.}$$

$$3t - 2 = 0 \quad \text{or} \quad 9t^2 + 6t + 4 = 0 \quad \text{Set each factor equal to 0.}$$

The first equation leads to a solution of $t = \frac{2}{3}$. The second equation does not factor, so we use the quadratic formula with $a = 9$, $b = 6$, and $c = 4$.

$$t = \frac{-6 \pm \sqrt{36 - 4(9)(4)}}{2(9)}$$

$$= \frac{-6 \pm \sqrt{36 - 144}}{18}$$

$$= \frac{-6 \pm \sqrt{-108}}{18}$$

$$= \frac{-6 \pm 6i\sqrt{3}}{18} \qquad \sqrt{-108} = i\sqrt{36 \cdot 3} = 6i\sqrt{3}$$

$$= \frac{6(-1 \pm i\sqrt{3})}{6 \cdot 3} \qquad \text{Factor 6 from the numerator and denominator.}$$

$$= \frac{-1 \pm i\sqrt{3}}{3} \qquad \text{Divide out common factor 6.}$$

The three solutions to our original equation are

$$\frac{2}{3}, \frac{-1 + i\sqrt{3}}{3}, \text{ and } \frac{-1 - i\sqrt{3}}{3}$$

B Applications

EXAMPLE 7 If an object is thrown downward with an initial velocity of 20 feet per second, the distance $s(t)$, in feet, it travels in t seconds is given by the function $s(t) = 20t + 16t^2$. How long does it take the object to fall 40 feet?

Solution We let $s(t) = 40$, and solve for t.

When → $s(t) = 40$

the function → $s(t) = 20t + 16t^2$

becomes → $40 = 20t + 16t^2$

$$16t^2 + 20t - 40 = 0$$

$$4t^2 + 5t - 10 = 0 \qquad \text{Divide by 4.}$$

Using the quadratic formula, we have

$$t = \frac{-5 \pm \sqrt{25 - 4(4)(-10)}}{2(4)}$$

$$= \frac{-5 \pm \sqrt{185}}{8}$$

$$t = \frac{-5 + \sqrt{185}}{8} \quad \text{or} \quad t = \frac{-5 - \sqrt{185}}{8}$$

The second solution is impossible because it is a negative number and time t must be positive. It takes

$$t = \frac{-5 + \sqrt{185}}{8} \quad \text{or approximately} \quad \frac{-5 + 13.60}{8} \approx 1.08 \text{ seconds}$$

for the object to fall 40 feet.

6. Solve $8t^3 - 27 = 0$.

7. An object thrown upward with an initial velocity of 32 feet per second rises and falls according to the equation $s = 32t - 16t^2$ where s is the height of the object above the ground at any time t. At what times t will the object be 12 feet above the ground?

Recall that the relationship between profit, revenue, and cost is given by the formula

$$P(x) = R(x) - C(x)$$

where $P(x)$ is the profit, $R(x)$ is the total revenue, and $C(x)$ is the total cost of producing and selling x items.

EXAMPLE 8 A company produces and sells copies of an accounting program for home computers. The total weekly cost (in dollars) to produce x copies of the program is $C(x) = 8x + 500$, and the weekly revenue for selling all x copies of the program is $R(x) = 35x - 0.1x^2$. How many programs must be sold each week for the weekly profit to be $1,200?

Solution Substituting the given expressions for $R(x)$ and $C(x)$ in the equation $P(x) = R(x) - C(x)$, we have a polynomial in x that represents the weekly profit $P(x)$.

$$P(x) = R(x) - C(x)$$

$$= 35x - 0.1x^2 - (8x + 500)$$

$$= 35x - 0.1x^2 - 8x - 500$$

$$= -500 + 27x - 0.1x^2$$

By setting this expression equal to 1,200, we have a quadratic equation to solve that gives us the number of programs x that need to be sold each week to bring in a profit of $1,200.

$$1,200 = -500 + 27x - 0.1x^2$$

We can write this equation in standard form by adding the opposite of each term on the right side of the equation to both sides of the equation. Doing so produces the following equation:

$$0.1x^2 - 27x + 1,700 = 0$$

Applying the quadratic formula to this equation with $a = 0.1$, $b = -27$, and $c = 1,700$, we have

$$x = \frac{27 \pm \sqrt{(-27)^2 - 4(0.1)(1,700)}}{2(0.1)}$$

$$= \frac{27 \pm \sqrt{729 - 680}}{0.2}$$

$$= \frac{27 \pm \sqrt{49}}{0.2}$$

$$= \frac{27 \pm 7}{0.2}$$

Writing this last expression as two separate expressions, we have our two solutions:

$$x = \frac{27 + 7}{0.2} \quad \text{or} \quad x = \frac{27 - 7}{0.2}$$

$$= \frac{34}{0.2} \qquad\qquad = \frac{20}{0.2}$$

$$= 170 \qquad\qquad\quad = 100$$

The weekly profit will be $1,200 if the company produces and sells 100 programs or 170 programs.

8. Use the information in Example 8 to find the number of programs the company must sell each week for its weekly profit to be $1,320.

Note What is interesting about this last example is that it has rational solutions, meaning it could have been solved by factoring. But looking back at the equation, factoring does not seem like a reasonable method of solution because the coefficients are either very large or very small. So there are times when using the quadratic formula is a faster method of solution, even though the equation you are solving is factorable.

Answer

8. 130 or 140 programs

USING TECHNOLOGY Graphing Calculators

More About Example 7

We can solve the problem discussed in Example 7 by graphing the function $Y_1 = 20X + 16X^2$ in a window with X from 0 to 2 (because X is taking the place of t and we know t is a positive quantity) and Y from 0 to 50 (because we are looking for X when Y_1 is 40). Graphing Y_1 gives a graph similar to the graph in Figure 1. Using the Zoom and Trace features at $Y_1 = 40$ gives us X = 1.08 to the nearest hundredth, matching the results we obtained by solving the original equation algebraically.

FIGURE 1

More About Example 8

To visualize the functions in Example 8, we set up our calculator this way:

$$Y_1 = 35X - .1X^2 \qquad \text{Revenue function}$$
$$Y_2 = 8X + 500 \qquad \text{Cost function}$$
$$Y_3 = Y_1 - Y_2 \qquad \text{Profit function}$$

Window: X from 0 to 350, Y from 0 to 3,500

Graphing these functions produces graphs similar to the ones shown in Figure 2. The lowest graph is the graph of the profit function. Using the Zoom and Trace features on the lowest graph at $Y_3 = 1,200$ produces two corresponding values of X, 170 and 100, which match the results in Example 8.

FIGURE 2

We will continue this discussion of the relationship between graphs of functions and solutions to equations in the Using Technology material in the next section.

GETTING READY FOR CLASS

After reading through the preceding section, respond in your own words and in complete sentences.

A. What is the quadratic formula?

B. Under what circumstances should the quadratic formula be applied?

C. When would the quadratic formula result in complex solutions?

D. When will the quadratic formula result in only one solution?

Vocabulary Review

Choose the correct words to fill in the blanks below.

rational quadratic standard form factored

1. An equation in _____, $ax^2 + bx + c = 0$, has two solutions that are always given by the formula $x = \frac{-b \pm \sqrt{b^2 - 4ac}}{2a}$.

2. The formula $x = \frac{-b \pm \sqrt{b^2 - 4ac}}{2a}$ is called the _____ formula.

3. If solutions to a quadratic equation turn out to be _____ numbers, then the original equation could have been solved by factoring.

4. Using the quadratic formula may be a faster way to finding a solution, even if the equation can be _____.

Problems

A Solve each equation. Use factoring or the quadratic formula, whichever is appropriate. (Try factoring first. If you have any difficulty factoring, then try the quadratic formula.)

1. $x^2 + 5x + 6 = 0$

2. $x^2 + 5x - 6 = 0$

3. $a^2 - 4a + 1 = 0$

4. $a^2 + 4a + 1 = 0$

5. $\frac{1}{6}x^2 - \frac{1}{2}x + \frac{1}{3} = 0$

6. $\frac{1}{6}x^2 + \frac{1}{2}x + \frac{1}{3} = 0$

7. $\frac{x^2}{2} + 1 = \frac{2x}{3}$

8. $\frac{x^2}{2} + \frac{2}{3} = -\frac{2x}{3}$

9. $y^2 - 5y = 0$

10. $2y^2 + 10y = 0$

11. $30x^2 + 40x = 0$

12. $50x^2 - 20x = 0$

13. $\frac{2t^2}{3} - t = -\frac{1}{6}$

14. $\frac{t^2}{3} - \frac{t}{2} = -\frac{3}{2}$

15. $0.01x^2 + 0.06x - 0.08 = 0$

16. $0.02x^2 - 0.03x + 0.05 = 0$

17. $2x + 3 = -2x^2$

18. $2x - 3 = 3x^2$

19. $100x^2 - 200x + 100 = 0$

20. $100x^2 - 600x + 900 = 0$

21. $\frac{1}{2}r^2 = \frac{1}{6}r - \frac{2}{3}$

22. $\frac{1}{4}r^2 = \frac{2}{5}r + \frac{1}{10}$

23. $(x - 3)(x - 5) = 1$

24. $(x - 3)(x + 1) = -6$

25. $(x + 3)^2 + (x - 8)(x - 1) = 16$ **26.** $(x - 4)^2 + (x + 2)(x + 1) = 9$ **27.** $\dfrac{x^2}{3} - \dfrac{5x}{6} = \dfrac{1}{2}$

28. $\dfrac{x^2}{6} + \dfrac{5}{6} = -\dfrac{x}{3}$ **29.** $\sqrt{x} = x - 1$ **30.** $\sqrt{x} = x - 2$

Multiply both sides of each equation by its LCD. Then solve the resulting equation.

31. $\dfrac{1}{x + 1} - \dfrac{1}{x} = \dfrac{1}{2}$ **32.** $\dfrac{1}{x + 1} + \dfrac{1}{x} = \dfrac{1}{3}$ **33.** $\dfrac{1}{y - 1} + \dfrac{1}{y + 1} = 1$

34. $\dfrac{2}{y + 2} + \dfrac{3}{y - 2} = 1$ **35.** $\dfrac{1}{x + 2} + \dfrac{1}{x + 3} = 1$ **36.** $\dfrac{1}{x + 3} + \dfrac{1}{x + 4} = 1$

37. $\dfrac{6}{r^2 - 1} - \dfrac{1}{2} = \dfrac{1}{r + 1}$ **38.** $2 + \dfrac{5}{r - 1} = \dfrac{12}{(r - 1)^2}$ **39.** $\dfrac{1}{x^2 - 4} + \dfrac{x}{x - 2} = 2$

40. $\dfrac{1}{x^2} - \dfrac{1}{x} = 2$ **41.** $2 - \dfrac{3}{x} = \dfrac{1}{x^2}$ **42.** $4 - \dfrac{1}{x} = \dfrac{2}{x^2}$

Solve each equation. In each case you will have three solutions.

43. $x^3 - 8 = 0$ **44.** $x^3 - 27 = 0$ **45.** $8a^3 + 27 = 0$

46. $27a^3 + 8 = 0$ **47.** $125t^3 - 1 = 0$ **48.** $64t^3 + 1 = 0$

Each of the following equations has three solutions. Look for the greatest common factor; then use the quadratic formula to find all solutions.

49. $2x^3 + 2x^2 + 3x = 0$ **50.** $6x^3 - 4x^2 + 6x = 0$ **51.** $3y^4 = 6y^3 - 6y^2$

52. $4y^4 = 16y^3 - 20y^2$ **53.** $6t^5 + 4t^4 = -2t^3$ **54.** $8t^5 + 2t^4 = -10t^3$

55. Which two of the expressions below are equivalent?

 a. $\dfrac{6 + 2\sqrt{3}}{4}$
 b. $\dfrac{3 + \sqrt{3}}{2}$
 c. $6 + \dfrac{\sqrt{3}}{2}$

56. Which two of the expressions below are equivalent?

 a. $\dfrac{8 - 4\sqrt{2}}{4}$
 b. $2 - 4\sqrt{3}$
 c. $2 - \sqrt{2}$

57. Solve $3x^2 - 5x = 0$

 a. by factoring.

 b. using the quadratic formula.

58. Solve $3x^2 + 23x - 70 = 0$

 a. by factoring.

 b. using the quadratic formula.

59. Can the equation $x^2 - 4x + 7 = 0$ be solved by factoring? Solve it.

60. Can the equation $x^2 = 5$ be solved by factoring? Solve it.

61. Is $x = -1 + i$ a solution to $x^2 + 2x = -2$?

62. Is $x = 2 + 2i$ a solution to $(x - 2)^2 = -4$?

Recall from section 2.4 that fixed points for a function are solutions to the equation $f(x) = x$. For each of the following functions, find the fixed points.

63. $f(x) = x^2 - 3x$
 64. $f(x) = x^2 + 5x$
 65. $f(x) = x(1.4 - x)$
 66. $f(x) = x(1.6 - x)$

67. $f(x) = x^2 + 5x + 2$
 68. $f(x) = x^2 - 4x + 3$
 69. $f(x) = \dfrac{x + 3}{x}$
 70. $f(x) = \dfrac{5 - x}{x}$

71. $f(x) = \dfrac{2x^2 - 6}{x}$
 72. $f(x) = \dfrac{x^2 + 5}{3x}$

B Applying the Concepts

73. Falling Object An object is thrown downward with an initial velocity of 5 feet per second. The relationship between the distance s it travels and time t is given by $s = 5t + 16t^2$. How long does it take the object to fall 74 feet?

74. Coin Toss A coin is tossed upward with an initial velocity of 32 feet per second from a height of 16 feet above the ground. The equation giving the object's height h at any time t is $h = 16 + 32t - 16t^2$. Does the object ever reach a height of 32 feet?

32 feet/sec

16 ft

75. Profit The total cost (in dollars) for a company to manufacture and sell x items per week is $C = 60x + 300$, whereas the revenue brought in by selling all x items is $R = 100x - 0.5x^2$. How many items must be sold to obtain a weekly profit of $300?

76. Profit Suppose a company manufactures and sells x picture frames each month with a total cost of $C = 1{,}200 + 3.5x$ dollars. If the revenue obtained by selling x frames is $R = 9x - 0.002x^2$, find the number of frames it must sell each month if its monthly profit is to be $2,300.

77. Photograph Cropping The following figure shows a photographic image on a 10.5-centimeter by 8.2-centimeter background. The overall area of the background is to be reduced to 80% of its original area by cutting off (cropping) equal strips on all four sides. What is the width of the strip that is cut from each side?

78. Area of a Garden A garden measures 20.3 meters by 16.4 meters. To double the area of the garden, strips of equal width are added to all four sides.

a. Draw a diagram that illustrates these conditions.

b. What are the new overall dimensions of the garden?

Image © Staci Truelson, 2011

79. Area and Perimeter A rectangle has a perimeter of 20 yards and an area of 15 square yards.

 a. Write two equations that state these facts in terms of the rectangle's length, l, and its width, w.

 b. Solve the two equations from part a to determine the actual length and width of the rectangle.

 c. Explain why two answers are possible to part b.

80. Population Size Writing in 1829, former President James Madison made some predictions about the growth of the population of the United States. The populations he predicted fit the equation

$$P(x) = 0.029x^2 - 1.39x + 42$$

where $P(x)$ is the population in millions of people x years from 1829.

 a. Use the equation to determine the approximate year President Madison would have predicted that the U.S. population would reach 100,000,000.

 b. If the U.S. population in 2010 was approximately 308,000,000, were President Madison's predictions accurate in the long term? Explain why or why not.

Getting Ready for the Next Section

Find the value of $b^2 - 4ac$ when

81. $a = 1, b = -3, c = -40$

82. $a = 2, b = 3, c = 4$

83. $a = 4, b = 12, c = 9$

84. $a = -3, b = 8, c = -1$

Solve.

85. $k^2 - 144 = 0$

86. $36 - 20k = 0$

Multiply.

87. $(x - 3)(x + 2)$

88. $(t - 5)(t + 5)$

89. $(x - 3)(x - 3)(x + 2)$

90. $(t - 5)(t + 5)(t - 3)$

Find the Mistake

Each sentence below contains a mistake. Circle the mistake and write the correct word or expression on the line provided.

1. Using the quadratic formula to solve a quadratic equation gives the solutions $x = -b \pm 4ac$.

2. Using the quadratic formula to solve $x^2 + 10x = 11$, $a = 1$, $b = 10$, and $c = 11$.

3. To solve $x^2 - 10x = -24$ using the quadratic formula, begin by finding the greatest common factor on both sides of the equals sign.

4. Solving $3x^2 - 2x - 4 = 0$ using the quadratic formula, we get $\frac{(1 \pm \sqrt{7})}{3}$ for our answer.

Chapter 13 Summary

The Square Root Property for Equations [13.1]

If $a^2 = b$, where b is a real number, then

$$a = \sqrt{b} \quad \text{or} \quad a = -\sqrt{b}$$

which can be written as $a = \pm\sqrt{b}$.

1. If $\quad (x - 3)^2 = 25$

then $\quad x - 3 = \pm 5$

$x = 3 \pm 5$

$x = 8 \text{ or } x = -2$

To Solve a Quadratic Equation by Completing the Square [13.1]

Step 1: Write the equation in the form $ax^2 + bx = c$.

Step 2: If $a \neq 1$, divide through by the constant a so the coefficient of x^2 is 1.

Step 3: Complete the square on the left side by adding the square of $\frac{1}{2}$ the coefficient of x to both sides.

Step 4: Write the left side of the equation as the square of a binomial. Simplify the right side if possible.

Step 5: Apply the square root property for equations, and solve as usual.

2. Solve $x^2 - 6x - 6 = 0$

$x^2 - 6x = 6$

$x^2 - 6x + 9 = 6 + 9$

$(x - 3)^2 = 15$

$x - 3 = \pm\sqrt{15}$

$x = 3 \pm \sqrt{15}$

The Quadratic Formula [13.2]

For any quadratic equation in the form $ax^2 + bx + c = 0$, $a \neq 0$, the two solutions are

$$x = \frac{-b \pm \sqrt{b^2 - 4ac}}{2a}$$

This equation is known as the *quadratic formula*.

3. If $2x^2 + 3x - 4 = 0$, then

$$x = \frac{-3 \pm \sqrt{9 - 4(2)(-4)}}{2(2)}$$

$$= \frac{-3 \pm \sqrt{41}}{4}$$

Solve each equation. [13.1, 13.2]

1. $(2x - 3)^2 = 36$

2. $(4x + 1)^2 = -12$

3. $y^2 + 8y + 16 = -25$

4. $(x + 4)(x + 6) = -3$

5. $64t^3 - 27 = 0$

6. $\dfrac{1}{a} - \dfrac{1}{2} = \dfrac{1}{a - 1}$

7. Solve the formula $49(r - 7)^2 = A$ for r. [13.1]

8. Solve $x^2 - 6x = -11$ by completing the square. [13.1]

9. If the length of the longest side of $30° - 60° - 90°$ triangle is 8 meters, find the lengths of the other two sides. [13.1]

60° 8 meters
30°

10. Projectile Motion An object projected upward with an initial velocity of 32 feet per second will rise and fall according to the equation $s(t) = 32t - 16t^2$, where s is its distance above the ground at time t. At what times will the object be 7 feet above the ground? [13.2]

11. Revenue The total weekly cost for a company to make x ceramic letter openers is given by the formula $C(x) = 9x + 100$. If the weekly revenue from selling all x letter openers is $R(x) = 30x - 0.2x^2$, how many letter openers must it sell a week to break even? [13.2]

Exponential and Logarithmic Functions

14

14.1 Exponential Functions and Applications

14.2 Inverse Functions

14.3 Logarithmic Functions and Applications

14.4 Properties of Logarithms

14.5 Common and Natural Logarithms with Applications

14.6 Exponential Equations, Change of Base, and Applications

Large earthquakes are often so destructive that their effects are difficult to miss (as the before and after photos above show of the 2010 earthquake in Haiti.) You may be surprised to know that thousands of earthquakes occur every day all over the world. Approximately 80 percent of these quakes occur along the rim of the Pacific Ocean, also known as the "Ring of Fire" for its large amount of volcanic activity. For some earthquakes to occur, large sections of the upper layers of Earth, called tectonic plates, collide and slide against each other. However, much of this sliding is usually gradual and goes unnoticed here on the surface.

The following table shows the number of earthquakes occurring from 2000 through most of 2010 based on their strength.

Magnitude	2000	2002	2004	2006	2008	2010
8.0–9.9	1	0	2	2	0	1
7.0–7.9	14	13	14	9	12	19
6.0–6.9	146	127	141	142	168	131
5.0–5.9	1,344	1,201	1,515	1,712	1,768	1,488
4.0–4.9	8,008	8,541	10,888	12,838	12,291	7,687
3.0–3.9	4,827	7,068	7,932	9,990	11,735	3,463
2.0–2.9	3,765	6,419	6,316	4,207	3,860	3,380
1.0–1.9	1,026	1,137	1,344	18	21	23
0.1–0.9	5	10	103	2	0	0
Total	22,256	27,454	31,194	29,568	31,777	16,217
Estimated Deaths	231	1,685	228,802	6,605	88,011	226,487

Source: earthquake.usgs.gov

In this chapter, we will discuss earthquakes and how they are measured. You will learn that the magnitude of an earthquake is measured on the Richter Scale, which is a logarithmic scale. As you work through this chapter, you will use mathematics to understand what that means and how to compare the strengths of earthquakes using logarithms.

OBJECTIVES

A Find function values for exponential functions.

B Graph exponential functions.

C Solve compound interest problems involving exponential functions.

KEY WORDS

exponential function

compound interest

natural exponential function

the number e

**Video Examples
Section 14.1**

To obtain an intuitive idea of how exponential functions behave, we can consider the heights attained by a bouncing ball. When a ball used in the game of racquetball is dropped from any height, the first bounce will reach a height that is $\frac{2}{3}$ of the original height. The second bounce will reach $\frac{2}{3}$ of the height of the first bounce, and so on, as shown in Figure 1.

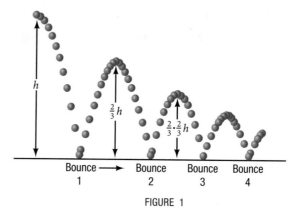

FIGURE 1

If the ball is initially dropped from a height of 1 meter, then during the first bounce it will reach a height of $\frac{2}{3}$ meter. The height of the second bounce will reach $\frac{2}{3}$ of the height reached on the first bounce. The maximum height of any bounce is $\frac{2}{3}$ of the height of the previous bounce.

$$\text{Initial height: } h = 1$$

$$\text{Bounce 1:} \quad h = \frac{2}{3}(1) = \frac{2}{3}$$

$$\text{Bounce 2:} \quad h = \frac{2}{3}\left(\frac{2}{3}\right) = \left(\frac{2}{3}\right)^2$$

$$\text{Bounce 3:} \quad h = \frac{2}{3}\left(\frac{2}{3}\right)^2 = \left(\frac{2}{3}\right)^3$$

$$\text{Bounce 4:} \quad h = \frac{2}{3}\left(\frac{2}{3}\right)^3 = \left(\frac{2}{3}\right)^4$$

$$\vdots \qquad \qquad \vdots$$

$$\text{Bounce } n: \quad h = \frac{2}{3}\left(\frac{2}{3}\right)^{n-1} = \left(\frac{2}{3}\right)^n$$

This last equation is exponential in form. We classify all exponential functions together with the following definition.

An *exponential function* is any function that can be written in the form

$$f(x) = b^x$$

where b is a positive real number other than 1.

Each of the following is an exponential function:

$$f(x) = 2^x \qquad g(x) = 3^x \qquad f(x) = \left(\frac{1}{4}\right)^x$$

A Evaluating Exponential Functions

The first step in becoming familiar with exponential functions is to find some values for specific exponential functions.

EXAMPLE 1 If the exponential functions $f(x)$ and $g(x)$ are defined by $f(x) = 2^x$ and $g(x) = 3^x$ then

$$f(0) = 2^0 = 1 \qquad\qquad g(0) = 3^0 = 1$$

$$f(1) = 2^1 = 2 \qquad\qquad g(1) = 3^1 = 3$$

$$f(2) = 2^2 = 4 \qquad\qquad g(2) = 3^2 = 9$$

$$f(3) = 2^3 = 8 \qquad\qquad g(3) = 3^3 = 27$$

$$f(-2) = 2^{-2} = \frac{1}{2^2} = \frac{1}{4} \qquad g(-2) = 3^{-2} = \frac{1}{3^2} = \frac{1}{9}$$

$$f(-3) = 2^{-3} = \frac{1}{2^3} = \frac{1}{8} \qquad g(-3) = 3^{-3} = \frac{1}{3^3} = \frac{1}{27}$$

Half-life is a term used in science and mathematics that quantifies the amount of decay or reduction in an element. The half-life of iodine-131 is 8 days, which means that every 8 days a sample of iodine-131 will decrease by half of its remaining amount. If we start with A_0 micrograms of iodine-131, then the function $A(t)$ represents the number of micrograms of iodine -131 in the sample after t days.

$$A(t) = A_0 \cdot 2^{-t/8}$$

EXAMPLE 2 A patient is administered a 1,200-microgram dose of iodine-131. How much iodine-131 will be in the patient's system after 10 days, and after 16 days?

Solution The initial amount of iodine-131 is $A_0 = 1,200$, so the function that gives the amount left in the patient's system after t days is

$$A(t) = 1,200 \cdot 2^{-t/8}$$

After 10 days, the amount left in the patient's system is

$$A(10) = 1,200 \cdot 2^{-10/8} = 1,200 \cdot 2^{-1.25} \approx 504.5 \text{ micrograms}$$

After 16 days, the amount left in the patient's system is

$$A(16) = 1,200 \cdot 2^{-16/8} = 1,200 \cdot 2^{-2} = 300 \text{ micrograms}$$

It is also important to note that the domain of an exponential function is the set of all real numbers, $(\infty, -\infty)$; whereas, the range is the set of all positive real numbers, $(0, \infty)$. This will become clearer as we start graphing these functions.

B Graphing Exponential Functions

3. Graph $f(x) = 3^x$.

EXAMPLE 3 Sketch the graph of the exponential function $f(x) = 2^x$.

Solution Using the results of Example 1, we produce the following table. Graphing the ordered pairs given in the table and connecting them with a smooth curve, we have the graph of $f(x) = 2^x$ shown in Figure 2.

x	f(x)
−3	$\frac{1}{8}$
−2	$\frac{1}{4}$
0	1
1	2
2	4
3	8

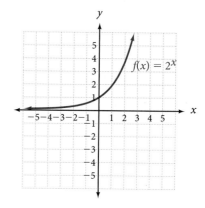

FIGURE 2

Notice that the graph does not cross the x-axis. It *approaches* the x-axis—in fact, we can get it as close to the x-axis as we want without it actually intersecting the x-axis. For the graph of $f(x) = 2^x$ to intersect the x-axis, we would have to find a value of x that would make $2^x = 0$. Because no such value of x exists, the graph of $f(x) = 2^x$ cannot intersect the x-axis.

4. Graph $f(x) = \left(\frac{1}{2}\right)^x$.

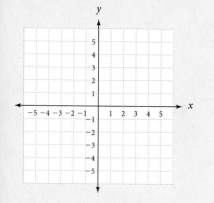

EXAMPLE 4 Sketch the graph of $f(x) = \left(\frac{1}{3}\right)^x$.

Solution The table beside Figure 3 gives some ordered pairs that satisfy the equation. Using the ordered pairs from the table, we have the graph shown in Figure 3.

x	f(x)
−3	27
−2	9
−1	3
0	1
1	$\frac{1}{3}$
2	$\frac{1}{9}$
3	$\frac{1}{27}$

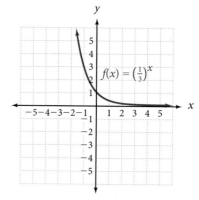

FIGURE 3

The graphs of all exponential functions have two things in common: (1) Each crosses the y-axis at $(0, 1)$ because $b^0 = 1$; and (2) none can cross the x-axis because $b^x = 0$ is impossible due to the restrictions on b.

Figures 4 and 5 show some families of exponential curves to help you become more familiar with them on an intuitive level.

FIGURE 4

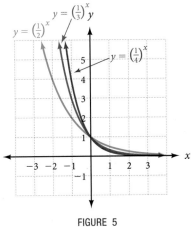

FIGURE 5

Note The use of parentheses around the fractional bases in Example 4 is important. The parentheses ensure that we raise the entire fraction to the given power, not just the numerator. In other words,

$$\left(\frac{1}{3}\right)^x \neq \frac{1^x}{3}$$

C Compound Interest

Among the many applications of exponential functions are the applications having to do with interest-bearing accounts including savings accounts and borrowed money accounts. Here are the details:

If P dollars are deposited in an account with annual interest rate r, compounded n times per year, then the amount of money in the account after t years is given by the formula

$$A(t) = P\left(1 + \frac{r}{n}\right)^{nt}$$

EXAMPLE 5 Suppose you borrow $500 from an account with an annual interest rate of 8% compounded quarterly. Find an equation that gives the amount of money owed after t years. Then find

a. The amount of money owed after 5 years.

b. The number of years it will take for the account balance owed to reach $1,000.

Solution First, we note that $P = 500$ and $r = 0.08$. Interest that is compounded quarterly is compounded four times a year, giving us $n = 4$. Substituting these numbers into the preceding formula, we have our function

$$A(t) = 500\left(1 + \frac{0.08}{4}\right)^{4t} = 500(1.02)^{4t}$$

a. To find the amount after 5 years, we let $t = 5$.

$$A(5) = 500(1.02)^{4 \cdot 5} = 500(1.02)^{20} \approx \$742.97$$

Our answer is found on a calculator, and then rounded to the nearest cent.

b. To see how long it will take for this account to total $1,000, we graph the equation $Y_1 = 500(1.02)^{4X}$ on a graphing calculator, and then look to see where it intersects the line $Y_2 = 1,000$. The two graphs are shown in Figure 6.

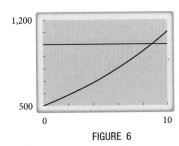

FIGURE 6

Using Zoom and Trace, or the Intersect function on the graphing calculator, we find that the two curves intersect at $X \approx 8.75$ and $Y = 1,000$. This means that our account balance will be $1,000 after the money has been borrowed for 8.75 years.

5. Repeat Example 5 if $600 is borrowed from an account that charges 6% annual interest, compounded monthly.

The Natural Exponential Function

A commonly occurring exponential function is based on a special number we denote with the letter e. The number e is a number like π. It is irrational and occurs in many formulas that describe the world around us. Like π, it can be approximated with a decimal number. Whereas π is approximately 3.1416, e is approximately 2.7183. (If you have a calculator with a key labeled e^x, you can use it to find e^1 to find a more accurate approximation to e.) We cannot give a more precise definition of the number e without using some of the topics taught in calculus. For the work we are going to do with the number e, we only need to know the following.

> **DEFINITION** number e
>
> The **number e** is an irrational number that is approximately 2.7183.

> **DEFINITION** natural exponential function
>
> The **natural exponential function** is any function that can be written in the form
>
> $$f(x) = e^x$$
>
> where e is the number e with an approximate value 2.7183.

Here are a table and graph (Figure 7) for the natural exponential function $f(x) = e^x$:

x	$f(x) = e^x$
-2	$f(-2) = e^{-2} = \frac{1}{e^2} \approx 0.135$
-1	$f(-1) = e^{-1} = \frac{1}{e} \approx 0.368$
0	$f(0) = e^0 = 1$
1	$f(1) = e^1 = e \approx 2.72$
2	$f(2) = e^2 \approx 7.39$
3	$f(3) = e^3 \approx 20.09$

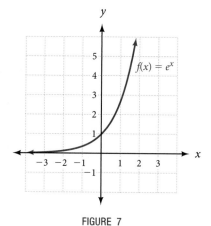

FIGURE 7

As you may have noticed, the domain and range of the natural exponential function is the same as the exponential function.

$$
\begin{array}{ccc}
 & Domain & Range \\
f(x) = b^x & (-\infty, \infty) & (-\infty, \infty) \\
f(x) = e^x & (-\infty, \infty) & (-\infty, \infty)
\end{array}
$$

One common application of natural exponential functions is with interest-bearing accounts. In Example 5, we worked with the formula

$$A(t) = P\left(1 + \frac{r}{n}\right)^{nt}$$

that gives the amount of money in an account if P dollars are deposited for t years at annual interest rate r, compounded n times per year. In Example 5, the number of compounding periods was four. What would happen if we let the number of compounding periods become larger and larger, so that we compounded the interest every day, then every hour,

then every second, and so on? If we take this as far as it can go, we end up compounding the interest every moment. When this happens, we have an account with interest that is compounded continuously, and the amount of money in such an account depends on the number e. Here are the details.

Continuously Compounded Interest If P dollars are deposited in an account with annual interest rate r, compounded continuously, then the amount of money in the account after t years is given by the formula

$$A(t) = Pe^{rt}$$

EXAMPLE 6 Suppose you deposit $500 in an account with an annual interest rate of 8% compounded continuously. Find an equation that gives the amount of money in the account after t years. Then find the amount of money in the account after 5 years.

Solution Because the interest is compounded continuously, we use the formula $A(t) = Pe^{rt}$. Substituting $P = 500$ and $r = 0.08$ into this formula, we have

$$A(t) = 500e^{0.08t}$$

After 5 years, this account will contain

$$A(5) = 500e^{0.08 \cdot 5} = 500e^{0.4} \approx \$745.91$$

to the nearest cent. Compare this result with the answer to Example 5a.

6. Suppose you deposit $600 in an account with an annual interest rate of 6% compounded continuously. How much money is in the account after 5 years?

GETTING READY FOR CLASS

After reading through the preceding section, respond in your own words and in complete sentences.

A. What is an exponential function?

B. In an exponential function, explain why the base b cannot equal 1.

C. Explain continuously compounded interest.

D. What is the special number e in an exponential function?

Vocabulary Review

Choose the correct words to fill in the blanks below.

number e exponential function x-axis $A(t) = Pe^{rt}$ $A(t) = P\left(1 + \dfrac{r}{n}\right)^{nt}$

1. An _____ is any function that can be written in the form $f(x) = b^x$ where b is a positive real number other than 1.

2. The graphs of all exponential functions have two things in common: each crosses the y-axis at $(0, 1)$, and none can cross the _____.

3. The _____ is an irrational number and is approximately 2.7183.

4. An account with an initial deposit P and an annual interest rate r, compounded n times per year, will have a balance after t years given by the function _____.

5. An account with an initial deposit P and an annual interest rate r, compounded continuously, will have a balance after t years given by the function _____.

Problems

A Let $f(x) = 3^x$ and $g(x) = \left(\dfrac{1}{2}\right)^x$, and evaluate each of the following.

1. $g(0)$

2. $f(0)$

3. $g(-1)$

4. $g(-4)$

5. $f(-3)$

6. $f(-1)$

7. $f(2) + g(-2)$

8. $f(2) - g(-2)$

Let $f(x) = 4^x$ and $g(x) = \left(\dfrac{1}{3}\right)^x$. Evaluate each of the following.

9. $f(-1) + g(1)$

10. $f(2) + g(-2)$

11. $\dfrac{f(-2)}{g(1)}$

12. $f(3) - f(2)$

13. $\dfrac{f(3) - f(2)}{3 - 2}$

14. $\dfrac{g(3) - g(2)}{3 - 2}$

15. $\dfrac{f(-1) - f(-2)}{-1 - (-2)}$

16. $\dfrac{g(-2) - g(-4)}{-2 - (-4)}$

Let $f(x) = 3^x$ and $g(x) = \left(\dfrac{1}{2}\right)^x$. Simplify each difference quotient.

17. $\dfrac{f(x) - f(2)}{x - 2}$

18. $\dfrac{g(x) - g(2)}{x - 2}$

19. $\dfrac{f(x) - f(3)}{x - 3}$

20. $\dfrac{g(x) - g(3)}{x - 3}$

21. $\dfrac{f(x + h) - f(x)}{h}$

22. $\dfrac{g(x + h) - g(x)}{h}$

B Graph each of the following exponential functions.

23. $f(x) = 4^x$

24. $f(x) = 2^{-x}$

25. $f(x) = 3^{-x}$

26. $f(x) = \left(\dfrac{1}{3}\right)^{-x}$

27. $g(x) = 2^{x+1}$

28. $g(x) = 2^{x-3}$

29. $g(x) = e^x$

30. $g(x) = e^{-x}$

31. $h(x) = \left(\dfrac{1}{3}\right)^x$

32. $h(x) = \left(\dfrac{1}{2}\right)^{-x}$

33. $h(x) = 3^{x+2}$

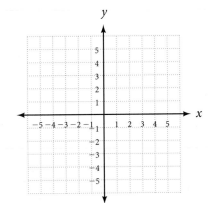

34. $h(x) = 2 \cdot 3^{-x}$

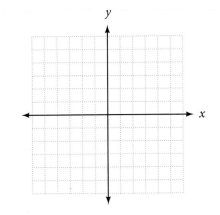

Graph each of the following functions on the same coordinate system for positive values of x only.

35. $y = 2x, y = x^2, y = 2^x$

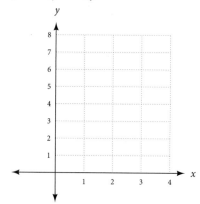

Based on your graph, which curve is growing faster?

36. $y = 3x, y = x^3, y = 3^x$

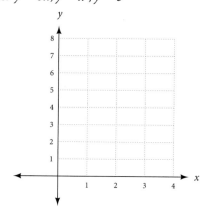

Based on your graph, which curve is growing faster?

37. On a graphing calculator, graph the family of curves $y = b^x, b = 2, 4, 6, 8$.

38. On a graphing calculator, graph the family of curves $y = b^x, b = \frac{1}{2}, \frac{1}{4}, \frac{1}{6}, \frac{1}{8}$.

Applying the Concepts

39. Bouncing Ball Suppose the ball mentioned in the introduction to this section is dropped from a height of 6 feet above the ground. Find an exponential function that gives the height h the ball will attain during the nth bounce. How high will it bounce on the fifth bounce?

40. Bouncing Ball A golf ball is manufactured so that if it is dropped from A feet above the ground onto a hard surface, the maximum height of each bounce will be one half of the height of the previous bounce. Find an exponential function that gives the height h the ball will attain during the nth bounce. If the ball is dropped from 10 feet above the ground onto a hard surface, how high will it bounce on the eighth bounce?

41. Exponential Decay Twinkies on the shelf of a convenience store lose their fresh tastiness over time. We say that the taste quality is 1 when the Twinkies are first put on the shelf at the store, and that the quality of tastiness declines according to the function $Q(t) = 0.85^t$ (t in days). Graph this function on a graphing calculator, and determine when the taste quality will be one half of its original value.

42. Exponential Growth Automobiles built before 1993 use Freon in their air conditioners. The federal government now prohibits the manufacture of Freon. Because the supply of Freon is decreasing, the price per pound is increasing exponentially. Current estimates put the formula for the price per pound of Freon at $p(t) = 1.89(1.25)^t$, where t is the number of years since 1990. Find the price of Freon in 1995 and 1990. How much will Freon cost in the year 2020?

43. Compound Interest Suppose you deposit $1,200 in an account with an annual interest rate of 6% compounded quarterly.

a. Find a function that gives the amount of money in the account after t years.

b. Find the amount of money in the account after 8 years.

c. How many years will it take for the account to contain $2,400?

d. If the interest were compounded continuously, how much money would the account contain after 8 years?

44. Compound Interest Suppose you deposit $500 in an account with an annual interest rate of 8% compounded monthly.

a. Find a function that gives the amount of money in the account after t years.

b. Find the amount of money in the account after 5 years.

c. How many years will it take for the account to contain $1,000?

d. If the interest were compounded continuously, how much money would the account contain after 5 years?

45. Student Loan Suppose you take out a student loan for $5,400. The loan will accrue interest at a rate of 5%, compounded quarterly, but you will not make any payments until after you graduate.

a. Find a function that gives the amount of money you owe when you graduate after t years.

b. Find the amount of money you owe if it takes you 5 years to graduate.

c. How long did it take you to graduate if you owe $8,000? Round to the nearest year.

d. If the interest were compounded continuously, how much money would you owe after 5 years?

46. Student Loan Suppose you take out a student loan for $2,300. The loan will accrue interest at a rate of 7%, compounded monthly, but you will not make any payments until after you graduate.

a. Find a function that gives the amount of money you owe when you graduate after t years.

b. Find the amount of money you owe if it takes you 6 years to graduate.

c. How long did it take you to graduate if you owe $4,000? Round to the nearest year.

d. If the interest were compounded continuously, how much money would you owe after 6 years?

47. Bacteria Growth Suppose it takes 12 hours for a certain strain of bacteria to reproduce by dividing in half. If 50 bacteria are present to begin with, then the total number present after x days will be $f(x) = 50 \cdot 4^x$. Find the total number present after 1 day, 2 days, and 3 days.

48. Bacteria Growth Suppose it takes 1 day for a certain strain of bacteria to reproduce by dividing in half. If 100 bacteria are present to begin with, then the total number present after x days will be $f(x) = 100 \cdot 2^x$. Find the total number present after 1 day, 2 days, 3 days, and 4 days. How many days must elapse before over 100,000 bacteria are present?

Declining-Balance Depreciation The declining-balance method of depreciation is an accounting method businesses use to deduct most of the cost of new equipment during the first few years of purchase. Unlike other methods, the declining-balance formula does not consider salvage value.

49. Value of a Crane The function

$$V(t) = 450,000 (1 - 0.30)^t$$

where V is value and t is time in years, can be used to find the value of a crane for the first 6 years of use.

a. What is the value of the crane after 3 years and 6 months?

b. State the domain of this function using interval notation.

c. Sketch the graph of this function.

d. State the range of this function using interval notation.

e. After how many years will the crane be worth only $85,000?

50. Value of a Printing Press The function

$$V(t) = 375,000(1 - 0.25)^t$$

where V is value and t is time in years, can be used to find the value of a printing press during the first 7 years of use.

a. What is the value of the printing press after 4 years and 9 months?

b. State the domain of this function using interval notation.

c. Sketch the graph of this function.

d. State the range of this function using interval notation.

e. After how many years will the printing press be worth only $65,000?

51. Value of a Painting Alana purchased a painting as an investment for $150. If the painting's value doubles every 3 years, then its value is given by the function

$$V(t) = 150 \cdot 2^{t/3} \text{ for } t \geq 0$$

where t is the number of years since it was purchased, and $V(t)$ is its value (in dollars) at that time. Graph this function.

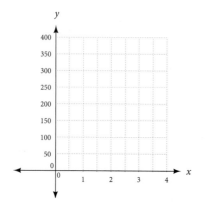

52. Value of a Painting Michael purchased a painting as an investment for $125. If the painting's value doubles every 5 years, then its value is given by the function

$$V(t) = 125 \cdot 2^{t/5} \text{ for } t \geq 0$$

where t is the number of years since it was purchased, and $V(t)$ is its value (in dollars) at that time. Graph this function.

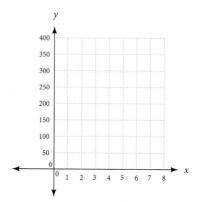

53. **Cost Increase** The cost of a can of Coca Cola in 1960 was $0.10. The exponential function that models the cost of a Coca Cola by year is given below, where t is the number of years since 1960.

$$C(t) = 0.10e^{0.0576t}$$

a. What was the expected cost of a can of Coca Cola in 1985?

b. What was the expected cost of a can of Coca Cola in 2000?

c. What was the expected cost of a can of Coca Cola in 2010?

d. What is the expected cost of a can of Coca Cola in 2050?

54. **Airline Travel** The number of airline passengers in 1990 was 466 million. The number of passengers traveling by airplane each year has increased exponentially according to the model, $P(t) = 466 \cdot 1.035^t$, where t is the number of years since 1990 (U.S. Census Bureau).

a. How many passengers traveled in 1997?

b. How many passengers will travel in 2015?

55. **Bankruptcy Model** In 1997, there were a total of 1,316,999 bankruptcies filed under the Bankruptcy Reform Act (Administrative Office of the U.S. Courts, Statistical Tables for the Federal Judiciary). The model for the number of bankruptcies filed is $B(t) = 0.798 \cdot 1.164^t$, where t is the number of years since 1994 and B is the number of bankruptcies filed in terms of millions. How close was the model in predicting the actual number of bankruptcies filed in 1997?

56. **Value of a Car** As a car ages, its value decreases. The value of a particular car with an original purchase price of $25,600 is modeled by the following function, where c is the value at time t. (Kelly Blue Book)

$$c(t) = 25,600(1 - 0.22)^t$$

a. What is the value of the car when it is 3 years old?

b. What is the total depreciation amount after 4 years?

57. **Bacteria Decay** You are conducting a biology experiment and begin with 5,000,000 cells, but some of those cells are dying each minute. The rate of death of the cells is modeled by the function $A(t) = A_0 \cdot e^{-.598t}$, where A_0 is the original number of cells, t is time in minutes, and A is the number of cells remaining after t minutes.

a. How may cells remain after 5 minutes?

b. How many cells remain after 10 minutes?

c. How many cells remain after 20 minutes?

58. **Health Care** In 1990, $699 billion were spent on health care expenditures. The amount of money, E, in billions spent on health care expenditures can be estimated using the function $E(t) = 78.16(1.11)^t$, where t is time in years since 1970 (U.S. Census Bureau).

a. How close was the estimate determined by the function in estimating the actual amount of money spent on health care expenditures in 1990?

b. What were the expected health care expenditures in 2008, 2009, and 2010?

Maintaining Your Skills

For each of the following relations, specify the domain and range; then indicate which are also functions.

59. $(-2, 6), (-2, 8), (2, 3)$

60. $(1, 2), (3, 4), (4, 1)$

State the domain for each of the following functions. Use interval notation.

61. $f(x) = \dfrac{-4}{x^2 + 2x - 35}$

62. $f(x) = \sqrt{3x + 1}$

If $f(x) = 2x^2 - 18$ and $g(x) = 2x - 6$, find

63. $f(0)$

64. $(g \circ f)(0)$

65. $\dfrac{g(x + h) - g(x)}{h}$

66. $\dfrac{g}{f(x)}$

The chart shows the increasing cost of Disneyland tickets from 1981 to 2011. Use it to answer the following questions.

67. Find the equation of the line that connects the first and the last points of the graph.

68. Use the equation found in problem 63 to find the cost of admission in 1990, 2000, and 2010.

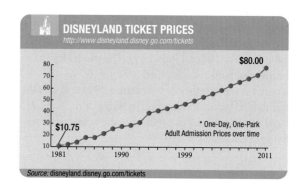

One Step Further

69. Reading Graphs The graphs of two exponential functions are given in Figures 8 and 9. Use the graphs to find the following:

a. $f(0)$ **b.** $f(-1)$ **c.** $f(1)$ **d.** $g(0)$ **e.** $g(1)$ **f.** $g(-1)$ **g.** $f(g(0))$ **h.** $g(f(0))$

FIGURE 8

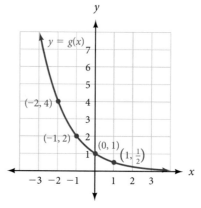

FIGURE 9

70. Analyzing Graphs The goal of this problem is to obtain a sense of the growth rate of an exponential function. We will compare the exponential function $y = 2^x$ to the familiar polynomial function $y = x^2$.

 a. On your calculator, graph $y = 2^x$ and $y = x^2$ using the window X: 0 to 3 and Y: 0 to 10. Which function appears to be growing faster as x grows larger?

b. Show algebraically that both graphs contain the points (2, 4) and (4, 16). You may also want to confirm this on your graph by enlarging the window.

c. On your calculator, graph $y = 2^x$ and $y = x^2$ with the window X: 0 to 15 and Y: 0 to 2,500. Which function dominates (grows the fastest) as x gets larger in the positive direction? Confirm your answer by evaluating $2^{(100)}$ and $(100)^2$.

Getting Ready for the Next Section

Solve each equation for y.

71. $x = 2y - 3$

72. $x = \dfrac{y + 7}{5}$

73. $x = y^2 - 3$

74. $x = (y + 4)^3$

75. $x = \dfrac{y - 4}{y - 2}$

76. $x = \dfrac{y + 5}{y - 3}$

77. $x = \sqrt{y - 3}$

78. $x = \sqrt{y} + 5$

Find the Mistake

Each problem below contains a mistake. Circle the mistake and write the correct number(s) or word(s) on the line provided.

1. The exponential function is any function written in the form $f(x) = x^b$. _____

2. The graph of $f(x) = 2^x$ gets very close to the y-axis but does not intersect it. _____

3. The special number e is a rational number. _____

4. To calculate annual interest rate compounded n times per year, use the formula $A(t) = Pe^{rt}$. _____

©iStockphoto.com/yellowcrestmedia

Note Recall that a relation has one or more outputs for every one input; whereas, a function has only one output for every input. Previously, we have used the vertical line test to tell if a graph is a function or a relation.

Practice Problems

1. If $f(x) = 4x + 1$, find the equation for the inverse of f.

Answer

1. $y = \frac{x-1}{4}$

Rocket Man Bob Maddox plans to strap himself to four rockets and blast himself into space. While he's raising money to build these rockets, he spends his free time building small pulse jet engines for bicycles. For power, the engines pulse 70 times per second. A single engine on a standard bicycle can propel the rider to a top speed of 50 miles per hour in 7 seconds. A bicycle with two engines attached can reach a top speed of 73 miles per hour. The function $f(m) = \frac{15m}{22}$ converts miles per hour, m, to feet per second. Based on our past knowledge of unit analysis, we can convert the bicycle's top speed with two engines from 73 miles per hour to approximately 49.77 feet per second. After reading this section, we will be able to use the inverse function to convert the feet per second quantity back to miles per hour.

A Finding the Inverse Relation

Suppose the function $f(x)$ is given by

$$f(x) = \{(1, 4), (2, 5), (3, 6), (4, 7)\}$$

The inverse of $f(x)$ is obtained by reversing the order of the coordinates in each ordered pair in $f(x)$. The inverse of $f(x)$ is the relation given by

$$g(x) = \{(4, 1), (5, 2), (6, 3), (7, 4)\}$$

It is obvious that the domain of $f(x)$ is now the range of $g(x)$, and the range of $f(x)$ is now the domain of $g(x)$. Every function (or relation) has an inverse that is obtained from the original function by interchanging the components of each ordered pair.

Suppose a function $f(x)$ is defined with an equation instead of a list of ordered pairs. We can obtain the equation of the inverse of $f(x)$ by interchanging the role of x and y in the equation for $f(x)$.

EXAMPLE 1 If the function $f(x)$ is defined by $f(x) = 2x - 3$, find the equation that represents the inverse of $f(x)$.

Solution Because the inverse of $f(x)$ is obtained by interchanging the components of all the ordered pairs belonging to $f(x)$, and each ordered pair in $f(x)$ satisfies the equation $y = 2x - 3$, we simply exchange x and y in the equation $y = 2x - 3$ to get the formula for the inverse of $f(x)$.

$$x = 2y - 3$$

We now solve this equation for y in terms of x.

$$x + 3 = 2y$$

$$\frac{x + 3}{2} = y$$

$$y = \frac{x + 3}{2}$$

The last line gives the equation that defines the inverse of $f(x)$. Let's compare the graphs of $f(x)$ and its inverse as given here. (See Figure 1.)

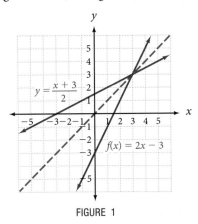

$y = \dfrac{x+3}{2}$

$f(x) = 2x - 3$

FIGURE 1

The graphs of $f(x)$ and its inverse have symmetry about the line $y = x$. This is a reasonable result since the one function was obtained from the other by interchanging x and y in the equation. The ordered pairs (a, b) and (b, a) always have symmetry about the line $y = x$.

B Graph a Function with its Inverse

EXAMPLE 2 Graph the function $y = x^2 - 2$ and its inverse. Give the equation for the inverse.

Solution We can obtain the graph of the inverse of $y = x^2 - 2$ by graphing $y = x^2 - 2$ by the usual methods, and then reflecting the graph about the line $y = x$. This reflection is graphically the equivalent to switching the domain and range.

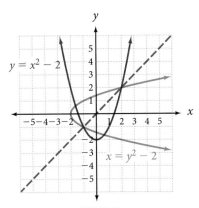

$y = x^2 - 2$

$x = y^2 - 2$

FIGURE 2

The equation that corresponds to the inverse of $y = x^2 - 2$ is obtained by interchanging x and y to get $x = y^2 - 2$.

We can solve the equation $x = y^2 - 2$ for y in terms of x as follows:

$$x = y^2 - 2$$
$$x + 2 = y^2$$
$$y = \pm\sqrt{x + 2}$$

Comparing the graphs from Examples 1 and 2, we observe that the inverse of a function is not always a function. In Example 1, both $f(x)$ and its inverse have graphs that are nonvertical straight lines and therefore both represent functions. In Example 2, the inverse of function $f(x)$ is not a function, since a vertical line crosses it in more than one place.

**Video Examples
Section 14.2**

2. Graph $y = x^2 + 1$ and its inverse. Give the equation for the inverse.

Answer

2. $y = \pm\sqrt{x - 1}$

One-to-One Functions

We can distinguish between those functions with inverses that are also functions and those functions with inverses that are not functions with the following definition:

> **DEFINITION** one-to-one functions
>
> A function is a **one-to-one function** if every element in the range comes from exactly one element in the domain.

This definition indicates that a one-to-one function will yield a set of ordered pairs in which no two different ordered pairs have the same second coordinates. For example, the function

$$f(x) = \{(2, 3), (-1, 3), (5, 8)\}$$

is not one-to-one because the element 3 in the range comes from both 2 and -1 in the domain. On the other hand, the function

$$g(x) = \{(5, 7), (3, -1), (4, 2)\}$$

is a one-to-one function because every element in the range comes from only one element in the domain.

Horizontal Line Test

If we have the graph of a function, then we can determine if the function is one-to-one with the following test. If a horizontal line crosses the graph of a function in more than one place, then the function is not a one-to-one function because the points at which the horizontal line crosses the graph will be points with the same y-coordinates, but different x-coordinates. Therefore, the function will have an element in the range (the y-coordinate) that comes from more than one element in the domain (the x-coordinates).

Of the functions we have covered previously, all the linear functions and exponential functions are one-to-one functions because no horizontal lines can be found that will cross their graphs in more than one place.

Functions Whose Inverses Are Also Functions

Because one-to-one functions do not repeat second coordinates, when we reverse the order of the ordered pairs in a one-to-one function, we obtain a relation in which no two ordered pairs have the same first coordinate — by definition, this relation must be a function. In other words, every one-to-one function has an inverse that is itself a function. Because of this, we can use function notation to represent that inverse.

Note The notation $f^{-1}(x)$ does not represent the reciprocal of $f(x)$. That is, the -1 in this notation is not an exponent. The notation $f^{-1}(x)$ is defined as representing the inverse function for a one-to-one function. In symbols,

$$f^{-1}(x) \neq \frac{1}{f(x)}$$

> **PROPERTY** Inverse Function Notation
>
> If $y = f(x)$ is a one-to-one function, then the inverse of $f(x)$ is also a function and can be denoted by $y = f^{-1}(x)$.

To illustrate, in Example 1 we found that the inverse of $f(x) = 2x - 3$ was the function $y = \frac{x+3}{2}$. We can write this inverse function with inverse function notation as

$$f^{-1}(x) = \frac{x+3}{2}$$

On the other hand, the inverse of the function in Example 2 is not itself a function, so we do not use the notation $f^{-1}(x)$ to represent it.

EXAMPLE 3 Find the inverse of $g(x) = \dfrac{x-4}{x-2}$.

Solution To find the inverse for g, we begin by replacing $g(x)$ with y to obtain

$$y = \frac{x-4}{x-2} \qquad \textit{Original function}$$

To find an equation for the inverse, we exchange x and y.

$$x = \frac{y-4}{y-2} \qquad \textit{Inverse of the original function}$$

To solve for y, we first multiply each side by $y - 2$ to obtain

$$x(y-2) = y-4$$

$$xy - 2x = y - 4 \qquad \textit{Distributive property}$$

$$xy - y = 2x - 4 \qquad \textit{Collect all terms containing y on the left side.}$$

$$y(x-1) = 2x - 4 \qquad \textit{Factor y from each term on the left side.}$$

$$y = \frac{2x-4}{x-1} \qquad \textit{Divide each side by x − 1.}$$

Because our original function is one-to-one, as verified by the graph in Figure 3, its inverse is also a function. Therefore, we can use inverse function notation to write

$$g^{-1}(x) = \frac{2x-4}{x-1}$$

FIGURE 3

EXAMPLE 4 Graph the function $y = 2^x$ and its inverse $x = 2^y$.

Solution We graphed $y = 2^x$ in the preceding section. We simply reflect its graph about the line $y = x$ to obtain the graph of its inverse $x = 2^y$.

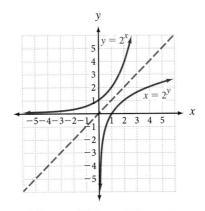

FIGURE 4

3. Find the inverse of $g(x) = \dfrac{x-3}{x+1}$.

4. Graph the function $y = 3^x$ and its inverse $x = 3^y$.

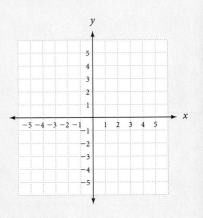

Answer

3. $g^{-1}(x) = \dfrac{3+x}{1-x}$

As you can see from the graph, $x = 2^y$ is a function. However, we have not developed the mathematical tools to solve this equation for y. Therefore, we are unable to use the inverse function notation to represent this function. In the next section, we will give a definition that solves this problem. For now, we simply leave the equation as $x = 2^y$.

Functions, Relations, and Inverses — A Summary

Here is a summary of some of the things we know about functions, relations, and their inverses:

1. Every function is a relation, but not every relation is a function.

2. Every function has an inverse, but only one-to-one functions have inverses that are also functions.

3. The domain of a function is the range of its inverse, and the range of a function is the domain of its inverse.

4. If $y = f(x)$ is a one-to-one function, then we can use the notation $y = f^{-1}(x)$ to represent its inverse function.

5. The graph of a function and its inverse have symmetry about the line $y = x$.

6. If (a, b) belongs to the function $f(x)$, then the point (b, a) belongs to its inverse.

GETTING READY FOR CLASS

After reading through the preceding section, respond in your own words and in complete sentences.

A. What is the inverse of a function?

B. What is the relationship between the graph of a function and the graph of its inverse?

C. What are one-to-one functions?

D. In words, explain inverse function notation.

Vocabulary Review

Choose the correct words to fill in the blanks below.

inverse function relation range one-to-one

1. Every function is a _____, but not every relation is a function.

2. A function is a one-to-one function if every element in the _____ comes from exactly one element in the domain.

3. A horizontal line test can help determine if a function is _____.

4. If $y = f(x)$ is a one-to-one function, then the _____ of $f(x)$ is also a function, given by $y = f^{-1}(x)$.

5. The graph of a _____ and its inverse have symmetry about the line $y = x$.

Problems

A For each of the following one-to-one functions, find the equation of the inverse. Write the inverse using the notation $f^{-1}(x)$.

1. $f(x) = 3x - 1$

2. $f(x) = 2x - 5$

3. $f(x) = x^3$

4. $f(x) = x^3 - 2$

5. $f(x) = \dfrac{x - 3}{x - 1}$

6. $f(x) = \dfrac{x - 2}{x - 3}$

7. $f(x) = \dfrac{x - 3}{4}$

8. $f(x) = \dfrac{x + 7}{2}$

9. $f(x) = \dfrac{1}{2}x - 3$

10. $f(x) = \dfrac{1}{3}x + 1$

11. $f(x) = \dfrac{2}{3}x - 3$

12. $f(x) = -\dfrac{1}{2}x + 4$

13. $f(x) = x^3 - 4$

14. $f(x) = -3x^3 + 2$

15. $f(x) = \dfrac{4x - 3}{2x + 1}$

16. $f(x) = \dfrac{3x - 5}{4x + 3}$

17. $f(x) = \dfrac{2x + 1}{3x + 1}$

18. $f(x) = \dfrac{3x + 2}{5x + 1}$

19. $f(x) = \sqrt[3]{3x - 2}$

20. $f(x) = \sqrt[3]{4x + 1}$

B For each of the following relations, sketch the graph of the relation and its inverse, and write an equation for the inverse.

21. $y = 2x - 1$

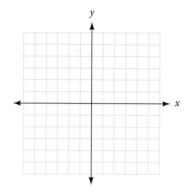

22. $y = 3x + 1$

23. $y = x^2 - 3$

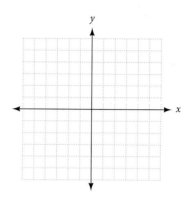

24. $y = x^2 + 1$

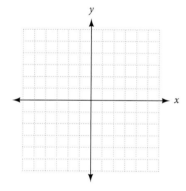

25. $y = x^2 - 2x - 3$

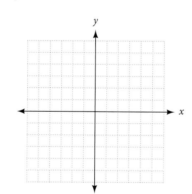

26. $y = x^2 + 2x - 3$

27. $y = 3^x$

28. $y = \left(\dfrac{1}{2}\right)^x$

29. $y = 4$

30. $y = -2$

31. $y = \frac{1}{2}x^3$

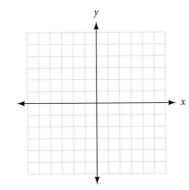

32. $y = x^3 - 2$

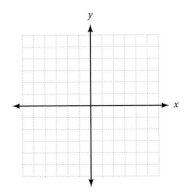

33. $y = \frac{1}{2}x + 2$

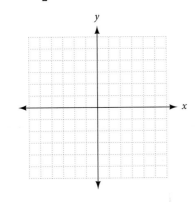

34. $y = \frac{1}{3}x - 1$

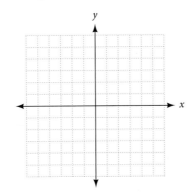

35. $y = \sqrt{x + 2}$

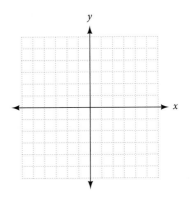

36. $y = \sqrt{x} + 2$

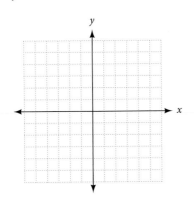

37. $y = \sqrt[3]{x} + 4$

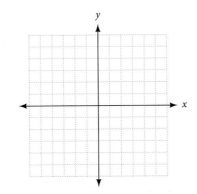

38. $y = \sqrt[3]{x} - 3$

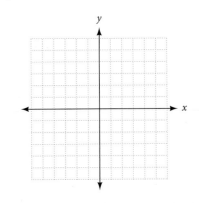

39. Determine if the following functions are one-to-one.

a.

b.

c.

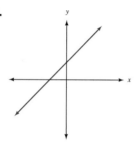

40. Could the following tables of values represent ordered pairs from one-to-one functions? Explain your answer.

a.

x	y
−2	5
−1	4
0	3
1	4
2	5

b.

x	y
1.5	0.1
2.0	0.2
2.5	0.3
3.0	0.4
3.5	0.5

For the given function $f(x)$, find the inverse $g(x)$ and determine whether $g(x)$ represents a function.

41. $f(x) = \{(1, 4), (2, 5), (3, 7)\}$

42. $f(x) = \{(-1, 4), (0, 3), (1, 4)\}$

43. $f(x) = \{(a, A), (b, B), (c, C)\}$

44. $f(x) = \{(d, 4), (f, 6), (g, 7), (D, 4)\}$

45. If $f(x) = 3x - 2$, then $f^{-1}(x) = \dfrac{x + 2}{3}$. Use these two functions to find

a. $f(2)$

b. $f^{-1}(2)$

c. $f(f^{-1}(2))$

d. $f^{-1}(f(2))$

46. If $f(x) = \frac{1}{2}x + 5$, then $f^{-1}(x) = 2x - 10$. Use these two functions to find

a. $f(-4)$

b. $f^{-1}(-4)$

c. $f(f^{-1}(-4))$

d. $f^{-1}(f(-4))$

47. Let $f(x) = \dfrac{1}{x}$, and find $f^{-1}(x)$.

48. Let $f(x) = \dfrac{a}{x}$, and find $f^{-1}(x)$. (a is a real number constant.)

Applying the Concepts

49. Inverse Functions in Words Inverses of one-to-one functions may also be found by *inverse reasoning*. For example, to find the inverse of $f(x) = 3x + 2$, first list, in order, the operations done to variable x:

a. Multiply by 3.

b. Add 2.

Then, to find the inverse, simply apply the inverse operations, in reverse order, to the variable x. That is:

c. Subtract 2.

d. Divide by 3.

The inverse function then becomes $f^{-1}(x) = \frac{x-2}{3}$. Use this method of inverse reasoning to find the inverse of the function $f(x) = \frac{x}{7} - 2$.

50. Inverse Functions in Words Refer to the method of inverse reasoning explained in Problem 49. Use inverse reasoning to find the following inverses:

a. $f(x) = 2x + 7$

b. $f(x) = \sqrt{x} - 9$

c. $f(x) = x^3 - 4$

d. $f(x) = \sqrt{x^3 - 4}$

51. Reading Tables Evaluate each of the following functions using the functions defined by Tables 1 and 2.

a. $f(g(-3))$

b. $g(f(-6))$

c. $g(f(2))$

d. $f(g(3))$

e. $f(g(-2))$

f. $g(f(3))$

Table 1	
x	**f(x)**
−6	3
2	−3
3	−2
6	4

Table 2	
x	**g(x)**
−3	2
−2	3
3	−6
4	6

g. What can you conclude about the relationship between functions $f(x)$ and $g(x)$?

52. Reading Tables Use the functions defined in Tables 1 and 2 in Problem 51 to answer the following questions.

a. What are the domain and range of $f(x)$?

b. What are the domain and range of $g(x)$?

c. How are the domain and range of $f(x)$ related to the domain and range of $g(x)$?

d. Is $f(x)$ a one-to-one function?

e. Is $g(x)$ a one-to-one function?

53. **Temperature** The function $C(F) = \frac{5(F-32)}{9}$ is used to convert the measurement of temperature from Fahrenheit to Celsius. In January 2010, the lowest recorded temperature was $-64°F$ in O'Brien, Alaska (USA Today).

 a. Use this function to determine the temperature in Celsius.

 b. Determine the inverse of this function.

 c. Use the inverse function to convert $0°C$ to Fahrenheit.

54. **Temperature** The function $F(C) = \frac{9}{5}C + 32$ is used to convert the measurement of temperature from Celsius to Fahrenheit. The lowest temperature recorded in Honolulu, Hawaii was $11°C$ in 1902 (The Weather Channel).

 a. Use this function to determine the temperature measured in Fahrenheit.

 b. Determine the inverse of this function.

 c. Use the inverse function to convert $32°F$ to Celsius.

55. Use the function given in problem 53 to find the temperature at which $C(F) = F$. What does this represent?

56. Use the function given in problem 54 to find the temperature at which $F(C) = C$. What does this represent?

57. **Social Security** A function that models the billions of dollars of Social Security payment (as shown in the chart) per year is $s(t) = 16t + 249.4$, where t is time in years since 1990 (U.S. Census Bureau).

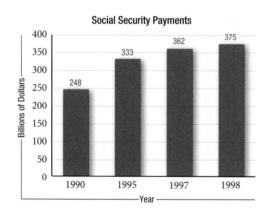

Social Security Payments

 a. Use the model to estimate the amount of Social Security payments paid in 2010.

 b. Write the inverse of the function.

 c. Using the inverse function, estimate the year in which payments reached $507 billion.

58. **Families** The function for the percentage of one-parent families (as shown in the following chart) is $f(x) = 0.417x + 24$, when x is the time in years since 1990 (U.S. Census Bureau).

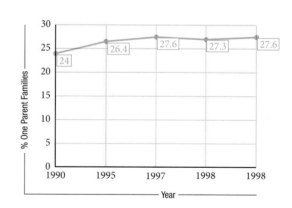

 a. Use the function to predict the percentage of families with one parent in the year 2010.

 b. Determine the inverse of the function, and estimate the year in which approximately 29% of the families were one-parent families.

59. Speed The fastest type of plane, a rocket plane, can travel at a speed of 4,520 miles per hour. The function $f(m) = \frac{22m}{15}$ converts miles per hour, m, to feet per second (World Book Encyclopedia).

 a. Use the function to convert the speed of the rocket plane to feet per second.

 b. Write the inverse of the function.

 c. Using the inverse function, convert 2 feet per second to miles per hour.

60. Speed A Lockheed SR-71A airplane set a world record (as reported by Air Force Armament Museum in 1996) with an absolute speed record of 2,193.167 miles per hour. The function $s(h) = 0.4468424h$ converts miles per hour, h, to meters per second, s (Air Force Armament Museum).

 a. What is the absolute speed of the Lockheed SR-71A in meters per second?

 b. What is the inverse of this function?

 c. Using the inverse function, determine the speed of an airplane in miles per hour that flies 150 meters per second.

Maintaining Your Skills

Solve each equation.

61. $(2x - 1)^2 = 25$

62. $(3x + 5)^2 = -12$

63. What number would you add to $x^2 - 10x$ to make it a perfect square trinomial?

64. What number would you add to $x^2 - 5x$ to make it a perfect square trinomial?

Solve by completing the square.

65. $x^2 - 10x + 8 = 0$
66. $x^2 - 5x + 4 = 0$
67. $3x^2 - 6x + 6 = 0$
68. $4x^2 - 16x - 8 = 0$

One Step Further

For each of the following functions, find $f^{-1}(x)$. Then show that $f(f^{-1}(x)) = x$.

69. $f(x) = 3x + 5$
70. $f(x) = 6 - 8x$
71. $f(x) = x^3 + 1$
72. $f(x) = x^3 - 8$

73. $f(x) = \dfrac{x - 4}{x - 2}$
74. $f(x) = \dfrac{x - 3}{x - 1}$
75. $f(x) = \dfrac{2x - 3}{x + 2}$
76. $f(x) = \dfrac{3x + 1}{2x - 1}$

77. Reading Graphs The graphs of a function and its inverse are shown in Figure 5. Use the graphs to find the following:

a. $f(0)$

b. $f(1)$

c. $f(2)$

d. $f^{-1}(1)$

e. $f^{-1}(2)$

f. $f^{-1}(5)$

g. $f^{-1}(f(2))$

h. $f(f^{-1}(5))$

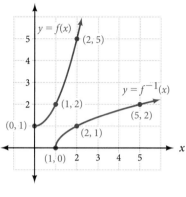

FIGURE 5

78. Domain From a Graph The function f is defined by the equation $f(x) = \sqrt{x + 4}$ for $x \geq -4$, meaning the domain is all x in the interval $[-4, \infty)$.

a. Graph the function f.

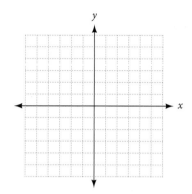

b. On the same set of axes, graph the line $y = x$ and the inverse function f^{-1}.

c. State the domain for the inverse function for f^{-1}.

d. Find the equation for $f^{-1}(x)$ and state its domain.

79. Currency Clark took a trip to Germany in November 2010. He had $1,000 USD of spending money for his trip. While in Germany he spent 694 EURO, the currency used for trade in Europe. The average conversion from USD to EURO at that time followed the function $E(D) = 0.73539D$, where E is the amount of EURO and D is the USD (ONADA FxConverter). How much money in EURO did Clark have left after his trip? USD?

80. Currency Lois left New York on October 21, 2010, to travel to Europe. On that day, she converted $500 USD in cash to EURO currency. The function for the conversion was $E(D) = 0.72088D$ where E is EURO and D is USD (ONADA FxConverter). On the date of her return trip, November 14, 2010, she had 86.50 EURO to convert to USD. On that date, the function for conversion was $E(D) = 0.73067D$.

a. How much money in EURO did Lois take with her to Europe?

b. How much money in USD did Lois return with to New York?

81. Assume that both $f(x)$ and $f^{-1}(x)$ are one-to-one functions. Given $f^{-1}(3) = -6$, solve the equation $f(3x + 2) = 3$.

82. Assume that both $f(x)$ and $f^{-1}(x)$ are one-to-one functions. Given $f(4) = -7$, solve the equation $f^{-1}(4x - 1) = 4$.

83. Let $f(x) = \dfrac{ax + b}{cx + d}$. Find $f^{-1}(x)$.

84. Let $f(x) = \dfrac{5x + 7}{6x - 5}$. Show that f is its own inverse.

85. Let $f(x) = \dfrac{ax + 5}{3x - a}$. Show that f is its own inverse.

86. Let $f(x) = \dfrac{ax + b}{cx - a}$. Show that f is its own inverse.

Getting Ready for the Next Section

Simplify.

87. 3^{-2}

88. 2^3

89. $\left(\dfrac{1}{2}\right)^4$

90. $\left(\dfrac{1}{3}\right)^{-3}$

Solve.

91. $2 = 3x$

92. $3 = 5x$

93. $4 = x^3$

94. $12 = x^2$

Fill in the blanks to make each statement true.

95. $8 = 2^{\square}$

96. $27 = 3^{\square}$

97. $10,000 = 10^{\square}$

98. $1,000 = 10^{\square}$

99. $81 = 3^{\square}$

100. $81 = 9^{\square}$

101. $6 = 6^{\square}$

102. $1 = 5^{\square}$

103. $27 = \left(\dfrac{1}{3}\right)^{\square}$

104. $16 = \left(\dfrac{1}{4}\right)^{\square}$

105. $\dfrac{1}{16} = 2^{\square}$

106. $\dfrac{1}{27} = 3^{\square}$

Find the Mistake

Each problem below contains a mistake. Circle the mistake and write the correct number(s) or word(s) on the line provided.

1. A function is one-to-one if every element in the domain comes from exactly one element in the range.

2. If $y = f(x)$ is a one-to-one function, then the inverse of $f(x)$ is also a relation and can be denoted by $y = f^{-1}(x)$.

3. The graph of a function and its inverse intersect the line $y = x$. _____

4. If (a, b) belongs to the function $f(x)$, then the point $(b, 0)$ belongs to its inverse. _____

A Convert between logarithmic form and exponential form.

B Use the definition of logarithms to solve simple logarithmic equations.

C Sketch the graph of logarithmic functions.

D Simplify expressions involving logarithms.

E Solve applications involving logarithms.

logarithm

logarithmic function

exponential form

logarithmic form

exponential function

**Video Examples
Section 14.3**

©iStockphoto.com/arindambanerjee

In January 2010, news organizations reported that an earthquake had rocked the tiny island nation of Haiti and caused massive destruction. They reported the strength of the earthquake as having measured 7.0 on the Richter scale. For comparison, Table 1 gives the Richter magnitude of a number of other earthquakes.

Table 1	Earthquakes	
Year	**Earthquake**	**Richter Magnitude**
1971	Los Angeles	6.6
1985	Mexico City	8.1
1989	San Francisco	7.1
1992	Kobe, Japan	7.2
1994	Northridge	6.6
1999	Armenia, Colombia	6.0
2004	Indian Ocean	9.1
2010	Haiti	7.0

Source: USGS

Although the sizes of the numbers in the table do not seem to be very different, the intensity of the earthquakes they measure can be very different. For example, the 1989 San Francisco earthquake was more than 10 times stronger than the 1999 earthquake in Colombia. The reason behind this is that the Richter scale is a *logarithmic scale*. In this section, we start our work with logarithms, which will give you an understanding of the Richter scale. Let's begin.

A Logarithmic Functions

As you know from your work in the previous sections, functions of the form

$$y = b^x \quad b > 0, b \neq 1$$

are called exponential functions. Because the equation of the inverse of a function can be obtained by exchanging x and y in the equation of the original function, the inverse of an exponential function must have the form

$$x = b^y \quad b > 0, b \neq 1$$

Now, this last equation is actually the equation of a logarithmic function, as the following definition indicates.

> **DEFINITION** logarithmic function
>
> The equation $y = \log_b x$ is read "y is the logarithm to the base b of x" and is equivalent to the equation
>
> *In symbols*: $\qquad\qquad x = b^y \qquad b > 0, b \neq 1$
>
> *In words*: $\qquad\qquad y$ is the number we raise b to in order to get x.

Notation When an equation is in the form $x = b^y$, it is said to be in *exponential form*. On the other hand, if an equation is in the form $y = \log_b x$, it is said to be in *logarithmic form*.

Here are some equivalent statements written in both forms:

Exponential Form		Logarithmic Form
$8 = 2^3$	\Leftrightarrow	$\log_2 8 = 3$
$25 = 5^2$	\Leftrightarrow	$\log_5 25 = 2$
$0.1 = 10^{-1}$	\Leftrightarrow	$\log_{10} 0.1 = -1$
$\frac{1}{8} = 2^{-3}$	\Leftrightarrow	$\log_2 \frac{1}{8} = -3$
$r = z^s$	\Leftrightarrow	$\log_z r = s$

B Logarithmic Equations

EXAMPLE 1 Solve $\log_3 x = -2$.

Solution In exponential form, the equation looks like this:

$$x = 3^{-2} \quad \text{or} \quad x = \frac{1}{9}$$

The solution is $\frac{1}{9}$.

Converting an equation in logarithmic form to its equivalent exponential form, as we did in Example 1, is the easiest way to isolate and solve for the variable x.

EXAMPLE 2 Solve $\log_x 4 = 3$.

Solution Again, we use the definition of logarithms to write the expression in exponential form.

$$4 = x^3$$

Taking the cube root of both sides, we have

$$\sqrt[3]{4} = \sqrt[3]{x^3}$$
$$x = \sqrt[3]{4}$$

The solution is $\sqrt[3]{4}$.

EXAMPLE 3 Solve $\log_8 4 = x$.

Solution We write the expression again in exponential form.

$$4 = 8^x$$

Because both 4 and 8 can be written as powers of 2, we write them in terms of powers of 2.

$$2^2 = (2^3)^x$$
$$2^2 = 2^{3x}$$

The only way the left and right sides of this last line can be equal is if the exponents are equal — that is, if

$$2 = 3x$$
$$x = \frac{2}{3}$$

The solution is $\frac{2}{3}$. We check as follows:

$$\log_8 4 = \frac{2}{3} \Leftrightarrow 4 = 8^{2/3}$$
$$4 = (\sqrt[3]{8})^2$$
$$4 = 2^2$$
$$4 = 4$$

The solution checks in the original equation.

C Graphing Logarithmic Functions

Graphing logarithmic functions can be done using the graphs of exponential functions and the fact that the graphs of inverse functions have symmetry about the line $y = x$. Here's an example to illustrate:

EXAMPLE 4 Graph the function $f(x) = \log_2 x$.

Solution Let $y = f(x)$. The equation $y = \log_2 x$ is, by definition, equivalent to the exponential equation

$$x = 2^y$$

which is the equation of the inverse of the function

$$y = 2^x$$

The graph of $y = 2^x$ was given in Figure 2 of Section 14.1. We simply reflect the graph of $y = 2^x$ about the line $y = x$ to get the graph of $x = 2^y$, which is also the graph of $y = \log_2 x$. (See Figure 1.)

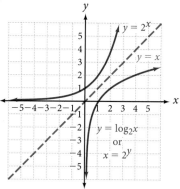

FIGURE 1

It is apparent from the graph that $y = \log_2 x$ is a function, because no vertical line will cross its graph in more than one place. The same is true for all logarithmic equations of the form $y = \log_b x$, where b is a positive number other than 1. Note also that the graph of $y = \log_b x$ will always appear to the right of the y-axis, meaning that x will always be positive in the equation $y = \log_b x$.

D Two Special Identities

If b is a positive real number other than 1, then each of the following is a consequence of the definition of a logarithm:

(1) $b^{\log_b x} = x$ where $x > 0$ and (2) $\log_b b^x = x$ where $x \in \mathbb{R}$.

The justifications for these identities are similar. Let's consider only the first one. Consider the expression

$$y = \log_b x$$

By definition, it is equivalent to

$$x = b^y$$

4. Graph the function $f(x) = \log_3 x$.

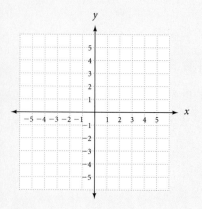

Note $x \in \mathbb{R}$ is read, "x is an element of the set of all real numbers."

Substituting $\log_b x$ for y in the last line gives us

$$x = b^{\log_b x}$$

The next examples in this section show how these two special properties can be used to simplify expressions involving logarithms.

EXAMPLE 5 Simplify the following logarithmic expressions.

a. $\log_2 8$ **b.** $\log_{10} 10{,}000$ **c.** $\log_b b$ **d.** $\log_b 1$ **e.** $\log_4 (\log_5 5)$

Solution

a. Substitute 2^3 for 8: $\log_2 8 = \log_2 2^3 = 3$

b. 10,000 can be written as 10^4: $\log_{10} 10{,}000 = \log_{10} 10^4 = 4$

c. Because $b^1 = b$, we have $\log_b b = \log_b b^1 = 1$

d. Because $1 = b^0$, we have $\log_b 1 = \log_b b^0 = 0$

e. Because $\log_5 5 = 1$, $\log_4 (\log_5 5) = \log_4 1 = 0$

Here is a summary of what we know about the logarithmic function $f(x) = \log_b x$, where b is a positive real number other than zero:

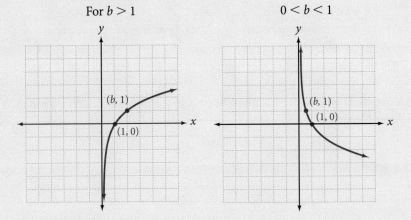

For $b > 1$ │ $0 < b < 1$

The function $f(x) = \log_b x$ is a one-to-one function. It has an x-intercept of $(1, 0)$, but it does not cross the y-axis. It also has a domain of $(0, \infty)$ and a range of $(-\infty, \infty)$.

Application

As we mentioned in the introduction to this section, one application of logarithms is in measuring the magnitude of an earthquake. If an earthquake has a shock wave T times greater than the smallest shock wave that can be measured on a seismograph, then the magnitude M of the earthquake, as measured on the Richter scale, is given by the formula

$$M = \log_{10} T$$

(When we talk about the size of a shock wave, we are talking about its amplitude. The amplitude of a wave is half the difference between its highest point and its lowest point.)

To illustrate the discussion, an earthquake that produces a shock wave that is 10,000 times greater than the smallest shock wave measurable on a seismograph will have a magnitude M on the Richter scale of

$$M = \log_{10} 10{,}000 = 4$$

5. Simplify.

 a. $\log_3 27$

 b. $\log_{10} 1{,}000$

 c. $\log_6 6$

 d. $\log_3 1$

 e. $\log_2 (\log_8 8)$

6. Repeat Example 6 if $M = 6$.

EXAMPLE 6 If an earthquake has a magnitude of $M = 5$ on the Richter scale, what can you say about the size of its shock wave?

Solution To answer this question, we put $M = 5$ into the formula $M = \log_{10} T$ to obtain

$$5 = \log_{10} T$$

Writing this expression in exponential form, we have

$$T = 10^5 = 100,000$$

We can say that an earthquake that measures 5 on the Richter scale has a shock wave 100,000 times greater than the smallest shock wave measurable on a seismograph.

From Example 6 and the discussion that preceded it, we find that an earthquake of magnitude 5 has a shock wave that is 10 times greater than an earthquake of magnitude 4, because 100,000 is 10 times 10,000.

GETTING READY FOR CLASS

After reading through the preceding section, respond in your own words and in complete sentences.

A. What is a logarithm?

B. What is the relationship between $y = 2^x$ and $y = \log_2 x$? How are their graphs related?

C. Will the graph of $y = \log_b x$ ever appear in the second or third quadrants? Explain why or why not.

D. How would you use the identity $b^{\log_b x} = x$ to simplify an equation involving a logarithm?

Answer

6. 1,000,000

Vocabulary Review

Match the following statements written in exponential form with the statement written in the correct logarithmic form.

1. $64 = 4^3$

2. $9 = 3^2$

3. $0.01 = 10^{-1}$

4. $\dfrac{1}{16} = 2^{-4}$

5. $z = x^y$

a. $\log_{10} 0.01 = -1$

b. $\log_4 64 = 3$

c. $\log_2 \dfrac{1}{16} = -4$

d. $\log_3 9 = 2$

e. $\log_x z = y$

Problems

A Write each of the following equations in logarithmic form.

1. $2^4 = 16$

2. $3^2 = 9$

3. $125 = 5^3$

4. $16 = 4^2$

5. $0.01 = 10^{-2}$

6. $0.001 = 10^{-3}$

7. $2^{-5} = \dfrac{1}{32}$

8. $4^{-2} = \dfrac{1}{16}$

9. $\left(\dfrac{1}{2}\right)^{-3} = 8$

10. $\left(\dfrac{1}{3}\right)^{-2} = 9$

11. $27 = 3^3$

12. $81 = 3^4$

Write each of the following equations in exponential form.

13. $\log_{10} 100 = 2$

14. $\log_2 8 = 3$

15. $\log_2 64 = 6$

16. $\log_2 32 = 5$

17. $\log_8 1 = 0$

18. $\log_9 9 = 1$

19. $\log_{10} 0.001 = -3$

20. $\log_{10} 0.0001 = -4$

21. $\log_6 36 = 2$

22. $\log_7 49 = 2$

23. $\log_5 \dfrac{1}{25} = -2$

24. $\log_3 \dfrac{1}{81} = -4$

B Solve each of the following equations for x.

25. $\log_3 x = 2$

26. $\log_4 x = 3$

27. $\log_5 x = -3$

28. $\log_2 x = -4$

29. $\log_2 16 = x$

30. $\log_3 27 = x$

31. $\log_8 2 = x$

32. $\log_{25} 5 = x$

33. $\log_x 4 = 2$

34. $\log_x 16 = 4$

35. $\log_x 5 = 3$

36. $\log_x 8 = 2$

37. $\log_5 25 = x$

38. $\log_5 x = -2$

39. $\log_x 36 = 2$

40. $\log_x \frac{1}{25} = 2$

41. $\log_8 4 = x$

42. $\log_{16} 8 = x$

43. $\log_9 \frac{1}{3} = x$

44. $\log_{27} 9 = x$

45. $\log_8 x = -2$

46. $\log_{36} \frac{1}{6} = x$

47. $\log_{16} 8 = x$

48. $\log_9 \frac{1}{27} = x$

C Sketch the graph of each of the following logarithmic functions.

49. $f(x) = \log_3 x$

50. $f(x) = \log_{1/2} x$

51. $f(x) = \log_{1/3} x$

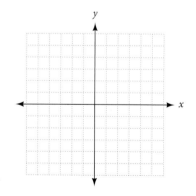

52. $f(x) = \log_4 x$

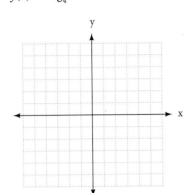

53. $f(x) = \log_5 x$

54. $f(x) = \log_{1/5} x$

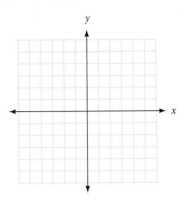

55. $f(x) = \log_{10} x$

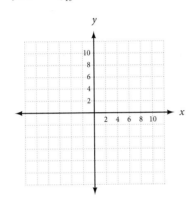

56. $f(x) = \log_{1/4} x$

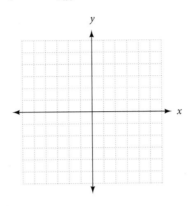

Each of the following graphs is a function of the form $f(x) = b^x$ or $f(x) = \log_b x$. Find the equation for each graph.

57.

58.

59.

60.

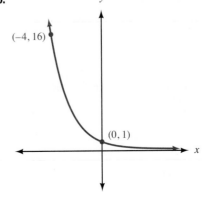

D Simplify each of the following.

61. $\log_2 16$

62. $\log_3 9$

63. $\log_{25} 125$

64. $\log_9 27$

65. $\log_{10} 1{,}000$

66. $\log_{10} 10{,}000$

67. $\log_3 3$

68. $\log_4 4$

69. $\log_5 1$

70. $\log_{10} 1$

71. $\log_{17} 1$

72. $\log_4 8$

73. $\log_{16} 4$

74. $\log_{10} 0.0001$

75. $\log_{100} 1000$

76. $\log_{32} 16$

77. $\log_3 (\log_2 8)$

78. $\log_5 (\log_{32} 2)$

79. $\log_{1/2} (\log_3 81)$

80. $\log_9 (\log_8 2)$

81. $\log_3 (\log_6 6)$

82. $\log_5 (\log_3 3)$

83. $\log_4 [\log_2 (\log_2 16)]$

84. $\log_4 [\log_3 (\log_2 8)]$

Applying the Concepts

85. **Metric System** The metric system uses logical and systematic prefixes for multiplication. For instance, to multiply a unit by 100, the prefix "hecto" is applied, so a hectometer is equal to 100 meters. For each of the prefixes in the following table find the logarithm, base 10, of the multiplying factor.

Prefix	Multiplying Factor	\log_{10} (Multiplying Factor)
Nano	0.000 000 001	
Micro	0.000 001	
Deci	0.1	
Giga	1,000,000,000	
Peta	1,000,000,000,000,000	

86. **Domain and Range** Use the graphs of $y = 2^x$ and $y = \log_2 x$ shown in Figure 1 of this section to find the domain and range for each function. Explain how the domain and range found for $y = 2^x$ relate to the domain and range found for $y = \log_2 x$.

87. Magnitude of an Earthquake Find the magnitude M of an earthquake with a shock wave that measures $T = 100$ on a seismograph.

88. Magnitude of an Earthquake Find the magnitude M of an earthquake with a shock wave that measures $T = 100,000$ on a seismograph.

89. Shock Wave If an earthquake has a magnitude of 8 on the Richter scale, how many times greater is its shock wave than the smallest shock wave measurable on a seismograph?

90. Shock Wave If the 1999 Colombia earthquake had a magnitude of 6 on the Richter scale, how many times greater was its shock wave than the smallest shock wave measurable on a seismograph?

Earthquake The table below categorizes earthquake by the magnitude and identifies the average annual occurrence.

Earthquakes		
Descriptor	Magnitude	Average Annual Occurrence
Great	≥8.0	1
Major	7–7.9	18
Strong	6–6.9	120
Moderate	5–5.9	800
Light	4–4.9	6,200
Minor	3–3.9	49,000
Very Minor	2–2.9	1,000 per day
Very Minor	1–1.9	8,000 per day

Source: *USGS*

91. What is the average number of earthquakes that occur per year when T is 100,000 or greater?

92. What is the average number of earthquakes that occur per year when T is 1,000,000 or greater?

Maintaining Your Skills

93. eBooks The chart shows the memory capacity on some consumer eReaders. If the average novel is 400 KB, find the number of novels that can fit on the following eReaders. Note: 1 MB = 1,024 KB and 1 GB = 1,024 MB.

94. iPads Using the chart, what percentage of iPad sales did not go to the seven states listed?

a. Amazon Kindle DX

b. Sony Ready PRS-600

c. Barnes and Noble Nook

Getting Ready for the Next Section

Simplify.

95. $8^{2/3}$

96. $27^{2/3}$

97. $16^{3/4}$

98. $81^{3/4}$

99. $(-27)^{2/3}$

100. $16^{-3/4}$

101. $27^{-2/3}$

102. $(-32)^{4/5}$

Solve.

103. $(x + 2)(x) = 2^3$

104. $(x + 3)(x) = 2^2$

105. $\dfrac{x-2}{x+1} = 9$

106. $\dfrac{x+1}{x-4} = 25$

Write in exponential form.

107. $\log_2 [(x + 2)(x)] = 3$

108. $\log_4 [x(x - 6)] = 2$

109. $\log_3\left(\dfrac{x-2}{x+1}\right) = 4$

110. $\log_3\left(\dfrac{x-1}{x-4}\right) = 2$

Find the Mistake

Each problem below contains a mistake. Circle the mistake and write the correct number(s) or word(s) on the line provided.

1. The expression $y = \log_b x$ is read "y is the logarithm to the base x of b." _____

2. The expression $\frac{1}{16} = 2^{-4}$ is equivalent to the exponential expression $\log_2 \frac{1}{16} = -4$. _____

3. The graph of $y = \log_2 x$ is a function because no horizontal line will cross its graph in more than one place.

4. A special identity for logarithms states that b raised to the power of $\log_b x$ is equal to b. _____

Landmark Review: Checking Your Progress

Let $f(x) = 3^x$ and $g(x) = \left(\frac{1}{3}\right)^x$, and evaluate each of the following.

1. $g(0)$

2. $f(-1)$

3. $f(-2) + g(2)$

4. $\frac{f(-3) - g(-2)}{-3-2}$

Graph each of the following exponential functions.

5. $f(x) = 3^x$

6. $g(x) = 4^{(x+1)}$

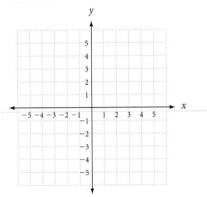

For each of the following functions, find the equation of the inverse.

7. $f(x) = 5x - 2$

8. $f(x) = 2x^3 - 2$

For each of the following relations, sketch the graph of the relation and its inverse, and write an equation for the inverse.

9. $y = 2x + 3$

10. $y = 2^x$

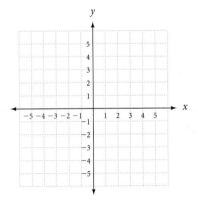

Write each of the following expressions in logarithmic form.

11. $4^2 = 16$

12. $2^4 = 16$

Write each of the following expressions in exponential form.

13. $\log(3)\frac{1}{9} = -2$

14. $\log(7)49 = 2$

Solve each of the following equations for x.

15. $\log(5)x = 2$

16. $\log(x)\frac{1}{64} = -2$

17. $\log(x)3 = 5$

18. $\log(25)\left(\frac{1}{5}\right) = x$

A Use the properties of logarithms to convert between expanded form and single logarithms.

B Use the properaties of logarithms to solve equations that contain logarithms.

product property for logarithms

quotient property for logarithms

power property for logarithms

Video Examples Section 14.4

14.4 Properties of Logarithms

Image © sxc.hu, JWilsher, 2008

Recall the pulse jet engines we discussed earlier in the chapter. A rider of the jet-powered bicycle must wear ear protection while the engine is going. The sound of the pulse jet can reach 150 decibels, which is enough to pop a person's eardrums without the protection.

A decibel is a measurement of sound intensity, more specifically the difference in intensity between two sounds. This difference is equal to ten times the logarithm of the ratio of the two sounds. The following table and bar chart compare the intensity of sounds with which we are familiar:

Decibels	Comparable to
10	Calm breathing
20	Whisper
60	Normal conversation
80	City traffic
90	Lawn mower
110	Amplified music
120	Snowmobile
140	Fireworks
180	Airplane takeoff

Source: *hearingaid.know.com*

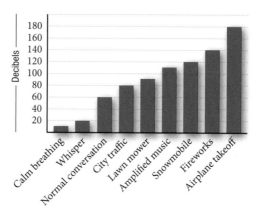

Chapter 14 Exponential and Logarithmic Functions

The precise definition for a *decibel* is

$$D = 10 \log_{10}\left(\frac{I}{I_0}\right)$$

where I is the intensity of the sound being measured, and I_0 is the intensity of the least audible sound. In this section, we will see that the preceding formula can also be written as

$$D = 10(\log_{10} I - \log_{10} I_0)$$

The rules we use to rewrite expressions containing logarithms are called the *properties of logarithms*. We will consider three of them here.

A Properties of Logarithms

For the following three properties, x, y, and b are all positive real numbers, $b \neq 1$, and r is any real number.

> **PROPERTY** Product Property for Logarithms
>
> $$\log_b(xy) = \log_b x + \log_b y$$
>
> *In words:* The logarithm of a **product** is the **sum** of the logarithms.

> **PROPERTY** Quotient Property for Logarithms
>
> $$\log_b\left(\frac{x}{y}\right) = \log_b x - \log_b y$$
>
> *In words:* The logarithm of a **quotient** is the **difference** of the logarithms.

> **PROPERTY** Power Property for Logarithms
>
> $$\log_b x^r = r \log_b x$$
>
> *In words:* The logarithm of a number raised to a **power** is the **product** of the power and the logarithm of the number.

Note Here are some common mistakes students make when working with properties of logarithms:

$$\log_b x + \log_b y \neq \log_b(x + y)$$
$$\log_b x - \log_b y \neq \log_b(x - y)$$
$$r\log_b x \neq (\log_b x)^r$$

Compare these errors with the correct properties to make sure you don't make the same mistakes.

Proof of the Product Property for Logarithms To prove the product property for logarithms, we simply apply the first identity for logarithms given in the preceding section.

$$b^{\log_b xy} = xy = (b^{\log_b x})(b^{\log_b y}) = b^{\log_b x + \log_b y}$$

Because the first and last expressions are equal and the bases are the same, the exponents $\log_b xy$ and $\log_b x + \log_b y$ must be equal. Therefore,

$$\log_b xy = \log_b x + \log_b y$$

The proofs of the quotient and power properties for logarithms proceed in much the same manner, so we will omit them here. The examples that follow show how the three properties can be used.

1. Expand, using the properties of logarithms.

$$\log_3 \frac{5a}{b}$$

2. Expand $\log_{10} \dfrac{x^2}{\sqrt[3]{y}}$.

Note When evaluating logarithmic expressions, it is generally helpful to change any radicals to equivalent exponents, as we have done in Example 2. However, we recommend converting final answers from exponential form to radical form, unless instructed otherwise, as we did in Example 3.

3. Write as a single logarithm.

$$3 \log_4 x + \log_4 y - 2 \log_4 z$$

EXAMPLE 1 Expand using the properties of logarithms: $\log_5 \dfrac{3xy}{z}$.

Solution Applying the quotient property for logarithms, we can write the quotient of $3xy$ and z in terms of a difference.

$$\log_5 \frac{3xy}{z} = \log_5 3xy - \log_5 z$$

Applying the product property for logarithms to the result $3xy$, we write it in terms of addition.

$$\log_5 \frac{3xy}{z} = \log_5 3 + \log_5 x + \log_5 y - \log_5 z$$

EXAMPLE 2 Expand using the properties of logarithms.

$$\log_2 \frac{x^4}{\sqrt{y} \cdot z^3}$$

Solution We write \sqrt{y} as $y^{1/2}$ and apply the properties.

$$\frac{\log_2 x^4}{\sqrt{y} \cdot z^3} = \log_2 \frac{x^4}{y^{1/2} z^3} \qquad \sqrt{y} = y^{1/2}$$

$$= \log_2 x^4 - \log_2(y^{1/2} \cdot z^3) \qquad \text{Quotient property for logarithms}$$

$$= \log_2 x^4 - (\log_2 y^{1/2} + \log_2 z^3) \qquad \text{Product property for logarithms}$$

$$= \log_2 x^4 - \log_2 y^{1/2} - \log_2 z^3 \qquad \text{Remove parentheses and distribute } -1.$$

$$= 4 \log_2 x - \frac{1}{2} \log_2 y - 3 \log_2 z \qquad \text{Power property for logarithms}$$

We can also use the three properties to write an expression in expanded form as just one logarithm.

EXAMPLE 3 Write the following as a single logarithm.

$$2 \log_{10} a + 3 \log_{10} b - \frac{1}{3} \log_{10} c$$

Solution We begin by applying the power property for logarithms.

$$2 \log_{10} a + 3 \log_{10} b - \frac{1}{3} \log_{10} c$$

$$= \log_{10} a^2 + \log_{10} b^3 - \log_{10} c^{1/3} \qquad \text{Power property for logarithms}$$

$$= \log_{10} (a^2 \cdot b^3) - \log_{10} c^{1/3} \qquad \text{Product property for logarithms}$$

$$= \log_{10} \frac{a^2 b^3}{c^{1/3}} \qquad \text{Quotient property for logarithms}$$

$$= \log_{10} \frac{a^2 b^3}{\sqrt[3]{c}} \qquad c^{1/3} = \sqrt[3]{c}$$

When evaluating logarithmic expressions, it is generally helpful to change any radicals to an equivalent exponent, as we have done in Example 2. However, we recommend converting final answers from exponential form to radical form, unless instructed otherwise, as we did in Example 3.

Answers

1. $\log_3 5 + \log_3 a - \log_3 b$

2. $2\log_{10} x - \frac{1}{3}\log_{10} y$

3. $\log_4 \frac{x^3 y}{z^2}$

B Equations with Logarithms

The properties of logarithms along with the definition of logarithms are useful in solving equations that involve logarithms.

EXAMPLE 4 Solve $\log_2(x + 2) + \log_2 x = 3$.

Solution Applying the product property for logarithms to the left side of the equation allows us to write it as a single logarithm.

$$\log_2(x + 2) + \log_2 x = 3$$

$$\log_2[(x + 2)(x)] = 3$$

The last line can be written in exponential form using the definition of logarithms:

$$(x + 2)(x) = 2^3$$

$$x^2 + 2x = 8$$

$$x^2 + 2x - 8 = 0$$

$$(x + 4)(x - 2) = 0$$

$$x + 4 = 0 \quad \text{or} \quad x - 2 = 0$$

$$x = -4 \qquad\qquad x = 2$$

In the previous section, we noted the fact that x in the expression $y = \log_b x$ cannot be a negative number. Because substitution of $x = -4$ into the original equation gives

$$\log_2(-2) + \log_2(-4) = 3$$

which contains logarithms of negative numbers, we cannot use -4 as a solution. The solution is 2.

●

4. Solve $\log_2(x + 3) + \log_2 x = 2$.

GETTING READY FOR CLASS

After reading through the preceding section, respond in your own words and in complete sentences.

A. Explain why the following statement is false: "The logarithm of a product is the product of the logarithms."

B. Explain the quotient property for logarithms in words.

C. Explain the difference between $\log_b m + \log_b n$ and $\log_b(m + n)$. Are they equivalent?

D. Explain the difference between $\log_b(mn)$ and $(\log_b m)(\log_b n)$. Are they equivalent?

Answer

4. 1

Vocabulary Review

Choose the correct words to fill in the blanks below.

quotient decibel power product

1. The definition of _____ is $D = 10 \log_{10}\left(\dfrac{I}{I_0}\right)$.

2. The _____ property for logarithms states that $\log_b(xy) = \log_b x + \log_b y$.

3. The _____ property for logarithms states that $\log_b\left(\dfrac{x}{y}\right) = \log_b x - \log_b y$.

4. The _____ property for logarithms states that $\log_b x^r = r \log_b x$.

Problems

A Use the three properties of logarithms given in this section to expand each expression as much as possible.

1. $\log_3 4x$

2. $\log_2 5x$

3. $\log_6 \dfrac{5}{x}$

4. $\log_3 \dfrac{x}{5}$

5. $\log_2 y^5$

6. $\log_7 y^3$

7. $\log_9 \sqrt[3]{z}$

8. $\log_8 \sqrt{z}$

9. $\log_6 x^2 y^4$

10. $\log_{10} x^2 y^4$

11. $\log_5 \sqrt{x} \cdot y^4$

12. $\log_8 \sqrt[3]{xy^6}$

13. $\log_b \dfrac{xy}{z}$

14. $\log_b \dfrac{3x}{y}$

15. $\log_{10} \dfrac{4}{xy}$

16. $\log_{10} \dfrac{5}{4y}$

17. $\log_{10} \dfrac{x^2 y}{\sqrt{z}}$

18. $\log_{10} \dfrac{\sqrt{x} \cdot y}{z^3}$

19. $\log_{10} \dfrac{x^3 \sqrt{y}}{z^4}$

20. $\log_{10} \dfrac{x^4 \sqrt[3]{y}}{\sqrt{z}}$

21. $\log_b \sqrt[3]{\dfrac{x^2 y}{z^4}}$

22. $\log_b \sqrt[4]{\dfrac{x^4 y^3}{z^5}}$

23. $\log_3 \sqrt[3]{\dfrac{x^2 y}{z^6}}$

24. $\log_8 \sqrt[4]{\dfrac{x^5 y^6}{z^3}}$

25. $\log_a \dfrac{4x^5}{9a^2}$

26. $\log_b \dfrac{16b^2}{25y^3}$

27. $\log_b \dfrac{9x^3}{64b^2}$

28. $\log_a \dfrac{27a^6}{16y^2}$

Write each expression as a single logarithm.

29. $\log_b x + \log_b z$

30. $\log_b x - \log_b z$

31. $2 \log_3 x - 3 \log_3 y$

32. $4 \log_2 x + 5 \log_2 y$

33. $\dfrac{1}{2} \log_{10} x + \dfrac{1}{3} \log_{10} y$

34. $\dfrac{1}{3} \log_{10} x - \dfrac{1}{4} \log_{10} y$

35. $3 \log_2 x + \dfrac{1}{2} \log_2 y - \log_2 z$

36. $2 \log_3 x + 3 \log_3 y - \log_3 z$

37. $\dfrac{1}{2} \log_2 x - 3 \log_2 y - 4 \log_2 z$

38. $3 \log_{10} x - \log_{10} y - \log_{10} z$

39. $\dfrac{3}{2} \log_{10} x - \dfrac{3}{4} \log_{10} y - \dfrac{4}{5} \log_{10} z$

40. $3 \log_{10} x - \dfrac{4}{3} \log_{10} y - 5 \log_{10} z$

41. $\dfrac{1}{2} \log_5 x + \dfrac{2}{3} \log_5 y - 4 \log_5 z$

42. $\dfrac{1}{4} \log_7 x + 5 \log_7 y - \dfrac{1}{3} \log_7 z$

43. $\log_3(x^2 - 16) - 2 \log_3(x + 4)$

44. $\log_4(x^2 - x - 6) - \log_4(x^2 - 9)$

B Solve each of the following equations.

45. $\log_2 x + \log_2 3 = 1$

46. $\log_3 x + \log_3 3 = 1$

47. $\log_3 x - \log_3 2 = 2$

48. $\log_3 x + \log_3 2 = 2$

49. $\log_3 x + \log_3(x - 2) = 1$

50. $\log_6 x + \log_6(x - 1) = 1$

51. $\log_3(x + 3) - \log_3(x - 1) = 1$

52. $\log_4(x - 2) - \log_4(x + 1) = 1$

53. $\log_2 x + \log_2(x - 2) = 3$

54. $\log_4 x + \log_4(x + 6) = 2$

55. $\log_8 x + \log_8(x - 3) = \dfrac{2}{3}$

56. $\log_{27} x + \log_{27}(x + 8) = \dfrac{2}{3}$

57. $\log_3(x + 2) - \log_3 x = 1$

58. $\log_2(x + 3) - \log_2(x - 3) = 2$

59. $\log_2(x + 1) + \log_2(x + 2) = 1$

60. $\log_3 x + \log_3(x + 6) = 3$

61. $\log_9 \sqrt{x} + \log_9 \sqrt{2x + 3} = \dfrac{1}{2}$

62. $\log_8 \sqrt{x} + \log_8 \sqrt{5x + 2} = \dfrac{2}{3}$

63. $4 \log_3 x - \log_3 x^2 = 6$

64. $9 \log_4 x - \log_4 x^3 = 12$

65. $\log_5 \sqrt{x} + \log_5 \sqrt{6x + 5} = 1$

66. $\log_2 \sqrt{x} + \log_2 \sqrt{6x + 5} = 1$

Applying the Concepts

67. Decibel Formula Use the properties of logarithms to rewrite the decibel formula $D = 10 \log_{10}\left(\frac{I}{I_0}\right)$ as

$$D = 10(\log_{10} I - \log_{10} I_0).$$

68. Decibel Formula In the decibel formula $D = 10 \log_{10}\left(\frac{I}{I_0}\right)$, the threshold of hearing, I_0, is

$$I_0 = 10^{-12} \text{ watts/meter}^2$$

Substitute 10^{-12} for I_0 in the decibel formula, then show that it simplifies to

$$D = 10(\log_{10} I + 12)$$

69. Finding Logarithms If $\log_{10} 8 = 0.903$ and $\log_{10} 5 = 0.699$, find the following without using a calculator.

a. $\log_{10} 40$

b. $\log_{10} 320$

c. $\log_{10} 1,600$

70. Matching Match each expression in the first column with an equivalent expression in the second column.

a. $\log_2(ab)$ i. b

b. $\log_2\left(\dfrac{a}{b}\right)$ ii. 2

c. $\log_5 a^b$ iii. $\log_2 a + \log_2 b$

d. $\log_a b^a$ iv. $\log_2 a - \log_2 b$

e. $\log_a a^b$ v. $a \log_a b$

f. $\log_3 9$ vi. $b \log_5 a$

71. Henderson–Hasselbalch Formula Doctors use the Henderson–Hasselbalch formula to calculate the pH of a person's blood. pH is a measure of the acidity and/or the alkalinity of a solution. This formula is represented as

$$\text{pH} = 6.1 + \log_{10}\left(\frac{x}{y}\right)$$

where x is the base concentration and y is the acidic concentration. Rewrite the Henderson–Hasselbalch formula so that the logarithm of a quotient is not involved.

72. Henderson–Hasselbalch Formula Refer to the information in the preceding problem about the Henderson–Hasselbalch formula. If most people have a blood pH of 7.4, use the Henderson–Hasselbalch formula to find the ratio of $\frac{x}{y}$ for an average person.

73. Food Processing The formula $M = 0.21(\log_{10} a - \log_{10} b)$ is used in the food processing industry to find the number of minutes M of heat processing a certain food should undergo at 250°F to reduce the probability of survival of *Clostridium botulinum* spores. The letter a represents the number of spores per can before heating, and b represents the number of spores per can after heating. Find M if $a = 1$ and $b = 10^{-12}$. Then find M using the same values for a and b in the formula $M = 0.21 \log_{10} \frac{a}{b}$.

74. Acoustic Powers The formula $N = \log_{10} \frac{P_1}{P_2}$ is used in radio electronics to find the ratio of the acoustic powers of two electric circuits in terms of their electric powers. Find N if P_1 is 100 and P_2 is 1. Then use the same two values of P_1 and P_2 to find N in the formula $N = \log_{10} P_1 - \log_{10} P_2$.

Getting Ready for the Next Section

Simplify.

75. 5^0

76. 4^1

77. $\log_3 3$

78. $\log_5 5$

79. $\log_b b^4$

80. $\log_a a^k$

81. $4^{\log_4 x}$

82. $3^{\log_3 a}$

Use a calculator to find each of the following. Write your answer in scientific notation with the first number in each answer rounded to the nearest tenth.

83. $10^{-5.6}$

84. $10^{-4.1}$

Divide and round to the nearest whole number

85. $\dfrac{2.00 \times 10^8}{3.96 \times 10^6}$

86. $\dfrac{3.25 \times 10^{12}}{1.72 \times 10^{10}}$

Find the Mistake

Each problem below contains a mistake. Circle the mistake and write the correct number(s) or word(s) on the line provided.

1. The logarithm of a product is the difference of the logarithms. _____

2. The quotient property for logarithms states that $\log_b \frac{x}{y} = \log_b x + \log_b y$. _____

3. The logarithm of a number raised to a power is the product of the number and the logarithm of the power.

4. To expand $\log_6 \frac{4xy}{z}$, apply the quotient property and the power property for logarithms. _____

14.5 Common and Natural Logarithms with Applications

OBJECTIVES

A Use a calculator to find common logarithms.

B Use a calculator to find a number given its common logarithm.

C Solve applications involving common logarithms.

D Simplify expressions containing natural logarithms.

KEY WORDS

common logarithm

characteristic

mantissa

antilogarithm

natural logarithm

Acid rain was first discovered in the 1960s by Gene Likens and his research team who studied the damage caused by acid rain to Hubbard Brook in New Hampshire. Acid rain is rain with a pH of 5.6 and below. pH is defined in terms of common logarithms — one of the topics we will present here. When you are finished with this section, you will have a more detailed knowledge of pH and acid rain.

Two kinds of logarithms occur more frequently than other logarithms. Logarithms with a base of 10 are very common because our number system is a base-10 number system. For this reason, we call base-10 logarithms *common logarithms*.

DEFINITION common logarithms

A *common logarithm* is a logarithm with a base of 10. Because common logarithms are used so frequently, it is customary, in order to save time, to omit notating the base. That is,

$$\log_{10} x = \log x$$

When the base is not shown, it is assumed to be 10.

A Common Logarithms

Common logarithms of powers of 10 are simple to evaluate. We need only recognize that $\log 10 = \log_{10} 10 = 1$ and apply the power property for logarithms $\log_b x^r = r \log_b x$.

$$
\begin{aligned}
\log 1{,}000 &= \log 10^3 &= &\quad 3 \log 10 = &\quad 3(1) = &\quad 3 \\
\log 100 &= \log 10^2 &= &\quad 2 \log 10 = &\quad 2(1) = &\quad 2 \\
\log 10 &= \log 10^1 &= &\quad 1 \log 10 = &\quad 1(1) = &\quad 1 \\
\log 1 &= \log 10^0 &= &\quad 0 \log 10 = &\quad 0(1) = &\quad 0 \\
\log 0.1 &= \log 10^{-1} &= &-1 \log 10 = &-1(1) = &-1 \\
\log 0.01 &= \log 10^{-2} &= &-2 \log 10 = &-2(1) = &-2 \\
\log 0.001 &= \log 10^{-3} &= &-3 \log 10 = &-3(1) = &-3
\end{aligned}
$$

To find common logarithms of numbers that are not powers of 10, we use a calculator with a $\boxed{\log}$ key.

Check the following logarithms to be sure you know how to use your calculator. (These answers have been rounded to the nearest ten-thousandth.)

$$\log 7.02 \approx 0.8463$$

$$\log 1.39 \approx 0.1430$$

$$\log 6.00 \approx 0.7782$$

$$\log 9.99 \approx 0.9996$$

**Video Examples
Section 14.5**

Practice Problems

1. Find log 27,600.

EXAMPLE 1 Use a calculator to find log 2,760.

Solution

$$\log 2{,}760 \approx 3.4409$$

To work this problem on a scientific calculator, we simply enter the number 2,760 and press the key labeled $\boxed{\log}$. On a graphing calculator we press the $\boxed{\log}$ key first, then 2,760.

 The 3 in the answer is called the *characteristic,* and the decimal part of the logarithm is called the *mantissa.*

2. Find log 0.00391.

EXAMPLE 2 Find log 0.0391.

Solution $\log 0.0391 \approx -1.4078$

3. Find log 0.00952.

EXAMPLE 3 Find log 0.00523.

Solution $\log 0.00523 \approx -2.2815$

B Antilogarithms

4. Find x if log $x = 3.9786$.

EXAMPLE 4 Find x if log $x = 3.8774$.

Solution We are looking for the number whose logarithm is 3.8774. On a scientific calculator, we enter 3.8774 and press the key labeled $\boxed{10^x}$. On a graphing calculator we press $\boxed{10^x}$ first, then 3.8774. The result is 7,540 to four significant digits. Here's why:

$$\text{If} \qquad \log x = 3.8774$$
$$\text{then} \qquad x = 10^{3.8774}$$
$$\approx 7{,}540$$

The number 7,540 is called the *antilogarithm* or just *antilog* of 3.8774. That is, 7,540 is the number whose logarithm is 3.8774.

5. Find x if log $x = -1.5901$.

EXAMPLE 5 Find x if log $x = -2.4179$.

Solution Using the $\boxed{10^x}$ key, the result is 0.00382.

$$\text{If} \qquad \log x = -2.4179$$
$$\text{then} \qquad x = 10^{-2.4179}$$
$$\approx 0.00382$$

The antilog of -2.4179 is 0.00382. That is, the logarithm of 0.00382 is -2.4179.

C Applications of Logarithms

In Section 8.3, we found that the magnitude M of an earthquake that produces a shock wave T times larger than the smallest shock wave that can be measured on a seismograph is given by the formula

$$M = \log_{10} T$$

We can rewrite this formula using our shorthand notation for common logarithms as

$$M = \log T$$

Answers

1. 4.4409
2. -2.4078
3. -2.0214
4. 9,519
5. 0.0257

EXAMPLE 6 The San Francisco earthquake of 1906 is estimated to have measured 8.3 on the Richter scale. The San Fernando earthquake of 1971 measured 6.6 on the Richter scale. Find T for each earthquake, and then give some indication of how much stronger the 1906 earthquake was than the 1971 earthquake.

Solution For the 1906 earthquake:

If $\log T = 8.3$, then $T = 2.00 \times 10^8$.

For the 1971 earthquake:

If $\log T = 6.6$, then $T = 3.98 \times 10^6$.

Dividing the two values of T and rounding our answer to the nearest whole number, we have

$$\frac{2.00 \times 10^8}{3.98 \times 10^6} \approx 50$$

The shock wave for the 1906 earthquake was approximately 50 times larger than the shock wave for the 1971 earthquake.

In chemistry, the pH of a solution is the measure of the acidity of the solution. The definition for pH involves common logarithms. Here it is:

$$\text{pH} = -\log[\text{H}^+]$$

where $[\text{H}^+]$ is the concentration of the hydrogen ion in moles per liter. The range for pH is from 0 to 14. Pure water, a neutral solution, has a pH of 7. An acidic solution, such as vinegar, will have a pH less than 7, and an alkaline solution, such as ammonia, has a pH above 7.

EXAMPLE 7 Normal rainwater has a pH of 5.6. What is the concentration of the hydrogen ion in normal rainwater?

Solution Substituting 5.6 for pH in the formula $\text{pH} = -\log[\text{H}^+]$, we have

$5.6 = -\log[\text{H}^+]$	Substitution
$\log[\text{H}^+] = -5.6$	Isolate the logarithm.
$[\text{H}^+] = 10^{-5.6}$	Write in exponential form.
$\approx 2.5 \times 10^{-6}$ moles per liter	Answer in scientific notation.

EXAMPLE 8 The concentration of the hydrogen ion in a sample of acid rain known to kill fish is 3.2×10^{-5} mole per liter. Find the pH of this acid rain to the nearest tenth.

Solution Substituting 3.2×10^{-5} for $[\text{H}^+]$ in the formula $\text{pH} = -\log[\text{H}^+]$, we have

$\text{pH} = -\log[3.2 \times 10^{-5}]$	Substitution
$\approx -(-4.5)$	Evaluate the logarithm.
≈ 4.5	Simplify.

D Natural Logarithms

Recall that the number e is an irrational number with an approximate value of 2.7183. Here is a logarithm that uses the number e.

> **DEFINITION** natural logarithms
>
> A *natural logarithm* is a logarithm with a base of e. The natural logarithm of x is denoted by $\ln x$. That is,
>
> $$\ln x = \log_e x$$

6. Find T if an earthquake measures 5.5 on the Richter scale.

Richter Scale

5.5

THE ACID SCALE

7. Find the concentration of the hydrogen ion in a can of cola if the pH is 4.1.

8. The concentration of hydrogen ions in a sample of acid rain is 1.8×10^{-5}. Find the pH. Round to the nearest tenth.

Answers

6. 3.16×10^5

7. 7.9×10^{-5}

8. 4.7

We can assume that all our properties of exponents and logarithms hold for expressions with a base of e, because e is a real number. Here are some examples intended to make you more familiar with the number e and natural logarithms.

9. Simplify each expression.
 a. $\ln e^2$
 b. $\ln e^4$
 c. $\ln e^{-2}$
 d. $\ln e^x$

EXAMPLE 9 Simplify each of the following expressions.

a. $e^0 = 1$

b. $e^1 = e$

c. $\ln e = 1$ In exponential form, $e^1 = e$.

d. $\ln 1 = 0$ In exponential form, $e^0 = 1$.

e. $\ln e^3 = 3$

f. $\ln e^{-4} = -4$

g. $\ln e^t = t$

10. Expand $\ln Pe^{rt}$.

EXAMPLE 10 Use the properties of logarithms to expand the expression $\ln (Ae^{5t})$.

Solution Because the properties of logarithms hold for natural logarithms, we have

$$\ln (Ae^{5t}) = \ln A + \ln e^{5t}$$
$$= \ln A + 5t \ln e$$
$$= \ln A + 5t \qquad \text{Because } \ln e = 1$$

11. If $\ln 5 = 1.6094$ and $\ln 7 = 1.9459$, find the following:
 a. $\ln 35$
 b. $\ln 0.2$
 c. $\ln 49$

EXAMPLE 11 If $\ln 2 = 0.6931$ and $\ln 3 = 1.0986$, find

a. $\ln 6$ **b.** $\ln 0.5$ **c.** $\ln 8$

Solution

a. Because $6 = 2 \cdot 3$, we have

$$\ln 6 = \ln 2 \cdot 3$$
$$= \ln 2 + \ln 3$$
$$= 0.6931 + 1.0986$$
$$= 1.7917$$

b. Writing 0.5 as $\frac{1}{2}$ and applying the quotient property for logarithms gives us

$$\ln 0.5 = \ln \frac{1}{2}$$
$$= \ln 1 - \ln 2$$
$$= 0 - 0.6931$$
$$= -0.6931$$

c. Writing 8 as 2^3 and applying the power property for logarithms, we have

$$\ln 8 = \ln 2^3$$
$$= 3 \ln 2$$
$$= 3(0.6931)$$
$$= 2.0793$$

GETTING READY FOR CLASS

After reading through the preceding section, respond in your own words and in complete sentences.

A. What is a common logarithm?

B. What is a natural logarithm?

C. Is e a rational number? Explain.

D. Find $\ln e$, and explain how you arrived at your answer.

Answers
9. a. 2 **b.** 4 **c.** -2 **d.** x
10. $\ln P + rt$
11. a. 3.5553 **b.** -1.6094 **c.** 3.8918

Vocabulary Review

Choose the correct words to fill in the blanks below.

| natural | common | characteristic | mantissa |

1. A _____ logarithm is a logarithm with a base of 10 and denoted as $\log_{10} x = \log x$.

2. When a logarithm is calculated as a decimal, the whole number part of the logarithm is called the _____ and the decimal part is called the _____.

3. A _____ logarithm is a logarithm with a base of e and denoted as $\ln x = \log_e x$.

Problems

A Find the following logarithms.

1. $\log 378$

2. $\log 426$

3. $\log 37.8$

4. $\log 42,600$

5. $\log 3,780$

6. $\log 0.4260$

7. $\log 0.0378$

8. $\log 0.0426$

9. $\log 37,800$

10. $\log 4,900$

11. $\log 600$

12. $\log 900$

13. $\log 2,010$

14. $\log 10,200$

15. $\log 0.00971$

16. $\log 0.0312$

17. $\log 0.0314$

18. $\log 0.00052$

19. $\log 0.399$

20. $\log 0.111$

B Find x in the following equations.

21. $\log x = 2.8802$

22. $\log x = 4.8802$

23. $\log x = -2.1198$

24. $\log x = -3.1198$

25. $\log x = 3.1553$

26. $\log x = 5.5911$

27. $\log x = -5.3497$

28. $\log x = -1.5670$

29. $\log x = -7.0372$

30. $\log x = -4.2000$

31. $\log x = 10$

32. $\log x = -1$

33. $\log x = -10$ **34.** $\log x = 1$ **35.** $\log x = 20$ **36.** $\log x = -20$

37. $\log x = -2$ **38.** $\log x = 4$ **39.** $\log x = \log_2 8$ **40.** $\log x = \log_3 9$

41. $\ln x = -1$ **42.** $\ln x = 4$ **43.** $\log x = 2 \log 5$ **44.** $\log x = -\log 4$

45. $\ln x = -3 \ln 2$ **46.** $\ln x = 5 \ln 3$ **47.** $\ln x = 4 \ln 3$ **48.** $\ln x = -2 \ln 5$

D Simplify each of the following expressions.

49. $\ln e$ **50.** $\ln 1$ **51.** $\ln e^5$ **52.** $\ln e^{-3}$

53. $\ln e^x$ **54.** $\ln e^y$ **55.** $\log 10{,}000$ **56.** $\log 0.001$

57. $\ln \dfrac{1}{e^3}$ **58.** $\ln \sqrt{e}$ **59.** $\log \sqrt{1000}$ **60.** $\log \sqrt[3]{10{,}000}$

Use the properties of logarithms to expand each of the following expressions.

61. $\ln (10e^{3t})$ **62.** $\ln (10e^{4t})$ **63.** $\ln (Ae^{-2t})$ **64.** $\ln (Ae^{-3t})$

65. $\log [100(1.01)^{3t}]$ **66.** $\log \left[\dfrac{1}{10} (1.5)^{t+2} \right]$ **67.** $\ln (Pe^{rt})$ **68.** $\ln \left(\dfrac{1}{2} e^{-kt} \right)$

69. $-\log (4.2 \times 10^{-3})$ **70.** $-\log (5.7 \times 10^{-10})$ **71.** $-\log (3.4 \times 10^4)$ **72.** $-\log (5.6 \times 10^8)$

If $\ln 2 = 0.6931$, $\ln 3 = 1.0986$, and $\ln 5 = 1.6094$, find each of the following.

73. $\ln 15$ **74.** $\ln 10$ **75.** $\ln \dfrac{1}{3}$ **76.** $\ln \dfrac{1}{5}$

77. $\ln 9$ **78.** $\ln 25$ **79.** $\ln 16$ **80.** $\ln 81$

Applying the Concepts

81. Atomic Bomb Tests The Bikini Atoll in the Pacific Ocean was used as a location for atomic bomb tests by the United States government in the 1950s. One such test resulted in an earthquake measurement of 5.0 on the Richter scale. Compare the 1906 San Francisco earthquake of estimated magnitude 8.3 on the Richter scale to this atomic bomb test. Use the shock wave T for purposes of comparison. The formula for determining the magnitude, M, of an earthquake on the Richter Scale is $M = \log_{10} T$, where T is the number of times the shockwave is greater than the smallest measurable shockwave.

82. Atomic Bomb Tests Today's nuclear weapons are 1,000 times more powerful than the atomic bombs tested in the Bikini Atoll mentioned in Problem 81. Use the shock wave T to determine the Richter scale measurement of a nuclear test today.

83. On March 11, 2011, one of the largest natural disasters in history occurred on the coast of Tohoku, Japan. An earthquake, which measured 9.0 on the Richter scale, and the resulting tsunami killed over 20,000 people. Below are the four largest earthquakes since 1900. For each, use the formula $T = 10^m$ to convert each magnitude to the level of energy of the quake.

84. For each of these earthquakes listed in problem 83, find how many times more energy they released compared to the 1906 San Francisco earthquake of magnitude 8.3.

Date	Location	Magnitude M	Energy Level T
5/22/1960	Chile	9.5	
3/28/1964	Alaska	9.2	
12/26/2004	Indonesia	9.1	
3/11/2011	Japan	9.0	

Source: *USGS*

85. Getting Close to e Use a calculator to complete the following table.

x	$(1 + x)^{1/x}$
1	
0.5	
0.1	
0.01	
0.001	
0.0001	
0.00001	

86. Getting Close to e Use a calculator to complete the following table.

x	$\left(1 + \dfrac{1}{x}\right)^x$
1	
10	
50	
100	
500	
1,000	
10,000	
1,000,000	

87. What number does the expression $(1 + x)^{1/x}$ seem to approach as x gets closer and closer to zero?

88. What number does the expression $\left(1 + \frac{1}{x}\right)^{x}$ seem to approach as x gets larger and larger?

89. University Enrollment The percentage of students enrolled in a university who are between the ages of 25 and 34 can be modeled by the formula $s = 5 \ln x$, where s is the percentage of students and x is the number of years since 1989. Predict the year in which approximately 15% of students enrolled in a university were between the ages of 25 and 34.

90. Memory A class of students take a test on the mathematics concept of solving quadratic equations. That class agrees to take a similar form of the test each month for the next 6 months to test their memory of the topic since instruction. The function of the average score earned each month on the test is $m(x) = 75 - 5\ln(x + 1)$, where x represents time in months. Complete the table to indicate the average score earned by the class at each month.

Time x	Score m
0	
1	
2	
3	
4	
5	
6	

Use the following figure to solve Problems 91–94.

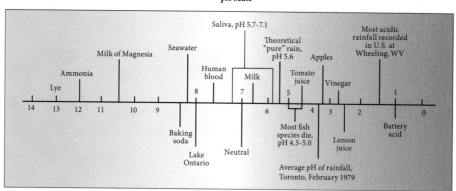

pH Scale

91. pH Find the pH of orange juice if the concentration of the hydrogen ion in the juice is $[H^+] = 6.50 \times 10^{-4}$.

92. pH Find the pH of milk if the concentration of the hydrogen ions in milk is $[H^+] = 1.88 \times 10^{-6}$.

93. pH Find the concentration of hydrogen ions in a glass of wine if the pH is 4.75.

94. pH Find the concentration of hydrogen ions in a bottle of vinegar if the pH is 5.75.

The Richter Scale Find the relative size T of the shock wave of earthquakes with the following magnitudes, as measured on the Richter scale.

95. 5.5 **96.** 6.6 **97.** 8.3 **98.** 8.7

99. Earthquake The chart below is a partial listing of earthquakes that were recorded in Canada during 2000. Complete the chart by computing the magnitude on the Richter Scale M or the number of times the associated shockwave is larger than the smallest measurable shockwave T.

Location	Date	Magnitude M	Shockwave T
Moresby Island	Jan. 23	4.0	
Vancouver Island	Apr. 30		1.99×10^5
Quebec City	June 29	3.2	
Mould Bay	Nov. 13	5.2	
St. Lawrence	Dec. 14		5.01×10^3

Source: *National Resources Canada, National Earthquake Hazards Program*

100. Earthquake On January 6, 2001, an earthquake with a magnitude of 7.7 on the Richter Scale hit southern India (*National Earthquake Information Center*). By what factor was this earthquake's shockwave greater than the smallest measurable shockwave?

Depreciation The annual rate of depreciation r on a car that is purchased for P dollars and is worth W dollars t years later can be found from the formula

$$\log(1 - r) = \frac{1}{t} \log \frac{W}{P}$$

101. Find the annual rate of depreciation on a car that is purchased for $9,000 and sold 5 years later for $4,500.

102. Find the annual rate of depreciation on a car that is purchased for $9,000 and sold 4 years later for $3,000.

Two cars depreciate in value according to the following depreciation tables. In each case, find the annual rate of depreciation.

103.

Age in years	Value in dollars
New	7,550
5	5,750

104.

Age in years	Value in dollars
New	7,550
3	5,750

Getting Ready for the Next Section

Solve.

105. $5(2x + 1) = 12$

106. $4(3x - 2) = 21$

107. $3^x = \dfrac{1}{27}$

108. $2^x = \dfrac{1}{8}$

Use a calculator to evaluate the following. Give answers to 4 decimal places.

109. $\dfrac{100{,}000}{32{,}000}$

110. $\dfrac{1.4982}{6.5681} + 3$

111. $\dfrac{1}{2}\left(\dfrac{-0.6931}{1.4289} + 3\right)$

112. $1 + \dfrac{0.04}{52}$

Use the power property to rewrite the following logarithms.

113. $\log 1.05^t$

114. $\log 1.033^t$

115. $\ln 1.045^{2t}$

116. $\ln 1.016^{10t}$

Use identities to simplify.

117. $\ln e^{0.05t}$

118. $\ln e^{-0.000121t}$

119. $\ln e^{-0.0042t}$

120. $\ln e^{0.0046t}$

Use a calculator to evaluate the following. Give answers to 2 decimal places.

121. $\dfrac{\ln 10}{0.0015}$

122. $\dfrac{\ln 2}{0.00125}$

123. $\dfrac{\ln 2}{\ln 1.043}$

124. $\dfrac{\ln 3}{\ln\left(1 + \dfrac{.08}{12}\right)}$

Find the Mistake

Each problem below contains a mistake. Circle the mistake and write the correct number(s) or word(s) on the line provided.

1. A common logarithm uses the expression $\log_{10} x$, which is equal to $\log x^{10}$. _____

2. When the base of a log is not shown, it is assumed to be 1. _____

3. When the logarithm of a whole number is approximated as a decimal, the decimal part of the answer is called the characteristic. _____

4. An antilogarithm is a logarithm with a base of e, denoted by $\ln x = \log_e x$. _____

OBJECTIVES

A Solve exponential equations.

B Use the change-of-base property to calculate logarithms.

C Solve application problems involving logarithmic or exponential equations.

KEY WORDS

doubling time

exponential equation

change-of-base property

For items involved in exponential growth, the time it takes for a quantity to double is called the *doubling time*. For example, if you invest $5,000 in an account that pays 5% annual interest, compounded quarterly, you may want to know how long it will take for your money to double in value. You can find this doubling time if you can solve the equation

$$10,000 = 5,000 \, (1.0125)^{4t}$$

You will see as you progress through this section that logarithms are the key to solving equations of this type.

A Exponential Equations

Logarithms are very important in solving equations in which the variable appears as an exponent. The equation

$$5^x = 12$$

is an example of one such equation. Equations of this form are called *exponential equations*. Because the quantities 5^x and 12 are equal, so are their common logarithms.

We begin our solution by taking the logarithm of both sides.

$$\log 5^x = \log 12$$

We now apply the power property for logarithms, $\log x^r = r \log x$, to turn x from an exponent into a coefficient.

$$x \log 5 = \log 12$$

Dividing both sides by log 5 gives us

$$x = \frac{\log 12}{\log 5}$$

If we want a decimal approximation to the solution, we can find log 12 and log 5 on a calculator and divide.

$$x \approx \frac{1.0792}{0.6990} \approx 1.5439$$

Note It is common to mistake the equation $y = x^2$ as an exponential equation. Yes, the equation has an exponent but it is not the variable appearing as such. The equation $y = 2^x$ is an exponential equation because the variable x appears as the exponent.

**Video Examples
Section 14.6**

The complete problem looks like this:

$$5^x = 12$$

$$\log 5^x = \log 12$$

$$x \log 5 = \log 12$$

$$x = \frac{\log 12}{\log 5}$$

$$\approx \frac{1.0792}{0.6990}$$

$$\approx 1.5439$$

Here is another example of solving an exponential equation using logarithms.

EXAMPLE 1 Solve $25^{2x+1} = 15$.

Solution Taking the logarithm of both sides and then writing the exponent $(2x + 1)$ as a coefficient, we proceed as follows:

$$25^{2x+1} = 15$$

$$\log 25^{2x+1} = \log 15 \qquad \textit{Take the log of both sides.}$$

$$(2x + 1)\log 25 = \log 15 \qquad \textit{Power property for logarithms}$$

$$2x + 1 = \frac{\log 15}{\log 25} \qquad \textit{Divide by log 25.}$$

$$2x = \frac{\log 15}{\log 25} - 1 \qquad \textit{Add } -1 \textit{ to both sides.}$$

$$x = \frac{1}{2}\left(\frac{\log 15}{\log 25} - 1\right) \qquad \textit{Multiply both sides by } \frac{1}{2}.$$

Using a calculator, we can write a decimal approximation to the answer.

$$x \approx \frac{1}{2}\left(\frac{1.1761}{1.3979} - 1\right)$$

$$\approx \frac{1}{2}(0.8413 - 1)$$

$$\approx \frac{1}{2}(-0.1587)$$

$$\approx -0.079$$

B Change of Base

There is a fourth property of logarithms we have not yet considered. This last property allows us to change from one base to another and is therefore called the *change-of-base property*.

> **PROPERTY** Change-of-Base Property for Logarithms
>
> If a and b are both positive numbers other than 1, and if $x > 0$, then
>
> $$\log_a x = \frac{\log_b x}{\log_b a}$$
>
> \uparrow Base a \uparrow Base b

The logarithm on the left side has a base of a, and both logarithms on the right side have a base of b. This allows us to change from base a to any other base b that is a positive number other than 1. Here is a proof of the change-of-base property for logarithms:

Practice Problems

1. Solve $12^{x+2} = 20$.

Answer

1. -0.7945 or -0.744 depending on whether you round your intermediate answers.

Proof We begin by writing the identity

$$a^{\log_a x} = x$$

Taking the logarithm base b of both sides and writing the exponent $\log_a x$ as a coefficient, we have

$$\log_b a^{\log_a x} = \log_b x$$

$$(\log_a x)(\log_b a) = \log_b$$

Dividing both sides by $\log_b a$, we have the desired result:

$$\frac{(\log_a x)(\log_b a)}{\log_b a} = \frac{\log_b x}{\log_b a}$$

$$\log_a x = \frac{\log_b x}{\log_b a}$$

We can use this property to find logarithms we could not otherwise compute on our calculators — that is, logarithms with bases other than 10 or e. The next example illustrates the use of this property.

EXAMPLE 2 Calculate $\log_8 24$.

Solution Because we do not have base-8 logarithms on our calculators, we can change this expression to an equivalent expression that contains only base-10 logarithms.

$$\log_8 24 = \frac{\log 24}{\log 8} \qquad \textit{Change-of-base property}$$

Don't be confused. We did not just drop the base, we changed to base 10. We could have written the last line like this:

$$\log_8 24 = \frac{\log_{10} 24}{\log_{10} 8}$$

From our calculators, we write

$$\log_8 24 \approx \frac{1.3802}{0.9031} \approx 1.5283$$

C Applications

If you invest P dollars in an account with an annual interest rate r that is compounded n times a year, then t years later the amount of money in that account will be

$$A = P\left(1 + \frac{r}{n}\right)^{nt}$$

EXAMPLE 3 How long does it take for $5,000 to double if it is deposited in an account that yields 5% interest compounded once a year?

Solution Substituting $P = 5,000$, $r = 0.05$, $n = 1$, and $A = 10,000$ into our formula, we have

$$10,000 = 5,000(1 + 0.05)^t$$

$$10,000 = 5,000(1.05)^t$$

$$2 = (1.05)^t \qquad \textit{Divide by 5,000.}$$

This is an exponential equation. We solve by taking the logarithm of both sides.

$$\log 2 = \log(1.05)^t$$

$$\log 2 = t \log 1.05$$

$$t = \frac{\log 2}{\log 1.05} \qquad \textit{Divide both sides by log 1.05.}$$

$$\approx 14.2$$

It takes a little over 14 years for $5,000 to double if it earns 5% interest per year, compounded once a year.

2. Calculate $\log_6 14$.

3. How long does it take for $5,000 to double if it is invested in an account that pays 11% interest compounded once a year?

Answers
2. 1.4728 or 1.4729
3. 6.64 years

4. How long will it take the city in Example 4 to grow to 75,000 people from its current population of 32,000?

EXAMPLE 4 Suppose that the population of a small city is 32,000 in the beginning of 2010. The city council assumes that the population size t years later can be estimated by the equation

$$P = 32{,}000e^{0.05t}$$

Approximately when will the city have a population of 50,000?

Solution We substitute 50,000 for P in the equation and solve for t.

$$50{,}000 = 32{,}000e^{0.05t}$$

$$1.5625 = e^{0.05t} \qquad \frac{50{,}000}{32{,}000} = 1.5625$$

To solve this equation for t, we can take the natural logarithm of each side.

$$\ln 1.5625 = \ln e^{0.05t}$$

$$= 0.05t \ln e \qquad \textit{Power property for logarithms}$$

$$= 0.05t \qquad \textit{Because } \ln e = 1$$

$$t = \frac{\ln 1.5625}{0.05} \qquad \textit{Divide each side by 0.05.}$$

$$\approx 8.93 \text{ years}$$

We can estimate that the population will reach 50,000 toward the end of 2018.

USING TECHNOLOGY Graphing Calculators

We can evaluate many logarithmic expressions on a graphing calculator by using the fact that logarithmic functions and exponential functions are inverses.

5. Evaluate the logarithmic expression $\log_5 8$ from the graph of an exponential function.

EXAMPLE 5 Evaluate the logarithmic expression $\log_3 7$ from the graph of an exponential function.

Solution First, we let $\log_3 7 = x$. Next, we write this expression in exponential form as $3^x = 7$. We can solve this equation graphically by finding the intersection of the graphs $Y_1 = 3^x$ and $Y_2 = 7$, as shown in Figure 1.

Using the calculator, we find the two graphs intersect at $(1.77, 7)$. Therefore, $\log_3 7 = 1.77$ to the nearest hundredth. We can check our work by evaluating the expression $3^{1.77}$ on our calculator with the key strokes

$$3 \boxed{\wedge} 1.77 \boxed{\text{ENTER}}$$

The result is 6.99 to the nearest hundredth, which seems reasonable since 1.77 is accurate to the nearest hundredth. To get a result closer to 7, we would need to find the intersection of the two graphs more accurately.

FIGURE 1

GETTING READY FOR CLASS

After reading through the preceding section, respond in your own words and in complete sentences.

A. What is an exponential equation?

B. How do logarithms help you solve exponential equations?

C. What is the change-of-base property?

D. Write an application modeled by the equation $A = 10{,}000\left(1 + \frac{0.08}{2}\right)^{(2)(5)}$.

Answers

4. Approximately 17.04 years

5. 1.29

Vocabulary Review

Choose the correct words to fill in the blanks below.

| variable | logarithms | exponential | change-of-base | doubling time |

1. When considering _____ growth, the time it takes for a quantity to double is called the _____.

2. An equation where the _____ appears as an exponent is called an exponential equation.

3. We use _____ to help solve exponential equations.

4. The _____ property states that $\log_a x = \dfrac{(\log_b s)}{(\log_b a)}$.

Problems

A Solve each exponential equation. Use a calculator to write the answer in decimal form. Round answers to four decimal places where appropriate.

1. $3^x = 5$

2. $4^x = 3$

3. $5^x = 3$

4. $3^x = 4$

5. $5^{-x} = 12$

6. $7^{-x} = 8$

7. $12^{-x} = 5$

8. $8^{-x} = 7$

9. $8^{x+1} = 4$

10. $9^{x+1} = 3$

11. $4^{x-1} = 4$

12. $3^{x-1} = 9$

13. $3^{2x+1} = 2$

14. $2^{2x+1} = 3$

15. $3^{1-2x} = 2$

16. $2^{1-2x} = 3$

17. $15^{3x-4} = 10$

18. $10^{3x-4} = 15$

19. $6^{5-2x} = 4$

20. $9^{7-3x} = 5$

21. $3^{-4x} = 81$

22. $2^{5x} = \dfrac{1}{16}$

23. $5^{3x-2} = 15$

24. $7^{4x+3} = 200$

25. $64^x = 16^{x-3}$

26. $81^x = 27^{x-4}$

27. $8^a = 16^{2a+5}$

28. $9^y = 81^{2y-3}$

29. $16^{x+1} = 8^{3x-2}$

30. $32^{x+3} = 4^{3x-2}$

31. $27^{3b-5} = 81^{2b-3}$

32. $16^{3x-3} = 128^{2x+4}$

33. $100e^{3t} = 250$

34. $150e^{0.065t} = 400$

35. $1200\left(1 + \dfrac{0.072}{4}\right)^{4t} = 25000$

36. $2700\left(1 + \dfrac{0.086}{12}\right)^{12t} = 10000$

37. $50e^{-0.0742t} = 32$

38. $19e^{-0.000243t} = 12$

B Use the change-of-base property to calculate each of the following logarithms. Round answers to four decimal places, where appropriate.

39. $\log_8 16$

40. $\log_9 27$

41. $\log_{16} 8$

42. $\log_{27} 9$

43. $\log_7 15$

44. $\log_3 12$

45. $\log_{15} 7$

46. $\log_{12} 3$

47. $\log_8 240$

48. $\log_6 180$

49. $\log_4 321$

50. $\log_5 462$

Find a decimal approximation to each of the following natural logarithms.

51. $\ln 345$

52. $\ln 3{,}450$

53. $\ln 0.345$

54. $\ln 0.0345$

55. $\ln 10$

56. $\ln 100$

57. $\ln 45{,}000$

58. $\ln 450{,}000$

Applying the Concepts

Use the formula $A = P\left(1 + \dfrac{r}{n}\right)^{nt}$ to solve the following compound interest problems. Round all answers to two decimal places.

59. Compound Interest How long will it take for $500 to double if it is invested at 6% annual interest compounded 2 times a year?

60. Compound Interest How long will it take for $500 to double if it is invested at 6% annual interest compounded 12 times a year?

61. Compound Interest How long will it take for $1,000 to triple if it is invested at 12% annual interest compounded 6 times a year?

62. Compound Interest How long will it take for $1,000 to become $4,000 if it is invested at 12% annual interest compounded 6 times a year?

63. Doubling Time How long does it take for an amount of money P to double itself if it is invested at 8% interest compounded 4 times a year?

64. Tripling Time How long does it take for an amount of money P to triple itself if it is invested at 8% interest compounded 4 times a year?

65. Tripling Time If a $25 investment is worth $75 today, how long ago must that $25 have been invested at 6% interest compounded twice a year?

66. Doubling Time If a $25 investment is worth $50 today, how long ago must that $25 have been invested at 6% interest compounded twice a year?

Recall from Section 14.1 that if P dollars are invested in an account with annual interest rate r, compounded continuously, then the amount of money in the account after t years is given by the formula

$$A(t) = Pe^{rt}$$

67. Continuously Compounded Interest Repeat Problem 51 if the interest is compounded continuously.

68. Continuously Compounded Interest Repeat Problem 54 if the interest is compounded continuously.

69. Continuously Compounded Interest How long will it take $500 to triple if it is invested at 6% annual interest, compounded continuously? Round your answer to two decimal places.

70. Continuously Compounded Interest How long will it take $500 to triple if it is invested at 12% annual interest, compounded continuously? Round your answer to two decimal places.

71. Continuously Compounded Interest How long will it take for $1,000 to be worth $2,500 at 8% interest, compounded continuously? Round your answer to two decimal places.

72. Continuously Compounded Interest How long will it take for $1,000 to be worth $5,000 at 8% interest, compounded continuously? Round your answer to two decimal places.

73. Exponential Growth Suppose that the population in a small city is 32,000 at the beginning of 2005 and that the city council assumes that the population size t years later can be estimated by the equation

$$P(t) = 32,000e^{0.05t}$$

In what month and year will the city have a population of 64,000?

74. Exponential Growth Suppose the population of a city is given by the equation

$$P(t) = 100,000e^{0.05t}$$

where t is the number of years from the present time. How large is the population now? ("Now" corresponds to a certain value of t. Once you realize what that value of t is, the problem becomes very simple.)

75. Airline Travel The number of airline passengers in 1990 was 466 million. The number of passengers traveling by airplane each year has increased exponentially according to the model, $P(t) = 466 \cdot 1.035^t$, where t is the number of years since 1990 (U.S. Census Bureau). In what year is it predicted that 900 million passengers will travel by airline?

76. Bankruptcy Model In 1997, there were a total of 1,316,999 bankruptcies filed under the Bankruptcy Reform Act. The model for the number of bankruptcies filed is $B(t) = 0.798 \cdot 1.164^t$, where t is the number of years since 1994 and B is the number of bankruptcies filed in terms of millions (Administrative Office of the U.S. Courts, *Statistical Tables for the Federal Judiciary*). In what year is it predicted that 12 million bankruptcies will be filed?

77. Health Care In 1990, $699 billion was spent on health care expenditures. The amount of money, E, in billions spent on health care expenditures can be estimated using the function $E(t) = 78.16(1.11)^t$, where t is time in years since 1970 (*U.S. Census Bureau*). In what year was it estimated that $800 billion would be spent on health care expenditures?

78. Value of a Car As a car ages, its value decreases. The value of a particular car with an original purchase price of $25,600 is modeled by the function $c(t) = 25,600(1 - 0.22)^t$, where c is the value at time t (Kelly Blue Book). How old is the car when its value is $10,000? Round your answer to two decimal places.

79. Compound Interest In 1991, the average cost of attending a public university through graduation was $20,972 (U.S. Department of Education, National Center for Educational Statistics). If John's parents deposited that amount in an account in 1991 at an interest rate of 7% compounded semi-annually, how long will it take for the money to double? Round your answer to two decimal places.

80. Carbon Dating Scientists use Carbon-14 dating to find the age of fossils and other artifacts. The amount of Carbon-14 in an organism will yield information concerning its age. A formula used in Carbon-14 dating is $A(t) = A_0 \cdot 2^{-t/5600}$, where A_0 is the amount of carbon originally in the organism, t is time in years, and A is the amount of carbon remaining after t years. Determine the number of years since an organism died if it originally contained 1,000 gram of Carbon-14 and it currently contains 600 gram of Carbon-14. Round to the nearest year.

81. Cost Increase The cost of a can of Coca Cola in 1960 was $0.10. The function that models the cost of a Coca Cola by year is $C(t) = 0.10e^{0.0576t}$, where t is the number of years since 1960. In what year was it expected that a can of Coca Cola will cost $1.00?

82. Online Banking Use The number of households using online banking services has increased from 754,000 in 1995 to 12,980,000 in 2000. The formula $H(t) = 0.76e^{0.55t}$ models the number of households H in millions when time is t years since 1995 (Home Banking Report). In what year was it estimated that 50,000,000 households would use online banking services?

Newtons Law of Cooling states that the temperature of an object after t minutes is given by the formula $T(t) = A + (T_o - A)e^{-kt}$, where A is the ambient (outside) temperature, T_o is the initial temperature of the object, and k is a constant. Use this formula to solve the following problems.

83. Coffee which is 200°F is poured into a coffee cup held in a room with temperature 75°F.

 a. If the coffee cools to 180°F in 5 minutes, find k.

 b. What will the coffee temperature be after 15 minutes using your value for k?

 c. When will the temperature of the coffee be 100°F using your value for k?

84. A cold soda which is 37°F is put on a table in a room with temperature 80°F

 a. If the soda heats to 45°F in 4 minutes, find k.

 b. What will the soda temperature be after 10 minutes using your value for k?

 c. When will the temperature of the soda be 70°F using your value for k?

Maintaining Your Skills

Find the vertex for each of the following parabolas, and then indicate if it is the highest or lowest point on the graph.

85. $y = 2x^2 + 8x - 15$ **86.** $y = 3x^2 - 9x - 10$ **87.** $y = 12x - 4x^2$ **88.** $y = 18x - 6x^2$

89. Maximum Height An object is projected into the air with an initial upward velocity of 64 feet per second. Its height h at any time t is given by the formula $h = 64t - 16t^2$. Find the time at which the object reaches its maximum height. Then find the maximum height.

90. Maximum Height An object is projected into the air with an initial upward velocity of 64 feet per second from the top of a building 40 feet high. If the height h of the object t seconds after it is projected into the air is $h = 40 + 64t - 16t^2$, find the time at which the object reaches its maximum height. Then find the maximum height it attains.

Find the Mistake

Each problem below contains a mistake. Circle the mistake and write the correct number(s) or word(s) on the line provided.

1. The equation $5^x = 12$ is an example of a logarithmic equation. _____

2. The change-of-base property for logarithms states that $\log_a x = \dfrac{\log_b a}{\log_b x}$. _____

3. For the change-of-base property to work, base a and base b must both be negative numbers. _____

4. To calculate $\log_9 36$, change the base to 10 and find the product of log 36 and log 9. _____

Chapter **14** Summary

EXAMPLES

1. For the exponential function $f(x) = 2^x$,

$$f(0) = 2^0 = 1$$
$$f(1) = 2^1 = 2$$
$$f(2) = 2^2 = 4$$
$$f(3) = 2^3 = 8$$

2. The function $f(x) = x^2$ is not one-to-one because 9, which is in the range, comes from both 3 and -3 in the domain.

3. The inverse of $f(x) = 2x - 3$ is

$$f^{-1}(x) = \frac{x + 3}{2}$$

4. The definition allows us to write expressions like

$$y = \log_3 27$$

equivalently in exponential form as

$$3^y = 27$$

which makes it apparent that y is 3.

5. Examples of the two special identities are

$$5^{\log_5 12} = 12$$

and

$$\log_8 8^3 = 3$$

6. We can rewrite the expression

$$\log_{10} \frac{45^6}{273}$$

using the properties of logarithms, as

$$6 \log_{10} 45 - \log_{10} 273$$

Exponential Functions [14.1]

Any function of the form

$$f(x) = b^x$$

where $b > 0$ and $b \neq 1$, is an *exponential function.*

One-to-One Functions [14.2]

A function is a *one-to-one function* if every element in the range comes from exactly one element in the domain.

Inverse Functions [14.2]

The *inverse* of a function is obtained by reversing the order of the coordinates of the ordered pairs belonging to the function. Only one-to-one functions have inverses that are also functions.

Definition of Logarithms [14.3]

If b is a positive number not equal to 1, then the expression

$$y = \log_b x$$

is equivalent to $x = b^y$; that is, in the expression $y = \log_b x$, y is the number to which we raise b in order to get x. Expressions written in the form $y = \log_b x$ are said to be in *logarithmic form.* Expressions like $x = b^y$ are in *exponential form.*

Two Special Identities [14.3]

For $b > 0$, $b \neq 1$, the following two expressions hold for all positive real numbers x:

1. $b^{\log_b x} = x$

2. $\log_b b^x = x$

Properties of Logarithms [14.4]

If x, y, and b are positive real numbers, $b \neq 1$, and r is any real number, then

1. $\log_b(xy) = \log_b x + \log_b y$ Product property for logarithms

2. $\log_b \left(\dfrac{x}{y} \right) = \log_b x - \log_b y$ Quotient property for logarithms

3. $\log_b x^r = r \log_b x$ Power property for logarithms

Common Logarithms [14.5]

Common logarithms are logarithms with a base of 10. To save time in writing, we omit the base when working with common logarithms; that is,

$$\log x = \log_{10} x$$

7. $\log_{10} 10{,}000 = \log 10{,}000$
$$= \log 10^4$$
$$= 4$$

Natural Logarithms [14.5]

Natural logarithms, written ln x, are logarithms with a base of e, where the number e is an irrational number (like the number π). A decimal approximation for e is 2.7183. All the properties of exponents and logarithms hold when the base is e.

8. $\ln e = 1$
$\ln 1 = 0$

Change of Base [14.6]

If x, a, and b are positive real numbers, $a \neq 1$ and $b \neq 1$, then

$$\log_a x = \frac{\log_b x}{\log_b a}$$

9. $\log_6 475 = \dfrac{\log 475}{\log 6}$
$$\approx \frac{2.6767}{0.7782}$$
$$\approx 3.44$$

> **COMMON MISTAKE**
>
> The most common mistakes that occur with logarithms come from trying to apply the properties of logarithms to situations in which they don't apply. For example, a very common mistake looks like this:
>
> $$\frac{\log 3}{\log 2} = \log 3 - \log 2 \qquad \text{Mistake}$$
>
> This is not a property of logarithms. To write the equation $\log 3 - \log 2$, we would have to start with
>
> $$\log \frac{3}{2} \qquad NOT \qquad \frac{\log 3}{\log 2}$$
>
> There is a difference.

CHAPTER 14 Test

Graph each exponential function. [14.1]

1. $f(x) = -3^x$

2. $g(x) = 2^{-x}$

Sketch the graph of each function and its inverse. Find $f^{-1}(x)$ for Problem 3. [14.2]

3. $f(x) = \frac{1}{2}x + \frac{3}{2}$

4. $f(x) = x^2 - 3$

Solve for x. [14.3]

5. $\log_5 x = 2$

6. $\log_x 7 = 3$

Graph each of the following. [14.3]

7. $y = \log_3 x$

8. $y = \log_{1/3} x$

Evaluate each of the following. Round to the nearest hundredth if necessary. [14.3, 14.4, 14.5]

9. $\log_9 3$

10. $\log_7 35$

11. $\log 14{,}500$

12. $\log 0.0203$

13. $\ln 56.3$

14. $\ln 0.034$

Use the properties of logarithms to expand each expression. [14.4]

15. $\log_7 \dfrac{4x^3}{y}$

16. $\log \dfrac{\sqrt[3]{x}}{y^2\sqrt{z}}$

Write each expression as a single logarithm. [14.4]

17. $\dfrac{1}{3}\log_2 x + 5\log_2 y$

18. $\dfrac{1}{4}\log_3 x - \log_3 y + 3\log_3 z$

Use a calculator to find x. Round to the nearest thousandth. [14.5]

19. $\log x = 5.3819$

20. $\log x = -2.4531$

Solve for x. Round to the nearest thousandth if necessary. [14.4, 14.6]

21. $4 = 3^x$

22. $9^{4x-1} = 27$

23. $\log_6 x + \log_6 3 = 1$

24. $\log_3 x + \log_3(x - 26) = 3$

25. **Chemistry** Use the formula $pH = -\log[H^+]$ to find the pH of a solution in which $[H^+] = 7.3 \times 10^{-6}$. [14.5]

26. **Compound Interest** If $300 is deposited in an account that earns 9% annual interest compounded twice a year, how much money will be in the account after 6 years? [14.1]

27. **Compound Interest** How long will it take $500 to become $2,500 if the $500 is deposited in an account that earns 6% annual interest compounded twice a year? [14.6]

Determine if the function is one-to-one. [14.2]

28. $f(x) = \{(4, 7), (5, 8), (7, 4), (8, 5)\}$

29. $g(x) = \{(1, 7), (2, 0), (3, 1), (4, 0)\}$

30.

31.

32.

33.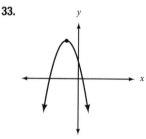

Answers to Odd-Numbered Problems

Chapter 0

Exercise Set 0.1

Vocabulary Review 1. place value **2.** condensed, expanded
3. hyphen **4.** comma **5.** set **6.** counting **7.** whole
Problems 1. 8 ones, 7 tens **3.** 5 ones, 4 tens
5. 8 ones, 4 tens, 3 hundreds **7.** 8 ones, 0 tens, 6 hundreds
9. 8 ones, 7 tens, 3 hundreds, 2 thousands **11.** 9 ones, 6 tens,
5 hundreds, 3 thousands, 7 ten thousands, 2 hundred thousands
13. Ten thousands **15.** Hundred millions **17.** Ones
19. Hundred thousands **21.** $600 + 50 + 8$ **23.** $60 + 8$
25. $4,000 + 500 + 80 + 7$ **27.** $30,000 + 2,000 + 600 + 70 + 4$
29. $3,000,000 + 400,000 + 60,000 + 2,000 + 500 + 70 + 7$
31. $400 + 7$ **33.** $30,000 + 60 + 8$ **35.** $3,000,000 + 4,000 + 8$
37. 547 **39.** 6,323 **41.** 27,035 **43.** 70,907
45. 180,643 **47.** 406,008 **49.** Twenty-nine
51. Forty **53.** Five hundred seventy-three
55. Seven hundred seven **57.** Seven hundred seventy
59. Twenty-three thousand, five hundred forty
61. Three thousand, four **63.** Three thousand, forty
65. One hundred four million, sixty-five thousand, seven hundred eighty
67. Five billion, three million, forty thousand, eight **69.** Two
million, five hundred forty-six thousand, seven hundred thirty-one
71. Twenty million, four hundred thirty-two thousand **73.** 325
75. 5,432 **77.** 86,762 **79.** 2,000,200 **81.** 2,002,200
83. Thousands **85.** $100,000 + 6,000 + 700 + 20 + 1$
87. Three million, fourteen thousand, five hundred seventy-two
89. 3 inches **91.** 5 inches **93.** 6 **95.** 3 **97.** 12

Find the Mistake 1. The place value of the 7 in the number 562,472
is tens. **2.** The number 12,789 written in expanded form is
$10,000 + 2,000 + 700 + 80 + 9$. **3.** The number 9,023,627,003
written in words is nine billion, twenty-three million, six hundred
twenty-seven thousand, three. **4.** Writing forty million, three
hundred forty-eight thousand, thirteen in digits gives 40,348,013.

Exercise Set 0.2

Vocabulary Review 1. addition **2.** place value **3.** carrying
4. sum **5.** zero **6.** commutative **7.** associative
8. solution, inspection **9.** perimeter
Problems 1. 15 **3.** 14 **5.** 24 **7.** 15 **9.** 20 **11.** 68
13. 98 **15.** 7,297 **17.** 6,487 **19.** 96 **21.** 7,449 **23.** 65
25. 102 **27.** 875 **29.** 829 **31.** 10,391 **33.** 16,204
35. 155,554 **37.** 111,110 **39.** 17,391 **41.** 14,892
43. 180 **45.** 2,220 **47.** 18,285 **49.** 774 **51.** 7,195

53.

First Number a	Second Number b	Their Sum $a + b$
61	38	99
63	36	99
65	34	99
67	32	99

55.

First Number a	Second Number b	Their Sum $a + b$
9	16	25
36	64	100
81	144	225
144	256	400

57. $9 + 5$ **59.** $8 + 3$ **61.** $4 + 6$ **63.** $1 + (2 + 3)$
65. $2 + (1 + 6)$ **67.** $(1 + 9) + 1$ **69.** $4 + (n + 1)$
71. $n = 4$ **73.** $n = 5$ **75.** $n = 8$ **77.** $n = 8$ **79.** $n = 9$
81. $n = 4$ **83.** The sum of 4 and 9 **85.** The sum of 8 and 1

87. The sum of 2 and 3 is 5. **89. a.** $5 + 2$ **b.** $8 + 3$
91. a. $m + 1$ **b.** $m + n$ **93.** 12 in. **95.** 16 ft
97. 26 yd **99.** 18 in. **101.** 34 gallons **103.** $349

Find the Mistake 1. To find the sum of 786 and 49, add the ones
place by writing the 5 and carrying the 1. **2.** The problem
$(12 + 7) + 3 = 12 + (7 + 3)$ uses the associative property of addition.
3. The solution for the equation $47 = x + 29$ is $x = 18$.
4. The perimeter of a rectangle with a length of 35 inches and a
width of 15 inches is 100 inches.

Exercise Set 0.3

Vocabulary Review 1. difference **2.** columns
3. borrowing **4.** estimation
Problems 1. 32 **3.** 22 **5.** 10 **7.** 111
9. The difference of 10 and 2. **11.** The difference of a and 6.
13. The difference of 8 and 2 is 6. **15.** $8 - 3$ **17.** $y - 9$
19. $3 - 2 = 1$ **21.** 312 **23.** 403 **25.** 1,111 **27.** 4,544
29. 15 **31.** 33 **33.** 5 **35.** 33 **37.** 95 **39.** 152
41. 274 **43.** 488 **45.** 538 **47.** 163 **49.** 1,610 **51.** 46,083
53.

First Number a	Second Number b	The Difference of a and b $a - b$
25	15	10
24	16	8
23	17	6
22	18	4

55.

First Number a	Second Number b	The Difference of a and b $a - b$
400	256	144
400	144	256
225	144	81
225	81	144

57. $n = 6$ **59.** $n = 19$ **61.** $n = 9$ **63.** $n = 13$ **65.** $n = 25$
67. $n = 11$ **69.** $n = 9$ **71.** $n = 1$ **73.** $255 **75.** 33 feet
77. 168 students **79.** $574 **81.** $350
83. a.

Year	Projected Sales (Millions)
2010	8.5
2011	19.4
2012	30.1

b. 21.6 million units

85. a.

Venue	Capacity
National Stadium	91,000
National Aquatic Center	17,000
National Indoor Stadium	18,000
Beijing Olympic Green Tennis Courts	17,400
Olympic Sports Center Stadium	38,000

b. 74,000 people

Find the Mistake 1. Translating "The difference of 22 and 3 is 19"
into symbols gives us $22 - 3 = 19$. **2.** To subtract 50 from
290, vertically align the digits wih the same place value. **3.** To
find the difference of 85 and 27, begin by borrowing 1 ten from the
tens column and subtracting 7 from 15. **4.** The solution for the
equation $108 = 130 - y$ is $y = 22$.

Landmark Review **1.** $500 + 40 + 9$

2. $1,000 + 400 + 90 + 3$ **3.** $60,000 + 200 + 40 + 3$
4. $30,000,000 + 400,000 + 3,0000 + 5$ **5.** 18 **6.** 31 **7.** 117
8. 7,650 **9.** 103,701 **10.** 402,162 **11.** 530 **12.** 1,235
13. 17 **14.** 22 **15.** 12 **16.** 102 **17.** $x = 5$

18. $n = 14$ **19.** $a = 7$ **20.** $y = 20$ **21.** $x = 16$
22. $n = 4$ **23.** $y = 5$ **24.** $x = 1$

Exercise Set 0.4

Vocabulary Review **1.** round **2.** less **3.** more
4. horizontal axis, vertical axis **5.** estimate
Problems **1.** 40 **3.** 50 **5.** 50 **7.** 80 **9.** 460 **11.** 470
13. 56,780 **15.** 4,500 **17.** 1,200 **19.** 500 **21.** 800
23. 900 **25.** 1,100 **27.** 5,000 **29.** 39,600
31. 5,000 **33.** 10,000 **35.** 1,000 **37.** 658,000
39. 608,000 **41.** 5,745,000

	Original Number	Rounded to the Nearest		
		Ten	Hundred	Thousand
43.	7,821	7,820	7,800	8,000
45.	5,999	6,000	6,000	6,000
47.	10,985	10,990	11,000	11,000
49.	99,999	100,000	100,000	100,000

51. $3,000,000 **53.** 27,380,000 views **55.** 8,500,000 viewers
57. $15,200 **59.** $31,000 **61.** 1,200 **63.** 1,900
65. 58,000 **67.** 11,900 **69.** 5,000 **71.** 96,000
73.

Calories Burned

Find the Mistake **1.** Rounding 12,456 to the nearest hundred gives us 12,500. **2.** To round 102,673 to the nearest ten, replace the 3 with a zero to get 102,670. **3.** To estimate the sum of 14,256 and 2,789 by rounding to the nearest hundred, add 14,300 and 2,800. **4.** Using the bar chart below, Student C's home is 8 more miles from school than Student A's home.

Exercise Set 0.5

Vocabulary Review **1.** addtion **2.** factors **3.** distributive
4. multiplication **5.** unchanged **6.** commutative
7. associative **8.** area
Problems **1.** 300 **3.** 600 **5.** 3,000 **7.** 5,000 **9.** 21,000
11. 81,000 **13.** The product of 6 and 7 **15.** The product of 2 and n
17. The product of 9 and 7 is 63. **19.** $7 \cdot n$ **21.** $6 \cdot 7 = 42$
23. $0 \cdot 6 = 0$ **25.** Products: $9 \cdot 7$ and 63 **27.** Products: $4(4)$ and 16
29. Factors: 2, 3, and 4 **31.** Factors: 2, 2, and 3 **33.** $9(5)$
35. $7 \cdot 6$ **37.** $(2 \cdot 7) \cdot 6$ **39.** $(3 \times 9) \times 1$ **41.** $7(2) + 7(3) = 35$
43. $9(4) + 9(7) = 99$ **45.** $3(x) + 3(1) = 3x + 3$
47. $2(x) + 2(5) = 2x + 10$ **49.** 100 **51.** 228 **53.** 36
55. 1,440 **57.** 950 **59.** 1,725 **61.** 121 **63.** 1,552
65. 4,200 **67.** 66,248 **69.** 279,200 **71.** 12,321
73. 106,400 **75.** 198,592 **77.** 612,928 **79.** 333,180
81. 18,053,805 **83.** 263,646,976
85.

First Number a	Second Number b	Their product ab
11	11	121
11	22	242
22	22	484
22	44	968

87.

First Number a	Second Number b	Their product ab
25	10	250
25	100	2,500
25	1,000	25,000
25	10,000	250,000

89.

First Number a	Second Number b	Their product ab
12	20	240
36	20	720
12	40	480
36	40	1,440

91. 25 in^2 **93.** 84 m^2 **95.** 192 cm^2 **97.** $n = 3$
99. $n = 9$ **101.** $n = 0$ **103.** $n = 12$ **105.** $n = 11$
107. $n = 9$ **109.** 2,860 miles **111.** 1,500 square feet without the garage, 2,067 square feet with the garage **113.** 2,081 calories
115. 560 calories **117.** Yes **119.** 8,000 **121.** 1,500,000
123. 1,400,000 **125.** 40 **127.** 54
Find the Mistake **1.** Factors are numbers that when multiplied together give a product. **2.** The distributive property is used to show that $16(5 + 9) = 16(5) + 16(9)$. **3.** The first step when multiplying 73 and 4 is to multiply 4 and 3 by writing down the 2 and carrying the 1. **4.** The area of a rectangle with a length of 35 inches and a width of 8 inches is 280 square inches.

Exercise Set 0.6

Vocabulary Review **1.** quotient **2.** dividend, divisor
3. remainder **4.** undefined
Problems **1.** $6 \div 3$ **3.** $45 \div 9$ **5.** $r \div s$ **7.** $20 \div 4 = 5$
9. $2 \cdot 3 = 6$ **11.** $9 \cdot 4 = 36$ **13.** $6 \cdot 8 = 48$ **15.** $7 \cdot 4 = 28$
17. 5 **19.** 8 **21.** Undefined **23.** 45 **25.** 23
27. 1,530 **29.** 1,350 **31.** 18,000 **33.** 16,680
35. 24,000 **37.** 45 **39.** 49 **41.** 432 **43.** 1,438
45. 705 **47.** 3,020 **49.** 61 R 4 **51.** 90 R 1 **53.** 13 R 7
55. 234 R 6 **57.** 402 R 4 **59.** 35 R 35
61.

First Number a	Second Number b	The Quotient of a and b $\frac{a}{b}$
100	25	4
100	26	3 R 22
100	27	3 R 19
100	28	3 R 16

63. a **65.** b **67.** 1 **69.** 2 **71.** 4 **73.** 6 **75.** $1,850
77. 16¢ **79.** 5 miles **81.** 6 glasses with 2 oz left over
83. 3 bottles **85.** $12 **87.** 665 mg **89.** 437
91. 3,247 **93.** 869 **95.** 5,684 **97.** 169 gal
Find the Mistake **1.** The division notation $\frac{10}{5}$ is equivalent to $5\overline{)10}$. **2.** The quotient of 198 and 11 is 18. **3.** To divide 1,640 by 12, first estimate the number of twelves we can subtract from 16. **4.** Dividing 2,380 by 13 gives a remainder of 1.

Exercise Set 0.7

Vocabulary Review **1.** multiplication **2.** factor
3. exponent, base **4.** zero **5.** grouping symbols
6. left, right **7.** before **8.** after

Problems **1.** Base 4, Exponent 5 **3.** Base 3, Exponent 6
5. Base 8, Exponent 2 **7.** Base 9, Exponent 1
9. Base 4, Exponent 0 **11.** 36 **13.** 8 **15.** 1 **17.** 1
19. 81 **21.** 10 **23.** 12 **25.** 1 **27.** 12 **29.** 100
31. 4 **33.** 43 **35.** 16 **37.** 84 **39.** 14 **41.** 74
43. 12,768 **45.** 104 **47.** 416 **49.** 66 **51.** 21
53. 7 **55.** 16 **57.** 84 **59.** 40 **61.** 41 **63.** 18
65. 405 **67.** 124 **69.** 11 **71.** 91 **73.** 7
75. $8(4 + 2) = 48$ **77.** $2(10 + 3) = 26$ **79.** $3(3 + 4) + 4 = 25$
81. $(20 \div 2) - 9 = 1$ **83.** $(8 \cdot 5) + (5 \cdot 4) = 60$ **85.** 247 calories
87. 407 calories **89.** Big Mac has 263 more calories which is
more than twice the calories. **91.** 4 **93.** 16
95. Answers will vary
Find the Mistake **1.** The expression 8^4 means that 8 is used as
a factor 4 times. **2.** The number 7 raised to the third power
equals $7^3 = 343$. **3.** Following the order of operations, work all
additions and subtractions after evaluating any exponents.
4. Simplifying $(4 + 1)^3 - (9 + 1)^2$ gives 25.

Chapter 0 Test

1. Thirty thousand, six hundred fifty-two **2.** 6,007,029
3. $200,000 + 80,000 + 5,000 + 600 + 30 + 4$ **4.** d **5.** g
6. c **7.** b **8.** 698 **9.** 24,603 **10.** $P = 62m$
11. 521 **12.** 4,787 **13.** 240,000 **14.** 810 **15.** 19,758
16. 72 **17.** 23 **18.** $2(13 + 4) = 34$ **19.** $(18 \div 6) + 12 = 15$
20. 40 **21.** 3 **22.** 80 **23.** $5x - 20$ **24.** 2,313,406;
Two million, three hundred thirteen thousand, four hundred six
25. 236,035; Two hundred thirty-six thousand, thirty-five

Chapter 1

Exercise Set 1.1

Vocabulary Review **1.** real number line **2.** coordinate
3. real **4.** equivalent **5.** nonzero **6.** absolute value
7. opposites **8.** algebraic expression, variables
Problems 1-7.

9. $\frac{18}{24}$ **11.** $\frac{12}{24}$ **13.** $\frac{15}{24}$ **15.** $\frac{36}{60}$ **17.** $\frac{22}{60}$ **19.** $\frac{50}{60}$
21. 6 **23.** 22 **25.** 3 **27.** 7 **29.** 3 **31.** 17
33. $-10, 10$ **35.** $-\frac{3}{4}, \frac{3}{4}$ **37.** $-\frac{11}{2}, \frac{11}{2}$ **39.** 3, 3
41. $\frac{2}{5}, \frac{2}{5}$ **43.** $-x, |x|$ **45.** $<$ **47.** $>$ **49.** $>$
51. $>$ **53.** $<$ **55.** $<$ **57.** $x + 5 = 14$
59. $5y < 30$ **61.** $5y \geq y - 16$ **63.** $\frac{x}{3} = x + 2$
65. 4 inches; 1 square inch **67.** 4.5 inches; 1.125 square inches
69. 10.25 centimeters; 5 square centimeters **71.** $-8, -2$
73. $-64°F; -54°F$ **75.** $-15°F$ **77.** -100 feet; -105 feet
79. 93.5 square inches, 39 inches **81.** 1,387 calories **83.** 654
more calories **85. a.** $263,800,000 **b.** False **c.** True
Find the Mistake **1.** Writing the fraction $\frac{4}{11}$ as an equivalent
fraction with a denominator of 33 gives us $\frac{12}{33}$. **2.** A real
number and its opposite will have the same absolute value.
3. In the algebraic expression $2x - y$, the x and the y are the
variables. **4.** The perimeter of a rectangle with a length of 32
and a width of 8 is 80.

Exercise Set 1.2

Vocabulary Review **1.** same **2.** different **3.** opposite
4. complementary **5.** supplementary

Problems **1.** 3 **3.** -7 **5.** -14 **7.** -3 **9.** -25
11. -12 **13.** -19 **15.** -25 **17.** -3 **19.** -6 **21.** 0
23. -10 **25.** -16 **27.** -12 **29.** -7 **31.** 35
33. 0 **35.** 9 **37.** -8 **39.** -4 **41.** -28 **43.** 6
45. 6 **47.** 8 **49.** 18 **51.** 10 **53.** 17 **55.** 13
57. 3 **59.** 15 **61.** 1 **63.** 1 **65.** 38 **67.** 68
69. $5 + 9 = 14$ **71.** $[-7 + (-5)] + 4 = -8$
73. $-7 - 4 = -11$ **75.** $[4 + (-5)] - 17 = -18$
77. $8 - 5 = 3$ **79.** $8 - (-5) = 13$ **81.** 3 **83.** 10
85. $35°$ **87.** $60°$ **89.** $73 + 10 - 8; 75°F$ **91.** 188 degrees
93. 265 degrees **95.** $2,000 **97.** 2006; $500
99. a. 5.7 million **b.** 8.5 million **c.** 41.3 million
Find the Mistake **1.** The difference of -22 and -17 is -5.
2. To subtract one real number from another, add its opposite.
3. To work the problem $4^3 - (-8 + 4) - 2$ using the order of
operations, we work the parentheses, then simplify the exponential
expression, and then subtract the 2. **4.** The supplement of an
angle $88°$ is 92.

Exercise Set 1.3

Vocabulary Review **1.** absolute values **2.** positive
3. negative **4.** zero
Problems **1.** -42 **3.** -16 **5.** 3 **7.** 121 **9.** 6 **11.** -60
13. 24 **15.** 49 **17.** -27 **19.** 6 **21.** 10 **23.** 9
25. 45 **27.** 14 **29.** -2 **31.** 216 **33.** -2 **35.** -18
37. 29 **39.** 38 **41.** -5 **43.** 37 **45.** 80 **47.** 17
49. 40 **51.** $-\frac{10}{21}$ **53.** -4 **55.** 1 **57.** $\frac{9}{16}$ **59.** 0 **61.** 0
63. 0 **65.** 0 **67.** 4 **69.** 24 **71.** 81 **73.** $\frac{9}{4}$ **75.** $1°F$
77. 275 calories **79.** $264.60 **81.** $848.20 was gained
Find the Mistake **1.** Multiplying a positive number by a
negative number will give a negative answer. **2.** Multiplying
two numbers with the same sign will give a positive answer.
3. Simplifying the expression $(-15)(-3)(6)(-2)$ gives us -540.
4. The number -12 raised to the third power is the opposite of the
number 12 raised to the third power.

Exercise Set 1.4

Vocabulary Review **1.** reciprocal **2.** positive **3.** negative
4. lowest terms **5.** zero
Problems **1.** -2 **3.** -3 **5.** $-\frac{1}{3}$ **7.** 3 **9.** $\frac{1}{7}$ **11.** 0
13. 9 **15.** -15 **17.** -36 **19.** $-\frac{1}{4}$ **21.** $\frac{16}{15}$ **23.** $\frac{4}{3}$
25. $-\frac{8}{13}$ **27.** -1 **29.** 1 **31.** $\frac{5}{9}$ **33.** $\frac{3}{5}$ **35.** $-\frac{5}{3}$
37. -2 **39.** -3 **41.** Undefined **43.** 0 **45.** 5
47. $-\frac{7}{3}$ **49.** -1 **51.** -7 **53.** $\frac{15}{17}$ **55.** $-\frac{32}{17}$ **57.** $\frac{1}{3}$
59. 1 **61.** 1 **63.** -2 **65.** $\frac{9}{7}$ **67.** $\frac{16}{11}$ **69.** -1
71. -1 **73. a.** 25 **b.** -25 **c.** -25 **d.** -25 **e.** 25
75. a. 10 **b.** 0 **c.** -100 **d.** -20 **77.** 3 **79.** -10
81. -3 **83.** -8 **85.** $350 **87.** Drops $3.5°F$ each hour
89. 14 blankets **91.** 48 bags **93.** 6 **95.** 28
97. a. $20,000 **b.** $50,000 **c.** Yes, the projected revenue for
5,000 e-mail addresses is $10,000, which is $5,000 more than the list
costs. **99.** $24.5°F/minute, 9.4°F/minute$
Find the Mistake **1.** The division problem $88 \div 4$ is the same as
$88 \cdot \left(\frac{1}{4}\right)$. **2.** Dividing two numbers with unlike signs gives a
negative answer. **3.** Dividing -55 by -5 will give us 11.
4. Simplifying $\frac{-3(6)}{9}$ gives us -2.

Landmark Review **1.** $-\frac{3}{4}, \frac{3}{4}$ **2.** 27, 27 **3.** 11 **4.** 9
5. $y + 15 = 27$ **6.** $3x < 24$ **7.** -39 **8.** -18

9. -3 **10.** -30 **11.** b **12.** a **13.** 25 **14.** 50
15. -3 **16.** Undefined **17.** -3 **18.** -4

Exercise Set 1.5

Vocabulary Review **1.** e **2.** b **3.** d **4.** h **5.** c
6. f **7.** j **8.** i **9.** a **10.** g
Problems **1.** $(4+2)+x = 6+x$ **3.** $x+(2+7) = x+9$
5. $(3 \cdot 5)x = 15x$ **7.** $(9 \cdot 6)y = 54y$ **9.** $\left(\frac{1}{2} \cdot 3\right)a = \frac{3}{2}a$
11. $\left(\frac{1}{3} \cdot 3\right)x = x$ **13.** $\left(\frac{1}{2} \cdot 2\right)y = y$ **15.** $\left(\frac{3}{4} \cdot \frac{4}{3}\right)x = x$
17. $\left(\frac{6}{5} \cdot \frac{5}{6}\right)a = a$ **19.** $(3 \cdot 5)xy = 15xy$ **21.** $(15 \cdot 2)xy = 30xy$
23. $8x+16$ **25.** $8x-16$ **27.** $4y+4$ **29.** $18x+15$
31. $6a+14$ **33.** $54y-72$ **35.** $x+2$ **37.** $-3x+12y$
39. $12x+18y$ **41.** $12a-8b$ **43.** $3x+2y$ **45.** $4a+25$
47. $-4a+11$ **49.** $6x+12$ **51.** $14x+38$ **53.** $8x-17$
55. $2x+1$ **57.** $6x-3$ **59.** $-2x+4$ **61.** $5x+10$
63. $6x+5$ **65.** $-3x-8$ **67.** $5x+6$ **69.** $6m-5$
71. $7+3x$ **73.** $3x-2y$ **75.** $0.09x+180$ **77.** $0.12x+60$
79. $a+1$ **81.** $1-a$ **83.** Commutative
85. Multiplicative inverse **87.** Commutative
89. Distributive **91.** Commutative, associative
93. Commutative, associative **95.** Commutative
97. Symmetric **99.** Commutative **101.** Additive inverse
103. Multiplicative inverse **105.** Symmetric **107.** $3x+6$
109. $9a+9b$ **111.** 0 **113.** 0 **115.** 10 **117.** 1
119. No **121.** $8 \div 4 \neq 4 \div 8$
123. $4(2+3) = 20$ $(4 \cdot 2)+(4 \cdot 3) = 20$

Find the Mistake **1.** To simplify the problem $8\left(\frac{4}{7}x\right)$, we used the associative property of multiplication. **2.** The distributive property states that multiplication distributes over addition. **3.** The multiplicative identity property states that $a(1) = a$ and $1(a) = a$. **4.** The expression $35\left(\frac{2}{3}y+18z\right)$ is equal to $35\left(\frac{2}{3}y\right)+35(18z)$ because of the distributive property.

Exercise Set 1.6

Vocabulary Review **1.** expression **2.** term **3.** similar
4. value, variable **5.** parentheses
Problems **1.** $-3x$ **3.** $-a$ **5.** $10x+2y$ **7.** $8a-2b$
9. $\frac{9}{2}x-3$ **11.** $a+5$ **13.** $5x-5$ **15.** $4a+2$
17. $-9x-2$ **19.** $0.75b-11$ **21.** $0.17x$ **23.** $-0.16x$
25. $10x-1$ **27** $21y+6$ **29.** $-6x+8$ **31.** $3a-2$
33. $-4x+26$ **35.** $y-10$ **37.** $-11x+4$ **39.** $2x-12$
41. $10a+33$ **43.** $1.1x-3.2$ **45.** $6y-46$ **47.** $-\frac{5}{2}x-\frac{7}{2}$
49. $2x$ **51.** $-2x-9$ **53.** $7x$ **55.** $10y$ **57.** $0.17x+180$
59. $0.02x+400$ **61.** 5 **63.** -9 **65.** 4 **67.** 4
69. -37 **71.** -41 **73.** 64 **75.** 64 **77.** 144
79. 144 **81.** 49 **83.** 40 **85.** 3 **87.** 0 **89.** 15
91. 6 **93. a.** 42°F **b.** 28°F **c.** -14°F **d.** -32°F
95. a. \$52 **b.** \$55 **c.** \$60 **97. a.** 20.7 lb/in² **b.** 24.7 lb/in²
c. 34.7 lb/in² **99.** $0.29H$; 2.32 hours **101.** $\frac{21}{2}a-5$

Find the Mistake **1.** Simplifying the expression $12z+9z-3z$ by combining similar terms gives us $18z$. **2.** The value of the expression $8y+\frac{1}{3}y$ when $y = 3$ is 25. **3.** To simplify the expression $-3(5x-4)-8x$, first apply distributive property and then combine similar terms. **4.** The value of the expression $(2x-4)^2$ when $x = \frac{1}{2}$ is different from the value when $x = 4$.

Chapter 1 Test

1. $15-x = 12$ **2.** $6a = 30$ **3.** $7+a > 15$ **4.** $\frac{12}{n} \leq 5$
5. -1 **6.** -11 **7.** $6, 6$ **8.** $-\frac{5}{3}, \frac{5}{3}$ **9.** -3 **10.** 12
11. -11 **12.** -7 **13.** -28 **14.** 24 **15.** -3
16. $-\frac{27}{8}$ **17.** -13 **18.** -10 **19.** -7 **20.** 62 **21.** 2
22. -6 **23.** d **24.** e **25.** a **26.** c **27.** $12+3x$
28. $-15y$ **29.** $-10x+15$ **30.** $-2x-4$ **31.** $-4n+10$
32. $-2a-16b$ **33.** $x+y$ **34.** $0.11n+240$
35. 123,760 dogs **36.** 82,446 dogs **37.** 4,771 dogs

Chapter 2

Exercise Set 2.1

Vocabulary Review **1.** fraction **2.** numerator, denominator
3. proper **4.** improper **5.** equivalent
Problems
1. 1 **3.** -2 **5.** x **7.** $-a$ **9.** 5 **11.** 1 **13.** -12
15.

Numerator a	Denominator b	Fraction $\frac{a}{b}$
3	5	$\frac{3}{5}$
-1	7	$-\frac{1}{7}$
$-x$	$-y$	$\frac{x}{y}$
$x+1$	x	$\frac{x+1}{x}$

17. $\frac{3}{4}, \frac{1}{2}, \frac{9}{10}$
19. True
21. False

23. – 31. Answers on number line below

33. $\frac{1}{20} < \frac{4}{25} < \frac{3}{10} < \frac{2}{5}$ **35.** $\frac{4}{6}$ **37.** $-\frac{5}{6}$ **39.** $\frac{8}{12}$
41. $-\frac{8}{12}$ **43.** $\frac{2x}{12x}$ **45.** $\frac{6x}{12x}$ **47.** $\frac{16}{8}$ **49.** $-\frac{40}{8}$
51. Answers will vary **53.** 3 **55.** -2 **57.** 37 **59.** $\frac{3}{4}$
61. $-\frac{43}{47}$ **63.** $\frac{4}{3}$ **65.** $-\frac{13}{17}$ **67. a.** $\frac{1}{2}$ **b.** $\frac{1}{2}$ **c.** $\frac{1}{4}$ **d.** $\frac{1}{4}$
69.

How often workers send non-work-related e-mail from the office	Fraction of respondents saying yes
Never	$\frac{4}{25}$
1 to 5 times a day	$\frac{47}{100}$
5 to 10 times a day	$\frac{8}{25}$
More than 10 times a day	$\frac{1}{20}$

71. $\frac{4}{5}$ **73.** $\frac{29}{43}$ **75.** $\frac{1,121}{1,791}$ **77.** $\frac{14}{50}$ **79.** d **81.** a
83. 108 **85.** 60 **87.** 4 **89.** 5 **91.** 7 **93.** 51

Find the Mistake **1.** For the fraction $\frac{21}{7}$, the denominator is 7.
2. The fraction $\frac{90}{15}$ is considered an improper fraction.
3. The multiplication property for fractions states that if the numerator and the denominator of a fraction are multiplied by the same nonzero number, the resulting fraction is equivalent to the original fraction. **4.** If we divide the numerator and denominator of the fraction $\frac{8}{12}$ by 4, then we get the equivalent fraction $\frac{2}{3}$.

Exercise Set 2.2

Vocabulary Review 1. prime **2.** remainder **3.** composite
4. lowest terms
Problems 1. Prime **3.** Composite; 3, 5, and 7 are factors
5. Composite; 3 is a factor **7.** Prime **9.** $2^2 \cdot 3$ **11.** 3^4
13. $5 \cdot 43$ **15.** $3 \cdot 5$ **17.** $\frac{1}{2}$ **19.** $\frac{2}{3}$ **21.** $\frac{4}{5}$ **23.** $\frac{9}{5}$
25. $\frac{7}{11}$ **27.** $\frac{3}{5}$ **29.** $\frac{1}{7}$ **31.** $\frac{7}{9}$ **33.** $\frac{7}{5}$ **35.** $\frac{5}{7}$
37. $\frac{3}{5}$ **39.** $\frac{5}{3}$ **41.** $\frac{8}{9}$ **43.** $\frac{42}{55}$ **45.** $\frac{17}{19}$ **47.** $\frac{14}{33}$
49. a. $\frac{2}{17}$ **b.** $\frac{3}{26}$ **c.** $\frac{1}{9}$ **d.** $\frac{3}{28}$ **e.** $\frac{2}{19}$ **51. a.** $\frac{1}{45}$ **b.** $\frac{1}{30}$ **c.** $\frac{1}{18}$ **d.** $\frac{1}{15}$ **e.** $\frac{1}{10}$
53. a. $\frac{1}{3}$ **b.** $\frac{5}{6}$ **c.** $\frac{1}{5}$ **55.** $\frac{9}{16}$

57. – 59.
$$\frac{1}{2} = \frac{2}{4} = \frac{4}{8} = \frac{8}{16}$$
$$\frac{5}{4} = \frac{10}{8} = \frac{20}{16}$$

61. $\frac{1}{4}$ **63.** $\frac{2}{3}$ **65.** $\frac{7}{15}$ **67.** $\frac{1}{4}$ **69.** 0 **71.** 3 **73.** 45
75. 25 **77.** $2^2 \cdot 3 \cdot 5$ **79.** $2^2 \cdot 3 \cdot 5$ **81.** 9 **83.** 25 **85.** b

Find the Mistake 1. The number 30 is a composite number because it has 10 as a divisor. **2.** The number 70 factored into a product of primes is $2 \cdot 5 \cdot 7$. **3.** When reducing the fraction $\frac{32}{48}$ to lowest terms, we divide out the common factor 2^4 to get $\frac{2}{3}$.
4. Reducing the fraction $\frac{112}{14}$ to lowest terms gives us $\frac{8}{1} = 1$.

Exercise Set 2.3

Vocabulary Review 1. product **2.** fractions
3. multiplication **4.** triangle
Problems 1. $\frac{8}{15}$ **3.** $\frac{7}{8}$ **5.** 1 **7.** $\frac{27}{4}$ **9.** 1 **11.** $\frac{1}{24}$
13. $\frac{24}{125}$ **15.** $\frac{105}{8}$

17.

First number	Second number	Their product
x	y	xy
$\frac{1}{2}$	$\frac{2}{3}$	$\frac{1}{3}$
$\frac{2}{3}$	$\frac{3}{4}$	$\frac{1}{2}$
$\frac{3}{4}$	$\frac{4}{5}$	$\frac{3}{5}$
$\frac{5}{a}$	$-\frac{a}{6}$	$-\frac{5}{6}$

19.

First number x	Second number y	Their product xy
$\frac{1}{2}$	30	15
$\frac{1}{5}$	30	6
$\frac{1}{6}$	30	5
$\frac{1}{15}$	30	2

21. $\frac{3}{5}$ **23.** 9 **25.** 1 **27.** -8 **29.** $-\frac{1}{15}$ **31.** $\frac{4}{9}$
33. $\frac{9}{16}$ **35.** $\frac{1}{4}$ **37.** $\frac{8}{27}$ **39.** $\frac{1}{2}$ **41.** $\frac{9}{100}$ **43.** 3
45. 24 **47.** 4 **49.** 9 **51.** $\frac{3}{10}$; numerator should be 3, not 4.
53. $\frac{1}{4}$; answer should be positive **55.** 14 **57.** -14
59. 5 **61.** -6 **63.** 133 in² **65.** $\frac{4}{9}$ ft²

67. 3 yd² **69.** 3,511 students **71.** 126,500 ft³
73. About 2,121 children **75.** $\frac{1}{27}$ **77.** $\frac{8}{27}$ **79.** 2
81. 3 **83.** 2 **85.** 5 **87.** 3 **89.** $\frac{4}{3}$ **91.** 3 **93.** $\frac{1}{7}$

Find the Mistake 1. To find the product of two fractions, multiply the numerators and multiply the denominators. **2.** To multiply $\frac{6}{7}$ by $\frac{12}{9}$, find the product of the numerators and divide it by the product of the denominators to get $\frac{72}{63}$. **3.** Simplifying $\left(\frac{5}{6}\right)^2 \cdot \frac{8}{9}$ gives $\frac{50}{81}$. **4.** The area of a triangle with a height of 14 inches and a base of 32 inches is 224 square inches.

Exercise Set 2.4

Vocabulary Review 1. product **2.** reciprocal **3.** divisor
4. fraction
Problems 1. $\frac{15}{4}$ **3.** $\frac{4}{3}$ **5.** 9 **7.** 200 **9.** $\frac{3}{8}$ **11.** -1
13. $\frac{49}{64}$ **15.** $\frac{3}{4}$ **17.** $\frac{15}{16}$ **19.** $\frac{1}{6}$ **21.** 6 **23.** $\frac{5}{18}$
25. $\frac{9}{2}$ **27.** $\frac{2}{9}$ **29.** 9 **31.** $\frac{4}{5}$ **33.** $-\frac{15}{2}$ **35.** 40 **37.** $\frac{7}{10}$
39. 13 **41.** 12 **43.** 186 **45.** 646 **47.** $\frac{3}{5}$ **49.** 40
51. $3 \cdot 5 = 15$; $3 \div \frac{1}{5} = 3 \cdot \frac{5}{1} = 15$ **53.** 14 blankets **55.** 48 bags
57. 6 **59.** $\frac{7}{16}$ **61.** 1,778 students **63.** 28 cartons

Find the Mistake 1. Two numbers whose product is 1 are said to be reciprocals. **2.** Dividing the fraction $\frac{12}{7}$ by $\frac{4}{9}$, is equivalent to $\frac{12}{7} \cdot \frac{9}{4}$. **3.** To work the problem $\frac{22}{5} \div \frac{10}{3}$, multiply the first fraction by the reciprocal of the second fraction. **4.** The quotient of $\frac{14}{11}$ and $\frac{32}{6}$ is $\frac{21}{88}$.

Exercise Set 2.5

Vocabulary Review 1. least common denominator
2. equivalent **3.** numerators **4.** lowest terms
Problems 1. $\frac{2}{3}$ **3.** $\frac{1}{4}$ **5.** $\frac{1}{2}$ **7.** $\frac{1}{3}$ **9.** $\frac{3}{2}$ **11.** $\frac{x+6}{2}$
13. $\frac{4}{5}$ **15.** $\frac{10}{3}$ **17.**

First Number a	Second Number b	The Sum of a and b $a + b$
$\frac{1}{2}$	$\frac{1}{3}$	$\frac{5}{6}$
$\frac{1}{3}$	$\frac{1}{4}$	$\frac{7}{12}$
$\frac{1}{4}$	$\frac{1}{5}$	$\frac{9}{20}$
$\frac{1}{5}$	$\frac{1}{6}$	$\frac{11}{30}$

19.

First Number a	Second Number b	The Sum of a and b $a + b$
$\frac{1}{12}$	$\frac{1}{2}$	$\frac{7}{12}$
$\frac{1}{12}$	$\frac{1}{3}$	$\frac{5}{12}$
$\frac{1}{12}$	$\frac{1}{4}$	$\frac{1}{3}$
$\frac{1}{12}$	$\frac{1}{6}$	$\frac{1}{4}$

21. $\frac{7}{9}$ **23.** $\frac{7}{3}$ **25.** $\frac{7}{4}$ **27.** $\frac{7}{6}$ **29.** $\frac{9}{20}$ **31.** $\frac{7}{10}$
33. $\frac{19}{24}$ **35.** $\frac{13}{60}$ **37.** $\frac{31}{100}$ **39.** $\frac{67}{144}$ **41.** $\frac{29}{35}$ **43.** $\frac{949}{1,260}$
45. $\frac{13}{420}$ **47.** $\frac{41}{24}$ **49.** $\frac{53}{60}$ **51.** $\frac{5}{4}$ **53.** $\frac{88}{9}$ **55.** $\frac{10}{3}$
57. $\frac{3}{4}$ **59.** $\frac{1}{4}$ **61.** -1 **63.** $-\frac{11}{15}$ **65.** 19 **67.** 3

69. 9 **71.** 4 **73.** $\frac{1}{4} < \frac{3}{8} < \frac{1}{2} < \frac{3}{4}$ **75.** $\frac{160}{63}$ **77.** $\frac{5}{8}$

79. $\frac{7}{3}$ **81.** 3 **83.** $\frac{9}{2}$ pints **85.** \$1,325 **87.** $\frac{7}{18}$

89.

Grade	Number of students	Fraction of students
A	5	$\frac{1}{8}$
B	8	$\frac{1}{5}$
C	20	$\frac{1}{2}$
Below C	7	$\frac{7}{40}$
Total	40	1

91. 10 lots

93. $\frac{3}{2}$ in.

95. $\frac{9}{5}$ ft **97.** 59

99. $\frac{16}{8}$ **101.** $\frac{8}{8}$

103. $\frac{11}{4}$ **105.** $\frac{17}{8}$

107. $\frac{9}{8}$ **109.** 2 R 3

111. 8 R 16

Find the Mistake 1. The fractions $\frac{a}{c}$ and $\frac{b}{c}$ can be added to become $\frac{a+b}{c}$ because they have a common denominator. **2.** Subtracting $\frac{12}{21}$ from $\frac{18}{21}$, gives us $\frac{6}{21} = \frac{2}{7}$. **3.** The least common denominator for a set of denominators is the smallest number that is exactly divisible by each denominator. **4.** The LCD for the fractions $\frac{4}{6}$, $\frac{2}{8}$ and $\frac{3}{4}$ is 24.

Exercise Set 2.6

Vocabulary Review 1. proper fraction **2.** addition
3. mixed number **4.** improper fraction **5.** denominator
6. remainder

Problems 1. $\frac{14}{3}$ **3.** $\frac{21}{4}$ **5.** $\frac{13}{8}$ **7.** $\frac{47}{3}$ **9.** $\frac{104}{21}$ **11.** $\frac{427}{33}$

13. $1\frac{1}{8}$ **15.** $-4\frac{3}{4}$ **17.** $4\frac{5}{6}$ **19.** $3\frac{1}{4}$ **21.** $4\frac{1}{27}$

23. $-28\frac{8}{15}$ **25. a.** $4\frac{13}{205}$ **b.** $\frac{38}{65}$ **27.** \$6 **29.** $\frac{71}{12}$

31. $\frac{1,526}{5}$¢ **33.** $\frac{11}{4}$ **35.** $\frac{37}{8}$ **37.** $\frac{14}{5}$ **39.** $\frac{9}{40}$ **41.** $-\frac{3}{8}$

43. $\frac{32}{35}$ **45.** $-\frac{4}{7}$ **47.** a

Find the Mistake 1. For mixed-number notation, writing the whole number next to the fraction implies addition. **2.** A shortcut for changing a mixed number to an improper fraction is to multiply the whole number by the denominator of the fraction, and then add the result to the numerator of the fraction. **3.** Changing $6\frac{4}{5}$ to an improper fraction gives us $\frac{34}{5}$. **4.** Writing $\frac{70}{12}$ as a mixed number gives us $5\frac{5}{6}$.

Exercise Set 2.7

Vocabulary Review 1. mixed number **2.** numerators
3. improper fractions **4.** divisor

Problems 1. $5\frac{1}{10}$ **3.** $13\frac{2}{3}$ **5.** $6\frac{93}{100}$ **7.** $5\frac{5}{6}$ **9.** $9\frac{3}{4}$ **11.** $3\frac{1}{5}$

13. $12\frac{1}{2}$ **15.** $9\frac{9}{20}$ **17.** $\frac{32}{45}$ **19.** $1\frac{2}{3}$ **21.** 4 **23.** $4\frac{3}{10}$

25. $\frac{1}{10}$ **27.** $-3\frac{1}{5}$ **29.** $2\frac{1}{8}$ **31.** $7\frac{1}{2}$ **33.** $\frac{11}{13}$

35. $5\frac{1}{2}$ cups **37.** $\frac{5}{6}$ cup **39.** $1\frac{1}{3}$ **41.** $2,441\frac{3}{5}$ cents

43. $163\frac{3}{4}$ mi **45.** $4\frac{1}{2}$ yd **47.** $2\frac{1}{4}$ yd^2 **49.** $\frac{3}{4} < \frac{5}{4} < 1\frac{1}{2} < 2\frac{1}{8}$

51. Can 1 contains $332\frac{1}{2}$ calories, whereas Can 2 contains 276 calories. Can 1 contains $56\frac{1}{2}$ more calories than Can 2.

53. Can 1 contains 1,325 milligrams of sodium, whereas Can 2 contains 1,095 milligrams of sodium. Can 1 contains 230 more milligrams of sodium than Can 2.

55. a. $\frac{10}{15}$ **b.** $\frac{3}{15}$ **c.** $\frac{9}{15}$ **d.** $\frac{5}{15}$ **57. a.** $\frac{5}{20}$ **b.** $\frac{12}{20}$ **c.** $\frac{18}{20}$ **d.** $\frac{2}{20}$

59. $\frac{13}{15}$ **61.** $\frac{14}{9} = 1\frac{5}{9}$ **63.** $\frac{3}{5}$ **65.** $\frac{1}{4}$ **67.** $2\frac{1}{4}$

69. $3\frac{1}{16}$ **71.** $2\frac{1}{4}$ ft^2 **73.** $3\frac{3}{8}$ ft^3

Find the Mistake 1. To multiply two mixed numbers, change to improper fractions, then multiply numerators and multiply denominators. **2.** Multiplying $4\frac{3}{8}$ and $9\frac{2}{7}$ gives us the mixed number $40\frac{5}{8}$. **3.** To divide the mixed numbers $12\frac{2}{5}$ and $3\frac{12}{5}$, change the mixed numbers to improper fractions and then multiply by the reciprocal of the divisor. **4.** The answer to the division problem $3\frac{9}{14} \div 2$ written as a mixed number is $1\frac{23}{28}$.

Landmark Review 1. 1 **2.** $\frac{1}{5}$ **3.** $\frac{19}{15}$ **4.** $\frac{5}{4}$

5. $\frac{73}{105}$ **6.** $\frac{2+x}{3}$ **7.** $\frac{41}{30} = 1\frac{11}{30}$ **8.** $\frac{1}{10}$ **9.** $\frac{29}{8}$ **10.** $\frac{14}{3}$

11. $\frac{21}{2}$ **12.** $\frac{5}{4}$ **13.** $4\frac{2}{3}$ **14.** $4\frac{3}{5}$ **15.** $3\frac{1}{2}$ **16.** $2\frac{8}{17}$

17. $23\frac{3}{8}$ **18.** $9\frac{4}{9}$ **19.** $37\frac{13}{24}$ **20.** $1\frac{1}{5}$ **21.** $2\frac{5}{16}$ **22.** $1\frac{47}{64}$

Exercise Set 2.8

Vocabulary Review 1. addition sign **2.** columns **3.** LCD
4. improper, proper **5.** borrow

Problems 1. $5\frac{4}{5}$ **3.** $12\frac{2}{5}$ **5.** $3\frac{4}{9}$ **7.** 12 **9.** $1\frac{3}{8}$ **11.** $14\frac{1}{6}$

13. $4\frac{1}{12}$ **15.** $2\frac{1}{12}$ **17.** $26\frac{7}{12}$ **19.** $11\frac{3}{8}$ **21.** 12 **23.** $2\frac{1}{2}$

25. $8\frac{6}{7}$ **27.** $3\frac{3}{8}$ **29.** $10\frac{4}{15}$ **31.** $2\frac{1}{15}$ **33.** 9 **35.** $18\frac{1}{10}$

37. 14 **39.** 17 **41.** $24\frac{1}{24}$ **43.** $27\frac{6}{7}$ **45.** $37\frac{3}{20}$ **47.** $33\frac{5}{24}$

49. $6\frac{1}{4}$ **51.** $9\frac{7}{10}$ **53.** $5\frac{1}{2}$ **55.** $\frac{2}{3}$ **57.** $1\frac{11}{12}$ **59.** $3\frac{11}{12}$

61. $5\frac{19}{20}$ **63.** $5\frac{1}{2}$ **65.** $\frac{13}{24}$ **67.** $10\frac{5}{12}$ **69.** $3\frac{1}{2}$ **71.** $5\frac{29}{40}$

73. $12\frac{1}{4}$ in. **75.** $\frac{1}{8}$ mi **77.** $31\frac{1}{6}$ in. **79.** NFL: $P = 306\frac{2}{3}$ yd, Canadian: $P = 350$ yd, Arena: $P = 156\frac{2}{3}$ yd **81. a.** $2\frac{1}{2}$ **b.** \$250

83. \$300 **85.** $4\frac{63}{64}$ **87.** 2 **89.** $\frac{11}{8} = 1\frac{3}{8}$ **91.** $3\frac{5}{8}$ **93.** $35\frac{3}{5}$ sec

Find the Mistake 1. To begin adding $3\frac{9}{14}$ and $5\frac{1}{3}$, write each mixed number with the addition sign and then apply the commutative and associative properties, such that $\left(3 + \frac{9}{14}\right) + \left(5 + \frac{1}{3}\right)$. **2.** The final answer for the problem $5\frac{2}{3} + 6\frac{7}{9}$ is $12\frac{4}{9}$. **3.** The first step when subtracting $8 - 3\frac{2}{7} = 4\frac{5}{7}$, is to borrow 1 from 8 in the form of $\frac{7}{7}$. **4.** Borrow 1 from the 5 in the form of $\frac{15}{15}$ so that $4\frac{25}{15} - 2\frac{12}{15} = 2\frac{13}{15}$.

Exercise Set 2.9

Vocabulary Review 1. complex fraction **2.** top, bottom
3. LCD **4.** whole numbers **5.** simplify

Problems 1. 7 **3.** 7 **5.** 2 **7.** 35 **9.** $\frac{7}{8}$ **11.** $8\frac{1}{3}$ **13.** $\frac{11}{36}$

15. $3\frac{2}{3}$ **17.** $6\frac{3}{8}$ **19.** $4\frac{5}{12}$ **21.** $\frac{8}{9}$ **23.** $\frac{1}{2}$ **25.** $1\frac{1}{10}$

27. 5 **29.** $\frac{3}{5}$ **31.** $\frac{7}{11}$ **33.** 5 **35.** $\frac{17}{28}$ **37.** $1\frac{7}{16}$ **39.** $\frac{13}{22}$

41. $\frac{5}{22}$ **43.** 14 **45.** $\frac{15}{16}$ **47.** $1\frac{5}{17}$ **49.** $\frac{3}{29}$ **51.** $1\frac{34}{67}$

53. $\frac{346}{441}$ **55.** $\frac{32}{33}$ **57.** $5\frac{2}{5}$ **59.** 8 **61.** $115\frac{2}{3}$ yd

Find the Mistake 1. The first step to solving the problem $\frac{1}{4} - \left(2\frac{3}{8} - 1\frac{5}{8}\right)^2$ is to subtract the fractions inside the parentheses before evaluating the exponent. **2.** To simplify $5\frac{2}{3} + \left(10\frac{1}{3} \cdot \frac{2}{3}\right)$ you must first change $10\frac{1}{3}$ into the improper fraction $\frac{31}{3}$ before multiplying by $\frac{2}{3}$. **3.** A complex fraction is a fraction in which a fraction or combination of fractions appear in the numerator and/or the denominator of the original fraction. **4.** To simplify the complex fraction $\frac{\frac{1}{6} + \frac{2}{3}}{\frac{5}{6} + \frac{5}{12}}$, multiply the top and bottom of the fraction by 12.

Chapter 2 Test

1. $\frac{1}{5}$ **2.** $\frac{1}{4}$ **3.** $\frac{15}{8}$ **4.** $\frac{1}{6}$ **5.** $-\frac{3}{10}$ **6.** $-\frac{1}{10}$ **7.** $\frac{3}{20}$
8. $\frac{4}{5}$ **9.** $-\frac{4}{5}$ **10.** $-\frac{1}{15}$ **11.** 6 **12.** $\frac{4}{9}$ **13.** $\frac{2}{5}$
14. $\frac{3}{14}$ **15.** $\frac{20}{11}$ **16.** $\frac{7}{18}$ **17.** $\frac{1}{5}$ **18.** $\frac{2}{3}$
19. $A = 15$ in^2 **20.** 10 dresses **21.** $\frac{1}{2}$ **22.** $\frac{3}{7}$
23. $\frac{45}{8} = 5\frac{5}{8}$ **24.** $1\frac{1}{2}$ **25.** $\frac{31}{24} = 1\frac{7}{24}$ **26.** $9\frac{19}{20}$
27. $\frac{5}{14}$ **28.** $3\frac{1}{2}$ **29.** $\frac{2}{9}$ **30.** $-1\frac{3}{8}$ **31.** $16\frac{4}{9}$ **32.** $5\frac{25}{72}$
33. $15\frac{7}{8}$ **34.** Perimeter $= 29\frac{1}{4}$ **35.** $14\frac{1}{2}$ **36.** $13\frac{1}{8}$
37. $\frac{5}{3} = 1\frac{2}{3}$ **38.** $\frac{3}{5}$ **39.** $\frac{3}{28}$ **40.** $1\frac{1}{13}$ **41.** $\frac{6}{100}$
42. $\frac{4}{200}$

Chapter 3

Exercise Set 3.1

Vocabulary Review **1.** fractional **2.** decimal point **3.** thousandths
4. tenths **5.** hundredths **6.** place value **7.** left **8.** right
Problems **1.** Three tenths **3.** Fifteen thousandths
5. Three and four tenths **7.** Fifty-two and seven tenths
9. $405\frac{36}{100}$ **11.** $9\frac{9}{1,000}$ **13.** $1\frac{234}{1,000}$ **15.** $\frac{305}{100,000}$ **17.** Tens
19. Tenths **21.** Hundred thousandths **23.** Ones **25.** Hundreds
27. 0.55 **29.** 6.9 **31.** 11.11 **33.** 100.02 **35.** 3,000.003

Rounded to the Nearest					
	Number	Whole	Tenth	Hundredth	Thousandth
37.	47.5479	48	47.5	47.55	47.548
39.	0.8175	1	0.8	0.82	0.818
41.	0.1562	0	0.2	0.16	0.156
43.	2,789.3241	2,789	2,789.3	2,789.32	2,789.324
45.	99.9999	100	100.0	100.00	100.000

47. Three and eleven hundredths; two and five tenths
49. 186,282.40 **51.** Fifteen hundredths
53.

Price Of 1 Gallon Of Regular Gasoline	
Date	Price (Dollars)
3/21/11	3.526
3/28/11	3.596
4/4/11	3.684
4/11/11	3.791

55. a. $<$ **b.** $>$
57. 0.002 0.005 0.02 0.025 0.05 0.052
59. 7.451 and 7.54 **61.** $\frac{1}{4}$
63. $\frac{1}{8}$ **65.** $\frac{5}{8}$ **67.** $\frac{7}{8}$

69. 9.99 **71.** 10.05 **73.** 0.05 **75.** 0.01 **77.** $6\frac{31}{100}$
79. $6\frac{23}{50}$ **81.** $18\frac{123}{1,000}$
Find the Mistake **1.** To move a place value from the tens column to the hundreds column, you must multiply by ten. **2.** The decimal 0.09 can be written as the fraction $\frac{9}{100}$. **3.** The decimal 142.9643 written as a mixed number is $142\frac{9,643}{10,000}$. **4.** Rounding the decimal 0.06479 to the nearest thousandth gives us 0.065.

Exercise Set 3.2

Vocabulary Review **1.** money **2.** columns **3.** decimal point
4. zeros **5.** value
Problems **1.** 6.19 **3.** 1.13 **5.** -6.29 **7.** 9.042

9. -8.021 **11.** 11.7843 **13.** 24.343 **15.** 24.111
17. 258.5414 **19.** 666.66 **21.** 11.11 **23.** 3.57
25. 4.22 **27.** 120.41 **29.** 44.933 **31.** 7.673 **33.** 530.865
35. 43.55 **37.** 5.918 **39.** 27.89 **41.** 35.64 **43.** 411.438
45. 6 **47.** 1 **49.** 3.1 **51.** 5.9 **53.** 4.17 **55.** 2.8
57. -11.41 **59.** -1.9 **61.** -17.5 **63.** 3.272 **65.** 4.001
67. $116.82 **69.** $1,571.10 **71.** 4.5 in. **73.** $5.43
75. 0.03 seconds **77. a.** Less **b.** Less **c.** 3%
79. 2 in. **81.** $3.25; three $1 bills and a quarter **83.** 3.25
85. $\frac{3}{100}$ **87.** $\frac{51}{10,000}$ **89.** $1\frac{1}{2}$ **91.** 1,400 **93.** $\frac{3}{20}$
95. $\frac{147}{1,000}$ **97.** 132,980 **99.** 2,115
101. a. 8 million **b.** Between 2009 and 2010 **c.** 625 million users
Find the Mistake **1.** To add 32.69 and 4.837, align the decimal point and add in columns. **2.** To add $0.004 + 5.06 + 32$ by first changing each decimal to a fraction would give us the problem $\frac{4}{1,000} + 5\frac{6}{100} + 32$. **3.** When subtracting $8.7 - 2.0163$, we make sure to keep the digits in the correct columns by writing 8.7 as 8.7000. **4.** Subtracting 4.367 from the sum of 12.1 and 0.036 gives us 7.769.

Exercise Set 3.3

Vocabulary Review multiply, answer, digits, decimal points
Problems **1.** 0.28 **3.** 0.028 **5.** 0.0027 **7.** 0.78 **9.** 0.792
11. 0.0156 **13.** 24.29821 **15.** 0.03 **17.** 187.85 **19.** 0.002
21. 27.96 **23.** 0.43 **25.** 49,940 **27.** 9,876,540 **29.** 1.89
31. 0.0025 **33.** 5.1106 **35.** 7.3485 **37.** 4.4 **39.** 2.074
41. 3.58 **43.** 187.4 **45.** 116.64 **47.** 20.75 **49.** 0.126
51. Moves it two places to the right **53.** $1,381.38 **55.** $7.10
57. $44.40 **59.** $293.04 **61.** 8,509 mm^2 **63.** 1.18 in^2
65. 1,879 **67.** 1,516 R 4 **69.** 298 **71.** 34.8
73. 49.896 **75.** 825 **77.** No
Find the Mistake **1.** To multiply 18.05 by 3.5, multiply as if the numbers were whole numbers and then place the decimal in the answer with three digits to its right. **2.** To estimate the answer for 24.9×7.3, round 24.9 to 25 and 7.3 to 7. **3.** To simplify $(8.43 + 1.002) - (0.05)(3.2)$, first work the operation inside the parentheses and find the product of 0.05 and 3.2 before subtracting.
4. Lucy pays $1.52 a pound for the first three pounds of candy she buys at a candy store, and pays $3.27 for each additional pound. To find how much she will pay if she buys 5.2 pounds of candy, we must solve the problem $1.52(3) + 3.27(2.2)$.

Exercise Set 3.4

Vocabulary Review **1.** long division **2.** above **3.** last
4. divisor, right
Problems **1.** 19.7 **3.** 6.2 **5.** 5.2 **7.** 11.04 **9.** 4.8
11. 9.7 **13.** 2.63 **15.** 4.24 **17.** 2.55 **19.** 1.35
21. 6.5 **23.** 9.9 **25.** 0.05 **27.** 89 **29.** 2.2 **31.** 4.6
33. 1.35 **35.** 16.97 **37.** 0.25 **39.** 2.71 **41.** 11.69
43. 3.98 **45.** 5.98 **47.** -4.24 **49.** 7.5 mi
51.

Rank	Name	Number of Tournaments	Average per Tournament
1.	Na Yeon Choi	23	$81,360
2.	Jiyai Shin	19	$93,850
3.	Cristie Kerr	21	$76,260
4.	Yani Tseng	19	$82,820
5.	Suzann Pettersen	19	$81,960

53. $5.65/hr **55.** 22.4 mi **57.** 5 hr **59.** 7 min **61.** 2.73

63. 0.13 **65.** 0.77778 **67.** 307.20607 **69.** 0.70945 **71.** $\frac{3}{4}$
73. $\frac{2}{3}$ **75.** $\frac{3}{8}$ **77.** $\frac{19}{50}$ **79.** $\frac{60}{100}$ **81.** $\frac{500}{100}$ **83.** $\frac{12}{15}$
85. $\frac{60}{15}$ **87.** $\frac{18}{15}$ **89.** 0.75 **91.** 0.875

Find the Mistake **1.** The answer to the problem $25\overline{)70.75}$ will have a decimal point placed with two digits to its right.
2. To work the problem $27.468 \div 8.4$, multiply both numbers by 10 and then divide. **3.** To divide 0.6778 by 0.54, multiply both numbers by 100 to move the decimal point two places to the right.
4. Samantha earns $10.16 an hour as a cashier. She received a paycheck for $309.88. To find out how many hours she worked, you must solve the problem $309.88 \div 10.16$.

Landmark Review **1.** One and fifteen hundredths
2. Forty-five and eight hundredths **3.** Five thousandths
4. Two hundred forty-five and one hundred fifty-seven thousandths
5. 0.0067 **6.** 5.6 **7.** 23.014 **8.** 2,013.15 **9.** 28.28
10. 9.150014 **11.** 124.15831 **12.** 11.799 **13.** 3.1 **14.** 7.07
15. 13.33 **16.** 78.37 **17.** 47.35 **18.** 0.00225 **19.** 20
20. 0.4 **21.** 5.16 **22.** 11.3505 **23.** 8.8 **24.** 33.46

Exercise Set 3.5

Vocabulary Review **1.** division **2.** repeats **3.** place values, reduce
Problems **1.** 0.125 **3.** 0.625
5.

Fraction	$\frac{1}{5}$	$\frac{2}{5}$	$\frac{3}{5}$	$\frac{4}{5}$	$\frac{5}{5}$
Decimal	0.2	0.4	0.6	0.8	1

7. 0.5 **9.** 0.56
11. 0.5625 **13.** −0.8125
15. 0.92 **17.** 0.27
19. 0.09 **21.** 0.28

23.

Decimal	0.125	0.250	0.375	0.500	0.625	0.750	0.875
Fraction	$\frac{1}{8}$	$\frac{1}{4}$	$\frac{3}{8}$	$\frac{1}{2}$	$\frac{5}{8}$	$\frac{3}{4}$	$\frac{7}{8}$

25. $\frac{3}{20}$ **27.** $\frac{2}{25}$ **29.** $\frac{3}{8}$ **31.** $5\frac{3}{5}$ **33.** $5\frac{3}{50}$ **35.** $1\frac{11}{50}$
37. 2.4 **39.** 3.98 **41.** 3.02 **43.** 0.3 **45.** 0.072 **47.** 0.8
49. 1 **51.** 0.25 **53.** $8.42 **55.** $38.66 **57.** 9 in.
59.

Change In Stock Price		
Date	Gain ($)	As a Decimal ($)
Monday	$\frac{3}{5}$	0.60
Tuesday	$\frac{1}{2}$	0.50
Wednesday	$\frac{1}{25}$	0.04
Thursday	$\frac{1}{5}$	0.20
Friday	$\frac{1}{10}$	0.10

61. 104.625 calories
63. $10.38
65. Yes **67.** 36
69. 25 **71.** 125
73. 9 **75.** $\frac{1}{81}$
77. $\frac{25}{36}$ **79.** 0.25
81. 1.44 **83.** 25
85. 100

Find the Mistake **1.** The correct way to write $\frac{6}{11}$ as a decimal is $0.\overline{54}$.
2. Writing 14.3 as a fraction gives us $14\frac{3}{10}$. **3.** The simplified answer to the problem $\frac{12}{45(0.256 + 0.14)}$ contains only fractions or only decimals. **4.** Simplifying the problem $\left(\frac{3}{2}\right)(0.5) + \left(\frac{1}{2}\right)^2(6.7)$ by first converting all decimals to fractions gives us $\left(\frac{3}{2}\right)\left(\frac{1}{2}\right) + \left(\frac{1}{2}\right)^2\left(\frac{67}{10}\right)$.

Exercise Set 3.6

Vocabulary Review **1.** square root **2.** radical sign
3. whole numbers **4.** irrational **5.** right **6.** hypotenuse
7. Pythagorean theorem **8.** spiral of roots
Problems **1.** 8 **3.** 9 **5.** 6 **7.** 5 **9.** 15 **11.** 48 **13.** 45
15. 48 **17.** 15 **19.** 1 **21.** 78 **23.** 9 **25.** $\frac{4}{7}$

27. $\frac{3}{4}$ **29.** $\frac{2}{3}$ **31.** $\frac{1}{4}$ **33.** False **35.** True **37.** 1.1180
39. 11.1803 **41.** 3.46 **43.** 11.18 **45.** 0.58 **47.** 0.58
49. 12.124 **51.** 9.327 **53.** 12.124 **55.** 12.124 **57.** 10 in.
59. 13 ft **61.** 8.06 km **63.** 17.49 m **65.** 30 ft **67.** 25 ft
71.

Height h(feet)	Distance d(miles)
10	4
50	9
90	12
130	14
170	16
190	17

73. a. 8.9 million
b. 8.25 million

Find the Mistake **1.** The square root of a positive number x is the number we square to get x. **2.** The square root of 225 is 15, and can be written in symbols as $\sqrt{225} = 15$. **3.** Simplifying the radical $\sqrt{\frac{196}{36}}$ gives us $\frac{14}{6}$ because $\left(\frac{14}{6}\right)^2 = \frac{14}{6} \cdot \frac{14}{6} = \frac{196}{36}$.
4. The Pythagorean theorem states that $c = \sqrt{a^2 + b^2}$.

Chapter 3 Test

1. Eleven and eight hundred nineteen thousandths **2.** Tenths
3. 73.0046 **4.** 100.91 **5.** 7.02 **6.** 11.724 **7.** 16.56
8. 5.84 **9.** 10.1 **10.** 7.89 **11.** −3.33 **12.** 0.22 **13.** $3.61
14. $18.10 **15.** $8.15 **16.** 0.68 **17.** $\frac{19}{50}$ **18.** 25.704
19. 2.38 **20.** 16.897 **21.** 0.47125 **22.** 8.92 **23.** 26.674
24. 10 **25.** 3 **26.** 15 **27.** 30 **28.** 11.18 in. **29.** 21.93 yds.
30. $44,033,333.33 per month **31.** $40,000,000 per month

Chapter 4

Exercise Set 4.1

Problems **1.** 750 tomatoes **3.** 36 ears of corn
5. 256.2 apples **7.** 210 apples **9.** 84.375 ears of corn
11. 93.75 ears of corn

Exercise Set 4.3

Problems **1.** 5 feet **3.** 10,912 yards **5.** 35 miles
7. 28.5 feet **9.** 6 miles **11.** 28 cm **13.** 0.4 km
15. 7,500 mm **17.** 0.00518 hm **19.** 630 m **21.** 6.21 mi
23. 68.6 m **25.** 170 cm. **27.** 77,200 m **29.** 1.22 km

Exercise Set 4.4

Problems **1.** 32 fl oz **3.** 1.75 qts **5.** 9 quarts, 1 pint
7. 2.25 gal **9.** 5 cups, 3 ounces **11.** 2,000 mL
13. 0.25 hL **15.** 40 L **17.** 0.580 kL **19.** 0.029 daL
21. 355 mL **23.** 13.2 L **25.** 2,590 gal **27.** 9.19 quarts
29. 126 cL

Exercise Set 4.5

Problems **1.** 7 pounds **3.** 205 oz **5.** 5 lb, 9.6 oz
7. 0.25 lb **9.** 4,300 lbs **11.** 250 dg
13. 0.0003 decagrams **15.** 0.0092 kg **17.** 9,300 mg
19. 23 cg **21.** 81.7 kg **23.** 1.54 tons **25.** 0.0176 oz
27. 2.47 kg **29.** 24,100 dg

Exercise Set 4.6

Problems **1.** 36.7° C **3.** 73.4° F **5.** −5° C **7.** 177° C
9. 32°F

Chapter 4 Test

Problems 1. 9.5 cm **2.** 56 oz **3.** 82.4° F **4.** 8.45 cups
5. Not possible **6.** 35,200 yds **7.** 1.5 qt **8.** 0.017g
9. 630 m **10.** 74.1 oz **11.** 218° C **12.** Not possible
13. 1.88 m **14.** 51.9 L **15.** 0.55 hm **16.** 0.5 gal
17. 32,400 lbs **18.** 2.57 kg **19.** 6.21 mi **20.** 230 mL

Chapter 5

Exercise Set 5.1

Vocabulary Review 1. solution set **2.** equivalent
3. addition **4.** subtraction
Problems 1. Yes **3.** Yes **5.** No **7.** Yes **9.** No
11. Yes **13.** 11 **15.** 4 **17.** $-\frac{3}{4}$ **19.** -5.8 **21.** -17
23. $-\frac{1}{8}$ **25.** -4 **27.** -3.6 **29.** 1 **31.** $-\frac{7}{45}$ **33.** 3
35. $\frac{11}{8}$ **37.** 21 **39.** 7 **41.** 3.5 **43.** 22 **45.** -2 **47.** -16
49. -3 **51.** 10 **53.** -12 **55.** 4 **57.** 2 **59.** -5
61. -1 **63.** -3 **65.** 8 **67.** -8 **69.** 2 **71.** 11
73. a. 280 tickets **b.** $2,100 **75.** 67° **77. a.** 225 **b.** $11,125

79.

n	t
32	72
50	90
44	84
96	136

81. $x + 55 + 55 = 180; 70°$ **83.** y
85. x **87.** 6 **89.** 6 **91.** -9
93. $-\frac{15}{8}$ **95.** 8 **97.** $-\frac{5}{4}$
99. $3x$ **101.** $-9x$

Find the Mistake 1. The number -2 is a solution to the equation $3x - 9 = 8x + 1$. **2.** Saying that the number 12 is a solution to the equation $\frac{1}{6}y + 3 = 5$ is a true statement. **3.** The addition property of equality states that $A + C = B + C$ if $A = B$.
4. To solve the equation $\frac{5}{6}b + 6 = 11$ for b, begin by adding -6 to both sides.

Exercise Set 5.2

Vocabulary Review 1. nonzero **2.** multiplication
3. division **4.** reciprocal
Problems 1. 2 **3.** 4 **5.** $-\frac{1}{2}$ **7.** -2 **9.** 3 **11.** 4
13. 0 **15.** 0 **17.** 6 **19.** -50 **21.** $\frac{3}{2}$ **23.** 12
25. -3 **27.** 32 **29.** -8 **31.** $\frac{1}{2}$ **33.** 4 **35.** 8
37. -12 **39.** -3 **41.** -4 **43.** 4 **45.** -15 **47.** $-\frac{1}{2}$
49. 3 **51.** 1 **53.** $\frac{1}{4}$ **55.** -3 **57.** 3 **59.** 2
61. $-\frac{3}{2}$ **63.** $-\frac{3}{2}$ **65.** 1 **67.** 1 **69.** -2 **71.** -2
73. 200 tickets **75.** $1,390.85 per month **77.** 31
79. 21 **81.** 20 songs **83. a.** 20 ft. **b.** 35 ft. **c.** 75 ft.
85. 55 hours **87.** 2 **89.** 6 **91.** 3,000 **93.** 650
95. $3x - 11$ **97.** $0.09x + 180$ **99.** $-6y + 4$ **101.** $4x - 11$
Find the Mistake 1. The multiplication property of equality states that multiplying both sides of an equation by the same nonzero number will not change the solution set. **2.** To solve $-\frac{2b}{9} = 3$ for b, we multiply both sides of the equation by the reciprocal of $-\frac{2}{9}$.
3. To begin solving the equation $\frac{5}{6}x + \frac{1}{2} = -\frac{1}{4}$, we can either add $-\frac{1}{2}$ to each side of the equation or multiply both sides by the LCD 12.
4. To solve the equation $-\frac{8}{9}y = -3$, we begin by multiplying both sides of the equation by 9.

Exercise Set 5.3

Vocabulary Review 1. a. distributive **b.** LCD, decimals **c.** similar
2. addition, constant **3.** multiplication **4.** solution

Problems 1. 3 **3.** -2 **5.** -1 **7.** 2 **9.** -4 **11.** -2
13. 0 **15.** 1 **17.** $\frac{1}{2}$ **19.** 7 **21.** 8 **23.** All real numbers
25. $\frac{3}{4}$ **27.** 75 **29.** 2 **31.** 6 **33.** 8 **35.** 0 **37.** $\frac{3}{7}$
39. No solution **41.** 1 **43.** -1 **45.** 6 **47.** $\frac{3}{4}$
49. 3 **51.** No solution **53.** 8 **55.** 6 **57.** -2
59. -2 **61.** 2 **63.** All real numbers **65.** 2 **67.** 20
69. 4,000 **71.** 700 **73.** 11 **75.** 7
77. a. $\frac{5}{4} = 1.25$ **b.** $\frac{15}{2} = 7.5$ **c.** $6x + 20$ **d.** 15 **e.** $4x - 20$ **f.** $\frac{45}{2} = 22.5$
79. $x = 3$ or 3 robot events were entered **81.** $4,200
83. 14 **85.** -3 **87.** $\frac{1}{4}$ **89.** $\frac{1}{3}$ **91.** $-\frac{3}{2}x + 3$
93. $\frac{5}{4}x - 5$ **95.** $4x - 3$ **97.** $8x + 3$
Find the Mistake 1. A linear equation in one variable is any equation that can be put in the form of $Ax + B = C$.
2. $12 + 6y - 4 = 3y$ is a linear equation simplified to $3y + 8 = 0$.
3. To solve the linear equation $3(2x + 7) = 2(x - 1)$, begin by applying the distributive property. **4.** To solve a linear equation, make sure to combine similar terms on each side.

Exercise Set 5.4

Vocabulary Review 1. variable **2.** formula **3.** isolate
4. complementary **5.** supplement
Problems 1. 100 feet **3.** 0 **5.** 2 **7.** 15 **9.** 10
11. -2 **13.** 1 **15. a.** 2 **b.** 4 **17. a.** 5 **b.** 18
19. $l = \frac{A}{w}$ **21.** $h = \frac{V}{lw}$ **23.** $a = P - b - c$
25. $x = 3y - 1$ **27.** $y = 3x + 6$ **29.** $y = -\frac{2}{3}x + 2$
31. $y = -2x - 5$ **33.** $y = -\frac{2}{3}x + 1$ **35.** $w = \frac{P - 2l}{2}$
37. $v = \frac{h - 16t^2}{t}$ **39.** $h = \frac{A - \pi r^2}{2\pi r}$ **41.** $t = \frac{I}{pr}$
43. a. $y = \frac{3}{5}x + 1$ **b.** $y = \frac{1}{2}x + 2$ **c.** $y = 4x + 3$
45. a. $y = -\frac{1}{2}x - 2$ **b.** $y = -3x + 2$ **c.** $y = -\frac{1}{3}x - 1$
47. $y = \frac{3}{7}x - 3$ **49.** $y = 2x + 8$ **51.** $60°; 150°$
53. $45°; 135°$ **55.** 10 **57.** 240 **59.** 25% **61.** 35%
63. 64 **65.** 2,000 **67.** $100°C$; yes **69.** $20°C$; yes
71. $C = \frac{5}{9}(F - 32)$ **73.** 46.3% **75.** 5.3% **77.** 7 meters
79. $\frac{3}{2}$ or 1.5 inches **81.** 132 feet **83.** $\frac{2}{9}$ centimeters

85.

Pitcher, Team	Rolaids Points
Heath Bell, San Diego	144
Brian Wilson, San Francisco	135
Carlos Marmol, Chicago	107
Billy Wagner, Atlanta	128
Francisco Cordero, Cincinnati	130

87. The sum of 4 and 1 is 5. **89.** The difference of 6 and 2 is 4.
91. The difference of a number and 5 is -12.
93. The sum of a number and 3 is four times the difference of that number and 3. **95.** $2(6 + 3) = 18$ **97.** $2(5) + 3 = 13$
99. $x + 5 = 13$ **101.** $5(x + 7) = 30$

Find the Mistake 1. A formula contains more than one variable.
2. To solve for a variable in a formula, make sure the variable is isolated on one side of the equation. **3.** To solve the formula $2ab = a + 8$ for b, divide each side of the equation by $2a$.
4. The question "What percent of 108 is 9?" can be translated into the equation $x \cdot 108 = 9$.

Landmark Review 1. 12 **2.** 25 **3.** 5.2 **4.** $\frac{13}{15}$ **5.** 4
6. 10 **7.** 4 **8.** 3 **9.** 3 **10.** 4 **11.** $\frac{A}{w} = l$
12. $P - a - b = c$ **13.** $18 - 3y = x$ **14.** $-\frac{4}{3}x + 3 = y$

Exercise Set 5.5

Vocabulary Review **1.** known **2.** unknown **3.** variable
4. equation **5.** sentence **6.** problem
Problems **1.** 8 **3.** 5 **5.** -1 **7.** 3 and 5 **9.** 6 and 14
11. Shelly is 39; Michele is 36 **13.** Evan is 11; Cody is 22
15. Barney is 27; Fred is 31 **17.** Lacy is 16; Jack is 32
19. Patrick is 18; Pat is 38 **21.** $s = 9$ inches **23.** $s = 15$ feet
25. 11 feet, 18 feet, 33 feet **27.** 26 feet, 13 feet, 14 feet
29. $l = 11$ inches; $w = 6$ inches **31.** $l = 25$ meters; $w = 9$ meters
33. $l = 15$ feet; $w = 3$ feet **35.** 9 dimes; 14 quarters
37. 12 quarters; 27 nickels **39.** 8 nickels; 17 dimes
41. 7 nickels; 10 dimes, 12 quarters **43.** 3 nickels; 9 dimes, 6 quarters
45. $5x$ **47.** $1.075x$ **49.** $0.09x + 180$
Find the Mistake **1.** False, $7 + \frac{m}{n} = n + 9$ **2.** True
3. False, $2(3x + 6) + 2x = 42$ **4.** False, $10(2x - 4) + 5x = 310$

Exercise Set 5.6

Vocabulary Review

1. consecutive **2.** even **3.** integers **4.** sum
Problems **1.** 5 and 6 **3.** -4 and -5 **5.** 13 and 15 **7.** 52 and 54
9. -14 and -16 **11.** 17, 19, and 21 **13.** 42, 44, and 46
15. \$4,000 at 8%, \$6,000 at 9% **17.** \$700 at 10%; \$1,200 at 12%
19. \$500 at 8%, \$1,000 at 9%, \$1,500 at 10% **21.** $45°, 45°, 90°$
23. $22.5°, 45°, 112.5°$ **25.** $53°, 90°$ **27.** $80°, 60°, 40°$
29. 16 adult and 22 children's tickets **31.** 16 minutes
33. 39 hours **35.** They are in offices 7329 and 7331.
37. Kendra is 8 years old and Marissa is 10 years old. **39.** Jeff
41. \$10.38 **43.** $l = 12$ meters; $w = 10$ meters **45.** $59°, 60°, 61°$
47. 18 cm, 24 cm, 36 cm **49.** \$54.00 **51.** Yes **53.** 5
55. a. 9 **b.** 3 **c.** -9 **d.** -3 **57. a.** -8 **b.** 8 **c.** 8 **d.** -8
59. -2.3125 **61.** $\frac{10}{3}$ **63.** $y = \frac{2}{3}x - 2$ **65.** $x = \frac{2}{3}y + \frac{5}{3}$
Find the Mistake **1.** True **2.** False, $0.07(4x) + 0.08x = 396$
3. True **4.** True

Exercise Set 5.7

Vocabulary Review **1.** inequalities **2.** positive **3.** negative
4. graph **5.** compound
Problems
1. $x < 12$

3. $a \leq 12$

5. $x > 13$

7. $y \geq 4$

9. $x > 9$

11. $x < -6$

13. $x < 2$

15. $a \leq 5$

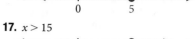

17. $x > 15$

19. $x < -3$

21. $x \leq 6$

23. $x \geq -50$

25. $y < -6$

27. $x < -4$

29. $x > -12$

31. $x < 6$

33. $y \geq -5$

35. $x < 3$

37. $x \leq 18$

39. $a < -20$

41. $y < 25$

43. $a \leq 3$

45. $x \geq \frac{15}{2}$

47. $x < -1$

49. $y \geq -2$

51. $x < -1$

53. $m \leq -6$

55. $x \leq -5$

57. $x \leq 3$

59. $x \leq \frac{25}{9}$

61. $y < -\frac{3}{2}x + 3$ **63.** $y < \frac{2}{5}x - 2$ **65.** $y \leq \frac{3}{7}x + 3$
67. $y \leq \frac{1}{2}x + 1$ **69. a.** 3 **b.** 2 **c.** No **d.** $x > 2$
71. $x < 3$ **73.** $x \leq 3$
75. $x \leq -7$ or $x \geq -3$

77. $x \leq -1$ or $x \geq \frac{3}{5}$

79. $x < -10$ or $x > 6$

81. $x < -3$ or $x \geq 2$

83. $3 \leq m \geq 7$

85. $-3 < a < 0$

87. $4 \leq a \leq 6$

89. $-4 < x < 2$

91. $-3 < x < 3$

93. $0 \leq x < 6$

95. $-\frac{2}{3} \leq x \leq 0$

97. $-\frac{3}{2} \leq x \leq \frac{17}{2}$

99. $\frac{3}{8} \leq x \leq \frac{3}{2}$

101. At least 291 **103.** $x < 2$ **105.** $x > -\frac{8}{3}$
107. $x \geq 6$; the width is at least 6 meters. **109.** $x > 6$; the shortest side is even and greater than 6 inches. **111.** $t \geq 100$
113. Lose money if they sell less than 200 tickets. Make a profit if they sell more than 200 tickets.
115. 8 **117.** 24 **119.** 25% **121.** 10% **123.** 80 **125.** 400
127. -5 **129.** 5 **131.** 7 **133.** 9 **135.** 6 **137.** $2x - 3$

Find the Mistake **1.** If the solution set to an inequality is all the real numbers greater than 9, the solution is $x > 9$. **2.** To include the number 14 in the solution to the inequality $x \geq 14$, we mark the number on the graph with a closed circle. **3.** The last step in solving an inequality is to graph the solution set.
4. Upon solving the inequality $-5y < 4x + 9$ for y, the inequality symbol is reversed.

Chapter 5 Test

1. No **2.** Yes **3.** Yes **4.** No **5.** 17 **6.** 5
7. 5.2 **8.** 3 **9.** -5 **10.** 1 **11.** 4 **12.** 1 **13.** 55
14. -3 **15.** $\frac{5}{2}$ **16.** 2 **17.** 5 **18.** -1 **19.** 8
20. $y = -\frac{1}{3}x + 2$ **21.** $a = \frac{x^2 - v^2}{2d}$ **22.** Paul is 36, Becca is 18
23. $l = 55$ cm, $w = 20$ cm **24.** 14 dimes, 6 nickels
25. $700 at 6% and $1,200 at 12% **26.** $x > 10$ **27.** $y \geq -4$
28. $x > -4$ **29.** $n \leq -2$ **30.** $x < 1$ **31.** $x \geq 2$
32. $x > -3$ **33.** $x \geq -1$ **34.** $x < -2$ **35.** 27.9%
36. 20.4%

Chapter 6

Exercise Set 6.1
Vocabulary Review **1.** base, exponent **2.** product
3. power **4.** distributive **5.** scientific notation
Problems **1.** 16 **3.** 0.09 **5.** 64 **7.** 25 **9.** -8
11. 81 **13.** $\frac{4}{9}$ **15.** $\frac{1}{16}$ **17.** $\frac{9}{16}$ **19.** $-\frac{1}{36}$
21. a.

Number x	1	2	3	4	5	6	7
Square x^2	1	4	9	16	25	36	49

b. Larger

23. x^9 **25.** y^{30} **27.** 2^{12} **29.** x^{28} **31.** x^{10} **33.** 5^{12}
35. y^9 **37.** 2^{50} **39.** a^{3x} **41.** b^{xy} **43.** $16x^2$ **45.** $32y^5$
47. $81x^4$ **49.** $0.25a^2b^2$ **51.** $64x^3y^3z^3$ **53.** $-64x^3y^3$ **55.** $8x^{12}$
57. $16a^6$ **59.** x^{14} **61.** a^{11} **63.** $128x^7$ **65.** $432x^{10}$
67. $16x^4y^6$ **69.** $\frac{8}{27}a^{12}b^{15}$

71.

Number x	-3	-2	-1	0	1	2	3
Square x^2	9	4	1	0	1	4	9

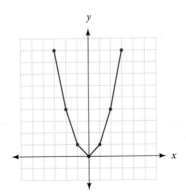

73.

Number x	-2.5	-1.5	-0.5	0	0.5	1.5	2.5
Square x^2	6.25	2.25	0.25	0	0.25	2.25	6.25

75. 4.32×10^4 **77.** 5.7×10^2 **79.** 2.38×10^5 **81.** 1.0×10^{10}

83. 2,490 **85.** 352 **87.** 28,000 **89.** $-310,000$

91.

Name	Location	Area (sq. mi.)	Scientific Notation
Caspian Sea	5 countries	152,239	1.52×10^5
Superior	US, Canada	31,820	3.18×10^4
Victoria	Tanzania, Uganda	26,828	2.68×10^4
Huron	US, Canada	23,010	2.30×10^4
Michigan	US	22,400	2.24×10^4

93. 27 inches3 **95.** 15.6 inches3 **97.** 36 inches3
99. Answers will vary **101.** 6.5×10^8 seconds **103.** \$740,000
105. \$180,000 **107.** \$30,957.48 **109.** 219 inches3
111. 182 inches3 **113.** -3 **115.** 11 **117.** -5
119. 5 **121.** 2 **123.** 6 **125.** 4 **127.** 3
Find the Mistake **1.** Simplifying the expression $\left(-\frac{12}{32}\right)^2$ will give a negative answer. **2.** You would use the product property for exponents to simplify the expression $x^8 \cdot x^4 \cdot x^{10}$. **3.** To simplify the expression $(x^3 \cdot y)^4 (y \cdot z^2)^8$, we must first use the distributive property of exponents. **4.** 5.83×10^5 in an example of a number written in scientific notation.

Exercise Set 6.2

Vocabulary Review **1.** negative **2.** quotient
3. expanded **4.** exponents
Problems **1.** $\frac{1}{9}$ **3.** $\frac{1}{36}$ **5.** $\frac{1}{64}$ **7.** $\frac{1}{125}$ **9.** $\frac{2}{x^3}$
11. $\frac{1}{8x^3}$ **13.** $\frac{1}{25y^2}$ **15.** $\frac{1}{100}$
17.

Number x	Square x^2	Power of 2 2^x
-3	9	$\frac{1}{8}$
-2	4	$\frac{1}{4}$
-1	1	$\frac{1}{2}$
0	0	1
1	1	2
2	4	4
3	9	8

19. $\frac{1}{25}$ **21.** x^6 **23.** 64 **25.** $8x^3$ **27.** 6^{10} **29.** $\frac{1}{6^{10}}$ **31.** $\frac{1}{2^8}$
33. 2^8 **35.** $27x^3$ **37.** $16x^2$ **39.** $\frac{x^3}{27}$ **41.** $\frac{16}{y^2}$ **43.** $-\frac{125}{x^3}$
45. $\frac{64}{y^2}$ **47.** 1 **49.** $-2y$ **51.** $-3ab^2$ **53.** 1 **55.** $81x^4y^4$
57. 1 **59.** $2a^2b$ **61.** $\frac{1}{49y^6}$ **63.** $\frac{1}{x^8}$ **65.** $\frac{1}{y^3}$ **67.** x^2
69. a^6 **71.** $\frac{1}{y^9}$ **73.** y^{40} **75.** $\frac{1}{x}$ **77.** x^9 **79.** $\frac{4b^8}{a^4}$
81. a^{16} **83.** $\frac{1}{a^4}$ **85.** $\frac{27}{16x^2y^4}$

87.

Number x	Power of x 2^x
-3	$\frac{1}{8}$
-2	$\frac{1}{4}$
-1	$\frac{1}{2}$
0	1
1	2
2	4
3	8

89. 4.8×10^{-3} **91.** 2.5×10^1 **93.** 9×10^{-6} **95.** -8.7×10^{-5}
97. 0.00423 **99.** 0.00008 **101.** 4.2 **103.** -0.0036
105.

Standard Form	Scientific Notation $n \times 10^r$
0.000357	3.57×10^{-4}
0.00357	3.57×10^{-3}
0.0357	3.57×10^{-2}
0.357	3.57×10^{-1}
3.57	3.57×10^0
35.7	3.57×10^1
357	3.57×10^2
3,570	3.57×10^3
35,700	3.57×10^4

107. 0.002 **109.** $4 \times 10^{-2}, 1.1 \times 10^{-1}, 1.4 \times 10^{-1}, 6.7 \times 10^{-1}$
111. 2.5×10^4 **113.** 2.35×10^5 **115.** 8.2×10^{-4}
117. 100 inches2, 400 inches2; 4 **119.** x^2; $4x^2$; 4
121. 216 inches3; 1,728 inches3; 8 **123.** x^3; $8x^3$; 8 **125.** 1,407
127. 13.5 **129.** 8 **131.** 26.52 **133.** 12 **135.** x^8
137. x **139.** $\frac{1}{y^2}$ **141.** 1 **143.** 340 **145.** 0.000035
Find the Mistake **1.** The negative exponent property says that $x^{-y} = \left(\frac{1}{x}\right)^y$. **2.** When using the quotient property of exponents to simplify the expression $\frac{y^{12}}{y^8}$, we get y^4. **3.** When using the expanded distributive property of exponents to simplify the expression $\left(\frac{x}{3}\right)^4$, we get $\frac{x^4}{81}$. **4.** The number 0.00238 written in scientific notation is 2.38×10^{-3}.

Exercise Set 6.3

Vocabulary Review **1.** constant **2.** terms **3.** coefficient
4. variable **5.** subtraction
Problems **1.** $12x^7$ **3.** $-16y^{11}$ **5.** $32x^2$ **7.** $200a^6$
9. $-24a^3b^3$ **11.** $24x^6y^8$ **13.** $3x$ **15.** $\frac{6}{y^3}$ **17.** $\frac{1}{2a}$
19. $-\frac{3a}{b^2}$ **21.** $\frac{x^2}{9z^2}$ **23.** $-\frac{y^2}{4x^3z}$
25.

a	b	ab	$\frac{a}{b}$	$\frac{b}{a}$
10	$5x$	$50x$	$\frac{2}{x}$	$\frac{x}{2}$
$20x^3$	$6x^2$	$120x^5$	$\frac{10x}{3}$	$\frac{3}{10x}$
$25x^5$	$5x^4$	$125x^9$	$5x$	$\frac{1}{5x}$
$3x^{-2}$	$3x^2$	9	$\frac{1}{x^4}$	x^4
$-2y^4$	$8y^7$	$-16y^{11}$	$-\frac{1}{4y^3}$	$-4y^3$

27. $4x^3$ **29.** $\frac{1}{b^2}$ **31.** $\frac{6y^{10}}{x^4}$ **33.** $-\frac{4x^2}{3y}$ **35.** 6×10^8
37. 1.75×10^{-1} **39.** 1.21×10^{-6} **41.** 4.2×10^3
43. 3×10^{10} **45.** 5×10^{-3} **47.** -5×10^{-4} **49.** 2×10^6
51. 1×10^1 **53.** 4.2×10^{-6} **55.** $8x^2$ **57.** $-11x^5$ **59.** 0
61. $4x^3$ **63.** $31ab^2$ **65.** $-11x^2y$
67.

a	b	ab	$a + b$
$5x$	$3x$	$15x^2$	$8x$
$4x^2$	$2x^2$	$8x^4$	$6x^2$
$3x^3$	$6x^3$	$18x^6$	$9x^3$
$2x^4$	$-3x^4$	$-6x^8$	$-x^4$
x^5	$7x^5$	$7x^{10}$	$8x^5$

69. $x^2y + x$ **71.** $x + y$ **73.** $x^2 - 4$ **75.** $x^2 - x - 6$
77. $x^2 - 5x$ **79.** $x^2 - 8x$ **81.** $9x^3$ **83.** $-20a^2$
85. $6x^5y^2$ **87.** -5 **89.** 6 **91.** 76 **93.** $6x^2$ **95.** $2x$

97. $-2x - 9$ **99.** 11 **101.** 36

Find the Mistake 1. The coefficient of the monomial $12x^4y^2z$ is 12.
2. Multiplying the monomials $-15x^3y$ and $2x^2$ gives us $-30x^5y$.
3. When dividing $28x^6y^2$ by $4x^2y$, write the problem as the separate fractions $\frac{28}{4} \cdot \frac{x^6}{x^2} \cdot \frac{y^2}{y}$. **4.** Monomials with similar terms have the same variable parts, but coefficients may differ.

Landmark Review 1. x^8 **2.** y^{27} **3.** x^6 **4.** a^{xy} **5.** $27a^3$
6. $256y^5$ **7.** 25 **8.** $27x^3$ **9.** $\frac{1}{4}$ **10.** $\frac{1}{a^2}$ **11.** $4xy$ **12.** 1
13. 3×10^2 **14.** 2×10^3 **15.** $18x^5y^4$ **16.** $\frac{a^2b}{2c^3}$

Exercise Set 6.4

Vocabulary Review 1. polynomial **2.** degree **3.** similar
4. opposite
Problems 1. Trinomial, 3 **3.** Trinomial, 3 **5.** Binomial, 1
7. Binomial, 2 **9.** Monomial, 2 **11.** Monomial, 0
13. $5x^2 + 5x + 9$ **15.** $5a^2 - 9a + 7$ **17.** $x^2 + 6x + 8$
19. $6x^2 - 13x + 5$ **21.** $x^2 - 9$ **23.** $3y^2 - 11y + 10$
25. $6x^3 + 5x^2 - 4x + 3$ **27.** $2x^2 - x + 1$ **29.** $2a^2 - 2a - 2$
31. $-\frac{1}{9}x^3 - \frac{2}{3}x^2 - \frac{5}{2}x + \frac{7}{4}$ **33.** $-4y^2 + 15y - 22$
35. $x^2 - 33x + 63$ **37.** $8y^2 + 4y + 26$ **39.** $75x^2 - 150x - 75$
41. $12x + 2$ **43.** 4 **45.** 18π inches3 **47.** 5 **49.** -6
51. $-20x^2$ **53.** $-21x$ **55.** $2x$ **57.** $-8x$ **59.** $6x - 18$
61. $-3xy + 12$
Find the Mistake 1. The degree of the polynomial $x^2 + 6 + 4x - x^7$ is 7. **2.** Adding $x^3 + 4x^2 - 8 + x^2$ gives us $x^3 + 5x^2 - 8$.
3. The first step when subtracting $(x - 5)^2$ from $(x + 5)^2$ is to apply the exponents to expand each expression. **4.** The value of the polynomial $7a^2 + 5a - 10$ when a is 4 is 122.

Exercise Set 6.5

Vocabulary Review 1. distributive **2.** term **3.** binomials
4. products **5.** column
Problems 1. $6x^2 + 2x$ **3.** $6x^4 - 4x^3 + 2x^2$ **5.** $2a^3b - 2a^2b^2 + 2ab$
7. $3y^4 + 9y^3 + 12y^2$ **9.** $8x^5y^2 + 12x^4y^3 + 32x^2y^4$
11. $x^2 + 7x + 12$ **13.** $x^2 + 7x + 6$ **15.** $x^2 + 2x + \frac{3}{4}$
17. $a^2 + 2a - 15$ **19.** $xy + bx - ay - ab$ **21.** $y^2 - \frac{16}{49}$
23. $3x^2 - 11x + 10$ **25.** $3a^2 - 16a - 12$ **27.** $6x^2 + 9x - 6$
29. $2ax - 8x - 3a + 12$ **31.** $36x^2 - 25$ **33.** $4x^2 + \frac{1}{2}x - \frac{3}{4}$
35. $3 - 7a - 6a^2$ **37.** $(x + 4)(x + 5) = x^2 + 4x + 5x + 20 = x^2 + 9x + 20$
39. $(2x + 1)(2x + 2) = 4x^2 + 6x + 2$ **41.** $a^3 + 7a^2 + 13a + 15$
43. $x^3 + 27$ **45.** $3x^3 - 23x^2 + 38x - 16$ **47.** $2x^4 - 7x^3 + 3x^2 - x + 3$
49. $20x^6 - 24x^3 - 32$ **51.** $28a^7 - 42a^4 - 32a^3 + 48$
53. $x^3 - 12x^2 + 47x - 60$ **55.** $6x^2 - 11x - 14$
57. $15x^2 + 19x - 4$ **59.** $x^2 - x - 20$ **61.** $6x + 19$
63. $x^2 + 4x$ **65.** $2x^2 - 8x + 6$ **67.** $a^2 - 4a + 8$
69. $A = x(3x + 2) = 3x^2 + 2x$ **71.** $A = x(x + 2) = x^2 + 2x$
73. a. $A = \pi(r + 2)^2 = \pi r^2 + 4\pi r + 4\pi$ **b.** $A = 4\pi(r + 1) = 4\pi r + 4\pi$
75. $R = 1{,}200p - 100p^2$ **77.** 169 **79.** $-10x$ **81.** 0 **83.** 0
85. $-12x + 16$ **87.** $x^2 + x - 2$ **89.** $x^2 + 6x + 9$ **91.** $25x^2 - 9$
Find the Mistake 1. To find the product of two binomials, multiply each term in the first by each term in the second.
2. When using the FOIL method to multiply the binomials $(9x - 2)$ and $(5x + 3)$, the product of the two outside terms is $27x$.
3. For the problem $(x^2 - 4)(3x^2 + 6)$, adding the products of the two outside terms with the two inside terms gives us $-6x^2$.
4. For the problem $4x^2(3x^2 + 4x + 9)$, the degree of the answer is 4.

Exercise Set 6.6

Vocabulary Review 1. sum **2.** square of a binomial
3. subtract **4.** difference of two squares
Problems 1. $x^2 - 4x + 4$ **3.** $a^2 + 6a + 9$ **5.** $x^2 - 10x + 25$

7. $a^2 - a + \frac{1}{4}$ **9.** $x^2 + 20x + 100$ **11.** $a^2 + 1.6a + 0.64$
13. $4x^2 - 4x + 1$ **15.** $16a^2 + 40a + 25$ **17.** $9x^2 - 12x + 4$
19. $9a^2 + 30ab + 25b^2$ **21.** $16x^2 - 40xy + 25y^2$
23. $49m^2 + 28mn + 4n^2$ **25.** $36x^2 - 120xy + 100y^2$
27. $x^4 + 10x^2 + 25$ **29.** $a^4 + 2a^2 + 1$ **31.** $x^6 - 8x^3 + 16$
33.

x	$(x + 3)^2$	$x^2 + 9$	$x^2 + 6x + 9$
1	16	10	16
2	25	13	25
3	36	18	36
4	49	25	49

35.

	a	1	3	3	4
	b	1	5	4	5
$(a + b)^2$		4	64	49	81
$a^2 + b^2$		2	34	25	41
$a^2 + ab + b^2$		3	49	37	61
$a^2 + 2ab + b^2$		4	64	49	81

37. $a^2 - 25$ **39.** $y^2 - 1$ **41.** $81 - x^2$ **43.** $4x^2 - 25$
45. $16x^2 - \frac{1}{9}$ **47.** $4a^2 - 49$ **49.** $36 - 49x^2$ **51.** $x^4 - 9$
53. $a^4 - 16$ **55.** $25y^8 - 64$ **57.** $2x^2 - 34$
59. $-12x^2 + 20x + 8$ **61.** $a^2 + 4a + 6$ **63.** $8x^3 + 36x^2 + 54x + 27$
65. $20x$ **67.** $-4xy$ **69.** $3x^2 + 2$ **71.** $8a^3 - 12a^2 + 6a - 1$
73. $(50 - 1)(50 + 1) = 2{,}500 - 1 = 2{,}499$ **75.** Both equal 25.
77. $x^2 + (x + 1)^2 = 2x^2 + 2x + 1$
79. $x^2 + (x + 1)^2 + (x + 2)^2 = 3x^2 + 6x + 5$
81. $a^2 + ab + ba + b^2 = a^2 + 2ab + b^2$
83. $A = (x - 2)^2 = x^2 - 4x + 4$ **85.** $2x^2$ **87.** x^2 **89.** $3x$ **91.** $3xy$
Find the Mistake 1. To find the square of the binomial $(5a^2 - 9a)$, we multiply $(5a^2 - 9a)$ by itself. **2.** The second term in a binomial square is twice the product of the two terms.
3. Simplifying $(7x - 3)^2$ gives us $49x^2 - 42x + 9$.
4. The problem $(4a - 6)(4a + 6) = 16a^2 - 36$ is an example of a difference of two squares.

Exercise Set 6.7

Vocabulary Review 1. division, distributive **2.** polynomial, term
Problems 1. $x - 2$ **3.** $3 - 2x^2$ **5.** $5xy - 2y$
7. $7x^4 - 6x^3 + 5x^2$ **9.** $10x^4 - 5x^2 + 1$ **11.** $-4x^2 + 3x - 1$
13. $-4a + 2$ **15.** $-8a^4 - 12a^3$ **17.** $-4b - 5a$
19. $-6a^2b + 3ab^2 - 7b^3$ **21.** $-\frac{a}{2} - b - \frac{b^2}{2a}$
23. $\frac{3}{2}a^3 + 4a^2 - 6a$ **25.** $3x + 4y$ **27.** $-y + 3$ **29.** $5y - 4$
31. $xy - x^2y^2$ **33.** $-1 + xy$ **35.** $-a + 1$ **37.** $x^2 - 3xy + y^2$
39. $2 - 3b + 5b$ **41.** $-2xy + 1$ **43.** $xy - \frac{1}{2}$ **45.** $\frac{1}{4x} - \frac{1}{2a} + \frac{3}{4}$
47. $\frac{4x^2}{3} + \frac{2}{3x} + \frac{1}{x^2}$ **49.** $3a^{3m} - 9a^m$ **51.** $2x^{4m} - 5x^{2m} + 7$
53. $3x^2 - x + 6$ **55.** 4 **57.** $x + 5$ **59.** Both equal 7.
61. $\frac{3(10) + 8}{2} = 19; 3(10) + 4 = 34$ **63.** $146\frac{20}{27}$ **65.** $2x + 5$
67. $x^2 - 3x$ **69.** $2x^3 - 10x^2$ **71.** $2x^3 - 3x^2$ **73.** $-3x + 21$
75. $-2x$ **77.** 2 **79.** 5 **81.** $-5x$
Find the Mistake 1. To divide a polynomial by a monomial, divide each term of the polynomial by the monomial.
2. Dividing $15x^3 - 6x$ by $-3x$ gives us $-5x^2 + 2$. **3.** The quotient of $12x^3y + 4x^4y^2 - x^3y^2$ and x^2y is $4x^2y - xy + 12$.
4. If you divide $9x^2 - 12x$ by $-3x$, the answer is $3x + 4$.

Exercise Set 6.8

Vocabulary Review 1. long divison **2.** estimate
3. opposite **4.** variable **5.** powers
Problems 1. $x - 2$ **3.** $a + 4$ **5.** $x - 3$ **7.** $x + 3$
9. $a - 5$ **11.** $x + 2 + \frac{2}{x + 3}$ **13.** $a - 2 + \frac{12}{a + 5}$

15. $x + 4 + \dfrac{9}{x-2}$ **17.** $x + 4 + \dfrac{-10}{x+1}$ **19.** $a + 1 + \dfrac{-1}{a+2}$

21. $x - 3 + \dfrac{17}{2x+4}$ **23.** $3a - 2 + \dfrac{7}{2a+3}$ **25.** $2a^2 - a - 3$

27. $4x^2 + 3x - \dfrac{1}{2} + \dfrac{-\frac{11}{2}}{2x-1}$ **29.** $x - 3$ **31.** $x^2 - x + 3$

33. $x^2 + 2x + 1$ **35.** $x^2 - x + 5$ **37.** $x^2 + x + 1$

39. $x^2 + 2x + 4$ **41.** $x^2 - 4x + 16$ **43.** 6.7 goals

45. 11.8 goals **47.** 47 **49.** 14 **51.** 70 **53.** 35 **55.** 5

57. 6,540 **59.** 1,760 **61.** 20 **63.** 63

Find the Mistake **1.** For a division of polynomials problem, you have found the answer when there is nothing left to bring down. **2.** To begin the problem $\dfrac{x^2 - 8x + 4}{x-4}$, we first estimate $x^2 \div x = x$. **3.** The quotient of $x^3 - 9x + 2$ and $x - 1$ has a remainder of -6. **4.** When dividing $9x^3 - 7x - 4$ by $3x - 2$, we first write the problem as $3x - 2\,\overline{)9x^3 + 0x^2 - 7x - 4}$.

Chapter 6 Test

1. -243 **2.** $\dfrac{4}{25}$ **3.** $72x^{10}$ **4.** 247,000 **5.** $-\dfrac{1}{9}$

6. $7a^2b^2$ **7.** $\dfrac{1}{a^3}$ **8.** $\dfrac{1}{x^3}$ **9.** 2.78×10^1 **10.** $\dfrac{5x^2y^2}{z^2}$

11. $9a^3$ **12.** $5x$ **13.** 1.6×10^4 **14.** $8x^2 - x - 12$

15. $x^2 + 7x$ **16.** $-6x - 10$ **17.** 11 **18.** $6a^4 + 10a^3 - 8a^2$

19. $x^2 - x - \dfrac{3}{4}$ **20.** $8x^2 - 2x - 6$ **21.** $x^2 - 9$

22. $x^4 - 18x^2 + 81$ **23.** $4a^2 - 4ab + b^2$ **24.** $3x^2 - 8y^2$

25. $a^4 - 3$ **26.** $3x - 2y$ **27.** $3x + 1$ **28.** $2x - \dfrac{1}{3x-1}$

29. 0.125 ft³ **30.** $V = w^3$ **31.** 7.83×10^8 **32.** 5.55×10^8

33. 2×10^{-2} **34.** 4×10^{-2}

Chapter 7

Exercise Set 7.1

Vocabulary Review **1.** monomial **2.** coefficient **3.** grouping, factors

Problems **1.** $5(3x + 5)$ **3.** $3(2a + 3)$ **5.** $4(x - 2y)$ **7.** $3(x^2 - 2x + 3)$ **9.** $3(a^2 - a + 20)$ **11.** $4(6y^2 - 13y + 9)$ **13.** $x^2(9 - 8x)$ **15.** $13a^2(1 - 2a)$ **17.** $7xy(3x - 4y)$ **19.** $11ab^2(2a - 1)$ **21.** $7x(x^2 + 3x + 4)$ **23.** $11(11y^4 + x^4)$ **25.** $25x^2(4x^2 - 2x + 1)$ **27.** $8(a^2 + 2b^2 + 4c^2)$ **29.** $4ab(a - 4b + 8ab)$ **31.** $6a^2b(ab + a - b)$ **33.** $11a^2b^2(11a - 2b + 3ab)$ **35.** $12x^2y^3(1 - 6x^3 - 3x^2y)$ **37.** $(x + 3)(y + 5)$ **39.** $(x + 2)(y + 6)$ **41.** $(a - 3)(b + 7)$ **43.** $(a - b)(x + y)$ **45.** $(2x - 5)(a + 3)$ **47.** $(b - 2)(3x - 4)$ **49.** $(x + 2)(x + a)$ **51.** $(x - b)(x - a)$ **53.** $(x + y)(a + b + c)$ **55.** $(3x + 2)(2x + 3)$ **57.** $(10x - 1)(2x + 5)$ **59.** $(4x + 5)(5x + 1)$ **61.** $(x + 2)(x^2 + 3)$ **63.** $(3x - 2)(2x^2 + 5)$ **65.** $(3x + 2)(x^2 - 5)$ **67.** 6 **69.** $3(4x^2 + 2x + 1)$ **71.** $A = 1{,}000(1 + r)$; \$1,120.00 **73. a.** $A = 1{,}000{,}000(1 + r)$ **b.** 1,300,000 **75.** $x^2 - 5x - 14$ **77.** $x^2 - x - 6$ **79.** $x^3 + 27$ **81.** $2x^3 + 9x^2 - 2x - 3$ **83.** $18x^7 - 12x^6 + 6x^5$ **85.** $x^2 + x + \dfrac{2}{9}$ **87.** $12x^2 - 10xy - 12y^2$ **89.** $81a^2 - 1$ **91.** $x^2 - 18x + 81$ **93.** $x^3 + 8$

Find the Mistake **1.** The greatest common factor for a polynomial is the monomial with the greatest coefficient and highest power of the variable that divides each term of the polynomial. **2.** The greatest common factor of the polynomial $12x^4 + 3x^3 - 6x$ is $3x$. **3.** Using the process of factoring by grouping, we can factor the polynomial $8xy + 2x - 12y - 3$ to get $2x(4y + 1) - 3(4y + 1) = (2x - 3)(4y + 1)$. **4.** To factor the polynomial $9xy - 27x - 2y + 6$ by grouping, factor $9x$ from the first two terms, and factor 2 from the last two terms to get $9x(y - 3) - 2(y - 3) = (9x - 2)(y - 3)$.

Exercise Set 7.2

Vocabulary Review coefficient, sum, product, leading, greatest common factor, trinomial

Problems **1.** $(x + 3)(x + 4)$ **3.** $(x + 1)(x + 2)$ **5.** $(a + 3)(a + 7)$ **7.** $(x - 2)(x - 5)$ **9.** $(y - 3)(y - 7)$ **11.** $(x - 4)(x + 3)$ **13.** $(y + 4)(y - 3)$ **15.** $(x + 7)(x - 2)$ **17.** $(r - 9)(r + 1)$ **19.** $(x - 6)(x + 5)$ **21.** $(a + 7)(a + 8)$ **23.** $(y + 6)\,(y - 7)$ **25.** $(x + 6)\,(x + 7)$ **27.** $(a - 9)(a + 7)$ **29.** $2(x + 1)(x + 2)$ **31.** $3(a + 4)(a - 5)$ **33.** $100(x - 2)(x - 3)$ **35.** $100(p - 5)(p - 8)$ **37.** $x^2(x + 3)(x - 4)$ **39.** $2r(r + 5)(r - 3)$ **41.** $2y^2(y + 1)(y - 4)$ **43.** $x^3(x + 2)(x + 2)$ **45.** $3y^2(y + 1)(y - 5)$ **47.** $4x^2(x - 4)(x - 9)$ **49.** $-2x(x - 3)(x - 2)$ **51.** $-7x^2(x + 2)(x + 1)$ **53.** $(x + 2y)(x + 3y)$ **55.** $(x - 4y)(x - 5y)$ **57.** $(a + 4b)(a - 2b)$ **59.** $(a - 5b)(a - 5b)$ **61.** $(a + 5b)(a + 5b)$ **63.** $(x - 6a)(x + 8a)$ **65.** $(x + 4b)(x - 9b)$ **67.** $2x(x - 3y)(x + y)$ **69.** $xy(x - 3y)(x - 4y)$ **71.** $-2a(a + 4b)(a - 2b)$ **73.** $(x^2 - 3)(x^2 - 2)$ **75.** $(x - 100)(x + 20)$ **77.** $\left(x - \dfrac{1}{2}\right)\left(x - \dfrac{1}{2}\right)$ **79.** $(x + 0.2)(x + 0.4)$ **81.** $x + 16$ **83.** $4x^2 - x - 3$ **85.** -2 **87.** -15 **89. a.** 5 GB **b.** 70 GB **c.** 240 GB **91.** $6a^2 + 13a + 2$ **93.** $6a^2 + 7a + 2$ **95.** $6a^2 + 8a + 2$ **97.** $4a^3 + 10a^2 - 24a$

Find the Mistake **1.** The leading coefficient of the polynomial $x^2 - 2x - 35$ is 1. **2.** Factoring the polynomial in the above sentence gives us $(x - 7)(x + 5)$. **3.** For the polynomial $8x^3 - 30x^2 - 108x$, the greatest common factor is $2x$. **4.** Factoring the polynomial in the above sentences gives us $2x(x - 6)(4x + 9)$.

Exercise Set 7.3

Vocabulary Review leading coefficient, trial and error, factors, products, multiply, polynomial

Problems **1.** $(2x + 1)(x + 3)$ **3.** $(2a - 3)(a + 1)$ **5.** $(3x + 5)(x - 1)$ **7.** $(3y + 1)(y - 5)$ **9.** $(2x + 3)(3x + 2)$ **11.** $(2x - 3y)(2x - 3y)$ **13.** $(4y + 1)(y - 3)$ **15.** $(4x - 5)(5x - 4)$ **17.** $(10a - b)(2a + 5b)$ **19.** $(4x - 5)(5x + 1)$ **21.** $(6m - 1)(2m + 3)$ **23.** $(4x + 5)\,(5x + 3)$ **25.** $(3a - 4b)(4a - 3b)$ **27.** $(3x - 7y)(x + 2y)$ **29.** $(2x + 5)(7x - 3)$ **31.** $(3x - 5)(2x - 11)$ **33.** $(5t - 19)(3t - 2)$ **35.** $(7t - 9)(4t + 3)$ **37.** $2(2x + 3)(x - 1)$ **39.** $2(4a - 3)(3a - 4)$ **41.** $x(5x - 4)(2x - 3)$ **43.** $x^2(3x + 2)(2x - 5)$ **45.** $2a(5a + 2)(a - 1)$ **47.** $3x(5x + 1)(x - 7)$ **49.** $5y(7y + 2)(y - 2)$ **51.** $a^2(5a + 1)(3a - 1)$ **53.** $3y(2x - 3)(4x + 5)$ **55.** $2y(2x - y)(3x - 7y)$ **57.** Both equal 25. **59.** $4x^2 - 9$ **61.** $x^4 - 81$ **63.** 1 **65.** -20

67.

Time t (seconds)	0	1	2	3	4
Height h (feet)	8	54	68	50	0

$h = 2(4 - t)(1 + 8t)$

69. a. $V = x \cdot (11 - 2x)(9 - 2x)$ **b.** 11 inches \times 9 inches **71.** $x^2 - 9$ **73.** $x^2 - 25$ **75.** $x^2 - 49$ **77.** $x^2 - 81$ **79.** $4x^2 - 9y^2$ **81.** $x^4 - 16$ **83.** $x^2 + 6x + 9$ **85.** $x^2 + 10x + 25$ **87.** $x^2 + 14x + 49$ **89.** $x^2 + 18x + 81$ **91.** $4x^2 + 12x + 9$ **93.** $16x^2 - 16xy + 4y^2$

Find the Mistake **1.** $(2x + 1)$ or $(3x + 2)$ is a factor of the polynomial $6x^2 + 7x + 12$. **2.** The first step to factoring $8a^3 - 4a^2 + 6a$ is to factor out the greatest common factor. **3.** All of the factors for $8a^3 - 4a^2 - 6a$ are $2a(4a - 1)(a + 3)$. **4.** The factors of the polynomial $40x^3z - 16x^2z - 12xz$ are $4xz(5x + 3)(2x - 2)$.

Landmark Review **1.** A greatest common factor. **2.** $2a(2a^2 + 4a - 5)$

3. $8xy(5x + 3y + 2xy)$ **4.** $(x - 1)(x - 8)$ **5.** $(x - 2)(y + 6)$
6. $(x + 2)(x - 2)(x - 7)$ **7.** $(x^2 + 6)(x - 8)$ **8.** $(x + 3)(x + 7)$
9. $(x + 4)(x - 7)$ **10.** $(x - 4)(x - 3)$ **11.** $(x - 4)(x + 6)$
12. $(3x + 2y)(2x - 3y)$ **13.** $(3a + 5)(a + 2)$ **14.** $(2y - 1)(3y - 2)$
15. $(5x - 6y)(4x - 5y)$

Exercise Set 7.4

Vocabulary Review 1. perfect squares, difference of two squares
2. sum **3.** trinomial **4.** square
Problems 1. $(x + 3)(x - 3)$ **3.** $(a + 6)(a - 6)$
5. $(x + 7)(x - 7)$ **7.** $4(a + 2)(a - 2)$ **9.** Cannot be factored.
11. $(5x + 13)(5x - 13)$ **13.** $(3a + 4b)(3a - 4b)$
15. $(3 + m)(3 - m)$ **17.** $(5 + 2x)(5 - 2x)$ **19.** $2(x + 3)(x - 3)$
21. $32(a + 2)(a - 2)$ **23.** $2y(2x + 3)(2x - 3)$
25. $(a^2 + b^2)(a + b)(a - b)$ **27.** $(4m^2 + 9)(2m + 3)(2m - 3)$
29. $3xy(x + 5y)(x - 5y)$ **31.** $-3x(x + 4)(x - 4)$ **33.** $(x - 1)^2$
35. $(x + 1)^2$ **37.** $(a - 5)^2$ **39.** $(y + 2)^2$ **41.** $(x - 2)^2$
43. $(m - 6)^2$ **45.** $(2a + 3)^2$ **47.** $(7x - 1)^2$ **49.** $(3y - 5)^2$
51. $(x + 5y)^2$ **53.** $(3a + b)^2$ **55.** $(2x - 3y)^2$ **57.** $3(a + 3)^2$
59. $2(x + 5y)^2$ **61.** $5x(x + 3y)^2$ **63.** $-4x(x - 3)^2$
65. $(x + 3 + y)(x + 3 - y)$ **67.** $(x + y + 3)(x + y - 3)$
69. 14 **71.** -10 **73.** 25 **75.** -16 **77.** 9
79. a. $x^2 - 16$ **b.** $(x + 4)(x - 4)$ **81.** $a^2 - b^2 = (a + b)(a - b)$
83. a. 1 **b.** 8 **c.** 27 **d.** 64 **e.** 125
85. a. $x^3 - x^2 + x$ **b.** $x^2 - x + 1$ **c.** $x^3 + 1$
87. a. $x^3 - 2x^2 + 4x$ **b.** $2x^2 - 4x + 8$ **c.** $x^3 + 8$
89. a. $x^3 - 3x^2 + 9x$ **b.** $3x^2 - 9x + 27$ **c.** $x^3 + 27$
91. a. $x^3 - 4x^2 + 16x$ **b.** $4x^2 - 16x + 64$ **c.** $x^3 + 64$
93. a. $x^3 - 5x^2 + 25x$ **b.** $5x^2 - 25x + 125$ **c.** $x^3 + 125$
Find the Mistake 1. The difference of two squares says that
$a^2 - b^2 = (a + b)(a - b)$. **2.** Factoring $81y^6 - 196$ gives us
$(9y^3 + 14)(9y^3 - 14)$. **3.** The polynomial $18x^2 - 84x + 98$ is an
example of a perfect square trinomial. **4.** The polynomial in the
above sentence factors to $2(3x - 7)^2$.

Exercise Set 7.5

Vocabulary Review 1. cubes **2.** difference **3.** factored **4.** sum
Problems 1. $(x - y)(x^2 + xy + y^2)$ **3.** $(a + 2)(a^2 - 2a + 4)$
5. $(3 + x)(9 - 3x + x^2)$ **7.** $(y - 1)(y^2 + y + 1)$
9. $(y - 4)(y^2 + 4y + 16)$ **11.** $(5h - t)(25h^2 + 5ht + t^2)$
13. $(x - 6)(x^2 + 6x + 36)$ **15.** $2(y - 3)(y^2 + 3y + 9)$
17. $2(a - 4b)(a^2 + 4ab + 16b^2)$ **19.** $2(x + 6y)(x^2 - 6xy + 36y^2)$
21. $10(a - 4b)(a^2 + 4ab + 16b^2)$ **23.** $10(r - 5)(r^2 + 5r + 25)$
25. $(4 + 3a)(16 - 12a + 9a^2)$ **27.** $(2x - 3y)(4x^2 + 6xy + 9y^2)$
29. $\left(t + \frac{1}{3}\right)\left(t^2 - \frac{1}{3}t + \frac{1}{9}\right)$ **31.** $\left(3x - \frac{1}{3}\right)\left(9x^2 + x + \frac{1}{9}\right)$
33. $(4a + 5b)(16a^2 - 20ab + 25b^2)$ **35.** $\left(\frac{1}{2}x - \frac{1}{3}y\right)\left(\frac{1}{4}x^2 + \frac{1}{6}xy + \frac{1}{9}y^2\right)$
37. $(a - b)(a^2 + ab + b^2)(a + b)(a^2 - ab + b^2)$
39. $(2x - y)(4x^2 + 2xy + y^2)(2x + y)(4x^2 - 2xy + y^2)$
41. $(x - 5y)(x^2 + 5xy + 25y^2)(x + 5y)(x^2 - 5xy + 25y^2)$
43. $(x^9 - 3)(x^{18} + 3x^9 + 9)$ **45.** 2 **47.** 125 **49.** $2x^5 - 8x^3$
51. $3x^4 - 18x^3 + 27x^2$ **53.** $y^3 + 25y$ **55.** $15a^2 - a - 2$
57. $4x^4 - 12x^3 - 40x^2$ **59.** $2ab^5 - 8ab^4 + 2ab^3$
Find the Mistake 1. The difference of two cubes says that
$a^3 - b^3 = (a - b)(a^2 + ab + b^2)$. **2.** The expression $a^3 + 8b^3$ is
an example of a sum of two cubes. **3.** The factors of the sum of
two cubes in the above sentence are $(a + 8b)$ and $(a^2 - 8ab + 64b^2)$.
4. Factoring $x^{12} - 343$ gives us $(x^4 - 7)(x^8 + 7x^4 + 49)$.

Exercise Set 7.6

Vocabulary Review 1. greatest common factor **2.** squares,
cubes **3.** perfect square trinomial, trial and error
4. grouping **5.** factors
Problems 1. $(x + 9)(x - 9)$ **3.** $(x + 5)(x - 3)$ **5.** $(x + 3)^2$
7. $(y - 5)^2$ **9.** $2ab(a^2 + 3a + 1)$ **11.** Cannot be factored.
13. $3(2a + 5)(2a - 5)$ **15.** $(3x - 2y)^2$ **17.** $4x(x^2 + 4y^2)$
19. $2y(y + 5)^2$ **21.** $a^4(a^2 + 4b^2)$ **23.** $(x + 4)(y + 3)$
25. $(x^2 + 4)(x + 2)(x - 2)$ **27.** $(x + 2)(y - 5)$
29. $5(a + b)^2$ **31.** Cannot be factored. **33.** $3(x + 2y)(x + 3y)$
35. $(2x + 19)(x - 2)$ **37.** $100(x - 2)(x - 1)$ **39.** $(x + 8)(x - 8)$
41. $(x + a)(x + 3)$ **43.** $a^5(7a + 3)(7a - 3)$
45. Cannot be factored. **47.** $a(5a + 1)(5a + 3)$ **49.** $(x + y)(a - b)$
51. $3a^2b(4a + 1)(4a - 1)$ **53.** $5x^2(2x + 3)(2x - 3)$
55. $(3x + 41y)(x - 2y)$ **57.** $2x^3(2x - 3)(4x - 5)$
59. $(2x + 3)(x + a)$ **61.** $(y^2 + 1)(y + 1)(y - 1)$
63. $3x^2y^2(2x + 3y)^2$ **65.** $-4x^5(x + 2)(x^2 - 2x + 4)$
67. $(x + 4)^2(x - 1)$ **69.** $(x + 2)^2(x^2 + 4x + 9)$
71. $4x(x + 1)(2x + 1)$ **73.** $(2x - 3)(4x^2 - 12x + 7)$
75. $(2001, 1)$, $(2005, 5.5)$, $(2006, 12)$, $(2007, 20)$, $(2008, 100)$,
$(2009, 175)$, $(2010, 400)$ **77.** 5 **79.** $-\frac{3}{2}$ **81.** $-\frac{3}{4}$
Find the Mistake 1. If a binomial is the sum of two squares, it
will not factor. **2.** To factor $14y^4 - 126y^2$, we first factor out
the greatest common factor $14y^2$. **3.** Once you factored out
the greatest common factor from the polynomial in the previous
sentence, you are left with a difference of two squares.
4. To factor the polynomial $9x^2 + 3xy - 4y + 16x^2$, group the first
two terms and the last two terms.

Exercise Set 7.7

Vocabulary Review 1. quadratic equation **2.** standard form
3. constant **4.** zero-factor property
List of Steps. 3, 4, 2, 5, 1
Problems 1. $-2, 1$ **3.** 4, 5 **5.** $0, -1, 3$ **7.** $-\frac{2}{3}, -\frac{3}{2}$
9. $0, -\frac{4}{3}, \frac{4}{3}$ **11.** $0, -\frac{1}{5}, -\frac{3}{5}$ **13.** $-1, -2$ **15.** 4, 5 **17.** 6, -4
19. 2, 3 **21.** -3 **23.** 4, -4 **25.** $\frac{3}{2}, -4$ **27.** $-\frac{2}{3}$ **29.** 5
31. 4, $-\frac{5}{2}$ **33.** $\frac{5}{3}, -4$ **35.** $\frac{7}{2}, -\frac{7}{2}$ **37.** $0, -6$ **39.** 0, 3
41. 0, 4 **43.** 0, 5 **45.** 2, 5 **47.** $\frac{1}{2}, -\frac{4}{3}$ **49.** 4, $-\frac{5}{2}$ **51.** 8, -10
53. 5, 8 **55.** 6, 8 **57.** -4 **59.** 5, 8 **61.** 6, -8 **63.** $0, -\frac{3}{2}, -4$
65. $0, 3, -\frac{5}{2}$ **67.** $0, \frac{1}{2}, -\frac{5}{2}$ **69.** $0, \frac{3}{5}, -\frac{3}{2}$ **71.** $\frac{1}{2}, \frac{3}{2}$
73. $-5, 4$ **75.** $-7, -6$ **77.** $-3, -1$ **79.** 2, 3 **81.** $-15, 10$
83. $-5, 3$ **85.** $-3, -2, 2$ **87.** $-4, -1, 4$ **89.** $x(x + 1) = 72$
91. $x(x + 2) = 99$ **93.** $x(x + 2) = 5[x + (x + 2)] - 10$
95. Bicycle $75, suit $15 **97.** House $2,400, lot $600
Find the Mistake 1. Standard form for a quadratic equation is
$a^2 + bx + c = 0$. **2.** The zero-factor property says that if
$(a + 8)(b - 3) = 0$, then $a + 8 = 0$ or $b - 3 = 0$. **3.** To solve
the equation $x^2 = -3x + 108$, we must first put the equation in
standard form. **4.** The solutions to the equation $3x^2 + 77 = 40x$
are $x = \frac{7}{3}$ and $x = 11$.

Exercise Set 7.8

Vocabulary Review 1. consecutive **2.** standard form
3. hypotenuse **4.** Pythagorean theorem
Problems 1. 8 and 10, or -10 and -8 **3.** 9 and 11, or -11 and -9
5. 8 and 10, or 0 and 2 **7.** 8 and 6 **9.** 2 and 12, or $-\frac{12}{5}$ and -10
11. 5 and 20, or 0 and 0 **13.** Width $= 3$ inches, length $= 4$ inches
15. 3 inches **17.** 6 inches and 8 inches **19.** 12 meters
21. 200 or 500 items **23.** 200 or 800 DVDs **25.** $4 or $8

27. $7 or $10 **29. a.** 5 feet **b.** 12 feet **31.** 6 inches by 12 inches
33. a. 25 seconds later

b.

t	h
0	100
5	1,680
10	2,460
15	2,440
20	1,620
25	0

35. $200x^{24}$ **37.** x^7
39. 8×10^1 **41.** $10ab^2$
43. $6x^4 + 6x^3 - 2x^2$
45. $9y^2 - 30y + 25$ **47.** $4a^2 - 49$

Find the Mistake 1. To find the two integers, you must solve for x in the equation $x^2 - 34x = -288$ or $x^2 - 34x + 288 = 0$.
2. The longest side of the rectangle measures 71 inches.
3. The Pythagorean theorem states that the square of the hypotenuse is equal to the sum of the squares of the legs. **4.** To find the lengths of the two legs, we must solve for x in the equation $x^2 + (2x + 2)^2 = 169$.

Chapter 7 Test

1. $6(x + 3)$ **2.** $4ab(3a - 6 + 2b)$ **3.** $(x - 2b)(x + 3a)$
4. $(3 - x)(5y - 4)$ **5.** $(x - 3)(x + 4)$ **6.** $(x - 7)(x + 3)$
7. $(x - 5)(x + 5)$ **8.** $(x - 2)(x + 2)(x^2 + 4)$ **9.** Does not factor
10. $2(3x - 4y)(3x + 4y)$ **11.** $(x^2 - 3)(x + 4)$ **12.** $(x - 3)(x + b)$
13. $2(2x - 5)(x + 1)$ **14.** $(4n - 3)(n + 4)$ **15.** $(3c - 2)(4c + 3)$
16. $3x(2x - 1)(2x + 3)$ **17.** $(x + 5y)(x^2 - 5xy + 25y^2)$
18. $2(3b - 4)(9b^2 + 12b + 16)$ **19.** $-3, 5$ **20.** $3, 4$ **21.** $-5, 5$
22. $-2, 7$ **23.** $-6, 5$ **24.** $-3, 0, 3$ **25.** $-4, \frac{3}{2}$ **26.** $-\frac{5}{3}, 6, 0$
27. 6 and 12 **28.** $-4, -2$ and $4, 6$ **29.** 4 ft by 13 ft
30. 6 ft, 8 ft **31.** 200 or 300 **32.** $3 or $5 **33. a.** 5 ft **b.** 12 ft
34. a. $t = 1, \frac{5}{8}$ sec **b.** $t = 2$ sec

Chapter 8

Exercise Set 8.1

Vocabulary Review 1. rational **2.** nonzero **3.** function
4. domain **5.** quotient
Problems 1. $\frac{x - 4}{6}$ **3.** $(a^2 + 9)(a + 3)$ **5.** $\frac{2y + 3}{y + 1}$ **7.** $\frac{x - 2}{x - 1}$
9. $\frac{x - 3}{x + 2}$ **11.** $\frac{x^2 - x + 1}{x - 1}$ **13.** $\frac{-4a}{3}$ **15.** $\frac{b - 1}{b + 1}$ **17.** $\frac{7x - 3}{7x + 5}$
19. $\frac{4x + 3}{4x - 3}$ **21.** $\frac{x + 5}{2x - 7}$ **23.** $-\frac{a + 5x}{3a + 5x}$ **25.** $\frac{a^2 - ab + b^2}{a - b}$
27. $\frac{2x - 2}{x}$ **29.** $\frac{x + 3}{y - 4}$ **31.** $x + 2$ **33.** $-\frac{3x + 2y}{3x + y}$
35. $\frac{x^2 + 2x + 4}{x + 2}$ **37.** $\frac{4x^2 + 6x + 9}{2x + 3}$ **39.** $\frac{1}{x^2 - 2x + 4}$ **41.** -1
43. $-(y + 6)$ **45.** $-\frac{3a + 1}{3a - 1}$ **47.** 3 **49.** $x + a$ **51.** $x + a + 3$
53. $\{x \mid x \ne 1\}$ **55.** $\{x \mid x \ne 2\}$ **57.** $\{t \mid t \ne 4, t \ne -4\}$
59. $g(0) = -3, g(-3) = 0, g(3) = 3, g(-1) = -1, g(1) =$ undefined
61. $h(0) = -3, h(-3) = 3, h(3) = 0, h(-1)$ is undefined, $h(1) = -1$
63. a. 4 **b.** 4 **65. a.** 5 **b.** 5 **67. a.** $x + a$ **b.** $2x + h$
69. a. $x + a$ **b.** $2x + h$ **71. a.** $x + a - 3$ **b.** $2x + h - 3$
73. a. $2x + 2a - 3$ **b.** $4x + 2h + 3$
75.

Weeks	Weight (lb)
x	W(x)
0	200
1	194.3
4	184
12	173.3
24	168

77. a. $C(n) = 1200 + 540n$
b. $\overline{C}(n) = \frac{1200 + 540n}{n}$
c. $660 **d.** $620 **e.** Decreasing
79. $\frac{2}{3}$ **81.** $20x^2y^2$
83. $(x + 2)(x - 2)$ **85.** $x^2(x - y)$
87. a. 2 **b.** -4 **c.** Undefined
d. 2 **e.** 1 **f.** -6 **g.** 4

Find the Mistake 1. The rational expression $\frac{8x^2 + 2x - 1}{16x^2 - 6x - 7}$ reduced to lowest terms is $\frac{4x - 1}{8x - 7}$. **2.** To reduce the rational expression $\frac{15x^2 - 16x + 4}{5x^2 - 17x + 6}$ to lowest terms, we must divide the numerator and denominator by the factor $5x - 2$. **3.** The rational function $f(x) = \frac{x^3 - 4x^2 + 4x + 16}{x^2 - 6x + 8}$ is undefined when $f(4)$ and $f(2)$. **4.** To find the difference quotient for $f(x) = x^2 + x - 7$, we use $f(a) = a^2 + a - 7$.

Exercise Set 8.2

Vocabulary Review 1. product, numerators **2.** quotient, reciprocal
Problems 1. $\frac{1}{6}$ **3.** $\frac{9}{4}$ **5.** $\frac{1}{2}$ **7.** $\frac{15y}{x^2}$ **9.** $\frac{b}{a}$ **11.** $\frac{2y^5}{z^3}$
13. $\frac{x + 3}{x + 2}$ **15.** $y + 1$ **17.** $\frac{3(x + 4)}{x - 2}$ **19.** $\frac{y^2}{xy + 1}$ **21.** $\frac{x^2 + 9}{x^2 - 9}$
23. $-\frac{20}{9}$ **25.** 1 **27.** $-\frac{x + 2y}{x - 2y}$ **29.** $\frac{9t^2 - 6t + 4}{4t^2 - 2t + 1}$
31. $\frac{x + 3}{x + 4}$ **33.** $\frac{a - b}{5}$ **35.** $\frac{5c - 1}{3c - 2}$ **37.** $-\frac{x + 2y}{x - 2y}$
39. 2 **41.** $x(x - 1)(x + 1)$ **43.** $\frac{(a + 4b)(a - 3b)}{(a - 4b)(a + 5b)}$
45. $\frac{2y - 1}{2y - 3}$ **47.** $\frac{(y - 2)(y + 1)}{(y + 2)(y - 1)}$ **49.** $\frac{x - 1}{x + 1}$ **51.** $\frac{x - 2}{x + 3}$
53. $\frac{w(y - 1)}{w - x}$ **55.** $\frac{(m + 2)(x + y)}{(2x + y)^2}$ **57.** $\frac{(1 - 2d)(d + c)}{d - c}$
59. $\frac{(r - s)^2(r + s)}{(r^2 + s^2)^2}$ **61.** $3x$ **63.** $2(x + 5)$ **65.** $x - 2$
67. $-(y - 4)$ or $4 - y$ **69.** $(a - 5)(a + 1)$
71. a. $\frac{5}{21}$ **b.** $\frac{5x + 3}{25x^2 + 15x + 9}$ **c.** $\frac{5x - 3}{25x^2 + 15x + 9}$ **d.** $\frac{5x + 3}{5x - 3}$
73. a. $\frac{(x + 2)^2}{(x - 1)^2}$ **b.** $(x - 3)^2$ **c.** $\frac{1}{(x - 3)^2}$
75. a.

Number of Copies	Price per Copy ($)
x	p(x)
1	20.33
10	9.33
20	6.40
50	4.00
100	3.05

b. $200
c. $305
d. $\frac{2x(x + 60)}{x + 5}$

77. Venus: 7,110 miles; Saturn: 75,050 miles **79.** $\frac{2}{3}$
81. $\frac{47}{105}$ **83.** $x - 7$ **85.** $-4x + 10$ **87.** $(x + 1)(x - 1)$
89. $2(x + 5)$ **91.** $(a - b)(a^2 + ab + b^2)$ **93.** $(2x + 3)(2x - 3)$
95. $\frac{x^4 - x^2y^2 + y^4}{x^2 + y^2}$ **97.** $\frac{(a - 1)(a + 5)}{3a^2 - 2a + 1}$ **99.** $-(p^2 - pq + q^2)$

Find the Mistake 1. Multiplying $\frac{x^2 - 4}{x - 2} \cdot \frac{x^2 + 2x - 8}{x^2 + 6x + 8}$ we get $x - 2$.
2. Multiplying $\frac{2x - 6}{4x^2 - 4x - 24} \cdot \frac{6x^2 + 2x - 20}{3x - 5}$ we get 1. **3.** To divide the rational expression $\frac{5x^2 + 21x - 20}{15x^2 - 17x + 4}$ by $\frac{x^2 - 25}{3x^2 - 4x + 1}$ we multiply by the reciprocal of the expression that follows the division symbol. **4.** Dividing $\frac{4x^2 + 27xy + 18y^2}{x^2 + 7xy + 6y^2} \div \frac{16x + 12y}{x^2 - y^2}$ to get $\frac{x - y}{4}$.

Exercise Set 8.3

Vocabulary Review 1. common **2.** denominator **3.** different
List of Steps: equivalent, numerators, lowest

Problems 1. 1 **3.** -1 **5.** $\frac{1}{x + y}$ **7.** 1 **9.** $\frac{a^2 + 2a - 3}{a^3}$
11. 1 **13.** $\frac{5}{4}$ **15.** $\frac{1}{3}$ **17.** $\frac{41}{24}$ **19.** $\frac{19}{144}$ **21.** $\frac{31}{24}$ **23.** $\frac{2}{3}$

25. a. $\frac{1}{16}$ **b.** $\frac{9}{4}$ **c.** $\frac{13}{24}$ **d.** $\frac{5x+15}{(x-3)^2}$ **e.** $\frac{x+3}{5}$ **f.** $\frac{x-2}{x-3}$

27. $f(x)-g(x)=\frac{1}{2}$ **29.** $f(x)-g(x)=\frac{1}{5}$ **31.** $\frac{x+3}{2(x+1)}$

33. $\frac{a-b}{a^2+ab+b^2}$ **35.** $\frac{x-3}{2x-1}$ **37.** $\frac{2y-3}{4y^2+6y+9}$

39. $\frac{2(2x-3)}{(x-3)(x-2)}$ **41.** $\frac{1}{2t-7}$ **43.** $\frac{4}{(a-3)(a+1)}$

45. $\frac{-4x^2}{(2x+1)(2x-1)(4x^2+2x+1)}$ **47.** $\frac{2}{(2x+3)(4x+3)}$ **49.** $\frac{a}{(a+4)(a+5)}$

51. $\frac{x+1}{(x-2)(x+3)}$ **53.** $\frac{x-1}{(x+1)(x+2)}$ **55.** $\frac{1}{(x+2)(x+1)}$

57. $\frac{1}{(x+2)(x+3)}$ **59.** $\frac{4x+5}{2x+1}$ **61.** $\frac{22-5t}{4-t}$ **63.** $\frac{2x^2+3x-4}{2x+3}$

65. $\frac{2x-3}{2x}$ **67.** $\frac{1}{2}$ **69.** $\frac{3}{x+4}$ **71.** $\frac{(2x+1)(x+5)}{(x-2)(x+1)(x+3)}$

73. $\frac{51}{10}$ **75.** 1 hour, 20 minutes **77.** $x+\frac{4}{x}=\frac{x^2+4}{x}$

79. $\frac{1}{x}+\frac{1}{x+1}=\frac{2x+1}{x(x+1)}$ **81.** $\frac{6}{5}$ **83.** $\frac{1}{9}$ **85.** $x+2$

87. $3-x$ **89.** $(x+2)(x-2)$ **91.** $(x-3)(x^2+3x+9)$ **93.** $\frac{x-1}{x+3}$

95. $\frac{a^2+ab+a-v}{u+v}$ **97.** $\frac{6x+5}{(4x-1)(3x-4)}$ **99.** $\frac{y(y^2+1)}{(y+1)^2(y-1)^2}$

Find the Mistake 1. If we add the rational expressions $\frac{x+5}{3x+9}$ and $\frac{4}{x^2-9}$ and reduce to the lowest terms, we get $\frac{x-1}{3x-9}$.

2. The least common denominator for the rational expressions $\frac{2x+3}{3x^2+6x-45}$ and $\frac{x+7}{3x^2+14x-5}$ is $(3x-1)(x+5)(3x-9)$.

3. If we subtract $\frac{5}{2x^2+x-3}$ from $\frac{2}{x^2-1}$, we get $\frac{-1}{(x+1)(2x+3)}$.

4. To subtract $\frac{x-4}{x^2-x-20}$ from $\frac{2x+5}{x^2-4x-5}$, make the expressions equivalent by first using LCD $(x+1)(x-5)(x+4)$.

Exercise Set 8.4
Vocabulary Review complex, quotient, numerator, LCD, reciprocal, denominator

Problems 1. $\frac{9}{8}$ **3.** $\frac{2}{15}$ **5.** $\frac{119}{20}$ **7.** $\frac{1}{x+1}$ **9.** $\frac{a+1}{a-1}$

11. $\frac{y-x}{y+x}$ **13.** $\frac{1}{(x+5)(x-2)}$ **15.** $\frac{1}{a^2-a+1}$ **17.** $\frac{x+3}{x+2}$

19. $\frac{a+3}{a-2}$ **21.** $\frac{x-3}{x+3}$ **23.** $\frac{a-1}{a+1}$ **25.** $-\frac{x}{3}$ **27.** $\frac{y^2+1}{2y}$

29. $\frac{-x^2+x-1}{x-1}$ **31.** $\frac{5}{3}$ **33.** $\frac{2x-1}{2x+3}$ **35.** $-\frac{1}{x(x+h)}$

37. $-\frac{1}{x^2+8x}$ **39.** $-\frac{12x+36}{x^2(x+6)^2}$ **41.** $\frac{3c+4a-2b}{5}$

43. $\frac{(t-4)(t+1)}{(t+6)(t-3)}$ **45.** $\frac{(5b-1)(b+5)}{2(2b-11)}$ **47.** $-\frac{3}{2x+14}$

49. $2m-9$ **51.** $-\frac{2}{x}$ **53.** $\frac{f(x)}{g(x)}=\frac{x-3}{x}$ **55.** $\frac{f(x)}{g(x)}=\frac{x+4}{x+2}$

57. a. $\frac{-4}{ax}$ **b.** $\frac{-1}{(x+1)(a+1)}$ **c.** $-\frac{a+x}{a^2x^2}$

59. a. As v approaches 0, the denominator approaches 1 **b.** $v=\frac{fs}{h}-s$

61. $xy-2x$ **63.** $3x-18$ **65.** ab **67.** $2t^2-3s$

69. $(y+5)(y-5)$ **71.** $x(a+b)$ **73.** 2 **75.** $-2,6$

Find the Mistake 1. Simplifying $\dfrac{\frac{2}{m}+\frac{4}{n}}{\frac{2}{m}-\frac{6}{n}}$ gives us $-\frac{2m+n}{3m-n}$.

2. If $f(x)=\frac{x+2}{x^2-16}$ and $g(x)=\frac{x^2-x-6}{x+4}$, then $\frac{f(x)}{g(x)}=\frac{1}{(x-4)(x-3)}$.

3. Simplifying $5-\frac{2}{x+\frac{1}{4}}$ gives us $\frac{20x-3}{4x+1}$.

4. The first step to simplifying $\frac{1-36x^{-2}}{1-9x^{-1}+18x^{-2}}$ is to write the expression with only positive exponents.

Landmark Review 1. $\frac{a+3}{2}$ **2.** $\frac{1}{y-3}$ **3.** $\frac{3x(x-5)}{x+2}$ **4.** $\frac{x+4}{5x-2}$

5. $\frac{a}{2}$ **6.** $\frac{x(x+4)}{x-1}$ **7.** $\frac{y(y-3)}{2(y+2)(y-5)}$ **8.** $2x^2+6$ **9.** 1

10. $\frac{4}{y-3}$ **11.** $\frac{a-3}{a-4}$ **12.** $x-4$ **13.** $\frac{x+4}{x+1}$ **14.** $\frac{x-2}{(x+3)(x+2)}$

15. $\frac{a-3}{a-4}$ **16.** $\frac{x+2y}{x+3y}$ **17.** $\frac{1}{(x+3)(4x-1)}$ **18.** $-\frac{2y}{y^2+1}$

Exercise Set 8.5
Vocabulary Review 1. rational **2.** extraneous **3.** function
4. asymptote

Problems 1. $-\frac{35}{3}$ **3.** $-\frac{18}{5}$ **5.** $\frac{36}{11}$ **7.** 2 **9.** 5 **11.** 2

13. $-3,4$ **15.** $1,-\frac{4}{3}$ **17.** No solution **19.** -4 **21.** $-\frac{1}{2},\frac{5}{3}$

23. $\frac{2}{3}$ **25.** -4 **27.** -6 **29.** -5 **31.** $\frac{53}{17}$ **33.** 2

35. No solution **37.** $\frac{22}{3}$ **39.** 18 **41.** No solution

43. a. $-\frac{9}{5},5$ **b.** $\frac{9}{2}$ **c.** No solution **45. a.** $\frac{1}{3}$ **b.** 3 **c.** 9 **d.** 4 **e.** $\frac{1}{3},3$

47. a. $\frac{6}{(x-4)(x+3)}$ **b.** $\frac{x-3}{x-4}$ **c.** 5 **49.** $x=\frac{ab}{a-b}$

51. $y=\frac{x-3}{x-1}$ **53.** $y=\frac{1-x}{3x-2}$

55.

$f(x)=\frac{1}{x-3}$

57.

$f(x)=\frac{4}{x+2}$

59.

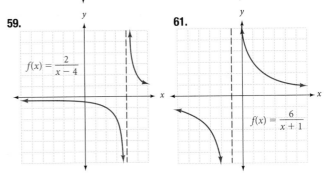

$f(x)=\frac{2}{x-4}$

61.

$f(x)=\frac{6}{x+1}$

63. Vertical asymptote: $x=\frac{5}{3}$; Domain: $\left\{x\,\middle|\,x\neq\frac{5}{3}\right\}$

65. Vertical asymptote: $x=\frac{3}{2}$; Domain: $\left\{x\,\middle|\,x\neq\frac{3}{2}\right\}$

67. Vertical asymptote: $x=\pm3$; Domain: $\{x\,|\,x\neq\pm3\}$

69. Vertical asymptote: $x=1,-3$; Domain: $\{x\,|\,x\neq1,-3\}$

71. Vertical asymptote: $x=1,2$; Domain: $\{x\,|\,x\neq1,2\}$

73. Vertical asymptote: $x=-\frac{1}{2},3$; Domain: $\left\{x\,\middle|\,x\neq-\frac{1}{2},3\right\}$

75. $\frac{24}{5}$ feet **77.** 2,358 **79.** 3 **81.** 12.3 **83.** 9, -1

85. 60 **87.** $-\frac{2}{3},-\frac{5}{2},\frac{5}{2}$ **89.** $-2,2,3$ **91.** $x=-\frac{a}{5}$

93. $v=\frac{16t^2+s}{t}$ **95.** $f=\frac{pg}{g-p}$

Find the Mistake 1. When we solve the rational equation $\frac{1}{x-3}-\frac{5}{x+3}=\frac{x}{x^2-9}$, we get $x=\frac{18}{5}$. **2.** To solve the rational equation $\frac{x+5}{x^2-5x}=\frac{20}{x^2-25}$, we must use the LCD $x(x-5)(x+5)$.

3. The graph for $f(x)=\frac{3}{x-1}$ will come very close but not touch the line $x=1$. **4.** The graph for $f(x)=\frac{2}{x+2}$ has a vertical asymptote of $x=-2$.

Exercise Set 8.6

Vocabulary Review **1.** blueprint **2.** LCD **3.** table **4.** asymptote

Problems **1.** $\frac{1}{5}$ and $\frac{3}{5}$ **3.** 3 or $\frac{1}{3}$ **5.** 3, 4 **7.** 3

9. a-b.

	d	r	t
Upstream	1.5	$5 - x$	$\frac{1.5}{5 - x}$
Downstream	3	$5 + x$	$\frac{3}{5 + x}$

c. They are the same. $\frac{1.5}{5 - x} = \frac{3}{5 + x}$

d. The speed of the current is $\frac{5}{3}$ mph. **11.** 6 mph

13. a-b.

	d	r	t
Train A	150	$x + 15$	$\frac{150}{x + 15}$
Train B	120	x	$\frac{120}{x}$

c. They are the same. $\frac{150}{x + 15} = \frac{120}{x}$

d. The speed of train A is 75 mph, and for train B it is 60 mph

15. 540 mph **17.** 54 mph **19.** 16 hours **21.** 15 hours

23. 5.25 minutes **25.** 1.52 minutes

27. $10 = \frac{1}{3}\left[\left(x + \frac{2}{3}x\right) + \frac{1}{3}\left(x + \frac{2}{3}x\right)\right]$; $x = \frac{27}{2}$

29. a. 30 grams **b.** 3.25 moles

31.

33.

35.

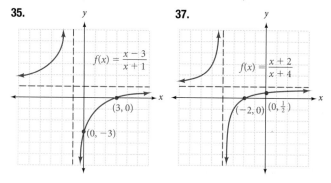

37.

39. 121,215 fans **41.** 2 **43.** $-2x^2y^2$ **45.** 185.12

47. $x - 3$ **49.** $4x^3 - 8x^2$ **51.** $4x^3 - 6x - 20$

53. $-3x + 9$ **55.** $(x + a)(x - a)$ **57.** $(x - 7y)(x + y)$

Find the Mistake **1.** The equation to find the numbers is $\frac{1}{x} + \frac{1}{3x} = 8$.
2. The rate equation $t = \frac{d}{r}$, is very useful for many applications of rational equations. **3.** To find the speed of the hikers, you must recognize that their travel distance is the same. **4.** To find how long it takes to finish together, you would use the equation $\frac{1}{4} - \frac{1}{3} = \frac{1}{x}$.

Chapter 8 Test

1. $x + y$ **2.** $\frac{2x - 3}{x - 1}$ **3.** 4.1 feet/second **4. a.** 8 **b.** 8

5. a. $2x + h + 1$ **b.** $x + a + 1$ **6.** $3(a + 2)$ **7.** $4(a - 2)$

8. $\frac{x + 1}{2}$ **9.** $\frac{27}{50}$ **10.** $\frac{17}{32}$ **11.** $\frac{1}{a + 5}$ **12.** $\frac{2(4x + 1)}{(x + 1)(2x - 1)}$

13. $\frac{x}{(x + 4)(x + 3)}$ **14.** $\frac{x + 2}{(x - 5)(x + 3)}$ **15.** $\frac{2a + 11}{2a + 13}$ **16.** $\frac{x - 5}{x + 2}$

17. $-\frac{10}{19}$ **18.** \varnothing **19.** $-\frac{12}{67}$ **20.** 7, 3 **21.** -2

22. 7 mph **23.** 10 hours

Chapter 9

Exercise Set 9.1

Vocabulary Review **1.** scatter diagram/plot **2.** line graph
3. ordered pair **4.** x–axis, y–axis **5.** quadrants **6.** origin

Problems

1.

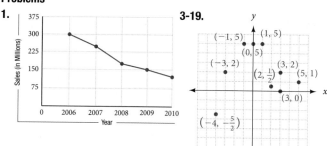

3-19.

21. $(-4, 4)$; QII **23.** $(-4, 2)$; QII **25.** $(-3, 0)$; no quadrant, x-axis

27. $(2, -2)$; QIV **29.** $(-5, -5)$; QIII

31. Yes **33.** No **35.** Yes **37.** No

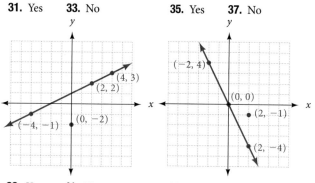

39. Yes **41.** No **43.** No **45.** No

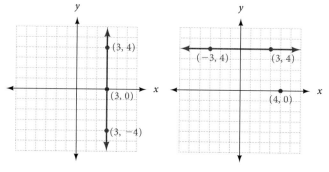

47. a. (5, 40), (10, 80), (20, 160), Answers may vary **b.** $320 **c.** 30 hours **d.** No, if she works 35 hours, she should be paid $280.

49. (1999, 70), (2001, 73), (2003, 68), (2005, 75), (2007, 80) , (2009, 73)

51. $A = (1, 2)$, $B = (6, 7)$ **53.** $A = (2, 2)$, $B = (2, 5)$, $C = (7, 5)$

55. a. -3 **b.** 6 **c.** 0 **4.** -4 **57. a.** 4 **b.** 2 **c.** -1 **d.** 9

59. -3 **61.** 2 **63.** 0 **65.** $y = -5x + 4$ **67.** $y = \frac{3}{2}x - 3$

Find the Mistake **1.** A scatter diagram uses dots to show data in the form of order pairs. **2.** In the ordered pair $(-3, 6)$, the x-coordinate is -3. **3.** To graph the ordered pair $(7, 3)$, start the origin, move 7 units to the right, and 3 units up. **4.** The ordered pair $(-2, 5)$ appears in the second quadrant.

Exercise Set 9.2

Vocabulary Review **1.** first **2.** second **3.** solution **4.** graphing **5.** linear **6.** ordered pairs **7.** graph **8.** straight line **9.** horizontal, vertical

Problems **1.** $(0, 6), (3, 0), (6, -6)$ **3.** $(0, 3), (4, 0), (-4, 6)$ **5.** $(1, 1), \left(\frac{3}{4}, 0\right), (5, 17)$ **7.** $(2, 13), (1, 6), (0, -1)$ **9.** $(-5, 4), (-5, -3), (-5, 0)$

11.

x	y
1	3
-3	-9
4	12
6	18

13.

x	y
0	0
$-\frac{1}{2}$	-2
-3	-12
3	12

15.

x	y
2	3
3	2
5	0
9	-4

17.

x	y
2	0
3	2
1	-2
-3	-10

19.

x	y
0	-1
-1	-7
-3	-19
$\frac{3}{2}$	8

21.

x	y
0	3
-2	7
2	-1
5	-7

23. $(0, -2)$ **25.** $(1, 5), (0, -2),$ and $(-2, -16)$ **27.** $(1, 6)$ and $(0, 0)$ **29.** $(2, -2)$ **31.** $(3, 0)$ and $(3, -3)$

33. $(0, 4), (2, 2), (4, 0)$

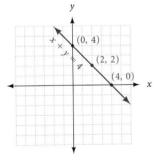

35. $(0, 3), (2, 1), (4, -1)$

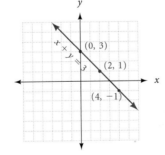

37. $(0, 0), (-2, -4), (2, 4)$

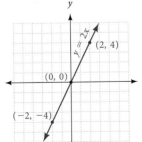

39. $(-3, -1), (0, 0), (3, 1)$

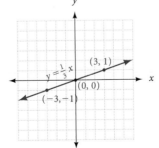

41. $(0, 1), (-1, -1), (1, 3)$

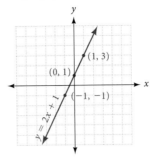

43. $(0, 4), (-1, 4), (2, 4)$

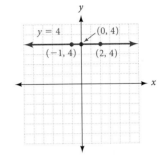

45. $(-2, 2), (0, 3), (2, 4)$

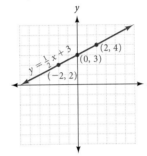

47. $(-3, 3), (0, 1), (3, -1)$

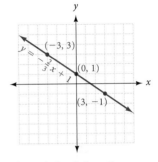

49. $(-1, 5), (0, 3), (1, 1)$

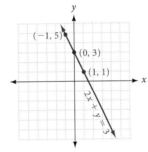

51. $(0, 3), (2, 0), (4, -3)$

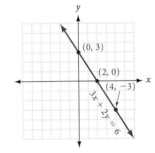

53. $(-2, 2), (0, 3), (2, 4)$

55.

57.

59.

61.

63.

65.

67.

69.

71.
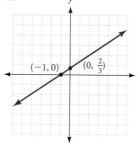

Problems

1.

Equation	x-intercept	y-intercept
$3x + 4y = 12$	4	3
$3x + 4y = 4$	$\frac{4}{3}$	1
$3x + 4y = 3$	1	$\frac{3}{4}$
$3x + 4y = 2$	$\frac{2}{3}$	$\frac{1}{2}$

3.

Equation	x-intercept	y-intercept
$x - 3y = 2$	2	$-\frac{2}{3}$
$y = \frac{1}{3}x - \frac{2}{3}$	2	$-\frac{2}{3}$
$x - 3y = 0$	0	0
$y = \frac{1}{3}x$	0	0

5.

7.

9.

11.

73.

Equation	H, V, and/or O
$x = 3$	V
$y = 3$	H
$y = 3x$	O
$y = 0$	O, H

75.

Equation	H, V, and/or O
$x = -\frac{3}{5}$	V
$y = -\frac{3}{5}$	H
$y = -\frac{3}{5}x$	O
$x = 0$	O, V

13.

15.

77.

x	y
−4	−3
−2	−2
0	−1
2	0
6	2

79. 12 inches

81. a. Yes **b.** No, she should earn $108 for working 9 hours. **c.** No, she should earn $84 for working 7 hours. **d.** Yes

83. a. $375,000 **b.** At the end of 6 years. **c.** No, the crane will be worth $195,000 after 9 years. **d.** $600,000

85. a. 2 **b.** 3 **87. a.** −4 **b.** 2

89. a. 6 **b.** 2

Find the Mistake **1.** A solution to a linear equation in two variable will be an ordered pair that makes the equation a true statement. **2.** A solution to the equation $3y = 6x + 9$ is (2, 7). **3.** To graph a linear equation in two variables, you must find two or more solutions. **4.** The line $y = \frac{1}{2}x$ intersects both axes, or passes through the origin.

Exercise Set 9.3

Vocabulary Review **1.** x-intercept **2.** y-intercept
3. x-axis **4.** y-axis

17.

19.

21.

23.

25.

27.

29.

31.

33.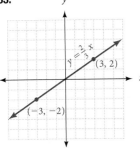

35. a. 0 **b.** $-\frac{3}{2}$ **c.** 1

d.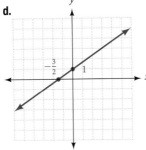

e. $y = \frac{2}{3}x + 1$

37. x-intercept = 3; y-intercept = 5

39. x-intercept = -1; y-intercept = -3

41.

43.

45.

47.

49.

51.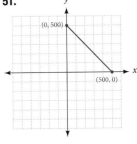

53. a. $10x + 12y = 480$
b. x-intercept = 48; y-intercept = 40
c.
d. 10 hours **e.** 18 hours

55. a. $50x + 60y = 450$
b. x-intercept = 9; y-intercept = 7.5
c.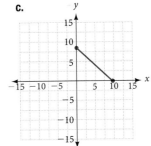
d. 5 hours **e.** 4.17 hours

57. a. $\frac{3}{2}$ **b.** $\frac{3}{2}$ **59. a.** $\frac{3}{2}$ **b.** $\frac{3}{2}$

Find the Mistake **1.** The y–intercept is where the graph of a line crosses the y–axis at the point $(0, y)$. **2.** The x–intercept for the linear equation $9x = \frac{1}{3}y + 5$ is $\left(\frac{5}{9}, 0\right)$. **3.** The y–intercept for the linear equation $2x + 7y = 3$ is $\left(0, \frac{3}{7}\right)$. **4.** The intercepts for the linear equation $15 = 3y - 5x$ are $(-3, 0)$ and $(0, 5)$.

Exercise Set 9.4
Vocabulary Review **1.** ratio **2.** vertical **3.** horizontal
4. slope **5.** zero **6.** undefined

Problems

1.

3.

5.

7.

9.

11.

13.

15.

17.

19.

21.

23.

25.

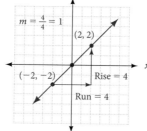

27. Slope = 3; y-intercept = 2

29. Slope = 2; y-intercept = -2

31.

33.

35.

37.

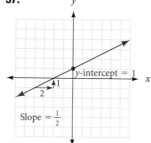

39. 6

41. All real numbers

43.

Equation	Slope
$x = 3$	Undefined
$y = 3$	0
$y = 3x$	3

45.

Equation	Slope
$y = -\frac{2}{3}$	0
$x = -\frac{2}{3}$	Undefined
$y = -\frac{2}{3}x$	$-\frac{2}{3}$

47. Slopes: A, 3.3; B, 3.1; C, 5.3, D, 1.9

49. Slopes: A, -50; B, -75, C, -25

51. a.

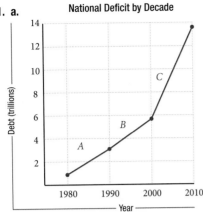

National Deficit by Decade

b. 0.23, 0.25, 0.79
c. The rate of change is increasing

53. 878 finishers. The number of finishers increased by 878 in one year.
55. a. 15,800 feet **b.** $-\frac{7}{100}$
57. 0.64% per year. Between 2003 and 2010 the percentage of computer users who use Macs has increased an average of 0.64% per year.
59. $y = 4x - 2$ **61.** $y = -\frac{3}{2}x + 3$ **63.** $y = \frac{1}{3}x - 2$
65. $y = -\frac{3}{4}x + 3$ **67.** $y = -\frac{2}{5}x - 2$ **69.** $y = -2x + 1$
71. $y = \frac{1}{2}x + 1$ **73.** $y = -\frac{3}{2}x - 1$

Find the Mistake **1.** A horizontal line has zero slope.
2. A line that rises from the left to the right has a positive slope.
3 The slope of the line that passes through the points $(-1, 6)$ and $(4, -9)$ is -3. **4.** Use the point $(2, -1)$ to graph a line with a y–intercept -6 and a slope $\frac{5}{2}$.

Landmark Review **1.** The origin **2.** 4 **3.** No
4. $(0, -3), \left(\frac{3}{4}, 0\right), (3, 9)$ **5.** $(-3, 3), (1, 9), \left(6, \frac{33}{2}\right)$
6. x-intercept $= 1$; y-intercept $= 1$
7. x-intercept $= 6$; y-intercept $= -3$ **8.** $\frac{1}{4}$ **9.** -2
10. No slope **11.** Rise **12.** There isn't one

Exercise Set 9.5

Vocabulary Review **1.** slope-intercept **2.** point-slope **3.** substitute
4. y–intercept **5.** slope **6.** parallel **7.** perpendicular
Problems **1.** $y = \frac{2}{3}x + 1$ **3.** $y = \frac{3}{2}x - 1$ **5.** $y = -\frac{2}{3}x + 3$
7. $y = 2x - 4$
9. $m = 2; b = 4$

11. $m = -3; b = 3$

13. $m = -\frac{3}{2}; b = 3$

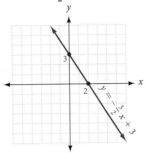

15. $m = \frac{4}{5}; b = -4$

17. $m = -\frac{2}{5}; b = -2$

19. $m = \frac{3}{2}; b = -3$

21. $y = 2x - 1$ **23.** $y = -\frac{1}{2}x - 1$ **25.** $y = \frac{3}{2}x - 6$
27. $y = -3x + 1$ **29.** $y = -4$ **31.** $x = -4$ **33.** $y = x - 2$
35. $y = 2x - 3$ **37.** $y = \frac{4}{3}x + 2$ **39.** $y = -\frac{2}{3}x - 3$
41. $y = 2$ **43.** $x = -5$ **45.** $m = 3, b = 3, y = 3x + 3$
47. $m = \frac{1}{4}, b = -1, y = \frac{1}{4}x - 1$
49. a. $-\frac{5}{2}$ **b.** $y = 2x + 6$
c. 6 **d.** 2
e.

51. $y = -\frac{2}{3}x + 2$
53. $y = -\frac{5}{2}x - 5$ **55.** $x = 3$
57. Any line given by $y = 4x + b$
59. $m = -1$ **61.** $m = -2$
63. a. $6,000
b. 3 years
c. Slope $= -3,000$
d. $3,000
e. $V = -3,000t + 21,000$

65. a. $3,000 **b.** 500 shirts **c.** Slope $= 2$ **d.** $2 **e.** $y = 2x + 1,000$
Find the Mistake **1.** The equation $y = -\frac{4}{7}x + 2$ is in slope–intercept form. **2.** For the equation $-8x + 2y = -4$, the slope is 4 and the y–intercept is -2. **3.** The line represented by the equation $(y - 2) = 6(x - 4)$ has a slope of 6. **4.** The lines represented by the equation $y = \frac{7}{2}x + \frac{1}{3}$ and $y = -\frac{2}{7}x - 5$ are perpendicular.

Exercise Set 9.6

Vocabulary Review **1.** linear inequality **2.** solution set
3. boundary **4.** solid **5.** broken

Problems

1.

3.

5.

7.

9.

11.

13.

15.

17.

19.

21.

23.

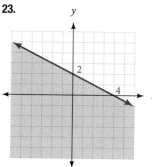

25. a. $y < \frac{8}{3}$ **b.** $y > -\frac{8}{3}$
c. $y = -\frac{4}{3}x + 4$
d.

27. a. $y = \frac{2}{5}x + 2$

b. $y < \frac{2}{5}x + 2$

c. $y > \frac{2}{5}x + 2$

29. $-6x + 11$ **31.** -8 **33.** -4 **35.** $w = \frac{P - 2l}{2}$

37.

39.

41. $y \geq \frac{3}{2}x - 6$ **43.** width = 2 inches; length = 11 inches

Find the Mistake **1.** The solution set for the linear inequality $y < 6x - \frac{3}{5}$ does not contain the points through which the boundary line passes. **2.** The boundary for the linear inequality $2 - 8y \geq x + 6$ is represented by a solid line. **3.** If a test point substituted into an original inequality results in a false statement, the solution set for the inequality lies on the opposite.
4. The solution set for the linear inequality $x < -8$, includes all points to the left of the line.

Chapter 9 Test

1-4.

5. $(0, -3), (2, 0), (4, 3), (-2, -6)$
6. $(0, 7), (4, -5)$

7.

8.

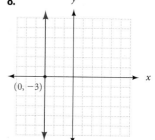

9. x-intercept $= 2$; y-intercept $= -4$

10. x-intercept $= -4$; y-intercept $= 6$

11. x-intercept $= -2$; y-intercept $= 4$

12. $-\frac{1}{2}$ **13.** $-\frac{8}{7}$ **14.** -3 **15.** Undefined **16.** 0

17. $y = -\frac{1}{2}x + 3$ **18.** $y = 3x - 5$ **19.** $y = -\frac{2}{3}x - 2$

20. $y = -3x - 6$ **21.** $y = \frac{3}{2}x + 6$ **22.** $y = -3x - 17$

23. $y = \frac{3}{2}x + 4$ **24.** $y = \frac{3}{2}x + 7$ **25.** $y = 2x - 6$

26.

27.

Chapter 10

Exercise Set 10.1

Vocabulary Review **1.** system **2.** ordered pairs **3.** axes

4. intersection

Problems

1.

3.

5.

7.

9.

11.

13.

15.

17.

19.

21.

23.

25.

No solution (inconsistent)

27.

Lines coincide (dependent)

29.

31.

33.

35. 33 $5 chips, 12 $25 chips

37. a. 25 hours **b.** Gigi's **c.** Marcy's
39. 1 **41.** 2 **43.** All real numbers **45.** $x = 3y - 1$
47. 1 **49.** 5 **51.** 34.5 **53.** 4 **55.** 33.95 **57.** 20
Find the Mistake 1. The point of intersection for two straight line graphs is a solution to both equations. **2.** The point of intersection for the equations $y = 6$ and $y = 4x + 2$ is $(1, 6)$.
3. The point of intersection for the equations $x = 5$ and $x = 9$ is not $(5, 9)$ because there is no point of intersection; the lines are parallel.
4. When lines coincide, we say that the equations are dependent.

Exercise Set 10.2

Vocabulary Review 1. equations **2.** substitute **3.** variables
4. check
Problems 1. $(4, 7)$ **3.** $(3, 17)$ **5.** $\left(\frac{3}{2}, 2\right)$ **7.** $(2, 4)$ **9.** $(0, 4)$
11. $(-1, 3)$ **13.** $(1, 1)$ **15.** $(2, -3)$ **17.** $\left(-2, \frac{3}{5}\right)$ **19.** $(-3, 5)$
21. No solution (inconsistent) **23.** Lines coincide (dependent)
25. $(3, 1)$ **27.** $\left(\frac{1}{2}, \frac{3}{4}\right)$ **29.** $(2, 6)$ **31.** $(4, 4)$ **33.** $(5, -2)$
35. $(18, 10)$ **37.** Lines coincide (dependent)
39. No solution (inconsistent) **41.** $(10, 12)$ **43.** 7.29 miles per hour
45. a. 560 miles **b.** Car **c.** Truck **d.** We are only working with positive numbers
47. $2x$ **49.** $7x$ **51.** $13x$ **53.** $3x - 2y$ **55.** $-12x + 8y$
57. $22x$ **59.** 1 **61.** 0 **63.** -5
Find the Mistake 1. Substitute the value for x from the first equation for x in the second equation. **2.** Add $3y$ to both sides of the first equation to isolate x. **3.** Substitute the value for y from the first equation for y in the second equation. **4.** Substitute the value for x from the first equation for x in the second equation.

Exercise Set 10.3

Vocabulary Review 1. variable **2.** multiplication property of equality
3. add **4.** solve **5.** substitute **6.** solution
Problems 1. $(2, 1)$ **3.** $(3, 7)$ **5.** $(2, -5)$ **7.** $(-1, 0)$
9. Lines coincide (dependent) **11.** No solution (inconsistent)
13. $(4, 8)$ **15.** $\left(\frac{1}{5}, 1\right)$ **17.** $(1, 0)$ **19.** $(3, 0)$ **21.** $(-1, -2)$
23. $\left(-5, \frac{3}{4}\right)$ **25.** $(-4, 5)$ **27.** $\left(-\frac{4}{3}, 1\right)$ **29.** $(-3, -10)$
31. $(3, 2)$ **33.** $\left(5, \frac{1}{3}\right)$ **35.** $\left(-2, \frac{2}{3}\right)$ **37.** $(2, 2)$
39. No solution (inconsistent) **41.** $(1, 1)$
43. No solution (inconsistent) **45.** $(10, 12)$
47. 1,478 adult tickets, 242 children tickets **49.** 3 and 23
51. 15 and 24 **53.** Length = 23 in.; width = 6 in.
55. 14 nickels and 10 dimes
Find the Mistake 1. Add the original equations to eliminate x and give a false statement, making the system inconsistent.
2. Multiply both sides of the first equation by -3 to eliminate x.
3. Add the original equations to eliminate x and solve for y.
4. When solving a system by the elimination method, if both

variables are eliminated and the resulting statement is true, then the lines coincide.
Landmark Review 1. $(-2, 1)$ **2.** $(-3, -4)$ **3.** $\left(\frac{3}{2}, 2\right)$
4. $\left(9, \frac{18}{5}\right)$ **5.** $(-1, 5)$ **6.** No solution (inconsistent)
7. $(5, -6)$ **8.** $(4, -3)$ **9.** $(-4, 5)$ **10.** $\left(4, \frac{1}{4}\right)$

Exercise Set 10.4

Vocabulary Review 1. triples **2.** dependent **3.** infinite
4. inconsistent
Problems 1. $(1, 2, 1)$ **3.** $(2, 1, 3)$ **5.** $(2, 0, 1)$
7. $\left(\frac{1}{2}, \frac{2}{3}, -\frac{1}{2}\right)$ **9.** Inconsistent (no solution) **11.** $(4, -3, -5)$
13. Dependent (parallel planes) **15.** $(4, -5, -3)$
17. Dependent (parallel planes) **19.** $\left(\frac{1}{2}, 1, 2\right)$ **21.** $\left(\frac{1}{2}, \frac{1}{3}, \frac{1}{4}\right)$
23. $\left(\frac{10}{3}, -\frac{5}{3}, -\frac{1}{3}\right)$ **25.** $\left(\frac{1}{4}, -\frac{1}{3}, \frac{1}{8}\right)$ **27.** $(6, 8, 12)$
29. $(-141, -210, -104)$ **31.** 4 amp, 3 amp, 1 amp
33. Suzie: 3 dozen; Minnie: 16 dozen; Bunny: 12 dozen
35. $2 + 3x$ **37.** $2x + 5y$ **39.** 6 **41.** $(-1, 5)$
43. 3, 7, 10 **45.** 3 ft, 6 ft, 11 ft
Find the Mistake 1. The solution to the following system is $(3, -3, -2)$.
2. It is easiest to solve the following system by eliminating z from each equation. **3.** The following system has no unique solution, dependent solution. **4.** The following system has solution: $(4, 3, 1)$.

Exercise Set 10.5

Vocabulary Review 1. common **2.** shaded **3.** second
4. first **5.** fourth **6.** third
Problems
1. **3.**

5. **7.**

9. **11.**

13.

15.

17.

19.
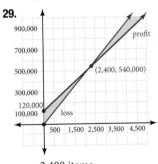

21. $x + y \leq 4; -x + y < 4$

23. $x + y \geq 4; -x + y < 4$
 b. 10 65-cent stamps

25. a.
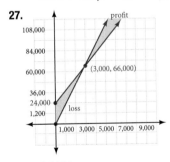

$0.55x + 0.65y \leq 40; x \geq 2y; x > 15, y \geq 0$

27.

3,000 items

29.

2,400 items

31.

6,500 items

33.

800 items

35. $p = \$420$; 1,680 units

37. $p = \$500$; 4,500 units

39. $p = \$600$; 9,600 units

41. a. $C(x) = 75{,}000 + 24x$,
 $R(x) = 40x$,
 $P(x) = 16x - 75{,}000$
b. $C(2{,}000) = \$123{,}000$,
 $R(2{,}000) = \$80{,}000$,
 $P(2{,}000) = -\$43{,}000$
c. $C(8{,}000) = \$267{,}000$,
 $R(8{,}000) = \$320{,}000$,
 $P(8{,}000) = \$53{,}000$
d. 4,688 items

43.
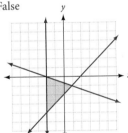

x-intercept $= 3$; y-intercept $= 6$; slope $= -2$

45.
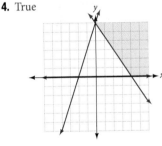

x-intercept $= -2$; no y-intercept; undefined

47. $y = -\frac{3}{7}x + \frac{5}{7}$ **49.** $x = 4$

51. Domain $= \mathbb{R}$; Range $= \{y \mid y \geq -9\}$; A function

53. -4 **55.** 4 **57.** $3x + 2$ **59.** $\dfrac{1}{x - 2}$

Find the Mistake **1.** The solution set for the system of inequalities $x \leq 0$ and $y \geq 0$ is all the points in the second quadrant only.
2. The solution set for the system of inequalities $x \leq 0$ and $y \leq 0$ is all the points in the third quadrant.
3. False

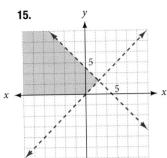

4. True

Chapter 10 Test
Problems **1.** $(0, 6)$ **2.** $(3, 2)$ **3.** $(7, 4)$ **4.** $(-2, 4)$
5. $(4, 0)$ **6.** $(-5, 3)$ **7.** $(-3, 4)$ **8.** $(11, 3)$ **9.** $(-4, 5)$
10. $(1, 2)$ **11.** $(-6, 3)$ **12.** Lines coincide **13.** $\left(1, \frac{2}{3}, \frac{2}{3}\right)$

14.

15.

16. $y < \frac{4}{3}x + 1, y \leq -\frac{4}{3}x + 1$ **17.** $y < \frac{4}{3}x + 1, y \geq -\frac{4}{3}x + 1$

Chapter 11

Exercise Set 11.1

Vocabulary Review **1.** linear inequality in one variable
2. addition property for inequalities **3.** set **4.** interval
5. multiplication property for inequalities

Problems

1. 0, $\frac{3}{2}$

3. 0, 4

5. -5, 0

7. 0, 4

9. -6, 0

11. 0, 4

13. -3, 0

15. -1, 0

17. -3, 0

19. 0, $\frac{7}{2}$

21. 0, 6

23. -52, 0

25. 0, $\frac{54}{7}$

27. -32, 0

29. 40

31. $(-\infty, -2]$ **33.** $[1, \infty)$ **35.** $(-\infty, 3)$ **37.** $(-\infty, -1]$
39. $[-17, \infty)$ **41.** $(-\infty, 10]$ **43.** $(1, \infty)$ **45.** $(435, \infty)$
47. $(-\infty, 900]$ **49. a.** 1 **b.** 16 **c.** no **d.** $x > 16$
51. $x \geq 6$; the width is at least 6 meters.
53. $x > 6$; the shortest side is even and greater than 6 inches.
55. $t \geq 100$ **57.** Lose money if they sell less than 200 tickets.
Make a profit if they sell more than 200 tickets.
59. a. $p \leq \$2.00$ **b.** $p < \$1.00$ **c.** $p > \$1.25$ **d.** $p \geq \$1.75$
61. a. 1983 and earlier **b.** 1991 and later **63.** 15.6 years
65. 19.6 years **67.** $x \geq 2$ **69.** $x < 4$ **71.** $-1 \leq x$
73. $x > -3$ **75.** $x < \frac{c-b}{a}$ **77.** $x < -\frac{a}{b}y + a$
79. $x > \frac{b-d}{c-a}$ **81.** $x < \frac{abc - acd}{bcd - abd}$
Find the Mistake **1.** False; The graph of the solution for the
inequality $-16 \leq 4x$ will use a bracket to show that -4 is included
in the solution. **2.** True **3.** False; $x > 3$ is a solution to
$-\frac{2}{3}x - 5 < -7$. **4.** False; The interval notation for the solution
set to $5 > \frac{1}{2} + 3(4x - 1)$ begins with a parentheses and ends with a
parentheses, $\left(\infty, \frac{5}{8}\right)$.

Exercise Set 11.2

Vocabulary Review **1.** set, elements **2.** subset **3.** empty set
4. union **5.** intersection **6.** set-builder **7.** compound
8. continued **9.** closed, open
Problems **1.** True **3.** False **5.** True **7.** True
9. $\{0, 1, 2, 3, 4, 5, 6\}$ **11.** $\{2, 4\}$ **13.** $\{0, 1, 2, 3, 4, 5, 6\}$
15. $\{0, 2\}$ **17.** $\{0, 1, 2, 3, 4, 5, 6, 7\}$ **19.** $\{1, 2, 4, 5\}$
21. a. $\{0, -1, -2\}$ **b.** $\{0, 1, 2\}$ **c.** $\{0, 1, 2, -1, -2\}$ **d.** $\{0\}$
23. a. $\{4, 5\}$ **b.** $\{-4, -5\}$ **c.** $\{4, 5, -4, 5\}$ **d.** \varnothing

25. **27.** $(-\infty, 1)$ **29.** $[3, \infty)$ **31.** $(-\infty, \infty)$

33. -1, 5 $(-\infty, -1)\cup(5, \infty)$

35. 1, 5 $(1, 5)$

37. 0, 7 $(0, 7]$

39. 2, 4 $(-\infty, 2)\cup(4, \infty)$

41. -1, 3 $(-1, 3)$

43. -3, -2 $(-3, -2]$

45. -3, 2 $[-3, 2)$

47. 3, 7 $[3, 7]$

49. -3, 0 $(-3, 0)$

51. 4, 6 $[4, 6]$

53. -4, 2 $(-4, 2)$

55. -3, 3 $(-3, 3)$

57. 0, 6 $(0, 6)$

59. $-\frac{2}{3}$, 0 $\left[-\frac{2}{3}, 0\right]$

61. $-\frac{3}{2}$, $\frac{17}{2}$ $\left[-\frac{3}{2}, \frac{17}{2}\right]$

63. $\frac{3}{8}$, $\frac{3}{2}$ $\left[\frac{3}{8}, \frac{3}{2}\right]$

65. $(-\infty, -7] \cup [-3, \infty)$ (number line: -7, -3)

67. $(-\infty, -1] \cup \left[\frac{3}{5}, \infty\right)$ (number line: -1, $\frac{3}{5}$)

69. $(-\infty, -10) \cup (6, \infty)$ (number line: -10, 6)

71. $(-\infty, -3) \cup [2, \infty)$ (number line: -3, 2)

73. $-2 < x \le 4$ **75.** $x < -4$ or $x \ge 1$

77. a. $35° \le C \le 45°$ **b.** $20° \le C \le 30°$ **c.** $-25° \le C \le -10°$
d. $-20° \le C \le -5°$ **79.** Adults: $0.61 \le r \le 0.83$; $61\% - 83\%$;
Juveniles: $0.06 \le r \le 0.20$; $6\% - 20\%$

81. -3 **83.** 15 **85.** No solution **87.** -1

89. $\frac{-c-b}{a} < x < \frac{c-b}{a}$ **91.** $\frac{-cd-b}{a} < x < \frac{cd-b}{a}$

Find the Mistake 1. The notation $x \in A$ is read "x is an element of set A." **2.** The union of two sets A and B is written $A \cup B$.
3. The interval $[-6, 8]$ for the continued inequality $-6 \le x \le 8$ is called a closed interval. **4.** To solve the compound inequality $-2 \le 4x - 3 \le 9$, add 3 from the first two parts of the inequality.

Exercise Set 11.3

Vocabulary Review 1. negative **2.** positive **3.** \Leftrightarrow **4.** same
Problems 1. $-4, 4$ **3.** $-2, 2$ **5.** \varnothing **7.** $-1, 1$ **9.** \varnothing
11. $\frac{17}{3}, \frac{7}{3}$ **13.** $-\frac{5}{2}, \frac{5}{6}$ **15.** $-1, 5$ **17.** \varnothing **19.** $20, -4$
21. $-4, 8$ **23.** $1, 4$ **25.** $-\frac{1}{7}, \frac{9}{7}$ **27.** $-3, 12$ **29.** $\frac{2}{3}, -\frac{10}{3}$
31. \varnothing **33.** $\frac{3}{2}, -1$ **35.** $5, 25$ **37.** $-30, 26$ **39.** $-12, 28$
41. $-2, 0$ **43.** $-\frac{1}{2}, \frac{7}{6}$ **45.** $0, 15$ **47.** $-\frac{23}{7}, -\frac{11}{7}$
49. $-5, \frac{3}{5}$ **51.** $1, \frac{1}{9}$ **53.** $-\frac{1}{2}$ **55.** 0 **57.** $-\frac{1}{6}, -\frac{7}{4}$
59. All real numbers **61.** All real numbers **63.** $-\frac{3}{10}, \frac{3}{2}$
65. $-\frac{1}{10}, -\frac{3}{5}$ **67. a.** $\frac{5}{4}$ **b.** $\frac{5}{4}$ **c.** 2 **d.** $\frac{1}{2}, 2$ **e.** $\frac{1}{3}, 4$
69. 1987 and 1995 **71.** $x < 4$ **73.** $-\frac{11}{3} \le a$ **75.** $t \le -\frac{3}{2}$
77. $x = a \pm b$ **79.** $x = \frac{-b \pm c}{a}$ **81.** $x = -\frac{a}{b}y \pm a$

Find the Mistake 1. Some absolute value equations have two solutions. **2.** There is one solution to the absolute value equation $|2x - 3| = |2x + 4|$. **3.** There is a solution of all real numbers for the absolute equation $|y - 7| = |7 - y|$. **4.** The absolute value equation $|6a - 4| = -9$ has no solution.

Exercise Set 11.4

Vocabulary Review 1. e **2.** f **3.** b **4.** d **5.** c **6.** a
Problems
1. (number line: -3, 3) $-3 < x < 3, (-3, 3)$

3. (number line: -2, 2) $x \le -2$ or $x \ge 2$, $(-\infty, -2] \cup [2, \infty)$

5. (number line: -3, 3) $-3 < x < 3, (-3, 3)$

7. (number line: -7, 7) $t < -7$ or $t > 7$, $(-\infty, -7] \cup [7, \infty)$

9. (number line) \varnothing

11. (number line) All real numbers, \mathbb{R}

13. (number line: -4, 10) $-4 < x < 10, (-4, 10)$

15. (number line: -9, -1) $a \le -9$ or $a \ge -1$, $(-\infty, -9] \cup [-1, \infty)$

17. (number line: 2, 8) $2 < x < 8, (2, 8)$

19. (number line) \varnothing

21. (number line: -1, 5) $-1 < x < 5, (-1, 5)$

23. (number line: -5, -1) $y \le -5$ or $y \ge -1$, $(-\infty, -5] \cup [-1, \infty)$

25. (number line: -5, 2) $k \le -5$ or $k \ge 2$, $(-\infty, -5] \cup [-2, \infty)$

27. (number line: -1, 7) $-1 < x < 7, (-1, 7)$

29. (number line: -2, 1) $a \le -2$ or $a \ge 1$, $(-\infty, -2] \cup [1, \infty)$

31. (number line: -6, $\frac{8}{3}$) $-6 < x < \frac{8}{3}, \left(-6, \frac{8}{3}\right)$

33. $[-2, 8]$ **35.** $\left(-2, \frac{4}{3}\right)$ **37.** $(-\infty, -5] \cup [-3, \infty)$
39. $\left(-\infty, -\frac{7}{2}\right) \cup \left(-\frac{3}{2}, \infty\right)$ **41.** $\left[-1, \frac{11}{5}\right]$ **43.** $\left(\frac{5}{3}, 3\right)$
45. $x < 2$ or $x > 8$ **47.** $x \le -3$ or $x \ge 12$ **49.** $x < 2$ or $x > 6$
51. $x < -6$ or $x > 18$ **53.** $0.99 < x < 1.01$
55. $x \le -\frac{3}{5}$ or $x \ge -\frac{2}{5}$ **57.** $\frac{5}{9} \le x \le \frac{7}{9}$ **59.** $x < -\frac{2}{3}$ or $x > 0$
61. $x \le \frac{2}{3}$ or $x \ge 2$ **63.** $-\frac{1}{6} \le x \le \frac{3}{2}$ **65.** $-0.05 < x < 0.25$
67. $\frac{1}{12} < x < \frac{5}{12}$ **69.** $|x| \le 4$ **71.** $|x - 5| \le 1$
73. a. 3 **b.** $-2, \frac{4}{5}$ **c.** no **d.** $x < -2$ or $x > \frac{4}{5}$
75. $-23 < v - 455 < 23$; $432 < v < 478$; blue
77. $x < a + b$ **79.** $x > \frac{b+c}{a}$ **81.** $x \le \frac{c-b}{a}$
83. $x > \frac{abc-a}{b}$ or $x < \frac{-abc-a}{b}$ **85.** $-2a < x < a$ **87.** $-a^2 \le x \le 3a^2$

Find the Mistake 1. False; The expression $|b| > 9$ is read "b is greater than 9 units from zero on the number line."
2. True **3.** False; When an absolute value expression is greater than a negative value, the solution set will be all real numbers.
4. True

Landmark Review

1. $x \ge -8$ (number line: -8)

2. $x > -6$ (number line: -6)

3. $x \le 28$ (number line: 28)

4. $x \le 10$ (number line: 10)

5. $\{0, 2, 4, 6, 1, 3, 5\}$ **6.** \varnothing **7.** $\{2, 4\}$ **8.** $\{0, 2, 4, 6, 1, 3, 5, 7\}$
9. $[16, 22]$ (number line: 16, 22)

10. $(10, 16)$ (number line: 10, 16)

11. $-\frac{14}{3}, 14$ **12.** $-6, -4$
13. $(-\infty, -3] \cup [1, \infty)$ (number line: -3, 1)

14. $(0, 2)$

Exercise Set 11.5

Vocabulary Review **1.** x-axis, y-axis **2.** x-coordinate, y-coordinate
3. origin **4.** x-intercept **5.** y-intercept **6.** vertical
7. horizontal

Problems 1.

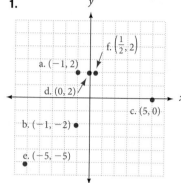

3. A. $(4, 1)$ **B.** $(-4, 3)$ **C.** $(-2, -5)$ **D.** $(2, -2)$ **E.** $(0, 5)$
F. $(-4, 0)$ **G.** $(1, 0)$
5. b **7.** b **9.** $y = x + 3$ **11.** $y = |x| - 3$
13. a. **b.**

c. **15. a.**

b. **c.**

17.

19. a.

b. **c.**

21. a. **b.**

c. **23.**

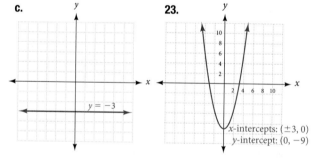

x-intercepts: $(\pm 3, 0)$
y-intercept: $(0, -9)$

25. **27.**

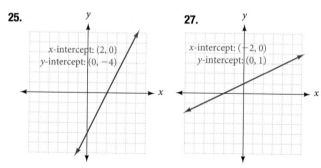

x-intercept: $(2, 0)$
y-intercept: $(0, -4)$

x-intercept: $(-2, 0)$
y-intercept: $(0, 1)$

29.

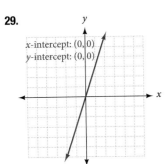

x-intercept: (0, 0)
y-intercept: (0, 0)

31.

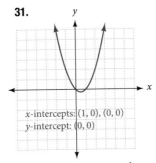

x-intercepts: (1, 0), (0, 0)
y-intercept: (0, 0)

33.

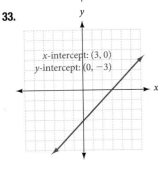

x-intercept: (3, 0)
y-intercept: (0, −3)

35. a. −7 **b.** −4 **c.** $-\frac{4}{3}$

d.

$4x + 12y = -16$

e. $y = -\frac{1}{3}x - \frac{4}{3}$

37. a. Yes **b.** No **c.** Yes

39.

Weekly Wages

41.

Light Intensity from a 100-watt Light Bulb

43.

Power Output from a Wind Turbine

45. a. 60 **b.** 70 **c.** 10 **d.** 6:30 & 7:00 **e.** about 22 minutes

47.

x	y
0	0
10	75
20	150

49.

x	y
0	0
$\frac{1}{2}$	3.75
1	7.5

Find the Mistake 1. False; On a rectangular coordinate system, the vertical axis is called the *y*-axis. **2.** True **3.** True **4.** True

Exercise Set 11.6

Vocabulary Review 1. domain, range **2.** left, right
3. function **4.** relation **5.** vertical line test
Problems

1.

x	y
−2	−8
0	0
2	8
5	20

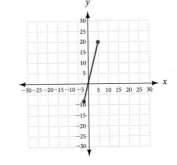

3.

x	y
0	100
5	80
10	60
15	40
20	20
25	0

5.

x	y
0	1600
40	1200
80	800
120	400
160	0

7.

x	y
0	40
25	102.5
50	165
75	227.5
100	290

9. Domain = {1, 3, 5, 7}; Range = {2, 4, 6, 8}; A function
11. Domain = {0, 1, 2, 3}; Range = {4, 5, 6}; A function

13. Domain = {a, b, c, d}; Range = {3, 4, 5}; A function
15. Domain = {a}; Range = {1, 2, 3, 4}; Not a function
17. Domain = {AAPL, DELL, MSFT, FB, GRPN};
Range = {$562, $12, $29, $32}
19. Domain = {Atlanta, Chicago, Boston, Houston};
Range = {Braves, White Sox, Cubs, Red Sox, Astros}
21. Yes **23.** No **25.** No **27.** Yes **29.** Yes **31.** Yes
33. Domain = {$x \mid -5 \le x \le 5$}; [$-5, 5$]
Range = {$y \mid 0 \le y \le 5$}, [0, 5]
35. Domain = {$x \mid -5 \le x \le 3$}; [$-5, 3$]
Range = {$y \mid y = 3$}, [3, 3]

37.

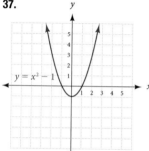

Domain = All real numbers;
Range = {$y \mid y \ge -1$}; A function

39.

Domain = All real numbers;
Range = {$y \mid y \ge 4$}; A function

41.

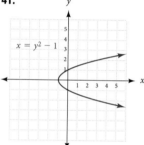

Domain = {$x \mid x \ge -1$};
Range = All real numbers;
Not a function

43.

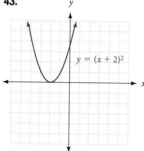

Domain = All real numbers;
Range = {$y \mid y \ge 0$}; A function

45.

Domain = {$x \mid x \ge 0$};
Range = All real numbers;
Not a function

47.

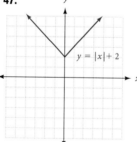

Domain = All real numbers;
Range = {$y \mid y \ge 2$}; A function

49. Domain = {2001, 2003, 2005, 2007, 2009};
Range = {5, 10, 40, 80, 160}

51. a. $y = 8.5x$ for $10 \le x \le 40$
b.

Table 4 Weekly Wages		
Hours Worked	Function Rule	Gross Pay ($)
x	$y = 8.5x$	y
10	$y = 8.5(10) = 85$	85
20	$y = 8.5(20) = 170$	170
30	$y = 8.5(30) = 255$	255
40	$y = 8.5(40) = 340$	340

c.

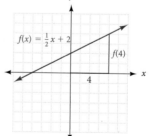

d. Domain = {$x \mid 10 \le x \le 40$}; Range = {$y \mid 85 \le y \le 340$}
e. Minimum = $85; Maximum = $340
53. a. Domain = {2008, 2009, 2010, 2011, 2012};
Range = {5.8, 5.2-5.5, 7.5, 9.7, 12.1} **b.** 2012 **c.** 2009
55. a. III **b.** I **c.** II **d.** IV **57. a.** 6 **b.** 8 **59. a.** 27 **b.** 6
61. 1 **63.** -3 **65.** $x = 0$ **67.** $-\frac{6}{5}$ **69.** $-\frac{35}{32}$
Find the Mistake 1. The set of all inputs for a function is called the
domain. **2.** For the relation (1, 2), (3, 4), (5, 6), (7, 8), the range is
{2, 4, 6, 8}. **3.** Graphing the relation $y = x^2 - 3$ will show that the
domain is all real numbers. **4.** If a vertical line crosses the graph
of a relation in more than one place, the relation cannot be a function.

Exercise Set 11.7
Vocabulary Review 1. independent **2.** dependent **3.** $f(x)$
4. function f
Problems 1. -1 **3.** -11 **5.** 2 **7.** 4 **9.** $a^2 + 3a + 4$
11. $2a + 7$ **13.** 1 **15.** -9 **17.** 8 **19.** 0 **21.** $3a^2 - 4a + 1$
23. $3a^2 + 8a + 5$ **25.** 4 **27.** 0 **29.** 2 **31.** 4
33. $x = \frac{1}{2}$ **35.** 24 **37.** -1 **39.** $2x^2 - 19x + 12$ **41.** 99
43. 28 **45.** 225 **47.** $\frac{3}{10}$ **49.** $\frac{2}{5}$ **51.** Undefined **53.** $\frac{1}{2}$
55. a. $a^2 - 7$ **b.** $a^2 - 6a + 5$ **c.** $x^2 - 2$ **d.** $x^2 + 4x$
e. $a^2 + 2ab + b^2 - 4$ **f.** $x^2 + 2xh + h^2 - 4$
57.

$f(x) = \frac{1}{2}x + 2$

$f(4)$

4

59. $x = 4$

61.

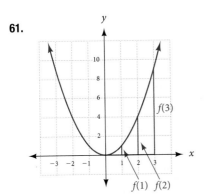

63. $V(3) = 300$, the painting is worth \$300 in 3 years; $V(6) = 600$, the painting is worth \$600 in 6 years.

65. a. True **b.** False **c.** True **d.** False **e.** True

67. a. \$5,625 **b.** \$1,500 **c.** $\{t \mid 0 \le t \le 5\}$

d.
Value of a Copy Machine

e. $\{V(t) \mid 1{,}500 \le V(t) \le 18{,}000\}$　　**f.** About 2.42 years

69. a.

d (feet)	$f(d)$ (KWH/year)
0	0
5	332
10	1328
15	2988
20	5312

Source: builditsolar.com/Projects/Wind/Wind.htm

b. $0 \le f(d) \le 5312$

71. $-0.1x^2 + 35x$ 　**73.** $4x^2 - 7x + 3$ 　**75.** $-0.1x^2 + 27x - 500$

77. $2x^2 + 8x + 8$ 　**79.** 10 　**81.** $2x^2 - 17x + 33$ 　**83.** $24x - 6$

Find the Mistake 　**1.** To evaluate the function $f(x) = 10x + 3$ for $f(1)$ we begin by replacing x with 1. 　**2.** For the average speed function $s(t) = \frac{60}{t}$, finding $s(6)$ shows an average speed of 10 miles per hour. 　**3.** If the volume of a ball is given by $V(r) = \frac{4}{3}\pi r^2$ where r is the radius, we see that the volume increases as the radius of the ball increases. 　**4.** If $f(x) = x^2$ and $g(x) = 2x + 4$, evaluating $f(g(2))$ is the same as finding $g(2)$ and using that value in $f(x) = x^2$ for x.

Chapter 11 Test

1. $[-4, \infty)$ 　**2.** $[-12, \infty)$ 　**3.** $(-\infty, 4)$ 　**4.** $(19, \infty)$

5. 10, 14 　**6.** $-21, -7$ 　**7.** $-6, 0$ 　**8.** \varnothing

9. $-\frac{5}{2} < y < -1$ 　**10.** $x \le -2$ or $x \ge 1$ 　**11.** All real numbers

12. \varnothing 　**13.** $3; -9; 3$ 　**14.** $8; 4; -\frac{1}{2}$ 　**15.** $10; -6; \frac{3}{5}$

16. -4; None; Undefined slope 　**17.** Domain = $\{0, 2, 6\}$; Range = $\{3, 5, -4\}$; A function 　**18.** Domain = All real numbers; Range = $\{y \mid y \ge 2\}$; A function 　**19.** Domain = $\{0, -6\}$; Range = $\{5, 7\}$ 　**20.** Domain = All real numbers; Range = All real numbers; A function

21. Domain = $\{x \mid -2 \le x \le 2\}$; Range = $\{y \mid -1 \le y \le 2\}$

22. Domain = $\{x \mid -2 \le x \le 2\}$; Range = $\{y \mid -3 \le y \le 3\}$

23. -16 　**24.** 11 　**25.** -12 　**26.** -9

Chapter 12

Exercise Set 12.1

Vocabulary Review 　**1.** positive 　**2.** cube 　**3.** radical 　**4.** radicand

5. index 　**6.** root 　**7.** rational 　**8.** exponent 　**9.** ratio

Problems 　**1.** 12 　**3.** Not a real number 　**5.** -7 　**7.** -3

9. 2 　**11.** Not a real number 　**13.** 0.2 　**15.** 0.2 　**17.** $6a^4$

19. $3a^4$ 　**21.** xy^2 　**23.** $2x^2y$ 　**25.** $2a^3b^5$ 　**27.** $-3x^3y^4$ 　**29.** $x^{4/5}$

31. $n^{2/3}$ 　**33.** $a^{3/6} = a^{1/2}$ 　**35.** $m^{5/8}$ 　**37.** 6 　**39.** -3 　**41.** 2

43. -2 　**45.** 2 　**47.** $\frac{9}{5}$ 　**49.** $\frac{4}{5}$ 　**51.** $\frac{2}{3}$ 　**53.** 9 　**55.** 125

57. 8 　**59.** 9 　**61.** $\frac{1}{3}$ 　**63.** $\frac{1}{27}$ 　**65.** $\frac{6}{5}$ 　**67.** $\frac{8}{27}$ 　**69.** 7 　**71.** $\frac{3}{4}$

73. $x^{4/5}$ 　**75.** a 　**77.** $\frac{1}{x^{2/5}}$ 　**79.** $x^{1/6}$ 　**81.** $a^{2/9}b^{1/6}c^{1/12}$ 　**83.** $x^{9/25}y^{1/2}z^{1/5}$

85. $\frac{b^{7/4}}{a^{1/8}}$ 　**87.** $y^{3/10}$ 　**89.** $\frac{1}{x^2y}$ 　**91.** $\frac{1}{a^2b^4}$ 　**93.** $\frac{s^{1/2}}{r^{20}}$ 　**95.** $10b^3$

97. $(9^{1/2} + 4^{1/2})^2 \overset{?}{=} 9 + 4$ 　**99.** $(a^{1/2})^{1/2} \overset{?}{=} a^{1/4}$ 　**101.** 25 mph
$\quad(3 + 2)^2 \overset{?}{=} 13$ 　　　　　$a^{1/4} = a^{1/4}$ 　**103.** 1.618
$\quad 5^2 \overset{?}{=} 13$
$\quad 25 \ne 13$

105. a. 420 picometers **b.** 594.0 pico meters **c.** 5.94×10^{-10} meters

107.

Pyramid Name	b	h	Slant height (ft)
Temple of Kukulkan	181 ft	98 ft	133
Temple IV at Tecal	118 ft	187 ft	196
Pyramid of the Sun	733 ft	246 ft	441
Great Pyramid of Khufu	755 ft	455 ft	591
Luxor Hotel	600 ft	350 ft	461
Transamerica Pyramid	175 ft	850 ft	854

Source: skyscrapercity.com

109. Transamerica 　**111.** 5 　**113.** 6 　**115.** $4x^2y$ 　**117.** $5y$

119. 3 　**121.** 2 　**123.** $2ab$ 　**125.** $5x^2y$ 　**127.** 25

129. $48x^4y^2$ 　**131.** $4x^6y^6$ 　**133.** $2xy^4$

Find the Mistake 　**1.** \sqrt{x} is the positive square number we square to get x. 　**2.** If we raise the fifth root of x to the fifth power, we get x. 　**3.** For the radical expression $\sqrt[4]{16}$, the 4 is the index and the 16 is the radicand. 　**4.** Using the rational exponent theorem, we can say that $16^{-3/2} = (16^{1/2})^{-3} = \frac{1}{64}$.

Exercise Set 12.2

Vocabulary Review 　**1.** product 　**2.** quotient 　**3.** simplified form

4. absolute value 　**5.** spiral of roots

Problems 　**1.** $2\sqrt{2}$ 　**3.** $7\sqrt{2}$ 　**5.** $12\sqrt{2}$ 　**7.** $4\sqrt{5}$

9. $4\sqrt{3}$ 　**11.** $15\sqrt{3}$ 　**13.** $3\sqrt[3]{2}$ 　**15.** $4\sqrt[3]{2}$ 　**17.** $6\sqrt[3]{2}$

19. $2\sqrt[5]{2}$ 　**21.** $3x\sqrt{2x}$ 　**23.** $2y\sqrt[4]{2y^3}$ 　**25.** $2xy^2\sqrt[4]{5xy}$

27. $4abc^2\sqrt{3b}$ 　**29.** $2bc\sqrt[3]{6a^2c}$ 　**31.** $2xy^2\sqrt[5]{2x^3y^2}$

33. $3xy^2z\sqrt[5]{x^2}$ 　**35.** $5xy^3\sqrt[3]{2x^2y}$ 　**37.** $-3x^3y^4\sqrt[4]{xy}$ 　**39.** $2\sqrt{3}$

41. $\sqrt{-20}$; not real number 　**43.** $\frac{\sqrt{11}}{2}$ 　**45.** $\frac{2\sqrt{3}}{3}$

47. $\frac{5\sqrt{6}}{6}$ 　**49.** $\frac{\sqrt{2}}{2}$ 　**51.** $\frac{\sqrt{5}}{5}$ 　**53.** $2\sqrt[3]{4}$ 　**55.** $\frac{2\sqrt[3]{3}}{3}$

57. $\frac{\sqrt[4]{24x^2}}{2x}$ 　**59.** $\frac{\sqrt[3]{8y^3}}{y}$ 　**61.** $\frac{\sqrt[3]{36xy^2}}{3y}$ 　**63.** $\frac{\sqrt[3]{6xy^2}}{3y}$ 　**65.** $\frac{\sqrt[4]{2x}}{2x}$

67. $\dfrac{\sqrt[4]{18x^2}}{2x}$ **69.** $\dfrac{3x\sqrt{15xy}}{5y}$ **71.** $\dfrac{5xy\sqrt{6xz}}{2z}$ **73.** $\dfrac{2ab\sqrt[3]{6ac^2}}{3c}$

75. $\dfrac{2xy^2\sqrt[3]{3z^2}}{3z}$ **77.** $5|x|$ **79.** $3|xy|\sqrt{3x}$ **81.** $|x-5|$

83. $|2x+3|$ **85.** $2|a(a+2)|$ **87.** $2|x|\sqrt{x-2}$

89. $\sqrt{9+16} \overset{?}{=} \sqrt{9}+\sqrt{16}; \sqrt{25} \overset{?}{=} 3+4; 5 \neq 7$

91. $5\sqrt{13}$ feet **93. a.** 89.4 miles **b.** 126.5 miles **c.** 154.9 miles

97. $a_{10}=\sqrt{11}; a_{100}=\sqrt{101}$ **99.** $7x$ **101.** $27xy^2$ **103.** $\frac{5}{6}x$

105. $-7(x+1)^2$ **107.** $3\sqrt{2}$ **109.** $5y\sqrt{3xy}$ **111.** $2a\sqrt[3]{ab^2}$

113. $2x^2y\sqrt[4]{2y}$

Find the Mistake 1. False: To write the radical expression $\sqrt{4x^7y^2}$ in simplified form, we apply the product property for radicals. **2.** False: The radical expression in statement 1 written in simplified form is $2x^3y\sqrt{x}$. **3.** False: To rationalize the denominator of the radical expression $\sqrt{\dfrac{8x^{4y}}{3x}}$, we use the quotient property for radicals and then multiply the numerator and denominator by $\sqrt{3x}$. **4.** False: For the spiral of roots, each new line segment drawn to the end of a diagonal is at a right angle to the previous line segment.

Exercise Set 12.3

Vocabulary Review simplified form, distributive, similar, radicand, radicals

Problems 1. $7\sqrt{5}$ **3.** $-x\sqrt{7}$ **5.** $\sqrt[3]{10}$ **7.** $9\sqrt[5]{6}$

9. 0 **11.** $\sqrt{5}$ **13.** $-32\sqrt{2}$ **15.** $-3x\sqrt{2}$

17. $5\sqrt{5y}+3\sqrt{2y}$ **19.** $-2\sqrt[3]{2}$ **21.** $8x\sqrt[3]{xy^2}$ **23.** $3a^2b\sqrt{3ab}$

25. $2ab\sqrt[3]{3a^2b}+3a^2b\sqrt[3]{3a^2b}$ **27.** $11ab\sqrt[3]{3a^2b}$ **29.** $10xy\sqrt[4]{3y}$

31. $\sqrt{2}$ **33.** $\dfrac{8\sqrt{5}}{15}$ **35.** $\dfrac{(x-1)\sqrt{x}}{x}$ **37.** $\dfrac{3\sqrt{2}}{2}$ **39.** $\dfrac{5\sqrt{6}}{6}$

41. $\dfrac{8\sqrt[3]{25}}{5}$ **43.** $\dfrac{7\sqrt{5}}{10}$ **45.** $\sqrt{12}\approx3.464; 2\sqrt{3}\approx2(1.732)=3.464$

47. $\sqrt{8}+\sqrt{18}\approx2.828+4.243=7.071; \sqrt{50}\approx7.071; \sqrt{26}\approx5.099$

49. $8\sqrt{2x}$ **51.** 5 **59.** $\sqrt{2}$

61. a. $\sqrt{2}{:}1\approx1.414{:}1$ **b.** $5{:}\sqrt{2}$ **c.** 5:4 **63.** 6

65. $4x^2+3xy-y^2$ **67.** x^2+6x+9 **69.** x^2-4 **71.** $6\sqrt{2}$

73. 6 **75.** $9x$ **77.** $48x$ **79.** $\dfrac{\sqrt{6}}{2}$ **81.** $\dfrac{\sqrt{3}}{2}$

Find the Mistake 1. False: When adding or subtracting radical expressions, put each in simplified form and apply the distributive property. **2.** True **3.** False: When combining $3\sqrt{5}+5\sqrt{3}-2\sqrt{5}$, the first and last radical are similar, so they can be combined to get $5\sqrt{3}+\sqrt{5}$. **4.** False: Combining $4\sqrt{8}+3\sqrt{32}, 20\sqrt{2}$.

Exercise Set 12.4

Vocabulary Review 1. polynomials **2.** conjugate **3.** radical **4.** denominator

Problems 1. $3\sqrt{2}$ **3.** $10\sqrt{21}$ **5.** 720 **7.** 54

9. $\sqrt{6}-9$ **11.** $24+6\sqrt[3]{4}$ **13.** $7+2\sqrt{6}$

15. $x+2\sqrt{x}-15$ **17.** $34+20\sqrt{3}$ **19.** $19+8\sqrt{3}$

21. $x-6\sqrt{x}+9$ **23.** $4a-12\sqrt{ab}+9b$ **25.** $x+4\sqrt{x-4}$

27. $x-6\sqrt{x-5}+4$ **29.** 1 **31.** $a-49$ **33.** $25-x$

35. $x-8$ **37.** $10+6\sqrt{3}$ **39.** $30\sqrt{3}-54$ **41.** $\dfrac{\sqrt{3}+1}{2}$

43. $\dfrac{5-\sqrt{5}}{4}$ **45.** $\dfrac{x+3\sqrt{x}}{x-9}$ **47.** $\dfrac{10+3\sqrt{5}}{11}$ **49.** $\dfrac{3\sqrt{x}+3\sqrt{y}}{x-y}$

51. $2+\sqrt{3}$ **53.** $\dfrac{11-4\sqrt{7}}{3}$ **55.** $\dfrac{a+2\sqrt{ab}+b}{a-b}$ **57.** $\dfrac{x+4\sqrt{x}+4}{x-4}$

59. $\dfrac{5-\sqrt{21}}{4}$ **61.** $\dfrac{\sqrt{x}-3x+2}{1-x}$ **63.** $\dfrac{6x-\sqrt{x}-1}{1-9x}$ **67.** $10\sqrt{3}$

69. $x+6\sqrt{x}+9$ **71.** 75 **73.** $x^2+2x\sqrt{3}+3$

75. $\dfrac{5\sqrt{2}}{4}$ second; $\frac{5}{2}$ second **81.** $t^2+10t+25$ **83.** x **85.** 7

87. $-4,-3$ **89.** $-6,-3$ **91.** $-5,-2$ **93.** Yes **95.** No

97. North America: 24.5 million; Asia: 7 million; Africa: 500,000

Find the Mistake 1. To multiply $\sqrt{5}(\sqrt{7}+6\sqrt{5})$, we use the distributive property to get our answer $30+\sqrt{35}$. **2.** To multiply $(2\sqrt{2}+\sqrt{7})(4\sqrt{2}-3\sqrt{7})$, we apply the FOIL method to get $-5-2\sqrt{14}$. **3.** To divide 4 by $(2\sqrt{5}-2)$, we rationalize the denominator by multiplying the numerator and the denominator by $2\sqrt{5}+2$. **4.** Dividing $\dfrac{(2\sqrt{3}-\sqrt{2})}{(4\sqrt{3}+\sqrt{2})}$ we get $\dfrac{(13-3\sqrt{6})}{23}$.

Landmark Review 1. 9 **2.** -9 **3.** Not a real number **4.** -3 **5.** x^2y^4 **6.** 8 **7.** $\frac{5}{4}$ **8.** 19 **9.** $2\sqrt{41}$ **10.** $3x^2y^3\sqrt{2x}$ **11.** $3x^2y^2z^3\sqrt[5]{2x^2z^3}$ **12.** $\dfrac{4\sqrt{5}}{5}$ **13.** $28\sqrt{6}$ **14.** $\dfrac{5\sqrt{2}}{6}$ **15.** $2\sqrt{3}$ **16.** 480 **17.** $30+11\sqrt{6}$ **18.** $6-2\sqrt{5}$ **19.** $\dfrac{1+\sqrt{3}}{2}$ **20.** $\dfrac{-7-3\sqrt{5}}{2}$

Exercise Set 12.5

Vocabulary Review 1. radical **2.** solutions **3.** extraneous **4.** original **5.** negative

Problems 1. 4 **3.** No solution **5.** 5 **7.** No solution

9. $\frac{39}{2}$ **11.** No solution **13.** 5 **15.** 3 **17.** $-\frac{32}{3}$

19. 3, 4 **21.** $-1,-2$ **23.** -1 **25.** No solution **27.** 7

29. 0, 3 **31.** -4 **33.** 8 **35.** 0 **37.** 9 **39.** 0 **41.** 8

43. No solution **45.** 0 **47.** 6 **49.** 3 **51.** 1 **53.** $\frac{16}{3}$

55. $h=100-16t^2$ **57.** $\frac{392}{121}\approx3.24$ feet **59.** $l=\frac{10}{3}(\sqrt{x+1})$

61. 5 meters **63.** 2,500 meters **65.** $0\le y\le100$

67.

69.

71.

73.

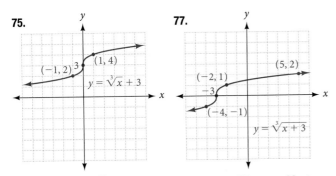

75. **77.**

79. 5 **81.** $2\sqrt{3}$ **83.** -1 **85.** 1 **87.** 4 **89.** 2
91. $10 - 2x$ **93.** $2 - 3x$ **95.** $6 + 7x - 20x^2$
97. $8x - 12x^2$ **99.** $4 + 12x + 9x^2$ **101.** $4 - 9x^2$
Find the Mistake **1.** Solving $\sqrt{6x + 1} = -5$ gives us an extraneous
solution. **2.** Solving $\sqrt[4]{3x - 8} = 2$ gives us a solution of $x = 8$.
3. Solving $\sqrt{x + 8} = x + 6$ gives a solution of $x = -4$
because $x = -7$ gives us a false statement when checked in the
original equation. **4.** The graphs of $y = \sqrt{x - 1}$ and $y = \sqrt{2x - 6}$
intersect at $(5, 2)$.

Exercise Set 12.6

Vocabulary Review **1.** number i **2.** complex number **3.** zero
4. equal **5.** imaginary **6.** conjugates
Problems **1.** $6i$ **3.** $-5i$ **5.** $6i\sqrt{2}$ **7.** $-2i\sqrt{3}$
9. 1 **11.** -1 **13.** $-i$ **15.** 1 **17.** $x = 3, y = -1$
19. $x = -2, y = -\frac{1}{2}$ **21.** $x = -8, y = -5$ **23.** $x = 7, y = \frac{1}{2}$
25. $x = \frac{3}{7}, y = \frac{2}{5}$ **27.** $5 + 9i$ **29.** $5 - i$ **31.** $2 - 4i$
33. $1 - 6i$ **35.** $2 + 2i$ **37.** $-1 - 7i$ **39.** $6 + 8i$ **41.** $2 - 24i$
43. $-15 + 12i$ **45.** $18 + 24i$ **47.** $10 + 11i$ **49.** $21 + 23i$
51. $-2 + 2i$ **53.** $2 - 11i$ **55.** $-21 + 20i$ **57.** $-2i$
59. $-7 - 24i$ **61.** 5 **63.** 40 **65.** 13 **67.** 164 **69.** 7
71. $-3 - 2i$ **73.** $-2 + 5i$ **75.** $\frac{8}{13} + \frac{12}{13}i$ **77.** $-\frac{18}{13} - \frac{12}{13}i$
79. $-\frac{5}{13} + \frac{12}{13}i$ **81.** $\frac{13}{15} - \frac{2}{5}i$ **83.** $-11 - 7i$ ohms
85. $-\frac{3}{2}$ **87.** $-3, \frac{1}{2}$ **89.** $\frac{5}{4}$ or $\frac{4}{5}$ **91.** $m = 0.23, \$8.90$
Find the Mistake **1.** The number i is such that $i = \sqrt{-1}$ is the
same as saying $i^2 = -1$. **2.** The form $a + bi$ is called standard
form for complex numbers and the bi is the imaginary part of
the number. **3.** The product of the complex numbers $(5 - 3i)$
and $(2 + 4i)$ is $22 + 14i$. **4.** The product of the two complex
conjugates $6 + 4i$ and $6 - 4i$ is $6^2 + 4^2 = 36 + 16 = 52$.

Chapter 12 Test

1. $\frac{1}{8}$ **2.** $\frac{8}{9}$ **3.** $\frac{1}{x^{11/21}}$ **4.** $\frac{a^{7/15}}{b^{12}}$ **5.** $5x^3y^{10}$ **6.** $3xy^5$

7. $\frac{2}{7a^2b^3}$ **8.** $x^{n - n^3}y^{n^3 - n}$ **9.** $3x^2y\sqrt{3xy}$ **10.** $4y^2\sqrt[3]{2x^2y}$

11. $\frac{\sqrt{15}}{5}$ **12.** $\frac{2ab^3\sqrt{14bc}}{7c}$ **13.** $7\sqrt{2} - 7$ **14.** $\frac{x + 2\sqrt{3x} + 3}{x - 3}$

15. $\frac{5 - x\sqrt{x}}{x}$ **16.** $\frac{17\sqrt{x^3 - 17}}{x^3 - 17}$ **17.** 0 **18.** $3a^2b\sqrt[4]{4}$

19. $4a^4 - 12a^2$ **20.** $4a^5 - 12a^{5/2} + 9$ **21.** $x + 4\sqrt{x} - 5$
22. $14 - 4\sqrt{6}$ **23.** 6, 2 does not check **24.** $-\frac{3}{2}$ **25.** 10

26. **27.**

28. E **29.** F **30.** A **31.** B **32.** D
33. C **34.** $x = -2; y = 1$ **35.** $7i$ **36.** $13 + i$
37. $-27 - 36i$ **38.** $\frac{5 - 12i}{13}$ **39.** $i^{36} = (i^2)^{18} = (-1)^{18} = 1$

Chapter 13

Exercise Set 13.1

Vocabulary Review 2, 1, 4, 3, 5
Problems **1.** ± 5 **3.** $\pm 3i$ **5.** $\pm\frac{\sqrt{3}}{2}$ **7.** $\pm 2i\sqrt{3}$
9. $\pm\frac{3\sqrt{5}}{2}$ **11.** $-2, 3$ **13.** $\frac{-3 \pm 3i}{2}$ **15.** $\frac{-2 \pm 2i\sqrt{2}}{5}$
17. $-4 \pm 3i\sqrt{3}$ **19.** $\frac{3 \pm 2i}{2}$ **21.** 36, 6 **23.** 4, 2
25. 25, 5 **27.** $\frac{25}{4}, \frac{5}{2}$ **29.** $\frac{49}{4}, \frac{7}{2}$ **31.** $\frac{1}{16}, \frac{1}{4}$ **33.** $\frac{1}{9}, \frac{1}{3}$
35. $-6, 2$ **37.** $-3, -9$ **39.** $1 \pm 2i$ **41.** $4 \pm \sqrt{15}$
43. $\frac{5 \pm \sqrt{37}}{2}$ **45.** $1 \pm \sqrt{5}$ **47.** $\frac{4 \pm \sqrt{13}}{3}$ **49.** $\frac{3 \pm i\sqrt{71}}{8}$
51. $\frac{-2 \pm \sqrt{7}}{3}$ **53.** $\frac{5 \pm \sqrt{47}}{2}$ **55.** $\frac{5 \pm i\sqrt{19}}{4}$ **57.** $\frac{15 \pm 2\sqrt{3}}{9}$
59. **a.** No **b.** $\pm 3i$ **61.** **a.** 0, 6 **b.** 0, 6 **63.** **a.** $-7, 5$ **b.** $-7, 5$
65. No **67.** **a.** $\frac{7}{5}$ **b.** 3 **c.** $\frac{7 \pm 2\sqrt{2}}{5}$ **d.** $\frac{71}{5}$ **e.** 3 **69.** $-1, 5$
71. $\frac{-3 \pm \sqrt{29}}{2}$ **73.** $\frac{7 \pm \sqrt{85}}{2}$ **75.** None **77.** $\frac{-3 \pm \sqrt{41}}{4}$
79. $\frac{-1 \pm 2\sqrt{3}}{2}$ **81.** $\frac{\sqrt{3}}{2}$ meter, 1 meter **83.** $\sqrt{2}$ inches
85. 781 feet **87.** 7.3% to the nearest tenth **89.** $20\sqrt{2} \approx 28$ feet
91. 169 **93.** 49 **95.** $\frac{85}{12}$ **97.** 1 **99.** $(3t - 2)(9t^2 + 6t + 4)$
101. $2(x + 3)(x^2 - 3x + 9)$ **103.** **a.** 1,006 **b.** 346

Find the Mistake **1.** The square root property for equations
says that if $a^2 = b$, then $a = \sqrt{b}$ or $a = -\sqrt{b}$. **2.** Solving
$(4x + 1)^2 = 16$ gives us two solutions, $\frac{3}{4}$ or $-\frac{3}{4}$. **3.** The first
step to solving a quadratic equation by completing the square is
to put the equation in the form $ax^2 + bx = c$. **4.** The method
of completing the square says that adding the square of half the
coefficient of the linear term to both sides will give a perfect square
trinomial.

Exercise Set 13.2

Vocabulary Review **1.** standard form **2.** quadratic
3. rational **4.** factored
Problems **1.** $-3, -2$ **3.** $2 \pm \sqrt{3}$ **5.** 1, 2 **7.** $\frac{2 \pm i\sqrt{14}}{3}$
9. 0, 5 **11.** $0, -\frac{4}{3}$ **13.** $\frac{3 \pm \sqrt{5}}{4}$ **15.** $-3 \pm \sqrt{17}$
17. $\frac{-1 \pm i\sqrt{5}}{2}$ **19.** 1 **21.** $\frac{1 \pm i\sqrt{47}}{6}$ **23.** $4 \pm \sqrt{2}$ **25.** $\frac{1}{2}, 1$
27. $-\frac{1}{2}, 3$ **29.** $\frac{3 \pm \sqrt{5}}{2}$ **31.** $\frac{-1 \pm i\sqrt{7}}{2}$ **33.** $1 \pm \sqrt{2}$
35. $\frac{-3 \pm \sqrt{5}}{2}$ **37.** $3, -5$ **39.** $1 \pm \sqrt{10}$ **41.** $\frac{3 \pm \sqrt{17}}{4}$
43. $2, -1 \pm i\sqrt{3}$ **45.** $-\frac{3}{2}, \frac{3 \pm 3i\sqrt{3}}{4}$ **47.** $\frac{1}{5}, \frac{-1 \pm i\sqrt{3}}{10}$
49. $0, \frac{-1 \pm i\sqrt{5}}{2}$ **51.** $0, 1 \pm i$ **53.** $0, \frac{-1 \pm i\sqrt{2}}{3}$
55. a and b **57.** **a.** $\frac{5}{3}, 0$ **b.** $\frac{5}{3}, 0$ **59.** No, $2 \pm i\sqrt{3}$

61. Yes **63.** 0, 4 **65.** 0, 0.4 **67.** $-2 \pm \sqrt{2}$

69. $\frac{1 \pm \sqrt{13}}{2}$ **71.** $\pm \sqrt{6}$ **73.** 2 seconds **75.** 20 or 60 items

77. 0.49 centimeter (8.86 cm is impossible)

79. a. $l + w = 10, lw = 15$ **b.** 8.16 yards, 1.84 yards **c.** Because either dimension (long or short) may be considered the length.

81. 169 **83.** 0 **85.** ± 12 **87.** $x^2 - x - 6$ **89.** $x^3 - 4x^2 - 3x + 18$

Find the Mistake **1.** Using the quadratic formula to solve a quadratic equation gives the solutions $x = \frac{-b \pm \sqrt{b^2 - 4ac}}{2a}$.

2. Using the quadratic formula to solve $x^2 + 10x = 11$, $a = 1$, $b = 10$, and $c = -11$. **3.** To solve $x^2 - 10x = -24$ using the quadratic formula, begin by writing the equation in standard form.

4. Solving $3x^2 - 2x - 4 = 0$ using the quadratic formula, we get $\frac{1 \pm \sqrt{13}}{7}$ for our answer.

Chapter 13 Test

1. $\frac{9}{2}, -\frac{3}{2}$ **2.** $\frac{-1 \pm 2i\sqrt{3}}{4}$ **3.** $-4 \pm 5i$ **4.** $-5 \pm i\sqrt{2}$

5. $\frac{3}{4}, \frac{-3 \pm 3i\sqrt{3}}{8}$ **6.** $\frac{1 \pm i\sqrt{7}}{2}$ **7.** $r = 7 \pm \frac{\sqrt{A}}{7}$ **8.** $3 \pm i\sqrt{2}$

9. 4 m, $4\sqrt{3}$ m **10.** $\frac{1}{4}$ second, $\frac{7}{4}$ seconds

11. 5 or 100 letter openers

Chapter 14

Exercise Set 14.1

Vocabulary Review **1.** exponential function **2.** x–axis

3. number e **4.** $A(t) = P\left(1 + \frac{r}{n}\right)^{nt}$ **5.** $A(t) = Pe^{rt}$

Problems **1.** 1 **3.** 2 **5.** $\frac{1}{27}$ **7.** 13 **9.** $\frac{7}{12}$ **11.** $\frac{3}{16}$

13. 48 **15.** $-\frac{35}{4}$ **17.** $\frac{3^x - 9}{x - 2}$ **19.** $\frac{3^x - 27}{x - 3}$ **21.** $\frac{3^x(3^h - 1)}{h}$

23.

25.

27.

29.

31.

33.

35. $y = 2x$

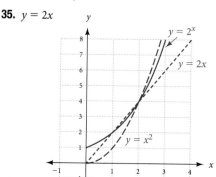

39. $h(n) = 6 \cdot \left(\frac{2}{3}\right)^n$; 5th bounce: $6\left(\frac{2}{3}\right)^5 \approx 0.79$ feet **41.** 4.27 days

43. a. $A(t) = 1,200\left(1 + \frac{.06}{4}\right)^{4t}$ **b.** \$1,932.39 **c.** About 11.64 years **d.** \$1,939.29

45. a. $A(t) = 5,400\left(1 + \frac{.05}{4}\right)^{4t}$ **b.** \$6,923.00 **c.** 8 years **d.** \$6,933.74

47. $f(1) = 200, f(2) = 800, f(3) = 3,200$

49. a. \$129,138.48 **b.** $[0, 6]$

c.

d. $[52,942.05, 450,000]$

e. After approximately 4 years and 8 months

51.

53. a. \$0.42 **b.** \$1.00 **c.** \$1.78 **d.** \$17.84

55. 1,258,525 bankruptcies, which is 58,474 less than the actual number.

57. a. 251,437 cells **b.** 12,644 cells **c.** 32 cells

59. $D = \{-2, 2\}, R = \{6, 8, 3\}$, Not a function

61. $(-\infty, -7) \cup (-7, 5) \cup (5, \infty)$ **63.** -18 **65.** 2

67. $y = 2.308x - 4562.06$

69. a. 1 **b.** $\frac{1}{3}$ **c.** 3 **d.** 1 **e.** $\frac{1}{2}$ **f.** 2 **g.** 3 **h.** $\frac{1}{2}$
71. $y = \frac{x+3}{2}$ **73.** $y = \pm\sqrt{x+3}$ **75.** $y = \frac{2x-4}{x-1}$ **77.** $y = x^2 + 3$
Find the Mistake **1.** The exponential function is any function written in the form $f(x) = b^x$. **2.** The graph of $f(x) = 2^x$ gets very close to the x–axis but does not intersect it. **3.** The special number e is an irrational number. **4.** To calculate annual interest rate compounded n times per year, use the formula $A(t) = P\left(1 + \frac{r}{n}\right)^{nt}$.

Exercise Set 14.2

Vocabulary Review **1.** relation **2.** range **3.** one-to-one
4. inverse **5.** function

Problems **1.** $f^{-1}(x) = \frac{x+1}{3}$ **3.** $f^{-1}(x) = \sqrt[3]{x}$
5. $f^{-1}(x) = \frac{x-3}{x-1}$ **7.** $f^{-1}(x) = 4x + 3$ **9.** $f^{-1}(x) = 2x + 6$
11. $f^{-1}(x) = \frac{3}{2}x + \frac{9}{2}$ **13.** $f^{-1}(x) = \sqrt[3]{x+4}$ **15.** $f^{-1}(x) = \frac{x+3}{4-2x}$
17. $f^{-1}(x) = \frac{1-x}{3x-2}$ **19.** $f^{-1}(x) = \frac{x^3+2}{3}$

21. **23.**

25. **27.**

29. **31.**

33. **35.**

37.

39. a. Yes **b.** No **c.** Yes
41. $g = \{(4, 1), (5, 2), (7, 3)\}$; Yes
43. $g = \{(A, a), (B, b), (C, c)\}$; Yes
45. a. 4 **b.** $\frac{4}{3}$ **c.** 2 **d.** 2
47. $f^{-1}(x) = \frac{1}{x}$

49. a. $3x$ **b.** $3x + 2$ **c.** $x - 2$
d. $\frac{x-2}{3}$; $f^{-1}(x) = 7(x + 2) = 7x + 14$
51. a. -3 **b.** -6 **c.** 2 **d.** 3 **e.** -2 **f.** 3 **g.** inverses
53. a. $-53.3°C$ **b.** $C^{-1}(F) = \frac{9}{5}F + 32$ **c.** $32°F$ **55.** -40
57. a. 569.4 **b.** $s^{-1}(t) = \frac{t-249.4}{16}$ **c.** 2006
59. a. 6629.33 ft/s **b.** $f^{-1}(m) = \frac{15m}{22}$ **c.** 1.36 mph
61. $-2, 3$ **63.** 25 **65.** $5 \pm \sqrt{17}$ **67.** $1 \pm i$ **69.** $f^{-1}(x) = \frac{x-5}{3}$
71. $f^{-1}(x) = \sqrt[3]{x-1}$ **73.** $f^{-1}(x) = \frac{2x-4}{x-1}$ **75.** $f^{-1}(x) = \frac{2x+3}{2-x}$
77. a. 1 **b.** 2 **c.** 5 **d.** 0 **e.** 1 **f.** 2 **g.** 2 **h.** 5
79. 41.39 EURO, $56.28 **81.** $x = -\frac{8}{3}$ **83.** $f^{-1}(x) = \frac{b-dx}{cx-a}$
85. $f^{-1}(x) = \frac{ax+5}{3x-a}$ **87.** $\frac{1}{9}$ **89.** $\frac{1}{16}$ **91.** $\frac{2}{3}$ **93.** $\sqrt[3]{4}$
95. 3 **97.** 4 **99.** 4 **101.** 1 **103.** -3 **105.** -4

Find the Mistake **1.** A function is a one-to-one function if every element in the range comes from exactly one element in the domain. **2.** If $y = f(x)$ is a one-to-one function, then the inverse of $f(x)$ is also a function and can be denoted by $y = f^{-1}(x)$. **3.** The graph of a function and its inverse have symmetry about the line $y = x$. **4.** If (a, b) belongs to the function $f(x)$, then the point (b, a) belongs to its inverse.

Exercise Set 14.3

Vocabulary Review **1.** b **2.** d **3.** a **4.** c **5.** e
Problems **1.** $\log_2 16 = 4$ **3.** $\log_5 125 = 3$ **5.** $\log_{10} 0.01 = -2$
7. $\log_2 \frac{1}{32} = -5$ **9.** $\log_{1/2} 8 = -3$ **11.** $\log_3 27 = 3$ **13.** $10^2 = 100$
15. $2^6 = 64$ **17.** $8^0 = 1$ **19.** $10^{-3} = 0.001$ **21.** $6^2 = 36$
23. $5^{-2} = \frac{1}{25}$ **25.** 9 **27.** $\frac{1}{125}$ **29.** 4 **31.** $\frac{1}{3}$ **33.** 2
35. $\sqrt[5]{5}$ **37.** 2 **39.** 6 **41.** $\frac{2}{3}$ **43.** $-\frac{1}{2}$ **45.** $\frac{1}{64}$ **47.** $\frac{3}{4}$

49. **51.**

53.

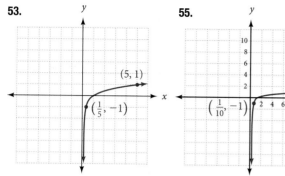

55.

57. $f(x) = 3^x$ **59.** $f(x) = \log_{1/3} x$ **61.** 4 **63.** $\frac{3}{2}$ **65.** 3 **67.** 1
69. 0 **71.** 0 **73.** $\frac{1}{2}$ **75.** $\frac{3}{2}$ **77.** 1 **79.** -2 **81.** 0 **83.** $\frac{1}{2}$

85.

Prefix	Multiplying Factor	\log_{10} (Multiplying Factor)
Nano	0.000 000 001	-9
Micro	0.000 001	-6
Deci	0.1	-1
Giga	1,000,000,000	9
Peta	1,000,000,000,000,000	15

87. 2 **89.** 10^8 times as large **91.** 939
93. a. 10,486 **b.** 1,311 **c.** 5,243 **95.** 4 **97.** 8 **99.** 9
101. $\frac{1}{9}$ **103.** $-4, 2$ **105.** $-\frac{11}{8}$ **107.** $2^3 = (x + 2)(x)$
109. $3^4 = \frac{x - 2}{x + 1}$

Find the Mistake 1. The expression $y = \log_b x$ is read "y is the logarithm to the base b of x." **2.** The expression $\frac{1}{16} = 2^{-4}$ is equivalent to the logarithmic expression $\log_2 \frac{1}{16} = -4$. **3.** The graph of $y = \log_2 x$ is a function because no vertical line will cross its graph in more than one place. **4.** A special identity for logarithms states that b raised to the power of $\log_b x$ is equal to x.

Landmark Review 1. 1 **2.** $\frac{1}{3}$ **3.** $\frac{2}{9}$ **4.** $-\frac{2}{27}$
5.

6.

7. $\frac{x + 2}{5}$ **8.** $\frac{\sqrt[3]{4x + 8}}{2}$
9.

11.

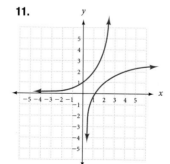

11. $\log_2 16 = 4$ **12.** $\log_4 16 = 2$ **13.** $3^{-2} = \frac{1}{9}$ **14.** $7^2 = 49$
15. 25 **16.** 8 **17.** $\sqrt[5]{3}$ **18.** $-\frac{1}{2}$

Exercise Set 14.4

Vocabulary Review 1. decibel **2.** product **3.** quotient **4.** power
Problems 1. $\log_3 4 + \log_3 x$ **3.** $\log_6 5 - \log_6 x$ **5.** $5 \log_2 y$
7. $\frac{1}{3} \log_9 z$ **9.** $2 \log_6 x + 4 \log_6 y$ **11.** $\frac{1}{2} \log_5 x + 4 \log_5 y$
13. $\log_b x + \log_b y - \log_b z$ **15.** $\log_{10} 4 - \log_{10} x - \log_{10} y$
17. $2 \log_{10} x + \log_{10} y - \frac{1}{2} \log_{10} z$ **19.** $3 \log_{10} x + \frac{1}{2} \log_{10} y - 4 \log_{10} z$
21. $\frac{2}{3} \log_b x + \frac{1}{3} \log_b y - \frac{4}{3} \log_b z$ **23.** $\frac{2}{3} \log_3 x + \frac{1}{3} \log_3 y - 2 \log_3 z$
25. $2 \log_a 2 + 5 \log_a x - 2 \log_a 3 - 2$ **27.** $2 \log_b 3 + 3 \log_b x - 6 \log_b 2 - 2$
29. $\log_b xz$ **31.** $\log_3 \frac{x^2}{y^3}$ **33.** $\log_{10} \left(\sqrt{x} \sqrt[3]{y} \right)$ **35.** $\log_2 \left(\frac{x^3 \sqrt{y}}{z} \right)$
37. $\log_2 \left(\frac{\sqrt{x}}{y^3 z^4} \right)$ **39.** $\log_{10} \left(\frac{x^{3/2}}{y^{3/4} z^{4/5}} \right)$ **41.** $\log_5 \left(\frac{\sqrt{x} \cdot \sqrt[3]{y^2}}{z^4} \right)$
43. $\log_3 \left(\frac{x - 4}{x + 4} \right)$ **45.** $\frac{2}{3}$ **47.** 18 **49.** 3 **51.** 3 **53.** 4
55. 4 **57.** 1 **59.** 0 **61.** $\frac{3}{2}$ **63.** 27 **65.** $\frac{5}{3}$
69. a. 1.602 **b.** 2.505 **c.** 3.204 **71.** $pH = 6.1 + \log_{10} x - \log_{10} y$
73. 2.52 **75.** 1 **77.** 1 **79.** 4 **81.** x **83.** 2.5×10^{-6} **85.** 51
Find the Mistake 1. The logarithm of a product is the sum of the logarithms. **2.** The quotient property for logarithms states that $\log_b \frac{x}{y} = \log_b x - \log_b y$. **3.** The logarithm of a number raised to a power is the product of the power and the logarithm of the number. **4.** To expand $\log_6 \frac{4xy}{z}$, apply the quotient property and the product property for logarithms.

Exercise Set 14.5

Vocabulary Review 1. common **2.** characteristic, mantissa
3. natural
Problems 1. 2.5775 **3.** 1.5775 **5.** 3.5775 **7.** -1.4225
9. 4.5775 **11.** 2.7782 **13.** 3.3032 **15.** -2.0128
17. -1.5031 **19.** -0.3990 **21.** 759 **23.** 0.00759
25. 1,430 **27.** 0.00000447 **29.** 0.0000000918 **31.** 10^{10}
33. 10^{-10} **35.** 10^{20} **37.** $\frac{1}{100}$ **39.** 1,000
41. 0.3679 or $\frac{1}{e}$ **43.** 25 **45.** $\frac{1}{8}$ **47.** 81 **49.** 1
51. 5 **53.** x **55.** 4 **57.** -3 **59.** $\frac{3}{2}$
61. $3t + \ln 10$ **63.** $-2t + \ln A$ **65.** $2 + 3t \log 1.01$
67. $rt + \ln P$ **69.** $3 - \log 4.2$ **71.** $-4 - \log 3.4$
73. 2.7080 **75.** -1.0986 **77.** 2.1972
79. 2.7724 **81.** San Francisco was approx. 2,000 times greater.
83.

Date	Location	Magnitude M	Energy Level T
5/22/1960	Chile	9.5	3.16×10^9
3/28/1964	Alaska	9.2	1.58×10^9
12/26/2004	Indonesia	9.1	1.26×10^9
3/11/2011	Japan	9.0	10^9

85.

x	$(1 + x)^{1/x}$
1	2
0.5	2.25
0.1	2.5937
0.01	2.7048
0.001	2.7169
0.0001	2.7181
0.00001	2.7183

87. e **89.** 2009
91. Approximately 3.19
93. 1.78×10^{-5}
95. 3.16×10^5
97. 2.00×10^8

99.

Location	Date	Magnitude M	Shockwave T
Moresby Island	Jan. 23	4.0	1.00×10^4
Vancouver Island	Apr. 30	5.3	1.99×10^5
Quebec City	June 29	3.2	1.58×10^3
Mould Bay	Nov. 13	5.2	1.58×10^5
St. Lawrence	Dec. 14	3.7	5.01×10^3

SOURCE: *National Resources Canada, National Earthquake Hazards Program.*

101. 12.9% **103.** 5.3% **105.** $\frac{7}{10}$ **107.** -3
109. 3.1250 **111.** 1.2575 **113.** $t \log 1.05$ **115.** $2t \ln 1.045$
117. $0.05t$ **119.** $-0.0042t$ **121.** 1535.06 **123.** 16.46

Find the Mistake **1.** A common logarithm uses the expression $\log_{10} x$, which is equal to $\log x$. **2.** When the base of a log is not shown, it is assumed to be 10. **3.** When the logarithm of a whole number is approximated as a decimal, the decimal part of the answer is called the mantissa. **4.** A natural logarithm is a logarithm with a base of e, denoted by $\ln x = \log_e x$.

Exercise Set 14.6

Vocabulary Review **1.** exponential, doubling time **2.** variable
3. logarithms **4.** change-of-base

Problems 1. 1.4650 **3.** 0.6826 **5.** -1.5440 **7.** -0.6477
9. $-\frac{1}{3}$ **11.** 2 **13.** -0.1845 **15.** 0.1845 **17.** 1.6168
19. 2.1131 **21.** -1 **23.** 1.2275 **25.** -6 **27.** -4 **29.** 2
31. 3 **33.** 0.3054 **35.** 42.5528 **37.** 6.0147 **39.** $\frac{4}{3}$
41. $\frac{3}{4}$ **43.** 1.3917 **45.** 0.7186 **47.** 2.6356 **49.** 4.1632
51. 5.8435 **53.** -1.0642 **55.** 2.3026 **57.** 10.7144
59. 11.72 years **61.** 9.25 years **63.** 8.75 years **65.** 18.58 years
67. 11.55 years **69.** 18.31 years **71.** 11.45 years
73. October 2018 **75.** 2009 **77.** 1992 **79.** 10.07 years
81. 2000 **83. a.** $K \approx 0.035$ **b.** 148.9°F **c.** 45.98 minutes
85. $(-2, -23)$, lowest **87.** $\left(\frac{3}{2}, 9\right)$, highest **89.** 2 seconds, 64 feet

Find the Mistake **1.** The equation $5^x = 12$ is an example of exponential equation. **2.** The change-of-base property for logarithms states that $\log_a x = \dfrac{\log_b x}{\log_b a}$. **3.** For the change-of-base property to work, base a and base b must both be positive numbers. **4.** To calculate $\log_9 36$, change the base to 10 and find the quotient of $\log 36$ and $\log 9$.

Chapter 14 Test

1.

2.

3.
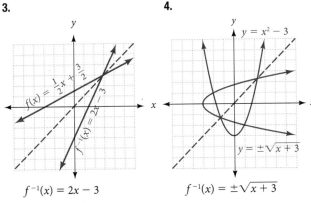

$f^{-1}(x) = 2x - 3$

4.

$f^{-1}(x) = \pm\sqrt{x + 3}$

5. 25 **6.** $\sqrt[3]{7}$

7.
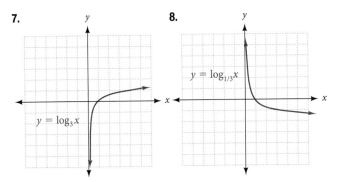

$y = \log_3 x$

8.

$y = \log_{1/3} x$

9. $\frac{1}{2}$ **10.** 1.83 **11.** 4.16 **12.** -1.69 **13.** 4.03 **14.** -3.38
15. $2\log_7 2 + 3\log_7 x - \log_7 y$ **16.** $\frac{1}{3}\log x - 2\log y - \frac{1}{2}\log z$
17. $\log_2 x^{1/3} y^5$ **18.** $\log_3 \dfrac{x^{1/4} z^3}{y}$ **19.** 240,935.059 **20.** 0.004
21. 1.262 **22.** $\frac{5}{8}$ **23.** 2 **24.** 27 **25.** 5.14
26. $508.76 **27.** 27.88 years **28.** Yes **29.** No **30.** Yes
31. Yes **32.** No **33.** No

Index

A

absolute value 95, 773
Addition
 associative property of 16
 commutative property of 15
 facts 12
 of decimals 255
 of fractions 197
 repeated 42
 with carrying 13
 with mixed numbers 223
 with table 12
 with whole numbers 13
 with zero 15
Addition property
 of zero 15
Addition Property for Inequalities 391, 752
Addition Property of Equality 327
Addition table 12
algebraic expression 98
analytic geometry 631, 792
and 836
antilog 988
antilogarithm 988
Area 48
 of a triangle 183
 rectangle 49
 square 49
Associative property
 of addition 16
 of multiplication 44
Associative Property of Addition 138
axes 790
Axis
 horizontal 37
 vertical 37

B

Bar chart 37, 789
Base 67, 410
Blueprint For Problem Solving 367
Borrowing 26

C

Carrying 13
Cartesian coordinate system 617
change-of-base property 998
characteristic 988
Chart
 bar 37
closed interval 767
coefficient 435, 716
common denominator 612
common logarithm 987
Commutative property
 of addition 15
 of multiplication 44, 137
complementary angles 111, 358

completing the square 910
complex conjugates 900
Complex fraction 233, 612
complex number 897, 906
Composite number 171
compound inequality 765
Compound Interest 941
conjugates 876
consecutive integers 378
continued inequalities 395, 767
Continuously Compounded Interest 943
coordinate 93
Counting numbers 6

D

decibel 979
Decimal numbers 247
 addition 256
 converting to fractions 283
 division 272
 estimating 265
 multiplication 264
 rounding 249
 subtraction 256
Decimal point 246
Denominator 161, 240
dependent 699, 749
dependent variable 824
Difference 24, 97, 979
difference quotient 550
Digit 2
Distributive property 44, 139
Dividend 57
Division, 338
 by one-digit numbers 58
 by two-digit numbers 59
 by zero 62
 long 59
 with fractions 189
 with remainders 61
 with whole numbers 58, 61
Division property
 for fractions 163
Divisor 57
domain 810, 811, 813, 837
doubling time 997

E

Equality 97
Equations 16
equivalent 94, 405
Equivalent 162
 fractions 162
equivalent equations 327
Estimating 28
 differences 36
 sums 36
Expanded form 3

exponent 410
exponential equations 997
exponential function 939, 942
Exponents 67
 one 68
 zero 68
extraneous solution 589, 884

F

factoring by grouping 484
Factors 43
finite set 764
formula 355
Fraction bar 57
Fractions 161
 addition 197
 comparing 164, 202
 complex 233
 converting to decimals 282
 dividing 189
 improper 161
 multiplying 180
 on the number line 161
 proper 161
 simplifying 180
 subtraction 197
 terms 161
function 811, 813, 837
function map 811
function notation 824

G

Geometric sequence 56
golden ratio 846
Grade point average (GPA) 275
graphing 629
greatest common factor 482
grouping 138

H

Height 17
horizontal asymptote 603
Horizontal axis 37
Horizontal Line Test 954
Hypotenuse 293

I

imaginary numbers 897
imaginary part 897, 906
Improper fraction 161
inconsistent 699, 749
independent variable 824
index 842, 905
Inequality 97
infinite set 764
input 810
Inspection 16
intercepts 797

intersection 764, 765
interval notation 753
Inverse Function 954
inverse reasoning 961
Irrational number 292

L

Large numbers 3
largest monomial 482
Least common denominator (LCD) 198, 570, 591
linear equation 344
linear equation in two variables 630
linear inequalities 752
line graph 616, 789
logarithmic scale 966
Long division 59
Lowest terms 172

M

mantissa 988
Mixed number 209
 addition 223
 division 217
 multiplication 216
 subtraction 225
monomial 435
Multiplication
 as repeated addition 42
 associative property of 44
 commutative property of 44
 with whole numbers 42
Multiplication property
 for fractions 162
 for inequalities 392, 754
 of equality 335
 of one 44
 of zero 44
Multiplication Property for Inequalities 392, 754
multiplication property of equality 335

N

natural logarithm 989
Natural numbers 6
negative square root 905
never 95
not 684
number e 942
number i 896
Number line 6
Numbers
 counting 6
 large 3
 natural 6
 rounding 34
 whole 6
 writing in digits 5
 writing in words 4
Numerator 161, 240
numerical coefficient 435

O

one-to-one function 954
open interval 766
opposites 96
or 836
ordered pair 627, 790
Order of operations 69
origin 618
output 810

P

paired data 789
percent 358
Perfect squares 291
Perimeter 17
 of a polygon 17
 of a rectangle 17
 of a square 17
 of a triangle 17
Place value 2, 246
point-slope form 673
Polygon 17
polynomial 443
positive square root 840, 841, 845, 905
Prime number 171
product 97, 979
Product 43
Proper fraction 161
properties of logarithms 979
Pythagorean theorem 293

Q

quadrants 618
quadratic equation. 523, 923
quadratic formula 923, 924
Quadratic Theorem 923
quotient 57, 97, 979

R

radical expression 842
radical sign 290, 842, 905
radicand 842, 905
range 810, 811, 813, 837
rational expression 546, 611
rational function 548
rationalizing the denominator 856
rational number 546, 611
real number line. 92
real numbers 93
real part 897, 906
Reciprocals 189, 241
Rectangle 17
 area 49
 perimeter 17
rectangular coordinate system 618, 789, 790
recursive 859
Reducing to lowest terms 171
relation 810, 837
Remainder 61
Repeated addition 42
revenue 451
Rhind Papyrus 608

R (continued)

Right triangle 293
rise 657
Rounding
 whole numbers 34
run 657

S

scatter diagram 616, 809
scientific notation 413
Sequence
 geometric 56
Set 6
set notation 753
similar (or like) terms. 147, 157, 437
similar radicals 866
simplified form 855, 905
slope 656, 657
Slope-Intercept Form 671
Solutions 16
solution set 326, 405, 725, 748
Spiral of roots 294
Square 17
 perimeter 17
Square inches 48
Square root 290
standard form 523, 630, 897
substitute 706
Subtraction
 of fractions 197
 with borrowing 26
 with mixed numbers 225
 with whole numbers 24
sum 14, 97, 979
Supplementary angles 111, 358
symmetric property 141
system of linear equations 696, 748

T

Table
 addition 12
term 147, 157
terms 435, 437
thirds 93, 161
three equal parts, 93
Triangle
 perimeter 17
 right 293

U

union 764, 765

V

variables 16, 97, 357
vertical asymptote 592
Vertical axis 37
Vertical Line Test 814
Vertical Translations 794

W

Whole numbers 6
 addition with 13
 division with 58, 61

multiplication with 42
rounding 34
subtraction with 24
writing in words 4
Width 17
Writing numbers
 in words 4
 with digits 5

X

x-axis 618, 790, 837
x-coordinate 617, 627, 790, 837
x-intercept 644, 797, 837

Y

y-axis 618, 790, 837
y-coordinate 617, 627, 790, 837
y-intercept 644, 797, 837

Z

Zero
 and addition 15
 and division 62
 and multiplication 44
 as an exponent 68
Zero-Factor Property 523